KU-755-118

Dedicated to
Marcia, Cherri, Robbie, Karen, and Teresa

Brief Contents

Contents

Appendix: An Introduction to PrecisionTree 21-34

Preface

This text is the 12th edition of *STATISTICS FOR BUSINESS AND ECONOMICS*. With this edition we welcome two eminent scholars to our author team: Jeffrey D. Camm of the University of Cincinnati and James J. Cochran of Louisiana Tech University. Both Jeff and Jim are accomplished teachers, researchers, and practitioners in the fields of statistics and business analytics. Jim is a fellow of the American Statistical Association. You can read more about their accomplishments in the About the Authors section which follows this preface. We believe that the addition of Jeff and Jim as our coauthors will both maintain and improve the effectiveness of *Statistics for Business and Economics*.

The purpose of *Statistics for Business and Economics* is to give students, primarily those in the fields of business administration and economics, a conceptual introduction to the field of statistics and its many applications. The text is applications oriented and written with the needs of the nonmathematician in mind; the mathematical prerequisite is knowledge of algebra.

Applications of data analysis and statistical methodology are an integral part of the organization and presentation of the text material. The discussion and development of each technique is presented in an application setting, with the statistical results providing insights to decisions and solutions to problems.

Although the book is applications oriented, we have taken care to provide sound methodological development and to use notation that is generally accepted for the topic being covered. Hence, students will find that this text provides good preparation for the study of more advanced statistical material. A bibliography to guide further study is included as an appendix.

The text introduces the student to the software packages of Minitab 16 and Microsoft® Office Excel 2010 and emphasizes the role of computer software in the application of statistical analysis. Minitab is illustrated as it is one of the leading statistical software packages for both education and statistical practice. Excel is not a statistical software package, but the wide availability and use of Excel make it important for students to understand the statistical capabilities of this package. Minitab and Excel procedures are provided in appendixes so that instructors have the flexibility of using as much computer emphasis as desired for the course. StatTools, a commercial Excel add-in developed by Palisade Corporation, extends the range of statistical options for Excel users. We show how to download and install StatTools in an appendix to Chapter 1, and most chapters include a chapter appendix that shows the steps required to accomplish a statistical procedure using StatTools. We have made the use of StatTools optional so that instructors who want to teach using only the standard tools available in Excel can do so.

Changes in the Twelfth Edition

We appreciate the acceptance and positive response to the previous editions *of Statistics for Business and Economics*. Accordingly, in making modifications for this new edition, we have maintained the presentation style and readability of those editions. There have been many changes made throughout the text to enhance its educational effectiveness. The most significant changes in the new edition are summarized here.

Content Revisions

- **Descriptive Statistics—Chapters 2 and 3.** We have significantly revised these chapters to incorporate new material on data visualization, best practices, and much more. Chapter 2 has been reorganized to include new material on side-by-side and stacked bar charts and a new section has been added on data visualization and best practices in creating effective displays. Chapter 3 now includes coverage of the geometric mean in the section on measures of location. The geometric mean has many applications in the computation of growth rates for financial assets, annual percentage rates, and so on. Chapter 3 also includes a new section on data dashboards and how summary statistics can be incorporated to enhance their effectiveness.

- **Discrete Probability Distributions—Chapter 5.** The introductory material in this chapter has been revised to explain better the role of probability distributions and to show how the material on assigning probabilities in Chapter 4 can be used to develop discrete probability distributions. We point out that the empirical discrete probability distribution is developed by using the relative frequency method to assign probabilities. At the request of many users, we have added a new section (Section 5.4) which covers bivariate discrete distributions and financial applications. We show how financial portfolios can be constructed and analyzed using these distributions.
- **Comparing Multiple Proportions, Tests of Independence, and Goodness of Fit—Chapter 12.** This chapter has undergone a major revision. We have added a new section on testing the equality of three or more population proportions. This section includes a procedure for making multiple comparison tests between all pairs of population proportions. The section on the test of independence has been rewritten to clarify that the test concerns the independence of two categorical variables. Revised appendixes with step-by-step instructions for Minitab, Excel, and StatTools are included.
- **New Case Problems.** We have added 8 new case problems to this edition; the total number of cases is 31. Three new descriptive statistics cases have been added to Chapters 2 and 3. Five new case problems involving regression appear in Chapters 14, 15, and 16. These case problems provide students with the opportunity to analyze larger data sets and prepare managerial reports based on the results of their analysis.
- **New Statistics in Practice Applications.** Each chapter begins with a Statistics in Practice vignette that describes an application of the statistical methodology to be covered in the chapter. New to this edition is a Statistics in Practice for Chapter 2 describing the use of data dashboards and data visualization at the Cincinnati Zoo. We have also added a new Statistics in Practice to Chapter 4 describing how a NASA team used probability to assist the rescue of 33 Chilean miners trapped by a cave-in.
- **New Examples and Exercises based on Real Data.** We continue to make a significant effort to update our text examples and exercises with the most current real data and referenced sources of statistical information. In this edition, we have added approximately 180 new examples and exercises based on real data and referenced sources. Using data from sources also used by *The Wall Street Journal, USA Today, Barron's,* and others, we have drawn from actual studies to develop explanations and to create exercises that demonstrate the many uses of statistics in business and economics. We believe that the use of real data helps generate more student interest in the material and enables the student to learn about both the statistical methodology and its application. The twelfth edition contains over 350 examples and exercises based on real data.

Features and Pedagogy

Authors Anderson, Sweeney, Williams, Camm, and Cochran have continued many of the features that appeared in previous editions. Important ones for students are noted here.

Methods Exercises and Applications Exercises

The end-of-section exercises are split into two parts, Methods and Applications. The Methods exercises require students to use the formulas and make the necessary computations.

The Applications exercises require students to use the chapter material in real-world situations. Thus, students first focus on the computational "nuts and bolts" and then move on to the subtleties of statistical application and interpretation.

Self-Test Exercises

Certain exercises are identified as "Self-Test Exercises." Completely worked-out solutions for these exercises are provided in Appendix D. Students can attempt the Self-Test Exercises and immediately check the solution to evaluate their understanding of the concepts presented in the chapter.

Margin Annotations and Notes and Comments

Margin annotations that highlight key points and provide additional insights for the student are a key feature of this text. These annotations, which appear in the margins, are designed to provide emphasis and enhance understanding of the terms and concepts being presented in the text.

At the end of many sections, we provide Notes and Comments designed to give the student additional insights about the statistical methodology and its application. Notes and Comments include warnings about or limitations of the methodology, recommendations for application, brief descriptions of additional technical considerations, and other matters.

Data Files Accompany the Text

Over 200 data files are available on the website that accompanies the text. The data sets are available in both Minitab and Excel formats. Webfile logos are used in the text to identify the data sets that are available on the website. Data sets for all case problems as well as data sets for larger exercises are included.

Acknowledgments

We would like to acknowledge the work of our reviewers, who provided comments and suggestions of ways to continue to improve our text. Thanks to

AbouEl-Makarim Aboueissa, University of Southern Maine

Kathleen Arano
Fort Hays State University

Musa Ayar
Uw-baraboo/Sauk County

Kathleen Burke
SUNY Cortland

YC Chang
University of Notre Dame

David Chen
Rosemont College and Saint Joseph's University

Margaret E. Cochran
Northwestern State University of Louisiana

Thomas A. Dahlstrom
Eastern University

Anne Drougas
Dominican University

Fesseha Gebremikael
Strayer University/ Calhoun Community College

Malcolm C. Gold
University of Wisconsin—Marshfield/Wood County

Joel Goldstein
Western Connecticut State University

Jim Grant
Lewis & Clark College

Reidar Hagtvedt
University of Alberta School of Business

Clifford B. Hawley
West Virginia University

Vance A. Hughey
Western Nevada College

Tony Hunnicutt
Ouachita Technical College

Stacey M. Jones
Albers School of Business and Economics, Seattle University

Dukpa Kim
University of Virginia

Rajaram Krishnan
Earlham College

Robert J. Lemke
Lake Forest College

Philip J. Mizzi
Arizona State University

Mehdi Mohaghegh
Norwich University

Mihail Motzev
Walla Walla University

Somnath Mukhopadhyay
The University of Texas at El Paso

Kenneth E. Murphy
Chapman University

Ogbonnaya John Nwoha
Grambling State University

Claudiney Pereira
Tulane University

J. G. Pitt
University of Toronto

Scott A. Redenius
Brandeis University

Sandra Robertson
Thomas Nelson Community College

Sunil Sapra
California State University, Los Angeles

Kyle Vann Scott
Snead State Community College

Rodney E. Stanley
Tennessee State University

Jennifer Strehler
Oakton Community College

Ronald Stunda
Valdosta State University

Cindy van Es
Cornell University

Jennifer VanGilder
Ursinus College

Jacqueline Wroughton
Northern Kentucky University

Dmitry Yarushkin
Grand View University

David Zimmer
Western Kentucky University

We continue to owe a debt to our many colleagues and friends for their helpful comments and suggestions in the development of this and earlier editions of our text. Among them are:

Mohammad Ahmadi
University of Tennessee at Chattanooga

Lari Arjomand
Clayton College and State University

Robert Balough
Clarion University

Philip Boudreaux
University of Louisiana

Mike Bourke
Houston Baptist University

James Brannon
University of Wisconsin—Oshkosh

John Bryant
University of Pittsburgh

Peter Bryant
University of Colorado

Terri L. Byczkowski
University of Cincinnati

Robert Carver
Stonehill College

Richard Claycombe
McDaniel College

Robert Cochran
University of Wyoming

Robert Collins
Marquette University

David W. Cravens
Texas Christian University

Tom Dahlstrom
Eastern College

Gopal Dorai
William Patterson University

Nicholas Farnum
California State University—Fullerton

Donald Gren
Salt Lake Community College

Paul Guy
California State University—Chico

Clifford Hawley
West Virginia University

Jim Hightower
California State University, Fullerton

Alan Humphrey
University of Rhode Island

Ann Hussein
Philadelphia College of Textiles and Science

C. Thomas Innis
University of Cincinnati

Ben Isselhardt
Rochester Institute of Technology

Jeffery Jarrett
University of Rhode Island

Ronald Klimberg
St. Joseph's University

David A. Kravitz
George Mason University

David Krueger
St. Cloud State University

John Leschke
University of Virginia

Martin S. Levy
University of Cincinnati

John S. Loucks
St. Edward's University

David Lucking-Reiley
Vanderbilt University

Bala Maniam
Sam Houston State University

Don Marx
University of Alaska, Anchorage

Tom McCullough
University of California—Berkeley

Ronald W. Michener
University of Virginia

Glenn Milligan
Ohio State University

Mitchell Muesham
Sam Houston State University

Roger Myerson
Northwestern University

Richard O'Connell
Miami University of Ohio

Alan Olinsky
Bryant College

Ceyhun Ozgur
Valparaiso University

Tom Pray
Rochester Institute of Technology

Harold Rahmlow
St. Joseph's University

H. V. Ramakrishna
Penn State University at Great Valley

Tom Ryan
Case Western Reserve University

Bill Seaver
University of Tennessee

Alan Smith
Robert Morris College

Willbann Terpening
Gonzaga University

Ted Tsukahara
St. Mary's College of California

Hroki Tsurumi
Rutgers University

David Tufte
University of New Orleans

Victor Ukpolo
Austin Peay State University

Ebenge Usip
Youngstown State University

Cindy Van Es
Cornell University

Jack Vaughn
University of Texas-El Paso

Andrew Welki
John Carroll University

Ari Wijetunga
Morehead State University

J. E. Willis
Louisiana State University

Mustafa Yilmaz
Northeastern University

Gary Yoshimoto
St. Cloud State University

Yan Yu
University of Cincinnati

Charles Zimmerman
Robert Morris College

We thank our associates from business and industry who supplied the Statistics in Practice features. We recognize them individually by a credit line in each of the articles. We are also indebted to our senior acquisitions editor, Charles McCormick Jr.; our developmental editor, Maggie Kubale; our content project manager, Tamborah Moore; our Project Manager at MPS Limited, Lynn Lustberg; our media editor, Chris Valentine; and others at Cengage South-Western for their editorial counsel and support during the preparation of this text.

David R. Anderson
Dennis J. Sweeney
Thomas A. Williams
Jeffrey D. Camm
James J. Cochran

About the Authors

David R. Anderson. David R. Anderson is Professor of Quantitative Analysis in the College of Business Administration at the University of Cincinnati. Born in Grand Forks, North Dakota, he earned his B.S., M.S., and Ph.D. degrees from Purdue University. Professor Anderson has served as Head of the Department of Quantitative Analysis and Operations Management and as Associate Dean of the College of Business Administration at the University of Cincinnati. In addition, he was the coordinator of the College's first Executive Program.

At the University of Cincinnati, Professor Anderson has taught introductory statistics for business students as well as graduate-level courses in regression analysis, multivariate analysis, and management science. He has also taught statistical courses at the Department of Labor in Washington, D.C. He has been honored with nominations and awards for excellence in teaching and excellence in service to student organizations.

Professor Anderson has coauthored 10 textbooks in the areas of statistics, management science, linear programming, and production and operations management. He is an active consultant in the field of sampling and statistical methods.

Dennis J. Sweeney. Dennis J. Sweeney is Professor of Quantitative Analysis and Founder of the Center for Productivity Improvement at the University of Cincinnati. Born in Des Moines, Iowa, he earned a B.S.B.A. degree from Drake University and his M.B.A. and D.B.A. degrees from Indiana University, where he was an NDEA Fellow. Professor Sweeney has worked in the management science group at Procter & Gamble and spent a year as a visiting professor at Duke University. Professor Sweeney served as Head of the Department of Quantitative Analysis and as Associate Dean of the College of Business Administration at the University of Cincinnati.

Professor Sweeney has published more than 30 articles and monographs in the area of management science and statistics. The National Science Foundation, IBM, Procter & Gamble, Federated Department Stores, Kroger, and Cincinnati Gas & Electric have funded his research, which has been published in *Management Science, Operations Research, Mathematical Programming, Decision Sciences,* and other journals.

Professor Sweeney has coauthored 10 textbooks in the areas of statistics, management science, linear programming, and production and operations management.

Thomas A. Williams. Thomas A. Williams is Professor of Management Science in the College of Business at Rochester Institute of Technology. Born in Elmira, New York, he earned his B.S. degree at Clarkson University. He did his graduate work at Rensselaer Polytechnic Institute, where he received his M.S. and Ph.D. degrees.

Before joining the College of Business at RIT, Professor Williams served for seven years as a faculty member in the College of Business Administration at the University of Cincinnati, where he developed the undergraduate program in Information Systems and then served as its coordinator. At RIT he was the first chairman of the Decision Sciences Department. He teaches courses in management science and statistics, as well as graduate courses in regression and decision analysis.

Professor Williams is the coauthor of 11 textbooks in the areas of management science, statistics, production and operations management, and mathematics. He has been a consultant for numerous *Fortune* 500 companies and has worked on projects ranging from the use of data analysis to the development of large-scale regression models.

Jeffrey D. Camm. Jeffrey D. Camm is Professor of Quantitative Analysis, Head of the Department of Operations, Business Analytics, and Information Systems and College of Business Research Fellow in the Carl H. Lindner College of Business at the University of Cincinnati, Born in Cincinnati, Ohio, he holds a B.S. from Xavier University and a Ph.D. from Clemson University. He has been at the University of Cincinnati since 1984 and has been a visiting scholar at Stanford University and a visiting professor of business administration at the Tuck School of Business at Dartmouth College.

Dr. Camm has published over 30 papers in the general area of optimization applied to problems in operations management. He has published his research in *Science, Management Science, Operations Research, Interfaces,* and other professional journals. At the University of Cincinnati, he was named the Dornoff Fellow of Teaching Excellence and he was the 2006 recipient of the INFORMS Prize for the Teaching of Operations Research Practice. A firm believer in practicing what he preaches, he has served as an operations research consultant to numerous companies and government agencies. From 2005 to 2010 he served as editor-in-chief of *Interfaces* and is currently on the editorial board of *INFORMS Transactions on Education.*

James J. Cochran. James J. Cochran is the Bank of Ruston Endowed Research Professor of Quantitative Analysis at Louisiana Tech University. Born in Dayton, Ohio, he earned his B.S., M.S., and M.B.A. degrees from Wright State University and a Ph.D. from the University of Cincinnati. He has been at Louisiana Tech University since 2000 and has been a visiting scholar at Stanford University, Universidad de Talca, and the University of South Africa.

Professor Cochran has published over two dozen papers in the development and application of operations research and statistical methods. He has published his research *in Management Science, The American Statistician, Communications in Statistics—Theory and Methods, European Journal of Operational Research, Journal of Combinatorial Optimization,* and other professional journals. He was the 2008 recipient of the INFORMS Prize for the Teaching of Operations Research Practice and the 2010 recipient of the Mu Sigma Rho Statistical Education Award. Professor Cochran was elected to the International Statistics Institute in 2005 and named a Fellow of the American Statistical Association in 2011. A strong advocate for effective operations research and statistics education as a means of improving the quality of applications to real problems, Professor Cochran has organized and chaired teaching effectiveness workshops in Montevideo, Uruguay; Cape Town, South Africa; Cartagena, Colombia; Jaipur, India; Buenos Aires, Argentina; and Nairobi, Kenya. He has served as an operations research consultant to numerous companies and not-for-profit organizations. He currently serves as editor-in-chief of *INFORMS Transactions on Education* and is on the editorial board of *Interfaces, the Journal of the Chilean Institute of Operations Research,* and *ORiON.*

CHAPTER 1

Data and Statistics

CONTENTS

STATISTICS *in* PRACTICE

*BLOOMBERG BUSINESSWEEK**
NEW YORK, NEW YORK

With a global circulation of more than 1 million, *Bloomberg Businessweek* is one of the most widely read business magazines in the world. Bloomberg's 1700 reporters in 145 service bureaus around the world enable *Bloomberg Businessweek* to deliver a variety of articles of interest to the global business and economic community. Along with feature articles on current topics, the magazine contains articles on international business, economic analysis, information processing, and science and technology. Information in the feature articles and the regular sections helps readers stay abreast of current developments and assess the impact of those developments on business and economic conditions.

Most issues of *Bloomberg Businessweek,* formerly *BusinessWeek,* provide an in-depth report on a topic of current interest. Often, the in-depth reports contain statistical facts and summaries that help the reader understand the business and economic information. For example, the cover story for the March 3, 2011 issue discussed the impact of businesses moving their most important work to cloud computing; the May 30, 2011 issue included a report on the crisis facing the U.S. Postal Service; and the August 1, 2011 issue contained a report on why the debt crisis is even worse than you think. In addition, *Bloomberg Businessweek* provides a variety of statistics about the state of the economy, including production indexes, stock prices, mutual funds, and interest rates.

Bloomberg Businessweek also uses statistics and statistical information in managing its own business. For example, an annual survey of subscribers helps the company learn about subscriber demographics, reading habits, likely purchases, lifestyles, and so on. *Bloomberg Businessweek* managers use statistical summaries from the survey to provide better services to subscribers and advertisers. One recent North American subscriber survey indicated that 90% of *Bloomberg Businessweek* subscribers use a personal computer at home and that 64% of *Bloomberg Businessweek* subscribers are involved with computer purchases at work. Such statistics alert *Bloomberg Businessweek* managers to subscriber interest in articles about new developments in computers. The results of the subscriber survey are also made available to potential advertisers. The high percentage of subscribers using personal computers at home and the high percentage of subscribers involved with computer purchases at work would be an incentive for a computer manufacturer to consider advertising in *Bloomberg Businessweek.*

In this chapter, we discuss the types of data available for statistical analysis and describe how the data are obtained. We introduce descriptive statistics and statistical inference as ways of converting data into meaningful and easily interpreted statistical information.

Bloomberg Businessweek uses statistical facts and summaries in many of its articles. © Kyodo/Photoshot.

*The authors are indebted to Charlene Trentham, Research Manager, for providing this Statistics in Practice.

Frequently, we see the following types of statements in newspapers and magazines:

- United States Department of Labor reported that the unemployment rate fell to 8.2%, the lowest in over three years (*The Washington Post,* April 6, 2012).
- Each American consumes an average of 23.2 quarts of ice cream, ice milk, sherbet, ices, and other commercially produced frozen dairy products per year (makeicecream.com website, April 2, 2012).

- The median selling price of a vacation home is $121,300 (@CNNMoney, March 29, 2012).
- The Wild Eagle rollercoaster at Dollywood in Pigeon Forge, Tennessee, reaches a maximum speed of 61 miles per hour (*USA Today* website, April 5, 2012).
- The number of registered users of Pinterest, a pinboard-style social photo sharing website, grew 85% between mid-January and mid-February (CNBC, March 29, 2012).
- The Pew Research Center reported that the United States median age of brides at the time of their first marriage is an all-time high of 26.5 years (*Significance,* February 2012).
- Canadians clocked an average of 45 hours online in the fourth quarter of 2011 (CBC News, March 2, 2012).
- The Federal Reserve reported that the average credit card debt is $5,204 per person (PRWeb website, April 5, 2012).

The numerical facts in the preceding statements (8.2%, 23.2, $121,300, 61, 85%, 26.5, 45, $5,204) are called **statistics**. In this usage, the term statistics refers to numerical facts such as averages, medians, percentages, and maximums that help us understand a variety of business and economic situations. However, as you will see, the field, or subject, of statistics involves much more than numerical facts. In a broader sense, statistics is the art and science of collecting, analyzing, presenting, and interpreting data. Particularly in business and economics, the information provided by collecting, analyzing, presenting, and interpreting data gives managers and decision makers a better understanding of the business and economic environment and thus enables them to make more informed and better decisions. In this text, we emphasize the use of statistics for business and economic decision making.

Chapter 1 begins with some illustrations of the applications of statistics in business and economics. In Section 1.2 we define the term *data* and introduce the concept of a data set. This section also introduces key terms such as *variables* and *observations,* discusses the difference between quantitative and categorical data, and illustrates the uses of cross-sectional and time series data. Section 1.3 discusses how data can be obtained from existing sources or through survey and experimental studies designed to obtain new data. The important role that the Internet now plays in obtaining data is also highlighted. The uses of data in developing descriptive statistics and in making statistical inferences are described in Sections 1.4 and 1.5. The last three sections of Chapter 1 provide the role of the computer in statistical analysis, an introduction to data mining, and a discussion of ethical guidelines for statistical practice. A chapter-ending appendix includes an introduction to the add-in StatTools which can be used to extend the statistical options for users of Microsoft Excel.

1.1 Applications in Business and Economics

In today's global business and economic environment, anyone can access vast amounts of statistical information. The most successful managers and decision makers understand the information and know how to use it effectively. In this section, we provide examples that illustrate some of the uses of statistics in business and economics.

Accounting

Public accounting firms use statistical sampling procedures when conducting audits for their clients. For instance, suppose an accounting firm wants to determine whether the amount of accounts receivable shown on a client's balance sheet fairly represents the actual amount of accounts receivable. Usually the large number of individual accounts receivable makes reviewing and validating every account too time-consuming and expensive. As common practice in such situations, the audit staff selects a subset of the accounts called a sample. After reviewing the accuracy of the sampled accounts, the auditors draw a conclusion as to whether the accounts receivable amount shown on the client's balance sheet is acceptable.

Finance

Financial analysts use a variety of statistical information to guide their investment recommendations. In the case of stocks, analysts review financial data such as price/earnings ratios and dividend yields. By comparing the information for an individual stock with information about the stock market averages, an analyst can begin to draw a conclusion as to whether the stock is a good investment. For example, *The Wall Street Journal* (March 19, 2012) reported that the average dividend yield for the S&P 500 companies was 2.2%. Microsoft showed a dividend yield of 2.42%. In this case, the statistical information on dividend yield indicates a higher dividend yield for Microsoft than the average dividend yield for the S&P 500 companies. This and other information about Microsoft would help the analyst make an informed buy, sell, or hold recommendation for Microsoft stock.

Marketing

Electronic scanners at retail checkout counters collect data for a variety of marketing research applications. For example, data suppliers such as ACNielsen and Information Resources, Inc., purchase point-of-sale scanner data from grocery stores, process the data, and then sell statistical summaries of the data to manufacturers. Manufacturers spend hundreds of thousands of dollars per product category to obtain this type of scanner data. Manufacturers also purchase data and statistical summaries on promotional activities such as special pricing and the use of in-store displays. Brand managers can review the scanner statistics and the promotional activity statistics to gain a better understanding of the relationship between promotional activities and sales. Such analyses often prove helpful in establishing future marketing strategies for the various products.

Production

Today's emphasis on quality makes quality control an important application of statistics in production. A variety of statistical quality control charts are used to monitor the output of a production process. In particular, an x-bar chart can be used to monitor the average output. Suppose, for example, that a machine fills containers with 12 ounces of a soft drink. Periodically, a production worker selects a sample of containers and computes the average number of ounces in the sample. This average, or x-bar value, is plotted on an x-bar chart. A plotted value above the chart's upper control limit indicates overfilling, and a plotted value below the chart's lower control limit indicates underfilling. The process is termed "in control" and allowed to continue as long as the plotted x-bar values fall between the chart's upper and lower control limits. Properly interpreted, an x-bar chart can help determine when adjustments are necessary to correct a production process.

Economics

Economists frequently provide forecasts about the future of the economy or some aspect of it. They use a variety of statistical information in making such forecasts. For instance, in forecasting inflation rates, economists use statistical information on such indicators as the Producer Price Index, the unemployment rate, and manufacturing capacity utilization. Often these statistical indicators are entered into computerized forecasting models that predict inflation rates.

Applications of statistics such as those described in this section are an integral part of this text. Such examples provide an overview of the breadth of statistical applications. To supplement these examples, practitioners in the fields of business and economics provided chapter-opening Statistics in Practice articles that introduce the material covered in each chapter. The Statistics in Practice applications show the importance of statistics in a wide variety of business and economic situations.

Information Systems

Information systems administrators are responsible for the day-to-day operation of an organization's computer networks. A variety of statistical information helps administrators assess the performance of computer networks, including local area networks (LANs), wide area networks (WANs), network segments, intranets, and other data communication systems. Statistics such as the mean number of users on the system, the proportion of time any component of the system is down, and the proportion of bandwidth utilized at various times of the day are examples of statistical information that help the system administrator better understand and manage the computer network.

1.2 Data

Data are the facts and figures collected, analyzed, and summarized for presentation and interpretation. All the data collected in a particular study are referred to as the **data set** for the study. Table 1.1 shows a data set containing information for 60 nations that participate in the World Trade Organization. The World Trade Organization encourages the free flow of international trade and provides a forum for resolving trade dispute.

Elements, Variables, and Observations

Elements are the entities on which data are collected. Each nation listed in Table 1.1 is an element with the nation or element name shown in the first column. With 60 nations, the data set contains 60 elements.

A **variable** is a characteristic of interest for the elements. The data set in Table 1.1 includes the following five variables:

- WTO Status: The nation's membership status in the World Trade Organization; this can be either as a member or an observer.
- Per Capita GDP ($): The total output of the nation divided by the number of people in the nation; this is commonly used to compare economic productivity of the nations.
- Trade Deficit ($1000s): The difference between total dollar value of the nation's imports and total dollar value of the nation's exports.
- Fitch Rating: The nation's sovereign credit rating as appraised by the Fitch Group[1]; the credit ratings range from a high of AAA to a low of F and can be modified by + or −.
- Fitch Outlook: An indication of the direction the credit rating is likely to move over the upcoming two years; the outlook can be negative, stable, or positive.

Measurements collected on each variable for every element in a study provide the data. The set of measurements obtained for a particular element is called an **observation**. Referring to Table 1.1, we see that the set of measurements for the first observation (Armenia) is

[1]The Fitch Group is one of three nationally recognized statistical rating organizations designated by the U.S. Securities and Exchange Commission. The other two are Standard and Poor's and Moody's investor service.

WEB file
Nations

Data sets such as Nations are available on the website for this text.

TABLE 1.1 DATA SET FOR 60 NATIONS IN THE WORLD TRADE ORGANIZATION

Nation	WTO Status	Per Capita GDP ($)	Trade Deficit ($1000s)	Fitch Rating	Fitch Outlook
Armenia	Member	5,400	2,673,359	BB−	Stable
Australia	Member	40,800	−33,304,157	AAA	Stable
Austria	Member	41,700	12,796,558	AAA	Stable
Azerbaijan	Observer	5,400	−16,747,320	BBB−	Positive
Bahrain	Member	27,300	3,102,665	BBB	Stable
Belgium	Member	37,600	−14,930,833	AA+	Negative
Brazil	Member	11,600	−29,796,166	BBB	Stable
Bulgaria	Member	13,500	4,049,237	BBB−	Positive
Canada	Member	40,300	−1,611,380	AAA	Stable
Cape Verde	Member	4,000	874,459	B+	Stable
Chile	Member	16,100	−14,558,218	A+	Stable
China	Member	8,400	−156,705,311	A+	Stable
Colombia	Member	10,100	−1,561,199	BBB−	Stable
Costa Rica	Member	11,500	5,807,509	BB+	Stable
Croatia	Member	18,300	8,108,103	BBB−	Negative
Cyprus	Member	29,100	6,623,337	BBB	Negative
Czech Republic	Member	25,900	−10,749,467	A+	Positive
Denmark	Member	40,200	−15,057,343	AAA	Stable
Ecuador	Member	8,300	1,993,819	B−	Stable
Egypt	Member	6,500	28,486,933	BB	Negative
El Salvador	Member	7,600	5,019,363	BB	Stable
Estonia	Member	20,200	802,234	A+	Stable
France	Member	35,000	118,841,542	AAA	Stable
Georgia	Member	5,400	4,398,153	B+	Positive
Germany	Member	37,900	−213,367,685	AAA	Stable
Hungary	Member	19,600	−9,421,301	BBB−	Negative
Iceland	Member	38,000	−504,939	BB+	Stable
Ireland	Member	39,500	−59,093,323	BBB+	Negative
Israel	Member	31,000	6,722,291	A	Stable
Italy	Member	30,100	33,568,668	A+	Negative
Japan	Member	34,300	31,675,424	AA	Negative
Kazakhstan	Observer	13,000	−33,220,437	BBB	Positive
Kenya	Member	1,700	9,174,198	B+	Stable
Latvia	Member	15,400	2,448,053	BBB−	Positive
Lebanon	Observer	15,600	13,715,550	B	Stable
Lithuania	Member	18,700	3,359,641	BBB	Positive
Malaysia	Member	15,600	−39,420,064	A−	Stable
Mexico	Member	15,100	1,288,112	BBB	Stable
Peru	Member	10,000	−7,888,993	BBB	Stable
Philippines	Member	4,100	15,667,209	BB+	Stable
Poland	Member	20,100	19,552,976	A−	Stable
Portugal	Member	23,200	21,060,508	BBB−	Negative
South Korea	Member	31,700	−37,509,141	A+	Stable
Romania	Member	12,300	13,323,709	BBB−	Stable
Russia	Observer	16,700	−151,400,000	BBB	Positive
Rwanda	Member	1,300	939,222	B	Stable
Serbia	Observer	10,700	8,275,693	BB−	Stable
Seychelles	Observer	24,700	666,026	B	Stable
Singapore	Member	59,900	−27,110,421	AAA	Stable
Slovakia	Member	23,400	−2,110,626	A+	Stable
Slovenia	Member	29,100	2,310,617	AA−	Negative
South Africa	Member	11,000	3,321,801	BBB+	Stable

Sweden	Member	40,600	−10,903,251	AAA	Stable
Switzerland	Member	43,400	−27,197,873	AAA	Stable
Thailand	Member	9,700	2,049,669	BBB	Stable
Turkey	Member	14,600	71,612,947	BB+	Positive
UK	Member	35,900	162,316,831	AAA	Negative
Uruguay	Member	15,400	2,662,628	BB	Positive
USA	Member	48,100	784,438,559	AAA	Stable
Zambia	Member	1,600	−1,805,198	B+	Stable

Member, 5,400, 2,673,359, BB−, and Stable. The set of measurements for the second observation (Australia) is Member, 40,800, −33,304,157, AAA, and Stable, and so on. A data set with 60 elements contains 60 observations.

Scales of Measurement

Data collection requires one of the following scales of measurement: nominal, ordinal, interval, or ratio. The scale of measurement determines the amount of information contained in the data and indicates the most appropriate data summarization and statistical analyses.

When the data for a variable consist of labels or names used to identify an attribute of the element, the scale of measurement is considered a **nominal scale.** For example, referring to the data in Table 1.1, the scale of measurement for the WTO Status variable is nominal because the data "member" and "observer" are labels used to identify the status category for the nation. In cases where the scale of measurement is nominal, a numerical code as well as a nonnumerical label may be used. For example, to facilitate data collection and to prepare the data for entry into a computer database, we might use a numerical code for WTO Status variable by letting 1 denote a member nation in the World Trade Organization and 2 denote an observer nation. The scale of measurement is nominal even though the data appear as numerical values.

The scale of measurement for a variable is considered an **ordinal scale** if the data exhibit the properties of nominal data and in addition, the order or rank of the data is meaningful. For example, referring to the data in Table 1.1, the scale of measurement for the Fitch Rating is ordinal because the rating labels which range from AAA to F can be rank ordered from best credit rating AAA to poorest credit rating F. The rating letters provide the labels similar to nominal data, but in addition, the data can also be ranked or ordered based on the credit rating, which makes the measurement scale ordinal. Ordinal data can also be recorded by a numerical code, for example, your class rank in school.

The scale of measurement for a variable is an **interval scale** if the data have all the properties of ordinal data and the interval between values is expressed in terms of a fixed unit of measure. Interval data are always numeric. College admission SAT scores are an example of interval-scaled data. For example, three students with SAT math scores of 620, 550, and 470 can be ranked or ordered in terms of best performance to poorest performance in math. In addition, the differences between the scores are meaningful. For instance, student 1 scored $620 - 550 = 70$ points more than student 2, while student 2 scored $550 - 470 = 80$ points more than student 3.

The scale of measurement for a variable is a **ratio scale** if the data have all the properties of interval data and the ratio of two values is meaningful. Variables such as distance, height, weight, and time use the ratio scale of measurement. This scale requires that a zero value be included to indicate that nothing exists for the variable at the zero point.

For example, consider the cost of an automobile. A zero value for the cost would indicate that the automobile has no cost and is free. In addition, if we compare the cost of $30,000 for one automobile to the cost of $15,000 for a second automobile, the ratio property shows that the first automobile is $30,000/$15,000 = 2 times, or twice, the cost of the second automobile.

Categorical and Quantitative Data

Data can be classified as either categorical or quantitative. Data that can be grouped by specific categories are referred to as **categorical data**. Categorical data use either the nominal or ordinal scale of measurement. Data that use numeric values to indicate how much or how many are referred to as **quantitative data**. Quantitative data are obtained using either the interval or ratio scale of measurement.

The statistical method appropriate for summarizing data depends upon whether the data are categorical or quantitative.

A **categorical variable** is a variable with categorical data, and a **quantitative variable** is a variable with quantitative data. The statistical analysis appropriate for a particular variable depends upon whether the variable is categorical or quantitative. If the variable is categorical, the statistical analysis is limited. We can summarize categorical data by counting the number of observations in each category or by computing the proportion of the observations in each category. However, even when the categorical data are identified by a numerical code, arithmetic operations such as addition, subtraction, multiplication, and division do not provide meaningful results. Section 2.1 discusses ways for summarizing categorical data.

Arithmetic operations provide meaningful results for quantitative variables. For example, quantitative data may be added and then divided by the number of observations to compute the average value. This average is usually meaningful and easily interpreted. In general, more alternatives for statistical analysis are possible when data are quantitative. Section 2.2 and Chapter 3 provide ways of summarizing quantitative data.

Cross-Sectional and Time Series Data

For purposes of statistical analysis, distinguishing between cross-sectional data and time series data is important. **Cross-sectional data** are data collected at the same or approximately the same point in time. The data in Table 1.1 are cross-sectional because they describe the five variables for the 60 World Trade Organization nations at the same point in time. **Time series data** are data collected over several time periods. For example, the time series in Figure 1.1 shows the U.S. average price per gallon of conventional regular gasoline between 2007 and 2012. Note that gasoline prices peaked in the summer of 2008 and then dropped sharply in the fall of 2008. Since 2008, the average price per gallon has continued to climb steadily, approaching an all-time high again in 2012.

Graphs of time series data are frequently found in business and economic publications. Such graphs help analysts understand what happened in the past, identify any trends over time, and project future values for the time series. The graphs of time series data can take on a variety of forms, as shown in Figure 1.2. With a little study, these graphs are usually easy to understand and interpret. For example, Panel (A) in Figure 1.2 is a graph that shows the Dow Jones Industrial Average Index from 2002 to 2012. In April 2002, the popular stock market index was near 10,000. Over the next five years the index rose to its all-time high of slightly over 14,000 in October 2007. However, notice the sharp decline in the time series after the high in 2007. By March 2009, poor economic conditions had caused the Dow Jones Industrial Average Index to return to the 7000 level. This was a scary and discouraging period for investors. However, by late 2009, the index was showing a recovery by reaching 10,000. The index has climbed steadily and was above 13,000 in early 2012.

FIGURE 1.1 U.S. AVERAGE PRICE PER GALLON FOR CONVENTIONAL REGULAR GASOLINE

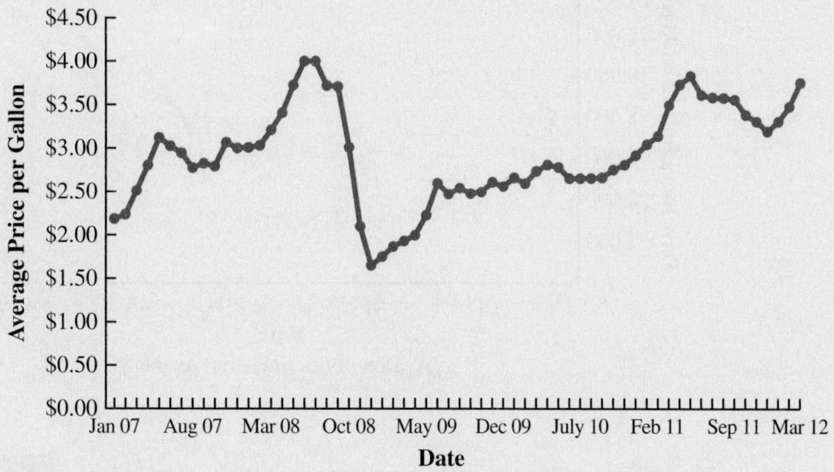

Source: Energy Information Administration, U.S. Department of Energy, March 2012.

The graph in Panel (B) shows the net income of McDonald's Inc. from 2005 to 2011. The declining economic conditions in 2008 and 2009 were actually beneficial to McDonald's as the company's net income rose to all-time highs. The growth in McDonald's net income showed that the company was thriving during the economic downturn as people were cutting back on the more expensive sit-down restaurants and seeking less-expensive alternatives offered by McDonald's. McDonald's net income continued to new all-time highs in 2010 and 2011.

Panel (C) shows the time series for the occupancy rate of hotels in South Florida over a one-year period. The highest occupancy rates, 95% and 98%, occur during the months of February and March when the climate of South Florida is attractive to tourists. In fact, January to April of each year is typically the high-occupancy season for South Florida hotels. On the other hand, note the low occupancy rates during the months of August to October, with the lowest occupancy rate of 50% occurring in September. High temperatures and the hurricane season are the primary reasons for the drop in hotel occupancy during this period.

NOTES AND COMMENTS

1. An observation is the set of measurements obtained for each element in a data set. Hence, the number of observations is always the same as the number of elements. The number of measurements obtained for each element equals the number of variables. Hence, the total number of data items can be determined by multiplying the number of observations by the number of variables.
2. Quantitative data may be discrete or continuous. Quantitative data that measure how many (e.g., number of calls received in 5 minutes) are discrete. Quantitative data that measure how much (e.g., weight or time) are continuous because no separation occurs between the possible data values.

FIGURE 1.2 A VARIETY OF GRAPHS OF TIME SERIES DATA

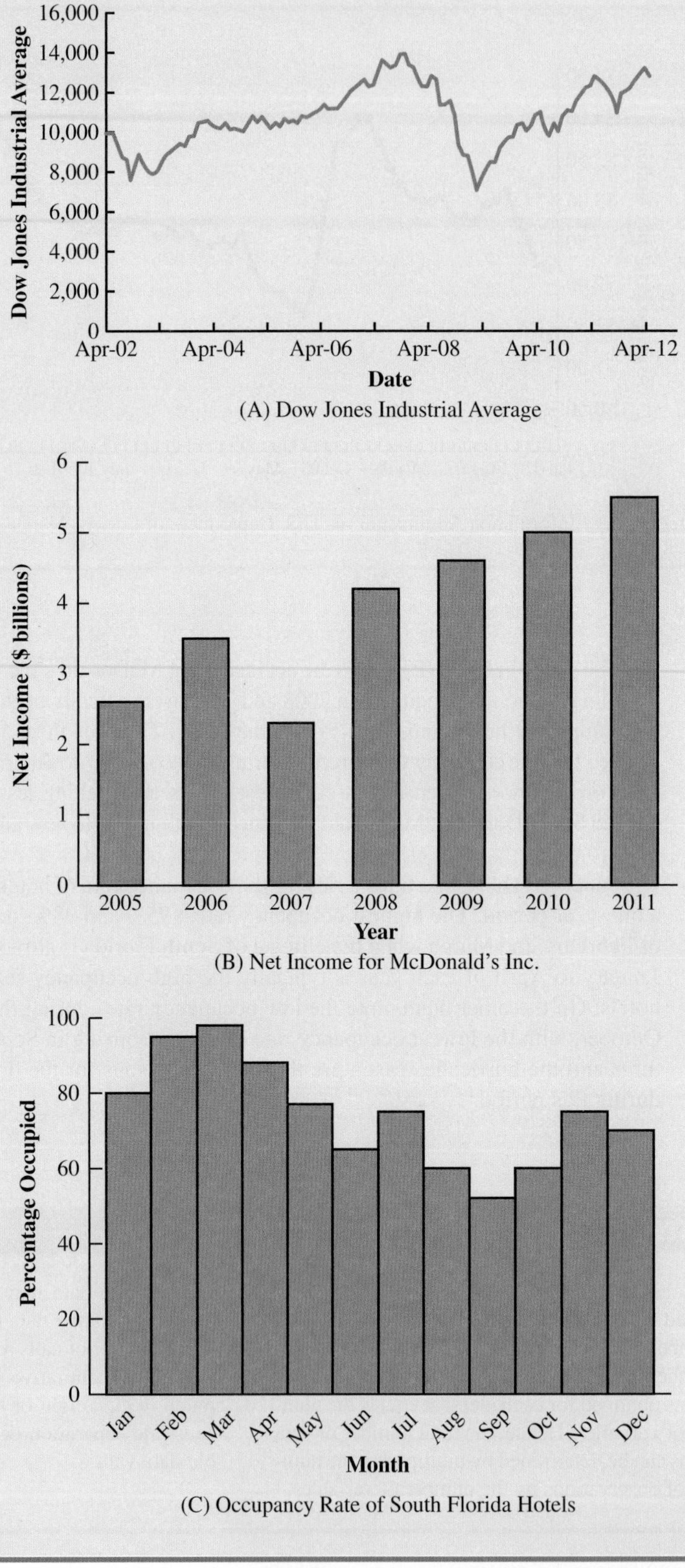

(A) Dow Jones Industrial Average

(B) Net Income for McDonald's Inc.

(C) Occupancy Rate of South Florida Hotels

1.3 Data Sources

Data can be obtained from existing sources or from surveys and experimental studies designed to collect new data.

Existing Sources

In some cases, data needed for a particular application already exist. Companies maintain a variety of databases about their employees, customers, and business operations. Data on employee salaries, ages, and years of experience can usually be obtained from internal personnel records. Other internal records contain data on sales, advertising expenditures, distribution costs, inventory levels, and production quantities. Most companies also maintain detailed data about their customers. Table 1.2 shows some of the data commonly available from internal company records.

Organizations that specialize in collecting and maintaining data make available substantial amounts of business and economic data. Companies access these external data sources through leasing arrangements or by purchase. Dun & Bradstreet, Bloomberg, and Dow Jones & Company are three firms that provide extensive business database services to clients. ACNielsen and Information Resources, Inc., built successful businesses collecting and processing data that they sell to advertisers and product manufacturers.

Data are also available from a variety of industry associations and special interest organizations. The Travel Industry Association of America maintains travel-related information such as the number of tourists and travel expenditures by states. Such data would be of interest to firms and individuals in the travel industry. The Graduate Management Admission Council maintains data on test scores, student characteristics, and graduate management education programs. Most of the data from these types of sources are available to qualified users at a modest cost.

The Internet is an important source of data and statistical information. Almost all companies maintain websites that provide general information about the company as well as data on sales, number of employees, number of products, product prices, and product specifications. In addition, a number of companies now specialize in making information available over the Internet. As a result, one can obtain access to stock quotes, meal prices at restaurants, salary data, and an almost infinite variety of information.

Government agencies are another important source of existing data. For instance, the U.S. Department of Labor maintains considerable data on employment rates, wage rates, size of the labor force, and union membership. Table 1.3 lists selected governmental agencies

TABLE 1.2 EXAMPLES OF DATA AVAILABLE FROM INTERNAL COMPANY RECORDS

Source	Some of the Data Typically Available
Employee records	Name, address, social security number, salary, number of vacation days, number of sick days, and bonus
Production records	Part or product number, quantity produced, direct labor cost, and materials cost
Inventory records	Part or product number, number of units on hand, reorder level, economic order quantity, and discount schedule
Sales records	Product number, sales volume, sales volume by region, and sales volume by customer type
Credit records	Customer name, address, phone number, credit limit, and accounts receivable balance
Customer profile	Age, gender, income level, household size, address, and preferences

TABLE 1.3 EXAMPLES OF DATA AVAILABLE FROM SELECTED GOVERNMENT AGENCIES

Government Agency	Some of the Data Available
Census Bureau	Population data, number of households, and household income
Federal Reserve Board	Data on the money supply, installment credit, exchange rates, and discount rates
Office of Management and Budget	Data on revenue, expenditures, and debt of the federal government
Department of Commerce	Data on business activity, value of shipments by industry, level of profits by industry, and growing and declining industries
Bureau of Labor Statistics	Consumer spending, hourly earnings, unemployment rate, safety records, and international statistics

and some of the data they provide. Most government agencies that collect and process data also make the results available through a website. Figure 1.3 shows the homepage for the U.S. Bureau of Labor Statistics website.

Statistical Studies

The largest experimental statistical study ever conducted is believed to be the 1954 Public Health Service experiment for the Salk polio vaccine. Nearly 2 million children in grades 1, 2, and 3 were selected from throughout the United States.

Sometimes the data needed for a particular application are not available through existing sources. In such cases, the data can often be obtained by conducting a statistical study. Statistical studies can be classified as either *experimental* or *observational.*

In an experimental study, a variable of interest is first identified. Then one or more other variables are identified and controlled so that data can be obtained about how they influence the variable of interest. For example, a pharmaceutical firm might be interested in conducting an experiment to learn how a new drug affects blood pressure. Blood pressure is the variable of interest in the study. The dosage level of the new drug is another variable that

FIGURE 1.3 U.S. BUREAU OF LABOR STATISTICS HOMEPAGE

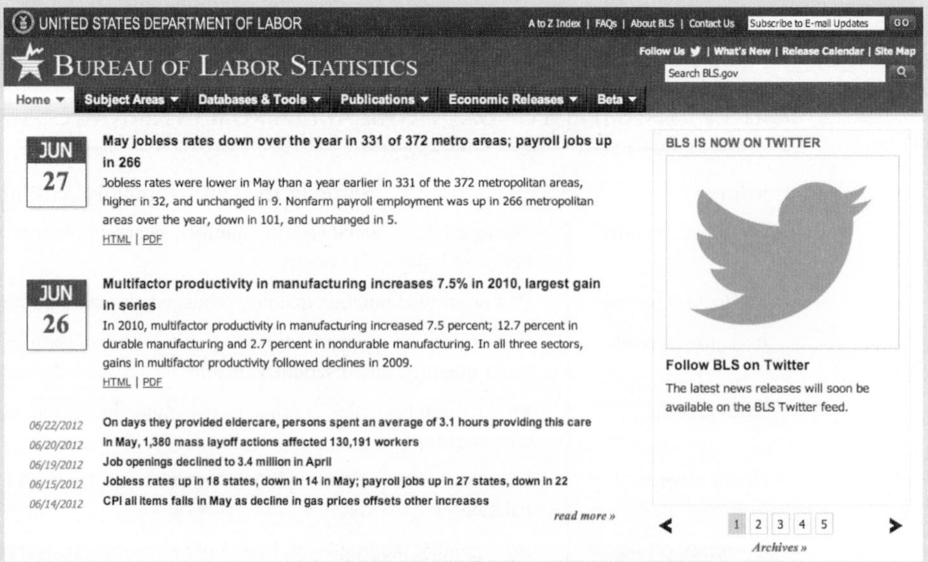

is hoped to have a causal effect on blood pressure. To obtain data about the effect of the new drug, researchers select a sample of individuals. The dosage level of the new drug is controlled, as different groups of individuals are given different dosage levels. Before and after data on blood pressure are collected for each group. Statistical analysis of the experimental data can help determine how the new drug affects blood pressure.

Studies of smokers and nonsmokers are observational studies because researchers do not determine or control who will smoke and who will not smoke.

Nonexperimental, or observational, statistical studies make no attempt to control the variables of interest. A survey is perhaps the most common type of observational study. For instance, in a personal interview survey, research questions are first identified. Then a questionnaire is designed and administered to a sample of individuals. Some restaurants use observational studies to obtain data about customer opinions on the quality of food, quality of service, atmosphere, and so on. A customer opinion questionnaire used by Chops City Grill in Naples, Florida, is shown in Figure 1.4. Note that the customers who fill out the questionnaire are asked to provide ratings for 12 variables, including overall experience, greeting by hostess, manager (table visit), overall service, and so on. The response categories of excellent, good, average, fair, and poor provide categorical data that enable Chops City Grill management to maintain high standards for the restaurant's food and service.

Anyone wanting to use data and statistical analysis as aids to decision making must be aware of the time and cost required to obtain the data. The use of existing data sources is desirable when data must be obtained in a relatively short period of time. If important data are not readily available from an existing source, the additional time and cost involved in obtaining the data must be taken into account. In all cases, the decision maker should

FIGURE 1.4 CUSTOMER OPINION QUESTIONNAIRE USED BY CHOPS CITY GRILL RESTAURANT IN NAPLES, FLORIDA

Chops
CITY GRILL

Date: ____________ Server Name: ____________

*O*ur customers are our top priority. Please take a moment to fill out our survey card, so we can better serve your needs. You may return this card to the front desk or return by mail. Thank you!

SERVICE SURVEY	Excellent	Good	Average	Fair	Poor
Overall Experience	❑	❑	❑	❑	❑
Greeting by Hostess	❑	❑	❑	❑	❑
Manager (Table Visit)	❑	❑	❑	❑	❑
Overall Service	❑	❑	❑	❑	❑
Professionalism	❑	❑	❑	❑	❑
Menu Knowledge	❑	❑	❑	❑	❑
Friendliness	❑	❑	❑	❑	❑
Wine Selection	❑	❑	❑	❑	❑
Menu Selection	❑	❑	❑	❑	❑
Food Quality	❑	❑	❑	❑	❑
Food Presentation	❑	❑	❑	❑	❑
Value for $ Spent	❑	❑	❑	❑	❑

What comments could you give us to improve our restaurant?

____

____

Thank you, we appreciate your comments. —The staff of Chops City Grill.

consider the contribution of the statistical analysis to the decision-making process. The cost of data acquisition and the subsequent statistical analysis should not exceed the savings generated by using the information to make a better decision.

Data Acquisition Errors

Managers should always be aware of the possibility of data errors in statistical studies. Using erroneous data can be worse than not using any data at all. An error in data acquisition occurs whenever the data value obtained is not equal to the true or actual value that would be obtained with a correct procedure. Such errors can occur in a number of ways. For example, an interviewer might make a recording error, such as a transposition in writing the age of a 24-year-old person as 42, or the person answering an interview question might misinterpret the question and provide an incorrect response.

Experienced data analysts take great care in collecting and recording data to ensure that errors are not made. Special procedures can be used to check for internal consistency of the data. For instance, such procedures would indicate that the analyst should review the accuracy of data for a respondent shown to be 22 years of age but reporting 20 years of work experience. Data analysts also review data with unusually large and small values, called outliers, which are candidates for possible data errors. In Chapter 3 we present some of the methods statisticians use to identify outliers.

Errors often occur during data acquisition. Blindly using any data that happen to be available or using data that were acquired with little care can result in misleading information and bad decisions. Thus, taking steps to acquire accurate data can help ensure reliable and valuable decision-making information.

1.4 Descriptive Statistics

Most of the statistical information in newspapers, magazines, company reports, and other publications consists of data that are summarized and presented in a form that is easy for the reader to understand. Such summaries of data, which may be tabular, graphical, or numerical, are referred to as **descriptive statistics.**

Refer to the data set in Table 1.1 showing data for 60 nations that participate in the World Trade Organization. Methods of descriptive statistics can be used to summarize these data. For example, consider the variable Fitch Outlook that indicates the direction the nation's credit rating is likely to move over the next two years. The Fitch Outlook is recorded as being negative, stable, or positive. A tabular summary of the data showing the number of nations with each of the Fitch Outlook ratings is shown in Table 1.4. A graphical summary of the same data, called a bar chart, is shown in Figure 1.5. These types of summaries make the data easier to interpret. Referring to Table 1.4 and Figure 1.5, we can see that the majority of Fitch Outlook credit ratings are stable, with 65% of the nations

TABLE 1.4 FREQUENCIES AND PERCENT FREQUENCIES FOR THE FITCH CREDIT RATING OUTLOOK OF 60 NATIONS

Fitch Outlook	Frequency	Percent Frequency (%)
Positive	10	16.7
Stable	39	65.0
Negative	11	18.3

FIGURE 1.5 BAR CHART FOR THE FITCH CREDIT RATING OUTLOOK FOR 60 NATIONS

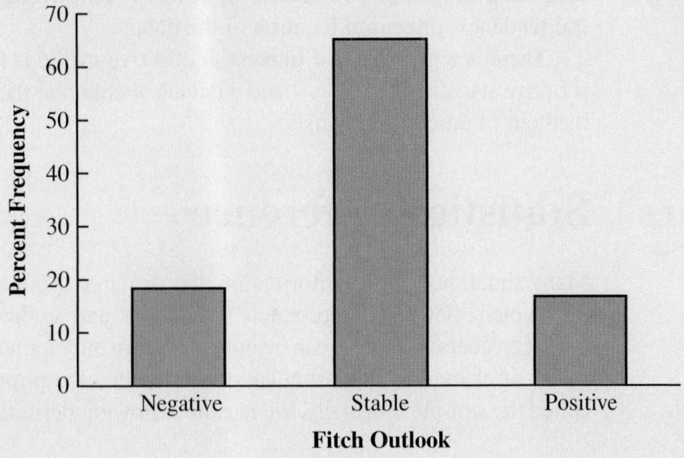

having this rating. Negative and positive outlook credit ratings are similar with slightly more nations having a negative outlook (18.3%) than a positive outlook (16.7%).

A graphical summary of the data for quantitative variable Per Capita GDP in Table 1.1, called a histogram, is provided in Figure 1.6. Using the histogram, it is easy to see that Per Capita GDP for the 60 nations ranges from \$0 to \$60,000, with the highest concentration between \$10,000 and \$20,000. Only one nation had a Per Capita GDP exceeding \$50,000.

In addition to tabular and graphical displays, numerical descriptive statistics are used to summarize data. The most common numerical measure is the average, or mean. Using

FIGURE 1.6 HISTOGRAM OF PER CAPITA GDP FOR 60 NATIONS

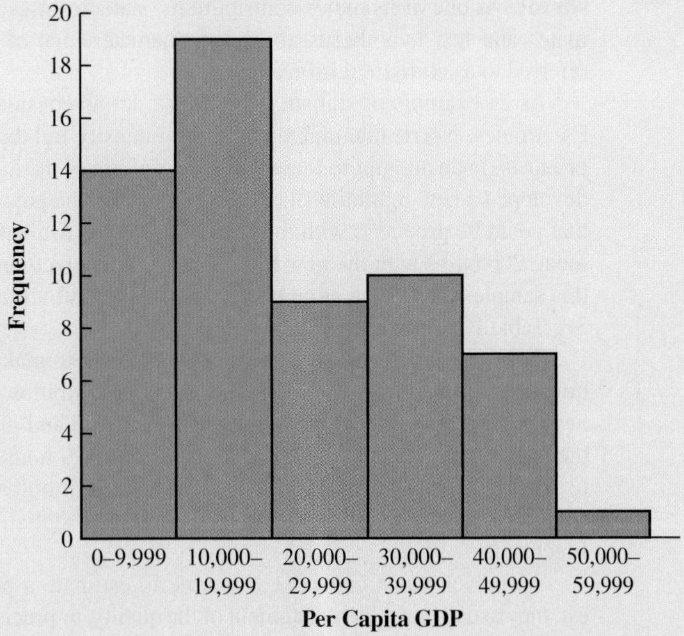

the data on Per Capita GDP for the 60 nations in Table 1.1, we can compute the average by adding Per Capita GDP for all 60 nations and dividing the total by 60. Doing so provides an average Per Capita GDP of $21,387. This average provides a measure of the central tendency, or central location of the data.

There is a great deal of interest in effective methods for developing and presenting descriptive statistics. Chapters 2 and 3 devote attention to the tabular, graphical, and numerical methods of descriptive statistics.

1.5 Statistical Inference

Many situations require information about a large group of elements (individuals, companies, voters, households, products, customers, and so on). But, because of time, cost, and other considerations, data can be collected from only a small portion of the group. The larger group of elements in a particular study is called the **population**, and the smaller group is called the **sample**. Formally, we use the following definitions.

POPULATION

A population is the set of all elements of interest in a particular study.

SAMPLE

A sample is a subset of the population.

The U.S. government conducts a census every 10 years. Market research firms conduct sample surveys every day.

The process of conducting a survey to collect data for the entire population is called a **census**. The process of conducting a survey to collect data for a sample is called a **sample survey**. As one of its major contributions, statistics uses data from a sample to make estimates and test hypotheses about the characteristics of a population through a process referred to as **statistical inference**.

As an example of statistical inference, let us consider the study conducted by Norris Electronics. Norris manufactures a high-intensity lightbulb used in a variety of electrical products. In an attempt to increase the useful life of the lightbulb, the product design group developed a new lightbulb filament. In this case, the population is defined as all lightbulbs that could be produced with the new filament. To evaluate the advantages of the new filament, 200 bulbs with the new filament were manufactured and tested. Data collected from this sample showed the number of hours each lightbulb operated before filament burnout. See Table 1.5.

Suppose Norris wants to use the sample data to make an inference about the average hours of useful life for the population of all lightbulbs that could be produced with the new filament. Adding the 200 values in Table 1.5 and dividing the total by 200 provides the sample average lifetime for the lightbulbs: 76 hours. We can use this sample result to estimate that the average lifetime for the lightbulbs in the population is 76 hours. Figure 1.7 provides a graphical summary of the statistical inference process for Norris Electronics.

Whenever statisticians use a sample to estimate a population characteristic of interest, they usually provide a statement of the quality, or precision, associated with the estimate. For the Norris example, the statistician might state that the point estimate of the average

TABLE 1.5 HOURS UNTIL BURNOUT FOR A SAMPLE OF 200 LIGHTBULBS FOR THE NORRIS ELECTRONICS EXAMPLE

WEB file
Norris

107	73	68	97	76	79	94	59	98	57
54	65	71	70	84	88	62	61	79	98
66	62	79	86	68	74	61	82	65	98
62	116	65	88	64	79	78	79	77	86
74	85	73	80	68	78	89	72	58	69
92	78	88	77	103	88	63	68	88	81
75	90	62	89	71	71	74	70	74	70
65	81	75	62	94	71	85	84	83	63
81	62	79	83	93	61	65	62	92	65
83	70	70	81	77	72	84	67	59	58
78	66	66	94	77	63	66	75	68	76
90	78	71	101	78	43	59	67	61	71
96	75	64	76	72	77	74	65	82	86
66	86	96	89	81	71	85	99	59	92
68	72	77	60	87	84	75	77	51	45
85	67	87	80	84	93	69	76	89	75
83	68	72	67	92	89	82	96	77	102
74	91	76	83	66	68	61	73	72	76
73	77	79	94	63	59	62	71	81	65
73	63	63	89	82	64	85	92	64	73

FIGURE 1.7 THE PROCESS OF STATISTICAL INFERENCE FOR THE NORRIS ELECTRONICS EXAMPLE

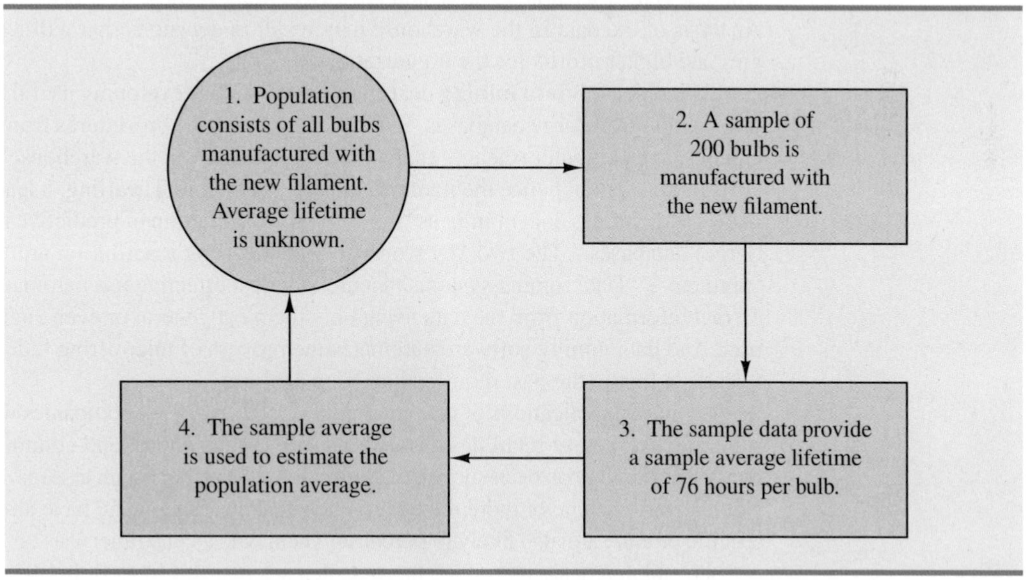

lifetime for the population of new lightbulbs is 76 hours with a margin of error of ±4 hours. Thus, an interval estimate of the average lifetime for all lightbulbs produced with the new filament is 72 hours to 80 hours. The statistician can also state how confident he or she is that the interval from 72 hours to 80 hours contains the population average.

1.6 Computers and Statistical Analysis

Statisticians frequently use computer software to perform the statistical computations required with large amounts of data. For example, computing the average lifetime for the 200 lightbulbs in the Norris Electronics example (see Table 1.5) would be quite tedious without a computer. To facilitate computer usage, many of the data sets in this book are available on the website that accompanies the text. The data files may be downloaded in either Minitab or Excel formats. In addition, the Excel add-in StatTools can be downloaded from the website. End-of-chapter appendixes cover the step-by-step procedures for using Minitab, Excel, and the Excel add-in StatTools to implement the statistical techniques presented in the chapter.

Minitab and Excel data sets and the Excel add-in StatTools are available on the website for this text.

1.7 Data Mining

With the aid of magnetic card readers, bar code scanners, and point-of-sale terminals, most organizations obtain large amounts of data on a daily basis. And, even for a small local restaurant that uses touch screen monitors to enter orders and handle billing, the amount of data collected can be substantial. For large retail companies, the sheer volume of data collected is hard to conceptualize, and figuring out how to effectively use these data to improve profitability is a challenge. Mass retailers such as Walmart capture data on 20 to 30 million transactions every day, telecommunication companies such as France Telecom and AT&T generate over 300 million call records per day, and Visa processes 6800 payment transactions per second or approximately 600 million transactions per day. Storing and managing the transaction data is a substantial undertaking.

The term *data warehousing* is used to refer to the process of capturing, storing, and maintaining the data. Computing power and data collection tools have reached the point where it is now feasible to store and retrieve extremely large quantities of data in seconds. Analysis of the data in the warehouse may result in decisions that will lead to new strategies and higher profits for the organization.

The subject of **data mining** deals with methods for developing useful decision-making information from large databases. Using a combination of procedures from statistics, mathematics, and computer science, analysts "mine the data" in the warehouse to convert it into useful information, hence the name *data mining*. Dr. Kurt Thearling, a leading practitioner in the field, defines data mining as "the automated extraction of predictive information from (large) databases." The two key words in Dr. Thearling's definition are "automated" and "predictive." Data mining systems that are the most effective use automated procedures to extract information from the data using only the most general or even vague queries by the user. And data mining software automates the process of uncovering hidden predictive information that in the past required hands-on analysis.

The major applications of data mining have been made by companies with a strong consumer focus, such as retail businesses, financial organizations, and communication companies. Data mining has been successfully used to help retailers such as Amazon and Barnes & Noble determine one or more related products that customers who have already purchased a specific product are also likely to purchase. Then, when a customer logs on to the company's website and purchases a product, the website uses pop-ups to alert the customer about additional products that the customer is likely to purchase. In another application, data mining may be used to identify customers who are likely to spend more than $20 on a particular shopping trip. These customers may then be identified as the ones to receive special e-mail or regular mail discount offers to encourage them to make their next shopping trip before the discount termination date.

Data mining is a technology that relies heavily on statistical methodology such as multiple regression, logistic regression, and correlation. But it takes a creative integration of all

Statistical methods play an important role in data mining, both in terms of discovering relationships in the data and predicting future outcomes. However, a thorough coverage of data mining and the use of statistics in data mining are outside the scope of this text.

these methods and computer science technologies involving artificial intelligence and machine learning to make data mining effective. A substantial investment in time and money is required to implement commercial data mining software packages developed by firms such as Oracle, Teradata, and SAS. The statistical concepts introduced in this text will be helpful in understanding the statistical methodology used by data mining software packages and enable you to better understand the statistical information that is developed.

Because statistical models play an important role in developing predictive models in data mining, many of the concerns that statisticians deal with in developing statistical models are also applicable. For instance, a concern in any statistical study involves the issue of model reliability. Finding a statistical model that works well for a particular sample of data does not necessarily mean that it can be reliably applied to other data. One of the common statistical approaches to evaluating model reliability is to divide the sample data set into two parts: a training data set and a test data set. If the model developed using the training data is able to accurately predict values in the test data, we say that the model is reliable. One advantage that data mining has over classical statistics is that the enormous amount of data available allows the data mining software to partition the data set so that a model developed for the training data set may be tested for reliability on other data. In this sense, the partitioning of the data set allows data mining to develop models and relationships and then quickly observe if they are repeatable and valid with new and different data. On the other hand, a warning for data mining applications is that with so much data available, there is a danger of overfitting the model to the point that misleading associations and cause/effect conclusions appear to exist. Careful interpretation of data mining results and additional testing will help avoid this pitfall.

1.8 Ethical Guidelines for Statistical Practice

Ethical behavior is something we should strive for in all that we do. Ethical issues arise in statistics because of the important role statistics plays in the collection, analysis, presentation, and interpretation of data. In a statistical study, unethical behavior can take a variety of forms including improper sampling, inappropriate analysis of the data, development of misleading graphs, use of inappropriate summary statistics, and/or a biased interpretation of the statistical results.

As you begin to do your own statistical work, we encourage you to be fair, thorough, objective, and neutral as you collect data, conduct analyses, make oral presentations, and present written reports containing information developed. As a consumer of statistics, you should also be aware of the possibility of unethical statistical behavior by others. When you see statistics in newspapers, on television, on the Internet, and so on, it is a good idea to view the information with some skepticism, always being aware of the source as well as the purpose and objectivity of the statistics provided.

The American Statistical Association, the nation's leading professional organization for statistics and statisticians, developed the report "Ethical Guidelines for Statistical Practice"[2] to help statistical practitioners make and communicate ethical decisions and assist students in learning how to perform statistical work responsibly. The report contains 67 guidelines organized into eight topic areas: Professionalism; Responsibilities to Funders, Clients, and Employers; Responsibilities in Publications and Testimony; Responsibilities to Research Subjects; Responsibilities to Research Team Colleagues; Responsibilities to Other Statisticians or Statistical Practitioners; Responsibilities Regarding Allegations of Misconduct; and Responsibilities of Employers Including Organizations, Individuals, Attorneys, or Other Clients Employing Statistical Practitioners.

---

[2]American Statistical Association "Ethical Guidelines for Statistical Practice," 1999.

One of the ethical guidelines in the professionalism area addresses the issue of running multiple tests until a desired result is obtained. Let us consider an example. In Section 1.5 we discussed a statistical study conducted by Norris Electronics involving a sample of 200 high-intensity lightbulbs manufactured with a new filament. The average lifetime for the sample, 76 hours, provided an estimate of the average lifetime for all lightbulbs produced with the new filament. However, consider this. Because Norris selected a sample of bulbs, it is reasonable to assume that another sample would have provided a different average lifetime.

Suppose Norris's management had hoped the sample results would enable them to claim that the average lifetime for the new lightbulbs was 80 hours or more. Suppose further that Norris's management decides to continue the study by manufacturing and testing repeated samples of 200 lightbulbs with the new filament until a sample mean of 80 hours or more is obtained. If the study is repeated enough times, a sample may eventually be obtained—by chance alone—that would provide the desired result and enable Norris to make such a claim. In this case, consumers would be misled into thinking the new product is better than it actually is. Clearly, this type of behavior is unethical and represents a gross misuse of statistics in practice.

Several ethical guidelines in the responsibilities and publications and testimony area deal with issues involving the handling of data. For instance, a statistician must account for all data considered in a study and explain the sample(s) actually used. In the Norris Electronics study the average lifetime for the 200 bulbs in the original sample is 76 hours; this is considerably less than the 80 hours or more that management hoped to obtain. Suppose now that after reviewing the results showing a 76 hour average lifetime, Norris discards all the observations with 70 or fewer hours until burnout, allegedly because these bulbs contain imperfections caused by startup problems in the manufacturing process. After discarding these lightbulbs, the average lifetime for the remaining lightbulbs in the sample turns out to be 82 hours. Would you be suspicious of Norris's claim that the lifetime for their lightbulbs is 82 hours?

If the Norris lightbulbs showing 70 or fewer hours until burnout were discarded to simply provide an average lifetime of 82 hours, there is no question that discarding the lightbulbs with 70 or fewer hours until burnout is unethical. But, even if the discarded lightbulbs contain imperfections due to startup problems in the manufacturing process—and, as a result, should not have been included in the analysis—the statistician who conducted the study must account for all the data that were considered and explain how the sample actually used was obtained. To do otherwise is potentially misleading and would constitute unethical behavior on the part of both the company and the statistician.

A guideline in the shared values section of the American Statistical Association report states that statistical practitioners should avoid any tendency to slant statistical work toward predetermined outcomes. This type of unethical practice is often observed when unrepresentative samples are used to make claims. For instance, in many areas of the country smoking is not permitted in restaurants. Suppose, however, a lobbyist for the tobacco industry interviews people in restaurants where smoking is permitted in order to estimate the percentage of people who are in favor of allowing smoking in restaurants. The sample results show that 90% of the people interviewed are in favor of allowing smoking in restaurants. Based upon these sample results, the lobbyist claims that 90% of all people who eat in restaurants are in favor of permitting smoking in restaurants. In this case we would argue that only sampling persons eating in restaurants that allow smoking has biased the results. If only the final results of such a study are reported, readers unfamiliar with the details of the study (i.e., that the sample was collected only in restaurants allowing smoking) can be misled.

The scope of the American Statistical Association's report is broad and includes ethical guidelines that are appropriate not only for a statistician, but also for consumers of statistical information. We encourage you to read the report to obtain a better perspective of ethical issues as you continue your study of statistics and to gain the background for determining how to ensure that ethical standards are met when you start to use statistics in practice.

Summary

Statistics is the art and science of collecting, analyzing, presenting, and interpreting data. Nearly every college student majoring in business or economics is required to take a course in statistics. We began the chapter by describing typical statistical applications for business and economics.

Data consist of the facts and figures that are collected and analyzed. Four scales of measurement used to obtain data on a particular variable include nominal, ordinal, interval, and ratio. The scale of measurement for a variable is nominal when the data are labels or names used to identify an attribute of an element. The scale is ordinal if the data demonstrate the properties of nominal data and the order or rank of the data is meaningful. The scale is interval if the data demonstrate the properties of ordinal data and the interval between values is expressed in terms of a fixed unit of measure. Finally, the scale of measurement is ratio if the data show all the properties of interval data and the ratio of two values is meaningful.

For purposes of statistical analysis, data can be classified as categorical or quantitative. Categorical data use labels or names to identify an attribute of each element. Categorical data use either the nominal or ordinal scale of measurement and may be nonnumeric or numeric. Quantitative data are numeric values that indicate how much or how many. Quantitative data use either the interval or ratio scale of measurement. Ordinary arithmetic operations are meaningful only if the data are quantitative. Therefore, statistical computations used for quantitative data are not always appropriate for categorical data.

In Sections 1.4 and 1.5 we introduced the topics of descriptive statistics and statistical inference. Descriptive statistics are the tabular, graphical, and numerical methods used to summarize data. The process of statistical inference uses data obtained from a sample to make estimates or test hypotheses about the characteristics of a population. The last three sections of the chapter provide information on the role of computers in statistical analysis, an introduction to the relative new field of data mining, and a summary of ethical guidelines for statistical practice.

Glossary

Statistics The art and science of collecting, analyzing, presenting, and interpreting data.

Data The facts and figures collected, analyzed, and summarized for presentation and interpretation.

Data set All the data collected in a particular study.

Elements The entities on which data are collected.

Variable A characteristic of interest for the elements.

Observation The set of measurements obtained for a particular element.

Nominal scale The scale of measurement for a variable when the data are labels or names used to identify an attribute of an element. Nominal data may be nonnumeric or numeric.

Ordinal scale The scale of measurement for a variable if the data exhibit the properties of nominal data and the order or rank of the data is meaningful. Ordinal data may be nonnumeric or numeric.

Interval scale The scale of measurement for a variable if the data demonstrate the properties of ordinal data and the interval between values is expressed in terms of a fixed unit of measure. Interval data are always numeric.

Ratio scale The scale of measurement for a variable if the data demonstrate all the properties of interval data and the ratio of two values is meaningful. Ratio data are always numeric.

Categorical data Labels or names used to identify an attribute of each element. Categorical data use either the nominal or ordinal scale of measurement and may be nonnumeric or numeric.
Quantitative data Numeric values that indicate how much or how many of something. Quantitative data are obtained using either the interval or ratio scale of measurement.
Categorical variable A variable with categorical data.
Quantitative variable A variable with quantitative data.
Cross-sectional data Data collected at the same or approximately the same point in time.
Time series data Data collected over several time periods.
Descriptive statistics Tabular, graphical, and numerical summaries of data.
Population The set of all elements of interest in a particular study.
Sample A subset of the population.
Census A survey to collect data on the entire population.
Sample survey A survey to collect data on a sample.
Statistical inference The process of using data obtained from a sample to make estimates or test hypotheses about the characteristics of a population.
Data mining The process of using procedures from statistics and computer science to extract useful information from extremely large databases.

Supplementary Exercises

1. Discuss the differences between statistics as numerical facts and statistics as a discipline or field of study.

SELF test

2. The U.S. Department of Energy provides fuel economy information for a variety of motor vehicles. A sample of 10 automobiles is shown in Table 1.6 (Fuel Economy website, February 22, 2008). Data show the size of the automobile (compact, midsize, or large), the number of cylinders in the engine, the city driving miles per gallon, the highway driving miles per gallon, and the recommended fuel (diesel, premium, or regular).
   a. How many elements are in this data set?
   b. How many variables are in this data set?
   c. Which variables are categorical and which variables are quantitative?
   d. What type of measurement scale is used for each of the variables?

SELF test

3. Refer to Table 1.6.
   a. What is the average miles per gallon for city driving?
   b. On average, how much higher is the miles per gallon for highway driving as compared to city driving?

TABLE 1.6 FUEL ECONOMY INFORMATION FOR 10 AUTOMOBILES

Car	Size	Cylinders	City MPG	Highway MPG	Fuel
Audi A8	Large	12	13	19	Premium
BMW 328Xi	Compact	6	17	25	Premium
Cadillac CTS	Midsize	6	16	25	Regular
Chrysler 300	Large	8	13	18	Premium
Ford Focus	Compact	4	24	33	Regular
Hyundai Elantra	Midsize	4	25	33	Regular
Jeep Grand Cherokee	Midsize	6	17	26	Diesel
Pontiac G6	Compact	6	15	22	Regular
Toyota Camry	Midsize	4	21	31	Regular
Volkswagen Jetta	Compact	5	21	29	Regular

TABLE 1.7 DATA FOR SEVEN COLLEGES AND UNIVERSITIES

School	State	Campus Setting	Endowment ($ billions)	% Applicants Admitted	NCAA Division
Amherst College	Massachusetts	Town: Fringe	1.7	18	III
Duke	North Carolina	City: Midsize	5.9	21	I-A
Harvard University	Massachusetts	City: Midsize	34.6	9	I-AA
Swarthmore College	Pennsylvania	Suburb: Large	1.4	18	III
University of Pennsylvania	Pennsylvania	City: Large	6.6	18	I-AA
Williams College	Massachusetts	Town: Fringe	1.9	18	III
Yale University	Connecticut	City: Midsize	22.5	9	I-AA

c. What percentage of the cars have four-cylinder engines?
d. What percentage of the cars use regular fuel?

4. Table 1.7 shows data for seven colleges and universities. The endowment (in billions of dollars) and the percentage of applicants admitted are shown (*USA Today,* February 3, 2008). The state each school is located in, the campus setting, and the NCAA Division for varsity teams were obtained from the National Center of Education Statistics website, February 22, 2008.
   a. How many elements are in the data set?
   b. How many variables are in the data set?
   c. Which of the variables are categorical and which are quantitative?

5. Consider the data set in Table 1.7
   a. Compute the average endowment for the sample.
   b. Compute the average percentage of applicants admitted.
   c. What percentage of the schools have NCAA Division III varsity teams?
   d. What percentage of the schools have a City: Midsize campus setting?

6. *Foreign Affairs* magazine conducted a survey to develop a profile of its subscribers (*Foreign Affairs* website, February 23, 2008). The following questions were asked.
   a. How many nights have you stayed in a hotel in the past 12 months?
   b. Where do you purchase books? Three options were listed: Bookstore, Internet, and Book Club.
   c. Do you own or lease a luxury vehicle? (Yes or No)
   d. What is your age?
   e. For foreign trips taken in the past three years, what was your destination? Seven international destinations were listed.

   Comment on whether each question provides categorical or quantitative data.

7. The Kroger Company is one of the largest grocery retailers in the United States with over 2000 grocery stores across the country. Kroger uses an online customer opinion questionnaire to obtain performance data about its products and services and learn about what motivates its customers (Kroger website, April 2012). In the survey, Kroger customers were asked if they would be willing to pay more for products that had each of the following four characteristics. The four questions were: Would you pay more for

   products that have a brand name?
   products that are environmentally friendly?
   products that are organic?
   products that have been recommended by others?

   For each question, the customers had the option of responding Yes if they would pay more or No if they would not pay more.

a. Are the data collected by Kroger in this example categorical or quantitative?
b. What measurement scale is used?

8. The *FinancialTimes*/Harris Poll is a monthly online poll of adults from six countries in Europe and the United States. A January poll included 1015 adults in the United States. One of the questions asked was, "How would you rate the Federal Bank in handling the credit problems in the financial markets?" Possible responses were Excellent, Good, Fair, Bad, and Terrible (Harris Interactive website, January 2008).
   a. What was the sample size for this survey?
   b. Are the data categorical or quantitative?
   c. Would it make more sense to use averages or percentages as a summary of the data for this question?
   d. Of the respondents in the United States, 10% said the Federal Bank is doing a good job. How many individuals provided this response?
9. The Commerce Department reported receiving the following applications for the Malcolm Baldrige National Quality Award: 23 from large manufacturing firms, 18 from large service firms, and 30 from small businesses.
   a. Is type of business a categorical or quantitative variable?
   b. What percentage of the applications came from small businesses?
10. The Bureau of Transportation Statistics Omnibus Household Survey is conducted annually and serves as an information source for the U.S. Department of Transportation. In one part of the survey the person being interviewed was asked to respond to the following statement: "Drivers of motor vehicles should be allowed to talk on a hand-held cell phone while driving." Possible responses were strongly agree, somewhat agree, somewhat disagree, and strongly disagree. Forty-four respondents said that they strongly agree with this statement, 130 said that they somewhat agree, 165 said they somewhat disagree, and 741 said they strongly disagree with this statement (Bureau of Transportation website, August 2010).
    a. Do the responses for this statement provide categorical or quantitative data?
    b. Would it make more sense to use averages or percentages as a summary of the responses for this statement?
    c. What percentage of respondents strongly agree with allowing drivers of motor vehicles to talk on a hand-held cell phone while driving?
    d. Do the results indicate general support for or against allowing drivers of motor vehicles to talk on a hand-held cell phone while driving?
11. J.D. Power and Associates conducts vehicle quality surveys to provide automobile manufacturers with consumer satisfaction information about their products (Vehicle Quality Survey, January 2010). Using a sample of vehicle owners from recent vehicle purchase records, the survey asks the owners a variety of questions about their new vehicles, such as those shown below. For each question, state whether the data collected are categorical or quantitative and indicate the measurement scale being used.
    a. What price did you pay for the vehicle?
    b. How did you pay for the vehicle? (Cash, Lease, or Finance)
    c. How likely would you be to recommend this vehicle to a friend? (Definitely Not, Probably Not, Probably Will, and Definitely Will)
    d. What is the current mileage?
    e. What is your overall rating of your new vehicle? A 10-point scale, ranging from 1 for unacceptable to 10 for truly exceptional, was used.
12. The Hawaii Visitors Bureau collects data on visitors to Hawaii. The following questions were among 16 asked in a questionnaire handed out to passengers during incoming airline flights.
    - This trip to Hawaii is my: 1st, 2nd, 3rd, 4th, etc.
    - The primary reason for this trip is: (10 categories, including vacation, convention, honeymoon)

- Where I plan to stay: (11 categories, including hotel, apartment, relatives, camping)
- Total days in Hawaii

a. What is the population being studied?
b. Is the use of a questionnaire a good way to reach the population of passengers on incoming airline flights?
c. Comment on each of the four questions in terms of whether it will provide categorical or quantitative data.

SELF test

13. Figure 1.8 provides a bar chart showing the amount of federal spending for the years 2004 to 2010 (Congressional Budget Office website, May 15, 2011).
    a. What is the variable of interest?
    b. Are the data categorical or quantitative?
    c. Are the data time series or cross-sectional?
    d. Comment on the trend in federal spending over time.

14. The following data show the number of rental cars in service for three rental car companies: Hertz, Avis, and Dollar. The data are for the years 2007–2010 and are in thousands of vehicles (Auto Rental News website, May 15, 2011).

	Cars in Service (1000s)			
Company	**2007**	**2008**	**2009**	**2010**
Hertz	327	311	286	290
Dollar	167	140	106	108
Avis	204	220	300	270

    a. Construct a time series graph for the years 2007 to 2010 showing the number of rental cars in service for each company. Show the time series for all three companies on the same graph.
    b. Comment on who appears to be the market share leader and how the market shares are changing over time.
    c. Construct a bar chart showing rental cars in service for 2010. Is this chart based on cross-sectional or time series data?

FIGURE 1.8 FEDERAL SPENDING

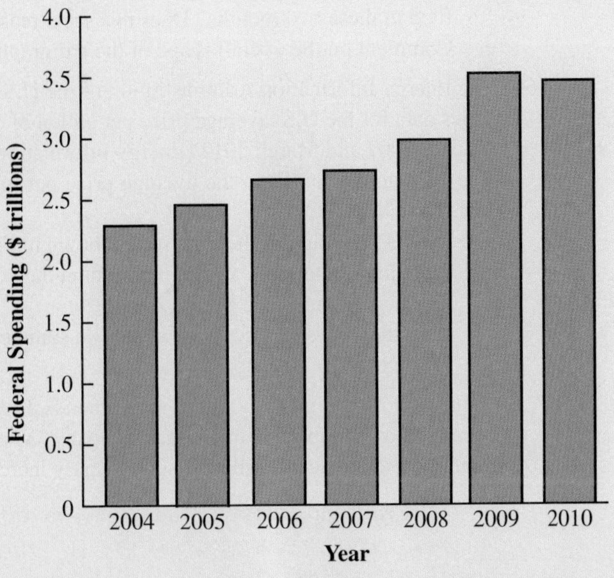

FIGURE 1.9 NUMBER OF RECREATIONAL BOATING ACCIDENTS

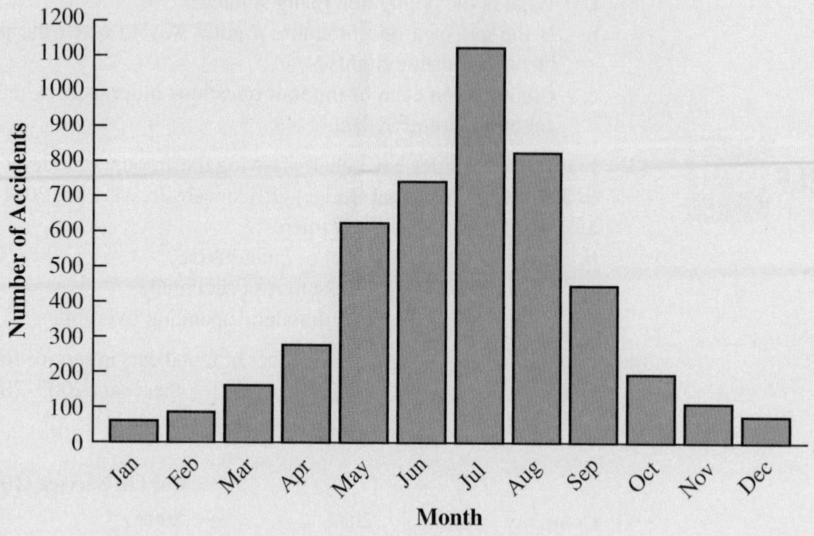

15. Every year, the U.S. Coast Guard collects data and compiles statistics on reported recreational boating accidents. These statistics are derived from accident reports that are filed by the owners/operators of recreational vessels involved in accidents. In 2009, 4730 recreational boating accident reports were filed. Figure 1.9 provides a bar chart summarizing the number of accident reports that were filed each month (U.S. Coast Guard's Boating Safety Division website, August 2010).
    a. Are the data categorical or quantitative?
    b. Are the data time series or cross-sectional?
    c. In what month were the most accident reports filed? Approximately how many?
    d. There were 61 accident reports filed in January and 76 accident reports filed in December. What percentage of the total number of accident reports for the year was filed in these two months? Does this seem reasonable?
    e. Comment on the overall shape of the bar graph.
16. The Energy Information Administration of the U.S. Department of Energy provided time series data for the U.S. average price per gallon of conventional regular gasoline between January 2007 and March 2012 (Energy Information Administration website, April 2012). Use the Internet to obtain the average price per gallon of conventional regular gasoline since March 2012.
    a. Extend the graph of the time series shown in Figure 1.1.
    b. What interpretations can you make about the average price per gallon of conventional regular gasoline since March 2012?
    c. Does the time series continue to show a summer increase in the average price per gallon? Explain.
17. A manager of a large corporation recommends a $10,000 raise be given to keep a valued subordinate from moving to another company. What internal and external sources of data might be used to decide whether such a salary increase is appropriate?

18. A random telephone survey of 1021 adults (aged 18 and older) was conducted by Opinion Research Corporation on behalf of CompleteTax, an online tax preparation and e-filing service. The survey results showed that 684 of those surveyed planned to file their taxes electronically (CompleteTax Tax Prep Survey 2010).
    a. Develop a descriptive statistic that can be used to estimate the percentage of all taxpayers who file electronically.
    b. The survey reported that the most frequently used method for preparing the tax return is to hire an accountant or professional tax preparer. If 60% of the people surveyed had their tax return prepared this way, how many people used an accountant or professional tax preparer?
    c. Other methods that the person filing the return often used include manual preparation, use of an online tax service, and use of a software tax program. Would the data for the method for preparing the tax return be considered categorical or quantitative?
19. A *Bloomberg Businessweek* North American subscriber study collected data from a sample of 2861 subscribers. Fifty-nine percent of the respondents indicated an annual income of $75,000 or more, and 50% reported having an American Express credit card.
    a. What is the population of interest in this study?
    b. Is annual income a categorical or quantitative variable?
    c. Is ownership of an American Express card a categorical or quantitative variable?
    d. Does this study involve cross-sectional or time series data?
    e. Describe any statistical inferences *Bloomberg Businessweek* might make on the basis of the survey.
20. A survey of 131 investment managers in *Barron's* Big Money poll revealed the following:
    - 43% of managers classified themselves as bullish or very bullish on the stock market.
    - The average expected return over the next 12 months for equities was 11.2%.
    - 21% selected health care as the sector most likely to lead the market in the next 12 months.
    - When asked to estimate how long it would take for technology and telecom stocks to resume sustainable growth, the managers' average response was 2.5 years.

    a. Cite two descriptive statistics.
    b. Make an inference about the population of all investment managers concerning the average return expected on equities over the next 12 months.
    c. Make an inference about the length of time it will take for technology and telecom stocks to resume sustainable growth.
21. A seven-year medical research study reported that women whose mothers took the drug DES during pregnancy were twice as likely to develop tissue abnormalities that might lead to cancer as were women whose mothers did not take the drug.
    a. This study compared two populations. What were the populations?
    b. Do you suppose the data were obtained in a survey or an experiment?
    c. For the population of women whose mothers took the drug DES during pregnancy, a sample of 3980 women showed that 63 developed tissue abnormalities that might lead to cancer. Provide a descriptive statistic that could be used to estimate the number of women out of 1000 in this population who have tissue abnormalities.
    d. For the population of women whose mothers did not take the drug DES during pregnancy, what is the estimate of the number of women out of 1000 who would be expected to have tissue abnormalities?
    e. Medical studies often use a relatively large sample (in this case, 3980). Why?

22. The Nielsen Company surveyed consumers in 47 markets from Europe, Asia-Pacific, the Americas, and the Middle East to determine which factors are most important in determining where they buy groceries. Using a scale of 1 (low) to 5 (high), the highest rated factor was *good value for money*, with an average point score of 4.32. The second highest rated factor was *better selection of high-quality brands and products*, with an average point score of 3.78, and the lowest rated factor was *uses recyclable bags and packaging*, with an average point score of 2.71 (Nielsen website, February 24, 2008). Suppose that you have been hired by a grocery store chain to conduct a similar study to determine what factors customers at the chain's stores in Charlotte, North Carolina, think are most important in determining where they buy groceries.
    a. What is the population for the survey that you will be conducting?
    b. How would you collect the data for this study?

23. Pew Research Center is a nonpartisan polling organization that provides information about issues, attitudes, and trends shaping America. In a recent poll, Pew researchers found that 47% of American adult respondents reported getting at least some local news on their cell phone or tablet computer (Pew Research website, May 14, 2011). Further findings showed that 42% of respondents who own cell phones or tablet computers use those devices to check local weather reports and 37% use the devices to find local restaurants or other businesses.
    a. One statistic concerned using cell phones or tablet computers for local news. What population is that finding applicable to?
    b. Another statistic concerned using cell phones or tablet computers to check local weather reports and to find local restaurants. What population is this finding applicable to?
    c. Do you think the Pew researchers conducted a census or a sample survey to obtain their results? Why?
    d. If you were a restaurant owner, would you find these results interesting? Why? How could you take advantage of this information?

24. A sample of midterm grades for five students showed the following results: 72, 65, 82, 90, 76. Which of the following statements are correct, and which should be challenged as being too generalized?
    a. The average midterm grade for the sample of five students is 77.
    b. The average midterm grade for all students who took the exam is 77.
    c. An estimate of the average midterm grade for all students who took the exam is 77.
    d. More than half of the students who take this exam will score between 70 and 85.
    e. If five other students are included in the sample, their grades will be between 65 and 90.

25. Table 1.8 shows a data set containing information for 25 of the shadow stocks tracked by the American Association of Individual Investors. Shadow stocks are common stocks of smaller companies that are not closely followed by Wall Street analysts. The data set is also on the website that accompanies the text in the file named Shadow02.
    a. How many variables are in the data set?
    b. Which of the variables are categorical and which are quantitative?
    c. For the Exchange variable, show the frequency and the percent frequency for AMEX, NYSE, and OTC. Construct a bar graph similar to Figure 1.5 for the Exchange variable.
    d. Show the frequency distribution for the Gross Profit Margin using the five intervals: 0–14.9, 15–29.9, 30–44.9, 45–59.9, and 60–74.9. Construct a histogram similar to Figure 1.6.
    e. What is the average price/earnings ratio?

TABLE 1.8 DATA SET FOR 25 SHADOW STOCKS

WEB file
Shadow02

Company	Exchange	Ticker Symbol	Market Cap ($ millions)	Price/ Earnings Ratio	Gross Profit Margin (%)
DeWolfe Companies	AMEX	DWL	36.4	8.4	36.7
North Coast Energy	OTC	NCEB	52.5	6.2	59.3
Hansen Natural Corp.	OTC	HANS	41.1	14.6	44.8
MarineMax, Inc.	NYSE	HZO	111.5	7.2	23.8
Nanometrics Incorporated	OTC	NANO	228.6	38.0	53.3
TeamStaff, Inc.	OTC	TSTF	92.1	33.5	4.1
Environmental Tectonics	AMEX	ETC	51.1	35.8	35.9
Measurement Specialties	AMEX	MSS	101.8	26.8	37.6
SEMCO Energy, Inc.	NYSE	SEN	193.4	18.7	23.6
Party City Corporation	OTC	PCTY	97.2	15.9	36.4
Embrex, Inc.	OTC	EMBX	136.5	18.9	59.5
Tech/Ops Sevcon, Inc.	AMEX	TO	23.2	20.7	35.7
ARCADIS NV	OTC	ARCAF	173.4	8.8	9.6
Qiao Xing Universal Tele.	OTC	XING	64.3	22.1	30.8
Energy West Incorporated	OTC	EWST	29.1	9.7	16.3
Barnwell Industries, Inc.	AMEX	BRN	27.3	7.4	73.4
Innodata Corporation	OTC	INOD	66.1	11.0	29.6
Medical Action Industries	OTC	MDCI	137.1	26.9	30.6
Instrumentarium Corp.	OTC	INMRY	240.9	3.6	52.1
Petroleum Development	OTC	PETD	95.9	6.1	19.4
Drexler Technology Corp.	OTC	DRXR	233.6	45.6	53.6
Gerber Childrenswear Inc.	NYSE	GCW	126.9	7.9	25.8
Gaiam, Inc.	OTC	GAIA	295.5	68.2	60.7
Artesian Resources Corp.	OTC	ARTNA	62.8	20.5	45.5
York Water Company	OTC	YORW	92.2	22.9	74.2

Appendix An Introduction to StatTools

StatTools is a professional add-in that expands the statistical capabilities available with Microsoft Excel. StatTools software can be downloaded from the website that accompanies this text.

Excel does not contain statistical functions or data analysis tools to perform all the statistical procedures discussed in the text. StatTools is a Microsoft Excel statistics add-in that extends the range of statistical and graphical options for Excel users. Most chapters include a chapter appendix that shows the steps required to accomplish a statistical procedure using StatTools. For those students who want to make more extensive use of the software, StatTools offers an excellent Help facility. The StatTools Help system includes detailed explanations of the statistical and data analysis options available, as well as descriptions and definitions of the types of output provided.

Getting Started with StatTools

StatTools software may be downloaded and installed on your computer by accessing the website that accompanies this text. After downloading and installing the software, perform the following steps to use StatTools as an Excel add-in.

Step 1. Click the **Start** button on the taskbar and then point to **All Programs**
Step 2. Point to the folder entitled **Palisade Decision Tools**
Step 3. Click **StatTools for Excel**

FIGURE 1.10 THE STATTOOLS—DATA SET MANAGER DIALOG BOX

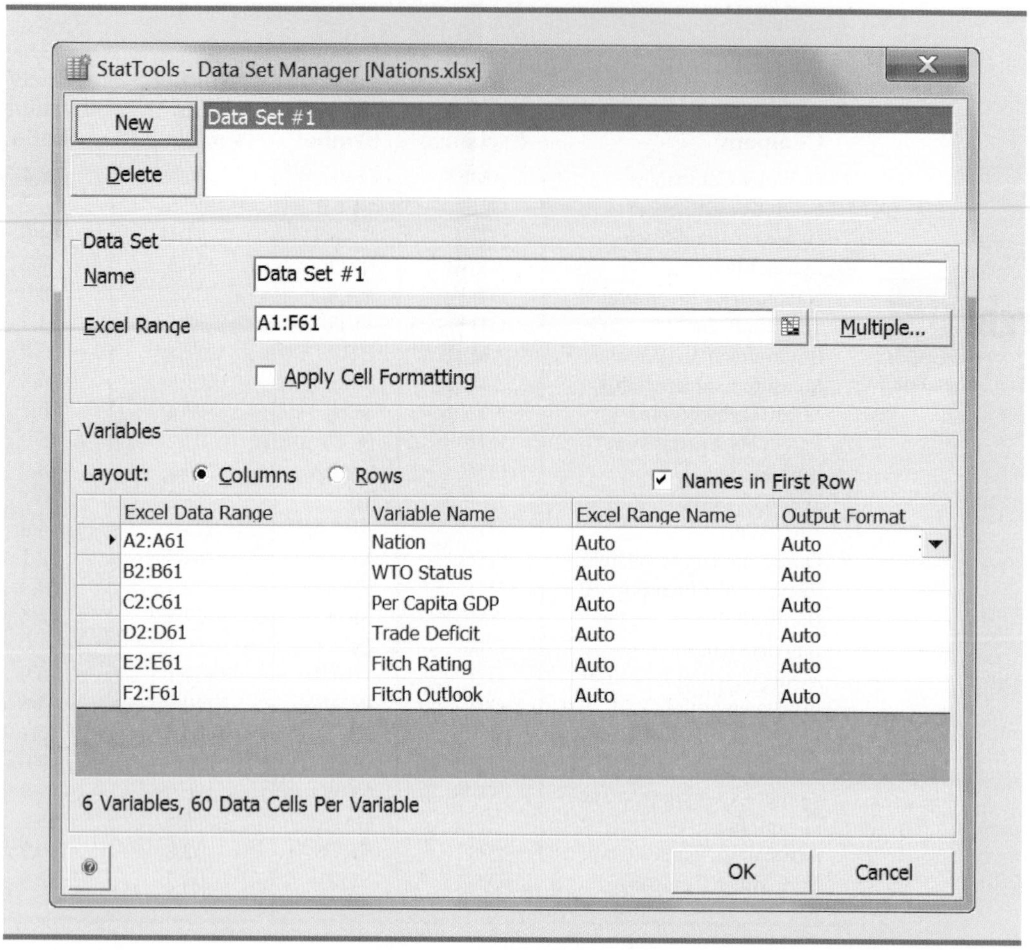

These steps will open Excel and add the StatTools tab next to the Add-Ins tab on the Excel Ribbon. Alternately, if you are already working in Excel, these steps will make StatTools available.

Using StatTools

WEB file
Nations

Before conducting any statistical analysis, we must create a StatTools data set using the StatTools Data Set Manager. Let us use the Excel worksheet for the 60 nations in the World Trade Organization data set in Table 1.1 to show how this is done. The following steps show how to create a StatTools data set for this application.

Step 1. Open the Excel file named Nations
Step 2. Select any cell in the data set (for example, cell A1)
Step 3. Click the **StatTools** tab on the Ribbon
Step 4. In the **Data** group, click **Data Set Manager**
Step 5. When StatTools asks if you want to add the range $A$1:$F$61 as a new StatTools data set, click **Yes**
Step 6. When the StatTools—Data Set Manager dialog box appears, click **OK**

Figure 1.10 shows the StatTools—Data Set Manager dialog box that appears in step 6. By default, the name of the new StatTools data set is Data Set #1. You can replace the name

FIGURE 1.11 THE STATTOOLS—APPLICATION SETTINGS DIALOG BOX

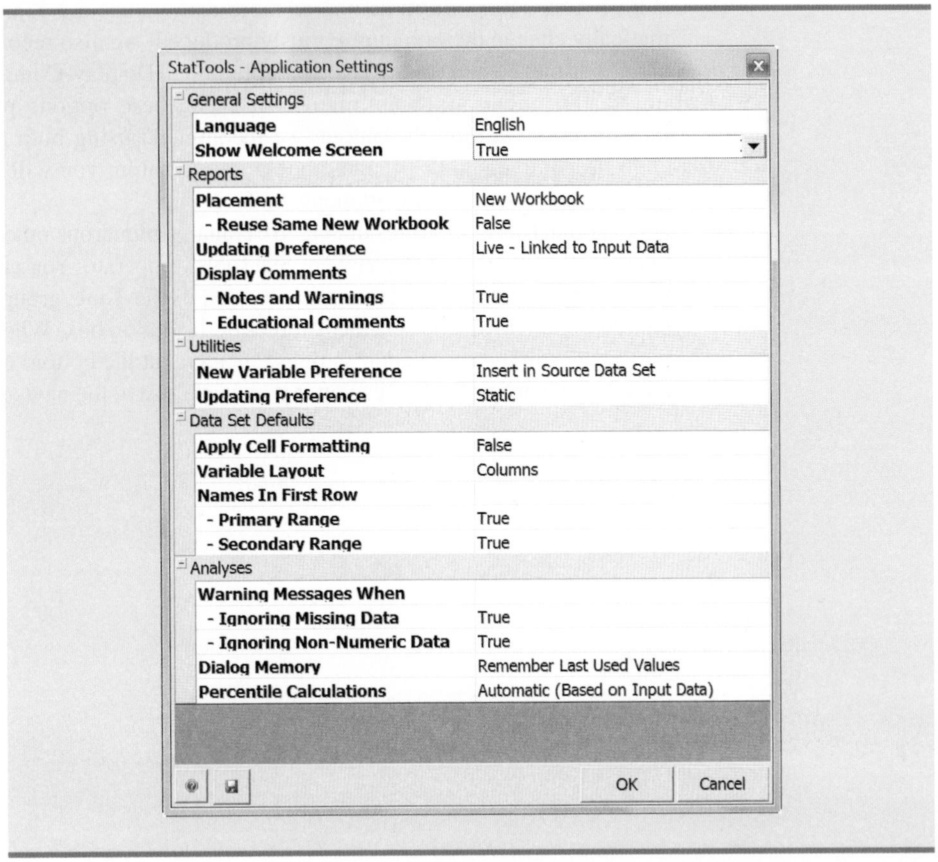

Data Set #1 in step 6 with a more descriptive name. And, if you select the Apply Cell Format option, the column labels will be highlighted in blue and the entire data set will have outside and inside borders. You can select the Data Set Manager at any time in your analysis to make these types of changes.

Recommended Application Settings

StatTools allows the user to specify some of the application settings that control such things as where statistical output is displayed and how calculations are performed. The following steps show how to access the StatTools—Application Settings dialog box.

Step 1. Click the **StatTools** tab on the Ribbon
Step 2. In the **Tools Group**, click **Utilities**
Step 3. Choose **Application Settings** from the list of options

Figure 1.11 shows that the StatTools—Application Settings dialog box has five sections: General Settings; Reports; Utilities; Data Set Defaults; and Analyses. Let us show how to make changes in the Reports section of the dialog box.

Figure 1.11 shows that the Placement option currently selected is **New Workbook**. Using this option, the StatTools output will be placed in a new workbook. But suppose you would like to place the StatTools output in the current (active) workbook. If you click the words **New Workbook**, a downward-pointing arrow will appear to the right. Clicking this arrow will display a list of all the placement options, including **Active Workbook**; we recommend using this option. Figure 1.11 also shows that the Updating

Preferences option in the Reports section is currently **Live—Linked to Input Data**. With live updating, anytime one or more data values are changed StatTools will automatically change the output previously produced; we also recommend using this option. Note that there are two options available under Display Comments: **Notes and Warnings** and **Educational Comments**. Because these options provide useful notes and information regarding the output, we recommend using both options. Thus, to include educational comments as part of the StatTools output you will have to change the value of False for Educational Comments to True.

The StatTools—Settings dialog box contains numerous other features that enable you to customize the way that you want StatTools to operate. You can learn more about these features by selecting the Help option located in the Tools group, or by clicking the Help icon located in the lower left-hand corner of the dialog box. When you have finished making changes in the application settings, click OK at the bottom of the dialog box and then click Yes when StatTools asks you if you want to save the new application settings.

CHAPTER 2

Descriptive Statistics: Tabular and Graphical Displays

CONTENTS

STATISTICS *in* PRACTICE

COLGATE-PALMOLIVE COMPANY*
NEW YORK, NEW YORK

The Colgate-Palmolive Company started as a small soap and candle shop in New York City in 1806. Today, Colgate-Palmolive employs more than 40,000 people working in more than 200 countries and territories around the world. Although best known for its brand names of Colgate, Palmolive, and Fab, the company also markets Mennen, Hill's Science Diet, and Hill's Prescription Diet products.

The Colgate-Palmolive Company uses statistics in its quality assurance program for home laundry detergent products. One concern is customer satisfaction with the quantity of detergent in a carton. Every carton in each size category is filled with the same amount of detergent by weight, but the volume of detergent is affected by the density of the detergent powder. For instance, if the powder density is on the heavy side, a smaller volume of detergent is needed to reach the carton's specified weight. As a result, the carton may appear to be underfilled when opened by the consumer.

To control the problem of heavy detergent powder, limits are placed on the acceptable range of powder density. Statistical samples are taken periodically, and the density of each powder sample is measured. Data summaries are then provided for operating personnel so that corrective action can be taken if necessary to keep the density within the desired quality specifications.

A frequency distribution for the densities of 150 samples taken over a one-week period and a histogram are shown in the accompanying table and figure. Density levels above .40 are unacceptably high. The frequency distribution and histogram show that the operation is meeting its quality guidelines with all of the densities less than or equal to .40. Managers viewing these statistical summaries would be pleased with the quality of the detergent production process.

In this chapter, you will learn about tabular and graphical methods of descriptive statistics such as frequency distributions, bar charts, histograms, stem-and-leaf displays, crosstabulations, and others. The goal of these methods is to summarize data so that the data can be easily understood and interpreted.

The Colgate-Palmolive Company uses statistical summaries to help maintain the quality of its products. © Kurt Brady/Alamy.

Frequency Distribution of Density Data

Density	Frequency
.29–.30	30
.31–.32	75
.33–.34	32
.35–.36	9
.37–.38	3
.39–.40	1
Total	150

Histogram of Density Data

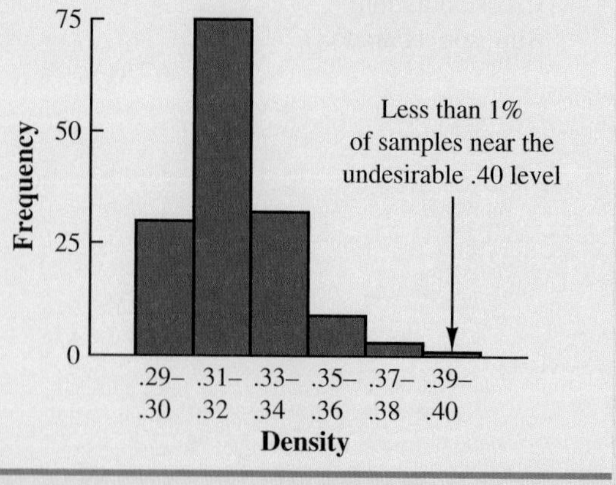

*The authors are indebted to William R. Fowle, Manager of Quality Assurance, Colgate-Palmolive Company, for providing this Statistics in Practice.

As indicated in Chapter 1, data can be classified as either categorical or quantitative. **Categorical data** use labels or names to identify categories of like items, and **quantitative data** are numerical values that indicate how much or how many. This chapter introduces the use of tabular and graphical displays for summarizing both categorical and quantitative data. Tabular and graphical displays can be found in annual reports, newspaper articles, and research studies. Everyone is exposed to these types of presentations. Hence, it is important to understand how they are constructed and how they should be interpreted.

We begin with a discussion of the use of tablular and graphical displays to summarize the data for a single variable. This is followed by a discussion of the use of tabular and graphical displays to summarize the data for two variables in a way that reveals the relationship between the two variables. **Data visualization** is a term often used to describe the use of graphical displays to summarize and present information about a data set. The last section of this chapter provides an introduction to data visualization and provides guidelines for creating effective graphical displays.

Modern statistical software packages provide extensive capabilities for summarizing data and preparing visual presentations. Minitab and Excel are two packages that are widely available. In the chapter appendixes, we show some of their capabilities.

2.1 Summarizing Data for a Categorical Variable

Frequency Distribution

We begin the discussion of how tabular and graphical displays can be used to summarize categorical data with the definition of a **frequency distribution**.

FREQUENCY DISTRIBUTION

A frequency distribution is a tabular summary of data showing the number (frequency) of observations in each of several nonoverlapping categories or classes.

Let us use the following example to demonstrate the construction and interpretation of a frequency distribution for categorical data. Coca-Cola, Diet Coke, Dr. Pepper, Pepsi, and Sprite are five popular soft drinks. Assume that the data in Table 2.1 show the soft drink selected in a sample of 50 soft drink purchases.

TABLE 2.1 DATA FROM A SAMPLE OF 50 SOFT DRINK PURCHASES

WEB file

SoftDrink

Coca-Cola	Coca-Cola	Coca-Cola	Sprite	Coca-Cola
Diet Coke	Dr. Pepper	Diet Coke	Dr. Pepper	Diet Coke
Pepsi	Sprite	Coca-Cola	Pepsi	Pepsi
Diet Coke	Coca-Cola	Sprite	Diet Coke	Pepsi
Coca-Cola	Diet Coke	Pepsi	Pepsi	Pepsi
Coca-Cola	Coca-Cola	Coca-Cola	Coca-Cola	Pepsi
Dr. Pepper	Coca-Cola	Coca-Cola	Coca-Cola	Coca-Cola
Diet Coke	Sprite	Coca-Cola	Coca-Cola	Dr. Pepper
Pepsi	Coca-Cola	Pepsi	Pepsi	Pepsi
Pepsi	Diet Coke	Coca-Cola	Dr. Pepper	Sprite

TABLE 2.2 FREQUENCY DISTRIBUTION OF SOFT DRINK PURCHASES

Soft Drink	Frequency
Coca-Cola	19
Diet Coke	8
Dr. Pepper	5
Pepsi	13
Sprite	5
Total	50

To develop a frequency distribution for these data, we count the number of times each soft drink appears in Table 2.1. Coca-Cola appears 19 times, Diet Coke appears 8 times, Dr. Pepper appears 5 times, Pepsi appears 13 times, and Sprite appears 5 times. These counts are summarized in the frequency distribution in Table 2.2.

This frequency distribution provides a summary of how the 50 soft drink purchases are distributed across the five soft drinks. This summary offers more insight than the original data shown in Table 2.1. Viewing the frequency distribution, we see that Coca-Cola is the leader, Pepsi is second, Diet Coke is third, and Sprite and Dr. Pepper are tied for fourth. The frequency distribution summarizes information about the popularity of the five soft drinks.

Relative Frequency and Percent Frequency Distributions

A frequency distribution shows the number (frequency) of observations in each of several nonoverlapping classes. However, we are often interested in the proportion, or percentage, of observations in each class. The *relative frequency* of a class equals the fraction or proportion of observations belonging to a class. For a data set with n observations, the relative frequency of each class can be determined as follows:

RELATIVE FREQUENCY

$$\text{Relative frequency of a class} = \frac{\text{Frequency of the class}}{n} \tag{2.1}$$

The *percent frequency* of a class is the relative frequency multiplied by 100.

A **relative frequency distribution** gives a tabular summary of data showing the relative frequency for each class. A **percent frequency distribution** summarizes the percent frequency of the data for each class. Table 2.3 shows a relative frequency distribution and a percent frequency distribution for the soft drink data. In Table 2.3 we see that the relative frequency for Coca-Cola is 19/50 = .38, the relative frequency for Diet Coke is 8/50 = .16, and so on. From the percent frequency distribution, we see that 38% of the purchases were Coca-Cola, 16% of the purchases were Diet Coke, and so on. We can also note that 38% + 26% + 16% = 80% of the purchases were for the top three soft drinks.

Bar Charts and Pie Charts

A **bar chart** is a graphical display for depicting categorical data summarized in a frequency, relative frequency, or percent frequency distribution. On one axis of the chart (usually the horizontal axis), we specify the labels that are used for the classes (categories). A frequency, relative frequency, or percent frequency scale can be used for the other axis of the chart

TABLE 2.3 RELATIVE FREQUENCY AND PERCENT FREQUENCY DISTRIBUTIONS OF SOFT DRINK PURCHASES

Soft Drink	Relative Frequency	Percent Frequency
Coca-Cola	.38	38
Diet Coke	.16	16
Dr. Pepper	.10	10
Pepsi	.26	26
Sprite	.10	10
Total	1.00	100

FIGURE 2.1 BAR CHART OF SOFT DRINK PURCHASES

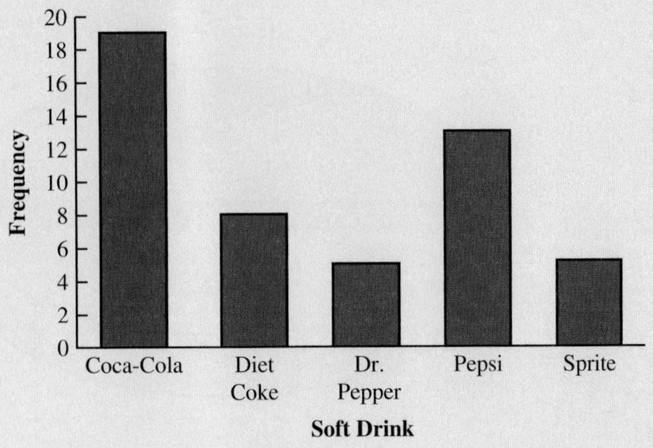

In quality control applications, bar charts are used to identify the most important causes of problems. When the bars are arranged in descending order of height from left to right with the most frequently occurring cause appearing first, the bar chart is called a pareto diagram. *This diagram is named for its founder, Vilfredo Pareto, an Italian economist.*

(usually the vertical axis). Then, using a bar of fixed width drawn above each class label, we extend the length of the bar until we reach the frequency, relative frequency, or percent frequency of the class. For categorical data, the bars should be separated to emphasize the fact that each category is separate. Figure 2.1 shows a bar chart of the frequency distribution for the 50 soft drink purchases. Note how the graphical display shows Coca-Cola, Pepsi, and Diet Coke to be the most preferred brands.

The **pie chart** provides another graphical display for presenting relative frequency and percent frequency distributions for categorical data. To construct a pie chart, we first draw a circle to represent all the data. Then we use the relative frequencies to subdivide the circle into sectors, or parts, that correspond to the relative frequency for each class. For example, because a circle contains 360 degrees and Coca-Cola shows a relative frequency of .38, the sector of the pie chart labeled Coca-Cola consists of .38(360) = 136.8 degrees. The sector of the pie chart labeled Diet Coke consists of .16(360) = 57.6 degrees. Similar calculations for the other classes yield the pie chart in Figure 2.2. The

FIGURE 2.2 PIE CHART OF SOFT DRINK PURCHASES

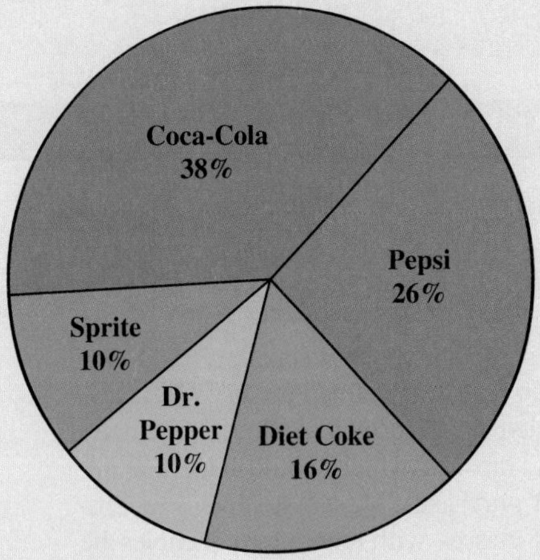

FIGURE 2.3 THREE-DIMENSIONAL PIE CHART OF SOFT DRINK PURCHASES

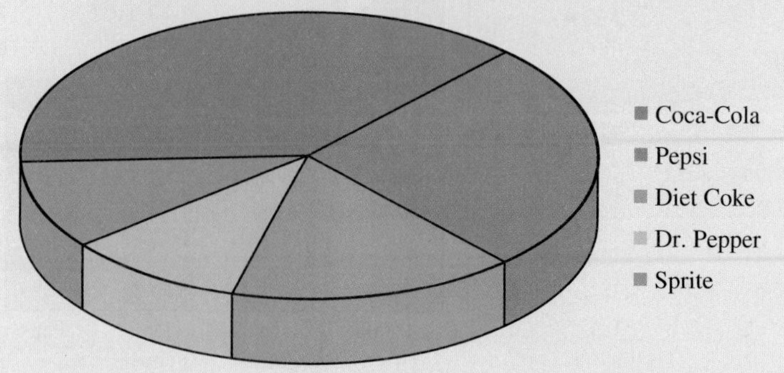

numerical values shown for each sector can be frequencies, relative frequencies, or percent frequencies.

Numerous options involving the use of colors, shading, legends, text font, and three-dimensional perspectives are available to enhance the visual appearance of bar and pie charts. When used carefully, such options can provide a more effective display. But, this is not always the case. For instance, consider the three-dimensional pie chart for the soft drink data shown in Figure 2.3. Compare it to the simpler presentation shown in Figure 2.2. The three-dimensional perspective adds no new understanding. In fact, because you have to view the three-dimensional pie chart in Figure 2.3 at an angle rather than straight overhead, it can be more difficult to visualize. The use of a legend in Figure 2.3 also forces your eyes to shift back and forth between the key and the chart. The simpler chart shown in Figure 2.2, which shows the percentages and classes directly on the pie, is more effective.

In general, pie charts are not the best way to present percentages for comparison. Research has shown that people are much better at accurately judging differences in length rather than differences in angles (or slices). When making such comparisons, we recommend you use a bar chart similar to Figure 2.1. In Section 2.5 we provide additional guidelines for creating effective visual displays.

NOTES AND COMMENTS

1. Often the number of classes in a frequency distribution is the same as the number of categories found in the data, as is the case for the soft drink purchase data in this section. The data involve only five soft drinks, and a separate frequency distribution class was defined for each one. Data that included all soft drinks would require many categories, most of which would have a small number of purchases. Most statisticians recommend that classes with smaller frequencies be grouped into an aggregate class called "other." Classes with frequencies of 5% or less would most often be treated in this fashion.
2. The sum of the frequencies in any frequency distribution always equals the number of observations. The sum of the relative frequencies in any relative frequency distribution always equals 1.00, and the sum of the percentages in a percent frequency distribution always equals 100.

Exercises

Methods

1. The response to a question has three alternatives: A, B, and C. A sample of 120 responses provides 60 A, 24 B, and 36 C. Show the frequency and relative frequency distributions.
2. A partial relative frequency distribution is given.

Class	Relative Frequency
A	.22
B	.18
C	.40
D	

 a. What is the relative frequency of class D?
 b. The total sample size is 200. What is the frequency of class D?
 c. Show the frequency distribution.
 d. Show the percent frequency distribution.

SELF test

3. A questionnaire provides 58 Yes, 42 No, and 20 no-opinion answers.
 a. In the construction of a pie chart, how many degrees would be in the section of the pie showing the Yes answers?
 b. How many degrees would be in the section of the pie showing the No answers?
 c. Construct a pie chart.
 d. Construct a bar chart.

Applications

WEB file

Syndicated

4. For the 2010–2011 viewing season, the top five syndicated programs were *Wheel of Fortune* (WoF), *Two and Half Men* (THM), *Jeopardy* (Jep), *Judge Judy* (JJ), and the *Oprah Winfrey Show* (OWS) (Nielsen Media Research website, April 16, 2012). Data indicating the preferred shows for a sample of 50 viewers follow.

WoF	Jep	JJ	Jep	THM
THM	WoF	OWS	Jep	THM
Jep	OWS	WoF	WoF	WoF
WoF	THM	OWS	THM	WoF
THM	JJ	JJ	Jep	THM
OWS	OWS	JJ	JJ	Jep
JJ	WoF	THM	WoF	WoF
THM	THM	WoF	JJ	JJ
Jep	THM	WoF	Jep	Jep
WoF	THM	OWS	OWS	Jep

 a. Are these data categorical or quantitative?
 b. Provide frequency and percent frequency distributions.
 c. Construct a bar chart and a pie chart.
 d. On the basis of the sample, which television show has the largest viewing audience? Which one is second?

5. In alphabetical order, the six most common last names in the United States are Brown, Johnson, Jones, Miller, Smith, and Williams (*The World Almanac,* 2012). Assume that a sample of 50 individuals with one of these last names provided the following data.

WEB file 2012Names

Brown	Williams	Williams	Williams	Brown
Smith	Jones	Smith	Johnson	Smith
Miller	Smith	Brown	Williams	Johnson
Johnson	Smith	Smith	Johnson	Brown
Williams	Miller	Johnson	Williams	Johnson
Williams	Johnson	Jones	Smith	Brown
Johnson	Smith	Smith	Brown	Jones
Jones	Jones	Smith	Smith	Miller
Miller	Jones	Williams	Miller	Smith
Jones	Johnson	Brown	Johnson	Miller

Summarize the data by constructing the following:

a. Relative and percent frequency distributions
b. A bar chart
c. A pie chart
d. Based on these data, what are the three most common last names?

6. Nielsen Media Research provided the list of the 25 top-rated single shows in television history (*The World Almanac,* 2012). The following data show the television network that produced each of these 25 top-rated shows.

WEB file 2012Networks

CBS	CBS	NBC	FOX	CBS
CBS	NBC	NBC	NBC	ABC
ABC	NBC	ABC	ABC	NBC
CBS	NBC	CBS	ABC	NBC
NBC	CBS	CBS	ABC	CBS

a. Construct a frequency distribution, percent frequency distribution, and bar chart for the data.
b. Which network or networks have done the best in terms of presenting top-rated television shows? Compare the performance of ABC, CBS, and NBC.

SELF test

7. The Canmark Research Center Airport Customer Satisfaction Survey uses an online questionnaire to provide airlines and airports with customer satisfaction ratings for all aspects of the customers' flight experience (airportsurvey website, July, 2012). After completing a flight, customers receive an e-mail asking them to go to the website and rate a variety of factors, including the reservation process, the check-in process, luggage policy, cleanliness of gate area, service by flight attendants, food/beverage selection, on-time arrival, and so on. A five-point scale, with Excellent (E), Very Good (V), Good (G), Fair (F), and Poor (P), is used to record customer ratings. Assume that passengers on a Delta Airlines flight from Myrtle Beach, South Carolina, to Atlanta, Georgia, provided the following ratings for the question, "Please rate the airline based on your overall experience with this flight." The sample ratings are shown below.

WEB file AirSurvey

E	E	G	V	V	E	V	V	V	E
E	G	V	E	E	V	E	E	E	V
V	V	V	F	V	E	V	E	G	E
G	E	V	E	V	E	V	V	V	V
E	E	V	V	E	P	E	V	P	V

a. Use a percent frequency distribution and a bar chart to summarize these data. What do these summaries indicate about the overall customer satisfaction with the Delta flight?
b. The online survey questionnaire enabled respondents to explain any aspect of the flight that failed to meet expectations. Would this be helpful information to a manager looking for ways to improve the overall customer satisfaction on Delta flights? Explain.

8. Data for a sample of 55 members of the Baseball Hall of Fame in Cooperstown, New York, are shown here. Each observation indicates the primary position played by the Hall of Famers: pitcher (P), catcher (H), 1st base (1), 2nd base (2), 3rd base (3), shortstop (S), left field (L), center field (C), and right field (R).

WEB file
BaseballHall

L	P	C	H	2	P	R	1	S	S	1	L	P	R	P
P	P	P	R	C	S	L	R	P	C	C	P	P	R	P
2	3	P	H	L	P	1	C	P	P	P	S	1	L	R
R	1	2	H	S	3	H	2	L	P					

a. Construct frequency and relative frequency distributions to summarize the data.
b. What position provides the most Hall of Famers?
c. What position provides the fewest Hall of Famers?
d. What outfield position (L, C, or R) provides the most Hall of Famers?
e. Compare infielders (1, 2, 3, and S) to outfielders (L, C, and R).

9. The Pew Research Center's Social & Demographic Trends project found that 46% of U.S. adults would rather live in a different type of community than the one where they are living now (Pew Research Center, January 29, 2009). The national survey of 2260 adults asked: "Where do you live now?" and "What do you consider to be the ideal community?" Response options were City (C), Suburb (S), Small Town (T), or Rural (R). A representative portion of this survey for a sample of 100 respondents is as follows.

WEB file
LivingArea

Where do you live now?

S	T	R	C	R	R	T	C	S	T	C	S	C	S	T
S	S	C	S	S	T	T	C	C	S	T	C	S	T	C
T	R	S	S	T	C	S	C	T	C	T	C	T	C	R
C	C	R	T	C	S	S	T	S	C	C	C	R	S	C
S	S	C	C	S	C	R	T	T	T	C	R	T	C	R
C	T	R	R	C	T	C	C	R	T	T	R	S	R	T
T	S	S	S	S	S	C	C	R	T					

What do you consider to be the ideal community?

S	C	R	R	R	S	T	S	S	T	T	S	C	S	T
C	C	R	T	R	S	T	T	S	S	C	C	T	T	S
S	R	C	S	C	C	S	C	R	C	T	S	R	R	R
C	T	S	T	T	T	R	R	S	C	C	R	R	S	S
S	T	C	T	T	C	R	T	T	T	C	T	T	R	R
C	S	R	T	C	T	C	C	T	T	T	R	C	R	T
T	C	S	S	C	S	T	S	S	R					

a. Provide a percent frequency distribution for each question.
b. Construct a bar chart for each question.
c. Where are most adults living now?
d. Where do most adults consider the ideal community?
e. What changes in living areas would you expect to see if people moved from where they currently live to their ideal community?

10. The *Financial Times*/Harris Poll is a monthly online poll of adults from six countries in Europe and the United States. The poll conducted in January 2008 included 1015 adults. One of the questions asked was, "How would you rate the Federal Bank in handling the credit problems in the financial markets?" Possible responses were Excellent, Good, Fair, Bad, and Terrible (Harris Interactive website, January 2008). The 1015 responses for this question can be found in the data file named FedBank.

WEB file
FedBank

a. Construct a frequency distribution.
b. Construct a percent frequency distribution.
c. Construct a bar chart for the percent frequency distribution.
d. Comment on how adults in the United States think the Federal Bank is handling the credit problems in the financial markets.

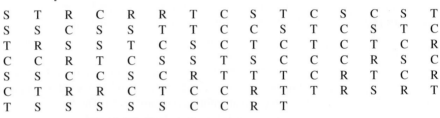

e. In Spain, 1114 adults were asked, "How would you rate the European Central Bank in handling the credit problems in the financial markets?" The percent frequency distribution obtained follows:

Rating	Percent Frequency
Excellent	0
Good	4
Fair	46
Bad	40
Terrible	10

Compare the results obtained in Spain with the results obtained in the United States.

2.2 Summarizing Data for a Quantitative Variable

Frequency Distribution

TABLE 2.4 YEAR-END AUDIT TIMES (IN DAYS)

12	14	19	18
15	15	18	17
20	27	22	23
22	21	33	28
14	18	16	13

WEB file

Audit

As defined in Section 2.1, a frequency distribution is a tabular summary of data showing the number (frequency) of observations in each of several nonoverlapping categories or classes. This definition holds for quantitative as well as categorical data. However, with quantitative data we must be more careful in defining the nonoverlapping classes to be used in the frequency distribution.

For example, consider the quantitative data in Table 2.4. These data show the time in days required to complete year-end audits for a sample of 20 clients of Sanderson and Clifford, a small public accounting firm. The three steps necessary to define the classes for a frequency distribution with quantitative data are

1. Determine the number of nonoverlapping classes.
2. Determine the width of each class.
3. Determine the class limits.

Let us demonstrate these steps by developing a frequency distribution for the audit time data in Table 2.4.

Number of classes Classes are formed by specifying ranges that will be used to group the data. As a general guideline, we recommend using between 5 and 20 classes. For a small number of data items, as few as five or six classes may be used to summarize the data. For a larger number of data items, a larger number of classes is usually required. The goal is to use enough classes to show the variation in the data, but not so many classes that some contain only a few data items. Because the number of data items in Table 2.4 is relatively small ($n = 20$), we chose to develop a frequency distribution with five classes.

Making the classes the same width reduces the chance of inappropriate interpretations by the user.

Width of the classes The second step in constructing a frequency distribution for quantitative data is to choose a width for the classes. As a general guideline, we recommend that the width be the same for each class. Thus the choices of the number of classes and the width of classes are not independent decisions. A larger number of classes means a smaller class width, and vice versa. To determine an approximate class width, we begin by identifying the largest and smallest data values. Then, with the desired number of classes specified, we can use the following expression to determine the approximate class width.

$$\text{Approximate class width} = \frac{\text{Largest data value} - \text{Smallest data value}}{\text{Number of classes}} \quad \textbf{(2.2)}$$

The approximate class width given by equation (2.2) can be rounded to a more convenient value based on the preference of the person developing the frequency distribution. For example, an approximate class width of 9.28 might be rounded to 10 simply because 10 is a more convenient class width to use in presenting a frequency distribution.

For the data involving the year-end audit times, the largest data value is 33 and the smallest data value is 12. Because we decided to summarize the data with five classes, using equation (2.2) provides an approximate class width of $(33 - 12)/5 = 4.2$. We therefore decided to round up and use a class width of five days in the frequency distribution.

No single frequency distribution is best for a data set. Different people may construct different, but equally acceptable, frequency distributions. The goal is to reveal the natural grouping and variation in the data.

In practice, the number of classes and the appropriate class width are determined by trial and error. Once a possible number of classes is chosen, equation (2.2) is used to find the approximate class width. The process can be repeated for a different number of classes. Ultimately, the analyst uses judgment to determine the combination of the number of classes and class width that provides the best frequency distribution for summarizing the data.

For the audit time data in Table 2.4, after deciding to use five classes, each with a width of five days, the next task is to specify the class limits for each of the classes.

Class limits Class limits must be chosen so that each data item belongs to one and only one class. The *lower class limit* identifies the smallest possible data value assigned to the class. The *upper class limit* identifies the largest possible data value assigned to the class. In developing frequency distributions for categorical data, we did not need to specify class limits because each data item naturally fell into a separate class. But with quantitative data, such as the audit times in Table 2.4, class limits are necessary to determine where each data value belongs.

TABLE 2.5
FREQUENCY DISTRIBUTION FOR THE AUDIT TIME DATA

Audit Time (days)	Frequency
10–14	4
15–19	8
20–24	5
25–29	2
30–34	1
Total	20

Using the audit time data in Table 2.4, we selected 10 days as the lower class limit and 14 days as the upper class limit for the first class. This class is denoted 10–14 in Table 2.5. The smallest data value, 12, is included in the 10–14 class. We then selected 15 days as the lower class limit and 19 days as the upper class limit of the next class. We continued defining the lower and upper class limits to obtain a total of five classes: 10–14, 15–19, 20–24, 25–29, and 30–34. The largest data value, 33, is included in the 30–34 class. The difference between the lower class limits of adjacent classes is the class width. Using the first two lower class limits of 10 and 15, we see that the class width is $15 - 10 = 5$.

With the number of classes, class width, and class limits determined, a frequency distribution can be obtained by counting the number of data values belonging to each class. For example, the data in Table 2.4 show that four values—12, 14, 14, and 13—belong to the 10–14 class. Thus, the frequency for the 10–14 class is 4. Continuing this counting process for the 15–19, 20–24, 25–29, and 30–34 classes provides the frequency distribution in Table 2.5. Using this frequency distribution, we can observe the following:

1. The most frequently occurring audit times are in the class of 15–19 days. Eight of the 20 audit times belong to this class.
2. Only one audit required 30 or more days.

Other conclusions are possible, depending on the interests of the person viewing the frequency distribution. The value of a frequency distribution is that it provides insights about the data that are not easily obtained by viewing the data in their original unorganized form.

Class midpoint In some applications, we want to know the midpoints of the classes in a frequency distribution for quantitative data. The **class midpoint** is the value halfway between the lower and upper class limits. For the audit time data, the five class midpoints are 12, 17, 22, 27, and 32.

Relative Frequency and Percent Frequency Distributions

We define the relative frequency and percent frequency distributions for quantitative data in the same manner as for categorical data. First, recall that the relative frequency is the proportion of

TABLE 2.6 RELATIVE FREQUENCY AND PERCENT FREQUENCY DISTRIBUTIONS FOR THE AUDIT TIME DATA

Audit Time (days)	Relative Frequency	Percent Frequency
10–14	.20	20
15–19	.40	40
20–24	.25	25
25–29	.10	10
30–34	.05	5
Total	1.00	100

the observations belonging to a class. With n observations,

$$\text{Relative frequency of class} = \frac{\text{Frequency of the class}}{n}$$

The percent frequency of a class is the relative frequency multiplied by 100.

Based on the class frequencies in Table 2.5 and with $n = 20$, Table 2.6 shows the relative frequency distribution and percent frequency distribution for the audit time data. Note that .40 of the audits, or 40%, required from 15 to 19 days. Only .05 of the audits, or 5%, required 30 or more days. Again, additional interpretations and insights can be obtained by using Table 2.6.

Dot Plot

One of the simplest graphical summaries of data is a **dot plot**. A horizontal axis shows the range for the data. Each data value is represented by a dot placed above the axis. Figure 2.4 is the dot plot for the audit time data in Table 2.4. The three dots located above 18 on the horizontal axis indicate that an audit time of 18 days occurred three times. Dot plots show the details of the data and are useful for comparing the distribution of the data for two or more variables.

Histogram

A common graphical display of quantitative data is a **histogram**. This graphical display can be prepared for data previously summarized in either a frequency, relative frequency, or percent frequency distribution. A histogram is constructed by placing the variable of interest on the horizontal axis and the frequency, relative frequency, or percent frequency on the vertical axis. The frequency, relative frequency, or percent frequency of each class is shown by drawing a rectangle whose base is determined by the class limits on the horizontal axis and whose height is the corresponding frequency, relative frequency, or percent frequency.

FIGURE 2.4 DOT PLOT FOR THE AUDIT TIME DATA

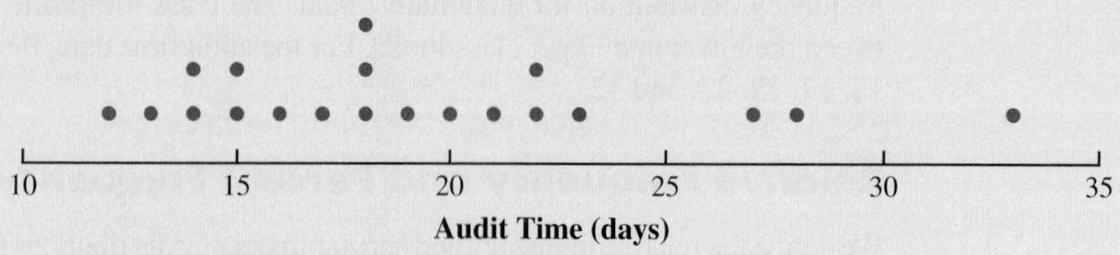

FIGURE 2.5 HISTOGRAM FOR THE AUDIT TIME DATA

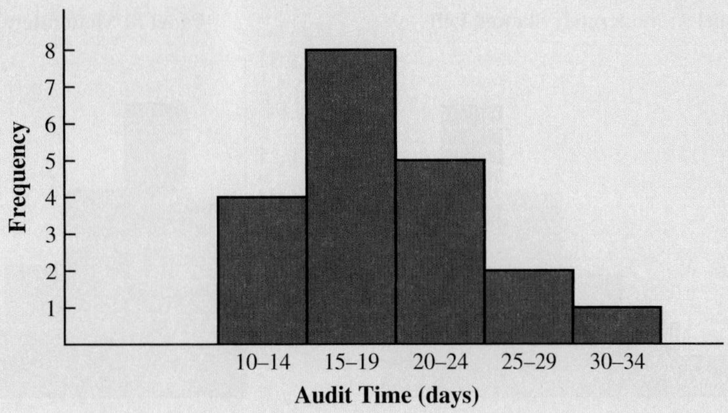

Figure 2.5 is a histogram for the audit time data. Note that the class with the greatest frequency is shown by the rectangle appearing above the class of 15–19 days. The height of the rectangle shows that the frequency of this class is 8. A histogram for the relative or percent frequency distribution of these data would look the same as the histogram in Figure 2.5 with the exception that the vertical axis would be labeled with relative or percent frequency values.

As Figure 2.5 shows, the adjacent rectangles of a histogram touch one another. Unlike a bar graph, a histogram contains no natural separation between the rectangles of adjacent classes. This format is the usual convention for histograms. Because the classes for the audit time data are stated as 10–14, 15–19, 20–24, 25–29, and 30–34, one-unit spaces of 14 to 15, 19 to 20, 24 to 25, and 29 to 30 would seem to be needed between the classes. These spaces are eliminated when constructing a histogram. Eliminating the spaces between classes in a histogram for the audit time data helps show that all values between the lower limit of the first class and the upper limit of the last class are possible.

One of the most important uses of a histogram is to provide information about the shape, or form, of a distribution. Figure 2.6 contains four histograms constructed from relative frequency distributions. Panel A shows the histogram for a set of data moderately skewed to the left. A histogram is said to be skewed to the left if its tail extends farther to the left. This histogram is typical for exam scores, with no scores above 100%, most of the scores above 70%, and only a few really low scores. Panel B shows the histogram for a set of data moderately skewed to the right. A histogram is said to be skewed to the right if its tail extends farther to the right. An example of this type of histogram would be for data such as housing prices; a few expensive houses create the skewness in the right tail.

Panel C shows a symmetric histogram. In a symmetric histogram, the left tail mirrors the shape of the right tail. Histograms for data found in applications are never perfectly symmetric, but the histogram for many applications may be roughly symmetric. Data for SAT scores, heights and weights of people, and so on lead to histograms that are roughly symmetric. Panel D shows a histogram highly skewed to the right. This histogram was constructed from data on the amount of customer purchases over one day at a women's apparel store. Data from applications in business and economics often lead to histograms that are skewed to the right. For instance, data on housing prices, salaries, purchase amounts, and so on often result in histograms skewed to the right.

FIGURE 2.6 HISTOGRAMS SHOWING DIFFERING LEVELS OF SKEWNESS

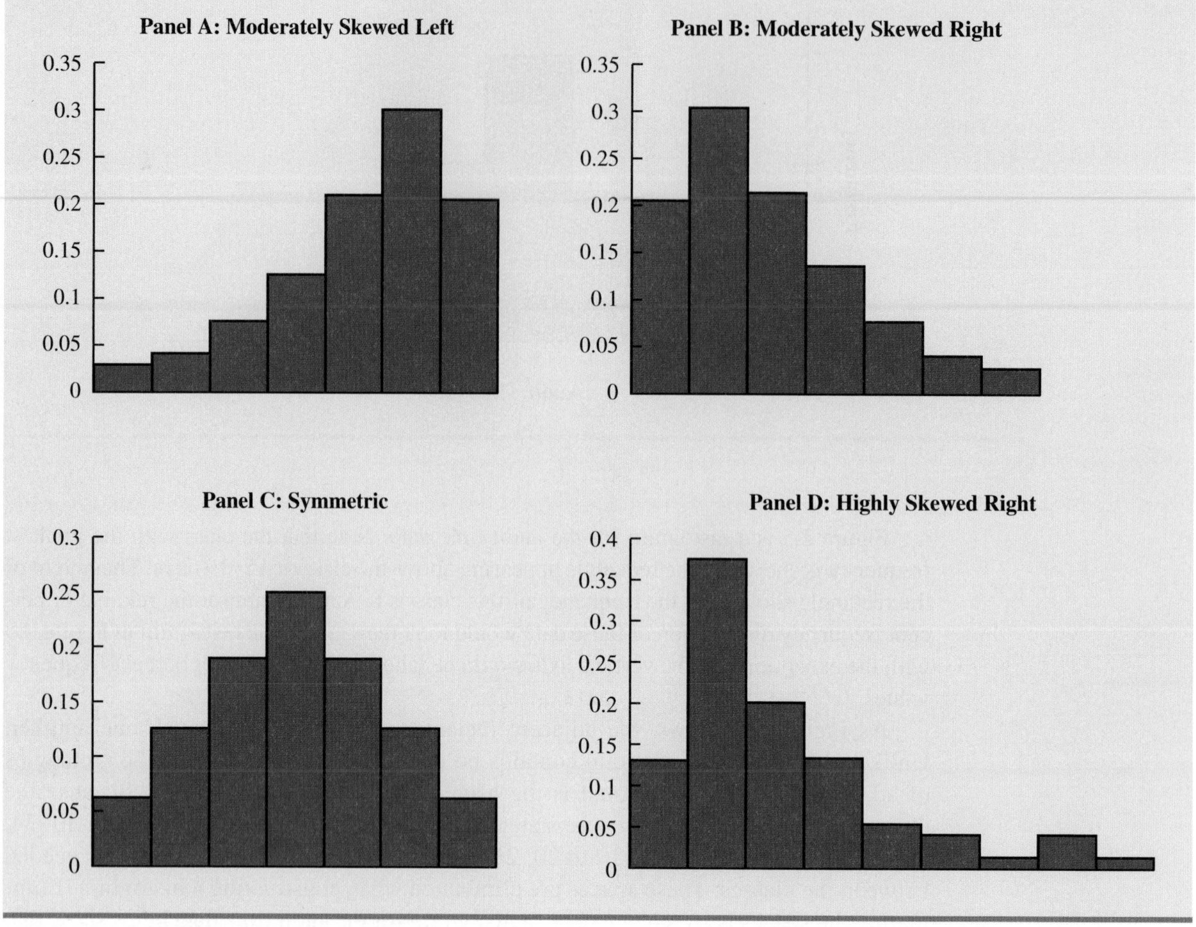

Cumulative Distributions

A variation of the frequency distribution that provides another tabular summary of quantitative data is the **cumulative frequency distribution**. The cumulative frequency distribution uses the number of classes, class widths, and class limits developed for the frequency distribution. However, rather than showing the frequency of each class, the cumulative frequency distribution shows the number of data items with values *less than or equal to the upper class limit* of each class. The first two columns of Table 2.7 provide the cumulative frequency distribution for the audit time data.

To understand how the cumulative frequencies are determined, consider the class with the description "less than or equal to 24." The cumulative frequency for this class is simply the sum of the frequencies for all classes with data values less than or equal to 24. For the frequency distribution in Table 2.5, the sum of the frequencies for classes 10–14, 15–19, and 20–24 indicates that $4 + 8 + 5 = 17$ data values are less than or equal to 24. Hence, the cumulative frequency for this class is 17. In addition, the cumulative frequency distribution in Table 2.7 shows that four audits were completed in 14 days or less and 19 audits were completed in 29 days or less.

TABLE 2.7 CUMULATIVE FREQUENCY, CUMULATIVE RELATIVE FREQUENCY, AND CUMULATIVE PERCENT FREQUENCY DISTRIBUTIONS FOR THE AUDIT TIME DATA

Audit Time (days)	Cumulative Frequency	Cumulative Relative Frequency	Cumulative Percent Frequency
Less than or equal to 14	4	.20	20
Less than or equal to 19	12	.60	60
Less than or equal to 24	17	.85	85
Less than or equal to 29	19	.95	95
Less than or equal to 34	20	1.00	100

As a final point, we note that a **cumulative relative frequency distribution** shows the proportion of data items, and a **cumulative percent frequency distribution** shows the percentage of data items with values less than or equal to the upper limit of each class. The cumulative relative frequency distribution can be computed either by summing the relative frequencies in the relative frequency distribution or by dividing the cumulative frequencies by the total number of items. Using the latter approach, we found the cumulative relative frequencies in column 3 of Table 2.7 by dividing the cumulative frequencies in column 2 by the total number of items ($n = 20$). The cumulative percent frequencies were again computed by multiplying the relative frequencies by 100. The cumulative relative and percent frequency distributions show that .85 of the audits, or 85%, were completed in 24 days or less, .95 of the audits, or 95%, were completed in 29 days or less, and so on.

Stem-and-Leaf Display

A **stem-and-leaf display** is a graphical display used to show simultaneously the rank order and shape of a distribution of data. To illustrate the use of a stem-and-leaf display, consider the data in Table 2.8. These data result from a 150-question aptitude test given to 50 individuals recently interviewed for a position at Haskens Manufacturing. The data indicate the number of questions answered correctly.

To develop a stem-and-leaf display, we first arrange the leading digits of each data value to the left of a vertical line. To the right of the vertical line, we record the last digit for each data value. Based on the top row of data in Table 2.8 (112, 72, 69, 97, and 107), the first

TABLE 2.8 NUMBER OF QUESTIONS ANSWERED CORRECTLY ON AN APTITUDE TEST

WEB file ApTest

112	72	69	97	107
73	92	76	86	73
126	128	118	127	124
82	104	132	134	83
92	108	96	100	92
115	76	91	102	81
95	141	81	80	106
84	119	113	98	75
68	98	115	106	95
100	85	94	106	119

five entries in constructing a stem-and-leaf display would be as follows:

```
 6 | 9
 7 | 2
 8 |
 9 | 7
10 | 7
11 | 2
12 |
13 |
14 |
```

For example, the data value 112 shows the leading digits 11 to the left of the line and the last digit 2 to the right of the line. Similarly, the data value 72 shows the leading digit 7 to the left of the line and last digit 2 to the right of the line. Continuing to place the last digit of each data value on the line corresponding to its leading digit(s) provides the following:

```
 6 | 9  8
 7 | 2  3  6  3  6  5
 8 | 6  2  3  1  1  0  4  5
 9 | 7  2  2  6  2  1  5  8  8  5  4
10 | 7  4  8  0  2  6  6  0  6
11 | 2  8  5  9  3  5  9
12 | 6  8  7  4
13 | 2  4
14 | 1
```

With this organization of the data, sorting the digits on each line into rank order is simple. Doing so provides the stem-and-leaf display shown here.

```
 6 | 8  9
 7 | 2  3  3  5  6  6
 8 | 0  1  1  2  3  4  5  6
 9 | 1  2  2  2  4  5  5  6  7  8  8
10 | 0  0  2  4  6  6  6  7  8
11 | 2  3  5  5  8  9  9
12 | 4  6  7  8
13 | 2  4
14 | 1
```

The numbers to the left of the vertical line (6, 7, 8, 9, 10, 11, 12, 13, and 14) form the *stem*, and each digit to the right of the vertical line is a *leaf*. For example, consider the first row with a stem value of 6 and leaves of 8 and 9.

```
6 | 8  9
```

This row indicates that two data values have a first digit of six. The leaves show that the data values are 68 and 69. Similarly, the second row

```
7 | 2  3  3  5  6  6
```

indicates that six data values have a first digit of seven. The leaves show that the data values are 72, 73, 73, 75, 76, and 76.

To focus on the shape indicated by the stem-and-leaf display, let us use a rectangle to contain the leaves of each stem. Doing so, we obtain the following:

6	8 9
7	2 3 3 5 6 6
8	0 1 1 2 3 4 5 6
9	1 2 2 2 4 5 5 6 7 8 8
10	0 0 2 4 6 6 6 7 8
11	2 3 5 5 8 9 9
12	4 6 7 8
13	2 4
14	1

Rotating this page counterclockwise onto its side provides a picture of the data that is similar to a histogram with classes of 60–69, 70–79, 80–89, and so on.

Although the stem-and-leaf display may appear to offer the same information as a histogram, it has two primary advantages.

1. The stem-and-leaf display is easier to construct by hand.
2. Within a class interval, the stem-and-leaf display provides more information than the histogram because the stem-and-leaf shows the actual data.

Just as a frequency distribution or histogram has no absolute number of classes, neither does a stem-and-leaf display have an absolute number of rows or stems. If we believe that our original stem-and-leaf display condensed the data too much, we can easily stretch the display by using two or more stems for each leading digit. For example, to use two stems for each leading digit, we would place all data values ending in 0, 1, 2, 3, and 4 in one row and all values ending in 5, 6, 7, 8, and 9 in a second row. The following stretched stem-and-leaf display illustrates this approach.

In a stretched stem-and-leaf display, whenever a stem value is stated twice, the first value corresponds to leaf values of 0–4, and the second value corresponds to leaf values of 5–9.

6	8 9
7	2 3 3
7	5 6 6
8	0 1 1 2 3 4
8	5 6
9	1 2 2 2 4
9	5 5 6 7 8 8
10	0 0 2 4
10	6 6 6 7 8
11	2 3
11	5 5 8 9 9
12	4
12	6 7 8
13	2 4
13	
14	1

Note that values 72, 73, and 73 have leaves in the 0–4 range and are shown with the first stem value of 7. The values 75, 76, and 76 have leaves in the 5–9 range and are shown with the second stem value of 7. This stretched stem-and-leaf display is similar to a frequency distribution with intervals of 65–69, 70–74, 75–79, and so on.

The preceding example showed a stem-and-leaf display for data with as many as three digits. Stem-and-leaf displays for data with more than three digits are possible. For example, consider the following data on the number of hamburgers sold by a fast-food restaurant for each of 15 weeks.

1565	1852	1644	1766	1888	1912	2044	1812
1790	1679	2008	1852	1967	1954	1733	

A stem-and-leaf display of these data follows.

Leaf unit = 10

15	6
16	4 7
17	3 6 9
18	1 5 5 8
19	1 5 6
20	0 4

A single digit is used to define each leaf in a stem-and-leaf display. The leaf unit indicates how to multiply the stem-and-leaf numbers in order to approximate the original data. Leaf units may be 100, 10, 1, 0.1, and so on.

Note that a single digit is used to define each leaf and that only the first three digits of each data value have been used to construct the display. At the top of the display we have specified Leaf unit = 10. To illustrate how to interpret the values in the display, consider the first stem, 15, and its associated leaf, 6. Combining these numbers, we obtain 156. To reconstruct an approximation of the original data value, we must multiply this number by 10, the value of the *leaf unit.* Thus, $156 \times 10 = 1560$ is an approximation of the original data value used to construct the stem-and-leaf display. Although it is not possible to reconstruct the exact data value from this stem-and-leaf display, the convention of using a single digit for each leaf enables stem-and-leaf displays to be constructed for data having a large number of digits. For stem-and-leaf displays where the leaf unit is not shown, the leaf unit is assumed to equal 1.

NOTES AND COMMENTS

1. A bar chart and a histogram are essentially the same thing; both are graphical presentations of the data in a frequency distribution. A histogram is just a bar chart with no separation between bars. For some discrete quantitative data, a separation between bars is also appropriate. Consider, for example, the number of classes in which a college student is enrolled. The data may only assume integer values. Intermediate values such as 1.5, 2.73, and so on are not possible. With continuous quantitative data, however, such as the audit times in Table 2.4, a separation between bars is not appropriate.
2. The appropriate values for the class limits with quantitative data depend on the level of accuracy of the data. For instance, with the audit time data of Table 2.4 the limits used were integer values. If the data were rounded to the nearest tenth of a day (e.g., 12.3, 14.4, and so on), then the limits would be stated in tenths of days. For instance, the first class would be 10.0–14.9. If the data were recorded to the nearest hundredth of a day (e.g., 12.34, 14.45, and so on), the limits would be stated in hundredths of days. For instance, the first class would be 10.00–14.99.
3. An *open-end* class requires only a lower class limit or an upper class limit. For example, in the audit time data of Table 2.4, suppose two of the audits had taken 58 and 65 days. Rather than continue with the classes of width 5 with classes 35–39, 40–44, 45–49, and so on, we could simplify the frequency distribution to show an open-end class of "35 or more." This class would have a frequency of 2. Most often the open-end class appears at the upper end of the distribution. Sometimes an open-end class appears at the lower end of the distribution, and occasionally such classes appear at both ends.
4. The last entry in a cumulative frequency distribution always equals the total number of observations. The last entry in a cumulative relative frequency distribution always equals 1.00 and the last entry in a cumulative percent frequency distribution always equals 100.

Exercises

Methods

11. Consider the following data.

WEB file
Frequency

14	21	23	21	16
19	22	25	16	16
24	24	25	19	16
19	18	19	21	12
16	17	18	23	25
20	23	16	20	19
24	26	15	22	24
20	22	24	22	20

a. Develop a frequency distribution using classes of 12–14, 15–17, 18–20, 21–23, and 24–26.
b. Develop a relative frequency distribution and a percent frequency distribution using the classes in part (a).

SELF test

12. Consider the following frequency distribution.

Class	Frequency
10–19	10
20–29	14
30–39	17
40–49	7
50–59	2

Construct a cumulative frequency distribution and a cumulative relative frequency distribution.

13. Construct a histogram for the data in exercise 12.
14. Consider the following data.

8.9	10.2	11.5	7.8	10.0	12.2	13.5	14.1	10.0	12.2
6.8	9.5	11.5	11.2	14.9	7.5	10.0	6.0	15.8	11.5

a. Construct a dot plot.
b. Construct a frequency distribution.
c. Construct a percent frequency distribution.

SELF test

15. Construct a stem-and-leaf display for the following data.

11.3	9.6	10.4	7.5	8.3	10.5	10.0
9.3	8.1	7.7	7.5	8.4	6.3	8.8

16. Construct a stem-and-leaf display for the following data. Use a leaf unit of 10.

1161	1206	1478	1300	1604	1725	1361	1422
1221	1378	1623	1426	1557	1730	1706	1689

Applications

SELF test

17. A doctor's office staff studied the waiting times for patients who arrive at the office with a request for emergency service. The following data with waiting times in minutes were collected over a one-month period.

2 5 10 12 4 4 5 17 11 8 9 8 12 21 6 8 7 13 18 3

Use classes of 0–4, 5–9, and so on in the following:
a. Show the frequency distribution.
b. Show the relative frequency distribution.

c. Show the cumulative frequency distribution.
d. Show the cumulative relative frequency distribution.
e. What proportion of patients needing emergency service wait 9 minutes or less?

18. A shortage of candidates has required school districts to pay higher salaries and offer extras to attract and retain school district superintendents. The following data show the annual base salary ($1000s) for superintendents in 20 districts in the greater Rochester, New York, area (*The Rochester Democrat and Chronicle,* February 10, 2008).

187	184	174	185
175	172	202	197
165	208	215	164
162	172	182	156
172	175	170	183

Use classes of 150–159, 160–169, and so on in the following.
a. Show the frequency distribution.
b. Show the percent frequency distribution.
c. Show the cumulative percent frequency distribution.
d. Develop a histogram for the annual base salary.
e. Do the data appear to be skewed? Explain.
f. What percentage of the superintendents make more than $200,000?

19. The Dow Jones Industrial Average (DJIA) underwent one of its infrequent reshufflings of companies when General Motors and Citigroup were replaced by Cisco Systems and Travelers (*The Wall Street Journal,* June 8, 2009). At the time, the prices per share for the 30 companies in the DJIA were as follows:

WEB file
DJIAPrices

Company	$/Share	Company	$/Share
3M	61	IBM	107
Alcoa	11	Intel	16
American Express	25	J.P. Morgan Chase	35
AT&T	24	Johnson & Johnson	56
Bank of America	12	Kraft Foods	27
Boeing	52	McDonald's	59
Caterpillar	38	Merck	26
Chevron	69	Microsoft	22
Cisco Systems	20	Pfizer	14
Coca-Cola	49	Procter & Gamble	53
DuPont	27	Travelers	43
ExxonMobil	72	United Technologies	56
General Electric	14	Verizon	29
Hewlett-Packard	37	Wal-Mart Stores	51
Home Depot	24	Walt Disney	25

a. What is the highest price per share? What is the lowest price per share?
b. Using a class width of 10, develop a frequency distribution for the data.
c. Prepare a histogram. Interpret the histogram, including a discussion of the general shape of the histogram, the midprice range, and the most frequent price range.
d. Use the *The Wall Street Journal* or another newspaper to find the current price per share for these companies. Prepare a histogram of the data and discuss any changes since June 2009. What company has had the largest increase in the price per share? What company has had the largest decrease in the price per share?

20. The London School of Economics and the Harvard Business School conducted a study of how chief executive officers (CEOs) spend their day. The study found that CEOs spend on average about 18 hours per week in meetings, not including conference calls, business

meals, and public events (*The Wall Street Journal,* February 14, 2012). Shown below is the time spent per week in meetings (hours) for a sample of 25 CEOs.

WEB file
CEOTime

14	15	18	23	15
19	20	13	15	23
23	21	15	20	21
16	15	18	18	19
19	22	23	21	12

a. What is the least amount of time spent per week on meetings? The highest?
b. Use a class width of two hours to prepare a frequency distribution and a percent frequency distribution for the data.
c. Prepare a histogram and comment on the shape of the distribution.

21. *Fortune* provides a list of America's largest corporations based on annual revenue. Shown below are the 50 largest corporations, with annual revenue expressed in billions of dollars (*CNN Money* website, January 15, 2010).

WEB file
LargeCorp

Corporation	Revenue	Corporation	Revenue
Amerisource Bergen	71	Lowe's	48
Archer Daniels Midland	70	Marathon Oil	74
AT&T	124	McKesson	102
Bank of America	113	Medco Health	51
Berkshire Hathaway	108	MetLife	55
Boeing	61	Microsoft	60
Cardinal Health	91	Morgan Stanley	62
Caterpillar	51	Pepsico	43
Chevron	263	Pfizer	48
Citigroup	112	Procter & Gamble	84
ConocoPhillips	231	Safeway	44
Costco Wholesale	72	Sears Holdings	47
CVS Caremark	87	State Farm Insurance	61
Dell	61	Sunoco	52
Dow Chemical	58	Target	65
ExxonMobil	443	Time Warner	47
Ford Motors	146	United Parcel Service	51
General Electric	149	United Technologies	59
Goldman Sachs	54	UnitedHealth Group	81
Hewlett-Packard	118	Valero Energy	118
Home Depot	71	Verizon	97
IBM	104	Walgreen	59
JPMorgan Chase	101	Walmart	406
Johnson & Johnson	64	WellPoint	61
Kroger	76	Wells Fargo	52

Summarize the data by constructing the following:

a. A frequency distribution (classes 0–49, 50–99, 100–149, and so on).
b. A relative frequency distribution.
c. A cumulative frequency distribution.
d. A cumulative relative frequency distribution.
e. What do these distributions tell you about the annual revenue of the largest corporations in America?
f. Show a histogram. Comment on the shape of the distribution.
g. What is the largest corporation in America and what is its annual revenue?

22. *Entrepreneur* magazine ranks franchises using performance measures such as growth rate, number of locations, startup costs, and financial stability. The number of locations for the top 20 U.S. franchises follow (*The World Almanac,* 2012).

WEB file Franchise

Franchise	No. U.S. Locations	Franchise	No. U.S. Locations
Hampton Inns	1864	Jan-Pro Franchising Intl. Inc.	12,394
ampm	3183	Hardee's	1901
McDonald's	32,805	Pizza Hut Inc.	13,281
7-Eleven Inc.	37,496	Kumon Math & Reading Centers	25,199
Supercuts	2130	Dunkin' Donuts	9947
Days Inn	1877	KFC Corp.	16,224
Vanguard Cleaning Systems	2155	Jazzercise Inc.	7683
		Anytime Fitness	1618
Servpro	1572	Matco Tools	1431
Subway	34,871	Stratus Building Solutions	5018
Denny's Inc.	1668		

Use classes 0–4999, 5000–9999, 10,000–14,999 and so forth to answer the following questions.

a. Construct a frequency distribution and a percent frequency distribution of the number of U.S. locations for these top-ranked franchises.
b. Construct a histogram of these data.
c. Comment on the shape of the distribution.

23. The *Nielsen Home Technology Report* provided information about home technology and its usage. The following data are the hours of personal computer usage during one week for a sample of 50 persons.

WEB file Computer

4.1	1.5	10.4	5.9	3.4	5.7	1.6	6.1	3.0	3.7
3.1	4.8	2.0	14.8	5.4	4.2	3.9	4.1	11.1	3.5
4.1	4.1	8.8	5.6	4.3	3.3	7.1	10.3	6.2	7.6
10.8	2.8	9.5	12.9	12.1	0.7	4.0	9.2	4.4	5.7
7.2	6.1	5.7	5.9	4.7	3.9	3.7	3.1	6.1	3.1

Summarize the data by constructing the following:

a. A frequency distribution (use a class width of three hours)
b. A relative frequency distribution
c. A histogram
d. Comment on what the data indicate about personal computer usage at home.

24. *Money* magazine listed top career opportunities for work that is enjoyable, pays well, and will still be around 10 years from now (*Money,* November 2009). Shown below are 20 top career opportunities, with the median pay and top pay for workers with two to seven years of experience in the field. Data are shown in thousands of dollars.

WEB file Careers

Career	Median Pay	Top Pay
Account Executive	81	157
Certified Public Accountant	74	138
Computer Security Consultant	100	138
Director of Communications	78	135
Financial Analyst	80	109
Finance Director	121	214
Financial Research Analyst	66	155
Hotel General Manager	77	146
Human Resources Manager	72	111
Investment Banking	106	221
IT Business Analyst	83	119
IT Project Manager	99	140

Career	Median Pay	Top Pay
Marketing Manager	77	126
Quality-Assurance Manager	80	122
Sales Representative	67	125
Senior Internal Auditor	76	106
Software Developer	79	116
Software Program Manager	110	152
Systems Engineer	87	130
Technical Writer	67	100

Develop a stem-and-leaf display for both the median pay and the top pay. Comment on what you learn about the pay for these careers.

25. A psychologist developed a new test of adult intelligence. The test was administered to 20 individuals, and the following data were obtained.

114	99	131	124	117	102	106	127	119	115
98	104	144	151	132	106	125	122	118	118

Construct a stem-and-leaf display for the data.

26. The 2011 Cincinnati Flying Pig Half-Marathon (13.1 miles) had 10,897 finishers (Cincinnati Flying Pig Marathon website). The following data show the ages for a sample of 40 half-marathoners.

WEB file
Marathon

49	33	40	37	56
44	46	57	55	32
50	52	43	64	40
46	24	30	37	43
31	43	50	36	61
27	44	35	31	43
52	43	66	31	50
72	26	59	21	47

a. Construct a stretched stem-and-leaf display.
b. What age group had the largest number of runners?
c. What age occurred most frequently?

2.3 Summarizing Data for Two Variables Using Tables

Thus far in this chapter, we have focused on using tabular and graphical displays to summarize the data for a single categorical or quantitative variable. Often a manager or decision maker needs to summarize the data for two variables in order to reveal the relationship—if any—between the variables. In this section, we show how to construct a tabular summary of the data for two variables.

Crosstabulation

A **crosstabulation** is a tabular summary of data for two variables. Although both variables can be either categorical or quantitative, crosstabulations in which one variable is categorical and the other variable is quantitative are just as common. We will illustrate this latter case by considering the following application based on data from Zagat's Restaurant Review. Data showing the quality rating and the typical meal price were collected for a sample of

WEB file

Restaurant

TABLE 2.9 QUALITY RATING AND MEAL PRICE DATA FOR 300 LOS ANGELES RESTAURANTS

Restaurant	Quality Rating	Meal Price ($)
1	Good	18
2	Very Good	22
3	Good	28
4	Excellent	38
5	Very Good	33
6	Good	28
7	Very Good	19
8	Very Good	11
9	Very Good	23
10	Good	13
.	.	.
.	.	.
.	.	.

300 restaurants in the Los Angeles area. Table 2.9 shows the data for the first 10 restaurants. Quality rating is a categorical variable with rating categories of good, very good, and excellent. Meal price is a quantitative variable that ranges from $10 to $49.

A crosstabulation of the data for this application is shown in Table 2.10. The labels shown in the margins of the table define the categories (classes) for the two variables. In the left margin, the row labels (good, very good, and excellent) correspond to the three rating categories for the quality rating variable. In the top margin, the column labels ($10–19, $20–29, $30–39, and $40–49) show that the meal price data have been grouped into four classes. Because each restaurant in the sample provides a quality rating and a meal price, each restaurant is associated with a cell appearing in one of the rows and one of the columns of the crosstabulation. For example, Table 2.9 shows restaurant 5 as having a very good quality rating and a meal price of $33. This restaurant belongs to the cell in row 2 and column 3 of the crosstabulation shown in Table 2.10. In constructing a crosstabulation, we simply count the number of restaurants that belong to each of the cells.

Grouping the data for a quantitative variable enables us to treat the quantitative variable as if it were a categorical variable when creating a crosstabulation.

Although four classes of the meal price variable were used to construct the crosstabulation shown in Table 2.10, the crosstabulation of quality rating and meal price could have been developed using fewer or more classes for the meal price variable. The issues involved in deciding how to group the data for a quantitative variable in a crosstabulation are similar to the issues involved in deciding the number of classes to use when constructing a frequency distribution for a quantitative variable. For this application, four classes of meal price were considered a reasonable number of classes to reveal any relationship between quality rating and meal price.

TABLE 2.10 CROSSTABULATION OF QUALITY RATING AND MEAL PRICE DATA FOR 300 LOS ANGELES RESTAURANTS

	Meal Price				
Quality Rating	$10–19	$20–29	$30–39	$40–49	Total
Good	42	40	2	0	84
Very Good	34	64	46	6	150
Excellent	2	14	28	22	66
Total	78	118	76	28	300

In reviewing Table 2.10, we see that the greatest number of restaurants in the sample (64) have a very good rating and a meal price in the $20–29 range. Only two restaurants have an excellent rating and a meal price in the $10–19 range. Similar interpretations of the other frequencies can be made. In addition, note that the right and bottom margins of the crosstabulation provide the frequency distributions for quality rating and meal price separately. From the frequency distribution in the right margin, we see that data on quality ratings show 84 restaurants with a good quality rating, 150 restaurants with a very good quality rating, and 66 restaurants with an excellent quality rating. Similarly, the bottom margin shows the frequency distribution for the meal price variable.

Dividing the totals in the right margin of the crosstabulation by the total for that column provides a relative and percent frequency distribution for the quality rating variable.

Quality Rating	Relative Frequency	Percent Frequency
Good	.28	28
Very Good	.50	50
Excellent	.22	22
Total	1.00	100

From the percent frequency distribution we see that 28% of the restaurants were rated good, 50% were rated very good, and 22% were rated excellent.

Dividing the totals in the bottom row of the crosstabulation by the total for that row provides a relative and percent frequency distribution for the meal price variable.

Meal Price	Relative Frequency	Percent Frequency
$10–19	.26	26
$20–29	.39	39
$30–39	.25	25
$40–49	.09	9
Total	1.00	100

Note that the sum of the values in the relative frequency column do not add exactly to 1.00 and the sum of the values in the percent frequency distribution do not add exactly to 100; the reason is that the values being summed are rounded. From the percent frequency distribution we see that 26% of the meal prices are in the lowest price class ($10–19), 39% are in the next higher class, and so on.

The frequency and relative frequency distributions constructed from the margins of a crosstabulation provide information about each of the variables individually, but they do not shed any light on the relationship between the variables. The primary value of a crosstabulation lies in the insight it offers about the relationship between the variables. A review of the crosstabulation in Table 2.10 reveals that restaurants with higher meal prices received higher quality ratings than restaurants with lower meal prices.

Converting the entries in a crosstabulation into row percentages or column percentages can provide more insight into the relationship between the two variables. For row percentages, the results of dividing each frequency in Table 2.10 by its corresponding row total are shown in Table 2.11. Each row of Table 2.11 is a percent frequency distribution of meal price for one of the quality rating categories. Of the restaurants with the lowest quality rating (good), we see that the greatest percentages are for the less expensive restaurants (50% have $10–19 meal prices and 47.6% have $20–29 meal prices). Of the restaurants with the highest quality rating (excellent), we see that the greatest

TABLE 2.11 ROW PERCENTAGES FOR EACH QUALITY RATING CATEGORY

	Meal Price				
Quality Rating	$10–19	$20–29	$30–39	$40–49	Total
Good	50.0	47.6	2.4	0.0	100
Very Good	22.7	42.7	30.6	4.0	100
Excellent	3.0	21.2	42.4	33.4	100

percentages are for the more expensive restaurants (42.4% have $30–39 meal prices and 33.4% have $40–49 meal prices). Thus, we continue to see that restaurants with higher meal prices received higher quality ratings.

Crosstabulations are widely used to investigate the relationship between two variables. In practice, the final reports for many statistical studies include a large number of crosstabulations. In the Los Angeles restaurant survey, the crosstabulation is based on one categorical variable (quality rating) and one quantitative variable (meal price). Crosstabulations can also be developed when both variables are categorical and when both variables are quantitative. When quantitative variables are used, however, we must first create classes for the values of the variable. For instance, in the restaurant example we grouped the meal prices into four classes ($10–19, $20–29, $30–39, and $40–49).

Simpson's Paradox

The data in two or more crosstabulations are often combined or aggregated to produce a summary crosstabulation showing how two variables are related. In such cases, conclusions drawn from two or more separate crosstabulations can be reversed when the data are aggregated into a single crosstabulation. The reversal of conclusions based on aggregate and unaggregated data is called **Simpson's paradox**. To provide an illustration of Simpson's paradox we consider an example involving the analysis of verdicts for two judges in two different courts.

Judges Ron Luckett and Dennis Kendall presided over cases in Common Pleas Court and Municipal Court during the past three years. Some of the verdicts they rendered were appealed. In most of these cases the appeals court upheld the original verdicts, but in some cases those verdicts were reversed. For each judge a crosstabulation was developed based upon two variables: Verdict (upheld or reversed) and Type of Court (Common Pleas and Municipal). Suppose that the two crosstabulations were then combined by aggregating the type of court data. The resulting aggregated crosstabulation contains two variables: Verdict (upheld or reversed) and Judge (Luckett or Kendall). This crosstabulation shows the number of appeals in which the verdict was upheld and the number in which the verdict was reversed for both judges. The following crosstabulation shows these results along with the column percentages in parentheses next to each value.

	Judge		
Verdict	Luckett	Kendall	Total
Upheld	129 (86%)	110 (88%)	239
Reversed	21 (14%)	15 (12%)	36
Total (%)	150 (100%)	125 (100%)	275

A review of the column percentages shows that 86% of the verdicts were upheld for Judge Luckett, while 88% of the verdicts were upheld for Judge Kendall. From this aggregated crosstabulation, we conclude that Judge Kendall is doing the better job because a greater percentage of Judge Kendall's verdicts are being upheld.

The following unaggregated crosstabulations show the cases tried by Judge Luckett and Judge Kendall in each court; column percentages are shown in parentheses next to each value.

Judge Luckett

Verdict	Common Pleas	Municipal Court	Total
Upheld	29 (91%)	100 (85%)	129
Reversed	3 (9%)	18 (15%)	21
Total (%)	32 (100%)	118 (100%)	150

Judge Kendall

Verdict	Common Pleas	Municipal Court	Total
Upheld	90 (90%)	20 (80%)	110
Reversed	10 (10%)	5 (20%)	15
Total (%)	100 (100%)	25 (100%)	125

From the crosstabulation and column percentages for Judge Luckett, we see that the verdicts were upheld in 91% of the Common Pleas Court cases and in 85% of the Municipal Court cases. From the crosstabulation and column percentages for Judge Kendall, we see that the verdicts were upheld in 90% of the Common Pleas Court cases and in 80% of the Municipal Court cases. Thus, when we unaggregate the data, we see that Judge Luckett has a better record because a greater percentage of Judge Luckett's verdicts are being upheld in both courts. This result contradicts the conclusion we reached with the aggregated data crosstabulation that showed Judge Kendall had the better record. This reversal of conclusions based on aggregated and unaggregated data illustrates Simpson's paradox.

The original crosstabulation was obtained by aggregating the data in the separate crosstabulations for the two courts. Note that for both judges the percentage of appeals that resulted in reversals was much higher in Municipal Court than in Common Pleas Court. Because Judge Luckett tried a much higher percentage of his cases in Municipal Court, the aggregated data favored Judge Kendall. When we look at the crosstabulations for the two courts separately, however, Judge Luckett shows the better record. Thus, for the original crosstabulation, we see that the *type of court* is a hidden variable that cannot be ignored when evaluating the records of the two judges.

Because of the possibility of Simpson's paradox, realize that the conclusion or interpretation may be reversed depending upon whether you are viewing unaggregated or aggregate crosstabulation data. Before drawing a conclusion, you may want to investigate whether the aggregate or unaggregate form of the crosstabulation provides the better insight and conclusion. Especially when the crosstabulation involves aggregated data, you should investigate whether a hidden variable could affect the results such that separate or unaggregated crosstabulations provide a different and possibly better insight and conclusion.

Exercises

Methods

SELF test

27. The following data are for 30 observations involving two categorical variables, x and y. The categories for x are A, B, and C; the categories for y are 1 and 2.

WEB file

Crosstab

Observation	x	y	Observation	x	y
1	A	1	16	B	2
2	B	1	17	C	1
3	B	1	18	B	1
4	C	2	19	C	1
5	B	1	20	B	1
6	C	2	21	C	2
7	B	1	22	B	1
8	C	2	23	C	2
9	A	1	24	A	1
10	B	1	25	B	1
11	A	1	26	C	2
12	B	1	27	C	2
13	C	2	28	A	1
14	C	2	29	B	1
15	C	2	30	B	2

a. Develop a crosstabulation for the data, with x as the row variable and y as the column variable.
b. Compute the row percentages.
c. Compute the column percentages.
d. What is the relationship, if any, between x and y?

28. The following observations are for two quantitative variables, x and y.

WEB file

Crosstab2

Observation	x	y	Observation	x	y
1	28	72	11	13	98
2	17	99	12	84	21
3	52	58	13	59	32
4	79	34	14	17	81
5	37	60	15	70	34
6	71	22	16	47	64
7	37	77	17	35	68
8	27	85	18	62	67
9	64	45	19	30	39
10	53	47	20	43	28

a. Develop a crosstabulation for the data, with x as the row variable and y as the column variable. For x use classes of 10–29, 30–49, and so on; for y use classes of 40–59, 60–79, and so on.
b. Compute the row percentages.
c. Compute the column percentages.
d. What is the relationship, if any, between x and y?

Applications

29. The following crosstabulation shows household income by educational level of the head of household (*Statistical Abstract of the United States: 2008*).
    a. Compute the row percentages and identify the percent frequency distributions of income for households in which the head is a high school graduate and in which the head holds a bachelor's degree.
    b. What percentage of households headed by high school graduates earn \$75,000 or more? What percentage of households headed by bachelor's degree recipients earn \$75,000 or more?

	Household Income ($1000s)					
Educational Level	**Under 25**	**25.0–49.9**	**50.0–74.9**	**75.0–99.9**	**100 or more**	**Total**
Not H.S. graduate	4207	3459	1389	539	367	9961
H.S. graduate	4917	6850	5027	2637	2668	22,099
Some college	2807	5258	4678	3250	4074	20,067
Bachelor's degree	885	2094	2848	2581	5379	13,787
Beyond bach. deg.	290	829	1274	1241	4188	7822
Total	13,106	18,490	15,216	10,248	16,676	73,736

c. Construct percent frequency histograms of income for households headed by persons with a high school degree and for those headed by persons with a bachelor's degree. Is any relationship evident between household income and educational level?

30. Refer again to the crosstabulation of household income by educational level shown in exercise 29.
   a. Compute column percentages and identify the percent frequency distributions displayed. What percentage of the heads of households did not graduate from high school?
   b. What percentage of the households earning $100,000 or more were headed by a person having schooling beyond a bachelor's degree? What percentage of the households headed by a person with schooling beyond a bachelor's degree earned over $100,000? Why are these two percentages different?
   c. Compare the percent frequency distributions for those households earning "Under 25," "100 or more," and for "Total." Comment on the relationship between household income and educational level of the head of household.

31. Recently, management at Oak Tree Golf Course received a few complaints about the condition of the greens. Several players complained that the greens are too fast. Rather than react to the comments of just a few, the Golf Association conducted a survey of 100 male and 100 female golfers. The survey results are summarized here.

Male Golfers

	Greens Condition	
Handicap	**Too Fast**	**Fine**
Under 15	10	40
15 or more	25	25

Female Golfers

	Greens Condition	
Handicap	**Too Fast**	**Fine**
Under 15	1	9
15 or more	39	51

   a. Combine these two crosstabulations into one with Male and Female as the row labels and Too Fast and Fine as the column labels. Which group shows the highest percentage saying that the greens are too fast?
   b. Refer to the initial crosstabulations. For those players with low handicaps (better players), which group (male or female) shows the highest percentage saying the greens are too fast?
   c. Refer to the initial crosstabulations. For those players with higher handicaps, which group (male or female) shows the highest percentage saying the greens are too fast?
   d. What conclusions can you draw about the preferences of men and women concerning the speed of the greens? Are the conclusions you draw from part (a) as compared with parts (b) and (c) consistent? Explain any apparent inconsistencies.

32. Table 2.12 shows a data set containing information for 45 mutual funds that are part of the *Morningstar Funds 500* for 2008. The data set includes the following five variables:

TABLE 2.12 FINANCIAL DATA FOR A SAMPLE OF 45 MUTUAL FUNDS

WEB file MutualFunds

Fund Name	Fund Type	Net Asset Value ($)	5-Year Average Return (%)	Expense Ratio (%)	Morningstar Rank
Amer Cent Inc & Growth Inv	DE	28.88	12.39	0.67	2-Star
American Century Intl. Disc	IE	14.37	30.53	1.41	3-Star
American Century Tax-Free Bond	FI	10.73	3.34	0.49	4-Star
American Century Ultra	DE	24.94	10.88	0.99	3-Star
Ariel	DE	46.39	11.32	1.03	2-Star
Artisan Intl Val	IE	25.52	24.95	1.23	3-Star
Artisan Small Cap	DE	16.92	15.67	1.18	3-Star
Baron Asset	DE	50.67	16.77	1.31	5-Star
Brandywine	DE	36.58	18.14	1.08	4-Star
Brown Cap Small	DE	35.73	15.85	1.20	4-Star
Buffalo Mid Cap	DE	15.29	17.25	1.02	3-Star
Delafield	DE	24.32	17.77	1.32	4-Star
DFA U.S. Micro Cap	DE	13.47	17.23	0.53	3-Star
Dodge & Cox Income	FI	12.51	4.31	0.44	4-Star
Fairholme	DE	31.86	18.23	1.00	5-Star
Fidelity Contrafund	DE	73.11	17.99	0.89	5-Star
Fidelity Municipal Income	FI	12.58	4.41	0.45	5-Star
Fidelity Overseas	IE	48.39	23.46	0.90	4-Star
Fidelity Sel Electronics	DE	45.60	13.50	0.89	3-Star
Fidelity Sh-Term Bond	FI	8.60	2.76	0.45	3-Star
Fidelity	DE	39.85	14.40	0.56	4-Star
FPA New Income	FI	10.95	4.63	0.62	3-Star
Gabelli Asset AAA	DE	49.81	16.70	1.36	4-Star
Greenspring	DE	23.59	12.46	1.07	3-Star
Janus	DE	32.26	12.81	0.90	3-Star
Janus Worldwide	IE	54.83	12.31	0.86	2-Star
Kalmar Gr Val Sm Cp	DE	15.30	15.31	1.32	3-Star
Managers Freemont Bond	FI	10.56	5.14	0.60	5-Star
Marsico 21st Century	DE	17.44	15.16	1.31	5-Star
Mathews Pacific Tiger	IE	27.86	32.70	1.16	3-Star
Meridan Value	DE	31.92	15.33	1.08	4-Star
Oakmark I	DE	40.37	9.51	1.05	2-Star
PIMCO Emerg Mkts Bd D	FI	10.68	13.57	1.25	3-Star
RS Value A	DE	26.27	23.68	1.36	4-Star
T. Rowe Price Latin Am.	IE	53.89	51.10	1.24	4-Star
T. Rowe Price Mid Val	DE	22.46	16.91	0.80	4-Star
Templeton Growth A	IE	24.07	15.91	1.01	3-Star
Thornburg Value A	DE	37.53	15.46	1.27	4-Star
USAA Income	FI	12.10	4.31	0.62	3-Star
Vanguard Equity-Inc	DE	24.42	13.41	0.29	4-Star
Vanguard Global Equity	IE	23.71	21.77	0.64	5-Star
Vanguard GNMA	FI	10.37	4.25	0.21	5-Star
Vanguard Sht-Tm TE	FI	15.68	2.37	0.16	3-Star
Vanguard Sm Cp Idx	DE	32.58	17.01	0.23	3-Star
Wasatch Sm Cp Growth	DE	35.41	13.98	1.19	4-Star

Fund Type: The type of fund, labeled DE (Domestic Equity), IE (International Equity), and FI (Fixed Income)

Net Asset Value ($): The closing price per share

5-Year Average Return (%): The average annual return for the fund over the past 5 years

Expense Ratio (%): The percentage of assets deducted each fiscal year for fund expenses

Morningstar Rank: The risk adjusted star rating for each fund; Morningstar ranks go from a low of 1-Star to a high of 5-Stars

a. Prepare a crosstabulation of the data on Fund Type (rows) and the average annual return over the past 5 years (columns). Use classes of 0–9.99, 10–19.99, 20–29.99, 30–39.99, 40–49.99, and 50–59.99 for the 5-Year Average Return (%).
b. Prepare a frequency distribution for the data on Fund Type.
c. Prepare a frequency distribution for the data on 5-Year Average Return (%).
d. How has the crosstabulation helped in preparing the frequency distributions in parts (b) and (c)?
e. What conclusions can you draw about the fund type and the average return over the past 5 years?

33. Refer to the data in Table 2.12.
a. Prepare a crosstabulation of the data on Fund Type (rows) and the Expense Ratio (%) (columns). Use classes of .25–.49, .50–.74, .75–.99, 1.00–1.24, and 1.25–1.49 for Expense Ratio (%).
b. Prepare a percent frequency distribution for Expense Ratio (%).
c. What conclusions can you draw about fund type and the expense ratio?

34. Refer to the data in Table 2.12.
a. Prepare a crosstabulation with 5-Year Average Return as the columns and Net Asset Value as the rows. Use classes starting with 0 in increments of 5 for 5-Year Average Return and classes starting at 0 with increments of 10 for Net Asset Value.
b. Comment on the relationship, if any, between the variables.

35. The U.S. Department of Energy's Fuel Economy Guide provides fuel efficiency data for cars and trucks (Fuel Economy website, September 8, 2012). A portion of the data for 149 compact, midsize, and large cars is shown in Table 2.13. The data set contains the following variables:

Size: Compact, Midsize, and Large

Displacement: Engine size in liters

Cylinders: Number of cylinders in the engine

TABLE 2.13 FUEL EFFICIENCY DATA FOR 311 CARS

WEB file
FuelData2012

Car	Size	Displacement	Cylinders	Drive	Fuel Type	City MPG	Hwy MPG
1	Compact	2.0	4	F	P	22	30
2	Compact	2.0	4	A	P	21	29
3	Compact	2.0	4	A	P	21	31
•	•	•	•	•	•	•	•
•	•	•	•	•	•	•	•
•	•	•	•	•	•	•	•
94	Midsize	3.5	6	A	R	17	25
95	Midsize	2.5	4	F	R	23	33
•	•	•	•	•	•	•	•
•	•	•	•	•	•	•	•
•	•	•	•	•	•	•	•
148	Large	6.7	12	R	P	11	18
149	Large	6.7	12	R	P	11	18

Drive: All wheel (A), front wheel (F), and rear wheel (R)

Fuel Type: Premium (P) or regular (R) fuel

City MPG: Fuel efficiency rating for city driving in terms of miles per gallon

Hwy MPG: Fuel efficiency rating for highway driving in terms of miles per gallon

The complete data set is contained in the file named FuelData2012.

a. Prepare a crosstabulation of the data on Size (rows) and Hwy MPG (columns). Use classes of 15–19, 20–24, 25–29, 30–34, 35–39, and 40–44 for Hwy MPG.
b. Comment on the relationship beween Size and Hwy MPG.
c. Prepare a crosstabulation of the data on Drive (rows) and City MPG (columns). Use classes of 10–14, 15–19, 20–24, 25–29, 30–34, and 35–39, and 40–44 for City MPG.
d. Comment on the relationship between Drive and City MPG.
e. Prepare a crosstabulation of the data on Fuel Type (rows) and City MPG (columns). Use classes of 10–14, 15–19, 20–24, 25–29, 30–34, 35–39, and 40–44 for City MPG.
f. Comment on the relationship between Fuel Type and City MPG.

2.4 Summarizing Data for Two Variables Using Graphical Displays

In the previous section we showed how a crosstabulation can be used to summarize the data for two variables and help reveal the relationship between the variables. In most cases, a graphical display is more useful for recognizing patterns and trends in the data.

In this section, we introduce a variety of graphical displays for exploring the relationships between two variables. Displaying data in creative ways can lead to powerful insights and allow us to make "common-sense inferences" based on our ability to visually compare, contrast, and recognize patterns. We begin with a discussion of scatter diagrams and trendlines.

Scatter Diagram and Trendline

A **scatter diagram** is a graphical display of the relationship between two quantitative variables, and a **trendline** is a line that provides an approximation of the relationship. As an illustration, consider the advertising/sales relationship for a stereo and sound equipment store in San Francisco. On 10 occasions during the past three months, the store used weekend television commercials to promote sales at its stores. The managers want to investigate whether a relationship exists between the number of commercials shown and sales at the store during the following week. Sample data for the 10 weeks with sales in hundreds of dollars are shown in Table 2.14.

TABLE 2.14 SAMPLE DATA FOR THE STEREO AND SOUND EQUIPMENT STORE

WEB file

Stereo

Week	Number of Commercials x	Sales ($100s) y
1	2	50
2	5	57
3	1	41
4	3	54
5	4	54
6	1	38
7	5	63
8	3	48
9	4	59
10	2	46

FIGURE 2.7 SCATTER DIAGRAM AND TRENDLINE FOR THE STEREO AND SOUND EQUIPMENT STORE

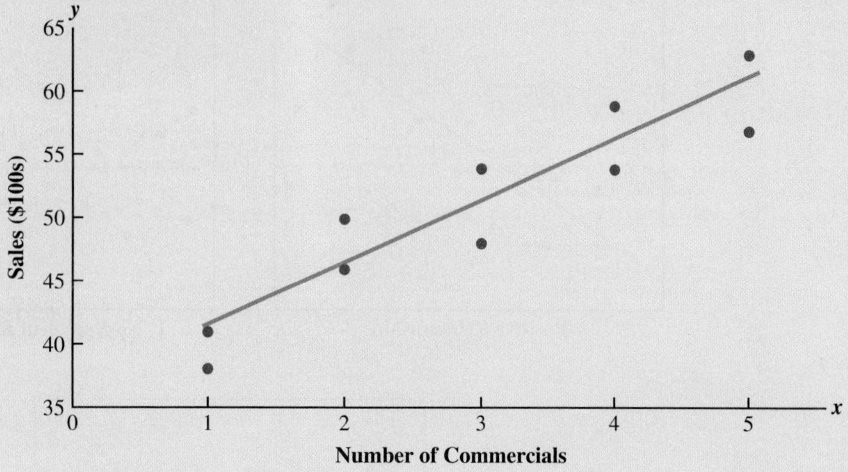

Figure 2.7 shows the scatter diagram and the trendline[1] for the data in Table 2.14. The number of commercials (x) is shown on the horizontal axis and the sales (y) are shown on the vertical axis. For week 1, $x = 2$ and $y = 50$. A point with those coordinates is plotted on the scatter diagram. Similar points are plotted for the other nine weeks. Note that during two of the weeks one commercial was shown, during two of the weeks two commercials were shown, and so on.

The scatter diagram in Figure 2.7 indicates a positive relationship between the number of commercials and sales. Higher sales are associated with a higher number of commercials. The relationship is not perfect in that all points are not on a straight line. However, the general pattern of the points and the trendline suggest that the overall relationship is positive.

Some general scatter diagram patterns and the types of relationships they suggest are shown in Figure 2.8. The top left panel depicts a positive relationship similar to the one for the number of commercials and sales example. In the top right panel, the scatter diagram shows no apparent relationship between the variables. The bottom panel depicts a negative relationship where y tends to decrease as x increases.

Side-by-Side and Stacked Bar Charts

In Section 2.1 we said that a bar chart is a graphical display for depicting categorical data summarized in a frequency, relative frequency, or percent frequency distribution. Side-by-side bar charts and stacked bar charts are extensions of basic bar charts that are used to display and compare two variables. By displaying two variables on the same chart, we may better understand the relationship between the variables.

A **side-by-side bar chart** is a graphical display for depicting multiple bar charts on the same display. To illustrate the construction of a side-by-side chart, recall the application involving the quality rating and meal price data for a sample of 300 restaurants located in the Los Angeles area. Quality rating is a categorical variable with rating categories of good, very good, and excellent. Meal price is a quantitative variable that ranges from $10 to $49. The crosstabulation displayed in Table 2.10 shows that the data for meal price were

[1]The equation of the trendline is $y = 36.15 + 4.95x$. The slope of the trendline is 4.95 and the y-intercept (the point where the trendline intersects the y-axis) is 36.15. We will discuss in detail the interpretation of the slope and y-intercept for a linear trendline in Chapter 14 when we study simple linear regression.

FIGURE 2.8 TYPES OF RELATIONSHIPS DEPICTED BY SCATTER DIAGRAMS

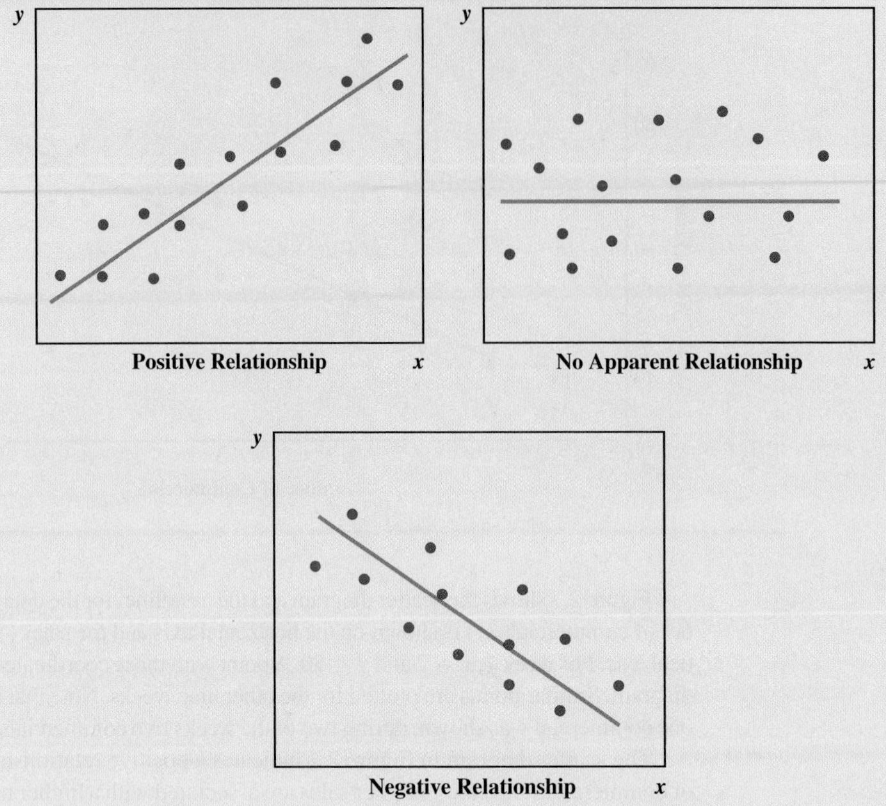

grouped into four classes: \$10–19, \$20–29, \$30–39, and \$40–49. We will use these classes to construct a side-by-side bar chart.

Figure 2.9 shows a side-by-side chart for the restaurant data. The color of each bar indicates the quality rating (blue = good, red = very good, and green = excellent). Each bar is constructed by extending the bar to the point on the vertical axis that represents the frequency

FIGURE 2.9 SIDE-BY-SIDE BAR CHART FOR THE QUALITY AND MEAL PRICE DATA

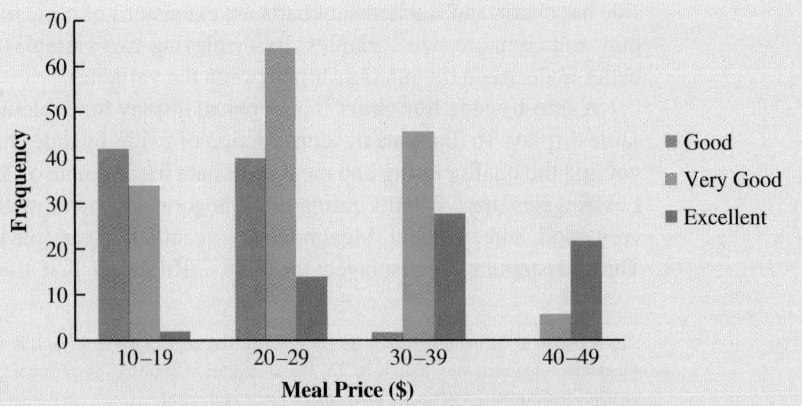

with which that quality rating occurred for each of the meal price categories. Placing each meal price category's quality rating frequency adjacent to one another allows us to quickly determine how a particular meal price category is rated. We see that the lowest meal price category ($10–$19) received mostly good and very good ratings, but very few excellent ratings. The highest price category ($40–49), however, shows a much different result. This meal price category received mostly excellent ratings, some very good ratings, but no good ratings.

Figure 2.9 also provides a good sense of the relationship between meal price and quality rating. Notice that as the price increases (left to right), the height of the blue bars decreases and the height of the green bars generally increases. This indicates that as price increases, the quality rating tends to be better. The very good rating, as expected, tends to be more prominent in the middle price categories as indicated by the dominance of the red bars in the middle of the chart.

Stacked bar charts are another way to display and compare two variables on the same display. A **stacked bar chart** is a bar chart in which each bar is broken into rectangular segments of a different color showing the relative frequency of each class in a manner similar to a pie chart. To illustrate a stacked bar chart we will use the quality rating and meal price data summarized in the crosstabulation shown in Table 2.10.

We can convert the frequency data in Table 2.10 into column percentages by dividing each element in a particular column by the total for that column. For instance, 42 of the 78 restaurants with a meal price in the $10–19 range had a good quality rating. In other words, (42/78)100 or 53.8% of the 78 restaurants had a good rating. Table 2.15 shows the column percentages for each meal price category. Using the data in Table 2.15 we constructed the stacked bar chart shown in Figure 2.10. Because the stacked bar chart is based on percentages, Figure 2.10 shows even more clearly than Figure 2.9 the

TABLE 2.15 COLUMN PERCENTAGES FOR EACH MEAL PRICE CATEGORY

	Meal Price			
Quality Rating	$10–19	$20–29	$30–39	$40–49
Good	53.8%	33.9%	2.6%	0.0%
Very Good	43.6	54.2	60.5	21.4
Excellent	2.6	11.9	36.8	78.6
Total	100.0%	100.0%	100.0%	100.0%

FIGURE 2.10 STACKED BAR CHART FOR QUALITY RATING AND MEAL PRICE DATA

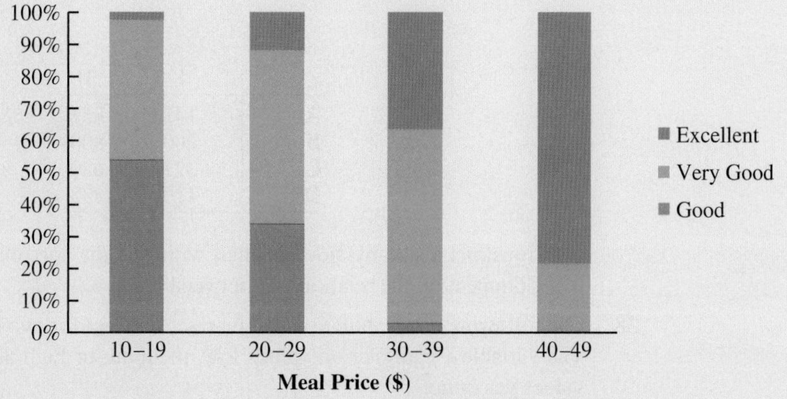

relationship between the variables. As we move from the low price category ($10–19) to the high price category ($40–49), the length of the blue bars decreases and the length of the green bars increases.

NOTES AND COMMENTS

1. A time series is a sequence of observations on a variable measured at successive points in time or over successive periods of time. A scatter diagram in which the value of time is shown on the horizontal axis and the time series values are shown on the vertical axis is referred to in time series analysis as a time series plot. We will discuss time series plots and how to analyze time series data in Chapter 17.
2. A stacked bar chart can also be used to display frequencies rather than percentage frequencies. In this case, the different color segments of each bar represent the contribution to the total for that bar, rather than the percentage contribution.

Exercises

Methods

SELF test

36. The following 20 observations are for two quantitative variables, *x* and *y*.

WEB file

Scatter

Observation	*x*	*y*	Observation	*x*	*y*
1	−22	22	11	−37	48
2	−33	49	12	34	−29
3	2	8	13	9	−18
4	29	−16	14	−33	31
5	−13	10	15	20	−16
6	21	−28	16	−3	14
7	−13	27	17	−15	18
8	−23	35	18	12	17
9	14	−5	19	−20	−11
10	3	−3	20	−7	−22

a. Develop a scatter diagram for the relationship between *x* and *y*.
b. What is the relationship, if any, between *x* and *y*?

37. Consider the following data on two categorical variables. The first variable, *x*, can take on values A, B, C, or D. The second variable, *y*, can take on values I or II. The following table gives the frequency with which each combination occurs.

	y	
x	I	II
A	143	857
B	200	800
C	321	679
D	420	580

a. Construct a side-by-side bar chart with *x* on the horizontal axis.
b. Comment on the relationship between *x* and *y*.

38. The following crosstabulation summarizes the data for two categorical variables, *x* and *y*. The variable *x* can take on values low, medium, or high and the variable *y* can take on values yes or no.

x	Yes	No	Total
	y		
Low	20	10	30
Medium	15	35	50
High	20	5	25
Total	55	50	105

a. Compute the row percentages.
b. Construct a stacked percent frequency bar chart with x on the horizontal axis.

Applications

39. A study on driving speed (miles per hour) and fuel efficiency (miles per gallon) for midsize automobiles resulted in the following data:

WEB file MPG

Driving Speed	30	50	40	55	30	25	60	25	50	55
Fuel Efficiency	28	25	25	23	30	32	21	35	26	25

a. Construct a scatter diagram with driving speed on the horizontal axis and fuel efficiency on the vertical axis.
b. Comment on any apparent relationship between these two variables.

WEB file Snow

40. The Current Results website lists the average annual high and low temperatures (degrees Fahrenheit) and average annual snowfall (inches) for fifty-one major U.S. cities, based on data from 1981 to 2010. The data are contained in the file *Snow*. For example, the average low temperature for Columbus, Ohio is 44 degrees and the average annual snowfall is 27.5 inches.
a. Construct a scatter diagram with the average annual low temperature on the horizontal axis and the average annual snowfall on the vertical axis.
b. Does there appear to be any relationship between these two variables?
c. Based on the scatter diagram, comment on any data points that seem to be unusual.

41. People often wait until middle age to worry about having a healthy heart. However, recent studies have shown that earlier monitoring of risk factors such as blood pressure can be very beneficial (*The Wall Street Journal,* January 10, 2012). Having higher than normal blood pressure, a condition known as hypertension, is a major risk factor for heart disease. Suppose a large sample of individuals of various ages and gender was selected and that each individual's blood pressure was measured to determine if they have hypertension. For the sample data, the following table shows the percentage of individuals with hypertension.

WEB file Hypertension

Age	Male	Female
20–34	11.00%	9.00%
35–44	24.00%	19.00%
45–54	39.00%	37.00%
55–64	57.00%	56.00%
65–74	62.00%	64.00%
75+	73.30%	79.00%

a. Develop a side-by-side bar chart with age on the horizontal axis, the percentage of individuals with hypertension on the vertical axis, and side-by-side bars based on gender.
b. What does the display you developed in part (a), indicate about hypertension and age?
c. Comment on differences by gender.

42. Smartphones are advanced mobile phones with Internet, photo, and music and video capability (The Pew Research Center, Internet & American Life Project, 2011). The following survey results show smartphone ownership by age.

WEB file
Smartphones

Age Category	Smartphone (%)	Other Cell Phone (%)	No Cell Phone (%)
18–24	49	46	5
25–34	58	35	7
35–44	44	45	11
45–54	28	58	14
55–64	22	59	19
65+	11	45	44

a. Construct a stacked bar chart to display the above survey data on type of mobile phone ownership. Use age category as the variable on the horizontal axis.
b. Comment on the relationship between age and smartphone ownership.
c. How would you expect the results of this survey to be different if conducted in 2021?

43. The Northwest regional manager of an outdoor equipment retailer conducted a study to determine how managers at three store locations are using their time. A summary of the results are shown in the following table.

WEB file
ManagerTime

	Percentage of Manager's Work Week Spent on			
Store Location	Meetings	Reports	Customers	Idle
Bend	18	11	52	19
Portland	52	11	24	13
Seattle	32	17	37	14

a. Create a stacked bar chart with store location on the horizontal axis and percentage of time spent on each task on the vertical axis.
b. Create a side-by-side bar chart with store location on the horizontal axis and side-by-side bars of the percentage of time spent on each task.
c. Which type of bar chart (stacked or side-by-side) do you prefer for these data? Why?

2.5 Data Visualization: Best Practices in Creating Effective Graphical Displays

Data visualization is a term used to describe the use of graphical displays to summarize and present information about a data set. The goal of data visualization is to communicate as effectively and clearly as possible, the key information about the data. In this section, we provide guidelines for creating an effective graphical display, discuss how to select an appropriate type of display given the purpose of the study, illustrate the use of data dashboards, and show how the Cincinnati Zoo and Botanical Garden uses data visualization techniques to improve decision making.

TABLE 2.16 PLANNED AND ACTUAL SALES BY SALES REGION ($1000s)

Sales Region	Planned Sales ($1000s)	Actual Sales ($1000s)
Northeast	540	447
Northwest	420	447
Southeast	575	556
Southwest	360	341

Creating Effective Graphical Displays

The data presented in Table 2.16 show the forecasted or planned value of sales ($1000s) and the actual value of sales ($1000s) by sales region in the United States for Gustin Chemical for the past year. Note that there are two quantitative variables (planned sales and actual sales) and one categorical variable (sales region). Suppose we would like to develop a graphical display that would enable management of Gustin Chemical to visualize how each sales region did relative to planned sales and simultaneously enable management to visualize sales performance across regions.

Figure 2.11 shows a side-by-side bar chart of the planned versus actual sales data. Note how this bar chart makes it very easy to compare the planned versus actual sales in a region, as well as across regions. This graphical display is simple, contains a title, is well labeled, and uses distinct colors to represent the two types of sales. Note also that the scale of the vertical axis begins at zero. The four sales regions are separated by space so that it is clear that they are distinct, whereas the planned versus actual sales values are side-by-side for easy comparison within each region. The side-by-side bar chart in Figure 2.11 makes it easy to see that the Southwest region is the lowest in both planned and actual sales and that the Northwest region slightly exceeded its planned sales.

Creating an effective graphical display is as much art as it is science. By following the general guidelines listed below you can increase the likelihood that your display will effectively convey the key information in the data.

- Give the display a clear and concise title.
- Keep the display simple. Do not use three dimensions when two dimensions are sufficient.

FIGURE 2.11 SIDE-BY-SIDE BAR CHART FOR PLANNED VERSUS ACTUAL SALES

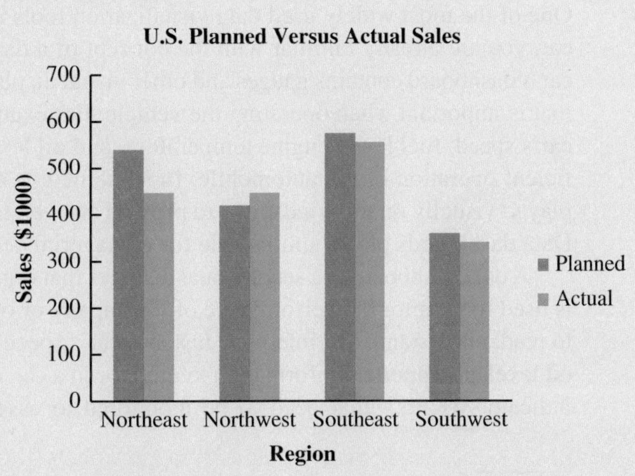

- Clearly label each axis and provide the units of measure.
- If color is used to distinguish categories, make sure the colors are distinct.
- If multiple colors or line types are used, use a legend to define how they are used and place the legend close to the representation of the data.

Choosing the Type of Graphical Display

In this chapter we discussed a variety of graphical displays, including bar charts, pie charts, dot plots, histograms, stem-and-leaf plots, scatter diagrams, side-by-side bar charts, and stacked bar charts. Each of these types of displays was developed for a specific purpose. In order to provide guidelines for choosing the appropriate type of graphical display, we now provide a summary of the types of graphical displays categorized by their purpose. We note that some types of graphical displays may be used effectively for multiple purposes.

Displays Used to Show the Distribution of Data

- Bar Chart—Used to show the frequency distribution and relative frequency distribution for categorical data
- Pie Chart—Used to show the relative frequency and percent frequency for categorical data
- Dot Plot—Used to show the distribution for quantitative data over the entire range of the data
- Histogram—Used to show the frequency distribution for quantitative data over a set of class intervals
- Stem-and-Leaf Display—Used to show both the rank order and shape of the distribution for quantitative data

Displays Used to Make Comparisons

- Side-by-Side Bar Chart—Used to compare two variables
- Stacked Bar Charts—Used to compare the relative frequency or percent frequency of two categorical variables

Displays Used to Show Relationships

- Scatter diagram—Used to show the relationship between two quantitative variables
- Trendline—Used to approximate the relationship of data in a scatter diagram

Data Dashboards

Data dashboards are also referred to as digital dashboards.

One of the most widely used data visualization tools is a **data dashboard**. If you drive a car, you are already familiar with the concept of a data dashboard. In an automobile, the car's dashboard contains gauges and other visual displays that provide the key information that is important when operating the vehicle. For example, the gauges used to display the car's speed, fuel level, engine temperature, and oil level are critical to ensure safe and efficient operation of the automobile. In some new vehicles, this information is even displayed visually on the windshield to provide an even more effective display for the driver. Data dashboards play a similar role for managerial decision making.

A data dashboard is a set of visual displays that organizes and presents information that is used to monitor the performance of a company or organization in a manner that is easy to read, understand, and interpret. Just as a car's speed, fuel level, engine temperature, and oil level are important information to monitor in a car, every business has key performance indicators (KPIs)[2] that need to be monitored to assess how a company is performing.

[2] Key performance indicators are sometimes referred to as Key Performance Metrics (KPMs).

Examples of KPIs are inventory on hand, daily sales, percentage of on-time deliveries, and sales revenue per quarter. A data dashboard should provide timely summary information (potentially from various sources) on KPIs that is important to the user, and it should do so in a manner that informs rather than overwhelms its user.

To illustrate the use of a data dashboard in decision making, we will discuss an application involving the Grogan Oil Company. Grogan has offices located in three cities in Texas: Austin (its headquarters), Houston, and Dallas. Grogan's Information Technology (IT) call center, located in the Austin office, handles calls from employees regarding computer-related problems involving software, Internet, and email issues. For example, if a Grogan employee in Dallas has a computer software problem, the employee can call the IT call center for assistance.

The data dashboard shown in Figure 2.12 was developed to monitor the performance of the call center. This data dashboard combines several displays to monitor the call center's KPIs. The data presented are for the current shift, which started at 8:00 A.M. The stacked bar chart in the upper left-hand corner shows the call volume for each type of problem (software, Internet, or email) over time. This chart shows that call volume is heavier during the first few hours of the shift, calls concerning email issues appear to decrease over time, and volume of calls regarding software issues are highest at midmorning. The pie chart in the upper

FIGURE 2.12 GROGAN OIL INFORMATION TECHNOLOGY CALL CENTER DATA DASHBOARD

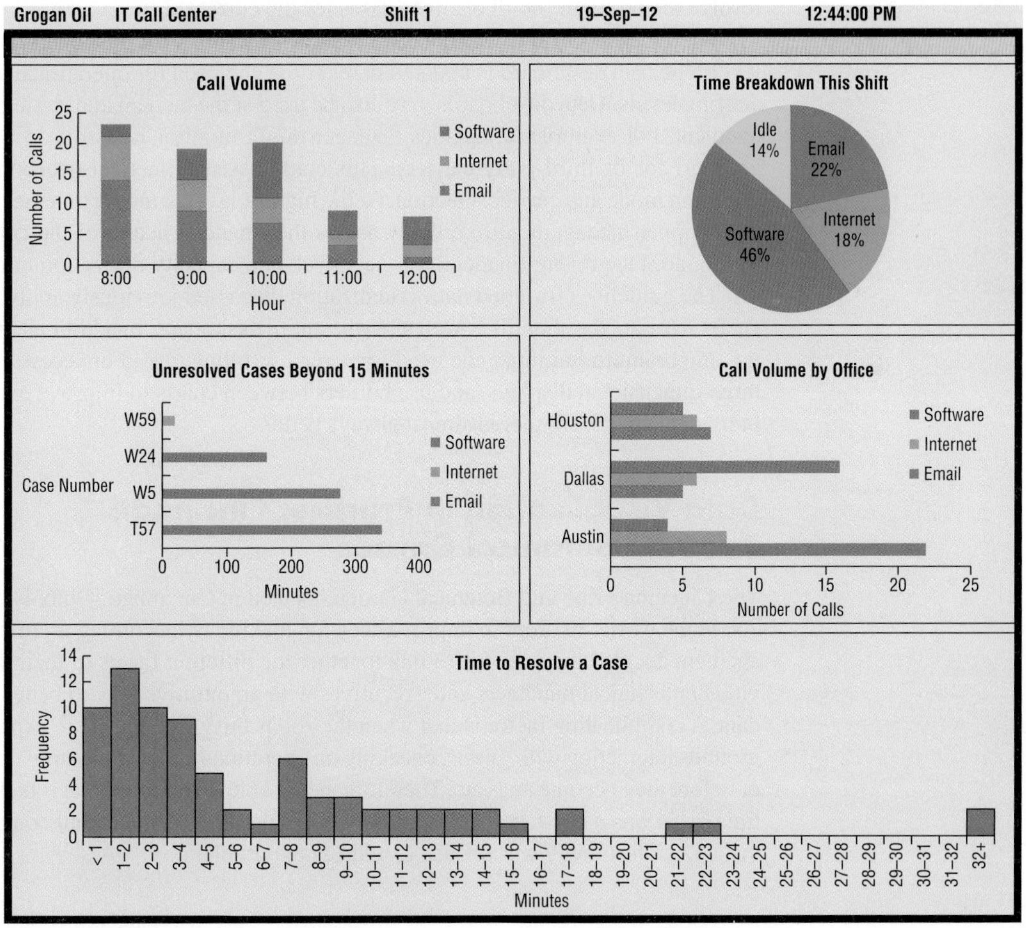

right-hand corner of the dashboard shows the percentage of time that call-center employees spent on each type of problem or not working on a call (idle). Both of these charts are important displays in determining optimal staffing levels. For instance, knowing the call mix and how stressed the system is—as measured by percentage of idle time—can help the IT manager make sure there are enough call center employees available with the right level of expertise.

The side-by-side bar chart below the pie chart shows the call volume by type of problem for each of Grogan's offices. This allows the IT manager to quickly identify if there is a particular type of problem by location. For example, it appears that the office in Austin is reporting a relatively high number of issues with email. If the source of the problem can be identified quickly, then the problem for many might be resolved quickly. Also, note that a relatively high number of software problems are coming from the Dallas office. The higher call volume in this case was simply due to the fact that the Dallas office is currently installing new software, and this has resulted in more calls to the IT call center. Because the IT manager was alerted to this by the Dallas office last week, the IT manager knew there would be an increase in calls coming from the Dallas office and was able to increase staffing levels to handle the expected increase in calls.

For each unresolved case that was received more than 15 minutes ago, the bar chart shown in the middle left-hand side of the data dashboard displays the length of time that each of these cases has been unresolved. This chart enables Grogan to quickly monitor the key problem cases and decide whether additional resources may be needed to resolve them. The worst case, T57, has been unresolved for over 300 minutes and is actually left over from the previous shift. Finally, the histogram at the bottom shows the distribution of the time to resolve the problem for all resolved cases for the current shift.

The Grogan Oil data dashboard illustrates the use of a dashboard at the operational level. The data dashboard is updated in real time and used for operational decisions such as staffing levels. Data dashboards may also be used at the tactical and strategic levels of management. For example, a logistics manager might monitor KPIs for on-time performance and cost for its third-party carriers. This could assist in tactical decisions such as transportation mode and carrier selection. At the highest level, a more strategic dashboard would allow upper management to quickly assess the financial health of the company by monitoring more aggregate financial, service level, and capacity utilization information.

The guidelines for good data visualization discussed previously apply to the individual charts in a data dashboard, as well as to the entire dashboard. In addition to those guidelines, it is important to minimize the need for screen scrolling, avoid unnecessary use of color or three-dimensional displays, and use borders between charts to improve readability. As with individual charts, simpler is almost always better.

Data Visualization in Practice: Cincinnati Zoo and Botanical Garden[3]

The Cincinnati Zoo and Botanical Garden, located in Cincinnati, Ohio, is the second oldest zoo in the world. In order to improve decision making by becoming more data-driven, management decided they needed to link together the different facets of their business and provide nontechnical managers and executives with an intuitive way to better understand their data. A complicating factor is that when the zoo is busy, managers are expected to be on the grounds interacting with guests, checking on operations, and anticipating issues as they arise or before they become an issue. Therefore, being able to monitor what is happening on a real-time basis was a key factor in deciding what to do. Zoo management concluded that a data visualization strategy was needed to address the problem.

[3] The authors are indebted to John Lucas of the Cincinnati Zoo and Botanical Garden for providing this application.

FIGURE 2.13 DATA DASHBOARD FOR THE CINCINNATI ZOO

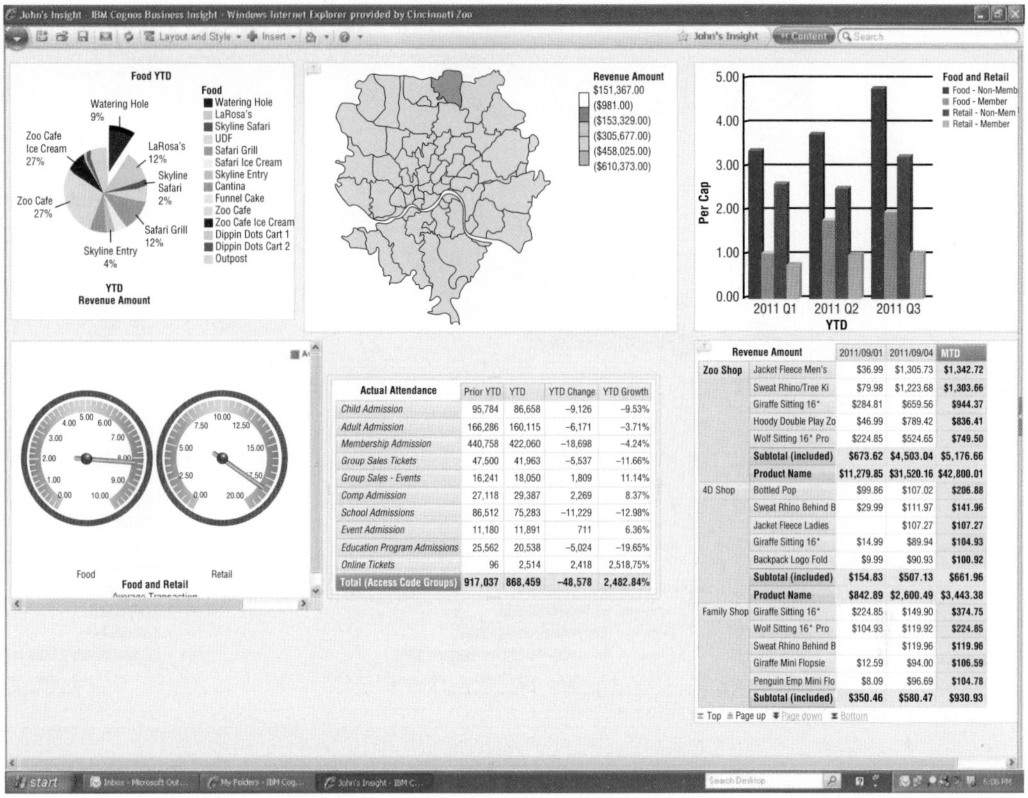

Because of its ease of use, real-time updating capability, and iPad compatibility, the Cincinnati Zoo decided to implement its data visualization strategy using IBM's Cognos advanced data visualization software. Using this software, the Cincinnati Zoo developed the data dashboard shown in Figure 2.13 to enable zoo management to track the following key performance indicators:

- Item Analysis (sales volumes and sales dollars by location within the zoo)
- Geo Analytics (using maps and displays of where the day's visitors are spending their time at the zoo)
- Customer Spending
- Cashier Sales Performance
- Sales and Attendance Data versus Weather Patterns
- Performance of the Zoo's Loyalty Rewards Program

An iPad mobile application was also developed to enable the zoo's managers to be out on the grounds and still see and anticipate what is occurring on a real-time basis. The Cincinnati Zoo's iPad data dashboard, shown in Figure 2.14, provides managers with access to the following information:

- Real-time attendance data, including what "types" of guests are coming to the zoo
- Real-time analysis showing which items are selling the fastest inside the zoo
- Real-time geographical representation of where the zoo's visitors live

FIGURE 2.14 THE CINCINNATI ZOO iPAD DATA DASHBOARD

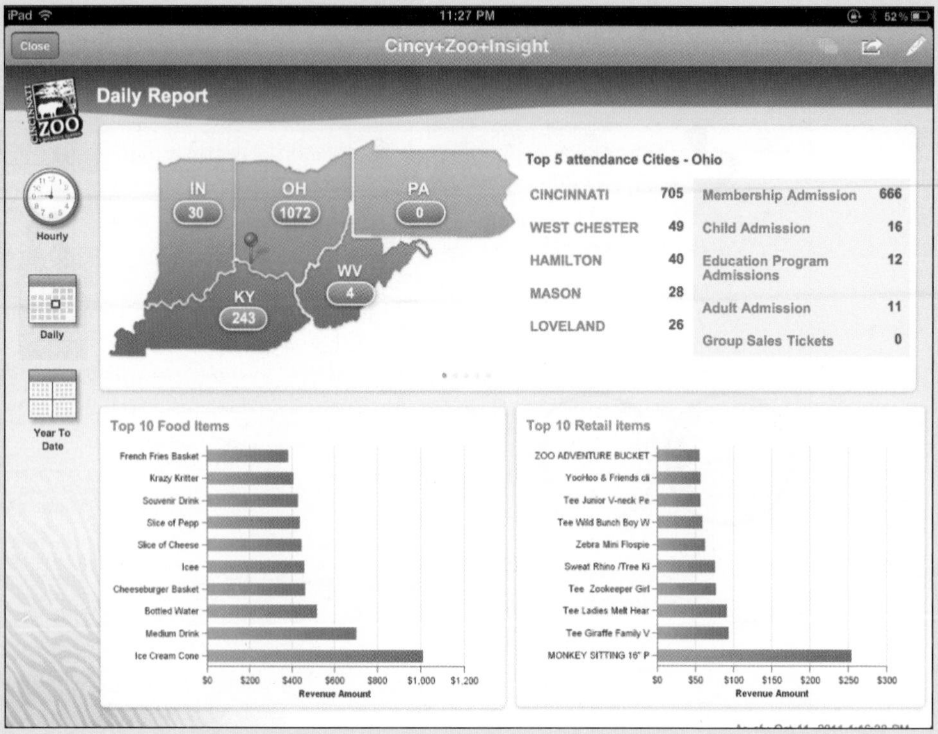

Having access to the data shown in Figures 2.13 and 2.14 allows the zoo managers to make better decisions on staffing levels within the zoo, which items to stock based upon weather and other conditions, and how to better target its advertising based on geodemographics.

The impact that data visualization has had on the zoo has been significant. Within the first year of use, the system has been directly responsible for revenue growth of over $500,000, increased visitation to the zoo, enhanced customer service, and reduced marketing costs.

NOTES AND COMMENTS

1. A variety of software is available for data visualization. Among the more popular packages are Cognos, JMP, Spotfire, and Tableau.
2. Radar charts and bubble charts are two other commonly used charts for displaying relationships between multiple variables. However, many experts in data visualization recommend against using these charts because they can be overcomplicated. Instead, the use of simpler displays such as bar charts and scatter diagrams is recommended.
3. A very powerful tool for visualizing geographic data is a Geographic Information System (GIS). A GIS uses color, symbols, and text on a map to help you understand how variables are distributed geographically. For example, a company interested in trying to locate a new distribution center might wish to better understand how the demand for its product varies throughout the United States. A GIS can be used to map the demand where red regions indicate high demand, blue lower demand, and no color for regions where the product is not sold. Locations closer to red high-demand regions might be good candidate sites for further consideration.

Summary

A set of data, even if modest in size, is often difficult to interpret directly in the form in which it is gathered. Tabular and graphical displays can be used to summarize and present data so that patterns are revealed and the data are more easily interpreted. Frequency distributions, relative frequency distributions, percent frequency distributions, bar charts, and pie charts were presented as tabular and graphical displays for summarizing the data for a single categorical variable. Frequency distributions, relative frequency distributions, percent frequency distributions, histograms, cumulative frequency distributions, cumulative relative frequency distributions, cumulative percent frequency distributions, and stem-and-leaf displays were presented as ways of summarizing the data for a single quantitative variable.

A crosstabulation was presented as a tabular display for summarizing the data for two variables and a scatter diagram was introduced as a graphical display for summarizing the data for two quantitative variables. We also showed that side-by-side bar charts and stacked bar charts are just extensions of basic bar charts that can be used to display and compare two categorical variables. Guidelines for creating effective graphical displays and how to choose the most appropriate type of display were discussed. Data dashboards were introduced to illustrate how a set of visual displays can be developed that organizes and presents information that is used to monitor a company's performance in a manner that is easy to read, understand, and interpret. Figure 2.15 provides a summary of the tabular and graphical methods presented in this chapter.

FIGURE 2.15 TABULAR AND GRAPHICAL DISPLAYS FOR SUMMARIZING DATA

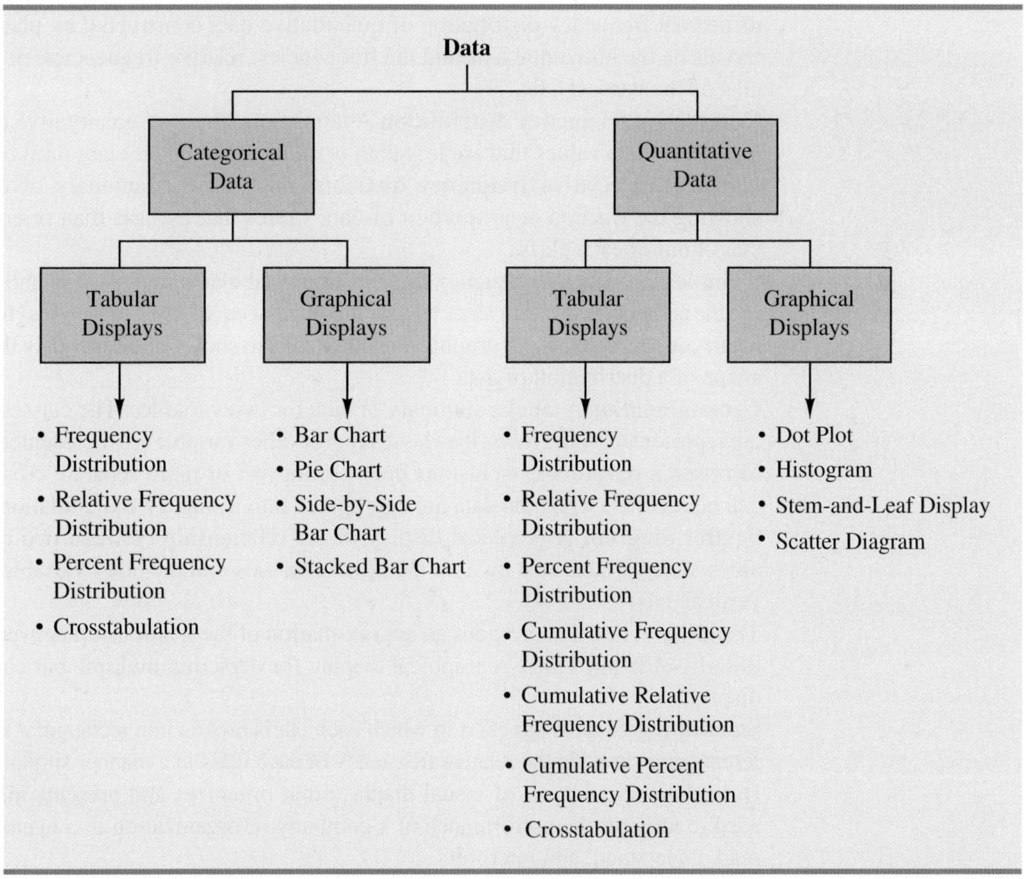

With large data sets, computer software packages are essential in constructing tabular and graphical summaries of data. In the chapter appendixes, we show how Minitab, Excel, and StatTools can be used for this purpose.

Glossary

Categorical data Labels or names used to identify categories of like items.

Quantitative data Numerical values that indicate how much or how many.

Data visualization A term used to describe the use of graphical displays to summarize and present information about a data set.

Frequency distribution A tabular summary of data showing the number (frequency) of observations in each of several nonoverlapping categories or classes.

Relative frequency distribution A tabular summary of data showing the fraction or proportion of observations in each of several nonoverlapping categories or classes.

Percent frequency distribution A tabular summary of data showing the percentage of observations in each of several nonoverlapping classes.

Bar chart A graphical device for depicting categorical data that have been summarized in a frequency, relative frequency, or percent frequency distribution.

Pie chart A graphical device for presenting data summaries based on subdivision of a circle into sectors that correspond to the relative frequency for each class.

Class midpoint The value halfway between the lower and upper class limits.

Dot plot A graphical device that summarizes data by the number of dots above each data value on the horizontal axis.

Histogram A graphical display of a frequency distribution, relative frequency distribution, or percent frequency distribution of quantitative data constructed by placing the class intervals on the horizontal axis and the frequencies, relative frequencies, or percent frequencies on the vertical axis.

Cumulative frequency distribution A tabular summary of quantitative data showing the number of data values that are less than or equal to the upper class limit of each class.

Cumulative relative frequency distribution A tabular summary of quantitative data showing the fraction or proportion of data values that are less than or equal to the upper class limit of each class.

Cumulative percent frequency distribution A tabular summary of quantitative data showing the percentage of data values that are less than or equal to the upper class limit of each class.

Stem-and-leaf display A graphical display used to show simultaneously the rank order and shape of a distribution of data.

Crosstabulation A tabular summary of data for two variables. The classes for one variable are represented by the rows; the classes for the other variable are represented by the columns.

Simpson's paradox Conclusions drawn from two or more separate crosstabulations that can be reversed when the data are aggregated into a single crosstabulation.

Scatter diagram A graphical display of the relationship between two quantitative variables. One variable is shown on the horizontal axis and the other variable is shown on the vertical axis.

Trendline A line that provides an approximation of the relationship between two variables.

Side-by-side bar chart A graphical display for depicting multiple bar charts on the same display.

Stacked bar chart A bar chart in which each bar is broken into rectangular segments of a different color showing the relative frequency of each class in a manner similar to a pie chart.

Data dashboard A set of visual displays that organizes and presents information that is used to monitor the performance of a company or organization in a manner that is easy to read, understand, and interpret.

Key Formulas

Relative Frequency

$$\frac{\text{Frequency of the class}}{n} \quad \textbf{(2.1)}$$

Approximate Class Width

$$\frac{\text{Largest data value} - \text{Smallest data value}}{\text{Number of classes}} \quad \textbf{(2.2)}$$

Supplementary Exercises

44. Approximately 1.5 million high school students take the SAT each year and nearly 80% of the college and universities without open admissions policies use SAT scores in making admission decisions (College Board, March 2009). The current version of the SAT includes three parts: reading comprehension, mathematics, and writing. A perfect combined score for all three parts is 2400. A sample of SAT scores for the combined three-part SAT are as follows:

WEB file
NewSAT

1665	1525	1355	1645	1780
1275	2135	1280	1060	1585
1650	1560	1150	1485	1990
1590	1880	1420	1755	1375
1475	1680	1440	1260	1730
1490	1560	940	1390	1175

a. Show a frequency distribution and histogram. Begin with the first class starting at 800 and use a class width of 200.
b. Comment on the shape of the distribution.
c. What other observations can be made about the SAT scores based on the tabular and graphical summaries?

45. The Pittsburgh Steelers defeated the Arizona Cardinals 27 to 23 in professional football's 43rd Super Bowl. With this win, its sixth championship, the Pittsburgh Steelers became the team with the most wins in the 43-year history of the event (*Tampa Tribune,* February 2,

WEB file
SuperBowl

Super Bowl	State	Won By Points	Super Bowl	State	Won By Points	Super Bowl	State	Won By Points
1	CA	25	16	MI	5	31	LA	14
2	FL	19	17	CA	10	32	CA	7
3	FL	9	18	FL	19	33	FL	15
4	LA	16	19	CA	22	34	GA	7
5	FL	3	20	LA	36	35	FL	27
6	FL	21	21	CA	19	36	LA	3
7	CA	7	22	CA	32	37	CA	27
8	TX	17	23	FL	4	38	TX	3
9	LA	10	24	LA	45	39	FL	3
10	FL	4	25	FL	1	40	MI	11
11	CA	18	26	MN	13	41	FL	12
12	LA	17	27	CA	35	42	AZ	3
13	FL	4	28	GA	17	43	FL	4
14	CA	12	29	FL	23			
15	LA	17	30	AZ	10			

2009). The Super Bowl has been played in eight different states: Arizona (AZ), California (CA), Florida (FL), Georgia (GA), Louisiana (LA), Michigan (MI), Minnesota (MN), and Texas (TX). Data in the following table show the state where the Super Bowls were played and the point margin of victory for the winning team.

a. Show a frequency distribution and bar chart for the state where the Super Bowl was played.
b. What conclusions can you draw from your summary in part (a)? What percentage of Super Bowls were played in the states of Florida or California? What percentage of Super Bowls were played in northern or cold-weather states?
c. Show a stretched stem-and-leaf display for the point margin of victory for the winning team. Show a histogram.
d. What conclusions can you draw from your summary in part (c)? What percentage of Super Bowls have been close games with the margin of victory less than 5 points? What percentage of Super Bowls have been won by 20 or more points?
e. The closest Super Bowl occurred when the New York Giants beat the Buffalo Bills. Where was this game played and what was the winning margin of victory? The biggest point margin in Super Bowl history occurred when the San Francisco 49ers beat the Denver Broncos. Where was this game played and what was the winning margin of victory?

46. Data showing the population by state in millions of people follow (*The World Almanac,* 2012).

WEB file
2012Population

State	Population	State	Population	State	Population
Alabama	4.8	Louisiana	4.5	Ohio	11.5
Alaska	0.7	Maine	1.3	Oklahoma	3.8
Arizona	6.4	Maryland	5.8	Oregon	4.3
Arkansas	2.9	Massachusetts	6.5	Pennsylvania	12.7
California	37.3	Michigan	9.9	Rhode Island	1.0
Colorado	5.0	Minnesota	5.3	South Carolina	4.6
Connecticut	3.6	Mississippi	3.0	South Dakota	0.8
Delaware	0.9	Missouri	6.0	Tennessee	6.3
Florida	18.8	Montana	0.9	Texas	25.1
Georgia	9.7	Nebraska	1.8	Utah	2.8
Hawaii	1.4	Nevada	2.7	Vermont	0.6
Idaho	1.6	New Hampshire	1.3	Virginia	8.0
Illinois	12.8	New Jersey	8.8	Washington	6.7
Indiana	6.5	New Mexico	2.0	West Virginia	1.9
Iowa	3.0	New York	19.4	Wisconsin	5.7
Kansas	2.9	North Carolina	9.5	Wyoming	0.6
Kentucky	4.3	North Dakota	0.7		

a. Develop a frequency distribution, a percent frequency distribution, and a histogram. Use a class width of 2.5 million.
b. Does there appear to be any skewness in the distribution? Explain.
c. What observations can you make about the population of the 50 states?

47. A startup company's ability to gain funding is a key to success. The funds raised (in millions of dollars) by 50 startup companies appear below (*The Wall Street Journal,* March 10, 2011).

WEB file
StartUps

81	61	103	166	168
80	51	130	77	78
69	119	81	60	20
73	50	110	21	60
192	18	54	49	63

91	272	58	54	40
47	24	57	78	78
154	72	38	131	52
48	118	40	49	55
54	112	129	156	31

a. Construct a stem-and-leaf display.
b. Comment on the display.

WEB file BBB

48. Consumer complaints are frequently reported to the Better Business Bureau. In 2011, the industries with the most complaints to the Better Business Bureau were banks; cable and satellite television companies; collection agencies; cellular phone providers; and new car dealerships (*USA Today,* April 16, 2012). The results for a sample of 200 complaints are contained in the file BBB.
    a. Show the frequency and percent frequency of complaints by industry.
    b. Construct a bar chart of the percent frequency distribution.
    c. Which industry had the highest number of complaints?
    d. Comment on the percentage frequency distribution for complaints.

49. Dividend yield is the annual dividend paid by a company expressed as a percentage of the price of the stock (Dividend/Stock Price × 100). The dividend yield for the Dow Jones Industrial Average companies is shown in Table 2.17 (*The Wall Street Journal,* June 8, 2009).
    a. Construct a frequency distribution and percent frequency distribution.
    b. Construct a histogram.
    c. Comment on the shape of the distribution.
    d. What do the tabular and graphical summaries indicate about the dividend yields among the Dow Jones Industrial Average companies?
    e. What company has the highest dividend yield? If the stock for this company currently sells for $14 per share and you purchase 500 shares, how much dividend income will this investment generate in one year?

50. One of the questions in a *Financial Times*/Harris Poll was, "How much do you favor or oppose a higher tax on higher carbon emission cars?" Possible responses were strongly favor, favor more than oppose, oppose more than favor, and strongly oppose. The following

TABLE 2.17 DIVIDEND YIELD FOR DOW JONES INDUSTRIAL AVERAGE COMPANIES

WEB file DYield

Company	Dividend Yield %	Company	Dividend Yield %
3M	3.6	IBM	2.1
Alcoa	1.3	Intel	3.4
American Express	2.9	J.P. Morgan Chase	0.5
AT&T	6.6	Johnson & Johnson	3.6
Bank of America	0.4	Kraft Foods	4.4
Boeing	3.8	McDonald's	3.4
Caterpillar	4.7	Merck	5.5
Chevron	3.9	Microsoft	2.5
Cisco Systems	0.0	Pfizer	4.2
Coca-Cola	3.3	Procter & Gamble	3.4
DuPont	5.8	Travelers	3.0
ExxonMobil	2.4	United Technologies	2.9
General Electric	9.2	Verizon	6.3
Hewlett-Packard	0.9	Wal-Mart Stores	2.2
Home Depot	3.9	Walt Disney	1.5

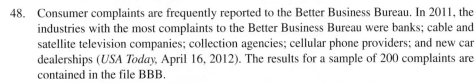

crosstabulation shows the responses obtained for 5372 adults surveyed in four countries in Europe and the United States (Harris Interactive website, February 27, 2008).

	Country					
Level of Support	Great Britain	Italy	Spain	Germany	United States	Total
Strongly favor	337	334	510	222	214	1617
Favor more than oppose	370	408	355	411	327	1871
Oppose more than favor	250	188	155	267	275	1135
Strongly oppose	130	115	89	211	204	749
Total	1087	1045	1109	1111	1020	5372

a. Construct a percent frequency distribution for the level of support variable. Do you think the results show support for a higher tax on higher carbon emission cars?
b. Construct a percent frequency distribution for the country variable.
c. Does the level of support among adults in the European countries appear to be different than the level of support among adults in the United States? Explain.

51. Western University has only one women's softball scholarship remaining for the coming year. The final two players that Western is considering are Allison Fealey and Emily Janson. The coaching staff has concluded that the speed and defensive skills are virtually identical for the two players, and that the final decision will be based on which player has the best batting average. Crosstabulations of each player's batting performance in their junior and senior years of high school are as follows:

	Allison Fealey	
Outcome	Junior	Senior
Hit	15	75
No Hit	25	175
Total At-Bats	40	250

	Emily Janson	
Outcome	Junior	Senior
Hit	70	35
No Hit	130	85
Total At Bats	200	120

A player's batting average is computed by dividing the number of hits a player has by the total number of at-bats. Batting averages are represented as a decimal number with three places after the decimal.

a. Calculate the batting average for each player in her junior year. Then calculate the batting average of each player in her senior year. Using this analysis, which player should be awarded the scholarship? Explain.
b. Combine or aggregate the data for the junior and senior years into one crosstabulation as follows:

	Player	
Outcome	Fealey	Janson
Hit		
No Hit		
Total At-Bats		

Calculate each player's batting average for the combined two years. Using this analysis, which player should be awarded the scholarship? Explain.
c. Are the recommendations you made in parts (a) and (b) consistent? Explain any apparent inconsistencies.

52. A survey of commercial buildings served by the Cincinnati Gas & Electric Company asked what main heating fuel was used and what year the building was constructed. A partial crosstabulation of the findings follows.

Year Constructed	Fuel Type				
	Electricity	**Natural Gas**	**Oil**	**Propane**	**Other**
1973 or before	40	183	12	5	7
1974–1979	24	26	2	2	0
1980–1986	37	38	1	0	6
1987–1991	48	70	2	0	1

a. Complete the crosstabulation by showing the row totals and column totals.
b. Show the frequency distributions for year constructed and for fuel type.
c. Prepare a crosstabulation showing column percentages.
d. Prepare a crosstabulation showing row percentages.
e. Comment on the relationship between year constructed and fuel type.

53. Table 2.18 shows a portion of the data for a sample of 103 private colleges and universities. The complete data set is contained in the file named Colleges. The data include the name of the college or university, the year the institution was founded, the tuition and fees (not including room and board) for the most recent academic year, and the percentage of full time, first-time bachelor's degree-seeking undergraduate students who obtain their degree in six years or less (The World Almanac, 2012)
   a. Construct a crosstabulation with Year Founded as the row variable and Tuition & Fees as the column variable. Use classes starting with 1600 and ending with 2000 in increments of 50 for Year Founded. For Tuition & Fees, use classes starting with 1 and ending 45000 in increments of 5000.
   b. Compute the row percentages for the crosstabulation in part (a).
   c. What relationship, if any, do you notice between Year Founded and Tuition & Fees?

54. Refer to the data set in Table 2.18.
   a. Construct a crosstabulation with Year Founded as the row variable and % Graduate as the column variable. Use classes starting with 1600 and ending with 2000 in increments of 50 for Year Founded. For % Graduate, use classes starting with 35% and ending with 100% in increments of 5%.
   b. Compute the row percentages for your crosstabulation in part (a).
   c. Comment on any relationship between the variables.

55. Refer to the data set in Table 2.18.
   a. Construct a scatter diagram to show the relationship between Year Founded and Tuition & Fees.
   b. Comment on any relationship between the variables.

TABLE 2.18 DATA FOR A SAMPLE OF PRIVATE COLLEGES AND UNIVERSITIES

WEB file
Colleges

School	Year Founded	Tuition & Fees	% Graduate
American University	1893	$36,697	79.00
Baylor University	1845	$29,754	70.00
Belmont University	1951	$23,680	68.00
·	·	·	·
·	·	·	·
·	·	·	·
Wofford College	1854	$31,710	82.00
Xavier University	1831	$29,970	79.00
Yale University	1701	$38,300	98.00

56. Refer to the data set in Table 2.18.
   a. Prepare a scatter diagram to show the relationship between Tuition & Fees and % Graduate.
   b. Comment on any relationship between the variables.
57. Google has changed its strategy with regard to how much and over which media it invests in advertising. The following table shows Google's marketing budget in millions of dollars for 2008 and 2011 (*The Wall Street Journal,* March 27, 2012).

	2008	2011
Internet	26.0	123.3
Newspaper, etc.	4.0	20.7
Television	0.0	69.3

   a. Construct a side-by-side bar chart with year as the variable on the horizontal axis. Comment on any trend in the display.
   b. Convert the above table to percentage allocation for each year. Construct a stacked bar chart with year as the variable on the horizontal axis.
   c. Is the display in part (a) or part (b) more insightful? Explain.
58. A zoo has categorized its visitors into three categories: member, school, and general. The member category refers to visitors who pay an annual fee to support the zoo. Members receive certain benefits such as discounts on merchandise and trips planned by the zoo. The school category includes faculty and students from day care and elementary and secondary schools; these visitors generally receive a discounted rate. The general category includes all other visitors. The zoo has been concerned about a recent drop in attendance. To help better understand attendance and membership, a zoo staff member has collected the following data:

WEB file
Zoo

	Attendance			
Visitor Category	2008	2009	2010	2011
General	153,713	158,704	163,433	169,106
Member	115,523	104,795	98,437	81,217
School	82,885	79,876	81,970	81,290
Total	352,121	343,375	343,840	331,613

   a. Construct a bar chart of total attendance over time. Comment on any trend in the data.
   b. Construct a side-by-side bar chart showing attendance by visitor category with year as the variable on the horizontal axis.
   c. Comment on what is happening to zoo attendance based on the charts from parts (a) and (b).

Case Problem 1 Pelican Stores

Pelican Stores, a division of National Clothing, is a chain of women's apparel stores operating throughout the country. The chain recently ran a promotion in which discount coupons were sent to customers of other National Clothing stores. Data collected for a sample of 100 in-store credit card transactions at Pelican Stores during one day while the promotion was running are contained in the file named PelicanStores. Table 2.19 shows a portion of the data set. The Proprietary Card method of payment refers to charges made using a National Clothing charge card. Customers who made a purchase using a discount coupon are referred to as promotional customers and customers who made a purchase but did not use a discount

TABLE 2.19 DATA FOR A SAMPLE OF 100 CREDIT CARD PURCHASES AT PELICAN STORES

Customer	Type of Customer	Items	Net Sales	Method of Payment	Gender	Marital Status	Age
1	Regular	1	39.50	Discover	Male	Married	32
2	Promotional	1	102.40	Proprietary Card	Female	Married	36
3	Regular	1	22.50	Proprietary Card	Female	Married	32
4	Promotional	5	100.40	Proprietary Card	Female	Married	28
5	Regular	2	54.00	MasterCard	Female	Married	34
.	.	.	.	.	.	.	.
.	.	.	.	.	.	.	.
.	.	.	.	.	.	.	.
96	Regular	1	39.50	MasterCard	Female	Married	44
97	Promotional	9	253.00	Proprietary Card	Female	Married	30
98	Promotional	10	287.59	Proprietary Card	Female	Married	52
99	Promotional	2	47.60	Proprietary Card	Female	Married	30
100	Promotional	1	28.44	Proprietary Card	Female	Married	44

WEB file

PelicanStores

coupon are referred to as regular customers. Because the promotional coupons were not sent to regular Pelican Stores customers, management considers the sales made to people presenting the promotional coupons as sales it would not otherwise make. Of course, Pelican also hopes that the promotional customers will continue to shop at its stores.

Most of the variables shown in Table 2.19 are self-explanatory, but two of the variables require some clarification.

Items	The total number of items purchased
Net Sales	The total amount ($) charged to the credit card

Pelican's management would like to use this sample data to learn about its customer base and to evaluate the promotion involving discount coupons.

Managerial Report

Use the tabular and graphical methods of descriptive statistics to help management develop a customer profile and to evaluate the promotional campaign. At a minimum, your report should include the following:

1. Percent frequency distribution for key variables.
2. A bar chart or pie chart showing the number of customer purchases attributable to the method of payment.
3. A crosstabulation of type of customer (regular or promotional) versus net sales. Comment on any similarities or differences present.
4. A scatter diagram to explore the relationship between net sales and customer age.

Case Problem 2 Motion Picture Industry

The motion picture industry is a competitive business. More than 50 studios produce a total of 300 to 400 new motion pictures each year, and the financial success of each motion picture varies considerably. The opening weekend gross sales ($millions), the total gross sales ($millions), the number of theaters the movie was shown in, and the number of weeks the motion picture was in release are common variables used to measure the success of a motion picture. Data collected for the top 100 motion pictures produced in 2011 are contained in the file named 2011Movies (Box Office Mojo, March 17, 2012). Table 2.20 shows the data for the first 10 motion pictures in this file.

TABLE 2.20 PERFORMANCE DATA FOR 10 MOTION PICTURES

WEB file 2011Movies

Motion Picture	Opening Gross Sales ($millions)	Total Gross Sales ($millions)	Number of Theaters	Weeks in Release
Harry Potter and the Deathly Hallows Part 2	169.19	381.01	4375	19
Transformers: Dark of the Moon	97.85	352.39	4088	15
The Twilight Saga: Breaking Dawn Part 1	138.12	281.29	4066	14
The Hangover Part II	85.95	254.46	3675	16
Pirates of the Caribbean: On Stranger Tides	90.15	241.07	4164	19
Fast Five	86.20	209.84	3793	15
Mission: Impossible—Ghost Protocol	12.79	208.55	3555	13
Cars 2	66.14	191.45	4115	25
Sherlock Holmes: A Game of Shadows	39.64	186.59	3703	13
Thor	65.72	181.03	3963	16

Managerial Report

Use the tabular and graphical methods of descriptive statistics to learn how these variables contribute to the success of a motion picture. Include the following in your report.

1. Tabular and graphical summaries for each of the four variables along with a discussion of what each summary tells us about the motion picture industry.
2. A scatter diagram to explore the relationship between Total Gross Sales and Opening Weekend Gross Sales. Discuss.
3. A scatter diagram to explore the relationship between Total Gross Sales and Number of Theaters. Discuss.
4. A scatter diagram to explore the relationship between Total Gross Sales and Number of Weeks in Release. Discuss.

Appendix 2.1 Using Minitab for Tabular and Graphical Presentations

Minitab offers extensive capabilities for constructing tabular and graphical summaries of data. In this appendix we show how Minitab can be used to construct several graphical summaries and the tabular summary of a crosstabulation. The graphical methods presented include the dot plot, the histogram, the stem-and-leaf display, and the scatter diagram.

Dot Plot

WEB file Audit

We use the audit time data in Table 2.4 to demonstrate. The data are in column C1 of a Minitab worksheet. The following steps will generate a dot plot.

Step 1. Select the **Graph** menu and choose **Dotplot**
Step 2. Select **One Y, Simple** and click **OK**

Step 3. When the Dotplot-One Y, Simple dialog box appears:
Enter C1 in the **Graph Variables** box
Click **OK**

Histogram

WEB file Audit

We show how to construct a histogram with frequencies on the vertical axis using the audit time data in Table 2.4. The data are in column C1 of a Minitab worksheet. The following steps will generate a histogram for audit times.

Step 1. Select the **Graph** menu
Step 2. Choose **Histogram**
Step 3. When the Histograms dialog box appears:
Choose **Simple** and click **OK**
Step 4. When the Histogram-Simple dialog box appears:
Enter C1 in the **Graph Variables** box
Click **OK**
Step 5. When the Histogram appears:*
Position the mouse pointer over any one of the bars
Double-click
Step 6. When the Edit Bars dialog box appears:
Click the **Binning** tab
Select **Cutpoint** for Interval Type
Select **Midpoint/Cutpoint positions** for Interval Definition
Enter 10:35/5 in the **Midpoint/Cutpoint positions** box
Click **OK**

Note that Minitab also provides the option of scaling the x-axis so that the numerical values appear at the midpoints of the histogram rectangles. If this option is desired, modify step 6 to include Select **Midpoint** for Interval Type and Enter 12:32/5 in the **Midpoint/Cutpoint positions** box. These steps provide the same histogram with the midpoints of the histogram rectangles labeled 12, 17, 22, 27, and 32.

Stem-and-Leaf Display

WEB file ApTest

We use the aptitude test data in Table 2.8 to demonstrate the construction of a stem-and-leaf display. The data are in column C1 of a Minitab worksheet. The following steps will generate the stretched stem-and-leaf display shown in Section 2.3.

Step 1. Select the **Graph** menu
Step 2. Choose **Stem-and-Leaf**
Step 3. When the Stem-and-Leaf dialog box appears:
Enter C1 in the **Graph Variables** box
Click **OK**

Scatter Diagram

WEB file Stereo

We use the stereo and sound equipment store data in Table 2.12 to demonstrate the construction of a scatter diagram. The weeks are numbered from 1 to 10 in column C1, the data for number of commercials are in column C2, and the data for sales are in column C3 of a Minitab worksheet. The following steps will generate the scatter diagram shown in Figure 2.7.

*Steps 5 and 6 are optional but are shown here to demonstrate user flexibility in displaying the histogram. The entry 10:35/5 in step 6 indicates that 10 is the starting value for the histogram, 35 is the ending value for the histogram, and 5 is the class width.

Step 1. Select the **Graph** menu
Step 2. Choose **Scatterplot**
Step 3. Select **Simple** and click **OK**
Step 4. When the Scatterplot-Simple dialog box appears:
Enter C3 under **Y variables** and C2 under **X variables**
Click **OK**

Crosstabulation

WEB file
Restaurant

We use the data from Zagat's restaurant review, part of which is shown in Table 2.9, to demonstrate. The restaurants are numbered from 1 to 300 in column C1 of the Minitab worksheet. The quality ratings are in column C2, and the meal prices are in column C3.

Minitab can only create a crosstabulation for qualitative variables and meal price is a quantitative variable. So we need to first code the meal price data by specifying the class to which each meal price belongs. The following steps will code the meal price data to create four classes of meal price in column C4: \$10–19, \$20–29, \$30–39, and \$40–49.

Step 1. Select the **Data** menu
Step 2. Choose **Code**
Step 3. Choose **Numeric to Text**
Step 4. When the Code-Numeric to Text dialog box appears:
Enter C3 in the **Code data from columns** box
Enter C4 in the **Store coded data in columns** box
Enter 10:19 in the first **Original values** box and \$10–19 in the adjacent **New** box
Enter 20:29 in the second **Original values** box and \$20–29 in the adjacent **New** box
Enter 30:39 in the third **Original values** box and \$30–39 in the adjacent **New** box
Enter 40:49 in the fourth **Original values** box and \$40–49 in the adjacent **New** box
Click **OK**

For each meal price in column C3 the associated meal price category will now appear in column C4. We can now develop a crosstabulation for quality rating and the meal price categories by using the data in columns C2 and C4. The following steps will create a crosstabulation containing the same information as shown in Table 2.10.

Step 1. Select the **Stat** menu
Step 2. Choose **Tables**
Step 3. Choose **Cross Tabulation and Chi-Square**
Step 4. When the Cross Tabulation and Chi-Square dialog box appears:
Enter C2 in the **For rows** box and C4 in the **For columns** box
Select **Counts** under Display
Click **OK**

Appendix 2.2 Using Excel for Tabular and Graphical Presentations

Excel offers extensive capabilities for constructing tabular and graphical summaries of data. In this appendix, we show how Excel can be used to construct a frequency distribution, bar chart, pie chart, histogram, scatter diagram, and crosstabulation. We will demonstrate three of Excel's most powerful tools for data analysis: chart tools, PivotChart Report, and PivotTable Report.

Frequency Distribution and Bar Chart for Categorical Data

In this section we show how Excel can be used to construct a frequency distribution and a bar chart for categorical data. We illustrate each using the data on soft drink purchases in Table 2.1.

Frequency distribution We begin by showing how the COUNTIF function can be used to construct a frequency distribution for the data in Table 2.1. Refer to Figure 2.16 as we describe the steps involved. The formula worksheet (showing the functions and formulas used) is set in the background, and the value worksheet (showing the results obtained using the functions and formulas) appears in the foreground.

WEB file SoftDrink

The label "Brand Purchased" and the data for the 50 soft drink purchases are in cells A1:A51. We also entered the labels "Soft Drink" and "Frequency" in cells C1:D1. The five soft drink names are entered into cells C2:C6. Excel's COUNTIF function can now be used to count the number of times each soft drink appears in cells A2:A51. The following steps are used.

Step 1. Select cell D2
Step 2. Enter =COUNTIF($A$2:$A$51,C2)
Step 3. Copy cell D2 to cells D3:D6

The formula worksheet in Figure 2.16 shows the cell formulas inserted by applying these steps. The value worksheet shows the values computed by the cell formulas. This worksheet shows the same frequency distribution that we developed in Table 2.2.

WEB file SoftDrink

Bar chart Here we show how Excel's chart tools can be used to construct a bar chart for the soft drink data. Refer to the frequency distribution shown in the value worksheet of Figure 2.16. The bar chart that we are going to develop is an extension of this worksheet. The worksheet and the bar chart developed are shown in Figure 2.17. The steps are as follows:

Step 1. Select cells C2:D6
Step 2. Click the **Insert** tab on the Ribbon
Step 3. In the **Charts** group, click **Column**
Step 4. When the list of column chart subtypes appears:
Go to the **2-D Column** section
Click **Clustered Column** (the leftmost chart)
Step 5. In the **Chart Layouts** group, click the **More** button (the downward-pointing arrow with a line over it) to display all the options
Step 6. Choose **Layout 9**
Step 7. Select the **Chart Title** and replace it with **Bar Chart of Soft Drink Purchases**
Step 8. Select the **Horizontal (Category) Axis Title** and replace it with **Soft Drink**
Step 9. Select the **Vertical (Value) Axis Title** and replace it with **Frequency**
Step 10. Right-click the **Series 1 Legend Entry**
Click **Delete**
Step 11. Right-click the vertical axis
Click **Format Axis**
Step 12. When the Format Axis dialog box appears:
Go to the **Axis Options** section
Select **Fixed** for **Major Unit** and enter 5.0 in the corresponding box
Click **Close**

The resulting bar chart is shown in Figure 2.17.*

---

*The bar chart in Figure 2.17 can be resized. Resizing an Excel chart is not difficult. First, select the chart. Sizing handles will appear on the chart border. Click on the sizing handles and drag them to resize the figure to your preference.

FIGURE 2.16 FREQUENCY DISTRIBUTION FOR SOFT DRINK PURCHASES CONSTRUCTED USING EXCEL'S COUNTIF FUNCTION

	A	B	C	D
1	**Brand Purchased**		**Soft Drink**	**Frequency**
2	Coca-Cola		Coca-Cola	=COUNTIF($A$2:$A$51,C2)
3	Diet Coke		Diet Coke	=COUNTIF($A$2:$A$51,C3)
4	Pepsi		Dr. Pepper	=COUNTIF($A$2:$A$51,C4)
5	Diet Coke		Pepsi	=COUNTIF($A$2:$A$51,C5)
6	Coca-Cola		Sprite	=COUNTIF($A$2:$A$51,C6)
7	Coca-Cola			
8	Dr. Pepper			
9	Diet Coke			
10	Pepsi			
11	Pepsi			
12	Coca-Cola			
13	Dr. Pepper			
14	Sprite			
15	Coca-Cola			
16	Diet Coke			
17	Coca-Cola			
18	Coca-Cola			
19	Sprite			
20	Coca-Cola			
50	Pepsi			
51	Sprite			

Note: Rows 21–49 are hidden.

	A	B	C	D
1	**Brand Purchased**		**Soft Drink**	**Frequency**
2	Coca-Cola		Coca-Cola	19
3	Diet Coke		Diet Coke	8
4	Pepsi		Dr. Pepper	5
5	Diet Coke		Pepsi	13
6	Coca-Cola		Sprite	5
7	Coca-Cola			
8	Dr. Pepper			
9	Diet Coke			
10	Pepsi			
11	Pepsi			
12	Coca-Cola			
13	Dr. Pepper			
14	Sprite			
15	Coca-Cola			
16	Diet Coke			
17	Coca-Cola			
18	Coca-Cola			
19	Sprite			
20	Coca-Cola			
50	Pepsi			
51	Sprite			

Excel can produce a pie chart for the soft drink data in a similar fashion. The major difference is that in step 3 you would click **Pie** in the **Charts** group. Several style pie charts are available. Likewise for comparing two variables, you can construct side-by-side or stacked bar charts. Once you have selected the cells containing the data for the two variables (step 1) following the steps as described (choosing **Clustered Column** from the **2-D Column** section) will create a side-by-side bar chart. To construct a stacked bar chart, the major difference is in step 4. Rather than selecting **Clustered Column**, you must select **Stacked Column** for a stacked bar chart of frequency or **100% Stacked Column** for a percent frequency stacked bar chart.

Frequency Distribution and Histogram for Quantitative Data

In a later section of this appendix we describe how to use Excel's PivotTable Report to construct a crosstabulation.

Excel's PivotTable Report is an interactive tool that allows you to quickly summarize data in a variety of ways, including developing a frequency distribution for quantitative data. Once a frequency distribution is created using the PivotTable Report, Excel's chart tools can then be used to construct the corresponding histogram. But, using Excel's PivotChart Report, we can construct a frequency distribution and a histogram simultaneously. We will illustrate this procedure using the audit time data in Table 2.4. The label "Audit Time" and the 20 audit

FIGURE 2.17 BAR CHART OF SOFT DRINK PURCHASES CONSTRUCTED USING EXCEL'S CHART TOOLS

	A	B	C	D	E	F	G
1	**Brand Purchased**		**Soft Drink**	**Frequency**			
2	Coca-Cola		Coca-Cola	19			
3	Diet Coke		Diet Coke	8			
4	Pepsi		Dr. Pepper	5			
5	Diet Coke		Pepsi	13			
6	Coca-Cola		Sprite	5			
7	Coca-Cola						
8	Dr. Pepper						
9	Diet Coke						
10	Pepsi						
11	Pepsi						
12	Coca-Cola						
13	Dr. Pepper						
14	Sprite						
15	Coca-Cola						
16	Diet Coke						
17	Coca-Cola						
18	Coca-Cola						
19	Sprite						
20	Coca-Cola						
50	Pepsi						
51	Sprite						

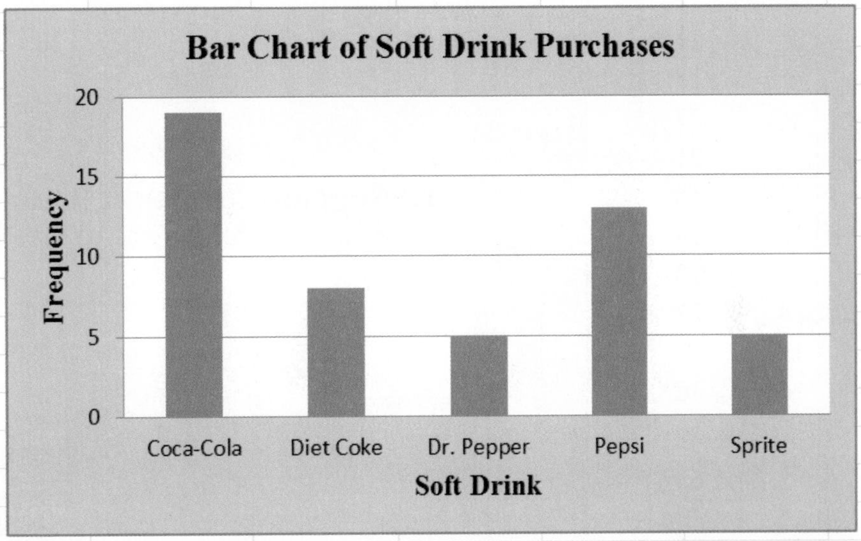

Note: Rows 21–49 are hidden.

WEB file

Audit

time values are entered into cells A1:A21 of an Excel worksheet. The following steps describe how to use Excel's PivotChart Report to construct a frequency distribution and a histogram for the audit time data. Refer to Figure 2.18 as we describe the steps involved.

Step 1. Click the **Insert** tab on the Ribbon
Step 2. In the **Tables** group, click the word **PivotTable**
Step 3. Choose **PivotChart** from the options that appear
Step 4. When the **Create PivotTable with PivotChart** dialog box appears,
Choose **Select a table or range**
Enter A1:A21 in the **Table/Range** box
Choose **Existing Worksheet** as the location for the PivotTable and PivotChart
Enter C1 in the **Location** box
Click **OK**
Step 5. In the **PivotTable Field List**, go to **Choose Fields to add to report**
Drag the **Audit Time** field to the **Axis Fields (Categories)** area
Drag the **Audit Time** field to the **Values** area
Step 6. Click **Sum of Audit Time** in the **Values** area
Step 7. Click **Value Field Settings** from the list of options that appears

FIGURE 2.18 USING EXCEL'S PIVOTCHART REPORT TO CONSTRUCT A FREQUENCY DISTRIBUTION AND HISTOGRAM FOR THE AUDIT TIME DATA

	A	B	C	D	E	F
1	**Audit Time**		**Row Labels**	**Count of Audit Time**		
2	12		10-14	4		
3	15		15-19	8		
4	20		20-24	5		
5	22		25-29	2		
6	14		30-34	1		
7	14		**Grand Total**	**20**		
8	15					
9	27					
10	21					
11	18					
12	19					
13	18					
14	22					
15	33					
16	16					
17	18					
18	17					
19	23					
20	28					
21	13					
22						
23						

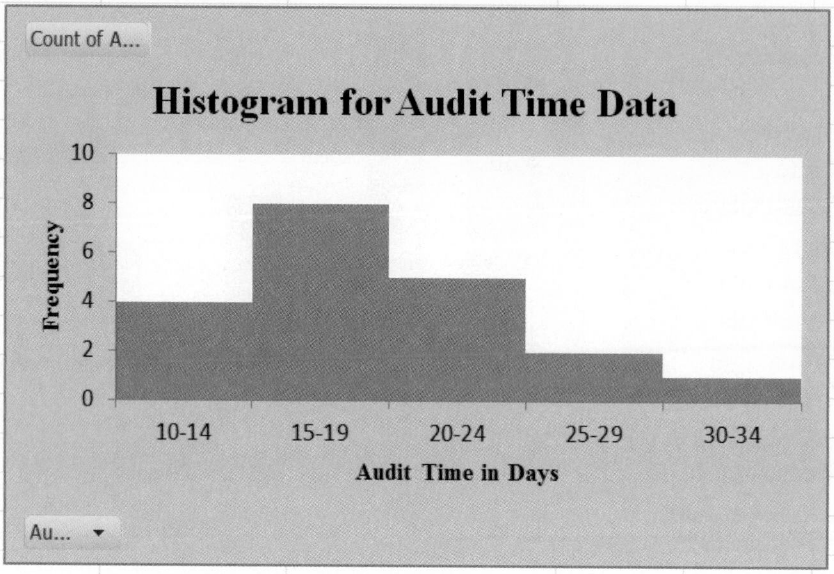

Step 8. When the Value Field Settings dialog appears,
Under **Summarize value field by**, choose **Count**
Click **OK**

Step 9. Close the **PivotTable Field List**

Step 10. Right-click cell C2 in the PivotTable report or any other cell containing an audit time

Step 11. Choose **Group** from the list of options that appears

Step 12. When the **Grouping** dialog box appears,
Enter 10 in the **Starting at** box
Enter 34 in the **Ending at** box
Enter 5 in the **By** box
Click **OK** (a PivotChart will appear)

Step 13. Click inside the resulting PivotChart

Step 14. Click the **Design** tab on the Ribbon

Step 15. In the **Chart Layouts** group, click the **More** button (the downward pointing arrow with a line over it) to display all the options

Step 16. Choose **Layout 8**

The PivotChart contains a Value field button in the upper left corner of the chart and an Axis field button in the lower left corner. To remove either or both of these buttons, right-click either button and select the option to remove that button or all field buttons shown on the chart.

Step 17. Select the **Chart Title** and replace it with **Histogram for Audit Time Data**

Step 18. Select the **Horizontal (Category) Axis Title** and replace it with **Audit Time in Days**

Step 19. Select the **Vertical (Value) Axis Title** and replace it with **Frequency**

Figure 2.18 shows the resulting PivotTable and PivotChart. We see that the PivotTable report provides the frequency distribution for the audit time data and the PivotChart provides the corresponding histogram. If desired, we can change the labels in any cell in the frequency distribution by selecting the cell and typing in the new label.

Crosstabulation

Excel's PivotTable Report provides an excellent way to summarize the data for two or more variables simultaneously. We will illustrate the use of Excel's PivotTable Report by showing how to develop a crosstabulation of quality ratings and meal prices for the sample of 300 Los Angeles restaurants. We will use the data in the file named Restaurant; the labels "Restaurant," "Quality Rating," and "Meal Price ($)" have been entered into cells A1:C1 of the worksheet as shown in Figure 2.19. The data for each of the restaurants in the sample have been entered into cells B2:C301.

In order to use the Pivot Table report to create a crosstabulation, we need to perform three tasks: Display the Initial PivotTable Field List and PivotTable Report; Set Up the PivotTable Field List; and Finalize the PivotTable Report. These tasks are described as follows.

Display the Initial PivotTable Field List and PivotTable Report: Three steps are needed to display the initial PivotTable Field List and PivotTable report.

Step 1. Click the **Insert** tab on the Ribbon

Step 2. In the **Tables** group, click the icon above the word PivotTable

FIGURE 2.19 EXCEL WORKSHEET CONTAINING RESTAURANT DATA

WEB file

Restaurant

Note: Rows 12–291 are hidden.

	A	B	C
1	Restaurant	Quality Rating	Meal Price ($)
2	1	Good	18
3	2	Very Good	22
4	3	Good	28
5	4	Excellent	38
6	5	Very Good	33
7	6	Good	28
8	7	Very Good	19
9	8	Very Good	11
10	9	Very Good	23
11	10	Good	13
292	291	Very Good	23
293	292	Very Good	24
294	293	Excellent	45
295	294	Good	14
296	295	Good	18
297	296	Good	17
298	297	Good	16
299	298	Good	15
300	299	Very Good	38
301	300	Very Good	31

FIGURE 2.20 INITIAL PIVOTTABLE FIELD LIST AND PIVOTTABLE FIELD REPORT FOR THE RESTAURANT DATA

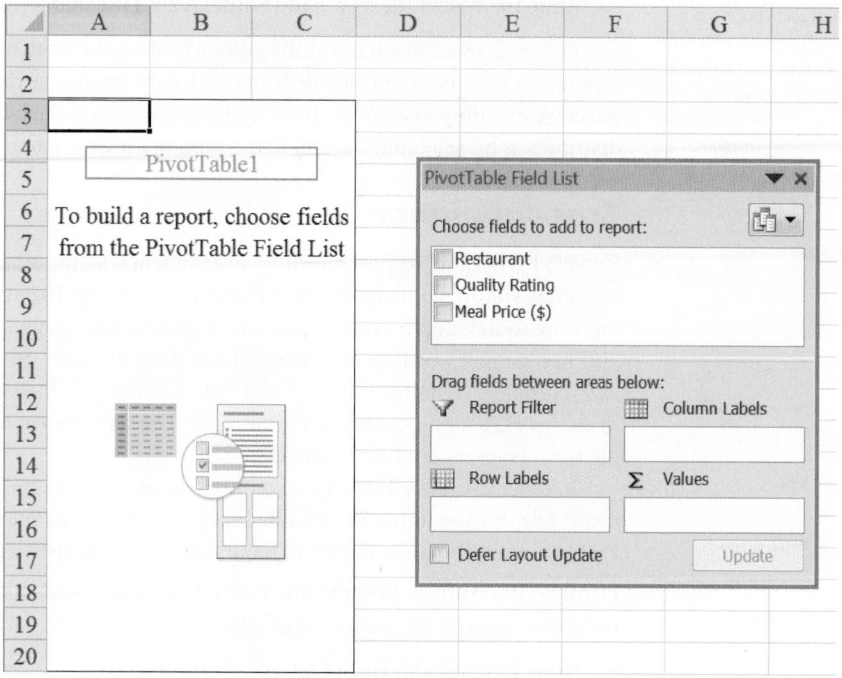

Step 3. When the **Create PivotTable** dialog box appears,
Choose **Select a Table or Range**
Enter A1:C301 in the **Table/Range** box
Choose **New Worksheet** as the location for the PivotTable Report
Click **OK**

The resulting initial PivotTable Field List and PivotTable Report are shown in Figure 2.20.

Set Up the PivotTable Field List: Each of the three columns in Figure 2.19 (labeled Restaurant, Quality Rating, and Meal Price ($)) is considered a field by Excel. Fields may be chosen to represent rows, columns, or values in the body of the PivotTable Report. The following steps show how to use Excel's PivotTable Field List to assign the Quality Rating field to the rows, the Meal Price ($) field to the columns, and the Restaurant field to the body of the PivotTable report.

Step 1. In the **PivotTable Field List,** go to **Choose Fields to add to report**
Drag the **Quality Rating** field to the **Row Labels** area
Drag the **Meal Price ($)** field to the **Column Labels** area
Drag the **Restaurant** field to the **Values** area
Step 2. Click on **Sum of Restaurant** in the **Values** area
Step 3. Click **Value Field Settings** from the list of options that appear
Step 4. When the Value Field Settings dialog appears,
Under **Summarize value field by**, choose **Count**
Click **OK**

Figure 2.21 shows the completed PivotTable Field List and a portion of the PivotTable worksheet as it now appears.

FIGURE 2.21 COMPLETED PIVOTTABLE FIELD LIST AND A PORTION OF THE PIVOTTABLE REPORT FOR THE RESTAURANT DATA (COLUMNS H:AK ARE HIDDEN)

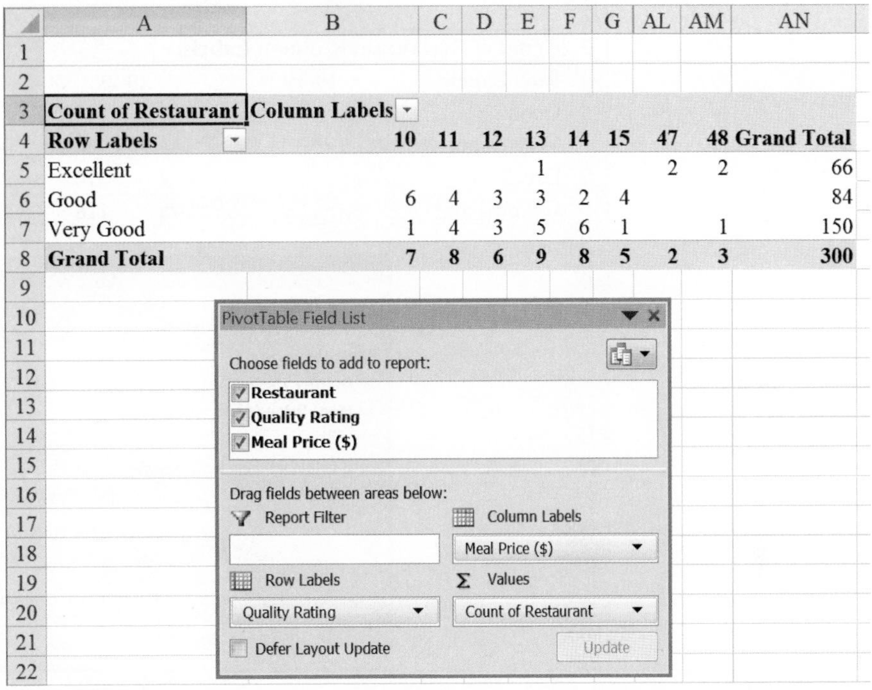

Finalize the PivotTable Report: To complete the PivotTable Report we need to group the columns representing meal prices and place the row labels for quality rating in the proper order. The following steps accomplish this.

Step 1. Right-click in cell B4 or any cell containing meal prices
Step 2. Choose **Group** from the list of options that appears
Step 3. When the **Grouping** dialog box appears,
Enter 10 in the **Starting at** box
Enter 49 in the **Ending at** box
Enter 10 in the **By** box
Click **OK**
Step 4. Right-click on **Excellent** in cell A5
Step 5. Choose **Move** and click **Move "Excellent" to End**

The final PivotTable Report is shown in Figure 2.22. Note that it provides the same information as the crosstabulation shown in Table 2.10.

Scatter Diagram

WEB file Stereo

We can use Excel's chart tools to construct a scatter diagram and a trend line for the stereo and sound equipment store data presented in Table 2.12. Refer to Figures 2.23 and 2.24 as we describe the steps involved. We will use the data in the file named Stereo; the labels Week, No. of Commercials, and Sales Volume have been entered into cells A1:C1 of the worksheet.

FIGURE 2.22 FINAL PIVOTTABLE REPORT FOR THE RESTAURANT DATA

	A	B	C	D	E	F
1						
2						
3	**Count of Restaurant**	**Column Labels**				
4	**Row Labels**	**10-19**	**20-29**	**30-39**	**40-49**	**Grand Total**
5	Good	42	40	2		84
6	Very Good	34	64	46	6	150
7	Excellent	2	14	28	22	66
8	**Grand Total**	**78**	**118**	**76**	**28**	**300**

FIGURE 2.23 SCATTER DIAGRAM FOR THE STEREO AND SOUND EQUIPMENT STORE USING EXCEL'S CHART TOOLS

	A	B	C
1	**Week**	**No. of Commercials**	**Sales Volume**
2	1	2	50
3	2	5	57
4	3	1	41
5	4	3	54
6	5	4	54
7	6	1	38
8	7	5	63
9	8	3	48
10	9	4	59
11	10	2	46

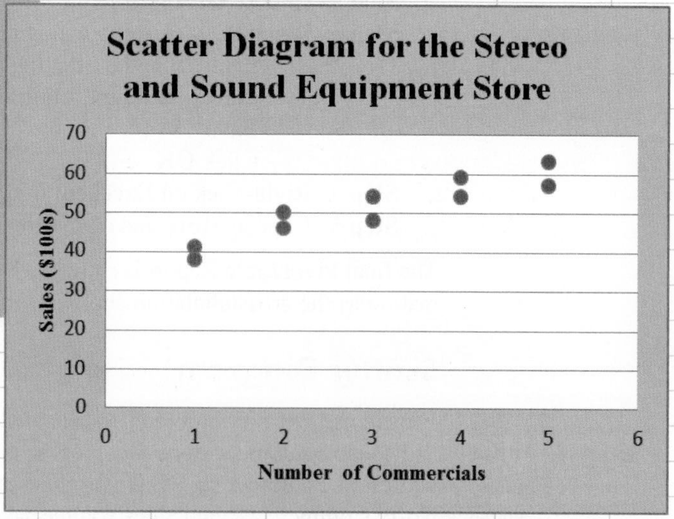

The data for each of the 10 weeks are entered into cells B2:C11. The following steps describe how to use Excel's chart tools to produce a scatter diagram for the data.

Step 1. Select cells B2:C11
Step 2. Click the **Insert** tab on the Ribbon
Step 3. In the **Charts** group, click **Scatter**
Step 4. When the list of scatter diagram subtypes appears, click **Scatter with only Markers** (the chart in the upper left corner)
Step 5. In the **Chart Layouts** group, click **Layout 1**
Step 6. Select the **Chart Title** and replace it with **Scatter Diagram for the Stereo and Sound Equipment Store**
Step 7. Select the **Horizontal (Value) Axis Title** and replace it with **Number of Commercials**
Step 8. Select the **Vertical (Value) Axis Title** and replace it with **Sales ($100s)**
Step 9. Right-click the **Series 1 Legend Entry** and click **Delete**

The worksheet displayed in Figure 2.23 shows the scatter diagram produced by Excel. The following steps describe how to add a trendline.

Step 1. Position the mouse pointer over any data point in the scatter diagram and right-click to display a list of options
Step 2. Choose **Add Trendline**
Step 3. When the **Format Trendline** dialog box appears,
Select **Trendline Options**
Choose **Linear** from the **Trend/Regression Type** list
Click **Close**

The worksheet displayed in Figure 2.24 shows the scatter diagram with the trendline added.

FIGURE 2.24 SCATTER DIAGRAM AND TRENDLINE FOR THE STEREO AND SOUND EQUIPMENT STORE USING EXCEL'S CHART TOOLS

	A	B	C
1	Week	No. of Commercials	Sales Volume
2	1	2	50
3	2	5	57
4	3	1	41
5	4	3	54
6	5	4	54
7	6	1	38
8	7	5	63
9	8	3	48
10	9	4	59
11	10	2	46

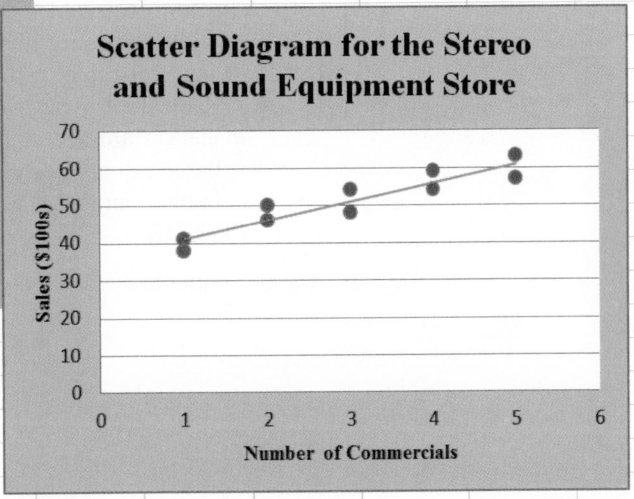

Appendix 2.3 Using StatTools for Tabular and Graphical Presentations

In this appendix we show how StatTools can be used to construct a histogram and a scatter diagram.

Histogram

We use the audit time data in Table 2.4 to illustrate. Begin by using the Data Set Manager to create a StatTools data set for these data using the procedure described in the appendix in Chapter 1. The following steps will generate a histogram.

WEB file
Audit

Step 1. Click the **StatTools** tab on the Ribbon
Step 2. In the **Analyses Group,** click **Summary Graphs**
Step 3. Choose the **Histogram** option
Step 4. When the StatTools-Histogram dialog box appears,
In the **Variables** section, select **Audit Time**
In the **Options** section,
Enter 5 in the **Number of Bins** box
Enter 9.5 in the **Histogram Minimum** box
Enter 34.5 in the **Histogram Maximum** box
Choose **Categorical** in the **X-Axis** box
Choose **Frequency** in the **Y-Axis** box
Click **OK**

A histogram for the audit time data similar to the histogram shown in Figure 2.5 will appear. The only difference is the histogram developed using StatTools shows the class midpoints on the horizontal axis.

WEB file
Stereo

Scatter Diagram

We use the stereo and sound equipment data in Table 2.12 to demonstrate the construction of a scatter diagram. Begin by using the Data Set Manager to create a StatTools data set for these data using the procedure described in the appendix in Chapter 1. The following steps will generate a scatter diagram.

Step 1. Click the **StatTools** tab on the Ribbon
Step 2. In the **Analyses Group,** click **Summary Graphs**
Step 3. Choose the **Scatterplot** option
Step 4. When the StatTools-Scatterplot dialog box appears,
In the **Variables** section,
In the column labeled **X**, select **No. of Commercials**
In the column labeled **Y**, select **Sales Volume**
Click **OK**

A scatter diagram similar to the one shown in Figure 2.23 will appear.

CHAPTER 3

Descriptive Statistics: Numerical Measures

CONTENTS

STATISTICS *in* PRACTICE

SMALL FRY DESIGN*
SANTA ANA, CALIFORNIA

Founded in 1997, Small Fry Design is a toy and accessory company that designs and imports products for infants. The company's product line includes teddy bears, mobiles, musical toys, rattles, and security blankets and features high-quality soft toy designs with an emphasis on color, texture, and sound. The products are designed in the United States and manufactured in China.

Small Fry Design uses independent representatives to sell the products to infant furnishing retailers, children's accessory and apparel stores, gift shops, upscale department stores, and major catalog companies. Currently, Small Fry Design products are distributed in more than 1000 retail outlets throughout the United States.

Cash flow management is one of the most critical activities in the day-to-day operation of this company. Ensuring sufficient incoming cash to meet both current and ongoing debt obligations can mean the difference between business success and failure. A critical factor in cash flow management is the analysis and control of accounts receivable. By measuring the average age and dollar value of outstanding invoices, management can predict cash availability and monitor changes in the status of accounts receivable. The company set the following goals: The average age for outstanding invoices should not exceed 45 days, and the dollar value of invoices more than 60 days old should not exceed 5% of the dollar value of all accounts receivable.

In a recent summary of accounts receivable status, the following descriptive statistics were provided for the age of outstanding invoices:

Mean	40 days
Median	35 days
Mode	31 days

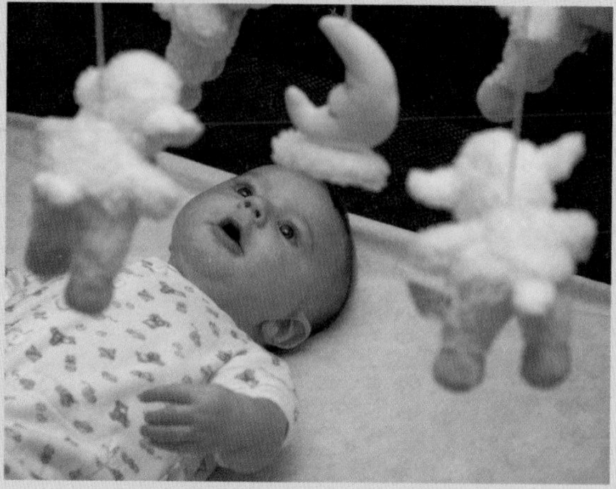

Small Fry Design uses descriptive statistics to monitor its accounts receivable and incoming cash flow. © Robert Dant/Alamy Limited.

Interpretation of these statistics shows that the mean or average age of an invoice is 40 days. The median shows that half of the invoices remain outstanding 35 days or more. The mode of 31 days, the most frequent invoice age, indicates that the most common length of time an invoice is outstanding is 31 days. The statistical summary also showed that only 3% of the dollar value of all accounts receivable was more than 60 days old. Based on the statistical information, management was satisfied that accounts receivable and incoming cash flow were under control.

In this chapter, you will learn how to compute and interpret some of the statistical measures used by Small Fry Design. In addition to the mean, median, and mode, you will learn about other descriptive statistics such as the range, variance, standard deviation, percentiles, and correlation. These numerical measures will assist in the understanding and interpretation of data.

*The authors are indebted to John A. McCarthy, President of Small Fry Design, for providing this Statistics in Practice.

In Chapter 2 we discussed tabular and graphical presentations used to summarize data. In this chapter, we present several numerical measures that provide additional alternatives for summarizing data.

We start by developing numerical summary measures for data sets consisting of a single variable. When a data set contains more than one variable, the same numerical measures can be computed separately for each variable. However, in the two-variable case, we will also develop measures of the relationship between the variables.

Numerical measures of location, dispersion, shape, and association are introduced. If the measures are computed for data from a sample, they are called **sample statistics**. If the measures are computed for data from a population, they are called **population parameters**. In statistical inference, a sample statistic is referred to as the **point estimator** of the corresponding population parameter. In Chapter 7 we will discuss in more detail the process of point estimation.

In the three chapter appendixes we show how Minitab, Excel, and StatTools can be used to compute the numerical measures described in the chapter.

3.1 Measures of Location

Mean

The mean is sometimes referred to as the arithmetic mean.

Perhaps the most important measure of location is the **mean**, or average value, for a variable. The mean provides a measure of central location for the data. If the data are for a sample, the mean is denoted by $\bar{x}$; if the data are for a population, the mean is denoted by the Greek letter μ.

In statistical formulas, it is customary to denote the value of variable x for the first observation by x_1, the value of variable x for the second observation by x_2, and so on. In general, the value of variable x for the ith observation is denoted by x_i. For a sample with n observations, the formula for the sample mean is as follows.

The sample mean $\bar{x}$ is a sample statistic.

SAMPLE MEAN

$$\bar{x} = \frac{\Sigma x_i}{n} \tag{3.1}$$

In the preceding formula, the numerator is the sum of the values of the n observations. That is,

$$\Sigma x_i = x_1 + x_2 + \cdots + x_n$$

The Greek letter Σ is the summation sign.

To illustrate the computation of a sample mean, let us consider the following class size data for a sample of five college classes.

46 54 42 46 32

We use the notation x_1, x_2, x_3, x_4, x_5 to represent the number of students in each of the five classes.

$$x_1 = 46 \qquad x_2 = 54 \qquad x_3 = 42 \qquad x_4 = 46 \qquad x_5 = 32$$

Hence, to compute the sample mean, we can write

$$\bar{x} = \frac{\Sigma x_i}{n} = \frac{x_1 + x_2 + x_3 + x_4 + x_5}{5} = \frac{46 + 54 + 42 + 46 + 32}{5} = 44$$

The sample mean class size is 44 students.

To provide a visual perspective of the mean and to show how it can be influenced by extreme values, consider the dot plot for the class size data shown in Figure 3.1. Treating the horizontal axis used to create the dot plot as a long narrow board in which each of the

FIGURE 3.1 THE MEAN AS THE CENTER OF BALANCE FOR THE DOT PLOT OF THE CLASSROOM SIZE DATA

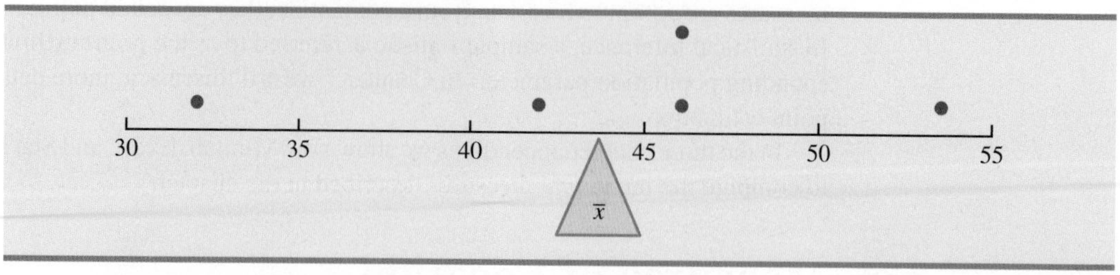

dots has the same fixed weight, the mean is the point at which we would place a fulcrum or pivot point under the board in order to balance the dot plot. This is the same principle by which a see-saw on a playground works, the only difference being that the see-saw is pivoted in the middle so that as one end goes up, the other end goes down. In the dot plot we are locating the pivot point based upon the location of the dots. Now consider what happens to the balance if we increase the largest value from 54 to 114. We will have to move the fulcrum under the new dot plot in a positive direction in order to reestablish balance. To determine how far we would have to shift the fulcrum, we simply compute the sample mean for the revised class size data.

$$\bar{x} = \frac{\Sigma x_i}{n} = \frac{x_1 + x_2 + x_3 + x_4 + x_5}{5} = \frac{46 + 114 + 42 + 46 + 32}{5} = \frac{280}{5} = 56$$

Thus, the mean for the revised class size data is 56, an increase of 12 students. In other words, we have to shift the balance point 12 units to the right to establish balance under the new dot plot.

Another illustration of the computation of a sample mean is given in the following situation. Suppose that a college placement office sent a questionnaire to a sample of business school graduates requesting information on monthly starting salaries. Table 3.1 shows the collected data. The mean monthly starting salary for the sample of 12 business college graduates is computed as

$$\bar{x} = \frac{\Sigma x_i}{n} = \frac{x_1 + x_2 + \cdots + x_{12}}{12}$$

$$= \frac{3850 + 3950 + \cdots + 3880}{12}$$

$$= \frac{47{,}280}{12} = 3940$$

TABLE 3.1 MONTHLY STARTING SALARIES FOR A SAMPLE OF 12 BUSINESS SCHOOL GRADUATES

WEB file

2012StartSalary

Graduate	Monthly Starting Salary ($)	Graduate	Monthly Starting Salary ($)
1	3850	7	3890
2	3950	8	4130
3	4050	9	3940
4	3880	10	4325
5	3755	11	3920
6	3710	12	3880

Equation (3.1) shows how the mean is computed for a sample with n observations. The formula for computing the mean of a population remains the same, but we use different notation to indicate that we are working with the entire population. The number of observations in a population is denoted by N and the symbol for a population mean is μ.

The sample mean $\bar{x}$ is a point estimator of the population mean μ.

POPULATION MEAN

$$\mu = \frac{\Sigma x_i}{N} \tag{3.2}$$

Weighted Mean

In the formulas for the sample mean and population mean, each x_i is given equal importance or weight. For instance, the formula for the sample mean can be written as follows:

$$\bar{x} = \frac{\Sigma x_i}{n} = \frac{1}{n}\left(\sum x_i\right) = \frac{1}{n}(x_1 + x_2 + \cdots + x_n) = \frac{1}{n}(x_1) + \frac{1}{n}(x_2) + \cdots + \frac{1}{n}(x_n)$$

This shows that each observation in the sample is given a weight of $1/n$. Although this practice is most common, in some instances the mean is computed by giving each observation a weight that reflects its relative importance. A mean computed in this manner is referred to as a **weighted mean**. The weighted mean is computed as follows:

WEIGHTED MEAN

$$\bar{x} = \frac{\Sigma w_i x_i}{\Sigma w_i} \tag{3.3}$$

where

$$w_i = \text{weight for observation } i$$

When the data are from a sample, equation (3.3) provides the weighted sample mean. If the data are from a population, μ replaces $\bar{x}$ and equation (3.3) provides the weighted population mean.

As an example of the need for a weighted mean, consider the following sample of five purchases of a raw material over the past three months.

Purchase	Cost per Pound ($)	Number of Pounds
1	3.00	1200
2	3.40	500
3	2.80	2750
4	2.90	1000
5	3.25	800

Note that the cost per pound varies from \$2.80 to \$3.40, and the quantity purchased varies from 500 to 2750 pounds. Suppose that a manager wanted to know the mean cost per pound of the raw material. Because the quantities ordered vary, we must use the formula for a weighted mean. The five cost-per-pound data values are $x_1 = 3.00$, $x_2 = 3.40$, $x_3 = 2.80$, $x_4 = 2.90$, and $x_5 = 3.25$. The weighted mean cost per pound is found by weighting each cost by its corresponding quantity. For this example, the weights are $w_1 = 1200$, $w_2 = 500$, $w_3 = 2750$, $w_4 = 1000$, and $w_5 = 800$. Based on equation (3.3), the weighted mean is calculated as follows:

$$\bar{x} = \frac{1200(3.00) + 500(3.40) + 2750(2.80) + 1000(2.90) + 800(3.25)}{1200 + 500 + 2750 + 1000 + 800}$$

$$= \frac{18{,}500}{6250} = 2.96$$

Thus, the weighted mean computation shows that the mean cost per pound for the raw material is \$2.96. Note that using equation (3.1) rather than the weighted mean formula in equation (3.3) would provide misleading results. In this case, the sample mean of the five cost-per-pound values is $(3.00 + 3.40 + 2.80 + 2.90 + 3.25)/5 = 15.35/5 = \$3.07$, which overstates the actual mean cost per pound purchased.

The choice of weights for a particular weighted mean computation depends upon the application. An example that is well known to college students is the computation of a grade point average (GPA). In this computation, the data values generally used are 4 for an A grade, 3 for a B grade, 2 for a C grade, 1 for a D grade, and 0 for an F grade. The weights are the number of credit hours earned for each grade. Exercise 16 at the end of this section provides an example of this weighted mean computation. In other weighted mean computations, quantities such as pounds, dollars, or volume are frequently used as weights. In any case, when observations vary in importance, the analyst must choose the weight that best reflects the importance of each observation in the determination of the mean.

Median

The **median** is another measure of central location. The median is the value in the middle when the data are arranged in ascending order (smallest value to largest value). With an odd number of observations, the median is the middle value. An even number of observations has no single middle value. In this case, we follow convention and define the median as the average of the values for the middle two observations. For convenience the definition of the median is restated as follows.

MEDIAN

Arrange the data in ascending order (smallest value to largest value).

(a) For an odd number of observations, the median is the middle value.
(b) For an even number of observations, the median is the average of the two middle values.

Let us apply this definition to compute the median class size for the sample of five college classes. Arranging the data in ascending order provides the following list.

32 42 46 46 54

Because $n = 5$ is odd, the median is the middle value. Thus the median class size is 46 students. Even though this data set contains two observations with values of 46, each observation is treated separately when we arrange the data in ascending order.

Suppose we also compute the median starting salary for the 12 business college graduates in Table 3.1. We first arrange the data in ascending order.

3710 3755 3850 3880 3880 3890 3920 3940 3950 4050 4130 4325

(3890 3920: Middle Two Values)

Because $n = 12$ is even, we identify the middle two values: 3890 and 3920. The median is the average of these values.

$$\text{Median} = \frac{3890 + 3920}{2} = 3905$$

The procedure we used to compute the median depends upon whether there is an odd number of observations or an even number of observations. Let us now describe a more conceptual and visual approach using the monthly starting salary for the 12 business college graduates. As before, we begin by arranging the data in ascending order.

3710 3755 3850 3880 3880 3890 3920 3940 3950 4050 4130 4325

Once the data are in ascending order, we trim pairs of extreme high and low values until no further pairs of values can be trimmed without completely eliminating all the data. For instance, after trimming the lowest observation (3710) and the highest observation (4325) we obtain a new data set with 10 observations.

~~3710~~ 3755 3850 3880 3880 3890 3920 3940 3950 4050 4130 ~~4325~~

We then trim the next lowest remaining value (3755) and the next highest remaining value (4130) to produce a new data set with eight observations.

~~3710~~ ~~3755~~ 3850 3880 3880 3890 3920 3940 3950 4050 ~~4130~~ ~~4325~~

Continuing this process we obtain the following results.

~~3710~~ ~~3755~~ ~~3850~~ 3880 3880 3890 3920 3940 3950 ~~4050~~ ~~4130~~ ~~4325~~

~~3710~~ ~~3755~~ ~~3850~~ ~~3880~~ 3880 3890 3920 3940 ~~3950~~ ~~4050~~ ~~4130~~ ~~4325~~

~~3710~~ ~~3755~~ ~~3850~~ ~~3880~~ ~~3880~~ 3890 3920 ~~3940~~ ~~3950~~ ~~4050~~ ~~4130~~ ~~4325~~

At this point no further trimming is possible without eliminating all the data. So, the median is just the average of the remaining two values. When there is an even number of observations, the trimming process will always result in two remaining values, and the average of these values will be the median. When there is an odd number of observations, the trimming process will always result in one final value, and this value will be the median. Thus, this method works whether the number of observations is odd or even.

The median is the measure of location most often reported for annual income and property value data because a few extremely large incomes or property values can inflate the mean. In such cases, the median is the preferred measure of central location.

Although the mean is the more commonly used measure of central location, in some situations the median is preferred. The mean is influenced by extremely small and large data values. For instance, suppose that the highest paid graduate (see Table 3.1) had a starting salary of $10,000 per month (maybe the individual's family owns the company). If we change the highest monthly starting salary in Table 3.1 from $4325 to $10,000 and recompute the mean, the sample mean changes from $3940 to $4413. The median of $3905, however, is unchanged, because $3890 and $3920 are still the middle two values. With the extremely high starting salary included, the median provides a better measure of central location than the mean. We can generalize to say that whenever a data set contains extreme values, the median is often the preferred measure of central location.

Geometric Mean

The **geometric mean** is a measure of location that is calculated by finding the nth root of the product of n values. The general formula for the geometric mean, denoted $\bar{x}_g$, follows.

GEOMETRIC MEAN

$$\bar{x}_g = \sqrt[n]{(x_1)(x_2)\cdots(x_n)} = [(x_1)(x_2)\cdots(x_n)]^{1/n} \tag{3.4}$$

The geometric mean is often used in analyzing growth rates in financial data. In these types of situations the arithmetic mean or average value will provide misleading results.

To illustrate the use of the geometric mean, consider Table 3.2 which shows the percentage annual returns, or growth rates, for a mutual fund over the past 10 years. Suppose we want to compute how much \$100 invested in the fund at the beginning of year 1 would be worth at the end of year 10. Let's start by computing the balance in the fund at the end of year 1. Because the percentage annual return for year 1 was −22.1%, the balance in the fund at the end of year 1 would be

$$\$100 - .221(\$100) = \$100(1 - .221) = \$100(.779) = \$77.90$$

The growth factor for each year is 1 plus .01 times the percentage return. A growth factor less than 1 indicates negative growth, while a growth factor greater than 1 indicates positive growth. The growth factor cannot be less than zero.

Note that .779 is identified as the growth factor for year 1 in Table 3.2. This result shows that we can compute the balance at the end of year 1 by multiplying the value invested in the fund at the beginning of year 1 times the growth factor for year 1.

The balance in the fund at the end of year 1, \$77.90, now becomes the beginning balance in year 2. So, with a percentage annual return for year 2 of 28.7%, the balance at the end of year 2 would be

$$\$77.90 + .287(\$77.90) = \$77.90(1 + .287) = \$77.90(1.287) = \$100.2573$$

Note that 1.287 is the growth factor for year 2. And, by substituting \$100(.779) for \$77.90 we see that the balance in the fund at the end of year 2 is

$$\$100(.779)(1.287) = \$100.2573$$

In other words, the balance at the end of year 2 is just the initial investment at the beginning of year 1 times the product of the first two growth factors. This result can be generalized to

TABLE 3.2 PERCENTAGE ANNUAL RETURNS AND GROWTH FACTORS FOR THE MUTUAL FUND DATA

Year	Return (%)	Growth Factor
1	−22.1	0.779
2	28.7	1.287
3	10.9	1.109
4	4.9	1.049
5	15.8	1.158
6	5.5	1.055
7	−37.0	0.630
8	26.5	1.265
9	15.1	1.151
10	2.1	1.021

show that the balance at the end of year 10 is the initial investment times the product of all 10 growth factors.

$$\$100[(.779)(1.287)(1.109)(1.049)(1.158)(1.055)(.630)(1.265)(1.151)(1.021)] =$$

$$\$100(1.334493) = \$133.4493$$

The nth root can be computed using most calculators or by using the POWER function in Excel. For instance, using Excel, the 10th root of 1.334493 = POWER (1.334493,1/10) or 1.029275.

So, a \$100 investment in the fund at the beginning of year 1 would be worth \$133.4493 at the end of year 10. Note that the product of the 10 growth factors is 1.334493. Thus, we can compute the balance at the end of year 10 for any amount of money invested at the beginning of year 1 by multiplying the value of the initial investment times 1.334493. For instance, an initial investment of \$2500 at the beginning of year 1 would be worth \$2500(1.334493) or approximately \$3336 at the end of year 10.

But, what was the mean percentage annual return or mean rate of growth for this investment over the 10-year period? Let us see how the geometric mean of the 10 growth factors can be used to answer to this question. Because the product of the 10 growth factors is 1.334493, the geometric mean is the 10th root of 1.334493 or

$$\bar{x}_g = \sqrt[10]{1.334493} = 1.029275$$

The geometric mean tells us that annual returns grew at an average annual rate of (1.029275 − 1)100% or 2.9275%. In other words, with an average annual growth rate of 2.9275%, a \$100 investment in the fund at the beginning of year 1 would grow to $\$100(1.029275)^{10} = \$133.4493$ at the end of 10 years.

It is important to understand that the arithmetic mean of the percentage annual returns does not provide the mean annual growth rate for this investment. The sum of the 10 annual percentage returns in Table 3.2 is 50.4. Thus, the arithmetic mean of the 10 percentage annual returns is 50.4/10 = 5.04%. A broker might try to convince you to invest in this fund by stating that the mean annual percentage return was 5.04%. Such a statement is not only misleading, it is inaccurate. A mean annual percentage return of 5.04% corresponds to an average growth factor of 1.0504. So, if the average growth factor were really 1.0504, \$100 invested in the fund at the beginning of year 1 would have grown to $\$100(1.0504)^{10} = \$163.51$ at the end of 10 years. But, using the 10 annual percentage returns in Table 3.2, we showed that an initial \$100 investment is worth \$133.45 at the end of 10 years. The broker's claim that the mean annual percentage return is 5.04% grossly overstates the true growth for this mutual fund. The problem is that the sample mean is only appropriate for an additive process. For a multiplicative process, such as applications involving growth rates, the geometric mean is the appropriate measure of location.

While the applications of the geometric mean to problems in finance, investments, and banking are particularly common, the geometric mean should be applied any time you want to determine the mean rate of change over several successive periods. Other common applications include changes in populations of species, crop yields, pollution levels, and birth and death rates. Also note that the geometric mean can be applied to changes that occur over any number of successive periods of any length. In addition to annual changes, the geometric mean is often applied to find the mean rate of change over quarters, months, weeks, and even days.

Mode

Another measure of location is the **mode**. The mode is defined as follows.

> MODE
>
> The mode is the value that occurs with greatest frequency.

To illustrate the identification of the mode, consider the sample of five class sizes. The only value that occurs more than once is 46. Because this value, occurring with a frequency of 2, has the greatest frequency, it is the mode. As another illustration, consider the sample of starting salaries for the business school graduates. The only monthly starting salary that occurs more than once is \$3880. Because this value has the greatest frequency, it is the mode.

Situations can arise for which the greatest frequency occurs at two or more different values. In these instances more than one mode exists. If the data contain exactly two modes, we say that the data are *bimodal*. If data contain more than two modes, we say that the data are *multimodal*. In multimodal cases the mode is almost never reported because listing three or more modes would not be particularly helpful in describing a location for the data.

Percentiles

A **percentile** provides information about how the data are spread over the interval from the smallest value to the largest value. For data that do not contain numerous repeated values, the pth percentile divides the data into two parts. Approximately p percent of the observations have values less than the pth percentile; approximately $(100 - p)$ percent of the observations have values greater than the pth percentile. The pth percentile is formally defined as follows.

PERCENTILE

The pth percentile is a value such that *at least p* percent of the observations are less than or equal to this value and *at least* $(100 - p)$ percent of the observations are greater than or equal to this value.

Colleges and universities frequently report admission test scores in terms of percentiles. For instance, suppose an applicant obtains a raw score of 54 on the verbal portion of an admission test. How this student performed in relation to other students taking the same test may not be readily apparent. However, if the raw score of 54 corresponds to the 70th percentile, we know that approximately 70% of the students scored lower than this individual and approximately 30% of the students scored higher than this individual.

The following procedure can be used to compute the pth percentile.

Following these steps makes it easy to calculate percentiles.

CALCULATING THE pTH PERCENTILE

Step 1. Arrange the data in ascending order (smallest value to largest value).
Step 2. Compute an index i

$$i = \left(\frac{p}{100}\right)n$$

where p is the percentile of interest and n is the number of observations.
Step 3. (a) If *i is not an integer, round up*. The next integer *greater* than i denotes the position of the pth percentile.
(b) If *i is an integer*, the pth percentile is the average of the values in positions i and $i + 1$.

As an illustration of this procedure, let us determine the 85th percentile for the starting salary data in Table 3.1.

Step 1. Arrange the data in ascending order.

3710 3755 3850 3880 3880 3890 3920 3940 3950 4050 4130 4325

Step 2.

$$i = \left(\frac{p}{100}\right)n = \left(\frac{85}{100}\right)12 = 10.2$$

Step 3. Because i is not an integer, *round up.* The position of the 85th percentile is the next integer greater than 10.2, the 11th position.

Returning to the data, we see that the 85th percentile is the data value in the 11th position, or 4130.

As another illustration of this procedure, let us consider the calculation of the 50th percentile for the starting salary data. Applying step 2, we obtain

$$i = \left(\frac{50}{100}\right)12 = 6$$

Because i is an integer, step 3(b) states that the 50th percentile is the average of the sixth and seventh data values; thus the 50th percentile is $(3890 + 3920)/2 = 3905$. Note that the *50th percentile is also the median.*

Quartiles

Quartiles are just specific percentiles; thus, the steps for computing percentiles can be applied directly in the computation of quartiles.

It is often desirable to divide data into four parts, with each part containing approximately one-fourth, or 25% of the observations. The division points are referred to as the **quartiles** and are defined as

Q_1 = first quartile, or 25th percentile

Q_2 = second quartile, or 50th percentile (also the median)

Q_3 = third quartile, or 75th percentile.

To compute the quartiles for the starting salary data, we begin by arranging the data in ascending order.

3710 3755 3850 3880 3880 3890 3920 3940 3950 4050 4130 4325

We already identified Q_2, the second quartile (median), as 3905. The computations of quartiles Q_1 and Q_3 require the use of the rule for finding the 25th and 75th percentiles. These calculations follow.

For Q_1,

$$i = \left(\frac{p}{100}\right)n = \left(\frac{25}{100}\right)12 = 3$$

Because i is an integer, step 3(b) indicates that the first quartile, or 25th percentile, is the average of the third and fourth data values; thus, $Q_1 = (3850 + 3880)/2 = 3865$.

For Q_3,

$$i = \left(\frac{p}{100}\right)n = \left(\frac{75}{100}\right)12 = 9$$

Again, because i is an integer, step 3(b) indicates that the third quartile, or 75th percentile, is the average of the ninth and tenth data values; thus, $Q_3 = (3950 + 4050)/2 = 4000$.

The quartiles divide the starting salary data into four parts, with each part containing 25% of the observations.

25% of the data			25% of the data			25% of the data			25% of the data		
3710	3755	3850	3880	3880	3890	3920	3940	3950	4050	4130	4325

$Q_1 = 3865$ $\quad$ $Q_2 = 3905$ (Median) $\quad$ $Q_3 = 4000$

We defined the quartiles as the 25th, 50th, and 75th percentiles. Thus, we computed the quartiles in the same way as percentiles. However, other conventions are sometimes used to compute quartiles, and the actual values reported for quartiles may vary slightly depending on the convention used. Nevertheless, the objective of all procedures for computing quartiles is to divide the data into four equal parts.

NOTES AND COMMENTS

1. It is better to use the median than the mean as a measure of central location when a data set contains extreme values. Another measure that is sometimes used when extreme values are present is the trimmed mean. The trimmed mean is obtained by deleting a percentage of the smallest and largest values from a data set and then computing the mean of the remaining values. For example, the 5% trimmed mean is obtained by removing the smallest 5% and the largest 5% of the data values and then computing the mean of the remaining values. Using the sample with $n = 12$ starting salaries, $0.05(12) = 0.6$. Rounding this value to 1 indicates that the 5% trimmed mean is obtained by removing the smallest data value and the largest data value and then computing the mean of the remaining 10 values. For the starting salary data, the 5% trimmed mean is 3924.50.

2. Other commonly used percentiles are the quintiles (the 20th, 40th, 60th, and 80th percentiles) and the deciles (the 10th, 20th, 30th, 40th, 50th, 60th, 70th, 80th, and 90th percentiles).

Exercises

Methods

1. Consider a sample with data values of 10, 20, 12, 17, and 16. Compute the mean and median.
2. Consider a sample with data values of 10, 20, 21, 17, 16, and 12. Compute the mean and median.

SELF test

3. Consider the following data and corresponding weights.

x_i	Weight (w_i)
3.2	6
2.0	3
2.5	2
5.0	8

a. Compute the weighted mean.
b. Compute the sample mean of the four data values without weighting. Note the difference in the results provided by the two computations.

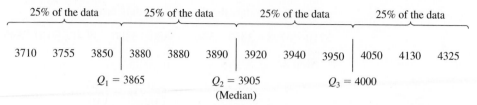

4. Consider the following data.

Period	Rate of Return (%)
1	−6.0
2	−8.0
3	−4.0
4	2.0
5	5.4

What is the mean growth rate over these five periods?

SELF test

5. Consider a sample with data values of 27, 25, 20, 15, 30, 34, 28, and 25. Compute the 20th, 25th, 65th, and 75th percentiles.

6. Consider a sample with data values of 53, 55, 70, 58, 64, 57, 53, 69, 57, 68, and 53. Compute the mean, median, and mode.

Applications

7. The average number of minutes Americans commute to work is 27.7 minutes (*Sterling's Best Places*, April 13, 2012). The average commute time in minutes for 48 cities are as follows:

WEB file
CommuteTime

Albuquerque	23.3	Jacksonville	26.2	Phoenix	28.3
Atlanta	28.3	Kansas City	23.4	Pittsburgh	25.0
Austin	24.6	Las Vegas	28.4	Portland	26.4
Baltimore	32.1	Little Rock	20.1	Providence	23.6
Boston	31.7	Los Angeles	32.2	Richmond	23.4
Charlotte	25.8	Louisville	21.4	Sacramento	25.8
Chicago	38.1	Memphis	23.8	Salt Lake City	20.2
Cincinnati	24.9	Miami	30.7	San Antonio	26.1
Cleveland	26.8	Milwaukee	24.8	San Diego	24.8
Columbus	23.4	Minneapolis	23.6	San Francisco	32.6
Dallas	28.5	Nashville	25.3	San Jose	28.5
Denver	28.1	New Orleans	31.7	Seattle	27.3
Detroit	29.3	New York	43.8	St. Louis	26.8
El Paso	24.4	Oklahoma City	22.0	Tucson	24.0
Fresno	23.0	Orlando	27.1	Tulsa	20.1
Indianapolis	24.8	Philadelphia	34.2	Washington, D.C.	32.8

a. What is the mean commute time for these 48 cities?
b. Compute the median commute time.
c. Compute the mode.
d. Compute the third quartile.

8. During the 2007–2008 NCAA college basketball season, men's basketball teams attempted an all-time high number of 3-point shots, averaging 19.07 shots per game (Associated Press Sports, January 24, 2009). In an attempt to discourage so many 3-point shots and encourage more inside play, the NCAA rules committee moved the 3-point line back from 19 feet, 9 inches to 20 feet, 9 inches at the beginning of the 2008–2009 basketball season. Shown in the following table are the 3-point shots taken and the 3-point shots made for a sample of 19 NCAA basketball games during the 2008–2009 season.

WEB file

3Points

3-Point Shots	Shots Made	3-Point Shots	Shots Made
23	4	17	7
20	6	19	10
17	5	22	7
18	8	25	11
13	4	15	6
16	4	10	5
8	5	11	3
19	8	25	8
28	5	23	7
21	7		

a. What is the mean number of 3-point shots taken per game?
b. What is the mean number of 3-point shots made per game?
c. Using the closer 3-point line, players were making 35.2% of their shots. What percentage of shots were players making from the new 3-point line?
d. What was the impact of the NCAA rules change that moved the 3-point line back to 20 feet, 9 inches for the 2008–2009 season? Would you agree with the Associated Press Sports article that stated, "Moving back the 3-point line hasn't changed the game dramatically"? Explain.

9. Endowment income is a critical part of the annual budgets at colleges and universities. A study by the National Association of College and University Business Officers reported that the 435 colleges and universities surveyed held a total of $413 billion in endowments. The 10 wealthiest universities are shown below (*The Wall Street Journal*, January 27, 2009). Amounts are in billion of dollars.

University	Endowment ($billion)	University	Endowment ($billion)
Columbia	7.2	Princeton	16.4
Harvard	36.6	Stanford	17.2
M.I.T.	10.1	Texas	16.1
Michigan	7.6	Texas A&M	6.7
Northwestern	7.2	Yale	22.9

a. What is the mean endowment for these universities?
b. What is the median endowment?
c. What is the mode endowment?
d. Compute the first and third quartiles.
e. What is the total endowment at these 10 universities? These universities represent 2.3% of the 435 colleges and universities surveyed. What percentage of the total $413 billion in endowments is held by these 10 universities?
f. *The Wall Street Journal* reported that over a recent five-month period, a downturn in the economy has caused endowments to decline 23%. What is the estimate of the dollar amount of the decline in the total endowments held by these 10 universities? Given this situation, what are some of the steps you would expect university administrators to be considering?

10. The cost of consumer purchases such as single-family housing, gasoline, Internet services, tax preparation, and hospitalization were provided in *The Wall-Street Journal* (January 2,

SELF test

WEB file

TaxCost

2007). Sample data typical of the cost of tax-return preparation by services such as H&R Block are shown below.

120	230	110	115	160
130	150	105	195	155
105	360	120	120	140
100	115	180	235	255

a. Compute the mean, median, and mode.
b. Compute the first and third quartiles.
c. Compute and interpret the 90th percentile.

11. According to the National Education Association (NEA), teachers generally spend more than 40 hours each week working on instructional duties (NEA website, April 2012). The following data show the number of hours worked per week for a sample of 13 high school science teachers and a sample of 11 high school English teachers.

High School Science Teachers: 53 56 54 54 55 58 49 61 54 54 52 53 54

High School English Teachers: 52 47 50 46 47 48 49 46 55 44 47

a. What is the median number of hours worked per week for the sample of 13 high school science teachers?
b. What is the median number of hours worked per week for the sample of 11 high school English teachers?
c. Which group has the highest median number of hours worked per week? What is the difference between the median number of hours worked per week?

WEB file

BigBangTheory

12. *The Big Bang Theory*, a situation comedy featuring Johnny Galecki, Jim Parsons, and Kaley Cuoco, is one of the most watched programs on network television. The first two episodes for the 2011–2012 season premiered on September 22, 2011; the first episode attracted 14.1 million viewers and the second episode attracted 14.7 million viewers. The following table shows the number of viewers in millions for the first 21 episodes of the 2011–2012 season (*The Big Bang Theory* website, April 17, 2012).

Air Date	Viewers (millions)	Air Date	Viewers (millions)
September 22, 2011	14.1	January 12, 2012	16.1
September 22, 2011	14.7	January 19, 2012	15.8
September 29, 2011	14.6	January 26, 2012	16.1
October 6, 2011	13.6	February 2, 2012	16.5
October 13, 2011	13.6	February 9, 2012	16.2
October 20, 2011	14.9	February 16, 2012	15.7
October 27, 2011	14.5	February 23, 2012	16.2
November 3, 2011	16.0	March 8, 2012	15.0
November 10, 2011	15.9	March 29, 2012	14.0
November 17, 2011	15.1	April 5, 2012	13.3
December 8, 2011	14.0		

a. Compute the minimum and maximum number of viewers.
b. Compute the mean, median, and mode.
c. Compute the first and third quartiles.
d. Has viewership grown or declined over the 2011–2012 season? Discuss.

13. In automobile mileage and gasoline-consumption testing, 13 automobiles were road tested for 300 miles in both city and highway driving conditions. The following data were recorded for miles-per-gallon performance.

City:	16.2	16.7	15.9	14.4	13.2	15.3	16.8	16.0	16.1	15.3	15.2	15.3	16.2
Highway:	19.4	20.6	18.3	18.6	19.2	17.4	17.2	18.6	19.0	21.1	19.4	18.5	18.7

Use the mean, median, and mode to make a statement about the difference in performance for city and highway driving.

WEB file

StateUnemp

14. The data contained in the file named StateUnemp show the unemployment rate in March 2011 and the unemployment rate in March 2012 for every state and the District of Columbia (Bureau of Labor Statistics website, April 20, 2012). To compare unemployment rates in March 2011 with unemployment rates in March 2012, compute the first quartile, the median, and the third quartile for the March 2011 unemployment data and the March 2012 unemployment data. What do these statistics suggest about the change in unemployment rates across the states?

15. Martinez Auto Supplies has retail stores located in eight cities in California. The price they charge for a particular product in each city varies because of differing competitive conditions. For instance, the price they charge for a case of a popular brand of motor oil in each city follows. Also shown are the number of cases that Martinez Auto sold last quarter in each city.

City	Price ($)	Sales (cases)
Bakersfield	34.99	501
Los Angeles	38.99	1425
Modesto	36.00	294
Oakland	33.59	882
Sacramento	40.99	715
San Diego	38.59	1088
San Francisco	39.59	1644
San Jose	37.99	819

Compute the average sales price per case for this product during the last quarter.

SELF test

16. The grade point average for college students is based on a weighted mean computation. For most colleges, the grades are given the following data values: A (4), B (3), C (2), D (1), and F (0). After 60 credit hours of course work, a student at State University earned 9 credit hours of A, 15 credit hours of B, 33 credit hours of C, and 3 credit hours of D.
    a. Compute the student's grade point average.
    b. Students at State University must maintain a 2.5 grade point average for their first 60 credit hours of course work in order to be admitted to the business college. Will this student be admitted?

17. Morningstar tracks the total return for a large number of mutual funds. The following table shows the total return and the number of funds for four categories of mutual funds (*Morningstar Funds 500*, 2008).

Type of Fund	Number of Funds	Total Return (%)
Domestic Equity	9191	4.65
International Equity	2621	18.15
Specialty Stock	1419	11.36
Hybrid	2900	6.75

a. Using the number of funds as weights, compute the weighted average total return for the mutual funds covered by Morningstar.
b. Is there any difficulty associated with using the "number of funds" as the weights in computing the weighted average total return for Morningstar in part (a)? Discuss. What else might be used for weights?
c. Suppose you had invested $10,000 in mutual funds at the beginning of 2007 and diversified the investment by placing $2000 in Domestic Equity funds, $4000 in International Equity funds, $3000 in Specialty Stock funds, and $1000 in Hybrid funds. What is the expected return on the portfolio?

18. Based on a survey of 425 master's programs in business administration, *U.S. News & World Report* ranked the Indiana University Kelley Business School as the 20th best business program in the country (*America's Best Graduate Schools,* 2009). The ranking was based in part on surveys of business school deans and corporate recruiters. Each survey respondent was asked to rate the overall academic quality of the master's program on a scale from 1 "marginal" to 5 "outstanding." Use the sample of responses shown below to compute the weighted mean score for the business school deans and the corporate recruiters. Discuss.

Quality Assessment	Business School Deans	Corporate Recruiters
5	44	31
4	66	34
3	60	43
2	10	12
1	0	0

19. Annual revenue for Corning Supplies grew by 5.5% in 2007; 1.1% in 2008; −3.5% in 2009; −1.1% in 2010; and 1.8% in 2011. What is the mean growth annual rate over this period?

20. Suppose that at the beginning of 2004 you invested $10,000 in the Stivers mutual fund and $5,000 in the Trippi mutual fund. The value of each investment at the end of each subsequent year is provided in the table below. Which mutual fund performed better?

Year	Stivers	Trippi
2004	11,000	5,600
2005	12,000	6,300
2006	13,000	6,900
2007	14,000	7,600
2008	15,000	8,500
2009	16,000	9,200
2010	17,000	9,900
2011	18,000	10,600

21. If an asset declines in value from $5,000 to $3,500 over nine years, what is the mean annual growth rate in the asset's value over these nine years?

22. The current value of a company is $25 million. If the value of the company six year ago was $10 million, what is the company's mean annual growth rate over the past six years?

3.2 Measures of Variability

The variability in the delivery time creates uncertainty for production scheduling. Methods in this section help measure and understand variability.

In addition to measures of location, it is often desirable to consider measures of variability, or dispersion. For example, suppose that you are a purchasing agent for a large manufacturing firm and that you regularly place orders with two different suppliers. After several months of operation, you find that the mean number of days required to fill orders is 10 days for both of the suppliers. The histograms summarizing the number of working days required to fill orders from the suppliers are shown in Figure 3.2. Although the mean number of days is 10 for both suppliers, do the two suppliers demonstrate the same degree of reliability in terms of making deliveries on schedule? Note the dispersion, or variability, in delivery times indicated by the histograms. Which supplier would you prefer?

For most firms, receiving materials and supplies on schedule is important. The 7- or 8-day deliveries shown for J.C. Clark Distributors might be viewed favorably; however, a few of the slow 13- to 15-day deliveries could be disastrous in terms of keeping a workforce busy and production on schedule. This example illustrates a situation in which the variability in the delivery times may be an overriding consideration in selecting a supplier. For most purchasing agents, the lower variability shown for Dawson Supply, Inc., would make Dawson the preferred supplier.

We turn now to a discussion of some commonly used measures of variability.

Range

The simplest measure of variability is the **range**.

RANGE

$$\text{Range} = \text{Largest value} - \text{Smallest value}$$

Let us refer to the data on starting salaries for business school graduates in Table 3.1. The largest starting salary is 4325 and the smallest is 3710. The range is $4325 - 3710 = 615$.

FIGURE 3.2 HISTORICAL DATA SHOWING THE NUMBER OF DAYS REQUIRED TO FILL ORDERS

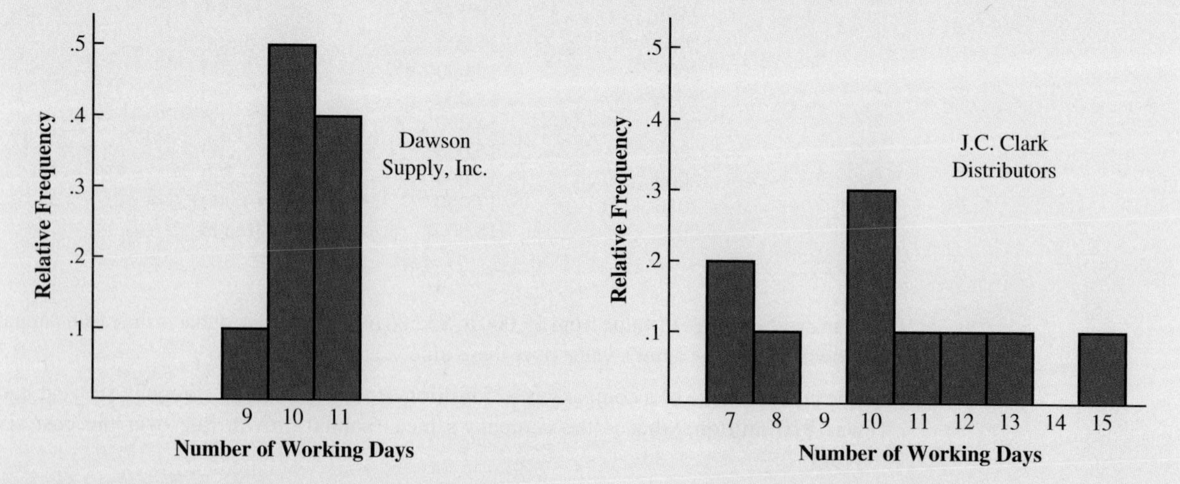

Although the range is the easiest of the measures of variability to compute, it is seldom used as the only measure. The reason is that the range is based on only two of the observations and thus is highly influenced by extreme values. Suppose the highest paid graduate received a starting salary of $10,000 per month. In this case, the range would be 10,000 − 3710 = 6290 rather than 615. This large value for the range would not be especially descriptive of the variability in the data because 11 of the 12 starting salaries are closely grouped between 3710 and 4130.

Interquartile Range

A measure of variability that overcomes the dependency on extreme values is the **interquartile range (IQR)**. This measure of variability is the difference between the third quartile, Q_3, and the first quartile, Q_1. In other words, the interquartile range is the range for the middle 50% of the data.

INTERQUARTILE RANGE

$$\text{IQR} = Q_3 - Q_1 \tag{3.5}$$

For the data on monthly starting salaries, the quartiles are $Q_3 = 4000$ and $Q_1 = 3865$. Thus, the interquartile range is 4000 − 3865 = 135.

Variance

The **variance** is a measure of variability that utilizes all the data. The variance is based on the difference between the value of each observation (x_i) and the mean. The difference between each x_i and the mean ($\bar{x}$ for a sample, μ for a population) is called a *deviation about the mean.* For a sample, a deviation about the mean is written $(x_i - \bar{x})$; for a population, it is written $(x_i - \mu)$. In the computation of the variance, the deviations about the mean are *squared.*

If the data are for a population, the average of the squared deviations is called the *population variance.* The population variance is denoted by the Greek symbol σ^2. For a population of N observations and with μ denoting the population mean, the definition of the population variance is as follows.

POPULATION VARIANCE

$$\sigma^2 = \frac{\Sigma(x_i - \mu)^2}{N} \tag{3.6}$$

In most statistical applications, the data being analyzed are for a sample. When we compute a sample variance, we are often interested in using it to estimate the population variance σ^2. Although a detailed explanation is beyond the scope of this text, it can be shown that if the sum of the squared deviations about the sample mean is divided by $n - 1$, and not n, the resulting sample variance provides an unbiased estimate of the population variance. For this reason, the *sample variance,* denoted by s^2, is defined as follows.

The sample variance s^2 is a point estimator of the population variance σ^2.

SAMPLE VARIANCE

$$s^2 = \frac{\Sigma(x_i - \bar{x})^2}{n - 1} \tag{3.7}$$

TABLE 3.3 COMPUTATION OF DEVIATIONS AND SQUARED DEVIATIONS ABOUT THE MEAN FOR THE CLASS SIZE DATA

Number of Students in Class (x_i)	Mean Class Size ($\bar{x}$)	Deviation About the Mean ($x_i - \bar{x}$)	Squared Deviation About the Mean $(x_i - \bar{x})^2$
46	44	2	4
54	44	10	100
42	44	−2	4
46	44	2	4
32	44	−12	144
		0	256
		$\Sigma(x_i - \bar{x})$	$\Sigma(x_i - \bar{x})^2$

To illustrate the computation of the sample variance, we will use the data on class size for the sample of five college classes as presented in Section 3.1. A summary of the data, including the computation of the deviations about the mean and the squared deviations about the mean, is shown in Table 3.3. The sum of squared deviations about the mean is $\Sigma(x_i - \bar{x})^2 = 256$. Hence, with $n - 1 = 4$, the sample variance is

$$s^2 = \frac{\Sigma(x_i - \bar{x})^2}{n - 1} = \frac{256}{4} = 64$$

Before moving on, let us note that the units associated with the sample variance often cause confusion. Because the values being summed in the variance calculation, $(x_i - \bar{x})^2$, are squared, the units associated with the sample variance are also *squared.* For instance, the sample variance for the class size data is $s^2 = 64$ (students)^2. The squared units associated with variance make it difficult to develop an intuitive understanding and interpretation of the numerical value of the variance. We recommend that you think of the variance as a measure useful in comparing the amount of variability for two or more variables. In a comparison of the variables, the one with the largest variance shows the most variability. Further interpretation of the value of the variance may not be necessary.

The variance is useful in comparing the variability of two or more variables.

As another illustration of computing a sample variance, consider the starting salaries listed in Table 3.1 for the 12 business school graduates. In Section 3.1, we showed that the sample mean starting salary was 3940. The computation of the sample variance ($s^2 = 27{,}440.91$) is shown in Table 3.4.

In Tables 3.3 and 3.4 we show both the sum of the deviations about the mean and the sum of the squared deviations about the mean. For any data set, the sum of the deviations about the mean will *always equal zero.* Note that in Tables 3.3 and 3.4, $\Sigma(x_i - \bar{x}) = 0$. The positive deviations and negative deviations cancel each other, causing the sum of the deviations about the mean to equal zero.

Standard Deviation

The **standard deviation** is defined to be the positive square root of the variance. Following the notation we adopted for a sample variance and a population variance, we use s to denote the sample standard deviation and σ to denote the population standard deviation. The standard deviation is derived from the variance in the following way.

TABLE 3.4 COMPUTATION OF THE SAMPLE VARIANCE FOR THE STARTING SALARY DATA

Monthly Salary (x_i)	Sample Mean ($\bar{x}$)	Deviation About the Mean ($x_i - \bar{x}$)	Squared Deviation About the Mean $(x_i - \bar{x})^2$
3850	3940	−90	8,100
3950	3940	10	100
4050	3940	110	12,100
3880	3940	−60	3,600
3755	3940	−185	34,225
3710	3940	−230	52,900
3890	3940	−50	2,500
4130	3940	190	36,100
3940	3940	0	0
4325	3940	385	148,225
3920	3940	−20	400
3880	3940	−60	3,600
		0	301,850
		$\Sigma(x_i - \bar{x})$	$\Sigma(x_i - \bar{x})^2$

Using equation (3.7),

$$s^2 = \frac{\Sigma(x_i - \bar{x})^2}{n - 1} = \frac{301{,}850}{11} = 27{,}440.91$$

The sample standard deviation s is a point estimator of the population standard deviation σ.

STANDARD DEVIATION

$$\text{Sample standard deviation} = s = \sqrt{s^2} \quad \textbf{(3.8)}$$

$$\text{Population standard deviation} = \sigma = \sqrt{\sigma^2} \quad \textbf{(3.9)}$$

Recall that the sample variance for the sample of class sizes in five college classes is $s^2 = 64$. Thus, the sample standard deviation is $s = \sqrt{64} = 8$. For the data on starting salaries, the sample standard deviation is $s = \sqrt{27{,}440.91} = 165.65$.

The standard deviation is easier to interpret than the variance because the standard deviation is measured in the same units as the data.

What is gained by converting the variance to its corresponding standard deviation? Recall that the units associated with the variance are squared. For example, the sample variance for the starting salary data of business school graduates is $s^2 = 27{,}440.91$ (dollars)^2. Because the standard deviation is the square root of the variance, the units of the variance, dollars squared, are converted to dollars in the standard deviation. Thus, the standard deviation of the starting salary data is \$165.65. In other words, the standard deviation is measured in the same units as the original data. For this reason the standard deviation is more easily compared to the mean and other statistics that are measured in the same units as the original data.

Coefficient of Variation

The coefficient of variation is a relative measure of variability; it measures the standard deviation relative to the mean.

In some situations we may be interested in a descriptive statistic that indicates how large the standard deviation is relative to the mean. This measure is called the **coefficient of variation** and is usually expressed as a percentage.

COEFFICIENT OF VARIATION

$$\left(\frac{\text{Standard deviation}}{\text{Mean}} \times 100\right)\% \tag{3.10}$$

For the class size data, we found a sample mean of 44 and a sample standard deviation of 8. The coefficient of variation is $[(8/44) \times 100]\% = 18.2\%$. In words, the coefficient of variation tells us that the sample standard deviation is 18.2% of the value of the sample mean. For the starting salary data with a sample mean of 3940 and a sample standard deviation of 165.65, the coefficient of variation, $[(165.65/3940) \times 100]\% = 4.2\%$, tells us the sample standard deviation is only 4.2% of the value of the sample mean. In general, the coefficient of variation is a useful statistic for comparing the variability of variables that have different standard deviations and different means.

NOTES AND COMMENTS

1. Statistical software packages and spreadsheets can be used to develop the descriptive statistics presented in this chapter. After the data are entered into a worksheet, a few simple commands can be used to generate the desired output. In three chapter-ending appendixes we show how Minitab, Excel, and StatTools can be used to develop descriptive statistics.
2. The standard deviation is a commonly used measure of the risk associated with investing in stock and stock funds (*Morningstar* website, July 21, 2012). It provides a measure of how monthly returns fluctuate around the long-run average return.
3. Rounding the value of the sample mean $\bar{x}$ and the values of the squared deviations $(x_i - \bar{x})^2$ may introduce errors when a calculator is used in the computation of the variance and standard deviation. To reduce rounding errors, we recommend carrying at least six significant digits during intermediate calculations. The resulting variance or standard deviati on can then be rounded to fewer digits.
4. An alternative formula for the computation of the sample variance is

$$s^2 = \frac{\Sigma x_i^2 - n\bar{x}^2}{n - 1}$$

where $\Sigma x_i^2 = x_1^2 + x_2^2 + \cdots + x_n^2$.

5. The mean absolute error (MAE) is another measure of variability that is computed by summing the absolute values of the deviations of the observations about the mean and dividing this sum by the number of observations. For a sample of size n, the MAE is computed as follows:

$$\text{MAE} = \frac{\Sigma|x_i - \bar{x}|}{n}$$

For the class size data presented in Section 3.1, $\bar{x} = 44$, $\Sigma|x_i - \bar{x}| = 28$, and the MAE $= 28/5 = 5.6$. You can learn more about the MAE and other measures of variability in Chapter 17.

Exercises

Methods

23. Consider a sample with data values of 10, 20, 12, 17, and 16. Compute the range and interquartile range.
24. Consider a sample with data values of 10, 20, 12, 17, and 16. Compute the variance and standard deviation.

SELF test

25. Consider a sample with data values of 27, 25, 20, 15, 30, 34, 28, and 25. Compute the range, interquartile range, variance, and standard deviation.

Applications

SELF test

26. A bowler's scores for six games were 182, 168, 184, 190, 170, and 174. Using these data as a sample, compute the following descriptive statistics:
    a. Range
    b. Variance
    c. Standard deviation
    d. Coefficient of variation
27. The results of a search to find the least expensive round-trip flights to Atlanta and Salt Lake City from 14 major U.S. cities are shown in the following table. The departure date was June 20, 2012, and the return date was June 27, 2012.

WEB file
Flights

	Round-Trip Cost ($)	
Departure City	**Atlanta**	**Salt Lake City**
Cincinnati	340.10	570.10
New York	321.60	354.60
Chicago	291.60	465.60
Denver	339.60	219.60
Los Angeles	359.60	311.60
Seattle	384.60	297.60
Detroit	309.60	471.60
Philadelphia	415.60	618.40
Washington, D.C.	293.60	513.60
Miami	249.60	523.20
San Francisco	539.60	381.60
Las Vegas	455.60	159.60
Phoenix	359.60	267.60
Dallas	333.90	458.60

   a. Compute the mean price for a round-trip flight into Atlanta and the mean price for a round-trip flight into Salt Lake City. Is Atlanta less expensive to fly into than Salt Lake City? If so, what could explain this difference?
   b. Compute the range, variance, and standard deviation for the two samples. What does this information tell you about the prices for flights into these two cities?
28. The Australian Open is the first of the four Grand Slam professional tennis events held each year. Victoria Azarenka beat Maria Sharapova to win the 2012 Australian Open women's title (*Washington Post,* January 27, 2012). During the tournament Ms. Azarenka's serve speed reached 178 kilometers per hour. A list of the 20 Women's Singles serve speed leaders for the 2012 Australian Open is provided below.

WEB file
AustralianOpen

Player	Serve Speed (km/h)	Player	Serve Speed (km/h)
S. Williams	191	G. Arn	179
S. Lisicki	190	V. Azarenka	178
M. Keys	187	A. Ivanovic	178
L. Hradecka	187	P. Kvitova	178
J. Gajdosova	187	M. Krajicek	178
J. Hampton	181	V. Dushevina	178
B. Mattek-Sands	181	S. Stosur	178
F. Schiavone	179	S. Cirstea	177
P. Parmentier	179	M. Barthel	177
N. Petrova	179	P. Ormaechea	177

   a. Compute the mean, variance, and standard deviation for the serve speeds.
   b. A similar sample of the 20 Women's Singles serve speed leaders for the 2011 Wimbledon tournament showed a sample mean serve speed of 182.5 kilometers per hour. The variance

and standard deviation were 33.3 and 5.77, respectively. Discuss any difference between the serve speeds in the Australian Open and the Wimbledon women's tournaments.

29. The *Los Angeles Times* regularly reports the air quality index for various areas of Southern California. A sample of air quality index values for Pomona provided the following data: 28, 42, 58, 48, 45, 55, 60, 49, and 50.
    a. Compute the range and interquartile range.
    b. Compute the sample variance and sample standard deviation.
    c. A sample of air quality index readings for Anaheim provided a sample mean of 48.5, a sample variance of 136, and a sample standard deviation of 11.66. What comparisons can you make between the air quality in Pomona and that in Anaheim on the basis of these descriptive statistics?

30. The following data were used to construct the histograms of the number of days required to fill orders for Dawson Supply, Inc., and J.C. Clark Distributors (see Figure 3.2).

Dawson Supply Days for Delivery:	11	10	9	10	11	11	10	11	10	10
Clark Distributors Days for Delivery:	8	10	13	7	10	11	10	7	15	12

Use the range and standard deviation to support the previous observation that Dawson Supply provides the more consistent and reliable delivery times.

31. The results of Accounting Principals' latest Workonomix survey indicate the average American worker spends $1092 on coffee annually (*The Consumerist,* January 20, 2012). To determine if there are any differences in coffee expenditures by age group, samples of 10 consumers were selected for three age groups (18–34, 35–44, and 45 and Older). The dollar amount each consumer in the sample spent last year on coffee is provided below.

WEB file

Coffee

18–34	35–44	45 and Older
1355	969	1135
115	434	956
1456	1792	400
2045	1500	1374
1621	1277	1244
994	1056	825
1937	1922	763
1200	1350	1192
1567	1586	1305
1390	1415	1510

a. Compute the mean, variance, and standard deviation for the each of these three samples.
b. What observations can be made based on these data?

WEB file

BackToSchool

32. The National Retail Federation reported that college freshman spend more on back-to-school items than any other college group (*USA Today,* August 4, 2006). Sample data comparing the back-to-school expenditures for 25 freshmen and 20 seniors are shown in the data file BackToSchool.
    a. What is the mean back-to-school expenditure for each group? Are the data consistent with the National Retail Federation's report?
    b. What is the range for the expenditures in each group?
    c. What is the interquartile range for the expenditures in each group?
    d. What is the standard deviation for expenditures in each group?
    e. Do freshmen or seniors have more variation in back-to-school expenditures?

33. Scores turned in by an amateur golfer at the Bonita Fairways Golf Course in Bonita Springs, Florida, during 2011 and 2012 are as follows:

2011 Season:	74	78	79	77	75	73	75	77
2012 Season:	71	70	75	77	85	80	71	79

a. Use the mean and standard deviation to evaluate the golfer's performance over the two-year period.
b. What is the primary difference in performance between 2011 and 2012? What improvement, if any, can be seen in the 2012 scores?

34. The following times were recorded by the quarter-mile and mile runners of a university track team (times are in minutes).

Quarter-Mile Times:	.92	.98	1.04	.90	.99
Mile Times:	4.52	4.35	4.60	4.70	4.50

After viewing this sample of running times, one of the coaches commented that the quarter-milers turned in the more consistent times. Use the standard deviation and the coefficient of variation to summarize the variability in the data. Does the use of the coefficient of variation indicate that the coach's statement should be qualified?

3.3 Measures of Distribution Shape, Relative Location, and Detecting Outliers

We have described several measures of location and variability for data. In addition, it is often important to have a measure of the shape of a distribution. In Chapter 2 we noted that a histogram provides a graphical display showing the shape of a distribution. An important numerical measure of the shape of a distribution is called **skewness**.

Distribution Shape

Figure 3.3 shows four histograms constructed from relative frequency distributions. The histograms in Panels A and B are moderately skewed. The one in Panel A is skewed to the left; its skewness is −.85. The histogram in Panel B is skewed to the right; its skewness is +.85. The histogram in Panel C is symmetric; its skewness is zero. The histogram in Panel D is highly skewed to the right; its skewness is 1.62. The formula used to compute skewness is somewhat complex.[1] However, the skewness can easily be computed using statistical software. For data skewed to the left, the skewness is negative; for data skewed to the right, the skewness is positive. If the data are symmetric, the skewness is zero.

For a symmetric distribution, the mean and the median are equal. When the data are positively skewed, the mean will usually be greater than the median; when the data are negatively skewed, the mean will usually be less than the median. The data used to construct the histogram in Panel D are customer purchases at a women's apparel store. The mean purchase amount is $77.60 and the median purchase amount is $59.70. The relatively few large purchase amounts tend to increase the mean, while the median remains unaffected by the large purchase amounts. The median provides the preferred measure of location when the data are highly skewed.

z-Scores

In addition to measures of location, variability, and shape, we are also interested in the relative location of values within a data set. Measures of relative location help us determine how far a particular value is from the mean.

[1]The formula for the skewness of sample data:

$$\text{Skewness} = \frac{n}{(n-1)(n-2)} \sum \left(\frac{x_i - \bar{x}}{s} \right)^3$$

FIGURE 3.3 HISTOGRAMS SHOWING THE SKEWNESS FOR FOUR DISTRIBUTIONS

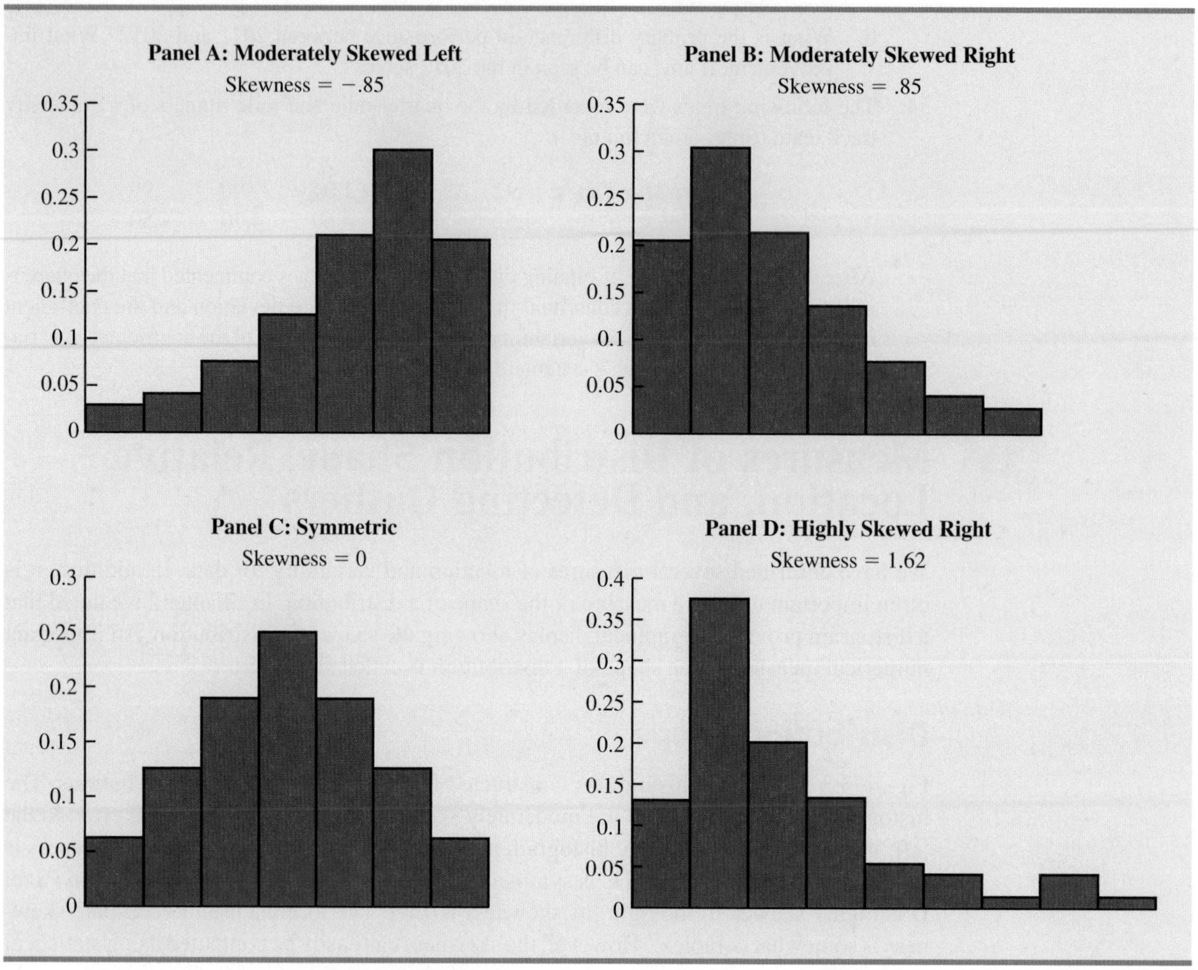

By using both the mean and standard deviation, we can determine the relative location of any observation. Suppose we have a sample of n observations, with the values denoted by $x_1, x_2, \ldots, x_n$. In addition, assume that the sample mean, $\bar{x}$, and the sample standard deviation, s, are already computed. Associated with each value, x_i, is another value called its **z-score**. Equation (3.11) shows how the z-score is computed for each x_i.

z-SCORE

$$z_i = \frac{x_i - \bar{x}}{s} \tag{3.11}$$

where

z_i = the z-score for x_i

$\bar{x}$ = the sample mean

s = the sample standard deviation

TABLE 3.5 z-SCORES FOR THE CLASS SIZE DATA

Number of Students in Class (x_i)	Deviation About the Mean ($x_i - \bar{x}$)	z-Score $\left(\frac{x_i - \bar{x}}{s}\right)$
46	2	2/8 = .25
54	10	10/8 = 1.25
42	−2	−2/8 = −.25
46	2	2/8 = .25
32	−12	−12/8 = −1.50

The z-score is often called the *standardized value*. The z-score, z_i, can be interpreted as the *number of standard deviations x_i is from the mean $\bar{x}$*. For example, $z_1 = 1.2$ would indicate that x_1 is 1.2 standard deviations greater than the sample mean. Similarly, $z_2 = -.5$ would indicate that x_2 is .5, or 1/2, standard deviation less than the sample mean. A z-score greater than zero occurs for observations with a value greater than the mean, and a z-score less than zero occurs for observations with a value less than the mean. A z-score of zero indicates that the value of the observation is equal to the mean.

The z-score for any observation can be interpreted as a measure of the relative location of the observation in a data set. Thus, observations in two different data sets with the same z-score can be said to have the same relative location in terms of being the same number of standard deviations from the mean.

The process of converting a value for a variable to a z-score is often referred to as a z transformation.

The z-scores for the class size data from Section 3.1 are computed in Table 3.5. Recall the previously computed sample mean, $\bar{x} = 44$, and sample standard deviation, $s = 8$. The z-score of -1.50 for the fifth observation shows it is farthest from the mean; it is 1.50 standard deviations below the mean. Figure 3.4 provides a dot plot of the class size data with a graphical representation of the associated z-scores on the axis below.

Chebyshev's Theorem

Chebyshev's theorem enables us to make statements about the proportion of data values that must be within a specified number of standard deviations of the mean.

FIGURE 3.4 DOT PLOT SHOWING CLASS SIZE DATA AND z-SCORES

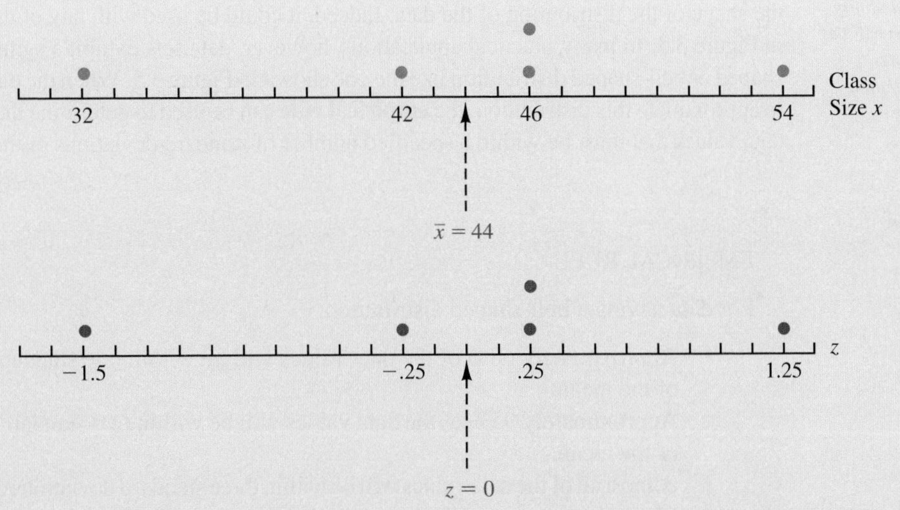

CHEBYSHEV'S THEOREM

At least $(1 - 1/z^2)$ of the data values must be within z standard deviations of the mean, where z is any value greater than 1.

Some of the implications of this theorem, with $z = 2$, 3, and 4 standard deviations, follow.

- At least .75, or 75%, of the data values must be within $z = 2$ standard deviations of the mean.
- At least .89, or 89%, of the data values must be within $z = 3$ standard deviations of the mean.
- At least .94, or 94%, of the data values must be within $z = 4$ standard deviations of the mean.

For an example using Chebyshev's theorem, suppose that the midterm test scores for 100 students in a college business statistics course had a mean of 70 and a standard deviation of 5. How many students had test scores between 60 and 80? How many students had test scores between 58 and 82?

For the test scores between 60 and 80, we note that 60 is two standard deviations below the mean and 80 is two standard deviations above the mean. Using Chebyshev's theorem, we see that at least .75, or at least 75%, of the observations must have values within two standard deviations of the mean. Thus, at least 75% of the students must have scored between 60 and 80.

Chebyshev's theorem requires $z > 1$; but z need not be an integer.

For the test scores between 58 and 82, we see that $(58 - 70)/5 = -2.4$ indicates 58 is 2.4 standard deviations below the mean and that $(82 - 70)/5 = +2.4$ indicates 82 is 2.4 standard deviations above the mean. Applying Chebyshev's theorem with $z = 2.4$, we have

$$\left(1 - \frac{1}{z^2}\right) = \left(1 - \frac{1}{(2.4)^2}\right) = .826$$

At least 82.6% of the students must have test scores between 58 and 82.

Empirical Rule

The empirical rule is based on the normal probability distribution, which will be discussed in Chapter 6. The normal distribution is used extensively throughout the text.

One of the advantages of Chebyshev's theorem is that it applies to any data set regardless of the shape of the distribution of the data. Indeed, it could be used with any of the distributions in Figure 3.3. In many practical applications, however, data sets exhibit a symmetric mound-shaped or bell-shaped distribution like the one shown in Figure 3.5. When the data are believed to approximate this distribution, the **empirical rule** can be used to determine the percentage of data values that must be within a specified number of standard deviations of the mean.

EMPIRICAL RULE

For data having a bell-shaped distribution:

- Approximately 68% of the data values will be within one standard deviation of the mean.
- Approximately 95% of the data values will be within two standard deviations of the mean.
- Almost all of the data values will be within three standard deviations of the mean.

FIGURE 3.5 A SYMMETRIC MOUND-SHAPED OR BELL-SHAPED DISTRIBUTION

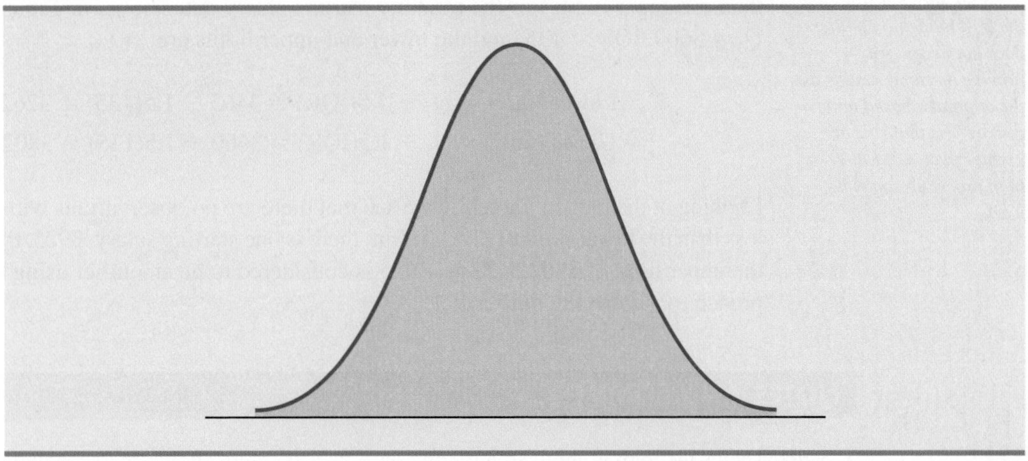

For example, liquid detergent cartons are filled automatically on a production line. Filling weights frequently have a bell-shaped distribution. If the mean filling weight is 16 ounces and the standard deviation is .25 ounces, we can use the empirical rule to draw the following conclusions.

- Approximately 68% of the filled cartons will have weights between 15.75 and 16.25 ounces (within one standard deviation of the mean).
- Approximately 95% of the filled cartons will have weights between 15.50 and 16.50 ounces (within two standard deviations of the mean).
- Almost all filled cartons will have weights between 15.25 and 16.75 ounces (within three standard deviations of the mean).

Detecting Outliers

Sometimes a data set will have one or more observations with unusually large or unusually small values. These extreme values are called **outliers**. Experienced statisticians take steps to identify outliers and then review each one carefully. An outlier may be a data value that has been incorrectly recorded. If so, it can be corrected before further analysis. An outlier may also be from an observation that was incorrectly included in the data set; if so, it can be removed. Finally, an outlier may be an unusual data value that has been recorded correctly and belongs in the data set. In such cases it should remain.

It is a good idea to check for outliers before making decisions based on data analysis. Errors are often made in recording data and entering data into the computer. Outliers should not necessarily be deleted, but their accuracy and appropriateness should be verified.

Standardized values (z-scores) can be used to identify outliers. Recall that the empirical rule allows us to conclude that for data with a bell-shaped distribution, almost all the data values will be within three standard deviations of the mean. Hence, in using z-scores to identify outliers, we recommend treating any data value with a z-score less than -3 or greater than $+3$ as an outlier. Such data values can then be reviewed for accuracy and to determine whether they belong in the data set.

Refer to the z-scores for the class size data in Table 3.5. The z-score of -1.50 shows the fifth class size is farthest from the mean. However, this standardized value is well within the -3 to $+3$ guideline for outliers. Thus, the z-scores do not indicate that outliers are present in the class size data.

Another approach to identifying outliers is based upon the values of the first and third quartiles (Q_1 and Q_3) and the interquartile range (IQR). Using this method, we first compute the following lower and upper limits:

$$\text{Lower Limit} = Q_1 - 1.5(\text{IQR})$$
$$\text{Upper Limit} = Q_3 + 1.5(\text{IQR})$$

The approach that uses the first and third quartiles and the IQR to identify outliers does not necessarily provide the same results as the approach based upon a z-score less than −3 or greater than +3. Either or both procedures may be used.

An observation is classified as an outlier if its value is less than the lower limit or greater than the upper limit. For the monthly starting salary data shown in Table 3.1, $Q_1 = 3465$, $Q_3 = 3600$, IQR $= 135$, and the lower and upper limits are

$$\text{Lower Limit} = Q_1 - 1.5(\text{IQR}) = 3465 - 1.5(135) = 3262.5$$
$$\text{Upper Limit} = Q_3 + 1.5(\text{IQR}) = 3600 + 1.5(135) = 3802.5$$

Looking at the data in Table 3.1 we see that there are no observations with a starting salary less than the lower limit of 3262.5. But, there is one starting salary, 3925, that is greater than the upper limit of 3802.5. Thus, 3925 is considered to be an outlier using this alternate approach to identifying outliers.

NOTES AND COMMENTS

1. Chebyshev's theorem is applicable for any data set and can be used to state the minimum number of data values that will be within a certain number of standard deviations of the mean. If the data are known to be approximately bell-shaped, more can be said. For instance, the empirical rule allows us to say that *approximately* 95% of the data values will be within two standard deviations of the mean; Chebyshev's theorem allows us to conclude only that at least 75% of the data values will be in that interval.

2. Before analyzing a data set, statisticians usually make a variety of checks to ensure the validity of data. In a large study it is not uncommon for errors to be made in recording data values or in entering the values into a computer. Identifying outliers is one tool used to check the validity of the data.

Exercises

Methods

35. Consider a sample with data values of 10, 20, 12, 17, and 16. Compute the z-score for each of the five observations.

36. Consider a sample with a mean of 500 and a standard deviation of 100. What are the z-scores for the following data values: 520, 650, 500, 450, and 280?

SELF test

37. Consider a sample with a mean of 30 and a standard deviation of 5. Use Chebyshev's theorem to determine the percentage of the data within each of the following ranges:
    a. 20 to 40
    b. 15 to 45
    c. 22 to 38
    d. 18 to 42
    e. 12 to 48

38. Suppose the data have a bell-shaped distribution with a mean of 30 and a standard deviation of 5. Use the empirical rule to determine the percentage of data within each of the following ranges:
    a. 20 to 40
    b. 15 to 45
    c. 25 to 35

Applications

SELF test

39. The results of a national survey showed that on average, adults sleep 6.9 hours per night. Suppose that the standard deviation is 1.2 hours.
    a. Use Chebyshev's theorem to calculate the percentage of individuals who sleep between 4.5 and 9.3 hours.

b. Use Chebyshev's theorem to calculate the percentage of individuals who sleep between 3.9 and 9.9 hours.

c. Assume that the number of hours of sleep follows a bell-shaped distribution. Use the empirical rule to calculate the percentage of individuals who sleep between 4.5 and 9.3 hours per day. How does this result compare to the value that you obtained using Chebyshev's theorem in part (a)?

40. The Energy Information Administration reported that the mean retail price per gallon of regular grade gasoline was \$3.43 (Energy Information Administration, July 2012). Suppose that the standard deviation was \$.10 and that the retail price per gallon has a bell-shaped distribution.

a. What percentage of regular grade gasoline sold between \$3.33 and \$3.53 per gallon?

b. What percentage of regular grade gasoline sold between \$3.33 and \$3.63 per gallon?

c. What percentage of regular grade gasoline sold for more than \$3.63 per gallon?

41. The national average for the math portion of the College Board's SAT test is 515 (*The World Almanac,* 2009). The College Board periodically rescales the test scores such that the standard deviation is approximately 100. Answer the following questions using a bell-shaped distribution and the empirical rule for the math test scores.

a. What percentage of students have an SAT math score greater than 615?

b. What percentage of students have an SAT math score greater than 715?

c. What percentage of students have an SAT math score between 415 and 515?

d. What percentage of students have an SAT math score between 315 and 615?

42. Many families in California are using backyard structures for home offices, art studios, and hobby areas as well as for additional storage. Suppose that the mean price for a customized wooden, shingled backyard structure is \$3100. Assume that the standard deviation is \$1200.

a. What is the z-score for a backyard structure costing \$2300?

b. What is the z-score for a backyard structure costing \$4900?

c. Interpret the z-scores in parts (a) and (b). Comment on whether either should be considered an outlier.

d. If the cost for a backyard shed-office combination built in Albany, California, is \$13,000, should this structure be considered an outlier? Explain.

43. Florida Power & Light (FP&L) Company has enjoyed a reputation for quickly fixing its electric system after storms. However, during the hurricane seasons of 2004 and 2005, a new reality was that the company's historical approach to emergency electric system repairs was no longer good enough (*The Wall Street Journal,* January 16, 2006). Data showing the days required to restore electric service after seven hurricanes during 2004 and 2005 follow.

Hurricane	Days to Restore Service
Charley	13
Frances	12
Jeanne	8
Dennis	3
Katrina	8
Rita	2
Wilma	18

Based on this sample of seven, compute the following descriptive statistics:

a. Mean, median, and mode

b. Range and standard deviation

c. Should Wilma be considered an outlier in terms of the days required to restore electric service?

d. The seven hurricanes resulted in 10 million service interruptions to customers. Do the statistics show that FP&L should consider updating its approach to emergency electric system repairs? Discuss.

44. A sample of 10 NCAA college basketball game scores provided the following data.

WEB file NCAA

Winning Team	Points	Losing Team	Points	Winning Margin
Arizona	90	Oregon	66	24
Duke	85	Georgetown	66	19
Florida State	75	Wake Forest	70	5
Kansas	78	Colorado	57	21
Kentucky	71	Notre Dame	63	8
Louisville	65	Tennessee	62	3
Oklahoma State	72	Texas	66	6
Purdue	76	Michigan State	70	6
Stanford	77	Southern Cal	67	10
Wisconsin	76	Illinois	56	20

a. Compute the mean and standard deviation for the points scored by the winning team.
b. Assume that the points scored by the winning teams for all NCAA games follow a bell-shaped distribution. Using the mean and standard deviation found in part (a), estimate the percentage of all NCAA games in which the winning team scores 84 or more points. Estimate the percentage of NCAA games in which the winning team scores more than 90 points.
c. Compute the mean and standard deviation for the winning margin. Do the data contain outliers? Explain.

45. The Associated Press Team Marketing Report listed the Dallas Cowboys as the team with the highest ticket prices in the National Football League (*USA Today,* October 20, 2009). Data showing the average ticket price for a sample of 14 teams in the National Football League are as follows.

WEB file NFLTickets

Team	Ticket Price	Team	Ticket Price
Atlanta Falcons	72	Green Bay Packers	63
Buffalo Bills	51	Indianapolis Colts	83
Carolina Panthers	63	New Orleans Saints	62
Chicago Bears	88	New York Jets	87
Cleveland Browns	55	Pittsburgh Steelers	67
Dallas Cowboys	160	Seattle Seahawks	61
Denver Broncos	77	Tennessee Titans	61

a. What is the mean ticket price?
b. The previous year, the mean ticket price was \$72.20. What was the percentage increase in the mean ticket price for the one-year period?
c. Compute the median ticket price.
d. Compute the first and third quartiles.
e. Compute the standard deviation.
f. What is the z-score for the Dallas Cowboys ticket price? Should this price be considered an outlier? Explain.

3.4 Five-Number Summaries and Box Plots

Summary statistics and easy-to-draw graphs based on summary statistics can be used to quickly summarize large quantities of data. In this section we show how five-number summaries and box plots can be developed to identify several characteristics of a large data set.

Five-Number Summary

In a **five-number summary**, five numbers are used to summarize the data:

1. Smallest value
2. First quartile (Q_1)
3. Median (Q_2)
4. Third quartile (Q_3)
5. Largest value

To develop a five-number summary we first arrange the data in ascending order. We then identify the smallest value, the three quartiles, and the largest value. For the monthly starting salary data in Table 3.1 we obtain the following results.

3710	3755	3850	|	3880	3880	3890	|	3920	3940	3950	|	4050	4130	4325
			$Q_1 = 3865$				$Q_2 = 3905$ (Median)				$Q_3 = 4000$			

We showed how to compute the median, 3905, and the quartiles, $Q_1 = 3865$ and $Q_3 = 4000$ in Section 3.1. Reviewing the data shows that the smallest value is 3710 and the largest value is 4325. Thus the five-number summary for the monthly starting salary data is 3710, 3865, 3905, 4000, 4325. Approximately one-fourth, or 25%, of the observations are between adjacent numbers in a five-number summary.

Box Plot

A **box plot** is a graphical summary of data that is based on a five-number summary. A key to the development of a box plot is the computation of the interquartile range, $\text{IQR} = Q_3 - Q_1$. Figure 3.6 shows a box plot for the monthly starting salary data. The steps used to construct the box plot follow.

Box plots provide a convenient visual display of several characteristics of a data set.

1. A box is drawn with the ends of the box located at the first and third quartiles. For the monthly starting salary data, $Q_1 = 3865$ and $Q_3 = 4000$. This box contains the middle 50% of the data.
2. A vertical line is drawn in the box at the location of the median (3905 for the monthly starting salary data).
3. By using the interquartile range, $\text{IQR} = Q_3 - Q_1$, *limits* are located at 1.5(IQR) below Q_1 and 1.5(IQR) above Q_3. For the monthly starting salary data, $\text{IQR} = Q_3 - Q_1 = 4000 - 3865 = 135$. Thus, the limits are $3865 - 1.5(135) = 3662.5$ and $4000 + 1.5(135) = 4202.5$. Data outside these limits are considered *outliers*.

FIGURE 3.6 BOX PLOT OF THE MONTHLY STARTING SALARY DATA WITH LINES SHOWING THE LOWER AND UPPER LIMITS

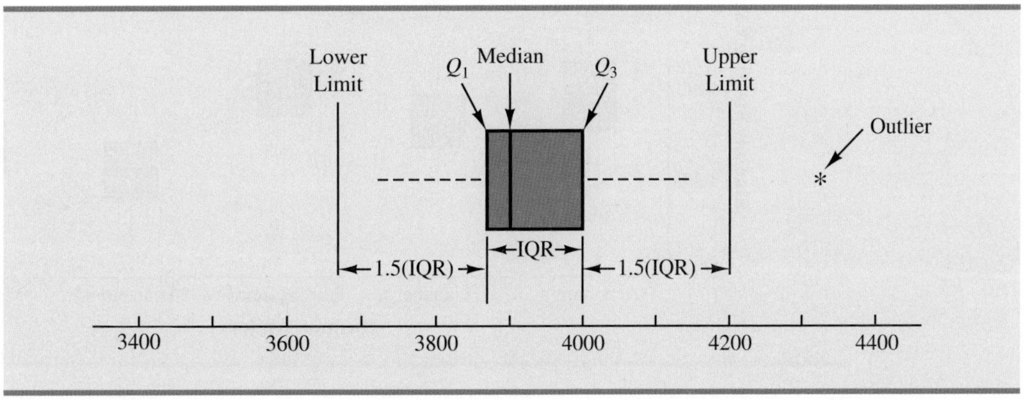

FIGURE 3.7 BOX PLOT OF THE MONTHLY STARTING SALARY DATA

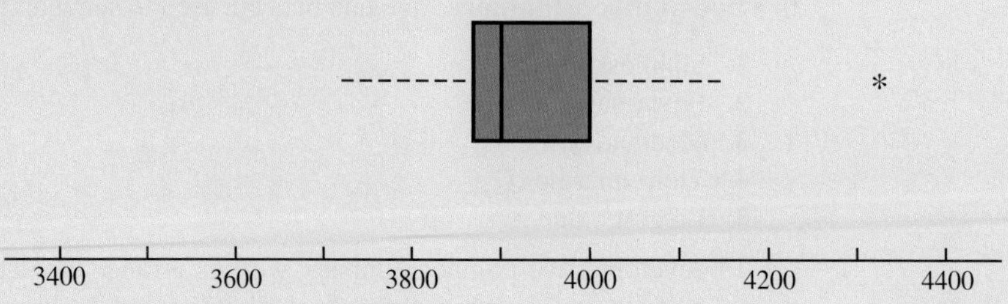

4. The dashed lines in Figure 3.6 are called *whiskers.* The whiskers are drawn from the ends of the box to the smallest and largest values *inside the limits* computed in step 3. Thus, the whiskers end at salary values of 3710 and 4130.
5. Finally, the location of each outlier is shown with the symbol *. In Figure 3.6 we see one outlier, 4325.

In Figure 3.6 we included lines showing the location of the upper and lower limits. These lines were drawn to show how the limits are computed and where they are located. Although the limits are always computed, generally they are not drawn on the box plots. Figure 3.7 shows the usual appearance of a box plot for the monthly starting salary data.

WEB file
MajorSalary

In order to compare monthly starting salaries for business school graduates by major, a sample of 111 recent graduates was selected. The major and the monthly starting salary were recorded for each graduate. Figure 3.8 shows the Minitab box plots for accounting, finance, information systems, management, and marketing majors. Note that the major is shown on the horizontal axis and each box plot is shown vertically above the corresponding major. Displaying box plots in this manner is an excellent graphical technique for making comparisons among two or more groups.

What observations can you make about monthly starting salaries by major using the box plots in Figure 3.8? Specifically, we note the following:

- The higher salaries are in accounting; the lower salaries are in management and marketing.

FIGURE 3.8 MINITAB BOX PLOTS OF MONTHLY STARTING SALARY BY MAJOR

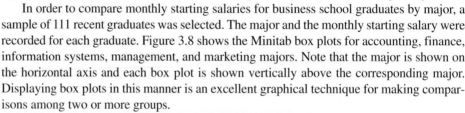

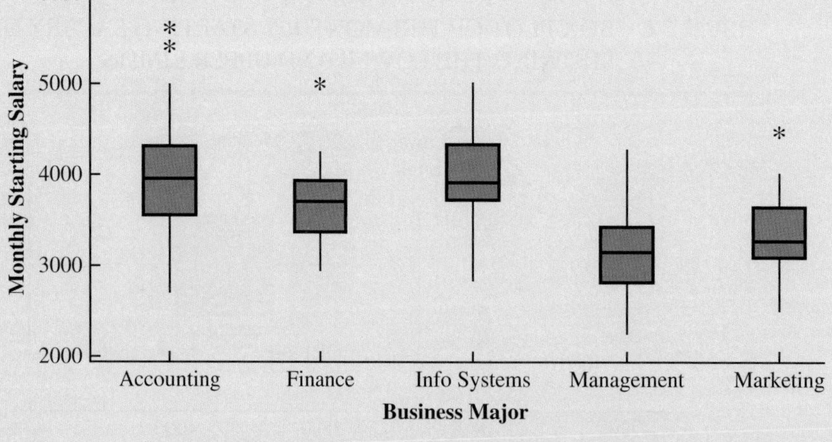

- Based on the medians, accounting and information systems have similar and higher median salaries. Finance is next with management and marketing showing lower median salaries.
- High salary outliers exist for accounting, finance, and marketing majors.
- Finance salaries appear to have the least variation, while accounting salaries appear to have the most variation.

Perhaps you can see additional interpretations based on these box plots.

NOTES AND COMMENTS

1. An advantage of the exploratory data analysis procedures is that they are easy to use; few numerical calculations are necessary. We simply sort the data values into ascending order and identify the five-number summary. The box plot can then be constructed. It is not necessary to compute the mean and the standard deviation for the data.

2. In Appendix 3.1, we show how to construct a box plot for the starting salary data using Minitab. The box plot obtained looks just like the one in Figure 3.7, but turned on its side.

Exercises

Methods

46. Consider a sample with data values of 27, 25, 20, 15, 30, 34, 28, and 25. Provide the five-number summary for the data.

47. Show the box plot for the data in exercise 36.

SELF test

48. Show the five-number summary and the box plot for the following data: 5, 15, 18, 10, 8, 12, 16, 10, 6.

49. A data set has a first quartile of 42 and a third quartile of 50. Compute the lower and upper limits for the corresponding box plot. Should a data value of 65 be considered an outlier?

Applications

50. Naples, Florida, hosts a half-marathon (13.1-mile race) in January each year. The event attracts top runners from throughout the United States as well as from around the world. In January 2009, 22 men and 31 women entered the 19–24 age class. Finish times in minutes are as follows (*Naples Daily News,* January 19, 2009). Times are shown in order of finish.

WEB file Runners

Finish	Men	Women	Finish	Men	Women	Finish	Men	Women
1	65.30	109.03	11	109.05	123.88	21	143.83	136.75
2	66.27	111.22	12	110.23	125.78	22	148.70	138.20
3	66.52	111.65	13	112.90	129.52	23		139.00
4	66.85	111.93	14	113.52	129.87	24		147.18
5	70.87	114.38	15	120.95	130.72	25		147.35
6	87.18	118.33	16	127.98	131.67	26		147.50
7	96.45	121.25	17	128.40	132.03	27		147.75
8	98.52	122.08	18	130.90	133.20	28		153.88
9	100.52	122.48	19	131.80	133.50	29		154.83
10	108.18	122.62	20	138.63	136.57	30		189.27
						31		189.28

a. George Towett of Marietta, Georgia, finished in first place for the men and Lauren Wald of Gainesville, Florida, finished in first place for the women. Compare the first-place finish times for men and women. If the 53 men and women runners had competed as one group, in what place would Lauren have finished?
b. What is the median time for men and women runners? Compare men and women runners based on their median times.
c. Provide a five-number summary for both the men and the women.
d. Are there outliers in either group?
e. Show the box plots for the two groups. Did men or women have the most variation in finish times? Explain.

SELF test

51. Annual sales, in millions of dollars, for 21 pharmaceutical companies follow.

8408	1374	1872	8879	2459	11413
608	14138	6452	1850	2818	1356
10498	7478	4019	4341	739	2127
3653	5794	8305			

a. Provide a five-number summary.
b. Compute the lower and upper limits.
c. Do the data contain any outliers?
d. Johnson & Johnson's sales are the largest on the list at $14,138 million. Suppose a data entry error (a transposition) had been made and the sales had been entered as $41,138 million. Would the method of detecting outliers in part (c) identify this problem and allow for correction of the data entry error?
e. Show a box plot.

52. *Consumer Reports* provided overall customer satisfaction scores for AT&T, Sprint, T-Mobile, and Verizon cell-phone services in major metropolitan areas throughout the United States. The rating for each service reflects the overall customer satisfaction considering a variety of factors such as cost, connectivity problems, dropped calls, static interference, and customer support. A satisfaction scale from 0 to 100 was used with 0 indicating completely dissatisfied and 100 indicating completely satisfied. The ratings for the four cell-phone services in 20 metropolitan areas are as shown (*Consumer Reports,* January 2009).

WEB file

CellService

Metropolitan Area	AT&T	Sprint	T-Mobile	Verizon
Atlanta	70	66	71	79
Boston	69	64	74	76
Chicago	71	65	70	77
Dallas	75	65	74	78
Denver	71	67	73	77
Detroit	73	65	77	79
Jacksonville	73	64	75	81
Las Vegas	72	68	74	81
Los Angeles	66	65	68	78
Miami	68	69	73	80
Minneapolis	68	66	75	77
Philadelphia	72	66	71	78
Phoenix	68	66	76	81
San Antonio	75	65	75	80
San Diego	69	68	72	79
San Francisco	66	69	73	75
Seattle	68	67	74	77
St. Louis	74	66	74	79
Tampa	73	63	73	79
Washington	72	68	71	76

a. Consider T-Mobile first. What is the median rating?
b. Develop a five-number summary for the T-Mobile service.
c. Are there outliers for T-Mobile? Explain.
d. Repeat parts (b) and (c) for the other three cell-phone services.
e. Show the box plots for the four cell-phone services on one graph. Discuss what a comparison of the box plots tells about the four services. Which service did *Consumer Reports* recommend as being best in terms of overall customer satisfaction?

53. The Philadelphia Phillies defeated the Tampa Bay Rays 4 to 3 to win the 2008 major league baseball World Series. Earlier in the major league baseball playoffs, the Philadelphia Phillies defeated the Los Angeles Dodgers to win the National League Championship, while the Tampa Bay Rays defeated the Boston Red Sox to win the American League Championship. The file MLBSalaries contains the salaries for the 28 players on each of these four teams (USA Today Salary Database, October 2008). The data, shown in thousands of dollars, have been ordered from the highest salary to the lowest salary for each team.

WEB file
MLBSalaries

a. Analyze the salaries for the World Champion Philadelphia Phillies. What is the total payroll for the team? What is the median salary? What is the five-number summary?
b. Were there salary outliers for the Philadelphia Phillies? If so, how many and what were the salary amounts?
c. What is the total payroll for each of the other three teams? Develop the five-number summary for each team and identify any outliers.
d. Show the box plots of the salaries for all four teams. What are your interpretations? Of these four teams, does it appear that the team with the higher salaries won the league championships and the World Series?

54. A listing of 46 mutual funds and their 12-month total return percentage is shown below (*Smart Money,* February 2004).
a. What are the mean and median return percentages for these mutual funds?
b. What are the first and third quartiles?
c. Provide a five-number summary.
d. Do the data contain any outliers? Show a box plot.

WEB file
Mutual

Mutual Fund	Return (%)	Mutual Fund	Return (%)
Alger Capital Appreciation	23.5	Nations Small Company	21.4
Alger LargeCap Growth	22.8	Nations SmallCap Index	24.5
Alger MidCap Growth	38.3	Nations Strategic Growth	10.4
Alger SmallCap	41.3	Nations Value Inv	10.8
AllianceBernstein Technology	40.6	One Group Diversified Equity	10.0
Federated American Leaders	15.6	One Group Diversified Int'l	10.9
Federated Capital Appreciation	12.4	One Group Diversified Mid Cap	15.1
Federated Equity-Income	11.5	One Group Equity Income	6.6
Federated Kaufmann	33.3	One Group Int'l Equity Index	13.2
Federated Max-Cap Index	16.0	One Group Large Cap Growth	13.6
Federated Stock	16.9	One Group Large Cap Value	12.8
Janus Adviser Int'l Growth	10.3	One Group Mid Cap Growth	18.7
Janus Adviser Worldwide	3.4	One Group Mid Cap Value	11.4
Janus Enterprise	24.2	One Group Small Cap Growth	23.6
Janus High-Yield	12.1	PBHG Growth	27.3
Janus Mercury	20.6	Putnam Europe Equity	20.4
Janus Overseas	11.9	Putnam Int'l Capital Opportunity	36.6
Janus Worldwide	4.1	Putnam International Equity	21.5
Nations Convertible Securities	13.6	Putnam Int'l New Opportunity	26.3
Nations Int'l Equity	10.7	Strong Advisor Mid Cap Growth	23.7
Nations LargeCap Enhd. Core	13.2	Strong Growth 20	11.7
Nations LargeCap Index	13.5	Strong Growth Inv	23.2
Nation MidCap Index	19.5	Strong Large Cap Growth	14.5

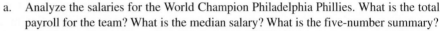

3.5 Measures of Association Between Two Variables

Thus far we have examined numerical methods used to summarize the data for *one variable at a time.* Often a manager or decision maker is interested in the *relationship between two variables.* In this section we present covariance and correlation as descriptive measures of the relationship between two variables.

We begin by reconsidering the application concerning a stereo and sound equipment store in San Francisco as presented in Section 2.4. The store's manager wants to determine the relationship between the number of weekend television commercials shown and the sales at the store during the following week. Sample data with sales expressed in hundreds of dollars are provided in Table 3.6. It shows 10 observations ($n = 10$), one for each week. The scatter diagram in Figure 3.9 shows a positive relationship, with higher sales (y) associated with a greater number of commercials (x). In fact, the scatter diagram suggests that a straight line could be used as an approximation of the relationship. In the following discussion, we introduce **covariance** as a descriptive measure of the linear association between two variables.

Covariance

For a sample of size n with the observations (x_1, y_1), (x_2, y_2), and so on, the sample covariance is defined as follows:

SAMPLE COVARIANCE

$$s_{xy} = \frac{\Sigma(x_i - \bar{x})(y_i - \bar{y})}{n - 1} \tag{3.12}$$

This formula pairs each x_i with a y_i. We then sum the products obtained by multiplying the deviation of each x_i from its sample mean $\bar{x}$ by the deviation of the corresponding y_i from its sample mean $\bar{y}$; this sum is then divided by $n - 1$.

To measure the strength of the linear relationship between the number of commercials x and the sales volume y in the stereo and sound equipment store problem, we use equation (3.12)

TABLE 3.6 SAMPLE DATA FOR THE STEREO AND SOUND EQUIPMENT STORE

WEB file Stereo

Week	Number of Commercials x	Sales Volume ($100s) y
1	2	50
2	5	57
3	1	41
4	3	54
5	4	54
6	1	38
7	5	63
8	3	48
9	4	59
10	2	46

FIGURE 3.9 SCATTER DIAGRAM FOR THE STEREO AND SOUND EQUIPMENT STORE

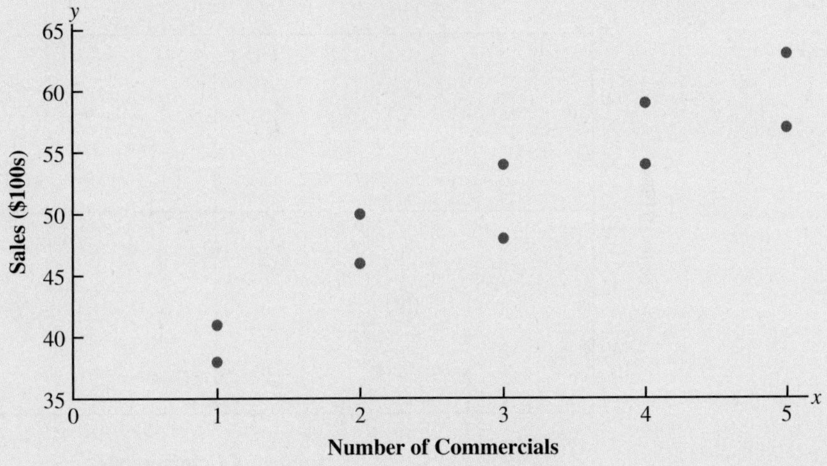

to compute the sample covariance. The calculations in Table 3.7 show the computation of $\Sigma(x_i - \bar{x})(y_i - \bar{y})$. Note that $\bar{x} = 30/10 = 3$ and $\bar{y} = 510/10 = 51$. Using equation (3.12), we obtain a sample covariance of

$$s_{xy} = \frac{\Sigma(x_i - \bar{x})(y_i - \bar{y})}{n - 1} = \frac{99}{9} = 11$$

The formula for computing the covariance of a population of size N is similar to equation (3.12), but we use different notation to indicate that we are working with the entire population.

POPULATION COVARIANCE

$$\sigma_{xy} = \frac{\Sigma(x_i - \mu_x)(y_i - \mu_y)}{N} \tag{3.13}$$

TABLE 3.7 CALCULATIONS FOR THE SAMPLE COVARIANCE

	x_i	y_i	$x_i - \bar{x}$	$y_i - \bar{y}$	$(x_i - \bar{x})(y_i - \bar{y})$
	2	50	−1	−1	1
	5	57	2	6	12
	1	41	−2	−10	20
	3	54	0	3	0
	4	54	1	3	3
	1	38	−2	−13	26
	5	63	2	12	24
	3	48	0	−3	0
	4	59	1	8	8
	2	46	−1	−5	5
Totals	30	510	0	0	99

$$s_{xy} = \frac{\Sigma(x_i - \bar{x})(y_i - \bar{y})}{n - 1} = \frac{99}{10 - 1} = 11$$

FIGURE 3.10 PARTITIONED SCATTER DIAGRAM FOR THE STEREO AND SOUND EQUIPMENT STORE

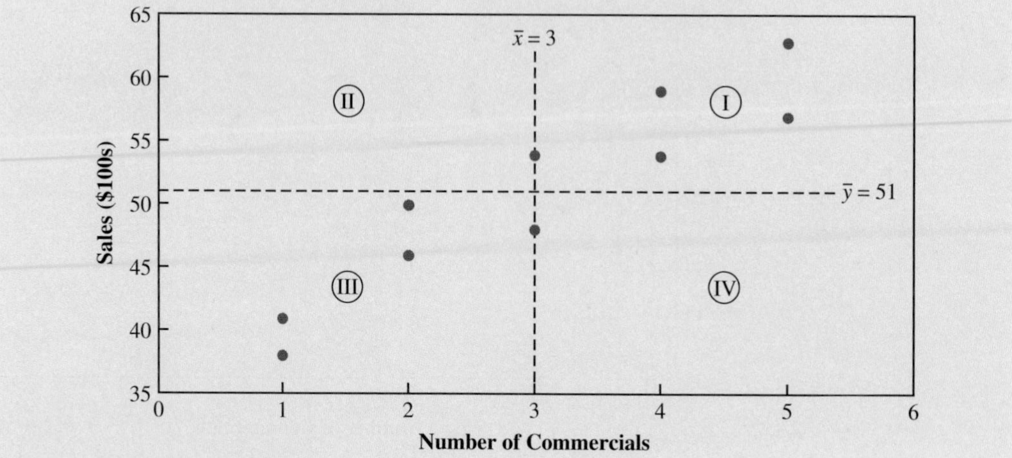

In equation (3.13) we use the notation μ_x for the population mean of the variable x and μ_y for the population mean of the variable y. The population covariance σ_{xy} is defined for a population of size N.

Interpretation of the Covariance

To aid in the interpretation of the sample covariance, consider Figure 3.10. It is the same as the scatter diagram of Figure 3.9 with a vertical dashed line at $\bar{x} = 3$ and a horizontal dashed line at $\bar{y} = 51$. The lines divide the graph into four quadrants. Points in quadrant I correspond to x_i greater than $\bar{x}$ and y_i greater than $\bar{y}$, points in quadrant II correspond to x_i less than $\bar{x}$ and y_i greater than $\bar{y}$, and so on. Thus, the value of $(x_i - \bar{x})(y_i - \bar{y})$ must be positive for points in quadrant I, negative for points in quadrant II, positive for points in quadrant III, and negative for points in quadrant IV.

The covariance is a measure of the linear association between two variables.

If the value of s_{xy} is positive, the points with the greatest influence on s_{xy} must be in quadrants I and III. Hence, a positive value for s_{xy} indicates a positive linear association between x and y; that is, as the value of x increases, the value of y increases. If the value of s_{xy} is negative, however, the points with the greatest influence on s_{xy} are in quadrants II and IV. Hence, a negative value for s_{xy} indicates a negative linear association between x and y; that is, as the value of x increases, the value of y decreases. Finally, if the points are evenly distributed across all four quadrants, the value of s_{xy} will be close to zero, indicating no linear association between x and y. Figure 3.11 shows the values of s_{xy} that can be expected with three different types of scatter diagrams.

Referring again to Figure 3.10, we see that the scatter diagram for the stereo and sound equipment store follows the pattern in the top panel of Figure 3.11. As we should expect, the value of the sample covariance indicates a positive linear relationship with $s_{xy} = 11$.

From the preceding discussion, it might appear that a large positive value for the covariance indicates a strong positive linear relationship and that a large negative value indicates a strong negative linear relationship. However, one problem with using covariance as a measure of the strength of the linear relationship is that the value of the covariance depends on the units of measurement for x and y. For example, suppose we are interested in the relationship between height x and weight y for individuals. Clearly the strength of the relationship should be the same whether we measure height in feet or inches. Measuring the height in

FIGURE 3.11 INTERPRETATION OF SAMPLE COVARIANCE

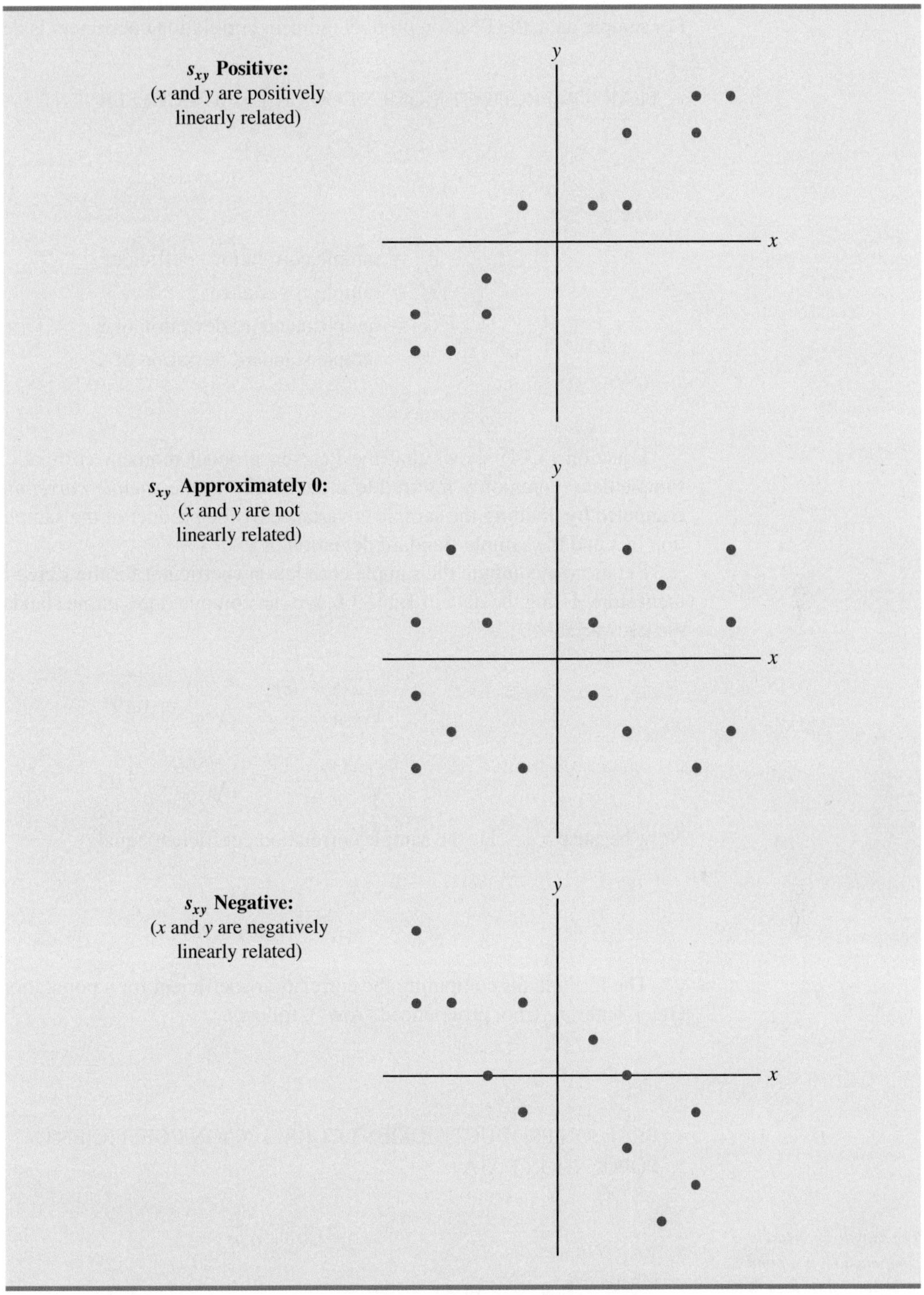

inches, however, gives us much larger numerical values for $(x_i - \bar{x})$ than when we measure height in feet. Thus, with height measured in inches, we would obtain a larger value for the numerator $\Sigma(x_i - \bar{x})(y_i - \bar{y})$ in equation (3.12)—and hence a larger covariance—when in fact the relationship does not change. A measure of the relationship between two variables that is not affected by the units of measurement for x and y is the **correlation coefficient**.

Correlation Coefficient

For sample data, the Pearson product moment correlation coefficient is defined as follows.

PEARSON PRODUCT MOMENT CORRELATION COEFFICIENT: SAMPLE DATA

$$r_{xy} = \frac{s_{xy}}{s_x s_y} \tag{3.14}$$

where

r_{xy} = sample correlation coefficient
s_{xy} = sample covariance
s_x = sample standard deviation of x
s_y = sample standard deviation of y

Equation (3.14) shows that the Pearson product moment correlation coefficient for sample data (commonly referred to more simply as the *sample correlation coefficient*) is computed by dividing the sample covariance by the product of the sample standard deviation of x and the sample standard deviation of y.

Let us now compute the sample correlation coefficient for the stereo and sound equipment store. Using the data in Table 3.6, we can compute the sample standard deviations for the two variables:

$$s_x = \sqrt{\frac{\Sigma(x_i - \bar{x})^2}{n-1}} = \sqrt{\frac{20}{9}} = 1.49$$

$$s_y = \sqrt{\frac{\Sigma(y_i - \bar{y})^2}{n-1}} = \sqrt{\frac{566}{9}} = 7.93$$

Now, because $s_{xy} = 11$, the sample correlation coefficient equals

$$r_{xy} = \frac{s_{xy}}{s_x s_y} = \frac{11}{(1.49)(7.93)} = .93$$

The formula for computing the correlation coefficient for a population, denoted by the Greek letter ρ_{xy} (rho, pronounced "row"), follows.

PEARSON PRODUCT MOMENT CORRELATION COEFFICIENT: POPULATION DATA

$$\rho_{xy} = \frac{\sigma_{xy}}{\sigma_x \sigma_y} \tag{3.15}$$

The sample correlation coefficient r_{xy} is a point estimator of the population correlation coefficient ρ_{xy}.

where

ρ_{xy} = population correlation coefficient
σ_{xy} = population covariance
σ_x = population standard deviation for x
σ_y = population standard deviation for y

FIGURE 3.12 SCATTER DIAGRAM DEPICTING A PERFECT POSITIVE LINEAR RELATIONSHIP

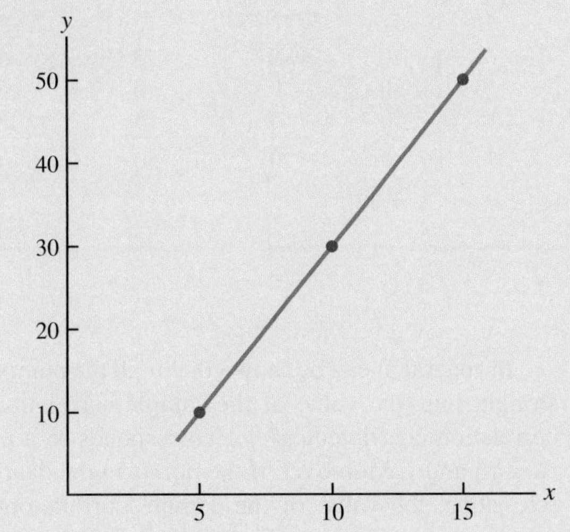

The sample correlation coefficient r_{xy} provides an estimate of the population correlation coefficient ρ_{xy}.

Interpretation of the Correlation Coefficient

First let us consider a simple example that illustrates the concept of a perfect positive linear relationship. The scatter diagram in Figure 3.12 depicts the relationship between x and y based on the following sample data.

x_i	y_i
5	10
10	30
15	50

The straight line drawn through each of the three points shows a perfect linear relationship between x and y. In order to apply equation (3.14) to compute the sample correlation we must first compute s_{xy}, s_x, and s_y. Some of the computations are shown in Table 3.8. Using the results in this table, we find

$$s_{xy} = \frac{\Sigma(x_i - \bar{x})(y_i - \bar{y})}{n - 1} = \frac{200}{2} = 100$$

$$s_x = \sqrt{\frac{\Sigma(x_i - \bar{x})^2}{n - 1}} = \sqrt{\frac{50}{2}} = 5$$

$$s_y = \sqrt{\frac{\Sigma(y_i - \bar{y})^2}{n - 1}} = \sqrt{\frac{800}{2}} = 20$$

$$r_{xy} = \frac{s_{xy}}{s_x s_y} = \frac{100}{5(20)} = 1$$

Thus, we see that the value of the sample correlation coefficient is 1.

TABLE 3.8 COMPUTATIONS USED IN CALCULATING THE SAMPLE CORRELATION COEFFICIENT

	x_i	y_i	$x_i - \bar{x}$	$(x_i - \bar{x})^2$	$y_i - \bar{y}$	$(y_i - \bar{y})^2$	$(x_i - \bar{x})(y_i - \bar{y})$
	5	10	−5	25	−20	400	100
	10	30	0	0	0	0	0
	15	50	5	25	20	400	100
Totals	30	90	0	50	0	800	200
	$\bar{x} = 10$	$\bar{y} = 30$					

The correlation coefficient ranges from −1 to +1. Values close to −1 or +1 indicate a strong linear relationship. The closer the correlation is to zero, the weaker the relationship.

In general, it can be shown that if all the points in a data set fall on a positively sloped straight line, the value of the sample correlation coefficient is +1; that is, a sample correlation coefficient of +1 corresponds to a perfect positive linear relationship between x and y. Moreover, if the points in the data set fall on a straight line having negative slope, the value of the sample correlation coefficient is −1; that is, a sample correlation coefficient of −1 corresponds to a perfect negative linear relationship between x and y.

Let us now suppose that a certain data set indicates a positive linear relationship between x and y but that the relationship is not perfect. The value of r_{xy} will be less than 1, indicating that the points in the scatter diagram are not all on a straight line. As the points deviate more and more from a perfect positive linear relationship, the value of r_{xy} becomes smaller and smaller. A value of r_{xy} equal to zero indicates no linear relationship between x and y, and values of r_{xy} near zero indicate a weak linear relationship.

For the data involving the stereo and sound equipment store, $r_{xy} = .93$. Therefore, we conclude that a strong positive linear relationship occurs between the number of commercials and sales. More specifically, an increase in the number of commercials is associated with an increase in sales.

In closing, we note that correlation provides a measure of linear association and not necessarily causation. A high correlation between two variables does not mean that changes in one variable will cause changes in the other variable. For example, we may find that the quality rating and the typical meal price of restaurants are positively correlated. However, simply increasing the meal price at a restaurant will not cause the quality rating to increase.

NOTES AND COMMENTS

1. Because the correlation coefficient measures only the strength of the linear relationship between two quantitative variables, it is possible for the correlation coefficient to be near zero, suggesting no linear relationship, when the relationship between the two variables is nonlinear. For example, the following scatter diagram shows the relationship between the amount spent by a small retail store for environmental control (heating and cooling) and the daily high outside temperature over 100 days.

   The sample correlation coefficient for these data is $r_{xy} = -.007$ and indicates there is no linear relationship between the two variables. However, the scatter diagram provides strong visual evidence of a nonlinear relationship. That is, we can see that as the daily high outside temperature increases, the money spent on environmental control first decreases as less heating is required and then increases as greater cooling is required.

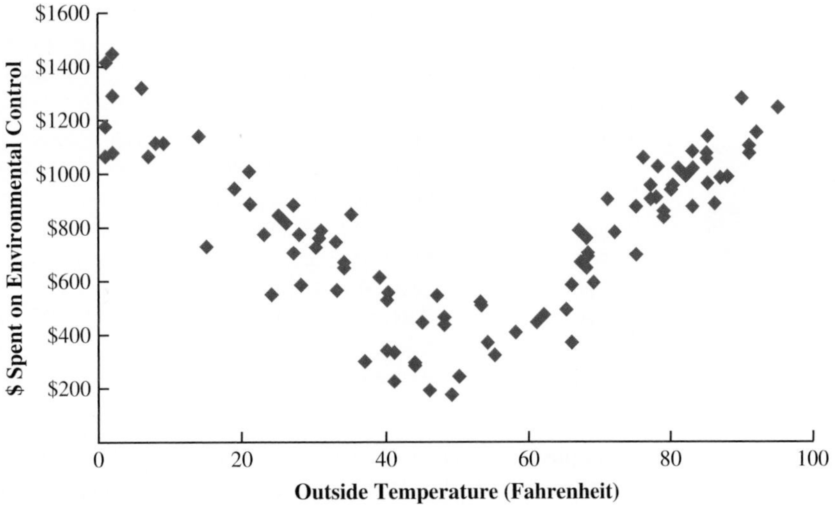

2. While the correlation coefficient is useful in assessing the relationship between two quantitative variables, other measures—such as the Spearman rank-correlation coefficient—can be used to assess a relationship between two variables when at least one of the variables is nominal or ordinal. We discuss the use of the Spearman rank-correlation coefficent in Chapter 18.

Exercises

Methods

SELF test

55. Five observations taken for two variables follow.

x_i	4	6	11	3	16
y_i	50	50	40	60	30

a. Develop a scatter diagram with x on the horizontal axis.
b. What does the scatter diagram developed in part (a) indicate about the relationship between the two variables?
c. Compute and interpret the sample covariance.
d. Compute and interpret the sample correlation coefficient.

56. Five observations taken for two variables follow.

x_i	6	11	15	21	27
y_i	6	9	6	17	12

a. Develop a scatter diagram for these data.
b. What does the scatter diagram indicate about a relationship between x and y?
c. Compute and interpret the sample covariance.
d. Compute and interpret the sample correlation coefficient.

Applications

57. Ten major college football bowl games were played in January 2010, with the University of Alabama beating the University of Texas 37 to 21 to become the national champion of college football. The results of the 10 bowl games follow (*USA Today,* January 8, 2010).

WEB file BowlGames

Bowl Game	Score	Predicted Point Margin	Actual Point Margin
Outback	Auburn 38 Northwestern 35	5	3
Gator	Florida State 33 West Virginia 21	1	12
Capital One	Penn State 19 LSU 17	3	2
Rose	Ohio State 26 Oregon 17	−2	9
Sugar	Florida 51 Cincinnati 24	14	27
Cotton	Mississippi State 21 Oklahoma State 7	3	14
Alamo	Texas Tech 41 Michigan State 31	9	10
Fiesta	Boise State 17 TCU 10	−4	7
Orange	Iowa 24 Georgia Tech 14	−3	10
Championship	Alabama 37 Texas 21	4	16

The predicted winning point margin was based on Las Vegas betting odds approximately one week before the bowl games were played. For example, Auburn was predicted to beat Northwestern in the Outback Bowl by five points. The actual winning point margin for Auburn was three points. A negative predicted winning point margin means that the team that won the bowl game was an underdog and expected to lose. For example, in the Rose Bowl, Ohio State was a two-point underdog to Oregon and ended up winning by nine points.

a. Develop a scatter diagram with predicted point margin on the horizontal axis.
b. What is the relationship between predicted and actual point margins?
c. Compute and interpret the sample covariance.
d. Compute the sample correlation coefficient. What does this value indicate about the relationship between the Las Vegas predicted point margin and the actual point margin in college football bowl games?

58. A department of transportation's study on driving speed and miles per gallon for midsize automobiles resulted in the following data:

Speed (Miles per Hour)	30	50	40	55	30	25	60	25	50	55
Miles per Gallon	28	25	25	23	30	32	21	35	26	25

Compute and interpret the sample correlation coefficient.

59. At the beginning of 2009, the economic downturn resulted in the loss of jobs and an increase in delinquent loans for housing. The national unemployment rate was 6.5% and the percentage of delinquent loans was 6.12% (*The Wall Street Journal,* January 27, 2009). In projecting where the real estate market was headed in the coming year, economists studied the relationship between the jobless rate and the percentage of delinquent loans. The expectation was that if the jobless rate continued to increase, there would also be an

WEB file Housing

Metro Area	Jobless Rate (%)	Delinquent Loan (%)	Metro Area	Jobless Rate (%)	Delinquent Loan (%)
Atlanta	7.1	7.02	New York	6.2	5.78
Boston	5.2	5.31	Orange County	6.3	6.08
Charlotte	7.8	5.38	Orlando	7.0	10.05
Chicago	7.8	5.40	Philadelphia	6.2	4.75
Dallas	5.8	5.00	Phoenix	5.5	7.22
Denver	5.8	4.07	Portland	6.5	3.79
Detroit	9.3	6.53	Raleigh	6.0	3.62
Houston	5.7	5.57	Sacramento	8.3	9.24
Jacksonville	7.3	6.99	St. Louis	7.5	4.40
Las Vegas	7.6	11.12	San Diego	7.1	6.91
Los Angeles	8.2	7.56	San Francisco	6.8	5.57
Miami	7.1	12.11	Seattle	5.5	3.87
Minneapolis	6.3	4.39	Tampa	7.5	8.42
Nashville	6.6	4.78			

increase in the percentage of delinquent loans. The data below show the jobless rate and the delinquent loan percentage for 27 major real estate markets.

a. Compute the correlation coefficient. Is there a positive correlation between the jobless rate and the percentage of delinquent housing loans? What is your interpretation?
b. Show a scatter diagram of the relationship between jobless rate and the percentage of delinquent housing loans.

60. The Dow Jones Industrial Average (DJIA) and the Standard & Poor's 500 Index (S&P 500) are both used to measure the performance of the stock market. The DJIA is based on the price of stocks for 30 large companies; the S&P 500 is based on the price of stocks for 500 companies. If both the DJIA and S&P 500 measure the performance of the stock market, how are they correlated? The following data show the daily percent increase or daily percent decrease in the DJIA and S&P 500 for a sample of nine days over a three-month period (*The Wall Street Journal,* January 15 to March 10, 2006).

WEB file StockMarket

DJIA	.20	.82	−.99	.04	−.24	1.01	.30	.55	−.25
S&P 500	.24	.19	−.91	.08	−.33	.87	.36	.83	−.16

a. Show a scatter diagram.
b. Compute the sample correlation coefficient for these data.
c. Discuss the association between the DJIA and S&P 500. Do you need to check both before having a general idea about the daily stock market performance?

61. The daily high and low temperatures for 14 cities around the world are shown (The Weather Channel, April 22, 2009).

WEB file WorldTemp

City	High	Low	City	High	Low
Athens	68	50	London	67	45
Beijing	70	49	Moscow	44	29
Berlin	65	44	Paris	69	44
Cairo	96	64	Rio de Janeiro	76	69
Dublin	57	46	Rome	69	51
Geneva	70	45	Tokyo	70	58
Hong Kong	80	73	Toronto	44	39

a. What is the sample mean high temperature?
b. What is the sample mean low temperature?
c. What is the correlation between the high and low temperatures? Discuss.

3.6 Data Dashboards: Adding Numerical Measures to Improve Effectiveness

In Section 2.5 we provided an introduction to data visualization, a term used to describe the use of graphical displays to summarize and present information about a data set. The goal of data visualization is to communicate key information about the data as effectively and clearly as possible. One of the most widely used data visualization tools is a data dashboard, a set of visual displays that organizes and presents information that is used to monitor the performance of a company or organization in a manner that is easy to read, understand, and interpret. In this section we extend the discussion of data dashboards to show how the addition of numerical measures can improve the overall effectiveness of the display.

The addition of numerical measures, such as the mean and standard deviation of key performance indicators (KPIs), to a data dashboard is critical because numerical measures often provide benchmarks or goals by which KPIs are evaluated. In addition, graphical displays that include numerical measures as components of the display are also frequently

FIGURE 3.13 INITIAL GROGAN OIL INFORMATION TECHNOLOGY CALL CENTER DATA DASHBOARD

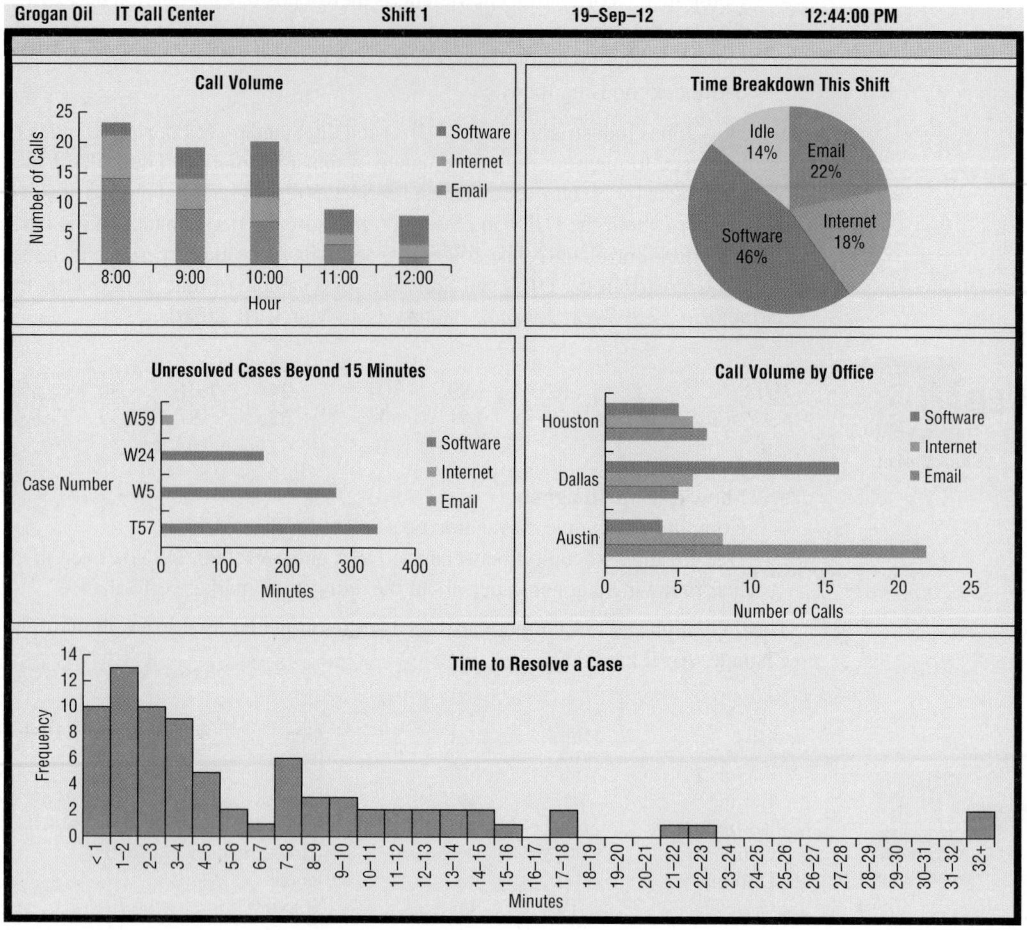

included in data dashboards. We must keep in mind that the purpose of a data dashboard is to provide information on the KPIs in a manner that is easy to read, understand, and interpret. Adding numerical measures and graphs that utilize numerical measures can help us accomplish these objectives.

To illustrate the use of numerical measures in a data dashboard, recall the Grogan Oil Company application that we used in Section 2.5 to introduce the concept of a data dashboard. Grogan Oil has offices located in three cities in Texas: Austin (its headquarters), Houston, and Dallas. Grogan's Information Technology (IT) call center, located in the Austin office, handles calls regarding computer-related problems (software, Internet, and email) from employees in the three offices. Figure 3.13 shows the data dashboard that Grogan developed to monitor the performance of the call center. The key components of this dashboard are as follows:

- The stacked bar chart in the upper left corner of the dashboard shows the call volume for each type of problem (software, Internet, or email) over time.
- The pie chart in the upper right corner of the dashboard shows the percentage of time that call center employees spent on each type of problem or not working on a call (idle).
- For each unresolved case that was received more than 15 minutes ago, the bar chart shown in the middle left portion of the dashboard shows the length of time that each of these cases has been unresolved.

- The bar chart in the middle right portion of the dashboard shows the call volume by office (Houston, Dallas, Austin) for each type of problem.
- The histogram at the bottom of the dashboard shows the distribution of the time to resolve a case for all resolved cases for the current shift.

In order to gain additional insight into the performance of the call center, Grogan's IT manager has decided to expand the current dashboard by adding box plots for the time required to resolve calls received for each type of problem (email, Internet, and software). In addition, a graph showing the time to resolve individual cases has been added in the lower left portion of the dashboard. Finally, the IT manager added a display of summary statistics for each type of problem and summary statistics for each of the first few hours of the shift. The updated dashboard is shown in Figure 3.14.

The IT call center has set a target performance level or benchmark of 10 minutes for the mean time to resolve a case. Furthermore, the center has decided it is undesirable for the time to resolve a case to exceed 15 minutes. To reflect these benchmarks, a black horizontal line at the mean target value of 10 minutes and a red horizontal line at the maximum acceptable level of 15 minutes have been added to both the graph showing the time to resolve cases and the box plots of the time required to resolve calls received for each type of problem.

The summary statistics in the dashboard in Figure 3.14 show that the mean time to resolve an email case is 4.6 minutes, the mean time to resolve an Internet case is 5.4 minutes, and the mean time to resolve a software case is 5.2 minutes. Thus, the mean time to resolve each type of case is better than the target mean (10 minutes).

Reviewing the box plots, we see that the box associated with the email cases is "larger" than the boxes associated with the other two types of cases. The summary statistics also show that the standard deviation of the time to resolve email cases is larger than the standard deviations of the times to resolve the other types of cases. This leads us to take a closer look at the email cases in the two new graphs. The box plot for the email cases has a whisker that extends beyond 15 minutes and an outlier well beyond 15 minutes. The graph of the time to resolve individual cases (in the lower left position of the dashboard) shows that this is because of two calls on email cases during the 9:00 hour that took longer than the target maximum time (15 minutes) to resolve. This analysis may lead the IT call center manager to further investigate why resolution times are more variable for email cases than for Internet or software cases. Based on this analysis, the IT manager may also decide to investigate the circumstances that led to inordinately long resolution times for the two email cases that took longer than 15 minutes to resolve.

The graph of the time to resolve individual cases also shows that most calls received during the first hour of the shift were resolved relatively quickly; the graph also shows that the time to resolve cases increased gradually throughout the morning. This could be due to a tendency for complex problems to arise later in the shift or possibly to the backlog of calls that accumulates over time. Although the summary statistics suggest that cases submitted during the 9:00 hour take the longest to resolve, the graph of time to resolve individual cases shows that two time-consuming email cases and one time-consuming software case were reported during that hour, and this may explain why the mean time to resolve cases during the 9.00 hour is larger than during any other hour of the shift. Overall, reported cases have generally been resolved in 15 minutes or less during this shift.

Drilling down refers to functionality in interactive data dashboards that allows the user to access information and analyses at an increasingly detailed level.

Dashboards such as the Grogan Oil data dashboard are often interactive. For instance, when a manager uses a mouse or a touch screen monitor to position the cursor over the display or point to something on the display, additional information, such as the time to resolve the problem, the time the call was received, and the individual and/or the location that reported the problem, may appear. Clicking on the individual item may also take the user to a new level of analysis at the individual case level.

FIGURE 3.14 UPDATED GROGAN OIL INFORMATION TECHNOLOGY CALL CENTER DATA DASHBOARD

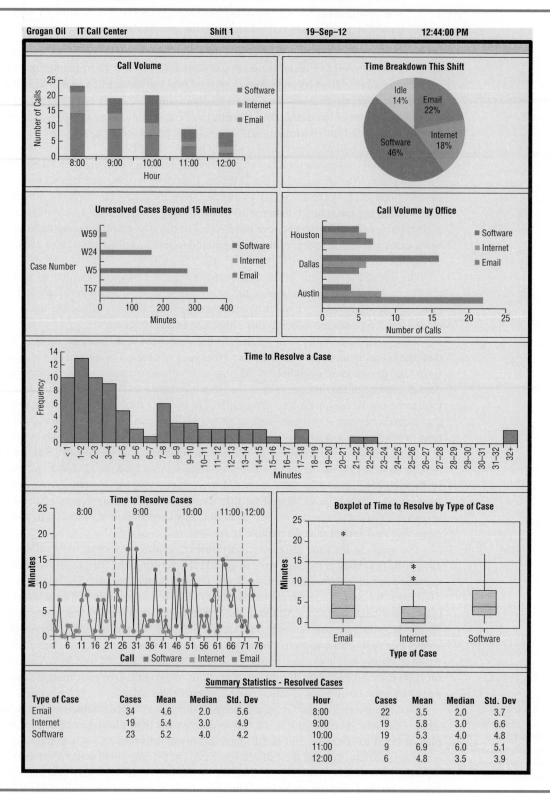

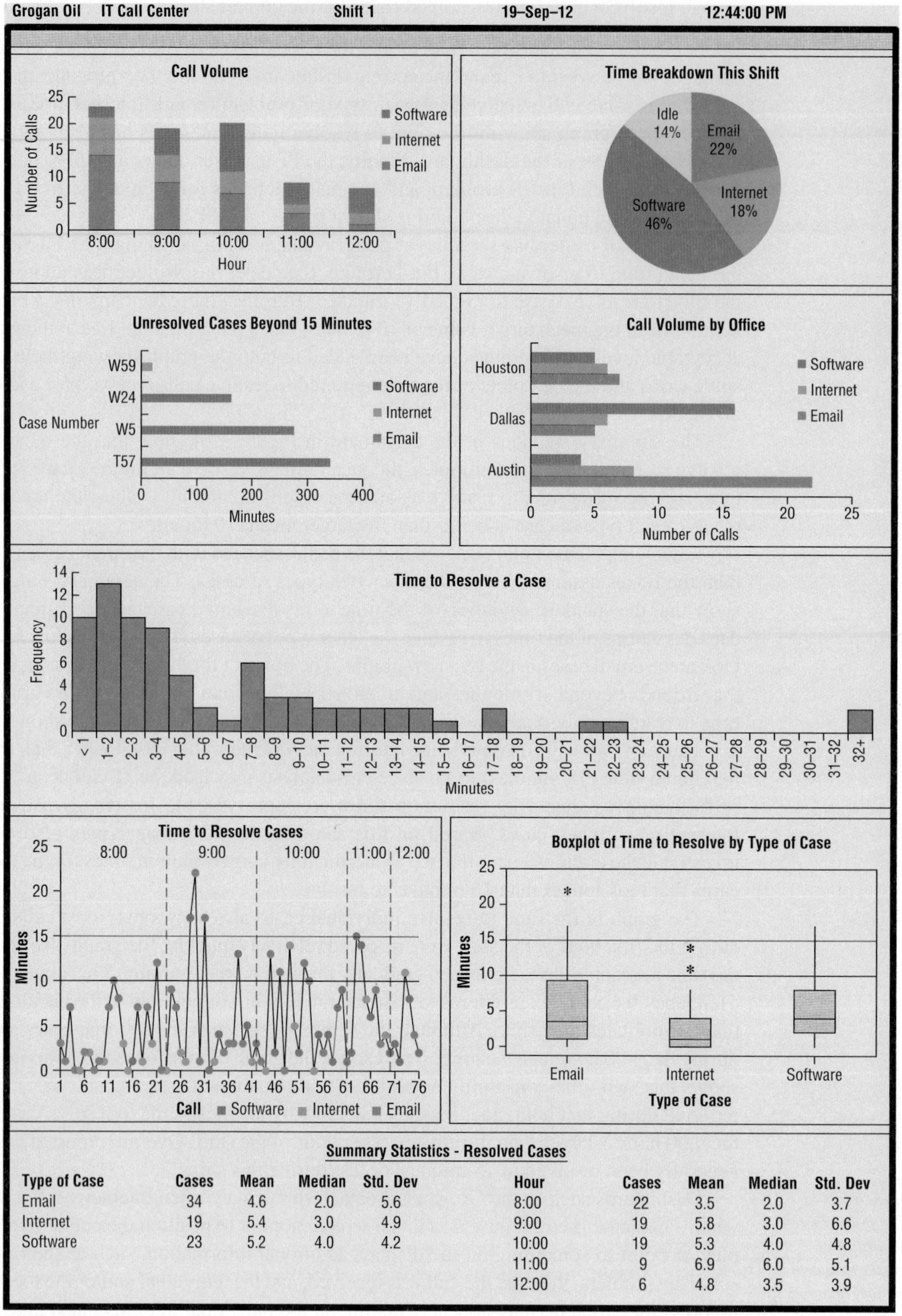

Summary Statistics - Resolved Cases

Type of Case	Cases	Mean	Median	Std. Dev
Email	34	4.6	2.0	5.6
Internet	19	5.4	3.0	4.9
Software	23	5.2	4.0	4.2

Hour	Cases	Mean	Median	Std. Dev
8:00	22	3.5	2.0	3.7
9:00	19	5.8	3.0	6.6
10:00	19	5.3	4.0	4.8
11:00	9	6.9	6.0	5.1
12:00	6	4.8	3.5	3.9

Summary

In this chapter we introduced several descriptive statistics that can be used to summarize the location, variability, and shape of a data distribution. Unlike the tabular and graphical displays introduced in Chapter 2, the measures introduced in this chapter summarize the data in terms of numerical values. When the numerical values obtained are for a sample, they are called sample statistics. When the numerical values obtained are for a population, they are called population parameters. Some of the notation used for sample statistics and population parameters follow.

In statistical inference, a sample statistic is referred to as a point estimator of the population parameter.

	Sample Statistic	Population Parameter
Mean	$\bar{x}$	μ
Variance	s^2	σ^2
Standard deviation	s	σ
Covariance	s_{xy}	σ_{xy}
Correlation	r_{xy}	ρ_{xy}

As measures of location, we defined the mean, median, mode, weighted mean, geometric mean, percentiles, and quartiles. Next, we presented the range, interquartile range, variance, standard deviation, and coefficient of variation as measures of variability or dispersion. Our primary measure of the shape of a data distribution was the skewness. Negative values of skewness indicate a data distribution skewed to the left, and positive values of skewness indicate a data distribution skewed to the right. We then described how the mean and standard deviation could be used, applying Chebyshev's theorem and the empirical rule, to provide more information about the distribution of data and to identify outliers.

In Section 3.4 we showed how to develop a five-number summary and a box plot to provide simultaneous information about the location, variability, and shape of the distribution. In Section 3.5 we introduced covariance and the correlation coefficient as measures of association between two variables. In the final section, we showed how adding numerical measures can improve the effectiveness of data dashboards.

The descriptive statistics we discussed can be developed using statistical software packages and spreadsheets. In the chapter-ending appendixes we show how to use Minitab, Excel, and StatTools to develop the descriptive statistics introduced in this chapter.

Glossary

Sample statistic A numerical value used as a summary measure for a sample (e.g., the sample mean, $\bar{x}$, the sample variance, s^2, and the sample standard deviation, s).

Population parameter A numerical value used as a summary measure for a population (e.g., the population mean, μ, the population variance, σ^2, and the population standard deviation, σ).

Point estimator A sample statistic, such as $\bar{x}$, s^2, and s, used to estimate the corresponding population parameter.

Mean A measure of central location computed by summing the data values and dividing by the number of observations.

Weighted mean The mean obtained by assigning each observation a weight that reflects its importance.

Median A measure of central location provided by the value in the middle when the data are arranged in ascending order.

Geometric mean A measure of location that is calculated by finding the nth root of the product of n values.

Mode A measure of location, defined as the value that occurs with greatest frequency.

Percentile A value such that at least p percent of the observations are less than or equal to this value and at least $(100 - p)$ percent of the observations are greater than or equal to this value. The 50th percentile is the median.

Quartiles The 25th, 50th, and 75th percentiles, referred to as the first quartile, the second quartile (median), and third quartile, respectively. The quartiles can be used to divide a data set into four parts, with each part containing approximately 25% of the data.

Range A measure of variability, defined to be the largest value minus the smallest value.

Interquartile range (IQR) A measure of variability, defined to be the difference between the third and first quartiles.

Variance A measure of variability based on the squared deviations of the data values about the mean.

Standard deviation A measure of variability computed by taking the positive square root of the variance.

Coefficient of variation A measure of relative variability computed by dividing the standard deviation by the mean and multiplying by 100.

Skewness A measure of the shape of a data distribution. Data skewed to the left result in negative skewness; a symmetric data distribution results in zero skewness; and data skewed to the right result in positive skewness.

z-score A value computed by dividing the deviation about the mean $(x_i - \bar{x})$ by the standard deviation s. A z-score is referred to as a standardized value and denotes the number of standard deviations x_i is from the mean.

Chebyshev's theorem A theorem that can be used to make statements about the proportion of data values that must be within a specified number of standard deviations of the mean.

Empirical rule A rule that can be used to compute the percentage of data values that must be within one, two, and three standard deviations of the mean for data that exhibit a bell-shaped distribution.

Outlier An unusually small or unusually large data value.

Five-number summary A technique that uses five numbers to summarize the data: smallest value, first quartile, median, third quartile, and largest value.

Box plot A graphical summary of data based on a five-number summary.

Covariance A measure of linear association between two variables. Positive values indicate a positive relationship; negative values indicate a negative relationship.

Correlation coefficient A measure of linear association between two variables that takes on values between -1 and $+1$. Values near $+1$ indicate a strong positive linear relationship; values near -1 indicate a strong negative linear relationship; and values near zero indicate the lack of a linear relationship.

Key Formulas

Sample Mean

$$\bar{x} = \frac{\Sigma x_i}{n} \tag{3.1}$$

Population Mean

$$\mu = \frac{\Sigma x_i}{N} \tag{3.2}$$

Weighted Mean

$$\bar{x} = \frac{\Sigma w_i x_i}{w_i} \tag{3.3}$$

Geometric Mean

$$\bar{x}_g = \sqrt[n]{(x_1)(x_2)\cdots(x_n)} = [(x_1)(x_2)\cdots(x_n)]^{1/n} \tag{3.4}$$

Interquartile Range

$$\text{IQR} = Q_3 - Q_1 \tag{3.5}$$

Population Variance

$$\sigma^2 = \frac{\Sigma(x_i - \mu)^2}{N} \tag{3.6}$$

Sample Variance

$$s^2 = \frac{\Sigma(x_i - \bar{x})^2}{n - 1} \tag{3.7}$$

Standard Deviation

$$\text{Sample standard deviation} = s = \sqrt{s^2} \tag{3.8}$$

$$\text{Population standard deviation} = \sigma = \sqrt{\sigma^2} \tag{3.9}$$

Coefficient of Variation

$$\left(\frac{\text{Standard deviation}}{\text{Mean}} \times 100\right)\% \tag{3.10}$$

***z*-Score**

$$z_i = \frac{x_i - \bar{x}}{s} \tag{3.11}$$

Sample Covariance

$$s_{xy} = \frac{\Sigma(x_i - \bar{x})(y_i - \bar{y})}{n - 1} \tag{3.12}$$

Population Covariance

$$\sigma_{xy} = \frac{\Sigma(x_i - \mu_x)(y_i - \mu_y)}{N} \tag{3.13}$$

Pearson Product Moment Correlation Coefficient: Sample Data

$$r_{xy} = \frac{s_{xy}}{s_x s_y} \tag{3.14}$$

Pearson Product Moment Correlation Coefficient: Population Data

$$\rho_{xy} = \frac{\sigma_{xy}}{\sigma_x \sigma_y} \tag{3.15}$$

Supplementary Exercises

62. The average number of times Americans dine out in a week fell from 4.0 in 2008 to 3.8 in 2012 (Zagat.com, April 1, 2012). The number of times a sample of 20 families dined out last week provides the following data.

6	1	5	3	7	3	0	3	1	3
4	1	2	4	1	0	5	6	3	1

a. Compute the mean and median.
b. Compute the first and third quartiles.
c. Compute the range and interquartile range.
d. Compute the variance and standard deviation.
e. The skewness measure for these data is 0.34. Comment on the shape of this distribution. Is it the shape you would expect? Why or why not?
f. Do the data contain outliers?

63. The U.S. Census Bureau provides statistics on family life in the United States, including the age at the time of first marriage, current marital status, and size of household (U.S. Census Bureau website, March 20, 2006). The following data show the age at the time of first marriage for a sample of men and a sample of women.

WEB file Ages

Men	26	23	28	25	27	30	26	35	28
	21	24	27	29	30	27	32	27	25
Women	20	28	23	30	24	29	26	25	
	22	22	25	23	27	26	19		

a. Determine the median age at the time of first marriage for men and women.
b. Compute the first and third quartiles for both men and women.
c. Twenty-five years ago the median age at the time of first marriage was 25 for men and 22 for women. What insight does this information provide about the decision of when to marry among young people today?

64. The average waiting time for a patient at an El Paso physician's office is just over 29 minutes, well above the national average of 21 minutes. In fact, El Paso has the longest physician's office waiting times in the United States (*El Paso Times,* January 8, 2012). In order to address the issue of long patient wait times, some physician's offices are using wait tracking systems to notify patients of expected wait times. Patients can adjust their arrival times based on this information and spend less time in waiting rooms. The following data show wait times (minutes) for a sample of patients at offices that do not have an office tracking system and wait times for a sample of patients at offices with an office tracking system.

WEB file WaitTracking

Without Wait Tracking System	With Wait Tracking System
24	31
67	11
17	14
20	18
31	12
44	37
12	9
23	13
16	12
37	15

a. What are the mean and median patient wait times for offices with a wait tracking system? What are the mean and median patient wait times for offices without a wait tracking system?
b. What are the variance and standard deviation of patient wait times for offices with a wait tracking system? What are the variance and standard deviation of patient wait times for visits to offices without a wait tracking system?
c. Do offices with a wait tracking system have shorter patient wait times than offices without a wait tracking system? Explain.
d. Considering only offices without a wait tracking system, what is the z-score for the tenth patient in the sample?
e. Considering only offices with a wait tracking system, what is the z-score for the sixth patient in the sample? How does this z-score compare with the z-score you calculated for part (d)?
f. Based on z-scores, do the data for offices without a wait tracking system contain any outliers? Based on z-scores, do the data for offices with a wait tracking system contain any outliers?

65. The U.S. Department of Education reports that about 50% of all college students use a student loan to help cover college expenses (National Center for Educational Studies, January 2006). A sample of students who graduated with student loan debt is shown here. The data, in thousands of dollars, show typical amounts of debt upon graduation.

10.1	14.8	5.0	10.2	12.4	12.2	2.0	11.5	17.8	4.0

a. For those students who use a student loan, what is the mean loan debt upon graduation?
b. What is the variance? Standard deviation?

66. Small business owners often look to payroll service companies to handle their employee payroll. Reasons are that small business owners face complicated tax regulations and penalties for employment tax errors are costly. According to the Internal Revenue Service, 26% of all small business employment tax returns contained errors that resulted in a tax penalty to the owner (*The Wall Street Journal,* January 30, 2006). The tax penalty for a sample of 20 small business owners follows:

WEB file
Penalty

820	270	450	1010	890	700	1350	350	300	1200
390	730	2040	230	640	350	420	270	370	620

a. What is the mean tax penalty for improperly filed employment tax returns?
b. What is the standard deviation?
c. Is the highest penalty, $2040, an outlier?
d. What are some of the advantages of a small business owner hiring a payroll service company to handle employee payroll services, including the employment tax returns?

67. Public transportation and the automobile are two methods an employee can use to get to work each day. Samples of times recorded for each method are shown. Times are in minutes.

Public Transportation:	28	29	32	37	33	25	29	32	41	34
Automobile:	29	31	33	32	34	30	31	32	35	33

a. Compute the sample mean time to get to work for each method.
b. Compute the sample standard deviation for each method.
c. On the basis of your results from parts (a) and (b), which method of transportation should be preferred? Explain.
d. Develop a box plot for each method. Does a comparison of the box plots support your conclusion in part (c)?

68. The National Association of Realtors reported the median home price in the United States and the increase in median home price over a five-year period (*The Wall Street Journal*, January 16, 2006). Use the sample home prices ($1000s) shown here to answer the following questions.

WEB file Homes

995.9	48.8	175.0	263.5	298.0	218.9	209.0
628.3	111.0	212.9	92.6	2325.0	958.0	212.5

a. What is the sample median home price?
b. In January 2001, the National Association of Realtors reported a median home price of $139,300 in the United States. What was the percentage increase in the median home price over the five-year period?
c. What are the first quartile and the third quartile for the sample data?
d. Provide a five-number summary for the home prices.
e. Do the data contain any outliers?
f. What is the mean home price for the sample? Why does the National Association of Realtors prefer to use the median home price in its reports?

69. The U.S. Census Bureau's American Community Survey reported the percentage of children under 18 years of age who had lived below the poverty level during the previous 12 months (U.S. Census Bureau website, August 2008). The region of the country, Northeast (NE), Southeast (SE), Midwest (MW), Southwest (SW), and West (W) and the percentage of children under 18 who had lived below the poverty level are shown for each state.
a. What is the median poverty level percentage for the 50 states?
b. What are the first and third quartiles? What is your interpretation of the quartiles?

WEB file PovertyLevel

State	Region	Poverty %	State	Region	Poverty %
Alabama	SE	23.0	Montana	W	17.3
Alaska	W	15.1	Nebraska	MW	14.4
Arizona	SW	19.5	Nevada	W	13.9
Arkansas	SE	24.3	New Hampshire	NE	9.6
California	W	18.1	New Jersey	NE	11.8
Colorado	W	15.7	New Mexico	SW	25.6
Connecticut	NE	11.0	New York	NE	20.0
Delaware	NE	15.8	North Carolina	SE	20.2
Florida	SE	17.5	North Dakota	MW	13.0
Georgia	SE	20.2	Ohio	MW	18.7
Hawaii	W	11.4	Oklahoma	SW	24.3
Idaho	W	15.1	Oregon	W	16.8
Illinois	MW	17.1	Pennsylvania	NE	16.9
Indiana	MW	17.9	Rhode Island	NE	15.1
Iowa	MW	13.7	South Carolina	SE	22.1
Kansas	MW	15.6	South Dakota	MW	16.8
Kentucky	SE	22.8	Tennessee	SE	22.7
Louisiana	SE	27.8	Texas	SW	23.9
Maine	NE	17.6	Utah	W	11.9
Maryland	NE	9.7	Vermont	NE	13.2
Massachusetts	NE	12.4	Virginia	SE	12.2
Michigan	MW	18.3	Washington	W	15.4
Minnesota	MW	12.2	West Virginia	SE	25.2
Mississippi	SE	29.5	Wisconsin	MW	14.9
Missouri	MW	18.6	Wyoming	W	12.0

c. Show a box plot for the data. Interpret the box plot in terms of what it tells you about the level of poverty for children in the United States. Are any states considered outliers? Discuss.
d. Identify the states in the lower quartile. What is your interpretation of this group and what region or regions are represented most in the lower quartile?

70. *Travel + Leisure* magazine presented its annual list of the 500 best hotels in the world (*Travel + Leisure,* January 2009). The magazine provides a rating for each hotel along with a brief description that includes the size of the hotel, amenities, and the cost per night for a double room. A sample of 12 of the top-rated hotels in the United States follows.

WEB file Travel

Hotel	Location	Rooms	Cost/Night
Boulders Resort & Spa	Phoenix, AZ	220	499
Disney's Wilderness Lodge	Orlando, FL	727	340
Four Seasons Hotel Beverly Hills	Los Angeles, CA	285	585
Four Seasons Hotel	Boston, MA	273	495
Hay-Adams	Washington, DC	145	495
Inn on Biltmore Estate	Asheville, NC	213	279
Loews Ventana Canyon Resort	Phoenix, AZ	398	279
Mauna Lani Bay Hotel	Island of Hawaii	343	455
Montage Laguna Beach	Laguna Beach, CA	250	595
Sofitel Water Tower	Chicago, IL	414	367
St. Regis Monarch Beach	Dana Point, CA	400	675
The Broadmoor	Colorado Springs, CO	700	420

a. What is the mean number of rooms?
b. What is the mean cost per night for a double room?
c. Develop a scatter diagram with the number of rooms on the horizontal axis and the cost per night on the vertical axis. Does there appear to be a relationship between the number of rooms and the cost per night? Discuss.
d. What is the sample correlation coefficient? What does it tell you about the relationship between the number of rooms and the cost per night for a double room? Does this appear reasonable? Discuss.

71. Morningstar tracks the performance of a large number of companies and publishes an evaluation of each. Along with a variety of financial data, Morningstar includes a Fair Value estimate for the price that should be paid for a share of the company's common stock. Data for 30 companies are available in the file named FairValue. The data include the Fair Value estimate per share of common stock, the most recent price per share, and the earning per share for the company (*Morningstar Stocks 500,* 2008).

WEB file FairValue

a. Develop a scatter diagram for the Fair Value and Share Price data with Share Price on the horizontal axis. What is the sample correlation coefficient, and what can you say about the relationship between the variables?
b. Develop a scatter diagram for the Fair Value and Earnings per Share data with Earnings per Share on the horizontal axis. What is the sample correlation coefficient, and what can you say about the relationship between the variables?

72. Does a major league baseball team's record during spring training indicate how the team will play during the regular season? Over the last six years, the correlation coefficient between a team's winning percentage in spring training and its winning percentage in the regular season is .18 (*The Wall Street Journal,* March 30, 2009). Shown are the winning percentages for the 14 American League teams during the 2008 season.

WEB file

SpringTraining

Team	Spring Training	Regular Season	Team	Spring Training	Regular Season
Baltimore Orioles	.407	.422	Minnesota Twins	.500	.540
Boston Red Sox	.429	.586	New York Yankees	.577	.549
Chicago White Sox	.417	.546	Oakland A's	.692	.466
Cleveland Indians	.569	.500	Seattle Mariners	.500	.377
Detroit Tigers	.569	.457	Tampa Bay Rays	.731	.599
Kansas City Royals	.533	.463	Texas Rangers	.643	.488
Los Angeles Angels	.724	.617	Toronto Blue Jays	.448	.531

a. What is the correlation coefficient between the spring training and the regular season winning percentages?
b. What is your conclusion about a team's record during spring training indicating how the team will play during the regular season? What are some of the reasons why this occurs? Discuss.

73. The days to maturity for a sample of five money market funds are shown here. The dollar amounts invested in the funds are provided. Use the weighted mean to determine the mean number of days to maturity for dollars invested in these five money market funds.

Days to Maturity	Dollar Value ($millions)
20	20
12	30
7	10
5	15
6	10

74. Automobiles traveling on a road with a posted speed limit of 55 miles per hour are checked for speed by a state police radar system. Following is a frequency distribution of speeds.

Speed (miles per hour)	Frequency
45–49	10
50–54	40
55–59	150
60–64	175
65–69	75
70–74	15
75–79	10
Total	475

a. What is the mean speed of the automobiles traveling on this road?
b. Compute the variance and the standard deviation.

75. The Panama Railroad Company was established in 1850 to construct a railroad across the isthmus that would allow fast and easy access between the Atlantic and Pacific Oceans. The following table (*The Big Ditch,* Mauer and Yu, 2011) provides annual returns for Panama Railroad stock from 1853 through 1880.

Year	Return on Panama Railroad Company Stock (%)
1853	−1
1854	−9
1855	19
1856	2
1857	3
1858	36
1859	21
1860	16
1861	−5
1862	43
1863	44
1864	48
1865	7
1866	11
1867	23
1868	20
1869	−11
1870	−51
1871	−42
1872	39
1873	42
1874	12
1875	26
1876	9
1877	−6
1878	25
1879	31
1880	30

WEB file

PanamaRailroad

a. Create a graph of the annual returns on the stock. The New York Stock Exchange earned an annual average return of 8.4% from 1853 through 1880. Can you tell from the graph if the Panama Railroad Company stock outperformed the New York Stock Exchange?
b. Calculate the mean annual return on Panama Railroad Company stock from 1853 through 1880. Did the stock outperform the New York Stock Exchange over the same period?

Case Problem 1 Pelican Stores

Pelican Stores, a division of National Clothing, is a chain of women's apparel stores operating throughout the country. The chain recently ran a promotion in which discount coupons were sent to customers of other National Clothing stores. Data collected for a sample of 100 in-store credit card transactions at Pelican Stores during one day while the promotion was running are contained in the file named PelicanStores. Table 3.9 shows a portion of the data set. The proprietary card method of payment refers to charges made using a National Clothing charge card. Customers who made a purchase using a discount coupon are referred to as promotional customers and customers who made a purchase but did not use a discount coupon are referred to as regular customers. Because the promotional coupons were not sent to regular Pelican Stores customers, management considers the sales made to people presenting the promotional coupons as sales it would not otherwise make. Of course, Pelican also hopes that the promotional customers will continue to shop at its stores.

TABLE 3.9 SAMPLE OF 100 CREDIT CARD PURCHASES AT PELICAN STORES

Customer	Type of Customer	Items	Net Sales	Method of Payment	Gender	Marital Status	Age
1	Regular	1	39.50	Discover	Male	Married	32
2	Promotional	1	102.40	Proprietary Card	Female	Married	36
3	Regular	1	22.50	Proprietary Card	Female	Married	32
4	Promotional	5	100.40	Proprietary Card	Female	Married	28
5	Regular	2	54.00	MasterCard	Female	Married	34
6	Regular	1	44.50	MasterCard	Female	Married	44
7	Promotional	2	78.00	Proprietary Card	Female	Married	30
8	Regular	1	22.50	Visa	Female	Married	40
9	Promotional	2	56.52	Proprietary Card	Female	Married	46
10	Regular	1	44.50	Proprietary Card	Female	Married	36
.	.	.	.	.	.	.	.
.	.	.	.	.	.	.	.
.	.	.	.	.	.	.	.
96	Regular	1	39.50	MasterCard	Female	Married	44
97	Promotional	9	253.00	Proprietary Card	Female	Married	30
98	Promotional	10	287.59	Proprietary Card	Female	Married	52
99	Promotional	2	47.60	Proprietary Card	Female	Married	30
100	Promotional	1	28.44	Proprietary Card	Female	Married	44

WEB file

PelicanStores

Most of the variables shown in Table 3.9 are self-explanatory, but two of the variables require some clarification.

Items	The total number of items purchased
Net Sales	The total amount ($) charged to the credit card

Pelican's management would like to use this sample data to learn about its customer base and to evaluate the promotion involving discount coupons.

Managerial Report

Use the methods of descriptive statistics presented in this chapter to summarize the data and comment on your findings. At a minimum, your report should include the following:

1. Descriptive statistics on net sales and descriptive statistics on net sales by various classifications of customers.
2. Descriptive statistics concerning the relationship between age and net sales.

Case Problem 2 Motion Picture Industry

The motion picture industry is a competitive business. More than 50 studios produce several hundred new motion pictures each year, and the financial success of the motion pictures varies considerably. The opening weekend gross sales, the total gross sales, the number of theaters the movie was shown in, and the number of weeks the motion picture was in release are common variables used to measure the success of a motion picture. Data on the top 100 grossing motion pictures released in 2011 (Box Office Mojo website, March 17, 2012) are contained in a file named 2011Movies. Table 3.10 shows the data for the first 10 motion pictures in this file. Note that some movies, such as *War Horse,* were released late in 2011 and continued to run in 2012.

TABLE 3.10 PERFORMANCE DATA FOR 10 MOTION PICTURES

Motion Picture	Opening Gross Sales ($millions)	Total Gross Sales ($millions)	Number of Theaters	Weeks in Release
Harry Potter and the Deathly Hallows Part 2	169.19	381.01	4375	19
Transformers: Dark of the Moon	97.85	352.39	4088	15
The Twilight Saga: Breaking Dawn Part 1	138.12	281.29	4066	14
The Hangover Part II	85.95	254.46	3675	16
Pirates of the Caribbean: On Stranger Tides	90.15	241.07	4164	19
Fast Five	86.20	209.84	3793	15
Mission: Impossible—Ghost Protocol	12.79	208.55	3555	13
Cars 2	66.14	191.45	4115	25
Sherlock Holmes: A Game of Shadows	39.64	186.59	3703	13
Thor	65.72	181.03	3963	16

WEB file 2011Movies

Managerial Report

Use the numerical methods of descriptive statistics presented in this chapter to learn how these variables contribute to the success of a motion picture. Include the following in your report:

1. Descriptive statistics for each of the four variables along with a discussion of what the descriptive statistics tell us about the motion picture industry.
2. What motion pictures, if any, should be considered high-performance outliers? Explain.
3. Descriptive statistics showing the relationship between total gross sales and each of the other variables. Discuss.

Case Problem 3 Business Schools of Asia-Pacific

WEB file Asian

The pursuit of a higher education degree in business is now international. A survey shows that more and more Asians choose the master of business administration (MBA) degree route to corporate success. As a result, the number of applicants for MBA courses at Asia-Pacific schools continues to increase.

Across the region, thousands of Asians show an increasing willingness to temporarily shelve their careers and spend two years in pursuit of a theoretical business qualification. Courses in these schools are notoriously tough and include economics, banking, marketing, behavioral sciences, labor relations, decision making, strategic thinking, business law, and more. The data set in Table 3.11 shows some of the characteristics of the leading Asia-Pacific business schools.

Managerial Report

Use the methods of descriptive statistics to summarize the data in Table 3.11. Discuss your findings.

1. Include a summary for each variable in the data set. Make comments and interpretations based on maximums and minimums, as well as the appropriate means and

TABLE 3.11 DATA FOR 25 ASIA-PACIFIC BUSINESS SCHOOLS

Business School	Full-Time Enrollment	Students per Faculty	Local Tuition ($)	Foreign Tuition ($)	Age	%Foreign	GMAT	English Test	Work Experience	Starting Salary ($)
Melbourne Business School	200	5	24,420	29,600	28	47	Yes	No	Yes	71,400
University of New South Wales (Sydney)	228	4	19,993	32,582	29	28	Yes	No	Yes	65,200
Indian Institute of Management (Ahmedabad)	392	5	4,300	4,300	22	0	No	No	No	7,100
Chinese University of Hong Kong	90	5	11,140	11,140	29	10	Yes	No	No	31,000
International University of Japan (Niigata)	126	4	33,060	33,060	28	60	Yes	Yes	No	87,000
Asian Institute of Management (Manila)	389	5	7,562	9,000	25	50	Yes	No	Yes	22,800
Indian Institute of Management (Bangalore)	380	5	3,935	16,000	23	1	Yes	No	No	7,500
National University of Singapore	147	6	6,146	7,170	29	51	Yes	Yes	Yes	43,300
Indian Institute of Management (Calcutta)	463	8	2,880	16,000	23	0	No	No	No	7,400
Australian National University (Canberra)	42	2	20,300	20,300	30	80	Yes	Yes	Yes	46,600
Nanyang Technological University (Singapore)	50	5	8,500	8,500	32	20	Yes	No	Yes	49,300
University of Queensland (Brisbane)	138	17	16,000	22,800	32	26	No	No	Yes	49,600
Hong Kong University of Science and Technology	60	2	11,513	11,513	26	37	Yes	No	Yes	34,000
Macquarie Graduate School of Management (Sydney)	12	8	17,172	19,778	34	27	No	No	Yes	60,100
Chulalongkorn University (Bangkok)	200	7	17,355	17,355	25	6	Yes	No	Yes	17,600
Monash Mt. Eliza Business School (Melbourne)	350	13	16,200	22,500	30	30	Yes	Yes	Yes	52,500
Asian Institute of Management (Bangkok)	300	10	18,200	18,200	29	90	No	Yes	Yes	25,000
University of Adelaide	20	19	16,426	23,100	30	10	No	No	Yes	66,000
Massey University (Palmerston North, New Zealand)	30	15	13,106	21,625	37	35	No	Yes	Yes	41,400
Royal Melbourne Institute of Technology Business Graduate School	30	7	13,880	17,765	32	30	No	Yes	Yes	48,900
Jamnalal Bajaj Institute of Management Studies (Mumbai)	240	9	1,000	1,000	24	0	No	No	Yes	7,000
Curtin Institute of Technology (Perth)	98	15	9,475	19,097	29	43	Yes	No	Yes	55,000
Lahore University of Management Sciences	70	14	11,250	26,300	23	2.5	No	No	No	7,500
Universiti Sains Malaysia (Penang)	30	5	2,260	2,260	32	15	No	Yes	Yes	16,000
De La Salle University (Manila)	44	17	3,300	3,600	28	3.5	Yes	No	Yes	13,100

proportions. What new insights do these descriptive statistics provide concerning Asia-Pacific business schools?

2. Summarize the data to compare the following:
   a. Any difference between local and foreign tuition costs.
   b. Any difference between mean starting salaries for schools requiring and not requiring work experience.
   c. Any difference between starting salaries for schools requiring and not requiring English tests.
3. Do starting salaries appear to be related to tuition?
4. Present any additional graphical and numerical summaries that will be beneficial in communicating the data in Table 3.11 to others.

Case Problem 4 Heavenly Chocolates Website Transactions

Heavenly Chocolates manufactures and sells quality chocolate products at its plant and retail store located in Saratoga Springs, New York. Two years ago the company developed a website and began selling its products over the Internet. Website sales have exceeded the company's expectations, and mangement is now considering stragegies to increase sales even further. To learn more about the website customers, a sample of 50 Heavenly Chocolate transactions was selected from the previous month's sales. Data showing the day of the week each transaction was made, the type of browser the customer used, the time spent on the website, the number of website pages viewed, and the amount spent by each of the 50 customers are contained in the file named Shoppers. A portion of the data are shown in Table 3.12.

Heavenly Chocolates would like to use the sample data to determine if online shoppers who spend more time and view more pages also spend more money during their visit to the website. The company would also like to investigate the effect that the day of the week and the type of browser have on sales.

WEB file
Shoppers

TABLE 3.12 A SAMPLE OF 50 HEAVENLY CHOCOLATES WEBSITE TRANSACTIONS

Customer	Day	Browser	Time (min)	Pages Viewed	Amount Spent ($)
1	Mon	Internet Explorer	12.0	4	54.52
2	Wed	Other	19.5	6	94.90
3	Mon	Internet Explorer	8.5	4	26.68
4	Tue	Firefox	11.4	2	44.73
5	Wed	Internet Explorer	11.3	4	66.27
6	Sat	Firefox	10.5	6	67.80
7	Sun	Internet Explorer	11.4	2	36.04
.	.	.	.	.	.
.	.	.	.	.	.
.	.	.	.	.	.
48	Fri	Internet Explorer	9.7	5	103.15
49	Mon	Other	7.3	6	52.15
50	Fri	Internet Explorer	13.4	3	98.75

Managerial Report

Use the methods of descriptive statistics to learn about the customers who visit the Heavenly Chocolates website. Include the following in your report.

1. Graphical and numerical summaries for the length of time the shopper spends on the website, the number of pages viewed, and the mean amount spent per transaction. Discuss what you learn about Heavenly Cholcolates' online shoppers from these numerical summaries.
2. Summarize the frequency, the total dollars spent, and the mean amount spent per transaction for each day of week. What observations can you make about Hevenly Chocolates' business based on the day of the week? Discuss.
3. Summarize the frequency, the total dollars spent, and the mean amount spent per transaction for each type of browser. What observations can you make about Heavenly Chocolate's business based on the type of browser? Discuss.
4. Develop a scatter diagram and compute the sample correlation coefficient to explore the relationship between the time spent on the website and the dollar amount spent. Use the horizontal axis for the time spent on the website. Discuss.
5. Develop a scatter diagram and compute the sample correlation coefficient to explore the relationship between the the number of website pages viewed and the amount spent. Use the horizontal axis for the number of website pages viewed. Discuss.
6. Develop a scatter diagram and compute the sample correlation coefficient to explore the relationship between the time spent on the website and the number of pages viewed. Use the horizontal axis to represent the number of pages viewed. Discuss.

Case Problem 5 African Elephant Populations

Although millions of elephants once roamed across Africa, by the mid-1980s elephant populations in African nations had been devastated by poaching. Elephants are important to African ecosystems. In tropical forests, elephants create clearings in the canopy that encourage new tree growth. In savannas, elephants reduce bush cover to create an environment that is favorable to browsing and grazing animals. In addition, the seeds of many plant species depend on passing through an elephant's digestive tract before germination.

The status of the elephant now varies greatly across the continent; in some nations, strong measures have been taken to effectively protect elephant populations, while in other nations the elephant populations remain in danger due to poaching for meat and ivory, loss of habitat, and conflict with humans. Table 3.13 shows elephant populations for several African nations in 1979, 1989, and 2007 (Lemieux and Clarke, "The International Ban on Ivory Sales and Its Effects on Elephant Poaching in Africa," *British Journal of Criminology*, 49(4), 2009).

The David Sheldrick Wildlife Trust was established in 1977 to honor the memory of naturalist David Leslie William Sheldrick who founded Warden of Tsavo East National Park in Kenya and headed the Planning Unit of the Wildlife Conservation and Management Department in that country. Management of the Sheldrick Trust would like to know what these data indicate about elephant populations in various African countries since 1979.

Managerial Report

Use methods of descriptive statistics to summarize the data and comment on changes in elephant populations in African nations since 1979. At a minimum your report should include the following.

1. The mean annual change in elephant population for each country in the 10 years from 1979 to 1989, and a discussion of which countries saw the largest changes in elephant population over this 10-year period.

TABLE 3.13 ELEPHANT POPULATIONS FOR SEVERAL AFRICAN NATIONS IN 1979, 1989, AND 2007

WEB file
AfricanElephants

	Elephant population		
Country	1979	1989	2007
Angola	12,400	12,400	2,530
Botswana	20,000	51,000	175,487
Cameroon	16,200	21,200	15,387
Cen African Rep	63,000	19,000	3,334
Chad	15,000	3,100	6,435
Congo	10,800	70,000	22,102
Dem Rep of Congo	377,700	85,000	23,714
Gabon	13,400	76,000	70,637
Kenya	65,000	19,000	31,636
Mozambique	54,800	18,600	26,088
Somalia	24,300	6,000	70
Sudan	134,000	4,000	300
Tanzania	316,300	80,000	167,003
Zambia	150,000	41,000	29,231
Zimbabwe	30,000	43,000	99,107

2. The mean annual change in elephant population for each country from 1989 to 2007, and a discussion of which countries saw the largest changes in elephant population over this 18-year period.
3. A comparison of your results from parts 1 and 2, and a discussion of the conclusions you can draw from this comparison.

Appendix 3.1 Descriptive Statistics Using Minitab

In this appendix, we describe how Minitab can be used to compute a variety of descriptive statistics and display box plots. We then show how Minitab can be used to obtain covariance and correlation measures for two variables.

Descriptive Statistics

WEB file
2012StartSalary

Table 3.1 provided the starting salaries for 12 business school graduates. These data are in column C2 of the file 2012StartSalary. The following steps can be used to generate descriptive statistics for the starting salary data.

Step 1. Select the **Stat** menu
Step 2. Choose **Basic Statistics**
Step 3. Choose **Display Descriptive Statistics**
Step 4. When the Display Descriptive Statistics dialog box appears:
Enter C2 in the **Variables** box
Click **OK**

Figure 3.15 shows the descriptive statistics obtained using Minitab. Definitions of the headings follow.

N	number of data values	Minimum	minimum data value
N*	number of missing data values	Q1	first quartile
Mean	mean	Median	median
SE Mean	standard error of mean	Q3	third quartile
StDev	standard deviation	Maximum	maximum data value

FIGURE 3.15 DESCRIPTIVE STATISTICS PROVIDED BY MINITAB

N	N*	Mean	SEMean	StDev
12	0	3940.0	47.8	165.7
Minimum	Q1	Median	Q3	Maximum
3710.0	4025.0	3905.0	4025.0	4325.0

The label SE Mean refers to the *standard error of the mean.* It is computed by dividing the standard deviation by the square root of *N*. The interpretation and use of this measure are discussed in Chapter 7 when we introduce the topics of sampling and sampling distributions.

The 10 descriptive statistics shown in Figure 3.15 are the default descriptive statistics selected automatically by Minitab. These descriptive statistics are of interest to the majority of users. However, Minitab provides 15 additional descriptive statistics that may be selected depending upon the preferences of the user. The variance, coefficient of variation, range, interquartile range, mode, and skewness are among the additional descriptive statistics available. To select one or more of these additional descriptive statistics, modify step 4 as follows.

Step 4. When the Display Descriptive Statistics dialog box appears:
Enter C2 in the **Variables** box
Click **Statistics**
Select the **descriptive statistics** you wish to obtain or
choose **All** to obtain all 25 descriptive statistics
Click **OK**
Click **OK**

Finally, note that Minitab's quartiles Q_1 = 3857.5 and Q_3 = 4025.0 are slightly different from the quartiles Q_1 = 3865 and Q_3 = 4000 computed in Section 3.1. The different conventions* used to identify the quartiles explain this variation. Hence, the values of Q_1 and Q_3 provided by one convention may not be identical to the values of Q_1 and Q_3 provided by another convention. Any differences tend to be negligible, however, and the results provided should not mislead the user in making the usual interpretations associated with quartiles.

Box Plot

The following steps use the file 2012StartSalary to generate the box plot for the starting salary data.

Step 1. Select the **Graph** menu
Step 2. Choose **Boxplot**
Step 3. Under the heading **OneY** select **Simple** and click **OK**
Step 4. When the Boxplot-One Y, Simple dialog box appears:
Enter C2 in the **Graph variables** box
Click **OK**

Covariance and Correlation

WEB file
Stereo

Table 3.6 provided the number of commercials and the sales volume for a stereo and sound equipment store. These data are available in the file Stereo, with the number of commercials

*With the n observations arranged in ascending order (smallest value to largest value), Minitab uses the positions given by $(n + 1)/4$ and $3(n + 1)/4$ to locate Q_1 and Q_3, respectively. When a position is fractional, Minitab interpolates between the two adjacent ordered data values to determine the corresponding quartile.

in column C2 and the sales volume in column C3. The following steps show how Minitab can be used to compute the covariance for the two variables.

Step 1. Select the **Stat** menu
Step 2. Choose **Basic Statistics**
Step 3. Choose **Covariance**
Step 4. When the Covariance dialog box appears:
Enter C2 C3 in the **Variables** box
Click **OK**

To obtain the correlation coefficient for the number of commercials and the sales volume, only one change is necessary in the preceding procedure. In step 3, choose the **Correlation** option.

Appendix 3.2 Descriptive Statistics Using Excel

Excel can be used to generate the descriptive statistics discussed in this chapter. We show how Excel can be used to generate several measures of location and variability for a single variable and to generate the covariance and correlation coefficient as measures of association between two variables.

Using Excel Functions

WEB file
2012StartSalary

Excel provides functions for computing the mean, median, mode, sample variance, and sample standard deviation. We illustrate the use of these Excel functions by computing the mean, median, mode, sample variance, and sample standard deviation for the starting salary data in Table 3.1. Refer to Figure 3.16 as we describe the steps involved. The data are entered in column B.

FIGURE 3.16 USING EXCEL FUNCTIONS FOR COMPUTING THE MEAN, MEDIAN, MODE, VARIANCE, AND STANDARD DEVIATION

	A	B	C	D	E
1	Graduate	Starting Salary		Mean	=AVERAGE(B2:B13)
2	1	3850		Median	=MEDIAN(B2:B13)
3	2	3950		Mode	=MODE.SNGL(B2:B13)
4	3	4050		Variance	=VAR.S(B2:B13)
5	4	3880		Standard Deviation	=STDEV.S(B2:B13)
6	5	3755			
7	6	3710			
8	7	3890			
9	8	4130			
10	9	3940			
11	10	4325			
12	11	3920			
13	12	3880			

	A	B	C	D	E
1	Graduate	Starting Salary		Mean	3940
2	1	3850		Median	3905
3	2	3950		Mode	3880
4	3	4050		Variance	27440.91
5	4	3880		Standard Deviation	165.65
6	5	3755			
7	6	3710			
8	7	3890			
9	8	4130			
10	9	3940			
11	10	4325			
12	11	3920			
13	12	3880			

WEB file

Stereo

To find the variance, standard deviation, and covariance for population data, follow the same steps but use the VAR.P, STDEV.P, and COV.P functions.

If the ***Analysis*** *tab doesn't appear on your ribbon or if the* ***Data Analysis*** *option doesn't appear in the* ***Data Analysis*** *tab, you need to activate the* ***Data Analysis ToolPak*** *by following these three steps:*

1. *Click the* ***File*** *tab, then click* ***Options****, and then click the* ***Add-Ins*** *category.*
2. *In the* ***Manage*** *box, click* ***Excel Add-ins****, and then click* ***Go****. The* ***Add-Ins*** *dialog box will then appear.*
3. *In the* ***Add-Ins available*** *box, select the check box next to the* ***Data Analysis ToolPak*** *add-in and click* ***OK****.*

The ***Analysis*** *tab will now be available with the* ***Data Analysis*** *option.*

Excel's AVERAGE function can be used to compute the mean by entering the following formula into cell E1:

=AVERAGE(B2:B13)

Similarly, the formulas =MEDIAN(B2:B13), =MODE.SNGL(B2:B13), =VAR.S(B2:B13), and =STDEV.S(B2:B13) are entered into cells E2:E5, respectively, to compute the median, mode, variance, and standard deviation for this sample. The worksheet in the foreground shows that the values computed using the Excel functions are the same as we computed earlier in the chapter.

Excel also provides functions that can be used to compute the sample covariance and the sample correlation coefficient. We show here how these functions can be used to compute the sample covariance and the sample correlation coefficient for the stereo and sound equipment store data in Table 3.6. Refer to Figure 3.17 as we present the steps involved.

Excel's sample covariance function, COVARIANCE.S, can be used to compute the sample covariance by entering the following formula into cell F1:

=COVARIANCE.S(B2:B11,C2:C11)

Similarly, the formula =CORREL(B2:B11,C2:C11) is entered into cell F2 to compute the sample correlation coefficient. The worksheet in the foreground shows the values computed using the Excel functions. Note that the value of the sample covariance (11) is the same as computed using equation (3.12). And the value of the sample correlation coefficient (.93) is the same as computed using equation (3.14).

Using Excel's Descriptive Statistics Tool

As we already demonstrated, Excel provides statistical functions to compute descriptive statistics for a data set. These functions can be used to compute one statistic at a time (e.g., mean, variance, etc.). Excel also provides a variety of Data Analysis Tools. One of these tools, called Descriptive Statistics, allows the user to compute a variety of descriptive statistics at once. We show here how it can be used to compute descriptive statistics for the starting salary data in Table 3.1.

FIGURE 3.17 USING EXCEL FUNCTIONS FOR COMPUTING THE COVARIANCE AND CORRELATION

	A	B	C	D	E	F	G
1	Week	Commercials	Sales Volume		Sample Covariance	=COVARIANCE.S(B2:B11,C2:C11)	
2	1	2	50		Sample Correlation	=CORREL(B2:B11,C2:C11)	
3	2	5	57				
4	3	1	41				
5	4	3	54				
6	5	4	54				
7	6	1	38				
8	7	5	63				
9	8	3	48				
10	9	4	59				
11	10	2	46				
12							

	A	B	C	D	E	F	G
1	Week	Commercials	Sales Volume		Sample Covariance	11	
2	1	2	50		Sample Correlation	0.9305	
3	2	5	57				
4	3	1	41				
5	4	3	54				
6	5	4	54				
7	6	1	38				
8	7	5	63				
9	8	3	48				
10	9	4	59				
11	10	2	46				
12							

FIGURE 3.18 EXCEL'S DESCRIPTIVE STATISTICS TOOL OUTPUT

	A	B	C	D	E
1	**Graduate**	**Starting Salary**		*Starting Salary*	
2	1	3850			
3	2	3950		**Mean**	3940
4	3	4050		Standard Error	47.82
5	4	3880		**Median**	3905
6	5	3755		**Mode**	3880
7	6	3710		**Standard Deviation**	165.65
8	7	3890		**Sample Variance**	27440.91
9	8	4130		Kurtosis	1.72
10	9	3940		Skewness	1.09
11	10	4325		**Range**	615
12	11	3920		**Minimum**	3710
13	12	3880		**Maximum**	4325
14				**Sum**	47280
15				**Count**	12

WEB file
2012StartSalary

Step 1. Click the **Data** tab on the Ribbon
Step 2. In the **Analysis** group, click **Data Analysis**
Step 3. When the Data Analysis dialog box appears:
Choose **Descriptive Statistics**
Click **OK**
Step 4. When the Descriptive Statistics dialog box appears:
Enter B1:B13 in the **Input Range** box
Select **Grouped By Columns**
Select **Labels in First Row**
Select **Output Range**
Enter D1 in the **Output Range** box (to identify the upper left-hand corner of the section of the worksheet where the descriptive statistics will appear)
Select **Summary statistics**
Click **OK**

Cells D1:E15 of Figure 3.18 show the descriptive statistics provided by Excel. The boldface entries are the descriptive statistics we covered in this chapter. The descriptive statistics that are not boldface are either covered subsequently in the text or discussed in more advanced texts.

Appendix 3.3 Descriptive Statistics Using StatTools

In this appendix, we describe how StatTools can be used to compute a variety of descriptive statistics and also display box plots. We then show how StatTools can be used to obtain covariance and correlation measures for two variables.

Descriptive Statistics

WEB file
2012StartSalary

We use the starting salary data in Table 3.1 to illustrate. Begin by using the Data Set Manager to create a StatTools data set for these data using the procedure described in the appendix in Chapter 1. The following steps will generate a variety of descriptive statistics.

Step 1. Click the **StatTools** tab on the Ribbon
Step 2. In the **Analyses Group,** click **Summary Statistics**
Step 3. Choose the **One-Variable Summary** option
Step 4. When the One-Variable Summary Statistics dialog box appears:
In the **Variables** section, select **Starting Salary**
Click **OK**

A variety of descriptive statistics as shown in Figure 3.18 will appear.

Box Plots

We use the starting salary data in Table 3.1 to illustrate. Begin by using the Data Set Manager to create a StatTools data set for these data using the procedure described in the appendix in Chapter 1. The following steps will create a box plot for these data.

WEB file
2012StartSalary

Step 1. Click the **StatTools** tab on the Ribbon
Step 2. In the **Analyses Group,** click **Summary Graphs**
Step 3. Choose the **Box-Whisker Plot** option
Step 4. When the StatTools-Box-Whisker Plot dialog box appears:
In the Variables section, select **Starting Salary**
Click **OK**

In its box plots, StatTools defines any value that is at least 1.5 IQR, but less than 3 IQR outside the box as a mild outlier, and any value that is at least 3 IRQ outside the box as an extreme outlier. The symbol □ is used to identify a mild outlier, the symbol ■ is used to signify an extreme outlier, and x is used to identify the mean.

Covariance and Correlation

We use the stereo and sound equipment data in Table 3.6 to demonstrate the computation of the sample covariance and the sample correlation coefficient. Begin by using the Data Set Manager to create a StatTools data set for these data using the procedure described in the appendix in Chapter 1. The following steps will provide the sample covariance and sample correlation coefficient.

WEB file
Stereo

Step 1. Click the **StatTools** tab on the Ribbon
Step 2. In the **Analyses Group,** click **Summary Statistics**
Step 3. Choose the **Correlation and Covariance** option
Step 4. When the StatTools-Correlation and Covariance dialog box appears:
In the **Variables** section
Select **No. of Commercials**
Select **Sales Volume**
In the **Tables to Create** section,
Select **Table of Correlations**
Select **Table of Covariances**
In the **Table Structure** section select **Symmetric**
Click **OK**

A table showing the correlation coefficient and the covariance will appear.

CHAPTER 4

Introduction to Probability

CONTENTS

STATISTICS *in* PRACTICE

PROBABILITY TO THE RESCUE*

On August 5, 2010, the San José copper and gold mine suffered a cave-in. Thirty-three men were trapped more than 2000 feet underground in the Atacama Desert near Copiapó, Chile. While most feared that these men's prospects were grim, several attempts were made to locate the miners and determine if they were still alive. Seventeen days later, rescuers reached the men with a 5½-inch borehole and ascertained that they were still alive.

While it was important to bring these men safely to the surface as quickly as possible, it was imperative that the rescue effort proceed cautiously. "The mine is old and there is concern of further collapses," Murray & Roberts Cementation Managing Director Henry Laas said in an interview with the *Santiago Times*. "The rescue methodology therefore has to be carefully designed and implemented."

NASA scientists based probabilities on similar circumstances experienced during space flights. © Hugo Infante/Government of Chile/Handout/ Reuters.

The Chilean government asked NASA to consult on the rescue operation. In response, NASA sent a four-person team consisting of an engineer (Clinton Cragg), two physicians (Michael Duncan and J. D. Polk), and a psychologist (Al Holland). When asked why a space agency was brought in to consult on the rescue of trapped miners, Duncan stated, "We brought our experience in vehicle design and long duration confinement to our Chilean counterparts."

The probability of failure was prominent in the thoughts of everyone involved. "We were thinking that the rescue vehicle would have to make over 40 round trips, so in consideration of the probability of part failures, we suggested the rescue team have three rigs and several sets of replacement parts available," said Cragg. "We also tried to increase the probability of success by placing spring-loaded rollers on the sides of the cage so the cage itself would not be damaged through direct contact with the rock wall as it moved through the rescue portal."

While it was important to consider probabilities of various events, precise estimates of probabilities could not be made. Probabilities could only be estimated subjectively based on similar circumstances NASA scientists had experienced during space flights. Duncan explained, "While we and the Chileans would have preferred to have precise estimates of various probabilities based on historical data, the uniqueness of the situation made this infeasible. For example, a miner had to stand virtually straight up in the cage on an ascent that was originally estimated to last two to four hours per miner, so we had to be concerned about fainting. All we could do was consider what we thought to be the facts and apply what we had learned from astronauts' experiences in their returns from short- and long-duration space missions." Duncan then concluded, "It actually took 15 minutes to bring the cage up from the bottom of the mine, so our estimates in this case were very conservative. Considering the risk involved, that is exactly what we wanted."

Ultimately the rescue approach designed by the Chileans in consultation with NASA was successful. On October 13 the last of the miners emerged; the 13-foot-long, 924-pound steel Fénix 2 rescue capsule withstood over 40 trips into the mine, and no miner fainted on ascent.

The use of subjective probabilities in unique situations is common for NASA. With limited space available on a space vehicle, assessing probability of failure for various components and the risks associated with these potential failures becomes critical to how NASA decides which spare components to include on a space flight. NASA also employs probability to estimate the likelihood of crew health and performance issues arising on space exploration missions.

*The authors are indebted to Dr. Michael Duncan and Clinton Cragg of NASA for providing input for this Statistics in Practice.

Managers often base their decisions on an analysis of uncertainties such as the following:

1. What are the chances that sales will decrease if we increase prices?
2. What is the likelihood a new assembly method will increase productivity?
3. How likely is it that the project will be finished on time?
4. What is the chance that a new investment will be profitable?

Some of the earliest work on probability originated in a series of letters between Pierre de Fermat and Blaise Pascal in the 1650s.

Probability is a numerical measure of the likelihood that an event will occur. Thus, probabilities can be used as measures of the degree of uncertainty associated with the four events previously listed. If probabilities are available, we can determine the likelihood of each event occurring.

Probability values are always assigned on a scale from 0 to 1. A probability near zero indicates an event is unlikely to occur; a probability near 1 indicates an event is almost certain to occur. Other probabilities between 0 and 1 represent degrees of likelihood that an event will occur. For example, if we consider the event "rain tomorrow," we understand that when the weather report indicates "a near-zero probability of rain," it means almost no chance of rain. However, if a .90 probability of rain is reported, we know that rain is likely to occur. A .50 probability indicates that rain is just as likely to occur as not. Figure 4.1 depicts the view of probability as a numerical measure of the likelihood of an event occurring.

4.1 Experiments, Counting Rules, and Assigning Probabilities

In discussing probability, we define an **experiment** as a process that generates well-defined outcomes. On any single repetition of an experiment, one and only one of the possible experimental outcomes will occur. Several examples of experiments and their associated outcomes follow.

Experiment	Experimental Outcomes
Toss a coin	Head, tail
Select a part for inspection	Defective, nondefective
Conduct a sales call	Purchase, no purchase
Roll a die	1, 2, 3, 4, 5, 6
Play a football game	Win, lose, tie

By specifying all possible experimental outcomes, we identify the **sample space** for an experiment.

FIGURE 4.1 PROBABILITY AS A NUMERICAL MEASURE OF THE LIKELIHOOD OF AN EVENT OCCURRING

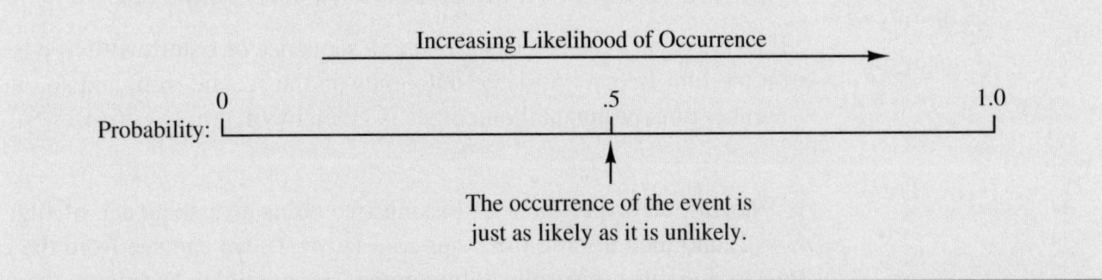

SAMPLE SPACE

The sample space for an experiment is the set of all experimental outcomes.

Experimental outcomes are also called sample points.

An experimental outcome is also called a **sample point** to identify it as an element of the sample space.

Consider the first experiment in the preceding table—tossing a coin. The upward face of the coin—a head or a tail—determines the experimental outcomes (sample points). If we let S denote the sample space, we can use the following notation to describe the sample space.

$$S = \{\text{Head, Tail}\}$$

The sample space for the second experiment in the table—selecting a part for inspection—can be described as follows:

$$S = \{\text{Defective, Nondefective}\}$$

Both of the experiments just described have two experimental outcomes (sample points). However, suppose we consider the fourth experiment listed in the table—rolling a die. The possible experimental outcomes, defined as the number of dots appearing on the upward face of the die, are the six points in the sample space for this experiment.

$$S = \{1, 2, 3, 4, 5, 6\}$$

Counting Rules, Combinations, and Permutations

Being able to identify and count the experimental outcomes is a necessary step in assigning probabilities. We now discuss three useful counting rules.

Multiple-step experiments The first counting rule applies to **multiple-step experiments**. Consider the experiment of tossing two coins. Let the experimental outcomes be defined in terms of the pattern of heads and tails appearing on the upward faces of the two coins. How many experimental outcomes are possible for this experiment? The experiment of tossing two coins can be thought of as a two-step experiment in which step 1 is the tossing of the first coin and step 2 is the tossing of the second coin. If we use H to denote a head and T to denote a tail, (H, H) indicates the experimental outcome with a head on the first coin and a head on the second coin. Continuing this notation, we can describe the sample space (S) for this coin-tossing experiment as follows:

$$S = \{(H, H), (H, T), (T, H), (T, T)\}$$

Thus, we see that four experimental outcomes are possible. In this case, we can easily list all the experimental outcomes.

The counting rule for multiple-step experiments makes it possible to determine the number of experimental outcomes without listing them.

COUNTING RULE FOR MULTIPLE-STEP EXPERIMENTS

If an experiment can be described as a sequence of k steps with n_1 possible outcomes on the first step, n_2 possible outcomes on the second step, and so on, then the total number of experimental outcomes is given by $(n_1)(n_2) \cdots (n_k)$.

Viewing the experiment of tossing two coins as a sequence of first tossing one coin ($n_1 = 2$) and then tossing the other coin ($n_2 = 2$), we can see from the counting rule that $(2)(2) = 4$ distinct experimental outcomes are possible. As shown, they are $S = \{(H, H),$

FIGURE 4.2 TREE DIAGRAM FOR THE EXPERIMENT OF TOSSING TWO COINS

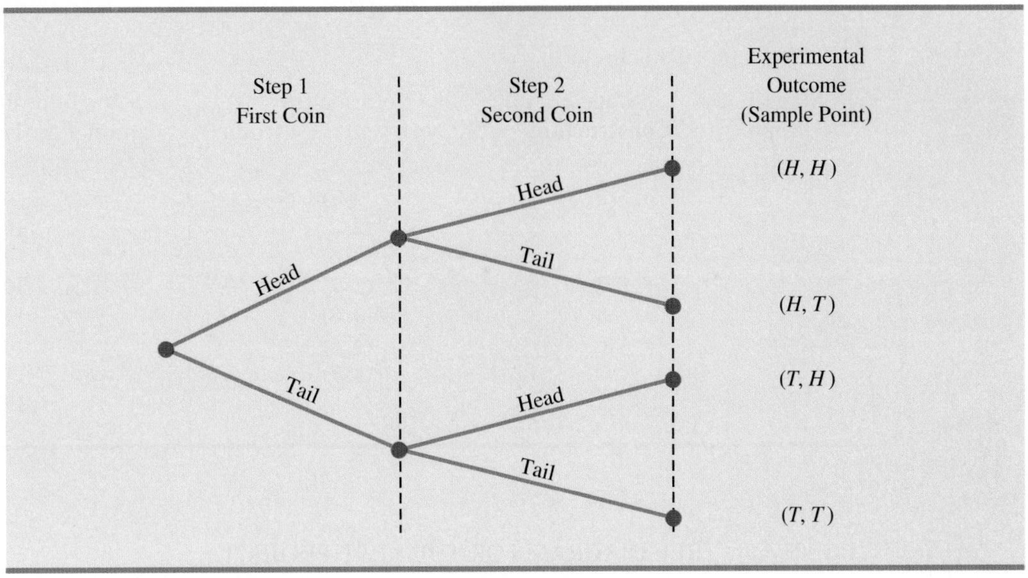

(H, T), (T, H), (T, T)}. The number of experimental outcomes in an experiment involving tossing six coins is (2)(2)(2)(2)(2)(2) = 64.

Without the tree diagram, one might think only three experimental outcomes are possible for two tosses of a coin: 0 heads, 1 head, and 2 heads.

A **tree diagram** is a graphical representation that helps in visualizing a multiple-step experiment. Figure 4.2 shows a tree diagram for the experiment of tossing two coins. The sequence of steps moves from left to right through the tree. Step 1 corresponds to tossing the first coin, and step 2 corresponds to tossing the second coin. For each step, the two possible outcomes are head or tail. Note that for each possible outcome at step 1 two branches correspond to the two possible outcomes at step 2. Each of the points on the right end of the tree corresponds to an experimental outcome. Each path through the tree from the leftmost node to one of the nodes at the right side of the tree corresponds to a unique sequence of outcomes.

Let us now see how the counting rule for multiple-step experiments can be used in the analysis of a capacity expansion project for the Kentucky Power & Light Company (KP&L). KP&L is starting a project designed to increase the generating capacity of one of its plants in northern Kentucky. The project is divided into two sequential stages or steps: stage 1 (design) and stage 2 (construction). Even though each stage will be scheduled and controlled as closely as possible, management cannot predict beforehand the exact time required to complete each stage of the project. An analysis of similar construction projects revealed possible completion times for the design stage of 2, 3, or 4 months and possible completion times for the construction stage of 6, 7, or 8 months. In addition, because of the critical need for additional electrical power, management set a goal of 10 months for the completion of the entire project.

Because this project has three possible completion times for the design stage (step 1) and three possible completion times for the construction stage (step 2), the counting rule for multiple-step experiments can be applied here to determine a total of (3)(3) = 9 experimental outcomes. To describe the experimental outcomes, we use a two-number notation; for instance, (2, 6) indicates that the design stage is completed in 2 months and the construction stage is completed in 6 months. This experimental outcome results in a total of 2 + 6 = 8 months to complete the entire project. Table 4.1 summarizes the nine experimental outcomes for the KP&L problem. The tree diagram in Figure 4.3 shows how the nine outcomes (sample points) occur.

The counting rule and tree diagram help the project manager identify the experimental outcomes and determine the possible project completion times. From the information in

TABLE 4.1 EXPERIMENTAL OUTCOMES (SAMPLE POINTS) FOR THE KP&L PROJECT

Completion Time (months)			
Stage 1 Design	Stage 2 Construction	Notation for Experimental Outcome	Total Project Completion Time (months)
2	6	(2, 6)	8
2	7	(2, 7)	9
2	8	(2, 8)	10
3	6	(3, 6)	9
3	7	(3, 7)	10
3	8	(3, 8)	11
4	6	(4, 6)	10
4	7	(4, 7)	11
4	8	(4, 8)	12

FIGURE 4.3 TREE DIAGRAM FOR THE KP&L PROJECT

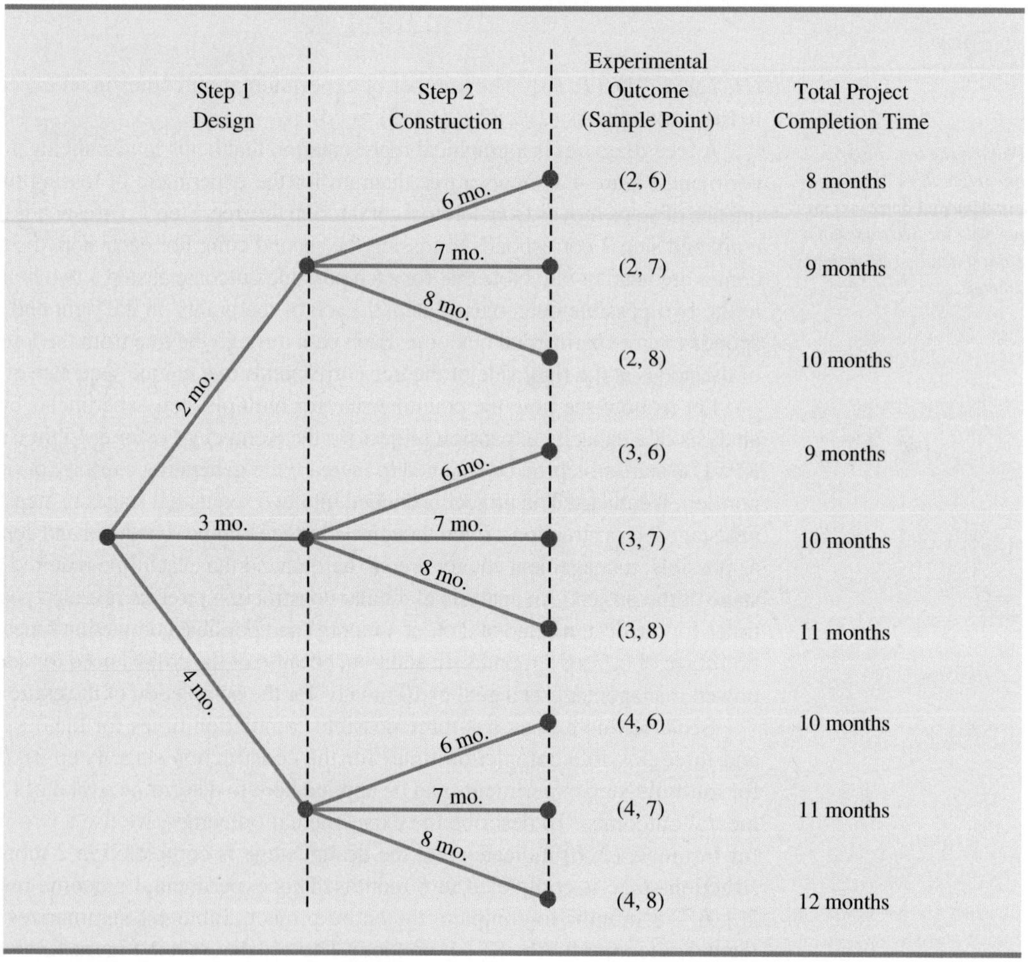

Figure 4.3, we see that the project will be completed in 8 to 12 months, with six of the nine experimental outcomes providing the desired completion time of 10 months or less. Even though identifying the experimental outcomes may be helpful, we need to consider how probability values can be assigned to the experimental outcomes before making an assessment of the probability that the project will be completed within the desired 10 months.

Combinations A second useful counting rule allows one to count the number of experimental outcomes when the experiment involves selecting n objects from a set of N objects. It is called the counting rule for **combinations**.

COUNTING RULE FOR COMBINATIONS

The number of combinations of N objects taken n at a time is

$$C_n^N = \binom{N}{n} = \frac{N!}{n!(N-n)!} \tag{4.1}$$

where

$$N! = N(N-1)(N-2)\cdots(2)(1)$$
$$n! = n(n-1)(n-2)\cdots(2)(1)$$

and, by definition,

$$0! = 1$$

In sampling from a finite population of size N, the counting rule for combinations is used to find the number of different samples of size n that can be selected.

The notation ! means *factorial;* for example, 5 factorial is $5! = (5)(4)(3)(2)(1) = 120$.

As an illustration of the counting rule for combinations, consider a quality control procedure in which an inspector randomly selects two of five parts to test for defects. In a group of five parts, how many combinations of two parts can be selected? The counting rule in equation (4.1) shows that with $N = 5$ and $n = 2$, we have

$$C_2^5 = \binom{5}{2} = \frac{5!}{2!(5-2)!} = \frac{(5)(4)(3)(2)(1)}{(2)(1)(3)(2)(1)} = \frac{120}{12} = 10$$

Thus, 10 outcomes are possible for the experiment of randomly selecting two parts from a group of five. If we label the five parts as A, B, C, D, and E, the 10 combinations or experimental outcomes can be identified as AB, AC, AD, AE, BC, BD, BE, CD, CE, and DE.

As another example, consider that the Florida lottery system uses the random selection of 6 integers from a group of 53 to determine the weekly winner. The counting rule for combinations, equation (4.1), can be used to determine the number of ways 6 different integers can be selected from a group of 53.

$$\binom{53}{6} = \frac{53!}{6!(53-6)!} = \frac{53!}{6!47!} = \frac{(53)(52)(51)(50)(49)(48)}{(6)(5)(4)(3)(2)(1)} = 22{,}957{,}480$$

The counting rule for combinations shows that the chance of winning the lottery is very unlikely.

The counting rule for combinations tells us that almost 23 million experimental outcomes are possible in the lottery drawing. An individual who buys a lottery ticket has 1 chance in 22,957,480 of winning.

Permutations A third counting rule that is sometimes useful is the counting rule for **permutations**. It allows one to compute the number of experimental outcomes when n objects are to be selected from a set of N objects where the order of selection is important. The same n objects selected in a different order are considered a different experimental outcome.

COUNTING RULE FOR PERMUTATIONS

The number of permutations of N objects taken n at a time is given by

$$P_n^N = n!\binom{N}{n} = \frac{N!}{(N-n)!} \tag{4.2}$$

The counting rule for permutations closely relates to the one for combinations; however, an experiment results in more permutations than combinations for the same number of objects because every selection of n objects can be ordered in $n!$ different ways.

As an example, consider again the quality control process in which an inspector selects two of five parts to inspect for defects. How many permutations may be selected? The counting rule in equation (4.2) shows that with $N = 5$ and $n = 2$, we have

$$P_2^5 = \frac{5!}{(5-2)!} = \frac{5!}{3!} = \frac{(5)(4)(3)(2)(1)}{(3)(2)(1)} = \frac{120}{6} = 20$$

Thus, 20 outcomes are possible for the experiment of randomly selecting two parts from a group of five when the order of selection must be taken into account. If we label the parts A, B, C, D, and E, the 20 permutations are AB, BA, AC, CA, AD, DA, AE, EA, BC, CB, BD, DB, BE, EB, CD, DC, CE, EC, DE, and ED.

Assigning Probabilities

Now let us see how probabilities can be assigned to experimental outcomes. The three approaches most frequently used are the classical, relative frequency, and subjective methods. Regardless of the method used, two **basic requirements for assigning probabilities** must be met.

BASIC REQUIREMENTS FOR ASSIGNING PROBABILITIES

1. The probability assigned to each experimental outcome must be between 0 and 1, inclusively. If we let E_i denote the ith experimental outcome and $P(E_i)$ its probability, then this requirement can be written as

$$0 \le P(E_i) \le 1 \text{ for all } i \tag{4.3}$$

2. The sum of the probabilities for all the experimental outcomes must equal 1.0. For n experimental outcomes, this requirement can be written as

$$P(E_1) + P(E_2) + \cdots + P(E_n) = 1 \tag{4.4}$$

The **classical method** of assigning probabilities is appropriate when all the experimental outcomes are equally likely. If n experimental outcomes are possible, a probability of $1/n$ is assigned to each experimental outcome. When using this approach, the two basic requirements for assigning probabilities are automatically satisfied.

For an example, consider the experiment of tossing a fair coin; the two experimental outcomes—head and tail—are equally likely. Because one of the two equally likely outcomes is a head, the probability of observing a head is 1/2, or .50. Similarly, the probability of observing a tail is also 1/2, or .50.

As another example, consider the experiment of rolling a die. It would seem reasonable to conclude that the six possible outcomes are equally likely, and hence each outcome is assigned a probability of 1/6. If $P(1)$ denotes the probability that one dot appears on the upward face of the die, then $P(1) = 1/6$. Similarly, $P(2) = 1/6$, $P(3) = 1/6$, $P(4) = 1/6$, $P(5) = 1/6$, and $P(6) = 1/6$. Note that these probabilities satisfy the two basic requirements of equations (4.3) and (4.4) because each of the probabilities is greater than or equal to zero and they sum to 1.0.

The **relative frequency method** of assigning probabilities is appropriate when data are available to estimate the proportion of the time the experimental outcome will occur if the experiment is repeated a large number of times. As an example, consider a study of waiting times in the X-ray department for a local hospital. A clerk recorded the number of patients waiting for service at 9:00 A.M. on 20 successive days and obtained the following results.

Number Waiting	Number of Days Outcome Occurred
0	2
1	5
2	6
3	4
4	3
Total	20

These data show that on 2 of the 20 days, zero patients were waiting for service; on 5 of the days, one patient was waiting for service; and so on. Using the relative frequency method, we would assign a probability of $2/20 = .10$ to the experimental outcome of zero patients waiting for service, $5/20 = .25$ to the experimental outcome of one patient waiting, $6/20 = .30$ to two patients waiting, $4/20 = .20$ to three patients waiting, and $3/20 = .15$ to four patients waiting. As with the classical method, using the relative frequency method automatically satisfies the two basic requirements of equations (4.3) and (4.4).

The **subjective method** of assigning probabilities is most appropriate when one cannot realistically assume that the experimental outcomes are equally likely and when little relevant data are available. When the subjective method is used to assign probabilities to the experimental outcomes, we may use any information available, such as our experience or intuition. After considering all available information, a probability value that expresses our *degree of belief* (on a scale from 0 to 1) that the experimental outcome will occur is specified. Because subjective probability expresses a person's degree of belief, it is personal. Using the subjective method, different people can be expected to assign different probabilities to the same experimental outcome.

The subjective method requires extra care to ensure that the two basic requirements of equations (4.3) and (4.4) are satisfied. Regardless of a person's degree of belief, the probability value assigned to each experimental outcome must be between 0 and 1, inclusive, and the sum of all the probabilities for the experimental outcomes must equal 1.0.

Consider the case in which Tom and Judy Elsbernd make an offer to purchase a house. Two outcomes are possible:

$$E_1 = \text{their offer is accepted}$$
$$E_2 = \text{their offer is rejected}$$

Judy believes that the probability their offer will be accepted is .8; thus, Judy would set $P(E_1) = .8$ and $P(E_2) = .2$. Tom, however, believes that the probability that their offer will be accepted is .6; hence, Tom would set $P(E_1) = .6$ and $P(E_2) = .4$. Note that Tom's probability estimate for E_1 reflects a greater pessimism that their offer will be accepted.

Bayes' theorem (see Section 4.5) provides a means for combining subjectively determined prior probabilities with probabilities obtained by other means to obtain revised, or posterior, probabilities.

Both Judy and Tom assigned probabilities that satisfy the two basic requirements. The fact that their probability estimates are different emphasizes the personal nature of the subjective method.

Even in business situations where either the classical or the relative frequency approach can be applied, managers may want to provide subjective probability estimates. In such cases, the best probability estimates often are obtained by combining the estimates from the classical or relative frequency approach with subjective probability estimates.

Probabilities for the KP&L Project

To perform further analysis on the KP&L project, we must develop probabilities for each of the nine experimental outcomes listed in Table 4.1. On the basis of experience and judgment, management concluded that the experimental outcomes were not equally likely. Hence, the classical method of assigning probabilities could not be used. Management then decided to conduct a study of the completion times for similar projects undertaken by KP&L over the past three years. The results of a study of 40 similar projects are summarized in Table 4.2.

After reviewing the results of the study, management decided to employ the relative frequency method of assigning probabilities. Management could have provided subjective probability estimates but felt that the current project was quite similar to the 40 previous projects. Thus, the relative frequency method was judged best.

In using the data in Table 4.2 to compute probabilities, we note that outcome (2, 6)—stage 1 completed in 2 months and stage 2 completed in 6 months—occurred six times in the 40 projects. We can use the relative frequency method to assign a probability of $6/40 = .15$ to this outcome. Similarly, outcome (2, 7) also occurred in six of the 40 projects, providing a $6/40 = .15$ probability. Continuing in this manner, we obtain the probability assignments for the sample points of the KP&L project shown in Table 4.3. Note that $P(2, 6)$ represents the probability of the sample point (2, 6), $P(2, 7)$ represents the probability of the sample point (2, 7), and so on.

TABLE 4.2 COMPLETION RESULTS FOR 40 KP&L PROJECTS

Completion Time (months)			
Stage 1 Design	**Stage 2 Construction**	**Sample Point**	**Number of Past Projects Having These Completion Times**
2	6	(2, 6)	6
2	7	(2, 7)	6
2	8	(2, 8)	2
3	6	(3, 6)	4
3	7	(3, 7)	8
3	8	(3, 8)	2
4	6	(4, 6)	2
4	7	(4, 7)	4
4	8	(4, 8)	6
		Total	40

TABLE 4.3 PROBABILITY ASSIGNMENTS FOR THE KP&L PROJECT BASED ON THE RELATIVE FREQUENCY METHOD

Sample Point	Project Completion Time	Probability of Sample Point
(2, 6)	8 months	$P(2, 6) = 6/40 = .15$
(2, 7)	9 months	$P(2, 7) = 6/40 = .15$
(2, 8)	10 months	$P(2, 8) = 2/40 = .05$
(3, 6)	9 months	$P(3, 6) = 4/40 = .10$
(3, 7)	10 months	$P(3, 7) = 8/40 = .20$
(3, 8)	11 months	$P(3, 8) = 2/40 = .05$
(4, 6)	10 months	$P(4, 6) = 2/40 = .05$
(4, 7)	11 months	$P(4, 7) = 4/40 = .10$
(4, 8)	12 months	$P(4, 8) = 6/40 = .15$
		Total 1.00

NOTES AND COMMENTS

1. In statistics, the notion of an experiment differs somewhat from the notion of an experiment in the physical sciences. In the physical sciences, researchers usually conduct an experiment in a laboratory or a controlled environment in order to learn about cause and effect. In statistical experiments, probability determines outcomes. Even though the experiment is repeated in exactly the same way, an entirely different outcome may occur. Because of this influence of probability on the outcome, the experiments of statistics are sometimes called *random experiments*.
2. When drawing a random sample without replacement from a population of size N, the counting rule for combinations is used to find the number of different samples of size n that can be selected.

Exercises

Methods

1. An experiment has three steps with three outcomes possible for the first step, two outcomes possible for the second step, and four outcomes possible for the third step. How many experimental outcomes exist for the entire experiment?

SELF test

2. How many ways can three items be selected from a group of six items? Use the letters A, B, C, D, E, and F to identify the items, and list each of the different combinations of three items.
3. How many permutations of three items can be selected from a group of six? Use the letters A, B, C, D, E, and F to identify the items, and list each of the permutations of items B, D, and F.
4. Consider the experiment of tossing a coin three times.
   a. Develop a tree diagram for the experiment.
   b. List the experimental outcomes.
   c. What is the probability for each experimental outcome?
5. Suppose an experiment has five equally likely outcomes: E_1, E_2, E_3, E_4, E_5. Assign probabilities to each outcome and show that the requirements in equations (4.3) and (4.4) are satisfied. What method did you use?

SELF test

6. An experiment with three outcomes has been repeated 50 times, and it was learned that E_1 occurred 20 times, E_2 occurred 13 times, and E_3 occurred 17 times. Assign probabilities to the outcomes. What method did you use?
7. A decision maker subjectively assigned the following probabilities to the four outcomes of an experiment: $P(E_1) = .10$, $P(E_2) = .15$, $P(E_3) = .40$, and $P(E_4) = .20$. Are these probability assignments valid? Explain.

Applications

8. In the city of Milford, applications for zoning changes go through a two-step process: a review by the planning commission and a final decision by the city council. At step 1 the planning commission reviews the zoning change request and makes a positive or negative recommendation concerning the change. At step 2 the city council reviews the planning commission's recommendation and then votes to approve or to disapprove the zoning change. Suppose the developer of an apartment complex submits an application for a zoning change. Consider the application process as an experiment.
   a. How many sample points are there for this experiment? List the sample points.
   b. Construct a tree diagram for the experiment.

SELF test

9. Simple random sampling uses a sample of size n from a population of size N to obtain data that can be used to make inferences about the characteristics of a population. Suppose that, from a population of 50 bank accounts, we want to take a random sample of four accounts in order to learn about the population. How many different random samples of four accounts are possible?

SELF test

10. Many students accumulate debt by the time they graduate from college. Shown in the following table is the percentage of graduates with debt and the average amount of debt for these graduates at four universities and four liberal arts colleges (*U.S. News and World Report, America's Best Colleges,* 2008).

University	% with Debt	Amount($)	College	% with Debt	Amount($)
Pace	72	32,980	Wartburg	83	28,758
Iowa State	69	32,130	Morehouse	94	27,000
Massachusetts	55	11,227	Wellesley	55	10,206
SUNY—Albany	64	11,856	Wofford	49	11,012

   a. If you randomly choose a graduate of Morehouse College, what is the probability that this individual graduated with debt?
   b. If you randomly choose one of these eight institutions for a follow-up study on student loans, what is the probability that you will choose an institution with more than 60% of its graduates having debt?
   c. If you randomly choose one of these eight institutions for a follow-up study on student loans, what is the probability that you will choose an institution whose graduates with debts have an average debt of more than $30,000?
   d. What is the probability that a graduate of Pace University does not have debt?
   e. For graduates of Pace University with debt, the average amount of debt is $32,980. Considering all graduates from Pace University, what is the average debt per graduate?

11. The National Occupant Protection Use Survey (NOPUS) was conducted to provide probability-based data on motorcycle helmet use in the United States. The survey was conducted by sending observers to randomly selected roadway sites where they collected data on motorcycle helmet use, including the number of motorcyclists wearing a Department of Transportation (DOT)-compliant helmet (National Highway Traffic Safety Administration website, January 7, 2010). Sample data consistent with the most recent NOPUS are shown below.

	Type of Helmet	
Region	DOT-Compliant	Noncompliant
Northeast	96	62
Midwest	86	43
South	92	49
West	76	16
Total	350	170

a. Use the sample data to compute an estimate of the probability that a motorcyclist wears a DOT-compliant helmet.
b. The probability that a motorcyclist wore a DOT-compliant helmet five years ago was .48, and last year this probability was .63. Would the National Highway Traffic Safety Administration be pleased with the most recent survey results?
c. What is the probability of DOT-compliant helmet use by region of the country? What region has the highest probability of DOT-compliant helmet use?

12. The Powerball lottery is played twice each week in 28 states, the Virgin Islands, and the District of Columbia. To play Powerball a participant must purchase a ticket and then select five numbers from the digits 1 through 55 and a Powerball number from the digits 1 through 42. To determine the winning numbers for each game, lottery officials draw five white balls out of a drum with 55 white balls, and one red ball out of a drum with 42 red balls. To win the jackpot, a participant's numbers must match the numbers on the five white balls in any order and the number on the red Powerball. Eight coworkers at the ConAgra Foods plant in Lincoln, Nebraska, claimed the record $365 million jackpot on February 18, 2006, by matching the numbers 15-17-43-44-49 and the Powerball number 29. A variety of other cash prizes are awarded each time the game is played. For instance, a prize of $200,000 is paid if the participant's five numbers match the numbers on the five white balls (Powerball website, March 19, 2006).
    a. Compute the number of ways the first five numbers can be selected.
    b. What is the probability of winning a prize of $200,000 by matching the numbers on the five white balls?
    c. What is the probability of winning the Powerball jackpot?

13. A company that manufactures toothpaste is studying five different package designs. Assuming that one design is just as likely to be selected by a consumer as any other design, what selection probability would you assign to each of the package designs? In an actual experiment, 100 consumers were asked to pick the design they preferred. The following data were obtained. Do the data confirm the belief that one design is just as likely to be selected as another? Explain.

Design	Number of Times Preferred
1	5
2	15
3	30
4	40
5	10

4.2 Events and Their Probabilities

In the introduction to this chapter we used the term *event* much as it would be used in everyday language. Then, in Section 4.1 we introduced the concept of an experiment and its associated experimental outcomes or sample points. Sample points and events provide the foundation for the study of probability. As a result, we must now introduce the formal definition of an **event** as it relates to sample points. Doing so will provide the basis for determining the probability of an event.

EVENT

An event is a collection of sample points.

For an example, let us return to the KP&L project and assume that the project manager is interested in the event that the entire project can be completed in 10 months or less. Referring to Table 4.3, we see that six sample points—(2, 6), (2, 7), (2, 8), (3, 6), (3, 7), and (4, 6)—provide a project completion time of 10 months or less. Let C denote the event that the project is completed in 10 months or less; we write

$$C = \{(2, 6), (2, 7), (2, 8), (3, 6), (3, 7), (4, 6)\}$$

Event C is said to occur if *any one* of these six sample points appears as the experimental outcome.

Other events that might be of interest to KP&L management include the following.

$$L = \text{The event that the project is completed in } \textit{less} \text{ than 10 months}$$
$$M = \text{The event that the project is completed in } \textit{more} \text{ than 10 months}$$

Using the information in Table 4.3, we see that these events consist of the following sample points.

$$L = \{(2, 6), (2, 7), (3, 6)\}$$
$$M = \{(3, 8), (4, 7), (4, 8)\}$$

A variety of additional events can be defined for the KP&L project, but in each case the event must be identified as a collection of sample points for the experiment.

Given the probabilities of the sample points shown in Table 4.3, we can use the following definition to compute the probability of any event that KP&L management might want to consider.

PROBABILITY OF AN EVENT

The probability of any event is equal to the sum of the probabilities of the sample points in the event.

Using this definition, we calculate the probability of a particular event by adding the probabilities of the sample points (experimental outcomes) that make up the event. We can now compute the probability that the project will take 10 months or less to complete. Because this event is given by $C = \{(2, 6), (2, 7), (2, 8), (3, 6), (3, 7), (4, 6)\}$, the probability of event C, denoted $P(C)$, is given by

$$P(C) = P(2, 6) + P(2, 7) + P(2, 8) + P(3, 6) + P(3, 7) + P(4, 6)$$

Refer to the sample point probabilities in Table 4.3; we have

$$P(C) = .15 + .15 + .05 + .10 + .20 + .05 = .70$$

Similarly, because the event that the project is completed in less than 10 months is given by $L = \{(2, 6), (2, 7), (3, 6)\}$, the probability of this event is given by

$$P(L) = P(2, 6) + P(2, 7) + P(3, 6)$$
$$= .15 + .15 + .10 = .40$$

Finally, for the event that the project is completed in more than 10 months, we have $M = \{(3, 8), (4, 7), (4, 8)\}$ and thus

$$P(M) = P(3, 8) + P(4, 7) + P(4, 8)$$
$$= .05 + .10 + .15 = .30$$

Using these probability results, we can now tell KP&L management that there is a .70 probability that the project will be completed in 10 months or less, a .40 probability that the project will be completed in less than 10 months, and a .30 probability that the project will be completed in more than 10 months. This procedure of computing event probabilities can be repeated for any event of interest to the KP&L management.

Any time that we can identify all the sample points of an experiment and assign probabilities to each, we can compute the probability of an event using the definition. However, in many experiments the large number of sample points makes the identification of the sample points, as well as the determination of their associated probabilities, extremely cumbersome, if not impossible. In the remaining sections of this chapter, we present some basic probability relationships that can be used to compute the probability of an event without knowledge of all the sample point probabilities.

NOTES AND COMMENTS

1. The sample space, S, is an event. Because it contains all the experimental outcomes, it has a probability of 1; that is, $P(S) = 1$.
2. When the classical method is used to assign probabilities, the assumption is that the experimental outcomes are equally likely. In such cases, the probability of an event can be computed by counting the number of experimental outcomes in the event and dividing the result by the total number of experimental outcomes.

Exercises

Methods

14. An experiment has four equally likely outcomes: E_1, E_2, E_3, and E_4.
    a. What is the probability that E_2 occurs?
    b. What is the probability that any two of the outcomes occur (e.g., E_1 or E_3)?
    c. What is the probability that any three of the outcomes occur (e.g., E_1 or E_2 or E_4)?

SELF test

15. Consider the experiment of selecting a playing card from a deck of 52 playing cards. Each card corresponds to a sample point with a 1/52 probability.
    a. List the sample points in the event an ace is selected.
    b. List the sample points in the event a club is selected.
    c. List the sample points in the event a face card (jack, queen, or king) is selected.
    d. Find the probabilities associated with each of the events in parts (a), (b), and (c).

16. Consider the experiment of rolling a pair of dice. Suppose that we are interested in the sum of the face values showing on the dice.
    a. How many sample points are possible? (*Hint:* Use the counting rule for multiple-step experiments.)
    b. List the sample points.
    c. What is the probability of obtaining a value of 7?
    d. What is the probability of obtaining a value of 9 or greater?
    e. Because each roll has six possible even values (2, 4, 6, 8, 10, and 12) and only five possible odd values (3, 5, 7, 9, and 11), the dice should show even values more often than odd values. Do you agree with this statement? Explain.
    f. What method did you use to assign the probabilities requested?

Applications

SELF test

17. Refer to the KP&L sample points and sample point probabilities in Tables 4.2 and 4.3.
    a. The design stage (stage 1) will run over budget if it takes 4 months to complete. List the sample points in the event the design stage is over budget.
    b. What is the probability that the design stage is over budget?
    c. The construction stage (stage 2) will run over budget if it takes 8 months to complete. List the sample points in the event the construction stage is over budget.
    d. What is the probability that the construction stage is over budget?
    e. What is the probability that both stages are over budget?

18. To investigate how often families eat at home, Harris Interactive surveyed 496 adults living with children under the age of 18 (*USA Today,* January 3, 2007). The survey results are shown in the following table.

Number of Family Meals per Week	Number of Survey Responses
0	11
1	11
2	30
3	36
4	36
5	119
6	114
7 or more	139

For a randomly selected family with children under the age of 18, compute the following.
    a. The probability the family eats no meals at home during the week.
    b. The probability the family eats at least four meals at home during the week.
    c. The probability the family eats two or fewer meals at home during the week.

19. Do you think the government protects investors adequately? This question was part of an online survey of investors under age 65 living in the United States and Great Britain (*Financial Times*/Harris Poll, October 1, 2009). The numbers of investors from the United States and Great Britain who answered Yes, No, or Unsure to this question are provided below.

Response	United States	Great Britain
Yes	187	197
No	334	411
Unsure	256	213

    a. Estimate the probability that an investor in the United States thinks the government is not protecting investors adequately.
    b. Estimate the probability that an investor in Great Britain thinks the government is not protecting investors adequately or is unsure the government is protecting investors adequately.
    c. For a randomly selected investor from these two countries, estimate the probability that the investor thinks the government is not protecting investors adequately.
    d. Based on the survey results, does there appear to be much difference between the perceptions of investors in the United States and investors in Great Britain regarding the issue of the government protecting investors adequately?

20. *Fortune* magazine publishes an annual list of the 500 largest companies in the United States. The following data show the five states with the largest number of *Fortune* 500 companies (*The New York Times Almanac,* 2006).

State	Number of Companies
New York	54
California	52
Texas	48
Illinois	33
Ohio	30

Suppose a *Fortune* 500 company is chosen for a follow-up questionnaire. What are the probabilities of the following events?

a. Let N be the event the company is headquartered in New York. Find $P(N)$.
b. Let T be the event the company is headquartered in Texas. Find $P(T)$.
c. Let B be the event the company is headquartered in one of these five states. Find $P(B)$.

21. Data on U.S. work-related fatalities by cause follow (*The World Almanac,* 2012).

Cause of Fatality	Number of Fatalities
Transportation incidents	1795
Assaults and violent acts	837
Contact with objects and equipment	741
Falls	645
Exposure to harmful substances or environments	404
Fires and explosions	113

Assume that a fatality will be randomly chosen from this population.

a. What is the probability the fatality resulted from a fall?
b. What is the probability the fatality resulted from a transportation incident?
c. What cause of fatality is least likely to occur? What is the probability the fatality resulted from this cause?

4.3 Some Basic Relationships of Probability

Complement of an Event

Given an event A, the **complement of** A is defined to be the event consisting of all sample points that are *not* in A. The complement of A is denoted by A^c. Figure 4.4 is a diagram, known as a **Venn diagram**, which illustrates the concept of a complement. The rectangular area represents the sample space for the experiment and as such contains all possible sample points. The circle represents event A and contains only the sample points that belong to A. The shaded region of the rectangle contains all sample points not in event A and is by definition the complement of A.

In any probability application, either event A or its complement A^c must occur. Therefore, we have

$$P(A) + P(A^c) = 1$$

FIGURE 4.4 COMPLEMENT OF EVENT A IS SHADED

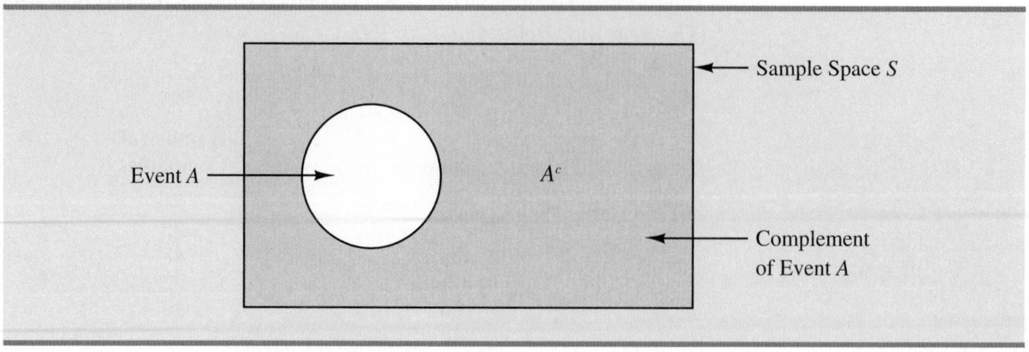

Solving for $P(A)$, we obtain the following result.

COMPUTING PROBABILITY USING THE COMPLEMENT

$$P(A) = 1 - P(A^c) \tag{4.5}$$

Equation (4.5) shows that the probability of an event A can be computed easily if the probability of its complement, $P(A^c)$, is known.

As an example, consider the case of a sales manager who, after reviewing sales reports, states that 80% of new customer contacts result in no sale. By allowing A to denote the event of a sale and A^c to denote the event of no sale, the manager is stating that $P(A^c) = .80$. Using equation (4.5), we see that

$$P(A) = 1 - P(A^c) = 1 - .80 = .20$$

We can conclude that a new customer contact has a .20 probability of resulting in a sale.

In another example, a purchasing agent states a .90 probability that a supplier will send a shipment that is free of defective parts. Using the complement, we can conclude that there is a $1 - .90 = .10$ probability that the shipment will contain defective parts.

Addition Law

The addition law is helpful when we are interested in knowing the probability that at least one of two events occurs. That is, with events A and B we are interested in knowing the probability that event A or event B or both occur.

Before we present the addition law, we need to discuss two concepts related to the combination of events: the *union* of events and the *intersection* of events. Given two events A and B, the **union of A and B** is defined as follows.

UNION OF TWO EVENTS

The *union* of A and B is the event containing *all* sample points belonging to A *or* B *or both.* The union is denoted by $A \cup B$.

The Venn diagram in Figure 4.5 depicts the union of events A and B. Note that the two circles contain all the sample points in event A as well as all the sample points in event B.

FIGURE 4.5 UNION OF EVENTS *A* AND *B* IS SHADED

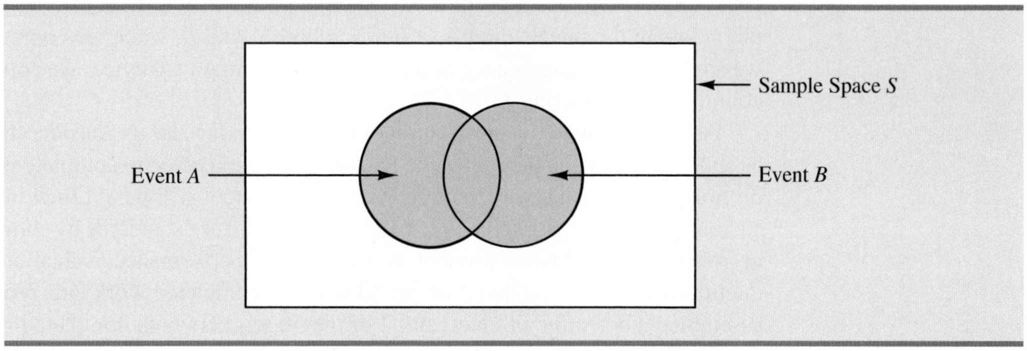

The fact that the circles overlap indicates that some sample points are contained in both *A* and *B*.

The definition of the **intersection of *A* and *B*** follows.

INTERSECTION OF TWO EVENTS

Given two events *A* and *B*, the *intersection* of *A* and *B* is the event containing the sample points belonging to *both A and B*. The intersection is denoted by $A \cap B$.

The Venn diagram depicting the intersection of events *A* and *B* is shown in Figure 4.6. The area where the two circles overlap is the intersection; it contains the sample points that are in both *A* and *B*.

Let us now continue with a discussion of the addition law. The **addition law** provides a way to compute the probability that event *A* or event *B* or both occur. In other words, the addition law is used to compute the probability of the union of two events. The addition law is written as follows.

ADDITION LAW

$$P(A \cup B) = P(A) + P(B) - P(A \cap B) \qquad \textbf{(4.6)}$$

FIGURE 4.6 INTERSECTION OF EVENTS *A* AND *B* IS SHADED

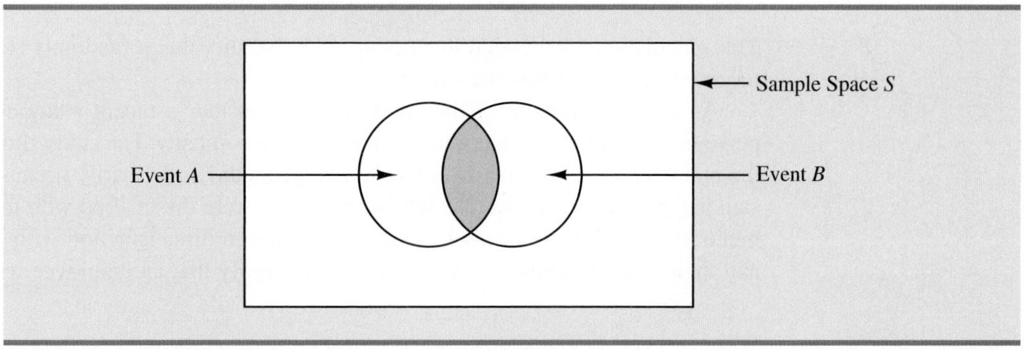

To understand the addition law intuitively, note that the first two terms in the addition law, $P(A) + P(B)$, account for all the sample points in $A \cup B$. However, because the sample points in the intersection $A \cap B$ are in both A and B, when we compute $P(A) + P(B)$, we are in effect counting each of the sample points in $A \cap B$ twice. We correct for this overcounting by subtracting $P(A \cap B)$.

As an example of an application of the addition law, let us consider the case of a small assembly plant with 50 employees. Each worker is expected to complete work assignments on time and in such a way that the assembled product will pass a final inspection. On occasion, some of the workers fail to meet the performance standards by completing work late or assembling a defective product. At the end of a performance evaluation period, the production manager found that 5 of the 50 workers completed work late, 6 of the 50 workers assembled a defective product, and 2 of the 50 workers both completed work late *and* assembled a defective product.

Let

$$L = \text{the event that the work is completed late}$$
$$D = \text{the event that the assembled product is defective}$$

The relative frequency information leads to the following probabilities.

$$P(L) = \frac{5}{50} = .10$$
$$P(D) = \frac{6}{50} = .12$$
$$P(L \cap D) = \frac{2}{50} = .04$$

After reviewing the performance data, the production manager decided to assign a poor performance rating to any employee whose work was either late or defective; thus the event of interest is $L \cup D$. What is the probability that the production manager assigned an employee a poor performance rating?

Note that the probability question is about the union of two events. Specifically, we want to know $P(L \cup D)$. Using equation (4.6), we have

$$P(L \cup D) = P(L) + P(D) - P(L \cap D)$$

Knowing values for the three probabilities on the right side of this expression, we can write

$$P(L \cup D) = .10 + .12 - .04 = .18$$

This calculation tells us that there is a .18 probability that a randomly selected employee received a poor performance rating.

As another example of the addition law, consider a recent study conducted by the personnel manager of a major computer software company. The study showed that 30% of the employees who left the firm within two years did so primarily because they were dissatisfied with their salary, 20% left because they were dissatisfied with their work assignments, and 12% of the former employees indicated dissatisfaction with *both* their salary and their work assignments. What is the probability that an employee who leaves within two years does so because of dissatisfaction with salary, dissatisfaction with the work assignment, or both?

Let

S = the event that the employee leaves because of salary

W = the event that the employee leaves because of work assignment

We have $P(S) = .30$, $P(W) = .20$, and $P(S \cap W) = .12$. Using equation (4.6), the addition law, we have

$$P(S \cup W) = P(S) + P(W) - P(S \cap W) = .30 + .20 - .12 = .38.$$

We find a .38 probability that an employee leaves for salary or work assignment reasons.

Before we conclude our discussion of the addition law, let us consider a special case that arises for **mutually exclusive events**.

MUTUALLY EXCLUSIVE EVENTS

Two events are said to be mutually exclusive if the events have no sample points in common.

Events A and B are mutually exclusive if, when one event occurs, the other cannot occur. Thus, a requirement for A and B to be mutually exclusive is that their intersection must contain no sample points. The Venn diagram depicting two mutually exclusive events A and B is shown in Figure 4.7. In this case $P(A \cap B) = 0$ and the addition law can be written as follows.

ADDITION LAW FOR MUTUALLY EXCLUSIVE EVENTS

$$P(A \cup B) = P(A) + P(B)$$

FIGURE 4.7 MUTUALLY EXCLUSIVE EVENTS

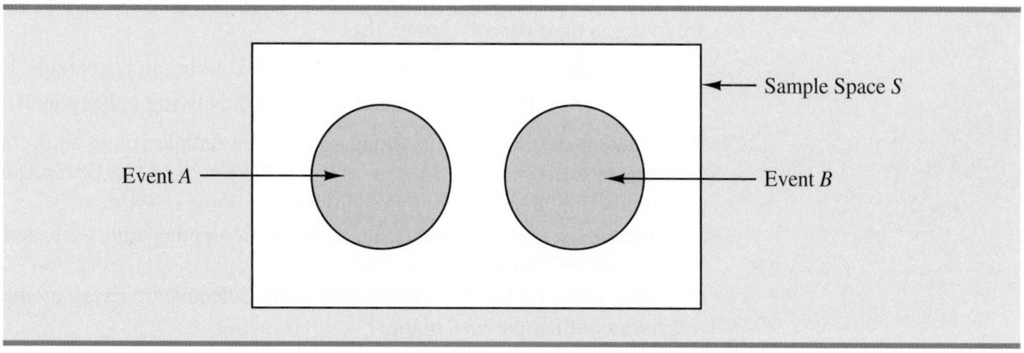

Exercises

Methods

22. Suppose that we have a sample space with five equally likely experimental outcomes: E_1, E_2, E_3, E_4, E_5. Let

$$A = \{E_1, E_2\}$$
$$B = \{E_3, E_4\}$$
$$C = \{E_2, E_3, E_5\}$$

a. Find $P(A)$, $P(B)$, and $P(C)$.
b. Find $P(A \cup B)$. Are A and B mutually exclusive?
c. Find A^c, C^c, $P(A^c)$, and $P(C^c)$.
d. Find $A \cup B^c$ and $P(A \cup B^c)$.
e. Find $P(B \cup C)$.

SELF test

23. Suppose that we have a sample space $S = \{E_1, E_2, E_3, E_4, E_5, E_6, E_7\}$, where $E_1, E_2, \ldots, E_7$ denote the sample points. The following probability assignments apply: $P(E_1) = .05$, $P(E_2) = .20$, $P(E_3) = .20$, $P(E_4) = .25$, $P(E_5) = .15$, $P(E_6) = .10$, and $P(E_7) = .05$. Let

$$A = \{E_1, E_4, E_6\}$$
$$B = \{E_2, E_4, E_7\}$$
$$C = \{E_2, E_3, E_5, E_7\}$$

a. Find $P(A)$, $P(B)$, and $P(C)$.
b. Find $A \cup B$ and $P(A \cup B)$.
c. Find $A \cap B$ and $P(A \cap B)$.
d. Are events A and C mutually exclusive?
e. Find B^c and $P(B^c)$.

Applications

24. Clarkson University surveyed alumni to learn more about what they think of Clarkson. One part of the survey asked respondents to indicate whether their overall experience at Clarkson fell short of expectations, met expectations, or surpassed expectations. The results showed that 4% of the respondents did not provide a response, 26% said that their experience fell short of expectations, and 65% of the respondents said that their experience met expectations.
   a. If we chose an alumnus at random, what is the probability that the alumnus would say their experience *surpassed* expectations?
   b. If we chose an alumnus at random, what is the probability that the alumnus would say their experience met or surpassed expectations?

25. The U.S. Census Bureau provides data on the number of young adults, ages 18–24, who are living in their parents' home.[1] Let

M = the event a male young adult is living in his parents' home
F = the event a female young adult is living in her parents' home

If we randomly select a male young adult and a female young adult, the Census Bureau data enable us to conclude $P(M) = .56$ and $P(F) = .42$ (*The World Almanac,* 2006). The probability that both are living in their parents' home is .24.
   a. What is the probability at least one of the two young adults selected is living in his or her parents' home?
   b. What is the probability both young adults selected are living on their own (neither is living in their parents' home)?

[1]The data include single young adults who are living in college dormitories because it is assumed these young adults will return to their parents' home when school is not in session.

26. Information about mutual funds provided by Morningstar Investment Research includes the type of mutual fund (Domestic Equity, International Equity, or Fixed Income) and the Morningstar rating for the fund. The rating is expressed from 1-star (lowest rating) to 5-star (highest rating). A sample of 25 mutual funds was selected from *Morningstar Funds 500* (2008). The following counts were obtained:
    - Sixteen mutual funds were Domestic Equity funds.
    - Thirteen mutual funds were rated 3-star or less.
    - Seven of the Domestic Equity funds were rated 4-star.
    - Two of the Domestic Equity funds were rated 5-star.

    Assume that one of these 25 mutual funds will be randomly selected in order to learn more about the mutual fund and its investment strategy.
    a. What is the probability of selecting a Domestic Equity fund?
    b. What is the probability of selecting a fund with a 4-star or 5-star rating?
    c. What is the probability of selecting a fund that is both a Domestic Equity fund *and* a fund with a 4-star or 5-star rating?
    d. What is the probability of selecting a fund that is a Domestic Equity fund *or* a fund with a 4-star or 5-star rating?

27. What NCAA college basketball conferences have the higher probability of having a team play in college basketball's national championship game? Over the last 20 years, the Atlantic Coast Conference (ACC) ranks first by having a team in the championship game 10 times. The Southeastern Conference (SEC) ranks second by having a team in the championship game 8 times. However, these two conferences have both had teams in the championship game only one time, when Arkansas (SEC) beat Duke (ACC) 76–70 in 1994 (NCAA website, April 2009). Use these data to estimate the following probabilities.
    a. What is the probability the ACC will have a team in the championship game?
    b. What is the probability the SEC will have team in the championship game?
    c. What is the probability the ACC and SEC will both have teams in the championship game?
    d. What is the probability at least one team from these two conferences will be in the championship game? That is, what is the probability a team from the ACC or SEC will play in the championship game?
    e. What is the probability that the championship game will not a have team from one of these two conferences?

SELF test

28. A survey of magazine subscribers showed that 45.8% rented a car during the past 12 months for business reasons, 54% rented a car during the past 12 months for personal reasons, and 30% rented a car during the past 12 months for both business and personal reasons.
    a. What is the probability that a subscriber rented a car during the past 12 months for business or personal reasons?
    b. What is the probability that a subscriber did not rent a car during the past 12 months for either business or personal reasons?

29. High school seniors with strong academic records apply to the nation's most selective colleges in greater numbers each year. Because the number of slots remains relatively stable, some colleges reject more early applicants. Suppose that for a recent admissions class, an Ivy League college received 2851 applications for early admission. Of this group, it admitted 1033 students early, rejected 854 outright, and deferred 964 to the regular admission pool for further consideration. In the past, this school has admitted 18% of the deferred early admission applicants during the regular admission process. Counting the students admitted early and the students admitted during the regular admission process, the total class size was 2375. Let E, R, and D represent the events that a student who applies for early admission is admitted early, rejected outright, or deferred to the regular admissions pool.
    a. Use the data to estimate $P(E)$, $P(R)$, and $P(D)$.
    b. Are events E and D mutually exclusive? Find $P(E \cap D)$.

c. For the 2375 students who were admitted, what is the probability that a randomly selected student was accepted during early admission?
d. Suppose a student applies for early admission. What is the probability that the student will be admitted for early admission or be deferred and later admitted during the regular admission process?

4.4 Conditional Probability

Often, the probability of an event is influenced by whether a related event already occurred. Suppose we have an event A with probability $P(A)$. If we obtain new information and learn that a related event, denoted by B, already occurred, we will want to take advantage of this information by calculating a new probability for event A. This new probability of event A is called a **conditional probability** and is written $P(A \mid B)$. We use the notation $\mid$ to indicate that we are considering the probability of event A *given* the condition that event B has occurred. Hence, the notation $P(A \mid B)$ reads "the probability of A given B."

As an illustration of the application of conditional probability, consider the situation of the promotion status of male and female officers of a major metropolitan police force in the eastern United States. The police force consists of 1200 officers, 960 men and 240 women. Over the past two years, 324 officers on the police force received promotions. The specific breakdown of promotions for male and female officers is shown in Table 4.4.

After reviewing the promotion record, a committee of female officers raised a discrimination case on the basis that 288 male officers had received promotions, but only 36 female officers had received promotions. The police administration argued that the relatively low number of promotions for female officers was due not to discrimination, but to the fact that relatively few females are members of the police force. Let us show how conditional probability could be used to analyze the discrimination charge.

Let

$$\begin{aligned} M &= \text{event an officer is a man} \\ W &= \text{event an officer is a woman} \\ A &= \text{event an officer is promoted} \\ A^c &= \text{event an officer is not promoted} \end{aligned}$$

Dividing the data values in Table 4.4 by the total of 1200 officers enables us to summarize the available information with the following probability values.

$P(M \cap A) = 288/1200 = .24$ probability that a randomly selected officer is a man *and* is promoted

$P(M \cap A^c) = 672/1200 = .56$ probability that a randomly selected officer is a man *and* is not promoted

TABLE 4.4 PROMOTION STATUS OF POLICE OFFICERS OVER THE PAST TWO YEARS

	Men	Women	Total
Promoted	288	36	324
Not Promoted	672	204	876
Total	960	240	1200

TABLE 4.5 JOINT PROBABILITY TABLE FOR PROMOTIONS

Joint probabilities appear in the body of the table.

	Men (M)	Women (W)	Total
Promoted (A)	.24	.03	.27
Not Promoted (A^c)	.56	.17	.73
Total	.80	.20	1.00

Marginal probabilities appear in the margins of the table.

$P(W \cap A) = 36/1200 = .03$ probability that a randomly selected officer is a woman *and* is promoted

$P(W \cap A^c) = 204/1200 = .17$ probability that a randomly selected officer is a woman *and* is not promoted

Because each of these values gives the probability of the intersection of two events, the probabilities are called **joint probabilities**. Table 4.5, which provides a summary of the probability information for the police officer promotion situation, is referred to as a *joint probability table.*

The values in the margins of the joint probability table provide the probabilities of each event separately. That is, $P(M) = .80$, $P(W) = .20$, $P(A) = .27$, and $P(A^c) = .73$. These probabilities are referred to as **marginal probabilities** because of their location in the margins of the joint probability table. We note that the marginal probabilities are found by summing the joint probabilities in the corresponding row or column of the joint probability table. For instance, the marginal probability of being promoted is $P(A) = P(M \cap A) + P(W \cap A) = .24 + .03 = .27$. From the marginal probabilities, we see that 80% of the force is male, 20% of the force is female, 27% of all officers received promotions, and 73% were not promoted.

Let us begin the conditional probability analysis by computing the probability that an officer is promoted given that the officer is a man. In conditional probability notation, we are attempting to determine $P(A \mid M)$. To calculate $P(A \mid M)$, we first realize that this notation simply means that we are considering the probability of the event A (promotion) given that the condition designated as event M (the officer is a man) is known to exist. Thus $P(A \mid M)$ tells us that we are now concerned only with the promotion status of the 960 male officers. Because 288 of the 960 male officers received promotions, the probability of being promoted given that the officer is a man is $288/960 = .30$. In other words, given that an officer is a man, that officer had a 30% chance of receiving a promotion over the past two years.

This procedure was easy to apply because the values in Table 4.4 show the number of officers in each category. We now want to demonstrate how conditional probabilities such as $P(A \mid M)$ can be computed directly from related event probabilities rather than the frequency data of Table 4.4.

We have shown that $P(A \mid M) = 288/960 = .30$. Let us now divide both the numerator and denominator of this fraction by 1200, the total number of officers in the study.

$$P(A \mid M) = \frac{288}{960} = \frac{288/1200}{960/1200} = \frac{.24}{.80} = .30$$

We now see that the conditional probability $P(A \mid M)$ can be computed as .24/.80. Refer to the joint probability table (Table 4.5). Note in particular that .24 is the joint probability of

A and M; that is, $P(A \cap M) = .24$. Also note that .80 is the marginal probability that a randomly selected officer is a man; that is, $P(M) = .80$. Thus, the conditional probability $P(A \mid M)$ can be computed as the ratio of the joint probability $P(A \cap M)$ to the marginal probability $P(M)$.

$$P(A \mid M) = \frac{P(A \cap M)}{P(M)} = \frac{.24}{.80} = .30$$

The fact that conditional probabilities can be computed as the ratio of a joint probability to a marginal probability provides the following general formula for conditional probability calculations for two events A and B.

CONDITIONAL PROBABILITY

$$P(A \mid B) = \frac{P(A \cap B)}{P(B)} \tag{4.7}$$

or

$$P(B \mid A) = \frac{P(A \cap B)}{P(A)} \tag{4.8}$$

The Venn diagram in Figure 4.8 is helpful in obtaining an intuitive understanding of conditional probability. The circle on the right shows that event B has occurred; the portion of the circle that overlaps with event A denotes the event $(A \cap B)$. We know that once event B has occurred, the only way that we can also observe event A is for the event $(A \cap B)$ to occur. Thus, the ratio $P(A \cap B)/P(B)$ provides the conditional probability that we will observe event A given that event B has already occurred.

Let us return to the issue of discrimination against the female officers. The marginal probability in row 1 of Table 4.5 shows that the probability of promotion of an officer is $P(A) = .27$ (regardless of whether that officer is male or female). However, the critical issue in the discrimination case involves the two conditional probabilities $P(A \mid M)$ and $P(A \mid W)$. That is, what is the probability of a promotion *given* that the officer is a man, and what is the probability of a promotion *given* that the officer is a woman? If these two probabilities are equal, a discrimination argument has no basis because the chances of a promotion are the same for male and female officers. However, a difference in the two conditional probabilities will support the position that male and female officers are treated differently in promotion decisions.

FIGURE 4.8 CONDITIONAL PROBABILITY $P(A \mid B) = P(A \cap B)/P(B)$

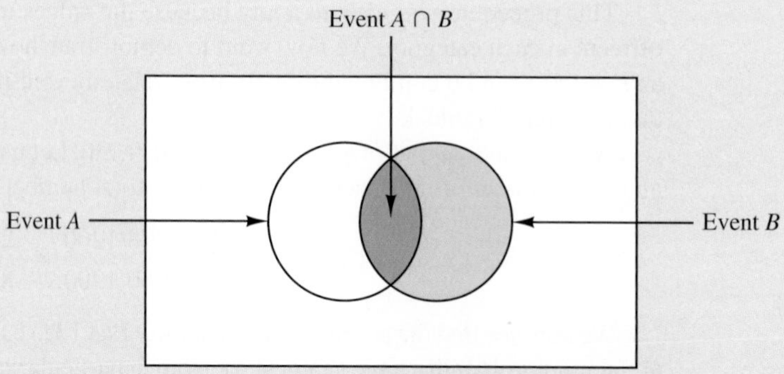

We already determined that $P(A \mid M) = .30$. Let us now use the probability values in Table 4.5 and the basic relationship of conditional probability in equation (4.7) to compute the probability that an officer is promoted given that the officer is a woman; that is, $P(A \mid W)$. Using equation (4.7), with W replacing B, we obtain

$$P(A \mid W) = \frac{P(A \cap W)}{P(W)} = \frac{.03}{.20} = .15$$

What conclusion do you draw? The probability of a promotion given that the officer is a man is .30, twice the .15 probability of a promotion given that the officer is a woman. Although the use of conditional probability does not in itself prove that discrimination exists in this case, the conditional probability values support the argument presented by the female officers.

Independent Events

In the preceding illustration, $P(A) = .27$, $P(A \mid M) = .30$, and $P(A \mid W) = .15$. We see that the probability of a promotion (event A) is affected or influenced by whether the officer is a man or a woman. Particularly, because $P(A \mid M) \neq P(A)$, we would say that events A and M are dependent events. That is, the probability of event A (promotion) is altered or affected by knowing that event M (the officer is a man) exists. Similarly, with $P(A \mid W) \neq P(A)$, we would say that events A and W are *dependent events.* However, if the probability of event A is not changed by the existence of event M—that is, $P(A \mid M) = P(A)$—we would say that events A and M are **independent events**. This situation leads to the following definition of the independence of two events.

INDEPENDENT EVENTS

Two events A and B are independent if

$$P(A \mid B) = P(A) \quad \textbf{(4.9)}$$

or

$$P(B \mid A) = P(B) \quad \textbf{(4.10)}$$

Otherwise, the events are dependent.

Multiplication Law

Whereas the addition law of probability is used to compute the probability of a union of two events, the multiplication law is used to compute the probability of the intersection of two events. The multiplication law is based on the definition of conditional probability. Using equations (4.7) and (4.8) and solving for $P(A \cap B)$, we obtain the **multiplication law**.

MULTIPLICATION LAW

$$P(A \cap B) = P(B)P(A \mid B) \quad \textbf{(4.11)}$$

or

$$P(A \cap B) = P(A)P(B \mid A) \quad \textbf{(4.12)}$$

To illustrate the use of the multiplication law, consider a newspaper circulation department where it is known that 84% of the households in a particular neighborhood subscribe to the daily edition of the paper. If we let D denote the event that a household subscribes to the daily edition, $P(D) = .84$. In addition, it is known that the probability that a household that already holds a daily subscription also subscribes to the Sunday edition (event S) is .75; that is, $P(S \mid D) = .75$. What is the probability that a household subscribes to both the Sunday and daily editions of the newspaper? Using the multiplication law, we compute the desired $P(S \cap D)$ as

$$P(S \cap D) = P(D)P(S \mid D) = .84(.75) = .63$$

We now know that 63% of the households subscribe to both the Sunday and daily editions.

Before concluding this section, let us consider the special case of the multiplication law when the events involved are independent. Recall that events A and B are independent whenever $P(A \mid B) = P(A)$ or $P(B \mid A) = P(B)$. Hence, using equations (4.11) and (4.12) for the special case of independent events, we obtain the following multiplication law.

MULTIPLICATION LAW FOR INDEPENDENT EVENTS

$$P(A \cap B) = P(A)P(B) \tag{4.13}$$

To compute the probability of the intersection of two independent events, we simply multiply the corresponding probabilities. Note that the multiplication law for independent events provides another way to determine whether A and B are independent. That is, if $P(A \cap B) = P(A)P(B)$, then A and B are independent; if $P(A \cap B) \neq P(A)P(B)$, then A and B are dependent.

As an application of the multiplication law for independent events, consider the situation of a service station manager who knows from past experience that 80% of the customers use a credit card when they purchase gasoline. What is the probability that the next two customers purchasing gasoline will each use a credit card? If we let

A = the event that the first customer uses a credit card

B = the event that the second customer uses a credit card

then the event of interest is $A \cap B$. Given no other information, we can reasonably assume that A and B are independent events. Thus,

$$P(A \cap B) = P(A)P(B) = (.80)(.80) = .64$$

To summarize this section, we note that our interest in conditional probability is motivated by the fact that events are often related. In such cases, we say the events are dependent and the conditional probability formulas in equations (4.7) and (4.8) must be used to compute the event probabilities. If two events are not related, they are independent; in this case neither event's probability is affected by whether the other event occurred.

NOTES AND COMMENTS

Do not confuse the notion of mutually exclusive events with that of independent events. Two events with nonzero probabilities cannot be both mutually exclusive and independent. If one mutually exclusive event is known to occur, the other cannot occur; thus, the probability of the other event occurring is reduced to zero. They are therefore dependent.

Exercises

Methods

30. Suppose that we have two events, A and B, with $P(A) = .50$, $P(B) = .60$, and $P(A \cap B) = .40$.

SELF test

   a. Find $P(A \mid B)$.
   b. Find $P(B \mid A)$.
   c. Are A and B independent? Why or why not?

31. Assume that we have two events, A and B, that are mutually exclusive. Assume further that we know $P(A) = .30$ and $P(B) = .40$.
   a. What is $P(A \cap B)$?
   b. What is $P(A \mid B)$?
   c. A student in statistics argues that the concepts of mutually exclusive events and independent events are really the same, and that if events are mutually exclusive they must be independent. Do you agree with this statement? Use the probability information in this problem to justify your answer.
   d. What general conclusion would you make about mutually exclusive and independent events given the results of this problem?

Applications

32. The automobile industry sold 657,000 vehicles in the United States during January 2009 (*The Wall Street Journal,* February 4, 2009). This volume was down 37% from January 2008 as economic conditions continued to decline. The Big Three U.S. automakers—General Motors, Ford, and Chrysler—sold 280,500 vehicles, down 48% from January 2008. A summary of sales by automobile manufacturer and type of vehicle sold is shown in the following table. Data are in thousands of vehicles. The non-U.S. manufacturers are led by Toyota, Honda, and Nissan. The category Light Truck includes pickup, minivan, SUV, and crossover models.

		Type of Vehicle	
		Car	**Light Truck**
Manufacturer	**U.S.**	87.4	193.1
	Non-U.S.	228.5	148.0

   a. Develop a joint probability table for these data and use the table to answer the remaining questions.
   b. What are the marginal probabilities? What do they tell you about the probabilities associated with the manufacturer and the type of vehicle sold?
   c. If a vehicle was manufactured by one of the U.S. automakers, what is the probability that the vehicle was a car? What is the probability it was a light truck?
   d. If a vehicle was not manufactured by one of the U.S. automakers, what is the probability that the vehicle was a car? What is the probability it was a light truck?
   e. If the vehicle was a light truck, what is the probability that it was manufactured by one of the U.S. automakers?
   f. What does the probability information tell you about sales?

SELF test

33. Students taking the Graduate Management Admissions Test (GMAT) were asked about their undergraduate major and intent to pursue their MBA as a full-time or part-time student. A summary of their responses follows.

		Undergraduate Major			
		Business	**Engineering**	**Other**	**Totals**
Intended Enrollment Status	**Full-Time**	352	197	251	800
	Part-Time	150	161	194	505
	Totals	502	358	445	1305

   a. Develop a joint probability table for these data.
   b. Use the marginal probabilities of undergraduate major (business, engineering, or other) to comment on which undergraduate major produces the most potential MBA students.

c. If a student intends to attend classes full-time in pursuit of an MBA degree, what is the probability that the student was an undergraduate engineering major?
d. If a student was an undergraduate business major, what is the probability that the student intends to attend classes full-time in pursuit of an MBA degree?
e. Let *A* denote the event that the student intends to attend classes full-time in pursuit of an MBA degree, and let *B* denote the event that the student was an undergraduate business major. Are events *A* and *B* independent? Justify your answer.

34. The U.S. Department of Transportation reported that during November, 83.4% of Southwest Airlines' flights, 75.1% of US Airways' flights, and 70.1% of JetBlue's flights arrived on time (*USA Today,* January 4, 2007). Assume that this on-time performance is applicable for flights arriving at concourse A of the Rochester International Airport, and that 40% of the arrivals at concourse A are Southwest Airlines flights, 35% are US Airways flights, and 25% are JetBlue flights.
    a. Develop a joint probability table with three rows (airlines) and two columns (on-time arrivals vs. late arrivals).
    b. An announcement has just been made that Flight 1424 will be arriving at gate 20 in concourse A. What is the most likely airline for this arrival?
    c. What is the probability that Flight 1424 will arrive on time?
    d. Suppose that an announcement is made saying that Flight 1424 will be arriving late. What is the most likely airline for this arrival? What is the least likely airline?

35. According to the Ameriprise Financial Money Across Generations study, 9 out of 10 parents with adult children ages 20 to 35 have helped their adult children with some type of financial assistance ranging from college, a car, rent, utilities, credit-card debt, and/or down payments for houses (*Money,* January 2009). The following table with sample data consistent with the study shows the number of times parents have given their adult children financial assistance to buy a car and to pay rent.

		Pay Rent	
		Yes	**No**
Buy a Car	**Yes**	56	52
	No	14	78

    a. Develop a joint probability table and use it to answer the remaining questions.
    b. Using the marginal probabilities for buy a car and pay rent, are parents more likely to assist their adult children with buying a car or paying rent? What is your interpretation of the marginal probabilities?
    c. If parents provided financial assistance to buy a car, what it the probability that the parents assisted with paying rent?
    d. If parents did not provide financial assistance to buy a car, what is the probability the parents assisted with paying rent?
    e. Is financial assistance to buy a car independent of financial assistance to pay rent? Use probabilities to justify your answer.
    f. What is the probability that parents provided financial assistance for their adult children by either helping buy a car or pay rent?

36. Jamal Crawford of the National Basketball Association's Portland Trail Blazers is the best free-throw shooter on the team, making 93% of his shots (ESPN website, April 5, 2012). Assume that late in a basketball game, Jamal Crawford is fouled and is awarded two shots.
    a. What is the probability that he will make both shots?
    b. What is the probability that he will make at least one shot?
    c. What is the probability that he will miss both shots?

d. Late in a basketball game, a team often intentionally fouls an opposing player in order to stop the game clock. The usual strategy is to intentionally foul the other team's worst free-throw shooter. Assume that the Portland Trail Blazers' center makes 58% of his free-throw shots. Calculate the probabilities for the center as shown in parts (a), (b), and (c), and show that intentionally fouling the Portland Trail Blazers' center is a better strategy than intentionally fouling Jamal Crawford. Assume as in parts (a), (b), and (c) that two shots will be awarded.

37. Visa Card USA studied how frequently young consumers, ages 18 to 24, use plastic (debit and credit) cards in making purchases (Associated Press, January 16, 2006). The results of the study provided the following probabilities.

- The probability that a consumer uses a plastic card when making a purchase is .37.
- Given that the consumer uses a plastic card, there is a .19 probability that the consumer is 18 to 24 years old.
- Given that the consumer uses a plastic card, there is a .81 probability that the consumer is more than 24 years old.

U.S. Census Bureau data show that 14% of the consumer population is 18 to 24 years old.

a. Given the consumer is 18 to 24 years old, what is the probability that the consumer uses a plastic card?
b. Given the consumer is over 24 years old, what is the probability that the consumer uses a plastic card?
c. What is the interpretation of the probabilities shown in parts (a) and (b)?
d. Should companies such as Visa, MasterCard, and Discover make plastic cards available to the 18 to 24 year old age group before these consumers have had time to establish a credit history? If no, why? If yes, what restrictions might the companies place on this age group?

38. Students in grades 3 through 8 in New York State are required to take a state mathematics exam. To meet the state's proficiency standards, a student must demonstrate an understanding of the mathematics expected at his or her grade level. The following data show the number of students tested in the New York City school system for grades 3 through 8 and the number who met and did not meet the proficiency standards on the exam (New York City Department of Education website, January 16, 2010).

	Met Proficiency Standards?	
Grade	**Yes**	**No**
3	47,401	23,975
4	35,020	34,740
5	36,062	33,540
6	36,361	32,929
7	40,945	29,768
8	40,720	31,931

a. Develop a joint probability table for these data.
b. What are the marginal probabilities? What do they tell about the probabilities of a student meeting or not meeting the proficiency standards on the exam?
c. If a randomly selected student is a third grader, what is the probability that the student met the proficiency standards? If the student is a fourth grader, what is the probability that the student met the proficiency standards?
d. If a randomly selected student is known to have met the proficiency standards on the exam, what is the probability that the student is a third grader? What is the probability if the student is a fourth grader?

4.5 Bayes' Theorem

In the discussion of conditional probability, we indicated that revising probabilities when new information is obtained is an important phase of probability analysis. Often, we begin the analysis with initial or **prior probability** estimates for specific events of interest. Then, from sources such as a sample, a special report, or a product test, we obtain additional information about the events. Given this new information, we update the prior probability values by calculating revised probabilities, referred to as **posterior probabilities. Bayes' theorem** provides a means for making these probability calculations. The steps in this probability revision process are shown in Figure 4.9.

As an application of Bayes' theorem, consider a manufacturing firm that receives shipments of parts from two different suppliers. Let A_1 denote the event that a part is from supplier 1 and A_2 denote the event that a part is from supplier 2. Currently, 65% of the parts purchased by the company are from supplier 1 and the remaining 35% are from supplier 2. Hence, if a part is selected at random, we would assign the prior probabilities $P(A_1) = .65$ and $P(A_2) = .35$.

The quality of the purchased parts varies with the source of supply. Historical data suggest that the quality ratings of the two suppliers are as shown in Table 4.6. If we let G denote the event that a part is good and B denote the event that a part is bad, the information in Table 4.6 provides the following conditional probability values.

$$P(G \mid A_1) = .98 \quad P(B \mid A_1) = .02$$
$$P(G \mid A_2) = .95 \quad P(B \mid A_2) = .05$$

The tree diagram in Figure 4.10 depicts the process of the firm receiving a part from one of the two suppliers and then discovering that the part is good or bad as a two-step experiment. We see that four experimental outcomes are possible; two correspond to the part being good and two correspond to the part being bad.

Each of the experimental outcomes is the intersection of two events, so we can use the multiplication rule to compute the probabilities. For instance,

$$P(A_1, G) = P(A_1 \cap G) = P(A_1)P(G \mid A_1)$$

FIGURE 4.9 PROBABILITY REVISION USING BAYES' THEOREM

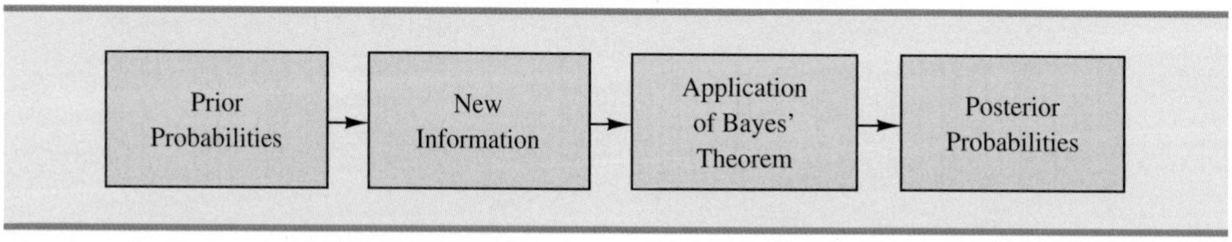

TABLE 4.6 HISTORICAL QUALITY LEVELS OF TWO SUPPLIERS

	Percentage Good Parts	Percentage Bad Parts
Supplier 1	98	2
Supplier 2	95	5

FIGURE 4.10 TREE DIAGRAM FOR TWO-SUPPLIER EXAMPLE

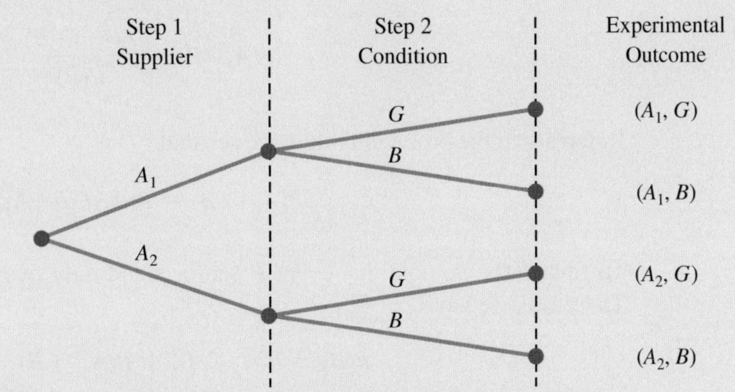

Note: Step 1 shows that the part comes from one of two suppliers, and step 2 shows whether the part is good or bad.

The process of computing these joint probabilities can be depicted in what is called a probability tree (see Figure 4.11). From left to right through the tree, the probabilities for each branch at step 1 are prior probabilities and the probabilities for each branch at step 2 are conditional probabilities. To find the probabilities of each experimental outcome, we simply multiply the probabilities on the branches leading to the outcome. Each of these joint probabilities is shown in Figure 4.11 along with the known probabilities for each branch.

Suppose now that the parts from the two suppliers are used in the firm's manufacturing process and that a machine breaks down because it attempts to process a bad part. Given the information that the part is bad, what is the probability that it came from supplier 1 and what is the probability that it came from supplier 2? With the information in the probability tree (Figure 4.11), Bayes' theorem can be used to answer these questions.

FIGURE 4.11 PROBABILITY TREE FOR TWO-SUPPLIER EXAMPLE

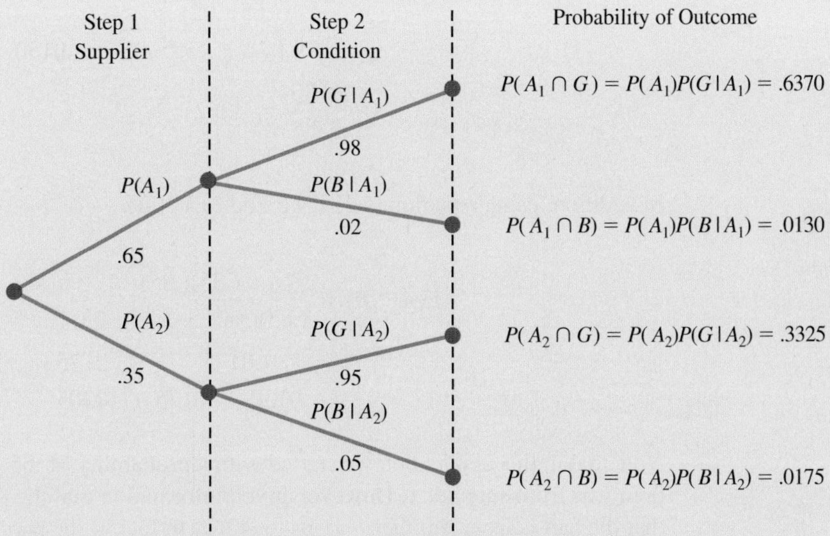

Letting B denote the event that the part is bad, we are looking for the posterior probabilities $P(A_1 \mid B)$ and $P(A_2 \mid B)$. From the law of conditional probability, we know that

$$P(A_1 \mid B) = \frac{P(A_1 \cap B)}{P(B)} \qquad \textbf{(4.14)}$$

Referring to the probability tree, we see that

$$P(A_1 \cap B) = P(A_1)P(B \mid A_1) \qquad \textbf{(4.15)}$$

To find $P(B)$, we note that event B can occur in only two ways: $(A_1 \cap B)$ and $(A_2 \cap B)$. Therefore, we have

$$\begin{aligned} P(B) &= P(A_1 \cap B) + P(A_2 \cap B) \\ &= P(A_1)P(B \mid A_1) + P(A_2)P(B \mid A_2) \end{aligned} \qquad \textbf{(4.16)}$$

Substituting from equations (4.15) and (4.16) into equation (4.14) and writing a similar result for $P(A_2 \mid B)$, we obtain Bayes' theorem for the case of two events.

The Reverend Thomas Bayes (1702–1761), a Presbyterian minister, is credited with the original work leading to the version of Bayes' theorem in use today.

BAYES' THEOREM (TWO-EVENT CASE)

$$P(A_1 \mid B) = \frac{P(A_1)P(B \mid A_1)}{P(A_1)P(B \mid A_1) + P(A_2)P(B \mid A_2)} \qquad \textbf{(4.17)}$$

$$P(A_2 \mid B) = \frac{P(A_2)P(B \mid A_2)}{P(A_1)P(B \mid A_1) + P(A_2)P(B \mid A_2)} \qquad \textbf{(4.18)}$$

Using equation (4.17) and the probability values provided in the example, we have

$$\begin{aligned} P(A_1 \mid B) &= \frac{P(A_1)P(B \mid A_1)}{P(A_1)P(B \mid A_1) + P(A_2)P(B \mid A_2)} \\ &= \frac{(.65)(.02)}{(.65)(.02) + (.35)(.05)} = \frac{.0130}{.0130 + .0175} \\ &= \frac{.0130}{.0305} = .4262 \end{aligned}$$

In addition, using equation (4.18), we find $P(A_2 \mid B)$.

$$\begin{aligned} P(A_2 \mid B) &= \frac{(.35)(.05)}{(.65)(.02) + (.35)(.05)} \\ &= \frac{.0175}{.0130 + .0175} = \frac{.0175}{.0305} = .5738 \end{aligned}$$

Note that in this application we started with a probability of .65 that a part selected at random was from supplier 1. However, given information that the part is bad, the probability that the part is from supplier 1 drops to .4262. In fact, if the part is bad, it has better than a 50–50 chance that it came from supplier 2; that is, $P(A_2 \mid B) = .5738$.

Bayes' theorem is applicable when the events for which we want to compute posterior probabilities are mutually exclusive and their union is the entire sample space.[2] For the case of n mutually exclusive events $A_1, A_2, \ldots, A_n$, whose union is the entire sample space, Bayes' theorem can be used to compute any posterior probability $P(A_i \mid B)$ as shown here.

BAYES' THEOREM

$$P(A_i \mid B) = \frac{P(A_i)P(B \mid A_i)}{P(A_1)P(B \mid A_1) + P(A_2)P(B \mid A_2) + \cdots + P(A_n)P(B \mid A_n)} \quad \textbf{(4.19)}$$

With prior probabilities $P(A_1), P(A_2), \ldots, P(A_n)$ and the appropriate conditional probabilities $P(B \mid A_1), P(B \mid A_2), \ldots, P(B \mid A_n)$, equation (4.19) can be used to compute the posterior probability of the events $A_1, A_2, \ldots, A_n$.

Tabular Approach

A tabular approach is helpful in conducting the Bayes' theorem calculations. Such an approach is shown in Table 4.7 for the parts supplier problem. The computations shown there are done in the following steps.

Step 1. Prepare the following three columns:
Column 1—The mutually exclusive events A_i for which posterior probabilities are desired
Column 2—The prior probabilities $P(A_i)$ for the events
Column 3—The conditional probabilities $P(B \mid A_i)$ of the new information B given each event

Step 2. In column 4, compute the joint probabilities $P(A_i \cap B)$ for each event and the new information B by using the multiplication law. These joint probabilities are found by multiplying the prior probabilities in column 2 by the corresponding conditional probabilities in column 3; that is, $P(A_i \cap B) = P(A_i)P(B \mid A_i)$.

Step 3. Sum the joint probabilities in column 4. The sum is the probability of the new information, $P(B)$. Thus we see in Table 4.7 that there is a .0130 probability that the part came from supplier 1 and is bad and a .0175 probability that the part came from supplier 2 and is bad. Because these are the only two ways in which a bad part can be obtained, the sum .0130 + .0175 shows an overall

TABLE 4.7 TABULAR APPROACH TO BAYES' THEOREM CALCULATIONS FOR THE TWO-SUPPLIER PROBLEM

(1) Events A_i	(2) Prior Probabilities $P(A_i)$	(3) Conditional Probabilities $P(B \mid A_i)$	(4) Joint Probabilities $P(A_i \cap B)$	(5) Posterior Probabilities $P(A_i \mid B)$
A_1	.65	.02	.0130	.0130/.0305 = .4262
A_2	.35	.05	.0175	.0175/.0305 = .5738
	1.00		$P(B)$ = .0305	1.0000

[2]If the union of events is the entire sample space, the events are said to be collectively exhaustive.

probability of .0305 of finding a bad part from the combined shipments of the two suppliers.

Step 4. In column 5, compute the posterior probabilities using the basic relationship of conditional probability.

$$P(A_i \mid B) = \frac{P(A_i \cap B)}{P(B)}$$

Note that the joint probabilities $P(A_i \cap B)$ are in column 4 and the probability $P(B)$ is the sum of column 4.

NOTES AND COMMENTS

1. Bayes' theorem is used extensively in decision analysis. The prior probabilities are often subjective estimates provided by a decision maker. Sample information is obtained and posterior probabilities are computed for use in choosing the best decision.
2. An event and its complement are mutually exclusive, and their union is the entire sample space. Thus, Bayes' theorem is always applicable for computing posterior probabilities of an event and its complement.

Exercises

Methods

SELF test

39. The prior probabilities for events A_1 and A_2 are $P(A_1) = .40$ and $P(A_2) = .60$. It is also known that $P(A_1 \cap A_2) = 0$. Suppose $P(B \mid A_1) = .20$ and $P(B \mid A_2) = .05$.
    a. Are A_1 and A_2 mutually exclusive? Explain.
    b. Compute $P(A_1 \cap B)$ and $P(A_2 \cap B)$.
    c. Compute $P(B)$.
    d. Apply Bayes' theorem to compute $P(A_1 \mid B)$ and $P(A_2 \mid B)$.
40. The prior probabilities for events A_1, A_2, and A_3 are $P(A_1) = .20$, $P(A_2) = .50$, and $P(A_3) = .30$. The conditional probabilities of event B given A_1, A_2, and A_3 are $P(B \mid A_1) = .50$, $P(B \mid A_2) = .40$, and $P(B \mid A_3) = .30$.
    a. Compute $P(B \cap A_1)$, $P(B \cap A_2)$, and $P(B \cap A_3)$.
    b. Apply Bayes' theorem, equation (4.19), to compute the posterior probability $P(A_2 \mid B)$.
    c. Use the tabular approach to applying Bayes' theorem to compute $P(A_1 \mid B)$, $P(A_2 \mid B)$, and $P(A_3 \mid B)$.

Applications

41. A consulting firm submitted a bid for a large research project. The firm's management initially felt they had a 50–50 chance of getting the project. However, the agency to which the bid was submitted subsequently requested additional information on the bid. Past experience indicates that for 75% of the successful bids and 40% of the unsuccessful bids the agency requested additional information.
    a. What is the prior probability of the bid being successful (that is, prior to the request for additional information)?
    b. What is the conditional probability of a request for additional information given that the bid will ultimately be successful?
    c. Compute the posterior probability that the bid will be successful given a request for additional information.

SELF test

42. A local bank reviewed its credit card policy with the intention of recalling some of its credit cards. In the past approximately 5% of cardholders defaulted, leaving the bank unable to collect the outstanding balance. Hence, management established a prior probability of .05 that any particular cardholder will default. The bank also found that the probability of missing a monthly payment is .20 for customers who do not default. Of course, the probability of missing a monthly payment for those who default is 1.
   a. Given that a customer missed one or more monthly payments, compute the posterior probability that the customer will default.
   b. The bank would like to recall its card if the probability that a customer will default is greater than .20. Should the bank recall its card if the customer misses a monthly payment? Why or why not?

43. Two Wharton professors analyzed 1,613,234 putts by golfers on the Professional Golfers Association (PGA) Tour and found that 983,764 of the putts were made and 629,470 of the putts were missed. Further analysis showed that for putts that were made, 64.0% of the time the player was attempting to make a par putt and 18.8% of the time the player was attempting to make a birdie putt. And, for putts that were missed, 20.3% of the time the player was attempting to make a par putt and 73.4% of the time the player was attempting to make a birdie putt. (D. G. Pope and M. E. Schweitzer, *Is Tiger Woods Loss Averse? Persistent Bias in the Face of Experience, Competition, and High Stakes,* June 2009, The Wharton School, University of Pennsylvania).
   a. What is the probability that a PGA Tour player makes a putt?
   b. Suppose that a PGA Tour player has a putt for par. What is the probability that the player will make the putt?
   c. Suppose that a PGA Tour player has a putt for birdie. What is the probability that the player will make the putt?
   d. Comment on the differences in the probabilities computed in parts (b) and (c).

44. According to the Open Doors 2011 Report, 9.5% of all full-time U.S. undergraduate students studied abroad during the 2009–2010 academic year (Institute of International Education, November 14, 2011). Assume that participation records show women make up 60% of the students who studied abroad during the 2009–2010 academic year, but women make up only 49% of the students who didn't participate.
   a. Let A_1 = the student studied abroad during the 2009–2010 academic year
      A_2 = the student did not study abroad during the 2009–2010 academic year
      W = the student is a female student
      Using the given information, what are the values for $P(A_1)$, $P(A_2)$, $P(W \mid A_1)$, and $P(W \mid A_2)$?
   b. What is the probability that a female student studied abroad during the 2009–2010 academic year?
   c. What is the probability that a male student studied abroad during the 2009–2010 academic year?
   d. Given the preceding results, what were the percentage of women and the percentage of men studying full-time during the 2009–2010 academic year?

45. In an article about investment alternatives, *Money* magazine reported that drug stocks provide a potential for long-term growth, with over 50% of the adult population of the United States taking prescription drugs on a regular basis. For adults age 65 and older, 82% take prescription drugs regularly. For adults age 18 to 64, 49% take prescription drugs regularly. The 18–64 age group accounts for 83.5% of the adult population (*Statistical Abstract of the United States*, 2008).
   a. What is the probability that a randomly selected adult is 65 or older?
   b. Given an adult takes prescription drugs regularly, what is the probability that the adult is 65 or older?

Summary

In this chapter we introduced basic probability concepts and illustrated how probability analysis can be used to provide helpful information for decision making. We described how probability can be interpreted as a numerical measure of the likelihood that an event will occur. In addition, we saw that the probability of an event can be computed either by summing the probabilities of the experimental outcomes (sample points) comprising the event or by using the relationships established by the addition, conditional probability, and multiplication laws of probability. For cases in which additional information is available, we showed how Bayes' theorem can be used to obtain revised or posterior probabilities.

Glossary

Probability A numerical measure of the likelihood that an event will occur.
Experiment A process that generates well-defined outcomes.
Sample space The set of all experimental outcomes.
Sample point An element of the sample space. A sample point represents an experimental outcome.
Multiple-step experiment An experiment that can be described as a sequence of steps. If a multiple-step experiment has k steps with n_1 possible outcomes on the first step, n_2 possible outcomes on the second step, and so on, the total number of experimental outcomes is given by $(n_1)(n_2) \cdots (n_k)$.
Tree diagram A graphical representation that helps in visualizing a multiple-step experiment.
Combination In an experiment we may be interested in determining the number of ways n objects may be selected from among N objects without regard to the *order in which the n objects are selected*. Each selection of n objects is called a combination and the total number of combinations of N objects taken n at a time is $C_n^N = \binom{N}{n} = \dfrac{N!}{n!(N-n)!}$ for $n = 0, 1, 2, \ldots, N$.
Permutation In an experiment we may be interested in determining the number of ways n objects may be selected from among N objects when the *order in which the n objects are selected* is important. Each ordering of n objects is called a permutation and the total number of permutations of N objects taken n at a time is $P_n^N = n!\binom{N}{n} = \dfrac{N!}{(N-n)!}$ for $n = 0, 1, 2, \ldots, N$.
Basic requirements for assigning probabilities Two requirements that restrict the manner in which probability assignments can be made: (1) for each experimental outcome E_i we must have $0 \le P(E_i) \le 1$; (2) considering all experimental outcomes, we must have $P(E_1) + P(E_2) + \cdots + P(E_n) = 1.0$.
Classical method A method of assigning probabilities that is appropriate when all the experimental outcomes are equally likely.
Relative frequency method A method of assigning probabilities that is appropriate when data are available to estimate the proportion of the time the experimental outcome will occur if the experiment is repeated a large number of times.
Subjective method A method of assigning probabilities on the basis of judgment.
Event A collection of sample points.
Complement of *A* The event consisting of all sample points that are not in A.
Venn diagram A graphical representation for showing symbolically the sample space and operations involving events in which the sample space is represented by a rectangle and events are represented as circles within the sample space.
Union of *A* and *B* The event containing all sample points belonging to A or B or both. The union is denoted $A \cup B$.

Intersection of A and B The event containing the sample points belonging to both A and B. The intersection is denoted $A \cap B$.
Addition law A probability law used to compute the probability of the union of two events. It is $P(A \cup B) = P(A) + P(B) - P(A \cap B)$. For mutually exclusive events, $P(A \cap B) = 0$; in this case the addition law reduces to $P(A \cup B) = P(A) + P(B)$.
Mutually exclusive events Events that have no sample points in common; that is, $A \cap B$ is empty and $P(A \cap B) = 0$.
Conditional probability The probability of an event given that another event already occurred. The conditional probability of A given B is $P(A \mid B) = P(A \cap B)/P(B)$.
Joint probability The probability of two events both occurring; that is, the probability of the intersection of two events.
Marginal probability The values in the margins of a joint probability table that provide the probabilities of each event separately.
Independent events Two events A and B where $P(A \mid B) = P(A)$ or $P(B \mid A) = P(B)$; that is, the events have no influence on each other.
Multiplication law A probability law used to compute the probability of the intersection of two events. It is $P(A \cap B) = P(B)P(A \mid B)$ or $P(A \cap B) = P(A)P(B \mid A)$. For independent events it reduces to $P(A \cap B) = P(A)P(B)$.
Prior probabilities Initial estimates of the probabilities of events.
Posterior probabilities Revised probabilities of events based on additional information.
Bayes' theorem A method used to compute posterior probabilities.

Key Formulas

Counting Rule for Combinations

$$C_n^N = \binom{N}{n} = \frac{N!}{n!(N-n)!} \tag{4.1}$$

Counting Rule for Permutations

$$P_n^N = n!\binom{N}{n} = \frac{N!}{(N-n)!} \tag{4.2}$$

Computing Probability Using the Complement

$$P(A) = 1 - P(A^c) \tag{4.5}$$

Addition Law

$$P(A \cup B) = P(A) + P(B) - P(A \cap B) \tag{4.6}$$

Conditional Probability

$$P(A \mid B) = \frac{P(A \cap B)}{P(B)} \tag{4.7}$$

$$P(B \mid A) = \frac{P(A \cap B)}{P(A)} \tag{4.8}$$

Multiplication Law

$$P(A \cap B) = P(B)P(A \mid B) \tag{4.11}$$

$$P(A \cap B) = P(A)P(B \mid A) \tag{4.12}$$

Multiplication Law for Independent Events

$$P(A \cap B) = P(A)P(B) \tag{4.13}$$

Bayes' Theorem

$$P(A_i \mid B) = \frac{P(A_i)P(B \mid A_i)}{P(A_1)P(B \mid A_1) + P(A_2)P(B \mid A_2) + \cdots + P(A_n)P(B \mid A_n)} \tag{4.19}$$

Supplementary Exercises

46. *The Wall Street Journal*/Harris Personal Finance poll asked 2082 adults if they owned a home (All Business website, January 23, 2008). A total of 1249 survey respondents answered Yes. Of the 450 respondents in the 18–34 age group, 117 responded Yes.
    a. What is the probability that a respondent to the poll owned a home?
    b. What is the probability that a respondent in the 18–34 age group owned a home?
    c. What is the probability that a respondent to the poll did not own a home?
    d. What is the probability that a respondent in the 18–34 age group did not own a home?

47. A financial manager made two new investments—one in the oil industry and one in municipal bonds. After a one-year period, each of the investments will be classified as either successful or unsuccessful. Consider the making of the two investments as an experiment.
    a. How many sample points exist for this experiment?
    b. Show a tree diagram and list the sample points.
    c. Let O = the event that the oil industry investment is successful and M = the event that the municipal bond investment is successful. List the sample points in O and in M.
    d. List the sample points in the union of the events $(O \cup M)$.
    e. List the sample points in the intersection of the events $(O \cap M)$.
    f. Are events O and M mutually exclusive? Explain.

48. Forty-three percent of Americans use social media and other websites to voice their opinions about television programs (*The Huffington Post,* November 23, 2011). Below are the results of a survey of 1400 individuals who were asked if they use social media and other websites to voice their opinions about television programs.

	Uses Social Media and Other Websites to Voice Opinions About Television Programs	Doesn't Use Social Media and Other Websites to Voice Opinions About Television Programs
Female	395	291
Male	323	355

    a. What is the probability a respondent is female?
    b. What is the conditional probability a respondent uses social media and other websites to voice opinions about television programs given the respondent is female?
    c. Let F denote the event that the respondent is female and A denote the event that the respondent uses social media and other websites to voice opinions about television programs. Are events F and A independent?

49. A study of 31,000 hospital admissions in New York State found that 4% of the admissions led to treatment-caused injuries. One-seventh of these treatment-caused injuries resulted in death, and one-fourth were caused by negligence. Malpractice claims were filed in one out of 7.5 cases involving negligence, and payments were made in one out of every two claims.
    a. What is the probability a person admitted to the hospital will suffer a treatment-caused injury due to negligence?

b. What is the probability a person admitted to the hospital will die from a treatment-caused injury?
c. In the case of a negligent treatment-caused injury, what is the probability a malpractice claim will be paid?

50. A telephone survey to determine viewer response to a new television show obtained the following data.

Rating	Frequency
Poor	4
Below average	8
Average	11
Above average	14
Excellent	13

a. What is the probability that a randomly selected viewer will rate the new show as average or better?
b. What is the probability that a randomly selected viewer will rate the new show below average or worse?

51. The following crosstabulation shows household income by educational level of the head of household (*Statistical Abstract of the United States*, 2008).

	Household Income ($1000s)					
Education Level	**Under 25**	**25.0–49.9**	**50.0–74.9**	**75.0–99.9**	**100 or more**	**Total**
Not H.S. Graduate	4207	3459	1389	539	367	9961
H.S. Graduate	4917	6850	5027	2637	2668	22,099
Some College	2807	5258	4678	3250	4074	20,067
Bachelor's Degree	885	2094	2848	2581	5379	13,787
Beyond Bach. Deg.	290	829	1274	1241	4188	7822
Total	13,106	18,490	15,216	10,248	16,676	73,736

a. Develop a joint probability table.
b. What is the probability of a head of household not being a high school graduate?
c. What is the probability of a head of household having a bachelor's degree or more education?
d. What is the probability of a household headed by someone with a bachelor's degree earning $100,000 or more?
e. What is the probability of a household having income below $25,000?
f. What is the probability of a household headed by someone with a bachelor's degree earning less than $25,000?
g. Is household income independent of educational level?

52. An MBA new-matriculants survey provided the following data for 2018 students.

		Applied to More Than One School	
		Yes	**No**
Age Group	**23 and under**	207	201
	24–26	299	379
	27–30	185	268
	31–35	66	193
	36 and over	51	169

a. For a randomly selected MBA student, prepare a joint probability table for the experiment consisting of observing the student's age and whether the student applied to one or more schools.
b. What is the probability that a randomly selected applicant is 23 or under?
c. What is the probability that a randomly selected applicant is older than 26?
d. What is the probability that a randomly selected applicant applied to more than one school?

53. Refer again to the data from the MBA new-matriculants survey in exercise 52.
a. Given that a person applied to more than one school, what is the probability that the person is 24–26 years old?
b. Given that a person is in the 36-and-over age group, what is the probability that the person applied to more than one school?
c. What is the probability that a person is 24–26 years old or applied to more than one school?
d. Suppose a person is known to have applied to only one school. What is the probability that the person is 31 or more years old?
e. Is the number of schools applied to independent of age? Explain.

54. In February 2012, the Pew Internet & American Life project conducted a survey that included several questions about how Internet users feel about search engines and other websites collecting information about them and using this information either to shape search results or target advertising to them (Pew Research Center, March 9, 2012). In one question, participants were asked, "If a search engine kept track of what you search for, and then used that information to personalize your future search results, how would you feel about that?" Respondents could indicate either "Would *not* be okay with it because you feel it is an invasion of your privacy" or "Would be *okay* with it, even if it means they are gathering information about you." Frequencies of responses by age group are summarized in the following table.

Age	Not Okay	Okay
18–29	0.1485	0.0604
30–49	0.2273	0.0907
50+	0.4008	0.0723

a. What is the probability a survey respondent will say she or he is *not okay* with this practice?
b. Given a respondent is 30–49 years old, what is the probability the respondent will say she or he is *okay* with this practice?
c. Given a respondent says she or he is *not okay* with this practice, what is the probability the respondent is 50+ years old?
d. Is the attitude about this practice independent of the age of the respondent? Why or why not?
e. Do attitudes toward this practice for respondents who are 18–29 years old and respondents who are 50+ years old differ?

55. A large consumer goods company ran a television advertisement for one of its soap products. On the basis of a survey that was conducted, probabilities were assigned to the following events.

B = individual purchased the product
S = individual recalls seeing the advertisement
$B \cap S$ = individual purchased the product and recalls seeing the advertisement

The probabilities assigned were $P(B) = .20$, $P(S) = .40$, and $P(B \cap S) = .12$.
a. What is the probability of an individual's purchasing the product given that the individual recalls seeing the advertisement? Does seeing the advertisement increase

the probability that the individual will purchase the product? As a decision maker, would you recommend continuing the advertisement (assuming that the cost is reasonable)?

b. Assume that individuals who do not purchase the company's soap product buy from its competitors. What would be your estimate of the company's market share? Would you expect that continuing the advertisement will increase the company's market share? Why or why not?

c. The company also tested another advertisement and assigned it values of $P(S) = .30$ and $P(B \cap S) = .10$. What is $P(B \mid S)$ for this other advertisement? Which advertisement seems to have had the bigger effect on customer purchases?

56. Cooper Realty is a small real estate company located in Albany, New York, specializing primarily in residential listings. They recently became interested in determining the likelihood of one of their listings being sold within a certain number of days. An analysis of company sales of 800 homes in previous years produced the following data.

		Days Listed Until Sold			
		Under 30	**31–90**	**Over 90**	Total
Initial Asking Price	**Under \$150,000**	50	40	10	100
	\$150,000–\$199,999	20	150	80	250
	\$200,000–\$250,000	20	280	100	400
	Over \$250,000	10	30	10	50
	Total	100	500	200	800

a. If A is defined as the event that a home is listed for more than 90 days before being sold, estimate the probability of A.

b. If B is defined as the event that the initial asking price is under \$150,000, estimate the probability of B.

c. What is the probability of $A \cap B$?

d. Assuming that a contract was just signed to list a home with an initial asking price of less than \$150,000, what is the probability that the home will take Cooper Realty more than 90 days to sell?

e. Are events A and B independent?

57. A company studied the number of lost-time accidents occurring at its Brownsville, Texas, plant. Historical records show that 6% of the employees suffered lost-time accidents last year. Management believes that a special safety program will reduce such accidents to 5% during the current year. In addition, it estimates that 15% of employees who had lost-time accidents last year will experience a lost-time accident during the current year.

a. What percentage of the employees will experience lost-time accidents in both years?

b. What percentage of the employees will suffer at least one lost-time accident over the two-year period?

58. A survey showed that 8% of Internet users age 18 and older report keeping a blog. Referring to the 18–29 age group as young adults, the survey showed that for bloggers 54% are young adults and for nonbloggers 24% are young adults (Pew Internet & American Life Project, July 19, 2006).

a. Develop a joint probability table for these data with two rows (bloggers vs. nonbloggers) and two columns (young adults vs. older adults).

b. What is the probability that an Internet user is a young adult?

c. What is the probability that an Internet user keeps a blog and is a young adult?

d. Suppose that in a follow-up phone survey we contact someone who is 24 years old. What is the probability that this person keeps a blog?

59. An oil company purchased an option on land in Alaska. Preliminary geologic studies assigned the following prior probabilities.

$$P(\text{high-quality oil}) = .50$$
$$P(\text{medium-quality oil}) = .20$$
$$P(\text{no oil}) = .30$$

a. What is the probability of finding oil?
b. After 200 feet of drilling on the first well, a soil test is taken. The probabilities of finding the particular type of soil identified by the test follow.

$$P(\text{soil} \mid \text{high-quality oil}) = .20$$
$$P(\text{soil} \mid \text{medium-quality oil}) = .80$$
$$P(\text{soil} \mid \text{no oil}) = .20$$

How should the firm interpret the soil test? What are the revised probabilities, and what is the new probability of finding oil?

60. The five most common words appearing in spam emails are *shipping!*, *today!*, *here!*, *available*, and *fingertips!* (Andy Greenberg, "The Most Common Words In Spam Email," *Forbes* website, March 17, 2010). Many spam filters separate spam from ham (email not considered to be spam) through application of Bayes' theorem. Suppose that for one email account, 1 in every 10 messages is spam and the proportions of spam messages that have the five most common words in spam email are given below.

shipping!	.051
today!	.045
here!	.034
available	.014
fingertips!	.014

Also suppose that the proportions of ham messages that have these words are

shipping!	.0015
today!	.0022
here!	.0022
available	.0041
fingertips!	.0011

a. If a message includes the word *shipping!*, what is the probability the message is spam? If a message includes the word *shipping!*, what is the probability the message is ham? Should messages that include the word *shipping!* be flagged as spam?
b. If a message includes the word *today!*, what is the probability the message is spam? If a message includes the word *here!*, what is the probability the message is spam? Which of these two words is a stronger indicator that a message is spam? Why?
c. If a message includes the word *available*, what is the probability the message is spam? If a message includes the word *fingertips!*, what is the probability the message is spam? Which of these two words is a stronger indicator that a message is spam? Why?
d. What insights do the results of parts (b) and (c) yield about what enables a spam filter that uses Bayes' theorem to work effectively?

Case Problem Hamilton County Judges

Hamilton County judges try thousands of cases per year. In an overwhelming majority of the cases disposed, the verdict stands as rendered. However, some cases are appealed, and of those appealed, some of the cases are reversed. Kristen DelGuzzi of *The Cincinnati Enquirer* conducted a study of cases handled by Hamilton County judges over a three-year period. Shown in Table 4.8 are the results for 182,908 cases handled (disposed) by

WEB file

Judge

TABLE 4.8 TOTAL CASES DISPOSED, APPEALED, AND REVERSED IN HAMILTON COUNTY COURTS

Common Pleas Court

Judge	Total Cases Disposed	Appealed Cases	Reversed Cases
Fred Cartolano	3037	137	12
Thomas Crush	3372	119	10
Patrick Dinkelacker	1258	44	8
Timothy Hogan	1954	60	7
Robert Kraft	3138	127	7
William Mathews	2264	91	18
William Morrissey	3032	121	22
Norbert Nadel	2959	131	20
Arthur Ney, Jr.	3219	125	14
Richard Niehaus	3353	137	16
Thomas Nurre	3000	121	6
John O'Connor	2969	129	12
Robert Ruehlman	3205	145	18
J. Howard Sundermann	955	60	10
Ann Marie Tracey	3141	127	13
Ralph Winkler	3089	88	6
Total	43,945	1762	199

Domestic Relations Court

Judge	Total Cases Disposed	Appealed Cases	Reversed Cases
Penelope Cunningham	2729	7	1
Patrick Dinkelacker	6001	19	4
Deborah Gaines	8799	48	9
Ronald Panioto	12,970	32	3
Total	30,499	106	17

Municipal Court

Judge	Total Cases Disposed	Appealed Cases	Reversed Cases
Mike Allen	6149	43	4
Nadine Allen	7812	34	6
Timothy Black	7954	41	6
David Davis	7736	43	5
Leslie Isaiah Gaines	5282	35	13
Karla Grady	5253	6	0
Deidra Hair	2532	5	0
Dennis Helmick	7900	29	5
Timothy Hogan	2308	13	2
James Patrick Kenney	2798	6	1
Joseph Luebbers	4698	25	8
William Mallory	8277	38	9
Melba Marsh	8219	34	7
Beth Mattingly	2971	13	1
Albert Mestemaker	4975	28	9
Mark Painter	2239	7	3
Jack Rosen	7790	41	13
Mark Schweikert	5403	33	6
David Stockdale	5371	22	4
John A. West	2797	4	2
Total	108,464	500	104

38 judges in Common Pleas Court, Domestic Relations Court, and Municipal Court. Two of the judges (Dinkelacker and Hogan) did not serve in the same court for the entire three-year period.

The purpose of the newspaper's study was to evaluate the performance of the judges. Appeals are often the result of mistakes made by judges, and the newspaper wanted to know which judges were doing a good job and which were making too many mistakes. You are called in to assist in the data analysis. Use your knowledge of probability and conditional probability to help with the ranking of the judges. You also may be able to analyze the likelihood of appeal and reversal for cases handled by different courts.

Managerial Report

Prepare a report with your rankings of the judges. Also, include an analysis of the likelihood of appeal and case reversal in the three courts. At a minimum, your report should include the following:

1. The probability of cases being appealed and reversed in the three different courts.
2. The probability of a case being appealed for each judge.
3. The probability of a case being reversed for each judge.
4. The probability of reversal given an appeal for each judge.
5. Rank the judges within each court. State the criteria you used and provide a rationale for your choice.

CHAPTER 5

Discrete Probability Distributions

CONTENTS

STATISTICS *in* PRACTICE

CITIBANK*
LONG ISLAND CITY, NEW YORK

Citibank, the retail banking division of Citigroup, offers a wide range of financial services including checking and saving accounts, loans and mortgages, insurance, and investment services. It delivers these services through a unique system referred to as Citibanking.

Citibank was one of the first banks in the United States to introduce automatic teller machines (ATMs). Citibank's ATMs, located in Citicard Banking Centers (CBCs), let customers do all of their banking in one place with the touch of a finger, 24 hours a day, 7 days a week. More than 150 different banking functions—from deposits to managing investments—can be performed with ease. Citibank customers use ATMs for 80% of their transactions.

Each Citibank CBC operates as a waiting line system with randomly arriving customers seeking service at one of the ATMs. If all ATMs are busy, the arriving customers wait in line. Periodic CBC capacity studies are used to analyze customer waiting times and to determine whether additional ATMs are needed.

Data collected by Citibank showed that the random customer arrivals followed a probability distribution known as the Poisson distribution. Using the Poisson distribution, Citibank can compute probabilities for the number of customers arriving at a CBC during any time period and make decisions concerning the number of ATMs needed. For example, let x = the number of customers arriving during a one-minute period. Assuming that a particular CBC has a mean arrival rate of two customers per minute, the following table shows the probabilities for the number of customers arriving during a one-minute period.

Each Citicard Banking Center operates as a waiting line system with randomly arriving customers seeking service at an ATM. © Jeff Greenberg/Alamy Limited.

x	**Probability**
0	.1353
1	.2707
2	.2707
3	.1804
4	.0902
5 or more	.0527

Discrete probability distributions, such as the one used by Citibank, are the topic of this chapter. In addition to the Poisson distribution, you will learn about the binomial and hypergeometric distributions and how they can be used to provide helpful probability information.

*The authors are indebted to Ms. Stacey Karter, Citibank, for providing this Statistics in Practice.

In this chapter we extend the study of probability by introducing the concepts of random variables and probability distributions. Random variables and probability distributions are models for populations of data. The values of what are called random variables represent the values of the data and the probability distribution provides either the probability of each data value or a rule for computing the probability of each data value or a set of data values. The focus of this chapter is on probability distributions for discrete data; that is, discrete probability distributions.

We will introduce two types of discrete probability distributions. The first type is a table with one column for the values of the random variable and a second column for the associated probabilities. We will see that the rules for assigning probabilities to experimental outcomes introduced in Chapter 4 are used to assign probabilities for such a distribution. The second type of discrete probability distribution uses a special mathematical function to compute the probabilities for each value of the random variable. We present three probability distributions of this type that are widely used in practice: the binomial, Poisson, and hypergeometric distributions.

5.1 Random Variables

In Chapter 4 we defined the concept of an experiment and its associated experimental outcomes. A random variable provides a means for describing experimental outcomes using numerical values. Random variables must assume numerical values.

Random variables must assume numerical values.

RANDOM VARIABLE

A **random variable** is a numerical description of the outcome of an experiment.

In effect, a random variable associates a numerical value with each possible experimental outcome. The particular numerical value of the random variable depends on the outcome of the experiment. A random variable can be classified as being either *discrete* or *continuous* depending on the numerical values it assumes.

Discrete Random Variables

A random variable that may assume either a finite number of values or an infinite sequence of values such as 0, 1, 2, . . . is referred to as a **discrete random variable**. For example, consider the experiment of an accountant taking the certified public accountant (CPA) examination. The examination has four parts. We can define a random variable as $x =$ the number of parts of the CPA examination passed. It is a discrete random variable because it may assume the finite number of values 0, 1, 2, 3, or 4.

As another example of a discrete random variable, consider the experiment of cars arriving at a tollbooth. The random variable of interest is $x =$ the number of cars arriving during a one-day period. The possible values for x come from the sequence of integers 0, 1, 2, and so on. Hence, x is a discrete random variable assuming one of the values in this infinite sequence.

Although the outcomes of many experiments can naturally be described by numerical values, others cannot. For example, a survey question might ask an individual to recall the message in a recent television commercial. This experiment would have two possible outcomes: The individual cannot recall the message and the individual can recall the message. We can still describe these experimental outcomes numerically by defining the discrete random variable x as follows: let $x = 0$ if the individual cannot recall the message and $x = 1$ if the individual can recall the message. The numerical values for this random variable are arbitrary (we could use 5 and 10), but they are acceptable in terms of the definition of a random variable—namely, x is a random variable because it provides a numerical description of the outcome of the experiment.

Table 5.1 provides some additional examples of discrete random variables. Note that in each example the discrete random variable assumes a finite number of values or an infinite sequence of values such as 0, 1, 2, ... These types of discrete random variables are discussed in detail in this chapter.

TABLE 5.1 EXAMPLES OF DISCRETE RANDOM VARIABLES

Experiment	Random Variable (x)	Possible Values for the Random Variable
Contact five customers	Number of customers who place an order	0, 1, 2, 3, 4, 5
Inspect a shipment of 50 radios	Number of defective radios	0, 1, 2, . . . , 49, 50
Operate a restaurant for one day	Number of customers	0, 1, 2, 3, . . .
Sell an automobile	Gender of the customer	0 if male; 1 if female

Continuous Random Variables

A random variable that may assume any numerical value in an interval or collection of intervals is called a **continuous random variable**. Experimental outcomes based on measurement scales such as time, weight, distance, and temperature can be described by continuous random variables. For example, consider an experiment of monitoring incoming telephone calls to the claims office of a major insurance company. Suppose the random variable of interest is x = the time between consecutive incoming calls in minutes. This random variable may assume any value in the interval $x \geq 0$. Actually, an infinite number of values are possible for x, including values such as 1.26 minutes, 2.751 minutes, 4.3333 minutes, and so on. As another example, consider a 90-mile section of interstate highway I-75 north of Atlanta, Georgia. For an emergency ambulance service located in Atlanta, we might define the random variable as x = number of miles to the location of the next traffic accident along this section of I-75. In this case, x would be a continuous random variable assuming any value in the interval $0 \leq x \leq 90$. Additional examples of continuous random variables are listed in Table 5.2. Note that each example describes a random variable that may assume any value in an interval of values. Continuous random variables and their probability distributions will be the topic of Chapter 6.

TABLE 5.2 EXAMPLES OF CONTINUOUS RANDOM VARIABLES

Experiment	Random Variable (x)	Possible Values for the Random Variable
Operate a bank	Time between customer arrivals in minutes	$x \geq 0$
Fill a soft drink can (max = 12.1 ounces)	Number of ounces	$0 \leq x \leq 12.1$
Construct a new library	Percentage of project complete after six months	$0 \leq x \leq 100$
Test a new chemical process	Temperature when the desired reaction takes place (min 150° F; max 212° F)	$150 \leq x \leq 212$

NOTES AND COMMENTS

One way to determine whether a random variable is discrete or continuous is to think of the values of the random variable as points on a line segment. Choose two points representing values of the random variable. If the entire line segment between the two points also represents possible values for the random variable, then the random variable is continuous.

Exercises

Methods

SELF test

1. Consider the experiment of tossing a coin twice.
   a. List the experimental outcomes.
   b. Define a random variable that represents the number of heads occurring on the two tosses.
   c. Show what value the random variable would assume for each of the experimental outcomes.
   d. Is this random variable discrete or continuous?
2. Consider the experiment of a worker assembling a product.
   a. Define a random variable that represents the time in minutes required to assemble the product.
   b. What values may the random variable assume?
   c. Is the random variable discrete or continuous?

Applications

SELF test

3. Three students scheduled interviews for summer employment at the Brookwood Institute. In each case the interview results in either an offer for a position or no offer. Experimental outcomes are defined in terms of the results of the three interviews.
   a. List the experimental outcomes.
   b. Define a random variable that represents the number of offers made. Is the random variable continuous?
   c. Show the value of the random variable for each of the experimental outcomes.
4. In January the U.S. unemployment rate dropped to 8.3% (U.S. Department of Labor website, February 10, 2012). The Census Bureau includes nine states in the Northeast region. Assume that the random variable of interest is the number of Northeastern states with an unemployment rate in January that was less than 8.3%. What values may this random variable assume?
5. To perform a certain type of blood analysis, lab technicians must perform two procedures. The first procedure requires either one or two separate steps, and the second procedure requires either one, two, or three steps.
   a. List the experimental outcomes associated with performing the blood analysis.
   b. If the random variable of interest is the total number of steps required to do the complete analysis (both procedures), show what value the random variable will assume for each of the experimental outcomes.
6. Listed is a series of experiments and associated random variables. In each case, identify the values that the random variable can assume and state whether the random variable is discrete or continuous.

Experiment	Random Variable (x)
a. Take a 20-question examination	Number of questions answered correctly
b. Observe cars arriving at a tollbooth for 1 hour	Number of cars arriving at tollbooth
c. Audit 50 tax returns	Number of returns containing errors
d. Observe an employee's work	Number of nonproductive hours in an eight-hour workday
e. Weigh a shipment of goods	Number of pounds

5.2 Developing Discrete Probability Distributions

The **probability distribution** for a random variable describes how probabilities are distributed over the values of the random variable. For a discrete random variable x, a **probability function**, denoted by $f(x)$, provides the probability for each value of the random variable. As such, you might suppose that the classical, subjective, and relative frequency methods of assigning probabilities introduced in Chapter 4 would be useful in developing discrete probability distributions. They are and in this section we show how. Application of this methodology leads to what we call tabular discrete probability distributions; that is, probability distributions that are presented in a table.

The classical method of assigning probabilities to values of a random variable is applicable when the experimental outcomes generate values of the random variable that are equally likely. For instance, consider the experiment of rolling a die and observing the number on the upward face. It must be one of the numbers 1, 2, 3, 4, 5, or 6 and each of these outcomes is equally likely. Thus, if we let x = number obtained on one roll of a die and $f(x)$ = the probability of x, the probability distribution of x is given in Table 5.3.

The subjective method of assigning probabilities can also lead to a table of values of the random variable together with the associated probabilities. With the subjective method the individual developing the probability distribution uses their best judgment to assign each probability. So, unlike probability distributions developed using the classical method, different people can be expected to obtain different probability distributions.

The relative frequency method of assigning probabilities to values of a random variable is applicable when reasonably large amounts of data are available. We then treat the data as if they were the population and use the relative frequency method to assign probabilities to the experimental outcomes. The use of the relative frequency method to develop discrete probability distributions leads to what is called an **empirical discrete distribution**. With the large amounts of data available today (e.g., scanner data, credit card data), this type of probability distribution is becoming more widely used in practice. Let us illustrate by considering the sale of automobiles at a dealership.

We will use the relative frequency method to develop a probability distribution for the number of cars sold per day at DiCarlo Motors in Saratoga, New York. Over the past 300 days, DiCarlo has experienced 54 days with no automobiles sold, 117 days with 1 automobile sold, 72 days with 2 automobiles sold, 42 days with 3 automobiles sold, 12 days with 4 automobiles sold, and 3 days with 5 automobiles sold. Suppose we consider the experiment of observing a day of operations at DiCarlo Motors and define the random variable of interest as x = the number of automobiles sold during a day. Using the relative frequencies

TABLE 5.3 PROBABILITY DISTRIBUTION FOR NUMBER OBTAINED ON ONE ROLL OF A DIE

Number Obtained x	**Probability of** x $f(x)$
1	1/6
2	1/6
3	1/6
4	1/6
5	1/6
6	1/6

TABLE 5.4 PROBABILITY DISTRIBUTION FOR THE NUMBER OF AUTOMOBILES SOLD DURING A DAY AT DICARLO MOTORS

x	$f(x)$
0	.18
1	.39
2	.24
3	.14
4	.04
5	.01
Total	1.00

to assign probabilities to the values of the random variable x, we can develop the probability distribution for x.

In probability function notation, $f(0)$ provides the probability of 0 automobiles sold, $f(1)$ provides the probability of 1 automobile sold, and so on. Because historical data show 54 of 300 days with 0 automobiles sold, we assign the relative frequency $54/300 = .18$ to $f(0)$, indicating that the probability of 0 automobiles being sold during a day is .18. Similarly, because 117 of 300 days had 1 automobile sold, we assign the relative frequency $117/300 = .39$ to $f(1)$, indicating that the probability of exactly 1 automobile being sold during a day is .39. Continuing in this way for the other values of the random variable, we compute the values for $f(2)$, $f(3)$, $f(4)$, and $f(5)$ as shown in Table 5.4.

A primary advantage of defining a random variable and its probability distribution is that once the probability distribution is known, it is relatively easy to determine the probability of a variety of events that may be of interest to a decision maker. For example, using the probability distribution for DiCarlo Motors as shown in Table 5.4, we see that the most probable number of automobiles sold during a day is 1 with a probability of $f(1) = .39$. In addition, there is an $f(3) + f(4) + f(5) = .14 + .04 + .01 = .19$ probability of selling 3 or more automobiles during a day. These probabilities, plus others the decision maker may ask about, provide information that can help the decision maker understand the process of selling automobiles at DiCarlo Motors.

In the development of a probability function for any discrete random variable, the following two conditions must be satisfied.

These conditions are the analogs to the two basic requirements for assigning probabilities to experimental outcomes presented in Chapter 4.

REQUIRED CONDITIONS FOR A DISCRETE PROBABILITY FUNCTION

$$f(x) \geq 0 \tag{5.1}$$

$$\Sigma f(x) = 1 \tag{5.2}$$

Table 5.4 shows that the probabilities for the random variable x satisfy equation (5.1); $f(x)$ is greater than or equal to 0 for all values of x. In addition, because the probabilities sum to 1, equation (5.2) is satisfied. Thus, the DiCarlo Motors probability function is a valid discrete probability function.

We can also show the DiCarlo Motors probability distribution graphically. In Figure 5.1 the values of the random variable x for DiCarlo Motors are shown on the horizontal axis and the probability associated with these values is shown on the vertical axis.

FIGURE 5.1 GRAPHICAL REPRESENTATION OF THE PROBABILITY DISTRIBUTION FOR THE NUMBER OF AUTOMOBILES SOLD DURING A DAY AT DICARLO MOTORS

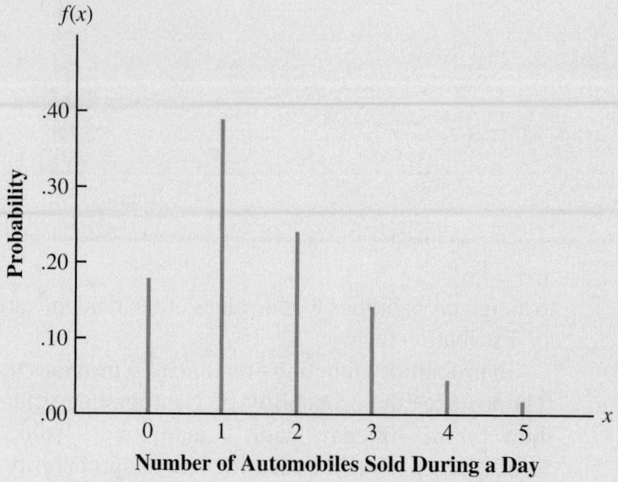

In addition to the probability distributions shown in tables, a formula that gives the probability function, $f(x)$, for every value of x is often used to describe probability distributions. The simplest example of a discrete probability distribution given by a formula is the **discrete uniform probability distribution**. Its probability function is defined by equation (5.3).

DISCRETE UNIFORM PROBABILITY FUNCTION

$$f(x) = 1/n \tag{5.3}$$

where

n = the number of values the random variable may assume

For example, consider again the experiment of rolling a die. We define the random variable x to be the number of dots on the upward face. For this experiment, $n = 6$ values are possible for the random variable; $x = 1, 2, 3, 4, 5, 6$. We showed earlier how the probability distribution for this experiment can be expressed as a table. Since the probabilities are equally likely, the discrete uniform probability function can also be used. The probability function for this discrete uniform random variable is

$$f(x) = 1/6 \qquad x = 1, 2, 3, 4, 5, 6$$

Several widely used discrete probability distributions are specified by formulas. Three important cases are the binomial, Poisson, and hypergeometric distributions; these distributions are discussed later in the chapter.

Exercises

Methods

SELF test

7. The probability distribution for the random variable x follows.

x	$f(x)$
20	.20
25	.15
30	.25
35	.40

a. Is this probability distribution valid? Explain.
b. What is the probability that $x = 30$?
c. What is the probability that x is less than or equal to 25?
d. What is the probability that x is greater than 30?

Applications

SELF test

8. The following data were collected by counting the number of operating rooms in use at Tampa General Hospital over a 20-day period: On three of the days only one operating room was used, on five of the days two were used, on eight of the days three were used, and on four days all four of the hospital's operating rooms were used.
   a. Use the relative frequency approach to construct an empirical discrete probability distribution for the number of operating rooms in use on any given day.
   b. Draw a graph of the probability distribution.
   c. Show that your probability distribution satisfies the required conditions for a valid discrete probability distribution.

9. For unemployed persons in the United States, the average number of months of unemployment at the end of December 2009 was approximately seven months (Bureau of Labor Statistics, January 2010). Suppose the following data are for a particular region in upstate New York. The values in the first column show the number of months unemployed and the values in the second column show the corresponding number of unemployed persons.

Months Unemployed	Number Unemployed
1	1029
2	1686
3	2269
4	2675
5	3487
6	4652
7	4145
8	3587
9	2325
10	1120

Let x be a random variable indicating the number of months a person is unemployed.
   a. Use the data to develop an empirical discrete probability distribution for x.
   b. Show that your probability distribution satisfies the conditions for a valid discrete probability distribution.

c. What is the probability that a person is unemployed for two months or less? Unemployed for more than two months?
d. What is the probability that a person is unemployed for more than six months?

10. The percent frequency distributions of job satisfaction scores for a sample of information systems (IS) senior executives and middle managers are as follows. The scores range from a low of 1 (very dissatisfied) to a high of 5 (very satisfied).

Job Satisfaction Score	IS Senior Executives (%)	IS Middle Managers (%)
1	5	4
2	9	10
3	3	12
4	42	46
5	41	28

a. Develop a probability distribution for the job satisfaction score of a senior executive.
b. Develop a probability distribution for the job satisfaction score of a middle manager.
c. What is the probability a senior executive will report a job satisfaction score of 4 or 5?
d. What is the probability a middle manager is very satisfied?
e. Compare the overall job satisfaction of senior executives and middle managers.

11. A technician services mailing machines at companies in the Phoenix area. Depending on the type of malfunction, the service call can take 1, 2, 3, or 4 hours. The different types of malfunctions occur at about the same frequency.
a. Develop a probability distribution for the duration of a service call.
b. Draw a graph of the probability distribution.
c. Show that your probability distribution satisfies the conditions required for a discrete probability function.
d. What is the probability a service call will take three hours?
e. A service call has just come in, but the type of malfunction is unknown. It is 3:00 P.M. and service technicians usually get off at 5:00 P.M. What is the probability the service technician will have to work overtime to fix the machine today?

12. The two largest cable providers are Comcast Cable Communications, with 21.5 million subscribers, and Time Warner Cable, with 11.0 million subscribers (*The New York Times Almanac*, 2007). Suppose that the management of Time Warner Cable subjectively assesses a probability distribution for the number of new subscribers next year in the state of New York as follows.

x	$f(x)$
100,000	.10
200,000	.20
300,000	.25
400,000	.30
500,000	.10
600,000	.05

a. Is this probability distribution valid? Explain.
b. What is the probability Time Warner will obtain more than 400,000 new subscribers?
c. What is the probability Time Warner will obtain fewer than 200,000 new subscribers?

13. A psychologist determined that the number of sessions required to obtain the trust of a new patient is either 1, 2, or 3. Let x be a random variable indicating the number of

sessions required to gain the patient's trust. The following probability function has been proposed.

$$f(x) = \frac{x}{6} \qquad \text{for } x = 1, 2, \text{ or } 3$$

a. Is this probability function valid? Explain.
b. What is the probability that it takes exactly 2 sessions to gain the patient's trust?
c. What is the probability that it takes at least 2 sessions to gain the patient's trust?

14. The following table is a partial probability distribution for the MRA Company's projected profits (x = profit in \$1000s) for the first year of operation (the negative value denotes a loss).

x	$f(x)$
−100	.10
0	.20
50	.30
100	.25
150	.10
200	

a. What is the proper value for $f(200)$? What is your interpretation of this value?
b. What is the probability that MRA will be profitable?
c. What is the probability that MRA will make at least \$100,000?

5.3 Expected Value and Variance

Expected Value

The **expected value**, or mean, of a random variable is a measure of the central location for the random variable. The formula for the expected value of a discrete random variable x follows.

The expected value is a weighted average of the values of the random variable where the weights are the probabilities.

EXPECTED VALUE OF A DISCRETE RANDOM VARIABLE

$$E(x) = \mu = \Sigma x f(x) \qquad (5.4)$$

Both the notations $E(x)$ and μ are used to denote the expected value of a random variable.

Equation (5.4) shows that to compute the expected value of a discrete random variable, we must multiply each value of the random variable by the corresponding probability $f(x)$ and then add the resulting products. Using the DiCarlo Motors automobile sales example from Section 5.2, we show the calculation of the expected value for the number of automobiles sold during a day in Table 5.5. The sum of the entries in the $xf(x)$ column shows that the expected value is 1.50 automobiles per day. We therefore know that although sales of 0, 1, 2, 3, 4, or 5 automobiles are possible on any one day, over time DiCarlo can anticipate selling an average of 1.50 automobiles per day. Assuming 30 days of operation during a month, we can use the expected value of 1.50 to forecast average monthly sales of 30(1.50) = 45 automobiles.

The expected value does not have to be a value the random variable can assume.

Variance

The expected value provides a measure of central tendency for a random variable, but we often also want a measure of variability, or dispersion. Just as we used the variance in Chapter 3 to summarize the variability in data, we now use **variance** to summarize the variability in the values of a random variable. The formula for the variance of a discrete random variable follows.

TABLE 5.5 CALCULATION OF THE EXPECTED VALUE FOR THE NUMBER OF AUTOMOBILES SOLD DURING A DAY AT DICARLO MOTORS

x	$f(x)$	$xf(x)$
0	.18	0(.18) = .00
1	.39	1(.39) = .39
2	.24	2(.24) = .48
3	.14	3(.14) = .42
4	.04	4(.04) = .16
5	.01	5(.01) = .05
		1.50

$E(x) = \mu = \Sigma xf(x)$

The variance is a weighted average of the squared deviations of a random variable from its mean. The weights are the probabilities.

VARIANCE OF A DISCRETE RANDOM VARIABLE

$$Var(x) = \sigma^2 = \Sigma(x - \mu)^2 f(x) \tag{5.5}$$

As equation (5.5) shows, an essential part of the variance formula is the deviation, $x - \mu$, which measures how far a particular value of the random variable is from the expected value, or mean, μ. In computing the variance of a random variable, the deviations are squared and then weighted by the corresponding value of the probability function. The sum of these weighted squared deviations for all values of the random variable is referred to as the *variance*. The notations $Var(x)$ and σ^2 are both used to denote the variance of a random variable.

The calculation of the variance for the probability distribution of the number of automobiles sold during a day at DiCarlo Motors is summarized in Table 5.6. We see that the variance is 1.25. The **standard deviation**, σ, is defined as the positive square root of the variance. Thus, the standard deviation for the number of automobiles sold during a day is

$$\sigma = \sqrt{1.25} = 1.118$$

The standard deviation is measured in the same units as the random variable ($\sigma = 1.118$ automobiles) and therefore is often preferred in describing the variability of a random variable. The variance σ^2 is measured in squared units and is thus more difficult to interpret.

TABLE 5.6 CALCULATION OF THE VARIANCE FOR THE NUMBER OF AUTOMOBILES SOLD DURING A DAY AT DICARLO MOTORS

x	$x - \mu$	$(x - \mu)^2$	$f(x)$	$(x - \mu)^2 f(x)$
0	0 − 1.50 = −1.50	2.25	.18	2.25(.18) = .4050
1	1 − 1.50 = −.50	.25	.39	.25(.39) = .0975
2	2 − 1.50 = .50	.25	.24	.25(.24) = .0600
3	3 − 1.50 = 1.50	2.25	.14	2.25(.14) = .3150
4	4 − 1.50 = 2.50	6.25	.04	6.25(.04) = .2500
5	5 − 1.50 = 3.50	12.25	.01	12.25(.01) = .1225
				1.2500

$\sigma^2 = \Sigma(x - \mu)^2 f(x)$

Exercises

Methods

15. The following table provides a probability distribution for the random variable x.

x	$f(x)$
3	.25
6	.50
9	.25

a. Compute $E(x)$, the expected value of x.
b. Compute σ^2, the variance of x.
c. Compute σ, the standard deviation of x.

SELF test

16. The following table provides a probability distribution for the random variable y.

y	$f(y)$
2	.20
4	.30
7	.40
8	.10

a. Compute $E(y)$.
b. Compute $Var(y)$ and σ.

Applications

17. The number of students taking the SAT has risen to an all-time high of more than 1.5 million (College Board, August 26, 2008). Students are allowed to repeat the test in hopes of improving the score that is sent to college and university admission offices. The number of times the SAT was taken and the number of students are as follows.

Number of Times	Number of Students
1	721,769
2	601,325
3	166,736
4	22,299
5	6730

a. Let x be a random variable indicating the number of times a student takes the SAT. Show the probability distribution for this random variable.
b. What is the probability that a student takes the SAT more than one time?
c. What is the probability that a student takes the SAT three or more times?
d. What is the expected value of the number of times the SAT is taken? What is your interpretation of the expected value?
e. What is the variance and standard deviation for the number of times the SAT is taken?

SELF test

18. The American Housing Survey reported the following data on the number of bedrooms in owner-occupied and renter-occupied houses in central cities (U.S. Census Bureau website, March 31, 2003).

Bedrooms	Number of Houses (1000s) Renter-Occupied	Owner-Occupied
0	547	23
1	5012	541
2	6100	3832
3	2644	8690
4 or more	557	3783

a. Define a random variable x = number of bedrooms in renter-occupied houses and develop a probability distribution for the random variable. (Let $x = 4$ represent 4 or more bedrooms.)
b. Compute the expected value and variance for the number of bedrooms in renter-occupied houses.
c. Define a random variable y = number of bedrooms in owner-occupied houses and develop a probability distribution for the random variable. (Let $y = 4$ represent 4 or more bedrooms.)
d. Compute the expected value and variance for the number of bedrooms in owner-occupied houses.
e. What observations can you make from a comparison of the number of bedrooms in renter-occupied versus owner-occupied homes?

19. The National Basketball Association (NBA) records a variety of statistics for each team. Two of these statistics are the percentage of field goals made by the team and the percentage of three-point shots made by the team. For a portion of the 2004 season, the shooting records of the 29 teams in the NBA showed the probability of scoring two points by making a field goal was .44, and the probability of scoring three points by making a three-point shot was .34 (NBA website, January 3, 2004).
a. What is the expected value of a two-point shot for these teams?
b. What is the expected value of a three-point shot for these teams?
c. If the probability of making a two-point shot is greater than the probability of making a three-point shot, why do coaches allow some players to shoot the three-point shot if they have the opportunity? Use expected value to explain your answer.

20. The probability distribution for damage claims paid by the Newton Automobile Insurance Company on collision insurance follows.

Payment ($)	Probability
0	.85
500	.04
1000	.04
3000	.03
5000	.02
8000	.01
10000	.01

a. Use the expected collision payment to determine the collision insurance premium that would enable the company to break even.
b. The insurance company charges an annual rate of $520 for the collision coverage. What is the expected value of the collision policy for a policyholder? (*Hint:* It is the expected payments from the company minus the cost of coverage.) Why does the policyholder purchase a collision policy with this expected value?

21. The following probability distributions of job satisfaction scores for a sample of information systems (IS) senior executives and middle managers range from a low of 1 (very dissatisfied) to a high of 5 (very satisfied).

Job Satisfaction Score	Probability IS Senior Executives	IS Middle Managers
1	.05	.04
2	.09	.10
3	.03	.12
4	.42	.46
5	.41	.28

a. What is the expected value of the job satisfaction score for senior executives?
b. What is the expected value of the job satisfaction score for middle managers?
c. Compute the variance of job satisfaction scores for executives and middle managers.
d. Compute the standard deviation of job satisfaction scores for both probability distributions.
e. Compare the overall job satisfaction of senior executives and middle managers.

22. The demand for a product of Carolina Industries varies greatly from month to month. The probability distribution in the following table, based on the past two years of data, shows the company's monthly demand.

Unit Demand	Probability
300	.20
400	.30
500	.35
600	.15

a. If the company bases monthly orders on the expected value of the monthly demand, what should Carolina's monthly order quantity be for this product?
b. Assume that each unit demanded generates \$70 in revenue and that each unit ordered costs \$50. How much will the company gain or lose in a month if it places an order based on your answer to part (a) and the actual demand for the item is 300 units?

23. The New York City Housing and Vacancy Survey showed a total of 59,324 rent-controlled housing units and 236,263 rent-stabilized units built in 1947 or later. For these rental units, the probability distributions for the number of persons living in the unit are given (U.S. Census Bureau website, January 12, 2004).

Number of Persons	Rent-Controlled	Rent-Stabilized
1	.61	.41
2	.27	.30
3	.07	.14
4	.04	.11
5	.01	.03
6	.00	.01

a. What is the expected value of the number of persons living in each type of unit?
b. What is the variance of the number of persons living in each type of unit?
c. Make some comparisons between the number of persons living in rent-controlled units and the number of persons living in rent-stabilized units.

24. The J. R. Ryland Computer Company is considering a plant expansion to enable the company to begin production of a new computer product. The company's president must determine whether to make the expansion a medium- or large-scale project. Demand for the new

product is uncertain, which for planning purposes may be low demand, medium demand, or high demand. The probability estimates for demand are .20, .50, and .30, respectively. Letting x and y indicate the annual profit in thousands of dollars, the firm's planners developed the following profit forecasts for the medium- and large-scale expansion projects.

		Medium-Scale Expansion Profit		Large-Scale Expansion Profit	
		x	$f(x)$	y	$f(y)$
	Low	50	.20	0	.20
Demand	**Medium**	150	.50	100	.50
	High	200	.30	300	.30

a. Compute the expected value for the profit associated with the two expansion alternatives. Which decision is preferred for the objective of maximizing the expected profit?
b. Compute the variance for the profit associated with the two expansion alternatives. Which decision is preferred for the objective of minimizing the risk or uncertainty?

5.4 Bivariate Distributions, Covariance, and Financial Portfolios

A probability distribution involving two random variables is called a **bivariate probability distribution**. In discussing bivariate probability distributions, it is useful to think of a bivariate experiment. Each outcome for a bivariate experiment consists of two values, one for each random variable. For example, consider the bivariate experiment of rolling a pair of dice. The outcome consists of two values, the number obtained with the first die and the number obtained with the second die. As another example, consider the experiment of observing the financial markets for a year and recording the percentage gain for a stock fund and a bond fund. Again, the experimental outcome provides a value for two random variables, the percent gain in the stock fund and the percent gain in the bond fund. When dealing with bivariate probability distributions, we are often interested in the relationship between the random variables. In this section, we introduce bivariate distributions and show how the covariance and correlation coefficient can be used as a measure of linear association between the random variables. We shall also see how bivariate probability distributions can be used to construct and analyze financial portfolios.

A Bivariate Empirical Discrete Probability Distribution

Recall that in Section 5.2 we developed an empirical discrete distribution for daily sales at the DiCarlo Motors automobile dealership in Saratoga, New York. DiCarlo has another dealership in Geneva, New York. Table 5.7 shows the number of cars sold at each of the dealerships over a 300-day period. The numbers in the bottom (total) row are the frequencies we used to develop an empirical probability distribution for daily sales at DiCarlo's Saratoga dealership in Section 5.2. The numbers in the right-most (total) column are the frequencies of daily sales for the Geneva dealership. Entries in the body of the table give the number of days the Geneva dealership had a level of sales indicated by the row, when the Saratoga dealership had the level of sales indicated by the column. For example, the entry of 33 in the Geneva dealership row labeled 1 and the Saratoga column labeled 2 indicates that for 33 days out of the 300, the Geneva dealership sold 1 car and the Saratoga dealership sold 2 cars.

TABLE 5.7 NUMBER OF AUTOMOBILES SOLD AT DICARLO'S SARATOGA AND GENEVA DEALERSHIPS OVER 300 DAYS

	Saratoga Dealership						
Geneva Dealership	**0**	**1**	**2**	**3**	**4**	**5**	**Total**
0	21	30	24	9	2	0	86
1	21	36	33	18	2	1	111
2	9	42	9	12	3	2	77
3	3	9	6	3	5	0	26
Total	54	117	72	42	12	3	300

Suppose we consider the bivariate experiment of observing a day of operations at DiCarlo Motors and recording the number of cars sold. Let us define x = number of cars sold at the Geneva dealership and y = the number of cars sold at the Saratoga dealership. We can now divide all of the frequencies in Table 5.7 by the number of observations (300) to develop a bivariate empirical discrete probability distribution for automobile sales at the two DiCarlo dealerships. Table 5.8 shows this bivariate discrete probability distribution. The probabilities in the lower margin provide the marginal distribution for the DiCarlo Motors Saratoga dealership. The probabilities in the right margin provide the marginal distribution for the DiCarlo Motors Geneva dealership.

The probabilities in the body of the table provide the bivariate probability distribution for sales at both dealerships. Bivariate probabilities are often called joint probabilities. We see that the joint probability of selling 0 automobiles at Geneva and 1 automobile at Saratoga on a typical day is $f(0, 1) = .1000$, the joint probability of selling 1 automobile at Geneva and 4 automobiles at Saratoga on a typical day is .0067, and so on. Note that there is one bivariate probability for each experimental outcome. With 4 possible values for x and 6 possible values for y, there are 24 experimental outcomes and bivariate probabilities.

Suppose we would like to know the probability distribution for total sales at both DiCarlo dealerships and the expected value and variance of total sales. We can define $s = x + y$ as total sales for DiCarlo Motors. Working with the bivariate probabilities in Table 5.8, we see that $f(s = 0) = .0700$, $f(s = 1) = .0700 + .1000 = .1700$, $f(s = 2) = .0300 + .1200 + .0800 = .2300$, and so on. We show the complete probability distribution for $s = x + y$ along with the computation of the expected value and variance in Table 5.9. The expected value is $E(s) = 2.6433$ and the variance is $Var(s) = 2.3895$.

With bivariate probability distributions, we often want to know the relationship between the two random variables. The covariance and/or correlation coefficient are good

TABLE 5.8 BIVARIATE EMPIRICAL DISCRETE PROBABILITY DISTRIBUTION FOR DAILY SALES AT DICARLO DEALERSHIPS IN SARATOGA AND GENEVA, NEW YORK

	Saratoga Dealership						
Geneva Dealership	**0**	**1**	**2**	**3**	**4**	**5**	**Total**
0	.0700	.1000	.0800	.0300	.0067	.0000	.2867
1	.0700	.1200	.1100	.0600	.0067	.0033	.3700
2	.0300	.1400	.0300	.0400	.0100	.0067	.2567
3	.0100	.0300	.0200	.0100	.0167	.0000	.0867
Total	.18	.39	.24	.14	.04	.01	1.0000

TABLE 5.9 CALCULATION OF THE EXPECTED VALUE AND VARIANCE FOR TOTAL DAILY SALES AT DICARLO MOTORS

s	$f(s)$	$sf(s)$	$s - E(s)$	$(s - E(s))^2$	$(s - E(s))^2 f(s)$
0	.0700	.0000	−2.6433	6.9872	.4891
1	.1700	.1700	−1.6433	2.7005	.4591
2	.2300	.4600	−0.6433	0.4139	.0952
3	.2900	.8700	0.3567	0.1272	.0369
4	.1267	.5067	1.3567	1.8405	.2331
5	.0667	.3333	2.3567	5.5539	.3703
6	.0233	.1400	3.3567	11.2672	.2629
7	.0233	.1633	4.3567	18.9805	.4429
8	.0000	.0000	5.3567	28.6939	.0000
		$E(s) = 2.6433$			$Var(s) = 2.3895$

measures of association between two random variables. We saw in Chapter 3 how to compute the covariance and correlation coefficient for sample data. The formula we will use for computing the covariance between two random variables x and y is given below.

COVARIANCE OF RANDOM VARIABLES x AND y[1]

$$\sigma_{xy} = [Var(x + y) - Var(x) - Var(y)]/2 \tag{5.6}$$

We have already computed $Var(s) = Var(x + y)$ and, in Section 5.2, we computed $Var(y)$. Now we need to compute $Var(x)$ before we can use equation (5.6) to compute the covariance of x and y. Using the probability distribution for x (the right margin of Table 5.8), we compute $E(x)$ and $Var(x)$ in Table 5.10.

We can now use equation (5.6) to compute the covariance of the random variables x and y.

$$\sigma_{xy} = [Var(x + y) - Var(x) - Var(y)]/2 = (2.3895 - .8696 - 1.25)/2 = .1350$$

A covariance of .1350 indicates that daily sales at DiCarlo's two dealerships have a positive relationship. To get a better sense of the strength of the relationship we can compute the correlation coefficient. The correlation coefficient for the two random variables x and y is given by equation (5.7).

TABLE 5.10 CALCULATION OF THE EXPECTED VALUE AND VARIANCE OF DAILY AUTOMOBILE SALES AT DICARLO MOTORS' GENEVA DEALERSHIP

x	$f(x)$	$xf(x)$	$x - E(x)$	$[(x - E(x)]^2$	$[x - E(x)]^2 f(x)$
0	.2867	.0000	−1.1435	1.3076	.3749
1	.3700	.3700	−.1435	0.0206	.0076
2	.2567	.5134	.8565	0.8565	.1883
3	.0867	.2601	1.8565	1.8565	.2988
		$E(x) = 1.1435$			$Var(x) = .8696$

[1]Another formula is often used to compute the covariance of x and y when $Var(x + y)$ is not known. It is $\sigma_{xy} = \sum_{i,j}[x_i - E(x_i)][y_j - E(y_j)]f(x_i,y_j)$.

CORRELATION BETWEEN RANDOM VARIABLES x AND y

$$\rho_{xy} = \frac{\sigma_{xy}}{\sigma_x \sigma_y} \tag{5.7}$$

From equation (5.7), we see that the correlation coefficient for two random variables is the covariance divided by the product of the standard deviations for the two random variables.

Let us compute the correlation coefficient between daily sales at the two DiCarlo dealerships. First we compute the standard deviations for sales at the Saratoga and Geneva dealerships by taking the square root of the variance.

$$\sigma_x = \sqrt{.8696} = .9325$$
$$\sigma_y = \sqrt{1.25} = 1.1180$$

Now we can compute the correlation coefficient as a measure of the linear association between the two random variables.

$$\rho_{xy} = \frac{\sigma_{xy}}{\sigma_x \sigma_y} = \frac{.1350}{(.9325)(1.1180)} = .1295$$

In Chapter 3 we defined the correlation coefficient as a measure of the linear association between two variables. Values near +1 indicate a strong positive linear relationship; values near −1 indicate a strong negative linear relationship; and values near zero indicate a lack of a linear relationship. This interpretation is also valid for random variables. The correlation coefficient of .1295 indicates there is a weak positive relationship between the random variables representing daily sales at the two DiCarlo dealerships. If the correlation coefficient had equaled zero, we would have concluded that daily sales at the two dealerships were independent.

Financial Applications

Let us now see how what we have learned can be useful in constructing financial portfolios that provide a good balance of risk and return. A financial advisor is considering four possible economic scenarios for the coming year and has developed a probability distribution showing the percent return, x, for investing in a large-cap stock fund and the percent return, y, for investing in a long-term government bond fund given each of the scenarios. The bivariate probability distribution for x and y is shown in Table 5.11. Table 5.11 is simply a list with a

TABLE 5.11 PROBABILITY DISTRIBUTION OF PERCENT RETURNS FOR INVESTING IN A LARGE-CAP STOCK FUND, x, AND INVESTING IN A LONG-TERM GOVERNMENT BOND FUND, y

Economic Scenario	Probability $f(x, y)$	Large-Cap Stock Fund (x)	Long-Term Government Bond Fund (y)
Recession	.10	−40	30
Weak growth	.25	5	5
Stable growth	.50	15	4
Strong growth	.15	30	2

separate row for each experimental outcome (economic scenario). Each row contains the joint probability for the experimental outcome and a value for each random variable. Since there are only 4 joint probabilities, the tabular form used in Table 5.11 is simpler than the one we used for DiCarlo Motors where there were (4)(6) = 24 joint probabilities.

Using the formula in Section 5.3 for computing the expected value of a single random variable, we can compute the expected percent return for investing in the stock fund, $E(x)$, and the expected percent return for investing in the bond fund, $E(y)$.

$$E(x) = .10(-40) + .25(5) + .5(15) + .15(30) = 9.25$$

$$E(y) = .10(30) + .25(5) + .5(4) + .15(2) = 6.55$$

Using this information, we might conclude that investing in the stock fund is a better investment. It has a higher expected return, 9.25%. But, financial analysts recommend that investors also consider the risk associated with an investment. The standard deviation of percent return is often used as a measure of risk. To compute the standard deviation, we must first compute the variance. Using the formula in Section 5.3 for computing the variance of a single random variable, we can compute the variance of the percent returns for the stock and bond fund investments.

$$Var(x) = .1(-40 - 9.25)^2 + .25(5 - 9.25)^2 + .50(15 - 9.25)^2 + .15(30 - 9.25)^2 = 328.1875$$

$$Var(y) = .1(30 - 6.55)^2 + .25(5 - 6.55)^2 + .50(4 - 6.55)^2 + .15(2 - 6.55)^2 = 61.9475$$

The standard deviation of the return from an investment in the stock fund is $\sigma_x = \sqrt{328.1875} = 18.1159\%$ and the standard deviation of the return from an investment in the bond fund is $\sigma_y = \sqrt{61.9475} = 7.8707\%$. So, we can conclude that investing in the bond fund is less risky. It has the smaller standard deviation. We have already seen that the stock fund offers a greater expected return, so if we want to choose between investing in either the stock fund or the bond fund it depends on our attitude toward risk and return. An aggressive investor might choose the stock fund because of the higher expected return; a conservative investor might choose the bond fund because of the lower risk. But, there are other options. What about the possibility of investing in a portfolio consisting of both an investment in the stock fund and an investment in the bond fund?

Suppose we would like to consider three alternatives: investing solely in the large-cap stock fund, investing solely in the long-term government bond fund, and splitting our funds equally between the stock fund and the bond fund (one-half in each). We have already computed the expected value and standard deviation for investing solely in the stock fund and the bond fund. Let us now evaluate the third alternative: constructing a portfolio by investing equal amounts in the large-cap stock fund and in the long-term government bond fund.

To evaluate this portfolio, we start by computing its expected return. We have previously defined x as the percent return from an investment in the stock fund and y as the percent return from an investment in the bond fund so the percent return for our portfolio is $r = .5x + .5y$. To find the expected return for a portfolio with one-half invested in the stock fund and one-half invested in the bond fund, we want to compute $E(r) = E(.5x + .5y)$. The expression $.5x + .5y$ is called a linear combination of the random variables x and y. Equation 5.8 provides an easy method for computing the expected value of a linear combination of the random variables x and y when we already know $E(x)$ and $E(y)$. In equation (5.8), a represents the coefficient of x and b represents the coefficient of y in the linear combination.

EXPECTED VALUE OF A LINEAR COMBINATION OF RANDOM VARIABLES x AND y

$$E(ax + by) = aE(x) + bE(y) \tag{5.8}$$

Since we have already computed $E(x) = 9.25$ and $E(y) = 6.55$, we can use equation (5.8) to compute the expected value of our portfolio.

$$E(.5x + .5y) = .5E(x) + .5E(y) = .5(9.25) + .5(6.55) = 7.9$$

We see that the expected return for investing in the portfolio is 7.9%. With \$100 invested, we would expect a return of \$100(.079) = \$7.90; with \$1000 invested we would expect a return of \$1000(.079) = \$79.00; and so on. But, what about the risk? As mentioned previously, financial analysts often use the standard deviation as a measure of risk.

Our portfolio is a linear combination of two random variables, so we need to be able to compute the variance and standard deviation of a linear combination of two random variables in order to assess the portfolio risk. When the covariance between two random variables is known, the formula given by equation (5.9) can be used to compute the variance of a linear combination of two random variables.

VARIANCE OF A LINEAR COMBINATION OF TWO RANDOM VARIABLES

$$Var(ax + by) = a^2Var(x) + b^2Var(y) + 2ab\sigma_{xy} \tag{5.9}$$

where σ_{xy} is the covariance of x and y

From equation (5.9), we see that both the variance of each random variable individually and the covariance between the random variables are needed to compute the variance of a linear combination of two random variables and hence the variance of our portfolio.

We computed $Var(x + y) = 119.46$ the same way we did for DiCarlo Motors in the previous subsection.

We have already computed the variance of each random variable individually: $Var(x) = 328.1875$ and $Var(y) = 61.9475$. Also, it can be shown that $Var(x + y) = 119.46$. So, using equation (5.6), the covariance of the random variables x and y is

$$\sigma_{xy} = [Var(x + y) - Var(x) - Var(y)]/2 = [119.46 - 328.1875 - 61.9475]/2 = -135.3375$$

A negative covariance between x and y, such as this, means that when x tends to be above its mean, y tends to be below its mean and vice versa.

We can now use equation (5.9) to compute the variance of return for our portfolio.

$$Var(.5x + .5y) = .5^2(328.1875) + .5^2(61.9475) + 2(.5)(.5)(-135.3375) = 29.865$$

The standard deviation of our portfolio is then given by $\sigma_{.5x+.5y} = \sqrt{29.865} = 5.4650\%$. This is our measure of risk for the portfolio consisting of investing 50% in the stock fund and 50% in the bond fund.

Perhaps we would now like to compare the three investment alternatives: investing solely in the stock fund, investing solely in the bond fund, or creating a portfolio by dividing our investment amount equally between the stock and bond funds. Table 5.12 shows the expected returns, variances, and standard deviations for each of the three alternatives.

TABLE 5.12 EXPECTED VALUES, VARIANCES, AND STANDARD DEVIATIONS FOR THREE INVESTMENT ALTERNATIVES

Investment Alternative	Expected Return (%)	Variance of Return	Standard Deviation of Return (%)
100% in Stock Fund	9.25	328.1875	18.1159
100% in Bond Fund	6.55	61.9475	7.8707
Portfolio (50% in Stock fund, 50% in Bond fund)	7.90	29.865	5.4650

Which of these alternatives would you prefer? The expected return is highest for investing 100% in the stock fund, but the risk is also highest. The standard deviation is 18.1159%. Investing 100% in the bond fund has a lower expected return, but a significantly smaller risk. Investing 50% in the stock fund and 50% in the bond fund (the portfolio) has an expected return that is halfway between that of the stock fund alone and the bond fund alone. But note that it has less risk than investing 100% in either of the individual funds. Indeed, it has both a higher return and less risk (smaller standard deviation) than investing solely in the bond fund. So we would say that investing in the portfolio dominates the choice of investing solely in the bond fund.

Whether you would choose to invest in the stock fund or the portfolio depends on your attitude toward risk. The stock fund has a higher expected return. But the portfolio has significantly less risk and also provides a fairly good return. Many would choose it. It is the negative covariance between the stock and bond funds that has caused the portfolio risk to be so much smaller than the risk of investing solely in either of the individual funds.

The portfolio analysis we just performed was for investing 50% in the stock fund and the other 50% in the bond fund. How would you calculate the expected return and the variance for other portfolios? Equations (5.8) and (5.9) can be used to make these calculations easily.

Suppose we wish to create a portfolio by investing 25% in the stock fund and 75% in the bond fund? What are the expected value and variance of this portfolio? The percent return for this portfolio is $r = .25x + .75y$, so we can use equation (5.8) to get the expected value of this portfolio:

$$E(.25x + .75y) = .25E(x) + .75E(y) = .25(9.25) + .75(6.55) = 7.225$$

Likewise, we may calculate the variance of the portfolio using equation (5.9):

$$\begin{aligned} Var(.25x + .75y) &= (.25)^2 Var(x) + (.75)^2\, Var(y) + 2(.25)(.75)\sigma_{xy} \\ &= .0625(328.1875) + (.5625)(61.9475) + (.375)(-135.3375) \\ &= 4.6056 \end{aligned}$$

The standard deviation of the new portfolio is $\sigma_{.25x+.75y} = \sqrt{4.6056} = 2.1461$.

Summary

We have introduced bivariate discrete probability distributions in this section. Since such distributions involve two random variables, we are often interested in a measure of association between the variables. The covariance and the correlation coefficient are the two measures we introduced and showed how to compute. A correlation coefficient near 1 or -1 indicates a strong correlation between the two random variables, a correlation coefficient near zero indicates a weak correlation between the variables. If two random variables are independent, the covariance and the correlation coefficient will equal zero.

We also showed how to compute the expected value and variance of linear combinations of random variables. From a statistical point of view, financial portfolios are linear combinations of random variables. They are actually a special kind of linear combination called a weighted average. The coefficients are nonnegative and add to 1. The portfolio example we presented showed how to compute the expected value and variance for a portfolio consisting of an investment in a stock fund and a bond fund. The same methodology can be used to compute the expected value and variance of a portfolio consisting of any two financial assets. It is the effect of covariance between the individual random variables on the variance of the portfolio that is the basis for much of the theory of reducing portfolio risk by diversifying across investment alternatives.

NOTES AND COMMENTS

1. Equations (5.8) and (5.9), along with their extensions to three or more random variables, are key building blocks in financial portfolio construction and analysis.
2. Equations (5.8) and (5.9) for computing the expected value and variance of a linear combination of two random variables can be extended to three or more random variables. The extension of equation (5.8) is straightforward; one more term is added for each additional random variable. The extension of equation (5.9) is more complicated because a separate term is needed for the covariance between all pairs of random variables. We leave these extensions to more advanced books.
3. The covariance term of equation (5.9) shows why negatively correlated random variables (investment alternatives) reduce the variance and, hence, the risk of a portfolio.

Exercises

Methods

SELF test

25. Given below is a bivariate distribution for the random variables x and y.

$f(x, y)$	x	y
.2	50	80
.5	30	50
.3	40	60

a. Compute the expected value and the variance for x and y.
b. Develop a probability distribution for $x + y$.
c. Using the result of part (b), compute $E(x + y)$ and $Var\ (x + y)$.
d. Compute the covariance and correlation for x and y. Are x and y positively related, negatively related, or unrelated?
e. Is the variance of the sum of x and y bigger, smaller, or the same as the sum of the individual variances? Why?

26. A person is interested in constructing a portfolio. Two stocks are being considered. Let x = percent return for an investment in stock 1, and y = percent return for an investment in stock 2. The expected return and variance for stock 1 are $E(x) = 8.45\%$ and $Var(x) = 25$. The expected return and variance for stock 2 are $E(y) = 3.20\%$ and $Var(y) = 1$. The covariance between the returns is $\sigma_{xy} = -3$.
a. What is the standard deviation for an investment in stock 1 and for an investment in stock 2? Using the standard deviation as a measure of risk, which of these stocks is the riskier investment?
b. What is the expected return and standard deviation, in dollars, for a person who invests \$500 in stock 1?
c. What is the expected percent return and standard deviation for a person who constructs a portfolio by investing 50% in each stock?
d. What is the expected percent return and standard deviation for a person who constructs a portfolio by investing 70% in stock 1 and 30% in stock 2?
e. Compute the correlation coefficient for x and y and comment on the relationship between the returns for the two stocks.

SELF test

27. The Chamber of Commerce in a Canadian city has conducted an evaluation of 300 restaurants in its metropolitan area. Each restaurant received a rating on a 3-point scale on typical meal price (1 least expensive to 3 most expensive) and quality (1 lowest quality to 3 greatest quality). A crosstabulation of the rating data is shown below. Forty-two of the restaurants

received a rating of 1 on quality and 1 on meal price, 39 of the restaurants received a rating of 1 on quality and 2 on meal price, and so on. Forty-eight of the restaurants received the highest rating of 3 on both quality and meal price.

	Meal Price (y)			
Quality (x)	**1**	**2**	**3**	**Total**
1	42	39	3	84
2	33	63	54	150
3	3	15	48	66
Total	78	117	105	300

a. Develop a bivariate probability distribution for quality and meal price of a randomly selected restaurant in this Canadian city. Let x = quality rating and y = meal price.
b. Compute the expected value and variance for quality rating, x.
c. Compute the expected value and variance for meal price, y.
d. The $Var(x + y) = 1.6691$. Compute the covariance of x and y. What can you say about the relationship between quality and meal price? Is this what you would expect?
e. Compute the correlation coefficient between quality and meal price? What is the strength of the relationship? Do you suppose it is likely to find a low cost restaurant in this city that is also high quality? Why or why not?

28. PortaCom has developed a design for a high-quality portable printer. The two key components of manufacturing cost are direct labor and parts. During a testing period, the company has developed prototypes and conducted extensive product tests with the new printer. PortaCom's engineers have developed the bivariate probability distribution shown below for the manufacturing costs. Parts cost (in dollars) per printer is represented by the random variable x and direct labor cost (in dollars) per printer is represented by the random variable y. Management would like to use this probability distribution to estimate manufacturing costs.

	Direct Labor (y)			
Parts (x)	**43**	**45**	**48**	**Total**
85	0.05	0.2	0.2	0.45
95	0.25	0.2	0.1	0.55
Total	0.30	0.4	0.3	1.00

a. Show the marginal distribution of direct labor cost and compute its expected value, variance, and standard deviation.
b. Show the marginal distribution of parts cost and compute its expected value, variance, and standard deviation.
c. Total manufacturing cost per unit is the sum of direct labor cost and parts cost. Show the probability distribution for total manufacturing cost per unit.
d. Compute the expected value, variance, and standard deviation of total manufacturing cost per unit.
e. Are direct labor and parts costs independent? Why or why not? If you conclude that they are not, what is the relationship between direct labor and parts cost?
f. PortaCom produced 1500 printers for its product introduction. The total manufacturing cost was $198,350. Is that about what you would expect? If it is higher or lower, what do you think may have caused it?

29. J.P. Morgan Asset Management publishes information about financial investments. Over the past 10 years, the expected return for the S&P 500 was 5.04% with a standard deviation of 19.45% and the expected return over that same period for a core bonds fund was 5.78% with a standard deviation of 2.13% (*J.P. Morgan Asset Management, Guide to the Markets*, 1st Quarter, 2012). The publication also reported that the correlation between the S&P 500 and core bonds is −.32. You are considering portfolio investments that are composed of an S&P 500 index fund and a core bonds fund.
    a. Using the information provided, determine the covariance between the S&P 500 and core bonds.
    b. Construct a portfolio that is 50% invested in an S&P 500 index fund and 50% in a core bonds fund. In percentage terms, what are the expected return and standard deviation for such a portfolio?
    c. Construct a portfolio that is 20% invested in an S&P 500 index fund and 80% invested in a core bonds fund. In percentage terms, what are the expected return and standard deviation for such a portfolio?
    d. Construct a portfolio that is 80% invested in an S&P 500 index fund and 20% invested in a core bonds fund. In percentage terms, what are the expected return and standard deviation for such a portfolio?
    e. Which of the portfolios in parts (b), (c), and (d) has the largest expected return? Which has the smallest standard deviation? Which of these portfolios is the best investment alternative?
    f. Discuss the advantages and disadvantages of investing in the three portfolios in parts (b), (c), and (d). Would you prefer investing all your money in the S&P 500 index, the core bonds fund, or one of the three portfolios? Why?
30. In addition to the information in exercise 29 on the S&P 500 and core bonds, J.P. Morgan Asset Management reported that the expected return for real estate investment trusts (REITs) was 13.07% with a standard deviation of 23.17% (*J.P. Morgan Asset Management, Guide to the Markets*, 1st Quarter, 2012). The correlation between the S&P 500 and REITs is .74 and the correlation between core bonds and REITs is −.04. You are considering portfolio investments that are composed of an S&P 500 index fund and REITs as well as portfolio investments composed of a core bonds fund and REITs.
    a. Using the information provided here and in exercise 29, determine the covariance between the S&P 500 and REITs and between core bonds and REITs.
    b. Construct a portfolio that is 50% invested in an S&P 500 fund and 50% invested in REITs. In percentage terms, what are the expected return and standard deviation for such a portfolio?
    c. Construct a portfolio that is 50% invested in a core bonds fund and 50% invested in REITs. In percentage terms, what are the expected return and standard deviation for such a portfolio?
    d. Construct a portfolio that is 80% invested in a core bonds fund and 20% invested in REITs. In percentage terms, what are the expected return and standard deviation for such a portfolio?
    e. Which of the portfolios in parts (b), (c), and (d) would you recommend to an aggressive investor? Which would you recommend to a conservative investor? Why?

5.5 Binomial Probability Distribution

The binomial probability distribution is a discrete probability distribution that has many applications. It is associated with a multiple-step experiment that we call the binomial experiment.

A Binomial Experiment

A **binomial experiment** exhibits the following four properties.

PROPERTIES OF A BINOMIAL EXPERIMENT

1. The experiment consists of a sequence of n identical trials.
2. Two outcomes are possible on each trial. We refer to one outcome as a *success* and the other outcome as a *failure.*
3. The probability of a success, denoted by p, does not change from trial to trial. Consequently, the probability of a failure, denoted by $1 - p$, does not change from trial to trial.
4. The trials are independent.

Jakob Bernoulli (1654–1705), the first of the Bernoulli family of Swiss mathematicians, published a treatise on probability that contained the theory of permutations and combinations, as well as the binomial theorem.

If properties 2, 3, and 4 are present, we say the trials are generated by a Bernoulli process. If, in addition, property 1 is present, we say we have a binomial experiment. Figure 5.2 depicts one possible sequence of successes and failures for a binomial experiment involving eight trials.

In a binomial experiment, our interest is in the *number of successes occurring in the n trials*. If we let x denote the number of successes occurring in the n trials, we see that x can assume the values of 0, 1, 2, 3, . . . , n. Because the number of values is finite, x is a *discrete* random variable. The probability distribution associated with this random variable is called the **binomial probability distribution**. For example, consider the experiment of tossing a coin five times and on each toss observing whether the coin lands with a head or a tail on its upward face. Suppose we want to count the number of heads appearing over the five tosses. Does this experiment show the properties of a binomial experiment? What is the random variable of interest? Note that:

1. The experiment consists of five identical trials; each trial involves the tossing of one coin.
2. Two outcomes are possible for each trial: a head or a tail. We can designate head a success and tail a failure.
3. The probability of a head and the probability of a tail are the same for each trial, with $p = .5$ and $1 - p = .5$.
4. The trials or tosses are independent because the outcome on any one trial is not affected by what happens on other trials or tosses.

FIGURE 5.2 ONE POSSIBLE SEQUENCE OF SUCCESSES AND FAILURES FOR AN EIGHT-TRIAL BINOMIAL EXPERIMENT

Property 1: The experiment consists of $n = 8$ identical trials.

Property 2: Each trial results in either success (S) or failure (F).

Trials →	1	2	3	4	5	6	7	8
Outcomes →	S	F	F	S	S	F	S	S

Thus, the properties of a binomial experiment are satisfied. The random variable of interest is x = the number of heads appearing in the five trials. In this case, x can assume the values of 0, 1, 2, 3, 4, or 5.

As another example, consider an insurance salesperson who visits 10 randomly selected families. The outcome associated with each visit is classified as a success if the family purchases an insurance policy and a failure if the family does not. From past experience, the salesperson knows the probability that a randomly selected family will purchase an insurance policy is .10. Checking the properties of a binomial experiment, we observe that:

1. The experiment consists of 10 identical trials; each trial involves contacting one family.
2. Two outcomes are possible on each trial: the family purchases a policy (success) or the family does not purchase a policy (failure).
3. The probabilities of a purchase and a nonpurchase are assumed to be the same for each sales call, with $p = .10$ and $1 - p = .90$.
4. The trials are independent because the families are randomly selected.

Because the four assumptions are satisfied, this example is a binomial experiment. The random variable of interest is the number of sales obtained in contacting the 10 families. In this case, x can assume the values of 0, 1, 2, 3, 4, 5, 6, 7, 8, 9, and 10.

Property 3 of the binomial experiment is called the *stationarity assumption* and is sometimes confused with property 4, independence of trials. To see how they differ, consider again the case of the salesperson calling on families to sell insurance policies. If, as the day wore on, the salesperson got tired and lost enthusiasm, the probability of success (selling a policy) might drop to .05, for example, by the tenth call. In such a case, property 3 (stationarity) would not be satisfied, and we would not have a binomial experiment. Even if property 4 held—that is, the purchase decisions of each family were made independently—it would not be a binomial experiment if property 3 was not satisfied.

In applications involving binomial experiments, a special mathematical formula, called the *binomial probability function,* can be used to compute the probability of x successes in the n trials. Using probability concepts introduced in Chapter 4, we will show in the context of an illustrative problem how the formula can be developed.

Martin Clothing Store Problem

Let us consider the purchase decisions of the next three customers who enter the Martin Clothing Store. On the basis of past experience, the store manager estimates the probability that any one customer will make a purchase is .30. What is the probability that two of the next three customers will make a purchase?

Using a tree diagram (Figure 5.3), we can see that the experiment of observing the three customers each making a purchase decision has eight possible outcomes. Using S to denote success (a purchase) and F to denote failure (no purchase), we are interested in experimental outcomes involving two successes in the three trials (purchase decisions). Next, let us verify that the experiment involving the sequence of three purchase decisions can be viewed as a binomial experiment. Checking the four requirements for a binomial experiment, we note that:

1. The experiment can be described as a sequence of three identical trials, one trial for each of the three customers who will enter the store.
2. Two outcomes—the customer makes a purchase (success) or the customer does not make a purchase (failure)—are possible for each trial.
3. The probability that the customer will make a purchase (.30) or will not make a purchase (.70) is assumed to be the same for all customers.
4. The purchase decision of each customer is independent of the decisions of the other customers.

FIGURE 5.3 TREE DIAGRAM FOR THE MARTIN CLOTHING STORE PROBLEM

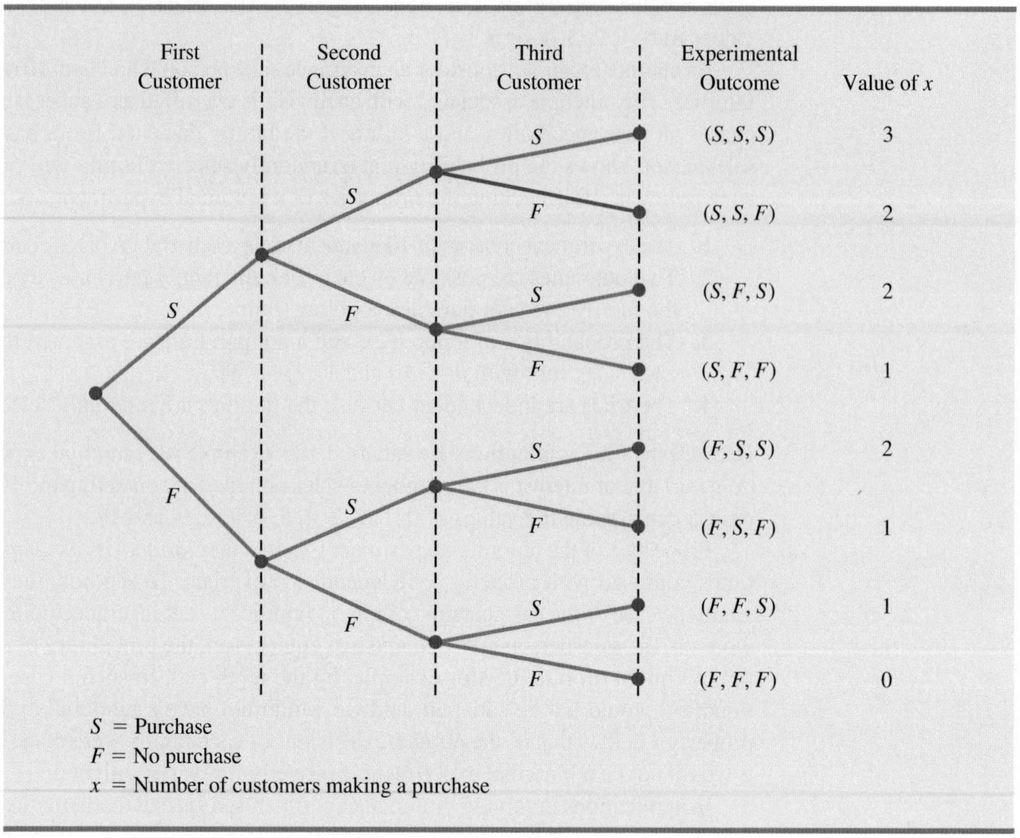

Hence, the properties of a binomial experiment are present.

The number of experimental outcomes resulting in exactly x successes in n trials can be computed using the following formula.[2]

NUMBER OF EXPERIMENTAL OUTCOMES PROVIDING EXACTLY x SUCCESSES IN n TRIALS

$$\binom{n}{x} = \frac{n!}{x!(n-x)!} \qquad \textbf{(5.10)}$$

where

$$n! = n(n-1)(n-2)\cdots(2)(1)$$

and, by definition,

$$0! = 1$$

Now let us return to the Martin Clothing Store experiment involving three customer purchase decisions. Equation (5.10) can be used to determine the number of experimental

[2]This formula, introduced in Chapter 4, determines the number of combinations of n objects selected x at a time. For the binomial experiment, this combinatorial formula provides the number of experimental outcomes (sequences of n trials) resulting in x successes.

outcomes involving two purchases; that is, the number of ways of obtaining $x = 2$ successes in the $n = 3$ trials. From equation (5.10) we have

$$\binom{n}{x} = \binom{3}{2} = \frac{3!}{2!(3-2)!} = \frac{(3)(2)(1)}{(2)(1)(1)} = \frac{6}{2} = 3$$

Equation (5.10) shows that three of the experimental outcomes yield two successes. From Figure 5.3 we see these three outcomes are denoted by (S, S, F), (S, F, S), and (F, S, S).

Using equation (5.10) to determine how many experimental outcomes have three successes (purchases) in the three trials, we obtain

$$\binom{n}{x} = \binom{3}{3} = \frac{3!}{3!(3-3)!} = \frac{3!}{3!0!} = \frac{(3)(2)(1)}{3(2)(1)(1)} = \frac{6}{6} = 1$$

From Figure 5.3 we see that the one experimental outcome with three successes is identified by (S, S, S).

We know that equation (5.10) can be used to determine the number of experimental outcomes that result in x successes in n trials. If we are to determine the probability of x successes in n trials, however, we must also know the probability associated with each of these experimental outcomes. Because the trials of a binomial experiment are independent, we can simply multiply the probabilities associated with each trial outcome to find the probability of a particular sequence of successes and failures.

The probability of purchases by the first two customers and no purchase by the third customer, denoted (S, S, F), is given by

$$pp(1-p)$$

With a .30 probability of a purchase on any one trial, the probability of a purchase on the first two trials and no purchase on the third is given by

$$(.30)(.30)(.70) = (.30)^2(.70) = .063$$

Two other experimental outcomes also result in two successes and one failure. The probabilities for all three experimental outcomes involving two successes follow.

Trial Outcomes				
1st Customer	**2nd Customer**	**3rd Customer**	**Experimental Outcome**	**Probability of Experimental Outcome**
Purchase	Purchase	No purchase	(S, S, F)	$pp(1-p) = p^2(1-p)$ $= (.30)^2(.70) = .063$
Purchase	No purchase	Purchase	(S, F, S)	$p(1-p)p = p^2(1-p)$ $= (.30)^2(.70) = .063$
No purchase	Purchase	Purchase	(F, S, S)	$(1-p)pp = p^2(1-p)$ $= (.30)^2(.70) = .063$

Observe that all three experimental outcomes with two successes have exactly the same probability. This observation holds in general. In any binomial experiment, all sequences of trial outcomes yielding x successes in n trials have the *same probability* of occurrence. The probability of each sequence of trials yielding x successes in n trials follows.

$$\text{Probability of a particular sequence of trial outcomes with } x \text{ successes in } n \text{ trials} = p^x(1-p)^{(n-x)} \qquad \textbf{(5.11)}$$

For the Martin Clothing Store, this formula shows that any experimental outcome with two successes has a probability of $p^2(1-p)^{(3-2)} = p^2(1-p)^1 = (.30)^2(.70)^1 = .063$.

Because equation (5.10) shows the number of outcomes in a binomial experiment with x successes and equation (5.11) gives the probability for each sequence involving x successes, we combine equations (5.10) and (5.11) to obtain the following **binomial probability function**.

BINOMIAL PROBABILITY FUNCTION

$$f(x) = \binom{n}{x} p^x(1-p)^{(n-x)} \qquad \textbf{(5.12)}$$

where

x = the number of successes
p = the probability of a success on one trial
n = the number of trials
$f(x)$ = the probability of x successes in n trials

$$\binom{n}{x} = \frac{n!}{x!(n-x)!}$$

For the binomial probability distribution, x is a discrete random variable with the probability function $f(x)$ applicable for values of $x = 0, 1, 2, \ldots, n$.

In the Martin Clothing Store example, let use equation (5.12) to compute the probability that no customer makes a purchase, exactly one customer makes a purchase, exactly two customers make a purchase, and all three customers make a purchase. The calculations are summarized in Table 5.13, which gives the probability distribution of the number of customers making a purchase. Figure 5.4 is a graph of this probability distribution.

The binomial probability function can be applied to *any* binomial experiment. If we are satisfied that a situation demonstrates the properties of a binomial experiment and if we

TABLE 5.13 PROBABILITY DISTRIBUTION FOR THE NUMBER OF CUSTOMERS MAKING A PURCHASE

x	$f(x)$
0	$\frac{3!}{0!3!}(.30)^0(.70)^3 = .343$
1	$\frac{3!}{1!2!}(.30)^1(.70)^2 = .441$
2	$\frac{3!}{2!1!}(.30)^2(.70)^1 = .189$
3	$\frac{3!}{3!0!}(.30)^3(.70)^0 = .027$
	1.000

FIGURE 5.4 GRAPHICAL REPRESENTATION OF THE PROBABILITY DISTRIBUTION FOR THE NUMBER OF CUSTOMERS MAKING A PURCHASE

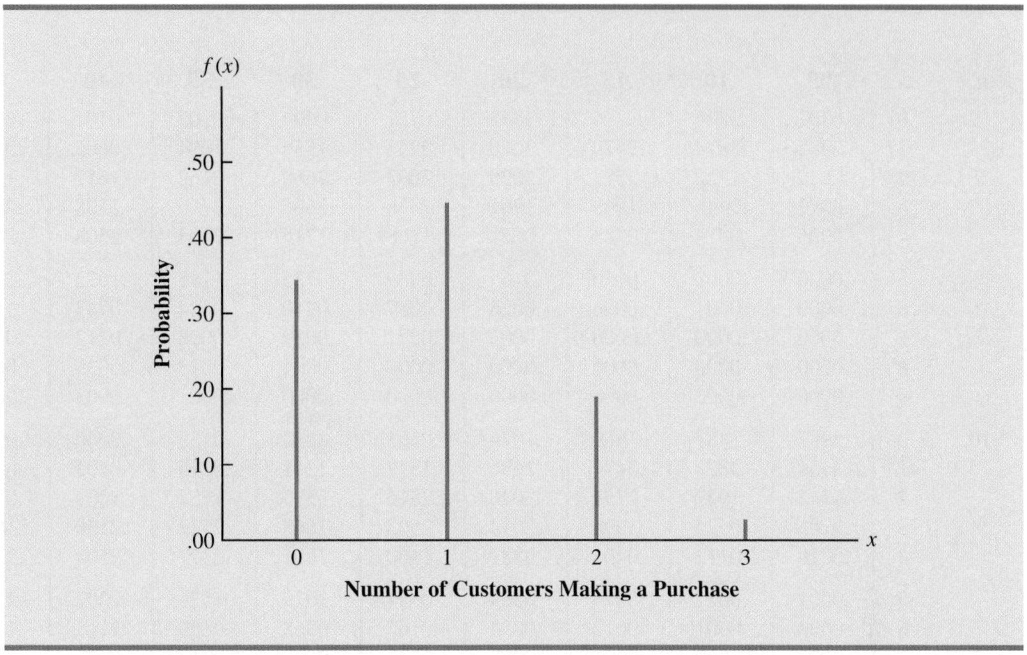

know the values of n and p, we can use equation (5.12) to compute the probability of x successes in the n trials.

If we consider variations of the Martin experiment, such as 10 customers rather than three entering the store, the binomial probability function given by equation (5.12) is still applicable. Suppose we have a binomial experiment with $n = 10$, $x = 4$, and $p = .30$. The probability of making exactly four sales to 10 customers entering the store is

$$f(4) = \frac{10!}{4!6!}(.30)^4(.70)^6 = .2001$$

Using Tables of Binomial Probabilities

Tables have been developed that give the probability of x successes in n trials for a binomial experiment. The tables are generally easy to use and quicker than equation (5.12). Table 5 of Appendix B provides such a table of binomial probabilities. A portion of this table appears in Table 5.14. To use this table, we must specify the values of n, p, and x for the binomial experiment of interest. In the example at the top of Table 5.14, we see that the probability of $x = 3$ successes in a binomial experiment with $n = 10$ and $p = .40$ is .2150. You can use equation (5.12) to verify that you would obtain the same answer if you used the binomial probability function directly.

Now let us use Table 5.14 to verify the probability of 4 successes in 10 trials for the Martin Clothing Store problem. Note that the value of $f(4) = .2001$ can be read directly from the table of binomial probabilities, with $n = 10$, $x = 4$, and $p = .30$.

Even though the tables of binomial probabilities are relatively easy to use, it is impossible to have tables that show all possible values of n and p that might be encountered in a binomial experiment. However, with today's calculators, using equation (5.12) to calculate the desired probability is not difficult, especially if the number of trials is not large. In the exercises, you should practice using equation (5.12) to compute the binomial probabilities unless the problem specifically requests that you use the binomial probability table.

TABLE 5.14 SELECTED VALUES FROM THE BINOMIAL PROBABILITY TABLE
EXAMPLE: $n = 10, x = 3, p = .40; f(3) = .2150$

					p						
n	x	.05	.10	.15	.20	.25	.30	.35	.40	.45	.50
9	0	.6302	.3874	.2316	.1342	.0751	.0404	.0207	.0101	.0046	.0020
	1	.2985	.3874	.3679	.3020	.2253	.1556	.1004	.0605	.0339	.0176
	2	.0629	.1722	.2597	.3020	.3003	.2668	.2162	.1612	.1110	.0703
	3	.0077	.0446	.1069	.1762	.2336	.2668	.2716	.2508	.2119	.1641
	4	.0006	.0074	.0283	.0661	.1168	.1715	.2194	.2508	.2600	.2461
	5	.0000	.0008	.0050	.0165	.0389	.0735	.1181	.1672	.2128	.2461
	6	.0000	.0001	.0006	.0028	.0087	.0210	.0424	.0743	.1160	.1641
	7	.0000	.0000	.0000	.0003	.0012	.0039	.0098	.0212	.0407	.0703
	8	.0000	.0000	.0000	.0000	.0001	.0004	.0013	.0035	.0083	.0176
	9	.0000	.0000	.0000	.0000	.0000	.0000	.0001	.0003	.0008	.0020
10	0	.5987	.3487	.1969	.1074	.0563	.0282	.0135	.0060	.0025	.0010
	1	.3151	.3874	.3474	.2684	.1877	.1211	.0725	.0403	.0207	.0098
	2	.0746	.1937	.2759	.3020	.2816	.2335	.1757	.1209	.0763	.0439
	3	.0105	.0574	.1298	.2013	.2503	.2668	.2522	**.2150**	.1665	.1172
	4	.0010	.0112	.0401	.0881	.1460	.2001	.2377	.2508	.2384	.2051
	5	.0001	.0015	.0085	.0264	.0584	.1029	.1536	.2007	.2340	.2461
	6	.0000	.0001	.0012	.0055	.0162	.0368	.0689	.1115	.1596	.2051
	7	.0000	.0000	.0001	.0008	.0031	.0090	.0212	.0425	.0746	.1172
	8	.0000	.0000	.0000	.0001	.0004	.0014	.0043	.0106	.0229	.0439
	9	.0000	.0000	.0000	.0000	.0000	.0001	.0005	.0016	.0042	.0098
	10	.0000	.0000	.0000	.0000	.0000	.0000	.0000	.0001	.0003	.0010

Statistical software packages such as Minitab and spreadsheet packages such as Excel also provide a capability for computing binomial probabilities. Consider the Martin Clothing Store example with $n = 10$ and $p = .30$. Figure 5.5 shows the binomial probabilities generated by Minitab for all possible values of x. Note that these values are the same as those found in the $p = .30$ column of Table 5.14. Appendix 5.1 gives the step-by-step procedure for using Minitab to generate the output in Figure 5.5. Appendix 5.2 describes how Excel can be used to compute binomial probabilities.

Expected Value and Variance for the Binomial Distribution

In Section 5.3 we provided formulas for computing the expected value and variance of a discrete random variable. In the special case where the random variable has a binomial distribution with a known number of trials n and a known probability of success p, the general formulas for the expected value and variance can be simplified. The results follow.

EXPECTED VALUE AND VARIANCE FOR THE BINOMIAL DISTRIBUTION

$$E(x) = \mu = np \quad \textbf{(5.13)}$$

$$Var(x) = \sigma^2 = np(1 - p) \quad \textbf{(5.14)}$$

FIGURE 5.5 MINITAB OUTPUT SHOWING BINOMIAL PROBABILITIES FOR THE MARTIN CLOTHING STORE PROBLEM

x	P(X = x)
0	0.0282
1	0.1211
2	0.2335
3	0.2668
4	0.2001
5	0.1029
6	0.0368
7	0.0090
8	0.0014
9	0.0001
10	0.0000

For the Martin Clothing Store problem with three customers, we can use equation (5.13) to compute the expected number of customers who will make a purchase.

$$E(x) = np = 3(.30) = .9$$

Suppose that for the next month the Martin Clothing Store forecasts 1000 customers will enter the store. What is the expected number of customers who will make a purchase? The answer is $\mu = np = (1000)(.3) = 300$. Thus, to increase the expected number of purchases, Martin's must induce more customers to enter the store and/or somehow increase the probability that any individual customer will make a purchase after entering.

For the Martin Clothing Store problem with three customers, we see that the variance and standard deviation for the number of customers who will make a purchase are

$$\sigma^2 = np(1 - p) = 3(.3)(.7) = .63$$
$$\sigma = \sqrt{.63} = .79$$

For the next 1000 customers entering the store, the variance and standard deviation for the number of customers who will make a purchase are

$$\sigma^2 = np(1 - p) = 1000(.3)(.7) = 210$$
$$\sigma = \sqrt{210} = 14.49$$

NOTES AND COMMENTS

1. The binomial table in Appendix B shows values of p up to and including $p = .95$. Some sources of the binomial table only show values of p up to and including $p = .50$. It would appear that such a table cannot be used when the probability of success exceeds $p = .50$. However, the table can be used by noting that the probability of $n - x$ failures is also the probability of x successes. Thus, when the probability of success is greater than $p = .50$, we can compute the probability of $n - x$ failures instead. The probability of failure, $1 - p$, will be less than .50 when $p > .50$.

2. Some sources present the binomial table in a cumulative form. In using such a table, one must subtract entries in the table to find the probability of exactly x success in n trials. For example, $f(2) = P(x \leq 2) - P(x \leq 1)$. The binomial table we provide in Appendix B provides $f(2)$ directly. To compute cumulative probabilities using the binomial table in Appendix B, sum the entries in the table. For example, to determine the cumulative probability $P(x \leq 2)$, compute the sum $f(0) + f(1) + f(2)$.

Exercises

Methods

SELF test

31. Consider a binomial experiment with two trials and $p = .4$.
    a. Draw a tree diagram for this experiment (see Figure 5.3).
    b. Compute the probability of one success, $f(1)$.
    c. Compute $f(0)$.
    d. Compute $f(2)$.
    e. Compute the probability of at least one success.
    f. Compute the expected value, variance, and standard deviation.
32. Consider a binomial experiment with $n = 10$ and $p = .10$.
    a. Compute $f(0)$.
    b. Compute $f(2)$.
    c. Compute $P(x \leq 2)$.
    d. Compute $P(x \geq 1)$.
    e. Compute $E(x)$.
    f. Compute $Var(x)$ and σ.
33. Consider a binomial experiment with $n = 20$ and $p = .70$.
    a. Compute $f(12)$.
    b. Compute $f(16)$.
    c. Compute $P(x \geq 16)$.
    d. Compute $P(x \leq 15)$.
    e. Compute $E(x)$.
    f. Compute $Var(x)$ and σ.

Applications

34. A Harris Interactive survey for InterContinental Hotels & Resorts asked respondents, "When traveling internationally, do you generally venture out on your own to experience culture, or stick with your tour group and itineraries?" The survey found that 23% of the respondents stick with their tour group (*USA Today,* January 21, 2004).
    a. In a sample of six international travelers, what is the probability that two will stick with their tour group?
    b. In a sample of six international travelers, what is the probability that at least two will stick with their tour group?
    c. In a sample of 10 international travelers, what is the probability that none will stick with the tour group?
35. In San Francisco, 30% of workers take public transportation daily (*USA Today,* December 21, 2005).
    a. In a sample of 10 workers, what is the probability that exactly 3 workers take public transportation daily?
    b. In a sample of 10 workers, what is the probability that at least 3 workers take public transportation daily?

SELF test

36. When a new machine is functioning properly, only 3% of the items produced are defective. Assume that we will randomly select two parts produced on the machine and that we are interested in the number of defective parts found.
    a. Describe the conditions under which this situation would be a binomial experiment.
    b. Draw a tree diagram similar to Figure 5.3 showing this problem as a two-trial experiment.
    c. How many experimental outcomes result in exactly one defect being found?
    d. Compute the probabilities associated with finding no defects, exactly one defect, and two defects.
37. A Randstad/Harris interactive survey reported that 25% of employees said their company is loyal to them (*USA Today,* November 11, 2009). Suppose 10 employees are selected randomly and will be interviewed about company loyalty.

a. Is the selection of 10 employees a binomial experiment? Explain.
b. What is the probability that none of the 10 employees will say their company is loyal to them?
c. What is the probability that 4 of the 10 employees will say their company is loyal to them?
d. What is the probability that at least 2 of the 10 employees will say their company is loyal to them?

38. Military radar and missile detection systems are designed to warn a country of an enemy attack. A reliability question is whether a detection system will be able to identify an attack and issue a warning. Assume that a particular detection system has a .90 probability of detecting a missile attack. Use the binomial probability distribution to answer the following questions.
a. What is the probability that a single detection system will detect an attack?
b. If two detection systems are installed in the same area and operate independently, what is the probability that at least one of the systems will detect the attack?
c. If three systems are installed, what is the probability that at least one of the systems will detect the attack?
d. Would you recommend that multiple detection systems be used? Explain.

39. Twelve of the top 20 finishers in the 2009 PGA Championship at Hazeltine National Golf Club in Chaska, Minnesota, used a Titleist brand golf ball (GolfBallTest website, November 12, 2009). Suppose these results are representative of the probability that a randomly selected PGA Tour player uses a Titleist brand golf ball. For a sample of 15 PGA Tour players, make the following calculations.
a. Compute the probability that exactly 10 of the 15 PGA Tour players use a Titleist brand golf ball.
b. Compute the probability that more than 10 of the 15 PGA Tour players use a Titleist brand golf ball.
c. For a sample of 15 PGA Tour players, compute the expected number of players who use a Titleist brand golf ball.
d. For a sample of 15 PGA Tour players, compute the variance and standard deviation of the number of players who use a Titleist brand golf ball.

40. The Census Bureau's Current Population Survey shows 28% of individuals, ages 25 and older, have completed four years of college (*The New York Times Almanac,* 2006). For a sample of 15 individuals, ages 25 and older, answer the following questions:
a. What is the probability 4 will have completed four years of college?
b. What is the probability 3 or more will have completed four years of college?

41. A university found that 20% of its students withdraw without completing the introductory statistics course. Assume that 20 students registered for the course.
a. Compute the probability that 2 or fewer will withdraw.
b. Compute the probability that exactly 4 will withdraw.
c. Compute the probability that more than 3 will withdraw.
d. Compute the expected number of withdrawals.

42. According to a survey conducted by TD Ameritrade, one out of four investors have exchange-traded funds in their portfolios (*USA Today,* January 11, 2007). Consider a sample of 20 investors.
a. Compute the probability that exactly 4 investors have exchange-traded funds in their portfolios.
b. Compute the probability that at least 2 of the investors have exchange-traded funds in their portfolios.
c. If you found that exactly 12 of the investors have exchange-traded funds in their portfolios, would you doubt the accuracy of the survey results?
d. Compute the expected number of investors who have exchange-traded funds in their portfolios.

43. Twenty-three percent of automobiles are not covered by insurance (CNN, February 23, 2006). On a particular weekend, 35 automobiles are involved in traffic accidents.
a. What is the expected number of these automobiles that are not covered by insurance?
b. What are the variance and standard deviation?

5.6 Poisson Probability Distribution

The Poisson probability distribution is often used to model random arrivals in waiting line situations.

In this section we consider a discrete random variable that is often useful in estimating the number of occurrences over a specified interval of time or space. For example, the random variable of interest might be the number of arrivals at a car wash in one hour, the number of repairs needed in 10 miles of highway, or the number of leaks in 100 miles of pipeline. If the following two properties are satisfied, the number of occurrences is a random variable described by the **Poisson probability distribution**.

PROPERTIES OF A POISSON EXPERIMENT

1. The probability of an occurrence is the same for any two intervals of equal length.
2. The occurrence or nonoccurrence in any interval is independent of the occurrence or nonoccurrence in any other interval.

The **Poisson probability function** is defined by equation (5.15).

Siméon Poisson taught mathematics at the Ecole Polytechnique in Paris from 1802 to 1808. In 1837, he published a work entitled, "Researches on the Probability of Criminal and Civil Verdicts," which includes a discussion of what later became known as the Poisson distribution.

POISSON PROBABILITY FUNCTION

$$f(x) = \frac{\mu^x e^{-\mu}}{x!} \tag{5.15}$$

where

$f(x)$ = the probability of x occurrences in an interval
μ = expected value or mean number of occurrences in an interval
e = 2.71828

For the Poisson probability distribution, x is a discrete random variable indicating the number of occurrences in the interval. Since there is no stated upper limit for the number of occurrences, the probability function $f(x)$ is applicable for values $x = 0, 1, 2, \ldots$ without limit. In practical applications, x will eventually become large enough so that $f(x)$ is approximately zero and the probability of any larger values of x becomes negligible.

An Example Involving Time Intervals

Bell Labs used the Poisson distribution to model the arrival of telephone calls.

Suppose that we are interested in the number of arrivals at the drive-up teller window of a bank during a 15-minute period on weekday mornings. If we can assume that the probability of a car arriving is the same for any two time periods of equal length and that the arrival or nonarrival of a car in any time period is independent of the arrival or nonarrival in any other time period, the Poisson probability function is applicable. Suppose these assumptions are satisfied and an analysis of historical data shows that the average number of cars arriving in a 15-minute period of time is 10; in this case, the following probability function applies.

$$f(x) = \frac{10^x e^{-10}}{x!}$$

The random variable here is x = number of cars arriving in any 15-minute period.

If management wanted to know the probability of exactly five arrivals in 15 minutes, we would set $x = 5$ and thus obtain

$$\begin{matrix}\text{Probability of exactly} \\ \text{5 arrivals in 15 minutes}\end{matrix} = f(5) = \frac{10^5 e^{-10}}{5!} = .0378$$

Although this probability was determined by evaluating the probability function with $\mu = 10$ and $x = 5$, it is often easier to refer to a table for the Poisson distribution. The table provides probabilities for specific values of x and μ. We included such a table as Table 7 of Appendix B. For convenience, we reproduced a portion of this table as Table 5.15. Note that to use the table of Poisson probabilities, we need know only the values of x and μ. From Table 5.15 we see that the probability of five arrivals in a 15-minute period is found by locating the value in the row of the table corresponding to $x = 5$ and the column of the table corresponding to $\mu = 10$. Hence, we obtain $f(5) = .0378$.

A property of the Poisson distribution is that the mean and variance are equal.

In the preceding example, the mean of the Poisson distribution is $\mu = 10$ arrivals per 15-minute period. A property of the Poisson distribution is that the mean of the distribution and the variance of the distribution are *equal.* Thus, the variance for the number of arrivals during 15-minute periods is $\sigma^2 = 10$. The standard deviation is $\sigma = \sqrt{10} = 3.16$.

Our illustration involves a 15-minute period, but other time periods can be used. Suppose we want to compute the probability of one arrival in a 3-minute period. Because 10 is the expected number of arrivals in a 15-minute period, we see that 10/15 = 2/3 is the expected number of arrivals in a 1-minute period and that (2/3)(3 minutes) = 2 is the expected number of arrivals in a 3-minute period. Thus, the probability of x arrivals in a 3-minute time period with $\mu = 2$ is given by the following Poisson probability function.

$$f(x) = \frac{2^x e^{-2}}{x!}$$

TABLE 5.15 SELECTED VALUES FROM THE POISSON PROBABILITY TABLES
EXAMPLE: $\mu = 10$, $x = 5$; $f(5) = .0378$

	μ									
x	**9.1**	**9.2**	**9.3**	**9.4**	**9.5**	**9.6**	**9.7**	**9.8**	**9.9**	**10**
0	.0001	.0001	.0001	.0001	.0001	.0001	.0001	.0001	.0001	.0000
1	.0010	.0009	.0009	.0008	.0007	.0007	.0006	.0005	.0005	.0005
2	.0046	.0043	.0040	.0037	.0034	.0031	.0029	.0027	.0025	.0023
3	.0140	.0131	.0123	.0115	.0107	.0100	.0093	.0087	.0081	.0076
4	.0319	.0302	.0285	.0269	.0254	.0240	.0226	.0213	.0201	.0189
5	.0581	.0555	.0530	.0506	.0483	.0460	.0439	.0418	.0398	**.0378**
6	.0881	.0851	.0822	.0793	.0764	.0736	.0709	.0682	.0656	.0631
7	.1145	.1118	.1091	.1064	.1037	.1010	.0982	.0955	.0928	.0901
8	.1302	.1286	.1269	.1251	.1232	.1212	.1191	.1170	.1148	.1126
9	.1317	.1315	.1311	.1306	.1300	.1293	.1284	.1274	.1263	.1251
10	.1198	.1210	.1219	.1228	.1235	.1241	.1245	.1249	.1250	.1251
11	.0991	.1012	.1031	.1049	.1067	.1083	.1098	.1112	.1125	.1137
12	.0752	.0776	.0799	.0822	.0844	.0866	.0888	.0908	.0928	.0948
13	.0526	.0549	.0572	.0594	.0617	.0640	.0662	.0685	.0707	.0729
14	.0342	.0361	.0380	.0399	.0419	.0439	.0459	.0479	.0500	.0521
15	.0208	.0221	.0235	.0250	.0265	.0281	.0297	.0313	.0330	.0347
16	.0118	.0127	.0137	.0147	.0157	.0168	.0180	.0192	.0204	.0217
17	.0063	.0069	.0075	.0081	.0088	.0095	.0103	.0111	.0119	.0128
18	.0032	.0035	.0039	.0042	.0046	.0051	.0055	.0060	.0065	.0071
19	.0015	.0017	.0019	.0021	.0023	.0026	.0028	.0031	.0034	.0037
20	.0007	.0008	.0009	.0010	.0011	.0012	.0014	.0015	.0017	.0019
21	.0003	.0003	.0004	.0004	.0005	.0006	.0006	.0007	.0008	.0009
22	.0001	.0001	.0002	.0002	.0002	.0002	.0003	.0003	.0004	.0004
23	.0000	.0001	.0001	.0001	.0001	.0001	.0001	.0001	.0002	.0002
24	.0000	.0000	.0000	.0000	.0000	.0000	.0000	.0001	.0001	.0001

The probability of one arrival in a 3-minute period is calculated as follows:

$$\text{Probability of exactly 1 arrival in 3 minutes} = f(1) = \frac{2^1 e^{-2}}{1!} = .2707$$

Earlier we computed the probability of five arrivals in a 15-minute period; it was .0378. Note that the probability of one arrival in a three-minute period (.2707) is not the same. When computing a Poisson probability for a different time interval, we must first convert the mean arrival rate to the time period of interest and then compute the probability.

An Example Involving Length or Distance Intervals

Let us illustrate an application not involving time intervals in which the Poisson distribution is useful. Suppose we are concerned with the occurrence of major defects in a highway one month after resurfacing. We will assume that the probability of a defect is the same for any two highway intervals of equal length and that the occurrence or nonoccurrence of a defect in any one interval is independent of the occurrence or nonoccurrence of a defect in any other interval. Hence, the Poisson distribution can be applied.

Suppose we learn that major defects one month after resurfacing occur at the average rate of two per mile. Let us find the probability of no major defects in a particular three-mile section of the highway. Because we are interested in an interval with a length of three miles, $\mu = (2 \text{ defects/mile})(3 \text{ miles}) = 6$ represents the expected number of major defects over the three-mile section of highway. Using equation (5.15), the probability of no major defects is $f(0) = 6^0 e^{-6}/0! = .0025$. Thus, it is unlikely that no major defects will occur in the three-mile section. In fact, this example indicates a $1 - .0025 = .9975$ probability of at least one major defect in the three-mile highway section.

Exercises

Methods

44. Consider a Poisson distribution with $\mu = 3$.
    a. Write the appropriate Poisson probability function.
    b. Compute $f(2)$.
    c. Compute $f(1)$.
    d. Compute $P(x \geq 2)$.

SELF test

45. Consider a Poisson distribution with a mean of two occurrences per time period.
    a. Write the appropriate Poisson probability function.
    b. What is the expected number of occurrences in three time periods?
    c. Write the appropriate Poisson probability function to determine the probability of x occurrences in three time periods.
    d. Compute the probability of two occurrences in one time period.
    e. Compute the probability of six occurrences in three time periods.
    f. Compute the probability of five occurrences in two time periods.

Applications

46. Phone calls arrive at the rate of 48 per hour at the reservation desk for Regional Airways.
    a. Compute the probability of receiving three calls in a 5-minute interval of time.
    b. Compute the probability of receiving exactly 10 calls in 15 minutes.
    c. Suppose no calls are currently on hold. If the agent takes 5 minutes to complete the current call, how many callers do you expect to be waiting by that time? What is the probability that none will be waiting?
    d. If no calls are currently being processed, what is the probability that the agent can take 3 minutes for personal time without being interrupted by a call?

47. During the period of time that a local university takes phone-in registrations, calls come in at the rate of one every two minutes.
    a. What is the expected number of calls in one hour?
    b. What is the probability of three calls in five minutes?
    c. What is the probability of no calls in a five-minute period?

SELF test

48. More than 50 million guests stay at bed and breakfasts (B&Bs) each year. The website for the Bed and Breakfast Inns of North America, which averages seven visitors per minute, enables many B&Bs to attract guests (*Time,* September 2001).
    a. Compute the probability of no website visitors in a one-minute period.
    b. Compute the probability of two or more website visitors in a one-minute period.
    c. Compute the probability of one or more website visitors in a 30-second period.
    d. Compute the probability of five or more website visitors in a one-minute period.

49. Airline passengers arrive randomly and independently at the passenger-screening facility at a major international airport. The mean arrival rate is 10 passengers per minute.
    a. Compute the probability of no arrivals in a one-minute period.
    b. Compute the probability that three or fewer passengers arrive in a one-minute period.
    c. Compute the probability of no arrivals in a 15-second period.
    d. Compute the probability of at least one arrival in a 15-second period.

50. An average of 15 aircraft accidents occur each year (*The World Almanac and Book of Facts,* 2004).
    a. Compute the mean number of aircraft accidents per month.
    b. Compute the probability of no accidents during a month.
    c. Compute the probability of exactly one accident during a month.
    d. Compute the probability of more than one accident during a month.

51. The National Safety Council (NSC) estimates that off-the-job accidents cost U.S. businesses almost $200 billion annually in lost productivity (National Safety Council, March 2006). Based on NSC estimates, companies with 50 employees are expected to average three employee off-the-job accidents per year. Answer the following questions for companies with 50 employees.
    a. What is the probability of no off-the-job accidents during a one-year period?
    b. What is the probability of at least two off-the-job accidents during a one-year period?
    c. What is the expected number of off-the-job accidents during six months?
    d. What is the probability of no off-the-job accidents during the next six months?

5.7 Hypergeometric Probability Distribution

The **hypergeometric probability distribution** is closely related to the binomial distribution. The two probability distributions differ in two key ways. With the hypergeometric distribution, the trials are not independent; and the probability of success changes from trial to trial.

In the usual notation for the hypergeometric distribution, r denotes the number of elements in the population of size N labeled success, and $N - r$ denotes the number of elements in the population labeled failure. The **hypergeometric probability function** is used to compute the probability that in a random selection of n elements, selected without replacement, we obtain x elements labeled success and $n - x$ elements labeled failure. For this outcome to occur, we must obtain x successes from the r successes in the population and $n - x$ failures from the $N - r$ failures. The following hypergeometric probability function provides $f(x)$, the probability of obtaining x successes in n trials.

HYPERGEOMETRIC PROBABILITY FUNCTION

$$f(x) = \frac{\binom{r}{x}\binom{N-r}{n-x}}{\binom{N}{n}} \qquad (5.16)$$

where

x = the number of successes
n = the number of trials
$f(x)$ = the probability of x successes in n trials
N = the number of elements in the population
r = the number of elements in the population labeled success

Note that $\binom{N}{n}$ represents the number of ways n elements can be selected from a population of size N; $\binom{r}{x}$ represents the number of ways that x successes can be selected from a total of r successes in the population; and $\binom{N-r}{n-x}$ represents the number of ways that $n - x$ failures can be selected from a total of $N - r$ failures in the population.

For the hypergeometric probability distribution, x is a discrete random variable and the probability function $f(x)$ given by equation (5.16) is usually applicable for values of $x = 0, 1, 2, \ldots, n$. However, only values of x where the number of observed successes is *less than or equal* to the number of successes in the population ($x \leq r$) and where the number of observed failures is *less than or equal to* the number of failures in the population ($n - x \leq N - r$) are valid. If these two conditions do not hold for one or more values of x, the corresponding $f(x) = 0$ indicating that the probability of this value of x is zero.

To illustrate the computations involved in using equation (5.16), let us consider the following quality control application. Electric fuses produced by Ontario Electric are packaged in boxes of 12 units each. Suppose an inspector randomly selects three of the 12 fuses in a box for testing. If the box contains exactly five defective fuses, what is the probability that the inspector will find exactly one of the three fuses defective? In this application, $n = 3$ and $N = 12$. With $r = 5$ defective fuses in the box the probability of finding $x = 1$ defective fuse is

$$f(1) = \frac{\binom{5}{1}\binom{7}{2}}{\binom{12}{3}} = \frac{\left(\frac{5!}{1!4!}\right)\left(\frac{7!}{2!5!}\right)}{\left(\frac{12!}{3!9!}\right)} = \frac{(5)(21)}{220} = .4773$$

Now suppose that we wanted to know the probability of finding *at least* 1 defective fuse. The easiest way to answer this question is to first compute the probability that the inspector does not find any defective fuses. The probability of $x = 0$ is

$$f(0) = \frac{\binom{5}{0}\binom{7}{3}}{\binom{12}{3}} = \frac{\left(\frac{5!}{0!5!}\right)\left(\frac{7!}{3!4!}\right)}{\left(\frac{12!}{3!9!}\right)} = \frac{(1)(35)}{220} = .1591$$

With a probability of zero defective fuses $f(0) = .1591$, we conclude that the probability of finding at least 1 defective fuse must be $1 - .1591 = .8409$. Thus, there is a reasonably high probability that the inspector will find at least 1 defective fuse.

The mean and variance of a hypergeometric distribution are as follows.

$$E(x) = \mu = n\left(\frac{r}{N}\right) \tag{5.17}$$

$$Var(x) = \sigma^2 = n\left(\frac{r}{N}\right)\left(1 - \frac{r}{N}\right)\left(\frac{N-n}{N-1}\right) \tag{5.18}$$

In the preceding example $n = 3$, $r = 5$, and $N = 12$. Thus, the mean and variance for the number of defective fuses are

$$\mu = n\left(\frac{r}{N}\right) = 3\left(\frac{5}{12}\right) = 1.25$$

$$\sigma^2 = n\left(\frac{r}{N}\right)\left(1 - \frac{r}{N}\right)\left(\frac{N-n}{N-1}\right) = 3\left(\frac{5}{12}\right)\left(1 - \frac{5}{12}\right)\left(\frac{12-3}{12-1}\right) = .60$$

The standard deviation is $\sigma = \sqrt{.60} = .77$.

NOTES AND COMMENTS

Consider a hypergeometric distribution with n trials. Let $p = (r/N)$ denote the probability of a success on the first trial. If the population size is large, the term $(N - n)/(N - 1)$ in equation (5.18) approaches 1. As a result, the expected value and variance can be written $E(x) = np$ and $Var(x) = np(1 - p)$. Note that these expressions are the same as the expressions used to compute the expected value and variance of a binomial distribution, as in equations (5.13) and (5.14). When the population size is large, a hypergeometric distribution can be approximated by a binomial distribution with n trials and a probability of success $p = (r/N)$.

Exercises

Methods

SELF test

52. Suppose $N = 10$ and $r = 3$. Compute the hypergeometric probabilities for the following values of n and x.
    a. $n = 4, x = 1$.
    b. $n = 2, x = 2$.
    c. $n = 2, x = 0$.
    d. $n = 4, x = 2$.
    e. $n = 4, x = 4$.
53. Suppose $N = 15$ and $r = 4$. What is the probability of $x = 3$ for $n = 10$?

Applications

54. In a survey conducted by the Gallup Organization, respondents were asked, "What is your favorite sport to watch?" Football and basketball ranked number one and two in terms of preference (Gallup website, January 3, 2004). Assume that in a group of 10 individuals, seven prefer football and three prefer basketball. A random sample of three of these individuals is selected.
    a. What is the probability that exactly two prefer football?
    b. What is the probability that the majority (either two or three) prefer football?

55. Blackjack, or twenty-one as it is frequently called, is a popular gambling game played in Las Vegas casinos. A player is dealt two cards. Face cards (jacks, queens, and kings) and tens have a point value of 10. Aces have a point value of 1 or 11. A 52-card deck contains 16 cards with a point value of 10 (jacks, queens, kings, and tens) and four aces.
    a. What is the probability that both cards dealt are aces or 10-point cards?
    b. What is the probability that both of the cards are aces?
    c. What is the probability that both of the cards have a point value of 10?
    d. A blackjack is a 10-point card and an ace for a value of 21. Use your answers to parts (a), (b), and (c) to determine the probability that a player is dealt blackjack. (*Hint:* Part (d) is not a hypergeometric problem. Develop your own logical relationship as to how the hypergeometric probabilities from parts (a), (b), and (c) can be combined to answer this question.)

SELF test

56. Axline Computers manufactures personal computers at two plants, one in Texas and the other in Hawaii. The Texas plant has 40 employees; the Hawaii plant has 20. A random sample of 10 employees is to be asked to fill out a benefits questionnaire.
    a. What is the probability that none of the employees in the sample work at the plant in Hawaii?
    b. What is the probability that 1 of the employees in the sample works at the plant in Hawaii?
    c. What is the probability that 2 or more of the employees in the sample work at the plant in Hawaii?
    d. What is the probability that 9 of the employees in the sample work at the plant in Texas?

57. The Zagat Restaurant Survey provides food, decor, and service ratings for some of the top restaurants across the United States. For 15 restaurants located in Boston, the average price of a dinner, including one drink and tip, was $48.60. You are leaving on a business trip to Boston and will eat dinner at three of these restaurants. Your company will reimburse you for a maximum of $50 per dinner. Business associates familiar with these restaurants have told you that the meal cost at one-third of these restaurants will exceed $50. Suppose that you randomly select three of these restaurants for dinner.
    a. What is the probability that none of the meals will exceed the cost covered by your company?
    b. What is the probability that one of the meals will exceed the cost covered by your company?
    c. What is the probability that two of the meals will exceed the cost covered by your company?
    d. What is the probability that all three of the meals will exceed the cost covered by your company?

58. The Troubled Asset Relief Program (TARP), passed by the U.S. Congress in October 2008, provided $700 billion in assistance for the struggling U.S. economy. Over $200 billion was given to troubled financial institutions with the hope that there would be an increase in lending to help jump-start the economy. But three months later, a Federal Reserve survey found that two-thirds of the banks that had received TARP funds had tightened terms for business loans (*The Wall Street Journal,* February 3, 2009). Of the 10 banks that were the biggest recipients of TARP funds, only 3 had actually increased lending during this period.

Increased Lending	**Decreased Lending**
BB&T	Bank of America
Sun Trust Banks	Capital One
U.S. Bancorp	Citigroup
	Fifth Third Bancorp
	J.P. Morgan Chase
	Regions Financial
	Wells Fargo

For the purposes of this exercise, assume that you will randomly select 3 of these 10 banks for a study that will continue to monitor bank lending practices. Let x be a random variable indicating the number of banks in the study that had increased lending.

a. What is $f(0)$? What is your interpretation of this value?
b. What is $f(3)$? What is your interpretation of this value?
c. Compute $f(1)$ and $f(2)$. Show the probability distribution for the number of banks in the study that had increased lending. What value of x has the highest probability?
d. What is the probability that the study will have at least one bank that had increased lending?
e. Compute the expected value, variance, and standard deviation for the random variable.

Summary

A random variable provides a numerical description of the outcome of an experiment. The probability distribution for a random variable describes how the probabilities are distributed over the values the random variable can assume. For any discrete random variable x, the probability distribution is defined by a probability function, denoted by $f(x)$, which provides the probability associated with each value of the random variable.

We introduced two types of discrete probability distributions. One type involved providing a list of the values of the random variable and the associated probabilities in a table. We showed how the relative frequency method of assigning probabilities could be used to develop empirical discrete probability distributions of this type. Bivariate empirical distributions were also discussed. With bivariate distributions, interest focuses on the relationship between two random variables. We showed how to compute the covariance and correlation coefficient as measures of such a relationship. We also showed how bivariate distributions involving market returns on financial assets could be used to create financial portfolios.

The second type of discrete probability distribution we discussed involved the use of a mathematical function to provide the probabilities for the random variable. The binomial, Poisson, and hypergeometric distributions discussed were all of this type. The binomial distribution can be used to determine the probability of x successes in n trials whenever the experiment has the following properties:

1. The experiment consists of a sequence of n identical trials.
2. Two outcomes are possible on each trial, one called success and the other failure.
3. The probability of a success p does not change from trial to trial. Consequently, the probability of failure, $1 - p$, does not change from trial to trial.
4. The trials are independent.

When the four properties hold, the binomial probability function can be used to determine the probability of obtaining x successes in n trials. Formulas were also presented for the mean and variance of the binomial distribution.

The Poisson distribution is used when it is desirable to determine the probability of obtaining x occurrences over an interval of time or space. The following assumptions are necessary for the Poisson distribution to be applicable.

1. The probability of an occurrence of the event is the same for any two intervals of equal length.
2. The occurrence or nonoccurrence of the event in any interval is independent of the occurrence or nonoccurrence of the event in any other interval.

A third discrete probability distribution, the hypergeometric, was introduced in Section 5.7. Like the binomial, it is used to compute the probability of x successes in n trials. But, in contrast to the binomial, the probability of success changes from trial to trial.

Glossary

Random variable A numerical description of the outcome of an experiment.

Discrete random variable A random variable that may assume either a finite number of values or an infinite sequence of values.

Continuous random variable A random variable that may assume any numerical value in an interval or collection of intervals.

Probability distribution A description of how the probabilities are distributed over the values of the random variable.

Probability function A function, denoted by $f(x)$, that provides the probability that x assumes a particular value for a discrete random variable.

Empirical discrete distribution A discrete probability distribution for which the relative frequency method is used to assign the probabilities.

Discrete uniform probability distribution A probability distribution for which each possible value of the random variable has the same probability.

Expected value A measure of the central location of a random variable.

Variance A measure of the variability, or dispersion, of a random variable.

Standard deviation The positive square root of the variance.

Bivariate probability distribution A probability distribution involving two random variables. A discrete bivariate probability distribution provides a probability for each pair of values that may occur for the two random variables.

Binomial experiment An experiment having the four properties stated at the beginning of Section 5.5.

Binomial probability distribution A probability distribution showing the probability of x successes in n trials of a binomial experiment.

Binomial probability function The function used to compute binomial probabilities.

Poisson probability distribution A probability distribution showing the probability of x occurrences of an event over a specified interval of time or space.

Poisson probability function The function used to compute Poisson probabilities.

Hypergeometric probability distribution A probability distribution showing the probability of x successes in n trials from a population with r successes and $N - r$ failures.

Hypergeometric probability function The function used to compute hypergeometric probabilities.

Key Formulas

Discrete Uniform Probability Function

$$f(x) = 1/n \tag{5.3}$$

Expected Value of a Discrete Random Variable

$$E(x) = \mu = \Sigma x f(x) \tag{5.4}$$

Variance of a Discrete Random Variable

$$Var(x) = \sigma^2 = \Sigma(x - \mu)^2 f(x) \tag{5.5}$$

Covariance of Random Variables x and y

$$\sigma_{xy} = [Var(x + y) - Var(x) - Var(y)]/2 \tag{5.6}$$

Correlation between Random Variables x and y

$$\rho_{xy} = \frac{\sigma_{xy}}{\sigma_x \sigma_y} \tag{5.7}$$

Expected Value of a Linear Combination of Random Variables x and y

$$E(ax + by) = aE(x) + bE(y) \tag{5.8}$$

Variance of a Linear Combination of Two Random Variables

$$Var(ax + by) = a^2 Var(x) + b^2 Var(y) + 2ab\sigma_{xy} \tag{5.9}$$

where σ_{xy} is the covariance of x and y

Number of Experimental Outcomes Providing Exactly x Successes in n Trials

$$\binom{n}{x} = \frac{n!}{x!(n - x)!} \tag{5.10}$$

Binomial Probability Function

$$f(x) = \binom{n}{x} p^x (1 - p)^{(n-x)} \tag{5.12}$$

Expected Value for the Binomial Distribution

$$E(x) = \mu = np \tag{5.13}$$

Variance for the Binomial Distribution

$$Var(x) = \sigma^2 = np(1 - p) \tag{5.14}$$

Poisson Probability Function

$$f(x) = \frac{\mu^x e^{-\mu}}{x!} \tag{5.15}$$

Hypergeometric Probability Function

$$f(x) = \frac{\binom{r}{x}\binom{N - r}{n - x}}{\binom{N}{n}} \tag{5.16}$$

Expected Value for the Hypergeometric Distribution

$$E(x) = \mu = n\left(\frac{r}{N}\right) \tag{5.17}$$

Variance for the Hypergeometric Distribution

$$Var(x) = \sigma^2 = n\left(\frac{r}{N}\right)\left(1 - \frac{r}{N}\right)\left(\frac{N - n}{N - 1}\right) \tag{5.18}$$

Supplementary Exercises

59. The *Barron's* Big Money Poll asked 131 investment managers across the United States about their short-term investment outlook (*Barron's,* October 28, 2002). Their responses showed 4% were very bullish, 39% were bullish, 29% were neutral, 21% were bearish, and 7% were very bearish. Let x be the random variable reflecting the level of optimism about the market. Set $x = 5$ for very bullish down through $x = 1$ for very bearish.
    a. Develop a probability distribution for the level of optimism of investment managers.
    b. Compute the expected value for the level of optimism.
    c. Compute the variance and standard deviation for the level of optimism.
    d. Comment on what your results imply about the level of optimism and its variability.

60. The American Association of Individual Investors publishes an annual guide to the top mutual funds (*The Individual Investor's Guide to the Top Mutual Funds,* 22e, American Association of Individual Investors, 2003). The total risk ratings for 29 categories of mutual funds are as follows.

Total Risk	Number of Fund Categories
Low	7
Below Average	6
Average	3
Above Average	6
High	7

    a. Let $x = 1$ for low risk up through $x = 5$ for high risk, and develop a probability distribution for level of risk.
    b. What are the expected value and variance for total risk?
    c. It turns out that 11 of the fund categories were bond funds. For the bond funds, seven categories were rated low and four were rated below average. Compare the total risk of the bond funds with the 18 categories of stock funds.

61. The budgeting process for a midwestern college resulted in expense forecasts for the coming year (in $ millions) of $9, $10, $11, $12, and $13. Because the actual expenses are unknown, the following respective probabilities are assigned: .3, .2, .25, .05, and .2.
    a. Show the probability distribution for the expense forecast.
    b. What is the expected value of the expense forecast for the coming year?
    c. What is the variance of the expense forecast for the coming year?
    d. If income projections for the year are estimated at $12 million, comment on the financial position of the college.

62. A bookstore at the Hartsfield-Jackson Airport in Atlanta sells reading materials (paperback books, newspapers, magazines) as well as snacks (peanuts, pretzels, candy, etc.). A point-of-sale terminal collects a variety of information about customer purchases. Shown below is a table showing the number of snack items and the number of items of reading material purchased by the most recent 600 customers.

		Reading Material		
		0	**1**	**2**
	0	0	60	18
Snacks	**1**	240	90	30
	2	120	30	12

a. Using the data in the table construct an empirical discrete bivariate probability distribution for x = number of snack items and y = number of reading materials in a randomly selected customer purchase. What is the probability of a customer purchase consisting of one item of reading materials and two snack items? What is the probability of a customer purchasing one snack item only? Why is the probability $f(x = 0, y = 0) = 0$?
b. Show the marginal probability distribution for the number of snack items purchased. Compute the expected value and variance.
c. What is the expected value and variance for the number of reading materials purchased by a customer?
d. Show the probability distribution for t = total number of items in a customer purchase. Compute its expected value and variance.
e. Compute the covariance and correlation coefficient between x and y. What is the relationship, if any, between the number of reading materials and number of snacks purchased on a customer visit?

63. The Knowles/Armitage (KA) group at Merrill Lynch advises clients on how to create a diversified investment portfolio. One of the investment alternatives they make available to clients is the All World Fund composed of global stocks with good dividend yields. One of their clients is interested in a portfolio consisting of investment in the All World Fund and a treasury bond fund. The expected percent return of an investment in the All World Fund is 7.80% with a standard deviation of 18.90%. The expected percent return of an investment in a treasury bond fund is 5.50% and the standard deviation is 4.60%. The covariance of an investment in the All World Fund with an investment in a treasury bond fund is −12.4.
    a. Which of the funds would be considered the more risky? Why?
    b. If KA recommends that the client invest 75% in the All World Fund and 25% in the treasury bond fund, what is the expected percent return and standard deviation for such a portfolio? What would be the expected return and standard deviation, in dollars, for a client investing $10,000 in such a portfolio?
    c. If KA recommends that the client invest 25% in the All World Fund and 75% in the treasury bond fund, what is the expected return and standard deviation for such a portfolio? What would be the expected return and standard deviation, in dollars, for a client investing $10,000 in such a portfolio?
    d. Which of the portfolios in parts (b) and (c) would you recommend for an aggressive investor? Which would you recommend for a conservative investor? Why?

64. A survey showed that the average commuter spends about 26 minutes on a one-way door-to-door trip from home to work. In addition, 5% of commuters reported a one-way commute of more than one hour (Bureau of Transportation Statistics website, January 12, 2004).
    a. If 20 commuters are surveyed on a particular day, what is the probability that 3 will report a one-way commute of more than one hour?
    b. If 20 commuters are surveyed on a particular day, what is the probability that none will report a one-way commute of more than one hour?
    c. If a company has 2000 employees, what is the expected number of employees who have a one-way commute of more than one hour?
    d. If a company has 2000 employees, what is the variance and standard deviation of the number of employees who have a one-way commute of more than one hour?

65. A political action group is planning to interview home owners to assess the impact caused by a recent slump in housing prices. According to a *Wall Street Journal*/Harris Interactive Personal Finance poll, 26% of individuals aged 18–34, 50% of individuals aged 35–44, and 88% of individuals aged 55 and over are home owners (All Business website, January 23, 2008).
    a. How many people from the 18–34 age group must be sampled to find an expected number of at least 20 home owners?
    b. How many people from the 35–44 age group must be sampled to find an expected number of at least 20 home owners?
    c. How many people from the 55 and over age group must be sampled to find an expected number of at least 20 home owners?

d. If the number of 18–34 year olds sampled is equal to the value identified in part (a), what is the standard deviation of the number who will be home owners?
e. If the number of 35–44 year olds sampled is equal to the value identified in part (b), what is the standard deviation of the number who will be home owners?

66. Many companies use a quality control technique called acceptance sampling to monitor incoming shipments of parts, raw materials, and so on. In the electronics industry, component parts are commonly shipped from suppliers in large lots. Inspection of a sample of n components can be viewed as the n trials of a binomial experiment. The outcome for each component tested (trial) will be that the component is classified as good or defective. Reynolds Electronics accepts a lot from a particular supplier if the defective components in the lot do not exceed 1%. Suppose a random sample of five items from a recent shipment is tested.
   a. Assume that 1% of the shipment is defective. Compute the probability that no items in the sample are defective.
   b. Assume that 1% of the shipment is defective. Compute the probability that exactly one item in the sample is defective.
   c. What is the probability of observing one or more defective items in the sample if 1% of the shipment is defective?
   d. Would you feel comfortable accepting the shipment if one item was found to be defective? Why or why not?

67. The unemployment rate in the state of Arizona is 4.1% (CNN Money website, May 2, 2007). Assume that 100 employable people in Arizona are selected randomly.
   a. What is the expected number of people who are unemployed?
   b. What are the variance and standard deviation of the number of people who are unemployed?

68. A poll conducted by Zogby International showed that of those Americans who said music plays a "very important" role in their lives, 30% said their local radio stations "always" play the kind of music they like (Zogby website, January 12, 2004). Suppose a sample of 800 people who say music plays an important role in their lives is taken.
   a. How many would you expect to say that their local radio stations always play the kind of music they like?
   b. What is the standard deviation of the number of respondents who think their local radio stations always play the kind of music they like?
   c. What is the standard deviation of the number of respondents who do not think their local radio stations always play the kind of music they like?

69. Cars arrive at a car wash randomly and independently; the probability of an arrival is the same for any two time intervals of equal length. The mean arrival rate is 15 cars per hour. What is the probability that 20 or more cars will arrive during any given hour of operation?

70. A new automated production process averages 1.5 breakdowns per day. Because of the cost associated with a breakdown, management is concerned about the possibility of having three or more breakdowns during a day. Assume that breakdowns occur randomly, that the probability of a breakdown is the same for any two time intervals of equal length, and that breakdowns in one period are independent of breakdowns in other periods. What is the probability of having three or more breakdowns during a day?

71. A regional director responsible for business development in the state of Pennsylvania is concerned about the number of small business failures. If the mean number of small business failures per month is 10, what is the probability that exactly 4 small businesses will fail during a given month? Assume that the probability of a failure is the same for any two months and that the occurrence or nonoccurrence of a failure in any month is independent of failures in any other month.

72. Customer arrivals at a bank are random and independent; the probability of an arrival in any one-minute period is the same as the probability of an arrival in any other one-minute period. Answer the following questions, assuming a mean arrival rate of three customers per minute.
   a. What is the probability of exactly three arrivals in a one-minute period?
   b. What is the probability of at least three arrivals in a one-minute period?

73. A deck of playing cards contains 52 cards, four of which are aces. What is the probability that the deal of a five-card hand provides
    a. A pair of aces?
    b. Exactly one ace?
    c. No aces?
    d. At least one ace?

74. *U.S. News & World Report*'s ranking of America's best graduate schools of business showed Harvard University and Stanford University in a tie for first place. In addition, 7 of the top 10 graduate schools of business showed students with an average undergraduate grade point average (GPA) of 3.50 or higher (*America's Best Graduate Schools, 2009 edition, U.S. News & World Report*). Suppose that we randomly select 2 of the top 10 graduate schools of business.
    a. What is the probability that exactly one school has students with an average undergraduate GPA of 3.50 or higher?
    b. What is the probability that both schools have students with an average undergraduate GPA of 3.50 or higher?
    c. What is the probability that neither school has students with an average undergraduate GPA of 3.50 or higher?

Appendix 5.1 Discrete Probability Distributions with Minitab

Statistical packages such as Minitab offer a relatively easy and efficient procedure for computing binomial probabilities. In this appendix, we show the step-by-step procedure for determining the binomial probabilities for the Martin Clothing Store problem in Section 5.4. Recall that the desired binomial probabilities are based on $n = 10$ and $p = .30$. Before beginning the Minitab routine, the user must enter the desired values of the random variable x into a column of the worksheet. We entered the values 0, 1, 2, . . . , 10 in column 1 (see Figure 5.5) to generate the entire binomial probability distribution. The Minitab steps to obtain the desired binomial probabilities follow.

Step 1. Select the **Calc** menu
Step 2. Choose **Probability Distributions**
Step 3. Choose **Binomial**
Step 4. When the Binomial Distribution dialog box appears:
Select **Probability**
Enter 10 in the **Number of trials** box
Enter .3 in the **Event probability** box
Enter C1 in the **Input column** box
Click **OK**

The Minitab output with the binomial probabilities will appear as shown in Figure 5.5.

Minitab provides Poisson and hypergeometric probabilities in a similar manner. For instance, to compute Poisson probabilities the only differences are in step 3, where the **Poisson** option would be selected, and step 4, where the **Mean** would be entered rather than the number of trials and the probability of success.

Appendix 5.2 Discrete Probability Distributions with Excel

Excel provides functions for computing probabilities for the binomial, Poisson, and hypergeometric distributions introduced in this chapter. The Excel function for computing binomial probabilities is BINOM.DIST. It has four arguments: x (the number of successes), n (the number of trials), p (the probability of success), and cumulative. FALSE is used for

FIGURE 5.6 EXCEL WORKSHEET FOR COMPUTING BINOMIAL PROBABILITIES

	A	B	C
1	Number of Trials (n)	10	
2	Probability of Success (p)	0.3	
3			
4		x	$f(x)$
5		0	=BINOM.DIST(B5,$B$1,$B$2,FALSE)
6		1	=BINOM.DIST(B6,$B$1,$B$2,FALSE)
7		2	=BINOM.DIST(B7,$B$1,$B$2,FALSE)
8		3	=BINOM.DIST(B8,$B$1,$B$2,FALSE)
9		4	=BINOM.DIST(B9,$B$1,$B$2,FALSE)
10		5	=BINOM.DIST(B10,$B$1,$B$2,FALSE)
11		6	=BINOM.DIST(B11,$B$1,$B$2,FALSE)
12		7	=BINOM.DIST(B12,$B$1,$B$2,FALSE)
13		8	=BINOM.DIST(B13,$B$1,$B$2,FALSE)
14		9	=BINOM.DIST(B14,$B$1,$B$2,FALSE)
15		10	=BINOM.DIST(B15,$B$1,$B$2,FALSE)

	A	B	C
1	Number of Trials (n)	10	
2	Probability of Success (p)	0.3	
3			
4		x	$f(x)$
5		0	0.0282
6		1	0.1211
7		2	0.2335
8		3	0.2668
9		4	0.2001
10		5	0.1029
11		6	0.0368
12		7	0.0090
13		8	0.0014
14		9	0.0001
15		10	0.0000

the fourth argument (cumulative) if we want the probability of x successes, and TRUE is used for the fourth argument if we want the cumulative probability of x or fewer successes. Here we show how to compute the probabilities of 0 through 10 successes for the Martin Clothing Store problem in Section 5.4 (see Figure 5.5).

As we describe the worksheet development, refer to Figure 5.6; the formula worksheet is set in the background, and the value worksheet appears in the foreground. We entered the number of trials (10) into cell B1, the probability of success into cell B2, and the values for the random variable into cells B5:B15. The following steps will generate the desired probabilities:

Step 1. Use the BINOM.DIST function to compute the probability of $x = 0$ by entering the following formula into cell C5:

=BINOM.DIST(B5,$B$1,$B$2,FALSE)

Step 2. Copy the formula in cell C5 into cells C6:C15.

The value worksheet in Figure 5.6 shows that the probabilities obtained are the same as in Figure 5.5. Poisson and hypergeometric probabilities can be computed in a similar fashion. The POISSON.DIST and HYPGEOM.DIST functions are used. Excel's Insert Function dialog box can help the user in entering the proper arguments for these functions (see Appendix E).

CHAPTER 6

Continuous Probability Distributions

CONTENTS

STATISTICS *in* PRACTICE

PROCTER & GAMBLE*
CINCINNATI, OHIO

Procter & Gamble (P&G) produces and markets such products as detergents, disposable diapers, over-the-counter pharmaceuticals, dentifrices, bar soaps, mouthwashes, and paper towels. Worldwide, it has the leading brand in more categories than any other consumer products company. Since its merger with Gillette, P&G also produces and markets razors, blades, and many other personal care products.

As a leader in the application of statistical methods in decision making, P&G employs people with diverse academic backgrounds: engineering, statistics, operations research, and business. The major quantitative technologies for which these people provide support are probabilistic decision and risk analysis, advanced simulation, quality improvement, and quantitative methods (e.g., linear programming, regression analysis, probability analysis).

The Industrial Chemicals Division of P&G is a major supplier of fatty alcohols derived from natural substances such as coconut oil and from petroleum-based derivatives. The division wanted to know the economic risks and opportunities of expanding its fatty-alcohol production facilities, so it called in P&G's experts in probabilistic decision and risk analysis to help. After structuring and modeling the problem, they determined that the key to profitability was the cost difference between the petroleum- and coconut-based raw materials. Future costs were unknown, but the analysts were able to approximate them with the following continuous random variables.

$x =$ the coconut oil price per pound of fatty alcohol

and

$y =$ the petroleum raw material price per pound of fatty alcohol

Because the key to profitability was the difference between these two random variables, a third random variable, $d = x - y$, was used in the analysis. Experts were interviewed to determine the probability distributions for x and y. In turn, this information was used to develop a probability distribution for the difference in prices d. This continuous probability distribution showed a .90 probability that the price difference would be \$.0655 or less and a .50 probability that the price difference would be \$.035 or less. In addition, there was only a .10 probability that the price difference would be \$.0045 or less.†

Procter & Gamble is a leader in the application of statistical methods in decision making. © John Sommers II/Reuters.

The Industrial Chemicals Division thought that being able to quantify the impact of raw material price differences was key to reaching a consensus. The probabilities obtained were used in a sensitivity analysis of the raw material price difference. The analysis yielded sufficient insight to form the basis for a recommendation to management.

The use of continuous random variables and their probability distributions was helpful to P&G in analyzing the economic risks associated with its fatty-alcohol production. In this chapter, you will gain an understanding of continuous random variables and their probability distributions, including one of the most important probability distributions in statistics, the normal distribution.

*The authors are indebted to Joel Kahn of Procter & Gamble for providing this Statistics in Practice.

†The price differences stated here have been modified to protect proprietary data.

In the preceding chapter we discussed discrete random variables and their probability distributions. In this chapter we turn to the study of continuous random variables. Specifically, we discuss three continuous probability distributions: the uniform, the normal, and the exponential.

A fundamental difference separates discrete and continuous random variables in terms of how probabilities are computed. For a discrete random variable, the probability function $f(x)$ provides the probability that the random variable assumes a particular value. With continuous random variables, the counterpart of the probability function is the **probability density function**, also denoted by $f(x)$. The difference is that the probability density function does not directly provide probabilities. However, the area under the graph of $f(x)$ corresponding to a given interval does provide the probability that the continuous random variable x assumes a value in that interval. So when we compute probabilities for continuous random variables we are computing the probability that the random variable assumes any value in an interval.

Because the area under the graph of $f(x)$ at any particular point is zero, one of the implications of the definition of probability for continuous random variables is that the probability of any particular value of the random variable is zero. In Section 6.1 we demonstrate these concepts for a continuous random variable that has a uniform distribution.

Much of the chapter is devoted to describing and showing applications of the normal distribution. The normal distribution is of major importance because of its wide applicability and its extensive use in statistical inference. The chapter closes with a discussion of the exponential distribution. The exponential distribution is useful in applications involving such factors as waiting times and service times.

6.1 Uniform Probability Distribution

Whenever the probability is proportional to the length of the interval, the random variable is uniformly distributed.

Consider the random variable x representing the flight time of an airplane traveling from Chicago to New York. Suppose the flight time can be any value in the interval from 120 minutes to 140 minutes. Because the random variable x can assume any value in that interval, x is a continuous rather than a discrete random variable. Let us assume that sufficient actual flight data are available to conclude that the probability of a flight time within any 1-minute interval is the same as the probability of a flight time within any other 1-minute interval contained in the larger interval from 120 to 140 minutes. With every 1-minute interval being equally likely, the random variable x is said to have a **uniform probability distribution**. The probability density function, which defines the uniform distribution for the flight-time random variable, is

$$f(x) = \begin{cases} 1/20 & \text{for } 120 \leq x \leq 140 \\ 0 & \text{elsewhere} \end{cases}$$

Figure 6.1 is a graph of this probability density function. In general, the uniform probability density function for a random variable x is defined by the following formula.

UNIFORM PROBABILITY DENSITY FUNCTION

$$f(x) = \begin{cases} \dfrac{1}{b - a} & \text{for } a \leq x \leq b \\ 0 & \text{elsewhere} \end{cases} \tag{6.1}$$

For the flight-time random variable, $a = 120$ and $b = 140$.

FIGURE 6.1 UNIFORM PROBABILITY DISTRIBUTION FOR FLIGHT TIME

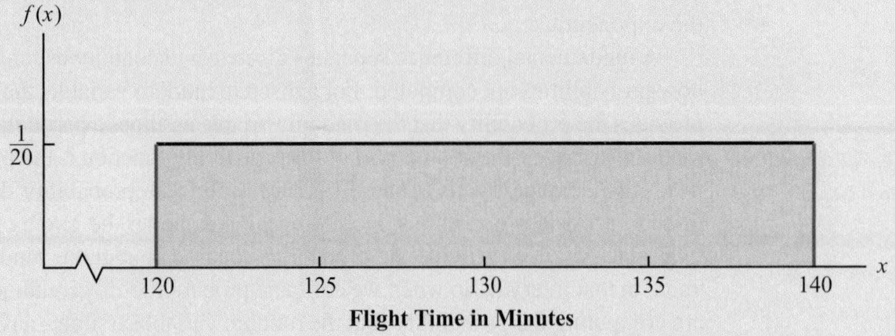

As noted in the introduction, for a continuous random variable, we consider probability only in terms of the likelihood that a random variable assumes a value within a specified interval. In the flight time example, an acceptable probability question is: What is the probability that the flight time is between 120 and 130 minutes? That is, what is $P(120 \leq x \leq 130)$? Because the flight time must be between 120 and 140 minutes and because the probability is described as being uniform over this interval, we feel comfortable saying $P(120 \leq x \leq 130) = .50$. In the following subsection we show that this probability can be computed as the area under the graph of $f(x)$ from 120 to 130 (see Figure 6.2).

Area as a Measure of Probability

Let us make an observation about the graph in Figure 6.2. Consider the area under the graph of $f(x)$ in the interval from 120 to 130. The area is rectangular, and the area of a rectangle is simply the width multiplied by the height. With the width of the interval equal to $130 - 120 = 10$ and the height equal to the value of the probability density function $f(x) = 1/20$, we have area = width $\times$ height $= 10(1/20) = 10/20 = .50$.

FIGURE 6.2 AREA PROVIDES PROBABILITY OF A FLIGHT TIME BETWEEN 120 AND 130 MINUTES

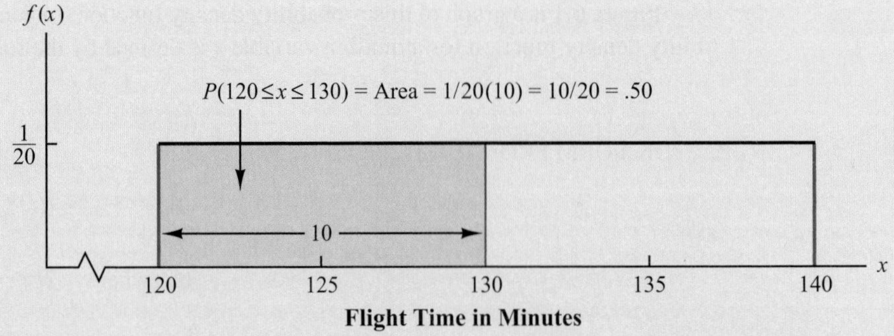

What observation can you make about the area under the graph of $f(x)$ and probability? They are identical! Indeed, this observation is valid for all continuous random variables. Once a probability density function $f(x)$ is identified, the probability that x takes a value between some lower value x_1 and some higher value x_2 can be found by computing the area under the graph of $f(x)$ over the interval from x_1 to x_2.

Given the uniform distribution for flight time and using the interpretation of area as probability, we can answer any number of probability questions about flight times. For example, what is the probability of a flight time between 128 and 136 minutes? The width of the interval is $136 - 128 = 8$. With the uniform height of $f(x) = 1/20$, we see that $P(128 \leq x \leq 136) = 8(1/20) = .40$.

Note that $P(120 \leq x \leq 140) = 20(1/20) = 1$; that is, the total area under the graph of $f(x)$ is equal to 1. This property holds for all continuous probability distributions and is the analog of the condition that the sum of the probabilities must equal 1 for a discrete probability function. For a continuous probability density function, we must also require that $f(x) \geq 0$ for all values of x. This requirement is the analog of the requirement that $f(x) \geq 0$ for discrete probability functions.

Two major differences stand out between the treatment of continuous random variables and the treatment of their discrete counterparts.

1. We no longer talk about the probability of the random variable assuming a particular value. Instead, we talk about the probability of the random variable assuming a value within some given interval.
2. The probability of a continuous random variable assuming a value within some given interval from x_1 to x_2 is defined to be the area under the graph of the probability density function between x_1 and x_2. Because a single point is an interval of zero width, this implies that the probability of a continuous random variable assuming any particular value exactly is zero. It also means that the probability of a continuous random variable assuming a value in any interval is the same whether or not the endpoints are included.

To see that the probability of any single point is 0, refer to Figure 6.2 and compute the probability of a single point, say, $x = 125$. $P(x = 125) = P(125 \leq x \leq 125) = 0(1/20) = 0$.

The calculation of the expected value and variance for a continuous random variable is analogous to that for a discrete random variable. However, because the computational procedure involves integral calculus, we leave the derivation of the appropriate formulas to more advanced texts.

For the uniform continuous probability distribution introduced in this section, the formulas for the expected value and variance are

$$E(x) = \frac{a + b}{2}$$

$$Var(x) = \frac{(b - a)^2}{12}$$

In these formulas, a is the smallest value and b is the largest value that the random variable may assume.

Applying these formulas to the uniform distribution for flight times from Chicago to New York, we obtain

$$E(x) = \frac{(120 + 140)}{2} = 130$$

$$Var(x) = \frac{(140 - 120)^2}{12} = 33.33$$

The standard deviation of flight times can be found by taking the square root of the variance. Thus, $\sigma = 5.77$ minutes.

NOTES AND COMMENTS

To see more clearly why the height of a probability density function is not a probability, think about a random variable with the following uniform probability distribution.

$$f(x) = \begin{cases} 2 & \text{for } 0 \leq x \leq .5 \\ 0 & \text{elsewhere} \end{cases}$$

The height of the probability density function, $f(x)$, is 2 for values of x between 0 and .5. However, we know probabilities can never be greater than 1. Thus, we see that $f(x)$ cannot be interpreted as the probability of x.

Exercises

Methods

SELF test

1. The random variable x is known to be uniformly distributed between 1.0 and 1.5.
   a. Show the graph of the probability density function.
   b. Compute $P(x = 1.25)$.
   c. Compute $P(1.0 \leq x \leq 1.25)$.
   d. Compute $P(1.20 < x < 1.5)$.
2. The random variable x is known to be uniformly distributed between 10 and 20.
   a. Show the graph of the probability density function.
   b. Compute $P(x < 15)$.
   c. Compute $P(12 \leq x \leq 18)$.
   d. Compute $E(x)$.
   e. Compute $Var(x)$.

Applications

3. Delta Airlines quotes a flight time of 2 hours, 5 minutes for its flights from Cincinnati to Tampa. Suppose we believe that actual flight times are uniformly distributed between 2 hours and 2 hours, 20 minutes.
   a. Show the graph of the probability density function for flight time.
   b. What is the probability that the flight will be no more than 5 minutes late?
   c. What is the probability that the flight will be more than 10 minutes late?
   d. What is the expected flight time?

SELF test

4. Most computer languages include a function that can be used to generate random numbers. In Excel, the RAND function can be used to generate random numbers between 0 and 1. If we let x denote a random number generated using RAND, then x is a continuous random variable with the following probability density function.

$$f(x) = \begin{cases} 1 & \text{for } 0 \leq x \leq 1 \\ 0 & \text{elsewhere} \end{cases}$$

   a. Graph the probability density function.
   b. What is the probability of generating a random number between .25 and .75?
   c. What is the probability of generating a random number with a value less than or equal to .30?
   d. What is the probability of generating a random number with a value greater than .60?
   e. Generate 50 random numbers by entering =RAND() into 50 cells of an Excel worksheet.
   f. Compute the mean and standard deviation for the random numbers in part (e).

5. The driving distance for the top 100 golfers on the PGA tour is between 284.7 and 310.6 yards (*Golfweek,* March 29, 2003). Assume that the driving distance for these golfers is uniformly distributed over this interval.
   a. Give a mathematical expression for the probability density function of driving distance.
   b. What is the probability the driving distance for one of these golfers is less than 290 yards?
   c. What is the probability the driving distance for one of these golfers is at least 300 yards?
   d. What is the probability the driving distance for one of these golfers is between 290 and 305 yards?
   e. How many of these golfers drive the ball at least 290 yards?
6. On average, 30-minute television sitcoms have 22 minutes of programming (CNBC, February 23, 2006). Assume that the probability distribution for minutes of programming can be approximated by a uniform distribution from 18 minutes to 26 minutes.
   a. What is the probability a sitcom will have 25 or more minutes of programming?
   b. What is the probability a sitcom will have between 21 and 25 minutes of programming?
   c. What is the probability a sitcom will have more than 10 minutes of commercials or other nonprogramming interruptions?
7. Suppose we are interested in bidding on a piece of land and we know one other bidder is interested.[1] The seller announced that the highest bid in excess of $10,000 will be accepted. Assume that the competitor's bid x is a random variable that is uniformly distributed between $10,000 and $15,000.
   a. Suppose you bid $12,000. What is the probability that your bid will be accepted?
   b. Suppose you bid $14,000. What is the probability that your bid will be accepted?
   c. What amount should you bid to maximize the probability that you get the property?
   d. Suppose you know someone who is willing to pay you $16,000 for the property. Would you consider bidding less than the amount in part (c)? Why or why not?

6.2 Normal Probability Distribution

Abraham de Moivre, a French mathematician, published The Doctrine of Chances *in 1733. He derived the normal distribution.*

The most important probability distribution for describing a continuous random variable is the **normal probability distribution**. The normal distribution has been used in a wide variety of practical applications in which the random variables are heights and weights of people, test scores, scientific measurements, amounts of rainfall, and other similar values. It is also widely used in statistical inference, which is the major topic of the remainder of this book. In such applications, the normal distribution provides a description of the likely results obtained through sampling.

Normal Curve

The form, or shape, of the normal distribution is illustrated by the bell-shaped normal curve in Figure 6.3. The probability density function that defines the bell-shaped curve of the normal distribution follows.

[1]This exercise is based on a problem suggested to us by Professor Roger Myerson of Northwestern University.

FIGURE 6.3 BELL-SHAPED CURVE FOR THE NORMAL DISTRIBUTION

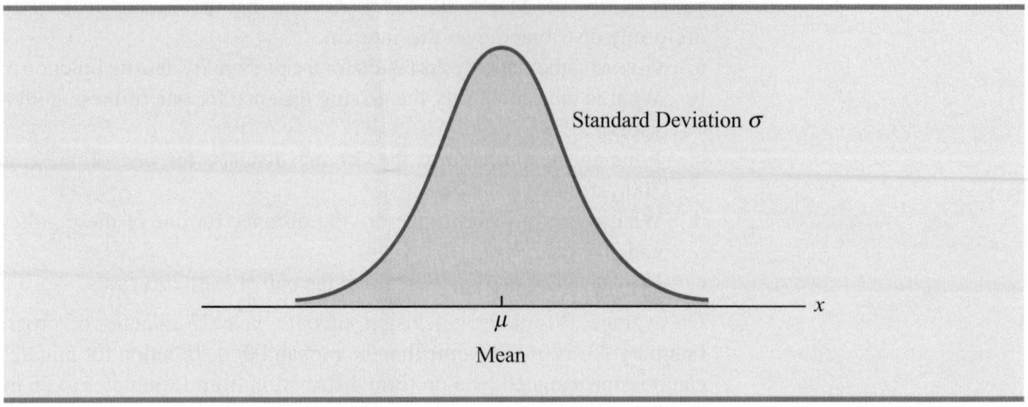

NORMAL PROBABILITY DENSITY FUNCTION

$$f(x) = \frac{1}{\sigma\sqrt{2\pi}} e^{-(x-\mu)^2/2\sigma^2} \tag{6.2}$$

where

μ = mean
σ = standard deviation
π = 3.14159
e = 2.71828

We make several observations about the characteristics of the normal distribution.

The normal curve has two parameters, μ and σ. They determine the location and shape of the normal distribution.

1. The entire family of normal distributions is differentiated by two parameters: the mean μ and the standard deviation σ.
2. The highest point on the normal curve is at the mean, which is also the median and mode of the distribution.
3. The mean of the distribution can be any numerical value: negative, zero, or positive. Three normal distributions with the same standard deviation but three different means (−10, 0, and 20) are shown here.

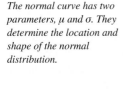

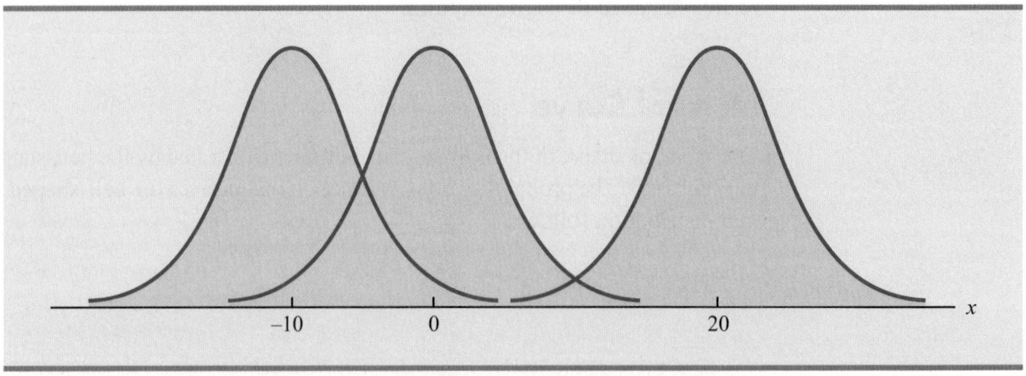

4. The normal distribution is symmetric, with the shape of the normal curve to the left of the mean a mirror image of the shape of the normal curve to the right of the mean. The tails of the normal curve extend to infinity in both directions and theoretically never touch the horizontal axis. Because it is symmetric, the normal distribution is not skewed; its skewness measure is zero.
5. The standard deviation determines how flat and wide the normal curve is. Larger values of the standard deviation result in wider, flatter curves, showing more variability in the data. Two normal distributions with the same mean but with different standard deviations are shown here.

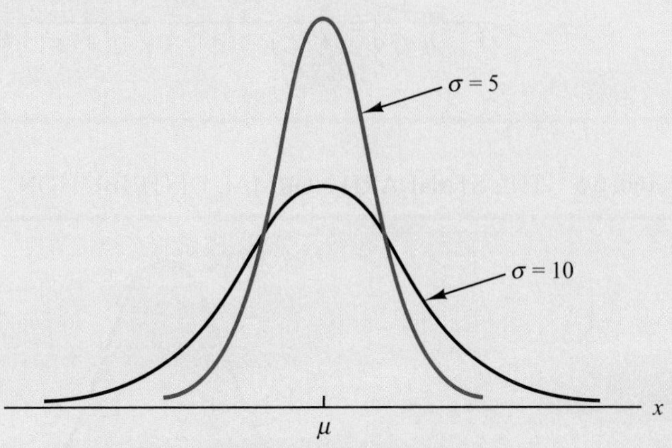

6. Probabilities for the normal random variable are given by areas under the normal curve. The total area under the curve for the normal distribution is 1. Because the distribution is symmetric, the area under the curve to the left of the mean is .50 and the area under the curve to the right of the mean is .50.
7. The percentage of values in some commonly used intervals are
   a. 68.3% of the values of a normal random variable are within plus or minus one standard deviation of its mean.
   b. 95.4% of the values of a normal random variable are within plus or minus two standard deviations of its mean.
   c. 99.7% of the values of a normal random variable are within plus or minus three standard deviations of its mean.

These percentages are the basis for the empirical rule introduced in Section 3.3.

Figure 6.4 shows properties (a), (b), and (c) graphically.

Standard Normal Probability Distribution

A random variable that has a normal distribution with a mean of zero and a standard deviation of one is said to have a **standard normal probability distribution**. The letter z is commonly used to designate this particular normal random variable. Figure 6.5 is the graph of the standard normal distribution. It has the same general appearance as other normal distributions, but with the special properties of $\mu = 0$ and $\sigma = 1$.

FIGURE 6.4 AREAS UNDER THE CURVE FOR ANY NORMAL DISTRIBUTION

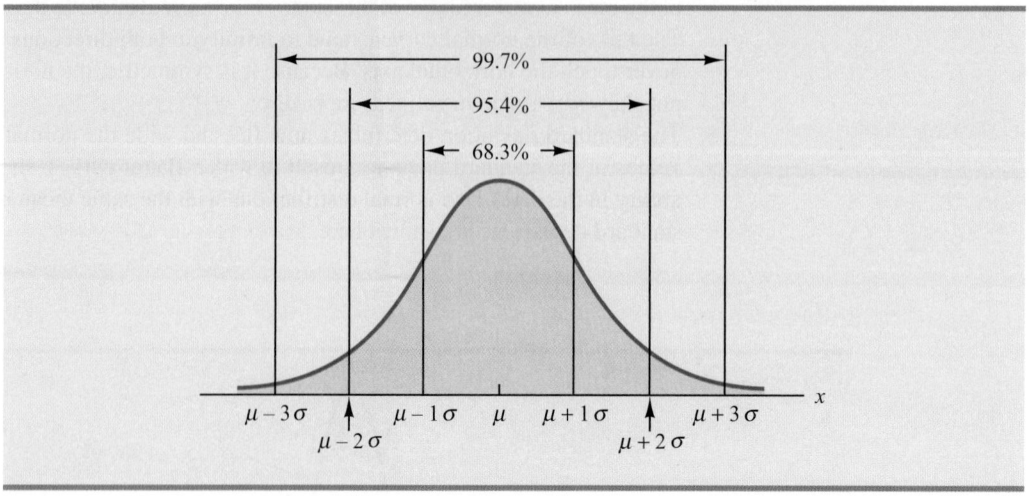

FIGURE 6.5 THE STANDARD NORMAL DISTRIBUTION

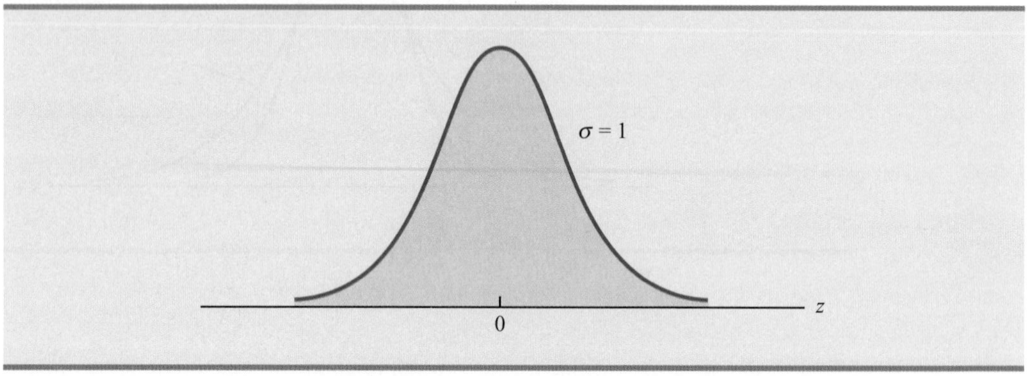

Because $\mu = 0$ and $\sigma = 1$, the formula for the standard normal probability density function is a simpler version of equation (6.2).

STANDARD NORMAL DENSITY FUNCTION

$$f(z) = \frac{1}{\sqrt{2\pi}} e^{-z^2/2}$$

As with other continuous random variables, probability calculations with any normal distribution are made by computing areas under the graph of the probability density function. Thus, to find the probability that a normal random variable is within any specific interval, we must compute the area under the normal curve over that interval.

For the normal probability density function, the height of the normal curve varies and more advanced mathematics is required to compute the areas that represent probability.

For the standard normal distribution, areas under the normal curve have been computed and are available in tables that can be used to compute probabilities. Such a table appears on the two pages inside the front cover of the text. The table on the left-hand page contains areas, or cumulative probabilities, for z values less than or equal to the mean of zero. The table on the right-hand page contains areas, or cumulative probabilities, for z values greater than or equal to the mean of zero.

The three types of probabilities we need to compute include (1) the probability that the standard normal random variable z will be less than or equal to a given value; (2) the probability that z will be between two given values; and (3) the probability that z will be greater than or equal to a given value. To see how the cumulative probability table for the standard normal distribution can be used to compute these three types of probabilities, let us consider some examples.

Because the standard normal random variable is continuous, $P(z \leq 1.00) = P(z < 1.00)$.

We start by showing how to compute the probability that z is less than or equal to 1.00; that is, $P(z \leq 1.00)$. This cumulative probability is the area under the normal curve to the left of $z = 1.00$ in the following graph.

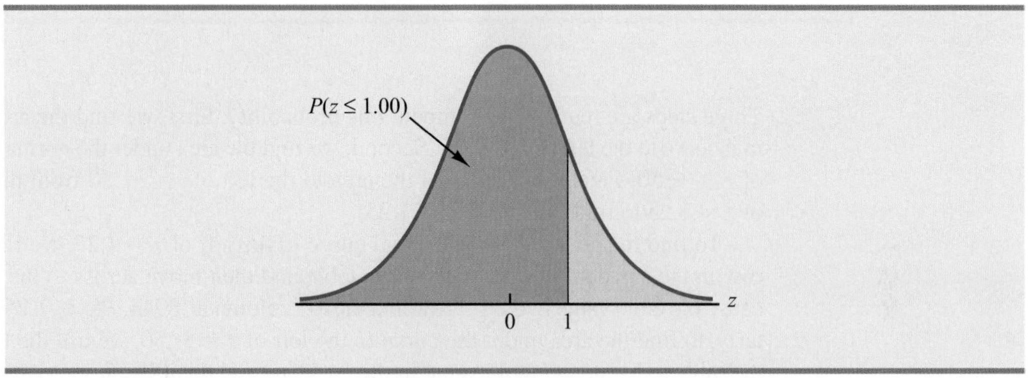

Refer to the right-hand page of the standard normal probability table inside the front cover of the text. The cumulative probability corresponding to $z = 1.00$ is the table value located at the intersection of the row labeled 1.0 and the column labeled .00. First we find 1.0 in the left column of the table and then find .00 in the top row of the table. By looking in the body of the table, we find that the 1.0 row and the .00 column intersect at the value of .8413; thus, $P(z \leq 1.00) = .8413$. The following excerpt from the probability table shows these steps.

z	**.00**	**.01**	**.02**
.			
.			
.			
.9	.8159	.8186	.8212
1.0	**.8413**	.8438	.8461
1.1	.8643	.8665	.8686
1.2	.8849	.8869	.8888
.			
.			
.			

$P(z \leq 1.00)$

To illustrate the second type of probability calculation we show how to compute the probability that z is in the interval between $-.50$ and 1.25; that is, $P(-.50 \leq z \leq 1.25)$. The following graph shows this area, or probability.

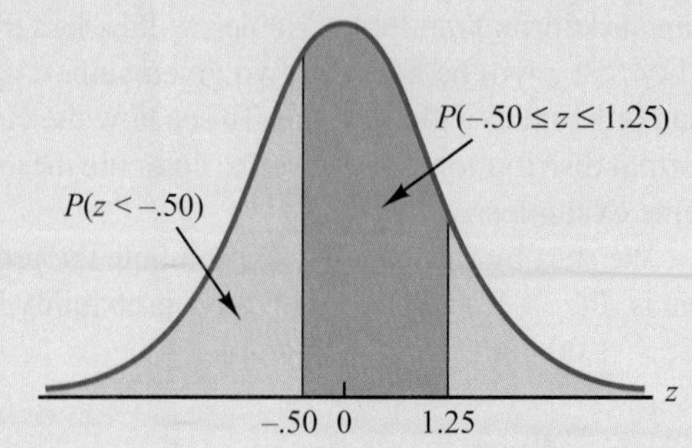

Three steps are required to compute this probability. First, we find the area under the normal curve to the left of $z = 1.25$. Second, we find the area under the normal curve to the left of $z = -.50$. Finally, we subtract the area to the left of $z = -.50$ from the area to the left of $z = 1.25$ to find $P(-.50 \leq z \leq 1.25)$.

To find the area under the normal curve to the left of $z = 1.25$, we first locate the 1.2 row in the standard normal probability table and then move across to the .05 column. Because the table value in the 1.2 row and the .05 column is .8944, $P(z \leq 1.25) = .8944$. Similarly, to find the area under the curve to the left of $z = -.50$, we use the left-hand page of the table to locate the table value in the $-.5$ row and the .00 column; with a table value of .3085, $P(z \leq -.50) = .3085$. Thus, $P(-.50 \leq z \leq 1.25) = P(z \leq 1.25) - P(z \leq -.50) = .8944 - .3085 = .5859$.

Let us consider another example of computing the probability that z is in the interval between two given values. Often it is of interest to compute the probability that a normal random variable assumes a value within a certain number of standard deviations of the mean. Suppose we want to compute the probability that the standard normal random variable is within one standard deviation of the mean; that is, $P(-1.00 \leq z \leq 1.00)$. To compute this probability we must find the area under the curve between -1.00 and 1.00. Earlier we found that $P(z \leq 1.00) = .8413$. Referring again to the table inside the front cover of the book, we find that the area under the curve to the left of $z = -1.00$ is .1587, so $P(z \leq -1.00) = .1587$. Therefore, $P(-1.00 \leq z \leq 1.00) = P(z \leq 1.00) - P(z \leq -1.00) = .8413 - .1587 = .6826$. This probability is shown graphically in the following figure.

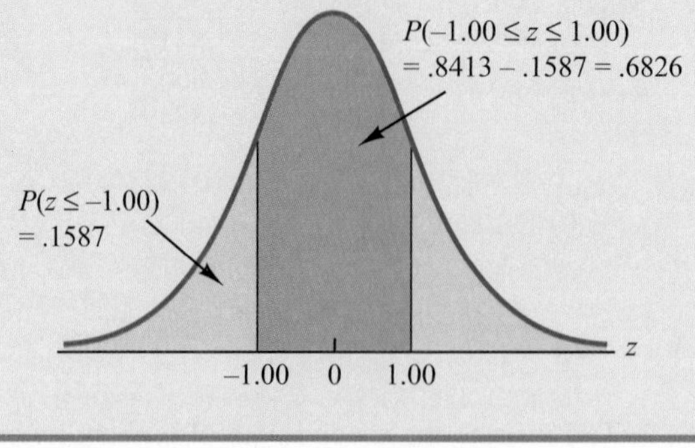

To illustrate how to make the third type of probability computation, suppose we want to compute the probability of obtaining a z value of at least 1.58; that is, $P(z \geq 1.58)$. The value in the $z = 1.5$ row and the .08 column of the cumulative normal table is .9429; thus, $P(z < 1.58) = .9429$. However, because the total area under the normal curve is 1, $P(z \geq 1.58) = 1 - .9429 = .0571$. This probability is shown in the following figure.

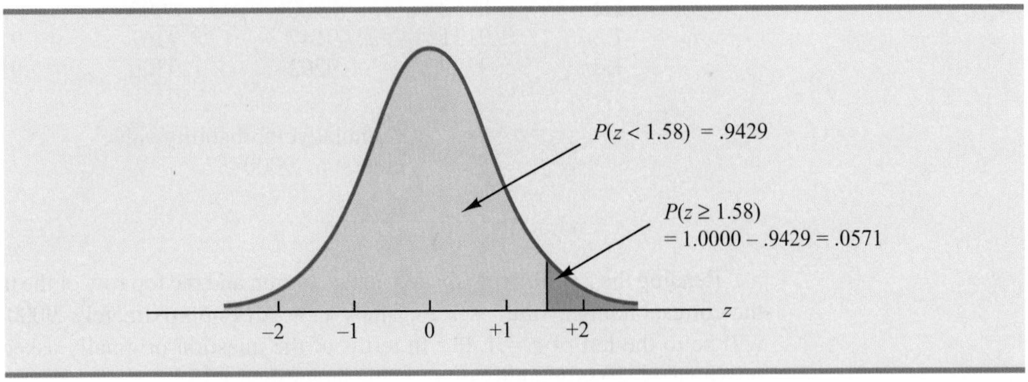

In the preceding illustrations, we showed how to compute probabilities given specified z values. In some situations, we are given a probability and are interested in working backward to find the corresponding z value. Suppose we want to find a z value such that the probability of obtaining a larger z value is .10. The following figure shows this situation graphically.

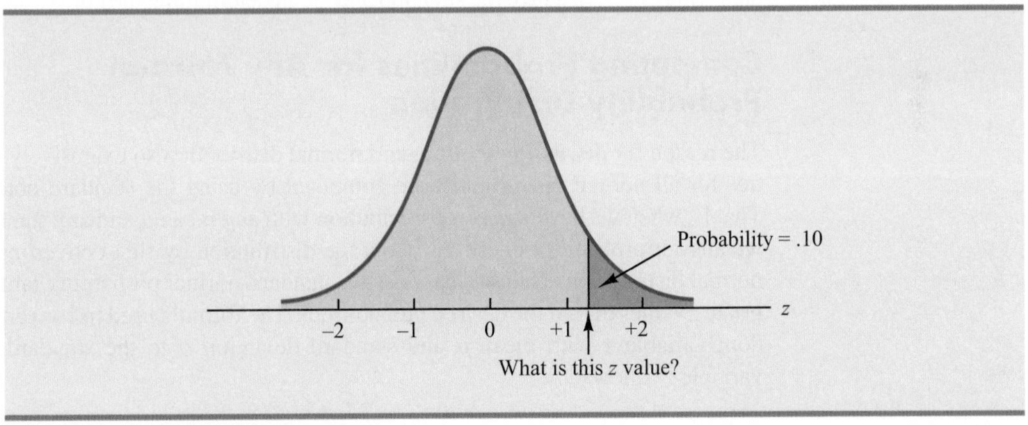

Given a probability, we can use the standard normal table in an inverse fashion to find the corresponding z value.

This problem is the inverse of those in the preceding examples. Previously, we specified the z value of interest and then found the corresponding probability, or area. In this example, we are given the probability, or area, and asked to find the corresponding z value. To do so, we use the standard normal probability table somewhat differently.

Recall that the standard normal probability table gives the area under the curve to the left of a particular z value. We have been given the information that the area in the upper tail of the curve is .10. Hence, the area under the curve to the left of the unknown z value must equal .9000. Scanning the body of the table, we find .8997 is the cumulative probability value closest to .9000. The section of the table providing this result follows.

z	.06	.07	.08	.09
•				
•				
•				
1.0	.8554	.8577	.8599	.8621
1.1	.8770	.8790	.8810	.8830
1.2	.8962	.8980	.8997	.9015
1.3	.9131	.9147	.9162	.9177
1.4	.9279	.9292	.9306	.9319
•				
•				
•				

Cumulative probability value closest to .9000

Reading the z value from the left-most column and the top row of the table, we find that the corresponding z value is 1.28. Thus, an area of approximately .9000 (actually .8997) will be to the left of $z = 1.28$.[2] In terms of the question originally asked, there is an approximately .10 probability of a z value larger than 1.28.

The examples illustrate that the table of cumulative probabilities for the standard normal probability distribution can be used to find probabilities associated with values of the standard normal random variable z. Two types of questions can be asked. The first type of question specifies a value, or values, for z and asks us to use the table to determine the corresponding areas or probabilities. The second type of question provides an area, or probability, and asks us to use the table to determine the corresponding z value. Thus, we need to be flexible in using the standard normal probability table to answer the desired probability question. In most cases, sketching a graph of the standard normal probability distribution and shading the appropriate area will help to visualize the situation and aid in determining the correct answer.

Computing Probabilities for Any Normal Probability Distribution

The reason for discussing the standard normal distribution so extensively is that probabilities for all normal distributions are computed by using the standard normal distribution. That is, when we have a normal distribution with any mean μ and any standard deviation σ, we answer probability questions about the distribution by first converting to the standard normal distribution. Then we can use the standard normal probability table and the appropriate z values to find the desired probabilities. The formula used to convert any normal random variable x with mean μ and standard deviation σ to the standard normal random variable z follows.

The formula for the standard normal random variable is similar to the formula we introduced in Chapter 3 for computing z-scores for a data set.

CONVERTING TO THE STANDARD NORMAL RANDOM VARIABLE

$$z = \frac{x - \mu}{\sigma} \tag{6.3}$$

[2] We could use interpolation in the body of the table to get a better approximation of the z value that corresponds to an area of .9000. Doing so to provide one more decimal place of accuracy would yield a z value of 1.282. However, in most practical situations, sufficient accuracy is obtained by simply using the table value closest to the desired probability.

A value of x equal to its mean μ results in $z = (\mu - \mu)/\sigma = 0$. Thus, we see that a value of x equal to its mean μ corresponds to $z = 0$. Now suppose that x is one standard deviation above its mean; that is, $x = \mu + \sigma$. Applying equation (6.3), we see that the corresponding z value is $z = [(\mu + \sigma) - \mu]/\sigma = \sigma/\sigma = 1$. Thus, an x value that is one standard deviation above its mean corresponds to $z = 1$. In other words, *we can interpret z as the number of standard deviations that the normal random variable x is from its mean μ.*

To see how this conversion enables us to compute probabilities for any normal distribution, suppose we have a normal distribution with $\mu = 10$ and $\sigma = 2$. What is the probability that the random variable x is between 10 and 14? Using equation (6.3), we see that at $x = 10$, $z = (x - \mu)/\sigma = (10 - 10)/2 = 0$ and that at $x = 14$, $z = (14 - 10)/2 = 4/2 = 2$. Thus, the answer to our question about the probability of x being between 10 and 14 is given by the equivalent probability that z is between 0 and 2 for the standard normal distribution. In other words, the probability that we are seeking is the probability that the random variable x is between its mean and two standard deviations above the mean. Using $z = 2.00$ and the standard normal probability table inside the front cover of the text, we see that $P(z \leq 2) = .9772$. Because $P(z \leq 0) = .5000$, we can compute $P(.00 \leq z \leq 2.00) = P(z \leq 2) - P(z \leq 0) = .9772 - .5000 = .4772$. Hence the probability that x is between 10 and 14 is .4772.

Grear Tire Company Problem

We turn now to an application of the normal probability distribution. Suppose the Grear Tire Company developed a new steel-belted radial tire to be sold through a national chain of discount stores. Because the tire is a new product, Grear's managers believe that the mileage guarantee offered with the tire will be an important factor in the acceptance of the product. Before finalizing the tire mileage guarantee policy, Grear's managers want probability information about x = number of miles the tires will last.

From actual road tests with the tires, Grear's engineering group estimated that the mean tire mileage is $\mu = 36{,}500$ miles and that the standard deviation is $\sigma = 5000$. In addition, the data collected indicate that a normal distribution is a reasonable assumption. What percentage of the tires can be expected to last more than 40,000 miles? In other words, what is the probability that the tire mileage, x, will exceed 40,000? This question can be answered by finding the area of the darkly shaded region in Figure 6.6.

FIGURE 6.6 GREAR TIRE COMPANY MILEAGE DISTRIBUTION

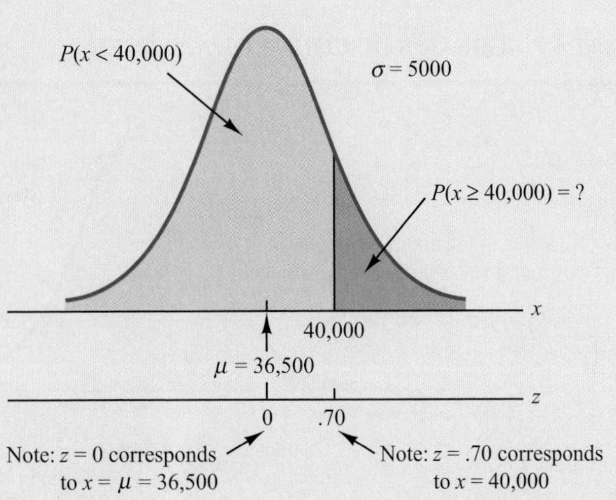

At $x = 40{,}000$, we have

$$z = \frac{x - \mu}{\sigma} = \frac{40{,}000 - 36{,}500}{5000} = \frac{3500}{5000} = .70$$

Refer now to the bottom of Figure 6.6. We see that a value of $x = 40{,}000$ on the Grear Tire normal distribution corresponds to a value of $z = .70$ on the standard normal distribution. Using the standard normal probability table, we see that the area under the standard normal curve to the left of $z = .70$ is .7580. Thus, $1.000 - .7580 = .2420$ is the probability that z will exceed .70 and hence x will exceed 40,000. We can conclude that about 24.2% of the tires will exceed 40,000 in mileage.

Let us now assume that Grear is considering a guarantee that will provide a discount on replacement tires if the original tires do not provide the guaranteed mileage. What should the guarantee mileage be if Grear wants no more than 10% of the tires to be eligible for the discount guarantee? This question is interpreted graphically in Figure 6.7.

According to Figure 6.7, the area under the curve to the left of the unknown guarantee mileage must be .10. So, we must first find the z value that cuts off an area of .10 in the left tail of a standard normal distribution. Using the standard normal probability table, we see that $z = -1.28$ cuts off an area of .10 in the lower tail. Hence, $z = -1.28$ is the value of the standard normal random variable corresponding to the desired mileage guarantee on the Grear Tire normal distribution. To find the value of x corresponding to $z = -1.28$, we have

The guarantee mileage we need to find is 1.28 standard deviations below the mean. Thus, $x = \mu - 1.28\sigma$.

$$z = \frac{x - \mu}{\sigma} = -1.28$$

$$x - \mu = -1.28\sigma$$

$$x = \mu - 1.28\sigma$$

With $\mu = 36{,}500$ and $\sigma = 5000$,

$$x = 36{,}500 - 1.28(5000) = 30{,}100$$

With the guarantee set at 30,000 miles, the actual percentage eligible for the guarantee will be 9.68%.

Thus, a guarantee of 30,100 miles will meet the requirement that approximately 10% of the tires will be eligible for the guarantee. Perhaps, with this information, the firm will set its tire mileage guarantee at 30,000 miles.

FIGURE 6.7 GREAR'S DISCOUNT GUARANTEE

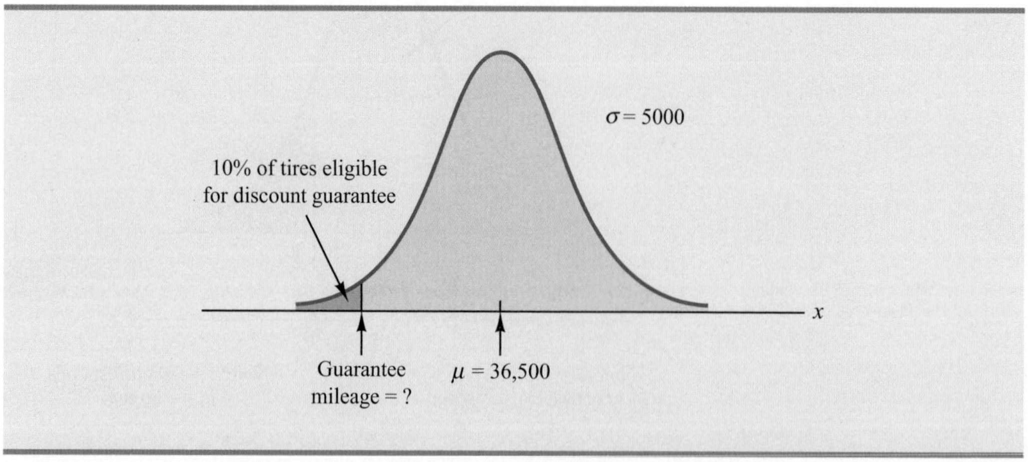

Again, we see the important role that probability distributions play in providing decision-making information. Namely, once a probability distribution is established for a particular application, it can be used to obtain probability information about the problem. Probability does not make a decision recommendation directly, but it provides information that helps the decision maker better understand the risks and uncertainties associated with the problem. Ultimately, this information may assist the decision maker in reaching a good decision.

EXERCISES

Methods

8. Using Figure 6.4 as a guide, sketch a normal curve for a random variable x that has a mean of $\mu = 100$ and a standard deviation of $\sigma = 10$. Label the horizontal axis with values of 70, 80, 90, 100, 110, 120, and 130.

9. A random variable is normally distributed with a mean of $\mu = 50$ and a standard deviation of $\sigma = 5$.
   a. Sketch a normal curve for the probability density function. Label the horizontal axis with values of 35, 40, 45, 50, 55, 60, and 65. Figure 6.4 shows that the normal curve almost touches the horizontal axis at three standard deviations below and at three standard deviations above the mean (in this case at 35 and 65).
   b. What is the probability the random variable will assume a value between 45 and 55?
   c. What is the probability the random variable will assume a value between 40 and 60?

10. Draw a graph for the standard normal distribution. Label the horizontal axis at values of $-3, -2, -1, 0, 1, 2$, and 3. Then use the table of probabilities for the standard normal distribution inside the front cover of the text to compute the following probabilities.
   a. $P(z \leq 1.5)$
   b. $P(z \leq 1)$
   c. $P(1 \leq z \leq 1.5)$
   d. $P(0 < z < 2.5)$

11. Given that z is a standard normal random variable, compute the following probabilities.
   a. $P(z \leq -1.0)$
   b. $P(z \geq -1)$
   c. $P(z \geq -1.5)$
   d. $P(-2.5 \leq z)$
   e. $P(-3 < z \leq 0)$

12. Given that z is a standard normal random variable, compute the following probabilities.
   a. $P(0 \leq z \leq .83)$
   b. $P(-1.57 \leq z \leq 0)$
   c. $P(z > .44)$
   d. $P(z \geq -.23)$
   e. $P(z < 1.20)$
   f. $P(z \leq -.71)$

SELF test

13. Given that z is a standard normal random variable, compute the following probabilities.
   a. $P(-1.98 \leq z \leq .49)$
   b. $P(.52 \leq z \leq 1.22)$
   c. $P(-1.75 \leq z \leq -1.04)$

14. Given that z is a standard normal random variable, find z for each situation.
   a. The area to the left of z is .9750.
   b. The area between 0 and z is .4750.
   c. The area to the left of z is .7291.
   d. The area to the right of z is .1314.
   e. The area to the left of z is .6700.
   f. The area to the right of z is .3300.

SELF test

15. Given that z is a standard normal random variable, find z for each situation.
   a. The area to the left of z is .2119.
   b. The area between $-z$ and z is .9030.
   c. The area between $-z$ and z is .2052.
   d. The area to the left of z is .9948.
   e. The area to the right of z is .6915.
16. Given that z is a standard normal random variable, find z for each situation.
   a. The area to the right of z is .01.
   b. The area to the right of z is .025.
   c. The area to the right of z is .05.
   d. The area to the right of z is .10.

Applications

17. For borrowers with good credit scores, the mean debt for revolving and installment accounts is \$15,015 (*BusinessWeek,* March 20, 2006). Assume the standard deviation is \$3540 and that debt amounts are normally distributed.
   a. What is the probability that the debt for a borrower with good credit is more than \$18,000?
   b. What is the probability that the debt for a borrower with good credit is less than \$10,000?
   c. What is the probability that the debt for a borrower with good credit is between \$12,000 and \$18,000?
   d. What is the probability that the debt for a borrower with good credit is no more than \$14,000?

SELF test

18. The average return for large-cap domestic stock funds over the three years 2009–2011 was 14.4% (*AAII Journal,* February, 2012). Assume the three-year returns were normally distributed across funds with a standard deviation of 4.4%.
   a. What is the probability an individual large-cap domestic stock fund had a three-year return of at least 20%?
   b. What is the probability an individual large-cap domestic stock fund had a three-year return of 10% or less?
   c. How big does the return have to be to put a domestic stock fund in the top 10% for the three-year period?
19. In an article about the cost of health care, *Money* magazine reported that a visit to a hospital emergency room for something as simple as a sore throat has a mean cost of \$328 (*Money,* January 2009). Assume that the cost for this type of hospital emergency room visit is normally distributed with a standard deviation of \$92. Answer the following questions about the cost of a hospital emergency room visit for this medical service.
   a. What is the probability that the cost will be more than \$500?
   b. What is the probability that the cost will be less than \$250?
   c. What is the probability that the cost will be between \$300 and \$400?
   d. If the cost to a patient is in the lower 8% of charges for this medical service, what was the cost of this patient's emergency room visit?
20. The average price for a gallon of gasoline in the United States is \$3.73 and in Russia it is \$3.40 (*Bloomberg Businessweek,* March 5–March 11, 2012). Assume these averages are the population means in the two countries and that the probability distributions are normally distributed with a standard deviation of \$.25 in the United States and a standard deviation of \$.20 in Russia.
   a. What is the probability that a randomly selected gas station in the United States charges less than \$3.50 per gallon?
   b. What percentage of the gas stations in Russia charge less than \$3.50 per gallon?
   c. What is the probability that a randomly selected gas station in Russia charged more than the mean price in the United States?
21. A person must score in the upper 2% of the population on an IQ test to qualify for membership in Mensa, the international high-IQ society. If IQ scores are normally distributed

with a mean of 100 and a standard deviation of 15, what score must a person have to qualify for Mensa?

22. Television viewing reached a new high when the Nielsen Company reported a mean daily viewing time of 8.35 hours per household (*USA Today,* November 11, 2009). Use a normal probability distribution with a standard deviation of 2.5 hours to answer the following questions about daily television viewing per household.
    a. What is the probability that a household views television between 5 and 10 hours a day?
    b. How many hours of television viewing must a household have in order to be in the top 3% of all television viewing households?
    c. What is the probability that a household views television more than 3 hours a day?

23. The time needed to complete a final examination in a particular college course is normally distributed with a mean of 80 minutes and a standard deviation of 10 minutes. Answer the following questions.
    a. What is the probability of completing the exam in one hour or less?
    b. What is the probability that a student will complete the exam in more than 60 minutes but less than 75 minutes?
    c. Assume that the class has 60 students and that the examination period is 90 minutes in length. How many students do you expect will be unable to complete the exam in the allotted time?

WEB file Volume

24. Trading volume on the New York Stock Exchange is heaviest during the first half hour (early morning) and last half hour (late afternoon) of the trading day. The early morning trading volumes (millions of shares) for 13 days in January and February are shown here (*Barron's,* January 23, 2006; February 13, 2006; and February 27, 2006).

214	163	265	194	180
202	198	212	201	
174	171	211	211	

The probability distribution of trading volume is approximately normal.
    a. Compute the mean and standard deviation to use as estimates of the population mean and standard deviation.
    b. What is the probability that, on a randomly selected day, the early morning trading volume will be less than 180 million shares?
    c. What is the probability that, on a randomly selected day, the early morning trading volume will exceed 230 million shares?
    d. How many shares would have to be traded for the early morning trading volume on a particular day to be among the busiest 5% of days?

25. According to the Sleep Foundation, the average night's sleep is 6.8 hours (*Fortune,* March 20, 2006). Assume the standard deviation is .6 hours and that the probability distribution is normal.
    a. What is the probability that a randomly selected person sleeps more than 8 hours?
    b. What is the probability that a randomly selected person sleeps 6 hours or less?
    c. Doctors suggest getting between 7 and 9 hours of sleep each night. What percentage of the population gets this much sleep?

6.3 Normal Approximation of Binomial Probabilities

In Section 5.5 we presented the discrete binomial distribution. Recall that a binomial experiment consists of a sequence of n identical independent trials with each trial having two possible outcomes, a success or a failure. The probability of a success on a trial is the same for all trials and is denoted by p. The binomial random variable is the number of successes in the n trials, and probability questions pertain to the probability of x successes in the n trials.

FIGURE 6.8 NORMAL APPROXIMATION TO A BINOMIAL PROBABILITY DISTRIBUTION WITH $n = 100$ AND $p = .10$ SHOWING THE PROBABILITY OF 12 ERRORS

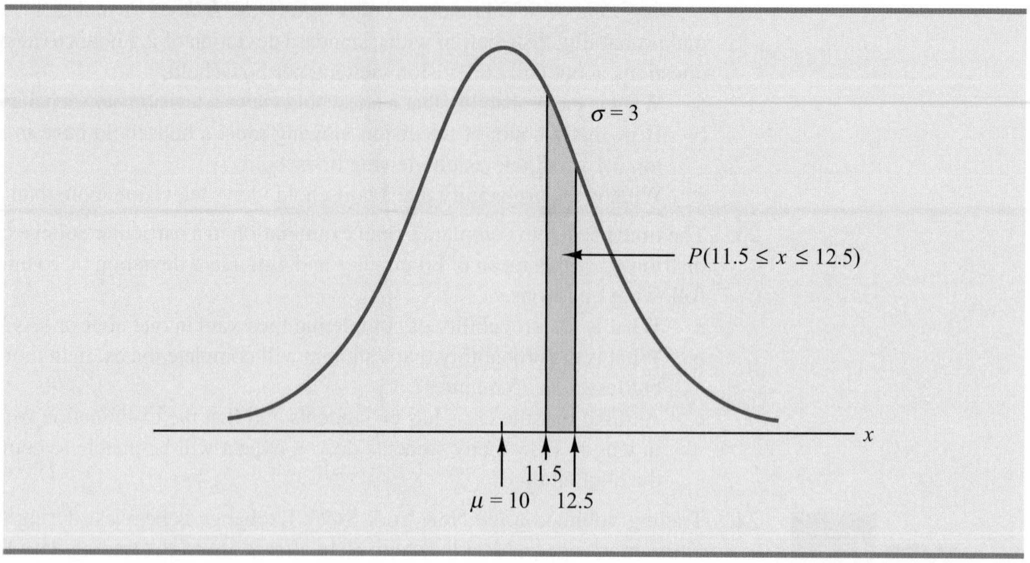

When the number of trials becomes large, evaluating the binomial probability function by hand or with a calculator is difficult. In cases where $np \geq 5$, and $n(1 - p) \geq 5$, the normal distribution provides an easy-to-use approximation of binomial probabilities. When using the normal approximation to the binomial, we set $\mu = np$ and $\sigma = \sqrt{np(1 - p)}$ in the definition of the normal curve.

Let us illustrate the normal approximation to the binomial by supposing that a particular company has a history of making errors in 10% of its invoices. A sample of 100 invoices has been taken, and we want to compute the probability that 12 invoices contain errors. That is, we want to find the binomial probability of 12 successes in 100 trials. In applying the normal approximation in this case, we set $\mu = np = (100)(.1) = 10$ and $\sigma = \sqrt{np(1 - p)} = \sqrt{(100)(.1)(.9)} = 3$. A normal distribution with $\mu = 10$ and $\sigma = 3$ is shown in Figure 6.8.

Recall that, with a continuous probability distribution, probabilities are computed as areas under the probability density function. As a result, the probability of any single value for the random variable is zero. Thus to approximate the binomial probability of 12 successes, we compute the area under the corresponding normal curve between 11.5 and 12.5. The .5 that we add and subtract from 12 is called a **continuity correction factor**. It is introduced because a continuous distribution is being used to approximate a discrete distribution. Thus, $P(x = 12)$ for the *discrete* binomial distribution is approximated by $P(11.5 \leq x \leq 12.5)$ for the *continuous* normal distribution.

Converting to the standard normal distribution to compute $P(11.5 \leq x \leq 12.5)$, we have

$$z = \frac{x - \mu}{\sigma} = \frac{12.5 - 10.0}{3} = .83 \quad \text{at } x = 12.5$$

and

$$z = \frac{x - \mu}{\sigma} = \frac{11.5 - 10.0}{3} = .50 \quad \text{at } x = 11.5$$

FIGURE 6.9 NORMAL APPROXIMATION TO A BINOMIAL PROBABILITY DISTRIBUTION WITH $n = 100$ AND $p = .10$ SHOWING THE PROBABILITY OF 13 OR FEWER ERRORS

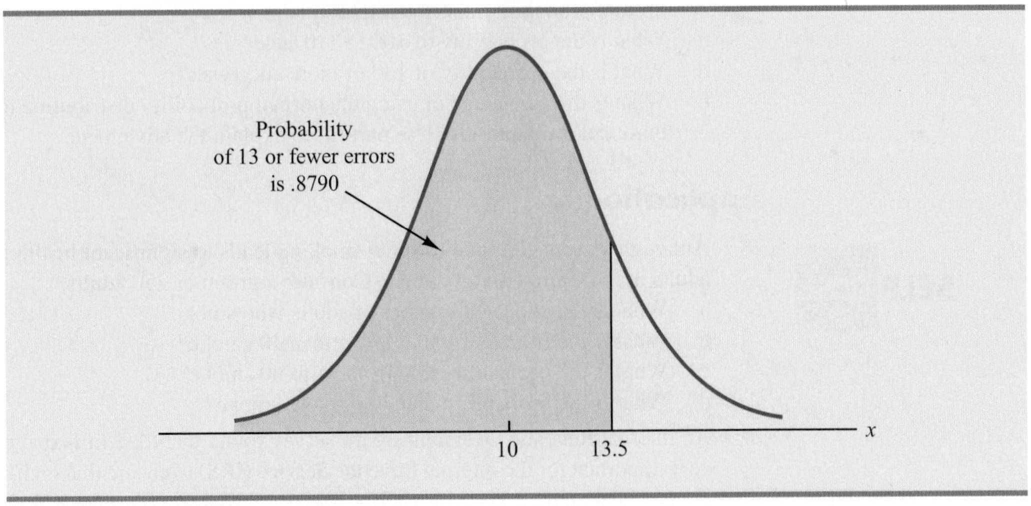

Using the standard normal probability table, we find that the area under the curve (in Figure 6.8) to the left of 12.5 is .7967. Similarly, the area under the curve to the left of 11.5 is .6915. Therefore, the area between 11.5 and 12.5 is .7967 − .6915 = .1052. The normal approximation to the probability of 12 successes in 100 trials is .1052.

For another illustration, suppose we want to compute the probability of 13 or fewer errors in the sample of 100 invoices. Figure 6.9 shows the area under the normal curve that approximates this probability. Note that the use of the continuity correction factor results in the value of 13.5 being used to compute the desired probability. The z value corresponding to $x = 13.5$ is

$$z = \frac{13.5 - 10.0}{3.0} = 1.17$$

The standard normal probability table shows that the area under the standard normal curve to the left of $z = 1.17$ is .8790. The area under the normal curve approximating the probability of 13 or fewer errors is given by the shaded portion of the graph in Figure 6.9.

Exercises

Methods

SELF test

26. A binomial probability distribution has $p = .20$ and $n = 100$.
    a. What are the mean and standard deviation?
    b. Is this situation one in which binomial probabilities can be approximated by the normal probability distribution? Explain.
    c. What is the probability of exactly 24 successes?
    d. What is the probability of 18 to 22 successes?
    e. What is the probability of 15 or fewer successes?

27. Assume a binomial probability distribution has $p = .60$ and $n = 200$.
   a. What are the mean and standard deviation?
   b. Is this situation one in which binomial probabilities can be approximated by the normal probability distribution? Explain.
   c. What is the probability of 100 to 110 successes?
   d. What is the probability of 130 or more successes?
   e. What is the advantage of using the normal probability distribution to approximate the binomial probabilities? Use part (d) to explain the advantage.

Applications

SELF test

28. Although studies continue to show smoking leads to significant health problems, 20% of adults in the United States smoke. Consider a group of 250 adults.
   a. What is the expected number of adults who smoke?
   b. What is the probability that fewer than 40 smoke?
   c. What is the probability that from 55 to 60 smoke?
   d. What is the probability that 70 or more smoke?
29. An Internal Revenue Oversight Board survey found that 82% of taxpayers said that it was very important for the Internal Revenue Service (IRS) to ensure that high-income tax payers do not cheat on their tax returns (*The Wall Street Journal,* February 11, 2009).
   a. For a sample of eight taxpayers, what is the probability that at least six taxpayers say that it is very important to ensure that high-income tax payers do not cheat on their tax returns? Use the binomial distribution probability function shown in Section 5.4 to answer this question.
   b. For a sample of 80 taxpayers, what is the probability that at least 60 taxpayers say that it is very important to ensure that high-income tax payers do not cheat on their tax returns? Use the normal approximation of the binomial distribution to answer this question.
   c. As the number of trails in a binomial distribution application becomes large, what is the advantage of using the normal approximation of the binomial distribution to compute probabilities?
   d. When the number of trials for a binominal distribution application becomes large, would developers of statistical software packages prefer to use the binomial distribution probability function shown in Section 5.5 or the normal approximation of the binomial distribution shown in Section 6.3? Explain.
30. Playing video and computer games is very popular. Over 70% of households play such games. Of those individuals who play video and computer games, 18% are under 18 years old, 53% are 18–59 years old, and 29% are over 59 years old (*The Wall Street Journal,* March 6, 2012).
   a. For a sample of 800 people who play these games, how many would you expect to be under 18 years of age?
   b. For a sample of 600 people who play these games, what is the probability that fewer than 100 will be under 18 years of age?
   c. For a sample of 800 people who play these games, what is the probability that 200 or more will be over 59 years of age?
31. A Bureau of National Affairs survey found that 79% of employers provide their workers with a two-day paid Thanksgiving holiday with workers off both Thursday and Friday (*USA Today,* November 12, 2009). Nineteen percent of employers provide a one-day paid holiday with workers off Thanksgiving Day. Two percent of employers do not provide a paid Thanksgiving holiday. Consider a sample of 120 employers.
   a. What is the probability that at least 85 of the employers provide a two-day paid Thanksgiving holiday?
   b. What is the probability that between 90 and 100 employers provide a two-day paid Thanksgiving holiday? That is, what is $P(90 \leq x \leq 100)$?
   c. What is the probability that less than 20 employers provide a one-day paid Thanksgiving holiday?

6.4 Exponential Probability Distribution

The **exponential probability distribution** may be used for random variables such as the time between arrivals at a car wash, the time required to load a truck, the distance between major defects in a highway, and so on. The exponential probability density function follows.

EXPONENTIAL PROBABILITY DENSITY FUNCTION

$$f(x) = \frac{1}{\mu} e^{-x/\mu} \quad \text{for } x \geq 0 \tag{6.4}$$

where μ = expected value or mean

As an example of the exponential distribution, suppose that x represents the loading time for a truck at the Schips loading dock and follows such a distribution. If the mean, or average, loading time is 15 minutes ($\mu = 15$), the appropriate probability density function for x is

$$f(x) = \frac{1}{15} e^{-x/15}$$

Figure 6.10 is the graph of this probability density function.

Computing Probabilities for the Exponential Distribution

In waiting line applications, the exponential distribution is often used for service time.

As with any continuous probability distribution, the area under the curve corresponding to an interval provides the probability that the random variable assumes a value in that interval. In the Schips loading dock example, the probability that loading a truck will take 6 minutes or less $P(x \leq 6)$ is defined to be the area under the curve in Figure 6.10 from $x = 0$ to $x = 6$. Similarly, the probability that the loading time will be 18 minutes or less $P(x \leq 18)$ is the area under the curve from $x = 0$ to $x = 18$. Note also that the probability that the loading time will be between 6 minutes and 18 minutes $P(6 \leq x \leq 18)$ is given by the area under the curve from $x = 6$ to $x = 18$.

FIGURE 6.10 EXPONENTIAL DISTRIBUTION FOR THE SCHIPS LOADING DOCK EXAMPLE

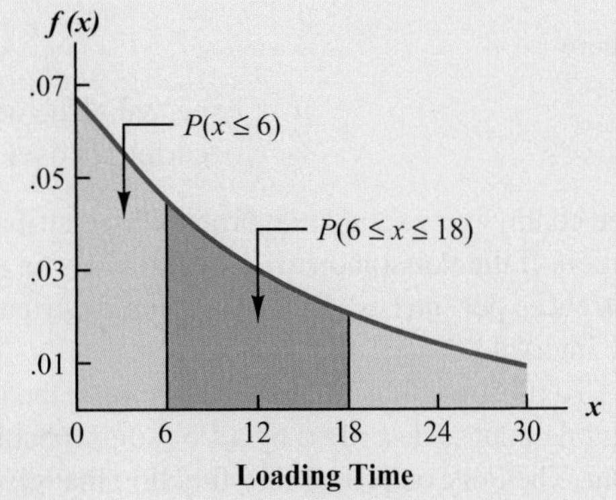

To compute exponential probabilities such as those just described, we use the following formula. It provides the cumulative probability of obtaining a value for the exponential random variable of less than or equal to some specific value denoted by x_0.

EXPONENTIAL DISTRIBUTION: CUMULATIVE PROBABILITIES

$$P(x \le x_0) = 1 - e^{-x_0/\mu} \tag{6.5}$$

For the Schips loading dock example, x = loading time in minutes and $\mu = 15$ minutes. Using equation (6.5)

$$P(x \le x_0) = 1 - e^{-x_0/15}$$

Hence, the probability that loading a truck will take 6 minutes or less is

$$P(x \le 6) = 1 - e^{-6/15} = .3297$$

Using equation (6.5), we calculate the probability of loading a truck in 18 minutes or less.

$$P(x \le 18) = 1 - e^{-18/15} = .6988$$

Thus, the probability that loading a truck will take between 6 minutes and 18 minutes is equal to $.6988 - .3297 = .3691$. Probabilities for any other interval can be computed similarly.

A property of the exponential distribution is that the mean and standard deviation are equal.

In the preceding example, the mean time it takes to load a truck is $\mu = 15$ minutes. A property of the exponential distribution is that the mean of the distribution and the standard deviation of the distribution are *equal.* Thus, the standard deviation for the time it takes to load a truck is $\sigma = 15$ minutes. The variance is $\sigma^2 = (15)^2 = 225$.

Relationship Between the Poisson and Exponential Distributions

In Section 5.6 we introduced the Poisson distribution as a discrete probability distribution that is often useful in examining the number of occurrences of an event over a specified interval of time or space. Recall that the Poisson probability function is

$$f(x) = \frac{\mu^x e^{-\mu}}{x!}$$

where

μ = expected value or mean number of occurrences over a specified interval

If arrivals follow a Poisson distribution, the time between arrivals must follow an exponential distribution.

The continuous exponential probability distribution is related to the discrete Poisson distribution. If the Poisson distribution provides an appropriate description of the number of occurrences per interval, the exponential distribution provides a description of the length of the interval between occurrences.

To illustrate this relationship, suppose the number of cars that arrive at a car wash during one hour is described by a Poisson probability distribution with a mean of 10 cars per hour. The Poisson probability function that gives the probability of x arrivals per hour is

$$f(x) = \frac{10^x e^{-10}}{x!}$$

Because the average number of arrivals is 10 cars per hour, the average time between cars arriving is

$$\frac{1 \text{ hour}}{10 \text{ cars}} = .1 \text{ hour/car}$$

Thus, the corresponding exponential distribution that describes the time between the arrivals has a mean of $\mu = .1$ hour per car; as a result, the appropriate exponential probability density function is

$$f(x) = \frac{1}{.1} e^{-x/.1} = 10e^{-10x}$$

NOTES AND COMMENTS

As we can see in Figure 6.10, the exponential distribution is skewed to the right. Indeed, the skewness measure for exponential distributions is 2. The exponential distribution gives us a good idea what a skewed distribution looks like.

Exercises

Methods

32. Consider the following exponential probability density function.

$$f(x) = \frac{1}{8} e^{-x/8} \quad \text{for } x \geq 0$$

a. Find $P(x \leq 6)$.
b. Find $P(x \leq 4)$.
c. Find $P(x \geq 6)$.
d. Find $P(4 \leq x \leq 6)$.

SELF test

33. Consider the following exponential probability density function.

$$f(x) = \frac{1}{3} e^{-x/3} \quad \text{for } x \geq 0$$

a. Write the formula for $P(x \leq x_0)$.
b. Find $P(x \leq 2)$.
c. Find $P(x \geq 3)$.
d. Find $P(x \leq 5)$.
e. Find $P(2 \leq x \leq 5)$.

Applications

34. Battery life between charges for the Motorola Droid Razr Maxx is 20 hours when the primary use is talk time (*The Wall Street Journal,* March 7, 2012). The battery life drops to 7 hours when the phone is primarily used for Internet applications over cellular. Assume that the battery life in both cases follows an exponential distribution.
   a. Show the probability density function for battery life for the Droid Razr Maxx phone when its primary use is talk time.
   b. What is the probability that the battery charge for a randomly selected Droid Razr Maxx phone will last no more than 15 hours when its primary use is talk time?
   c. What is the probability that the battery charge for a randomly selected Droid Razr Maxx phone will last more than 20 hours when its primary use is talk time?

d. What is the probability that the battery charge for a randomly selected Droid Razr Maxx phone will last no more than 5 hours when its primary use is Internet applications?

SELF test

35. The time between arrivals of vehicles at a particular intersection follows an exponential probability distribution with a mean of 12 seconds.
   a. Sketch this exponential probability distribution.
   b. What is the probability that the arrival time between vehicles is 12 seconds or less?
   c. What is the probability that the arrival time between vehicles is 6 seconds or less?
   d. What is the probability of 30 or more seconds between vehicle arrivals?

36. Comcast Corporation is the largest cable television company, the second largest Internet service provider, and the fourth largest telephone service provider in the United States. Generally known for quality and reliable service, the company periodically experiences unexpected service interruptions. On January 14, 2009, such an interruption occurred for the Comcast customers living in southwest Florida. When customers called the Comcast office, a recorded message told them that the company was aware of the service outage and that it was anticipated that service would be restored in two hours. Assume that two hours is the mean time to do the repair and that the repair time has an exponential probability distribution.
   a. What is the probability that the cable service will be repaired in one hour or less?
   b. What is the probability that the repair will take between one hour and two hours?
   c. For a customer who calls the Comcast office at 1:00 P.M., what is the probability that the cable service will not be repaired by 5:00 P.M.?

37. Collina's Italian Café in Houston, Texas, advertises that carryout orders take about 25 minutes (Collina's website, February 27, 2008). Assume that the time required for a carryout order to be ready for customer pickup has an exponential distribution with a mean of 25 minutes.
   a. What is the probability than a carryout order will be ready within 20 minutes?
   b. If a customer arrives 30 minutes after placing an order, what is the probability that the order will not be ready?
   c. A particular customer lives 15 minutes from Collina's Italian Café. If the customer places a telephone order at 5:20 P.M., what is the probability that the customer can drive to the café, pick up the order, and return home by 6:00 P.M.?

38. Do interruptions while you are working reduce your productivity? According to a University of California–Irvine study, businesspeople are interrupted at the rate of approximately 5½ times per hour (*Fortune,* March 20, 2006). Suppose the number of interruptions follows a Poisson probability distribution.
   a. Show the probability distribution for the time between interruptions.
   b. What is the probability a businessperson will have no interruptions during a 15-minute period?
   c. What is the probability that the next interruption will occur within 10 minutes for a particular businessperson?

Summary

This chapter extended the discussion of probability distributions to the case of continuous random variables. The major conceptual difference between discrete and continuous probability distributions involves the method of computing probabilities. With discrete distributions, the probability function $f(x)$ provides the probability that the random variable x assumes various values. With continuous distributions, the probability density function $f(x)$ does not provide probability values directly. Instead, probabilities are given by areas under the curve or graph of the probability density function $f(x)$. Because the area under the curve above a single point is zero, we observe that the probability of any particular value is zero for a continuous random variable.

Three continuous probability distributions—the uniform, normal, and exponential distributions—were treated in detail. The normal distribution is used widely in statistical inference and will be used extensively throughout the remainder of the text.

Glossary

Probability density function A function used to compute probabilities for a continuous random variable. The area under the graph of a probability density function over an interval represents probability.

Uniform probability distribution A continuous probability distribution for which the probability that the random variable will assume a value in any interval is the same for each interval of equal length.

Normal probability distribution A continuous probability distribution. Its probability density function is bell-shaped and determined by its mean μ and standard deviation σ.

Standard normal probability distribution A normal distribution with a mean of zero and a standard deviation of one.

Continuity correction factor A value of .5 that is added to or subtracted from a value of x when the continuous normal distribution is used to approximate the discrete binomial distribution.

Exponential probability distribution A continuous probability distribution that is useful in computing probabilities for the time it takes to complete a task.

Key Formulas

Uniform Probability Density Function

$$f(x) = \begin{cases} \dfrac{1}{b - a} & \text{for } a \le x \le b \\ 0 & \text{elsewhere} \end{cases} \tag{6.1}$$

Normal Probability Density Function

$$f(x) = \frac{1}{\sigma\sqrt{2\pi}} e^{-(x-\mu)^2/2\sigma^2} \tag{6.2}$$

Converting to the Standard Normal Random Variable

$$z = \frac{x - \mu}{\sigma} \tag{6.3}$$

Exponential Probability Density Function

$$f(x) = \frac{1}{\mu} e^{-x/\mu} \qquad \text{for } x \ge 0 \tag{6.4}$$

Exponential Distribution: Cumulative Probabilities

$$P(x \le x_0) = 1 - e^{-x_0/\mu} \tag{6.5}$$

Supplementary Exercises

39. A business executive, transferred from Chicago to Atlanta, needs to sell her house in Chicago quickly. The executive's employer has offered to buy the house for \$210,000, but the offer expires at the end of the week. The executive does not currently have a better offer but can afford to leave the house on the market for another month. From conversations with her realtor, the executive believes the price she will get by leaving the house on the market for another month is uniformly distributed between \$200,000 and \$225,000.

a. If she leaves the house on the market for another month, what is the mathematical expression for the probability density function of the sales price?
b. If she leaves it on the market for another month, what is the probability she will get at least $215,000 for the house?
c. If she leaves it on the market for another month, what is the probability she will get less than $210,000?
d. Should the executive leave the house on the market for another month? Why or why not?

40. The NCAA estimates that the yearly value of a full athletic scholarship at in-state public universities is $19,000 (*The Wall Street Journal,* March 12, 2012). Assume the scholarship value is normally distributed with a standard deviation of $2100.
a. For the 10% of athletic scholarships of least value, how much are they worth?
b. What percentage of athletic scholarships are valued at $22,000 or more?
c. For the 3% of athletic scholarships that are most valuable, how much are they worth?

41. Motorola used the normal distribution to determine the probability of defects and the number of defects expected in a production process. Assume a production process produces items with a mean weight of 10 ounces. Calculate the probability of a defect and the expected number of defects for a 1000-unit production run in the following situations.
a. The process standard deviation is .15, and the process control is set at plus or minus one standard deviation. Units with weights less than 9.85 or greater than 10.15 ounces will be classified as defects.
b. Through process design improvements, the process standard deviation can be reduced to .05. Assume the process control remains the same, with weights less than 9.85 or greater than 10.15 ounces being classified as defects.
c. What is the advantage of reducing process variation, thereby causing process control limits to be at a greater number of standard deviations from the mean?

42. During early 2012, economic hardship was stretching the limits of France's welfare system. One indicator of the level of hardship was the increase in the number of people bringing items to a Paris pawnbroker; the number of people bringing items to the pawnbroker had increased to 658 per day (*Bloomberg Businessweek,* March 5–March 11, 2012). Assume the number of people bringing items to the pawnshop per day in 2012 is normally distributed with a mean of 658.
a. Suppose you learn that on 3% of the days, 610 or fewer people brought items to the pawnshop. What is the standard deviation of the number of people bringing items to the pawnshop per day?
b. On any given day, what is the probability that between 600 and 700 people bring items to the pawnshop?
c. How many people bring items to the pawnshop on the busiest 3% of days?

43. *Condé Nast Traveler* publishes a Gold List of the top hotels all over the world. The Broadmoor Hotel in Colorado Springs contains 700 rooms and is on the 2004 Gold List (*Condé Nast Traveler,* January 2004). Suppose Broadmoor's marketing group forecasts a mean demand of 670 rooms for the coming weekend. Assume that demand for the upcoming weekend is normally distributed with a standard deviation of 30.
a. What is the probability all the hotel's rooms will be rented?
b. What is the probability 50 or more rooms will not be rented?
c. Would you recommend the hotel consider offering a promotion to increase demand? What considerations would be important?

44. Ward Doering Auto Sales is considering offering a special service contract that will cover the total cost of any service work required on leased vehicles. From experience, the company manager estimates that yearly service costs are approximately normally distributed, with a mean of $150 and a standard deviation of $25.
a. If the company offers the service contract to customers for a yearly charge of $200, what is the probability that any one customer's service costs will exceed the contract price of $200?
b. What is Ward's expected profit per service contract?

45. Is lack of sleep causing traffic fatalities? A study conducted under the auspices of the National Highway Traffic Safety Administration found that the average number of fatal crashes caused by drowsy drivers each year was 1550 (*BusinessWeek,* January 26, 2004). Assume the annual number of fatal crashes per year is normally distributed with a standard deviation of 300.
    a. What is the probability of fewer than 1000 fatal crashes in a year?
    b. What is the probability the number of fatal crashes will be between 1000 and 2000 for a year?
    c. For a year to be in the upper 5% with respect to the number of fatal crashes, how many fatal crashes would have to occur?

46. Assume that the test scores from a college admissions test are normally distributed, with a mean of 450 and a standard deviation of 100.
    a. What percentage of the people taking the test score between 400 and 500?
    b. Suppose someone receives a score of 630. What percentage of the people taking the test score better? What percentage score worse?
    c. If a particular university will not admit anyone scoring below 480, what percentage of the persons taking the test would be acceptable to the university?

47. According to Salary Wizard, the average base salary for a brand manager in Houston, Texas, is $88,592 and the average base salary for a brand manager in Los Angeles, California, is $97,417 (Salary Wizard website, February 27, 2008). Assume that salaries are normally distributed, the standard deviation for brand managers in Houston is $19,900, and the standard deviation for brand managers in Los Angeles is $21,800.
    a. What is the probability that a brand manager in Houston has a base salary in excess of $100,000?
    b. What is the probability that a brand manager in Los Angeles has a base salary in excess of $100,000?
    c. What is the probability that a brand manager in Los Angeles has a base salary of less than $75,000?
    d. How much would a brand manager in Los Angeles have to make in order to have a higher salary than 99% of the brand managers in Houston?

48. A machine fills containers with a particular product. The standard deviation of filling weights is known from past data to be .6 ounce. If only 2% of the containers hold less than 18 ounces, what is the mean filling weight for the machine? That is, what must μ equal? Assume the filling weights have a normal distribution.

49. Consider a multiple-choice examination with 50 questions. Each question has four possible answers. Assume that a student who has done the homework and attended lectures has a 75% probability of answering any question correctly.
    a. A student must answer 43 or more questions correctly to obtain a grade of A. What percentage of the students who have done their homework and attended lectures will obtain a grade of A on this multiple-choice examination?
    b. A student who answers 35 to 39 questions correctly will receive a grade of C. What percentage of students who have done their homework and attended lectures will obtain a grade of C on this multiple-choice examination?
    c. A student must answer 30 or more questions correctly to pass the examination. What percentage of the students who have done their homework and attended lectures will pass the examination?
    d. Assume that a student has not attended class and has not done the homework for the course. Furthermore, assume that the student will simply guess at the answer to each question. What is the probability that this student will answer 30 or more questions correctly and pass the examination?

50. A blackjack player at a Las Vegas casino learned that the house will provide a free room if play is for four hours at an average bet of $50. The player's strategy provides a probability of .49 of winning on any one hand, and the player knows that there are 60 hands per hour. Suppose the player plays for four hours at a bet of $50 per hand.

a. What is the player's expected payoff?
b. What is the probability the player loses $1000 or more?
c. What is the probability the player wins?
d. Suppose the player starts with $1500. What is the probability of going broke?

51. The Information Systems Audit and Control Association surveyed office workers to learn about the anticipated usage of office computers for personal holiday shopping (*USA Today*, November 11, 2009). Assume that the number of hours a worker spends doing holiday shopping on an office computer follows an exponential distribution.
 a. The study reported that there is a .53 probability that a worker uses an office computer for holiday shopping 5 hours or less. Is the mean time spent using an office computer for holiday shopping closest to 5.8, 6.2, 6.6, or 7 hours?
 b. Using the mean time from part (a), what is the probability that a worker uses an office computer for holiday shopping more than 10 hours?
 c. What is the probability that a worker uses an office computer for holiday shopping between 4 and 8 hours?

52. The website for the Bed and Breakfast Inns of North America gets approximately seven visitors per minute (*Time*, September 2001). Suppose the number of website visitors per minute follows a Poisson probability distribution.
 a. What is the mean time between visits to the website?
 b. Show the exponential probability density function for the time between website visits.
 c. What is the probability no one will access the website in a 1-minute period?
 d. What is the probability no one will access the website in a 12-second period?

53. The American Community Survey showed that residents of New York City have the longest travel times to get to work compared to residents of other cities in the United States (U.S. Census Bureau website, August 2008). According to the latest statistics available, the average travel time to work for residents of New York City is 38.3 minutes.
 a. Assume the exponential probability distribution is applicable and show the probability density function for the travel time to work for a resident of this city.
 b. What is the probability it will take a resident of this city between 20 and 40 minutes to travel to work?
 c. What is the probability it will take a resident of this city more than one hour to travel to work?

54. The time (in minutes) between telephone calls at an insurance claims office has the following exponential probability distribution.

$$f(x) = .50e^{-.50x} \qquad \text{for } x \geq 0$$

 a. What is the mean time between telephone calls?
 b. What is the probability of having 30 seconds or less between telephone calls?
 c. What is the probability of having 1 minute or less between telephone calls?
 d. What is the probability of having 5 or more minutes without a telephone call?

Case Problem Specialty Toys

Specialty Toys, Inc., sells a variety of new and innovative children's toys. Management learned that the preholiday season is the best time to introduce a new toy, because many families use this time to look for new ideas for December holiday gifts. When Specialty discovers a new toy with good market potential, it chooses an October market entry date.

In order to get toys in its stores by October, Specialty places one-time orders with its manufacturers in June or July of each year. Demand for children's toys can be highly volatile. If a new toy catches on, a sense of shortage in the marketplace often increases the demand to high levels and large profits can be realized. However, new toys can also flop, leaving Specialty stuck with high levels of inventory that must be sold at reduced prices. The most important question the company faces is deciding how many units of a new toy should be purchased to meet anticipated sales demand. If too few are purchased, sales will be lost; if too many are purchased, profits will be reduced because of low prices realized in clearance sales.

For the coming season, Specialty plans to introduce a new product called Weather Teddy. This variation of a talking teddy bear is made by a company in Taiwan. When a child presses Teddy's hand, the bear begins to talk. A built-in barometer selects one of five responses that predict the weather conditions. The responses range from "It looks to be a very nice day! Have fun" to "I think it may rain today. Don't forget your umbrella." Tests with the product show that, even though it is not a perfect weather predictor, its predictions are surprisingly good. Several of Specialty's managers claimed Teddy gave predictions of the weather that were as good as many local television weather forecasters.

As with other products, Specialty faces the decision of how many Weather Teddy units to order for the coming holiday season. Members of the management team suggested order quantities of 15,000, 18,000, 24,000, or 28,000 units. The wide range of order quantities suggested indicates considerable disagreement concerning the market potential. The product management team asks you for an analysis of the stock-out probabilities for various order quantities, an estimate of the profit potential, and to help make an order quantity recommendation. Specialty expects to sell Weather Teddy for $24 based on a cost of $16 per unit. If inventory remains after the holiday season, Specialty will sell all surplus inventory for $5 per unit. After reviewing the sales history of similar products, Specialty's senior sales forecaster predicted an expected demand of 20,000 units with a .95 probability that demand would be between 10,000 units and 30,000 units.

Managerial Report

Prepare a managerial report that addresses the following issues and recommends an order quantity for the Weather Teddy product.

1. Use the sales forecaster's prediction to describe a normal probability distribution that can be used to approximate the demand distribution. Sketch the distribution and show its mean and standard deviation.
2. Compute the probability of a stock-out for the order quantities suggested by members of the management team.
3. Compute the projected profit for the order quantities suggested by the management team under three scenarios: worst case in which sales = 10,000 units, most likely case in which sales = 20,000 units, and best case in which sales = 30,000 units.
4. One of Specialty's managers felt that the profit potential was so great that the order quantity should have a 70% chance of meeting demand and only a 30% chance of any stock-outs. What quantity would be ordered under this policy, and what is the projected profit under the three sales scenarios?
5. Provide your own recommendation for an order quantity and note the associated profit projections. Provide a rationale for your recommendation.

Appendix 6.1 Continuous Probability Distributions with Minitab

Let us demonstrate the Minitab procedure for computing continuous probabilities by referring to the Grear Tire Company problem where tire mileage was described by a normal distribution with $\mu = 36,500$ and $\sigma = 5000$. One question asked was: What is the probability that the tire mileage will exceed 40,000 miles?

For continuous probability distributions, Minitab provides the cumulative probability that the random variable takes on a value less than or equal to a specified constant. For the Grear tire mileage question, Minitab can be used to determine the cumulative probability that the tire mileage will be less than or equal to 40,000 miles. After obtaining the cumulative probability, we must subtract it from 1 to determine the probability that the tire mileage will exceed 40,000 miles.

Prior to using Minitab to compute a probability, we must enter the specified constant into a column of the worksheet. For the Grear tire mileage question we enter 40,000 into column C1 of the Minitab worksheet. The steps in using Minitab to compute the cumulative probability follow.

Step 1. Select the **Calc** menu
Step 2. Choose **Probability Distributions**
Step 3. Choose **Normal**
Step 4. When the Normal Distribution dialog box appears:
Select **Cumulative probability**
Enter 36500 in the **Mean** box
Enter 5000 in the **Standard deviation** box
Enter C1 in the **Input column** box (the column containing 40,000)
Click **OK**

Minitab shows that this probability is .7580. Because we are interested in the probability that the tire mileage will be greater than 40,000, the desired probability is $1 - .7580 = .2420$.

A second question in the Grear Tire Company problem was: What mileage guarantee should Grear set to ensure that no more than 10% of the tires qualify for the guarantee? Here we are given a probability and want to find the corresponding value for the random variable. Minitab uses an inverse calculation routine to find the value of the random variable associated with a given cumulative probability. First, we must enter the cumulative probability into a column of the Minitab worksheet. In this case, the desired cumulative probability of .10 is entered into column C1. Then, the first three steps of the Minitab procedure are as already listed. In step 4, we select **Inverse cumulative probability** instead of **Cumulative probability** and complete the remaining parts of the step. Minitab then displays the mileage guarantee of 30,092 miles.

Minitab is capable of computing probabilities for other continuous probability distributions, including the exponential probability distribution. To compute exponential probabilities, follow the procedure shown previously for the normal probability distribution and choose the **Exponential** option in step 3. Step 4 is as shown, with the exception that entering the standard deviation is not required. Output for cumulative probabilities and inverse cumulative probabilities is identical to that described for the normal probability distribution.

Appendix 6.2 Continuous Probability Distributions with Excel

Excel provides the capability for computing probabilities for several continuous probability distributions, including the normal and exponential probability distributions. In this appendix, we describe how Excel can be used to compute probabilities for any normal distribution. The procedures for the exponential and other continuous distributions are similar to the one we describe for the normal distribution.

Let us return to the Grear Tire Company problem where the tire mileage was described by a normal distribution with $\mu = 36{,}500$ and $\sigma = 5000$. Assume we are interested in the probability that tire mileage will exceed 40,000 miles.

Excel's NORM.DIST function can be used to compute cumulative probabilities for a normal distribution. The general form of the function is NORM.DIST(x,μ,σ,cumulative). For the fourth argument, TRUE is specified if a cumulative probability is desired. Thus, to compute the cumulative probability that the tire mileage will be less than or equal to 40,000 miles we would enter the following formula into any cell of an Excel worksheet:

=NORM.DIST(40000,36500,5000,TRUE)

At this point, .7580 will appear in the cell where the formula was entered, indicating that the probability of tire mileage being less than or equal to 40,000 miles is .7580. Therefore, the probability that tire mileage will exceed 40,000 miles is $1 - .7580 = .2420$.

Excel's NORM.INV function uses an inverse computation to find the x value corresponding to a given cumulative probability. For instance, suppose we want to find the guaranteed mileage Grear should offer so that no more than 10% of the tires will be eligible for the guarantee. We would enter the following formula into any cell of an Excel worksheet:

=NORM.INV(.1,36500,5000)

At this point, 30092 will appear in the cell where the formula was entered, indicating that the probability of a tire lasting 30,092 miles or less is .10.

The Excel function for computing exponential probabilities is EXPON.DIST. This function requires three inputs: x, the value of the variable; lambda, which is $1/\mu$, and TRUE if you would like the cumulative probability. For example, consider an exponential probability distribution with mean $\mu = 15$. The probability that the exponential variable is less than or equal to 6 can be computed by the Excel formula

=EXPON.DIST(6,1/15,TRUE)

If you need help inserting functions in a worksheet, Excel's Insert Function dialog box may be used. See Appendix E.

CHAPTER 7

Sampling and Sampling Distributions

CONTENTS

STATISTICS *in* PRACTICE

MEADWESTVACO CORPORATION*
STAMFORD, CONNECTICUT

MeadWestvaco Corporation, a leading producer of packaging, coated and specialty papers, and specialty chemicals, employs more than 17,000 people. It operates worldwide in 30 countries and serves customers located in approximately 100 countries. MeadWestvaco's internal consulting group uses sampling to provide a variety of information that enables the company to obtain significant productivity benefits and remain competitive.

For example, MeadWestvaco maintains large woodland holdings, which supply the trees, or raw material, for many of the company's products. Managers need reliable and accurate information about the timberlands and forests to evaluate the company's ability to meet its future raw material needs. What is the present volume in the forests? What is the past growth of the forests? What is the projected future growth of the forests? With answers to these important questions MeadWestvaco's managers can develop plans for the future, including long-term planting and harvesting schedules for the trees.

How does MeadWestvaco obtain the information it needs about its vast forest holdings? Data collected from sample plots throughout the forests are the basis for learning about the population of trees owned by the company. To identify the sample plots, the timberland holdings are first divided into three sections based on location and types of trees. Using maps and random numbers, MeadWestvaco analysts identify random samples of 1/5- to 1/7-acre plots in each section of the forest. MeadWestvaco foresters collect data from these sample plots to learn about the forest population.

*The authors are indebted to Dr. Edward P. Winkofsky for providing this Statistics in Practice.

Random sampling of its forest holdings enables MeadWestvaco Corporation to meet future raw material needs. © Robert Crum/Shutterstock.com.

Foresters throughout the organization participate in the field data collection process. Periodically, two-person teams gather information on each tree in every sample plot. The sample data are entered into the company's continuous forest inventory (CFI) computer system. Reports from the CFI system include a number of frequency distribution summaries containing statistics on types of trees, present forest volume, past forest growth rates, and projected future forest growth and volume. Sampling and the associated statistical summaries of the sample data provide the reports essential for the effective management of MeadWestvaco's forests and timberlands.

In this chapter you will learn about simple random sampling and the sample selection process. In addition, you will learn how statistics such as the sample mean and sample proportion are used to estimate the population mean and population proportion. The important concept of a sampling distribution is also introduced.

In Chapter 1 we presented the following definitions of an element, a population, and a sample.

- An *element* is the entity on which data are collected.
- A *population* is the collection of all the elements of interest.
- A *sample* is a subset of the population.

The reason we select a sample is to collect data to make an inference and answer research questions about a population.

Let us begin by citing two examples in which sampling was used to answer a research question about a population.

1. Members of a political party in Texas were considering supporting a particular candidate for election to the U.S. Senate, and party leaders wanted to estimate the proportion of registered voters in the state favoring the candidate. A sample of 400 registered voters in Texas was selected and 160 of the 400 voters indicated a preference for the candidate. Thus, an estimate of the proportion of the population of registered voters favoring the candidate is 160/400 = .40.
2. A tire manufacturer is considering producing a new tire designed to provide an increase in mileage over the firm's current line of tires. To estimate the mean useful life of the new tires, the manufacturer produced a sample of 120 tires for testing. The test results provided a sample mean of 36,500 miles. Hence, an estimate of the mean useful life for the population of new tires was 36,500 miles.

A sample mean provides an estimate of a population mean, and a sample proportion provides an estimate of a population proportion. With estimates such as these, some estimation error can be expected. This chapter provides the basis for determining how large that error might be.

It is important to realize that sample results provide only *estimates* of the values of the corresponding population characteristics. We do not expect exactly .40, or 40%, of the population of registered voters to favor the candidate, nor do we expect the sample mean of 36,500 miles to exactly equal the mean mileage for the population of all new tires produced. The reason is simply that the sample contains only a portion of the population. Some sampling error is to be expected. With proper sampling methods, the sample results will provide "good" estimates of the population parameters. But how good can we expect the sample results to be? Fortunately, statistical procedures are available for answering this question.

Let us define some of the terms used in sampling. The **sampled population** is the population from which the sample is drawn, and a **frame** is a list of the elements that the sample will be selected from. In the first example, the sampled population is all registered voters in Texas, and the frame is a list of all the registered voters. Because the number of registered voters in Texas is a finite number, the first example is an illustration of sampling from a finite population. In Section 7.2, we discuss how a simple random sample can be selected when sampling from a finite population.

The sampled population for the tire mileage example is more difficult to define because the sample of 120 tires was obtained from a production process at a particular point in time. We can think of the sampled population as the conceptual population of all the tires that could have been made by the production process at that particular point in time. In this sense the sampled population is considered infinite, making it impossible to construct a frame to draw the sample from. In Section 7.2, we discuss how to select a random sample in such a situation.

In this chapter, we show how simple random sampling can be used to select a sample from a finite population and describe how a random sample can be taken from an infinite population that is generated by an ongoing process. We then show how data obtained from a sample can be used to compute estimates of a population mean, a population standard deviation, and a population proportion. In addition, we introduce the important concept of a sampling distribution. As we will show, knowledge of the appropriate sampling distribution enables us to make statements about how close the sample estimates are to the corresponding population parameters. The last section discusses some alternatives to simple random sampling that are often employed in practice.

7.1 The Electronics Associates Sampling Problem

The director of personnel for Electronics Associates, Inc. (EAI), has been assigned the task of developing a profile of the company's 2500 managers. The characteristics to be identified include the mean annual salary for the managers and the proportion of managers having completed the company's management training program.

WEB file
EAI

Using the 2500 managers as the population for this study, we can find the annual salary and the training program status for each individual by referring to the firm's personnel records. The data set containing this information for all 2500 managers in the population is in the file named EAI.

Using the EAI data and the formulas presented in Chapter 3, we computed the population mean and the population standard deviation for the annual salary data.

$$\text{Population mean: } \mu = \$51{,}800$$
$$\text{Population standard deviation: } \sigma = \$4000$$

The data for the training program status show that 1500 of the 2500 managers completed the training program.

Numerical characteristics of a population are called **parameters**. Letting p denote the proportion of the population that completed the training program, we see that $p = 1500/2500 = .60$. The population mean annual salary ($\mu = \$51{,}800$), the population standard deviation of annual salary ($\sigma = \$4000$), and the population proportion that completed the training program ($p = .60$) are parameters of the population of EAI managers.

Often the cost of collecting information from a sample is substantially less than from a population, especially when personal interviews must be conducted to collect the information.

Now, suppose that the necessary information on all the EAI managers was not readily available in the company's database. The question we now consider is how the firm's director of personnel can obtain estimates of the population parameters by using a sample of managers rather than all 2500 managers in the population. Suppose that a sample of 30 managers will be used. Clearly, the time and the cost of developing a profile would be substantially less for 30 managers than for the entire population. If the personnel director could be assured that a sample of 30 managers would provide adequate information about the population of 2500 managers, working with a sample would be preferable to working with the entire population. Let us explore the possibility of using a sample for the EAI study by first considering how we can identify a sample of 30 managers.

7.2 Selecting a Sample

In this section we describe how to select a sample. We first describe how to sample from a finite population and then describe how to select a sample from an infinite population.

Sampling from a Finite Population

Other methods of probability sampling are described in Section 7.8

Statisticians recommend selecting a probability sample when sampling from a finite population because a probability sample allows them to make valid statistical inferences about the population. The simplest type of probability sample is one in which each sample of size n has the same probability of being selected. It is called a simple random sample. A simple random sample of size n from a finite population of size N is defined as follows.

SIMPLE RANDOM SAMPLE (FINITE POPULATION)

A **simple random sample** of size n from a finite population of size N is a sample selected such that each possible sample of size n has the same probability of being selected.

We describe how Excel, Minitab, and StatTools can be used to generate a simple random sample in the chapter appendices.

One procedure for selecting a simple random sample from a finite population is to use a table of random numbers to choose the elements for the sample one at a time in such a way that, at each step, each of the elements remaining in the population has the same probability of being selected. Sampling n elements in this way will satisfy the definition of a simple random sample from a finite population.

To select a simple random sample from the finite population of EAI managers, we first construct a frame by assigning each manager a number. For example, we can assign the

TABLE 7.1 RANDOM NUMBERS

63271	59986	71744	51102	15141	80714	58683	93108	13554	79945
88547	09896	95436	79115	08303	01041	20030	63754	08459	28364
55957	57243	83865	09911	19761	66535	40102	26646	60147	15702
46276	87453	44790	67122	45573	84358	21625	16999	13385	22782
55363	07449	34835	15290	76616	67191	12777	21861	68689	03263
69393	92785	49902	58447	42048	30378	87618	26933	40640	16281
13186	29431	88190	04588	38733	81290	89541	70290	40113	08243
17726	28652	56836	78351	47327	18518	92222	55201	27340	10493
36520	64465	05550	30157	82242	29520	69753	72602	23756	54935
81628	36100	39254	56835	37636	02421	98063	89641	64953	99337
84649	48968	75215	75498	49539	74240	03466	49292	36401	45525
63291	11618	12613	75055	43915	26488	41116	64531	56827	30825
70502	53225	03655	05915	37140	57051	48393	91322	25653	06543
06426	24771	59935	49801	11082	66762	94477	02494	88215	27191
20711	55609	29430	70165	45406	78484	31639	52009	18873	96927
41990	70538	77191	25860	55204	73417	83920	69468	74972	38712
72452	36618	76298	26678	89334	33938	95567	29380	75906	91807
37042	40318	57099	10528	09925	89773	41335	96244	29002	46453
53766	52875	15987	46962	67342	77592	57651	95508	80033	69828
90585	58955	53122	16025	84299	53310	67380	84249	25348	04332
32001	96293	37203	64516	51530	37069	40261	61374	05815	06714
62606	64324	46354	72157	67248	20135	49804	09226	64419	29457
10078	28073	85389	50324	14500	15562	64165	06125	71353	77669
91561	46145	24177	15294	10061	98124	75732	00815	83452	97355
13091	98112	53959	79607	52244	63303	10413	63839	74762	50289

managers the numbers 1 to 2500 in the order that their names appear in the EAI personnel file. Next, we refer to the table of random numbers shown in Table 7.1. Using the first row of the table, each digit, 6, 3, 2, . . . , is a random digit having an equal chance of occurring. Because the largest number in the population list of EAI managers, 2500, has four digits, we will select random numbers from the table in sets or groups of four digits. Even though we may start the selection of random numbers anywhere in the table and move systematically in a direction of our choice, we will use the first row of Table 7.1 and move from left to right. The first 7 four-digit random numbers are

The random numbers in the table are shown in groups of five for readability.

6327 1599 8671 7445 1102 1514 1807

Because the numbers in the table are random, these four-digit numbers are equally likely.

We can now use these four-digit random numbers to give each manager in the population an equal chance of being included in the random sample. The first number, 6327, is greater than 2500. It does not correspond to one of the numbered managers in the population, and hence is discarded. The second number, 1599, is between 1 and 2500. Thus the first manager selected for the random sample is number 1599 on the list of EAI managers. Continuing this process, we ignore the numbers 8671 and 7445 before identifying managers number 1102, 1514, and 1807 to be included in the random sample. This process continues until the simple random sample of 30 EAI managers has been obtained.

In implementing this simple random sample selection process, it is possible that a random number used previously may appear again in the table before the complete sample of 30 EAI managers has been selected. Because we do not want to select a manager more than one time, any previously used random numbers are ignored because the corresponding manager is already included in the sample. Selecting a sample in this manner is referred to as **sampling without replacement**. If we selected a sample such that previously used random numbers are acceptable and specific managers could be included in the sample two or more times, we would be **sampling with replacement**. Sampling with replacement is a valid way of identifying a simple random sample. However, sampling without replacement is the sampling procedure used most often in practice. When we refer to simple random sampling, we will assume the sampling is without replacement.

Sampling from an Infinite Population

Sometimes we want to select a sample from a population, but the population is infinitely large or the elements of the population are being generated by an ongoing process for which there is no limit on the number of elements that can be generated. Thus, it is not possible to develop a list of all the elements in the population. This is considered the infinite population case. With an infinite population, we cannot select a simple random sample because we cannot construct a frame consisting of all the elements. In the infinite population case, statisticians recommend selecting what is called a random sample.

RANDOM SAMPLE (INFINITE POPULATION)

A **random sample** of size n from an infinite population is a sample selected such that the following conditions are satisfied.

1. Each element selected comes from the same population.
2. Each element is selected independently.

Care and judgment must be exercised in implementing the selection process for obtaining a random sample from an infinite population. Each case may require a different selection procedure. Let us consider two examples to see what we mean by the conditions (1) each element selected comes from the same population and (2) each element is selected independently.

A common quality control application involves a production process where there is no limit on the number of elements that can be produced. The conceptual population we are sampling from is all the elements that could be produced (not just the ones that are produced) by the ongoing production process. Because we cannot develop a list of all the elements that could be produced, the population is considered infinite. To be more specific, let us consider a production line designed to fill boxes of a breakfast cereal with a mean weight of 24 ounces of breakfast cereal per box. Samples of 12 boxes filled by this process are periodically selected by a quality control inspector to determine if the process is operating properly or if, perhaps, a machine malfunction has caused the process to begin underfilling or overfilling the boxes.

With a production operation such as this, the biggest concern in selecting a random sample is to make sure that condition 1, the sampled elements are selected from the same population, is satisfied. To ensure that this condition is satisfied, the boxes must be selected at approximately the same point in time. This way the inspector avoids the possibility of selecting some boxes when the process is operating properly and other boxes when the process is not operating properly and is underfilling or overfilling the boxes. With a production process such as this, the second condition, each element is selected independently, is satisfied by designing the production process so that each box of cereal is filled independently. With this assumption, the quality control inspector only needs to worry about satisfying the same population condition.

As another example of selecting a random sample from an infinite population, consider the population of customers arriving at a fast-food restaurant. Suppose an employee is asked to select and interview a sample of customers in order to develop a profile of customers who visit the restaurant. The customer arrival process is ongoing and there is no way to obtain a list of all customers in the population. So, for practical purposes, the population for this ongoing process is considered infinite. As long as a sampling procedure is designed so that all the elements in the sample are customers of the restaurant and they are selected independently, a random sample will be obtained. In this case, the employee collecting the sample needs to select the sample from people who come into the restaurant and make a purchase to ensure that the same population condition is satisfied. If, for instance, the employee selected someone for the sample who came into the restaurant just to use the restroom, that person would not be a customer and the same population condition would be violated. So, as long as the interviewer selects the sample from people making a purchase at the restaurant, condition 1 is satisfied. Ensuring that the customers are selected independently can be more difficult.

The purpose of the second condition of the random sample selection procedure (each element is selected independently) is to prevent selection bias. In this case, selection bias would occur if the interviewer were free to select customers for the sample arbitrarily. The interviewer might feel more comfortable selecting customers in a particular age group and might avoid customers in other age groups. Selection bias would also occur if the interviewer selected a group of five customers who entered the restaurant together and asked all of them to participate in the sample. Such a group of customers would be likely to exhibit similar characteristics, which might provide misleading information about the population of customers. Selection bias such as this can be avoided by ensuring that the selection of a particular customer does not influence the selection of any other customer. In other words, the elements (customers) are selected independently.

McDonald's, the fast-food restaurant leader, implemented a random sampling procedure for this situation. The sampling procedure was based on the fact that some customers presented discount coupons. Whenever a customer presented a discount coupon, the next customer served was asked to complete a customer profile questionnaire. Because arriving customers presented discount coupons randomly and independently of other customers, this sampling procedure ensured that customers were selected independently. As a result, the sample satisfied the requirements of a random sample from an infinite population.

Situations involving sampling from an infinite population are usually associated with a process that operates over time. Examples include parts being manufactured on a production line, repeated experimental trials in a laboratory, transactions occurring at a bank, telephone calls arriving at a technical support center, and customers entering a retail store. In each case, the situation may be viewed as a process that generates elements from an infinite population. As long as the sampled elements are selected from the same population and are selected independently, the sample is considered a random sample from an infinite population.

NOTES AND COMMENTS

1. In this section we have been careful to define two types of samples: a simple random sample from a finite population and a random sample from an infinite population. In the remainder of the text, we will generally refer to both of these as either a *random sample* or simply a *sample*. We will not make a distinction of the sample being a "simple" random sample unless it is necessary for the exercise or discussion.
2. Statisticians who specialize in sample surveys from finite populations use sampling methods that provide probability samples. With a probability sample, each possible sample has a known probability of selection and a random process is used to select the elements for the sample. Simple random sampling is one of these methods. In Section 7.8, we describe some other probability sampling methods: stratified random sampling, cluster sampling, and systematic sampling. We use the term "simple" in simple random sampling to clarify that this is the probability sampling method that assures each sample of size n has the same probability of being selected.
3. The number of different simple random samples of size n that can be selected from a finite population of size N is

$$\frac{N!}{n!(N - n)!}$$

In this formula, $N!$ and $n!$ are the factorial formulas discussed in Chapter 4. For the EAI problem with $N = 2500$ and $n = 30$, this expression can be used to show that approximately 2.75×10^{69} different simple random samples of 30 EAI managers can be obtained.

Exercises

Methods

SELF test

1. Consider a finite population with five elements labeled A, B, C, D, and E. Ten possible simple random samples of size 2 can be selected.
   a. List the 10 samples beginning with AB, AC, and so on.
   b. Using simple random sampling, what is the probability that each sample of size 2 is selected?
   c. Assume random number 1 corresponds to A, random number 2 corresponds to B, and so on. List the simple random sample of size 2 that will be selected by using the random digits 8 0 5 7 5 3 2.
2. Assume a finite population has 350 elements. Using the last three digits of each of the following five-digit random numbers (e.g., 601, 022, 448, . . .), determine the first four elements that will be selected for the simple random sample.

   98601 73022 83448 02147 34229 27553 84147 93289 14209

Applications

SELF test

3. *Fortune* publishes data on sales, profits, assets, stockholders' equity, market value, and earnings per share for the 500 largest U.S. industrial corporations (*Fortune* 500, 2012). Assume that you want to select a simple random sample of 10 corporations from the *Fortune* 500 list. Use the last three digits in column 9 of Table 7.1, beginning with 554. Read down the column and identify the numbers of the 10 corporations that would be selected.
4. The 10 most active stocks on the New York Stock Exchange on March 6, 2006, are shown here (*The Wall Street Journal,* March 7, 2006).

AT&T	Lucent	Nortel	Qwest	Bell South
Pfizer	Texas Instruments	Gen. Elect.	iShrMSJpn	LSI Logic

   Exchange authorities decided to investigate trading practices using a sample of three of these stocks.
   a. Beginning with the first random digit in column 6 of Table 7.1, read down the column to select a simple random sample of three stocks for the exchange authorities.
   b. Using the information in the third Note and Comment, determine how many different simple random samples of size 3 can be selected from the list of 10 stocks.
5. A student government organization is interested in estimating the proportion of students who favor a mandatory "pass-fail" grading policy for elective courses. A list of names and addresses of the 645 students enrolled during the current quarter is available from the registrar's office. Using three-digit random numbers in row 10 of Table 7.1 and moving across the row from left to right, identify the first 10 students who would be selected using simple random sampling. The three-digit random numbers begin with 816, 283, and 610.
6. The *County and City Data Book,* published by the Census Bureau, lists information on 3139 counties throughout the United States. Assume that a national study will collect data from 30 randomly selected counties. Use four-digit random numbers from the last column of Table 7.1 to identify the numbers corresponding to the first five counties selected for the sample. Ignore the first digits and begin with the four-digit random numbers 9945, 8364, 5702, and so on.

7. Assume that we want to identify a simple random sample of 12 of the 372 doctors practicing in a particular city. The doctors' names are available from a local medical organization. Use the eighth column of five-digit random numbers in Table 7.1 to identify the 12 doctors for the sample. Ignore the first two random digits in each five-digit grouping of the random numbers. This process begins with random number 108 and proceeds down the column of random numbers.

8. The following stocks make up the Dow Jones Industrial Average (*Barron's,* July 30, 2012).

1. 3M	11. Disney	21. McDonald's
2. AT&T	12. DuPont	22. Merck
3. Alcoa	13. ExxonMobil	23. Microsoft
4. American Express	14. General Electric	24. J.P. Morgan
5. Bank of America	15. Hewlett-Packard	25. Pfizer
6. Boeing	16. Home Depot	26. Procter & Gamble
7. Caterpillar	17. IBM	27. Travelers
8. Chevron	18. Intel	28. United Technologies
9. Cisco Systems	19. Johnson & Johnson	29. Verizon
10. Coca-Cola	20. Kraft Foods	30. Wal-Mart

Suppose you would like to select a sample of six of these companies to conduct an in-depth study of management practices. Use the first two digits in each row of the ninth column of Table 7.1 to select a simple random sample of six companies.

9. *The Wall Street Journal* provides the net asset value, the year-to-date percent return, and the three-year percent return for 555 mutual funds (*The Wall Street Journal,* April 25, 2003). Assume that a simple random sample of 12 of the 555 mutual funds will be selected for a follow-up study on the size and performance of mutual funds. Use the fourth column of the random numbers in Table 7.1, beginning with 51102, to select the simple random sample of 12 mutual funds. Begin with mutual fund 102 and use the *last* three digits in each row of the fourth column for your selection process. What are the numbers of the 12 mutual funds in the simple random sample?

10. Indicate which of the following situations involve sampling from a finite population and which involve sampling from an infinite population. In cases where the sampled population is finite, describe how you would construct a frame.
    a. Obtain a sample of licensed drivers in the state of New York.
    b. Obtain a sample of boxes of cereal produced by the Breakfast Choice company.
    c. Obtain a sample of cars crossing the Golden Gate Bridge on a typical weekday.
    d. Obtain a sample of students in a statistics course at Indiana University.
    e. Obtain a sample of the orders that are processed by a mail-order firm.

7.3 Point Estimation

Now that we have described how to select a simple random sample, let us return to the EAI problem. A simple random sample of 30 managers and the corresponding data on annual salary and management training program participation are as shown in Table 7.2. The notation x_1, x_2, and so on is used to denote the annual salary of the first manager in the sample, the annual salary of the second manager in the sample, and so on. Participation in the management training program is indicated by Yes in the management training program column.

To estimate the value of a population parameter, we compute a corresponding characteristic of the sample, referred to as a **sample statistic**. For example, to estimate the population mean μ and the population standard deviation σ for the annual salary of EAI managers, we use the data in Table 7.2 to calculate the corresponding sample statistics: the

TABLE 7.2 ANNUAL SALARY AND TRAINING PROGRAM STATUS FOR A SIMPLE RANDOM SAMPLE OF 30 EAI MANAGERS

Annual Salary ($)	Management Training Program	Annual Salary ($)	Management Training Program
x_1 = 49,094.30	Yes	x_{16} = 51,766.00	Yes
x_2 = 53,263.90	Yes	x_{17} = 52,541.30	No
x_3 = 49,643.50	Yes	x_{18} = 44,980.00	Yes
x_4 = 49,894.90	Yes	x_{19} = 51,932.60	Yes
x_5 = 47,621.60	No	x_{20} = 52,973.00	Yes
x_6 = 55,924.00	Yes	x_{21} = 45,120.90	Yes
x_7 = 49,092.30	Yes	x_{22} = 51,753.00	Yes
x_8 = 51,404.40	Yes	x_{23} = 54,391.80	No
x_9 = 50,957.70	Yes	x_{24} = 50,164.20	No
x_{10} = 55,109.70	Yes	x_{25} = 52,973.60	No
x_{11} = 45,922.60	Yes	x_{26} = 50,241.30	No
x_{12} = 57,268.40	No	x_{27} = 52,793.90	No
x_{13} = 55,688.80	Yes	x_{28} = 50,979.40	Yes
x_{14} = 51,564.70	No	x_{29} = 55,860.90	Yes
x_{15} = 56,188.20	No	x_{30} = 57,309.10	No

sample mean and the sample standard deviation s. Using the formulas for a sample mean and a sample standard deviation presented in Chapter 3, the sample mean is

$$\bar{x} = \frac{\Sigma x_i}{n} = \frac{1{,}554{,}420}{30} = \$51{,}814$$

and the sample standard deviation is

$$s = \sqrt{\frac{\Sigma(x_i - \bar{x})^2}{n - 1}} = \sqrt{\frac{325{,}009{,}260}{29}} = \$3348$$

To estimate p, the proportion of managers in the population who completed the management training program, we use the corresponding sample proportion $\bar{p}$. Let x denote the number of managers in the sample who completed the management training program. The data in Table 7.2 show that $x = 19$. Thus, with a sample size of $n = 30$, the sample proportion is

$$\bar{p} = \frac{x}{n} = \frac{19}{30} = .63$$

By making the preceding computations, we perform the statistical procedure called *point estimation.* We refer to the sample mean $\bar{x}$ as the **point estimator** of the population mean μ, the sample standard deviation s as the point estimator of the population standard deviation σ, and the sample proportion $\bar{p}$ as the point estimator of the population proportion p. The numerical value obtained for $\bar{x}$, s, or $\bar{p}$ is called the **point estimate**. Thus, for the simple random sample of 30 EAI managers shown in Table 7.2, \$51,814 is the point estimate of μ, \$3348 is the point estimate of σ, and .63 is the point estimate of p. Table 7.3 summarizes the sample results and compares the point estimates to the actual values of the population parameters.

As is evident from Table 7.3, the point estimates differ somewhat from the corresponding population parameters. This difference is to be expected because a sample, and not a census of the entire population, is being used to develop the point estimates. In the next chapter, we will show how to construct an interval estimate in order to provide information about how close the point estimate is to the population parameter.

TABLE 7.3 SUMMARY OF POINT ESTIMATES OBTAINED FROM A SIMPLE RANDOM SAMPLE OF 30 EAI MANAGERS

Population Parameter	Parameter Value	Point Estimator	Point Estimate
μ = Population mean annual salary	\$51,800	$\bar{x}$ = Sample mean annual salary	\$51,814
σ = Population standard deviation for annual salary	\$4000	s = Sample standard deviation for annual salary	\$3348
p = Population proportion having completed the management training program	.60	$\bar{p}$ = Sample proportion having completed the management training program	.63

Practical Advice

The subject matter of most of the rest of the book is concerned with statistical inference. Point estimation is a form of statistical inference. We use a sample statistic to make an inference about a population parameter. When making inferences about a population based on a sample, it is important to have a close correspondence between the sampled population and the target population. The **target population** is the population we want to make inferences about, while the sampled population is the population from which the sample is actually taken. In this section, we have described the process of drawing a simple random sample from the population of EAI managers and making point estimates of characteristics of that same population. So the sampled population and the target population are identical, which is the desired situation. But in other cases, it is not as easy to obtain a close correspondence between the sampled and target populations.

Consider the case of an amusement park selecting a sample of its customers to learn about characteristics such as age and time spent at the park. Suppose all the sample elements were selected on a day when park attendance was restricted to employees of a single company. Then the sampled population would be composed of employees of that company and members of their families. If the target population we wanted to make inferences about were typical park customers over a typical summer, then we might encounter a significant difference between the sampled population and the target population. In such a case, we would question the validity of the point estimates being made. Park management would be in the best position to know whether a sample taken on a particular day was likely to be representative of the target population.

In summary, whenever a sample is used to make inferences about a population, we should make sure that the study is designed so that the sampled population and the target population are in close agreement. Good judgment is a necessary ingredient of sound statistical practice.

Exercises

Methods

SELF test

11. The following data are from a simple random sample.

5 8 10 7 10 14

a. What is the point estimate of the population mean?
b. What is the point estimate of the population standard deviation?

12. A survey question for a sample of 150 individuals yielded 75 Yes responses, 55 No responses, and 20 No Opinions.
a. What is the point estimate of the proportion in the population who respond Yes?
b. What is the point estimate of the proportion in the population who respond No?

Applications

SELF test

13. A sample of 5 months of sales data provided the following information:

Month:	1	2	3	4	5
Units Sold:	94	100	85	94	92

a. Develop a point estimate of the population mean number of units sold per month.
b. Develop a point estimate of the population standard deviation.

WEB file

MutualFund

14. *BusinessWeek* published information on 283 equity mutual funds (*BusinessWeek,* January 26, 2004). A sample of 40 of those funds is contained in the data set MutualFund. Use the data set to answer the following questions.
a. Develop a point estimate of the proportion of the *BusinessWeek* equity funds that are load funds.
b. Develop a point estimate of the proportion of funds that are classified as high risk.
c. Develop a point estimate of the proportion of funds that have a below-average risk rating.

15. Many drugs used to treat cancer are expensive. *BusinessWeek* reported on the cost per treatment of Herceptin, a drug used to treat breast cancer (*BusinessWeek,* January 30, 2006). Typical treatment costs (in dollars) for Herceptin are provided by a simple random sample of 10 patients.

4376	5578	2717	4920	4495
4798	6446	4119	4237	3814

a. Develop a point estimate of the mean cost per treatment with Herceptin.
b. Develop a point estimate of the standard deviation of the cost per treatment with Herceptin.

16. A sample of 426 U.S. adults age 50 and older were asked how important a variety of issues were in choosing whom to vote for in the 2012 presidential election (*AARP Bulletin,* March 2012).
a. What is the sampled population for this study?
b. Social Security and Medicare was cited as "very important" by 350 respondents. Estimate the proportion of the population of U.S. adults age 50 and over who believe this issue is very important.
c. Education was cited as "very important" by 74% of the respondents. Estimate the number of respondents who believe this issue is very important.
d. Job Growth was cited as "very important" by 354 respondents. Estimate the proportion of U.S. adults age 50 and over who believe job growth is very important.
e. What is the target population for the inferences being made in parts (b) and (d)? Is it the same as the sampled population you identified in part (a)? Suppose you later learn that the sample was restricted to members of the American Association of Retired People (AARP). Would you still feel the inferences being made in parts (b) and (d) are valid? Why or why not?

17. The American Association of Individual Investors (AAII) polls its subscribers on a weekly basis to determine the number who are bullish, bearish, or neutral on the short-term prospects for the stock market. Their findings for the week ending March 2, 2006, are consistent with the following sample results (AAII website, March 7, 2006).

Bullish 409 Neutral 299 Bearish 291

Develop a point estimate of the following population parameters.
a. The proportion of all AAII subscribers who are bullish on the stock market.
b. The proportion of all AAII subscribers who are neutral on the stock market.
c. The proportion of all AAII subscribers who are bearish on the stock market.
d. What is the sampled population? What is the target population for parts (a), (b), and (c)? Would you be comfortable extending these results to the target population of all investors?

7.4 Introduction to Sampling Distributions

In the preceding section we said that the sample mean $\bar{x}$ is the point estimator of the population mean μ, and the sample proportion $\bar{p}$ is the point estimator of the population proportion p. For the simple random sample of 30 EAI managers shown in Table 7.2, the point estimate of μ is $\bar{x} = \$51{,}814$ and the point estimate of p is $\bar{p} = .63$. Suppose we select another simple random sample of 30 EAI managers and obtain the following point estimates:

$$\text{Sample mean: } \bar{x} = \$52{,}670$$
$$\text{Sample proportion: } \bar{p} = .70$$

Note that different values of $\bar{x}$ and $\bar{p}$ were obtained. Indeed, a second simple random sample of 30 EAI managers cannot be expected to provide the same point estimates as the first sample.

The ability to understand the material in subsequent chapters depends heavily on the ability to understand and use the sampling distributions presented in this chapter.

Now, suppose we repeat the process of selecting a simple random sample of 30 EAI managers over and over again, each time computing the values of $\bar{x}$ and $\bar{p}$. Table 7.4 contains a portion of the results obtained for 500 simple random samples, and Table 7.5 shows the frequency and relative frequency distributions for the 500 $\bar{x}$ values. Figure 7.1 shows the relative frequency histogram for the $\bar{x}$ values.

In Chapter 5 we defined a random variable as a numerical description of the outcome of an experiment. If we consider the process of selecting a simple random sample as an

TABLE 7.4 VALUES OF $\bar{x}$ AND $\bar{p}$ FROM 500 SIMPLE RANDOM SAMPLES OF 30 EAI MANAGERS

Sample Number	Sample Mean ($\bar{x}$)	Sample Proportion ($\bar{p}$)
1	51,814	.63
2	52,670	.70
3	51,780	.67
4	51,588	.53
.	.	.
.	.	.
.	.	.
500	51,752	.50

TABLE 7.5 FREQUENCY AND RELATIVE FREQUENCY DISTRIBUTIONS OF $\bar{x}$ FROM 500 SIMPLE RANDOM SAMPLES OF 30 EAI MANAGERS

Mean Annual Salary ($)	Frequency	Relative Frequency
49,500.00–49,999.99	2	.004
50,000.00–50,499.99	16	.032
50,500.00–50,999.99	52	.104
51,000.00–51,499.99	101	.202
51,500.00–51,999.99	133	.266
52,000.00–52,499.99	110	.220
52,500.00–52,999.99	54	.108
53,000.00–53,499.99	26	.052
53,500.00–53,999.99	6	.012
Totals	500	1.000

FIGURE 7.1 RELATIVE FREQUENCY HISTOGRAM OF $\bar{x}$ VALUES FROM 500 SIMPLE RANDOM SAMPLES OF SIZE 30 EACH

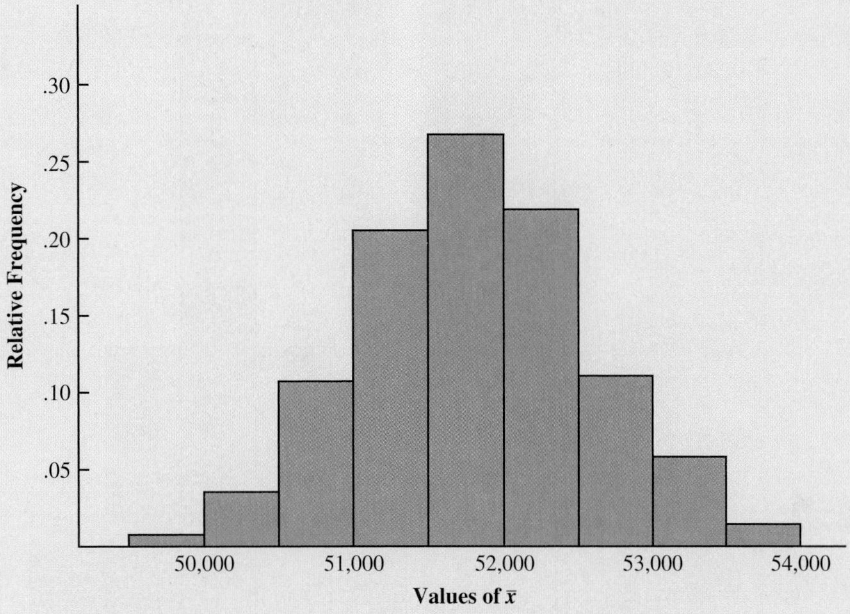

experiment, the sample mean $\bar{x}$ is the numerical description of the outcome of the experiment. Thus, the sample mean $\bar{x}$ is a random variable. As a result, just like other random variables, $\bar{x}$ has a mean or expected value, a standard deviation, and a probability distribution. Because the various possible values of $\bar{x}$ are the result of different simple random samples, the probability distribution of $\bar{x}$ is called the **sampling distribution** of $\bar{x}$. Knowledge of this sampling distribution and its properties will enable us to make probability statements about how close the sample mean $\bar{x}$ is to the population mean μ.

Let us return to Figure 7.1. We would need to enumerate every possible sample of 30 managers and compute each sample mean to completely determine the sampling distribution of $\bar{x}$. However, the histogram of 500 $\bar{x}$ values gives an approximation of this sampling distribution. From the approximation we observe the bell-shaped appearance of the distribution. We note that the largest concentration of the $\bar{x}$ values and the mean of the 500 $\bar{x}$ values is near the population mean $\mu = \$51,800$. We will describe the properties of the sampling distribution of $\bar{x}$ more fully in the next section.

The 500 values of the sample proportion $\bar{p}$ are summarized by the relative frequency histogram in Figure 7.2. As in the case of $\bar{x}$, $\bar{p}$ is a random variable. If every possible sample of size 30 were selected from the population and if a value of $\bar{p}$ were computed for each sample, the resulting probability distribution would be the sampling distribution of $\bar{p}$. The relative frequency histogram of the 500 sample values in Figure 7.2 provides a general idea of the appearance of the sampling distribution of $\bar{p}$.

In practice, we select only one simple random sample from the population. We repeated the sampling process 500 times in this section simply to illustrate that many different samples are possible and that the different samples generate a variety of values for the sample statistics $\bar{x}$ and $\bar{p}$. The probability distribution of any particular sample statistic is called the sampling distribution of the statistic. In Section 7.5 we show the characteristics of the sampling distribution of $\bar{x}$. In Section 7.6 we show the characteristics of the sampling distribution of $\bar{p}$.

FIGURE 7.2 RELATIVE FREQUENCY HISTOGRAM OF $\bar{p}$ VALUES FROM 500 SIMPLE RANDOM SAMPLES OF SIZE 30 EACH

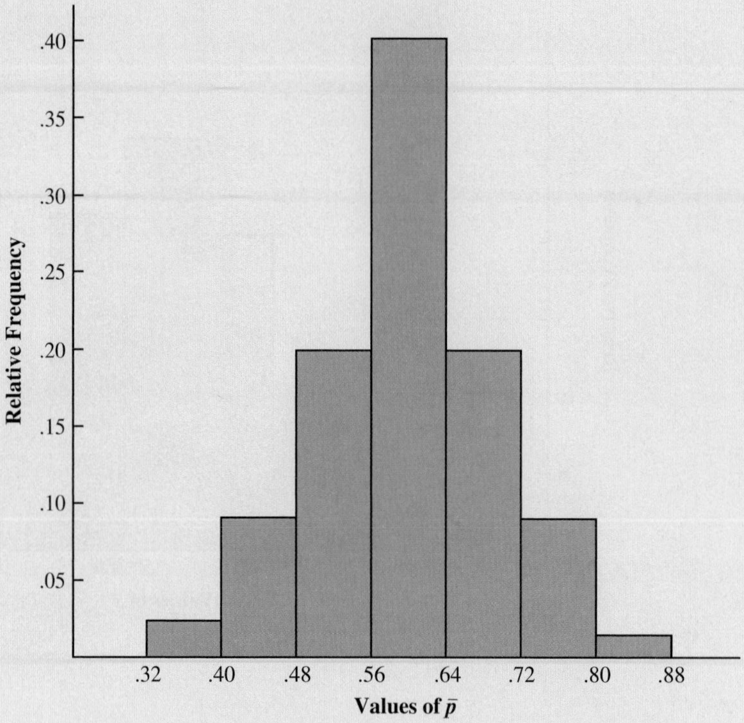

7.5 Sampling Distribution of $\bar{x}$

In the previous section we said that the sample mean $\bar{x}$ is a random variable and its probability distribution is called the sampling distribution of $\bar{x}$.

SAMPLING DISTRIBUTION OF $\bar{x}$

The sampling distribution of $\bar{x}$ is the probability distribution of all possible values of the sample mean $\bar{x}$.

This section describes the properties of the sampling distribution of $\bar{x}$. Just as with other probability distributions we studied, the sampling distribution of $\bar{x}$ has an expected value or mean, a standard deviation, and a characteristic shape or form. Let us begin by considering the mean of all possible $\bar{x}$ values, which is referred to as the expected value of $\bar{x}$.

Expected Value of $\bar{x}$

In the EAI sampling problem we saw that different simple random samples result in a variety of values for the sample mean $\bar{x}$. Because many different values of the random variable $\bar{x}$ are possible, we are often interested in the mean of all possible values of $\bar{x}$ that can be generated by the various simple random samples. The mean of the $\bar{x}$ random variable is the expected value of $\bar{x}$. Let $E(\bar{x})$ represent the expected value of $\bar{x}$ and μ represent the mean of the population from which we are selecting a simple random sample. It can be shown that with simple random sampling, $E(\bar{x})$ and μ are equal.

The expected value of $\bar{x}$ equals the mean of the population from which the sample is selected.

EXPECTED VALUE OF $\bar{x}$

$$E(\bar{x}) = \mu \qquad (7.1)$$

where

$$E(\bar{x}) = \text{the expected value of } \bar{x}$$
$$\mu = \text{the population mean}$$

This result shows that with simple random sampling, the expected value or mean of the sampling distribution of $\bar{x}$ is equal to the mean of the population. In Section 7.1 we saw that the mean annual salary for the population of EAI managers is $\mu = \$51{,}800$. Thus, according to equation (7.1), the mean of all possible sample means for the EAI study is also \$51,800.

When the expected value of a point estimator equals the population parameter, we say the point estimator is **unbiased**. Thus, equation (7.1) shows that $\bar{x}$ is an unbiased estimator of the population mean μ.

Standard Deviation of $\bar{x}$

Let us define the standard deviation of the sampling distribution of $\bar{x}$. We will use the following notation.

$$\sigma_{\bar{x}} = \text{the standard deviation of } \bar{x}$$
$$\sigma = \text{the standard deviation of the population}$$
$$n = \text{the sample size}$$
$$N = \text{the population size}$$

It can be shown that the formula for the standard deviation of $\bar{x}$ depends on whether the population is finite or infinite. The two formulas for the standard deviation of $\bar{x}$ follow.

STANDARD DEVIATION OF $\bar{x}$

Finite Population	*Infinite Population*	
$\sigma_{\bar{x}} = \sqrt{\frac{N-n}{N-1}}\left(\frac{\sigma}{\sqrt{n}}\right)$	$\sigma_{\bar{x}} = \frac{\sigma}{\sqrt{n}}$	(7.2)

In comparing the two formulas in (7.2), we see that the factor $\sqrt{(N-n)/(N-1)}$ is required for the finite population case but not for the infinite population case. This factor is commonly referred to as the **finite population correction factor**. In many practical sampling situations, we find that the population involved, although finite, is "large," whereas the sample size is relatively "small." In such cases the finite population correction factor $\sqrt{(N-n)/(N-1)}$ is close to 1. As a result, the difference between the values of the standard deviation of $\bar{x}$ for the finite and infinite population cases becomes negligible. Then, $\sigma_{\bar{x}} = \sigma/\sqrt{n}$ becomes a good approximation to the standard deviation of $\bar{x}$ even though the population is finite. This observation leads to the following general guideline, or rule of thumb, for computing the standard deviation of $\bar{x}$.

USE THE FOLLOWING EXPRESSION TO COMPUTE THE STANDARD DEVIATION OF $\bar{x}$

$$\sigma_{\bar{x}} = \frac{\sigma}{\sqrt{n}} \tag{7.3}$$

whenever

1. The population is infinite; or
2. The population is finite *and* the sample size is less than or equal to 5% of the population size; that is, $n/N \le .05$.

Problem 21 shows that when $n/N \le .05$, the finite population correction factor has little effect on the value of $\sigma_{\bar{x}}$.

The term standard error is used throughout statistical inference to refer to the standard deviation of a point estimator.

In cases where $n/N > .05$, the finite population version of formula (7.2) should be used in the computation of $\sigma_{\bar{x}}$. Unless otherwise noted, throughout the text we will assume that the population size is "large," $n/N \le .05$, and expression (7.3) can be used to compute $\sigma_{\bar{x}}$.

To compute $\sigma_{\bar{x}}$, we need to know σ, the standard deviation of the population. To further emphasize the difference between $\sigma_{\bar{x}}$ and σ, we refer to the standard deviation of $\bar{x}$, $\sigma_{\bar{x}}$, as the **standard error** of the mean. In general, the term *standard error* refers to the standard deviation of a point estimator. Later we will see that the value of the standard error of the mean is helpful in determining how far the sample mean may be from the population mean. Let us now return to the EAI example and compute the standard error of the mean associated with simple random samples of 30 EAI managers.

In Section 7.1 we saw that the standard deviation of annual salary for the population of 2500 EAI managers is $\sigma = 4000$. In this case, the population is finite, with $N = 2500$. However, with a sample size of 30, we have $n/N = 30/2500 = .012$. Because the sample size is less than 5% of the population size, we can ignore the finite population correction factor and use equation (7.3) to compute the standard error.

$$\sigma_{\bar{x}} = \frac{\sigma}{\sqrt{n}} = \frac{4000}{\sqrt{30}} = 730.3$$

Form of the Sampling Distribution of $\bar{x}$

The preceding results concerning the expected value and standard deviation for the sampling distribution of $\bar{x}$ are applicable for any population. The final step in identifying the characteristics of the sampling distribution of $\bar{x}$ is to determine the form or shape of the sampling distribution. We will consider two cases: (1) The population has a normal distribution; and (2) the population does not have a normal distribution.

Population has a normal distribution. In many situations it is reasonable to assume that the population from which we are selecting a random sample has a normal, or nearly normal, distribution. When the population has a normal distribution, the sampling distribution of $\bar{x}$ is normally distributed for any sample size.

Population does not have a normal distribution. When the population from which we are selecting a random sample does not have a normal distribution, the **central limit theorem** is helpful in identifying the shape of the sampling distribution of $\bar{x}$. A statement of the central limit theorem as it applies to the sampling distribution of $\bar{x}$ follows.

CENTRAL LIMIT THEOREM

In selecting random samples of size n from a population, the sampling distribution of the sample mean $\bar{x}$ can be approximated by a *normal distribution* as the sample size becomes large.

Figure 7.3 shows how the central limit theorem works for three different populations; each column refers to one of the populations. The top panel of the figure shows that none of the populations are normally distributed. Population I follows a uniform distribution. Population II is often called the rabbit-eared distribution. It is symmetric, but the more likely

FIGURE 7.3 ILLUSTRATION OF THE CENTRAL LIMIT THEOREM FOR THREE POPULATIONS

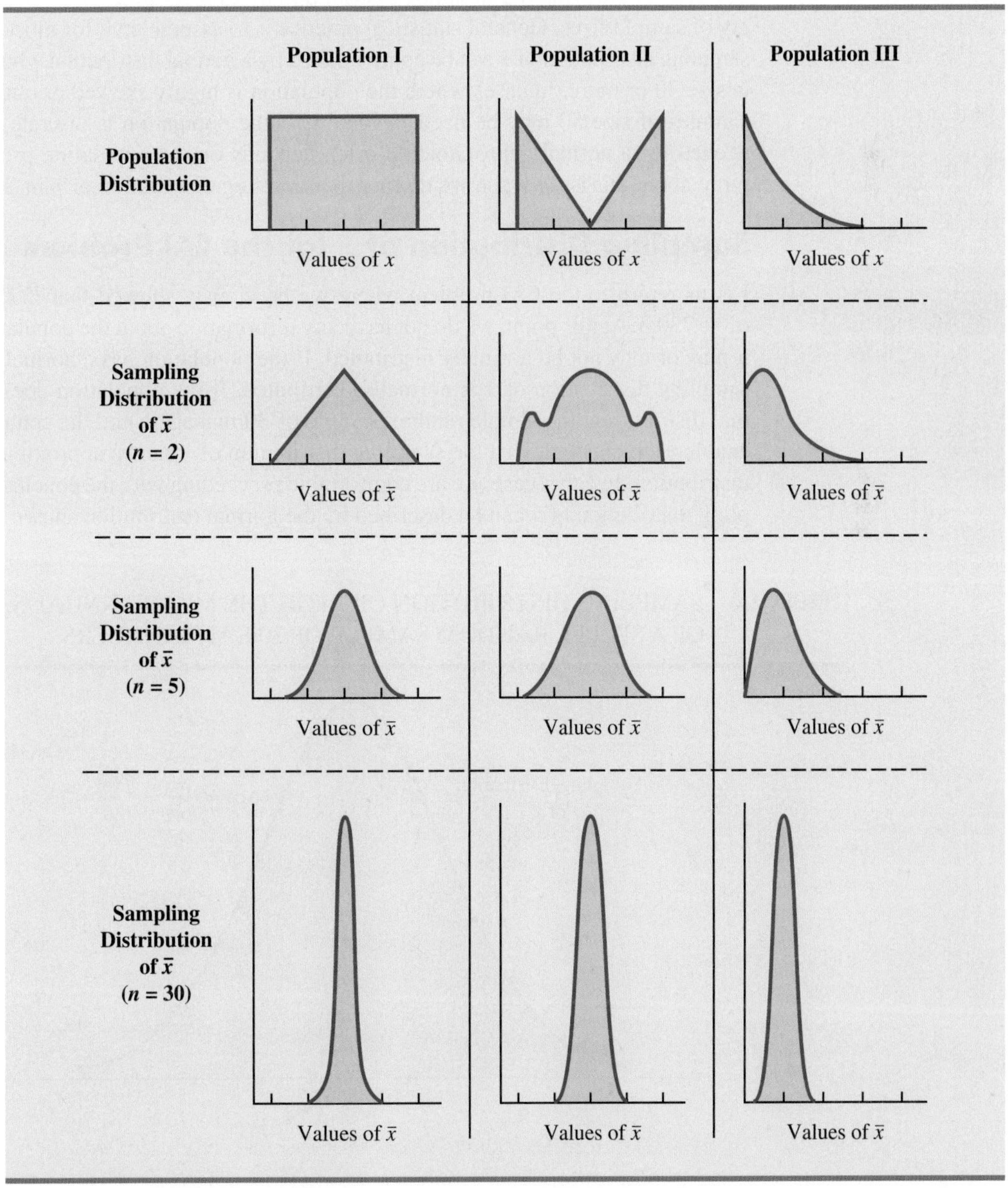

values fall in the tails of the distribution. Population III is shaped like the exponential distribution; it is skewed to the right.

The bottom three panels of Figure 7.3 show the shape of the sampling distribution for samples of size $n = 2$, $n = 5$, and $n = 30$. When the sample size is 2, we see that the shape of each sampling distribution is different from the shape of the corresponding population distribution. For samples of size 5, we see that the shapes of the sampling distributions for populations I and II begin to look similar to the shape of a normal distribution. Even though the shape of the sampling distribution for population III begins to look similar to the shape of a normal distribution, some skewness to the right is still present. Finally, for samples of size 30, the shapes of each of the three sampling distributions are approximately normal.

From a practitioner standpoint, we often want to know how large the sample size needs to be before the central limit theorem applies and we can assume that the shape of the sampling distribution is approximately normal. Statistical researchers have investigated this question by studying the sampling distribution of $\bar{x}$ for a variety of populations and a variety of sample sizes. General statistical practice is to assume that, for most applications, the sampling distribution of $\bar{x}$ can be approximated by a normal distribution whenever the sample is size 30 or more. In cases where the population is highly skewed or outliers are present, samples of size 50 may be needed. Finally, if the population is discrete, the sample size needed for a normal approximation often depends on the population proportion. We say more about this issue when we discuss the sampling distribution of $\bar{p}$ in Section 7.6.

Sampling Distribution of $\bar{x}$ for the EAI Problem

Let us return to the EAI problem where we previously showed that $E(\bar{x}) = \$51{,}800$ and $\sigma_{\bar{x}} = 730.3$. At this point, we do not have any information about the population distribution; it may or may not be normally distributed. If the population has a normal distribution, the sampling distribution of $\bar{x}$ is normally distributed. If the population does not have a normal distribution, the simple random sample of 30 managers and the central limit theorem enable us to conclude that the sampling distribution of $\bar{x}$ can be approximated by a normal distribution. In either case, we are comfortable proceeding with the conclusion that the sampling distribution of $\bar{x}$ can be described by the normal distribution shown in Figure 7.4.

FIGURE 7.4 SAMPLING DISTRIBUTION OF $\bar{x}$ FOR THE MEAN ANNUAL SALARY OF A SIMPLE RANDOM SAMPLE OF 30 EAI MANAGERS

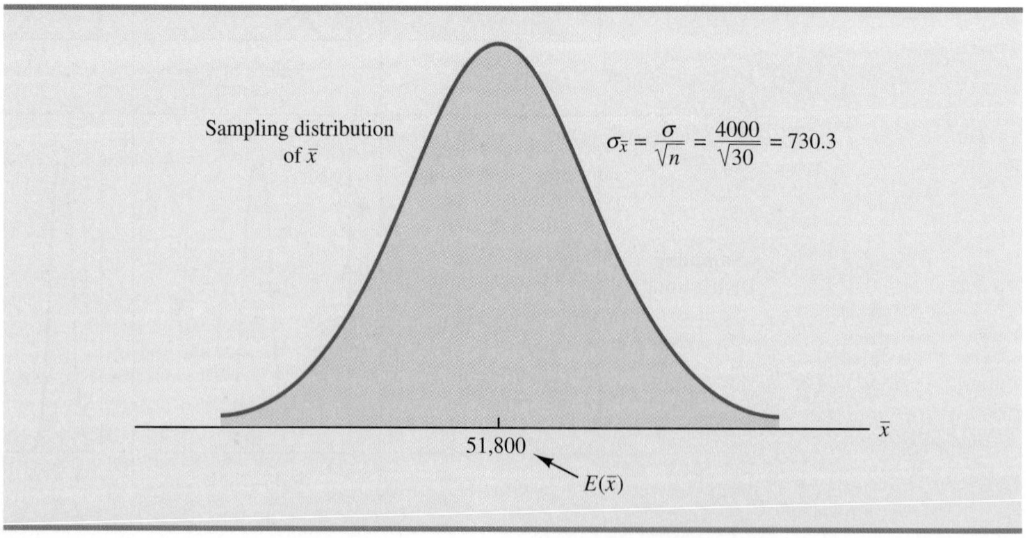

Practical Value of the Sampling Distribution of $\bar{x}$

Whenever a simple random sample is selected and the value of the sample mean is used to estimate the value of the population mean μ, we cannot expect the sample mean to exactly equal the population mean. The practical reason we are interested in the sampling distribution of $\bar{x}$ is that it can be used to provide probability information about the difference between the sample mean and the population mean. To demonstrate this use, let us return to the EAI problem.

Suppose the personnel director believes the sample mean will be an acceptable estimate of the population mean if the sample mean is within \$500 of the population mean. However, it is not possible to guarantee that the sample mean will be within \$500 of the population mean. Indeed, Table 7.5 and Figure 7.1 show that some of the 500 sample means differed by more than \$2000 from the population mean. So we must think of the personnel director's request in probability terms. That is, the personnel director is concerned with the following question: What is the probability that the sample mean computed using a simple random sample of 30 EAI managers will be within \$500 of the population mean?

Because we have identified the properties of the sampling distribution of $\bar{x}$ (see Figure 7.4), we will use this distribution to answer the probability question. Refer to the sampling distribution of $\bar{x}$ shown again in Figure 7.5. With a population mean of \$51,800, the personnel director wants to know the probability that $\bar{x}$ is between \$51,300 and \$52,300. This probability is given by the darkly shaded area of the sampling distribution shown in Figure 7.5. Because the sampling distribution is normally distributed, with mean 51,800 and standard error of the mean 730.3, we can use the standard normal probability table to find the area or probability.

We first calculate the z value at the upper endpoint of the interval (52,300) and use the table to find the area under the curve to the left of that point (left tail area). Then we compute the z value at the lower endpoint of the interval (51,300) and use the table to find the area under the curve to the left of that point (another left tail area). Subtracting the second tail area from the first gives us the desired probability.

At $\bar{x} = 52{,}300$, we have

$$z = \frac{52{,}300 - 51{,}800}{730.30} = .68$$

FIGURE 7.5 PROBABILITY OF A SAMPLE MEAN BEING WITHIN \$500 OF THE POPULATION MEAN FOR A SIMPLE RANDOM SAMPLE OF 30 EAI MANAGERS

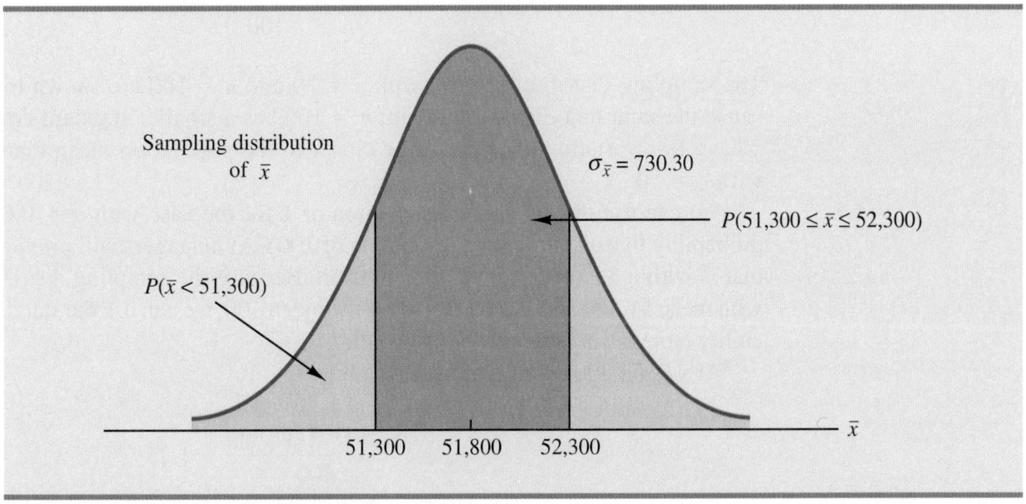

Referring to the standard normal probability table, we find a cumulative probability (area to the left of $z = .68$) of .7517.

At $\bar{x} = 51{,}300$, we have

$$z = \frac{51{,}300 - 51{,}800}{730.30} = -.68$$

The area under the curve to the left of $z = -.68$ is .2483. Therefore, $P(51{,}300 \le \bar{x} \le 52{,}300) = P(z \le .68) - P(z < -.68) = .7517 - .2483 = .5034$.

The sampling distribution of $\bar{x}$ can be used to provide probability information about how close the sample mean $\bar{x}$ is to the population mean μ.

The preceding computations show that a simple random sample of 30 EAI managers has a .5034 probability of providing a sample mean $\bar{x}$ that is within \$500 of the population mean. Thus, there is a $1 - .5034 = .4966$ probability that the difference between $\bar{x}$ and $\mu = \$51{,}800$ will be more than \$500. In other words, a simple random sample of 30 EAI managers has roughly a 50–50 chance of providing a sample mean within the allowable \$500. Perhaps a larger sample size should be considered. Let us explore this possibility by considering the relationship between the sample size and the sampling distribution of $\bar{x}$.

Relationship Between the Sample Size and the Sampling Distribution of $\bar{x}$

Suppose that in the EAI sampling problem we select a simple random sample of 100 EAI managers instead of the 30 originally considered. Intuitively, it would seem that with more data provided by the larger sample size, the sample mean based on $n = 100$ should provide a better estimate of the population mean than the sample mean based on $n = 30$. To see how much better, let us consider the relationship between the sample size and the sampling distribution of $\bar{x}$.

First note that $E(\bar{x}) = \mu$ regardless of the sample size. Thus, the mean of all possible values of $\bar{x}$ is equal to the population mean μ regardless of the sample size n. However, note that the standard error of the mean, $\sigma_{\bar{x}} = \sigma/\sqrt{n}$, is related to the square root of the sample size. Whenever the sample size is increased, the standard error of the mean $\sigma_{\bar{x}}$ decreases. With $n = 30$, the standard error of the mean for the EAI problem is 730.3. However, with the increase in the sample size to $n = 100$, the standard error of the mean is decreased to

$$\sigma_{\bar{x}} = \frac{\sigma}{\sqrt{n}} = \frac{4000}{\sqrt{100}} = 400$$

The sampling distributions of $\bar{x}$ with $n = 30$ and $n = 100$ are shown in Figure 7.6. Because the sampling distribution with $n = 100$ has a smaller standard error, the values of $\bar{x}$ have less variation and tend to be closer to the population mean than the values of $\bar{x}$ with $n = 30$.

We can use the sampling distribution of $\bar{x}$ for the case with $n = 100$ to compute the probability that a simple random sample of 100 EAI managers will provide a sample mean that is within \$500 of the population mean. Because the sampling distribution is normal, with mean 51,800 and standard error of the mean 400, we can use the standard normal probability table to find the area or probability.

At $\bar{x} = 52{,}300$ (see Figure 7.7), we have

$$z = \frac{52{,}300 - 51{,}800}{400} = 1.25$$

FIGURE 7.6 A COMPARISON OF THE SAMPLING DISTRIBUTIONS OF $\bar{x}$ FOR SIMPLE RANDOM SAMPLES OF $n = 30$ AND $n = 100$ EAI MANAGERS

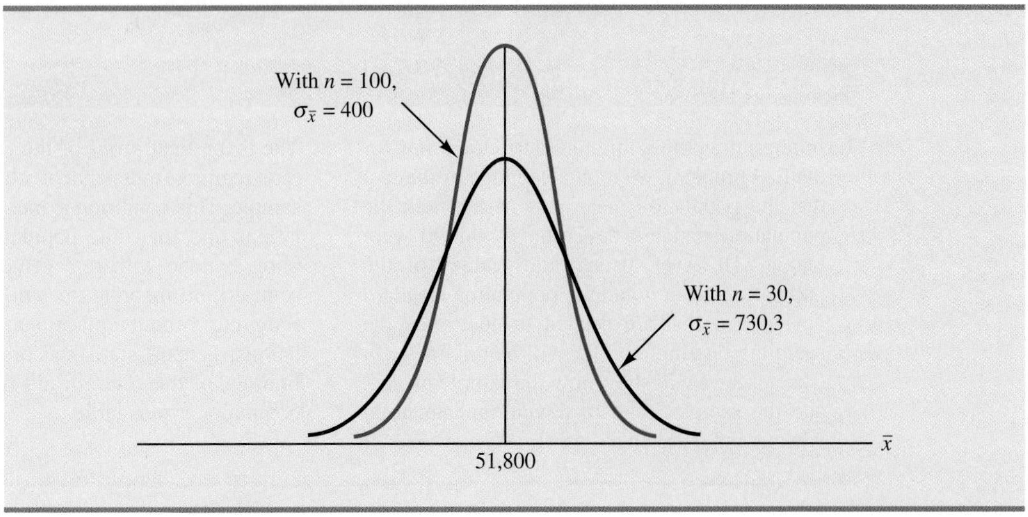

Referring to the standard normal probability table, we find a cumulative probability corresponding to $z = 1.25$ of .8944.

At $\bar{x} = 51{,}300$, we have

$$z = \frac{51{,}300 - 51{,}800}{400} = -1.25$$

The cumulative probability corresponding to $z = -1.25$ is .1056. Therefore, $P(51{,}300 \le \bar{x} \le 52{,}300) = P(z \le 1.25) - P(z \le -1.25) = .8944 - .1056 = .7888$. Thus, by increasing the sample size from 30 to 100 EAI managers, we increase the probability of obtaining a sample mean within \$500 of the population mean from .5034 to .7888.

FIGURE 7.7 PROBABILITY OF A SAMPLE MEAN BEING WITHIN \$500 OF THE POPULATION MEAN FOR A SIMPLE RANDOM SAMPLE OF 100 EAI MANAGERS

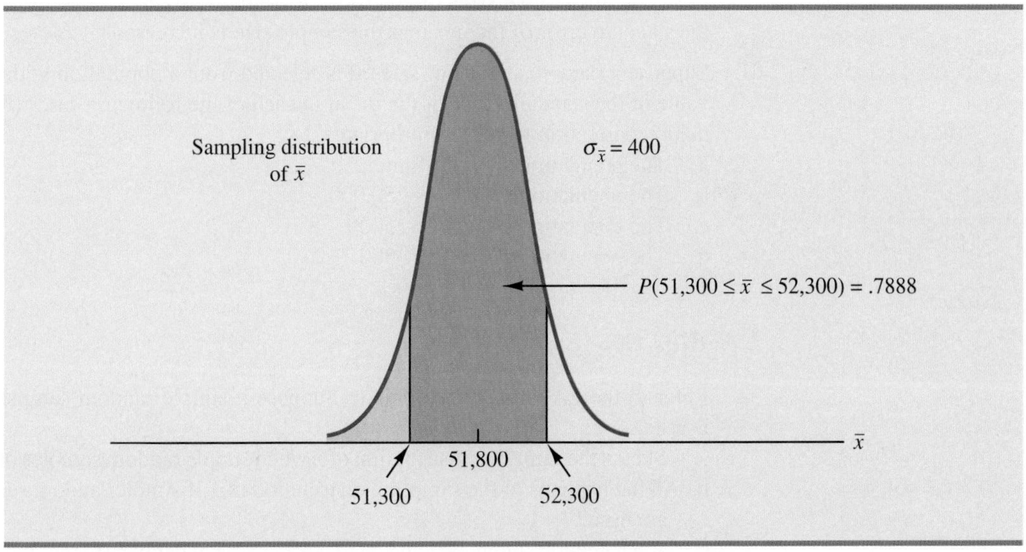

The important point in this discussion is that as the sample size is increased, the standard error of the mean decreases. As a result, the larger sample size provides a higher probability that the sample mean is within a specified distance of the population mean.

NOTES AND COMMENTS

1. In presenting the sampling distribution of $\bar{x}$ for the EAI problem, we took advantage of the fact that the population mean $\mu = 51{,}800$ and the population standard deviation $\sigma = 4000$ were known. However, usually the values of the population mean μ and the population standard deviation σ that are needed to determine the sampling distribution of $\bar{x}$ will be unknown. In Chapter 8 we will show how the sample mean $\bar{x}$ and the sample standard deviation s are used when μ and σ are unknown.

2. The theoretical proof of the central limit theorem requires independent observations in the sample. This condition is met for infinite populations and for finite populations where sampling is done with replacement. Although the central limit theorem does not directly address sampling without replacement from finite populations, general statistical practice applies the findings of the central limit theorem when the population size is large.

Exercises

Methods

18. A population has a mean of 200 and a standard deviation of 50. A sample of size 100 will be taken and the sample mean $\bar{x}$ will be used to estimate the population mean.
    a. What is the expected value of $\bar{x}$?
    b. What is the standard deviation of $\bar{x}$?
    c. Show the sampling distribution of $\bar{x}$.
    d. What does the sampling distribution of $\bar{x}$ show?

SELF test

19. A population has a mean of 200 and a standard deviation of 50. Suppose a sample of size 100 is selected and $\bar{x}$ is used to estimate μ.
    a. What is the probability that the sample mean will be within ± 5 of the population mean?
    b. What is the probability that the sample mean will be within ± 10 of the population mean?

20. Assume the population standard deviation is $\sigma = 25$. Compute the standard error of the mean, $\sigma_{\bar{x}}$, for sample sizes of 50, 100, 150, and 200. What can you say about the size of the standard error of the mean as the sample size is increased?

21. Suppose a random sample of size 50 is selected from a population with $\sigma = 10$. Find the value of the standard error of the mean in each of the following cases (use the finite population correction factor if appropriate).
    a. The population size is infinite.
    b. The population size is $N = 50{,}000$.
    c. The population size is $N = 5000$.
    d. The population size is $N = 500$.

Applications

22. Refer to the EAI sampling problem. Suppose a simple random sample of 60 managers is used.
    a. Sketch the sampling distribution of $\bar{x}$ when simple random samples of size 60 are used.
    b. What happens to the sampling distribution of $\bar{x}$ if simple random samples of size 120 are used?
    c. What general statement can you make about what happens to the sampling distribution of $\bar{x}$ as the sample size is increased? Does this generalization seem logical? Explain.

SELF test

23. In the EAI sampling problem (see Figure 7.5), we showed that for $n = 30$, there was .5034 probability of obtaining a sample mean within $\pm$\$500 of the population mean.
    a. What is the probability that $\bar{x}$ is within \$500 of the population mean if a sample of size 60 is used?
    b. Answer part (a) for a sample of size 120.

24. *Barron's* reported that the average number of weeks an individual is unemployed is 17.5 weeks (*Barron's,* February 18, 2008). Assume that for the population of all unemployed individuals the population mean length of unemployment is 17.5 weeks and that the population standard deviation is 4 weeks. Suppose you would like to select a sample of 50 unemployed individuals for a follow-up study.
    a. Show the sampling distribution of $\bar{x}$, the sample mean average for a sample of 50 unemployed individuals.
    b. What is the probability that a simple random sample of 50 unemployed individuals will provide a sample mean within 1 week of the population mean?
    c. What is the probability that a simple random sample of 50 unemployed individuals will provide a sample mean within 1/2 week of the population mean?

25. The College Board reported the following mean scores for the three parts of the Scholastic Aptitude Test (SAT) (*The World Almanac,* 2009):

Critical Reading	502
Mathematics	515
Writing	494

Assume that the population standard deviation on each part of the test is $\sigma = 100$.
    a. What is the probability a sample of 90 test takers will provide a sample mean test score within 10 points of the population mean of 502 on the Critical Reading part of the test?
    b. What is the probability a sample of 90 test takers will provide a sample mean test score within 10 points of the population mean of 515 on the Mathematics part of the test? Compare this probability to the value computed in part (a).
    c. What is the probability a sample of 100 test takers will provide a sample mean test score within 10 of the population mean of 494 on the writing part of the test? Comment on the differences between this probability and the values computed in parts (a) and (b).

26. The mean annual cost of automobile insurance is \$939 (CNBC, February 23, 2006). Assume that the standard deviation is $\sigma = \$245$.
    a. What is the probability that a sample of automobile insurance policies will have a sample mean within \$25 of the population mean for each of the following sample sizes: 30, 50, 100, and 400?
    b. What is the advantage of a larger sample size when attempting to estimate the population mean?

27. The Economic Policy Institute periodically issues reports on wages of entry level workers. The institute reported that entry level wages for male college graduates were \$21.68 per hour and for female college graduates were \$18.80 per hour in 2011 (Economic Policy Institute website, March 30, 2012). Assume the standard deviation for male graduates is \$2.30, and for female graduates it is \$2.05.
    a. What is the probability that a sample of 50 male graduates will provide a sample mean within \$.50 of the population mean, \$21.68?
    b. What is the probability that a sample of 50 female graduates will provide a sample mean within \$.50 of the population mean, \$18.80?
    c. In which of the preceding two cases, part (a) or part (b), do we have a higher probability of obtaining a sample estimate within \$.50 of the population mean? Why?
    d. What is the probability that a sample of 120 female graduates will provide a sample mean more than \$.30 below the population mean?

28. The average score for male golfers is 95 and the average score for female golfers is 106 (*Golf Digest,* April 2006). Use these values as the population means for men and women and assume that the population standard deviation is $\sigma = 14$ strokes for both. A sample of 30 male golfers and another sample of 45 female golfers will be taken.
    a. Show the sampling distribution of $\bar{x}$ for male golfers.
    b. What is the probability that the sample mean is within 3 strokes of the population mean for the sample of male golfers?
    c. What is the probability that the sample mean is within 3 strokes of the population mean for the sample of female golfers?
    d. In which case, part (b) or part (c), is the probability of obtaining a sample mean within 3 strokes of the population mean higher? Why?

29. The mean preparation fee H&R Block charged retail customers last year was \$183 (*The Wall Street Journal,* March 7, 2012). Use this price as the population mean and assume the population standard deviation of preparation fees is \$50.
    a. What is the probability that the mean price for a sample of 30 H&R Block retail customers is within \$8 of the population mean?
    b. What is the probability that the mean price for a sample of 50 H&R Block retail customers is within \$8 of the population mean?
    c. What is the probability that the mean price for a sample of 100 H&R Block retail customers is within \$8 of the population mean?
    d. Which, if any, of the sample sizes in parts (a), (b), and (c) would you recommend to have at least a .95 probability that the sample mean is within \$8 of the population mean?

30. To estimate the mean age for a population of 4000 employees, a simple random sample of 40 employees is selected.
    a. Would you use the finite population correction factor in calculating the standard error of the mean? Explain.
    b. If the population standard deviation is $\sigma = 8.2$ years, compute the standard error both with and without the finite population correction factor. What is the rationale for ignoring the finite population correction factor whenever $n/N \leq .05$?
    c. What is the probability that the sample mean age of the employees will be within ± 2 years of the population mean age?

7.6 Sampling Distribution of $\bar{p}$

The sample proportion $\bar{p}$ is the point estimator of the population proportion p. The formula for computing the sample proportion is

$$\bar{p} = \frac{x}{n}$$

where

x = the number of elements in the sample that possess the characteristic of interest

n = sample size

As noted in Section 7.4, the sample proportion $\bar{p}$ is a random variable and its probability distribution is called the sampling distribution of $\bar{p}$.

SAMPLING DISTRIBUTION OF $\bar{p}$

The sampling distribution of $\bar{p}$ is the probability distribution of all possible values of the sample proportion $\bar{p}$.

To determine how close the sample proportion $\bar{p}$ is to the population proportion p, we need to understand the properties of the sampling distribution of $\bar{p}$: the expected value of $\bar{p}$, the standard deviation of $\bar{p}$, and the shape or form of the sampling distribution of $\bar{p}$.

Expected Value of $\bar{p}$

The expected value of $\bar{p}$, the mean of all possible values of $\bar{p}$, is equal to the population proportion p.

EXPECTED VALUE OF $\bar{p}$

$$E(\bar{p}) = p \tag{7.4}$$

where

$$E(\bar{p}) = \text{the expected value of } \bar{p}$$
$$p = \text{the population proportion}$$

Because $E(\bar{p}) = p$, $\bar{p}$ is an unbiased estimator of p. Recall from Section 7.1 we noted that $p = .60$ for the EAI population, where p is the proportion of the population of managers who participated in the company's management training program. Thus, the expected value of $\bar{p}$ for the EAI sampling problem is .60.

Standard Deviation of $\bar{p}$

Just as we found for the standard deviation of $\bar{x}$, the standard deviation of $\bar{p}$ depends on whether the population is finite or infinite. The two formulas for computing the standard deviation of $\bar{p}$ follow.

STANDARD DEVIATION OF $\bar{p}$

Finite Population *Infinite Population*

$$\sigma_{\bar{p}} = \sqrt{\frac{N-n}{N-1}}\sqrt{\frac{p(1-p)}{n}} \qquad \sigma_{\bar{p}} = \sqrt{\frac{p(1-p)}{n}} \tag{7.5}$$

Comparing the two formulas in (7.5), we see that the only difference is the use of the finite population correction factor $\sqrt{(N-n)/(N-1)}$.

As was the case with the sample mean $\bar{x}$, the difference between the expressions for the finite population and the infinite population becomes negligible if the size of the finite population is large in comparison to the sample size. We follow the same rule of thumb that we recommended for the sample mean. That is, if the population is finite with $n/N \le .05$, we will use $\sigma_{\bar{p}} = \sqrt{p(1-p)/n}$. However, if the population is finite with $n/N > .05$, the finite population correction factor should be used. Again, unless specifically noted, throughout the text we will assume that the population size is large in relation to the sample size and thus the finite population correction factor is unnecessary.

In Section 7.5 we used the term standard error of the mean to refer to the standard deviation of $\bar{x}$. We stated that in general the term standard error refers to the standard deviation of a point estimator. Thus, for proportions we use *standard error of the proportion* to refer to the standard deviation of $\bar{p}$. Let us now return to the EAI example and compute the standard error of the proportion associated with simple random samples of 30 EAI managers.

For the EAI study we know that the population proportion of managers who participated in the management training program is $p = .60$. With $n/N = 30/2500 = .012$, we can ignore the finite population correction factor when we compute the standard error of the proportion. For the simple random sample of 30 managers, $\sigma_{\bar{p}}$ is

$$\sigma_{\bar{p}} = \sqrt{\frac{p(1-p)}{n}} = \sqrt{\frac{.60(1-.60)}{30}} = .0894$$

Form of the Sampling Distribution of $\bar{p}$

Now that we know the mean and standard deviation of the sampling distribution of $\bar{p}$, the final step is to determine the form or shape of the sampling distribution. The sample proportion is $\bar{p} = x/n$. For a simple random sample from a large population, the value of x is a binomial random variable indicating the number of elements in the sample with the characteristic of interest. Because n is a constant, the probability of x/n is the same as the binomial probability of x, which means that the sampling distribution of $\bar{p}$ is also a discrete probability distribution and that the probability for each value of x/n is the same as the probability of x.

In Chapter 6 we also showed that a binomial distribution can be approximated by a normal distribution whenever the sample size is large enough to satisfy the following two conditions:

$$np \geq 5 \quad \text{and} \quad n(1-p) \geq 5$$

Assuming these two conditions are satisfied, the probability distribution of x in the sample proportion, $\bar{p} = x/n$, can be approximated by a normal distribution. And because n is a constant, the sampling distribution of $\bar{p}$ can also be approximated by a normal distribution. This approximation is stated as follows:

> The sampling distribution of $\bar{p}$ can be approximated by a normal distribution whenever $np \geq 5$ and $n(1-p) \geq 5$.

In practical applications, when an estimate of a population proportion is desired, we find that sample sizes are almost always large enough to permit the use of a normal approximation for the sampling distribution of $\bar{p}$.

Recall that for the EAI sampling problem we know that the population proportion of managers who participated in the training program is $p = .60$. With a simple random sample of size 30, we have $np = 30(.60) = 18$ and $n(1-p) = 30(.40) = 12$. Thus, the sampling distribution of $\bar{p}$ can be approximated by a normal distribution shown in Figure 7.8.

Practical Value of the Sampling Distribution of $\bar{p}$

The practical value of the sampling distribution of $\bar{p}$ is that it can be used to provide probability information about the difference between the sample proportion and the population proportion. For instance, suppose that in the EAI problem the personnel director wants to know the probability of obtaining a value of $\bar{p}$ that is within .05 of the population proportion of EAI managers who participated in the training program. That is, what is the probability of obtaining a sample with a sample proportion $\bar{p}$ between .55 and .65? The darkly shaded

FIGURE 7.8 SAMPLING DISTRIBUTION OF $\bar{p}$ FOR THE PROPORTION OF EAI MANAGERS WHO PARTICIPATED IN THE MANAGEMENT TRAINING PROGRAM

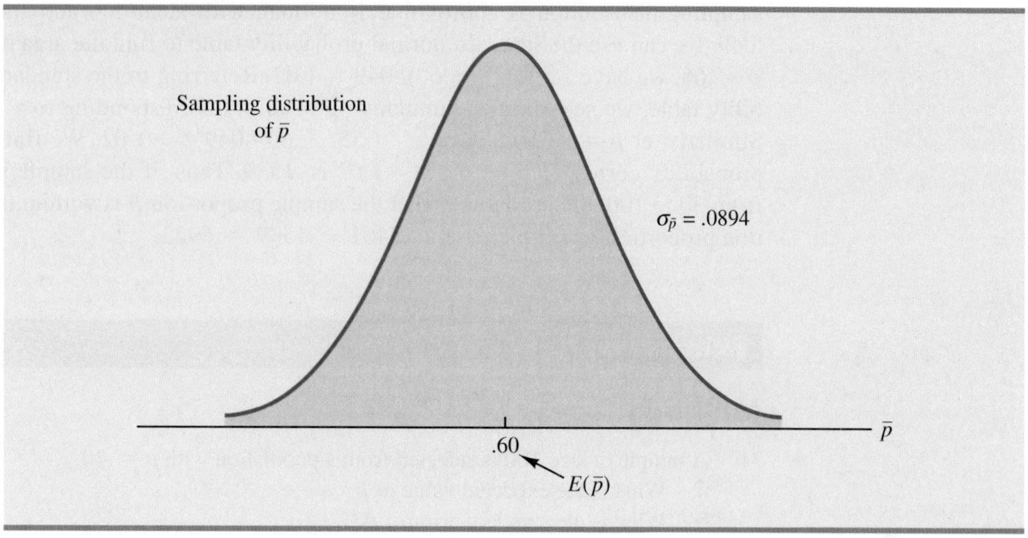

area in Figure 7.9 shows this probability. Using the fact that the sampling distribution of $\bar{p}$ can be approximated by a normal distribution with a mean of .60 and a standard error of the proportion of $\sigma_{\bar{p}} = .0894$, we find that the standard normal random variable corresponding to $\bar{p} = .65$ has a value of $z = (.65 - .60)/.0894 = .56$. Referring to the standard normal probability table, we see that the cumulative probability corresponding to $z = .56$ is .7123. Similarly, at $\bar{p} = .55$, we find $z = (.55 - .60)/.0894 = -.56$. From the standard normal probability table, we find the cumulative probability corresponding to $z = -.56$ is .2877. Thus, the probability of selecting a sample that provides a sample proportion $\bar{p}$ within .05 of the population proportion p is given by $.7123 - .2877 = .4246$.

If we consider increasing the sample size to $n = 100$, the standard error of the proportion becomes

$$\sigma_{\bar{p}} = \sqrt{\frac{.60(1 - .60)}{100}} = .049$$

FIGURE 7.9 PROBABILITY OF OBTAINING $\bar{p}$ BETWEEN .55 AND .65

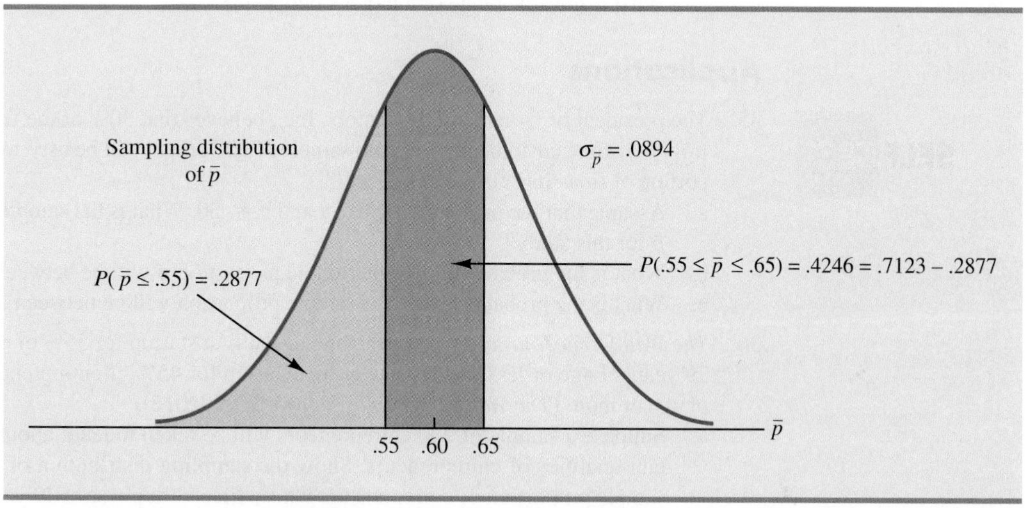

With a sample size of 100 EAI managers, the probability of the sample proportion having a value within .05 of the population proportion can now be computed. Because the sampling distribution is approximately normal, with mean .60 and standard deviation .049, we can use the standard normal probability table to find the area or probability. At $\bar{p} = .65$, we have $z = (.65 - .60)/.049 = 1.02$. Referring to the standard normal probability table, we see that the cumulative probability corresponding to $z = 1.02$ is .8461. Similarly, at $\bar{p} = .55$, we have $z = (.55 - .60)/.049 = -1.02$. We find the cumulative probability corresponding to $z = -1.02$ is .1539. Thus, if the sample size is increased from 30 to 100, the probability that the sample proportion $\bar{p}$ is within .05 of the population proportion p will increase to $.8461 - .1539 = .6922$.

Exercises

Methods

31. A sample of size 100 is selected from a population with $p = .40$.
    a. What is the expected value of $\bar{p}$?
    b. What is the standard error of $\bar{p}$?
    c. Show the sampling distribution of $\bar{p}$.
    d. What does the sampling distribution of $\bar{p}$ show?

SELF test

32. A population proportion is .40. A sample of size 200 will be taken and the sample proportion $\bar{p}$ will be used to estimate the population proportion.
    a. What is the probability that the sample proportion will be within ±.03 of the population proportion?
    b. What is the probability that the sample proportion will be within ±.05 of the population proportion?
33. Assume that the population proportion is .55. Compute the standard error of the proportion, $\sigma_{\bar{p}}$, for sample sizes of 100, 200, 500, and 1000. What can you say about the size of the standard error of the proportion as the sample size is increased?
34. The population proportion is .30. What is the probability that a sample proportion will be within ±.04 of the population proportion for each of the following sample sizes?
    a. $n = 100$
    b. $n = 200$
    c. $n = 500$
    d. $n = 1000$
    e. What is the advantage of a larger sample size?

Applications

SELF test

35. The president of Doerman Distributors, Inc., believes that 30% of the firm's orders come from first-time customers. A random sample of 100 orders will be used to estimate the proportion of first-time customers.
    a. Assume that the president is correct and $p = .30$. What is the sampling distribution of $\bar{p}$ for this study?
    b. What is the probability that the sample proportion $\bar{p}$ will be between .20 and .40?
    c. What is the probability that the sample proportion will be between .25 and .35?
36. *The Wall Street Journal* reported that the age at first startup for 55% of entrepreneurs was 29 years of age or less and the age at first startup for 45% of entrepreneurs was 30 years of age or more (*The Wall Street Journal,* March 19, 2012).
    a. Suppose a sample of 200 entrepreneurs will be taken to learn about the most important qualities of entrepreneurs. Show the sampling distribution of $\bar{p}$ where $\bar{p}$ is the sample proportion of entrepreneurs whose first startup was at 29 years of age or less.

b. What is the probability that the sample proportion in part (a) will be within $\pm.05$ of its population proportion?
c. Suppose a sample of 200 entrepreneurs will be taken to learn about the most important qualities of entrepreneurs. Show the sampling distribution of $\bar{p}$ where $\bar{p}$ is now the sample proportion of entrepreneurs whose first startup was at 30 years of age or more.
d. What is the probability that the sample proportion in part (c) will be within $\pm.05$ of its population proportion?
e Is the probability different in parts (b) and (d)? Why?
f. Answer part (b) for a sample of size 400. Is the probability smaller? Why?

37. People end up tossing 12% of what they buy at the grocery store (*Reader's Digest,* March 2009). Assume this is the true population proportion and that you plan to take a sample survey of 540 grocery shoppers to further investigate their behavior.
    a. Show the sampling distribution of $\bar{p}$, the proportion of groceries thrown out by your sample respondents.
    b. What is the probability that your survey will provide a sample proportion within $\pm.03$ of the population proportion?
    c. What is the probability that your survey will provide a sample proportion within $\pm.015$ of the population proportion?

38. Forty-two percent of primary care doctors think their patients receive unnecessary medical care (*Reader's Digest,* December 2011/January 2012).
    a. Suppose a sample of 300 primary care doctors were taken. Show the sampling distribution of the proportion of the doctors who think their patients receive unnecessary medical care.
    b. What is the probability that the sample proportion will be within $\pm.03$ of the population proportion?
    c. What is the probability that the sample proportion will be within $\pm.05$ of the population proportion?
    d. What would be the effect of taking a larger sample on the probabilities in parts (b) and (c)? Why?

39. In 2008 the Better Business Bureau settled 75% of complaints they received (*USA Today,* March 2, 2009). Suppose you have been hired by the Better Business Bureau to investigate the complaints they received this year involving new car dealers. You plan to select a sample of new car dealer complaints to estimate the proportion of complaints the Better Business Bureau is able to settle. Assume the population proportion of complaints settled for new car dealers is .75, the same as the overall proportion of complaints settled in 2008.
    a. Suppose you select a sample of 450 complaints involving new car dealers. Show the sampling distribution of $\bar{p}$.
    b. Based upon a sample of 450 complaints, what is the probability that the sample proportion will be within .04 of the population proportion?
    c. Suppose you select a sample of 200 complaints involving new car dealers. Show the sampling distribution of $\bar{p}$.
    d. Based upon the smaller sample of only 200 complaints, what is the probability that the sample proportion will be within .04 of the population proportion?
    e. As measured by the increase in probability, how much do you gain in precision by taking the larger sample in part (b)?

40. The Grocery Manufacturers of America reported that 76% of consumers read the ingredients listed on a product's label. Assume the population proportion is $p = .76$ and a sample of 400 consumers is selected from the population.
    a. Show the sampling distribution of the sample proportion $\bar{p}$ where $\bar{p}$ is the proportion of the sampled consumers who read the ingredients listed on a product's label.

b. What is the probability that the sample proportion will be within ±.03 of the population proportion?
c. Answer part (b) for a sample of 750 consumers.

41. The Food Marketing Institute shows that 17% of households spend more than \$100 per week on groceries. Assume the population proportion is $p = .17$ and a sample of 800 households will be selected from the population.
   a. Show the sampling distribution of $\bar{p}$, the sample proportion of households spending more than \$100 per week on groceries.
   b. What is the probability that the sample proportion will be within ±.02 of the population proportion?
   c. Answer part (b) for a sample of 1600 households.

7.7 Properties of Point Estimators

In this chapter we showed how sample statistics such as a sample mean $\bar{x}$, a sample standard deviation s, and a sample proportion $\bar{p}$ can be used as point estimators of their corresponding population parameters μ, σ, and p. It is intuitively appealing that each of these sample statistics is the point estimator of its corresponding population parameter. However, before using a sample statistic as a point estimator, statisticians check to see whether the sample statistic demonstrates certain properties associated with good point estimators. In this section we discuss three properties of good point estimators: unbiased, efficiency, and consistency.

Because several different sample statistics can be used as point estimators of different population parameters, we use the following general notation in this section.

$$\theta = \text{the population parameter of interest}$$
$$\hat{\theta} = \text{the sample statistic or point estimator of } \theta$$

The notation θ is the Greek letter theta, and the notation $\hat{\theta}$ is pronounced "theta-hat." In general, θ represents any population parameter such as a population mean, population standard deviation, population proportion, and so on; $\hat{\theta}$ represents the corresponding sample statistic such as the sample mean, sample standard deviation, and sample proportion.

Unbiased

If the expected value of the sample statistic is equal to the population parameter being estimated, the sample statistic is said to be an *unbiased estimator* of the population parameter.

UNBIASED

The sample statistic $\hat{\theta}$ is an unbiased estimator of the population parameter θ if

$$E(\hat{\theta}) = \theta$$

where

$$E(\hat{\theta}) = \text{the expected value of the sample statistic } \hat{\theta}$$

FIGURE 7.10 EXAMPLES OF UNBIASED AND BIASED POINT ESTIMATORS

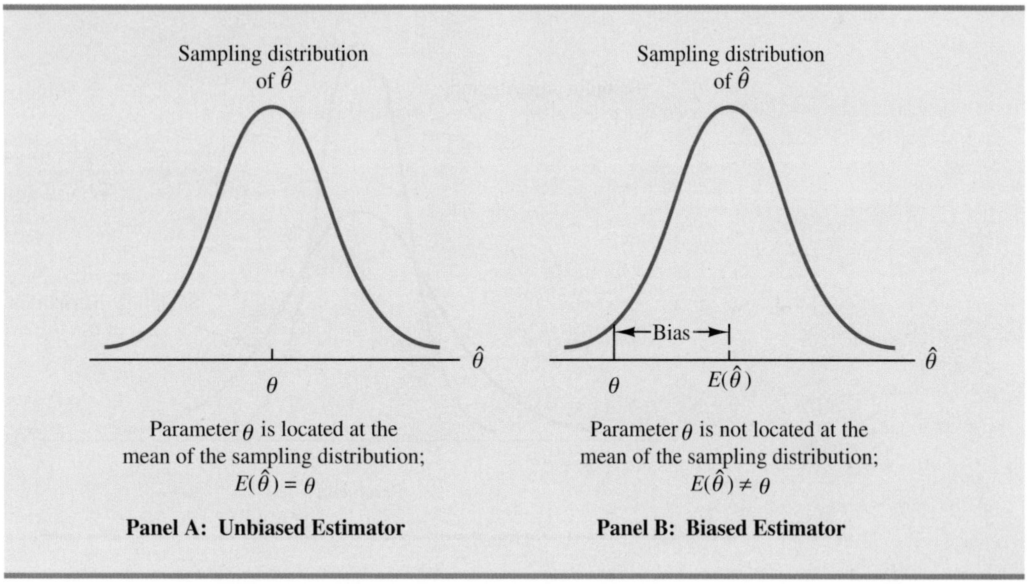

Hence, the expected value, or mean, of all possible values of an unbiased sample statistic is equal to the population parameter being estimated.

Figure 7.10 shows the cases of unbiased and biased point estimators. In the illustration showing the unbiased estimator, the mean of the sampling distribution is equal to the value of the population parameter. The estimation errors balance out in this case, because sometimes the value of the point estimator $\hat{\theta}$ may be less than θ and other times it may be greater than θ. In the case of a biased estimator, the mean of the sampling distribution is less than or greater than the value of the population parameter. In the illustration in Panel B of Figure 7.10, $E(\hat{\theta})$ is greater than θ; thus, the sample statistic has a high probability of overestimating the value of the population parameter. The amount of the bias is shown in the figure.

In discussing the sampling distributions of the sample mean and the sample proportion, we stated that $E(\bar{x}) = \mu$ and $E(\bar{p}) = p$. Thus, both $\bar{x}$ and $\bar{p}$ are unbiased estimators of their corresponding population parameters μ and p.

In the case of the sample standard deviation s and the sample variance s^2, it can be shown that $E(s^2) = \sigma^2$. Thus, we conclude that the sample variance s^2 is an unbiased estimator of the population variance σ^2. In fact, when we first presented the formulas for the sample variance and the sample standard deviation in Chapter 3, $n - 1$ rather than n was used in the denominator. The reason for using $n - 1$ rather than n is to make the sample variance an unbiased estimator of the population variance.

Efficiency

When sampling from a normal population, the standard error of the sample mean is less than the standard error of the sample median. Thus, the sample mean is more efficient than the sample median.

Assume that a simple random sample of n elements can be used to provide two unbiased point estimators of the same population parameter. In this situation, we would prefer to use the point estimator with the smaller standard error, because it tends to provide estimates closer to the population parameter. The point estimator with the smaller standard error is said to have greater **relative efficiency** than the other.

Figure 7.11 shows the sampling distributions of two unbiased point estimators, $\hat{\theta}_1$ and $\hat{\theta}_2$. Note that the standard error of $\hat{\theta}_1$ is less than the standard error of $\hat{\theta}_2$; thus,

FIGURE 7.11 SAMPLING DISTRIBUTIONS OF TWO UNBIASED POINT ESTIMATORS

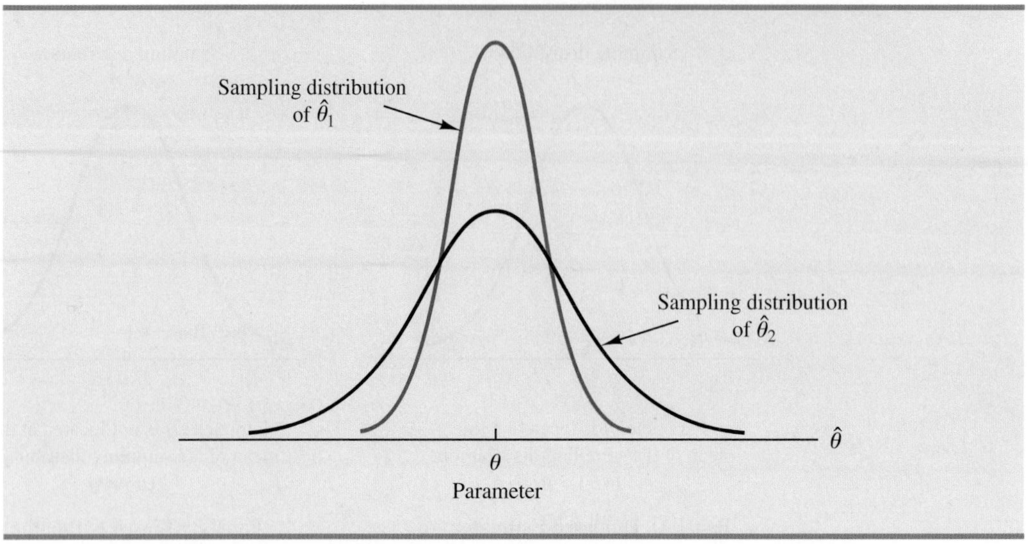

values of $\hat{\theta}_1$ have a greater chance of being close to the parameter θ than do values of $\hat{\theta}_2$. Because the standard error of point estimator $\hat{\theta}_1$ is less than the standard error of point estimator $\hat{\theta}_2$, $\hat{\theta}_1$ is relatively more efficient than $\hat{\theta}_2$ and is the preferred point estimator.

Consistency

A third property associated with good point estimators is **consistency**. Loosely speaking, a point estimator is consistent if the values of the point estimator tend to become closer to the population parameter as the sample size becomes larger. In other words, a large sample size tends to provide a better point estimate than a small sample size. Note that for the sample mean $\bar{x}$, we showed that the standard error of $\bar{x}$ is given by $\sigma_{\bar{x}} = \sigma/\sqrt{n}$. Because $\sigma_{\bar{x}}$ is related to the sample size such that larger sample sizes provide smaller values for $\sigma_{\bar{x}}$, we conclude that a larger sample size tends to provide point estimates closer to the population mean μ. In this sense, we can say that the sample mean $\bar{x}$ is a consistent estimator of the population mean μ. Using a similar rationale, we can also conclude that the sample proportion $\bar{p}$ is a consistent estimator of the population proportion p.

NOTES AND COMMENTS

In Chapter 3 we stated that the mean and the median are two measures of central location. In this chapter we discussed only the mean. The reason is that in sampling from a normal population, where the population mean and population median are identical, the standard error of the median is approximately 25% larger than the standard error of the mean. Recall that in the EAI problem where $n = 30$, the standard error of the mean is $\sigma_{\bar{x}} = 730.3$. The standard error of the median for this problem would be $1.25 \times (730.3) = 913$. As a result, the sample mean is more efficient and will have a higher probability of being within a specified distance of the population mean.

7.8 Other Sampling Methods

This section provides a brief introduction to survey sampling methods other than simple random sampling.

We described simple random sampling as a procedure for sampling from a finite population and discussed the properties of the sampling distributions of $\bar{x}$ and $\bar{p}$ when simple random sampling is used. Other methods such as stratified random sampling, cluster sampling, and systematic sampling provide advantages over simple random sampling in some of these situations. In this section we briefly introduce these alternative sampling methods. A more in-depth treatment is provided in Chapter 22, which is located on the website that accompanies the text.

Stratified Random Sampling

Stratified random sampling works best when the variance among elements in each stratum is relatively small.

In **stratified random sampling**, the elements in the population are first divided into groups called *strata,* such that each element in the population belongs to one and only one stratum. The basis for forming the strata, such as department, location, age, industry type, and so on, is at the discretion of the designer of the sample. However, the best results are obtained when the elements within each stratum are as much alike as possible. Figure 7.12 is a diagram of a population divided into H strata.

After the strata are formed, a simple random sample is taken from each stratum. Formulas are available for combining the results for the individual stratum samples into one estimate of the population parameter of interest. The value of stratified random sampling depends on how homogeneous the elements are within the strata. If elements within strata are alike, the strata will have low variances. Thus relatively small sample sizes can be used to obtain good estimates of the strata characteristics. If strata are homogeneous, the stratified random sampling procedure provides results just as precise as those of simple random sampling by using a smaller total sample size.

Cluster Sampling

Cluster sampling works best when each cluster provides a small-scale representation of the population.

In **cluster sampling**, the elements in the population are first divided into separate groups called *clusters*. Each element of the population belongs to one and only one cluster (see Figure 7.13). A simple random sample of the clusters is then taken. All elements within each sampled cluster form the sample. Cluster sampling tends to provide the best results when the elements within the clusters are not alike. In the ideal case, each cluster is a representative small-scale version of the entire population. The value of cluster sampling depends on how representative each cluster is of the entire population. If all clusters are alike in this regard, sampling a small number of clusters will provide good estimates of the population parameters.

FIGURE 7.12 DIAGRAM FOR STRATIFIED RANDOM SAMPLING

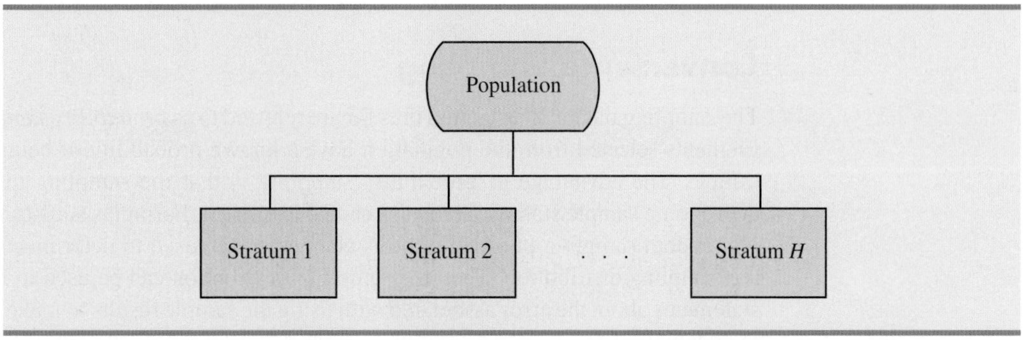

FIGURE 7.13 DIAGRAM FOR CLUSTER SAMPLING

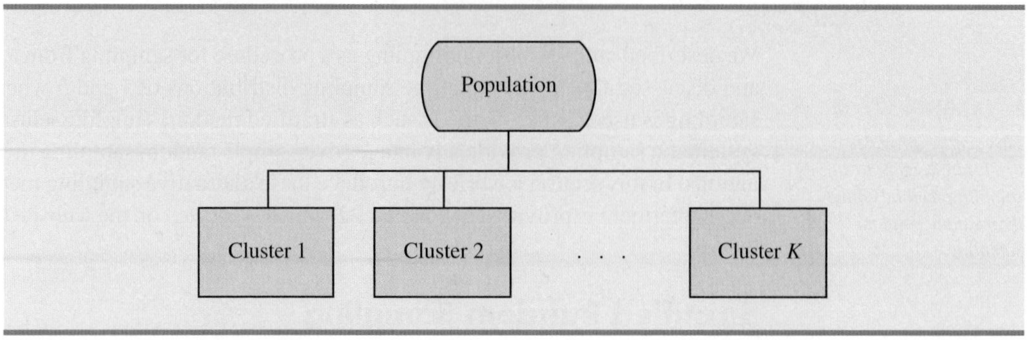

One of the primary applications of cluster sampling is area sampling, where clusters are city blocks or other well-defined areas. Cluster sampling generally requires a larger total sample size than either simple random sampling or stratified random sampling. However, it can result in cost savings because of the fact that when an interviewer is sent to a sampled cluster (e.g., a city-block location), many sample observations can be obtained in a relatively short time. Hence, a larger sample size may be obtainable with a significantly lower total cost.

Systematic Sampling

In some sampling situations, especially those with large populations, it is time-consuming to select a simple random sample by first finding a random number and then counting or searching through the list of the population until the corresponding element is found. An alternative to simple random sampling is **systematic sampling**. For example, if a sample size of 50 is desired from a population containing 5000 elements, we will sample one element for every 5000/50 = 100 elements in the population. A systematic sample for this case involves selecting randomly one of the first 100 elements from the population list. Other sample elements are identified by starting with the first sampled element and then selecting every 100th element that follows in the population list. In effect, the sample of 50 is identified by moving systematically through the population and identifying every 100th element after the first randomly selected element. The sample of 50 usually will be easier to identify in this way than it would be if simple random sampling were used. Because the first element selected is a random choice, a systematic sample is usually assumed to have the properties of a simple random sample. This assumption is especially applicable when the list of elements in the population is a random ordering of the elements.

Convenience Sampling

The sampling methods discussed thus far are referred to as *probability sampling* techniques. Elements selected from the population have a known probability of being included in the sample. The advantage of probability sampling is that the sampling distribution of the appropriate sample statistic generally can be identified. Formulas such as the ones for simple random sampling presented in this chapter can be used to determine the properties of the sampling distribution. Then the sampling distribution can be used to make probability statements about the error associated with using the sample results to make inferences about the population.

Convenience sampling is a *nonprobability sampling* technique. As the name implies, the sample is identified primarily by convenience. Elements are included in the sample without prespecified or known probabilities of being selected. For example, a professor conducting research at a university may use student volunteers to constitute a sample simply because they are readily available and will participate as subjects for little or no cost. Similarly, an inspector may sample a shipment of oranges by selecting oranges haphazardly from among several crates. Labeling each orange and using a probability method of sampling would be impractical. Samples such as wildlife captures and volunteer panels for consumer research are also convenience samples.

Convenience samples have the advantage of relatively easy sample selection and data collection; however, it is impossible to evaluate the "goodness" of the sample in terms of its representativeness of the population. A convenience sample may provide good results or it may not; no statistically justified procedure allows a probability analysis and inference about the quality of the sample results. Sometimes researchers apply statistical methods designed for probability samples to a convenience sample, arguing that the convenience sample can be treated as though it were a probability sample. However, this argument cannot be supported, and we should be cautious in interpreting the results of convenience samples that are used to make inferences about populations.

Judgment Sampling

One additional nonprobability sampling technique is **judgment sampling**. In this approach, the person most knowledgeable on the subject of the study selects elements of the population that he or she feels are most representative of the population. Often this method is a relatively easy way of selecting a sample. For example, a reporter may sample two or three senators, judging that those senators reflect the general opinion of all senators. However, the quality of the sample results depends on the judgment of the person selecting the sample. Again, great caution is warranted in drawing conclusions based on judgment samples used to make inferences about populations.

NOTES AND COMMENTS

We recommend using probability sampling methods when sampling from finite populations: simple random sampling, stratified random sampling, cluster sampling, or systematic sampling. For these methods, formulas are available for evaluating the "goodness" of the sample results in terms of the closeness of the results to the population parameters being estimated. An evaluation of the goodness cannot be made with convenience or judgment sampling. Thus, great care should be used in interpreting the results based on nonprobability sampling methods.

Summary

In this chapter we presented the concepts of sampling and sampling distributions. We demonstrated how a simple random sample can be selected from a finite population and how a random sample can be collected from an infinite population. The data collected from such samples can be used to develop point estimates of population parameters. Because different samples provide different values for the point estimators, point estimators such as $\bar{x}$ and $\bar{p}$ are random variables. The probability distribution of such a random variable is called a sampling distribution. In particular, we described the sampling distributions of the sample mean $\bar{x}$ and the sample proportion $\bar{p}$.

In considering the characteristics of the sampling distributions of $\bar{x}$ and $\bar{p}$, we stated that $E(\bar{x}) = \mu$ and $E(\bar{p}) = p$. Thus $\bar{x}$ and $\bar{p}$ are unbiased estimators. After developing the standard deviation or standard error formulas for these estimators, we described the conditions necessary for the sampling distributions of $\bar{x}$ and $\bar{p}$ to follow a normal distribution. Other sampling methods including stratified random sampling, cluster sampling, systematic sampling, convenience sampling, and judgment sampling were discussed.

Glossary

Sampled population The population from which the sample is taken.
Frame A listing of the elements the sample will be selected from.
Parameter A numerical characteristic of a population, such as a population mean μ, a population standard deviation σ, a population proportion p, and so on.
Simple random sample A simple random sample of size n from a finite population of size N is a sample selected such that each possible sample of size n has the same probability of being selected.
Sampling without replacement Once an element has been included in the sample, it is removed from the population and cannot be selected a second time.
Sampling with replacement Once an element has been included in the sample, it is returned to the population. A previously selected element can be selected again and therefore may appear in the sample more than once.
Random sample A random sample from an infinite population is a sample selected such that the following conditions are satisfied: (1) Each element selected comes from the same population; (2) each element is selected independently.
Sample statistic A sample characteristic, such as a sample mean $\bar{x}$, a sample standard deviation s, a sample proportion $\bar{p}$, and so on. The value of the sample statistic is used to estimate the value of the corresponding population parameter.
Point estimator The sample statistic, such as $\bar{x}$, s, or $\bar{p}$, that provides the point estimate of the population parameter.
Point estimate The value of a point estimator used in a particular instance as an estimate of a population parameter.
Target population The population for which statistical inferences such as point estimates are made. It is important for the target population to correspond as closely as possible to the sampled population.
Sampling distribution A probability distribution consisting of all possible values of a sample statistic.
Unbiased A property of a point estimator that is present when the expected value of the point estimator is equal to the population parameter it estimates.
Finite population correction factor The term $\sqrt{(N - n)/(N - 1)}$ that is used in the formulas for $\sigma_{\bar{x}}$ and $\sigma_{\bar{p}}$ whenever a finite population, rather than an infinite population, is being sampled. The generally accepted rule of thumb is to ignore the finite population correction factor whenever $n/N \leq .05$.
Standard error The standard deviation of a point estimator.
Central limit theorem A theorem that enables one to use the normal probability distribution to approximate the sampling distribution of $\bar{x}$ whenever the sample size is large.
Relative efficiency Given two unbiased point estimators of the same population parameter, the point estimator with the smaller standard error is more efficient.
Consistency A property of a point estimator that is present whenever larger sample sizes tend to provide point estimates closer to the population parameter.
Stratified random sampling A probability sampling method in which the population is first divided into strata and a simple random sample is then taken from each stratum.

Cluster sampling A probability sampling method in which the population is first divided into clusters and then a simple random sample of the clusters is taken.
Systematic sampling A probability sampling method in which we randomly select one of the first k elements and then select every kth element thereafter.
Convenience sampling A nonprobability method of sampling whereby elements are selected for the sample on the basis of convenience.
Judgment sampling A nonprobability method of sampling whereby elements are selected for the sample based on the judgment of the person doing the study.

Key Formulas

Expected Value of $\bar{x}$

$$E(\bar{x}) = \mu \tag{7.1}$$

Standard Deviation of $\bar{x}$ (Standard Error)

Finite Population *Infinite Population*

$$\sigma_{\bar{x}} = \sqrt{\frac{N-n}{N-1}}\left(\frac{\sigma}{\sqrt{n}}\right) \qquad \sigma_{\bar{x}} = \frac{\sigma}{\sqrt{n}} \tag{7.2}$$

Expected Value of $\bar{p}$

$$E(\bar{p}) = p \tag{7.4}$$

Standard Deviation of $\bar{p}$ (Standard Error)

Finite Population *Infinite Population*

$$\sigma_{\bar{p}} = \sqrt{\frac{N-n}{N-1}}\sqrt{\frac{p(1-p)}{n}} \qquad \sigma_{\bar{p}} = \sqrt{\frac{p(1-p)}{n}} \tag{7.5}$$

Supplementary Exercises

42. *U.S. News & World Report* publishes comprehensive information on America's best colleges (*America's Best Colleges,* 2009 ed.). Among other things, they provide a listing of their 133 best national universities. You would like to take a sample of these universities for a follow-up study on their students. Begin at the bottom of the third column of random digits in Table 7.1. Ignoring the first two digits in each five-number group and using the three-digit random numbers beginning with 959 read *up* the column to identify the number (from 1 to 133) of the first seven universities to be included in a simple random sample. Continue by starting at the bottom of the fourth and fifth columns and reading up if necessary.

43. The latest available data showed health expenditures were \$8086 per person in the United States or 17.6% of gross domestic product (Centers for Medicare & Medicaid Services website, April 1, 2012). Use \$8086 as the population mean and suppose a survey research firm will take a sample of 100 people to investigate the nature of their health expenditures. Assume the population standard deviation is \$2500.
    a. Show the sampling distribution of the mean amount of health care expenditures for a sample of 100 people.

b. What is the probability the sample mean will be within ± \$200 of the population mean?

c. What is the probability the sample mean will be greater than \$9000? If the survey research firm reports a sample mean greater than \$9000, would you question whether the firm followed correct sampling procedures? Why or why not?

44. Foot Locker uses sales per square foot as a measure of store productivity. Sales are currently running at an annual rate of \$406 per square foot (*The Wall Street Journal,* March 7, 2012). You have been asked by management to conduct a study of a sample of 64 Foot Locker stores. Assume the standard deviation in annual sales per square foot for the population of all 3400 Foot Locker stores is \$80.
    a. Show the sampling distribution of $\bar{x}$, the sample mean annual sales per square foot for a sample of 64 Foot Locker stores.
    b. What is the probability that the sample mean will be within \$15 of the population mean?
    c. Suppose you find a sample mean of \$380. What is the probability of finding a sample mean of \$380 or less? Would you consider such a sample to be an unusually low performing group of stores?

45. The mean television viewing time for Americans is 15 hours per week (*Money,* November 2003). Suppose a sample of 60 Americans is taken to further investigate viewing habits. Assume the population standard deviation for weekly viewing time is $\sigma = 4$ hours.
    a. What is the probability the sample mean will be within 1 hour of the population mean?
    b. What is the probability the sample mean will be within 45 minutes of the population mean?

46. After deducting grants based on need, the average cost to attend the University of Southern California (USC) is \$27,175 (*U.S. News & World Report, America's Best Colleges,* 2009 ed.). Assume the population standard deviation is \$7400. Suppose that a random sample of 60 USC students will be taken from this population.
    a. What is the value of the standard error of the mean?
    b. What is the probability that the sample mean will be more than \$27,175?
    c. What is the probability that the sample mean will be within \$1000 of the population mean?
    d. How would the probability in part (c) change if the sample size were increased to 100?

47. Three firms carry inventories that differ in size. Firm A's inventory contains 2000 items, firm B's inventory contains 5000 items, and firm C's inventory contains 10,000 items. The population standard deviation for the cost of the items in each firm's inventory is $\sigma = 144$. A statistical consultant recommends that each firm take a sample of 50 items from its inventory to provide statistically valid estimates of the average cost per item. Managers of the small firm state that because it has the smallest population, it should be able to make the estimate from a much smaller sample than that required by the larger firms. However, the consultant states that to obtain the same standard error and thus the same precision in the sample results, all firms should use the same sample size regardless of population size.
    a. Using the finite population correction factor, compute the standard error for each of the three firms given a sample of size 50.
    b. What is the probability that for each firm the sample mean $\bar{x}$ will be within ± 25 of the population mean μ?

48. A researcher reports survey results by stating that the standard error of the mean is 20. The population standard deviation is 500.
    a. How large was the sample used in this survey?
    b. What is the probability that the point estimate was within ± 25 of the population mean?

49. A production process is checked periodically by a quality control inspector. The inspector selects simple random samples of 30 finished products and computes the sample mean product weights $\bar{x}$. If test results over a long period of time show that 5% of the $\bar{x}$ values are over 2.1 pounds and 5% are under 1.9 pounds, what are the mean and the standard deviation for the population of products produced with this process?

50. About 28% of private companies are owned by women (*The Cincinnati Enquirer*, January 26, 2006). Answer the following questions based on a sample of 240 private companies.
    a. Show the sampling distribution of $\bar{p}$, the sample proportion of companies that are owned by women.
    b. What is the probability the sample proportion will be within ±.04 of the population proportion?
    c. What is the probability the sample proportion will be within ±.02 of the population proportion?

51. A market research firm conducts telephone surveys with a 40% historical response rate. What is the probability that in a new sample of 400 telephone numbers, at least 150 individuals will cooperate and respond to the questions? In other words, what is the probability that the sample proportion will be at least 150/400 = .375?

52. Advertisers contract with Internet service providers and search engines to place ads on websites. They pay a fee based on the number of potential customers who click on their ad. Unfortunately, click fraud—the practice of someone clicking on an ad solely for the purpose of driving up advertising revenue—has become a problem. Forty percent of advertisers claim they have been a victim of click fraud (*BusinessWeek,* March 13, 2006). Suppose a simple random sample of 380 advertisers will be taken to learn more about how they are affected by this practice.
    a. What is the probability that the sample proportion will be within ±.04 of the population proportion experiencing click fraud?
    b. What is the probability that the sample proportion will be greater than .45?

53. The proportion of individuals insured by the All-Driver Automobile Insurance Company who received at least one traffic ticket during a five-year period is .15.
    a. Show the sampling distribution of $\bar{p}$ if a random sample of 150 insured individuals is used to estimate the proportion having received at least one ticket.
    b. What is the probability that the sample proportion will be within ±.03 of the population proportion?

54. Lori Jeffrey is a successful sales representative for a major publisher of college textbooks. Historically, Lori obtains a book adoption on 25% of her sales calls. Viewing her sales calls for one month as a sample of all possible sales calls, assume that a statistical analysis of the data yields a standard error of the proportion of .0625.
    a. How large was the sample used in this analysis? That is, how many sales calls did Lori make during the month?
    b. Let $\bar{p}$ indicate the sample proportion of book adoptions obtained during the month. Show the sampling distribution of $\bar{p}$.
    c. Using the sampling distribution of $\bar{p}$, compute the probability that Lori will obtain book adoptions on 30% or more of her sales calls during a one-month period.

Appendix 7.1 The Expected Value and Standard Deviation of $\bar{x}$

In this appendix we present the mathematical basis for the expressions for $E(\bar{x})$, the expected value of $\bar{x}$ as given by equation (7.1), and $\sigma_{\bar{x}}$, the standard deviation of $\bar{x}$ as given by equation (7.2).

Expected Value of $\bar{x}$

Assume a population with mean μ and variance σ^2. A simple random sample of size n is selected with individual observations denoted $x_1, x_2, \ldots, x_n$. A sample mean $\bar{x}$ is computed as follows.

$$\bar{x} = \frac{\Sigma x_i}{n}$$

With repeated simple random samples of size n, $\bar{x}$ is a random variable that assumes different numerical values depending on the specific n items selected. The expected value of the random variable $\bar{x}$ is the mean of all possible $\bar{x}$ values.

$$\begin{aligned}\text{Mean of } \bar{x} = E(\bar{x}) &= E\left(\frac{\Sigma x_i}{n}\right)\\ &= \frac{1}{n}[E(x_1 + x_2 + \cdots + x_n)]\\ &= \frac{1}{n}[E(x_1) + E(x_2) + \cdots + E(x_n)]\end{aligned}$$

For any x_i we have $E(x_i) = \mu$; therefore we can write

$$\begin{aligned}E(\bar{x}) &= \frac{1}{n}(\mu + \mu + \cdots + \mu)\\ &= \frac{1}{n}(n\mu) = \mu\end{aligned}$$

This result shows that the mean of all possible $\bar{x}$ values is the same as the population mean μ. That is, $E(\bar{x}) = \mu$.

Standard Deviation of $\bar{x}$

Again assume a population with mean μ, variance σ^2, and a sample mean given by

$$\bar{x} = \frac{\Sigma x_i}{n}$$

With repeated simple random samples of size n, we know that $\bar{x}$ is a random variable that takes different numerical values depending on the specific n items selected. What follows is the derivation of the expression for the standard deviation of the $\bar{x}$ values, $\sigma_{\bar{x}}$, for the case of an infinite population. The derivation of the expression for $\sigma_{\bar{x}}$ for a finite population when sampling is done without replacement is more difficult and is beyond the scope of this text.

Returning to the infinite population case, recall that a simple random sample from an infinite population consists of observations $x_1, x_2, \ldots, x_n$ that are independent. The following two expressions are general formulas for the variance of random variables.

$$Var(ax) = a^2\, Var(x)$$

where a is a constant and x is a random variable, and

$$Var(x + y) = Var(x) + Var(y)$$

where x and y are *independent* random variables. Using the two preceding equations, we can develop the expression for the variance of the random variable $\bar{x}$ as follows.

$$Var(\bar{x}) = Var\left(\frac{\Sigma x_i}{n}\right) = Var\left(\frac{1}{n}\Sigma x_i\right)$$

Then, with $1/n$ a constant, we have

$$\begin{aligned} Var(\bar{x}) &= \left(\frac{1}{n}\right)^2 Var(\Sigma x_i) \\ &= \left(\frac{1}{n}\right)^2 Var(x_1 + x_2 + \cdots + x_n) \end{aligned}$$

In the infinite population case, the random variables $x_1, x_2, \ldots, x_n$ are independent, which enables us to write

$$Var(\bar{x}) = \left(\frac{1}{n}\right)^2 [Var(x_1) + Var(x_2) + \cdots + Var(x_n)]$$

For any x_i, we have $Var(x_i) = \sigma^2$; therefore we have

$$Var(\bar{x}) = \left(\frac{1}{n}\right)^2 (\sigma^2 + \sigma^2 + \cdots + \sigma^2)$$

With n values of σ^2 in this expression, we have

$$Var(\bar{x}) = \left(\frac{1}{n}\right)^2 (n\sigma^2) = \frac{\sigma^2}{n}$$

Taking the square root provides the formula for the standard deviation of $\bar{x}$.

$$\sigma_{\bar{x}} = \sqrt{Var(\bar{x})} = \frac{\sigma}{\sqrt{n}}$$

Appendix 7.2 Random Sampling with Minitab

If a list of the elements in a population is available in a Minitab file, Minitab can be used to select a simple random sample. For example, a list of the top 100 metropolitan areas in the United States and Canada is provided in column 1 of the data set MetAreas (*Places Rated Almanac—The Millennium Edition 2000*). Column 2 contains the overall rating of each metropolitan area. The first 10 metropolitan areas in the data set and their corresponding ratings are shown in Table 7.6.

Suppose that you would like to select a simple random sample of 30 metropolitan areas in order to do an in-depth study of the cost of living in the United States and Canada. The following steps can be used to select the sample.

Step 1. Select the **Calc** pull-down menu
Step 2. Choose **Random Data**
Step 3. Choose **Sample From Columns**

TABLE 7.6 OVERALL RATING FOR THE FIRST 10 METROPOLITAN AREAS IN THE DATA SET METAREAS

Metropolitan Area	Rating
Albany, NY	64.18
Albuquerque, NM	66.16
Appleton, WI	60.56
Atlanta, GA	69.97
Austin, TX	71.48
Baltimore, MD	69.75
Birmingham, AL	69.59
Boise City, ID	68.36
Boston, MA	68.99
Buffalo, NY	66.10

WEB file
MetAreas

Step 4. When the Sample From Columns dialog box appears:
Enter 30 in the **Number of rows to sample** box
Enter C1 C2 in the **From columns** box below
Enter C3 C4 in the **Store samples in** box
Step 5. Click **OK**

The random sample of 30 metropolitan areas appears in columns C3 and C4.

Appendix 7.3 Random Sampling with Excel

If a list of the elements in a population is available in an Excel file, Excel can be used to select a simple random sample. For example, a list of the top 100 metropolitan areas in the United States and Canada is provided in column A of the data set MetAreas (*Places Rated Almanac—The Millennium Edition 2000*). Column B contains the overall rating of each metropolitan area. The first 10 metropolitan areas in the data set and their corresponding ratings are shown in Table 7.6. Assume that you would like to select a simple random sample of 30 metropolitan areas in order to do an in-depth study of the cost of living in the United States and Canada.

The rows of any Excel data set can be placed in a random order by adding an extra column to the data set and filling the column with random numbers using the =RAND() function. Then, using Excel's sort ascending capability on the random number column, the rows of the data set will be reordered randomly. The random sample of size n appears in the first n rows of the reordered data set.

In the MetAreas data set, labels are in row 1 and the 100 metropolitan areas are in rows 2 to 101. The following steps can be used to select a simple random sample of 30 metropolitan areas.

Step 1. Enter =RAND() in cell C2
Step 2. Copy cell C2 to cells C3:C101
Step 3. Select any cell in Column C
Step 4. Click the **Home** tab on the Ribbon
Step 5. In the **Editing** group, click **Sort & Filter**
Step 6. Click **Sort Smallest to Largest**

The random sample of 30 metropolitan areas appears in rows 2 to 31 of the reordered data set. The random numbers in column C are no longer necessary and can be deleted if desired.

Appendix 7.4 Random Sampling with StatTools

WEB file
MetAreas

If a list of the elements in a population is available in an Excel file, StatTools Random Sample Utility can be used to select a simple random sample. For example, a list of the top 100 metropolitan areas in the United States and Canada is provided in column A of the data set MetAreas (*Places Rated Almanac—The Millennium Edition 2000*). Column B contains the overall rating of each metropolitan area. Assume that you would like to select a simple random sample of 30 metropolitan areas in order to do an in-depth study of the cost of living in the United States and Canada.

Begin by using the Data Set Manager to create a StatTools data set for these data using the procedure described in the appendix to Chapter 1. The following steps will generate a simple random sample of 30 metropolitan areas.

Step 1. Click the **StatTools** tab on the Ribbon
Step 2. In the **Data Group** click **Data Utilities**
Step 3. Choose the **Random Sample** option
Step 4. When the StatTools—Random Sample Utility dialog box appears:
- In the **Variables** section:
  - Select **Metropolitan Area**
  - Select **Rating**
- In the **Options** section:
  - Enter 1 in the **Number of Samples** box
  - Enter 30 in the **Sample Size** box
- Click **OK**

The random sample of 30 metropolitan areas will appear in columns A and B of the worksheet entitled Random Sample.

CHAPTER 8

Interval Estimation

CONTENTS

STATISTICS *in* PRACTICE

FOOD LION*
SALISBURY, NORTH CAROLINA

Founded in 1957 as Food Town, Food Lion is one of the largest supermarket chains in the United States, with 1300 stores in 11 Southeastern and Mid-Atlantic states. The company sells more than 24,000 different products and offers nationally and regionally advertised brand-name merchandise, as well as a growing number of high-quality private label products manufactured especially for Food Lion. The company maintains its low price leadership and quality assurance through operating efficiencies such as standard store formats, innovative warehouse design, energy-efficient facilities, and data synchronization with suppliers. Food Lion looks to a future of continued innovation, growth, price leadership, and service to its customers.

Being in an inventory-intense business, Food Lion made the decision to adopt the LIFO (last-in, first-out) method of inventory valuation. This method matches current costs against current revenues, which minimizes the effect of radical price changes on profit and loss results. In addition, the LIFO method reduces net income thereby reducing income taxes during periods of inflation.

Food Lion establishes a LIFO index for each of seven inventory pools: Grocery, Paper/Household, Pet Supplies, Health & Beauty Aids, Dairy, Cigarette/Tobacco, and Beer/Wine. For example, a LIFO index of 1.008 for the Grocery pool would indicate that the company's grocery inventory value at current costs reflects a 0.8% increase due to inflation over the most recent one-year period.

A LIFO index for each inventory pool requires that the year-end inventory count for each product be valued at the current year-end cost and at the preceding year-end cost. To avoid excessive time and expense associated with counting the inventory in all 1200 store locations, Food Lion selects a random sample of 50 stores. Year-end physical inventories are taken in each of the sample stores. The current-year and preceding-year costs for each item are then used to construct the required LIFO indexes for each inventory pool.

For a recent year, the sample estimate of the LIFO index for the Health & Beauty Aids inventory pool was 1.015. Using a 95% confidence level, Food Lion computed a margin of error of .006 for the sample estimate. Thus, the interval from 1.009 to 1.021 provided a 95% confidence interval estimate of the population LIFO index. This level of precision was judged to be very good.

In this chapter you will learn how to compute the margin of error associated with sample estimates. You will also learn how to use this information to construct and interpret interval estimates of a population mean and a population proportion.

As an inventory-intense business, Food Lion adopted the LIFO method of inventory valuation. © Bloomberg/Getty Images.

*The authors are indebted to Keith Cunningham, Tax Director, and Bobby Harkey, Staff Tax Accountant, at Food Lion for providing this Statistics in Practice.

In Chapter 7, we stated that a point estimator is a sample statistic used to estimate a population parameter. For instance, the sample mean $\bar{x}$ is a point estimator of the population mean μ and the sample proportion $\bar{p}$ is a point estimator of the population proportion p. Because a point estimator cannot be expected to provide the exact value of the population parameter, an **interval estimate** is often computed by adding and subtracting a value, called the **margin of error**, to the point estimate. The general form of an interval estimate is as follows:

$$\text{Point estimate} \pm \text{Margin of error}$$

The purpose of an interval estimate is to provide information about how close the point estimate, provided by the sample, is to the value of the population parameter.

In this chapter we show how to compute interval estimates of a population mean μ and a population proportion p. The general form of an interval estimate of a population mean is

$$\bar{x} \pm \text{Margin of error}$$

Similarly, the general form of an interval estimate of a population proportion is

$$\bar{p} \pm \text{Margin of error}$$

The sampling distributions of $\bar{x}$ and $\bar{p}$ play key roles in computing these interval estimates.

8.1 Population Mean: σ Known

In order to develop an interval estimate of a population mean, either the population standard deviation σ or the sample standard deviation s must be used to compute the margin of error. In most applications σ is not known, and s is used to compute the margin of error. In some applications, large amounts of relevant historical data are available and can be used to estimate the population standard deviation prior to sampling. Also, in quality control applications where a process is assumed to be operating correctly, or "in control," it is appropriate to treat the population standard deviation as known. We refer to such cases as the **σ known** case. In this section we introduce an example in which it is reasonable to treat σ as known and show how to construct an interval estimate for this case.

Each week Lloyd's Department Store selects a simple random sample of 100 customers in order to learn about the amount spent per shopping trip. With x representing the amount spent per shopping trip, the sample mean $\bar{x}$ provides a point estimate of μ, the mean amount spent per shopping trip for the population of all Lloyd's customers. Lloyd's has been using the weekly survey for several years. Based on the historical data, Lloyd's now assumes a known value of $\sigma = \$20$ for the population standard deviation. The historical data also indicate that the population follows a normal distribution.

WEB file Lloyd's

During the most recent week, Lloyd's surveyed 100 customers ($n = 100$) and obtained a sample mean of $\bar{x} = \$82$. The sample mean amount spent provides a point estimate of the population mean amount spent per shopping trip, μ. In the discussion that follows, we show how to compute the margin of error for this estimate and develop an interval estimate of the population mean.

Margin of Error and the Interval Estimate

In Chapter 7 we showed that the sampling distribution of $\bar{x}$ can be used to compute the probability that $\bar{x}$ will be within a given distance of μ. In the Lloyd's example, the historical data show that the population of amounts spent is normally distributed with a standard deviation of $\sigma = 20$. So, using what we learned in Chapter 7, we can conclude that the sampling distribution of $\bar{x}$ follows a normal distribution with a standard error of $\sigma_{\bar{x}} = \sigma/\sqrt{n} = 20/\sqrt{100} = 2$. This sampling distribution is shown in Figure 8.1.[1] Because

[1]We use the fact that the population of amounts spent has a normal distribution to conclude that the sampling distribution of $\bar{x}$ has a normal distribution. If the population did not have a normal distribution, we could rely on the central limit theorem and the sample size of $n = 100$ to conclude that the sampling distribution of $\bar{x}$ is approximately normal. In either case, the sampling distribution of $\bar{x}$ would appear as shown in Figure 8.1.

FIGURE 8.1 SAMPLING DISTRIBUTION OF THE SAMPLE MEAN AMOUNT SPENT FROM SIMPLE RANDOM SAMPLES OF 100 CUSTOMERS

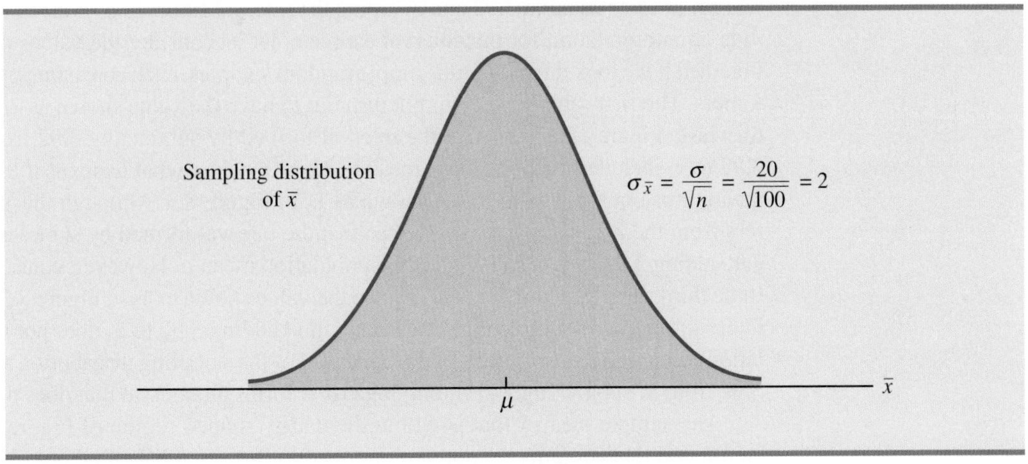

the sampling distribution shows how values of $\bar{x}$ are distributed around the population mean μ, the sampling distribution of $\bar{x}$ provides information about the possible differences between $\bar{x}$ and μ.

Using the standard normal probability table, we find that 95% of the values of any normally distributed random variable are within ± 1.96 standard deviations of the mean. Thus, when the sampling distribution of $\bar{x}$ is normally distributed, 95% of the $\bar{x}$ values must be within $\pm 1.96\sigma_{\bar{x}}$ of the mean μ. In the Lloyd's example we know that the sampling distribution of $\bar{x}$ is normally distributed with a standard error of $\sigma_{\bar{x}} = 2$. Because $\pm 1.96\sigma_{\bar{x}} = 1.96(2) = 3.92$, we can conclude that 95% of all $\bar{x}$ values obtained using a sample size of $n = 100$ will be within ± 3.92 of the population mean μ. See Figure 8.2.

FIGURE 8.2 SAMPLING DISTRIBUTION OF $\bar{x}$ SHOWING THE LOCATION OF SAMPLE MEANS THAT ARE WITHIN 3.92 OF μ

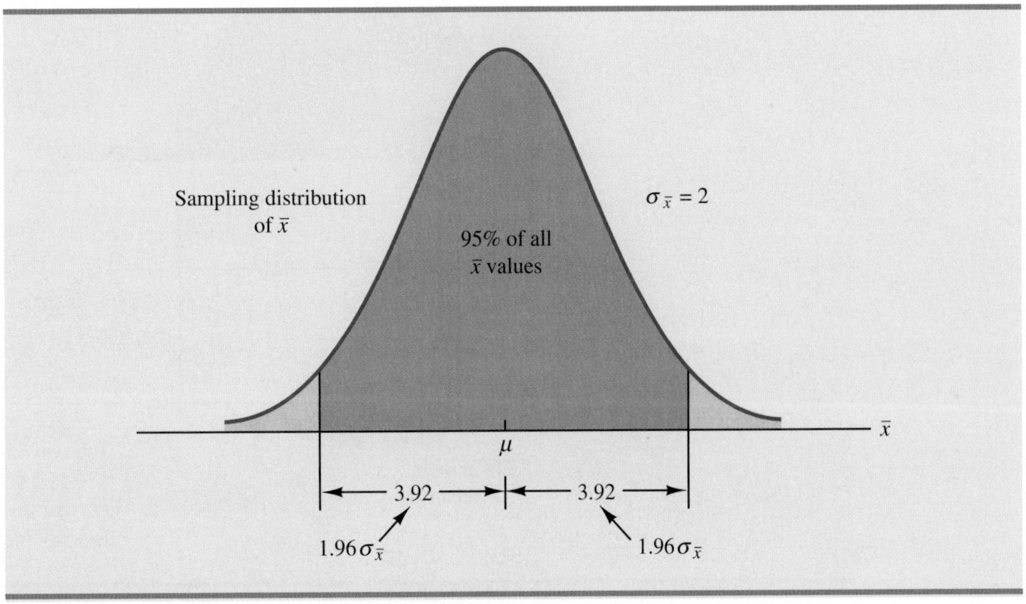

In the introduction to this chapter we said that the general form of an interval estimate of the population mean μ is $\bar{x} \pm$ margin of error. For the Lloyd's example, suppose we set the margin of error equal to 3.92 and compute the interval estimate of μ using $\bar{x} \pm 3.92$. To provide an interpretation for this interval estimate, let us consider the values of $\bar{x}$ that could be obtained if we took three *different* simple random samples, each consisting of 100 Lloyd's customers. The first sample mean might turn out to have the value shown as $\bar{x}_1$ in Figure 8.3. In this case, Figure 8.3 shows that the interval formed by subtracting 3.92 from $\bar{x}_1$ and adding 3.92 to $\bar{x}_1$ includes the population mean μ. Now consider what happens if the second sample mean turns out to have the value shown as $\bar{x}_2$ in Figure 8.3. Although this sample mean differs from the first sample mean, we see that the interval formed by subtracting 3.92 from $\bar{x}_2$ and adding 3.92 to $\bar{x}_2$ also includes the population mean μ. However, consider what happens if the third sample mean turns out to have the value shown as $\bar{x}_3$ in Figure 8.3. In this case, the interval formed by subtracting 3.92 from $\bar{x}_3$ and adding 3.92 to $\bar{x}_3$ does not include the population mean μ. Because $\bar{x}_3$ falls in the upper tail of the sampling distribution and is farther than 3.92 from μ, subtracting and adding 3.92 to $\bar{x}_3$ forms an interval that does not include μ.

Any sample mean $\bar{x}$ that is within the darkly shaded region of Figure 8.3 will provide an interval that contains the population mean μ. Because 95% of all possible sample means are in the darkly shaded region, 95% of all intervals formed by subtracting 3.92 from $\bar{x}$ and adding 3.92 to $\bar{x}$ will include the population mean μ.

Recall that during the most recent week, the quality assurance team at Lloyd's surveyed 100 customers and obtained a sample mean amount spent of $\bar{x} = 82$. Using $\bar{x} \pm 3.92$ to

FIGURE 8.3 INTERVALS FORMED FROM SELECTED SAMPLE MEANS AT LOCATIONS $\bar{x}_1$, $\bar{x}_2$, AND $\bar{x}_3$

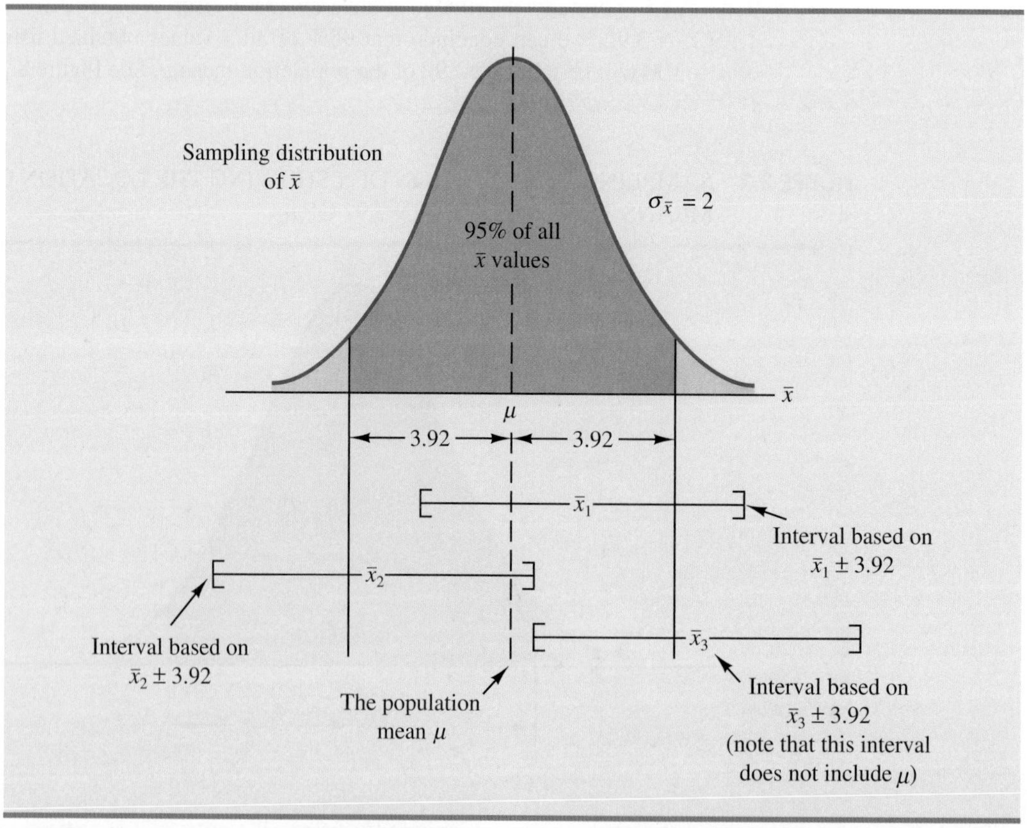

construct the interval estimate, we obtain 82 ± 3.92. Thus, the specific interval estimate of μ based on the data from the most recent week is $82 - 3.92 = 78.08$ to $82 + 3.92 = 85.92$. Because 95% of all the intervals constructed using $\bar{x} \pm 3.92$ will contain the population mean, we say that we are 95% confident that the interval 78.08 to 85.92 includes the population mean μ. We say that this interval has been established at the 95% **confidence level**. The value .95 is referred to as the **confidence coefficient**, and the interval 78.08 to 85.92 is called the 95% **confidence interval**.

This discussion provides insight as to why the interval is called a 95% confidence interval.

With the margin of error given by $z_{\alpha/2}(\sigma/\sqrt{n})$, the general form of an interval estimate of a population mean for the σ known case follows.

INTERVAL ESTIMATE OF A POPULATION MEAN: σ KNOWN

$$\bar{x} \pm z_{\alpha/2}\frac{\sigma}{\sqrt{n}} \tag{8.1}$$

where $(1 - \alpha)$ is the confidence coefficient and $z_{\alpha/2}$ is the z value providing an area of $\alpha/2$ in the upper tail of the standard normal probability distribution.

Let us use expression (8.1) to construct a 95% confidence interval for the Lloyd's example. For a 95% confidence interval, the confidence coefficient is $(1 - \alpha) = .95$ and thus, $\alpha = .05$. Using the standard normal probability table, an area of $\alpha/2 = .05/2 = .025$ in the upper tail provides $z_{.025} = 1.96$. With the Lloyd's sample mean $\bar{x} = 82$, $\sigma = 20$, and a sample size $n = 100$, we obtain

$$82 \pm 1.96\frac{20}{\sqrt{100}}$$

$$82 \pm 3.92$$

Thus, using expression (8.1), the margin of error is 3.92 and the 95% confidence interval is $82 - 3.92 = 78.08$ to $82 + 3.92 = 85.92$.

Although a 95% confidence level is frequently used, other confidence levels such as 90% and 99% may be considered. Values of $z_{\alpha/2}$ for the most commonly used confidence levels are shown in Table 8.1. Using these values and expression (8.1), the 90% confidence interval for the Lloyd's example is

$$82 \pm 1.645\frac{20}{\sqrt{100}}$$

$$82 \pm 3.29$$

TABLE 8.1 VALUES OF $z_{\alpha/2}$ FOR THE MOST COMMONLY USED CONFIDENCE LEVELS

Confidence Level	α	$\alpha/2$	$z_{\alpha/2}$
90%	.10	.05	1.645
95%	.05	.025	1.960
99%	.01	.005	2.576

Thus, at 90% confidence, the margin of error is 3.29 and the confidence interval is 82 − 3.29 = 78.71 to 82 + 3.29 = 85.29. Similarly, the 99% confidence interval is

$$82 \pm 2.576 \frac{20}{\sqrt{100}}$$

$$82 \pm 5.15$$

Thus, at 99% confidence, the margin of error is 5.15 and the confidence interval is 82 − 5.15 = 76.85 to 82 + 5.15 = 87.15.

Comparing the results for the 90%, 95%, and 99% confidence levels, we see that in order to have a higher degree of confidence, the margin of error and thus the width of the confidence interval must be larger.

Practical Advice

If the population follows a normal distribution, the confidence interval provided by expression (8.1) is exact. In other words, if expression (8.1) were used repeatedly to generate 95% confidence intervals, exactly 95% of the intervals generated would contain the population mean. If the population does not follow a normal distribution, the confidence interval provided by expression (8.1) will be approximate. In this case, the quality of the approximation depends on both the distribution of the population and the sample size.

In most applications, a sample size of $n \geq 30$ is adequate when using expression (8.1) to develop an interval estimate of a population mean. If the population is not normally distributed but is roughly symmetric, sample sizes as small as 15 can be expected to provide good approximate confidence intervals. With smaller sample sizes, expression (8.1) should only be used if the analyst believes, or is willing to assume, that the population distribution is at least approximately normal.

NOTES AND COMMENTS

1. The interval estimation procedure discussed in this section is based on the assumption that the population standard deviation σ is known. By σ known we mean that historical data or other information are available that permit us to obtain a good estimate of the population standard deviation prior to taking the sample that will be used to develop an estimate of the population mean. So technically we don't mean that σ is actually known with certainty. We just mean that we obtained a good estimate of the standard deviation prior to sampling and thus we won't be using the same sample to estimate both the population mean and the population standard deviation.
2. The sample size n appears in the denominator of the interval estimation expression (8.1). Thus, if a particular sample size provides too wide an interval to be of any practical use, we may want to consider increasing the sample size. With n in the denominator, a larger sample size will provide a smaller margin of error, a narrower interval, and greater precision. The procedure for determining the size of a simple random sample necessary to obtain a desired precision is discussed in Section 8.3.

Exercises

Methods

1. A simple random sample of 40 items resulted in a sample mean of 25. The population standard deviation is $\sigma = 5$.
   a. What is the standard error of the mean, $\sigma_{\bar{x}}$?
   b. At 95% confidence, what is the margin of error?

SELF test

2. A simple random sample of 50 items from a population with $\sigma = 6$ resulted in a sample mean of 32.
   a. Provide a 90% confidence interval for the population mean.
   b. Provide a 95% confidence interval for the population mean.
   c. Provide a 99% confidence interval for the population mean.
3. A simple random sample of 60 items resulted in a sample mean of 80. The population standard deviation is $\sigma = 15$.
   a. Compute the 95% confidence interval for the population mean.
   b. Assume that the same sample mean was obtained from a sample of 120 items. Provide a 95% confidence interval for the population mean.
   c. What is the effect of a larger sample size on the interval estimate?
4. A 95% confidence interval for a population mean was reported to be 152 to 160. If $\sigma = 15$, what sample size was used in this study?

Applications

SELF test

WEB file Houston

5. Data were collected on the amount spent by 64 customers for lunch at a major Houston restaurant. These data are contained in the file named Houston. Based upon past studies the population standard deviation is known with $\sigma = \$6$.
   a. At 99% confidence, what is the margin of error?
   b. Develop a 99% confidence interval estimate of the mean amount spent for lunch.

WEB file Nielsen

6. Nielsen Media Research conducted a study of household television viewing times during the 8 P.M. to 11 P.M. time period. The data contained in the file named Nielsen are consistent with the findings reported (*The World Almanac,* 2003). Based upon past studies the population standard deviation is assumed known with $\sigma = 3.5$ hours. Develop a 95% confidence interval estimate of the mean television viewing time per week during the 8 P.M. to 11 P.M. time period.
7. *The Wall Street Journal* reported that automobile crashes cost the United States \$162 billion annually (*The Wall Street Journal,* March 5, 2008). The average cost per person for crashes in the Tampa, Florida, area was reported to be \$1599. Suppose this average cost was based on a sample of 50 persons who had been involved in car crashes and that the population standard deviation is $\sigma = \$600$. What is the margin of error for a 95% confidence interval? What would you recommend if the study required a margin of error of \$150 or less?
8. Studies show that massage therapy has a variety of health benefits and it is not too expensive (*The Wall Street Journal,* March 13, 2012). A sample of 10 typical one-hour massage therapy sessions showed an average charge of \$59. The population standard deviation for a one-hour session is $\sigma = \$5.50$.
   a. What assumptions about the population should we be willing to make if a margin of error is desired?
   b. Using 95% confidence, what is the margin of error?
   c. Using 99% confidence, what is the margin of error?

WEB file TaxReturn

9. AARP reported on a study conducted to learn how long it takes individuals to prepare their federal income tax return (*AARP Bulletin,* April 2008). The data contained in the file named TaxReturn are consistent with the study results. These data provide the time in hours required for 40 individuals to complete their federal income tax returns. Using past years' data, the population standard deviation can be assumed known with $\sigma = 9$ hours. What is the 95% confidence interval estimate of the mean time it takes an individual to complete a federal income tax return?
10. *Playbill* magazine reported that the mean annual household income of its readers is \$119,155 (*Playbill,* January 2006). Assume this estimate of the mean annual household income is based on a sample of 80 households, and based on past studies, the population standard deviation is known to be $\sigma = \$30,000$.

a. Develop a 90% confidence interval estimate of the population mean.
b. Develop a 95% confidence interval estimate of the population mean.
c. Develop a 99% confidence interval estimate of the population mean.
d. Discuss what happens to the width of the confidence interval as the confidence level is increased. Does this result seem reasonable? Explain.

8.2 Population Mean: σ Unknown

When developing an interval estimate of a population mean we usually do not have a good estimate of the population standard deviation either. In these cases, we must use the same sample to estimate both μ and σ. This situation represents the **σ unknown** case. When s is used to estimate σ, the margin of error and the interval estimate for the population mean are based on a probability distribution known as the ***t* distribution**. Although the mathematical development of the t distribution is based on the assumption of a normal distribution for the population we are sampling from, research shows that the t distribution can be successfully applied in many situations where the population deviates significantly from normal. Later in this section we provide guidelines for using the t distribution if the population is not normally distributed.

William Sealy Gosset, writing under the name "Student," is the founder of the t distribution. Gosset, an Oxford graduate in mathematics, worked for the Guinness Brewery in Dublin, Ireland. He developed the t distribution while working on small-scale materials and temperature experiments.

The t distribution is a family of similar probability distributions, with a specific t distribution depending on a parameter known as the **degrees of freedom**. The t distribution with one degree of freedom is unique, as is the t distribution with two degrees of freedom, with three degrees of freedom, and so on. As the number of degrees of freedom increases, the difference between the t distribution and the standard normal distribution becomes smaller and smaller. Figure 8.4 shows t distributions with 10 and 20 degrees of freedom and their relationship to the standard normal probability distribution. Note that a t distribution with more degrees of freedom exhibits less variability and more

FIGURE 8.4 COMPARISON OF THE STANDARD NORMAL DISTRIBUTION WITH t DISTRIBUTIONS HAVING 10 AND 20 DEGREES OF FREEDOM

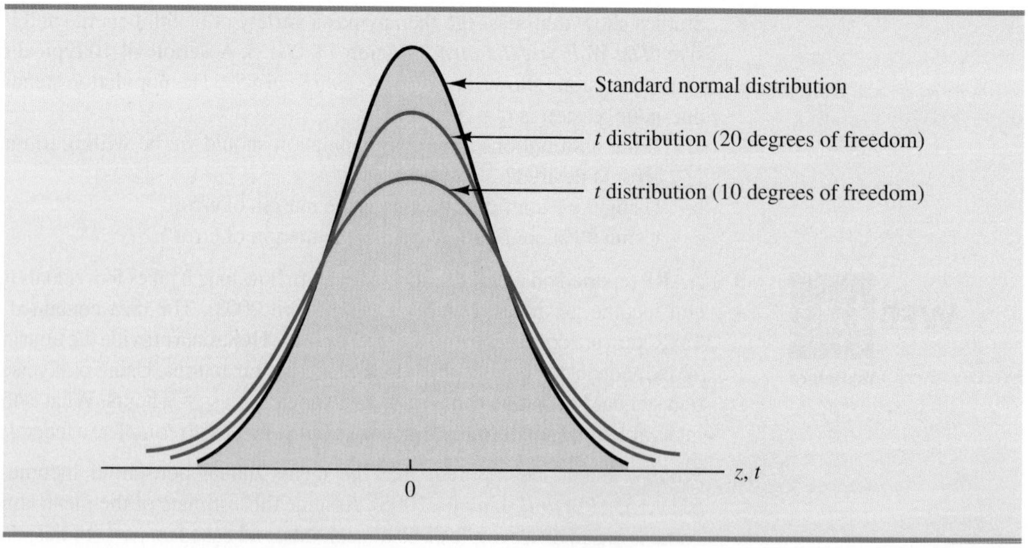

closely resembles the standard normal distribution. Note also that the mean of the t distribution is zero.

We place a subscript on t to indicate the area in the upper tail of the t distribution. For example, just as we used $z_{.025}$ to indicate the z value providing a .025 area in the upper tail of a standard normal distribution, we will use $t_{.025}$ to indicate a .025 area in the upper tail of a t distribution. In general, we will use the notation $t_{\alpha/2}$ to represent a t value with an area of $\alpha/2$ in the upper tail of the t distribution. See Figure 8.5.

As the degrees of freedom increase, the t distribution approaches the standard normal distribution.

Table 2 in Appendix B contains a table for the t distribution. A portion of this table is shown in Table 8.2. Each row in the table corresponds to a separate t distribution with the degrees of freedom shown. For example, for a t distribution with 9 degrees of freedom, $t_{.025} = 2.262$. Similarly, for a t distribution with 60 degrees of freedom, $t_{.025} = 2.000$. As the degrees of freedom continue to increase, $t_{.025}$ approaches $z_{.025} = 1.96$. In fact, the standard normal distribution z values can be found in the infinite degrees of freedom row (labeled ∞) of the t distribution table. If the degrees of freedom exceed 100, the infinite degrees of freedom row can be used to approximate the actual t value; in other words, for more than 100 degrees of freedom, the standard normal z value provides a good approximation to the t value.

Margin of Error and the Interval Estimate

In Section 8.1 we showed that an interval estimate of a population mean for the σ known case is

$$\bar{x} \pm z_{\alpha/2}\frac{\sigma}{\sqrt{n}}$$

To compute an interval estimate of μ for the σ unknown case, the sample standard deviation s is used to estimate σ, and $z_{\alpha/2}$ is replaced by the t distribution value $t_{\alpha/2}$. The margin

FIGURE 8.5 t DISTRIBUTION WITH $\alpha/2$ AREA OR PROBABILITY IN THE UPPER TAIL

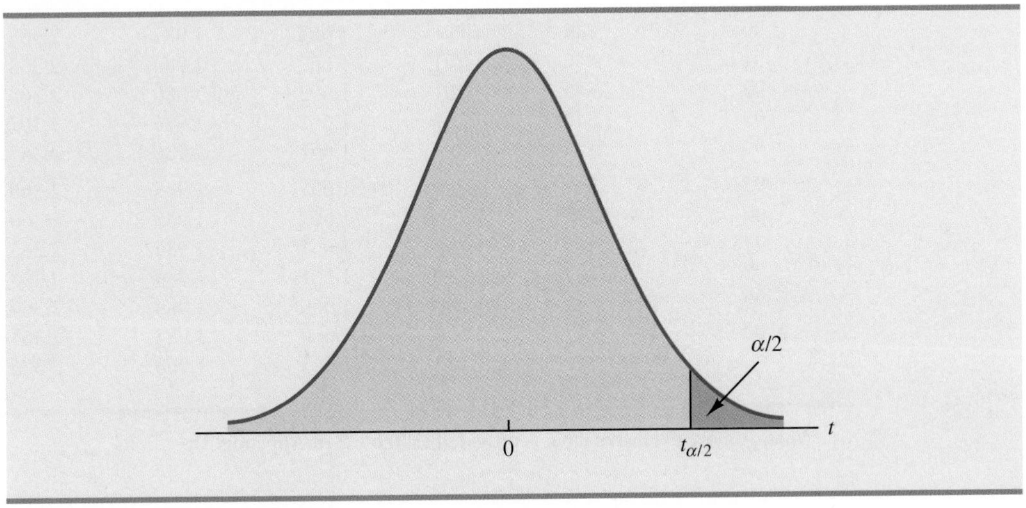

TABLE 8.2 SELECTED VALUES FROM THE t DISTRIBUTION TABLE*

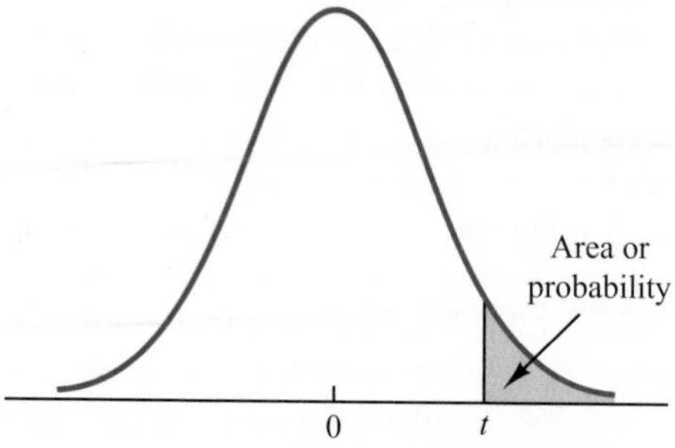

Degrees of Freedom	Area in Upper Tail					
	.20	**.10**	**.05**	**.025**	**.01**	**.005**
1	1.376	3.078	6.314	12.706	31.821	63.656
2	1.061	1.886	2.920	4.303	6.965	9.925
3	.978	1.638	2.353	3.182	4.541	5.841
4	.941	1.533	2.132	2.776	3.747	4.604
5	.920	1.476	2.015	2.571	3.365	4.032
6	.906	1.440	1.943	2.447	3.143	3.707
7	.896	1.415	1.895	2.365	2.998	3.499
8	.889	1.397	1.860	2.306	2.896	3.355
9	.883	1.383	1.833	2.262	2.821	3.250
⋮	⋮	⋮	⋮	⋮	⋮	⋮
60	.848	1.296	1.671	2.000	2.390	2.660
61	.848	1.296	1.670	2.000	2.389	2.659
62	.847	1.295	1.670	1.999	2.388	2.657
63	.847	1.295	1.669	1.998	2.387	2.656
64	.847	1.295	1.669	1.998	2.386	2.655
65	.847	1.295	1.669	1.997	2.385	2.654
66	.847	1.295	1.668	1.997	2.384	2.652
67	.847	1.294	1.668	1.996	2.383	2.651
68	.847	1.294	1.668	1.995	2.382	2.650
69	.847	1.294	1.667	1.995	2.382	2.649
⋮	⋮	⋮	⋮	⋮	⋮	⋮
90	.846	1.291	1.662	1.987	2.368	2.632
91	.846	1.291	1.662	1.986	2.368	2.631
92	.846	1.291	1.662	1.986	2.368	2.630
93	.846	1.291	1.661	1.986	2.367	2.630
94	.845	1.291	1.661	1.986	2.367	2.629
95	.845	1.291	1.661	1.985	2.366	2.629
96	.845	1.290	1.661	1.985	2.366	2.628
97	.845	1.290	1.661	1.985	2.365	2.627
98	.845	1.290	1.661	1.984	2.365	2.627
99	.845	1.290	1.660	1.984	2.364	2.626
100	.845	1.290	1.660	1.984	2.364	2.626
∞	.842	1.282	1.645	1.960	2.326	2.576

**Note:* A more extensive table is provided as Table 2 of Appendix B.

of error is then given by $t_{\alpha/2} s/\sqrt{n}$. With this margin of error, the general expression for an interval estimate of a population mean when σ is unknown follows.

INTERVAL ESTIMATE OF A POPULATION MEAN: σ UNKNOWN

$$\bar{x} \pm t_{\alpha/2} \frac{s}{\sqrt{n}} \tag{8.2}$$

where s is the sample standard deviation, $(1 - \alpha)$ is the confidence coefficient, and $t_{\alpha/2}$ is the t value providing an area of $\alpha/2$ in the upper tail of the t distribution with $n - 1$ degrees of freedom.

The reason the number of degrees of freedom associated with the t value in expression (8.2) is $n - 1$ concerns the use of s as an estimate of the population standard deviation σ. The expression for the sample standard deviation is

$$s = \sqrt{\frac{\Sigma(x_i - \bar{x})^2}{n - 1}}$$

Degrees of freedom refer to the number of independent pieces of information that go into the computation of $\Sigma(x_i - \bar{x})^2$. The n pieces of information involved in computing $\Sigma(x_i - \bar{x})^2$ are as follows: $x_1 - \bar{x}, x_2 - \bar{x}, \ldots, x_n - \bar{x}$. In Section 3.2 we indicated that $\Sigma(x_i - \bar{x}) = 0$ for any data set. Thus, only $n - 1$ of the $x_i - \bar{x}$ values are independent; that is, if we know $n - 1$ of the values, the remaining value can be determined exactly by using the condition that the sum of the $x_i - \bar{x}$ values must be 0. Thus, $n - 1$ is the number of degrees of freedom associated with $\Sigma(x_i - \bar{x})^2$ and hence the number of degrees of freedom for the t distribution in expression (8.2).

To illustrate the interval estimation procedure for the σ unknown case, we will consider a study designed to estimate the mean credit card debt for the population of U.S. households. A sample of $n = 70$ households provided the credit card balances shown in Table 8.3. For this situation, no previous estimate of the population standard deviation σ is available. Thus, the sample data must be used to estimate both the population mean and the population standard deviation. Using the data in Table 8.3, we compute the sample mean $\bar{x} = \$9312$ and the sample standard deviation $s = \$4007$. With 95% confidence and $n - 1 = 69$ degrees of

TABLE 8.3 CREDIT CARD BALANCES FOR A SAMPLE OF 70 HOUSEHOLDS

9430	14661	7159	9071	9691	11032
7535	12195	8137	3603	11448	6525
4078	10544	9467	16804	8279	5239
5604	13659	12595	13479	5649	6195
5179	7061	7917	14044	11298	12584
4416	6245	11346	6817	4353	15415
10676	13021	12806	6845	3467	15917
1627	9719	4972	10493	6191	12591
10112	2200	11356	615	12851	9743
6567	10746	7117	13627	5337	10324
13627	12744	9465	12557	8372	
18719	5742	19263	6232	7445	

WEB file

NewBalance

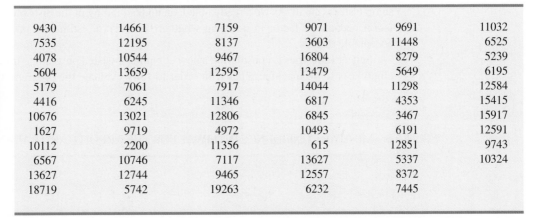

freedom, Table 8.2 can be used to obtain the appropriate value for $t_{.025}$. We want the t value in the row with 69 degrees of freedom, and the column corresponding to .025 in the upper tail. The value shown is $t_{.025} = 1.995$.

We use expression (8.2) to compute an interval estimate of the population mean credit card balance.

$$9312 \pm 1.995 \frac{4007}{\sqrt{70}}$$

$$9312 \pm 955$$

The point estimate of the population mean is \$9312, the margin of error is \$955, and the 95% confidence interval is 9312 − 955 = \$8357 to 9312 + 955 = \$10,267. Thus, we are 95% confident that the mean credit card balance for the population of all households is between \$8357 and \$10,267.

The procedures used by Minitab, Excel and StatTools to develop confidence intervals for a population mean are described in Appendixes 8.1, 8.2 and 8.3. For the household credit card balances study, the results of the Minitab interval estimation procedure are shown in Figure 8.6. The sample of 70 households provides a sample mean credit card balance of \$9312, a sample standard deviation of \$4007, a standard error of the mean of \$479, and a 95% confidence interval of \$8357 to \$10,267.

Practical Advice

If the population follows a normal distribution, the confidence interval provided by expression (8.2) is exact and can be used for any sample size. If the population does not follow a normal distribution, the confidence interval provided by expression (8.2) will be approximate. In this case, the quality of the approximation depends on both the distribution of the population and the sample size.

Larger sample sizes are needed if the distribution of the population is highly skewed or includes outliers.

In most applications, a sample size of $n \geq 30$ is adequate when using expression (8.2) to develop an interval estimate of a population mean. However, if the population distribution is highly skewed or contains outliers, most statisticians would recommend increasing the sample size to 50 or more. If the population is not normally distributed but is roughly symmetric, sample sizes as small as 15 can be expected to provide good approximate confidence intervals. With smaller sample sizes, expression (8.2) should only be used if the analyst believes, or is willing to assume, that the population distribution is at least approximately normal.

Using a Small Sample

In the following example we develop an interval estimate for a population mean when the sample size is small. As we already noted, an understanding of the distribution of the population becomes a factor in deciding whether the interval estimation procedure provides acceptable results.

Scheer Industries is considering a new computer-assisted program to train maintenance employees to do machine repairs. In order to fully evaluate the program, the director of

FIGURE 8.6 MINITAB CONFIDENCE INTERVAL FOR THE CREDIT CARD BALANCE SURVEY

Variable	N	Mean	StDev	SE Mean	95% CI
NewBalance	70	9312	4007	479	(8357, 10267)

TABLE 8.4 TRAINING TIME IN DAYS FOR A SAMPLE OF 20 SCHEER INDUSTRIES EMPLOYEES

WEB file Scheer

52	59	54	42
44	50	42	48
55	54	60	55
44	62	62	57
45	46	43	56

manufacturing requested an estimate of the population mean time required for maintenance employees to complete the computer-assisted training.

A sample of 20 employees is selected, with each employee in the sample completing the training program. Data on the training time in days for the 20 employees are shown in Table 8.4. A histogram of the sample data appears in Figure 8.7. What can we say about the distribution of the population based on this histogram? First, the sample data do not support the conclusion that the distribution of the population is normal, yet we do not see any evidence of skewness or outliers. Therefore, using the guidelines in the previous subsection, we conclude that an interval estimate based on the t distribution appears acceptable for the sample of 20 employees.

We continue by computing the sample mean and sample standard deviation as follows.

$$\bar{x} = \frac{\Sigma x_i}{n} = \frac{1030}{20} = 51.5 \text{ days}$$

$$s = \sqrt{\frac{\Sigma(x_i - \bar{x})^2}{n - 1}} = \sqrt{\frac{889}{20 - 1}} = 6.84 \text{ days}$$

FIGURE 8.7 HISTOGRAM OF TRAINING TIMES FOR THE SCHEER INDUSTRIES SAMPLE

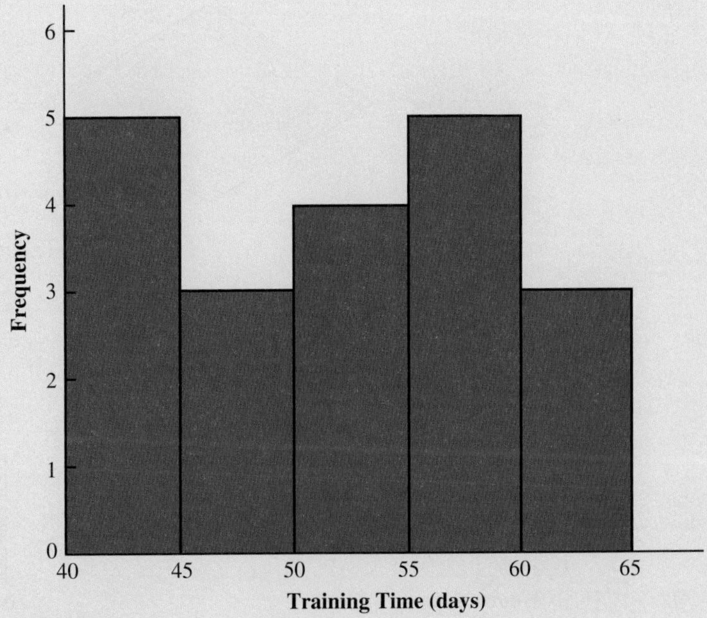

For a 95% confidence interval, we use Table 2 of Appendix B and $n - 1 = 19$ degrees of freedom to obtain $t_{.025} = 2.093$. Expression (8.2) provides the interval estimate of the population mean.

$$51.5 \pm 2.093\left(\frac{6.84}{\sqrt{20}}\right)$$

$$51.5 \pm 3.2$$

The point estimate of the population mean is 51.5 days. The margin of error is 3.2 days and the 95% confidence interval is $51.5 - 3.2 = 48.3$ days to $51.5 + 3.2 = 54.7$ days.

Using a histogram of the sample data to learn about the distribution of a population is not always conclusive, but in many cases it provides the only information available. The histogram, along with judgment on the part of the analyst, can often be used to decide whether expression (8.2) can be used to develop the interval estimate.

Summary of Interval Estimation Procedures

We provided two approaches to developing an interval estimate of a population mean. For the σ known case, σ and the standard normal distribution are used in expression (8.1) to compute the margin of error and to develop the interval estimate. For the σ unknown case, the sample standard deviation s and the t distribution are used in expression (8.2) to compute the margin of error and to develop the interval estimate.

A summary of the interval estimation procedures for the two cases is shown in Figure 8.8. In most applications, a sample size of $n \geq 30$ is adequate. If the population has a normal or approximately normal distribution, however, smaller sample sizes may be used.

FIGURE 8.8 SUMMARY OF INTERVAL ESTIMATION PROCEDURES FOR A POPULATION MEAN

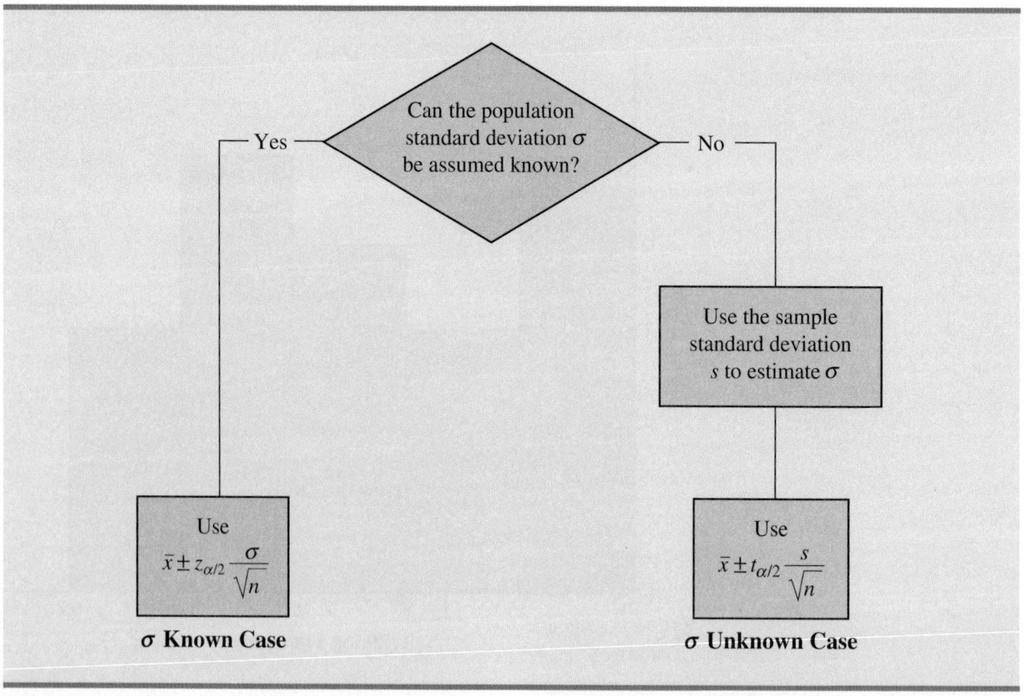

For the σ unknown case a sample size of $n \geq 50$ is recommended if the population distribution is believed to be highly skewed or has outliers.

NOTES AND COMMENTS

1. When σ is known, the margin of error, $z_{\alpha/2}(\sigma/\sqrt{n})$, is fixed and is the same for all samples of size n. When σ is unknown, the margin of error, $t_{\alpha/2}(s/\sqrt{n})$, varies from sample to sample. This variation occurs because the sample standard deviation s varies depending upon the sample selected. A large value for s provides a larger margin of error, while a small value for s provides a smaller margin of error.
2. What happens to confidence interval estimates when the population is skewed? Consider a population that is skewed to the right with large data values stretching the distribution to the right. When such skewness exists, the sample mean $\bar{x}$ and the sample standard deviation s are positively correlated. Larger values of s tend to be associated with larger values of $\bar{x}$. Thus, when $\bar{x}$ is larger than the population mean, s tends to be larger than σ. This skewness causes the margin of error, $t_{\alpha/2}(s/\sqrt{n})$, to be larger than it would be with σ known. The confidence interval with the larger margin of error tends to include the population mean μ more often than it would if the true value of σ were used. But when $\bar{x}$ is smaller than the population mean, the correlation between $\bar{x}$ and s causes the margin of error to be small. In this case, the confidence interval with the smaller margin of error tends to miss the population mean more than it would if we knew σ and used it. For this reason, we recommend using larger sample sizes with highly skewed population distributions.

Exercises

Methods

11. For a t distribution with 16 degrees of freedom, find the area, or probability, in each region.
    a. To the right of 2.120
    b. To the left of 1.337
    c. To the left of -1.746
    d. To the right of 2.583
    e. Between -2.120 and 2.120
    f. Between -1.746 and 1.746
12. Find the t value(s) for each of the following cases.
    a. Upper tail area of .025 with 12 degrees of freedom
    b. Lower tail area of .05 with 50 degrees of freedom
    c. Upper tail area of .01 with 30 degrees of freedom
    d. Where 90% of the area falls between these two t values with 25 degrees of freedom
    e. Where 95% of the area falls between these two t values with 45 degrees of freedom

SELF test

13. The following sample data are from a normal population: 10, 8, 12, 15, 13, 11, 6, 5.
    a. What is the point estimate of the population mean?
    b. What is the point estimate of the population standard deviation?
    c. With 95% confidence, what is the margin of error for the estimation of the population mean?
    d. What is the 95% confidence interval for the population mean?
14. A simple random sample with $n = 54$ provided a sample mean of 22.5 and a sample standard deviation of 4.4.
    a. Develop a 90% confidence interval for the population mean.
    b. Develop a 95% confidence interval for the population mean.

c. Develop a 99% confidence interval for the population mean.
d. What happens to the margin of error and the confidence interval as the confidence level is increased?

Applications

SELF test

15. Sales personnel for Skillings Distributors submit weekly reports listing the customer contacts made during the week. A sample of 65 weekly reports showed a sample mean of 19.5 customer contacts per week. The sample standard deviation was 5.2. Provide 90% and 95% confidence intervals for the population mean number of weekly customer contacts for the sales personnel.

WEB file
CorporateBonds

16. A sample containing years to maturity and yield for 40 corporate bonds are contained in the data file named CorporateBonds (*Barron's,* April 2, 2012).
    a. What is the sample mean years to maturity for corporate bonds and what is the sample standard deviation?
    b. Develop a 95% confidence interval for the population mean years to maturity.
    c. What is the sample mean yield on corporate bonds and what is the sample standard deviation?
    d. Develop a 95% confidence interval for the population mean yield on corporate bonds.

17. The International Air Transport Association surveys business travelers to develop quality ratings for transatlantic gateway airports. The maximum possible rating is 10. Suppose a simple random sample of 50 business travelers is selected and each traveler is asked to provide a rating for the Miami International Airport. The ratings obtained from the sample of 50 business travelers follow.

WEB file
Miami

6	4	6	8	7	7	6	3	3	8	10	4	8
7	8	7	5	9	5	8	4	3	8	5	5	4
4	4	8	4	5	6	2	5	9	9	8	4	8
9	9	5	9	7	8	3	10	8	9	6		

Develop a 95% confidence interval estimate of the population mean rating for Miami.

WEB file
JobSearch

18. Older people often have a hard time finding work. AARP reported on the number of weeks it takes a worker aged 55 plus to find a job. The data on number of weeks spent searching for a job contained in the file JobSearch are consistent with the AARP findings (*AARP Bulletin,* April 2008).
    a. Provide a point estimate of the population mean number of weeks it takes a worker aged 55 plus to find a job.
    b. At 95% confidence, what is the margin of error?
    c. What is the 95% confidence interval estimate of the mean?
    d. Discuss the degree of skewness found in the sample data. What suggestion would you make for a repeat of this study?

19. The average cost per night of a hotel room in New York City is \$273 (*SmartMoney,* March 2009). Assume this estimate is based on a sample of 45 hotels and that the sample standard deviation is \$65.
    a. With 95% confidence, what is the margin of error?
    b. What is the 95% confidence interval estimate of the population mean?
    c. Two years ago the average cost of a hotel room in New York City was \$229. Discuss the change in cost over the two-year period.

WEB file
Program

20. Is your favorite TV program often interrupted by advertising? CNBC presented statistics on the average number of programming minutes in a half-hour sitcom (CNBC, February 23, 2006). The following data (in minutes) are representative of their findings.

21.06	22.24	20.62
21.66	21.23	23.86
23.82	20.30	21.52
21.52	21.91	23.14
20.02	22.20	21.20
22.37	22.19	22.34
23.36	23.44	

Assume the population is approximately normal. Provide a point estimate and a 95% confidence interval for the mean number of programming minutes during a half-hour television sitcom.

WEB file
Alcohol

21. Consumption of alcoholic beverages by young women of drinking age in the United Kingdom, the United States, and Europe was reported (*The Wall Street Journal,* February 15, 2006). Data (annual consumption in liters) consistent with the findings reported in *The Wall Street Journal* article are shown for a sample of 20 European young women.

266	82	199	174	97
170	222	115	130	169
164	102	113	171	0
93	0	93	110	130

Assuming the population is roughly symmetric, construct a 95% confidence interval for the mean annual consumption of alcoholic beverages by European young women.

22. Disney's *Hannah Montana: The Movie* opened on Easter weekend in April 2009. Over the three-day weekend, the movie became the number-one box office attraction (*The Wall Street Journal,* April 13, 2009). The ticket sales revenue in dollars for a sample of 25 theaters is as follows.

WEB file
TicketSales

20,200	10,150	13,000	11,320	9700
8350	7300	14,000	9940	11,200
10,750	6240	12,700	7430	13,500
13,900	4200	6750	6700	9330
13,185	9200	21,400	11,380	10,800

a. What is the 95% confidence interval estimate for the mean ticket sales revenue per theater? Interpret this result.
b. Using the movie ticket price of \$7.16 per ticket, what is the estimate of the mean number of customers per theater?
c. The movie was shown in 3118 theaters. Estimate the total number of customers who saw *Hannah Montana: The Movie* and the total box office ticket sales for the three-day weekend.

8.3 Determining the Sample Size

If a desired margin of error is selected prior to sampling, the procedures in this section can be used to determine the sample size necessary to satisfy the margin of error requirement.

In providing practical advice in the two preceding sections, we commented on the role of the sample size in providing good approximate confidence intervals when the population is not normally distributed. In this section, we focus on another aspect of the sample size issue. We describe how to choose a sample size large enough to provide a desired margin of error. To understand how this process works, we return to the σ known case presented in Section 8.1. Using expression (8.1), the interval estimate is

$$\bar{x} \pm z_{\alpha/2}\frac{\sigma}{\sqrt{n}}$$

The quantity $z_{\alpha/2}(\sigma/\sqrt{n})$ is the margin of error. Thus, we see that $z_{\alpha/2}$, the population standard deviation σ, and the sample size n combine to determine the margin of error. Once we select a confidence coefficient $1 - \alpha$, $z_{\alpha/2}$ can be determined. Then, if we have a value for σ, we can determine the sample size n needed to provide any desired margin of error. Development of the formula used to compute the required sample size n follows.

Let E = the desired margin of error:

$$E = z_{\alpha/2} \frac{\sigma}{\sqrt{n}}$$

Solving for $\sqrt{n}$, we have

$$\sqrt{n} = \frac{z_{\alpha/2}\sigma}{E}$$

Squaring both sides of this equation, we obtain the following expression for the sample size.

Equation (8.3) can be used to provide a good sample size recommendation. However, judgment on the part of the analyst should be used to determine whether the final sample size should be adjusted upward.

SAMPLE SIZE FOR AN INTERVAL ESTIMATE OF A POPULATION MEAN

$$n = \frac{(z_{\alpha/2})^2\sigma^2}{E^2} \tag{8.3}$$

This sample size provides the desired margin of error at the chosen confidence level.

In equation (8.3), E is the margin of error that the user is willing to accept, and the value of $z_{\alpha/2}$ follows directly from the confidence level to be used in developing the interval estimate. Although user preference must be considered, 95% confidence is the most frequently chosen value ($z_{.025} = 1.96$).

Finally, use of equation (8.3) requires a value for the population standard deviation σ. However, even if σ is unknown, we can use equation (8.3) provided we have a preliminary or *planning value* for σ. In practice, one of the following procedures can be chosen.

A planning value for the population standard deviation σ must be specified before the sample size can be determined. Three methods of obtaining a planning value for σ are discussed here.

1. Use the estimate of the population standard deviation computed from data of previous studies as the planning value for σ.
2. Use a pilot study to select a preliminary sample. The sample standard deviation from the preliminary sample can be used as the planning value for σ.
3. Use judgment or a "best guess" for the value of σ. For example, we might begin by estimating the largest and smallest data values in the population. The difference between the largest and smallest values provides an estimate of the range for the data. Finally, the range divided by 4 is often suggested as a rough approximation of the standard deviation and thus an acceptable planning value for σ.

Let us demonstrate the use of equation (8.3) to determine the sample size by considering the following example. A previous study that investigated the cost of renting automobiles in the United States found a mean cost of approximately $55 per day for renting a midsize automobile. Suppose that the organization that conducted this study would like to conduct a new study in order to estimate the population mean daily rental cost for a midsize automobile in the United States. In designing the new study, the project director specifies that the population mean daily rental cost be estimated with a margin of error of $2 and a 95% level of confidence.

The project director specified a desired margin of error of $E = 2$, and the 95% level of confidence indicates $z_{.025} = 1.96$. Thus, we only need a planning value for the population standard deviation σ in order to compute the required sample size. At this point, an analyst reviewed the sample data from the previous study and found that the sample standard deviation for the daily rental cost was $9.65. Using 9.65 as the planning value for σ, we obtain

Equation (8.3) provides the minimum sample size needed to satisfy the desired margin of error requirement. If the computed sample size is not an integer, rounding up to the next integer value will provide a margin of error slightly smaller than required.

$$n = \frac{(z_{\alpha/2})^2\sigma^2}{E^2} = \frac{(1.96)^2(9.65)^2}{2^2} = 89.43$$

Thus, the sample size for the new study needs to be at least 89.43 midsize automobile rentals in order to satisfy the project director's \$2 margin-of-error requirement. In cases where the computed n is not an integer, we round up to the next integer value; hence, the recommended sample size is 90 midsize automobile rentals.

Exercises

Methods

23. How large a sample should be selected to provide a 95% confidence interval with a margin of error of 10? Assume that the population standard deviation is 40.

SELF test

24. The range for a set of data is estimated to be 36.
    a. What is the planning value for the population standard deviation?
    b. At 95% confidence, how large a sample would provide a margin of error of 3?
    c. At 95% confidence, how large a sample would provide a margin of error of 2?

Applications

SELF test

25. Refer to the Scheer Industries example in Section 8.2. Use 6.84 days as a planning value for the population standard deviation.
    a. Assuming 95% confidence, what sample size would be required to obtain a margin of error of 1.5 days?
    b. If the precision statement was made with 90% confidence, what sample size would be required to obtain a margin of error of 2 days?

26. The U.S. Energy Information Administration (US EIA) reported that the average price for a gallon of regular gasoline is \$3.94 (US EIA website, April 6, 2012). The US EIA updates its estimates of average gas prices on a weekly basis. Assume the standard deviation is \$.25 for the price of a gallon of regular gasoline and recommend the appropriate sample size for the US EIA to use if they wish to report each of the following margins of error at 95% confidence.
    a. The desired margin of error is \$.10.
    b. The desired margin of error is \$.07.
    c. The desired margin of error is \$.05.

27. Annual starting salaries for college graduates with degrees in business administration are generally expected to be between \$30,000 and \$45,000. Assume that a 95% confidence interval estimate of the population mean annual starting salary is desired. What is the planning value for the population standard deviation? How large a sample should be taken if the desired margin of error is
    a. \$500?
    b. \$200?
    c. \$100?
    d. Would you recommend trying to obtain the \$100 margin of error? Explain.

28. An online survey by ShareBuilder, a retirement plan provider, and Harris Interactive reported that 60% of female business owners are not confident they are saving enough for retirement (*SmallBiz*, Winter 2006). Suppose we would like to do a follow-up study to determine how much female business owners are saving each year toward retirement and want to use \$100 as the desired margin of error for an interval estimate of the population mean. Use \$1100 as a planning value for the standard deviation and recommend a sample size for each of the following situations.
    a. A 90% confidence interval is desired for the mean amount saved.
    b. A 95% confidence interval is desired for the mean amount saved.

c. A 99% confidence interval is desired for the mean amount saved.

d. When the desired margin of error is set, what happens to the sample size as the confidence level is increased? Would you recommend using a 99% confidence interval in this case? Discuss.

29. The travel-to-work time for residents of the 15 largest cities in the United States is reported in the *2003 Information Please Almanac.* Suppose that a preliminary simple random sample of residents of San Francisco is used to develop a planning value of 6.25 minutes for the population standard deviation.

a. If we want to estimate the population mean travel-to-work time for San Francisco residents with a margin of error of 2 minutes, what sample size should be used? Assume 95% confidence.

b. If we want to estimate the population mean travel-to-work time for San Francisco residents with a margin of error of 1 minute, what sample size should be used? Assume 95% confidence.

30. There has been a trend toward less driving in the last few years, especially by young people. From 2001 to 2009 the annual vehicle miles traveled by people from 16 to 34 years of age decreased from 10,300 to 7900 miles per person (U.S. PIRG and Education Fund website, April 6, 2012). Assume the standard deviation was 2000 miles in 2009. Suppose you would like to conduct a survey to develop a 95% confidence interval estimate of the annual vehicle-miles per person for people 16 to 34 years of age at the current time. A margin of error of 100 miles is desired. How large a sample should be used for the current survey?

8.4 Population Proportion

In the introduction to this chapter we said that the general form of an interval estimate of a population proportion p is

$$\bar{p} \pm \text{Margin of error}$$

The sampling distribution of $\bar{p}$ plays a key role in computing the margin of error for this interval estimate.

In Chapter 7 we said that the sampling distribution of $\bar{p}$ can be approximated by a normal distribution whenever $np \geq 5$ and $n(1 - p) \geq 5$. Figure 8.9 shows the normal approximation

FIGURE 8.9 NORMAL APPROXIMATION OF THE SAMPLING DISTRIBUTION OF $\bar{p}$

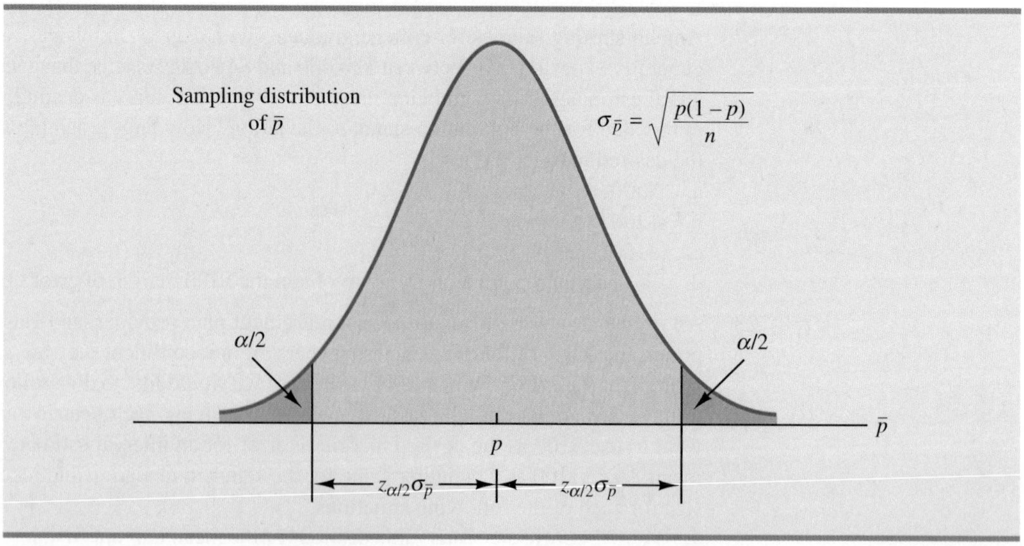

of the sampling distribution of $\bar{p}$. The mean of the sampling distribution of $\bar{p}$ is the population proportion p, and the standard error of $\bar{p}$ is

$$\sigma_{\bar{p}} = \sqrt{\frac{p(1 - p)}{n}} \tag{8.4}$$

Because the sampling distribution of $\bar{p}$ is normally distributed, if we choose $z_{\alpha/2}\sigma_{\bar{p}}$ as the margin of error in an interval estimate of a population proportion, we know that $100(1 - \alpha)\%$ of the intervals generated will contain the true population proportion. But $\sigma_{\bar{p}}$ cannot be used directly in the computation of the margin of error because p will not be known; p is what we are trying to estimate. So $\bar{p}$ is substituted for p and the margin of error for an interval estimate of a population proportion is given by

$$\text{Margin of error} = z_{\alpha/2}\sqrt{\frac{\bar{p}(1 - \bar{p})}{n}} \tag{8.5}$$

With this margin of error, the general expression for an interval estimate of a population proportion is as follows.

When developing confidence intervals for proportions, the quantity $z_{\alpha/2}\sqrt{\bar{p}(1 - \bar{p})/n}$ provides the margin of error.

INTERVAL ESTIMATE OF A POPULATION PROPORTION

$$\bar{p} \pm z_{\alpha/2}\sqrt{\frac{\bar{p}(1 - \bar{p})}{n}} \tag{8.6}$$

where $1 - \alpha$ is the confidence coefficient and $z_{\alpha/2}$ is the z value providing an area of $\alpha/2$ in the upper tail of the standard normal distribution.

WEB file

TeeTimes

The following example illustrates the computation of the margin of error and interval estimate for a population proportion. A national survey of 900 women golfers was conducted to learn how women golfers view their treatment at golf courses in the United States. The survey found that 396 of the women golfers were satisfied with the availability of tee times. Thus, the point estimate of the proportion of the population of women golfers who are satisfied with the availability of tee times is 396/900 = .44. Using expression (8.6) and a 95% confidence level,

$$\bar{p} \pm z_{\alpha/2}\sqrt{\frac{\bar{p}(1 - \bar{p})}{n}}$$

$$.44 \pm 1.96\sqrt{\frac{.44(1 - .44)}{900}}$$

$$.44 \pm .0324$$

Thus, the margin of error is .0324 and the 95% confidence interval estimate of the population proportion is .4076 to .4724. Using percentages, the survey results enable us to state with 95% confidence that between 40.76% and 47.24% of all women golfers are satisfied with the availability of tee times.

Determining the Sample Size

Let us consider the question of how large the sample size should be to obtain an estimate of a population proportion at a specified level of precision. The rationale for the sample size determination in developing interval estimates of p is similar to the rationale used in Section 8.3 to determine the sample size for estimating a population mean.

Previously in this section we said that the margin of error associated with an interval estimate of a population proportion is $z_{\alpha/2}\sqrt{\bar{p}(1-\bar{p})/n}$. The margin of error is based on the value of $z_{\alpha/2}$, the sample proportion $\bar{p}$, and the sample size n. Larger sample sizes provide a smaller margin of error and better precision.

Let E denote the desired margin of error.

$$E = z_{\alpha/2}\sqrt{\frac{\bar{p}(1-\bar{p})}{n}}$$

Solving this equation for n provides a formula for the sample size that will provide a margin of error of size E.

$$n = \frac{(z_{\alpha/2})^2\bar{p}(1-\bar{p})}{E^2}$$

Note, however, that we cannot use this formula to compute the sample size that will provide the desired margin of error because $\bar{p}$ will not be known until after we select the sample. What we need, then, is a planning value for $\bar{p}$ that can be used to make the computation. Using p^* to denote the planning value for $\bar{p}$, the following formula can be used to compute the sample size that will provide a margin of error of size E.

SAMPLE SIZE FOR AN INTERVAL ESTIMATE OF A POPULATION PROPORTION

$$n = \frac{(z_{\alpha/2})^2 p^*(1-p^*)}{E^2} \qquad \textbf{(8.7)}$$

In practice, the planning value p^* can be chosen by one of the following procedures.

1. Use the sample proportion from a previous sample of the same or similar units.
2. Use a pilot study to select a preliminary sample. The sample proportion from this sample can be used as the planning value, p^*.
3. Use judgment or a "best guess" for the value of p^*.
4. If none of the preceding alternatives apply, use a planning value of $p^* = .50$.

Let us return to the survey of women golfers and assume that the company is interested in conducting a new survey to estimate the current proportion of the population of women golfers who are satisfied with the availability of tee times. How large should the sample be if the survey director wants to estimate the population proportion with a margin of error of .025 at 95% confidence? With $E = .025$ and $z_{\alpha/2} = 1.96$, we need a planning value p^* to answer the sample size question. Using the previous survey result of $\bar{p} = .44$ as the planning value p^*, equation (8.7) shows that

$$n = \frac{(z_{\alpha/2})^2 p^*(1-p^*)}{E^2} = \frac{(1.96)^2(.44)(1-.44)}{(.025)^2} = 1514.5$$

TABLE 8.5 SOME POSSIBLE VALUES FOR $p^*(1 - p^*)$

p^*	$p^*(1 - p^*)$	
.10	(.10)(.90) = .09	
.30	(.30)(.70) = .21	
.40	(.40)(.60) = .24	
.50	(.50)(.50) = .25	← Largest value for $p^*(1 - p^*)$
.60	(.60)(.40) = .24	
.70	(.70)(.30) = .21	
.90	(.90)(.10) = .09	

Thus, the sample size must be at least 1514.5 women golfers to satisfy the margin of error requirement. Rounding up to the next integer value indicates that a sample of 1515 women golfers is recommended to satisfy the margin of error requirement.

The fourth alternative suggested for selecting a planning value p^* is to use $p^* = .50$. This value of p^* is frequently used when no other information is available. To understand why, note that the numerator of equation (8.7) shows that the sample size is proportional to the quantity $p^*(1 - p^*)$. A larger value for the quantity $p^*(1 - p^*)$ will result in a larger sample size. Table 8.5 gives some possible values of $p^*(1 - p^*)$. Note that the largest value of $p^*(1 - p^*)$ occurs when $p^* = .50$. Thus, in case of any uncertainty about an appropriate planning value, we know that $p^* = .50$ will provide the largest sample size recommendation. In effect, we play it safe by recommending the largest necessary sample size. If the sample proportion turns out to be different from the .50 planning value, the margin of error will be smaller than anticipated. Thus, in using $p^* = .50$, we guarantee that the sample size will be sufficient to obtain the desired margin of error.

In the survey of women golfers example, a planning value of $p^* = .50$ would have provided the sample size

$$n = \frac{(z_{\alpha/2})^2 p^*(1 - p^*)}{E^2} = \frac{(1.96)^2(.50)(1 - .50)}{(.025)^2} = 1536.6$$

Thus, a slightly larger sample size of 1537 women golfers would be recommended.

NOTES AND COMMENTS

The desired margin of error for estimating a population proportion is almost always .10 or less. In national public opinion polls conducted by organizations such as Gallup and Harris, a .03 or .04 margin of error is common. With such margins of error, equation (8.7) will almost always provide a sample size that is large enough to satisfy the requirements of $np \geq 5$ and $n(1 - p) \geq 5$ for using a normal distribution as an approximation for the sampling distribution of $\bar{x}$.

Exercises

Methods

SELF test

31. A simple random sample of 400 individuals provides 100 Yes responses.
    a. What is the point estimate of the proportion of the population that would provide Yes responses?
    b. What is your estimate of the standard error of the proportion, $\sigma_{\bar{p}}$?
    c. Compute the 95% confidence interval for the population proportion.

32. A simple random sample of 800 elements generates a sample proportion $\bar{p} = .70$.
    a. Provide a 90% confidence interval for the population proportion.
    b. Provide a 95% confidence interval for the population proportion.
33. In a survey, the planning value for the population proportion is $p^* = .35$. How large a sample should be taken to provide a 95% confidence interval with a margin of error of .05?
34. At 95% confidence, how large a sample should be taken to obtain a margin of error of .03 for the estimation of a population proportion? Assume that past data are not available for developing a planning value for p^*.

Applications

SELF test

35. The Consumer Reports National Research Center conducted a telephone survey of 2000 adults to learn about the major economic concerns for the future (*Consumer Reports,* January 2009). The survey results showed that 1760 of the respondents think the future health of Social Security is a major economic concern.
    a. What is the point estimate of the population proportion of adults who think the future health of Social Security is a major economic concern.
    b. At 90% confidence, what is the margin of error?
    c. Develop a 90% confidence interval for the population proportion of adults who think the future health of Social Security is a major economic concern.
    d. Develop a 95% confidence interval for this population proportion.
36. According to statistics reported on CNBC, a surprising number of motor vehicles are not covered by insurance (CNBC, February 23, 2006). Sample results, consistent with the CNBC report, showed 46 of 200 vehicles were not covered by insurance.
    a. What is the point estimate of the proportion of vehicles not covered by insurance?
    b. Develop a 95% confidence interval for the population proportion.
37. Towers Perrin, a New York human resources consulting firm, conducted a survey of 1100 employees at medium-sized and large companies to determine how dissatisfied employees were with their jobs (*The Wall Street Journal,* January 29, 2003). Representative data are shown in the file JobSatisfaction. A response of Yes indicates the employee strongly disliked the current work experience.

WEB file
JobSatisfaction

    a. What is the point estimate of the proportion of the population of employees who strongly dislike their current work experience?
    b. At 95% confidence, what is the margin of error?
    c. What is the 95% confidence interval for the proportion of the population of employees who strongly dislike their current work experience?
    d. Towers Perrin estimates that it costs employers one-third of an hourly employee's annual salary to find a successor and as much as 1.5 times the annual salary to find a successor for a highly compensated employee. What message did this survey send to employers?
38. According to Thomson Financial, through January 25, 2006, the majority of companies reported profits had beaten estimates (*BusinessWeek,* February 6, 2006). A sample of 162 companies showed 104 beat estimates, 29 matched estimates, and 29 fell short.
    a. What is the point estimate of the proportion that fell short of estimates?
    b. Determine the margin of error and provide a 95% confidence interval for the proportion that beat estimates.
    c. How large a sample is needed if the desired margin of error is .05?

SELF test

39. The percentage of people not covered by health care insurance in 2003 was 15.6% (*Statistical Abstract of the United States*, 2006). A congressional committee has been charged with conducting a sample survey to obtain more current information.
    a. What sample size would you recommend if the committee's goal is to estimate the current proportion of individuals without health care insurance with a margin of error of .03? Use a 95% confidence level.
    b. Repeat part (a) using a 99% confidence level.

40. For many years businesses have struggled with the rising cost of health care. But recently, the increases have slowed due to less inflation in health care prices and employees paying for a larger portion of health care benefits. A recent Mercer survey showed that 52% of U.S. employers were likely to require higher employee contributions for health care coverage in 2009 (*BusinessWeek,* February 16, 2009). Suppose the survey was based on a sample of 800 companies. Compute the margin of error and a 95% confidence interval for the proportion of companies likely to require higher employee contributions for health care coverage in 2009.

41. Fewer young people are driving. In 1983, 87% of 19-year-olds had a driver's license. Twenty-five years later that percentage had dropped to 75% (University of Michigan Transportation Research Institute website, April 7, 2012). Suppose these results are based on a random sample of 1200 19-year-olds in 1983 and again in 2008.
    a. At 95% confidence, what is the margin of error and the interval estimate of the number of 19-year-old drivers in 1983?
    b. At 95% confidence, what is the margin of error and the interval estimate of the number of 19-year-old drivers in 2008?
    c. Is the margin of error the same in parts (a) and (b)? Why or why not?

42. A poll for the presidential campaign sampled 491 potential voters in June. A primary purpose of the poll was to obtain an estimate of the proportion of potential voters who favored each candidate. Assume a planning value of $p^* = .50$ and a 95% confidence level.
    a. For $p^* = .50$, what was the planned margin of error for the June poll?
    b. Closer to the November election, better precision and smaller margins of error are desired. Assume the following margins of error are requested for surveys to be conducted during the presidential campaign. Compute the recommended sample size for each survey.

Survey	Margin of Error
September	.04
October	.03
Early November	.02
Pre-Election Day	.01

43. A Phoenix Wealth Management/Harris Interactive survey of 1500 individuals with net worth of $1 million or more provided a variety of statistics on wealthy people (*BusinessWeek,* September 22, 2003). The previous three-year period had been bad for the stock market, which motivated some of the questions asked.
    a. The survey reported that 53% of the respondents lost 25% or more of their portfolio value over the past three years. Develop a 95% confidence interval for the proportion of wealthy people who lost 25% or more of their portfolio value over the past three years.
    b. The survey reported that 31% of the respondents feel they have to save more for retirement to make up for what they lost. Develop a 95% confidence interval for the population proportion.
    c. Five percent of the respondents gave $25,000 or more to charity over the previous year. Develop a 95% confidence interval for the proportion who gave $25,000 or more to charity.
    d. Compare the margin of error for the interval estimates in parts (a), (b), and (c). How is the margin of error related to $\bar{p}$? When the same sample is being used to estimate a variety of proportions, which of the proportions should be used to choose the planning value p^*? Why do you think $p^* = .50$ is often used in these cases?

Summary

In this chapter we presented methods for developing interval estimates of a population mean and a population proportion. A point estimator may or may not provide a good estimate of a population parameter. The use of an interval estimate provides a measure of the precision of an estimate. Both the interval estimate of the population mean and the population proportion are of the form: point estimate ± margin of error.

We presented interval estimates for a population mean for two cases. In the σ known case, historical data or other information is used to develop an estimate of σ prior to taking a sample. Analysis of new sample data then proceeds based on the assumption that σ is known. In the σ unknown case, the sample data are used to estimate both the population mean and the population standard deviation. The final choice of which interval estimation procedure to use depends upon the analyst's understanding of which method provides the best estimate of σ.

In the σ known case, the interval estimation procedure is based on the assumed value of σ and the use of the standard normal distribution. In the σ unknown case, the interval estimation procedure uses the sample standard deviation s and the t distribution. In both cases the quality of the interval estimates obtained depends on the distribution of the population and the sample size. If the population is normally distributed the interval estimates will be exact in both cases, even for small sample sizes. If the population is not normally distributed, the interval estimates obtained will be approximate. Larger sample sizes will provide better approximations, but the more highly skewed the population is, the larger the sample size needs to be to obtain a good approximation. Practical advice about the sample size necessary to obtain good approximations was included in Sections 8.1 and 8.2. In most cases a sample of size 30 or more will provide good approximate confidence intervals.

The general form of the interval estimate for a population proportion is $\bar{p}$ ± margin of error. In practice the sample sizes used for interval estimates of a population proportion are generally large. Thus, the interval estimation procedure is based on the standard normal distribution.

Often a desired margin of error is specified prior to developing a sampling plan. We showed how to choose a sample size large enough to provide the desired precision.

Glossary

Interval estimate An estimate of a population parameter that provides an interval believed to contain the value of the parameter. For the interval estimates in this chapter, it has the form: point estimate ± margin of error.

Margin of error The ± value added to and subtracted from a point estimate in order to develop an interval estimate of a population parameter.

σ known The case when historical data or other information provides a good value for the population standard deviation prior to taking a sample. The interval estimation procedure uses this known value of σ in computing the margin of error.

Confidence level The confidence associated with an interval estimate. For example, if an interval estimation procedure provides intervals such that 95% of the intervals formed using the procedure will include the population parameter, the interval estimate is said to be constructed at the 95% confidence level.

Confidence coefficient The confidence level expressed as a decimal value. For example, .95 is the confidence coefficient for a 95% confidence level.

Confidence interval Another name for an interval estimate.

σ unknown The more common case when no good basis exists for estimating the population standard deviation prior to taking the sample. The interval estimation procedure uses the sample standard deviation s in computing the margin of error.

t distribution A family of probability distributions that can be used to develop an interval estimate of a population mean whenever the population standard deviation σ is unknown and is estimated by the sample standard deviation s.

Degrees of freedom A parameter of the t distribution. When the t distribution is used in the computation of an interval estimate of a population mean, the appropriate t distribution has $n - 1$ degrees of freedom, where n is the size of the sample.

Key Formulas

Interval Estimate of a Population Mean: σ Known

$$\bar{x} \pm z_{\alpha/2} \frac{\sigma}{\sqrt{n}} \tag{8.1}$$

Interval Estimate of a Population Mean: σ Unknown

$$\bar{x} \pm t_{\alpha/2} \frac{s}{\sqrt{n}} \tag{8.2}$$

Sample Size for an Interval Estimate of a Population Mean

$$n = \frac{(z_{\alpha/2})^2\sigma^2}{E^2} \tag{8.3}$$

Interval Estimate of a Population Proportion

$$\bar{p} \pm z_{\alpha/2}\sqrt{\frac{\bar{p}(1-\bar{p})}{n}} \tag{8.6}$$

Sample Size for an Interval Estimate of a Population Proportion

$$n = \frac{(z_{\alpha/2})^2 p^*(1-p^*)}{E^2} \tag{8.7}$$

Supplementary Exercises

44. A sample survey of 54 discount brokers showed that the mean price charged for a trade of 100 shares at $50 per share was $33.77 (*AAII Journal,* February 2006). The survey is conducted annually. With the historical data available, assume a known population standard deviation of $15.
    a. Using the sample data, what is the margin of error associated with a 95% confidence interval?
    b. Develop a 95% confidence interval for the mean price charged by discount brokers for a trade of 100 shares at $50 per share.
45. A survey conducted by the American Automobile Association showed that a family of four spends an average of $215.60 per day while on vacation. Suppose a sample of 64 families of four vacationing at Niagara Falls resulted in a sample mean of $252.45 per day and a sample standard deviation of $74.50.
    a. Develop a 95% confidence interval estimate of the mean amount spent per day by a family of four visiting Niagara Falls.
    b. Based on the confidence interval from part (a), does it appear that the population mean amount spent per day by families visiting Niagara Falls differs from the mean reported by the American Automobile Association? Explain.
46. The 92 million Americans of age 50 and over control 50 percent of all discretionary income (*AARP Bulletin,* March 2008). AARP estimated that the average annual expenditure on restaurants and carryout food was $1873 for individuals in this age group. Suppose this estimate is based on a sample of 80 persons and that the sample standard deviation is $550.
    a. At 95% confidence, what is the margin of error?
    b. What is the 95% confidence interval for the population mean amount spent on restaurants and carryout food?
    c. What is your estimate of the total amount spent by Americans of age 50 and over on restaurants and carryout food?
    d. If the amount spent on restaurants and carryout food is skewed to the right, would you expect the median amount spent to be greater or less than $1873?

47. Many stock market observers say that when the P/E ratio for stocks gets over 20 the market is overvalued. The P/E ratio is the stock price divided by the most recent 12 months of earnings. Suppose you are interested in seeing whether the current market is overvalued and would also like to know what proportion of companies pay dividends. A random sample of 30 companies listed on the New York Stock Exchange (NYSE) is provided (*Barron's,* January 19, 2004).

WEB file NYSEStocks

Company	Dividend	P/E Ratio	Company	Dividend	P/E Ratio
Albertsons	Yes	14	NY Times A	Yes	25
BRE Prop	Yes	18	Omnicare	Yes	25
CityNtl	Yes	16	PallCp	Yes	23
DelMonte	No	21	PubSvcEnt	Yes	11
EnrgzHldg	No	20	SensientTch	Yes	11
Ford Motor	Yes	22	SmtProp	Yes	12
Gildan A	No	12	TJX Cos	Yes	21
HudsnUtdBcp	Yes	13	Thomson	Yes	30
IBM	Yes	22	USB Hldg	Yes	12
JeffPilot	Yes	16	US Restr	Yes	26
KingswayFin	No	6	Varian Med	No	41
Libbey	Yes	13	Visx	No	72
MasoniteIntl	No	15	Waste Mgt	No	23
Motorola	Yes	68	Wiley A	Yes	21
Ntl City	Yes	10	Yum Brands	No	18

a. What is a point estimate of the P/E ratio for the population of stocks listed on the New York Stock Exchange? Develop a 95% confidence interval.
b. Based on your answer to part (a), do you believe that the market is overvalued?
c. What is a point estimate of the proportion of companies on the NYSE that pay dividends? Is the sample size large enough to justify using the normal distribution to construct a confidence interval for this proportion? Why or why not?

WEB file Flights

48. US Airways conducted a number of studies that indicated a substantial savings could be obtained by encouraging Dividend Miles frequent flyer customers to redeem miles and schedule award flights online (*US Airways Attaché,* February 2003). One study collected data on the amount of time required to redeem miles and schedule an award flight over the telephone. A sample showing the time in minutes required for each of 150 award flights scheduled by telephone is contained in the data set Flights. Use Minitab or Excel to help answer the following questions.
a. What is the sample mean number of minutes required to schedule an award flight by telephone?
b. What is the 95% confidence interval for the population mean time to schedule an award flight by telephone?
c. Assume a telephone ticket agent works 7.5 hours per day. How many award flights can one ticket agent be expected to handle a day?
d. Discuss why this information supported US Airways' plans to use an online system to reduce costs.

WEB file Standing

49. A recent article reported that there are approximately 11 minutes of actual playing time in a typical National Football League (NFL) game (*The Wall Street Journal,* January 15, 2010). The article included information about the amount of time devoted to replays, the amount of time devoted to commercials, and the amount of time the players spend standing around between plays. Data consistent with the findings published in *The Wall Street Journal* are in the file named Standing. These data provide the amount of time players spend standing around between plays for a sample of 60 NFL games.
a. Use the Standing data set to develop a point estimate of the number of minutes during an NFL game that players are standing around between plays. Compare this to the actual playing time reported in the article. Are you surprised?
b. What is the sample standard deviation?

c. Develop a 95% confidence interval for the number of minutes players spend standing around between plays.

50. Mileage tests are conducted for a particular model of automobile. If a 98% confidence interval with a margin of error of 1 mile per gallon is desired, how many automobiles should be used in the test? Assume that preliminary mileage tests indicate the standard deviation is 2.6 miles per gallon.

51. In developing patient appointment schedules, a medical center wants to estimate the mean time that a staff member spends with each patient. How large a sample should be taken if the desired margin of error is two minutes at a 95% level of confidence? How large a sample should be taken for a 99% level of confidence? Use a planning value for the population standard deviation of eight minutes.

52. Annual salary plus bonus data for chief executive officers are presented in the *BusinessWeek* Annual Pay Survey. A preliminary sample showed that the standard deviation is $675 with data provided in thousands of dollars. How many chief executive officers should be in a sample if we want to estimate the population mean annual salary plus bonus with a margin of error of $100,000? (*Note:* The desired margin of error would be $E = 100$ if the data are in thousands of dollars.) Use 95% confidence.

53. The National Center for Education Statistics reported that 47% of college students work to pay for tuition and living expenses. Assume that a sample of 450 college students was used in the study.
    a. Provide a 95% confidence interval for the population proportion of college students who work to pay for tuition and living expenses.
    b. Provide a 99% confidence interval for the population proportion of college students who work to pay for tuition and living expenses.
    c. What happens to the margin of error as the confidence is increased from 95% to 99%?

54. A *USA Today*/CNN/Gallup survey of 369 working parents found 200 who said they spend too little time with their children because of work commitments.
    a. What is the point estimate of the proportion of the population of working parents who feel they spend too little time with their children because of work commitments?
    b. At 95% confidence, what is the margin of error?
    c. What is the 95% confidence interval estimate of the population proportion of working parents who feel they spend too little time with their children because of work commitments?

55. Which would be hardest for you to give up: Your computer or your television? In a recent survey of 1677 U.S. Internet users, 74% of the young tech elite (average age of 22) say their computer would be very hard to give up (*PC Magazine,* February 3, 2004). Only 48% say their television would be very hard to give up.
    a. Develop a 95% confidence interval for the proportion of the young tech elite that would find it very hard to give up their computer.
    b. Develop a 99% confidence interval for the proportion of the young tech elite that would find it very hard to give up their television.
    c. In which case, part (a) or part (b), is the margin of error larger? Explain why.

56. Cincinnati/Northern Kentucky International Airport had the second highest on-time arrival rate for 2005 among the nation's busiest airports (*The Cincinnati Enquirer,* February 3, 2006). Assume the findings were based on 455 on-time arrivals out of a sample of 550 flights.
    a. Develop a point estimate of the on-time arrival rate (proportion of flights arriving on time) for the airport.
    b. Construct a 95% confidence interval for the on-time arrival rate of the population of all flights at the airport during 2005.

57. The *2003 Statistical Abstract of the United States* reported the percentage of people 18 years of age and older who smoke. Suppose that a study designed to collect new data on smokers and nonsmokers uses a preliminary estimate of the proportion who smoke of .30.
    a. How large a sample should be taken to estimate the proportion of smokers in the population with a margin of error of .02? Use 95% confidence.

b. Assume that the study uses your sample size recommendation in part (a) and finds 520 smokers. What is the point estimate of the proportion of smokers in the population?
c. What is the 95% confidence interval for the proportion of smokers in the population?

58. A well-known bank credit card firm wishes to estimate the proportion of credit card holders who carry a nonzero balance at the end of the month and incur an interest charge. Assume that the desired margin of error is .03 at 98% confidence.
    a. How large a sample should be selected if it is anticipated that roughly 70% of the firm's card holders carry a nonzero balance at the end of the month?
    b. How large a sample should be selected if no planning value for the proportion could be specified?
59. Workers in several industries were surveyed to determine the proportion of workers who feel their industry is understaffed. In the government sector, 37% of the respondents said they were understaffed, in the health care sector 33% said they were understaffed, and in the education sector 28% said they were understaffed (*USA Today,* January 11, 2010). Suppose that 200 workers were surveyed in each industry.
    a. Construct a 95% confidence interval for the proportion of workers in each of these industries who feel their industry is understaffed.
    b. Assuming the same sample size will be used in each industry, how large would the sample need to be to ensure that the margin of error is .05 or less for each of the three confidence intervals?
60. Although airline schedules and cost are important factors for business travelers when choosing an airline carrier, a *USA Today* survey found that business travelers list an airline's frequent flyer program as the most important factor. From a sample of $n = 1993$ business travelers who responded to the survey, 618 listed a frequent flyer program as the most important factor.
    a. What is the point estimate of the proportion of the population of business travelers who believe a frequent flyer program is the most important factor when choosing an airline carrier?
    b. Develop a 95% confidence interval estimate of the population proportion.
    c. How large a sample would be required to report the margin of error of .01 at 95% confidence? Would you recommend that *USA Today* attempt to provide this degree of precision? Why or why not?

Case Problem 1 Young Professional Magazine

Young Professional magazine was developed for a target audience of recent college graduates who are in their first 10 years in a business/professional career. In its two years of publication, the magazine has been fairly successful. Now the publisher is interested in expanding the magazine's advertising base. Potential advertisers continually ask about the demographics and interests of subscribers to *Young Professional*. To collect this information, the magazine commissioned a survey to develop a profile of its subscribers. The survey results will be used to help the magazine choose articles of interest and provide advertisers with a profile of subscribers. As a new employee of the magazine, you have been asked to help analyze the survey results.

Some of the survey questions follow:

WEB file
Professional

1. What is your age?
2. Are you: Male_________ Female___________
3. Do you plan to make any real estate purchases in the next two years? Yes______ No______
4. What is the approximate total value of financial investments, exclusive of your home, owned by you or members of your household?
5. How many stock/bond/mutual fund transactions have you made in the past year?

TABLE 8.6 PARTIAL SURVEY RESULTS FOR *YOUNG PROFESSIONAL* MAGAZINE

Age	Gender	Real Estate Purchases	Value of Investments($)	Number of Transactions	Broadband Access	Household Income($)	Children
38	Female	No	12200	4	Yes	75200	Yes
30	Male	No	12400	4	Yes	70300	Yes
41	Female	No	26800	5	Yes	48200	No
28	Female	Yes	19600	6	No	95300	No
31	Female	Yes	15100	5	No	73300	Yes
⋮	⋮	⋮	⋮	⋮	⋮	⋮	⋮

6. Do you have broadband access to the Internet at home? Yes______ No______
7. Please indicate your total household income last year. ___________
8. Do you have children? Yes______ No______

The file entitled Professional contains the responses to these questions. Table 8.6 shows the portion of the file pertaining to the first five survey respondents.

Managerial Report

Prepare a managerial report summarizing the results of the survey. In addition to statistical summaries, discuss how the magazine might use these results to attract advertisers. You might also comment on how the survey results could be used by the magazine's editors to identify topics that would be of interest to readers. Your report should address the following issues, but do not limit your analysis to just these areas.

1. Develop appropriate descriptive statistics to summarize the data.
2. Develop 95% confidence intervals for the mean age and household income of subscribers.
3. Develop 95% confidence intervals for the proportion of subscribers who have broadband access at home and the proportion of subscribers who have children.
4. Would *Young Professional* be a good advertising outlet for online brokers? Justify your conclusion with statistical data.
5. Would this magazine be a good place to advertise for companies selling educational software and computer games for young children?
6. Comment on the types of articles you believe would be of interest to readers of *Young Professional*.

Case Problem 2 Gulf Real Estate Properties

Gulf Real Estate Properties, Inc., is a real estate firm located in southwest Florida. The company, which advertises itself as "expert in the real estate market," monitors condominium sales by collecting data on location, list price, sale price, and number of days it takes to sell each unit. Each condominium is classified as *Gulf View* if it is located directly on the Gulf of Mexico or *No Gulf View* if it is located on the bay or a golf course, near but not on the Gulf. Sample data from the Multiple Listing Service in Naples, Florida, provided recent sales data for 40 Gulf View condominiums and 18 No Gulf View condominiums. Prices are in thousands of dollars. The data are shown in Table 8.7.

Managerial Report

1. Use appropriate descriptive statistics to summarize each of the three variables for the 40 Gulf View condominiums.

TABLE 8.7 SALES DATA FOR GULF REAL ESTATE PROPERTIES

WEB file

GulfProp

Gulf View Condominiums			No Gulf View Condominiums		
List Price	Sale Price	Days to Sell	List Price	Sale Price	Days to Sell
495.0	475.0	130	217.0	217.0	182
379.0	350.0	71	148.0	135.5	338
529.0	519.0	85	186.5	179.0	122
552.5	534.5	95	239.0	230.0	150
334.9	334.9	119	279.0	267.5	169
550.0	505.0	92	215.0	214.0	58
169.9	165.0	197	279.0	259.0	110
210.0	210.0	56	179.9	176.5	130
975.0	945.0	73	149.9	144.9	149
314.0	314.0	126	235.0	230.0	114
315.0	305.0	88	199.8	192.0	120
885.0	800.0	282	210.0	195.0	61
975.0	975.0	100	226.0	212.0	146
469.0	445.0	56	149.9	146.5	137
329.0	305.0	49	160.0	160.0	281
365.0	330.0	48	322.0	292.5	63
332.0	312.0	88	187.5	179.0	48
520.0	495.0	161	247.0	227.0	52
425.0	405.0	149			
675.0	669.0	142			
409.0	400.0	28			
649.0	649.0	29			
319.0	305.0	140			
425.0	410.0	85			
359.0	340.0	107			
469.0	449.0	72			
895.0	875.0	129			
439.0	430.0	160			
435.0	400.0	206			
235.0	227.0	91			
638.0	618.0	100			
629.0	600.0	97			
329.0	309.0	114			
595.0	555.0	45			
339.0	315.0	150			
215.0	200.0	48			
395.0	375.0	135			
449.0	425.0	53			
499.0	465.0	86			
439.0	428.5	158			

2. Use appropriate descriptive statistics to summarize each of the three variables for the 18 No Gulf View condominiums.
3. Compare your summary results. Discuss any specific statistical results that would help a real estate agent understand the condominium market.
4. Develop a 95% confidence interval estimate of the population mean sales price and population mean number of days to sell for Gulf View condominiums. Interpret your results.

5. Develop a 95% confidence interval estimate of the population mean sales price and population mean number of days to sell for No Gulf View condominiums. Interpret your results.
6. Assume the branch manager requested estimates of the mean selling price of Gulf View condominiums with a margin of error of \$40,000 and the mean selling price of No Gulf View condominiums with a margin of error of \$15,000. Using 95% confidence, how large should the sample sizes be?
7. Gulf Real Estate Properties just signed contracts for two new listings: a Gulf View condominium with a list price of \$589,000 and a No Gulf View condominium with a list price of \$285,000. What is your estimate of the final selling price and number of days required to sell each of these units?

Case Problem 3 Metropolitan Research, Inc.

Metropolitan Research, Inc., a consumer research organization, conducts surveys designed to evaluate a wide variety of products and services available to consumers. In one particular study, Metropolitan looked at consumer satisfaction with the performance of automobiles produced by a major Detroit manufacturer. A questionnaire sent to owners of one of the manufacturer's full-sized cars revealed several complaints about early transmission problems. To learn more about the transmission failures, Metropolitan used a sample of actual transmission repairs provided by a transmission repair firm in the Detroit area. The following data show the actual number of miles driven for 50 vehicles at the time of transmission failure.

WEB file

Auto

85,092	32,609	59,465	77,437	32,534	64,090	32,464	59,902
39,323	89,641	94,219	116,803	92,857	63,436	65,605	85,861
64,342	61,978	67,998	59,817	101,769	95,774	121,352	69,568
74,276	66,998	40,001	72,069	25,066	77,098	69,922	35,662
74,425	67,202	118,444	53,500	79,294	64,544	86,813	116,269
37,831	89,341	73,341	85,288	138,114	53,402	85,586	82,256
77,539	88,798						

Managerial Report

1. Use appropriate descriptive statistics to summarize the transmission failure data.
2. Develop a 95% confidence interval for the mean number of miles driven until transmission failure for the population of automobiles with transmission failure. Provide a managerial interpretation of the interval estimate.
3. Discuss the implication of your statistical findings in terms of the belief that some owners of the automobiles experienced early transmission failures.
4. How many repair records should be sampled if the research firm wants the population mean number of miles driven until transmission failure to be estimated with a margin of error of 5000 miles? Use 95% confidence.
5. What other information would you like to gather to evaluate the transmission failure problem more fully?

Appendix 8.1 Interval Estimation with Minitab

We describe the use of Minitab in constructing confidence intervals for a population mean and a population proportion.

Population Mean: σ Known

WEB file

Lloyd's

We illustrate interval estimation using the Lloyd's example in Section 8.1. The amounts spent per shopping trip for the sample of 100 customers are in column C1 of a Minitab worksheet. The population standard deviation $\sigma = 20$ is assumed known. The following steps can be used to compute a 95% confidence interval estimate of the population mean.

Step 1. Select the **Stat** menu
Step 2. Choose **Basic Statistics**
Step 3. Choose **1-Sample Z**
Step 4. When the 1-Sample Z dialog box appears:
  Enter C1 in the **Samples in columns** box
  Enter 20 in the **Standard deviation** box
Step 5. Click **OK**

The Minitab default is a 95% confidence level. In order to specify a different confidence level such as 90%, add the following to step 4.

Select **Options**
When the 1-Sample Z-Options dialog box appears:
  Enter 90 in the **Confidence level** box
Click **OK**

Population Mean: σ Unknown

WEB file

NewBalance

We illustrate interval estimation using the data in Table 8.3 showing the credit card balances for a sample of 70 households. The data are in column C1 of a Minitab worksheet. In this case the population standard deviation σ will be estimated by the sample standard deviation s. The following steps can be used to compute a 95% confidence interval estimate of the population mean.

Step 1. Select the **Stat** menu
Step 2. Choose **Basic Statistics**
Step 3. Choose **1-Sample t**
Step 4. When the 1-Sample t dialog box appears:
  Enter C1 in the **Samples in columns** box
Step 5. Click **OK**

The Minitab default is a 95% confidence level. In order to specify a different confidence level such as 90%, add the following to step 4.

Select **Options**
When the 1-Sample t-Options dialog box appears:
  Enter 90 in the **Confidence level** box
Click **OK**

Population Proportion

WEB file

TeeTimes

We illustrate interval estimation using the survey data for women golfers presented in Section 8.4. The data are in column C1 of a Minitab worksheet. Individual responses are recorded as Yes if the golfer is satisfied with the availability of tee times and No otherwise. The following steps can be used to compute a 95% confidence interval estimate of the proportion of women golfers who are satisfied with the availability of tee times.

Step 1. Select the **Stat** menu
Step 2. Choose **Basic Statistics**
Step 3. Choose **1 Proportion**

Step 4. When the 1 Proportion dialog box appears:
Enter C1 in the **Samples in columns** box
Step 5. Select **Options**
Step 6. When the 1 Proportion-Options dialog box appears:
Select **Use test and interval based on normal distribution**
Click **OK**
Step 7. Click **OK**

The Minitab default is a 95% confidence level. In order to specify a different confidence level such as 90%, enter 90 in the **Confidence Level** box when the 1 Proportion-Options dialog box appears in step 6.

Note: Minitab's 1 Proportion routine uses an alphabetical ordering of the responses and selects the *second response* for the population proportion of interest. In the women golfers example, Minitab used the alphabetical ordering No-Yes and then provided the confidence interval for the proportion of Yes responses. Because Yes was the response of interest, the Minitab output was fine. However, if Minitab's alphabetical ordering does not provide the response of interest, select any cell in the column and use the sequence: Editor > Column > Value Order. It will provide you with the option of entering a user-specified order, but you must list the response of interest second in the define-an-order box.

Appendix 8.2 Interval Estimation Using Excel

We describe the use of Excel in constructing confidence intervals for a population mean and a population proportion.

Population Mean: σ Known

WEB file
Lloyd's

We illustrate interval estimation using the Lloyd's example in Section 8.1. The population standard deviation $\sigma = 20$ is assumed known. The amounts spent for the sample of 100 customers are in column A of an Excel worksheet. Excel's AVERAGE and CONFIDENCE.NORM functions can be used to compute the point estimate and the margin of error for an estimate of the population mean.

Step 1. Select cell C1 and enter the Excel formula =AVERAGE(A2:A101)
Step 2. Select cell C2 and enter the Excel formula =CONFIDENCE.NORM(.05,20,100)

The three inputs of the CONFIDENCE.NORM function are

Alpha = 1 − confidence coefficient = 1 − .95 = .05
The population standard deviation = 20
The sample size = 100

The point estimate of the population mean (82) in cell C1 and the margin of error (3.92) in cell C2 allow the confidence interval for the population mean to be easily computed.

Population Mean: σ Unknown

WEB file
NewBalance

We illustrate interval estimation using the data in Table 8.3, which show the credit card balances for a sample of 70 households. The data are in column A of an Excel worksheet. The following steps can be used to compute the point estimate and the margin of error for an interval estimate of a population mean. We will use Excel's Descriptive Statistics Tool described in Chapter 3.

FIGURE 8.10 INTERVAL ESTIMATION OF THE POPULATION MEAN CREDIT CARD BALANCE USING EXCEL

	A	B	C	D	E	F
1	**NewBalance**		*NewBalance*			
2	9430				Point Estimate	
3	7535		Mean	9312		
4	4078		Standard Error	478.9281		
5	5604		Median	9466		
6	5179		Mode	13627		
7	4416		Standard Deviation	4007		
8	10676		Sample Variance	16056048		
9	1627		Kurtosis	-0.2960		
10	10112		Skewness	0.1879		
11	6567		Range	18648		
12	13627		Minimum	615		
13	18719		Maximum	19263		
14	14661		Sum	651840		
15	12195		Count	70	Margin of Error	
16	10544		Confidence Level(95.0%)	955		
17	13659					
70	9743					
71	10324					
72						

Note: Rows 18 to 69 are hidden.

Step 1. Click the **Data** tab on the Ribbon
Step 2. In the **Analysis** group, click **Data Analysis**
Step 3. Choose **Descriptive Statistics** from the list of Analysis Tools
Step 4. When the Descriptive Statistics dialog box appears:
Enter A1:A71 in the **Input Range** box
Select **Grouped by Columns**
Select **Labels in First Row**
Select **Output Range**
Enter C1 in the Output Range box
Select **Summary Statistics**
Select **Confidence Level for Mean**
Enter 95 in the Confidence Level for Mean box
Click **OK**

The summary statistics will appear in columns C and D. The point estimate of the population mean appears in cell D3. The margin of error, labeled "Confidence Level(95.0%)," appears in cell D16. The point estimate (\$9312) and the margin of error (\$955) allow the confidence interval for the population mean to be easily computed. The output from this Excel procedure is shown in Figure 8.10.

FIGURE 8.11 EXCEL TEMPLATE FOR INTERVAL ESTIMATION OF A POPULATION PROPORTION

	A	B	C	D	E
1	**Response**		**Interval Estimate of a Population Proportion**		
2	Yes				
3	No		**Sample Size**	=COUNTA(A2:A901)	
4	Yes		**Response of Interest**	Yes	
5	Yes		**Count for Response**	=COUNTIF(A2:A901,D4)	
6	No		**Sample Proportion**	=D5/D3	
7	No				
8	No		**Confidence Coefficient**	0.95	
9	Yes		***z* Value**	=NORM.S.INV(0.5+D8/2)	
10	Yes				
11	Yes		**Standard Error**	=SQRT(D6*(1-D6)/D3)	
12	No		**Margin of Error**	=D9*D11	
13	No				
14	Yes		**Point Estimate**	=D6	
15	No		**Lower Limit**	=D14-D12	
16	No		**Upper Limit**	=D14+D12	
17	Yes				
18	No				
901	Yes				
902					

WEB file

TeeTimes

	A	B	C	D	E	F	G
1	**Response**		**Interval Estimate of a Population Proportion**				
2	Yes						
3	No		**Sample Size**	900	Enter the response of interest		
4	Yes		**Response of Interest**	Yes			
5	Yes		**Count for Response**	396			
6	No		**Sample Proportion**	0.4400			
7	No				Enter the confidence coefficient		
8	No		**Confidence Coefficient**	0.95			
9	Yes		***z* Value**	1.960			
10	Yes						
11	Yes		**Standard Error**	0.0165			
12	No		**Margin of Error**	0.0324			
13	No						
14	Yes		**Point Estimate**	0.4400			
15	No		**Lower Limit**	0.4076			
16	No		**Upper Limit**	0.4724			
17	Yes						
18	No						
901	Yes						
902							

Note: Rows 19 to 900 are hidden.

Population Proportion

WEB file

Interval p

We illustrate interval estimation using the survey data for women golfers presented in Section 8.4. The data are in column A of an Excel worksheet. Individual responses are recorded as Yes if the golfer is satisfied with the availability of tee times and No otherwise. Excel does not offer a built-in routine to handle the estimation of a population proportion; however, it is relatively easy to develop an Excel template that can be used for this purpose. The template shown in Figure 8.11 provides the 95% confidence interval estimate of the proportion of women golfers who are satisfied with the availability of tee times. Note that the background worksheet in Figure 8.11 shows the cell formulas that provide the interval estimation results shown in the foreground worksheet. The following steps are necessary to use the template for this data set.

Step 1. Enter the data range A2:A901 into the =COUNTA cell formula in cell D3
Step 2. Enter Yes as the response of interest in cell D4
Step 3. Enter the data range A2:A901 into the =COUNTIF cell formula in cell D5
Step 4. Enter .95 as the confidence coefficient in cell D8

The template automatically provides the confidence interval in cells D15 and D16.

This template can be used to compute the confidence interval for a population proportion for other applications. For instance, to compute the interval estimate for a new data set, enter the new sample data into column A of the worksheet and then make the changes to the four cells as shown. If the new sample data have already been summarized, the sample data do not have to be entered into the worksheet. In this case, enter the sample size into cell D3 and the sample proportion into cell D6; the worksheet template will then provide the confidence interval for the population proportion. The worksheet in Figure 8.11 is available in the file Interval p on the website that accompanies this book.

Appendix 8.3 Interval Estimation with StatTools

In this appendix we show how StatTools can be used to develop an interval estimate of a population mean for the σ unknown case and determine the sample size needed to provide a desired margin of error.

Population Mean: σ Unknown Case

In this case the population standard deviation σ will be estimated by the sample standard deviation s. We use the credit card balance data in Table 8.3 to illustrate. Begin by using the Data Set Manager to create a StatTools data set for these data using the procedure described in the appendix to Chapter 1. The following steps can be used to compute a 95% confidence interval estimate of the population mean.

WEB file
NewBalance

Step 1. Click the **StatTools** tab on the Ribbon
Step 2. In the **Analyses** group, click **Statistical Inference**
Step 3. Choose the **Confidence Interval** option
Step 4. Choose Mean/Std. Deviation
Step 5. When the StatTools-Confidence Interval for Mean/Std. Deviation dialog box appears:
For **Analysis Type** choose **One-Sample Analysis**
In the **Variables** section, select **NewBalance**
In the **Confidence Intervals to Calculate** section:
Select the **For the Mean** option
Select 95% for the **Confidence Level**
Click **OK**

Some descriptive statistics and the confidence interval will appear.

Determining the Sample Size

In Section 8.3 we showed how to determine the sample size needed to provide a desired margin of error. The example used involved a study designed to estimate the population mean daily rental cost for a midsize automobile in the United States. The project director specified that the population mean daily rental cost be estimated with a margin of error of

$2 and a 95% level of confidence. Sample data from a previous study provided a sample standard deviation of $9.65; this value was used as the planning value for the population standard deviation. The following steps can be used to compute the recommended sample size required to provide a 95% confidence interval estimate of the population mean with a margin of error of $2.

The half-length of interval is the margin of error.

Step 1. Click the **StatTools** tab on the Ribbon
Step 2. In the **Analyses** group, click **Statistical Inference**
Step 3. Choose the **Sample Size Selection** option
Step 4. When the StatTools-Sample Size Selection dialog box appears:
In the **Parameter to Estimate** section, select **Mean**
In the **Confidence Interval Specification** section:
Select **95%** for the **Confidence Level**
Enter **2** in the **Half-Length of Interval** box
Enter **9.65** in the **Estimated Std Dev** box
Click **OK**

The output showing a recommended sample size of 90 will appear.

Population Proportion

We illustrate using the survey data for women golfers presented in Section 8.4. Begin by using the Data Set Manager to create a StatTools data set for these data using the procedure described in the appendix to Chapter 1. The following steps can be used to compute a 95% confidence interval estimate of the population mean.

WEB file
Tee Times

Step 1. Click the **StatTools** tab on the Ribbon
Step 2. In the **Analyses** group, click **Statistical Inference**
Step 3. Choose **Confidence Interval**
Step 4. Choose **Proportion**
Step 5. When the StatTools-Confidence Interval for Proportion dialog box appears:
For **Analysis Type** choose **One-Sample Analysis**
In the **Variables** section, select **Response**
In the **Categories to Analyze** section, select **Yes**
In the **Options** section, enter 95% in the **Confidence Level** box
Click **OK**

Some descriptive statistics and the confidence interval will appear.

StatTools also provides a capability for determining the appropriate sample size to provide a desired margin of error. The steps are similar to those for determining the sample size in the previous subsection.

CHAPTER 9

Hypothesis Tests

CONTENTS

STATISTICS *in* PRACTICE

JOHN MORRELL & COMPANY*
CINCINNATI, OHIO

John Morrell & Company, which began in England in 1827, is considered the oldest continuously operating meat manufacturer in the United States. It is a wholly owned and independently managed subsidiary of Smithfield Foods, Smithfield, Virginia. John Morrell & Company offers an extensive product line of processed meats and fresh pork to consumers under 13 regional brands including John Morrell, E-Z-Cut, Tobin's First Prize, Dinner Bell, Hunter, Kretschmar, Rath, Rodeo, Shenson, Farmers Hickory Brand, Iowa Quality, and Peyton's. Each regional brand enjoys high brand recognition and loyalty among consumers.

Market research at Morrell provides management with up-to-date information on the company's various products and how the products compare with competing brands of similar products. A recent study compared a Beef Pot Roast made by Morrell to similar beef products from two major competitors. In the three-product comparison test, a sample of consumers was used to indicate how the products rated in terms of taste, appearance, aroma, and overall preference.

One research question concerned whether the Beef Pot Roast made by Morrell was the preferred choice of more than 50% of the consumer population. Letting p indicate the population proportion preferring Morrell's product, the hypothesis test for the research question is as follows:

$$H_0: p \leq .50$$
$$H_a: p > .50$$

The null hypothesis H_0 indicates the preference for Morrell's product is less than or equal to 50%. If the sample data support rejecting H_0 in favor of the alternative hypothesis H_a, Morrell will draw the research conclusion that in a three-product comparison, their Beef Pot Roast is preferred by more than 50% of the consumer population.

Market research at Morrell provides up-to-date information on their various products and how they compare with competing brands.
© Jeff Greenberg/Alamy Limited.

In an independent taste test study using a sample of 224 consumers in Cincinnati, Milwaukee, and Los Angeles, 150 consumers selected the Beef Pot Roast made by Morrell as the preferred product. Using statistical hypothesis testing procedures, the null hypothesis H_0 was rejected. The study provided statistical evidence supporting H_a and the conclusion that the Morrell product is preferred by more than 50% of the consumer population.

The point estimate of the population proportion was $\bar{p} = 150/224 = .67$. Thus, the sample data provided support for a food magazine advertisement showing that in a three-product taste comparison, Beef Pot Roast made by Morrell was "preferred 2 to 1 over the competition."

In this chapter we will discuss how to formulate hypotheses and how to conduct tests like the one used by Morrell. Through the analysis of sample data, we will be able to determine whether a hypothesis should or should not be rejected.

*The authors are indebted to Marty Butler, Vice President of Marketing, John Morrell, for providing this Statistics in Practice.

In Chapters 7 and 8 we showed how a sample could be used to develop point and interval estimates of population parameters. In this chapter we continue the discussion of statistical inference by showing how hypothesis testing can be used to determine whether a statement about the value of a population parameter should or should not be rejected.

In hypothesis testing we begin by making a tentative assumption about a population parameter. This tentative assumption is called the **null hypothesis** and is denoted by H_0. We then define another hypothesis, called the **alternative hypothesis**, which is the opposite of what is stated in the null hypothesis. The alternative hypothesis is denoted by H_a.

The hypothesis testing procedure uses data from a sample to test the two competing statements indicated by H_0 and H_a.

This chapter shows how hypothesis tests can be conducted about a population mean and a population proportion. We begin by providing examples that illustrate approaches to developing null and alternative hypotheses.

9.1 Developing Null and Alternative Hypotheses

It is not always obvious how the null and alternative hypotheses should be formulated. Care must be taken to structure the hypotheses appropriately so that the hypothesis testing conclusion provides the information the researcher or decision maker wants. The context of the situation is very important in determining how the hypotheses should be stated. All hypothesis testing applications involve collecting a sample and using the sample results to provide evidence for drawing a conclusion. Good questions to consider when formulating the null and alternative hypotheses are, What is the purpose of collecting the sample? What conclusions are we hoping to make?

Learning to correctly formulate hypotheses will take some practice. Expect some initial confusion over the proper choice of the null and alternative hypotheses. The examples in this section are intended to provide guidelines.

In the chapter introduction, we stated that the null hypothesis H_0 is a tentative assumption about a population parameter such as a population mean or a population proportion. The alternative hypothesis H_a is a statement that is the opposite of what is stated in the null hypothesis. In some situations it is easier to identify the alternative hypothesis first and then develop the null hypothesis. In other situations it is easier to identify the null hypothesis first and then develop the alternative hypothesis. We will illustrate these situations in the following examples.

The Alternative Hypothesis as a Research Hypothesis

Many applications of hypothesis testing involve an attempt to gather evidence in support of a research hypothesis. In these situations, it is often best to begin with the alternative hypothesis and make it the conclusion that the researcher hopes to support. Consider a particular automobile that currently attains a fuel efficiency of 24 miles per gallon in city driving. A product research group has developed a new fuel injection system designed to increase the miles-per-gallon rating. The group will run controlled tests with the new fuel injection system looking for statistical support for the conclusion that the new fuel injection system provides more miles per gallon than the current system.

Several new fuel injection units will be manufactured, installed in test automobiles, and subjected to research-controlled driving conditions. The sample mean miles per gallon for these automobiles will be computed and used in a hypothesis test to determine if it can be concluded that the new system provides more than 24 miles per gallon. In terms of the population mean miles per gallon μ, the research hypothesis $\mu > 24$ becomes the alternative hypothesis. Since the current system provides an average or mean of 24 miles per gallon, we will make the tentative assumption that the new system is not any better than the current system and choose $\mu \leq 24$ as the null hypothesis. The null and alternative hypotheses are:

$$H_0: \mu \leq 24$$
$$H_a: \mu > 24$$

If the sample results lead to the conclusion to reject H_0, the inference can be made that $H_a: \mu > 24$ is true. The researchers have the statistical support to state that the new fuel injection system increases the mean number of miles per gallon. The production of automobiles with the new fuel injection system should be considered. However, if the sample results lead to the conclusion that H_0 cannot be rejected, the researchers cannot conclude

The conclusion that the research hypothesis is true is made if the sample data provide sufficient evidence to show that the null hypothesis can be rejected.

that the new fuel injection system is better than the current system. Production of automobiles with the new fuel injection system on the basis of better gas mileage cannot be justified. Perhaps more research and further testing can be conducted.

Successful companies stay competitive by developing new products, new methods, new systems, and the like, that are better than what is currently available. Before adopting something new, it is desirable to conduct research to determine if there is statistical support for the conclusion that the new approach is indeed better. In such cases, the research hypothesis is stated as the alternative hypothesis. For example, a new teaching method is developed that is believed to be better than the current method. The alternative hypothesis is that the new method is better. The null hypothesis is that the new method is no better than the old method. A new sales force bonus plan is developed in an attempt to increase sales. The alternative hypothesis is that the new bonus plan increases sales. The null hypothesis is that the new bonus plan does not increase sales. A new drug is developed with the goal of lowering blood pressure more than an existing drug. The alternative hypothesis is that the new drug lowers blood pressure more than the existing drug. The null hypothesis is that the new drug does not provide lower blood pressure than the existing drug. In each case, rejection of the null hypothesis H_0 provides statistical support for the research hypothesis. We will see many examples of hypothesis tests in research situations such as these throughout this chapter and in the remainder of the text.

The Null Hypothesis as an Assumption to Be Challenged

Of course, not all hypothesis tests involve research hypotheses. In the following discussion we consider applications of hypothesis testing where we begin with a belief or an assumption that a statement about the value of a population parameter is true. We will then use a hypothesis test to challenge the assumption and determine if there is statistical evidence to conclude that the assumption is incorrect. In these situations, it is helpful to develop the null hypothesis first. The null hypothesis H_0 expresses the belief or assumption about the value of the population parameter. The alternative hypothesis H_a is that the belief or assumption is incorrect.

As an example, consider the situation of a manufacturer of soft drink products. The label on a soft drink bottle states that it contains 67.6 fluid ounces. We consider the label correct provided the population mean filling weight for the bottles is *at least* 67.6 fluid ounces. Without any reason to believe otherwise, we would give the manufacturer the benefit of the doubt and assume that the statement provided on the label is correct. Thus, in a hypothesis test about the population mean fluid weight per bottle, we would begin with the assumption that the label is correct and state the null hypothesis as $\mu \geq 67.6$. The challenge to this assumption would imply that the label is incorrect and the bottles are being underfilled. This challenge would be stated as the alternative hypothesis $\mu < 67.6$. Thus, the null and alternative hypotheses are:

$$H_0: \mu \geq 67.6$$
$$H_a: \mu < 67.6$$

A manufacturer's product information is usually assumed to be true and stated as the null hypothesis. The conclusion that the information is incorrect can be made if the null hypothesis is rejected.

A government agency with the responsibility for validating manufacturing labels could select a sample of soft drinks bottles, compute the sample mean filling weight, and use the sample results to test the preceding hypotheses. If the sample results lead to the conclusion to reject H_0, the inference that H_a: $\mu < 67.6$ is true can be made. With this statistical support, the agency is justified in concluding that the label is incorrect and underfilling of the bottles is occurring. Appropriate action to force the manufacturer to comply with labeling standards would be considered. However, if the sample results indicate H_0 cannot be rejected, the assumption that the manufacturer's labeling is correct cannot be rejected. With this conclusion, no action would be taken.

Let us now consider a variation of the soft drink bottle filling example by viewing the same situation from the manufacturer's point of view. The bottle-filling operation has been designed to fill soft drink bottles with 67.6 fluid ounces as stated on the label. The company does not want to underfill the containers because that could result in an underfilling complaint from customers or, perhaps, a government agency. However, the company does not want to overfill containers either because putting more soft drink than necessary into the containers would be an unnecessary cost. The company's goal would be to adjust the bottle-filling operation so that the population mean filling weight per bottle is 67.6 fluid ounces as specified on the label.

Although this is the company's goal, from time to time any production process can get out of adjustment. If this occurs in our example, underfilling or overfilling of the soft drink bottles will occur. In either case, the company would like to know about it in order to correct the situation by readjusting the bottle-filling operation to the designed 67.6 fluid ounces. In a hypothesis testing application, we would again begin with the assumption that the production process is operating correctly and state the null hypothesis as $\mu = 67.6$ fluid ounces. The alternative hypothesis that challenges this assumption is that $\mu \neq 67.6$, which indicates either overfilling or underfilling is occurring. The null and alternative hypotheses for the manufacturer's hypothesis test are:

$$H_0\colon \mu = 67.6$$
$$H_a\colon \mu \neq 67.6$$

Suppose that the soft drink manufacturer uses a quality control procedure to periodically select a sample of bottles from the filling operation and computes the sample mean filling weight per bottle. If the sample results lead to the conclusion to reject H_0, the inference is made that $H_a\colon \mu \neq 67.6$ is true. We conclude that the bottles are not being filled properly and the production process should be adjusted to restore the population mean to 67.6 fluid ounces per bottle. However, if the sample results indicate H_0 cannot be rejected, the assumption that the manufacturer's bottle filling operation is functioning properly cannot be rejected. In this case, no further action would be taken and the production operation would continue to run.

The two preceding forms of the soft drink manufacturing hypothesis test show that the null and alternative hypotheses may vary depending upon the point of view of the researcher or decision maker. To correctly formulate hypotheses it is important to understand the context of the situation and structure the hypotheses to provide the information the researcher or decision maker wants.

Summary of Forms for Null and Alternative Hypotheses

The hypothesis tests in this chapter involve two population parameters: the population mean and the population proportion. Depending on the situation, hypothesis tests about a population parameter may take one of three forms: two use inequalities in the null hypothesis; the third uses an equality in the null hypothesis. For hypothesis tests involving a population mean, we let μ_0 denote the hypothesized value and we must choose one of the following three forms for the hypothesis test.

The three possible forms of hypotheses H_0 and H_a are shown here. Note that the equality always *appears in the null hypothesis H_0.*

$$H_0\colon \mu \geq \mu_0 \qquad H_0\colon \mu \leq \mu_0 \qquad H_0\colon \mu = \mu_0$$
$$H_a\colon \mu < \mu_0 \qquad H_a\colon \mu > \mu_0 \qquad H_a\colon \mu \neq \mu_0$$

For reasons that will be clear later, the first two forms are called one-tailed tests. The third form is called a two-tailed test.

In many situations, the choice of H_0 and H_a is not obvious and judgment is necessary to select the proper form. However, as the preceding forms show, the equality part of the expression (either $\geq$, $\leq$, or $=$) *always* appears in the null hypothesis. In selecting the proper form of H_0 and H_a, keep in mind that the alternative hypothesis is often what the test is attempting to establish. Hence, asking whether the user is looking for evidence to support $\mu < \mu_0$, $\mu > \mu_0$, or $\mu \neq \mu_0$ will help determine H_a. The following exercises are designed to provide practice in choosing the proper form for a hypothesis test involving a population mean.

Exercises

1. The manager of the Danvers-Hilton Resort Hotel stated that the mean guest bill for a weekend is \$600 or less. A member of the hotel's accounting staff noticed that the total charges for guest bills have been increasing in recent months. The accountant will use a sample of future weekend guest bills to test the manager's claim.
   a. Which form of the hypotheses should be used to test the manager's claim? Explain.

$$H_0: \mu \geq 600 \qquad H_0: \mu \leq 600 \qquad H_0: \mu = 600$$
$$H_a: \mu < 600 \qquad H_a: \mu > 600 \qquad H_a: \mu \neq 600$$

   b. What conclusion is appropriate when H_0 cannot be rejected?
   c. What conclusion is appropriate when H_0 can be rejected?

SELF test

2. The manager of an automobile dealership is considering a new bonus plan designed to increase sales volume. Currently, the mean sales volume is 14 automobiles per month. The manager wants to conduct a research study to see whether the new bonus plan increases sales volume. To collect data on the plan, a sample of sales personnel will be allowed to sell under the new bonus plan for a one-month period.
   a. Develop the null and alternative hypotheses most appropriate for this situation.
   b. Comment on the conclusion when H_0 cannot be rejected.
   c. Comment on the conclusion when H_0 can be rejected.
3. A production line operation is designed to fill cartons with laundry detergent to a mean weight of 32 ounces. A sample of cartons is periodically selected and weighed to determine whether underfilling or overfilling is occurring. If the sample data lead to a conclusion of underfilling or overfilling, the production line will be shut down and adjusted to obtain proper filling.
   a. Formulate the null and alternative hypotheses that will help in deciding whether to shut down and adjust the production line.
   b. Comment on the conclusion and the decision when H_0 cannot be rejected.
   c. Comment on the conclusion and the decision when H_0 can be rejected.
4. Because of high production-changeover time and costs, a director of manufacturing must convince management that a proposed manufacturing method reduces costs before the new method can be implemented. The current production method operates with a mean cost of \$220 per hour. A research study will measure the cost of the new method over a sample production period.
   a. Develop the null and alternative hypotheses most appropriate for this study.
   b. Comment on the conclusion when H_0 cannot be rejected.
   c. Comment on the conclusion when H_0 can be rejected.

9.2 Type I and Type II Errors

The null and alternative hypotheses are competing statements about the population. Either the null hypothesis H_0 is true or the alternative hypothesis H_a is true, but not both. Ideally the hypothesis testing procedure should lead to the acceptance of H_0 when H_0 is true and the

TABLE 9.1 ERRORS AND CORRECT CONCLUSIONS IN HYPOTHESIS TESTING

		Population Condition	
		H_0 True	H_a True
Conclusion	Accept H_0	Correct Conclusion	Type II Error
	Reject H_0	Type I Error	Correct Conclusion

rejection of H_0 when H_a is true. Unfortunately, the correct conclusions are not always possible. Because hypothesis tests are based on sample information, we must allow for the possibility of errors. Table 9.1 illustrates the two kinds of errors that can be made in hypothesis testing.

The first row of Table 9.1 shows what can happen if the conclusion is to accept H_0. If H_0 is true, this conclusion is correct. However, if H_a is true, we make a **Type II error**; that is, we accept H_0 when it is false. The second row of Table 9.1 shows what can happen if the conclusion is to reject H_0. If H_0 is true, we make a **Type I error**; that is, we reject H_0 when it is true. However, if H_a is true, rejecting H_0 is correct.

Recall the hypothesis testing illustration discussed in Section 9.1 in which an automobile product research group developed a new fuel injection system designed to increase the miles-per-gallon rating of a particular automobile. With the current model obtaining an average of 24 miles per gallon, the hypothesis test was formulated as follows.

$$H_0: \mu \leq 24$$
$$H_a: \mu > 24$$

The alternative hypothesis, $H_a: \mu > 24$, indicates that the researchers are looking for sample evidence to support the conclusion that the population mean miles per gallon with the new fuel injection system is greater than 24.

In this application, the Type I error of rejecting H_0 when it is true corresponds to the researchers claiming that the new system improves the miles-per-gallon rating ($\mu > 24$) when in fact the new system is not any better than the current system. In contrast, the Type II error of accepting H_0 when it is false corresponds to the researchers concluding that the new system is not any better than the current system ($\mu \leq 24$) when in fact the new system improves miles-per-gallon performance.

For the miles-per-gallon rating hypothesis test, the null hypothesis is $H_0: \mu \leq 24$. Suppose the null hypothesis is true as an equality; that is, $\mu = 24$. The probability of making a Type I error when the null hypothesis is true as an equality is called the **level of significance**. Thus, for the miles-per-gallon rating hypothesis test, the level of significance is the probability of rejecting $H_0: \mu \leq 24$ when $\mu = 24$. Because of the importance of this concept, we now restate the definition of level of significance.

LEVEL OF SIGNIFICANCE

The level of significance is the probability of making a Type I error when the null hypothesis is true as an equality.

The Greek symbol α (alpha) is used to denote the level of significance, and common choices for α are .05 and .01.

In practice, the person responsible for the hypothesis test specifies the level of significance. By selecting α, that person is controlling the probability of making a Type I error. If the cost of making a Type I error is high, small values of α are preferred. If the cost of making a Type I error is not too high, larger values of α are typically used. Applications of hypothesis testing that only control for the Type I error are called *significance tests.* Many applications of hypothesis testing are of this type.

If the sample data are consistent with the null hypothesis H_0, we will follow the practice of concluding "do not reject H_0." This conclusion is preferred over "accept H_0," because the conclusion to accept H_0 puts us at risk of making a Type II error.

Although most applications of hypothesis testing control for the probability of making a Type I error, they do not always control for the probability of making a Type II error. Hence, if we decide to accept H_0, we cannot determine how confident we can be with that decision. Because of the uncertainty associated with making a Type II error when conducting significance tests, statisticians usually recommend that we use the statement "do not reject H_0" instead of "accept H_0." Using the statement "do not reject H_0" carries the recommendation to withhold both judgment and action. In effect, by not directly accepting H_0, the statistician avoids the risk of making a Type II error. Whenever the probability of making a Type II error has not been determined and controlled, we will not make the statement "accept H_0." In such cases, only two conclusions are possible: *do not reject* H_0 or *reject* H_0.

Although controlling for a Type II error in hypothesis testing is not common, it can be done. In Sections 9.7 and 9.8 we will illustrate procedures for determining and controlling the probability of making a Type II error. If proper controls have been established for this error, action based on the "accept H_0" conclusion can be appropriate.

NOTES AND COMMENTS

Walter Williams, syndicated columnist and professor of economics at George Mason University, points out that the possibility of making a Type I or a Type II error is always present in decision making (*The Cincinnati Enquirer,* August 14, 2005). He notes that the Food and Drug Administration (FDA) runs the risk of making these errors in its drug approval process. The FDA must either approve a new drug or not approve it. Thus the FDA runs the risk of making a Type I error by approving a new drug that is not safe and effective, or making a Type II error by failing to approve a new drug that is safe and effective. Regardless of the decision made, the possibility of making a costly error cannot be eliminated.

Exercises

SELF test

5. Duke Energy reported that the cost of electricity for an efficient home in a particular neighborhood of Cincinnati, Ohio, was \$104 per month (*Home Energy Report,* Duke Energy, March 2012). A researcher believes that the cost of electricity for a comparable neighborhood in Chicago, Illinois, is higher. A sample of homes in this Chicago neighborhood will be taken and the sample mean monthly cost of electricity will be used to test the following null and alternative hypotheses.

$$H_0: \mu \leq 104$$
$$H_a: \mu > 104$$

   a. Assume the sample data led to rejection of the null hypothesis. What would be your conclusion about the cost of electricity in the Chicago neighborhood?
   b. What is the Type I error in this situation? What are the consequences of making this error?
   c. What is the Type II error in this situation? What are the consequences of making this error?

6. The label on a 3-quart container of orange juice states that the orange juice contains an average of 1 gram of fat or less. Answer the following questions for a hypothesis test that could be used to test the claim on the label.
   a. Develop the appropriate null and alternative hypotheses.
   b. What is the Type I error in this situation? What are the consequences of making this error?
   c. What is the Type II error in this situation? What are the consequences of making this error?
7. Carpetland salespersons average $8000 per week in sales. Steve Contois, the firm's vice president, proposes a compensation plan with new selling incentives. Steve hopes that the results of a trial selling period will enable him to conclude that the compensation plan increases the average sales per salesperson.
   a. Develop the appropriate null and alternative hypotheses.
   b. What is the Type I error in this situation? What are the consequences of making this error?
   c. What is the Type II error in this situation? What are the consequences of making this error?
8. Suppose a new production method will be implemented if a hypothesis test supports the conclusion that the new method reduces the mean operating cost per hour.
   a. State the appropriate null and alternative hypotheses if the mean cost for the current production method is $220 per hour.
   b. What is the Type I error in this situation? What are the consequences of making this error?
   c. What is the Type II error in this situation? What are the consequences of making this error?

9.3 Population Mean: σ Known

In Chapter 8 we said that the σ known case corresponds to applications in which historical data and/or other information are available that enable us to obtain a good estimate of the population standard deviation prior to sampling. In such cases the population standard deviation can, for all practical purposes, be considered known. In this section we show how to conduct a hypothesis test about a population mean for the σ known case.

The methods presented in this section are exact if the sample is selected from a population that is normally distributed. In cases where it is not reasonable to assume the population is normally distributed, these methods are still applicable if the sample size is large enough. We provide some practical advice concerning the population distribution and the sample size at the end of this section.

One-Tailed Test

One-tailed tests about a population mean take one of the following two forms.

Lower Tail Test	Upper Tail Test
$H_0: \mu \geq \mu_0$	$H_0: \mu \leq \mu_0$
$H_a: \mu < \mu_0$	$H_a: \mu > \mu_0$

Let us consider an example involving a lower tail test.

The Federal Trade Commission (FTC) periodically conducts statistical studies designed to test the claims that manufacturers make about their products. For example, the label on a large can of Hilltop Coffee states that the can contains 3 pounds of coffee. The FTC knows that Hilltop's production process cannot place exactly 3 pounds of coffee in each can, even if the mean filling weight for the population of all cans filled is 3 pounds per can. However, as long as the population mean filling weight is at least 3 pounds per can, the rights of consumers will be protected. Thus, the FTC interprets the label information on a large can of coffee as a claim by Hilltop that the population mean filling weight is at least 3 pounds per can. We will show how the FTC can check Hilltop's claim by conducting a lower tail hypothesis test.

The first step is to develop the null and alternative hypotheses for the test. If the population mean filling weight is at least 3 pounds per can, Hilltop's claim is correct. This establishes the null hypothesis for the test. However, if the population mean weight is less than 3 pounds per can, Hilltop's claim is incorrect. This establishes the alternative hypothesis. With μ denoting the population mean filling weight, the null and alternative hypotheses are as follows:

$$H_0: \mu \geq 3$$
$$H_a: \mu < 3$$

Note that the hypothesized value of the population mean is $\mu_0 = 3$.

If the sample data indicate that H_0 cannot be rejected, the statistical evidence does not support the conclusion that a label violation has occurred. Hence, no action should be taken against Hilltop. However, if the sample data indicate H_0 can be rejected, we will conclude that the alternative hypothesis, H_a: $\mu < 3$, is true. In this case a conclusion of underfilling and a charge of a label violation against Hilltop would be justified.

Suppose a sample of 36 cans of coffee is selected and the sample mean $\bar{x}$ is computed as an estimate of the population mean μ. If the value of the sample mean $\bar{x}$ is less than 3 pounds, the sample results will cast doubt on the null hypothesis. What we want to know is how much less than 3 pounds must $\bar{x}$ be before we would be willing to declare the difference significant and risk making a Type I error by falsely accusing Hilltop of a label violation. A key factor in addressing this issue is the value the decision maker selects for the level of significance.

As noted in the preceding section, the level of significance, denoted by α, is the probability of making a Type I error by rejecting H_0 when the null hypothesis is true as an equality. The decision maker must specify the level of significance. If the cost of making a Type I error is high, a small value should be chosen for the level of significance. If the cost is not high, a larger value is more appropriate. In the Hilltop Coffee study, the director of the FTC's testing program made the following statement: "If the company is meeting its weight specifications at $\mu = 3$, I do not want to take action against them. But, I am willing to risk a 1% chance of making such an error." From the director's statement, we set the level of significance for the hypothesis test at $\alpha = .01$. Thus, we must design the hypothesis test so that the probability of making a Type I error when $\mu = 3$ is .01.

For the Hilltop Coffee study, by developing the null and alternative hypotheses and specifying the level of significance for the test, we carry out the first two steps required in conducting every hypothesis test. We are now ready to perform the third step of hypothesis testing: collect the sample data and compute the value of what is called a test statistic.

Test statistic For the Hilltop Coffee study, previous FTC tests show that the population standard deviation can be assumed known with a value of $\sigma = .18$. In addition, these tests also show that the population of filling weights can be assumed to have a normal distribution. From the study of sampling distributions in Chapter 7 we know that if the population from which we are sampling is normally distributed, the sampling distribution of $\bar{x}$ will also be normally distributed. Thus, for the Hilltop Coffee study, the sampling distribution of $\bar{x}$ is normally distributed. With a known value of $\sigma = .18$ and a sample size of $n = 36$, Figure 9.1 shows the sampling distribution of $\bar{x}$ when the null hypothesis is true as an equality; that is, when $\mu = \mu_0 = 3$.[1] Note that the standard error of $\bar{x}$ is given by $\sigma_{\bar{x}} = \sigma/\sqrt{n} = .18/\sqrt{36} = .03$.

The standard error of $\bar{x}$ is the standard deviation of the sampling distribution of $\bar{x}$.

Because the sampling distribution of $\bar{x}$ is normally distributed, the sampling distribution of

$$z = \frac{\bar{x} - \mu_0}{\sigma_{\bar{x}}} = \frac{\bar{x} - 3}{.03}$$

[1]In constructing sampling distributions for hypothesis tests, it is assumed that H_0 is satisfied as an equality.

FIGURE 9.1 SAMPLING DISTRIBUTION OF $\bar{x}$ FOR THE HILLTOP COFFEE STUDY WHEN THE NULL HYPOTHESIS IS TRUE AS AN EQUALITY ($\mu = 3$)

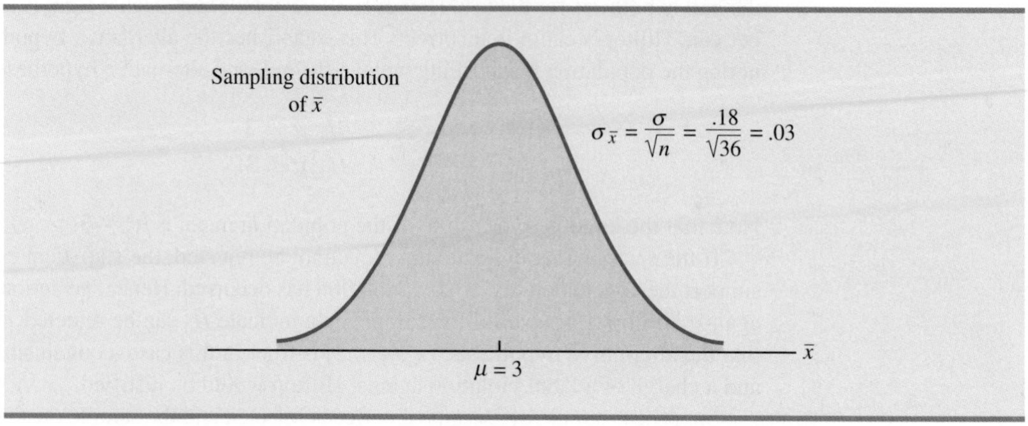

is a standard normal distribution. A value of $z = -1$ means that the value of $\bar{x}$ is one standard error below the hypothesized value of the mean, a value of $z = -2$ means that the value of $\bar{x}$ is two standard errors below the hypothesized value of the mean, and so on. We can use the standard normal probability table to find the lower tail probability corresponding to any z value. For instance, the lower tail area at $z = -3.00$ is .0013. Hence, the probability of obtaining a value of z that is three or more standard errors below the mean is .0013. As a result, the probability of obtaining a value of $\bar{x}$ that is 3 or more standard errors below the hypothesized population mean $\mu_0 = 3$ is also .0013. Such a result is unlikely if the null hypothesis is true.

For hypothesis tests about a population mean in the σ known case, we use the standard normal random variable z as a **test statistic** to determine whether $\bar{x}$ deviates from the hypothesized value of μ enough to justify rejecting the null hypothesis. With $\sigma_{\bar{x}} = \sigma/\sqrt{n}$, the test statistic is as follows.

TEST STATISTIC FOR HYPOTHESIS TESTS ABOUT A POPULATION MEAN: σ KNOWN

$$z = \frac{\bar{x} - \mu_0}{\sigma/\sqrt{n}} \tag{9.1}$$

The key question for a lower tail test is, How small must the test statistic z be before we choose to reject the null hypothesis? Two approaches can be used to answer this question: the p-value approach and the critical value approach.

p-value approach The p-value approach uses the value of the test statistic z to compute a probability called a **p-value**.

A small p-value indicates the value of the test statistic is unusual given the assumption that H_0 is true.

p-VALUE

A p-value is a probability that provides a measure of the evidence against the null hypothesis provided by the sample. Smaller p-values indicate more evidence against H_0.

The p-value is used to determine whether the null hypothesis should be rejected.

Let us see how the p-value is computed and used. The value of the test statistic is used to compute the p-value. The method used depends on whether the test is a lower tail, an upper tail, or a two-tailed test. For a lower tail test, the p-value is the probability of obtaining a value for the test statistic as small as or smaller than that provided by the sample. Thus, to compute the p-value for the lower tail test in the σ known case, we use the standard normal distribution to find the probability that z is less than or equal to the value of the test statistic. After computing the p-value, we must then decide whether it is small enough to reject the null hypothesis; as we will show, this decision involves comparing the p-value to the level of significance.

WEB file
Coffee

Let us now compute the p-value for the Hilltop Coffee lower tail test. Suppose the sample of 36 Hilltop coffee cans provides a sample mean of $\bar{x} = 2.92$ pounds. Is $\bar{x} = 2.92$ small enough to cause us to reject H_0? Because this is a lower tail test, the p-value is the area under the standard normal curve for values of $z \leq$ the value of the test statistic. Using $\bar{x} = 2.92$, $\sigma = .18$, and $n = 36$, we compute the value of the test statistic z.

$$z = \frac{\bar{x} - \mu_0}{\sigma/\sqrt{n}} = \frac{2.92 - 3}{.18/\sqrt{36}} = -2.67$$

Thus, the p-value is the probability that z is less than or equal to -2.67 (the lower tail area corresponding to the value of the test statistic).

Using the standard normal probability table, we find that the lower tail area at $z = -2.67$ is .0038. Figure 9.2 shows that $\bar{x} = 2.92$ corresponds to $z = -2.67$ and a p-value $=$ .0038. This p-value indicates a small probability of obtaining a sample mean of $\bar{x} = 2.92$ (and a test statistic of -2.67) or smaller when sampling from a population with $\mu = 3$. This

FIGURE 9.2 p-VALUE FOR THE HILLTOP COFFEE STUDY WHEN $\bar{x} = 2.92$ AND $z = -2.67$

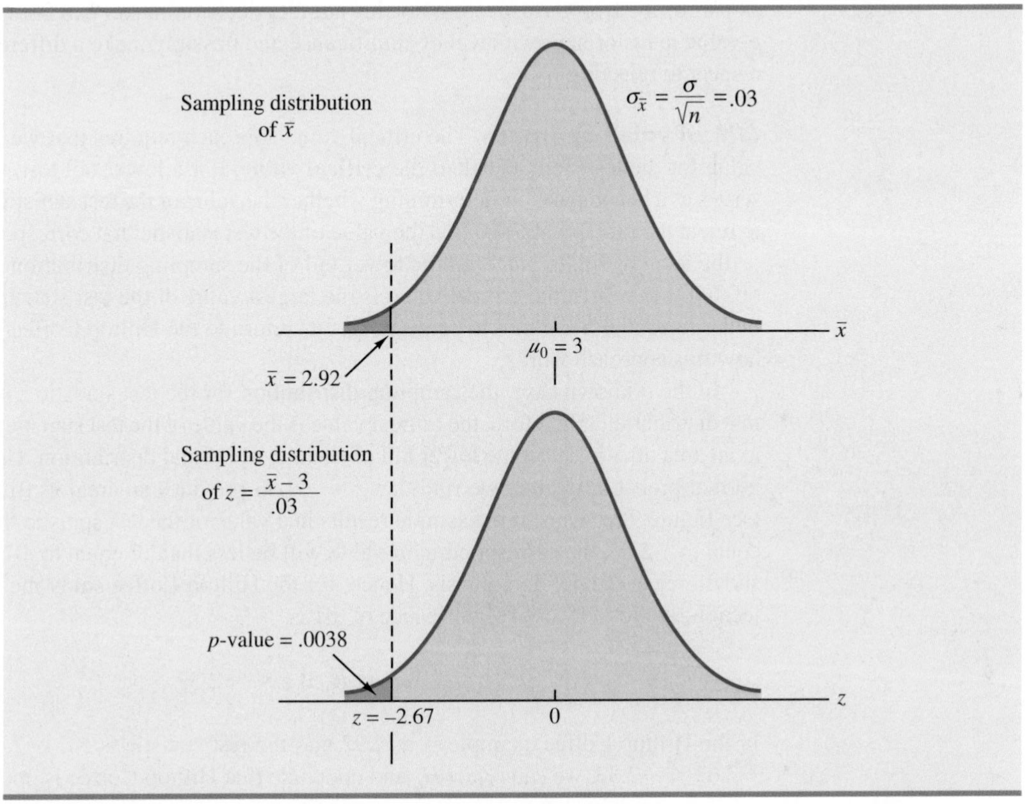

p-value does not provide much support for the null hypothesis, but is it small enough to cause us to reject H_0? The answer depends upon the level of significance for the test.

As noted previously, the director of the FTC's testing program selected a value of .01 for the level of significance. The selection of $\alpha = .01$ means that the director is willing to tolerate a probability of .01 of rejecting the null hypothesis when it is true as an equality ($\mu_0 = 3$). The sample of 36 coffee cans in the Hilltop Coffee study resulted in a p-value = .0038, which means that the probability of obtaining a value of $\bar{x} = 2.92$ or less when the null hypothesis is true as an equality is .0038. Because .0038 is less than or equal to $\alpha = .01$, we reject H_0. Therefore, we find sufficient statistical evidence to reject the null hypothesis at the .01 level of significance.

We can now state the general rule for determining whether the null hypothesis can be rejected when using the p-value approach. For a level of significance α, the rejection rule using the p-value approach is as follows:

REJECTION RULE USING p-VALUE

$$\text{Reject } H_0 \text{ if } p\text{-value} \leq \alpha$$

In the Hilltop Coffee test, the p-value of .0038 resulted in the rejection of the null hypothesis. Although the basis for making the rejection decision involves a comparison of the p-value to the level of significance specified by the FTC director, the observed p-value of .0038 means that we would reject H_0 for any value of $\alpha \geq .0038$. For this reason, the p-value is also called the *observed level of significance.*

Different decision makers may express different opinions concerning the cost of making a Type I error and may choose a different level of significance. By providing the p-value as part of the hypothesis testing results, another decision maker can compare the reported p-value to his or her own level of significance and possibly make a different decision with respect to rejecting H_0.

Critical value approach The critical value approach requires that we first determine a value for the test statistic called the **critical value**. For a lower tail test, the critical value serves as a benchmark for determining whether the value of the test statistic is small enough to reject the null hypothesis. It is the value of the test statistic that corresponds to an area of α (the level of significance) in the lower tail of the sampling distribution of the test statistic. In other words, the critical value is the largest value of the test statistic that will result in the rejection of the null hypothesis. Let us return to the Hilltop Coffee example and see how this approach works.

In the σ known case, the sampling distribution for the test statistic z is a standard normal distribution. Therefore, the critical value is the value of the test statistic that corresponds to an area of $\alpha = .01$ in the lower tail of a standard normal distribution. Using the standard normal probability table, we find that $z = -2.33$ provides an area of .01 in the lower tail (see Figure 9.3). Thus, if the sample results in a value of the test statistic that is less than or equal to -2.33, the corresponding p-value will be less than or equal to .01; in this case, we should reject the null hypothesis. Hence, for the Hilltop Coffee study the critical value rejection rule for a level of significance of .01 is

$$\text{Reject } H_0 \text{ if } z \leq -2.33$$

In the Hilltop Coffee example, $\bar{x} = 2.92$ and the test statistic is $z = -2.67$. Because $z = -2.67 < -2.33$, we can reject H_0 and conclude that Hilltop Coffee is underfilling cans.

FIGURE 9.3 CRITICAL VALUE = −2.33 FOR THE HILLTOP COFFEE HYPOTHESIS TEST

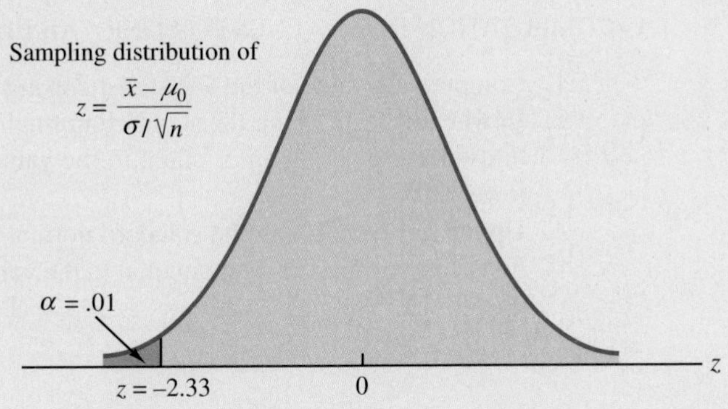

We can generalize the rejection rule for the critical value approach to handle any level of significance. The rejection rule for a lower tail test follows.

REJECTION RULE FOR A LOWER TAIL TEST: CRITICAL VALUE APPROACH

$$\text{Reject } H_0 \text{ if } z \leq -z_\alpha$$

where $-z_\alpha$ is the critical value; that is, the z value that provides an area of α in the lower tail of the standard normal distribution.

Summary The p-value approach to hypothesis testing and the critical value approach will always lead to the same rejection decision; that is, whenever the p-value is less than or equal to α, the value of the test statistic will be less than or equal to the critical value. The advantage of the p-value approach is that the p-value tells us *how* significant the results are (the observed level of significance). If we use the critical value approach, we only know that the results are significant at the stated level of significance.

At the beginning of this section, we said that one-tailed tests about a population mean take one of the following two forms:

Lower Tail Test	Upper Tail Test
$H_0: \mu \geq \mu_0$	$H_0: \mu \leq \mu_0$
$H_a: \mu < \mu_0$	$H_a: \mu > \mu_0$

We used the Hilltop Coffee study to illustrate how to conduct a lower tail test. We can use the same general approach to conduct an upper tail test. The test statistic z is still computed using equation (9.1). But, for an upper tail test, the p-value is the probability of obtaining a value for the test statistic as large as or larger than that provided by the sample. Thus, to compute the p-value for the upper tail test in the σ known case, we must use the standard normal distribution to find the probability that z is greater than or equal to the value of the test statistic. Using the critical value approach causes us to reject the null hypothesis if the value of the test statistic is greater than or equal to the critical value z_α; in other words, we reject H_0 if $z \geq z_\alpha$.

Let us summarize the steps involved in computing p-values for one-tailed hypothesis tests.

COMPUTATION OF p-VALUES FOR ONE-TAILED TESTS

1. Compute the value of the test statistic using equation (9.1).
2. **Lower tail test:** Using the standard normal distribution, compute the probability that z is less than or equal to the value of the test statistic (area in the lower tail).
3. **Upper tail test:** Using the standard normal distribution, compute the probability that z is greater than or equal to the value of the test statistic (area in the upper tail).

Two-Tailed Test

In hypothesis testing, the general form for a **two-tailed test** about a population mean is as follows:

$$H_0: \mu = \mu_0$$
$$H_a: \mu \neq \mu_0$$

In this subsection we show how to conduct a two-tailed test about a population mean for the σ known case. As an illustration, we consider the hypothesis testing situation facing MaxFlight, Inc.

The U.S. Golf Association (USGA) establishes rules that manufacturers of golf equipment must meet if their products are to be acceptable for use in USGA events. MaxFlight uses a high-technology manufacturing process to produce golf balls with a mean driving distance of 295 yards. Sometimes, however, the process gets out of adjustment and produces golf balls with a mean driving distance different from 295 yards. When the mean distance falls below 295 yards, the company worries about losing sales because the golf balls do not provide as much distance as advertised. When the mean distance passes 295 yards, MaxFlight's golf balls may be rejected by the USGA for exceeding the overall distance standard concerning carry and roll.

MaxFlight's quality control program involves taking periodic samples of 50 golf balls to monitor the manufacturing process. For each sample, a hypothesis test is conducted to determine whether the process has fallen out of adjustment. Let us develop the null and alternative hypotheses. We begin by assuming that the process is functioning correctly; that is, the golf balls being produced have a mean distance of 295 yards. This assumption establishes the null hypothesis. The alternative hypothesis is that the mean distance is not equal to 295 yards. With a hypothesized value of $\mu_0 = 295$, the null and alternative hypotheses for the MaxFlight hypothesis test are as follows:

$$H_0: \mu = 295$$
$$H_a: \mu \neq 295$$

If the sample mean $\bar{x}$ is significantly less than 295 yards or significantly greater than 295 yards, we will reject H_0. In this case, corrective action will be taken to adjust the manufacturing process. On the other hand, if $\bar{x}$ does not deviate from the hypothesized mean $\mu_0 = 295$ by a significant amount, H_0 will not be rejected and no action will be taken to adjust the manufacturing process.

The quality control team selected $\alpha = .05$ as the level of significance for the test. Data from previous tests conducted when the process was known to be in adjustment show that

FIGURE 9.4 SAMPLING DISTRIBUTION OF $\bar{x}$ FOR THE MAXFLIGHT HYPOTHESIS TEST

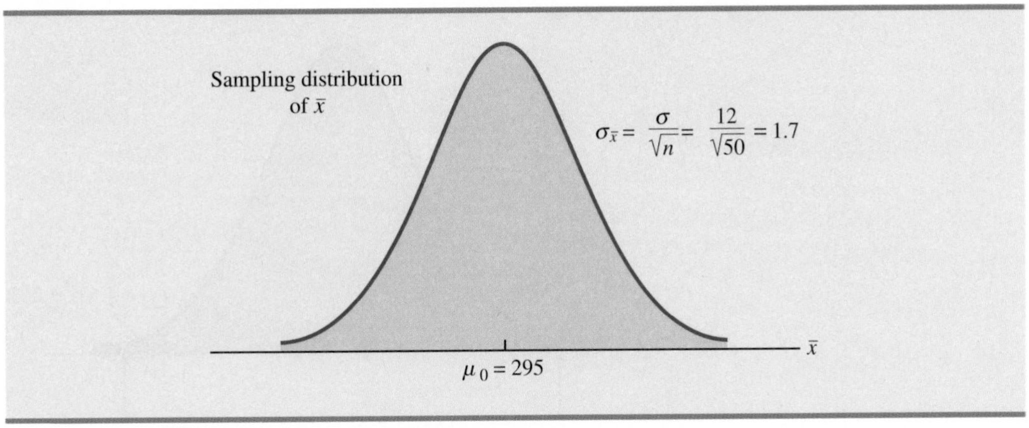

the population standard deviation can be assumed known with a value of $\sigma = 12$. Thus, with a sample size of $n = 50$, the standard error of $\bar{x}$ is

$$\sigma_{\bar{x}} = \frac{\sigma}{\sqrt{n}} = \frac{12}{\sqrt{50}} = 1.7$$

Because the sample size is large, the central limit theorem (see Chapter 7) allows us to conclude that the sampling distribution of $\bar{x}$ can be approximated by a normal distribution. Figure 9.4 shows the sampling distribution of $\bar{x}$ for the MaxFlight hypothesis test with a hypothesized population mean of $\mu_0 = 295$.

WEB file
GolfTest

Suppose that a sample of 50 golf balls is selected and that the sample mean is $\bar{x} = 297.6$ yards. This sample mean provides support for the conclusion that the population mean is larger than 295 yards. Is this value of $\bar{x}$ enough larger than 295 to cause us to reject H_0 at the .05 level of significance? In the previous section we described two approaches that can be used to answer this question: the *p*-value approach and the critical value approach.

p-value approach Recall that the *p*-value is a probability used to determine whether the null hypothesis should be rejected. For a two-tailed test, values of the test statistic in *either* tail provide evidence against the null hypothesis. For a two-tailed test, the *p*-value is the probability of obtaining a value for the test statistic *as unlikely as or more unlikely than* that provided by the sample. Let us see how the *p*-value is computed for the MaxFlight hypothesis test.

First we compute the value of the test statistic. For the σ known case, the test statistic z is a standard normal random variable. Using equation (9.1) with $\bar{x} = 297.6$, the value of the test statistic is

$$z = \frac{\bar{x} - \mu_0}{\sigma/\sqrt{n}} = \frac{297.6 - 295}{12/\sqrt{50}} = 1.53$$

Now to compute the *p*-value we must find the probability of obtaining a value for the test statistic *at least as unlikely as* $z = 1.53$. Clearly values of $z \geq 1.53$ are *at least as unlikely.* But, because this is a two-tailed test, values of $z \leq -1.53$ are also *at least as unlikely as* the value of the test statistic provided by the sample. In Figure 9.5, we see that the two-tailed *p*-value in this case is given by $P(z \leq -1.53) + P(z \geq 1.53)$. Because the normal curve is symmetric, we can compute this probability by finding the upper tail area at $z = 1.53$ and doubling it. The table for the standard normal distribution shows that $P(z < 1.53) = .9370$. Thus, the upper tail area is $P(z \geq 1.53) = 1.0000 - .9370 = .0630$. Doubling this, we find the *p*-value for the MaxFlight two-tailed hypothesis test is *p*-value $= 2(.0630) = .1260$.

FIGURE 9.5 p-VALUE FOR THE MAXFLIGHT HYPOTHESIS TEST

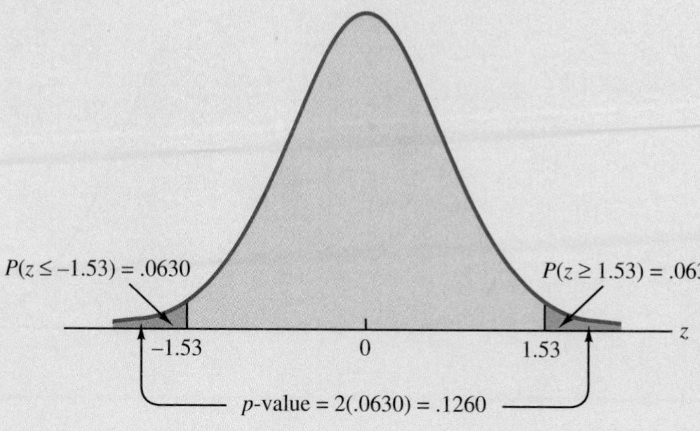

Next we compare the p-value to the level of significance to see whether the null hypothesis should be rejected. With a level of significance of $\alpha = .05$, we do not reject H_0 because the p-value $= .1260 > .05$. Because the null hypothesis is not rejected, no action will be taken to adjust the MaxFlight manufacturing process.

The computation of the p-value for a two-tailed test may seem a bit confusing as compared to the computation of the p-value for a one-tailed test. But it can be simplified by following three steps.

COMPUTATION OF p-VALUES FOR TWO-TAILED TESTS

1. Compute the value of the test statistic using equation (9.1).
2. If the value of the test statistic is in the upper tail, compute the probability that z is greater than or equal to the value of the test statistic (the upper tail area). If the value of the test statistic is in the lower tail, compute the probability that z is less than or equal to the value of the test statistic (the lower tail area).
3. Double the probability (or tail area) from step 2 to obtain the p-value.

Critical value approach Before leaving this section, let us see how the test statistic z can be compared to a critical value to make the hypothesis testing decision for a two-tailed test. Figure 9.6 shows that the critical values for the test will occur in both the lower and upper tails of the standard normal distribution. With a level of significance of $\alpha = .05$, the area in each tail corresponding to the critical values is $\alpha/2 = .05/2 = .025$. Using the standard normal probability table, we find the critical values for the test statistic are $-z_{.025} = -1.96$ and $z_{.025} = 1.96$. Thus, using the critical value approach, the two-tailed rejection rule is

$$\text{Reject } H_0 \text{ if } z \leq -1.96 \text{ or if } z \geq 1.96$$

Because the value of the test statistic for the MaxFlight study is $z = 1.53$, the statistical evidence will not permit us to reject the null hypothesis at the .05 level of significance.

Summary and Practical Advice

We presented examples of a lower tail test and a two-tailed test about a population mean. Based upon these examples, we can now summarize the hypothesis testing procedures about a population mean for the σ known case as shown in Table 9.2. Note that μ_0 is the hypothesized value of the population mean.

FIGURE 9.6 CRITICAL VALUES FOR THE MAXFLIGHT HYPOTHESIS TEST

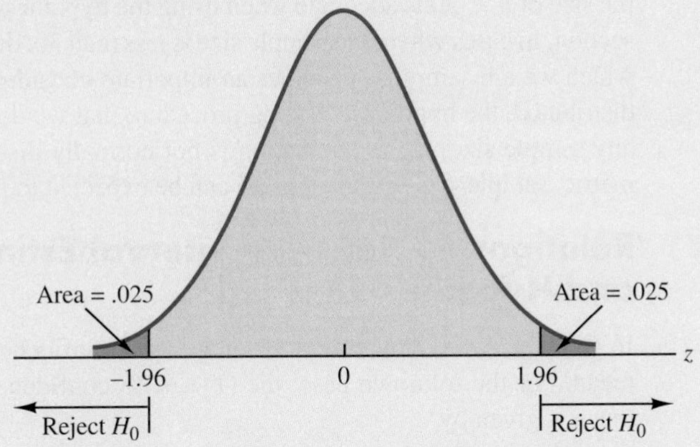

TABLE 9.2 SUMMARY OF HYPOTHESIS TESTS ABOUT A POPULATION MEAN: σ KNOWN CASE

	Lower Tail Test	Upper Tail Test	Two-Tailed Test
Hypotheses	$H_0: \mu \geq \mu_0$ $H_a: \mu < \mu_0$	$H_0: \mu \leq \mu_0$ $H_a: \mu > \mu_0$	$H_0: \mu = \mu_0$ $H_a: \mu \neq \mu_0$
Test Statistic	$z = \frac{\bar{x} - \mu_0}{\sigma/\sqrt{n}}$	$z = \frac{\bar{x} - \mu_0}{\sigma/\sqrt{n}}$	$z = \frac{\bar{x} - \mu_0}{\sigma/\sqrt{n}}$
Rejection Rule: *p*-Value Approach	Reject H_0 if p-value $\leq \alpha$	Reject H_0 if p-value $\leq \alpha$	Reject H_0 if p-value $\leq \alpha$
Rejection Rule: Critical Value Approach	Reject H_0 if $z \leq -z_\alpha$	Reject H_0 if $z \geq z_\alpha$	Reject H_0 if $z \leq -z_{\alpha/2}$ or if $z \geq z_{\alpha/2}$

The hypothesis testing steps followed in the two examples presented in this section are common to every hypothesis test.

STEPS OF HYPOTHESIS TESTING

Step 1. Develop the null and alternative hypotheses.
Step 2. Specify the level of significance.
Step 3. Collect the sample data and compute the value of the test statistic.

p-Value Approach

Step 4. Use the value of the test statistic to compute the p-value.
Step 5. Reject H_0 if the p-value $\leq \alpha$.
Step 6. Interpret the statistical conclusion in the context of the application.

Critical Value Approach

Step 4. Use the level of significance to determine the critical value and the rejection rule.
Step 5. Use the value of the test statistic and the rejection rule to determine whether to reject H_0.
Step 6. Interpret the statistical conclusion in the context of the application.

Practical advice about the sample size for hypothesis tests is similar to the advice we provided about the sample size for interval estimation in Chapter 8. In most applications, a sample size of $n \geq 30$ is adequate when using the hypothesis testing procedure described in this section. In cases where the sample size is less than 30, the distribution of the population from which we are sampling becomes an important consideration. If the population is normally distributed, the hypothesis testing procedure that we described is exact and can be used for any sample size. If the population is not normally distributed but is at least roughly symmetric, sample sizes as small as 15 can be expected to provide acceptable results.

Relationship Between Interval Estimation and Hypothesis Testing

In Chapter 8 we showed how to develop a confidence interval estimate of a population mean. For the σ known case, the $(1 - \alpha)\%$ confidence interval estimate of a population mean is given by

$$\bar{x} \pm z_{\alpha/2} \frac{\sigma}{\sqrt{n}}$$

In this chapter, we showed that a two-tailed hypothesis test about a population mean takes the following form:

$$H_0: \mu = \mu_0$$
$$H_a: \mu \neq \mu_0$$

where μ_0 is the hypothesized value for the population mean.

Suppose that we follow the procedure described in Chapter 8 for constructing a $100(1 - \alpha)\%$ confidence interval for the population mean. We know that $100(1 - \alpha)\%$ of the confidence intervals generated will contain the population mean and $100\alpha\%$ of the confidence intervals generated will not contain the population mean. Thus, if we reject H_0 whenever the confidence interval does not contain μ_0, we will be rejecting the null hypothesis when it is true ($\mu = \mu_0$) with probability α. Recall that the level of significance is the probability of rejecting the null hypothesis when it is true. So constructing a $100(1 - \alpha)\%$ confidence interval and rejecting H_0 whenever the interval does not contain μ_0 is equivalent to conducting a two-tailed hypothesis test with α as the level of significance. The procedure for using a confidence interval to conduct a two-tailed hypothesis test can now be summarized.

A CONFIDENCE INTERVAL APPROACH TO TESTING A HYPOTHESIS OF THE FORM

$$H_0: \mu = \mu_0$$
$$H_a: \mu \neq \mu_0$$

1. Select a simple random sample from the population and use the value of the sample mean $\bar{x}$ to develop the confidence interval for the population mean μ.

$$\bar{x} \pm z_{\alpha/2} \frac{\sigma}{\sqrt{n}}$$

2. If the confidence interval contains the hypothesized value μ_0, do not reject H_0. Otherwise, reject[2] H_0.

For a two-tailed hypothesis test, the null hypothesis can be rejected if the confidence interval does not include μ_0.

[2] To be consistent with the rule for rejecting H_0 when the p-value $\leq \alpha$, we would also reject H_0 using the confidence interval approach if μ_0 happens to be equal to one of the end points of the $100(1 - \alpha)\%$ confidence interval.

Let us illustrate by conducting the MaxFlight hypothesis test using the confidence interval approach. The MaxFlight hypothesis test takes the following form:

$$H_0: \mu = 295$$
$$H_a: \mu \neq 295$$

To test these hypotheses with a level of significance of $\alpha = .05$, we sampled 50 golf balls and found a sample mean distance of $\bar{x} = 297.6$ yards. Recall that the population standard deviation is $\sigma = 12$. Using these results with $z_{.025} = 1.96$, we find that the 95% confidence interval estimate of the population mean is

$$\bar{x} \pm z_{.025}\frac{\sigma}{\sqrt{n}}$$
$$297.6 \pm 1.96\frac{12}{\sqrt{50}}$$
$$297.6 \pm 3.3$$

or

$$294.3 \text{ to } 300.9$$

This finding enables the quality control manager to conclude with 95% confidence that the mean distance for the population of golf balls is between 294.3 and 300.9 yards. Because the hypothesized value for the population mean, $\mu_0 = 295$, is in this interval, the hypothesis testing conclusion is that the null hypothesis, H_0: $\mu = 295$, cannot be rejected.

Note that this discussion and example pertain to two-tailed hypothesis tests about a population mean. However, the same confidence interval and two-tailed hypothesis testing relationship exists for other population parameters. The relationship can also be extended to one-tailed tests about population parameters. Doing so, however, requires the development of one-sided confidence intervals, which are rarely used in practice.

NOTES AND COMMENTS

We have shown how to use p-values. The smaller the p-value the greater the evidence against H_0 and the more the evidence in favor of H_a. Here are some guidelines statisticians suggest for interpreting small p-values.

- Less than .01—Overwhelming evidence to conclude H_a is true.
- Between .01 and .05—Strong evidence to conclude H_a is true.
- Between .05 and .10—Weak evidence to conclude H_a is true.
- Greater than .10—Insufficient evidence to conclude H_a is true.

Exercises

Note to Student: Some of the exercises that follow ask you to use the p-value approach and others ask you to use the critical value approach. Both methods will provide the same hypothesis testing conclusion. We provide exercises with both methods to give you practice using both. In later sections and in following chapters, we will generally emphasize the p-value approach as the preferred method, but you may select either based on personal preference.

Methods

9. Consider the following hypothesis test:

$$H_0: \mu \geq 20$$
$$H_a: \mu < 20$$

A sample of 50 provided a sample mean of 19.4. The population standard deviation is 2.
a. Compute the value of the test statistic.
b. What is the p-value?
c. Using $\alpha = .05$, what is your conclusion?
d. What is the rejection rule using the critical value? What is your conclusion?

SELF test

10. Consider the following hypothesis test:

$$H_0: \mu \leq 25$$
$$H_a: \mu > 25$$

A sample of 40 provided a sample mean of 26.4. The population standard deviation is 6.
a. Compute the value of the test statistic.
b. What is the p-value?
c. At $\alpha = .01$, what is your conclusion?
d. What is the rejection rule using the critical value? What is your conclusion?

SELF test

11. Consider the following hypothesis test:

$$H_0: \mu = 15$$
$$H_a: \mu \neq 15$$

A sample of 50 provided a sample mean of 14.15. The population standard deviation is 3.
a. Compute the value of the test statistic.
b. What is the p-value?
c. At $\alpha = .05$, what is your conclusion?
d. What is the rejection rule using the critical value? What is your conclusion?

12. Consider the following hypothesis test:

$$H_0: \mu \geq 80$$
$$H_a: \mu < 80$$

A sample of 100 is used and the population standard deviation is 12. Compute the p-value and state your conclusion for each of the following sample results. Use $\alpha = .01$.
a. $\bar{x} = 78.5$
b. $\bar{x} = 77$
c. $\bar{x} = 75.5$
d. $\bar{x} = 81$

13. Consider the following hypothesis test:

$$H_0: \mu \leq 50$$
$$H_a: \mu > 50$$

A sample of 60 is used and the population standard deviation is 8. Use the critical value approach to state your conclusion for each of the following sample results. Use $\alpha = .05$.
a. $\bar{x} = 52.5$
b. $\bar{x} = 51$
c. $\bar{x} = 51.8$

14. Consider the following hypothesis test:

$$H_0: \mu = 22$$
$$H_a: \mu \neq 22$$

A sample of 75 is used and the population standard deviation is 10. Compute the p-value and state your conclusion for each of the following sample results. Use $\alpha = .01$.

a. $\bar{x} = 23$
b. $\bar{x} = 25.1$
c. $\bar{x} = 20$

Applications

SELF test

15. Individuals filing federal income tax returns prior to March 31 received an average refund of \$1056. Consider the population of "last-minute" filers who mail their tax return during the last five days of the income tax period (typically April 10 to April 15).
    a. A researcher suggests that a reason individuals wait until the last five days is that on average these individuals receive lower refunds than do early filers. Develop appropriate hypotheses such that rejection of H_0 will support the researcher's contention.
    b. For a sample of 400 individuals who filed a tax return between April 10 and 15, the sample mean refund was \$910. Based on prior experience a population standard deviation of $\sigma = \$1600$ may be assumed. What is the p-value?
    c. At $\alpha = .05$, what is your conclusion?
    d. Repeat the preceding hypothesis test using the critical value approach.

16. In a study entitled How Undergraduate Students Use Credit Cards, it was reported that undergraduate students have a mean credit card balance of \$3173 (*Sallie Mae*, April 2009). This figure was an all-time high and had increased 44% over the previous five years. Assume that a current study is being conducted to determine if it can be concluded that the mean credit card balance for undergraduate students has continued to increase compared to the April 2009 report. Based on previous studies, use a population standard deviation $\sigma = \$1000$.
    a. State the null and alternative hypotheses.
    b. What is the p-value for a sample of 180 undergraduate students with a sample mean credit card balance of \$3325?
    c. Using a .05 level of significance, what is your conclusion?

17. The mean hourly wage for employees in goods-producing industries is currently \$24.57 (Bureau of Labor Statistics website, April, 12, 2012). Suppose we take a sample of employees from the manufacturing industry to see if the mean hourly wage differs from the reported mean of \$24.57 for the goods-producing industries.
    a. State the null and alternative hypotheses we should use to test whether the population mean hourly wage in the manufacturing industry differs from the population mean hourly wage in the goods-producing industries.
    b. Suppose a sample of 30 employees from the manufacturing industry showed a sample mean of \$23.89 per hour. Assume a population standard deviation of \$2.40 per hour and compute the p-value.
    c. With $\alpha = .05$ as the level of significance, what is your conclusion?
    d. Repeat the preceding hypothesis test using the critical value approach.

18. The average annual total return for U.S. Diversified Equity mutual funds from 1999 to 2003 was 4.1% (*BusinessWeek,* January 26, 2004). A researcher would like to conduct a

hypothesis test to see whether the returns for mid-cap growth funds over the same period are significantly different from the average for U.S. Diversified Equity funds.

a. Formulate the hypotheses that can be used to determine whether the mean annual return for mid-cap growth funds differ from the mean for U.S. Diversified Equity funds.
b. A sample of 40 mid-cap growth funds provides a mean return of $\bar{x} = 3.4\%$. Assume the population standard deviation for mid-cap growth funds is known from previous studies to be $\sigma = 2\%$. Use the sample results to compute the test statistic and p-value for the hypothesis test.
c. At $\alpha = .05$, what is your conclusion?

19. The Internal Revenue Service (IRS) provides a toll-free help line for taxpayers to call in and get answers to questions as they prepare their tax returns. In recent years, the IRS has been inundated with taxpayer calls and has redesigned its phone service as well as posted answers to frequently asked questions on its website (*The Cincinnati Enquirer*, January 7, 2010). According to a report by a taxpayer advocate, callers using the new system can expect to wait on hold for an unreasonably long time of 12 minutes before being able to talk to an IRS employee. Suppose you select a sample of 50 callers after the new phone service has been implemented; the sample results show a mean waiting time of 10 minutes before an IRS employee comes on the line. Based upon data from past years, you decide it is reasonable to assume that the standard deviation of waiting time is 8 minutes. Using your sample results, can you conclude that the actual mean waiting time turned out to be significantly less than the 12-minute claim made by the taxpayer advocate? Use $\alpha = .05$.

20. For the United States, the mean monthly Internet bill is \$32.79 per household (CNBC, January 18, 2006). A sample of 50 households in a southern state showed a sample mean of \$30.63. Use a population standard deviation of $\sigma = \$5.60$.
    a. Formulate hypotheses for a test to determine whether the sample data support the conclusion that the mean monthly Internet bill in the southern state is less than the national mean of \$32.79.
    b. What is the value of the test statistic?
    c. What is the p-value?
    d. At $\alpha = .01$, what is your conclusion?

WEB file

Fowle

21. Fowle Marketing Research, Inc., bases charges to a client on the assumption that telephone surveys can be completed in a mean time of 15 minutes or less. If a longer mean survey time is necessary, a premium rate is charged. A sample of 35 surveys provided the survey times shown in the file named Fowle. Based upon past studies, the population standard deviation is assumed known with $\sigma = 4$ minutes. Is the premium rate justified?
    a. Formulate the null and alternative hypotheses for this application.
    b. Compute the value of the test statistic.
    c. What is the p-value?
    d. At $\alpha = .01$, what is your conclusion?

22. CCN and ActMedia provided a television channel targeted to individuals waiting in supermarket checkout lines. The channel showed news, short features, and advertisements. The length of the program was based on the assumption that the population mean time a shopper stands in a supermarket checkout line is 8 minutes. A sample of actual waiting times will be used to test this assumption and determine whether actual mean waiting time differs from this standard.
    a. Formulate the hypotheses for this application.
    b. A sample of 120 shoppers showed a sample mean waiting time of 8.4 minutes. Assume a population standard deviation of $\sigma = 3.2$ minutes. What is the p-value?
    c. At $\alpha = .05$, what is your conclusion?
    d. Compute a 95% confidence interval for the population mean. Does it support your conclusion?

9.4 Population Mean: σ Unknown

In this section we describe how to conduct hypothesis tests about a population mean for the σ unknown case. Because the σ unknown case corresponds to situations in which an estimate of the population standard deviation cannot be developed prior to sampling, the sample must be used to develop an estimate of both μ and σ. Thus, to conduct a hypothesis test about a population mean for the σ unknown case, the sample mean $\bar{x}$ is used as an estimate of μ and the sample standard deviation s is used as an estimate of σ.

The steps of the hypothesis testing procedure for the σ unknown case are the same as those for the σ known case described in Section 9.3. But, with σ unknown, the computation of the test statistic and p-value is a bit different. Recall that for the σ known case, the sampling distribution of the test statistic has a standard normal distribution. For the σ unknown case, however, the sampling distribution of the test statistic follows the t distribution; it has slightly more variability because the sample is used to develop estimates of both μ and σ.

In Section 8.2 we showed that an interval estimate of a population mean for the σ unknown case is based on a probability distribution known as the t distribution. Hypothesis tests about a population mean for the σ unknown case are also based on the t distribution. For the σ unknown case, the test statistic has a t distribution with $n - 1$ degrees of freedom.

TEST STATISTIC FOR HYPOTHESIS TESTS ABOUT A POPULATION MEAN: σ UNKNOWN

$$t = \frac{\bar{x} - \mu_0}{s/\sqrt{n}} \tag{9.2}$$

In Chapter 8 we said that the t distribution is based on an assumption that the population from which we are sampling has a normal distribution. However, research shows that this assumption can be relaxed considerably when the sample size is large enough. We provide some practical advice concerning the population distribution and sample size at the end of the section.

One-Tailed Test

WEB file AirRating

Let us consider an example of a one-tailed test about a population mean for the σ unknown case. A business travel magazine wants to classify transatlantic gateway airports according to the mean rating for the population of business travelers. A rating scale with a low score of 0 and a high score of 10 will be used, and airports with a population mean rating greater than 7 will be designated as superior service airports. The magazine staff surveyed a sample of 60 business travelers at each airport to obtain the ratings data. The sample for London's Heathrow Airport provided a sample mean rating of $\bar{x} = 7.25$ and a sample standard deviation of $s = 1.052$. Do the data indicate that Heathrow should be designated as a superior service airport?

We want to develop a hypothesis test for which the decision to reject H_0 will lead to the conclusion that the population mean rating for the Heathrow Airport is *greater* than 7. Thus, an upper tail test with H_a: $\mu > 7$ is required. The null and alternative hypotheses for this upper tail test are as follows:

$$H_0: \mu \leq 7$$
$$H_a: \mu > 7$$

We will use $\alpha = .05$ as the level of significance for the test.

Using equation (9.2) with $\bar{x} = 7.25$, $\mu_0 = 7$, $s = 1.052$, and $n = 60$, the value of the test statistic is

$$t = \frac{\bar{x} - \mu_0}{s/\sqrt{n}} = \frac{7.25 - 7}{1.052/\sqrt{60}} = 1.84$$

The sampling distribution of t has $n - 1 = 60 - 1 = 59$ degrees of freedom. Because the test is an upper tail test, the p-value is $P(t \geq 1.84)$, that is, the upper tail area corresponding to the value of the test statistic.

The t distribution table provided in most textbooks will not contain sufficient detail to determine the exact p-value, such as the p-value corresponding to $t = 1.84$. For instance, using Table 2 in Appendix B, the t distribution with 59 degrees of freedom provides the following information.

Area in Upper Tail	**.20**	**.10**	**.05**	**.025**	**.01**	**.005**
***t* Value (59 *df*)**	.848	1.296	1.671	2.001	2.391	2.662

$t = 1.84$ (arrow pointing between 1.671 and 2.001)

We see that $t = 1.84$ is between 1.671 and 2.001. Although the table does not provide the exact p-value, the values in the "Area in Upper Tail" row show that the p-value must be less than .05 and greater than .025. With a level of significance of $\alpha = .05$, this placement is all we need to know to make the decision to reject the null hypothesis and conclude that Heathrow should be classified as a superior service airport.

Appendix F shows how to compute p-values using Excel or Minitab.

Because it is cumbersome to use a t table to compute p-values, and only approximate values are obtained, we show how to compute the exact p-value using Excel or Minitab. The directions can be found in Appendix F at the end of this text. Using Excel or Minitab with $t = 1.84$ provides the upper tail p-value of .0354 for the Heathrow Airport hypothesis test. With $.0354 < .05$, we reject the null hypothesis and conclude that Heathrow should be classified as a superior service airport.

The decision whether to reject the null hypothesis in the σ unknown case can also be made using the critical value approach. The critical value corresponding to an area of $\alpha = .05$ in the upper tail of a t distribution with 59 degrees of freedom is $t_{.05} = 1.671$. Thus the rejection rule using the critical value approach is to reject H_0 if $t \geq 1.671$. Because $t = 1.84 > 1.671$, H_0 is rejected. Heathrow should be classified as a superior service airport.

Two-Tailed Test

To illustrate how to conduct a two-tailed test about a population mean for the σ unknown case, let us consider the hypothesis testing situation facing Holiday Toys. The company manufactures and distributes its products through more than 1000 retail outlets. In planning production levels for the coming winter season, Holiday must decide how many units of each product to produce prior to knowing the actual demand at the retail level. For this year's most important new toy, Holiday's marketing director is expecting demand to average 40 units per retail outlet. Prior to making the final production decision based upon this estimate, Holiday decided to survey a sample of 25 retailers in order to develop more information about the demand for the new product. Each retailer was provided with information about the features of the new toy along with the cost and the suggested selling price. Then each retailer was asked to specify an anticipated order quantity.

With μ denoting the population mean order quantity per retail outlet, the sample data will be used to conduct the following two-tailed hypothesis test:

$$H_0: \mu = 40$$
$$H_a: \mu \neq 40$$

If H_0 cannot be rejected, Holiday will continue its production planning based on the marketing director's estimate that the population mean order quantity per retail outlet will be $\mu = 40$ units. However, if H_0 is rejected, Holiday will immediately reevaluate its production plan for the product. A two-tailed hypothesis test is used because Holiday wants to reevaluate the production plan if the population mean quantity per retail outlet is less than anticipated or greater than anticipated. Because no historical data are available (it's a new product), the population mean μ and the population standard deviation must both be estimated using $\bar{x}$ and s from the sample data.

WEB file
Orders

The sample of 25 retailers provided a mean of $\bar{x} = 37.4$ and a standard deviation of $s = 11.79$ units. Before going ahead with the use of the t distribution, the analyst constructed a histogram of the sample data in order to check on the form of the population distribution. The histogram of the sample data showed no evidence of skewness or any extreme outliers, so the analyst concluded that the use of the t distribution with $n - 1 = 24$ degrees of freedom was appropriate. Using equation (9.2) with $\bar{x} = 37.4$, $\mu_0 = 40$, $s = 11.79$, and $n = 25$, the value of the test statistic is

$$t = \frac{\bar{x} - \mu_0}{s/\sqrt{n}} = \frac{37.4 - 40}{11.79/\sqrt{25}} = -1.10$$

Because we have a two-tailed test, the p-value is two times the area under the curve of the t distribution for $t \leq -1.10$. Using Table 2 in Appendix B, the t distribution table for 24 degrees of freedom provides the following information.

Area in Upper Tail	**.20**	**.10**	**.05**	**.025**	**.01**	**.005**
***t*-Value (24 *df*)**	.857	1.318	1.711	2.064	2.492	2.797

$t = 1.10$

The t distribution table only contains positive t values. Because the t distribution is symmetric, however, the upper tail area at $t = 1.10$ is the same as the lower tail area at $t = -1.10$. We see that $t = 1.10$ is between 0.857 and 1.318. From the "Area in Upper Tail" row, we see that the area in the upper tail at $t = 1.10$ is between .20 and .10. When we double these amounts, we see that the p-value must be between .40 and .20. With a level of significance of $\alpha = .05$, we now know that the p-value is greater than α. Therefore, H_0 cannot be rejected. Sufficient evidence is not available to conclude that Holiday should change its production plan for the coming season.

Appendix F shows how the p-value for this test can be computed using Excel or Minitab. The p-value obtained is .2822. With a level of significance of $\alpha = .05$, we cannot reject H_0 because $.2822 > .05$.

The test statistic can also be compared to the critical value to make the two-tailed hypothesis testing decision. With $\alpha = .05$ and the t distribution with 24 degrees of freedom, $-t_{.025} = -2.064$ and $t_{.025} = 2.064$ are the critical values for the two-tailed test. The rejection rule using the test statistic is

$$\text{Reject } H_0 \text{ if } t \leq -2.064 \text{ or if } t \geq 2.064$$

TABLE 9.3 SUMMARY OF HYPOTHESIS TESTS ABOUT A POPULATION MEAN: σ UNKNOWN CASE

	Lower Tail Test	Upper Tail Test	Two-Tailed Test
Hypotheses	$H_0: \mu \geq \mu_0$ $H_a: \mu < \mu_0$	$H_0: \mu \leq \mu_0$ $H_a: \mu > \mu_0$	$H_0: \mu = \mu_0$ $H_a: \mu \neq \mu_0$
Test Statistic	$t = \dfrac{\bar{x} - \mu_0}{s/\sqrt{n}}$	$t = \dfrac{\bar{x} - \mu_0}{s/\sqrt{n}}$	$t = \dfrac{\bar{x} - \mu_0}{s/\sqrt{n}}$
Rejection Rule: *p*-Value Approach	Reject H_0 if p-value $\leq \alpha$	Reject H_0 if p-value $\leq \alpha$	Reject H_0 if p-value $\leq \alpha$
Rejection Rule: Critical Value Approach	Reject H_0 if $t \leq -t_\alpha$	Reject H_0 if $t \geq t_\alpha$	Reject H_0 if $t \leq -t_{\alpha/2}$ or if $t \geq t_{\alpha/2}$

Based on the test statistic $t = -1.10$, H_0 cannot be rejected. This result indicates that Holiday should continue its production planning for the coming season based on the expectation that $\mu = 40$.

Summary and Practical Advice

Table 9.3 provides a summary of the hypothesis testing procedures about a population mean for the σ unknown case. The key difference between these procedures and the ones for the σ known case is that s is used, instead of σ, in the computation of the test statistic. For this reason, the test statistic follows the t distribution.

The applicability of the hypothesis testing procedures of this section is dependent on the distribution of the population being sampled from and the sample size. When the population is normally distributed, the hypothesis tests described in this section provide exact results for any sample size. When the population is not normally distributed, the procedures are approximations. Nonetheless, we find that sample sizes of 30 or greater will provide good results in most cases. If the population is approximately normal, small sample sizes (e.g., $n < 15$) can provide acceptable results. If the population is highly skewed or contains outliers, sample sizes approaching 50 are recommended.

Exercises

Methods

23. Consider the following hypothesis test:

$$H_0: \mu \leq 12$$
$$H_a: \mu > 12$$

A sample of 25 provided a sample mean $\bar{x} = 14$ and a sample standard deviation $s = 4.32$.

a. Compute the value of the test statistic.
b. Use the t distribution table (Table 2 in Appendix B) to compute a range for the p-value.
c. At $\alpha = .05$, what is your conclusion?
d. What is the rejection rule using the critical value? What is your conclusion?

SELF test

24. Consider the following hypothesis test:

$$H_0: \mu = 18$$
$$H_a: \mu \neq 18$$

A sample of 48 provided a sample mean $\bar{x} = 17$ and a sample standard deviation $s = 4.5$.

a. Compute the value of the test statistic.
b. Use the t distribution table (Table 2 in Appendix B) to compute a range for the p-value.
c. At $\alpha = .05$, what is your conclusion?
d. What is the rejection rule using the critical value? What is your conclusion?

25. Consider the following hypothesis test:

$$H_0: \mu \geq 45$$
$$H_a: \mu < 45$$

A sample of 36 is used. Identify the p-value and state your conclusion for each of the following sample results. Use $\alpha = .01$.

a. $\bar{x} = 44$ and $s = 5.2$
b. $\bar{x} = 43$ and $s = 4.6$
c. $\bar{x} = 46$ and $s = 5.0$

26. Consider the following hypothesis test:

$$H_0: \mu = 100$$
$$H_a: \mu \neq 100$$

A sample of 65 is used. Identify the p-value and state your conclusion for each of the following sample results. Use $\alpha = .05$.

a. $\bar{x} = 103$ and $s = 11.5$
b. $\bar{x} = 96.5$ and $s = 11.0$
c. $\bar{x} = 102$ and $s = 10.5$

Applications

SELF test

27. The Employment and Training Administration reported that the U.S. mean unemployment insurance benefit was \$238 per week (*The World Almanac,* 2003). A researcher in the state of Virginia anticipated that sample data would show evidence that the mean weekly unemployment insurance benefit in Virginia was below the national average.

a. Develop appropriate hypotheses such that rejection of H_0 will support the researcher's contention.
b. For a sample of 100 individuals, the sample mean weekly unemployment insurance benefit was \$231 with a sample standard deviation of \$80. What is the p-value?
c. At $\alpha = .05$, what is your conclusion?
d. Repeat the preceding hypothesis test using the critical value approach.

28. A shareholders' group, in lodging a protest, claimed that the mean tenure for a chief executive office (CEO) was at least nine years. A survey of companies reported in *The Wall Street Journal* found a sample mean tenure of $\bar{x} = 7.27$ years for CEOs with a standard deviation of $s = 6.38$ years (*The Wall Street Journal*, January 2, 2007).

a. Formulate hypotheses that can be used to challenge the validity of the claim made by the shareholders' group.
b. Assume 85 companies were included in the sample. What is the p-value for your hypothesis test?
c. At $\alpha = .01$, what is your conclusion?

WEB file

Administrator

29. The national mean annual salary for a school administrator is \$90,000 a year (*The Cincinnati Enquirer,* April 7, 2012). A school official took a sample of 25 school administrators in the state of Ohio to learn about salaries in that state to see if they differed from the national average.
    a. Formulate hypotheses that can be used to determine whether the population mean annual administrator salary in Ohio differs from the national mean of \$90,000.
    b. The sample data for 25 Ohio administrators is contained in the file named Administrator. What is the p-value for your hypothesis test in part (a)?
    c. At $\alpha = .05$, can your null hypothesis be rejected? What is your conclusion?
    d. Repeat the preceding hypothesis test using the critical value approach.

WEB file

ChildCare

30. The time married men with children spend on child care averages 6.4 hours per week (*Time,* March 12, 2012). You belong to a professional group on family practices that would like to do its own study to determine if the time married men in your area spend on child care per week differs from the reported mean of 6.4 hours per week. A sample of 40 married couples will be used with the data collected showing the hours per week the husband spends on child care. The sample data are contained in the file ChildCare.
    a. What are the hypotheses if your group would like to determine if the population mean number of hours married men are spending in child care differs from the mean reported by *Time* in your area?
    b. What is the sample mean and the p-value?
    c. Select your own level of significance. What is your conclusion?
31. The Coca-Cola Company reported that the mean per capita annual sales of its beverages in the United States was 423 eight-ounce servings (Coca-Cola Company website, February 3, 2009). Suppose you are curious whether the consumption of Coca-Cola beverages is higher in Atlanta, Georgia, the location of Coca-Cola's corporate headquarters. A sample of 36 individuals from the Atlanta area showed a sample mean annual consumption of 460.4 eight-ounce servings with a standard deviation of $s = 101.9$ ounces. Using $\alpha = .05$, do the sample results support the conclusion that mean annual consumption of Coca-Cola beverage products is higher in Atlanta?

WEB file

UsedCars

32. According to the National Automobile Dealers Association, the mean price for used cars is \$10,192. A manager of a Kansas City used car dealership reviewed a sample of 50 recent used car sales at the dealership in an attempt to determine whether the population mean price for used cars at this particular dealership differed from the national mean. The prices for the sample of 50 cars are shown in the file named UsedCars.
    a. Formulate the hypotheses that can be used to determine whether a difference exists in the mean price for used cars at the dealership.
    b. What is the p-value?
    c. At $\alpha = .05$, what is your conclusion?
33. Annual per capita consumption of milk is 21.6 gallons (*Statistical Abstract of the United States: 2006*). Being from the Midwest, you believe milk consumption is higher there and wish to support your opinion. A sample of 16 individuals from the midwestern town of Webster City showed a sample mean annual consumption of 24.1 gallons with a standard deviation of $s = 4.8$.
    a. Develop a hypothesis test that can be used to determine whether the mean annual consumption in Webster City is higher than the national mean.
    b. What is a point estimate of the difference between mean annual consumption in Webster City and the national mean?
    c. At $\alpha = .05$, test for a significant difference. What is your conclusion?
34. Joan's Nursery specializes in custom-designed landscaping for residential areas. The estimated labor cost associated with a particular landscaping proposal is based on the number of plantings of trees, shrubs, and so on to be used for the project. For cost-estimating purposes, managers use two hours of labor time for the planting of a

medium-sized tree. Actual times from a sample of 10 plantings during the past month follow (times in hours).

1.7	1.5	2.6	2.2	2.4	2.3	2.6	3.0	1.4	2.3

With a .05 level of significance, test to see whether the mean tree-planting time differs from two hours.

a. State the null and alternative hypotheses.
b. Compute the sample mean.
c. Compute the sample standard deviation.
d. What is the p-value?
e. What is your conclusion?

9.5 Population Proportion

In this section we show how to conduct a hypothesis test about a population proportion p. Using p_0 to denote the hypothesized value for the population proportion, the three forms for a hypothesis test about a population proportion are as follows.

$$\begin{array}{lll} H_0: p \geq p_0 & H_0: p \leq p_0 & H_0: p = p_0 \\ H_a: p < p_0 & H_a: p > p_0 & H_a: p \neq p_0 \end{array}$$

The first form is called a lower tail test, the second form is called an upper tail test, and the third form is called a two-tailed test.

Hypothesis tests about a population proportion are based on the difference between the sample proportion $\bar{p}$ and the hypothesized population proportion p_0. The methods used to conduct the hypothesis test are similar to those used for hypothesis tests about a population mean. The only difference is that we use the sample proportion and its standard error to compute the test statistic. The p-value approach or the critical value approach is then used to determine whether the null hypothesis should be rejected.

Let us consider an example involving a situation faced by Pine Creek golf course. Over the past year, 20% of the players at Pine Creek were women. In an effort to increase the proportion of women players, Pine Creek implemented a special promotion designed to attract women golfers. One month after the promotion was implemented, the course manager requested a statistical study to determine whether the proportion of women players at Pine Creek had increased. Because the objective of the study is to determine whether the proportion of women golfers increased, an upper tail test with H_a: $p > .20$ is appropriate. The null and alternative hypotheses for the Pine Creek hypothesis test are as follows:

$$\begin{array}{l} H_0: p \leq .20 \\ H_a: p > .20 \end{array}$$

If H_0 can be rejected, the test results will give statistical support for the conclusion that the proportion of women golfers increased and the promotion was beneficial. The course manager specified that a level of significance of $\alpha = .05$ be used in carrying out this hypothesis test.

The next step of the hypothesis testing procedure is to select a sample and compute the value of an appropriate test statistic. To show how this step is done for the Pine Creek upper tail test, we begin with a general discussion of how to compute the value of the test statistic for any form of a hypothesis test about a population proportion. The sampling distribution of $\bar{p}$, the point estimator of the population parameter p, is the basis for developing the test statistic.

When the null hypothesis is true as an equality, the expected value of $\bar{p}$ equals the hypothesized value p_0; that is, $E(\bar{p}) = p_0$. The standard error of $\bar{p}$ is given by

$$\sigma_{\bar{p}} = \sqrt{\frac{p_0(1 - p_0)}{n}}$$

In Chapter 7 we said that if $np \geq 5$ and $n(1 - p) \geq 5$, the sampling distribution of $\bar{p}$ can be approximated by a normal distribution.[3] Under these conditions, which usually apply in practice, the quantity

$$z = \frac{\bar{p} - p_0}{\sigma_{\bar{p}}} \tag{9.3}$$

has a standard normal probability distribution. With $\sigma_{\bar{p}} = \sqrt{p_0(1 - p_0)/n}$, the standard normal random variable z is the test statistic used to conduct hypothesis tests about a population proportion.

TEST STATISTIC FOR HYPOTHESIS TESTS ABOUT A POPULATION PROPORTION

$$z = \frac{\bar{p} - p_0}{\sqrt{\dfrac{p_0(1 - p_0)}{n}}} \tag{9.4}$$

WEB file WomenGolf

We can now compute the test statistic for the Pine Creek hypothesis test. Suppose a random sample of 400 players was selected, and that 100 of the players were women. The proportion of women golfers in the sample is

$$\bar{p} = \frac{100}{400} = .25$$

Using equation (9.4), the value of the test statistic is

$$z = \frac{\bar{p} - p_0}{\sqrt{\dfrac{p_0(1 - p_0)}{n}}} = \frac{.25 - .20}{\sqrt{\dfrac{.20(1 - .20)}{400}}} = \frac{.05}{.02} = 2.50$$

Because the Pine Creek hypothesis test is an upper tail test, the p-value is the probability that z is greater than or equal to $z = 2.50$; that is, it is the upper tail area corresponding to $z \geq 2.50$. Using the standard normal probability table, we find that the area to the left of $z = 2.50$ is .9938. Thus, the p-value for the Pine Creek test is $1.0000 - .9938 = .0062$. Figure 9.7 shows this p-value calculation.

Recall that the course manager specified a level of significance of $\alpha = .05$. A p-value $= .0062 < .05$ gives sufficient statistical evidence to reject H_0 at the .05 level of significance. Thus, the test provides statistical support for the conclusion that the special promotion increased the proportion of women players at the Pine Creek golf course.

[3]In most applications involving hypothesis tests of a population proportion, sample sizes are large enough to use the normal approximation. The exact sampling distribution of $\bar{p}$ is discrete with the probability for each value of $\bar{p}$ given by the binomial distribution. So hypothesis testing is a bit more complicated for small samples when the normal approximation cannot be used.

FIGURE 9.7 CALCULATION OF THE p-VALUE FOR THE PINE CREEK HYPOTHESIS TEST

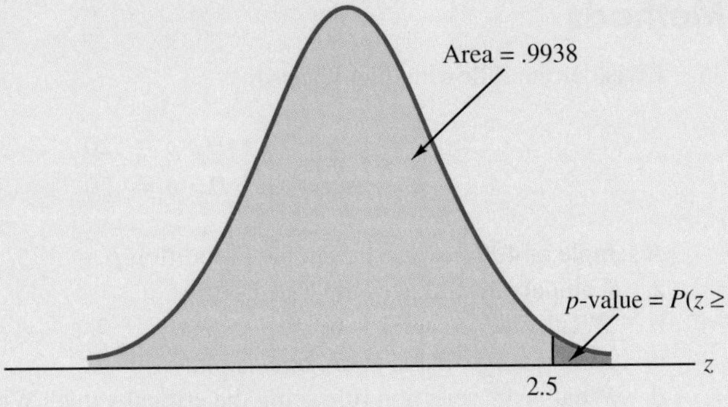

The decision whether to reject the null hypothesis can also be made using the critical value approach. The critical value corresponding to an area of .05 in the upper tail of a normal probability distribution is $z_{.05} = 1.645$. Thus, the rejection rule using the critical value approach is to reject H_0 if $z \geq 1.645$. Because $z = 2.50 > 1.645$, H_0 is rejected.

Again, we see that the p-value approach and the critical value approach lead to the same hypothesis testing conclusion, but the p-value approach provides more information. With a p-value $= .0062$, the null hypothesis would be rejected for any level of significance greater than or equal to .0062.

Summary

The procedure used to conduct a hypothesis test about a population proportion is similar to the procedure used to conduct a hypothesis test about a population mean. Although we only illustrated how to conduct a hypothesis test about a population proportion for an upper tail test, similar procedures can be used for lower tail and two-tailed tests. Table 9.4 provides a summary of the hypothesis tests about a population proportion. We assume that $np \geq 5$ and $n(1 - p) \geq 5$; thus the normal probability distribution can be used to approximate the sampling distribution of $\bar{p}$.

TABLE 9.4 SUMMARY OF HYPOTHESIS TESTS ABOUT A POPULATION PROPORTION

	Lower Tail Test	Upper Tail Test	Two-Tailed Test
Hypotheses	$H_0: p \geq p_0$ $H_a: p < p_0$	$H_0: p \leq p_0$ $H_a: p > p_0$	$H_0: p = p_0$ $H_a: p \neq p_0$
Test Statistic	$z = \dfrac{\bar{p} - p_0}{\sqrt{\dfrac{p_0(1 - p_0)}{n}}}$	$z = \dfrac{\bar{p} - p_0}{\sqrt{\dfrac{p_0(1 - p_0)}{n}}}$	$z = \dfrac{\bar{p} - p_0}{\sqrt{\dfrac{p_0(1 - p_0)}{n}}}$
Rejection Rule: p-Value Approach	Reject H_0 if p-value $\leq \alpha$	Reject H_0 if p-value $\leq \alpha$	Reject H_0 if p-value $\leq \alpha$
Rejection Rule: Critical Value Approach	Reject H_0 if $z \leq -z_\alpha$	Reject H_0 if $z \geq z_\alpha$	Reject H_0 if $z \leq -z_{\alpha/2}$ or if $z \geq z_{\alpha/2}$

Exercises

Methods

35. Consider the following hypothesis test:

$$H_0: p = .20$$
$$H_a: p \neq .20$$

A sample of 400 provided a sample proportion $\bar{p} = .175$.

a. Compute the value of the test statistic.
b. What is the p-value?
c. At $\alpha = .05$, what is your conclusion?
d. What is the rejection rule using the critical value? What is your conclusion?

SELF test

36. Consider the following hypothesis test:

$$H_0: p \geq .75$$
$$H_a: p < .75$$

A sample of 300 items was selected. Compute the p-value and state your conclusion for each of the following sample results. Use $\alpha = .05$.

a. $\bar{p} = .68$
b. $\bar{p} = .72$
c. $\bar{p} = .70$
d. $\bar{p} = .77$

Applications

37. A study found that, in 2005, 12.5% of U.S. workers belonged to unions (*The Wall Street Journal,* January 21, 2006). Suppose a sample of 400 U.S. workers is collected in 2006 to determine whether union efforts to organize have increased union membership.
    a. Formulate the hypotheses that can be used to determine whether union membership increased in 2006.
    b. If the sample results show that 52 of the workers belonged to unions, what is the p-value for your hypothesis test?
    c. At $\alpha = .05$, what is your conclusion?

SELF test

38. A study by *Consumer Reports* showed that 64% of supermarket shoppers believe supermarket brands to be as good as national name brands. To investigate whether this result applies to its own product, the manufacturer of a national name-brand ketchup asked a sample of shoppers whether they believed that supermarket ketchup was as good as the national brand ketchup.
    a. Formulate the hypotheses that could be used to determine whether the percentage of supermarket shoppers who believe that the supermarket ketchup was as good as the national brand ketchup differed from 64%.
    b. If a sample of 100 shoppers showed 52 stating that the supermarket brand was as good as the national brand, what is the p-value?
    c. At $\alpha = .05$, what is your conclusion?
    d. Should the national brand ketchup manufacturer be pleased with this conclusion? Explain.

WEB file
AgeGroup

39. According to the Pew Internet & American Life Project, 75% of American adults use the Internet (Pew Internet website, April 19, 2008). The Pew project authors also reported on the percentage of Americans who use the Internet by age group. The data in the file AgeGroup are consistent with their findings. These data were obtained from a sample of 100 Internet users in the 30–49 age group and 200 Internet users in the 50–64 age group. A Yes indicates the survey repondent had used the Internet; a No indicates the survey repondent had not.

a. Formulate hypotheses that could be used to determine whether the percentage of Internet users in the two age groups differs from the overall average of 75%
b. Estimate the proportion of Internet users in the 30–49 age group. Does this proportion differ significantly from the overall proportion of .75? Use $\alpha = .05$
c. Estimate the proportion of Internet users in the 50–64 age group. Does this proportion differ significantly from the overall proportion of .75? Use $\alpha = .05$
d. Would you expect the proportion of users in the 18–29 age group to be larger or smaller than the proportion for the 30–49 age group? Support you conclusion with the results obtained in parts (b) and (c).

40. In 2008, 46% of business owners gave a holiday gift to their employees. A 2009 survey of business owners indicated that 35% planned to provide a holiday gift to their employees (Radio WEZV, Myrtle Beach, SC, November 11, 2009). Suppose the survey results are based on a sample of 60 business owners.
a. How many business owners in the survey planned to provide a holiday gift to their employees in 2009?
b. Suppose the business owners in the sample did as they planned. Compute the p-value for a hypothesis test that can be used to determine if the proportion of business owners providing holiday gifts had decreased from the 2008 level.
c. Using a .05 level of significance, would you conclude that the proportion of business owners providing gifts decreased? What is the smallest level of significance for which you could draw such a conclusion?

41. Speaking to a group of analysts in January 2006, a brokerage firm executive claimed that at least 70% of investors are currently confident of meeting their investment objectives. A UBS Investor Optimism Survey, conducted over the period January 2 to January 15, found that 67% of investors were confident of meeting their investment objectives (CNBC, January 20, 2006).
a. Formulate the hypotheses that can be used to test the validity of the brokerage firm executive's claim.
b. Assume the UBS Investor Optimism Survey collected information from 300 investors. What is the p-value for the hypothesis test?
c. At $\alpha = .05$, should the executive's claim be rejected?

42. According to the University of Nevada Center for Logistics Management, 6% of all merchandise sold in the United States gets returned (*BusinessWeek,* January 15, 2007). A Houston department store sampled 80 items sold in January and found that 12 of the items were returned.
a. Construct a point estimate of the proportion of items returned for the population of sales transactions at the Houston store.
b. Construct a 95% confidence interval for the porportion of returns at the Houston store.
c. Is the proportion of returns at the Houston store significantly different from the returns for the nation as a whole? Provide statistical support for your answer.

WEB file
Eagle

43. Eagle Outfitters is a chain of stores specializing in outdoor apparel and camping gear. They are considering a promotion that involves mailing discount coupons to all their credit card customers. This promotion will be considered a success if more than 10% of those receiving the coupons use them. Before going national with the promotion, coupons were sent to a sample of 100 credit card customers.
a. Develop hypotheses that can be used to test whether the population proportion of those who will use the coupons is sufficient to go national.
b. The file Eagle contains the sample data. Develop a point estimate of the population proportion.
c. Use $\alpha = .05$ to conduct your hypothesis test. Should Eagle go national with the promotion?

WEB file
Drowsy

44. In a cover story, *BusinessWeek* published information about sleep habits of Americans (*BusinessWeek,* January 26, 2004). The article noted that sleep deprivation causes a number of problems, including highway deaths. Fifty-one percent of adult drivers admit to

driving while drowsy. A researcher hypothesized that this issue was an even bigger problem for night shift workers.

a. Formulate the hypotheses that can be used to help determine whether more than 51% of the population of night shift workers admit to driving while drowsy.
b. A sample of 400 night shift workers identified those who admitted to driving while drowsy. See the Drowsy file. What is the sample proportion? What is the p-value?
c. At $\alpha = .01$, what is your conclusion?

45. Many investors and financial analysts believe the Dow Jones Industrial Average (DJIA) provides a good barometer of the overall stock market. On January 31, 2006, 9 of the 30 stocks making up the DJIA increased in price (*The Wall Street Journal,* February 1, 2006). On the basis of this fact, a financial analyst claims we can assume that 30% of the stocks traded on the New York Stock Exchange (NYSE) went up the same day.
   a. Formulate null and alternative hypotheses to test the analyst's claim.
   b. A sample of 50 stocks traded on the NYSE that day showed that 24 went up. What is your point estimate of the population proportion of stocks that went up?
   c. Conduct your hypothesis test using $\alpha = .01$ as the level of significance. What is your conclusion?

9.6 Hypothesis Testing and Decision Making

In the previous sections of this chapter we have illustrated hypothesis testing applications that are considered significance tests. After formulating the null and alternative hypotheses, we selected a sample and computed the value of a test statistic and the associated p-value. We then compared the p-value to a controlled probability of a Type I error, α, which is called the level of significance for the test. If p-value $\leq \alpha$, we made the conclusion "reject H_0" and declared the results significant; otherwise, we made the conclusion "do not reject H_0." With a significance test, we control the probability of making the Type I error, but not the Type II error. Thus, we recommended the conclusion "do not reject H_0" rather than "accept H_0" because the latter puts us at risk of making the Type II error of accepting H_0 when it is false. With the conclusion "do not reject H_0," the statistical evidence is considered inconclusive and is usually an indication to postpone a decision or action until further research and testing can be undertaken.

However, if the purpose of a hypothesis test is to make a decision when H_0 is true and a different decision when H_a is true, the decision maker may want to, and in some cases be forced to, take action with both the conclusion *do not reject* H_0 and the conclusion *reject* H_0. If this situation occurs, statisticians generally recommend controlling the probability of making a Type II error. With the probabilities of both the Type I and Type II error controlled, the conclusion from the hypothesis test is either to *accept* H_0 or *reject* H_0. In the first case, H_0 is concluded to be true, while in the second case, H_a is concluded true. Thus, a decision and appropriate action can be taken when either conclusion is reached.

A good illustration of hypothesis testing for decision making is lot-acceptance sampling, a topic we will discuss in more depth in Chapter 20. For example, a quality control manager must decide to accept a shipment of batteries from a supplier or to return the shipment because of poor quality. Assume that design specifications require batteries from the supplier to have a mean useful life of at least 120 hours. To evaluate the quality of an incoming shipment, a sample of 36 batteries will be selected and tested. On the basis of the sample, a decision must be made to accept the shipment of batteries or to return it to the supplier because of poor quality. Let μ denote the mean number of hours of useful life for batteries in the shipment. The null and alternative hypotheses about the population mean follow.

$$H_0: \mu \geq 120$$
$$H_a: \mu < 120$$

If H_0 is rejected, the alternative hypothesis is concluded to be true. This conclusion indicates that the appropriate action is to return the shipment to the supplier. However, if H_0 is not rejected, the decision maker must still determine what action should be taken. Thus, without directly concluding that H_0 is true, but merely by not rejecting it, the decision maker will have made the decision to accept the shipment as being of satisfactory quality.

In such decision-making situations, it is recommended that the hypothesis testing procedure be extended to control the probability of making a Type II error. Because a decision will be made and action taken when we do not reject H_0, knowledge of the probability of making a Type II error will be helpful. In Sections 9.7 and 9.8 we explain how to compute the probability of making a Type II error and how the sample size can be adjusted to help control the probability of making a Type II error.

9.7 Calculating the Probability of Type II Errors

In this section we show how to calculate the probability of making a Type II error for a hypothesis test about a population mean. We illustrate the procedure by using the lot-acceptance example described in Section 9.6. The null and alternative hypotheses about the mean number of hours of useful life for a shipment of batteries are H_0: $\mu \geq 120$ and H_a: $\mu < 120$. If H_0 is rejected, the decision will be to return the shipment to the supplier because the mean hours of useful life are less than the specified 120 hours. If H_0 is not rejected, the decision will be to accept the shipment.

Suppose a level of significance of $\alpha = .05$ is used to conduct the hypothesis test. The test statistic in the σ known case is

$$z = \frac{\bar{x} - \mu_0}{\sigma/\sqrt{n}} = \frac{\bar{x} - 120}{\sigma/\sqrt{n}}$$

Based on the critical value approach and $z_{.05} = 1.645$, the rejection rule for the lower tail test is

$$\text{Reject } H_0 \text{ if } z \leq -1.645$$

Suppose a sample of 36 batteries will be selected and based upon previous testing the population standard deviation can be assumed known with a value of $\sigma = 12$ hours. The rejection rule indicates that we will reject H_0 if

$$z = \frac{\bar{x} - 120}{12/\sqrt{36}} \leq -1.645$$

Solving for $\bar{x}$ in the preceding expression indicates that we will reject H_0 if

$$\bar{x} \leq 120 - 1.645\left(\frac{12}{\sqrt{36}}\right) = 116.71$$

Rejecting H_0 when $\bar{x} \leq 116.71$ means that we will make the decision to accept the shipment whenever

$$\bar{x} > 116.71$$

With this information, we are ready to compute probabilities associated with making a Type II error. First, recall that we make a Type II error whenever the true shipment mean is less than 120 hours and we make the decision to accept H_0: $\mu \geq 120$. Hence, to compute the probability of making a Type II error, we must select a value of μ less than 120 hours.

FIGURE 9.8 PROBABILITY OF A TYPE II ERROR WHEN $\mu = 112$

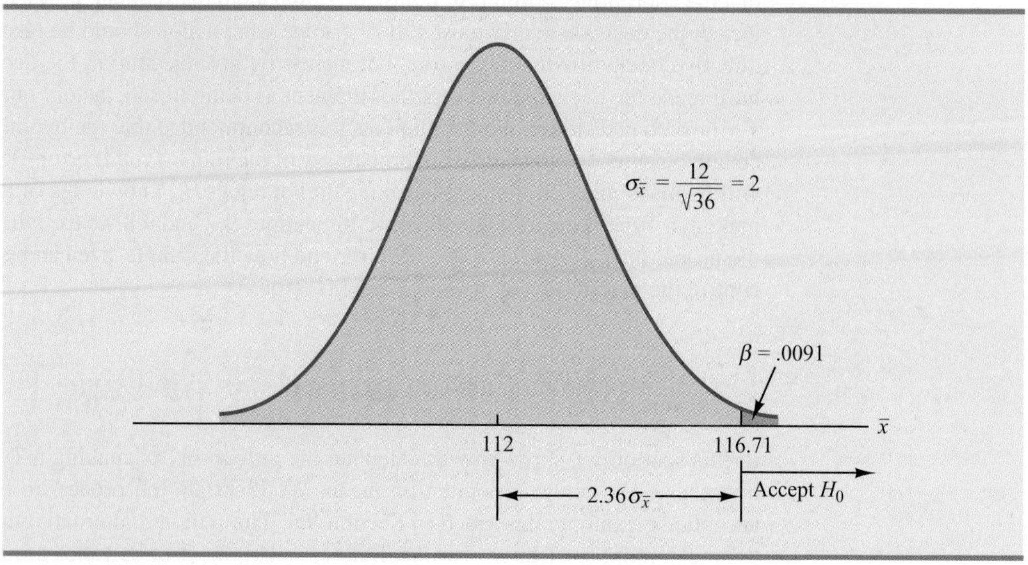

For example, suppose the shipment is considered to be of poor quality if the batteries have a mean life of $\mu = 112$ hours. If $\mu = 112$ is really true, what is the probability of accepting H_0: $\mu \geq 120$ and hence committing a Type II error? Note that this probability is the probability that the sample mean $\bar{x}$ is greater than 116.71 when $\mu = 112$.

Figure 9.8 shows the sampling distribution of $\bar{x}$ when the mean is $\mu = 112$. The shaded area in the upper tail gives the probability of obtaining $\bar{x} > 116.71$. Using the standard normal distribution, we see that at $\bar{x} = 116.71$

$$z = \frac{\bar{x} - \mu}{\sigma/\sqrt{n}} = \frac{116.71 - 112}{12/\sqrt{36}} = 2.36$$

The standard normal probability table shows that with $z = 2.36$, the area in the upper tail is $1.0000 - .9909 = .0091$. Thus, .0091 is the probability of making a Type II error when $\mu = 112$. Denoting the probability of making a Type II error as β, we see that when $\mu = 112$, $\beta = .0091$. Therefore, we can conclude that if the mean of the population is 112 hours, the probability of making a Type II error is only .0091.

We can repeat these calculations for other values of μ less than 120. Doing so will show a different probability of making a Type II error for each value of μ. For example, suppose the shipment of batteries has a mean useful life of $\mu = 115$ hours. Because we will accept H_0 whenever $\bar{x} > 116.71$, the z value for $\mu = 115$ is given by

$$z = \frac{\bar{x} - \mu}{\sigma/\sqrt{n}} = \frac{116.71 - 115}{12/\sqrt{36}} = .86$$

From the standard normal probability table, we find that the area in the upper tail of the standard normal distribution for $z = .86$ is $1.0000 - .8051 = .1949$. Thus, the probability of making a Type II error is $\beta = .1949$ when the true mean is $\mu = 115$.

In Table 9.5 we show the probability of making a Type II error for a variety of values of μ less than 120. Note that as μ increases toward 120, the probability of making a Type II error increases toward an upper bound of .95. However, as μ decreases to values farther below 120, the probability of making a Type II error diminishes. This pattern is what we

TABLE 9.5 PROBABILITY OF MAKING A TYPE II ERROR FOR THE LOT-ACCEPTANCE HYPOTHESIS TEST

Value of μ	$z = \dfrac{116.71 - \mu}{12/\sqrt{36}}$	Probability of a Type II Error (β)	Power ($1 - \beta$)
112	2.36	.0091	.9909
114	1.36	.0869	.9131
115	.86	.1949	.8051
116.71	.00	.5000	.5000
117	−.15	.5596	.4404
118	−.65	.7422	.2578
119.999	−1.645	.9500	.0500

As Table 9.5 shows, the probability of a Type II error depends on the value of the population mean μ. For values of μ near μ_0, the probability of making the Type II error can be high.

should expect. When the true population mean μ is close to the null hypothesis value of $\mu = 120$, the probability is high that we will make a Type II error. However, when the true population mean μ is far below the null hypothesis value of $\mu = 120$, the probability is low that we will make a Type II error.

The probability of correctly rejecting H_0 when it is false is called the **power** of the test. For any particular value of μ, the power is $1 - \beta$; that is, the probability of correctly rejecting the null hypothesis is 1 minus the probability of making a Type II error. Values of power are also listed in Table 9.5. On the basis of these values, the power associated with each value of μ is shown graphically in Figure 9.9. Such a graph is called a **power curve**. Note that the power curve extends over the values of μ for which the null hypothesis is false. The height of the power curve at any value of μ indicates the probability of correctly rejecting H_0 when H_0 is false.[4]

FIGURE 9.9 POWER CURVE FOR THE LOT-ACCEPTANCE HYPOTHESIS TEST

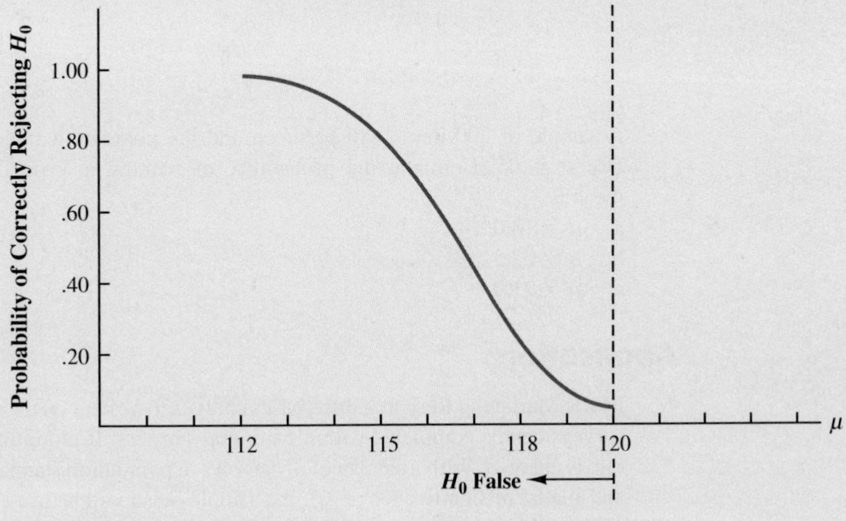

[4]Another graph, called the operating characteristic curve, is sometimes used to provide information about the probability of making a Type II error. The operating characteristic curve shows the probability of accepting H_0 and thus provides β for the values of μ where the null hypothesis is false. The probability of making a Type II error can be read directly from this graph.

In summary, the following step-by-step procedure can be used to compute the probability of making a Type II error in hypothesis tests about a population mean.

1. Formulate the null and alternative hypotheses.
2. Use the level of significance α and the critical value approach to determine the critical value and the rejection rule for the test.
3. Use the rejection rule to solve for the value of the sample mean corresponding to the critical value of the test statistic.
4. Use the results from step 3 to state the values of the sample mean that lead to the acceptance of H_0. These values define the acceptance region for the test.
5. Use the sampling distribution of $\bar{x}$ for a value of μ satisfying the alternative hypothesis, and the acceptance region from step 4, to compute the probability that the sample mean will be in the acceptance region. This probability is the probability of making a Type II error at the chosen value of μ.

Exercises

Methods

SELF test

46. Consider the following hypothesis test.

$$H_0: \mu \geq 10$$
$$H_a: \mu < 10$$

The sample size is 120 and the population standard deviation is assumed known with $\sigma = 5$. Use $\alpha = .05$.

a. If the population mean is 9, what is the probability that the sample mean leads to the conclusion *do not reject* H_0?
b. What type of error would be made if the actual population mean is 9 and we conclude that H_0: $\mu \geq 10$ is true?
c. What is the probability of making a Type II error if the actual population mean is 8?

47. Consider the following hypothesis test.

$$H_0: \mu = 20$$
$$H_a: \mu \neq 20$$

A sample of 200 items will be taken and the population standard deviation is $\sigma = 10$. Use $\alpha = .05$. Compute the probability of making a Type II error if the population mean is:

a. $\mu = 18.0$
b. $\mu = 22.5$
c. $\mu = 21.0$

Applications

48. Fowle Marketing Research, Inc., bases charges to a client on the assumption that telephone surveys can be completed within 15 minutes or less. If more time is required, a premium rate is charged. With a sample of 35 surveys, a population standard deviation of 4 minutes, and a level of significance of .01, the sample mean will be used to test the null hypothesis H_0: $\mu \leq 15$.

a. What is your interpretation of the Type II error for this problem? What is its impact on the firm?
b. What is the probability of making a Type II error when the actual mean time is $\mu = 17$ minutes?

c. What is the probability of making a Type II error when the actual mean time is $\mu = 18$ minutes?
d. Sketch the general shape of the power curve for this test.

SELF test

49. A consumer research group is interested in testing an automobile manufacturer's claim that a new economy model will travel at least 25 miles per gallon of gasoline (H_0: $\mu \geq 25$).
    a. With a .02 level of significance and a sample of 30 cars, what is the rejection rule based on the value of $\bar{x}$ for the test to determine whether the manufacturer's claim should be rejected? Assume that σ is 3 miles per gallon.
    b. What is the probability of committing a Type II error if the actual mileage is 23 miles per gallon?
    c. What is the probability of committing a Type II error if the actual mileage is 24 miles per gallon?
    d. What is the probability of committing a Type II error if the actual mileage is 25.5 miles per gallon?
50. *Young Adult* magazine states the following hypotheses about the mean age of its subscribers.

$$H_0: \mu = 28$$
$$H_a: \mu \neq 28$$

    a. What would it mean to make a Type II error in this situation?
    b. The population standard deviation is assumed known at $\sigma = 6$ years and the sample size is 100. With $\alpha = .05$, what is the probability of accepting H_0 for μ equal to 26, 27, 29, and 30?
    c. What is the power at $\mu = 26$? What does this result tell you?
51. A production line operation is tested for filling weight accuracy using the following hypotheses.

Hypothesis	**Conclusion and Action**
H_0: $\mu = 16$	Filling okay; keep running
H_a: $\mu \neq 16$	Filling off standard; stop and adjust machine

The sample size is 30 and the population standard deviation is $\sigma = .8$. Use $\alpha = .05$.
    a. What would a Type II error mean in this situation?
    b. What is the probability of making a Type II error when the machine is overfilling by .5 ounces?
    c. What is the power of the statistical test when the machine is overfilling by .5 ounces?
    d. Show the power curve for this hypothesis test. What information does it contain for the production manager?
52. Refer to exercise 48. Assume the firm selects a sample of 50 surveys and repeat parts (b) and (c). What observation can you make about how increasing the sample size affects the probability of making a Type II error?
53. Sparr Investments, Inc., specializes in tax-deferred investment opportunities for its clients. Recently Sparr offered a payroll deduction investment program for the employees of a particular company. Sparr estimates that the employees are currently averaging \$100 or less per month in tax-deferred investments. A sample of 40 employees will be used to test Sparr's hypothesis about the current level of investment activity among the population of employees. Assume the employee monthly tax-deferred investment amounts have a standard deviation of \$75 and that a .05 level of significance will be used in the hypothesis test.
    a. What is the Type II error in this situation?
    b. What is the probability of the Type II error if the actual mean employee monthly investment is \$120?
    c. What is the probability of the Type II error if the actual mean employee monthly investment is \$130?
    d. Assume a sample size of 80 employees is used and repeat parts (b) and (c).

9.8 Determining the Sample Size for a Hypothesis Test About a Population Mean

Assume that a hypothesis test is to be conducted about the value of a population mean. The level of significance specified by the user determines the probability of making a Type I error for the test. By controlling the sample size, the user can also control the probability of making a Type II error. Let us show how a sample size can be determined for the following lower tail test about a population mean.

$$H_0: \mu \geq \mu_0$$
$$H_a: \mu < \mu_0$$

The upper panel of Figure 9.10 is the sampling distribution of $\bar{x}$ when H_0 is true with $\mu = \mu_0$. For a lower tail test, the critical value of the test statistic is denoted $-z_\alpha$. In the upper panel of the figure the vertical line, labeled c, is the corresponding value of $\bar{x}$. Note that, if we reject H_0 when $\bar{x} \leq c$, the probability of a Type I error will be α. With z_α representing the z value corresponding to an area of α in the upper tail of the standard normal distribution, we compute c using the following formula:

$$c = \mu_0 - z_\alpha \frac{\sigma}{\sqrt{n}} \tag{9.5}$$

FIGURE 9.10 DETERMINING THE SAMPLE SIZE FOR SPECIFIED LEVELS OF THE TYPE I (α) AND TYPE II (β) ERRORS

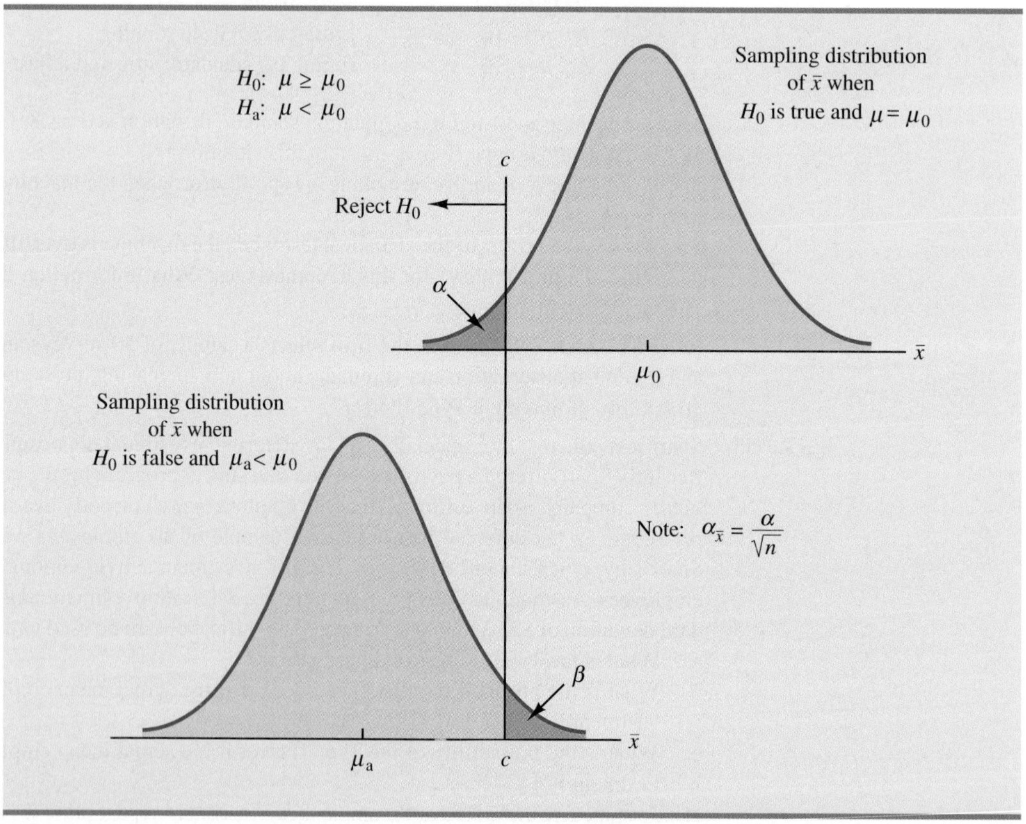

The lower panel of Figure 9.10 is the sampling distribution of $\bar{x}$ when the alternative hypothesis is true with $\mu = \mu_a < \mu_0$. The shaded region shows β, the probability of a Type II error that the decision maker will be exposed to if the null hypothesis is accepted when $\bar{x} > c$. With z_β representing the z value corresponding to an area of β in the upper tail of the standard normal distribution, we compute c using the following formula:

$$c = \mu_a + z_\beta \frac{\sigma}{\sqrt{n}} \tag{9.6}$$

Now what we want to do is to select a value for c so that when we reject H_0 and accept H_a, the probability of a Type I error is equal to the chosen value of α and the probability of a Type II error is equal to the chosen value of β. Therefore, both equations (9.5) and (9.6) must provide the same value for c, and the following equation must be true.

$$\mu_0 - z_\alpha \frac{\sigma}{\sqrt{n}} = \mu_a + z_\beta \frac{\sigma}{\sqrt{n}}$$

To determine the required sample size, we first solve for the $\sqrt{n}$ as follows.

$$\mu_0 - \mu_a = z_\alpha \frac{\sigma}{\sqrt{n}} + z_\beta \frac{\sigma}{\sqrt{n}}$$

$$\mu_0 - \mu_a = \frac{(z_\alpha + z_\beta)\sigma}{\sqrt{n}}$$

and

$$\sqrt{n} = \frac{(z_\alpha + z_\beta)\sigma}{(\mu_0 - \mu_a)}$$

Squaring both sides of the expression provides the following sample size formula for a one-tailed hypothesis test about a population mean.

SAMPLE SIZE FOR A ONE-TAILED HYPOTHESIS TEST ABOUT A POPULATION MEAN

$$n = \frac{(z_\alpha + z_\beta)^2\sigma^2}{(\mu_0 - \mu_a)^2} \tag{9.7}$$

where

z_α = z value providing an area of α in the upper tail of a standard normal distribution
z_β = z value providing an area of β in the upper tail of a standard normal distribution
σ = the population standard deviation
μ_0 = the value of the population mean in the null hypothesis
μ_a = the value of the population mean used for the Type II error

Note: In a two-tailed hypothesis test, use (9.7) with $z_{\alpha/2}$ replacing z_α.

Although the logic of equation (9.7) was developed for the hypothesis test shown in Figure 9.10, it holds for any one-tailed test about a population mean. In a two-tailed hypothesis test about a population mean, $z_{\alpha/2}$ is used instead of z_α in equation (9.7).

Let us return to the lot-acceptance example from Sections 9.6 and 9.7. The design specification for the shipment of batteries indicated a mean useful life of at least 120 hours for the batteries. Shipments were rejected if H_0: $\mu \geq 120$ was rejected. Let us assume that the quality control manager makes the following statements about the allowable probabilities for the Type I and Type II errors.

Type I error statement: If the mean life of the batteries in the shipment is $\mu = 120$, I am willing to risk an $\alpha = .05$ probability of rejecting the shipment.

Type II error statement: If the mean life of the batteries in the shipment is 5 hours under the specification (i.e., $\mu = 115$), I am willing to risk a $\beta = .10$ probability of accepting the shipment.

These statements are based on the judgment of the manager. Someone else might specify different restrictions on the probabilities. However, statements about the allowable probabilities of both errors must be made before the sample size can be determined.

In the example, $\alpha = .05$ and $\beta = .10$. Using the standard normal probability distribution, we have $z_{.05} = 1.645$ and $z_{.10} = 1.28$. From the statements about the error probabilities, we note that $\mu_0 = 120$ and $\mu_a = 115$. Finally, the population standard deviation was assumed known at $\sigma = 12$. By using equation (9.7), we find that the recommended sample size for the lot-acceptance example is

$$n = \frac{(1.645 + 1.28)^2(12)^2}{(120 - 115)^2} = 49.3$$

Rounding up, we recommend a sample size of 50.

Because both the Type I and Type II error probabilities have been controlled at allowable levels with $n = 50$, the quality control manager is now justified in using the *accept* H_0 and *reject* H_0 statements for the hypothesis test. The accompanying inferences are made with allowable probabilities of making Type I and Type II errors.

We can make three observations about the relationship among α, β, and the sample size n.

1. Once two of the three values are known, the other can be computed.
2. For a given level of significance α, increasing the sample size will reduce β.
3. For a given sample size, decreasing α will increase β, whereas increasing α will decrease β.

The third observation should be kept in mind when the probability of a Type II error is not being controlled. It suggests that one should not choose unnecessarily small values for the level of significance α. For a given sample size, choosing a smaller level of significance means more exposure to a Type II error. Inexperienced users of hypothesis testing often think that smaller values of α are always better. They are better if we are concerned only about making a Type I error. However, smaller values of α have the disadvantage of increasing the probability of making a Type II error.

Exercises

Methods

SELF test

54. Consider the following hypothesis test.

$$H_0: \mu \geq 10$$
$$H_a: \mu < 10$$

The sample size is 120 and the population standard deviation is 5. Use $\alpha = .05$. If the actual population mean is 9, the probability of a Type II error is .2912. Suppose the researcher wants to reduce the probability of a Type II error to .10 when the actual population mean is 9. What sample size is recommended?

55. Consider the following hypothesis test.

$$H_0\colon \mu = 20$$
$$H_a\colon \mu \neq 20$$

The population standard deviation is 10. Use $\alpha = .05$. How large a sample should be taken if the researcher is willing to accept a .05 probability of making a Type II error when the actual population mean is 22?

Applications

56. Suppose the project director for the Hilltop Coffee study (see Section 9.3) asked for a .10 probability of claiming that Hilltop was not in violation when it really was underfilling by 1 ounce ($\mu_a = 2.9375$ pounds). What sample size would have been recommended?

SELF test

57. A special industrial battery must have a life of at least 400 hours. A hypothesis test is to be conducted with a .02 level of significance. If the batteries from a particular production run have an actual mean use life of 385 hours, the production manager wants a sampling procedure that only 10% of the time would show erroneously that the batch is acceptable. What sample size is recommended for the hypothesis test? Use 30 hours as an estimate of the population standard deviation.

58. *Young Adult* magazine states the following hypotheses about the mean age of its subscribers.

$$H_0\colon \mu = 28$$
$$H_a\colon \mu \neq 28$$

If the manager conducting the test will permit a .15 probability of making a Type II error when the true mean age is 29, what sample size should be selected? Assume $\sigma = 6$ and a .05 level of significance.

59. An automobile mileage study tested the following hypotheses.

Hypothesis	Conclusion
$H_0\colon \mu \geq 25$ mpg	Manufacturer's claim supported
$H_a\colon \mu < 25$ mpg	Manufacturer's claim rejected; average mileage per gallon less than stated

For $\sigma = 3$ and a .02 level of significance, what sample size would be recommended if the researcher wants an 80% chance of detecting that μ is less than 25 miles per gallon when it is actually 24?

Summary

Hypothesis testing is a statistical procedure that uses sample data to determine whether a statement about the value of a population parameter should or should not be rejected. The hypotheses are two competing statements about a population parameter. One statement is called the null hypothesis (H_0), and the other statement is called the alternative hypothesis (H_a). In Section 9.1 we provided guidelines for developing hypotheses for situations frequently encountered in practice.

Whenever historical data or other information provides a basis for assuming that the population standard deviation is known, the hypothesis testing procedure for the population mean is based on the standard normal distribution. Whenever σ is unknown, the sample standard deviation s is used to estimate σ and the hypothesis testing procedure is based on the t distribution. In both cases, the quality of results depends on both the form of the population distribution and the sample size. If the population has a normal distribution, both hypothesis testing procedures are applicable, even with small sample sizes. If the population is not normally distributed, larger sample sizes are needed. General guidelines about the sample size were provided in Sections 9.3 and 9.4. In the case of hypothesis tests about a population proportion, the hypothesis testing procedure uses a test statistic based on the standard normal distribution.

In all cases, the value of the test statistic can be used to compute a p-value for the test. A p-value is a probability used to determine whether the null hypothesis should be rejected. If the p-value is less than or equal to the level of significance α, the null hypothesis can be rejected.

Hypothesis testing conclusions can also be made by comparing the value of the test statistic to a critical value. For lower tail tests, the null hypothesis is rejected if the value of the test statistic is less than or equal to the critical value. For upper tail tests, the null hypothesis is rejected if the value of the test statistic is greater than or equal to the critical value. Two-tailed tests consist of two critical values: one in the lower tail of the sampling distribution and one in the upper tail. In this case, the null hypothesis is rejected if the value of the test statistic is less than or equal to the critical value in the lower tail or greater than or equal to the critical value in the upper tail.

Extensions of hypothesis testing procedures to include an analysis of the Type II error were also presented. In Section 9.7 we showed how to compute the probability of making a Type II error. In Section 9.8 we showed how to determine a sample size that will control for the probability of making both a Type I error and a Type II error.

Glossary

Null hypothesis The hypothesis tentatively assumed true in the hypothesis testing procedure.

Alternative hypothesis The hypothesis concluded to be true if the null hypothesis is rejected.

Type II error The error of accepting H_0 when it is false.

Type I error The error of rejecting H_0 when it is true.

Level of significance The probability of making a Type I error when the null hypothesis is true as an equality.

One-tailed test A hypothesis test in which rejection of the null hypothesis occurs for values of the test statistic in one tail of its sampling distribution.

Test statistic A statistic whose value helps determine whether a null hypothesis should be rejected.

***p*-value** A probability that provides a measure of the evidence against the null hypothesis given by the sample. Smaller p-values indicate more evidence against H_0. For a lower tail test, the p-value is the probability of obtaining a value for the test statistic as small as or smaller than that provided by the sample. For an upper tail test, the p-value is the probability of obtaining a value for the test statistic as large as or larger than that provided by the sample. For a two-tailed test, the p-value is the probability of obtaining a value for the test statistic at least as unlikely as or more unlikely than that provided by the sample.

Critical value A value that is compared with the test statistic to determine whether H_0 should be rejected.

Two-tailed test A hypothesis test in which rejection of the null hypothesis occurs for values of the test statistic in either tail of its sampling distribution.

Power The probability of correctly rejecting H_0 when it is false.

Power curve A graph of the probability of rejecting H_0 for all possible values of the population parameter not satisfying the null hypothesis. The power curve provides the probability of correctly rejecting the null hypothesis.

Key Formulas

Test Statistic for Hypothesis Tests About a Population Mean: σ Known

$$z = \frac{\bar{x} - \mu_0}{\sigma/\sqrt{n}} \quad (9.1)$$

Test Statistic for Hypothesis Tests About a Population Mean: σ Unknown

$$t = \frac{\bar{x} - \mu_0}{s/\sqrt{n}} \quad (9.2)$$

Test Statistic for Hypothesis Tests About a Population Proportion

$$z = \frac{\bar{p} - p_0}{\sqrt{\dfrac{p_0(1 - p_0)}{n}}} \quad (9.4)$$

Sample Size for a One-Tailed Hypothesis Test About a Population Mean

$$n = \frac{(z_\alpha + z_\beta)^2\sigma^2}{(\mu_0 - \mu_a)^2} \quad (9.7)$$

In a two-tailed test, replace z_α with $z_{\alpha/2}$.

Supplementary Exercises

60. A production line operates with a mean filling weight of 16 ounces per container. Overfilling or underfilling presents a serious problem and when detected requires the operator to shut down the production line to readjust the filling mechanism. From past data, a population standard deviation $\sigma = .8$ ounces is assumed. A quality control inspector selects a sample of 30 items every hour and at that time makes the decision of whether to shut down the line for readjustment. The level of significance is $\alpha = .05$.
    a. State the hypothesis test for this quality control application.
    b. If a sample mean of $\bar{x} = 16.32$ ounces were found, what is the p-value? What action would you recommend?
    c. If a sample mean of $\bar{x} = 15.82$ ounces were found, what is the p-value? What action would you recommend?
    d. Use the critical value approach. What is the rejection rule for the preceding hypothesis testing procedure? Repeat parts (b) and (c). Do you reach the same conclusion?

61. At Western University the historical mean of scholarship examination scores for freshman applications is 900. A historical population standard deviation $\sigma = 180$ is assumed known. Each year, the assistant dean uses a sample of applications to determine whether the mean examination score for the new freshman applications has changed.
    a. State the hypotheses.
    b. What is the 95% confidence interval estimate of the population mean examination score if a sample of 200 applications provided a sample mean $\bar{x} = 935$?
    c. Use the confidence interval to conduct a hypothesis test. Using $\alpha = .05$, what is your conclusion?
    d. What is the p-value?

62. *Playbill* is a magazine distributed around the country to people attending musicals and other theatrical productions. The mean annual household income for the population of *Playbill* readers is $119,155 (*Playbill,* January 2006). Assume the standard deviation is

σ = \$20,700. A San Francisco civic group has asserted that the mean for theatergoers in the Bay Area is higher. A sample of 60 theater attendees in the Bay Area showed a sample mean household income of \$126,100.

a. Develop hypotheses that can be used to determine whether the sample data support the conclusion that theater attendees in the Bay Area have a higher mean household income than that for all *Playbill* readers.
b. What is the *p*-value based on the sample of 60 theater attendees in the Bay Area?
c. Use $\alpha = .01$ as the level of significance. What is your conclusion?

63. On Friday, Wall Street traders were anxiously awaiting the federal government's release of numbers on the January increase in nonfarm payrolls. The early consensus estimate among economists was for a growth of 250,000 new jobs (CNBC, February 3, 2006). However, a sample of 20 economists taken Thursday afternoon provided a sample mean of 266,000 with a sample standard deviation of 24,000. Financial analysts often call such a sample mean, based on late-breaking news, the *whisper number*. Treat the "consensus estimate" as the population mean. Conduct a hypothesis test to determine whether the whisper number justifies a conclusion of a statistically significant increase in the consensus estimate of economists. Use $\alpha = .01$ as the level of significance.

WEB file
FirstBirth

64. Data released by the National Center for Health Statistics showed that the mean age at which women had their first child was 25.0 in 2006 (*The Wall Street Journal,* February 4, 2009). The reporter, Sue Shellenbarger, noted that this was the first decrease in the average age at which women had their first child in several years. A recent sample of 42 women provided the data in the website file named FirstBirth concerning the age at which these women had their first child. Do the data indicate a change from 2006 in the mean age at which women had their first child? Use $\alpha = .05$.

WEB file
WeeklyPay

65. A recent issue of the *AARP Bulletin* reported that the average weekly pay for a woman with a high school school degree is \$520 (*AARP Bulletin*, January–February, 2010). Suppose you would like to determine if the average weekly pay for all working women is significantly greater than that for women with a high school degree. Data providing the weekly pay for a sample of 50 working women are available in the file named WeeklyPay. These data are consistent with the findings reported in the AARP article.

a. State the hypotheses that should be used to test whether the mean weekly pay for all women is significantly greater than the mean weekly pay for women with a high school degree.
b. Use the data in the file named WeeklyPay to compute the sample mean, the test statistic, and the *p*-value.
c. Use $\alpha = .05$. What is your conclusion?
d. Repeat the hypothesis test using the critical value approach.

66. The chamber of commerce of a Florida Gulf Coast community advertises that area residential property is available at a mean cost of \$125,000 or less per lot. Suppose a sample of 32 properties provided a sample mean of \$130,000 per lot and a sample standard deviation of \$12,500. Use a .05 level of significance to test the validity of the advertising claim.

67. In Hamilton County, Ohio, the mean number of days needed to sell a house is 86 days (Cincinnati Multiple Listing Service, April, 2012). Data for the sale of 40 houses in a nearby county showed a sample mean of 80 days with a sample standard deviation of 20 days. Conduct a hypothesis test to determine whether the mean number of days until a house is sold is different than the Hamilton County mean of 86 days in the nearby county. Use $\alpha = .05$ for the level of significance, and state your conclusion.

68. On December 25, 2009, an airline passenger was subdued while attempting to blow up a Northwest Airlines flight headed for Detroit, Michigan. The passenger had smuggled explosives hidden in his underwear past a metal detector at an airport screening facility. As a result, the Transportation Security Administration (TSA) proposed installing full-body scanners to replace the metal detectors at the nation's largest airports. This proposal resulted in strong objections from privacy advocates who considered the scanners an invasion of privacy.

On January 5–6, 2010, *USA Today* conducted a poll of 542 adults to learn what proportion of airline travelers approved of using full-body scanners (*USA Today,* January 11, 2010). The poll results showed that 455 of the respondents felt that full-body scanners would improve airline security and 423 indicated that they approved of using the devices.

a. Conduct a hypothesis test to determine if the results of the poll justify concluding that over 80% of airline travelers feel that the use of full-body scanners will improve airline security. Use $\alpha = .05$.

b. Suppose the TSA will go forward with the installation and mandatory use of full-body scanners if over 75% of airline travelers approve of using the devices. You have been told to conduct a statistical analysis using the poll results to determine if the TSA should require mandatory use of the full-body scanners. Because this is viewed as a very sensitive decision, use $\alpha = .01$. What is your recommendation?

69. An airline promotion to business travelers is based on the assumption that two-thirds of business travelers use a laptop computer on overnight business trips.
    a. State the hypotheses that can be used to test the assumption.
    b. What is the sample proportion from an American Express sponsored survey that found 355 of 546 business travelers use a laptop computer on overnight business trips?
    c. What is the p-value?
    d. Use $\alpha = .05$. What is your conclusion?

70. Virtual call centers are staffed by individuals working out of their homes. Most home agents earn \$10 to \$15 per hour without benefits versus \$7 to \$9 per hour with benefits at a traditional call center (*BusinessWeek,* January 23, 2006). Regional Airways is considering employing home agents, but only if a level of customer satisfaction greater than 80% can be maintained. A test was conducted with home service agents. In a sample of 300 customers, 252 reported that they were satisfied with service.
    a. Develop hypotheses for a test to determine whether the sample data support the conclusion that customer service with home agents meets the Regional Airways criterion.
    b. What is your point estimate of the percentage of satisfied customers?
    c. What is the p-value provided by the sample data?
    d. What is your hypothesis testing conclusion? Use $\alpha = .05$ as the level of significance.

71. During the 2004 election year, new polling results were reported daily. In an IBD/TIPP poll of 910 adults, 503 respondents reported that they were optimistic about the national outlook, and President Bush's leadership index jumped 4.7 points to 55.3 (*Investor's Business Daily,* January 14, 2004).
    a. What is the sample proportion of respondents who are optimistic about the national outlook?
    b. A campaign manager wants to claim that this poll indicates that the majority of adults are optimistic about the national outlook. Construct a hypothesis test so that rejection of the null hypothesis will permit the conclusion that the proportion optimistic is greater than 50%.
    c. Use the polling data to compute the p-value for the hypothesis test in part (b). Explain to the manager what this p-value means about the level of significance of the results.

72. A radio station in Myrtle Beach announced that at least 90% of the hotels and motels would be full for the Memorial Day weekend. The station advised listeners to make reservations in advance if they planned to be in the resort over the weekend. On Saturday night a sample of 58 hotels and motels showed 49 with a no-vacancy sign and 9 with vacancies. What is your reaction to the radio station's claim after seeing the sample evidence? Use $\alpha = .05$ in making the statistical test. What is the p-value?

73. According to the federal government, 24% of workers covered by their company's health care plan were not required to contribute to the premium (*Statistical Abstract of the United*

States: 2006). A recent study found that 81 out of 400 workers sampled were not required to contribute to their company's health care plan.

a. Develop hypotheses that can be used to test whether the percent of workers not required to contribute to their company's health care plan has declined.
b. What is a point estimate of the proportion receiving free company-sponsored health care insurance?
c. Has a statistically significant decline occurred in the proportion of workers receiving free company-sponsored health care insurance? Use $\alpha = .05$.

74. Shorney Construction Company bids on projects assuming that the mean idle time per worker is 72 or fewer minutes per day. A sample of 30 construction workers will be used to test this assumption. Assume that the population standard deviation is 20 minutes.
    a. State the hypotheses to be tested.
    b. What is the probability of making a Type II error when the population mean idle time is 80 minutes?
    c. What is the probability of making a Type II error when the population mean idle time is 75 minutes?
    d. What is the probability of making a Type II error when the population mean idle time is 70 minutes?
    e. Sketch the power curve for this problem.
75. A federal funding program is available to low-income neighborhoods. To qualify for the funding, a neighborhood must have a mean household income of less than $15,000 per year. Neighborhoods with mean annual household income of $15,000 or more do not qualify. Funding decisions are based on a sample of residents in the neighborhood. A hypothesis test with a .02 level of significance is conducted. If the funding guidelines call for a maximum probability of .05 of not funding a neighborhood with a mean annual household income of $14,000, what sample size should be used in the funding decision study? Use $\sigma = \$4000$ as a planning value.
76. H_0: $\mu = 120$ and H_a: $\mu \neq 120$ are used to test whether a bath soap production process is meeting the standard output of 120 bars per batch. Use a .05 level of significance for the test and a planning value of 5 for the standard deviation.
    a. If the mean output drops to 117 bars per batch, the firm wants to have a 98% chance of concluding that the standard production output is not being met. How large a sample should be selected?
    b. With your sample size from part (a), what is the probability of concluding that the process is operating satisfactorily for each of the following actual mean outputs: 117, 118, 119, 121, 122, and 123 bars per batch? That is, what is the probability of a Type II error in each case?

Case Problem 1 Quality Associates, Inc.

Quality Associates, Inc., a consulting firm, advises its clients about sampling and statistical procedures that can be used to control their manufacturing processes. In one particular application, a client gave Quality Associates a sample of 800 observations taken during a time in which that client's process was operating satisfactorily. The sample standard deviation for these data was .21; hence, with so much data, the population standard deviation was assumed to be .21. Quality Associates then suggested that random samples of size 30 be taken periodically to monitor the process on an ongoing basis. By analyzing the new samples, the client could quickly learn whether the process was operating satisfactorily. When the process was not operating satisfactorily, corrective action could be taken to eliminate the problem. The design specification indicated the mean for the process should be 12. The hypothesis test suggested by Quality Associates follows.

$$H_0: \mu = 12$$
$$H_a: \mu \neq 12$$

Corrective action will be taken any time H_0 is rejected.

The following samples were collected at hourly intervals during the first day of operation of the new statistical process control procedure. These data are available in the data set Quality.

WEB file
Quality

Sample 1	Sample 2	Sample 3	Sample 4
11.55	11.62	11.91	12.02
11.62	11.69	11.36	12.02
11.52	11.59	11.75	12.05
11.75	11.82	11.95	12.18
11.90	11.97	12.14	12.11
11.64	11.71	11.72	12.07
11.80	11.87	11.61	12.05
12.03	12.10	11.85	11.64
11.94	12.01	12.16	12.39
11.92	11.99	11.91	11.65
12.13	12.20	12.12	12.11
12.09	12.16	11.61	11.90
11.93	12.00	12.21	12.22
12.21	12.28	11.56	11.88
12.32	12.39	11.95	12.03
11.93	12.00	12.01	12.35
11.85	11.92	12.06	12.09
11.76	11.83	11.76	11.77
12.16	12.23	11.82	12.20
11.77	11.84	12.12	11.79
12.00	12.07	11.60	12.30
12.04	12.11	11.95	12.27
11.98	12.05	11.96	12.29
12.30	12.37	12.22	12.47
12.18	12.25	11.75	12.03
11.97	12.04	11.96	12.17
12.17	12.24	11.95	11.94
11.85	11.92	11.89	11.97
12.30	12.37	11.88	12.23
12.15	12.22	11.93	12.25

Managerial Report

1. Conduct a hypothesis test for each sample at the .01 level of significance and determine what action, if any, should be taken. Provide the test statistic and *p*-value for each test.
2. Compute the standard deviation for each of the four samples. Does the assumption of .21 for the population standard deviation appear reasonable?
3. Compute limits for the sample mean $\bar{x}$ around $\mu = 12$ such that, as long as a new sample mean is within those limits, the process will be considered to be operating satisfactorily. If $\bar{x}$ exceeds the upper limit or if $\bar{x}$ is below the lower limit, corrective action will be taken. These limits are referred to as upper and lower control limits for quality control purposes.
4. Discuss the implications of changing the level of significance to a larger value. What mistake or error could increase if the level of significance is increased?

Case Problem 2 Ethical Behavior of Business Students at Bayview University

During the global recession of 2008 and 2009, there were many accusations of unethical behavior by Wall Street executives, financial managers, and other corporate officers. At that time, an article appeared that suggested that part of the reason for such unethical business behavior may stem from the fact that cheating has become more prevalent among business students (*Chronicle of Higher Education,* February 10, 2009). The article reported that 56 percent of business students admitted to cheating at some time during their academic career as compared to 47 percent of nonbusiness students.

Cheating has been a concern of the dean of the College of Business at Bayview University for several years. Some faculty members in the college believe that cheating is more widespread at Bayview than at other universities, while other faculty members think that cheating is not a major problem in the college. To resolve some of these issues, the dean commissioned a study to assess the current ethical behavior of business students at Bayview. As part of this study, an anonymous exit survey was administered to a sample of 90 business students from this year's graduating class. Responses to the following questions were used to obtain data regarding three types of cheating.

During your time at Bayview, did you ever present work copied off the Internet as your own?

Yes ________ No ________

During your time at Bayview, did you ever copy answers off another student's exam?

Yes ________ No ________

During your time at Bayview, did you ever collaborate with other students on projects that were supposed to be completed individually?

Yes ________ No ________

Any student who answered Yes to one or more of these questions was considered to have been involved in some type of cheating. A portion of the data collected follows. The complete data set is in the file named Bayview.

WEB file
Bayview

Student	Copied from Internet	Copied on Exam	Collaborated on Individual Project	Gender
1	No	No	No	Female
2	No	No	No	Male
3	Yes	No	Yes	Male
4	Yes	Yes	No	Male
5	No	No	Yes	Male
6	Yes	No	No	Female
.	.	.	.	.
.	.	.	.	.
.	.	.	.	.
88	No	No	No	Male
89	No	Yes	Yes	Male
90	No	No	No	Female

Managerial Report

Prepare a report for the dean of the college that summarizes your assessment of the nature of cheating by business students at Bayview University. Be sure to include the following items in your report.

1. Use descriptive statistics to summarize the data and comment on your findings.
2. Develop 95% confidence intervals for the proportion of all students, the proportion of male students, and the proportion of female students who were involved in some type of cheating.
3. Conduct a hypothesis test to determine if the proportion of business students at Bayview University who were involved in some type of cheating is less than that of business students at other institutions as reported by the *Chronicle of Higher Education*.
4. Conduct a hypothesis test to determine if the proportion of business students at Bayview University who were involved in some form of cheating is less than that of nonbusiness students at other institutions as reported by the *Chronicle of Higher Education*.
5. What advice would you give to the dean based upon your analysis of the data?

Appendix 9.1 Hypothesis Testing with Minitab

We describe the use of Minitab to conduct hypothesis tests about a population mean and a population proportion.

Population Mean: σ Known

We illustrate using the MaxFlight golf ball distance example in Section 9.3. The data are in column C1 of a Minitab worksheet. The population standard deviation $\sigma = 12$ is assumed known and the level of significance is $\alpha = .05$. The following steps can be used to test the hypothesis H_0: $\mu = 295$ versus H_a: $\mu \neq 295$.

WEB file

GolfTest

Step 1. Select the **Stat** menu
Step 2. Choose **Basic Statistics**
Step 3. Choose **1-Sample Z**
Step 4. When the 1-Sample Z dialog box appears:
    Enter C1 in the **Samples in columns** box
    Enter 12 in the **Standard deviation** box
    Select **Perform Hypothesis Test**
    Enter 295 in the **Hypothesized mean** box
    Select **Options**
Step 5. When the 1-Sample Z-Options dialog box appears:
    Enter 95 in the **Confidence level** box*
    Select **not equal** in the **Alternative** box
    Click **OK**
Step 6. Click **OK**

In addition to the hypothesis testing results, Minitab provides a 95% confidence interval for the population mean.

*Minitab provides both hypothesis testing and interval estimation results simultaneously. The user may select any confidence level for the interval estimate of the population mean: 95% confidence is suggested here.

The procedure can be easily modified for a one-tailed hypothesis test by selecting the less than or greater than option in the **Alternative** box in step 5.

Population Mean: σ Unknown

WEB file

AirRating

The ratings that 60 business travelers gave for Heathrow Airport are entered in column C1 of a Minitab worksheet. The level of significance for the test is $\alpha = .05$, and the population standard deviation σ will be estimated by the sample standard deviation s. The following steps can be used to test the hypothesis H_0: $\mu \leq 7$ against H_a: $\mu > 7$.

Step 1. Select the **Stat** menu
Step 2. Choose **Basic Statistics**
Step 3. Choose **1-Sample t**
Step 4. When the 1-Sample t dialog box appears:
Enter C1 in the **Samples in columns** box
Select **Perform Hypothesis Test**
Enter 7 in the **Hypothesized mean** box
Select **Options**
Step 5. When the 1-Sample t-Options dialog box appears:
Enter 95 in the **Confidence level** box
Select **greater than** in the **Alternative** box
Click **OK**
Step 6. Click **OK**

The Heathrow Airport rating study involved a greater than alternative hypothesis. The preceding steps can be easily modified for other hypothesis tests by selecting the less than or not equal options in the **Alternative** box in step 5.

Population Proportion

WEB file

WomenGolf

We illustrate using the Pine Creek golf course example in Section 9.5. The data with responses Female and Male are in column C1 of a Minitab worksheet. Minitab uses an alphabetical ordering of the responses and selects the *second response* for the population proportion of interest. In this example, Minitab uses the alphabetical ordering Female-Male to provide results for the population proportion of Male responses. Because Female is the response of interest, we change Minitab's ordering as follows: Select any cell in the column and use the sequence: Editor > Column > Value Order. Then choose the option of entering a user-specified order. Enter Male-Female in the **Define-an-order** box and click OK. Minitab's 1 Proportion routine will then provide the hypothesis test results for the population proportion of female golfers. We proceed as follows:

Step 1. Select the **Stat** menu
Step 2. Choose **Basic Statistics**
Step 3. Choose **1 Proportion**
Step 4. When the 1 Proportion dialog box appears:
Enter C1 in the **Samples in Columns** box
Select **Perform Hypothesis Test**
Enter .20 in the **Hypothesized proportion** box
Select **Options**
Step 5. When the 1 Proportion-Options dialog box appears:
Enter 95 in the **Confidence level** box
Select greater than in the **Alternative** box
Select **Use test and interval based on normal distribution**
Click **OK**
Step 6. Click **OK**

Appendix 9.2 Hypothesis Testing with Excel

Excel does not provide built-in routines for the hypothesis tests presented in this chapter. To handle these situations, we present Excel worksheets that we designed to use as templates for testing hypotheses about a population mean and a population proportion. The worksheets are easy to use and can be modified to handle any sample data. The worksheets are available on the website that accompanies this book.

Population Mean: σ Known

We illustrate using the MaxFlight golf ball distance example in Section 9.3. The data are in column A of an Excel worksheet. The population standard deviation $\sigma = 12$ is assumed known and the level of significance is $\alpha = .05$. The following steps can be used to test the hypothesis H_0: $\mu = 295$ versus H_a: $\mu \neq 295$.

WEB file

Hyp Sigma Known

Refer to Figure 9.11 as we describe the procedure. The worksheet in the background shows the cell formulas used to compute the results shown in the foreground worksheet. The data are entered into cells A2:A51. The following steps are necessary to use the template for this data set.

Step 1. Enter the data range A2:A51 into the =COUNT cell formula in cell D4
Step 2. Enter the data range A2:A51 into the =AVERAGE cell formula in cell D5
Step 3. Enter the population standard deviation $\sigma = 12$ into cell D7
Step 4. Enter the hypothesized value for the population mean 295 into cell D8

The remaining cell formulas automatically provide the standard error, the value of the test statistic z, and three p-values. Because the alternative hypothesis ($\mu_0 \neq 295$) indicates a two-tailed test, the p-value (Two Tail) in cell D15 is used to make the rejection decision. With p-value $= .1255 > \alpha = .05$, the null hypothesis cannot be rejected. The p-values in cells D13 or D14 would be used if the hypotheses involved a one-tailed test.

This template can be used to make hypothesis testing computations for other applications. For instance, to conduct a hypothesis test for a new data set, enter the new sample data into column A of the worksheet. Modify the formulas in cells D4 and D5 to correspond to the new data range. Enter the population standard deviation into cell D7 and the hypothesized value for the population mean into cell D8 to obtain the results. If the new sample data have already been summarized, the new sample data do not have to be entered into the worksheet. In this case, enter the sample size into cell D4, the sample mean into cell D5, the population standard deviation into cell D7, and the hypothesized value for the population mean into cell D8 to obtain the results. The worksheet in Figure 9.11 is available in the file Hyp Sigma Known on the website that accompanies this book.

Population Mean: σ Unknown

We illustrate using the Heathrow Airport rating example in Section 9.4. The data are in column A of an Excel worksheet. The population standard deviation σ is unknown and will be estimated by the sample standard deviation s. The level of significance is $\alpha = .05$. The following steps can be used to test the hypothesis H_0: $\mu \leq 7$ versus H_a: $\mu > 7$.

WEB file

Hyp Sigma Unknown

Refer to Figure 9.12 as we describe the procedure. The background worksheet shows the cell formulas used to compute the results shown in the foreground version of the worksheet. The data are entered into cells A2:A61. The following steps are necessary to use the template for this data set.

Step 1. Enter the data range A2:A61 into the =COUNT cell formula in cell D4
Step 2. Enter the data range A2:A61 into the =AVERAGE cell formula in cell D5
Step 3. Enter the data range A2:A61 into the =STDEV cell formula in cell D6
Step 4. Enter the hypothesized value for the population mean 7 into cell D8

FIGURE 9.11 EXCEL WORKSHEET FOR HYPOTHESIS TESTS ABOUT A POPULATION MEAN WITH σ KNOWN

	A	B	C	D	E
1	**Yards**		**Hypothesis Test about a Population Mean:**		
2	303		**σ Known Case**		
3	282				
4	289		**Sample Size**	=COUNT(A2:A51)	
5	298		**Sample Mean**	=AVERAGE(A2:A51)	
6	283				
7	317		**Population Standard Deviation**	12	
8	297		**Hypothesized Value**	295	
9	308				
10	317		**Standard Error**	=D7/SQRT(D4)	
11	293		**Test Statistic z**	=(D5-D8)/D10	
12	284				
13	290		***p*-value (Lower Tail)**	=NORM.S.DIST(D11,TRUE)	
14	304		***p*-value (Upper Tail)**	=1-D13	
15	290		***p*-value (Two Tail)**	=2*(MIN(D13,D14))	
16	311				
50	301				
51	292				
52					

	A	B	C	D	E
1	**Yards**		**Hypothesis Test about a Population Mean:**		
2	303		**σ Known Case**		
3	282				
4	289		**Sample Size**	50	
5	298		**Sample Mean**	297.6	
6	283				
7	317		**Population Standard Deviation**	12	
8	297		**Hypothesized Value**	295	
9	308				
10	317		**Standard Error**	1.70	
11	293		**Test Statistic z**	1.53	
12	284				
13	290		***p*-value (Lower Tail)**	0.9372	
14	304		***p*-value (Upper Tail)**	0.0628	
15	290		***p*-value (Two Tail)**	0.1255	
16	311				
50	301				
51	292				
52					

Note: Rows 17 to 49 are hidden.

The remaining cell formulas automatically provide the standard error, the value of the test statistic t, the number of degrees of freedom, and three p-values. Because the alternative hypothesis ($\mu > 7$) indicates an upper tail test, the p-value (Upper Tail) in cell D15 is used to make the decision. With p-value $= .0353 < \alpha = .05$, the null hypothesis is rejected. The p-values in cells D14 or D16 would be used if the hypotheses involved a lower tail test or a two-tailed test.

This template can be used to make hypothesis testing computations for other applications. For instance, to conduct a hypothesis test for a new data set, enter the new sample data into column A of the worksheet and modify the formulas in cells D4, D5, and D6 to

FIGURE 9.12 EXCEL WORKSHEET FOR HYPOTHESIS TESTS ABOUT A POPULATION MEAN WITH σ UNKNOWN

	A	B	C	D	E
1	**Rating**		**Hypothesis Test About a Population Mean**		
2	5		**With σ Unknown**		
3	7				
4	8		**Sample Size**	=COUNT(A2:A61)	
5	7		**Sample Mean**	=AVERAGE(A2:A61)	
6	8		**Sample Std. Deviation**	=STDEV(A2:A61)	
7	8				
8	8		**Hypothesized Value**	7	
9	7				
10	8		**Standard Error**	=D6/SQRT(D4)	
11	10		**Test Statistic *t***	=(D5-D8)/D10	
12	6		**Degrees of Freedom**	=D4-1	
13	7				
14	8		***p*-value (Lower Tail)**	=T.DIST(D11,D12,TRUE)	
15	8		***p*-value (Upper Tail)**	=1-D14	
16	9		***p*-value (Two Tail)**	=2*(MIN(D14,D15))	
17	7				
59	7				
60	7				
61	8				
62					

	A	B	C	D	E	F
1	**Rating**		**Hypothesis Test About a Population Mean**			
2	5		**With σ Unknown**			
3	7					
4	8		**Sample Size**	60		
5	7		**Sample Mean**	7.25		
6	8		**Sample Std. Deviation**	1.05		
7	8					
8	8		**Hypothesized Value**	7		
9	7					
10	8		**Standard Error**	0.136		
11	10		**Test Statistic *t***	1.841		
12	6		**Degrees of Freedom**	59		
13	7					
14	8		***p*-value (Lower Tail)**	0.9647		
15	8		***p*-value (Upper Tail)**	0.0353		
16	9		***p*-value (Two Tail)**	0.0706		
17	7					
59	7					
60	7					
61	8					
62						

Note: Rows 18 to 58 are hidden.

correspond to the new data range. Enter the hypothesized value for the population mean into cell D8 to obtain the results. If the new sample data have already been summarized, the new sample data do not have to be entered into the worksheet. In this case, enter the sample size into cell D4, the sample mean into cell D5, the sample standard deviation into cell D6, and the hypothesized value for the population mean into cell D8 to obtain the results. The worksheet in Figure 9.12 is available in the file Hyp Sigma Unknown on the website that accompanies this book.

FIGURE 9.13 EXCEL WORKSHEET FOR HYPOTHESIS TESTS ABOUT A POPULATION PROPORTION

	A	B	C	D	E
1	Golfer		Hypothesis Test about a Population Proportion		
2	Female				
3	Male		Sample Size	=COUNTA(A2:A401)	
4	Female		Response of Interest	Female	
5	Male		Count for Response	=COUNTIF(A2:A903,D4)	
6	Male		Sample Proportion	=D5/D3	
7	Female				
8	Male		Hypothesized Value	0.2	
9	Male				
10	Female		Standard Error	=SQRT(D8*(1-D8)/D3)	
11	Male		Test Statistic *z*	=(D6-D8)/D10	
12	Male				
13	Male		*p*-value (Lower Tail)	=NORM.S.DIST(D11,TRUE)	
14	Male		*p*-value (Upper Tail)	=1-D13	
15	Male		*p*-value (TwoTail)	=2*MIN(D13,D14)	
16	Female				
400	Male				
401	Male				
402					

	A	B	C	D	E	F
1	Golfer		Hypothesis Test about a Population Proportion			
2	Female					
3	Male		Sample Size	400		
4	Female		Response of Interest	Female		
5	Male		Count for Response	100		
6	Male		Sample Proportion	0.25		
7	Female					
8	Male		Hypothesized Value	0.20		
9	Male					
10	Female		Standard Error	0.02		
11	Male		Test Statistic *z*	2.5000		
12	Male					
13	Male		*p*-value (Lower Tail)	0.9938		
14	Male		*p*-value (Upper Tail)	0.0062		
15	Male		*p*-value (TwoTail)	0.0124		
16	Female					
400	Male					
401	Male					
402						

Note: Rows 17 to 399 are hidden.

Population Proportion

WEB file

Hypothesis p

We illustrate using the Pine Creek golf course survey data presented in Section 9.5. The data of Male or Female golfer are in column A of an Excel worksheet. Refer to Figure 9.13 as we describe the procedure. The background worksheet shows the cell formulas used to compute the results shown in the foreground worksheet. The data are entered into cells A2:A401. The following steps can be used to test the hypothesis H_0: $p \leq .20$ versus H_a: $p > .20$.

Step 1. Enter the data range A2:A401 into the =COUNTA cell formula in cell D3
Step 2. Enter Female as the response of interest in cell D4
Step 3. Enter the data range A2:A401 into the =COUNTIF cell formula in cell D5
Step 4. Enter the hypothesized value for the population proportion .20 into cell D8

The remaining cell formulas automatically provide the standard error, the value of the test statistic z, and three p-values. Because the alternative hypothesis ($p > .20$) indicates an upper tail test, the p-value (Upper Tail) in cell D14 is used to make the decision. With p-value $= .0062 < \alpha = .05$, the null hypothesis is rejected. The p-values in cells D13 or D15 would be used if the hypothesis involved a lower tail test or a two-tailed test.

This template can be used to make hypothesis testing computations for other applications. For instance, to conduct a hypothesis test for a new data set, enter the new sample data into column A of the worksheet. Modify the formulas in cells D3 and D5 to correspond to the new data range. Enter the response of interest into cell D4 and the hypothesized value for the population proportion into cell D8 to obtain the results. If the new sample data have already been summarized, the new sample data do not have to be entered into the worksheet. In this case, enter the sample size into cell D3, the sample proportion into cell D6, and the hypothesized value for the population proportion into cell D8 to obtain the results. The worksheet in Figure 9.13 is available in the file Hypothesis p on the website that accompanies this book.

Appendix 9.3 Hypothesis Testing with StatTools

In this appendix we show how StatTools can be used to conduct hypothesis tests about a population mean for the σ unknown case

Population Mean: σ Unknown Case

WEB file
AirRating

In this case the population standard deviation σ will be estimated by the sample standard deviation s. We use the example discussed in Section 9.4 involving ratings that 60 business travelers gave for Heathrow Airport.

Begin by using the Data Set Manager to create a StatTools data set for these data using the procedure described in the appendix in Chapter 1. The following steps can be used to test the hypothesis H_0: $\mu \leq 7$ against H_a: $\mu > 7$.

Step 1. Click the **StatTools** tab on the Ribbon
Step 2. In the **Analyses** group, click **Statistical Inference**
Step 3. Choose the **Hypothesis Test** option
Step 4. Choose Mean/Std. Deviation
Step 5. When the StatTools-Hypothesis Test for Mean/Std. Deviation dialog box appears:
For **Analysis Type,** choose **One-Sample Analysis**
In the **Variables** section, select **Rating**
In the **Hypothesis Tests to Perform** section:
Select the **Mean** option
Enter 7 in the **Null Hypothesis Value** box
Select **Greater Than Null Value (One-Tailed Test)** in the **Alternative Hypothesis Type** box
If selected, remove the check in the **Standard Deviation** box
Click **OK**

The results from the hypothesis test will appear. They include the p-value and the value of the test statistic.

Population Proportion

We illustrate using the Pine Creek golf course example in Section 9.5. Begin by using the Data Set Manager to create a StatTools data set for these data using the procedure described in the appendix to Chapter 1. The following steps can be used to conduct a hypothesis test of the population proportion.

WEB file

WomenGolf

Step 1. Click the **StatTools** tab on the Ribbon
Step 2. In the **Analyses** group, click **Statistical Inference**
Step 3. Choose **Hypothesis Test**
Step 4. Choose **Proportion**
Step 5. When the StatTools-Hypothesis Test for Proportion dialog box appears:
    For **Analysis Type** choose **One-Sample Analysis**
    In the **Variables** section, select **Golfer**
    In the **Categories to Analyze** section, select **Female**
    In the **Hypotheses About Proportion** section:
        Enter .20 in the **Null Hypothesis Value** box
        In the **Alternaive Hypothesis Type** box, choose **Greater Than Null Value (One-Tailed Test)**
    Click **OK**

The results from the hypothesis test will appear. They include the *p*-value and the value of the test statistic.

CHAPTER 10

Inference About Means and Proportions with Two Populations

CONTENTS

STATISTICS *in* PRACTICE

U.S. FOOD AND DRUG ADMINISTRATION
WASHINGTON, D.C.

It is the responsibility of the U.S. Food and Drug Administration (FDA), through its Center for Drug Evaluation and Research (CDER), to ensure that drugs are safe and effective. But CDER does not do the actual testing of new drugs itself. It is the responsibility of the company seeking to market a new drug to test it and submit evidence that it is safe and effective. CDER statisticians and scientists then review the evidence submitted.

Companies seeking approval of a new drug conduct extensive statistical studies to support their application. The testing process in the pharmaceutical industry usually consists of three stages: (1) preclinical testing, (2) testing for long-term usage and safety, and (3) clinical efficacy testing. At each successive stage, the chance that a drug will pass the rigorous tests decreases; however, the cost of further testing increases dramatically. Industry surveys indicate that on average the research and development for one new drug costs $250 million and takes 12 years. Hence, it is important to eliminate unsuccessful new drugs in the early stages of the testing process, as well as to identify promising ones for further testing.

Statistics plays a major role in pharmaceutical research, where government regulations are stringent and rigorously enforced. In preclinical testing, a two- or three-population statistical study typically is used to determine whether a new drug should continue to be studied in the long-term usage and safety program. The populations may consist of the new drug, a control, and a standard drug. The preclinical testing process begins when a new drug is sent to the pharmacology group for evaluation of efficacy—the capacity of the drug to produce the desired effects. As part of the process, a statistician is asked to design an experiment that can be used to test the new drug. The design must specify the sample size and the statistical methods of analysis. In a two-population study, one sample is used to obtain data on the efficacy of the new drug (population 1) and a second sample is used to obtain data on the efficacy of a standard drug (population 2). Depending on the intended use, the new and standard drugs are tested in such disciplines as neurology, cardiology, and immunology. In most studies, the statistical method involves hypothesis testing for the difference between the means of the new drug population and the standard drug population. If a new drug lacks efficacy or produces undesirable effects in comparison with the standard drug, the new drug is rejected and withdrawn from further testing. Only new drugs that show promising comparisons with the standard drugs are forwarded to the long-term usage and safety testing program.

Statistical methods are used to test and develop new drugs. © Lisa S./Shutterstock.com.

Further data collection and multipopulation studies are conducted in the long-term usage and safety testing program and in the clinical testing programs. The FDA requires that statistical methods be defined prior to such testing to avoid data-related biases. In addition, to avoid human biases, some of the clinical trials are double or triple blind. That is, neither the subject nor the investigator knows what drug is administered to whom. If the new drug meets all requirements in relation to the standard drug, a new drug application (NDA) is filed with the FDA. The application is rigorously scrutinized by statisticians and scientists at the agency.

In this chapter you will learn how to construct interval estimates and make hypothesis tests about means and proportions with two populations. Techniques will be presented for analyzing independent random samples as well as matched samples.

In Chapters 8 and 9 we showed how to develop interval estimates and conduct hypothesis tests for situations involving a single population mean and a single population proportion. In this chapter we continue our discussion of statistical inference by showing how interval estimates and hypothesis tests can be developed for situations involving two populations when the difference between the two population means or the two population proportions is of prime importance. For example, we may want to develop an interval estimate of the difference between the mean starting salary for a population of men and the mean starting salary for a population of women or conduct a hypothesis test to determine whether any difference is present between the proportion of defective parts in a population of parts produced by supplier A and the proportion of defective parts in a population of parts produced by supplier B. We begin our discussion of statistical inference about two populations by showing how to develop interval estimates and conduct hypothesis tests about the difference between the means of two populations when the standard deviations of the two populations are assumed known.

10.1 Inferences About the Difference Between Two Population Means: σ_1 and σ_2 Known

Letting μ_1 denote the mean of population 1 and μ_2 denote the mean of population 2, we will focus on inferences about the difference between the means: $\mu_1 - \mu_2$. To make an inference about this difference, we select a simple random sample of n_1 units from population 1 and a second simple random sample of n_2 units from population 2. The two samples, taken separately and independently, are referred to as **independent simple random samples**. In this section, we assume that information is available such that the two population standard deviations, σ_1 and σ_2, can be assumed known prior to collecting the samples. We refer to this situation as the σ_1 and σ_2 known case. In the following example we show how to compute a margin of error and develop an interval estimate of the difference between the two population means when σ_1 and σ_2 are known.

Interval Estimation of $\mu_1 - \mu_2$

Greystone Department Stores, Inc., operates two stores in Buffalo, New York: One is in the inner city and the other is in a suburban shopping center. The regional manager noticed that products that sell well in one store do not always sell well in the other. The manager believes this situation may be attributable to differences in customer demographics at the two locations. Customers may differ in age, education, income, and so on. Suppose the manager asks us to investigate the difference between the mean ages of the customers who shop at the two stores.

Let us define population 1 as all customers who shop at the inner-city store and population 2 as all customers who shop at the suburban store.

μ_1 = mean of population 1 (i.e., the mean age of all customers who shop at the inner-city store)

μ_2 = mean of population 2 (i.e., the mean age of all customers who shop at the suburban store)

The difference between the two population means is $\mu_1 - \mu_2$.

To estimate $\mu_1 - \mu_2$, we will select a simple random sample of n_1 customers from population 1 and a simple random sample of n_2 customers from population 2. We then compute the two sample means.

$\bar{x}_1$ = sample mean age for the simple random sample of n_1 inner-city customers

$\bar{x}_2$ = sample mean age for the simple random sample of n_2 suburban customers

The point estimator of the difference between the two population means is the difference between the two sample means.

POINT ESTIMATOR OF THE DIFFERENCE BETWEEN TWO POPULATION MEANS

$$\bar{x}_1 - \bar{x}_2 \tag{10.1}$$

Figure 10.1 provides an overview of the process used to estimate the difference between two population means based on two independent simple random samples.

The standard error of $\bar{x}_1 - \bar{x}_2$ is the standard deviation of the sampling distribution of $\bar{x}_1 - \bar{x}_2$.

As with other point estimators, the point estimator $\bar{x}_1 - \bar{x}_2$ has a standard error that describes the variation in the sampling distribution of the estimator. With two independent simple random samples, the standard error of $\bar{x}_1 - \bar{x}_2$ is as follows:

STANDARD ERROR OF $\bar{x}_1 - \bar{x}_2$

$$\sigma_{\bar{x}_1 - \bar{x}_2} = \sqrt{\frac{\sigma_1^2}{n_1} + \frac{\sigma_2^2}{n_2}} \tag{10.2}$$

If both populations have a normal distribution, or if the sample sizes are large enough that the central limit theorem enables us to conclude that the sampling distributions of $\bar{x}_1$ and $\bar{x}_2$ can be approximated by a normal distribution, the sampling distribution of $\bar{x}_1 - \bar{x}_2$ will have a normal distribution with mean given by $\mu_1 - \mu_2$.

As we showed in Chapter 8, an interval estimate is given by a point estimate $\pm$ a margin of error. In the case of estimation of the difference between two population means, an interval estimate will take the following form:

$$\bar{x}_1 - \bar{x}_2 \pm \text{Margin of error}$$

FIGURE 10.1 ESTIMATING THE DIFFERENCE BETWEEN TWO POPULATION MEANS

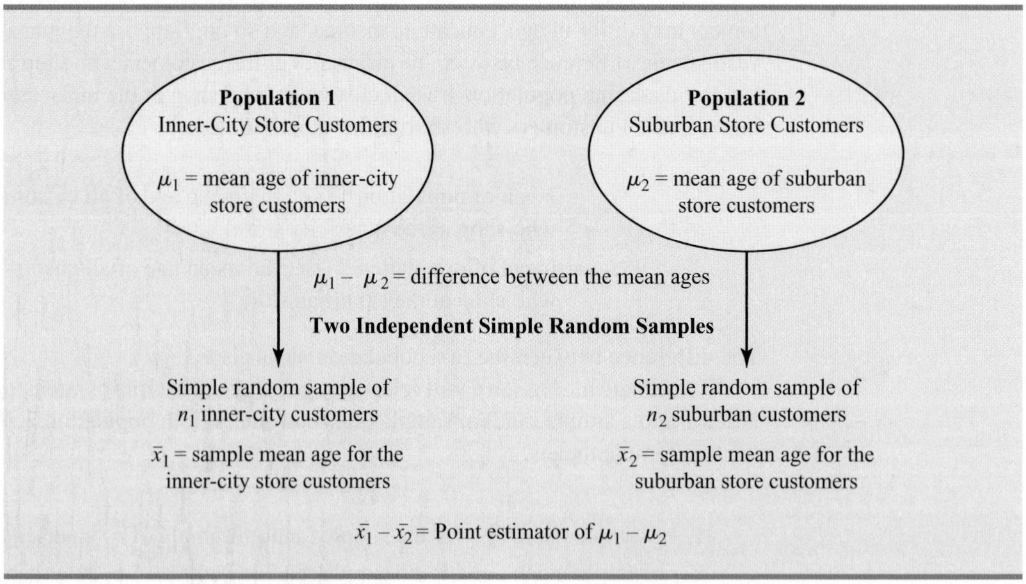

With the sampling distribution of $\bar{x}_1 - \bar{x}_2$ having a normal distribution, we can write the margin of error as follows:

The margin of error is given by multiplying the standard error by $z_{\alpha/2}$.

$$\text{Margin of error} = z_{\alpha/2}\sigma_{\bar{x}_1-\bar{x}_2} = z_{\alpha/2}\sqrt{\frac{\sigma_1^2}{n_1} + \frac{\sigma_2^2}{n_2}} \qquad \textbf{(10.3)}$$

Thus the interval estimate of the difference between two population means is as follows:

INTERVAL ESTIMATE OF THE DIFFERENCE BETWEEN TWO POPULATION MEANS: σ_1 AND σ_2 KNOWN

$$\bar{x}_1 - \bar{x}_2 \pm z_{\alpha/2}\sqrt{\frac{\sigma_1^2}{n_1} + \frac{\sigma_2^2}{n_2}} \qquad \textbf{(10.4)}$$

where $1 - \alpha$ is the confidence coefficient.

Let us return to the Greystone example. Based on data from previous customer demographic studies, the two population standard deviations are known with $\sigma_1 = 9$ years and $\sigma_2 = 10$ years. The data collected from the two independent simple random samples of Greystone customers provided the following results.

	Inner City Store	**Suburban Store**
Sample Size	$n_1 = 36$	$n_2 = 49$
Sample Mean	$\bar{x}_1 = 40$ years	$\bar{x}_2 = 35$ years

Using expression (10.1), we find that the point estimate of the difference between the mean ages of the two populations is $\bar{x}_1 - \bar{x}_2 = 40 - 35 = 5$ years. Thus, we estimate that the customers at the inner-city store have a mean age five years greater than the mean age of the suburban store customers. We can now use expression (10.4) to compute the margin of error and provide the interval estimate of $\mu_1 - \mu_2$. Using 95% confidence and $z_{\alpha/2} = z_{.025} = 1.96$, we have

$$\bar{x}_1 - \bar{x}_2 \pm z_{\alpha/2}\sqrt{\frac{\sigma_1^2}{n_1} + \frac{\sigma_2^2}{n_2}}$$

$$40 - 35 \pm 1.96\sqrt{\frac{9^2}{36} + \frac{10^2}{49}}$$

$$5 \pm 4.06$$

Thus, the margin of error is 4.06 years and the 95% confidence interval estimate of the difference between the two population means is $5 - 4.06 = .94$ years to $5 + 4.06 = 9.06$ years.

Hypothesis Tests About $\mu_1 - \mu_2$

Let us consider hypothesis tests about the difference between two population means. Using D_0 to denote the hypothesized difference between μ_1 and μ_2, the three forms for a hypothesis test are as follows:

$$H_0\colon \mu_1 - \mu_2 \ge D_0 \qquad H_0\colon \mu_1 - \mu_2 \le D_0 \qquad H_0\colon \mu_1 - \mu_2 = D_0$$
$$H_a\colon \mu_1 - \mu_2 < D_0 \qquad H_a\colon \mu_1 - \mu_2 > D_0 \qquad H_a\colon \mu_1 - \mu_2 \ne D_0$$

In many applications, $D_0 = 0$. Using the two-tailed test as an example, when $D_0 = 0$ the null hypothesis is H_0: $\mu_1 - \mu_2 = 0$. In this case, the null hypothesis is that μ_1 and μ_2 are equal. Rejection of H_0 leads to the conclusion that H_a: $\mu_1 - \mu_2 \neq 0$ is true; that is, μ_1 and μ_2 are not equal.

The steps for conducting hypothesis tests presented in Chapter 9 are applicable here. We must choose a level of significance, compute the value of the test statistic and find the p-value to determine whether the null hypothesis should be rejected. With two independent simple random samples, we showed that the point estimator $\bar{x}_1 - \bar{x}_2$ has a standard error $\sigma_{\bar{x}_1 - \bar{x}_2}$ given by expression (10.2) and, when the sample sizes are large enough, the distribution of $\bar{x}_1 - \bar{x}_2$ can be described by a normal distribution. In this case, the test statistic for the difference between two population means when σ_1 and σ_2 are known is as follows.

TEST STATISTIC FOR HYPOTHESIS TESTS ABOUT $\mu_1 - \mu_2$: σ_1 AND σ_2 KNOWN

$$z = \frac{(\bar{x}_1 - \bar{x}_2) - D_0}{\sqrt{\dfrac{\sigma_1^2}{n_1} + \dfrac{\sigma_2^2}{n_2}}} \tag{10.5}$$

Let us demonstrate the use of this test statistic in the following hypothesis testing example.

As part of a study to evaluate differences in education quality between two training centers, a standardized examination is given to individuals who are trained at the centers. The difference between the mean examination scores is used to assess quality differences between the centers. The population means for the two centers are as follows.

μ_1 = the mean examination score for the population of individuals trained at center A

μ_2 = the mean examination score for the population of individuals trained at center B

We begin with the tentative assumption that no difference exists between the training quality provided at the two centers. Hence, in terms of the mean examination scores, the null hypothesis is that $\mu_1 - \mu_2 = 0$. If sample evidence leads to the rejection of this hypothesis, we will conclude that the mean examination scores differ for the two populations. This conclusion indicates a quality differential between the two centers and suggests that a follow-up study investigating the reason for the differential may be warranted. The null and alternative hypotheses for this two-tailed test are written as follows.

$$H_0: \mu_1 - \mu_2 = 0$$
$$H_a: \mu_1 - \mu_2 \neq 0$$

The standardized examination given previously in a variety of settings always resulted in an examination score standard deviation near 10 points. Thus, we will use this information to assume that the population standard deviations are known with $\sigma_1 = 10$ and $\sigma_2 = 10$. An $\alpha = .05$ level of significance is specified for the study.

WEB file
ExamScores

Independent simple random samples of $n_1 = 30$ individuals from training center A and $n_2 = 40$ individuals from training center B are taken. The respective sample means are $\bar{x}_1 = 82$ and $\bar{x}_2 = 78$. Do these data suggest a significant difference between the population means at the two training centers? To help answer this question, we compute the test statistic using equation (10.5).

$$z = \frac{(\bar{x}_1 - \bar{x}_2) - D_0}{\sqrt{\dfrac{\sigma_1^2}{n_1} + \dfrac{\sigma_2^2}{n_2}}} = \frac{(82 - 78) - 0}{\sqrt{\dfrac{10^2}{30} + \dfrac{10^2}{40}}} = 1.66$$

Next let us compute the p-value for this two-tailed test. Because the test statistic z is in the upper tail, we first compute the area under the curve to the right of $z = 1.66$. Using the standard normal distribution table, the area to the left of $z = 1.66$ is .9515. Thus, the area in the upper tail of the distribution is $1.0000 - .9515 = .0485$. Because this test is a two-tailed test, we must double the tail area: p-value $= 2(.0485) = .0970$. Following the usual rule to reject H_0 if p-value $\leq \alpha$, we see that the p-value of .0970 does not allow us to reject H_0 at the .05 level of significance. The sample results do not provide sufficient evidence to conclude the training centers differ in quality.

In this chapter we will use the p-value approach to hypothesis testing as described in Chapter 9. However, if you prefer, the test statistic and the critical value rejection rule may be used. With $\alpha = .05$ and $z_{\alpha/2} = z_{.025} = 1.96$, the rejection rule employing the critical value approach would be reject H_0 if $z \leq -1.96$ or if $z \geq 1.96$. With $z = 1.66$, we reach the same do not reject H_0 conclusion.

In the preceding example, we demonstrated a two-tailed hypothesis test about the difference between two population means. Lower tail and upper tail tests can also be considered. These tests use the same test statistic as given in equation (10.5). The procedure for computing the p-value and the rejection rules for these one-tailed tests are the same as those presented in Chapter 9.

Practical Advice

In most applications of the interval estimation and hypothesis testing procedures presented in this section, random samples with $n_1 \geq 30$ and $n_2 \geq 30$ are adequate. In cases where either or both sample sizes are less than 30, the distributions of the populations become important considerations. In general, with smaller sample sizes, it is more important for the analyst to be satisfied that it is reasonable to assume that the distributions of the two populations are at least approximately normal.

Exercises

Methods

SELF test

1. The following results come from two independent random samples taken of two populations.

Sample 1	Sample 2
$n_1 = 50$	$n_2 = 35$
$\bar{x}_1 = 13.6$	$\bar{x}_2 = 11.6$
$\sigma_1 = 2.2$	$\sigma_2 = 3.0$

a. What is the point estimate of the difference between the two population means?
b. Provide a 90% confidence interval for the difference between the two population means.
c. Provide a 95% confidence interval for the difference between the two population means.

SELF test

2. Consider the following hypothesis test.

$$H_0: \mu_1 - \mu_2 \leq 0$$
$$H_a: \mu_1 - \mu_2 > 0$$

The following results are for two independent samples taken from the two populations.

Sample 1	Sample 2
$n_1 = 40$	$n_2 = 50$
$\bar{x}_1 = 25.2$	$\bar{x}_2 = 22.8$
$\sigma_1 = 5.2$	$\sigma_2 = 6.0$

a. What is the value of the test statistic?
b. What is the p-value?
c. With $\alpha = .05$, what is your hypothesis testing conclusion?

3. Consider the following hypothesis test.

$$H_0: \mu_1 - \mu_2 = 0$$
$$H_a: \mu_1 - \mu_2 \neq 0$$

The following results are for two independent samples taken from the two populations.

Sample 1	Sample 2
$n_1 = 80$	$n_2 = 70$
$\bar{x}_1 = 104$	$\bar{x}_2 = 106$
$\sigma_1 = 8.4$	$\sigma_2 = 7.6$

a. What is the value of the test statistic?
b. What is the p-value?
c. With $\alpha = .05$, what is your hypothesis testing conclusion?

Applications

SELF test

4. *Condé Nast Traveler* conducts an annual survey in which readers rate their favorite cruise ship. All ships are rated on a 100-point scale, with higher values indicating better service. A sample of 37 ships that carry fewer than 500 passengers resulted in an average rating of 85.36, and a sample of 44 ships that carry 500 or more passengers provided an average rating of 81.40 (*Condé Nast Traveler,* February 2008). Assume that the population standard deviation is 4.55 for ships that carry fewer than 500 passengers and 3.97 for ships that carry 500 or more passengers.
   a. What is the point estimate of the difference between the population mean rating for ships that carry fewer than 500 passengers and the population mean rating for ships that carry 500 or more passengers?
   b. At 95% confidence, what is the margin of error?
   c. What is a 95% confidence interval estimate of the difference between the population mean ratings for the two sizes of ships?
5. The average expenditure on Valentine's Day was expected to be \$100.89 (*USA Today,* February 13, 2006). Do male and female consumers differ in the amounts they spend? The average expenditure in a sample survey of 40 male consumers was \$135.67, and the average expenditure in a sample survey of 30 female consumers was \$68.64. Based on past surveys, the standard deviation for male consumers is assumed to be \$35, and the standard deviation for female consumers is assumed to be \$20.

a. What is the point estimate of the difference between the population mean expenditure for males and the population mean expenditure for females?
b. At 99% confidence, what is the margin of error?
c. Develop a 99% confidence interval for the difference between the two population means.

6. Suppose that you are responsible for making arrangements for a business convention. Because of budget cuts due to the recent recession, you have been charged with choosing a city for the convention that has the least expensive hotel rooms. You have narrowed your choices to Atlanta and Houston. The file named Hotel contains samples of prices for rooms in Atlanta and Houston that are consistent with the results reported by Smith Travel Research (*SmartMoney,* March 2009). Because considerable historical data on the prices of rooms in both cities are available, the population standard deviations for the prices can be assumed to be \$20 in Atlanta and \$25 in Houston. Based on the sample data, can you conclude that the mean price of a hotel room in Atlanta is lower than one in Houston?

WEB file
Hotel

7. *Consumer Reports* uses a survey of readers to obtain customer satisfaction ratings for the nation's largest retailers (*Consumer Reports,* March 2012). Each survey respondent is asked to rate a specified retailer in terms of six factors: quality of products, selection, value, checkout efficiency, service, and store layout. An overall satisfaction score summarizes the rating for each respondent with 100 meaning the respondent is completely satisfied in terms of all six factors. Sample data representative of independent samples of Target and Walmart customers are shown below.

Target	Walmart
$n_1 = 25$	$n_2 = 30$
$\bar{x}_1 = 79$	$\bar{x}_2 = 71$

a. Formulate the null and alternative hypotheses to test whether there is a difference between the population mean customer satisfaction scores for the two retailers.
b. Assume that experience with the *Consumer Reports* satisfaction rating scale indicates that a population standard deviation of 12 is a reasonable assumption for both retailers. Conduct the hypothesis test and report the p-value. At a .05 level of significance what is your conclusion?
c. Which retailer, if either, appears to have the greater customer satisfaction? Provide a 95% confidence interval for the difference between the population mean customer satisfaction scores for the two retailers.

8. Will improving customer service result in higher stock prices for the companies providing the better service? "When a company's satisfaction score has improved over the prior year's results and is above the national average (currently 75.7), studies show its shares have a good chance of outperforming the broad stock market in the long run" (*Businessweek,* March 2, 2009). The following satisfaction scores of three companies for the 4th quarters of 2007 and 2008 were obtained from the American Customer Satisfaction Index. Assume that the scores are based on a poll of 60 customers from each company. Because the polling has been done for several years, the standard deviation can be assumed to equal 6 points in each case.

Company	2007 Score	2008 Score
Rite Aid	73	76
Expedia	75	77
J.C. Penney	77	78

a. For Rite Aid, is the increase in the satisfaction score from 2007 to 2008 statistically significant? Use $\alpha = .05$. What can you conclude?
b. Can you conclude that the 2008 score for Rite Aid is above the national average of 75.7? Use $\alpha = .05$.
c. For Expedia, is the increase from 2007 to 2008 statistically significant? Use $\alpha = .05$.
d. When conducting a hypothesis test with the values given for the standard deviation, sample size, and α, how large must the increase from 2007 to 2008 be for it to be statistically significant?
e. Use the result of part (d) to state whether the increase for J.C. Penney from 2007 to 2008 is statistically significant.

10.2 Inferences About the Difference Between Two Population Means: σ_1 and σ_2 Unknown

In this section we extend the discussion of inferences about the difference between two population means to the case when the two population standard deviations, σ_1 and σ_2, are unknown. In this case, we will use the sample standard deviations, s_1 and s_2, to estimate the unknown population standard deviations. When we use the sample standard deviations, the interval estimation and hypothesis testing procedures will be based on the t distribution rather than the standard normal distribution.

Interval Estimation of $\mu_1 - \mu_2$

In the following example we show how to compute a margin of error and develop an interval estimate of the difference between two population means when σ_1 and σ_2 are unknown. Clearwater National Bank is conducting a study designed to identify differences between checking account practices by customers at two of its branch banks. A simple random sample of 28 checking accounts is selected from the Cherry Grove Branch and an independent simple random sample of 22 checking accounts is selected from the Beechmont Branch. The current checking account balance is recorded for each of the checking accounts. A summary of the account balances follows:

WEB file
CheckAcct

	Cherry Grove	Beechmont
Sample Size	$n_1 = 28$	$n_2 = 22$
Sample Mean	$\bar{x}_1 = \$1025$	$\bar{x}_2 = \$910$
Sample Standard Deviation	$s_1 = \$150$	$s_2 = \$125$

Clearwater National Bank would like to estimate the difference between the mean checking account balance maintained by the population of Cherry Grove customers and the population of Beechmont customers. Let us develop the margin of error and an interval estimate of the difference between these two population means.

In Section 10.1, we provided the following interval estimate for the case when the population standard deviations, σ_1 and σ_2, are known.

$$\bar{x}_1 - \bar{x}_2 \pm z_{\alpha/2}\sqrt{\frac{\sigma_1^2}{n_1} + \frac{\sigma_2^2}{n_2}}$$

When σ_1 and σ_2 are estimated by s_1 and s_2, the t distribution is used to make inferences about the difference between two population means.

With σ_1 and σ_2 unknown, we will use the sample standard deviations s_1 and s_2 to estimate σ_1 and σ_2 and replace $z_{\alpha/2}$ with $t_{\alpha/2}$. As a result, the interval estimate of the difference between two population means is given by the following expression:

INTERVAL ESTIMATE OF THE DIFFERENCE BETWEEN TWO POPULATION MEANS: σ_1 AND σ_2 UNKNOWN

$$\bar{x}_1 - \bar{x}_2 \pm t_{\alpha/2}\sqrt{\frac{s_1^2}{n_1} + \frac{s_2^2}{n_2}} \tag{10.6}$$

where $1 - \alpha$ is the confidence coefficient.

In this expression, the use of the t distribution is an approximation, but it provides excellent results and is relatively easy to use. The only difficulty that we encounter in using expression (10.6) is determining the appropriate degrees of freedom for $t_{\alpha/2}$. Statistical software packages compute the appropriate degrees of freedom automatically. The formula used is as follows:

DEGREES OF FREEDOM: t DISTRIBUTION WITH TWO INDEPENDENT RANDOM SAMPLES

$$df = \frac{\left(\frac{s_1^2}{n_1} + \frac{s_2^2}{n_2}\right)^2}{\frac{1}{n_1 - 1}\left(\frac{s_1^2}{n_1}\right)^2 + \frac{1}{n_2 - 1}\left(\frac{s_2^2}{n_2}\right)^2} \tag{10.7}$$

Let us return to the Clearwater National Bank example and show how to use expression (10.6) to provide a 95% confidence interval estimate of the difference between the population mean checking account balances at the two branch banks. The sample data show $n_1 = 28$, $\bar{x}_1 = \$1025$, and $s_1 = \$150$ for the Cherry Grove branch, and $n_2 = 22$, $\bar{x}_2 = \$910$, and $s_2 = \$125$ for the Beechmont branch. The calculation for degrees of freedom for $t_{\alpha/2}$ is as follows:

$$df = \frac{\left(\frac{s_1^2}{n_1} + \frac{s_2^2}{n_2}\right)^2}{\frac{1}{n_1 - 1}\left(\frac{s_1^2}{n_1}\right)^2 + \frac{1}{n_2 - 1}\left(\frac{s_2^2}{n_2}\right)^2} = \frac{\left(\frac{150^2}{28} + \frac{125^2}{22}\right)^2}{\frac{1}{28 - 1}\left(\frac{150^2}{28}\right)^2 + \frac{1}{22 - 1}\left(\frac{125^2}{22}\right)^2} = 47.8$$

We round the noninteger degrees of freedom *down* to 47 to provide a larger t-value and a more conservative interval estimate. Using the t distribution table with 47 degrees of freedom, we find $t_{.025} = 2.012$. Using expression (10.6), we develop the 95% confidence interval estimate of the difference between the two population means as follows.

$$\bar{x}_1 - \bar{x}_2 \pm t_{.025}\sqrt{\frac{s_1^2}{n_1} + \frac{s_2^2}{n_2}}$$

$$1025 - 910 \pm 2.012\sqrt{\frac{150^2}{28} + \frac{125^2}{22}}$$

$$115 \pm 78$$

The point estimate of the difference between the population mean checking account balances at the two branches is \$115. The margin of error is \$78, and the 95% confidence interval estimate of the difference between the two population means is 115 − 78 = \$37 to 115 + 78 = \$193.

This suggestion should help if you are using equation (10.7) to calculate the degrees of freedom by hand.

The computation of the degrees of freedom (equation (10.7)) is cumbersome if you are doing the calculation by hand, but it is easily implemented with a computer software package. However, note that the expressions s_1^2/n_1 and s_2^2/n_2 appear in both expression (10.6) and equation (10.7). These values only need to be computed once in order to evaluate both (10.6) and (10.7).

Hypothesis Tests About $\mu_1 - \mu_2$

Let us now consider hypothesis tests about the difference between the means of two populations when the population standard deviations σ_1 and σ_2 are unknown. Letting D_0 denote the hypothesized difference between μ_1 and μ_2, Section 10.1 showed that the test statistic used for the case where σ_1 and σ_2 are known is as follows.

$$z = \frac{(\bar{x}_1 - \bar{x}_2) - D_0}{\sqrt{\dfrac{\sigma_1^2}{n_1} + \dfrac{\sigma_2^2}{n_2}}}$$

The test statistic, z, follows the standard normal distribution.

When σ_1 and σ_2 are unknown, we use s_1 as an estimator of σ_1 and s_2 as an estimator of σ_2. Substituting these sample standard deviations for σ_1 and σ_2 provides the following test statistic when σ_1 and σ_2 are unknown.

TEST STATISTIC FOR HYPOTHESIS TESTS ABOUT $\mu_1 - \mu_2$: σ_1 AND σ_2 UNKNOWN

$$t = \frac{(\bar{x}_1 - \bar{x}_2) - D_0}{\sqrt{\dfrac{s_1^2}{n_1} + \dfrac{s_2^2}{n_2}}} \tag{10.8}$$

The degrees of freedom for t are given by equation (10.7).

Let us demonstrate the use of this test statistic in the following hypothesis testing example.

Consider a new computer software package developed to help systems analysts reduce the time required to design, develop, and implement an information system. To evaluate the benefits of the new software package, a random sample of 24 systems analysts is selected. Each analyst is given specifications for a hypothetical information system. Then 12 of the analysts are instructed to produce the information system by using current technology. The other 12 analysts are trained in the use of the new software package and then instructed to use it to produce the information system.

This study involves two populations: a population of systems analysts using the current technology and a population of systems analysts using the new software package. In terms of the time required to complete the information system design project, the population means are as follow.

μ_1 = the mean project completion time for systems analysts using the current technology

μ_2 = the mean project completion time for systems analysts using the new software package

The researcher in charge of the new software evaluation project hopes to show that the new software package will provide a shorter mean project completion time. Thus, the researcher is looking for evidence to conclude that μ_2 is less than μ_1; in this case, the

TABLE 10.1 COMPLETION TIME DATA AND SUMMARY STATISTICS FOR THE SOFTWARE TESTING STUDY

WEB file
SoftwareTest

	Current Technology	New Software
	300	274
	280	220
	344	308
	385	336
	372	198
	360	300
	288	315
	321	258
	376	318
	290	310
	301	332
	283	263
Summary Statistics		
Sample size	$n_1 = 12$	$n_2 = 12$
Sample mean	$\bar{x}_1 = 325$ hours	$\bar{x}_2 = 286$ hours
Sample standard deviation	$s_1 = 40$	$s_2 = 44$

difference between the two population means, $\mu_1 - \mu_2$, will be greater than zero. The research hypothesis $\mu_1 - \mu_2 > 0$ is stated as the alternative hypothesis. Thus, the hypothesis test becomes

$$H_0: \mu_1 - \mu_2 \leq 0$$
$$H_a: \mu_1 - \mu_2 > 0$$

We will use $\alpha = .05$ as the level of significance.

Suppose that the 24 analysts complete the study with the results shown in Table 10.1. Using the test statistic in equation (10.8), we have

$$t = \frac{(\bar{x}_1 - \bar{x}_2) - D_0}{\sqrt{\dfrac{s_1^2}{n_1} + \dfrac{s_2^2}{n_2}}} = \frac{(325 - 286) - 0}{\sqrt{\dfrac{40^2}{12} + \dfrac{44^2}{12}}} = 2.27$$

Computing the degrees of freedom using equation (10.7), we have

$$df = \frac{\left(\dfrac{s_1^2}{n_1} + \dfrac{s_2^2}{n_2}\right)^2}{\dfrac{1}{n_1 - 1}\left(\dfrac{s_1^2}{n_1}\right)^2 + \dfrac{1}{n_2 - 1}\left(\dfrac{s_2^2}{n_2}\right)^2} = \frac{\left(\dfrac{40^2}{12} + \dfrac{44^2}{12}\right)^2}{\dfrac{1}{12 - 1}\left(\dfrac{40^2}{12}\right)^2 + \dfrac{1}{12 - 1}\left(\dfrac{44^2}{12}\right)^2} = 21.8$$

Rounding down, we will use a t distribution with 21 degrees of freedom. This row of the t distribution table is as follows:

Area in Upper Tail	**.20**	**.10**	**.05**	**.025**	**.01**	**.005**
***t*-Value (21 *df*)**	0.859	1.323	1.721	2.080	2.518	2.831

$t = 2.27$

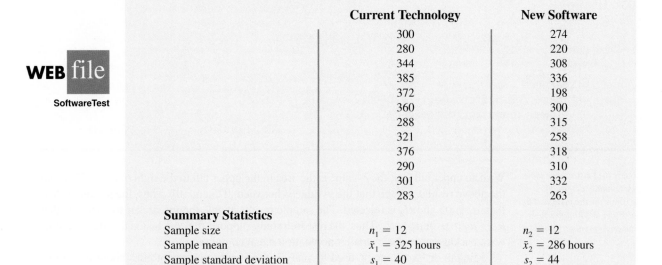

FIGURE 10.2 MINITAB OUTPUT FOR THE HYPOTHESIS TEST OF THE CURRENT AND NEW SOFTWARE TECHNOLOGY

```
Two-sample T for Current vs New

          N      Mean     StDev   SE Mean
Current  12     325.0      40.0        12
New      12     286.0      44.0        13

Difference = mu Current - mu New
Estimate for difference:  39.0000
95% lower bound for difference = 9.5
T-Test of difference = 0 (vs >):  T-Value = 2.27  P-Value = 0.017  DF = 21
```

Using the t distribution table, we can only determine a range for the p-value. Use of Excel or Minitab shows the exact p-value = .017.

With an upper tail test, the *p*-value is the area in the upper tail to the right of $t = 2.27$. From the above results, we see that the *p*-value is between .025 and .01. Thus, the *p*-value is less than $\alpha = .05$ and H_0 is rejected. The sample results enable the researcher to conclude that $\mu_1 - \mu_2 > 0$, or $\mu_1 > \mu_2$. Thus, the research study supports the conclusion that the new software package provides a smaller population mean completion time.

Minitab or Excel can be used to analyze data for testing hypotheses about the difference between two population means. The Minitab output comparing the current and new software technology is shown in Figure 10.2. The last line of the output shows $t = 2.27$ and *p*-value = .017. Note that Minitab used equation (10.7) to compute 21 degrees of freedom for this analysis.

Practical Advice

Whenever possible, equal sample sizes, $n_1 = n_2$, are recommended.

The interval estimation and hypothesis testing procedures presented in this section are robust and can be used with relatively small sample sizes. In most applications, equal or nearly equal sample sizes such that the total sample size $n_1 + n_2$ is at least 20 can be expected to provide very good results even if the populations are not normal. Larger sample sizes are recommended if the distributions of the populations are highly skewed or contain outliers. Smaller sample sizes should only be used if the analyst is satisfied that the distributions of the populations are at least approximately normal.

NOTES AND COMMENTS

Another approach used to make inferences about the difference between two population means when σ_1 and σ_2 are unknown is based on the assumption that the two population standard deviations are *equal* ($\sigma_1 = \sigma_2 = \sigma$). Under this assumption, the two sample standard deviations are combined to provide the following *pooled sample variance:*

$$s_p^2 = \frac{(n_1 - 1)s_1^2 + (n_2 - 1)s_2^2}{n_1 + n_2 - 2}$$

The *t* test statistic becomes

$$t = \frac{(\bar{x}_1 - \bar{x}_2) - D_0}{s_p\sqrt{\dfrac{1}{n_1} + \dfrac{1}{n_2}}}$$

and has $n_1 + n_2 - 2$ degrees of freedom. At this point, the computation of the *p*-value and the interpretation of the sample results are identical to the procedures discussed earlier in this section.

A difficulty with this procedure is that the assumption that the two population standard deviations are equal is usually difficult to verify. Unequal population standard deviations are frequently encountered. Using the pooled procedure may not provide satisfactory results, especially if the sample sizes n_1 and n_2 are quite different.

The *t* procedure that we presented in this section does not require the assumption of equal population standard deviations and can be applied whether the population standard deviations are equal or not. It is a more general procedure and is recommended for most applications.

Exercises

Methods

SELF test

9. The following results are for independent random samples taken from two populations.

Sample 1	Sample 2
$n_1 = 20$	$n_2 = 30$
$\bar{x}_1 = 22.5$	$\bar{x}_2 = 20.1$
$s_1 = 2.5$	$s_2 = 4.8$

a. What is the point estimate of the difference between the two population means?
b. What is the degrees of freedom for the t distribution?
c. At 95% confidence, what is the margin of error?
d. What is the 95% confidence interval for the difference between the two population means?

SELF test

10. Consider the following hypothesis test.

$$H_0: \mu_1 - \mu_2 = 0$$
$$H_a: \mu_1 - \mu_2 \neq 0$$

The following results are from independent samples taken from two populations.

Sample 1	Sample 2
$n_1 = 35$	$n_2 = 40$
$\bar{x}_1 = 13.6$	$\bar{x}_2 = 10.1$
$s_1 = 5.2$	$s_2 = 8.5$

a. What is the value of the test statistic?
b. What is the degrees of freedom for the t distribution?
c. What is the p-value?
d. At $\alpha = .05$, what is your conclusion?

11. Consider the following data for two independent random samples taken from two normal populations.

Sample 1	10	7	13	7	9	8
Sample 2	8	7	8	4	6	9

a. Compute the two sample means.
b. Compute the two sample standard deviations.
c. What is the point estimate of the difference between the two population means?
d. What is the 90% confidence interval estimate of the difference between the two population means?

Applications

SELF test

12. The U.S. Department of Transportation provides the number of miles that residents of the 75 largest metropolitan areas travel per day in a car. Suppose that for a simple random sample of 50 Buffalo residents the mean is 22.5 miles a day and the standard deviation is

8.4 miles a day, and for an independent simple random sample of 40 Boston residents the mean is 18.6 miles a day and the standard deviation is 7.4 miles a day.

a. What is the point estimate of the difference between the mean number of miles that Buffalo residents travel per day and the mean number of miles that Boston residents travel per day?
b. What is the 95% confidence interval for the difference between the two population means?

WEB file

CollegeCosts

13. The average annual cost (including tuition, room, board, books, and fees) to attend a public college takes nearly a third of the annual income of a typical family with college-age children (*Money,* April 2012). At private colleges, the average annual cost is equal to about 60% of the typical family's income. The following random samples show the annual cost of attending private and public colleges. Data are in thousands of dollars.

Private Colleges

52.8	43.2	45.0	33.3	44.0	
30.6	45.8	37.8	50.5	42.0	

Public Colleges

20.3	22.0	28.2	15.6	24.1	28.5
22.8	25.8	18.5	25.6	14.4	21.8

a. Compute the sample mean and sample standard deviation for private and public colleges.
b. What is the point estimate of the difference between the two population means? Interpret this value in terms of the annual cost of attending private and public colleges.
c. Develop a 95% confidence interval of the difference between the mean annual cost of attending private and pubic colleges.

14. Are nursing salaries in Tampa, Florida, lower than those in Dallas, Texas? Salary data show staff nurses in Tampa earn less than staff nurses in Dallas (*The Tampa Tribune,* January 15, 2007). Suppose that in a follow-up study of 40 staff nurses in Tampa and 50 staff nurses in Dallas you obtain the following results.

Tampa	Dallas
$n_1 = 40$	$n_2 = 50$
$\bar{x}_1 = \$56,100$	$\bar{x}_2 = \$59,400$
$s_1 = \$6000$	$s_2 = \$7000$

a. Formulate hypothesis so that, if the null hypothesis is rejected, we can conclude that salaries for staff nurses in Tampa are significantly lower than for those in Dallas. Use $\alpha = .05$.
b. What is the value of the test statistic?
c. What is the p-value?
d. What is your conclusion?

15. Commercial real estate prices and rental rates suffered substantial declines in 2008 and 2009 (*Newsweek,* July 27, 2009). These declines were particularly severe in Asia; annual lease rates in Tokyo, Hong Kong, and Singapore declined by 40% or more. Even with such large declines, annual lease rates in Asia were still higher than those in many cities in Europe. Annual lease rates for a sample of 30 commercial properties in Hong Kong showed a mean of $1,114 per square meter with a standard deviation of $230.

Annual lease rates for a sample of 40 commercial properties in Paris showed a mean lease rate of \$989 per square meter with a standard deviation of \$195.

a. On the basis of the sample results, can we conclude that the mean annual lease rate is higher in Hong Kong than in Paris? Develop appropriate null and alternative hypotheses.
b. Use $\alpha = .01$. What is your conclusion?

WEB file
SATMath

16. The College Board provided comparisons of Scholastic Aptitude Test (SAT) scores based on the highest level of education attained by the test taker's parents. A research hypothesis was that students whose parents had attained a higher level of education would on average score higher on the SAT. The overall mean SAT math score was 514 (College Board website, January 8, 2012). SAT math scores for independent samples of students follow. The first sample shows the SAT math test scores for students whose parents are college graduates with a bachelor's degree. The second sample shows the SAT math test scores for students whose parents are high school graduates but do not have a college degree.

Student's Parents			
College Grads		**High School Grads**	
485	487	442	492
534	533	580	478
650	526	479	425
554	410	486	485
550	515	528	390
572	578	524	535
497	448		
592	469		

a. Formulate the hypotheses that can be used to determine whether the sample data support the hypothesis that students show a higher population mean math score on the SAT if their parents attained a higher level of education.
b. What is the point estimate of the difference between the means for the two populations?
c. Compute the p-value for the hypothesis test.
d. At $\alpha = .05$, what is your conclusion?

17. Periodically, Merrill Lynch customers are asked to evaluate Merrill Lynch financial consultants and services. Higher ratings on the client satisfaction survey indicate better service, with 7 the maximum service rating. Independent samples of service ratings for two financial consultants are summarized here. Consultant A has 10 years of experience, whereas consultant B has 1 year of experience. Use $\alpha = .05$ and test to see whether the consultant with more experience has the higher population mean service rating.

Consultant A	Consultant B
$n_1 = 16$	$n_2 = 10$
$\bar{x}_1 = 6.82$	$\bar{x}_2 = 6.25$
$s_1 = .64$	$s_2 = .75$

a. State the null and alternative hypotheses.
b. Compute the value of the test statistic.
c. What is the p-value?
d. What is your conclusion?

WEB file

AirDelay

18. Researchers at Purdue University and Wichita State University found that airlines are doing a better job of getting passengers to their destinations on time (Associated Press, April 2, 2012). AirTran Airways and Southwest Airlines were among the leaders in on-time arrivals with both having 88% of their flights arriving on time. But for the 12% of flights that were delayed, how many minutes were these flights late? Sample data showing the number of minutes that delayed flights were late are provided in the file named AirDelay. Data are shown for both airlines.
    a. Formulate the hypotheses that can be used to test for a difference between the population mean minutes late for delayed flights by these two airlines.
    b. What is the sample mean number of minutes late for delayed flights for each of these two airlines?
    c. Using a .05 level of significance, what is the p-value and what is your conclusion?

10.3 Inferences About the Difference Between Two Population Means: Matched Samples

Suppose employees at a manufacturing company can use two different methods to perform a production task. To maximize production output, the company wants to identify the method with the smaller population mean completion time. Let μ_1 denote the population mean completion time for production method 1 and μ_2 denote the population mean completion time for production method 2. With no preliminary indication of the preferred production method, we begin by tentatively assuming that the two production methods have the same population mean completion time. Thus, the null hypothesis is H_0: $\mu_1 - \mu_2 = 0$. If this hypothesis is rejected, we can conclude that the population mean completion times differ. In this case, the method providing the smaller mean completion time would be recommended. The null and alternative hypotheses are written as follows.

$$H_0: \mu_1 - \mu_2 = 0$$
$$H_a: \mu_1 - \mu_2 \neq 0$$

In choosing the sampling procedure that will be used to collect production time data and test the hypotheses, we consider two alternative designs. One is based on independent samples and the other is based on **matched samples**.

1. *Independent sample design:* A simple random sample of workers is selected and each worker in the sample uses method 1. A second independent simple random sample of workers is selected and each worker in this sample uses method 2. The test of the difference between population means is based on the procedures in Section 10.2.
2. *Matched sample design:* One simple random sample of workers is selected. Each worker first uses one method and then uses the other method. The order of the two methods is assigned randomly to the workers, with some workers performing method 1 first and others performing method 2 first. Each worker provides a pair of data values, one value for method 1 and another value for method 2.

In the matched sample design the two production methods are tested under similar conditions (i.e., with the same workers); hence this design often leads to a smaller sampling error than the independent sample design. The primary reason is that in a matched sample design, variation between workers is eliminated because the same workers are used for both production methods.

Let us demonstrate the analysis of a matched sample design by assuming it is the method used to test the difference between population means for the two production methods.

TABLE 10.2 TASK COMPLETION TIMES FOR A MATCHED SAMPLE DESIGN

Worker	Completion Time for Method 1 (minutes)	Completion Time for Method 2 (minutes)	Difference in Completion Times (d_i)
1	6.0	5.4	.6
2	5.0	5.2	−.2
3	7.0	6.5	.5
4	6.2	5.9	.3
5	6.0	6.0	.0
6	6.4	5.8	.6

WEB file

Matched

A random sample of six workers is used. The data on completion times for the six workers are given in Table 10.2. Note that each worker provides a pair of data values, one for each production method. Also note that the last column contains the difference in completion times d_i for each worker in the sample.

The key to the analysis of the matched sample design is to realize that we consider only the column of differences. Therefore, we have six data values (.6, −.2, .5, .3, .0, and .6) that will be used to analyze the difference between population means of the two production methods.

Let μ_d = the mean of the *difference* in values for the population of workers. With this notation, the null and alternative hypotheses are rewritten as follows.

$$H_0: \mu_d = 0$$
$$H_a: \mu_d \neq 0$$

If H_0 is rejected, we can conclude that the population mean completion times differ.

Other than the use of the d notation, the formulas for the sample mean and sample standard deviation are the same ones used previously in the text.

The d notation is a reminder that the matched sample provides *difference* data. The sample mean and sample standard deviation for the six difference values in Table 10.2 follow.

$$\bar{d} = \frac{\Sigma d_i}{n} = \frac{1.8}{6} = .30$$

$$s_d = \sqrt{\frac{\Sigma(d_i - \bar{d})^2}{n - 1}} = \sqrt{\frac{.56}{5}} = .335$$

It is not necessary to make the assumption that the population has a normal distribution if the sample size is large. Sample size guidelines for using the t distribution were presented in Chapters 8 and 9.

With the small sample of $n = 6$ workers, we need to make the assumption that the population of differences has a normal distribution. This assumption is necessary so that we may use the t distribution for hypothesis testing and interval estimation procedures. Based on this assumption, the following test statistic has a t distribution with $n - 1$ degrees of freedom.

TEST STATISTIC FOR HYPOTHESIS TESTS INVOLVING MATCHED SAMPLES

$$t = \frac{\bar{d} - \mu_d}{s_d/\sqrt{n}} \qquad (10.9)$$

Once the difference data are computed, the t distribution procedure for matched samples is the same as the one-population estimation and hypothesis testing procedures described in Chapters 8 and 9.

Let us use equation (10.9) to test the hypotheses H_0: $\mu_d = 0$ and H_a: $\mu_d \neq 0$, using $\alpha = .05$. Substituting the sample results $\bar{d} = .30$, $s_d = .335$, and $n = 6$ into equation (10.9), we compute the value of the test statistic.

$$t = \frac{\bar{d} - \mu_d}{s_d/\sqrt{n}} = \frac{.30 - 0}{.335/\sqrt{6}} = 2.20$$

Now let us compute the p-value for this two-tailed test. Because $t = 2.20 > 0$, the test statistic is in the upper tail of the t distribution. With $t = 2.20$, the area in the upper tail to the right of the test statistic can be found by using the t distribution table with degrees of freedom $= n - 1 = 6 - 1 = 5$. Information from the 5 degrees of freedom row of the t distribution table is as follows:

Area in Upper Tail	**.20**	**.10**	**.05**	**.025**	**.01**	**.005**
***t*-Value (5 *df*)**	0.920	1.476	2.015	2.571	3.365	4.032

$t = 2.20$ (between 2.015 and 2.571)

Thus, we see that the area in the upper tail is between .05 and .025. Because this test is a two-tailed test, we double these values to conclude that the p-value is between .10 and .05. This p-value is greater than $\alpha = .05$. Thus, the null hypothesis H_0: $\mu_d = 0$ is not rejected. Using Excel or Minitab and the data in Table 10.2, we find the exact p-value $= .080$.

In addition we can obtain an interval estimate of the difference between the two population means by using the single population methodology of Chapter 8. At 95% confidence, the calculation follows.

$$\bar{d} \pm t_{.025}\frac{s_d}{\sqrt{n}}$$

$$.3 \pm 2.571\left(\frac{.335}{\sqrt{6}}\right)$$

$$.3 \pm .35$$

Thus, the margin of error is .35 and the 95% confidence interval for the difference between the population means of the two production methods is −.05 minutes to .65 minutes.

NOTES AND COMMENTS

1. In the example presented in this section, workers performed the production task with first one method and then the other method. This example illustrates a matched sample design in which each sampled element (worker) provides a pair of data values. It is also possible to use different but "similar" elements to provide the pair of data values. For example, a worker at one location could be matched with a similar worker at another location (similarity based on age, education, gender, experience, etc.). The pairs of workers would provide the difference data that could be used in the matched sample analysis.
2. A matched sample procedure for inferences about two population means generally provides better precision than the independent sample approach; therefore it is the recommended design. However, in some applications the matching cannot be achieved, or perhaps the time and cost associated with matching are excessive. In such cases, the independent sample design should be used.

Exercises

Methods

SELF test

19. Consider the following hypothesis test.

$$H_0: \mu_d \leq 0$$
$$H_a: \mu_d > 0$$

The following data are from matched samples taken from two populations.

	Population	
Element	1	2
1	21	20
2	28	26
3	18	18
4	20	20
5	26	24

a. Compute the difference value for each element.
b. Compute $\bar{d}$.
c. Compute the standard deviation s_d.
d. Conduct a hypothesis test using $\alpha = .05$. What is your conclusion?

20. The following data are from matched samples taken from two populations.

	Population	
Element	1	2
1	11	8
2	7	8
3	9	6
4	12	7
5	13	10
6	15	15
7	15	14

a. Compute the difference value for each element.
b. Compute $\bar{d}$.
c. Compute the standard deviation s_d.
d. What is the point estimate of the difference between the two population means?
e. Provide a 95% confidence interval for the difference between the two population means.

Applications

SELF test

21. A market research firm used a sample of individuals to rate the purchase potential of a particular product before and after the individuals saw a new television commercial about the product. The purchase potential ratings were based on a 0 to 10 scale, with higher values indicating a higher purchase potential. The null hypothesis stated that the mean rating "after" would be less than or equal to the mean rating "before." Rejection of this hypothesis would show that the commercial improved the mean purchase potential rating. Use $\alpha = .05$ and the following data to test the hypothesis and comment on the value of the commercial.

	Purchase Rating			Purchase Rating	
Individual	After	Before	Individual	After	Before
1	6	5	5	3	5
2	6	4	6	9	8
3	7	7	7	7	5
4	4	3	8	6	6

WEB file

StockPrices

22. The price per share of stock for a sample of 25 companies was recorded at the beginning of 2012 and then again at the end of the 1st quarter of 2012 (*The Wall Street Journal,* April 2, 2012). How stocks perform during the 1st quarter is an indicator of what is ahead for the stock market and the economy. Use the sample data in the file entitled StockPrices to answer the following.
   a. Let d_i denote the change in price per share for company i where d_i = 1st quarter of 2012 price per share minus the beginning of 2012 price per share. Use the sample mean of these values to estimate the dollar amount a share of stock has changed during the 1st quarter.
   b. What is the 95% confidence interval estimate of the population mean change in the price per share of stock during the first quarter? Interpret this result.

23. Bank of America's Consumer Spending Survey collected data on annual credit card charges in seven different categories of expenditures: transportation, groceries, dining out, household expenses, home furnishings, apparel, and entertainment. Using data from a sample of 42 credit card accounts, assume that each account was used to identify the annual credit card charges for groceries (population 1) and the annual credit card charges for dining out (population 2). Using the difference data, the sample mean difference was $\bar{d}$ = \$850, and the sample standard deviation was s_d = \$1123.
   a. Formulate the null and alternative hypotheses to test for no difference between the population mean credit card charges for groceries and the population mean credit card charges for dining out.
   b. Use a .05 level of significance. Can you conclude that the population means differ? What is the p-value?
   c. Which category, groceries or dining out, has a higher population mean annual credit card charge? What is the point estimate of the difference between the population means? What is the 95% confidence interval estimate of the difference between the population means?

WEB file

BusinessTravel

24. The Global Business Travel Association reported the domestic airfare for business travel for the current year and the previous year (*INC. Magazine,* February 2012). Below is a sample of 12 flights with their domestic airfares shown for both years.

Current Year	Previous Year	Current Year	Previous Year
345	315	635	585
526	463	710	650
420	462	605	545
216	206	517	547
285	275	570	508
405	432	610	580

   a. Formulate the hypotheses and test for a significant increase in the mean domestic airfare for business travel for the one-year period. What is the p-value? Using a .05 level of significance, what is your conclusion?
   b. What is the sample mean domestic airfare for business travel for each year?
   c. What is the percentage change in the airfare for the one-year period?

25. The College Board SAT college entrance exam consists of three parts: math, writing, and critical reading (*The World Almanac,* 2012). Sample data showing the math and writing scores for a sample of 12 students who took the SAT follow.

WEB file

TestScores

Student	Math	Writing	Student	Math	Writing
1	540	474	7	480	430
2	432	380	8	499	459
3	528	463	9	610	615
4	574	612	10	572	541
5	448	420	11	390	335
6	502	526	12	593	613

a. Use a .05 level of significance and test for a difference between the population mean for the math scores and the population mean for the writing scores. What is the p-value and what is your conclusion?

b. What is the point estimate of the difference between the mean scores for the two tests? What are the estimates of the population mean scores for the two tests? Which test reports the higher mean score?

26. Scores in the first and fourth (final) rounds for a sample of 20 golfers who competed in PGA tournaments are shown in the following table (*Golfweek,* February 14, 2009, and February 28, 2009). Suppose you would like to determine if the mean score for the first round of a PGA Tour event is significantly different than the mean score for the fourth and final round. Does the pressure of playing in the final round cause scores to go up? Or does the increased player concentration cause scores to come down?

WEB file

GolfScores

Player	First Round	Final Round	Player	First Round	Final Round
Michael Letzig	70	72	Aron Price	72	72
Scott Verplank	71	72	Charles Howell	72	70
D. A. Points	70	75	Jason Dufner	70	73
Jerry Kelly	72	71	Mike Weir	70	77
Soren Hansen	70	69	Carl Pettersson	68	70
D. J. Trahan	67	67	Bo Van Pelt	68	65
Bubba Watson	71	67	Ernie Els	71	70
Reteif Goosen	68	75	Cameron Beckman	70	68
Jeff Klauk	67	73	Nick Watney	69	68
Kenny Perry	70	69	Tommy Armour III	67	71

a. Use $\alpha = .10$ to test for a statistically significantly difference between the population means for first- and fourth-round scores. What is the p-value? What is your conclusion?

b. What is the point estimate of the difference between the two population means? For which round is the population mean score lower?

c. What is the margin of error for a 90% confidence interval estimate for the difference between the population means? Could this confidence interval have been used to test the hypothesis in part (a)? Explain.

27. A manufacturer produces both a deluxe and a standard model of an automatic sander designed for home use. Selling prices obtained from a sample of retail outlets follow.

	Model Price ($)			Model Price ($)	
Retail Outlet	Deluxe	Standard	Retail Outlet	Deluxe	Standard
1	39	27	5	40	30
2	39	28	6	39	34
3	45	35	7	35	29
4	38	30			

a. The manufacturer's suggested retail prices for the two models show a \$10 price differential. Use a .05 level of significance and test that the mean difference between the prices of the two models is \$10.
b. What is the 95% confidence interval for the difference between the mean prices of the two models?

10.4 Inferences About the Difference Between Two Population Proportions

Letting p_1 denote the proportion for population 1 and p_2 denote the proportion for population 2, we next consider inferences about the difference between the two population proportions: $p_1 - p_2$. To make an inference about this difference, we will select two independent random samples consisting of n_1 units from population 1 and n_2 units from population 2.

Interval Estimation of $p_1 - p_2$

In the following example, we show how to compute a margin of error and develop an interval estimate of the difference between two population proportions.

A tax preparation firm is interested in comparing the quality of work at two of its regional offices. By randomly selecting samples of tax returns prepared at each office and verifying the sample returns' accuracy, the firm will be able to estimate the proportion of erroneous returns prepared at each office. Of particular interest is the difference between these proportions.

p_1 = proportion of erroneous returns for population 1 (office 1)
p_2 = proportion of erroneous returns for population 2 (office 2)
$\bar{p}_1$ = sample proportion for a simple random sample from population 1
$\bar{p}_2$ = sample proportion for a simple random sample from population 2

The difference between the two population proportions is given by $p_1 - p_2$. The point estimator of $p_1 - p_2$ is as follows.

POINT ESTIMATOR OF THE DIFFERENCE BETWEEN TWO POPULATION PROPORTIONS

$$\bar{p}_1 - \bar{p}_2 \tag{10.10}$$

Thus, the point estimator of the difference between two population proportions is the difference between the sample proportions of two independent simple random samples.

As with other point estimators, the point estimator $\bar{p}_1 - \bar{p}_2$ has a sampling distribution that reflects the possible values of $\bar{p}_1 - \bar{p}_2$ if we repeatedly took two independent random samples. The mean of this sampling distribution is $p_1 - p_2$ and the standard error of $\bar{p}_1 - \bar{p}_2$ is as follows:

STANDARD ERROR OF $\bar{p}_1 - \bar{p}_2$

$$\sigma_{\bar{p}_1 - \bar{p}_2} = \sqrt{\frac{p_1(1 - p_1)}{n_1} + \frac{p_2(1 - p_2)}{n_2}} \tag{10.11}$$

If the sample sizes are large enough that n_1p_1, $n_1(1 - p_1)$, n_2p_2, and $n_2(1 - p_2)$ are all greater than or equal to 5, the sampling distribution of $\bar{p}_1 - \bar{p}_2$ can be approximated by a normal distribution.

As we showed previously, an interval estimate is given by a point estimate ± a margin of error. In the estimation of the difference between two population proportions, an interval estimate will take the following form:

$$\bar{p}_1 - \bar{p}_2 \pm \text{Margin of error}$$

With the sampling distribution of $\bar{p}_1 - \bar{p}_2$ approximated by a normal distribution, we would like to use $z_{\alpha/2}\sigma_{\bar{p}_1 - \bar{p}_2}$ as the margin of error. However, $\sigma_{\bar{p}_1 - \bar{p}_2}$ given by equation (10.11) cannot be used directly because the two population proportions, p_1 and p_2, are unknown. Using the sample proportion $\bar{p}_1$ to estimate p_1 and the sample proportion $\bar{p}_2$ to estimate p_2, the margin of error is as follows.

$$\text{Margin of error} = z_{\alpha/2}\sqrt{\frac{\bar{p}_1(1 - \bar{p}_1)}{n_1} + \frac{\bar{p}_2(1 - \bar{p}_2)}{n_2}} \qquad \textbf{(10.12)}$$

The general form of an interval estimate of the difference between two population proportions is as follows.

INTERVAL ESTIMATE OF THE DIFFERENCE BETWEEN TWO POPULATION PROPORTIONS

$$\bar{p}_1 - \bar{p}_2 \pm z_{\alpha/2}\sqrt{\frac{\bar{p}_1(1 - \bar{p}_1)}{n_1} + \frac{\bar{p}_2(1 - \bar{p}_2)}{n_2}} \qquad \textbf{(10.13)}$$

where $1 - \alpha$ is the confidence coefficient.

Returning to the tax preparation example, we find that independent simple random samples from the two offices provide the following information.

Office 1	**Office 2**
$n_1 = 250$	$n_2 = 300$
Number of returns with errors = 35	Number of returns with errors = 27

WEB file
TaxPrep

The sample proportions for the two offices follow.

$$\bar{p}_1 = \frac{35}{250} = .14$$

$$\bar{p}_2 = \frac{27}{300} = .09$$

The point estimate of the difference between the proportions of erroneous tax returns for the two populations is $\bar{p}_1 - \bar{p}_2 = .14 - .09 = .05$. Thus, we estimate that office 1 has a .05, or 5%, greater error rate than office 2.

Expression (10.13) can now be used to provide a margin of error and interval estimate of the difference between the two population proportions. Using a 90% confidence interval with $z_{\alpha/2} = z_{.05} = 1.645$, we have

$$\bar{p}_1 - \bar{p}_2 \pm z_{\alpha/2}\sqrt{\frac{\bar{p}_1(1-\bar{p}_1)}{n_1} + \frac{\bar{p}_2(1-\bar{p}_2)}{n_2}}$$

$$.14 - .09 \pm 1.645\sqrt{\frac{.14(1-.14)}{250} + \frac{.09(1-.09)}{300}}$$

$$.05 \pm .045$$

Thus, the margin of error is .045, and the 90% confidence interval is .005 to .095.

Hypothesis Tests About $p_1 - p_2$

Let us now consider hypothesis tests about the difference between the proportions of two populations. We focus on tests involving no difference between the two population proportions. In this case, the three forms for a hypothesis test are as follows:

All hypotheses considered use 0 as the difference of interest.

$$H_0: p_1 - p_2 \geq 0 \qquad H_0: p_1 - p_2 \leq 0 \qquad H_0: p_1 - p_2 = 0$$
$$H_a: p_1 - p_2 < 0 \qquad H_a: p_1 - p_2 > 0 \qquad H_a: p_1 - p_2 \neq 0$$

When we assume H_0 is true as an equality, we have $p_1 - p_2 = 0$, which is the same as saying that the population proportions are equal, $p_1 = p_2$.

We will base the test statistic on the sampling distribution of the point estimator $\bar{p}_1 - \bar{p}_2$. In equation (10.11), we showed that the standard error of $\bar{p}_1 - \bar{p}_2$ is given by

$$\sigma_{\bar{p}_1-\bar{p}_2} = \sqrt{\frac{p_1(1-p_1)}{n_1} + \frac{p_2(1-p_2)}{n_2}}$$

Under the assumption H_0 is true as an equality, the population proportions are equal and $p_1 = p_2 = p$. In this case, $\sigma_{\bar{p}_1-\bar{p}_2}$ becomes

STANDARD ERROR OF $\bar{p}_1 - \bar{p}_2$ WHEN $p_1 = p_2 = p$

$$\sigma_{\bar{p}_1-\bar{p}_2} = \sqrt{\frac{p(1-p)}{n_1} + \frac{p(1-p)}{n_2}} = \sqrt{p(1-p)\left(\frac{1}{n_1} + \frac{1}{n_2}\right)} \qquad \textbf{(10.14)}$$

With p unknown, we pool, or combine, the point estimators from the two samples ($\bar{p}_1$ and $\bar{p}_2$) to obtain a single point estimator of p as follows:

POOLED ESTIMATOR OF p WHEN $p_1 = p_2 = p$

$$\bar{p} = \frac{n_1\bar{p}_1 + n_2\bar{p}_2}{n_1 + n_2} \qquad \textbf{(10.15)}$$

This **pooled estimator of p** is a weighted average of $\bar{p}_1$ and $\bar{p}_2$.

Substituting $\bar{p}$ for p in equation (10.14), we obtain an estimate of the standard error of $\bar{p}_1 - \bar{p}_2$. This estimate of the standard error is used in the test statistic. The general form of the test statistic for hypothesis tests about the difference between two population proportions is the point estimator divided by the estimate of $\sigma_{\bar{p}_1-\bar{p}_2}$.

TEST STATISTIC FOR HYPOTHESIS TESTS ABOUT $p_1 - p_2$

$$z = \frac{(\bar{p}_1 - \bar{p}_2)}{\sqrt{\bar{p}(1 - \bar{p})\left(\frac{1}{n_1} + \frac{1}{n_2}\right)}} \tag{10.16}$$

This test statistic applies to large sample situations where n_1p_1, $n_1(1 - p_1)$, n_2p_2, and $n_2(1 - p_2)$ are all greater than or equal to 5.

Let us return to the tax preparation firm example and assume that the firm wants to use a hypothesis test to determine whether the error proportions differ between the two offices. A two-tailed test is required. The null and alternative hypotheses are as follows:

$$H_0: p_1 - p_2 = 0$$
$$H_a: p_1 - p_2 \neq 0$$

If H_0 is rejected, the firm can conclude that the error rates at the two offices differ. We will use $\alpha = .10$ as the level of significance.

The sample data previously collected showed $\bar{p}_1 = .14$ for the $n_1 = 250$ returns sampled at office 1 and $\bar{p}_2 = .09$ for the $n_2 = 300$ returns sampled at office 2. We continue by computing the pooled estimate of p.

$$\bar{p} = \frac{n_1\bar{p}_1 + n_2\bar{p}_2}{n_1 + n_2} = \frac{250(.14) + 300(.09)}{250 + 300} = .1127$$

Using this pooled estimate and the difference between the sample proportions, the value of the test statistic is as follows.

$$z = \frac{(\bar{p}_1 - \bar{p}_2)}{\sqrt{\bar{p}(1 - \bar{p})\left(\frac{1}{n_1} + \frac{1}{n_2}\right)}} = \frac{(.14 - .09)}{\sqrt{.1127(1 - .1127)\left(\frac{1}{250} + \frac{1}{300}\right)}} = 1.85$$

In computing the p-value for this two-tailed test, we first note that $z = 1.85$ is in the upper tail of the standard normal distribution. Using $z = 1.85$ and the standard normal distribution table, we find the area in the upper tail is $1.0000 - .9678 = .0322$. Doubling this area for a two-tailed test, we find the p-value $= 2(.0322) = .0644$. With the p-value less than $\alpha = .10$, H_0 is rejected at the .10 level of significance. The firm can conclude that the error rates differ between the two offices. This hypothesis testing conclusion is consistent with the earlier interval estimation results that showed the interval estimate of the difference between the population error rates at the two offices to be .005 to .095, with Office 1 having the higher error rate.

Exercises

Methods

SELF test

28. Consider the following results for independent samples taken from two populations.

Sample 1	Sample 2
$n_1 = 400$	$n_2 = 300$
$\bar{p}_1 = .48$	$\bar{p}_2 = .36$

a. What is the point estimate of the difference between the two population proportions?
b. Develop a 90% confidence interval for the difference between the two population proportions.
c. Develop a 95% confidence interval for the difference between the two population proportions.

SELF test

29. Consider the hypothesis test

$$H_0: p_1 - p_2 \leq 0$$
$$H_a: p_1 - p_2 > 0$$

The following results are for independent samples taken from the two populations.

Sample 1	Sample 2
$n_1 = 200$	$n_2 = 300$
$\bar{p}_1 = .22$	$\bar{p}_2 = .16$

a. What is the p-value?
b. With $\alpha = .05$, what is your hypothesis testing conclusion?

Applications

SELF test

30. A *Businessweek*/Harris survey asked senior executives at large corporations their opinions about the economic outlook for the future. One question was, "Do you think that there will be an increase in the number of full-time employees at your company over the next 12 months?" In the current survey, 220 of 400 executives answered Yes, while in a previous year survey, 192 of 400 executives had answered Yes. Provide a 95% confidence interval estimate for the difference between the proportions at the two points in time. What is your interpretation of the interval estimate?

31. The Professional Golf Association (PGA) measured the putting accuracy of professional golfers playing on the PGA Tour and the best amateur golfers playing in the World Amateur Championship (*Golf Magazine,* January 2007). A sample of 1075 6-foot putts by professional golfers found 688 made puts. A sample of 1200 6-foot putts by amateur golfers found 696 made putts.
a. Estimate the proportion of made 6-foot putts by professional golfers. Estimate the proportion of made 6-foot putts by amateur golfers. Which group had a better putting accuracy?
b. What is the point estimate of the difference between the proportions of the two populations? What does this estimate tell you about the percentage of putts made by the two groups of golfers?
c. What is the 95% confidence interval for the difference between the two population proportions? Interpret his confidence interval in terms of the percentage of putts made by the two groups of golfers.

32. An American Automobile Association (AAA) study investigated the question of whether a man or a woman was more likely to stop and ask for directions (AAA, January 2006). The situation referred to in the study stated the following: "If you and your spouse are driving together and become lost, would you stop and ask for directions?" A sample representative of the data used by AAA showed 300 of 811 women said that they would stop and ask for directions, while 255 of 750 men said that they would stop and ask for directions.
a. The AAA research hypothesis was that women would be more likely to say that they would stop and ask for directions. Formulate the null and alternative hypotheses for this study.
b. What is the percentage of women who indicated that they would stop and ask for directions?

c. What is the percentage of men who indicated that they would stop and ask for directions?
d. At $\alpha = .05$, test the hypothesis. What is the p-value, and what conclusion would you expect AAA to draw from this study?

33. Chicago O'Hare and Atlanta Hartsfield-Jackson are the two busiest airports in the United States. The congestion often leads to delayed flight arrivals as well as delayed flight departures. The Bureau of Transportation tracks the on-time and delayed performance at major airports (*Travel & Leisure,* November 2006). A flight is considered delayed if it is more than 15 minutes behind schedule. The following sample data show the delayed departures at Chicago O'Hare and Atlanta Hartsfield-Jackson airports.

	Chicago O'Hare	**Atlanta Hartsfield-Jackson**
Flights	900	1200
Delayed Departures	252	312

a. State the hypotheses that can be used to determine whether the population proportions of delayed departures differ at these two airports.
b. What is the point estimate of the proportion of flights that have delayed departures at Chicago O'Hare?
c. What is the point estimate of the proportion of flights that have delayed departures at Atlanta Hartsfield-Jackson?
d. What is the p-value for the hypothesis test? What is your conclusion?

34. *Businessweek* reported that there seems to be a difference by age group in how well people like life in Russia (*Businessweek,* March 10, 2008). The following sample data are consistent with the *Businessweek* findings and show the responses by age group to the question: "Do you like life in Russia?"

	Russian Age Group	
	17–26	40 and over
Sample	300	260
Responded Yes	192	117

a. What is the point estimate of the proportion of Russians aged 17 to 26 who like life in Russia?
b. What is the point estimate of the proportion of Russians aged 40 and over who like life in Russia?
c. Provide a 95% confidence interval estimate of the difference between the proportion of young Russians aged 17 to 26 and older Russians aged 40 and over who like life in Russia.

35. In a test of the quality of two television commercials, each commercial was shown in a separate test area six times over a one-week period. The following week a telephone survey was conducted to identify individuals who had seen the commercials. Those individuals were asked to state the primary message in the commercials. The following results were recorded.

	Commercial A	**Commercial B**
Number Who Saw Commercial	150	200
Number Who Recalled Message	63	60

a. Use $\alpha = .05$ and test the hypothesis that there is no difference in the recall proportions for the two commercials.
b. Compute a 95% confidence interval for the difference between the recall proportions for the two populations.

36. Winter visitors are extremely important to the economy of Southwest Florida. Hotel occupancy is an often-reported measure of visitor volume and visitor activity (*Naples Daily News,* March 22, 2012). Hotel occupancy data for February in two consecutive years are as follows.

	Current Year	Previous Year
Occupied Rooms	1470	1458
Total Rooms	1750	1800

a. Formulate the hypothesis test that can be used to determine if there has been an increase in the proportion of rooms occupied over the one-year period.
b. What is the estimated proportion of hotel rooms occupied each year?
c. Using a .05 level of significance, what is your hypothesis test conclusion? What is the p-value?
d. What is the 95% confidence interval estimate of the change in occupancy for the one-year period? Do you think area officials would be pleased with the results?

37. The Adecco Workplace Insights Survey sampled men and women workers and asked if they expected to get a raise or promotion this year (*USA Today,* February 16, 2012). Suppose the survey sampled 200 men and 200 women. If 104 of the men replied Yes and 74 of the women replied Yes, are the results statistically significant in that you can conclude a greater proportion of men are expecting to get a raise or a promotion this year?
a. State the hypothesis test in terms of the population proportion of men and the population proportion of women?
b. What is the sample proportion for men? For women?
c. Use a .01 level of significance. What is the p-value and what is your conclusion?

Summary

In this chapter we discussed procedures for developing interval estimates and conducting hypothesis tests involving two populations. First, we showed how to make inferences about the difference between two population means when independent simple random samples are selected. We first considered the case where the population standard deviations, σ_1 and σ_2, could be assumed known. The standard normal distribution z was used to develop the interval estimate and served as the test statistic for hypothesis tests. We then considered the case where the population standard deviations were unknown and estimated by the sample standard deviations s_1 and s_2. In this case, the t distribution was used to develop the interval estimate and served as the test statistic for hypothesis tests.

Inferences about the difference between two population means were then discussed for the matched sample design. In the matched sample design each element provides a pair of data values, one from each population. The difference between the paired data values is then used in the statistical analysis. The matched sample design is generally preferred to the independent sample design because the matched-sample procedure often improves the precision of the estimate.

Finally, interval estimation and hypothesis testing about the difference between two population proportions were discussed. Statistical procedures for analyzing the difference between two population proportions are similar to the procedures for analyzing the difference between two population means.

Glossary

Independent simple random samples Samples selected from two populations in such a way that the elements making up one sample are chosen independently of the elements making up the other sample.
Matched samples Samples in which each data value of one sample is matched with a corresponding data value of the other sample.
Pooled estimator of p An estimator of a population proportion obtained by computing a weighted average of the point estimators obtained from two independent samples.

Key Formulas

Point Estimator of the Difference Between Two Population Means

$$\bar{x}_1 - \bar{x}_2 \tag{10.1}$$

Standard Error of $\bar{x}_1 - \bar{x}_2$

$$\sigma_{\bar{x}_1 - \bar{x}_2} = \sqrt{\frac{\sigma_1^2}{n_1} + \frac{\sigma_2^2}{n_2}} \tag{10.2}$$

Interval Estimate of the Difference Between Two Population Means: σ_1 and σ_2 Known

$$\bar{x}_1 - \bar{x}_2 \pm z_{\alpha/2}\sqrt{\frac{\sigma_1^2}{n_1} + \frac{\sigma_2^2}{n_2}} \tag{10.4}$$

Test Statistic for Hypothesis Tests About $\mu_1 - \mu_2$: σ_1 and σ_2 Known

$$z = \frac{(\bar{x}_1 - \bar{x}_2) - D_0}{\sqrt{\dfrac{\sigma_1^2}{n_1} + \dfrac{\sigma_2^2}{n_2}}} \tag{10.5}$$

Interval Estimate of the Difference Between Two Population Means: σ_1 and σ_2 Unknown

$$\bar{x}_1 - \bar{x}_2 \pm t_{\alpha/2}\sqrt{\frac{s_1^2}{n_1} + \frac{s_2^2}{n_2}} \tag{10.6}$$

Degrees of Freedom: t Distribution with Two Independent Random Samples

$$df = \frac{\left(\dfrac{s_1^2}{n_1} + \dfrac{s_2^2}{n_2}\right)^2}{\dfrac{1}{n_1 - 1}\left(\dfrac{s_1^2}{n_1}\right)^2 + \dfrac{1}{n_2 - 1}\left(\dfrac{s_2^2}{n_2}\right)^2} \tag{10.7}$$

Test Statistic for Hypothesis Tests About $\mu_1 - \mu_2$: σ_1 and σ_2 Unknown

$$t = \frac{(\bar{x}_1 - \bar{x}_2) - D_0}{\sqrt{\dfrac{s_1^2}{n_1} + \dfrac{s_2^2}{n_2}}} \tag{10.8}$$

Test Statistic for Hypothesis Tests Involving Matched Samples

$$t = \frac{\bar{d} - \mu_d}{s_d/\sqrt{n}} \tag{10.9}$$

Point Estimator of the Difference Between Two Population Proportions

$$\bar{p}_1 - \bar{p}_2 \tag{10.10}$$

Standard Error of $\bar{p}_1 - \bar{p}_2$

$$\sigma_{\bar{p}_1 - \bar{p}_2} = \sqrt{\frac{p_1(1 - p_1)}{n_1} + \frac{p_2(1 - p_2)}{n_2}} \tag{10.11}$$

Interval Estimate of the Difference Between Two Population Proportions

$$\bar{p}_1 - \bar{p}_2 \pm z_{\alpha/2}\sqrt{\frac{\bar{p}_1(1 - \bar{p}_1)}{n_1} + \frac{\bar{p}_2(1 - \bar{p}_2)}{n_2}} \tag{10.13}$$

Standard Error of $\bar{p}_1 - \bar{p}_2$ when $p_1 = p_2 = p$

$$\sigma_{\bar{p}_1 - \bar{p}_2} = \sqrt{p(1 - p)\left(\frac{1}{n_1} + \frac{1}{n_2}\right)} \tag{10.14}$$

Pooled Estimator of p when $p_1 = p_2 = p$

$$\bar{p} = \frac{n_1\bar{p}_1 + n_2\bar{p}_2}{n_1 + n_2} \tag{10.15}$$

Test Statistic for Hypothesis Tests About $p_1 - p_2$

$$z = \frac{(\bar{p}_1 - \bar{p}_2)}{\sqrt{\bar{p}(1 - \bar{p})\left(\frac{1}{n_1} + \frac{1}{n_2}\right)}} \tag{10.16}$$

Supplementary Exercises

38. Safegate Foods, Inc., is redesigning the checkout lanes in its supermarkets throughout the country and is considering two designs. Tests on customer checkout times conducted at two stores where the two new systems have been installed result in the following summary of the data.

System A	System B
$n_1 = 120$	$n_2 = 100$
$\bar{x}_1 = 4.1$ minutes	$\bar{x}_2 = 3.4$ minutes
$\sigma_1 = 2.2$ minutes	$\sigma_2 = 1.5$ minutes

Test at the .05 level of significance to determine whether the population mean checkout times of the two systems differ. Which system is preferred?

WEB file HomePrices

39. Home values tend to increase over time under normal conditions, but the recession of 2008 and 2009 has reportedly caused the sales price of existing homes to fall nationwide (*Businessweek,* March 9, 2009). You would like to see if the data support this conclusion. The file HomePrices contains data on 30 existing home sales in 2006 and 40 existing home sales in 2009.
    a. Provide a point estimate of the difference between the population mean prices for the two years.
    b. Develop a 99% confidence interval estimate of the difference between the resale prices of houses in 2006 and 2009.
    c. Would you feel justified in concluding that resale prices of existing homes have declined from 2006 to 2009? Why or why not?

40. Mutual funds are classified as *load* or *no-load* funds. Load funds require an investor to pay an initial fee based on a percentage of the amount invested in the fund. The no-load funds do not require this initial fee. Some financial advisors argue that the load mutual funds may be worth the extra fee because these funds provide a higher mean rate of return than the no-load mutual funds. A sample of 30 load mutual funds and a sample of 30 no-load mutual funds were selected. Data were collected on the annual return for the funds over a five-year period. The data are contained in the data set Mutual. The data for the first five load and first five no-load mutual funds are as follows.

WEB file Mutual

Mutual Funds—Load	Return	Mutual Funds—No Load	Return
American National Growth	15.51	Amana Income Fund	13.24
Arch Small Cap Equity	14.57	Berger One Hundred	12.13
Bartlett Cap Basic	17.73	Columbia International Stock	12.17
Calvert World International	10.31	Dodge & Cox Balanced	16.06
Colonial Fund A	16.23	Evergreen Fund	17.61

a. Formulate H_0 and H_a such that rejection of H_0 leads to the conclusion that the load mutual funds have a higher mean annual return over the five-year period.
b. Use the 60 mutual funds in the data set Mutual to conduct the hypothesis test. What is the p-value? At $\alpha = .05$, what is your conclusion?

41. The National Association of Home Builders provided data on the cost of the most popular home remodeling projects. Sample data on cost in thousands of dollars for two types of remodeling projects are as follows.

Kitchen	Master Bedroom	Kitchen	Master Bedroom
25.2	18.0	23.0	17.8
17.4	22.9	19.7	24.6
22.8	26.4	16.9	21.0
21.9	24.8	21.8	
19.7	26.9	23.6	

a. Develop a point estimate of the difference between the population mean remodeling costs for the two types of projects.
b. Develop a 90% confidence interval for the difference between the two population means.

42. In early 2009, the economy was experiencing a recession. But how was the recession affecting the stock market? Shown are data from a sample of 15 companies. Shown for each company is the price per share of stock on January 1 and April 30 (*The Wall Street Journal,* May 1, 2009).

WEB file

PriceChange

Company	January 1 ($)	April 30 ($)
Applied Materials	10.13	12.21
Bank of New York	28.33	25.48
Chevron	73.97	66.10
Cisco Systems	16.30	19.32
Coca-Cola	45.27	43.05
Comcast	16.88	15.46
Ford Motors	2.29	5.98
General Electric	16.20	12.65
Johnson & Johnson	59.83	52.36
JP Morgan Chase	31.53	33.00
Microsoft	19.44	20.26
Oracle	17.73	19.34
Pfizer	17.71	13.36
Philip Morris	43.51	36.18
Procter & Gamble	61.82	49.44

a. What is the change in the mean price per share of stock over the four-month period?
b. Provide a 90% confident interval estimate of the change in the mean price per share of stock. Interpret the results.
c. What was the percentage change in the mean price per share of stock over the four-month period?
d. If this same percentage change were to occur for the next four months and again for the four months after that, what would be the mean price per share of stock at the end of the year 2009?

43. Country Financial, a financial services company, uses surveys of adults age 18 and older to determine if personal financial fitness is changing over time (*USA Today,* April 4, 2012). In February 2012, a sample of 1000 adults showed 410 indicating that their financial security was more than fair. In February 2010, a sample of 900 adults showed 315 indicating that their financial security was more than fair.
a. State the hypotheses that can be used to test for a significant difference between the population proportions for the two years.
b. What is the sample proportion indicating that their financial security was more than fair in 2012? In 2010?
c. Conduct the hypothesis test and compute the p-value. At a .05 level of significance, what is your conclusion?
d. What is the 95% confidence interval estimate of the difference between the two population proportions?

44. A large automobile insurance company selected samples of single and married male policyholders and recorded the number who made an insurance claim over the preceding three-year period.

Single Policyholders	Married Policyholders
$n_1 = 400$	$n_2 = 900$
Number making claims = 76	Number making claims = 90

a. Use $\alpha = .05$. Test to determine whether the claim rates differ between single and married male policyholders.
b. Provide a 95% confidence interval for the difference between the proportions for the two populations.

45. Medical tests were conducted to learn about drug-resistant tuberculosis. Of 142 cases tested in New Jersey, 9 were found to be drug-resistant. Of 268 cases tested in Texas, 5 were found

to be drug-resistant. Do these data suggest a statistically significant difference between the proportions of drug-resistant cases in the two states? Use a .02 level of significance. What is the p-value, and what is your conclusion?

WEB file Occupancy

46. Vacation occupancy rates were expected to be up during March 2008 in Myrtle Beach, South Carolina (*The Sun News,* February 29, 2008). Data in the file Occupancy will allow you to replicate the findings presented in the newspaper. The data show units rented and not rented for a random sample of vacation properties during the first week of March 2007 and March 2008.
   a. Estimate the proportion of units rented during the first week of March 2007 and the first week of March 2008.
   b. Provide a 95% confidence interval for the difference in proportions.
   c. On the basis of your findings, does it appear March rental rates for 2008 will be up from those a year earlier?
47. For the week ended January 15, 2009, the bullish sentiment of individual investors was 27.6% (*AAII Journal,* February 2009). The bullish sentiment was reported to be 48.7% one week earlier and 39.7% one month earlier. The sentiment measures are based on a poll conducted by the American Assocation of Individual Investors. Assume that each of the bullish sentiment measures was based on a sample size of 240.
   a. Develop a 95% confidence interval for the difference between the bullish sentiment measures for the most recent two weeks.
   b. Develop hypotheses so that rejection of the null hypothesis will allow us to conclude that the most recent bullish sentiment is weaker than that of one month ago.
   c. Conduct a hypotheses test of part (b) using $\alpha = .01$. What is your conclusion?

Case Problem Par, Inc.

Par, Inc., is a major manufacturer of golf equipment. Management believes that Par's market share could be increased with the introduction of a cut-resistant, longer-lasting golf ball. Therefore, the research group at Par has been investigating a new golf ball coating designed to resist cuts and provide a more durable ball. The tests with the coating have been promising.

One of the researchers voiced concern about the effect of the new coating on driving distances. Par would like the new cut-resistant ball to offer driving distances comparable to those of the current-model golf ball. To compare the driving distances for the two balls, 40 balls of both the new and current models were subjected to distance tests. The testing was performed with a mechanical hitting machine so that any difference between the mean distances for the two models could be attributed to a difference in the two models. The results of the tests, with distances measured to the nearest yard, follow. These data are available on the website that accompanies the text.

WEB file Golf

Model		Model		Model		Model	
Current	New	Current	New	Current	New	Current	New
264	277	270	272	263	274	281	283
261	269	287	259	264	266	274	250
267	263	289	264	284	262	273	253
272	266	280	280	263	271	263	260
258	262	272	274	260	260	275	270
283	251	275	281	283	281	267	263
258	262	265	276	255	250	279	261
266	289	260	269	272	263	274	255
259	286	278	268	266	278	276	263
270	264	275	262	268	264	262	279

Managerial Report

1. Formulate and present the rationale for a hypothesis test that Par could use to compare the driving distances of the current and new golf balls.
2. Analyze the data to provide the hypothesis testing conclusion. What is the p-value for your test? What is your recommendation for Par, Inc.?
3. Provide descriptive statistical summaries of the data for each model.
4. What is the 95% confidence interval for the population mean driving distance of each model, and what is the 95% confidence interval for the difference between the means of the two populations?
5. Do you see a need for larger sample sizes and more testing with the golf balls? Discuss.

Appendix 10.1 Inferences About Two Populations Using Minitab

We describe the use of Minitab to develop interval estimates and conduct hypothesis tests about the difference between two population means and the difference between two population proportions. Minitab provides both interval estimation and hypothesis testing results within the same module. Thus, the Minitab procedure is the same for both types of inferences. In the examples that follow, we will demonstrate interval estimation and hypothesis testing for the same two samples. We note that Minitab does not provide a routine for inferences about the difference between two population means when the population standard deviations σ_1 and σ_2 are known.

Difference Between Two Population Means: σ_1 and σ_2 Unknown

WEB file
CheckAcct

We will use the data for the checking account balances example presented in Section 10.2. The checking account balances at the Cherry Grove branch are in column C1, and the checking account balances at the Beechmont branch are in column C2. In this example, we will use the Minitab 2-Sample t procedure to provide a 95% confidence interval estimate of the difference between population means for the checking account balances at the two branch banks. The output of the procedure also provides the p-value for the hypothesis test: H_0: $\mu_1 - \mu_2 = 0$ versus H_a: $\mu_1 - \mu_2 \neq 0$. The following steps are necessary to execute the procedure:

Step 1. Select the **Stat** menu
Step 2. Choose **Basic Statistics**
Step 3. Choose **2-Sample t**
Step 4. When the 2-Sample t (Test and Confidence Interval) dialog box appears:
Select **Samples in different columns**
Enter C1 in the **First** box
Enter C2 in the **Second** box
Select **Options**
Step 5. When the 2-Sample t - Options dialog box appears:
Enter 95 in the **Confidence level** box
Enter 0 in the **Test difference** box
Enter not equal in the **Alternative** box
Click **OK**
Step 6. Click **OK**

The 95% confidence interval estimate is \$37 to \$193, as described in Section 10.2. The p-value = .005 shows the null hypothesis of equal population means can be rejected at the $\alpha = .01$ level of significance. In other applications, step 5 may be modified to provide different confidence levels, different hypothesized values, and different forms of the hypotheses.

Difference Between Two Population Means with Matched Samples

WEB file
Matched

We use the data on production times in Table 10.2 to illustrate the matched-sample procedure. The completion times for method 1 are entered into column C1 and the completion times for method 2 are entered into column C2. The Minitab steps for a matched sample are as follows:

Step 1. Select the **Stat** menu
Step 2. Choose **Basic Statistics**
Step 3. Choose **Paired t**
Step 4. When the Paired t (Test and Confidence Interval) dialog box appears:
Select **Samples in columns**
Enter C1 in the **First sample** box
Enter C2 in the **Second sample** box
Select **Options**
Step 5. When the Paired t - Options dialog box appears:
Enter 95 in the **Confidence level**
Enter 0 in the **Test mean** box
Enter not equal in the **Alternative** box
Click **OK**
Step 6. Click **OK**

The 95% confidence interval estimate is −.05 to .65, as described in Section 10.3. The p-value = .08 shows that the null hypothesis of no difference in completion times cannot be rejected at $\alpha = .05$. Step 5 may be modified to provide different confidence levels, different hypothesized values, and different forms of the hypothesis.

Difference Between Two Population Proportions

WEB file
TaxPrep

We will use the data on tax preparation errors presented in Section 10.4. The sample results for 250 tax returns prepared at Office 1 are in column C_1 and the sample results for 300 tax returns prepared at Office 2 are in column C_2. Yes denotes an error was found in the tax return and No indicates no error was found. The procedure we describe provides both a 95% confidence interval estimate of the difference between the two population proportions and hypothesis testing results for H_0: $p_1 - p_2 = 0$ and H_a: $p_1 - p_2 \neq 0$.

Step 1. Select the **Stat** menu
Step 2. Choose **Basic Statistics**
Step 3. Choose **2 Proportions**
Step 4. When the 2 Proportions (Test and Confidence Interval) dialog box appears:
Select **Samples in different columns**
Enter C1 in the **First** box
Enter C2 in the **Second** box
Select **Options**
Step 5. When the 2 Proportions-Options dialog box appears:
Enter 90 in the **Confidence level** box
Enter 0 in the **Test difference** box
Enter not equal in the **Alternative** box
Select **Use pooled estimate of p for test**
Click **OK**
Step 6. Click **OK**

The 90% confidence interval estimate is .005 to .095, as described in Section 10.4. The p-value = .065 shows the null hypothesis of no difference in error rates can be rejected at

$\alpha = .10$. Step 5 may be modified to provide different confidence levels, different hypothesized values, and different forms of the hypotheses.

In the tax preparation example, the data are categorical. Yes and No are used to indicate whether an error is present. In modules involving proportions, Minitab calculates proportions for the response coming second in alphabetic order. Thus, in the tax preparation example, Minitab computes the proportion of Yes responses, which is the proportion we wanted.

If Minitab's alphabetical ordering does not compute the proportion for the response of interest, we can fix it. Select any cell in the data column, go to the Minitab menu bar, and select Editor > Column > Value Order. This sequence will provide the option of entering a user-specified order. Simply make sure that the response of interest is listed second in the define-an-order box. Minitab's 2 Proportion routine will then provide the confidence interval and hypothesis testing results for the population proportion of interest.

Finally, we note that Minitab's 2 Proportion routine uses a computational procedure different from the procedure described in the text. Thus, the Minitab output can be expected to provide slightly different interval estimates and slightly different *p*-values. However, results from the two methods should be close and are expected to provide the same interpretation and conclusion.

Appendix 10.2 Inferences About Two Populations Using Excel

We describe the use of Excel to conduct hypothesis tests about the difference between two population means.* We begin with inferences about the difference between the means of two populations when the population standard deviations σ_1 and σ_2 are known.

Difference Between Two Population Means: σ_1 and σ_2 Known

WEB file
ExamScores

We will use the examination scores for the two training centers discussed in Section 10.1. The label Center A is in cell A1 and the label Center B is in cell B1. The examination scores for Center A are in cells A2:A31 and examination scores for Center B are in cells B2:B41. The population standard deviations are assumed known with $\sigma_1 = 10$ and $\sigma_2 = 10$. The Excel routine will request the input of variances which are $\sigma_1^2 = 100$ and $\sigma_2^2 = 100$. The following steps can be used to conduct a hypothesis test about the difference between the two population means.

Step 1. Click the **Data** tab on the Ribbon
Step 2. In the **Analysis** group, click **Data Analysis**
Step 3. When the Data Analysis dialog box appears:
  Choose **z-Test: Two Sample for Means**
Step 4. When the z-Test: Two Sample for Means dialog box appears:
  Enter A1:A31 in the **Variable 1 Range** box
  Enter B1:B41 in the **Variable 2 Range** box
  Enter 0 in the **Hypothesized Mean Difference** box
  Enter 100 in the **Variable 1 Variance (known)** box
  Enter 100 in the **Variable 2 Variance (known)** box
  Select **Labels**
  Enter .05 in the **Alpha** box
  Select **Output Range** and enter C1 in the box
  Click **OK**

---

*Excel's data analysis tools provide hypothesis testing procedures for the difference between two population means. No routines are available for interval estimation of the difference between two population means nor for inferences about the difference between two population proportions.

The two-tailed p-value is denoted P(Z<=z) two-tail. Its value of .0977 does not allow us to reject the null hypothesis at $\alpha = .05$.

Difference Between Two Population Means: σ_1 and σ_2 Unknown

WEB file
SoftwareTest

We use the data for the software testing study in Table 10.1. The data are already entered into an Excel worksheet with the label Current in cell A1 and the label New in cell B1. The completion times for the current technology are in cells A2:A13, and the completion times for the new software are in cells B2:B13. The following steps can be used to conduct a hypothesis test about the difference between two population means with σ_1 and σ_2 unknown.

Step 1. Click the **Data** tab on the Ribbon
Step 2. In the **Analysis** group, click **Data Analysis**
Step 3. When the Data Analysis dialog box appears:
Choose **t-Test: Two Sample Assuming Unequal Variances**
Step 4. When the t-Test: Two Sample Assuming Unequal Variances dialog box appears:
Enter A1:A13 in the **Variable 1 Range** box
Enter B1:B13 in the **Variable 2 Range** box
Enter 0 in the **Hypothesized Mean Difference** box
Select **Labels**
Enter .05 in the **Alpha** box
Select **Output Range** and enter C1 in the box
Click **OK**

The appropriate p-value is denoted P(T<=t) one-tail. Its value of .017 allows us to reject the null hypothesis at $\alpha = .05$.

Difference Between Two Population Means with Matched Samples

WEB file
Matched

We use the matched-sample completion times in Table 10.2 to illustrate. The data are entered into a worksheet with the label Method 1 in cell A1 and the label Method 2 in cell B2. The completion times for method 1 are in cells A2:A7 and the completion times for method 2 are in cells B2:B7. The Excel procedure uses the steps previously described for the t-Test except the user chooses the **t-Test: Paired Two Sample for Means** data analysis tool in step 3. The variable 1 range is A1:A7 and the variable 2 range is B1:B7.

The appropriate p-value is denoted P(T<=t) two-tail. Its value of .08 does not allow us to reject the null hypothesis at $\alpha = .05$.

Appendix 10.3 Inferences About Two Populations Using StatTools

In this appendix we show how StatTools can be used to develop interval estimates and conduct hypothesis tests about the difference between two population means for the σ_1 and σ_2 unknown case.

Interval Estimation of μ_1 and μ_2

WEB file
CheckAcct

We will use the data for the checking account balances example presented in Section 10.2. Begin by using the Data Set Manager to create a StatTools data set for these data using the procedure described in the appendix in Chapter 1. The following steps can be used to compute a 95% confidence interval estimate of the difference between the two population means.

Step 1. Click the **StatTools** tab on the Ribbon
Step 2. In the **Analyses** group, click **Statistical Inference**
Step 3. Select the **Confidence Interval** option
Step 4. Choose Mean/Std. Deviation
Step 5. When the StatTools-Confidence Interval for Mean/Std. Deviation dialog box appears:
For **Analysis Type**, choose **Two-Sample Analysis**
In the **Variables** section,
Select **Cherry Grove**
Select **Beechmont**
In the **Confidence Intervals to Calculate** section,
Select the **For the Difference of Means** option
Select 95% for the **Confidence Level**
Click **OK**

Because the sample size for Cherry Grove ($n_1 = 28$) differs from the sample size for Beechmont ($n_2 = 22$), StatTools will inform you of this difference after you click OK. A dialog box will appear saying "The variable Beechmont contains missing data. This analysis will ignore the missing data." Click OK. A Choose Variable Ordering dialog box then appears, indicating that the analysis will compare the difference between the Cherry Grove data set and the Beechmont data set. Click OK and the StatTools interval estimation output will appear.

Hypothesis Tests About μ_1 and μ_2

WEB file
SoftwareTest

We will use the software evaluation example and the completion time data presented in Table 10.1. Begin by using the Data Set Manager to create a StatTools data set for these data using the procedure described in the appendix in Chapter 1. The following steps can be used to test the hypothesis: $H_0: \mu_1 - \mu_2 \leq 0$ against $H_a: \mu_1 - \mu_2 > 0$.

Step 1. Click the **StatTools** tab on the Ribbon
Step 2. In the **Analyses** group, click **Statistical Inference**
Step 3. Select the **Hypothesis Test** option
Step 4. Choose Mean/Std. Deviation
Step 5. When the StatTools-Hypothesis Test for Mean/Std. Deviation dialog box appears:
For **Analysis Type**, choose **Two-Sample Analysis**
In the **Variables** section,
Select **Current**
Select **New**
In the **Hypothesis Test to Perform** section,
Select **Difference of Means**
Enter 0 in the **Null Hypothesis Value** box
Select **Greater Than Null Value (One-Tailed Test)** in the **Alternative Hypothesis Type** box
Click **OK**
When the Choose Variable Ordering dialog box appears, click **OK**

The results of the hypothesis test will then appear.

Inferences About the Difference Between Two Population Means: Matched Samples

WEB file
Matched

StatTools can be used to develop interval estimates and conduct hypothesis tests for the difference between population means for the matched samples case. We will use the matched-sample completion times in Table 10.2 to illustrate.

Begin by using the Data Set Manager to create a StatTools data set for these data using the procedure described in the appendix in Chapter 1. The following steps can be used to compute a 95% confidence interval estimate of the difference between the population mean completion times.

Step 1. Click the **StatTools** tab on the Ribbon
Step 2. In the **Analyses** group, click **Statistical Inference**
Step 3. Select the **Confidence Interval** option
Step 4. Choose Mean/Std. Deviation
Step 5. When the StatTools-Confidence Interval for Mean/Std. Deviation dialog box appears:
For **Analysis Type**, choose **Paired-Sample Analysis**
In the **Variables** section,
Select **Method 1**
Select **Method 2**
In the **Confidence Intervals to Calculate** section,
Select the **For the Difference of Means** option
Select 95% for the **Confidence Level**
If selected, remove the check in the **For the Standard Deviation box**
Click **OK**
When the Choose Variable Ordering dialog box appears, click **OK**

The confidence interval will appear.

Conducting hypothesis tests for the matched samples case is very similar to conducting hypothesis tests for the difference in two means shown previously. Choose the Hypothesis Test option in step 3. Then when the Hypothesis Test for Mean/Std. Deviation dialog box appears in step 5, describe the type of hypothesis test desired.

CHAPTER 11

Inferences About Population Variances

CONTENTS

STATISTICS *in* PRACTICE

U.S. GOVERNMENT ACCOUNTABILITY OFFICE*
WASHINGTON, D.C.

The U.S. Government Accountability Office (GAO) is an independent, nonpolitical audit organization in the legislative branch of the federal government. GAO evaluators determine the effectiveness of current and proposed federal programs. To carry out their duties, evaluators must be proficient in records review, legislative research, and statistical analysis techniques.

In one case, GAO evaluators studied a Department of Interior program established to help clean up the nation's rivers and lakes. As part of this program, federal grants were made to small cities throughout the United States. Congress asked the GAO to determine how effectively the program was operating. To do so, the GAO examined records and visited the sites of several waste treatment plants.

One objective of the GAO audit was to ensure that the effluent (treated sewage) at the plants met certain standards. Among other things, the audits reviewed sample data on the oxygen content, the pH level, and the amount of suspended solids in the effluent. A requirement of the program was that a variety of tests be taken daily at each plant and that the collected data be sent periodically to the state engineering department. The GAO's investigation of the data showed whether various characteristics of the effluent were within acceptable limits.

For example, the mean or average pH level of the effluent was examined carefully. In addition, the variance in the reported pH levels was reviewed. The following hypothesis test was conducted about the variance in pH level for the population of effluent.

$$H_0\colon \sigma^2 = \sigma_0^2$$
$$H_a\colon \sigma^2 \neq \sigma_0^2$$

In this test, σ_0^2 is the population variance in pH level expected at a properly functioning plant. In one particular plant, the null hypothesis was rejected. Further analysis showed that this plant had a variance in pH level that was significantly less than normal.

Effluent at this facility must fall within a statistically determined pH range. © ZUMA Wire Service/Alamy Limited.

The auditors visited the plant to examine the measuring equipment and to discuss their statistical findings with the plant manager. The auditors found that the measuring equipment was not being used because the operator did not know how to work it. Instead, the operator had been told by an engineer what level of pH was acceptable and had simply recorded similar values without actually conducting the test. The unusually low variance in this plant's data resulted in rejection of H_0. The GAO suspected that other plants might have similar problems and recommended an operator training program to improve the data collection aspect of the pollution control program.

In this chapter you will learn how to conduct statistical inferences about the variances of one and two populations. Two new distributions, the chi-square distribution and the F distribution, will be introduced and used to make interval estimates and hypothesis tests about population variances.

*The authors thank Mr. Art Foreman and Mr. Dale Ledman of the U.S. Government Accountability Office for providing this Statistics in Practice.

In the preceding four chapters we examined methods of statistical inference involving population means and population proportions. In this chapter we expand the discussion to situations involving inferences about population variances. As an example of a case in which a variance can provide important decision-making information, consider the production process of filling containers with a liquid detergent product. The filling mechanism for the process is adjusted so that the mean filling weight is 16 ounces per container. Although a mean of 16 ounces is desired, the variance of the filling weights is also critical.

That is, even with the filling mechanism properly adjusted for the mean of 16 ounces, we cannot expect every container to have exactly 16 ounces. By selecting a sample of containers, we can compute a sample variance for the number of ounces placed in a container. This value will serve as an estimate of the variance for the population of containers being filled by the production process. If the sample variance is modest, the production process will be continued. However, if the sample variance is excessive, overfilling and underfilling may be occurring even though the mean is correct at 16 ounces. In this case, the filling mechanism will be readjusted in an attempt to reduce the filling variance for the containers.

In many manufacturing applications, controlling the process variance is extremely important in maintaining quality.

In the first section we consider inferences about the variance of a single population. Subsequently, we will discuss procedures that can be used to make inferences about the variances of two populations.

11.1 Inferences About a Population Variance

The sample variance

$$s^2 = \frac{\Sigma(x_i - \bar{x})^2}{n - 1} \tag{11.1}$$

is the point estimator of the population variance σ^2. In using the sample variance as a basis for making inferences about a population variance, the sampling distribution of the quantity $(n - 1)s^2/\sigma^2$ is helpful. This sampling distribution is described as follows.

SAMPLING DISTRIBUTION OF $(n - 1)s^2/\sigma^2$

Whenever a simple random sample of size n is selected from a normal population, the sampling distribution of

$$\frac{(n - 1)s^2}{\sigma^2} \tag{11.2}$$

has a chi-square distribution with $n - 1$ degrees of freedom.

The chi-square distribution is based on sampling from a normal population.

Figure 11.1 shows some possible forms of the sampling distribution of $(n - 1)s^2/\sigma^2$.

Since the sampling distribution of $(n - 1)s^2/\sigma^2$ is known to have a chi-square distribution whenever a simple random sample of size n is selected from a normal population, we can use the chi-square distribution to develop interval estimates and conduct hypothesis tests about a population variance.

Interval Estimation

To show how the chi-square distribution can be used to develop a confidence interval estimate of a population variance σ^2, suppose that we are interested in estimating the population variance for the production filling process mentioned at the beginning of this chapter. A sample of 20 containers is taken, and the sample variance for the filling quantities is found to be $s^2 = .0025$. However, we know we cannot expect the variance of a sample of 20 containers to provide the exact value of the variance for the population of containers filled by the production process. Hence, our interest will be in developing an interval estimate for the population variance.

FIGURE 11.1 EXAMPLES OF THE SAMPLING DISTRIBUTION OF $(n - 1)s^2/\sigma^2$ (A CHI-SQUARE DISTRIBUTION)

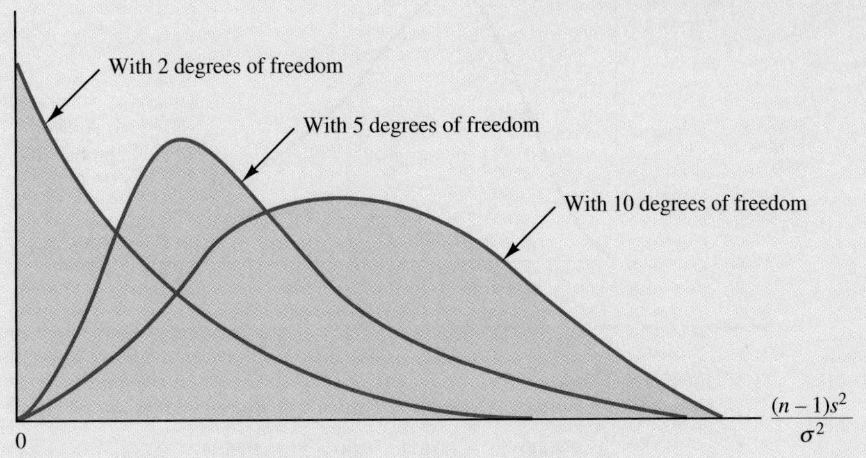

We will use the notation χ^2_α to denote the value for the chi-square distribution that provides an area or probability of α to the *right* of the χ^2_α value. For example, in Figure 11.2 the chi-square distribution with 19 degrees of freedom is shown with $\chi^2_{.025} = 32.852$ indicating that 2.5% of the chi-square values are to the right of 32.852, and $\chi^2_{.975} = 8.907$ indicating that 97.5% of the chi-square values are to the right of 8.907. Tables of areas or probabilities are readily available for the chi-square distribution. Refer to Table 11.1 and verify that these chi-square values with 19 degrees of freedom (19th row of the table) are correct. Table 3 of Appendix B provides a more extensive table of chi-square values.

From the graph in Figure 11.2 we see that .95, or 95%, of the chi-square values are between $\chi^2_{.975}$ and $\chi^2_{.025}$. That is, there is a .95 probability of obtaining a χ^2 value such that

$$\chi^2_{.975} \leq \chi^2 \leq \chi^2_{.025}$$

FIGURE 11.2 A CHI-SQUARE DISTRIBUTION WITH 19 DEGREES OF FREEDOM

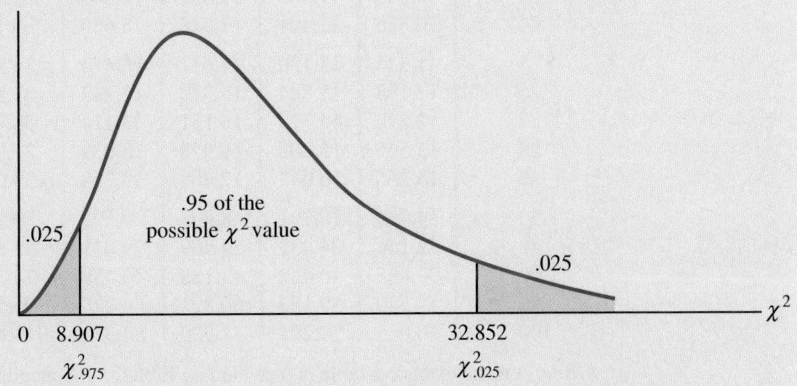

TABLE 11.1 SELECTED VALUES FROM THE CHI-SQUARE DISTRIBUTION TABLE*

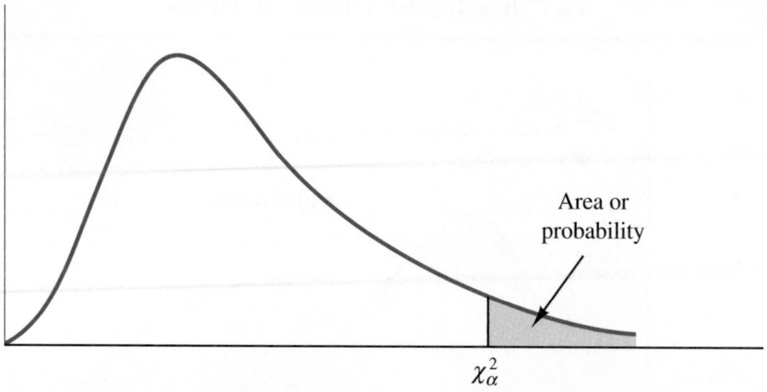

Degrees of Freedom	Area in Upper Tail .99	.975	.95	.90	.10	.05	.025	.01
1	.000	.001	.004	.016	2.706	3.841	5.024	6.635
2	.020	.051	.103	.211	4.605	5.991	7.378	9.210
3	.115	.216	.352	.584	6.251	7.815	9.348	11.345
4	.297	.484	.711	1.064	7.779	9.488	11.143	13.277
5	.554	.831	1.145	1.610	9.236	11.070	12.832	15.086
6	.872	1.237	1.635	2.204	10.645	12.592	14.449	16.812
7	1.239	1.690	2.167	2.833	12.017	14.067	16.013	18.475
8	1.647	2.180	2.733	3.490	13.362	15.507	17.535	20.090
9	2.088	2.700	3.325	4.168	14.684	16.919	19.023	21.666
10	2.558	3.247	3.940	4.865	15.987	18.307	20.483	23.209
11	3.053	3.816	4.575	5.578	17.275	19.675	21.920	24.725
12	3.571	4.404	5.226	6.304	18.549	21.026	23.337	26.217
13	4.107	5.009	5.892	7.041	19.812	22.362	24.736	27.688
14	4.660	5.629	6.571	7.790	21.064	23.685	26.119	29.141
15	5.229	6.262	7.261	8.547	22.307	24.996	27.488	30.578
16	5.812	6.908	7.962	9.312	23.542	26.296	28.845	32.000
17	6.408	7.564	8.672	10.085	24.769	27.587	30.191	33.409
18	7.015	8.231	9.390	10.865	25.989	28.869	31.526	34.805
19	7.633	8.907	10.117	11.651	27.204	30.144	32.852	36.191
20	8.260	9.591	10.851	12.443	28.412	31.410	34.170	37.566
21	8.897	10.283	11.591	13.240	29.615	32.671	35.479	38.932
22	9.542	10.982	12.338	14.041	30.813	33.924	36.781	40.289
23	10.196	11.689	13.091	14.848	32.007	35.172	38.076	41.638
24	10.856	12.401	13.848	15.659	33.196	36.415	39.364	42.980
25	11.524	13.120	14.611	16.473	34.382	37.652	40.646	44.314
26	12.198	13.844	15.379	17.292	35.563	38.885	41.923	45.642
27	12.878	14.573	16.151	18.114	36.741	40.113	43.195	46.963
28	13.565	15.308	16.928	18.939	37.916	41.337	44.461	48.278
29	14.256	16.047	17.708	19.768	39.087	42.557	45.722	49.588
30	14.953	16.791	18.493	20.599	40.256	43.773	46.979	50.892
40	22.164	24.433	26.509	29.051	51.805	55.758	59.342	63.691
60	37.485	40.482	43.188	46.459	74.397	79.082	83.298	88.379
80	53.540	57.153	60.391	64.278	96.578	101.879	106.629	112.329
100	70.065	74.222	77.929	82.358	118.498	124.342	129.561	135.807

*Note: A more extensive table is provided as Table 3 of Appendix B.

We stated in expression (11.2) that $(n - 1)s^2/\sigma^2$ follows a chi-square distribution; therefore we can substitute $(n - 1)s^2/\sigma^2$ for χ^2 and write

$$\chi^2_{.975} \leq \frac{(n-1)s^2}{\sigma^2} \leq \chi^2_{.025} \tag{11.3}$$

In effect, expression (11.3) provides an interval estimate in that .95, or 95%, of all possible values for $(n - 1)s^2/\sigma^2$ will be in the interval $\chi^2_{.975}$ to $\chi^2_{.025}$. We now need to do some algebraic manipulations with expression (11.3) to develop an interval estimate for the population variance σ^2. Working with the leftmost inequality in expression (11.3), we have

$$\chi^2_{.975} \leq \frac{(n-1)s^2}{\sigma^2}$$

Thus

$$\sigma^2\chi^2_{.975} \leq (n-1)s^2$$

or

$$\sigma^2 \leq \frac{(n-1)s^2}{\chi^2_{.975}} \tag{11.4}$$

Performing similar algebraic manipulations with the rightmost inequality in expression (11.3) gives

$$\frac{(n-1)s^2}{\chi^2_{.025}} \leq \sigma^2 \tag{11.5}$$

The results of expressions (11.4) and (11.5) can be combined to provide

$$\frac{(n-1)s^2}{\chi^2_{.025}} \leq \sigma^2 \leq \frac{(n-1)s^2}{\chi^2_{.975}} \tag{11.6}$$

Because expression (11.3) is true for 95% of the $(n - 1)s^2/\sigma^2$ values, expression (11.6) provides a 95% confidence interval estimate for the population variance σ^2.

Let us return to the problem of providing an interval estimate for the population variance of filling quantities. Recall that the sample of 20 containers provided a sample variance of $s^2 = .0025$. With a sample size of 20, we have 19 degrees of freedom. As shown in Figure 11.2, we have already determined that $\chi^2_{.975} = 8.907$ and $\chi^2_{.025} = 32.852$. Using these values in expression (11.6) provides the following interval estimate for the population variance.

$$\frac{(19)(.0025)}{32.852} \leq \sigma^2 \leq \frac{(19)(.0025)}{8.907}$$

or

$$.0014 \leq \sigma^2 \leq .0053$$

A confidence interval for a population standard deviation can be found by computing the square roots of the lower limit and upper limit of the confidence interval for the population variance.

Taking the square root of these values provides the following 95% confidence interval for the population standard deviation.

$$.0380 \leq \sigma \leq .0730$$

Thus, we illustrated the process of using the chi-square distribution to establish interval estimates of a population variance and a population standard deviation. Note specifically that because $\chi^2_{.975}$ and $\chi^2_{.025}$ were used, the interval estimate has a .95 confidence coefficient. Extending expression (11.6) to the general case of any confidence coefficient, we have the following interval estimate of a population variance.

INTERVAL ESTIMATE OF A POPULATION VARIANCE

$$\frac{(n-1)s^2}{\chi^2_{\alpha/2}} \leq \sigma^2 \leq \frac{(n-1)s^2}{\chi^2_{(1-\alpha/2)}} \tag{11.7}$$

where the χ^2 values are based on a chi-square distribution with $n - 1$ degrees of freedom and where $1 - \alpha$ is the confidence coefficient.

Hypothesis Testing

Using σ_0^2 to denote the hypothesized value for the population variance, the three forms for a hypothesis test about a population variance are as follows:

$$\begin{aligned} &H_0\colon \sigma^2 \geq \sigma_0^2 \qquad &&H_0\colon \sigma^2 \leq \sigma_0^2 \qquad &&H_0\colon \sigma^2 = \sigma_0^2 \\ &H_a\colon \sigma^2 < \sigma_0^2 &&H_a\colon \sigma^2 > \sigma_0^2 &&H_a\colon \sigma^2 \neq \sigma_0^2 \end{aligned}$$

These three forms are similar to the three forms that we used to conduct one-tailed and two-tailed hypothesis tests about population means and proportions in Chapters 9 and 10.

The procedure for conducting a hypothesis test about a population variance uses the hypothesized value for the population variance σ_0^2 and the sample variance s^2 to compute the value of a χ^2 test statistic. Assuming that the population has a normal distribution, the test statistic is as follows:

TEST STATISTIC FOR HYPOTHESIS TESTS ABOUT A POPULATION VARIANCE

$$\chi^2 = \frac{(n-1)s^2}{\sigma_0^2} \tag{11.8}$$

where χ^2 has a chi-square distribution with $n - 1$ degrees of freedom.

After computing the value of the χ^2 test statistic, either the p-value approach or the critical value approach may be used to determine whether the null hypothesis can be rejected.

Let us consider the following example. The St. Louis Metro Bus Company wants to promote an image of reliability by encouraging its drivers to maintain consistent schedules. As a standard policy the company would like arrival times at bus stops to have low variability. In terms of the variance of arrival times, the company standard specifies an arrival time variance of 4 or less when arrival times are measured in minutes. The following hypothesis test is formulated to help the company determine whether the arrival time population variance is excessive.

$$\begin{aligned} &H_0\colon \sigma^2 \leq 4 \\ &H_a\colon \sigma^2 > 4 \end{aligned}$$

In tentatively assuming H_0 is true, we are assuming that the population variance of arrival times is within the company guideline. We reject H_0 if the sample evidence indicates that the population variance exceeds the guideline. In this case, follow-up steps should be taken to reduce the population variance. We conduct the hypothesis test using a level of significance of $\alpha = .05$.

WEB file

BusTimes

Suppose that a random sample of 24 bus arrivals taken at a downtown intersection provides a sample variance of $s^2 = 4.9$. Assuming that the population distribution of arrival times is approximately normal, the value of the test statistic is as follows.

$$\chi^2 = \frac{(n-1)s^2}{\sigma_0^2} = \frac{(24-1)(4.9)}{4} = 28.18$$

The chi-square distribution with $n - 1 = 24 - 1 = 23$ degrees of freedom is shown in Figure 11.3. Because this is an upper tail test, the area under the curve to the right of the test statistic $\chi^2 = 28.18$ is the p-value for the test.

Like the t distribution table, the chi-square distribution table does not contain sufficient detail to enable us to determine the p-value exactly. However, we can use the chi-square distribution table to obtain a range for the p-value. For example, using Table 11.1, we find the following information for a chi-square distribution with 23 degrees of freedom.

Area in Upper Tail	**.10**	**.05**	**.025**	**.01**
χ^2 Value (23 *df*)	32.007	35.172	38.076	41.638

$\chi^2 = 28.18$

Because $\chi^2 = 28.18$ is less than 32.007, the area in upper tail (the p-value) is greater than .10. With the p-value $> \alpha = .05$, we cannot reject the null hypothesis. The sample does not support the conclusion that the population variance of the arrival times is excessive.

Because of the difficulty of determining the exact p-value directly from the chi-square distribution table, a computer software package such as Minitab or Excel is helpful. Appendix F, at the back of the book, describes how to compute p-values. In the appendix, we show that the exact p-value corresponding to $\chi^2 = 28.18$ is .2091.

As with other hypothesis testing procedures, the critical value approach can also be used to draw the hypothesis testing conclusion. With $\alpha = .05$, $\chi^2_{.05}$ provides the critical value for

FIGURE 11.3 CHI-SQUARE DISTRIBUTION FOR THE ST. LOUIS METRO BUS EXAMPLE

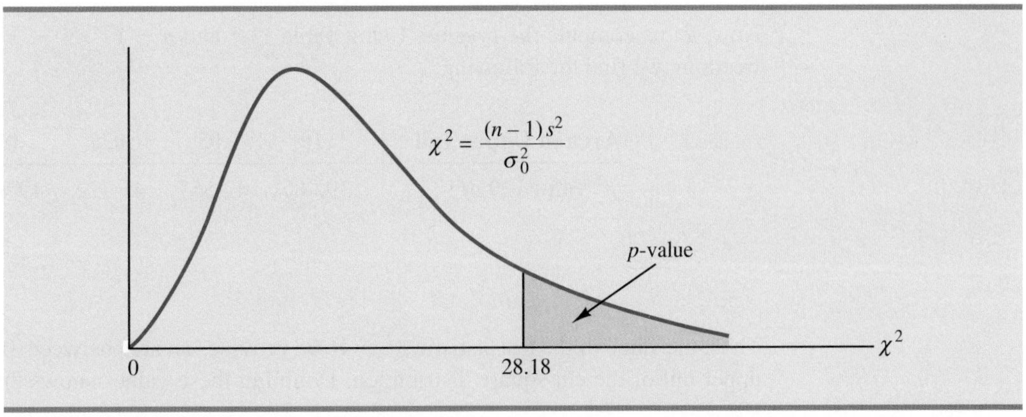

the upper tail hypothesis test. Using Table 11.1 and 23 degrees of freedom, $\chi^2_{.05} = 35.172$. Thus, the rejection rule for the bus arrival time example is as follows:

$$\text{Reject } H_0 \text{ if } \chi^2 \geq 35.172$$

Because the value of the test statistic is $\chi^2 = 28.18$, we cannot reject the null hypothesis.

In practice, upper tail tests as presented here are the most frequently encountered tests about a population variance. In situations involving arrival times, production times, filling weights, part dimensions, and so on, low variances are desirable, whereas large variances are unacceptable. With a statement about the maximum allowable population variance, we can test the null hypothesis that the population variance is less than or equal to the maximum allowable value against the alternative hypothesis that the population variance is greater than the maximum allowable value. With this test structure, corrective action will be taken whenever rejection of the null hypothesis indicates the presence of an excessive population variance.

As we saw with population means and proportions, other forms of hypothesis tests can be developed. Let us demonstrate a two-tailed test about a population variance by considering a situation faced by a bureau of motor vehicles. Historically, the variance in test scores for individuals applying for driver's licenses has been $\sigma^2 = 100$. A new examination with new test questions has been developed. Administrators of the bureau of motor vehicles would like the variance in the test scores for the new examination to remain at the historical level. To evaluate the variance in the new examination test scores, the following two-tailed hypothesis test has been proposed.

$$H_0\colon \sigma^2 = 100$$
$$H_a\colon \sigma^2 \neq 100$$

Rejection of H_0 will indicate that a change in the variance has occurred and suggest that some questions in the new examination may need revision to make the variance of the new test scores similar to the variance of the old test scores. A sample of 30 applicants for driver's licenses will be given the new version of the examination. We will use a level of significance $\alpha = .05$ to conduct the hypothesis test.

The sample of 30 examination scores provided a sample variance $s^2 = 162$. The value of the chi-square test statistic is as follows:

$$\chi^2 = \frac{(n-1)s^2}{\sigma_0^2} = \frac{(30-1)(162)}{100} = 46.98$$

Now, let us compute the p-value. Using Table 11.1 and $n - 1 = 30 - 1 = 29$ degrees of freedom, we find the following.

Area in Upper Tail	**.10**	**.05**	**.025**	**.01**
χ^2 Value (29 *df*)	39.087	42.557	45.722	49.588

$\chi^2 = 46.98$ (falls between 45.722 and 49.588)

Thus, the value of the test statistic $\chi^2 = 46.98$ provides an area between .025 and .01 in the upper tail of the chi-square distribution. Doubling these values shows that the two-tailed

TABLE 11.2 SUMMARY OF HYPOTHESIS TESTS ABOUT A POPULATION VARIANCE

	Lower Tail Test	Upper Tail Test	Two-Tailed Test
Hypotheses	$H_0: \sigma^2 \geq \sigma_0^2$ $H_a: \sigma^2 < \sigma_0^2$	$H_0: \sigma^2 \leq \sigma_0^2$ $H_a: \sigma^2 > \sigma_0^2$	$H_0: \sigma^2 = \sigma_0^2$ $H_a: \sigma^2 \neq \sigma_0^2$
Test Statistic	$\chi^2 = \frac{(n-1)s^2}{\sigma_0^2}$	$\chi^2 = \frac{(n-1)s^2}{\sigma_0^2}$	$\chi^2 = \frac{(n-1)s^2}{\sigma_0^2}$
Rejection Rule: *p*-value Approach	Reject H_0 if p-value $\leq \alpha$	Reject H_0 if p-value $\leq \alpha$	Reject H_0 if p-value $\leq \alpha$
Rejection Rule: Critical Value Approach	Reject H_0 if $\chi^2 \leq \chi^2_{(1-\alpha)}$	Reject H_0 if $\chi^2 \geq \chi^2_{\alpha}$	Reject H_0 if $\chi^2 \leq \chi^2_{(1-\alpha/2)}$ or if $\chi^2 \geq \chi^2_{\alpha/2}$

p-value is between .05 and .02. Excel or Minitab can be used to show the exact p-value = .0374. With p-value $\leq \alpha = .05$, we reject H_0 and conclude that the new examination test scores have a population variance different from the historical variance of $\sigma^2 = 100$. A summary of the hypothesis testing procedures for a population variance is shown in Table 11.2.

Exercises

Methods

1. Find the following chi-square distribution values from Table 11.1 or Table 3 of Appendix B.
   a. $\chi^2_{.05}$ with $df = 5$
   b. $\chi^2_{.025}$ with $df = 15$
   c. $\chi^2_{.975}$ with $df = 20$
   d. $\chi^2_{.01}$ with $df = 10$
   e. $\chi^2_{.95}$ with $df = 18$

SELF test

2. A sample of 20 items provides a sample standard deviation of 5.
   a. Compute the 90% confidence interval estimate of the population variance.
   b. Compute the 95% confidence interval estimate of the population variance.
   c. Compute the 95% confidence interval estimate of the population standard deviation.
3. A sample of 16 items provides a sample standard deviation of 9.5. Test the following hypotheses using $\alpha = .05$. What is your conclusion? Use both the p-value approach and the critical value approach.

$$H_0: \sigma^2 \leq 50$$
$$H_a: \sigma^2 > 50$$

Applications

4. The variance in drug weights is critical in the pharmaceutical industry. For a specific drug, with weights measured in grams, a sample of 18 units provided a sample variance of $s^2 = .36$.
   a. Construct a 90% confidence interval estimate of the population variance for the weight of this drug.
   b. Construct a 90% confidence interval estimate of the population standard deviation.

5. John Calipari, head basketball coach for the 2012 national champion University of Kentucky Wildcats, is the highest paid coach in college basketball with an annual salary of $5.4 million (*USA Today,* March 29, 2012). The sample below shows the head basketball coach's salary for a sample of 10 schools playing NCAA Division I basketball. Salary data are in millions of dollars.

University	Coach's Salary	University	Coach's Salary
Indiana	2.2	Syracuse	1.5
Xavier	.5	Murray State	.2
Texas	2.4	Florida State	1.5
Connecticut	2.7	South Dakota State	.1
West Virginia	2.0	Vermont	.2

a. Use the sample mean for the 10 schools to estimate the population mean annual salary for head basketball coaches at colleges and universities playing NCAA Division I basketball.
b. Use the data to estimate the population standard deviation for the annual salary for head basketball coaches.
c. What is the 95% confidence interval for the population variance?
d. What is the 95% confidence interval for the population standard deviation?

WEB file
Halloween

6. Americans spend nearly $7 billion on Halloween costumes and decorations (*The Wall Street Journal,* October 27, 2011). Sample data showing the amount, in dollars, 16 adults spent on a Halloween costume are as follows.

12	69	22	64
33	36	31	44
52	16	13	98
45	32	63	26

a. What is the estimate of the population mean amount adults spend on a Halloween costume?
b. What is the sample standard deviation?
c. Provide a 95% confidence interval estimate of the population standard deviation for the amount adults spend on a Halloween costume?

7. To analyze the risk, or volatility, associated with investing in General Electric common stock, a sample of the eight quarterly percent total returns was identified as shown below (Charles Schwab website, January 2012). The percent total return includes the stock price change plus the dividend payment for the quarter.

20.0	−20.5	12.2	12.6	10.5	−5.8	−18.7	15.3

a. What is the value of the sample mean? What is its interpretation?
b. Compute the sample variance and sample standard deviation as measures of volatility for the quarterly return for General Electric?
c. Construct a 95% confidence interval for the population variance?
d. Construct a 95% confidence interval for the population standard deviation?

8. March 4, 2009, was one of the few good days for the stock market in early 2009. The Dow Jones Industrial Average went up 149.82 points (*The Wall Street Journal,* March 5, 2009). The following table shows the stock price changes for a sample of 12 companies on that day.

a. Compute the sample variance for the daily price change.

WEB file PriceChange

Company	Price Change ($)	Company	Price Change ($)
Aflac	0.81	John.&John.	1.46
Bank of Am.	−0.05	Loews Cp	0.92
Cablevision	0.41	Nokia	0.21
Diageo	1.32	SmpraEngy	0.97
Flour Cp	2.37	Sunoco	0.52
Goodrich	0.3	Tyson Food	0.12

b. Compute the sample standard deviation for the price change.
c. Provide 95% confidence interval estimates of the population variance and the population standard deviation.

SELF test

9. An automotive part must be machined to close tolerances to be acceptable to customers. Production specifications call for a maximum variance in the lengths of the parts of .0004. Suppose the sample variance for 30 parts turns out to be $s^2 = .0005$. Use $\alpha = .05$ to test whether the population variance specification is being violated.

WEB file Costco

10. *Consumer Reports* uses a 100-point customer satisfaction score to rate the nation's major chain stores. Assume that from past experience with the satisfaction rating score, a population standard deviation of $\sigma = 12$ is expected. In 2012, Costco with its 432 warehouses in 40 states was the only chain store to earn an outstanding rating for overall quality (*Consumer Reports,* March 2012). A sample of 15 Costco customer satisfaction scores follows.

95	90	83	75	95
98	80	83	82	93
86	80	94	64	62

a. What is the sample mean customer satisfaction score for Costco?
b. What is the sample variance?
c. What is the sample standard deviation?
d. Construct a hypothesis test to determine whether the population standard deviation of $\sigma = 12$ should be rejected for Costco. With a .05 level of significance, what is your conclusion?

11. At the end of 2008, the variance in the semiannual yields of overseas government bond was $\sigma^2 = .70$. A group of bond investors met at that time to discuss future trends in overseas bond yields. Some expected the variability in overseas bond yields to increase and others took the opposite view. The following table shows the semiannual yields for 12 overseas countries as of March 6, 2009 (*Barron's,* March 9, 2009).

WEB file Yields

Country	Yield (%)	Country	Yield (%)
Australia	3.98	Italy	4.51
Belgium	3.78	Japan	1.32
Canada	2.95	Netherlands	3.53
Denmark	3.55	Spain	3.90
France	3.44	Sweden	2.48
Germany	3.08	U.K.	3.76

a. Compute the mean, variance, and standard deviation of the overseas bond yields as of March 6, 2009.
b. Develop hypotheses to test whether the sample data indicate that the variance in bond yields has changed from that at the end of 2008.
c. Use $\alpha = .05$ to conduct the hypothesis test formulated in part (b). What is your conclusion?

12. A *Fortune* study found that the variance in the number of vehicles owned or leased by subscribers to *Fortune* magazine is .94. Assume a sample of 12 subscribers to another

magazine provided the following data on the number of vehicles owned or leased: 2, 1, 2, 0, 3, 2, 2, 1, 2, 1, 0, and 1.

a. Compute the sample variance in the number of vehicles owned or leased by the 12 subscribers.
b. Test the hypothesis H_0: $\sigma^2 = .94$ to determine whether the variance in the number of vehicles owned or leased by subscribers of the other magazine differs from $\sigma^2 = .94$ for *Fortune*. At a .05 level of significance, what is your conclusion?

11.2 Inferences About Two Population Variances

In some statistical applications we may want to compare the variances in product quality resulting from two different production processes, the variances in assembly times for two assembly methods, or the variances in temperatures for two heating devices. In making comparisons about the two population variances, we will be using data collected from two independent random samples, one from population 1 and another from population 2. The two sample variances s_1^2 and s_2^2 will be the basis for making inferences about the two population variances σ_1^2 and σ_2^2. Whenever the variances of two normal populations are equal ($\sigma_1^2 = \sigma_2^2$), the sampling distribution of the ratio of the two sample variances s_1^2/s_2^2 is as follows.

SAMPLING DISTRIBUTION OF s_1^2/s_2^2 WHEN $\sigma_1^2 = \sigma_2^2$

Whenever independent simple random samples of sizes n_1 and n_2 are selected from two normal populations with equal variances, the sampling distribution of

$$\frac{s_1^2}{s_2^2} \tag{11.9}$$

has an F distribution with $n_1 - 1$ degrees of freedom for the numerator and $n_2 - 1$ degrees of freedom for the denominator; s_1^2 is the sample variance for the random sample of n_1 items from population 1, and s_2^2 is the sample variance for the random sample of n_2 items from population 2.

The F distribution is based on sampling from two normal populations.

Figure 11.4 is a graph of the F distribution with 20 degrees of freedom for both the numerator and denominator. As indicated by this graph, the F distribution is not symmetric,

FIGURE 11.4 F DISTRIBUTION WITH 20 DEGREES OF FREEDOM FOR THE NUMERATOR AND 20 DEGREES OF FREEDOM FOR THE DENOMINATOR

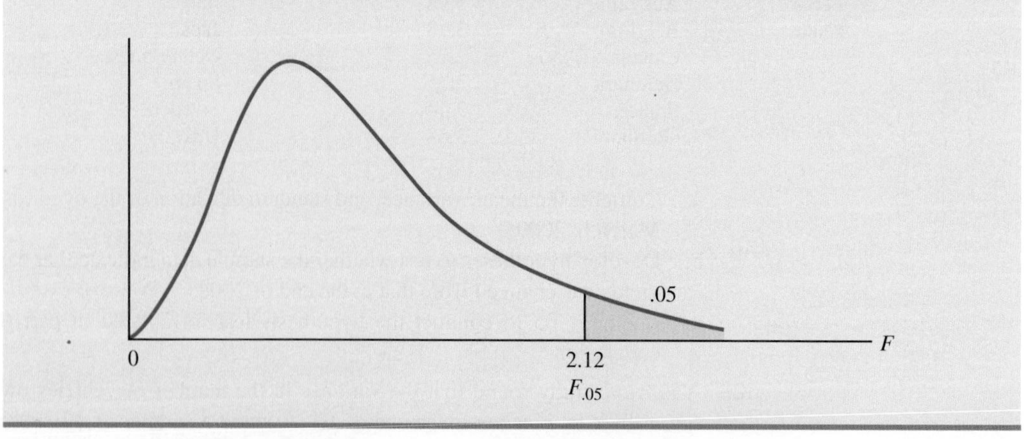

and the F values can never be negative. The shape of any particular F distribution depends on its numerator and denominator degrees of freedom.

We will use F_α to denote the value of F that provides an area or probability of α in the upper tail of the distribution. For example, as noted in Figure 11.4, $F_{.05}$ denotes the upper tail area of .05 for an F distribution with 20 degrees of freedom for the numerator and 20 degrees of freedom for the denominator. The specific value of $F_{.05}$ can be found by referring to the F distribution table, a portion of which is shown in Table 11.3. Using 20 degrees of freedom for the numerator, 20 degrees of freedom for the denominator, and the row corresponding to an area of .05 in the upper tail, we find $F_{.05} = 2.12$. Note that the

TABLE 11.3 SELECTED VALUES FROM THE *F* DISTRIBUTION TABLE*

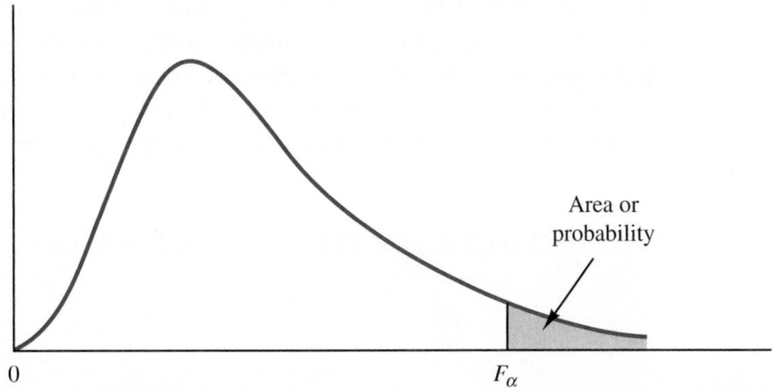

Denominator Degrees of Freedom	Area in Upper Tail	Numerator Degrees of Freedom 10	15	20	25	30
10	.10	2.32	2.24	2.20	2.17	2.16
	.05	2.98	2.85	2.77	2.73	2.70
	.025	3.72	3.52	3.42	3.35	3.31
	.01	4.85	4.56	4.41	4.31	4.25
15	.10	2.06	1.97	1.92	1.89	1.87
	.05	2.54	2.40	2.33	2.28	2.25
	.025	3.06	2.86	2.76	2.69	2.64
	.01	3.80	3.52	3.37	3.28	3.21
20	.10	1.94	1.84	1.79	1.76	1.74
	.05	2.35	2.20	2.12	2.07	2.04
	.025	2.77	2.57	2.46	2.40	2.35
	.01	3.37	3.09	2.94	2.84	2.78
25	.10	1.87	1.77	1.72	1.68	1.66
	.05	2.24	2.09	2.01	1.96	1.92
	.025	2.61	2.41	2.30	2.23	2.18
	.01	3.13	2.85	2.70	2.60	2.54
30	.10	1.82	1.72	1.67	1.63	1.61
	.05	2.16	2.01	1.93	1.88	1.84
	.025	2.51	2.31	2.20	2.12	2.07
	.01	2.98	2.70	2.55	2.45	2.39

**Note:* A more extensive table is provided as Table 4 of Appendix B.

table can be used to find F values for upper tail areas of .10, .05, .025, and .01. See Table 4 of Appendix B for a more extensive table for the F distribution.

Let us show how the F distribution can be used to conduct a hypothesis test about the variances of two populations. We begin with a test of the equality of two population variances. The hypotheses are stated as follows.

$$H_0: \sigma_1^2 = \sigma_2^2$$
$$H_a: \sigma_1^2 \neq \sigma_2^2$$

We make the tentative assumption that the population variances are equal. If H_0 is rejected, we will draw the conclusion that the population variances are not equal.

The procedure used to conduct the hypothesis test requires two independent random samples, one from each population. The two sample variances are then computed. We refer to the population providing the *larger* sample variance as population 1. Thus, a sample size of n_1 and a sample variance of s_1^2 correspond to population 1, and a sample size of n_2 and a sample variance of s_2^2 correspond to population 2. Based on the assumption that both populations have a normal distribution, the ratio of sample variances provides the following F test statistic.

TEST STATISTIC FOR HYPOTHESIS TESTS ABOUT POPULATION VARIANCES WITH $\sigma_1^2 = \sigma_2^2$

$$F = \frac{s_1^2}{s_2^2} \tag{11.10}$$

Denoting the population with the larger sample variance as population 1, the test statistic has an F distribution with $n_1 - 1$ degrees of freedom for the numerator and $n_2 - 1$ degrees of freedom for the denominator.

Because the F test statistic is constructed with the larger sample variance s_1^2 in the numerator, the value of the test statistic will be in the upper tail of the F distribution. Therefore, the F distribution table as shown in Table 11.3 and in Table 4 of Appendix B need only provide upper tail areas or probabilities. If we did not construct the test statistic in this manner, lower tail areas or probabilities would be needed. In this case, additional calculations or more extensive F distribution tables would be required. Let us now consider an example of a hypothesis test about the equality of two population variances.

Dullus County Schools is renewing its school bus service contract for the coming year and must select one of two bus companies, the Milbank Company or the Gulf Park Company. We will use the variance of the arrival or pickup/delivery times as a primary measure of the quality of the bus service. Low variance values indicate the more consistent and higher-quality service. If the variances of arrival times associated with the two services are equal, Dullus School administrators will select the company offering the better financial terms. However, if the sample data on bus arrival times for the two companies indicate a significant difference between the variances, the administrators may want to give special consideration to the company with the better or lower variance service. The appropriate hypotheses follow.

$$H_0: \sigma_1^2 = \sigma_2^2$$
$$H_a: \sigma_1^2 \neq \sigma_2^2$$

If H_0 can be rejected, the conclusion of unequal service quality is appropriate. We will use a level of significance of $\alpha = .10$ to conduct the hypothesis test.

WEB file SchoolBus

A sample of 26 arrival times for the Milbank service provides a sample variance of 48 and a sample of 16 arrival times for the Gulf Park service provides a sample variance of 20. Because the Milbank sample provided the larger sample variance, we will denote Milbank as population 1. Using equation (11.10), we find the value of the test statistic:

$$F = \frac{s_1^2}{s_2^2} = \frac{48}{20} = 2.40$$

The corresponding F distribution has $n_1 - 1 = 26 - 1 = 25$ numerator degrees of freedom and $n_2 - 1 = 16 - 1 = 15$ denominator degrees of freedom.

As with other hypothesis testing procedures, we can use the p-value approach or the critical value approach to obtain the hypothesis testing conclusion. Table 11.3 shows the following areas in the upper tail and corresponding F values for an F distribution with 25 numerator degrees of freedom and 15 denominator degrees of freedom.

Area in Upper Tail	**.10**	**.05**	**.025**	**.01**
***F* Value ($df_1 = 25, df_2 = 15$)**	1.89	2.28	2.69	3.28

$F = 2.40$ (between 2.28 and 2.69)

Because $F = 2.40$ is between 2.28 and 2.69, the area in the upper tail of the distribution is between .05 and .025. For this two-tailed test, we double the upper tail area, which results in a p-value between .10 and .05. Because we selected $\alpha = .10$ as the level of significance, the p-value $< \alpha = .10$. Thus, the null hypothesis is rejected. This finding leads to the conclusion that the two bus services differ in terms of pickup/delivery time variances. The recommendation is that the Dullus County School administrators give special consideration to the better or lower variance service offered by the Gulf Park Company.

We can use Excel or Minitab to show that the test statistic $F = 2.40$ provides a two-tailed p-value $= .0811$. With $.0811 < \alpha = .10$, the null hypothesis of equal population variances is rejected.

To use the critical value approach to conduct the two-tailed hypothesis test at the $\alpha = .10$ level of significance, we would select critical values with an area of $\alpha/2 = .10/2 = .05$ in each tail of the distribution. Because the value of the test statistic computed using equation (11.10) will always be in the upper tail, we only need to determine the upper tail critical value. From Table 11.3, we see that $F_{.05} = 2.28$. Thus, even though we use a two-tailed test, the rejection rule is stated as follows.

$$\text{Reject } H_0 \text{ if } F \geq 2.28$$

Because the test statistic $F = 2.40$ is greater than 2.28, we reject H_0 and conclude that the two bus services differ in terms of pickup/delivery time variances.

One-tailed tests involving two population variances are also possible. In this case, we use the F distribution to determine whether one population variance is significantly greater than the other. A one-tailed hypothesis test about two population variances will always be formulated as an *upper tail* test:

A one-tailed hypothesis test about two population variances can always be formulated as an upper tail test. This approach eliminates the need for lower tail F values.

$$H_0: \sigma_1^2 \leq \sigma_2^2$$
$$H_a: \sigma_1^2 > \sigma_2^2$$

This form of the hypothesis test always places the p-value and the critical value in the upper tail of the F distribution. As a result, only upper tail F values will be needed, simplifying both the computations and the table for the F distribution.

Let us demonstrate the use of the F distribution to conduct a one-tailed test about the variances of two populations by considering a public opinion survey. Samples of 31 men and 41 women will be used to study attitudes about current political issues. The researcher conducting the study wants to test to see whether the sample data indicate that women show a greater variation in attitude on political issues than men. In the form of the one-tailed hypothesis test given previously, women will be denoted as population 1 and men will be denoted as population 2. The hypothesis test will be stated as follows.

$$H_0: \sigma^2_{\text{women}} \leq \sigma^2_{\text{men}}$$
$$H_a: \sigma^2_{\text{women}} > \sigma^2_{\text{men}}$$

A rejection of H_0 gives the researcher the statistical support necessary to conclude that women show a greater variation in attitude on political issues.

With the sample variance for women in the numerator and the sample variance for men in the denominator, the F distribution will have $n_1 - 1 = 41 - 1 = 40$ numerator degrees of freedom and $n_2 - 1 = 31 - 1 = 30$ denominator degrees of freedom. We will use a level of significance $\alpha = .05$ to conduct the hypothesis test. The survey results provide a sample variance of $s_1^2 = 120$ for women and a sample variance of $s_2^2 = 80$ for men. The test statistic is as follows.

$$F = \frac{s_1^2}{s_2^2} = \frac{120}{80} = 1.50$$

Referring to Table 4 in Appendix B, we find that an F distribution with 40 numerator degrees of freedom and 30 denominator degrees of freedom has $F_{.10} = 1.57$. Because the test statistic $F = 1.50$ is less than 1.57, the area in the upper tail must be greater than .10. Thus, we can conclude that the p-value is greater than .10. Using Excel or Minitab provides a p-value $= .1256$. Because the p-value $> \alpha = .05$, H_0 cannot be rejected. Hence, the sample results do not support the conclusion that women show greater variation in attitude on political issues than men. Table 11.4 provides a summary of hypothesis tests about two population variances.

TABLE 11.4 SUMMARY OF HYPOTHESIS TESTS ABOUT TWO POPULATION VARIANCES

	Upper Tail Test	**Two-Tailed Test**
Hypotheses	$H_0: \sigma_1^2 \leq \sigma_2^2$ $H_a: \sigma_1^2 > \sigma_2^2$	$H_0: \sigma_1^2 = \sigma_2^2$ $H_a: \sigma_1^2 \neq \sigma_2^2$ Note: Population 1 has the larger sample variance
Test Statistic	$F = \frac{s_1^2}{s_2^2}$	$F = \frac{s_1^2}{s_2^2}$
Rejection Rule: p-value	Reject H_0 if p-value $\leq \alpha$	Reject H_0 if p-value $\leq \alpha$
Rejection Rule: Critical Value Approach	Reject H_0 if $F \geq F_\alpha$	Reject H_0 if $F \geq F_{\alpha/2}$

NOTES AND COMMENTS

Research confirms the fact that the F distribution is sensitive to the assumption of normal populations. The F distribution should not be used unless it is reasonable to assume that both populations are at least approximately normally distributed.

Exercises

Methods

13. Find the following F distribution values from Table 4 of Appendix B.
    a. $F_{.05}$ with degrees of freedom 5 and 10
    b. $F_{.025}$ with degrees of freedom 20 and 15
    c. $F_{.01}$ with degrees of freedom 8 and 12
    d. $F_{.10}$ with degrees of freedom 10 and 20
14. A sample of 16 items from population 1 has a sample variance $s_1^2 = 5.8$ and a sample of 21 items from population 2 has a sample variance $s_2^2 = 2.4$. Test the following hypotheses at the .05 level of significance.

$$H_0: \sigma_1^2 \leq \sigma_2^2$$
$$H_a: \sigma_1^2 > \sigma_2^2$$

    a. What is your conclusion using the p-value approach?
    b. Repeat the test using the critical value approach.

SELF test

15. Consider the following hypothesis test.

$$H_0: \sigma_1^2 = \sigma_2^2$$
$$H_a: \sigma_1^2 \neq \sigma_2^2$$

    a. What is your conclusion if $n_1 = 21$, $s_1^2 = 8.2$, $n_2 = 26$, and $s_2^2 = 4.0$? Use $\alpha = .05$ and the p-value approach.
    b. Repeat the test using the critical value approach.

Applications

16. Investors commonly use the standard deviation of the monthly percentage return for a mutual fund as a measure of the risk for the fund; in such cases, a fund that has a larger standard deviation is considered more risky than a fund with a lower standard deviation. The standard deviation for the American Century Equity Growth fund and the standard deviation for the Fidelity Growth Discovery fund were recently reported to be 15.0% and 18.9%, respectively (*The Top Mutual Funds*, AAII, 2009). Assume that each of these standard deviations is based on a sample of 60 months of returns. Do the sample results support the conclusion that the Fidelity fund has a larger population variance than the American Century fund? Which fund is more risky?

SELF test

17. Most individuals are aware of the fact that the average annual repair cost for an automobile depends on the age of the automobile. A researcher is interested in finding out whether the variance of the annual repair costs also increases with the age of the automobile. A sample of 26 automobiles 4 years old showed a sample standard deviation for annual repair costs of \$170 and a sample of 25 automobiles 2 years old showed a sample standard deviation for annual repair costs of \$100.
    a. State the null and alternative versions of the research hypothesis that the variance in annual repair costs is larger for the older automobiles.
    b. At a .01 level of significance, what is your conclusion? What is the p-value? Discuss the reasonableness of your findings.

18. Data were collected on the top 1000 financial advisers by *Barron's* (*Barron's*, February 9, 2009). Merrill Lynch had 239 people on the list and Morgan Stanley had 121 people on the list. A sample of 16 of the Merrill Lynch advisers and 10 of the Morgan Stanley advisers showed that the advisers managed many very large accounts with a large variance in the total amount of funds managed. The standard deviation of the amount managed by the Merrill Lynch advisers was $s_1 = \$587$ million. The standard deviation of the amount managed by the Morgan Stanley advisers was $s_2 = \$489$ million. Conduct a hypothesis test at $\alpha = .10$ to determine if there is a significant difference in the population variances for the amounts managed by the two companies. What is your conclusion about the variability in the amount of funds managed by advisers from the two firms?

19. The variance in a production process is an important measure of the quality of the process. A large variance often signals an opportunity for improvement in the process by finding ways to reduce the process variance. Conduct a statistical test to determine whether there is a significant difference between the variances in the bag weights for two machines. Use a .05 level of significance. What is your conclusion? Which machine, if either, provides the greater opportunity for quality improvements?

WEB file Bags

Machine 1	2.95	3.45	3.50	3.75	3.48	3.26	3.33	3.20
	3.16	3.20	3.22	3.38	3.90	3.36	3.25	3.28
	3.20	3.22	2.98	3.45	3.70	3.34	3.18	3.35
	3.12							
Machine 2	3.22	3.30	3.34	3.28	3.29	3.25	3.30	3.27
	3.38	3.34	3.35	3.19	3.35	3.05	3.36	3.28
	3.30	3.28	3.30	3.20	3.16	3.33		

20. On the basis of data provided by a Romac salary survey, the variance in annual salaries for seniors in public accounting firms is approximately 2.1 and the variance in annual salaries for managers in public accounting firms is approximately 11.1. The salary data were provided in thousands of dollars. Assuming that the salary data were based on samples of 25 seniors and 26 managers, test the hypothesis that the population variances in the salaries are equal. At a .05 level of significance, what is your conclusion?

21. Many smartphones, especially those of the LTE-enabled persuasion, have earned a bad rap for exceptionally poor battery life. Battery life between charges for the Motorola Droid Razr Max averages 20 hours when the primary use is talk time and 7 hours when the primary use is Internet applications (*The Wall Street Journal,* March 7, 2012). Since the mean hours for talk time usage is greater than the mean hours for Internet usage, the question was raised as to whether the variance in hours of usage is also greater when the primary use is talk time. Sample data showing battery hours of use for the two applications follows.

WEB file BatteryTime

Primary Use: Talking

35.8	22.2	4.0	32.6	8.5	42.5
8.0	3.8	30.0	12.8	10.3	35.5

Primary Use: Internet

14.0	12.5	16.4	1.9	9.9
5.4	1.0	15.2	4.0	4.7

a. Formulate hypotheses about the two population variances that can be used to determine if the population variance in battery hours of use is greater for the talk time application.
b. What are the standard deviations of battery hours of use for the two samples?
c. Conduct the hypothesis test and compute the p-value. Using a .05 level of significance, what is your conclusion?

22. A research hypothesis is that the variance of stopping distances of automobiles on wet pavement is substantially greater than the variance of stopping distances of automobiles on dry pavement. In the research study, 16 automobiles traveling at the same speeds are tested for stopping distances on wet pavement and then tested for stopping distances on dry pavement. On wet pavement, the standard deviation of stopping distances is 32 feet. On dry pavement, the standard deviation is 16 feet.
    a. At a .05 level of significance, do the sample data justify the conclusion that the variance in stopping distances on wet pavement is greater than the variance in stopping distances on dry pavement? What is the *p*-value?
    b. What are the implications of your statistical conclusions in terms of driving safety recommendations?

Summary

In this chapter we presented statistical procedures that can be used to make inferences about population variances. In the process we introduced two new probability distributions: the chi-square distribution and the F distribution. The chi-square distribution can be used as the basis for interval estimation and hypothesis tests about the variance of a normal population.

We illustrated the use of the F distribution in hypothesis tests about the variances of two normal populations. In particular, we showed that with independent simple random samples of sizes n_1 and n_2 selected from two normal populations with equal variances $\sigma_1^2 = \sigma_2^2$, the sampling distribution of the ratio of the two sample variances s_1^2/s_2^2 has an F distribution with $n_1 - 1$ degrees of freedom for the numerator and $n_2 - 1$ degrees of freedom for the denominator.

Key Formulas

Interval Estimate of a Population Variance

$$\frac{(n-1)s^2}{\chi^2_{\alpha/2}} \le \sigma^2 \le \frac{(n-1)s^2}{\chi^2_{(1-\alpha/2)}} \tag{11.7}$$

Test Statistic for Hypothesis Tests About a Population Variance

$$\chi^2 = \frac{(n-1)s^2}{\sigma_0^2} \tag{11.8}$$

Test Statistic for Hypothesis Tests About Population Variances with $\sigma_1^2 = \sigma_2^2$

$$F = \frac{s_1^2}{s_2^2} \tag{11.10}$$

Supplementary Exercises

23. Because of staffing decisions, managers of the Gibson-Marimont Hotel are interested in the variability in the number of rooms occupied per day during a particular season of the year. A sample of 20 days of operation shows a sample mean of 290 rooms occupied per day and a sample standard deviation of 30 rooms.
    a. What is the point estimate of the population variance?
    b. Provide a 90% confidence interval estimate of the population variance.
    c. Provide a 90% confidence interval estimate of the population standard deviation.

24. Initial public offerings (IPOs) of stocks are on average underpriced. The standard deviation measures the dispersion, or variation, in the underpricing-overpricing indicator. A sample of 13 Canadian IPOs that were subsequently traded on the Toronto Stock Exchange had a standard deviation of 14.95. Develop a 95% confidence interval estimate of the population standard deviation for the underpricing-overpricing indicator.

25. The estimated daily living costs for an executive traveling to various major cities follow. The estimates include a single room at a four-star hotel, beverages, breakfast, taxi fares, and incidental costs.

WEB file Travel

City	Daily Living Cost ($)	City	Daily Living Cost ($)
Bangkok	242.87	Mexico City	212.00
Bogotá	260.93	Milan	284.08
Cairo	194.19	Mumbai	139.16
Dublin	260.76	Paris	436.72
Frankfurt	355.36	Rio de Janeiro	240.87
Hong Kong	346.32	Seoul	310.41
Johannesburg	165.37	Tel Aviv	223.73
Lima	250.08	Toronto	181.25
London	326.76	Warsaw	238.20
Madrid	283.56	Washington, D.C.	250.61

a. Compute the sample mean.
b. Compute the sample standard deviation.
c. Compute a 95% confidence interval for the population standard deviation.

26. Part variability is critical in the manufacturing of ball bearings. Large variances in the size of the ball bearings cause bearing failure and rapid wearout. Production standards call for a maximum variance of .0001 when the bearing sizes are measured in inches. A sample of 15 bearings shows a sample standard deviation of .014 inches.
a. Use $\alpha = .10$ to determine whether the sample indicates that the maximum acceptable variance is being exceeded.
b. Compute the 90% confidence interval estimate of the variance of the ball bearings in the population.

27. The filling variance for boxes of cereal is designed to be .02 or less. A sample of 41 boxes of cereal shows a sample standard deviation of .16 ounces. Use $\alpha = .05$ to determine whether the variance in the cereal box fillings is exceeding the design specification.

28. City Trucking, Inc., claims consistent delivery times for its routine customer deliveries. A sample of 22 truck deliveries shows a sample variance of 1.5. Test to determine whether H_0: $\sigma^2 \leq 1$ can be rejected. Use $\alpha = .10$.

29. A sample of 9 days over the past six months showed that a dentist treated the following numbers of patients: 22, 25, 20, 18, 15, 22, 24, 19, and 26. If the number of patients seen per day is normally distributed, would an analysis of these sample data reject the hypothesis that the variance in the number of patients seen per day is equal to 10? Use a .10 level of significance. What is your conclusion?

30. A sample standard deviation for the number of passengers taking a particular airline flight is 8. A 95% confidence interval estimate of the population standard deviation is 5.86 passengers to 12.62 passengers.
a. Was a sample size of 10 or 15 used in the statistical analysis?
b. Suppose the sample standard deviation of $s = 8$ was based on a sample of 25 flights. What change would you expect in the confidence interval for the population standard deviation? Compute a 95% confidence interval estimate of σ with a sample size of 25.

31. Is there any difference in the variability in golf scores for players on the LPGA Tour (the women's professional golf tour) and players on the PGA Tour (the men's professional golf tour)? A sample of 20 tournament scores from LPGA events showed a standard deviation of 2.4623 strokes, and a sample of 30 tournament scores from PGA events showed a standard deviation of 2.2118 (*Golfweek*, February 7, 2009, and March 7, 2009). Conduct a hypothesis test for equal population variances to determine if there is any statistically significant difference in the variability of golf scores for male and female professional golfers. Use $\alpha = .10$. What is your conclusion?

32. The grade point averages of 352 students who completed a college course in financial accounting have a standard deviation of .940. The grade point averages of 73 students who dropped out of the same course have a standard deviation of .797. Do the data indicate a difference between the variances of grade point averages for students who completed a financial accounting course and students who dropped out? Use a .05 level of significance. *Note:* $F_{.025}$ with 351 and 72 degrees of freedom is 1.466.

33. The accounting department analyzes the variance of the weekly unit costs reported by two production departments. A sample of 16 cost reports for each of the two departments shows cost variances of 2.3 and 5.4, respectively. Is this sample sufficient to conclude that the two production departments differ in terms of unit cost variance? Use $\alpha = .10$.

34. Two new assembly methods are tested and the variances in assembly times are reported. Use $\alpha = .10$ and test for equality of the two population variances.

	Method A	Method B
Sample Size	$n_1 = 31$	$n_2 = 25$
Sample Variation	$s_1^2 = 25$	$s_2^2 = 12$

Case Problem Air Force Training Program

An Air Force introductory course in electronics uses a personalized system of instruction whereby each student views a videotaped lecture and then is given a programmed instruction text. The students work independently with the text until they have completed the training and passed a test. Of concern is the varying pace at which the students complete this portion of their training program. Some students are able to cover the programmed instruction text relatively quickly, whereas other students work much longer with the text and require additional time to complete the course. The fast students wait until the slow students complete the introductory course before the entire group proceeds together with other aspects of their training.

A proposed alternative system involves use of computer-assisted instruction. In this method, all students view the same videotaped lecture and then each is assigned to a computer terminal for further instruction. The computer guides the student, working independently, through the self-training portion of the course.

To compare the proposed and current methods of instruction, an entering class of 122 students was assigned randomly to one of the two methods. One group of 61 students used the current programmed-text method and the other group of 61 students used the proposed computer-assisted method. The time in hours was recorded for each student in the study. The following data are provided in the data set Training.

Course Completion Times (hours) for Current Training Method

76	76	77	74	76	74	74	77	72	78	73
78	75	80	79	72	69	79	72	70	70	81
76	78	72	82	72	73	71	70	77	78	73
79	82	65	77	79	73	76	81	69	75	75
77	79	76	78	76	76	73	77	84	74	74
69	79	66	70	74	72					

WEB file

Training

Course Completion Times (hours) for Proposed Computer-Assisted Method

74	75	77	78	74	80	73	73	78	76	76
74	77	69	76	75	72	75	72	76	72	77
73	77	69	77	75	76	74	77	75	78	72
77	78	78	76	75	76	76	75	76	80	77
76	75	73	77	77	77	79	75	75	72	82
76	76	74	72	78	71					

Managerial Report

1. Use appropriate descriptive statistics to summarize the training time data for each method. What similarities or differences do you observe from the sample data?
2. Use the methods of Chapter 10 to comment on any difference between the population means for the two methods. Discuss your findings.
3. Compute the standard deviation and variance for each training method. Conduct a hypothesis test about the equality of population variances for the two training methods. Discuss your findings.
4. What conclusion can you reach about any differences between the two methods? What is your recommendation? Explain.
5. Can you suggest other data or testing that might be desirable before making a final decision on the training program to be used in the future?

Appendix 11.1 Population Variances with Minitab

Here we describe how to use Minitab to conduct a hypothesis test involving two population variances.

WEB file

SchoolBus

We will use the data for the Dullus County School bus study in Section 11.2. The arrival times for Milbank appear in column C1, and the arrival times for Gulf Park appear in column C2. The following Minitab procedure can be used to conduct the hypothesis test H_0: $\sigma_1^2 = \sigma_2^2$ and H_a: $\sigma_1^2 \neq \sigma_2^2$.

Step 1. Select the **Stat** menu
Step 2. Choose **Basic Statistics**
Step 3. Choose **2-Variances**
Step 4. When the 2-Variances (Test and Confidence Interval) dialog box appears:
Select **Samples in different columns** in the **Data** box
Enter C1 in the **First** box
Enter C2 in the **Second** box
Select **Options**

Step 5. When the 2-Variances-Options dialog box appears:
Enter 95 in the **Confidence level** box
Select **Variance 1/Variance 2** in the **Hypothesized ratio** box
Enter 1 in the **Value** box
Select **not equal** in the **Alternative** box
Click **OK**
Click **OK**

The F test results show the test statistic $F = 2.40$ and the p-value $= .081$. The procedure may be used for either one-tailed or two-tailed tests and the output provides confidence intervals for the ratio of the variances as well as the ratio of the standard deviations.

Appendix 11.2 Population Variances with Excel

Here we describe how to use Excel to conduct a hypothesis test involving two population variances.

WEB file SchoolBus

We will use the data for the Dullus County School bus study in Section 11.2. The Excel worksheet has the label Milbank in cell A1 and the label Gulf Park in cell B1. The times for the Milbank sample are in cells A2:A27 and the times for the Gulf Park sample are in cells B2:B17. The steps to conduct the hypothesis test H_0: $\sigma_1^2 = \sigma_2^2$ and H_a: $\sigma_1^2 \neq \sigma_2^2$ are as follows:

Step 1. Click the **Data** tab on the Ribbon
Step 2. In the **Analysis** group, click **Data Analysis**
Step 3. When the Data Analysis dialog box appears:
Choose **F-Test Two-Sample for Variances**
Step 4. When the F-Test Two Sample for Variances dialog box appears:
Enter A1:A27 in the **Variable 1 Range** box
Enter B1:B17 in the **Variable 2 Range** box
Select **Labels**
Enter .05 in the **Alpha** box
(*Note:* This Excel procedure uses alpha as the area in the upper tail.)
Select **Output Range** and enter C1 in the box
Click **OK**

The output P(F<=f) one-tail = .0405 is the one-tailed area associated with the test statistic $F = 2.40$. Thus, the two-tailed p-value is 2(.0405) = .081. If the hypothesis test had been a one-tailed test, the one-tailed area in the cell labeled P(F<=f) one-tail provides the information necessary to determine the p-value for the test.

Appendix 11.3 Single Population Standard Deviation with StatTools

WEB file BusTimes

In this appendix we show how StatTools can be used to conduct hypothesis tests about a population standard deviation. StatTools conducts hypothesis tests on the population standard deviation, not on the population variance directly. We use the example discussed in Section 11.1 involving bus arrival times at a downtown intersection to illustrate.

Begin by using the Data Set Manager to create a StatTools data set for the BusTimes data using the procedure described in the appendix in Chapter 1. The following steps can be used to test the hypothesis $H_0 : \sigma \leq 2$ against $H_a : \sigma > 2$.

Step 1. Click the **StatTools** tab on the Ribbon
Step 2. In the **Analyses** group, click **Statistical Inference**

Step 3. Choose the **Hypothesis Test** option
Step 4. Choose **Mean/Std. Deviation**
Step 5. When the StatTools-Hypothesis Test for Mean/Std. Deviation dialog box appears:
For **Analysis Type,** choose **One-Sample Analysis**
In the variables section, select **Times**
In the **Hypothesis Tests to Perform** section:
Remove the check mark from the **Mean** box
Select the **Standard Deviation** option
Enter 2 in the **Null Hypothesis Value** box
Select **Greater Than Null Value (One-Tailed Test)** in the **Alternative Hypothesis Type** box
Click **OK**

The results from the hypothesis test will appear. They include the *p*-value and the value of the χ^2 test statistic.

CHAPTER 12

Comparing Multiple Proportions, Test of Independence and Goodness of Fit

CONTENTS

STATISTICS *in* PRACTICE

UNITED WAY*
ROCHESTER, NEW YORK

United Way of Greater Rochester is a nonprofit organization dedicated to improving the quality of life for all people in the seven counties it serves by meeting the community's most important human care needs.

The annual United Way/Red Cross fund-raising campaign funds hundreds of programs offered by more than 200 service providers. These providers meet a wide variety of human needs—physical, mental, and social—and serve people of all ages, backgrounds, and economic means.

The United Way of Greater Rochester decided to conduct a survey to learn more about community perceptions of charities. Focus-group interviews were held with professional, service, and general worker groups to obtain preliminary information on perceptions. The information obtained was then used to help develop the questionnaire for the survey. The questionnaire was pretested, modified, and distributed to 440 individuals.

A variety of descriptive statistics, including frequency distributions and crosstabulations, were provided from the data collected. An important part of the analysis involved the use of chi-square tests of independence. One use of such statistical tests was to determine whether perceptions of administrative expenses were independent of the occupation of the respondent.

The hypotheses for the test of independence were:

H_0: Perception of United Way administrative expenses is independent of the occupation of the respondent.

H_a: Perception of United Way administrative expenses is not independent of the occupation of the respondent.

Two questions in the survey provided categorical data for the statistical test. One question obtained data on perceptions of the percentage of funds going to administrative expenses (up to 10%, 11–20%, and 21% or more). The other question asked for the occupation of the respondent.

The test of independence led to rejection of the null hypothesis and to the conclusion that perception of United Way administrative expenses is not independent of the occupation of the respondent. Actual administrative expenses were less than 9%, but 35% of the respondents perceived that administrative expenses were 21% or more. Hence, many respondents had inaccurate perceptions of administrative expenses. In this group, production-line, clerical, sales, and professional-technical employees had the more inaccurate perceptions.

The community perceptions study helped United Way of Rochester develop adjustments to its programs and fund-raising activities. In this chapter, you will learn how tests, such as described here, are conducted.

United Way programs meet the needs of children as well as adults. © Jim West/Alamy.

*The authors are indebted to Dr. Philip R. Tyler, marketing consultant to the United Way, for providing this Statistics in Practice.

In Chapters 9, 10, and 11 we introduced methods of statistical inference for hypothesis tests about the means, proportions, and variances of one and two populations. In this chapter, we introduce three additional hypothesis-testing procedures that expand our capacity for making statistical inferences about populations.

The test statistic used in conducting the hypothesis tests in this chapter is based on the chi-square (χ^2) distribution. In all cases, the data are categorical. These chi-square tests are versatile and expand hypothesis testing with the following applications.

1. Testing the equality of population proportions for three or more populations
2. Testing the independence of two categorical variables
3. Testing whether a probability distribution for a population follows a specific historical or theoretical probability distribution

We begin by considering hypothesis tests for the equality of population proportions for three or more populations.

12.1 Testing the Equality of Population Proportions for Three or More Populations

In Section 10.2 we introduced methods of statistical inference for population proportions with two populations where the hypothesis test conclusion was based on the standard normal (z) test statistic. We now show how the chi-square (χ^2) test statistic can be used to make statistical inferences about the equality of population proportions for three or more populations. Using the notation

$$p_1 = \text{population proportion for population 1}$$
$$p_2 = \text{population proportion for population 2}$$

and

$$p_k = \text{population proportion for population } k$$

the hypotheses for the equality of population proportions for $k \geq 3$ populations are as follows:

$$H_0\text{: } p_1 = p_2 = \cdots = p_k$$
$$H_a\text{: Not all population proportions are equal}$$

If the sample data and the chi-square test computations indicate H_0 cannot be rejected, we cannot detect a difference among the k population proportions. However, if the sample data and the chi-square test computations indicate H_0 can be rejected, we have the statistical evidence to conclude that not all k population proportions are equal; that is, one or more population proportions differ from the other population proportions. Further analyses can be done to conclude which population proportion or proportions are significantly different from others. Let us demonstrate this chi-square test by considering an application.

Organizations such as J.D. Power and Associates use the proportion of owners likely to repurchase a particular automobile as an indication of customer loyalty for the automobile. An automobile with a greater proportion of owners likely to repurchase is concluded to have greater customer loyalty. Suppose that in a particular study we want to compare the customer loyalty for three automobiles: Chevrolet Impala, Ford Fusion, and Honda Accord. The current owners of each of the three automobiles form the three populations for the study. The three population proportions of interest are as follows:

p_1 = proportion likely to repurchase an Impala for the population of Chevrolet Impala owners

p_2 = proportion likely to repurchase a Fusion for the population of Ford Fusion owners

p_3 = proportion likely to repurchase an Accord for the population of Honda Accord owners

TABLE 12.1 SAMPLE RESULTS OF LIKELY TO REPURCHASE FOR THREE POPULATIONS OF AUTOMOBILE OWNERS (OBSERVED FREQUENCIES)

WEB file
AutoLoyalty

		Automobile Owners			
		Chevrolet Impala	**Ford Fusion**	**Honda Accord**	**Total**
Likely to Repurchase	**Yes**	69	120	123	312
	No	56	80	52	188
	Total	125	200	175	500

The hypotheses are stated as follows:

$$H_0: p_1 = p_2 = p_3$$
$$H_a: \text{Not all population proportions are equal}$$

In studies such as these, we often use the same sample size for each population. We have chosen different sample sizes in this example to show that the chi-square test is not restricted to equal sample sizes for each of the k populations.

To conduct this hypothesis test we begin by taking a sample of owners from each of the three populations. Thus we will have a sample of Chevrolet Impala owners, a sample of Ford Fusion owners, and a sample of Honda Accord owners. Each sample provides categorical data indicating whether the respondents are likely or not likely to repurchase the automobile. The data for samples of 125 Chevrolet Impala owners, 200 Ford Fusion owners, and 175 Honda Accord owners are summarized in the tabular format shown in Table 12.1. This table has two rows for the responses Yes and No and three columns, one corresponding to each of the populations. The observed frequencies are summarized in the six cells of the table corresponding to each combination of the likely to repurchase responses and the three populations.

Using Table 12.1, we see that 69 of the 125 Chevrolet Impala owners indicated that they were likely to repurchase a Chevrolet Impala. One hundred and twenty of the 200 Ford Fusion owners and 123 of the 175 Honda Accord owners indicated that they were likely to repurchase their current automobile. Also, across all three samples, 312 of the 500 owners in the study indicated that they were likely to repurchase their current automobile. The question now is how do we analyze the data in Table 12.1 to determine if the hypothesis H_0: $p_1 = p_2 = p_3$ should be rejected?

The data in Table 12.1 are the *observed frequencies* for each of the six cells that represent the six combinations of the likely to repurchase response and the owner population. If we can determine the *expected frequencies under the assumption H_0 is true*, we can use the chi-square test statistic to determine whether there is a significant difference between the observed and expected frequencies. If a significant difference exists between the observed and expected frequencies, the hypothesis H_0 can be rejected and there is evidence that not all the population proportions are equal.

Expected frequencies for the six cells of the table are based on the following rationale. First, we assume that the null hypothesis of equal population proportions is true. Then we note that in the entire sample of 500 owners, a total of 312 owners indicated that they were likely to repurchase their current automobile. Thus, 312/500 = .624 is the overall sample proportion of owners indicating they are likely to repurchase their current automobile. If H_0: $p_1 = p_2 = p_3$ is true, .624 would be the best estimate of the proportion responding likely to repurchase for each of the automobile owner populations. So if the assumption of H_0 is true, we would expect .624 of the 125 Chevrolet Impala owners, or .624(125) = 78 owners to indicate they are likely to repurchase the Impala. Using the .624 overall sample proportion, we would expect .624(200) = 124.8 of the 200 Ford Fusion owners and .624(175) = 109.2

of the Honda Accord owners to respond that they are likely to repurchase their respective model of automobile.

Let us generalize the approach to computing expected frequencies by letting e_{ij} denote the expected frequency for the cell in row i and column j of the table. With this notation, now reconsider the expected frequency calculation for the response of likely to repurchase Yes (row 1) for Chevrolet Impala owners (column 1), that is, the expected frequency e_{11}.

Note that 312 is the total number of Yes responses (row 1 total), 175 is the total sample size for Chevrolet Impala owners (column 1 total), and 500 is the total sample size. Following the logic in the preceding paragraph, we can show

$$e_{11} = \left(\frac{\text{Row 1 Total}}{\text{Total Sample Size}}\right)(\text{Column 1 Total}) = \left(\frac{312}{500}\right)125 = (.624)125 = 78$$

Starting with the first part of the above expression, we can write

$$e_{11} = \frac{(\text{Row 1 Total})(\text{Column 1 Total})}{\text{Total Sample Size}}$$

Generalizing this expression shows that the following formula can be used to provide the expected frequencies under the assumption H_0 is true.

EXPECTED FREQUENCIES UNDER THE ASSUMPTION H_0 IS TRUE

$$e_{ij} = \frac{(\text{Row } i \text{ Total})(\text{Column } j \text{ Total})}{\text{Total Sample Size}} \tag{12.1}$$

Using equation (12.1), we see that the expected frequency of Yes responses (row 1) for Honda Accord owners (column 3) would be e_{13} = (Row 1 Total)(Column 3 Total)/(Total Sample Size) = (312)(175)/500 = 109.2. Use equation (12.1) to verify the other expected frequencies are as shown in Table 12.2.

The test procedure for comparing the observed frequencies of Table 12.1 with the expected frequencies of Table 12.2 involves the computation of the following chi-square statistic:

CHI-SQUARE TEST STATISTIC

$$\chi^2 = \sum_i \sum_j \frac{(f_{ij} - e_{ij})^2}{e_{ij}} \tag{12.2}$$

where

f_{ij} = observed frequency for the cell in row i and column j

e_{ij} = expected frequency for the cell in row i and column j under the assumption H_0 is true

Note: In a chi-square test involving the equality of k population proportions, the above test statistic has a chi-square distribution with $k - 1$ degrees of freedom provided the expected frequency is 5 *or more* for each cell.

TABLE 12.2 EXPECTED FREQUENCIES FOR LIKELY TO REPURCHASE FOR THREE POPULATIONS OF AUTOMOBILE OWNERS IF H_0 IS TRUE

		Automobile Owners			
		Chevrolet Impala	**Ford Fusion**	**Honda Accord**	**Total**
Likely to Repurchase	**Yes**	78	124.8	109.2	312
	No	47	75.2	65.8	188
	Total	125	200	175	500

Reviewing the expected frequencies in Table 12.2, we see that the expected frequency is at least five for each cell in the table. We therefore proceed with the computation of the chi-square test statistic. The calculations necessary to compute the value of the test statistic are shown in Table 12.3. In this case, we see that the value of the test statistic is $\chi^2 = 7.89$.

In order to understand whether or not $\chi^2 = 7.89$ leads us to reject H_0: $p_1 = p_2 = p_3$, you will need to understand and refer to values of the chi-square distribution. Table 12.4 shows the general shape of the chi-square distribution, but note that the shape of a specific chi-square distribution depends upon the number of degrees of freedom. The table shows the upper tail areas of .10, .05, .025, .01, and .005 for chi-square distributions with up to 15 degrees of freedom. This version of the chi-square table will enable you to conduct the hypothesis tests presented in this chapter.

Since the expected frequencies shown in Table 12.2 are based on the assumption that H_0: $p_1 = p_2 = p_3$ is true, observed frequencies, f_{ij}, that are in agreement with expected frequencies, e_{ij}, provide small values of $(f_{ij} - e_{ij})^2$ in equation (12.2). If this is the case, the value of the chi-square test statistic will be relatively small and H_0 cannot be rejected. On the other hand, if the differences between the observed and expected frequencies are *large*, values of $(f_{ij} - e_{ij})^2$ and the computed value of the test statistic will be large. In this case, the null hypothesis of equal population proportions can be rejected. Thus a chi-square test for equal population proportions will always be an upper tail test with rejection of H_0 occurring when the test statistic is in the upper tail of the chi-square distribution.

The chi-square test presented in this section is always a one-tailed test with the rejection of H_0 occurring in the upper tail of the chi-square distribution.

We can use the upper tail area of the appropriate chi-square distribution and the p-value approach to determine whether the null hypothesis can be rejected. In the automobile brand loyalty study, the three owner populations indicate that the appropriate chi-square

TABLE 12.3 COMPUTATION OF THE CHI-SQUARE TEST STATISTIC FOR THE TEST OF EQUAL POPULATION PROPORTIONS

Likely to Repurchase?	**Automobile Owner**	**Observed Frequency (f_{ij})**	**Expected Frequency (e_{ij})**	**Difference ($f_{ij} - e_{ij}$)**	**Squared Difference $(f_{ij} - e_{ij})^2$**	**Squared Difference Divided by Expected Frequency $(f_{ij} - e_{ij})^2/e_{ij}$**
Yes	Impala	69	78.0	−9.0	81.00	1.04
Yes	Fusion	120	124.8	−4.8	23.04	0.18
Yes	Accord	123	109.2	13.8	190.44	1.74
No	Impala	56	47.0	9.0	81.00	1.72
No	Fusion	80	75.2	4.8	23.04	0.31
No	Accord	52	65.8	−13.8	190.44	2.89
	Total	500	500			$\chi^2 = 7.89$

TABLE 12.4 SELECTED VALUES OF THE CHI-SQUARE DISTRIBUTION

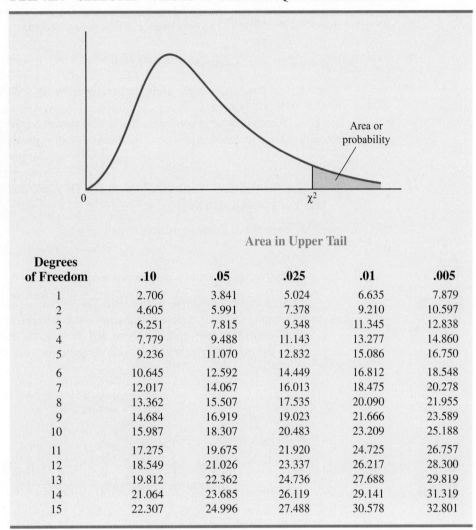

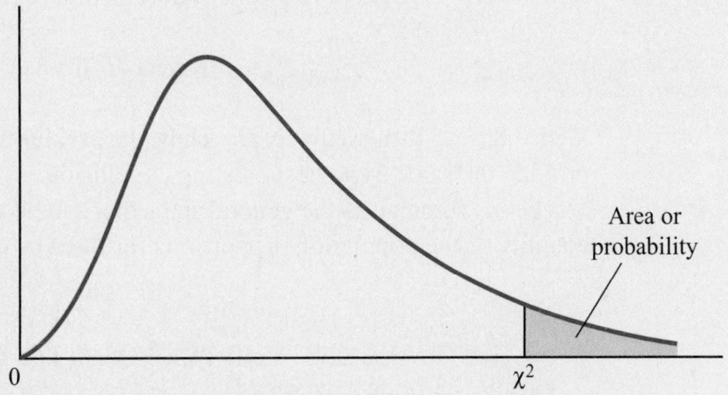

Degrees of Freedom	Area in Upper Tail .10	.05	.025	.01	.005
1	2.706	3.841	5.024	6.635	7.879
2	4.605	5.991	7.378	9.210	10.597
3	6.251	7.815	9.348	11.345	12.838
4	7.779	9.488	11.143	13.277	14.860
5	9.236	11.070	12.832	15.086	16.750
6	10.645	12.592	14.449	16.812	18.548
7	12.017	14.067	16.013	18.475	20.278
8	13.362	15.507	17.535	20.090	21.955
9	14.684	16.919	19.023	21.666	23.589
10	15.987	18.307	20.483	23.209	25.188
11	17.275	19.675	21.920	24.725	26.757
12	18.549	21.026	23.337	26.217	28.300
13	19.812	22.362	24.736	27.688	29.819
14	21.064	23.685	26.119	29.141	31.319
15	22.307	24.996	27.488	30.578	32.801

distribution has $k - 1 = 3 - 1 = 2$ degrees of freedom. Using row two of the chi-square distribution table, we have the following:

Area in Upper Tail	**.10**	**.05**	**.025**	**.01**	**.005**
χ^2 Value (2 *df*)	4.605	5.991	7.378	9.210	10.597

$\chi^2 = 7.89$ (between 7.378 and 9.210)

We see the upper tail area at $\chi^2 = 7.89$ is between .025 and .01. Thus, the corresponding upper tail area or p-value must be between .025 and .01. With p-value $\leq .05$, we reject H_0 and conclude that the three population proportions are not all equal and thus there is a difference in brand loyalties among the Chevrolet Impala, Ford Fusion, and Honda Accord owners. Minitab or Excel procedures provided in Appendix F can be used to show $\chi^2 = 7.89$ with 2 degrees of freedom yields a p-value $= .0193$.

Instead of using the p-value, we could use the critical value approach to draw the same conclusion. With $\alpha = .05$ and 2 degrees of freedom, the critical value for the chi-square test statistic is $\chi^2 = 5.991$. The upper tail rejection region becomes

$$\text{Reject } H_0 \text{ if } \chi^2 \geq 5.991$$

With $7.89 \geq 5.991$, we reject H_0. Thus, the p-value approach and the critical value approach provide the same hypothesis-testing conclusion.

Let us summarize the general steps that can be used to conduct a chi-square test for the equality of the population proportions for three or more populations.

A CHI-SQUARE TEST FOR THE EQUALITY OF POPULATION PROPORTIONS FOR $k \geq 3$ POPULATIONS

1. State the null and alternative hypotheses

   H_0: $p_1 = p_2 = \cdots = p_k$
   H_a: Not all population proportions are equal

2. Select a random sample from each of the populations and record the observed frequencies, f_{ij}, in a table with 2 rows and k columns
3. Assume the null hypothesis is true and compute the expected frequencies, e_{ij}
4. If the expected frequency, e_{ij}, is 5 or more for each cell, compute the test statistic:

$$\chi^2 = \sum_i \sum_j \frac{(f_{ij} - e_{ij})^2}{e_{ij}}$$

5. Rejection rule:

   p-value approach: Reject H_0 if p-value $\leq \alpha$
   Critical value approach: Reject H_0 if $\chi^2 \geq \chi^2_\alpha$

   where the chi-square distribution has $k - 1$ degrees of freedom and α is the level of significance for the test.

A Multiple Comparison Procedure

We have used a chi-square test to conclude that the population proportions for the three populations of automobile owners are not all equal. Thus, some differences among the population proportions exist and the study indicates that customer loyalties are not all the same for the Chevrolet Impala, Ford Fusion, and Honda Accord owners. To identify where the differences between population proportions exist, we can begin by computing the three sample proportions as follows:

Brand Loyalty Sample Proportions

Chevrolet Impala	$\bar{p}_1 = 69/125 = .5520$
Ford Fusion	$\bar{p}_2 = 120/200 = .6000$
Honda Accord	$\bar{p}_3 = 123/175 = .7029$

Since the chi-square test indicated that not all population proportions are equal, it is reasonable for us to proceed by attempting to determine where differences among the population proportions exist. For this we will rely on a multiple comparison procedure that can be used to conduct statistical tests between all pairs of population proportions. In the following, we discuss a multiple comparison procedure known as the **Marascuilo procedure**. This is a relatively straightforward procedure for making pairwise comparisons of all pairs of population proportions. We will demonstrate the computations required by this multiple comparison test procedure for the automobile customer loyalty study.

We begin by computing the absolute value of the pairwise difference between sample proportions for each pair of populations in the study. In the three-population automobile brand loyalty study we compare populations 1 and 2, populations 1 and 3, and then populations 2 and 3 using the sample proportions as follows:

Chevrolet Impala and Ford Fusion

$$|\bar{p}_1 - \bar{p}_2| = |.5520 - .6000| = .0480$$

Chevrolet Impala and Honda Accord

$$|\bar{p}_1 - \bar{p}_3| = |.5520 - .7029| = .1509$$

Ford Fusion and Honda Accord

$$|\bar{p}_2 - \bar{p}_3| = |.6000 - .7029| = .1029$$

In a second step, we select a level of significance and compute the corresponding critical value for each pairwise comparison using the following expression.

CRITICAL VALUES FOR THE MARASCUILO PAIRWISE COMPARISON PROCEDURE FOR k POPULATION PROPORTIONS

For each pairwise comparison compute a critical value as follows:

$$CV_{ij} = \sqrt{\chi^2_\alpha}\sqrt{\frac{\bar{p}_i(1 - \bar{p}_i)}{n_i} + \frac{\bar{p}_j(1 - \bar{p}_j)}{n_j}} \qquad \textbf{(12.3)}$$

where

χ^2_α = chi-square with a level of significance α and $k - 1$ degrees of freedom

$\bar{p}_i$ and $\bar{p}_j$ = sample proportions for populations i and j

n_i and n_j = sample sizes for populations i and j

Using the chi-square distribution in Table 12.4, $k - 1 = 3 - 1 = 2$ degrees of freedom, and a .05 level of significance, we have $\chi^2_{.05} = 5.991$. Now using the sample proportions $\bar{p}_1 = .5520$, $\bar{p}_2 = .6000$, and $\bar{p}_3 = .7029$, the critical values for the three pairwise comparison tests are as follows:

Chevrolet Impala and Ford Fusion

$$CV_{12} = \sqrt{5.991}\sqrt{\frac{.5520(1 - .5520)}{125} + \frac{.6000(1 - .6000)}{200}} = .1380$$

Chevrolet Impala and Honda Accord

$$CV_{13} = \sqrt{5.991}\sqrt{\frac{.5520(1 - .5520)}{125} + \frac{.7029(1 - .7029)}{175}} = .1379$$

TABLE 12.5 PAIRWISE COMPARISON TESTS FOR THE AUTOMOBILE BRAND LOYALTY STUDY

Pairwise Comparison	$\lvert\bar{p}_i - \bar{p}_j\rvert$	CV_{ij}	Significant if $\lvert\bar{p}_i - \bar{p}_j\rvert > CV_{ij}$
Chevrolet Impala vs. Ford Fusion	.0480	.1380	Not significant
Chevrolet Impala vs. Honda Accord	.1509	.1379	Significant
Ford Fusion vs. Honda Accord	.1029	.1198	Not significant

Ford Fusion and Honda Accord

$$CV_{23} = \sqrt{5.991}\sqrt{\frac{.6000(1 - .6000)}{200} + \frac{.7029(1 - .7029)}{175}} = .1198$$

If the absolute value of any pairwise sample proportion difference $\lvert\bar{p}_i - \bar{p}_j\rvert$ exceeds its corresponding critical value, CV_{ij}, the pairwise difference is significant at the .05 level of significance and we can conclude that the two corresponding population proportions are different. The final step of the pairwise comparison procedure is summarized in Table 12.5.

The conclusion from the pairwise comparison procedure is that the only significant difference in customer loyalty occurs between the Chevrolet Impala and the Honda Accord. Our sample results indicate that the Honda Accord had a greater population proportion of owners who say they are likely to repurchase the Honda Accord. Thus, we can conclude that the Honda Accord ($\bar{p}_3 = .7029$) has a greater customer loyalty than the Chevrolet Impala ($\bar{p}_1 = .5520$).

The results of the study are inconclusive as to the comparative loyalty of the Ford Fusion. While the Ford Fusion did not show significantly different results when compared to the Chevrolet Impala or Honda Accord, a larger sample may have revealed a significant difference between Ford Fusion and the other two automobiles in terms of customer loyalty. It is not uncommon for a multiple comparison procedure to show significance for some pairwise comparisons and yet not show significance for other pairwise comparisons in the study.

NOTES AND COMMENTS

1. In Chapter 10, we used the standard normal distribution and the z test statistic to conduct hypothesis tests about the proportions of two populations. However, the chi-square test introduced in this section can also be used to conduct the hypothesis test that the proportions of two populations are equal. The results will be the same under both test procedures and the value of the test statistic χ^2 will be equal to the square of the value of the test statistic z. An advantage of the methodology in Chapter 10 is that it can be used for either a one-tailed or a two-tailed hypothesis about the proportions of two populations whereas the chi-square test in this section can be used only for two-tailed tests. Exercise 12.6 will give you a chance to use the chi-square test for the hypothesis that the proportions of two populations are equal.
2. Each of the k populations in this section had two response outcomes, Yes or No. In effect, each population had a binomial distribution with parameter p the population proportion of Yes responses. An extension of the chi-square procedure in this section applies when each of the k populations has three or more possible responses. In this case, each population is said to have a multinomial distribution. The chi-square calculations for the expected frequencies, e_{ij}, and the test statistic, χ^2, are the same as shown in expressions (12.1) and (12.2). The only difference is that the null hypothesis assumes that the multinomial distribution for the response variable is the same for all populations. With r responses for each of the k populations, the chi-square test statistic has $(r - 1)(k - 1)$ degrees of freedom. Exercise 12.8 will give you a chance to use the chi-square test to compare three populations with multinomial distributions.

Exercises

Methods

SELF test

1. Use the sample data below to test the hypotheses

$$H_0: p_1 = p_2 = p_3$$
$$H_a: \text{Not all population proportions are equal}$$

where p_i is the population proportion of Yes responses for population i. Using a .05 level of significance, what is the p-value and what is your conclusion?

	Populations		
Response	**1**	**2**	**3**
Yes	150	150	96
No	100	150	104

SELF test

2. Reconsider the observed frequencies in exercise 1
   a. Compute the sample proportion for each population.
   b. Use the multiple comparison procedure to determine which population proportions differ significantly. Use a .05 level of significance.

Applications

3. The sample data below represent the number of late and on time flights for Delta, United, and US Airways (*Bureau of Transportation Statistics,* March 2012).

	Airline		
Flight	**Delta**	**United**	**US Airways**
Late	39	51	56
On Time	261	249	344

   a. Formulate the hypotheses for a test that will determine if the population proportion of late flights is the same for all three airlines.
   b. Conduct the hypothesis test with a .05 level of significance. What is the p-value and what is your conclusion?
   c. Compute the sample proportion of late flights for each airline. What is the overall proportion of late flights for the three airlines?

SELF test

4. Benson Manufacturing is considering ordering electronic components from three different suppliers. The suppliers may differ in terms of quality in that the proportion or percentage of defective components may differ among the suppliers. To evaluate the proportion of defective components for the suppliers, Benson has requested a sample shipment of 500 components from each supplier. The number of defective components and the number of good components found in each shipment are as follows.

	Supplier		
Component	**A**	**B**	**C**
Defective	15	20	40
Good	485	480	460

a. Formulate the hypotheses that can be used to test for equal proportions of defective components provided by the three suppliers.
b. Using a .05 level of significance, conduct the hypothesis test. What is the p-value and what is your conclusion?
c. Conduct a multiple comparison test to determine if there is an overall best supplier or if one supplier can be eliminated because of poor quality.

5. Kate Sanders, a researcher in the department of biology at IPFW University, studied the effect of agriculture contaminants on the stream fish population in Northeastern Indiana (April 2012). Specially designed traps collected samples of fish at each of four stream locations. A research question was, Did the differences in agricultural contaminants found at the four locations alter the proportion of the fish population by gender? Observed frequencies were as follows.

	Stream Locations			
Gender	A	B	C	D
Male	49	44	49	39
Female	41	46	36	44

a. Focusing on the proportion of male fish at each location, test the hypothesis that the population proportions are equal for all four locations. Use a .05 level of significance. What is the p-value and what is your conclusion?
b. Does it appear that differences in agricultural contaminants found at the four locations altered the fish population by gender?

Exercise 6 shows a chi-square test can be used when the hypothesis is about the equality of two population proportions.

6. A tax preparation firm is interested in comparing the quality of work at two of its regional offices. The observed frequencies showing the number of sampled returns with errors and the number of sampled returns that were correct are as follows.

	Regional Office	
Return	**Office 1**	**Office 2**
Error	35	27
Correct	215	273

a. What are the sample proportions of returns with errors at the two offices?
b. Use the chi-square test procedure to see if there is a significant difference between the population proportion of error rates for the two offices. Test the null hypothesis H_0: $p_1 = p_2$ with a .10 level of significance. What is the p-value and what is your conclusion? *Note*: We generally use the chi-square test of equal proportions when there are three or more populations, but this example shows that the same chi-square test can be used for testing equal proportions with two populations.
c. In the Section 10.2, a z test was used to conduct the above test. Either a χ^2 test statistic or a z test statistic may be used to test the hypothesis. However, when we want to make inferences about the proportions for two populations, we generally prefer the z test statistic procedure. Refer to the Notes and Comments at the end of this section and comment on why the z test statistic provides the user with more options for inferences about the proportions of two populations.

7. Social networking is becoming more and more popular around the world. Pew Research Center used a survey of adults in several countries to determine the percentage of adults who use social networking sites (*USA Today,* February 8, 2012). Assume that the results for surveys in Great Britain, Israel, Russia, and United States are as follows.

Use Social Networking Sites	Country			
	Great Britain	Israel	Russia	United States
Yes	344	265	301	500
No	456	235	399	500

a. Conduct a hypothesis test to determine whether the proportion of adults using social networking sites is equal for all four countries. What is the p-value? Using a .05 level of significance, what is your conclusion?

b. What are the sample proportions for each of the four countries? Which country has the largest proportion of adults using social networking sites?

c. Using a .05 level of significance, conduct multiple pairwise comparison tests among the four countries. What is your conclusion?

Exercise 8 shows a chi-square test can also be used for multiple population tests when the categorical response variable has three or more outcomes.

8. A manufacturer is considering purchasing parts from three different suppliers. The parts received from the suppliers are classified as having a minor defect, having a major defect, or being good. Test results from samples of parts received from each of the three suppliers are shown below. Note that any test with these data is no longer a test of proportions for the three supplier populations because the categorical response variable has three outcomes: minor defect, major defect, and good.

Part Tested	Supplier		
	A	B	C
Minor Defect	15	13	21
Major Defect	5	11	5
Good	130	126	124

Using the data above, conduct a hypothesis test to determine if the distribution of defects is the same for the three suppliers. Use the chi-square test calculations as presented in this section with the exception that a table with r rows and c columns results in a chi-square test statistic with $(r - 1)(c - 1)$ degrees of freedom. Using a .05 level of significance, what is the p-value and what is your conclusion.

12.2 Test of Independence

An important application of a chi-square test involves using sample data to test for the independence of two categorical variables. For this test we take one sample from a population and record the observations for two categorical variables. We will summarize the data by counting the number of responses for each combination of a category for variable 1 and a category for variable 2. The null hypothesis for this test is that the two categorical variables are independent. Thus, the test is referred to as a **test of independence**. We will illustrate this test with the following example.

A beer industry association conducts a survey to determine the preferences of beer drinkers for light, regular, and dark beers. A sample of 200 beer drinkers is taken with each person in the sample asked to indicate a preference for one of the three types of beers: light, regular, or dark. At the end of the survey questionnaire, the respondent is asked to provide information on a variety of demographics including gender: male or female. A research question of interest to the association is whether preference for the three types of beer is independent of the gender of the beer drinker. If the two categorical variables, beer preference and gender, are independent, beer preference does not depend on gender and the preference for light, regular, and dark beer can be expected to be the same for male and female beer drinkers. However, if the test conclusion is that the two categorical variables are not independent, we have evidence that beer preference is associated or dependent upon the gender of the beer drinker. As a result, we can expect beer preferences to differ for male and female beer drinkers. In this case, a beer manufacturer could use this information to customize its promotions and advertising for the different target markets of male and female beer drinkers.

The hypotheses for this test of independence are as follows:

H_0: Beer preference is independent of gender

H_a: Beer preference is not independent of gender

The sample data will be summarized in a two-way table with beer preferences of light, regular, and dark as one of the variables and gender of male and female as the other variable. Since an objective of the study is to determine if there is difference between the beer preferences for male and female beer drinkers, we consider gender an explanatory variable and follow the usual practice of making the explanatory variable the column variable in the data tabulation table. The beer preference is the categorical response variable and is shown as the row variable. The sample results of the 200 beer drinkers in the study are summarized in Table 12.6.

The sample data are summarized based on the combination of beer preference and gender for the individual respondents. For example, 51 individuals in the study were males who preferred light beer, 56 individuals in the study were males who preferred regular beer, and so on. Let us now analyze the data in the table and test for independence of beer preference and gender.

First of all, since we selected a sample of beer drinkers, summarizing the data for each variable separately will provide some insights into the characteristics of the beer drinker population. For the categorical variable gender, we see 132 of the 200 in the sample were male. This gives us the estimate that 132/200 = .66, or 66%, of the beer drinker population is male. Similarly we estimate that 68/200 = .34, or 34%, of the beer drinker population is female. Thus male beer drinkers appear to outnumber female beer drinkers approximately 2 to 1. Sample proportions or percentages for the three types of beer are

Prefer Light Beer	90/200 = .450, or 45.0%
Prefer Regular Beer	77/200 = .385, or 38.5%
Prefer Dark Beer	33/200 = .165, or 16.5%

Across all beer drinkers in the sample, light beer is preferred most often and dark beer is preferred least often.

Let us now conduct the chi-square test to determine if beer preference and gender are independent. The computations and formulas used are the same as those used for the

WEB file

BeerPreference

TABLE 12.6 SAMPLE RESULTS FOR BEER PREFERENCES OF MALE AND FEMALE BEER DRINKERS (OBSERVED FREQUENCIES)

		Gender		
		Male	**Female**	**Total**
Beer Preference	**Light**	51	39	90
	Regular	56	21	77
	Dark	25	8	33
	Total	132	68	200

TABLE 12.7 EXPECTED FREQUENCIES IF BEER PREFERENCE IS INDEPENDENT OF THE GENDER OF THE BEER DRINKER

		Gender		
		Male	Female	Total
Beer Preference	Light	59.40	30.60	90
	Regular	50.82	26.18	77
	Dark	21.78	11.22	33
	Total	132	68	200

chi-square test in Section 12.1. Utilizing the observed frequencies in Table 12.6 for row i and column j, f_{ij}, we compute the expected frequencies, e_{ij}, under the assumption that the beer preferences and gender are independent. The computation of the expected frequencies follows the same logic and formula used in Section 12.1. Thus the expected frequency for row i and column j is given by

$$e_{ij} = \frac{(\text{Row } i \text{ Total})(\text{Column } j \text{ Total})}{\text{Sample Size}} \tag{12.4}$$

For example, $e_{11} = (90)(132)/200 = 59.40$ is the expected frequency for male beer drinkers who would prefer light beer if beer preference is independent of gender. Show that equation (12.4) can be used to find the other expected frequencies shown in Table 12.7.

Following the chi-square test procedure discussed in Section 12.1, we use the following expression to compute the value of the chi-square test statistic.

$$\chi^2 = \sum_i \sum_j \frac{(f_{ij} - e_{ij})^2}{e_{ij}} \tag{12.5}$$

With r rows and c columns in the table, the chi-square distribution will have $(r - 1)(c - 1)$ degrees of freedom provided the expected frequency is at least 5 for each cell. Thus, in this application we will use a chi-square distribution with $(3 - 1)(2 - 1) = 2$ degrees of freedom. The complete steps to compute the chi-square test statistic are summarized in Table 12.8.

We can use the upper tail area of the chi-square distribution with 2 degrees of freedom and the p-value approach to determine whether the null hypothesis that beer preference

TABLE 12.8 COMPUTATION OF THE CHI-SQUARE TEST STATISTIC FOR THE TEST OF INDEPENDENCE BETWEEN BEER PREFERENCE AND GENDER

Beer Preference	Gender	Observed Frequency f_{ij}	Expected Frequency e_{ij}	Difference $(f_{ij} - e_{ij})$	Squared Difference $(f_{ij} - e_{ij})^2$	Squared Difference Divided by Expected Frequency $(f_{ij} - e_{ij})^2/e_{ij}$
Light	Male	51	59.40	−8.40	70.56	1.19
Light	Female	39	30.60	8.40	70.56	2.31
Regular	Male	56	50.82	5.18	26.83	.53
Regular	Female	21	26.18	−5.18	26.83	1.02
Dark	Male	25	21.78	3.22	10.37	.48
Dark	Female	8	11.22	−3.22	10.37	.92
	Total	200	200			$\chi^2 = 6.45$

is independent of gender can be rejected. Using row two of the chi-square distribution table shown in Table 12.4, we have the following:

Area in Upper Tail	**.10**	**.05**	**.025**	**.01**	**.005**
χ^2 Value (2 *df*)	4.605	5.991	7.378	9.210	10.597

$\chi^2 = 6.45$

Thus, we see the upper tail area at $\chi^2 = 6.45$ is between .05 and .025, and so the corresponding upper tail area or *p*-value must be between .05 and .025. With *p*-value $\leq .05$, we reject H_0 and conclude that beer preference is not independent of the gender of the beer drinker. Stated another way, the study shows that beer preference can be expected to differ for male and female beer drinkers. Minitab or Excel procedures provided in Appendix F can be used to show $\chi^2 = 6.45$ with two degrees of freedom yields a *p*-value $= .0398$.

Instead of using the *p*-value, we could use the critical value approach to draw the same conclusion. With $\alpha = .05$ and 2 degrees of freedom, the critical value for the chi-square test statistic is $\chi^2_{.05} = 5.991$. The upper tail rejection region becomes

$$\text{Reject } H_0 \text{ if } \geq 5.991$$

With $6.45 \geq 5.991$, we reject H_0. Again we see that the *p*-value approach and the critical value approach provide the same conclusion.

While we now have evidence that beer preference and gender are not independent, we will need to gain additional insight from the data to assess the nature of the association between these two variables. One way to do this is to compute the probability of the beer preference responses for males and females separately. These calculations are as follows:

Beer Preference	**Male**	**Female**
Light	51/132 = .3864, or 38.64%	39/68 = .5735, or 57.35%
Regular	56/132 = .4242, or 42.42%	21/68 = .3088, or 30.88%
Dark	25/132 = .1894, or 18.94%	8/68 = .1176, or 11.76%

The bar chart for male and female beer drinkers of the three kinds of beer is shown in Figure 12.1.

FIGURE 12.1 BAR CHART COMPARISON OF BEER PREFERENCE BY GENDER

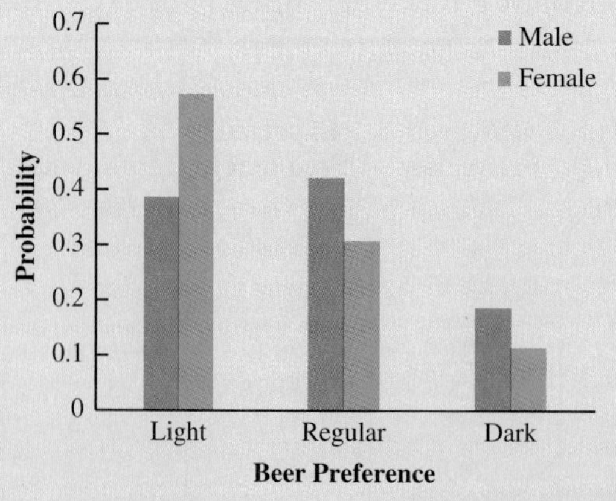

What observations can you make about the association between beer preference and gender? For female beer drinkers in the sample, the highest preference is for light beer at 57.35%. For male beer drinkers in the sample, regular beer is most frequently preferred at 42.42%. While female beer drinkers have a higher preference for light beer than males, male beer drinkers have a higher preference for both regular beer and dark beer. Data visualization through bar charts such as shown in Figure 12.1 is helpful in gaining insight as to how two categorical variables are associated.

Before we leave this discussion, we summarize the steps for a test of independence.

The expected frequencies must all be 5 or more for the chi-square test to be valid.

This chi-square test is also a one-tailed test with rejection of H_0 occurring in the upper tail of a chi-square distribution with $(r - 1)(c - 1)$ degrees of freedom.

CHI-SQUARE TEST FOR INDEPENDENCE OF TWO CATEGORICAL VARIABLES

1. State the null and alternative hypotheses.

   H_0: The two categorical variables are independent
   H_a: The two categorical variables are not independent

2. Select a random sample from the population and collect data for both variables for every element in the sample. Record the observed frequencies, f_{ij}, in a table with r rows and c columns.
3. Assume the null hypothesis is true and compute the expected frequencies, e_{ij}
4. If the expected frequency, e_{ij}, is 5 or more for each cell, compute the test statistic:

$$\chi^2 = \sum_i \sum_j \frac{(f_{ij} - e_{ij})^2}{e_{ij}}$$

5. Rejection rule:

   p-value approach: Reject H_0 if p-value $\leq \alpha$
   Critical value approach: Reject H_0 if $\chi^2 \geq \chi^2_\alpha$

   where the chi-square distribution has $(r - 1)(c - 1)$ degrees of freedom and α is the level of significance for the test.

Finally, if the null hypothesis of independence is rejected, summarizing the probabilities as shown in the above example will help the analyst determine where the association or dependence exists for the two categorical variables.

Exercises

Methods

SELF test

9. The following table contains observed frequencies for a sample of 200. Test for independence of the row and column variables using $\alpha = .05$.

	Column Variable		
Row Variable	**A**	**B**	**C**
P	20	44	50
Q	30	26	30

10. The following table contains observed frequencies for a sample of 240. Test for independence of the row and column variables using $\alpha = .05$.

	Column Variable		
Row Variable	A	B	C
P	20	30	20
Q	30	60	25
R	10	15	30

Applications

SELF test

11. A *Bloomberg Businessweek* subscriber study asked, "In the past 12 months, when traveling for business, what type of airline ticket did you purchase most often?" A second question asked if the type of airline ticket purchased most often was for domestic or international travel. Sample data obtained are shown in the following table.

	Type of Flight	
Type of Ticket	Domestic	International
First class	29	22
Business class	95	121
Economy class	518	135

a. Using a .05 level of significance, is the type of ticket purchased independent of the type of flight? What is your conclusion?
b. Discuss any dependence that exists between the type of ticket and type of flight.

WEB file
WorkforcePlan

12. A Deloitte employment survey asked a sample of human resource executives how their company planned to change its workforce over the next 12 months (*INC. Magazine,* February 2012). A categorical response variable showed three options: The company plans to hire and add to the number of employees, the company plans no change in the number of employees, or the company plans to lay off and reduce the number of employees. Another categorical variable indicated if the company was private or public. Sample data for 180 companies are summarized as follows.

	Company	
Employment Plan	Private	Public
Add Employees	37	32
No Change	19	34
Lay Off Employees	16	42

a. Conduct a test of independence to determine if the employment plan for the next 12 months is independent of the type of company. At a .05 level of significance, what is your conclusion?
b. Discuss any differences in the employment plans for private and public companies over the next 12 months.

13. Health insurance benefits vary by the size of the company (*Atlanta Business Chronicle,* December 31, 2010). The sample data below show the number of companies providing health insurance for small, medium, and large companies. For purposes of this study, small companies are companies that have fewer than 100 employees. Medium-sized companies have 100 to 999 employees, and large companies have 1000 or more employees. The

questionnaire sent to 225 employees asked whether or not the employee had health insurance and then asked the employee to indicate the size of the company.

	Size of the Company		
Health Insurance	**Small**	**Medium**	**Large**
Yes	36	65	88
No	14	10	12

a. Conduct a test of independence to determine whether health insurance coverage is independent of the size of the company. What is the *p*-value? Using a .05 level of significance, what is your conclusion?
b. A newspaper article indicated employees of small companies are more likely to lack health insurance coverage. Use percentages based on the above data to support this conclusion.

WEB file
AutoQuality

14. A vehicle quality survey asked new owners a variety of questions about their recently purchased automobile (J.D. Power and Associates, March 2012). One question asked for the owner's rating of the vehicle using categorical responses of average, outstanding, and exceptional. Another question asked for the owner's education level with the categorical responses some high school, high school graduate, some college, and college graduate. Assume the sample data below are for 500 owners who had recently purchased an automobile.

	Education			
Quality Rating	**Some HS**	**HS Grad**	**Some College**	**College Grad**
Average	35	30	20	60
Outstanding	45	45	50	90
Exceptional	20	25	30	50

a. Use a .05 level of significance and a test of independence to determine if a new owner's vehicle quality rating is independent of the owner's education. What is the *p*-value and what is your conclusion?
b. Use the overall percentage of average, outstanding, and exceptional ratings to comment upon how new owners rate the quality of their recently purchased automobiles.

15. *The Wall Street Journal* Corporate Perceptions Study 2011 surveyed readers and asked how each rated the quality of management and the reputation of the company for over 250 worldwide corporations. Both the quality of management and the reputation of the company were rated on an excellent, good, and fair categorical scale. Assume the sample data for 200 respondents below applies to this study.

	Reputation of Company		
Quality of Management	**Excellent**	**Good**	**Fair**
Excellent	40	25	5
Good	35	35	10
Fair	25	10	15

a. Use a .05 level of significance and test for independence of the quality of management and the reputation of the company. What is the *p*-value and what is your conclusion?
b. If there is a dependence or association between the two ratings, discuss and use probabilities to justify your answer.

16. As the price of oil rises, there is increased worldwide interest in alternate sources of energy. A *Financial Times*/Harris Poll surveyed people in six countries to assess attitudes toward a variety of alternate forms of energy (Harris Interactive website, February 27, 2008). The data in the following table are a portion of the poll's findings concerning whether people favor or oppose the building of new nuclear power plants.

	Country					
Response	**Great Britain**	**France**	**Italy**	**Spain**	**Germany**	**United States**
Strongly favor	141	161	298	133	128	204
Favor more than oppose	348	366	309	222	272	326
Oppose more than favor	381	334	219	311	322	316
Strongly oppose	217	215	219	443	389	174

a. How large was the sample in this poll?
b. Conduct a hypothesis test to determine whether people's attitude toward building new nuclear power plants is independent of country. What is your conclusion?
c. Using the percentage of respondents who "strongly favor" and "favor more than oppose," which country has the most favorable attitude toward building new nuclear power plants? Which country has the least favorable attitude?

17. The National Sleep Foundation used a survey to determine whether hours of sleep per night are independent of age (*Newsweek*, January 19, 2004). A sample of individuals was asked to indicate the number of hours of sleep per night with categorical options: fewer than 6 hours, 6 to 6.9 hours, 7 to 7.9 hours, and 8 hours or more. Later in the survey, the individuals were asked to indicate their age with categorical options: age 39 or younger and age 40 or older. Sample data follow.

	Age Group	
Hours of Sleep	**39 or younger**	**40 or older**
Fewer than 6	38	36
6 to 6.9	60	57
7 to 7.9	77	75
8 or more	65	92

a. Conduct a test of independence to determine whether hours of sleep are independent of age. Using a .05 level of significance, what is the p-value and what is your conclusion?
b. What is your estimate of the percentages of individuals who sleep fewer than 6 hours, 6 to 6.9 hours, 7 to 7.9 hours, and 8 hours or more per night?

18. On a syndicated television show the two hosts often create the impression that they strongly disagree about which movies are best. Each movie review is categorized as Pro ("thumbs up"), Con ("thumbs down"), or Mixed. The results of 160 movie ratings by the two hosts are shown here.

	Host B		
Host A	**Con**	**Mixed**	**Pro**
Con	24	8	13
Mixed	8	13	11
Pro	10	9	64

Use a test of independence with a .01 level of significance to analyze the data. What is your conclusion?

12.3 Goodness of Fit Test

In this section we use a chi-square test to determine whether a population being sampled has a specific probability distribution. We first consider a population with a historical multinomial probability distribution and use a goodness of fit test to determine if new sample data indicate there has been a change in the population distribution compared to the historical distribution. We then consider a situation where an assumption is made that a population has a normal probability distribution. In this case, we use a goodness of fit test to determine if sample data indicate that the assumption of a normal probability distribution is or is not appropriate. Both tests are referred to as **goodness of fit tests**.

Multinomial Probability Distribution

The multinomial probability distribution is an extension of the binomial probability distribution to the case where there are three or more outcomes per trial.

With a **multinomial probability distribution**, each element of a population is assigned to one and only one of three or more categories. As an example, consider the market share study being conducted by Scott Marketing Research. Over the past year, market shares for a certain product have stabilized at 30% for company A, 50% for company B, and 20% for company C. Since each customer is classified as buying from one of these companies, we have a multinomial probability distribution with three possible outcomes. The probability for each of the three outcomes is as follows.

p_{A} = probability a customer purchases the company A product

p_{B} = probability a customer purchases the company B product

p_{C} = probability a customer purchases the company C product

The sum of the probabilities for a multinomial probability distribution equal 1.

Using the historical market shares, we have multinomial probability distribution with $p_{\text{A}} = .30$, $p_{\text{B}} = .50$, and $p_{\text{C}} = .20$.

Company C plans to introduce a "new and improved" product to replace its current entry in the market. Company C has retained Scott Marketing Research to determine whether the new product will alter or change the market shares for the three companies. Specifically, the Scott Marketing Research study will introduce a sample of customers to the new company C product and then ask the customers to indicate a preference for the company A product, the company B product, or the new company C product. Based on the sample data, the following hypothesis test can be used to determine if the new company C product is likely to change the historical market shares for the three companies.

H_0: $p_{\text{A}} = .30$, $p_{\text{B}} = .50$, and $p_{\text{C}} = .20$

H_a: The population proportions are not $p_{\text{A}} = .30$, $p_{\text{B}} = .50$, and $p_{\text{C}} = .20$

The null hypothesis is based on the historical multinomial probability distribution for the market shares. If sample results lead to the rejection of H_0, Scott Marketing Research will have evidence to conclude that the introduction of the new company C product will change the market shares.

Let us assume that the market research firm has used a consumer panel of 200 customers. Each customer was asked to specify a purchase preference among the three alternatives: company A's product, company B's product, and company C's new product. The 200 responses are summarized here.

Observed Frequency		
Company A's Product	**Company B's Product**	**Company C's New Product**
48	98	54

We now can perform a goodness of fit test that will determine whether the sample of 200 customer purchase preferences is consistent with the null hypothesis. Like other chi-square tests, the goodness of fit test is based on a comparison of observed frequencies with the expected frequencies under the assumption that the null hypothesis is true. Hence, the next step is to compute expected purchase preferences for the 200 customers under the assumption that H_0: $p_A = .30$, $p_B = .50$, and $p_C = .20$ is true. Doing so provides the expected frequencies as follows.

Expected Frequency		
Company A's Product	**Company B's Product**	**Company C's New Product**
200(.30) = 60	200(.50) = 100	200(.20) = 40

Note that the expected frequency for each category is found by multiplying the sample size of 200 by the hypothesized proportion for the category.

The goodness of fit test now focuses on the differences between the observed frequencies and the expected frequencies. Whether the differences between the observed and expected frequencies are "large" or "small" is a question answered with the aid of the following chi-square test statistic.

TEST STATISTIC FOR GOODNESS OF FIT

$$\chi^2 = \sum_{i=1}^{k} \frac{(f_i - e_i)^2}{e_i} \tag{12.6}$$

where

f_i = observed frequency for category i
e_i = expected frequency for category i
k = the number of categories

Note: The test statistic has a chi-square distribution with $k - 1$ degrees of freedom provided that the expected frequencies are 5 *or more* for all categories.

Let us continue with the Scott Marketing Research example and use the sample data to test the hypothesis that the multinomial population has the market share proportions $p_A = .30$, $p_B = .50$, and $p_C = .20$. We will use an $\alpha = .05$ level of significance. We proceed by using the observed and expected frequencies to compute the value of the test statistic. With the expected frequencies all 5 or more, the computation of the chi-square test statistic is shown in Table 12.9. Thus, we have $\chi^2 = 7.34$.

The test for goodness of fit is always a one-tailed test with the rejection occurring in the upper tail of the chi-square distribution.

We will reject the null hypothesis if the differences between the observed and expected frequencies are large. Thus the test of goodness of fit will always be an upper tail test. We can use the upper tail area for the test statistic and the p-value approach to determine whether the null hypothesis can be rejected. With $k - 1 = 3 - 1 = 2$ degrees of freedom,

TABLE 12.9 COMPUTATION OF THE CHI-SQUARE TEST STATISTIC FOR THE SCOTT MARKETING RESEARCH MARKET SHARE STUDY

Category	Hypothesized Proportion	Observed Frequency (f_i)	Expected Frequency (e_i)	Difference ($f_i - e_i$)	Squared Difference $(f_i - e_i)^2$	Squared Difference Divided by Expected Frequency $(f_i - e_i)^2/e_i$
Company A	.30	48	60	−12	144	2.40
Company B	.50	98	100	−2	4	0.04
Company C	.20	54	40	14	196	4.90
Total		200				$\chi^2 = 7.34$

row two of the chi-square distribution table in Table 12.4 provides the following:

Area in Upper Tail	**.10**	**.05**	**.025**	**.01**	**.005**
χ^2 Value (2 *df*)	4.605	5.991	7.378	9.210	10.597

$\chi^2 = 7.34$ (between 5.991 and 7.378)

The test statistic $\chi^2 = 7.34$ is between 5.991 and 7.378. Thus, the corresponding upper tail area or *p*-value must be between .05 and .025. With *p*-value $\leq .05$, we reject H_0 and conclude that the introduction of the new product by company C will alter the historical market shares. Minitab or Excel procedures provided in Appendix F can be used to show $\chi^2 = 7.34$ provides a *p*-value = .0255.

Instead of using the *p*-value, we could use the critical value approach to draw the same conclusion. With $\alpha = .05$ and 2 degrees of freedom, the critical value for the test statistic is $\chi^2_{.05} = 5.991$. The upper tail rejection rule becomes

$$\text{Reject } H_0 \text{ if } \chi^2 \geq 5.991$$

With $7.34 > 5.991$, we reject H_0. The *p*-value approach and critical value approach provide the same hypothesis testing conclusion.

Now that we have concluded the introduction of a new company C product will alter the market shares for the three companies, we are interested in knowing more about how the market shares are likely to change. Using the historical market shares and the sample data, we summarize the data as follows:

Company	Historical Market Share (%)	Sample Data Market Share (%)
A	30	48/200 = .24, or 24
B	50	98/200 = .49, or 49
C	20	54/200 = .27, or 27

The historical market shares and the sample market shares are compared in the bar chart shown in Figure 12.2. This data visualization process shows that the new product will likely increase the market share for company C. Comparisons for the other two companies indicate that company C's gain in market share will hurt company A more than company B.

FIGURE 12.2 BAR CHART OF MARKET SHARES BY COMPANY BEFORE AND AFTER THE NEW PRODUCT FOR COMPANY C

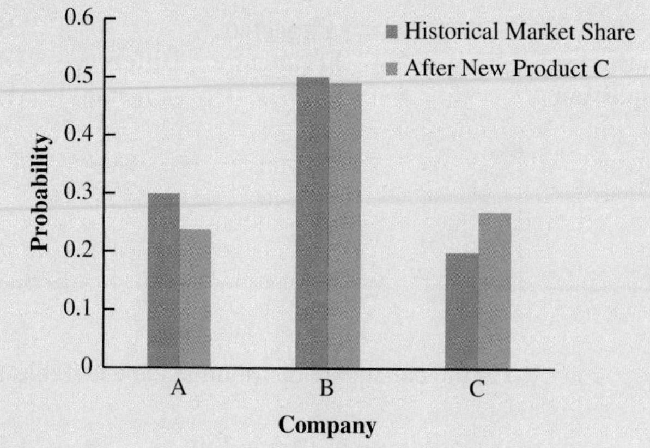

Let us summarize the steps that can be used to conduct a goodness of fit test for a hypothesized multinomial population distribution.

MULTINOMIAL PROBABILITY DISTRIBUTION GOODNESS OF FIT TEST

1. State the null and alternative hypotheses.

   H_0: The population follows a multinomial probability distribution with specified probabilities for each of the k categories

   H_a: The population does not follow a multinomial distribution with the specified probabilities for each of the k categories

2. Select a random sample and record the observed frequencies f_i for each category.
3. Assume the null hypothesis is true and determine the expected frequency e_i in each category by multiplying the category probability by the sample size.
4. If the expected frequency e_i is at least 5 for each category, compute the value of the test statistic.

$$\chi^2 = \sum_{i=1}^{k} \frac{(f_i - e_i)^2}{e_i}$$

5. Rejection rule:

p-value approach:	Reject H_0 if p-value $\leq \alpha$
Critical value approach:	Reject H_0 if $\chi^2 \geq \chi^2_\alpha$

   where α is the level of significance for the test and there are $k - 1$ degrees of freedom.

Normal Probability Distribution

The goodness of fit test for a normal probability distribution is also based on the use of the chi-square distribution. In particular, observed frequencies for several categories of sample data are compared to expected frequencies under the assumption that the population has a

TABLE 12.10 CHEMLINE EMPLOYEE APTITUDE TEST SCORES FOR 50 RANDOMLY CHOSEN JOB APPLICANTS

71	66	61	65	54	93
60	86	70	70	73	73
55	63	56	62	76	54
82	79	76	68	53	58
85	80	56	61	61	64
65	62	90	69	76	79
77	54	64	74	65	65
61	56	63	80	56	71
79	84				

normal probability distribution. Because the normal probability distribution is continuous, we must modify the way the categories are defined and how the expected frequencies are computed. Let us demonstrate the goodness of fit test for a normal distribution by considering the job applicant test data for Chemline, Inc., shown in Table 12.10.

Chemline hires approximately 400 new employees annually for its four plants located throughout the United States. The personnel director asks whether a normal distribution applies for the population of test scores. If such a distribution can be used, the distribution would be helpful in evaluating specific test scores; that is, scores in the upper 20%, lower 40%, and so on, could be identified quickly. Hence, we want to test the null hypothesis that the population of test scores has a normal distribution.

Let us first use the data in Table 12.10 to develop estimates of the mean and standard deviation of the normal distribution that will be considered in the null hypothesis. We use the sample mean $\bar{x}$ and the sample standard deviation s as point estimators of the mean and standard deviation of the normal distribution. The calculations follow.

$$\bar{x} = \frac{\Sigma x_i}{n} = \frac{3421}{50} = 68.42$$

$$s = \sqrt{\frac{\Sigma(x_i - \bar{x})^2}{n-1}} = \sqrt{\frac{5310.0369}{49}} = 10.41$$

WEB file Chemline

Using these values, we state the following hypotheses about the distribution of the job applicant test scores.

H_0: The population of test scores has a normal distribution with mean 68.42 and standard deviation 10.41

H_a: The population of test scores does not have a normal distribution with mean 68.42 and standard deviation 10.41

The hypothesized normal distribution is shown in Figure 12.3.

With the continuous normal probability distribution, we must use a different procedure for defining the categories. We need to define the categories in terms of *intervals* of test scores.

With a continuous probability distribution, establish intervals such that each interval has an expected frequency of five or more.

Recall the rule of thumb for an expected frequency of at least five in each interval or category. We define the categories of test scores such that the expected frequencies will be at least five for each category. With a sample size of 50, one way of establishing categories

FIGURE 12.3 HYPOTHESIZED NORMAL DISTRIBUTION OF TEST SCORES FOR THE CHEMLINE JOB APPLICANTS

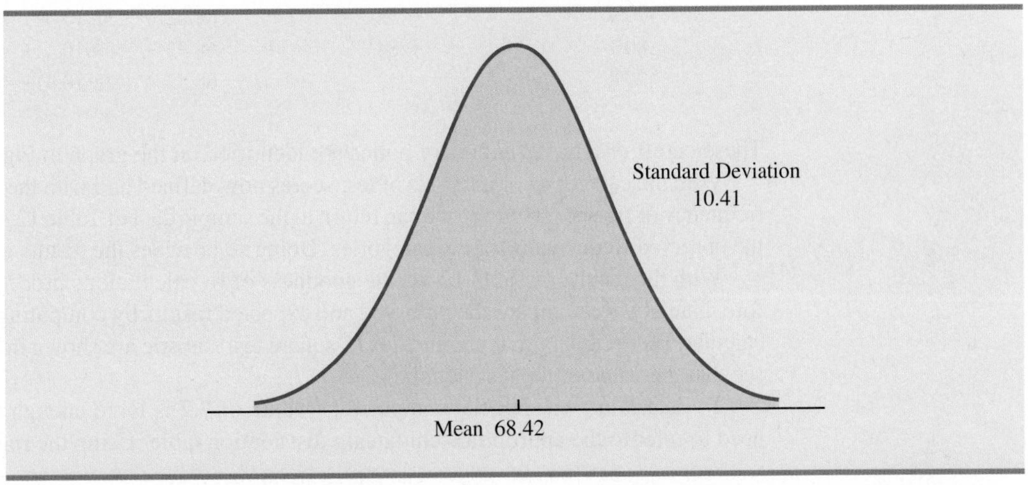

FIGURE 12.4 NORMAL DISTRIBUTION FOR THE CHEMLINE EXAMPLE WITH 10 EQUAL-PROBABILITY INTERVALS

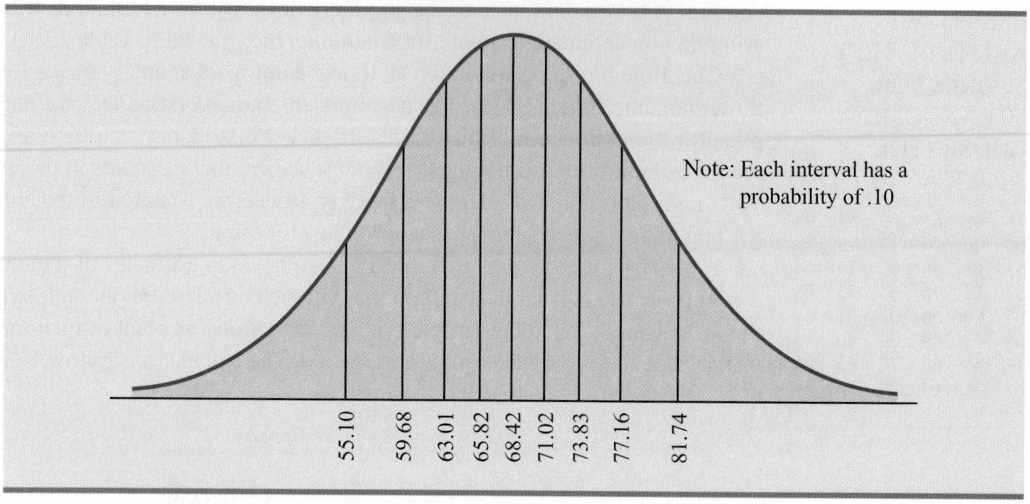

is to divide the normal probability distribution into 10 equal-probability intervals (see Figure 12.4). With a sample size of 50, we would expect five outcomes in each interval or category, and the rule of thumb for expected frequencies would be satisfied.

Let us look more closely at the procedure for calculating the category boundaries. When the normal probability distribution is assumed, the standard normal probability tables can be used to determine these boundaries. First consider the test score cutting off the lowest 10% of the test scores. From the table for the standard normal distribution we find that the z value for this test score is -1.28. Therefore, the test score of $x = 68.42 - 1.28(10.41) = 55.10$ provides this cutoff value for the lowest 10% of the scores. For the lowest 20%, we find $z = -.84$, and thus $x = 68.42 - .84(10.41) = 59.68$. Working through the normal distribution in that way provides the following test score values.

Percentage	z	Test Score
10%	−1.28	68.42 − 1.28(10.41) = 55.10
20%	−.84	68.42 − .84(10.41) = 59.68
30%	−.52	68.42 − .52(10.41) = 63.01
40%	−.25	68.42 − .25(10.41) = 65.82
50%	.00	68.42 + 0(10.41) = 68.42
60%	+.25	68.42 + .25(10.41) = 71.02
70%	+.52	68.42 + .52(10.41) = 73.83
80%	+.84	68.42 + .84(10.41) = 77.16
90%	+1.28	68.42 + 1.28(10.41) = 81.74

These cutoff or interval boundary points are identified on the graph in Figure 12.4.

With the categories or intervals of test scores now defined and with the known expected frequency of five per category, we can return to the sample data of Table 12.10 and determine the observed frequencies for the categories. Doing so provides the results in Table 12.11.

With the results in Table 12.11, the goodness of fit calculations proceed exactly as before. Namely, we compare the observed and expected results by computing a χ^2 value. The calculations necessary to compute the chi-square test statistic are shown in Table 12.12. We see that the value of the test statistic is $\chi^2 = 7.2$.

To determine whether the computed χ^2 value of 7.2 is large enough to reject H_0, we need to refer to the appropriate chi-square distribution table. Using the rule for computing

TABLE 12.11 OBSERVED AND EXPECTED FREQUENCIES FOR CHEMLINE JOB APPLICANT TEST SCORES

Test Score Interval	Observed Frequency (f_i)	Expected Frequency (e_i)
Less than 55.10	5	5
55.10 to 59.68	5	5
59.68 to 63.01	9	5
63.01 to 65.82	6	5
65.82 to 68.42	2	5
68.42 to 71.02	5	5
71.02 to 73.83	2	5
73.83 to 77.16	5	5
77.16 to 81.74	5	5
81.74 and over	6	5
Total	50	50

TABLE 12.12 COMPUTATION OF THE CHI-SQUARE TEST STATISTIC FOR THE CHEMLINE JOB APPLICANT EXAMPLE

Test Score Interval	Observed Frequency (f_i)	Expected Frequency (e_i)	Difference ($f_i - e_i$)	Squared Difference $(f_i - e_i)^2$	Squared Difference Divided by Expected Frequency $(f_i - e_i)^2/e_i$
Less than 55.10	5	5	0	0	0.0
55.10 to 59.68	5	5	0	0	0.0
59.68 to 63.01	9	5	4	16	3.2
63.01 to 65.82	6	5	1	1	0.2
65.82 to 68.42	2	5	−3	9	1.8
68.42 to 71.02	5	5	0	0	0.0
71.02 to 73.83	2	5	−3	9	1.8
73.83 to 77.16	5	5	0	0	0.0
77.16 to 81.74	5	5	0	0	0.0
81.74 and over	6	5	1	1	0.2
Total	50	50			$\chi^2 = 7.2$

the number of degrees of freedom for the goodness of fit test, we have $k - p - 1 = 10 - 2 - 1 = 7$ degrees of freedom based on $k = 10$ categories and $p = 2$ parameters (mean and standard deviation) estimated from the sample data.

Estimating the two parameters of the normal distribution will cause a loss of two degrees of freedom in the χ^2 test.

Suppose that we test the null hypothesis that the distribution for the test scores is a normal distribution with a .10 level of significance. To test this hypothesis, we need to determine the p-value for the test statistic $\chi^2 = 7.2$ by finding the area in the upper tail of a chi-square distribution with 7 degrees of freedom. Using row seven of Table 12.4, we find that $\chi^2 = 7.2$ provides an area in the upper tail greater than .10. Thus, we know that the p-value is greater than .10. Minitab or Excel procedures in Appendix F can be used to show $\chi^2 = 7.2$ provides a p-value $= .4084$. With p-value $>.10$, the hypothesis that the probability distribution for the Chemline job applicant test scores is a normal probability distribution cannot be rejected. The normal probability distribution may be applied to assist in the interpretation of test scores. A summary of the goodness fit test for a normal probability distribution follows.

NORMAL PROBABILITY DISTRIBUTION GOODNESS OF FIT TEST

1. State the null and alternative hypotheses.

H_0: The population has a normal probability distribution
H_a: The population does not have a normal probability distribution

2. Select a random sample and
  a. Compute the sample mean and sample standard deviation.
  b. Define k intervals of values so that the expected frequency is at least five for each interval. Using equal probability intervals is a good approach.
  c. Record the observed frequency of data values f_i in each interval defined.

3. Compute the expected number of occurrences e_i for each interval of values defined in step 2(b). Multiply the sample size by the probability of a normal random variable being in the interval.

4. Compute the value of the test statistic.

$$\chi^2 = \sum_{i=1}^{k} \frac{(f_i - e_i)^2}{e_i}$$

5. Rejection rule:

p-value approach:	Reject H_0 if p-value $\leq \alpha$
Critical value approach:	Reject H_0 if $\chi^2 \geq \chi^2_\alpha$

where α is the level of significance. The degrees of freedom $= k - p - 1$, where p is the number of parameters of the distribution estimated by the sample. In step 2a, the sample is used to estimate the mean and standard deviation. Thus, $p = 2$ and the degrees of freedom $= k - 2 - 1 = k - 3$.

Exercises

Methods

SELF test

19. Test the following hypotheses by using the χ^2 goodness of fit test.

H_0: $p_A = .40$, $p_B = .40$, and $p_C = .20$
H_a: The population proportions are not $p_A = .40$, $p_B = .40$, and $p_C = .20$

A sample of size 200 yielded 60 in category A, 120 in category B, and 20 in category C. Use $\alpha = .01$ and test to see whether the proportions are as stated in H_0.

a. Use the p-value approach.
b. Repeat the test using the critical value approach.

SELF test

20. The following data are believed to have come from a normal distribution. Use the goodness of fit test and $\alpha = .05$ to test this claim.

17	23	22	24	19	23	18	22	20	13	11	21	18	20	21
21	18	15	24	23	23	43	29	27	26	30	28	33	23	29

Applications

SELF test

21. During the first 13 weeks of the television season, the Saturday evening 8:00 P.M. to 9:00 P.M. audience proportions were recorded as ABC 29%, CBS 28%, NBC 25%, and independents 18%. A sample of 300 homes two weeks after a Saturday night schedule revision yielded the following viewing audience data: ABC 95 homes, CBS 70 homes, NBC 89 homes, and independents 46 homes. Test with $\alpha = .05$ to determine whether the viewing audience proportions changed.

WEB file

M&M

22. Mars, Inc. manufactures M&M's, one of the most popular candy treats in the world. The milk chocolate candies come in a variety of colors including blue, brown, green, orange, red, and yellow (M&M website, March 2012). The overall proportions for the colors are .24 blue, .13 brown, .20 green, .16 orange, .13 red, and .14 yellow. In a sampling study, several bags of M&M milk chocolates were opened and the following color counts were obtained.

Blue	**Brown**	**Green**	**Orange**	**Red**	**Yellow**
105	72	89	84	70	80

Use a .05 level of significance and the sample data to test the hypothesis that the overall proportions for the colors are as stated above. What is your conclusion?

23. *The Wall Street Journal's* Shareholder Scoreboard tracks the performance of 1000 major U.S. companies. The performance of each company is rated based on the annual total return, including stock price changes and the reinvestment of dividends. Ratings are assigned by dividing all 1000 companies into five groups from A (top 20%), B (next 20%), to E (bottom 20%). Shown here are the one-year ratings for a sample of 60 of the largest companies. Do the largest companies differ in performance from the performance of the 1000 companies in the Shareholder Scoreboard? Use $\alpha = .05$.

A	**B**	**C**	**D**	**E**
5	8	15	20	12

24. The National Highway Traffic Safety Administration reported the percentage of traffic accidents occurring each day of the week (*Time,* March 12, 2012). Assume that a sample of 420 accidents provided the following data.

Sunday	**Monday**	**Tuesday**	**Wednesday**	**Thursday**	**Friday**	**Saturday**
66	50	53	47	55	69	80

a. Conduct a hypothesis test to determine if the proportion of traffic accidents is the same for each day of the week. What is the p-value? Using a .05 level of significance, what is your conclusion?

b. Compute the percentage of traffic accidents occurring on each day of the week. What day has the highest percentage of traffic accidents? Does this seem reasonable? Discuss.

25. Use $\alpha = .01$ and conduct a goodness of fit test to see whether the following sample appears to have been selected from a normal probability distribution.

55	86	94	58	55	95	55	52	69	95	90	65	87	50	56
55	57	98	58	79	92	62	59	88	65					

After you complete the goodness of fit calculations, construct a histogram of the data. Does the histogram representation support the conclusion reached with the goodness of fit test? (*Note:* $\bar{x} = 71$ and $s = 17$.)

WEB file

Demand

26. The weekly demand for a product is believed to be normally distributed. Use a goodness of fit test and the following data to test this assumption. Use $\alpha = .10$. The sample mean is 24.5 and the sample standard deviation is 3.

18	20	22	27	22
25	22	27	25	24
26	23	20	24	26
27	25	19	21	25
26	25	31	29	25
25	28	26	28	24

Summary

In this chapter we have introduced hypothesis tests for the following applications.

1. Testing the equality of population proportions for three or more populations.
2. Testing the independence of two categorical variables.
3. Testing whether a probability distribution for a population follows a specific historical or theoretical probability distribution.

All tests apply to categorical variables and all tests use a chi-square (χ^2) test statistic that is based on the differences between observed frequencies and expected frequencies. In each case, expected frequencies are computed under the assumption that the null hypothesis is true. These chi-square tests are upper tailed tests. Large differences between observed and expected frequencies provide a large value for the chi-square test statistic and indicate that the null hypothesis should be rejected.

The test for the equality of population proportions for three or more populations is based on independent random samples selected from each of the populations. The sample data show the counts for each of two categorical responses for each population. The null hypothesis is that the population proportions are equal. Rejection of the null hypothesis supports the conclusion that the population proportions are not all equal.

The test of independence between two categorical variables uses one sample from a population with the data showing the counts for each combination of two categorical variables. The null hypothesis is that the two variables are independent and the test is referred to as a test of independence. If the null hypothesis is rejected, there is statistical evidence of an association or dependency between the two variables.

The goodness of fit test is used to test the hypothesis that a population has a specific historical or theoretical probability distribution. We showed applications for populations with a multinomial probability distribution and with a normal probability distribution. Since the normal probability distribution applies to continuous data, intervals of data values were established to create the categories for the categorical variable required for the goodness of fit test.

Glossary

Marascuilo procedure A multiple comparison procedure that can be used to test for a significant difference between pairs of population proportions. This test can be helpful in identifying differences between pairs of population proportions whenever the hypothesis of equal population proportions has been rejected.

Test of independence A chi-square test that can be used to test for the independence between two categorical variables. If the hypothesis of independence is rejected, it can be concluded that the categorical variables are associated or dependent.

Goodness of fit test A chi-square test that can be used to test that a population probability distribution has a specific historical or theoretical probability distribution. This test was demonstrated for both a multinomial probability distribution and a normal probability distribution.

Multinomial probability distribution A probability distribution where each outcome belongs to one of three or more categories. The multinomial probability distribution extends the binomial probability from two to three or more outcomes per trial.

Key Formulas

Expected Frequencies Under the Assumption H_0 Is True

$$e_{ij} = \frac{(\text{Row } i \text{ Total})(\text{Column } j \text{ Total})}{\text{Sample Size}} \tag{12.1}$$

Chi-Square Test Statistic

$$\chi^2 = \sum_i \sum_j \frac{(f_{ij} - e_{ij})^2}{e_{ij}} \tag{12.2}$$

Critical Values for the Marascuilo Pairwise Comparison Procedure

$$CV_{ij} = \sqrt{\chi^2_\alpha}\sqrt{\frac{\bar{p}_i(1 - \bar{p}_i)}{n_i} + \frac{\bar{p}_j(1 - \bar{p}_j)}{n_j}} \tag{12.3}$$

Chi-Square Test Statistic for the Goodness of Fit Test

$$\chi^2 = \sum_i \frac{(f_i - e_i)^2}{e_i} \tag{12.6}$$

Supplementary Exercises

27. In a quality control test of parts manufactured at Dabco Corporation, an engineer sampled parts produced on the first, second, and third shifts. The research study was designed to determine if the population proportion of good parts was the same for all three shifts. Sample data follow.

	Production Shift		
Quality	**First**	**Second**	**Third**
Good	285	368	176
Defective	15	32	24

   a. Using a .05 level of significance, conduct a hypothesis test to determine if the population proportion of good parts is the same for all three shifts. What is the p-value and what is your conclusion?
   b. If the conclusion is that the population proportions are not all equal, use a multiple comparison procedure to determine how the shifts differ in terms of quality. What shift or shifts need to improve the quality of parts produced?

28. Phoenix Marketing International identified Bridgeport, Connecticut, Los Alamos, New Mexico, Naples, Florida and Washington D.C. as the four U.S. cities with the highest percentage of

millionaires (*USA Today,* December 7, 2011). Data consistent with that study show the following number of millionaires for samples of individuals from each of the four cities.

	City			
Millionaire	**Bridgeport**	**Los Alamos**	**Naples**	**Washington DC**
Yes	44	35	36	34
No	456	265	364	366

a. What is the estimate of the percentage of millionaires in each of these cities?
b. Using a .05 level of significance, test for the equality of the population proportion of millionaires for these four cities. What is the p-value and what is your conclusion?

WEB file

BothWork

29. Samples taken in three cities, Anchorage, Atlanta, and Minneapolis, were used to learn about the proportion of married couples where both husband and wife are in the workforce (*USA Today,* January 15, 2006).

	City		
In Workforce	**Anchorage**	**Atlanta**	**Minneapolis**
Both Work	57	70	63
Only One	33	50	27

a. Conduct a hypothesis test to determine if the population proportion of married couples with both husband and wife in the workforce is the same for the three cities. Using a .05 level of significance, what is the p-value and what is your conclusion?
b. Using these three samples, what is an estimate of the proportion of married couples with both husband and wife in the workforce?

30. A Pew Research Center survey asked respondents if they would rather live in a place with a slower pace of life or a place with a faster pace of life (*USA Today,* February 13, 2009). The survey also asked the respondent's gender. Consider the following sample data.

	Gender	
Preferred Pace of Life	**Male**	**Female**
Slower	230	218
No Preference	20	24
Faster	90	48

a. Is the preferred pace of life independent of gender? Using a .05 level of significance, what is the p-value and what is your conclusion?
b. Discuss any differences between the preferences of men and women.

31. Bara Research Group conducted a survey about church attendance. The survey respondents were asked about their church attendance and asked to indicate their age. Use the sample data to determine whether church attendance is independent of age. Using a .05 level of significance, what is the p-value and what is your conclusion? What conclusion can you draw about church attendance as individuals grow older?

	Age			
Church Attendance	**20 to 29**	**30 to 39**	**40 to 49**	**50 to 59**
Yes	31	63	94	72
No	69	87	106	78

WEB file
Ambulance

32. An ambulance service responds to emergency calls for two counties in Virginia. One county is an urban county and the other is a rural county. A sample of 471 ambulance calls over the past two years showed the county and the day of the week for each emergency call. Data are as follows.

	Day of Week						
County	Sun	Mon	Tue	Wed	Thu	Fri	Sat
Urban	61	48	50	55	63	73	43
Rural	7	9	16	13	9	14	10

Test for independence of the county and the day of the week. Using a .05 level of significance, what is the *p*-value and what is your conclusion?

33. Based on sales over a six-month period, the five top-selling compact cars are Chevy Cruze, Ford Focus, Hyundai Elantra, Honda Civic, and Toyota Corolla (*Motor Trend,* November 2, 2011). Based on total sales, the market shares for these five compact cars were Chevy Cruze 24%, Ford Focus 21%, Hyundai Elantra 19%, Honda Civic 18%, and Toyota Corolla 17%. A sample of 400 compact car sales in Chicago showed the following number of vehicles sold.

Chevy Cruze	108
Ford Focus	92
Hyundai Elantra	64
Honda Civic	84
Toyota Corolla	52

Use a goodness of fit test to determine if the sample data indicate that the market shares for the five compact cars in Chicago are different than the market shares reported by *Motor Trend.* Using a .05 level of significance, what is the *p*-value and what is your conclusion? What market share differences, if any, exist in Chicago?

WEB file
Grades

34. A random sample of final examination grades for a college course follows.

55	85	72	99	48	71	88	70	59	98	80	74	93	85	74
82	90	71	83	60	95	77	84	73	63	72	95	79	51	85
76	81	78	65	75	87	86	70	80	64					

Use $\alpha = .05$ and test to determine whether a normal probability distribution should be rejected as being representative of the population distribution of grades.

35. A salesperson makes four calls per day. A sample of 100 days gives the following frequencies of sales volumes.

Number of Sales	Observed Frequency (days)
0	30
1	32
2	25
3	10
4	3
Total	100

Records show sales are made to 30% of all sales calls. Assuming independent sales calls, the number of sales per day should follow a binomial probability distribution. The binomial probability function presented in Chapter 5 is

$$f(x) = \frac{n!}{x!(n - x)!} p^x (1 - p)^{n-x}$$

For this exercise, assume that the population has a binomial probability distribution with $n = 4$, $p = .30$, and $x = 0, 1, 2, 3$, and 4.

a. Compute the expected frequencies for $x = 0, 1, 2, 3$, and 4 by using the binomial probability function. Combine categories if necessary to satisfy the requirement that the expected frequency is five or more for all categories.
b. Use the goodness of fit test to determine whether the assumption of a binomial probability distribution should be rejected. Use $\alpha = .05$. Because no parameters of the binomial probability distribution were estimated from the sample data, the degrees of freedom are $k - 1$ when k is the number of categories.

Case Problem A Bipartisan Agenda for Change

In a study conducted by Zogby International for the *Democrat and Chronicle,* more than 700 New Yorkers were polled to determine whether the New York state government works. Respondents surveyed were asked questions involving pay cuts for state legislators, restrictions on lobbyists, term limits for legislators, and whether state citizens should be able to put matters directly on the state ballot for a vote. The results regarding several proposed reforms had broad support, crossing all demographic and political lines.

Suppose that a follow-up survey of 100 individuals who live in the western region of New York was conducted. The party affiliation (Democrat, Independent, Republican) of each individual surveyed was recorded, as well as their responses to the following three questions.

1. Should legislative pay be cut for every day the state budget is late?
   Yes ____ No ____
2. Should there be more restrictions on lobbyists?
   Yes ____ No ____
3. Should there be term limits requiring that legislators serve a fixed number of years?
   Yes ____ No ____

WEB file NYReform

The responses were coded using 1 for a Yes response and 2 for a No response. The complete data set is available in the file named NYReform.

Managerial Report

1. Use descriptive statistics to summarize the data from this study. What are your preliminary conclusions about the independence of the response (Yes or No) and party affiliation for each of the three questions in the survey?
2. With regard to question 1, test for the independence of the response (Yes and No) and party affiliation. Use $\alpha = .05$.
3. With regard to question 2, test for the independence of the response (Yes and No) and party affiliation. Use $\alpha = .05$.
4. With regard to question 3, test for the independence of the response (Yes and No) and party affiliation. Use $\alpha = .05$.
5. Does it appear that there is broad support for change across all political lines? Explain.

Appendix 12.1 Chi-Square Tests Using Minitab

Test the Equality of Population Proportions and Test of Independence

The Minitab procedure is identical for both of these applications. We will describe the procedure for the following situations.

1. A data set shows the responses for each element in the sample.
2. A tabular summary of the data shows the observed frequencies for the response categories.

WEB file
AutoLoyalty

We begin with the automobile loyalty example presented in Section 12.1. Responses for a sample of 500 automobile owners is contained in the web file AutoLoyalty. Column C1 shows the population the owner belongs to (Chevrolet Impala, Ford Fusion, or Honda Accord) and column C2 contains the likely to purchase response (Yes or No). The Minitab steps to conduct a chi-square test using the data set follow.

Step 1. Select the **Stat** menu
Step 2. Select **Tables**
Step 3. Choose **Cross Tabulation and Chi-Square**
Step 4. When then Cross Tabulation and Chi-Square dialog box appears:
Enter C2 in the **For Rows** box
Enter C1 in the **For Columns** box
Under the **Display** options, select **Counts**
Select **Chi-Square**
Step 5. When the Cross Tabulation—Chi Square dialog box appears:
Select **Chi-Square analysis**
Click **OK**
Step 6. Click **OK**

The output shows both a tabular summary of the data and the chi-square test results.

This procedure can also be used for a test of independence. Use the web file BeerPreference to conduct the test for the example in Section 12.2.

Next let us show how to conduct this test if a tabular summary of the data showing observed frequencies has already been obtained. We begin with a new Minitab worksheet and label the columns C1 to C3 with the titles of the three populations: Chevrolet Impala, Ford Fusion, and Honda Accord. Then we enter the observed frequencies of the Yes and No responses for each population in its corresponding column. Thus, we enter 69 and 56 in column 1, enter 120 and 80 in column 2, and enter 123 and 52 in column 3. The Minitab steps for this test are as follows.

Step 1. Select the **Stat** menu
Step 2. Select **Tables**
Step 3. Choose **Chi-Square Test (Two-Way Table in Worksheet)**
Step 4. When the Chi-Square Test dialog box appears:
Enter C1-C3 in the **Columns containing the table** box
Click **OK**

Goodness of Fit Test

In order to use Minitab to conduct a goodness of fit test, the user must first obtain a sample from the population and determine the observed frequency for each of k categories. Under the assumption that the hypothesized population distribution is true, the user must also determine the hypothesized or expected proportion for each of the k categories. Using a new

Minitab worksheet, the observed frequencies are entered in column C1 and the corresponding hypothesized proportions are entered in column C2.

Using the Scott Marketing Research example presented in Section 12.3, the sample of 200 customer preferences for products A, B, and C provided observed frequencies of 48, 98, and 54. These frequencies are entered in column C1. Using historical market share data, the hypothesized proportions, .30, .50 and .20, are entered in column C2. The Minitab steps for the goodness of fit test for this multinomial probability distribution follow.

Step 1. Select the **Stat** menu
Step 2. Select **Tables**
Step 3. Choose **Chi-Square Goodness of Fit Test (One Variable)**
Step 4. When the Chi-Square Goodness of Fit Test dialog box appears:
Select **Observed counts**
Enter C1 in the **Observed counts** box
Select **Specific proportions**
Enter C2 in the **Specific proportions** box
Click **OK**

If in any application of the goodness of fit test the null hypothesis is *equal* proportions for the k categories, column C2 is not necessary. In this case, the user can select **Equal proportions** rather than **Specific proportions** in step 4.

Appendix 12.2 Chi-Square Tests Using Excel

The Excel procedure for tests for the equality of population proportions, tests of independence, and goodness of fit tests are basically the same as all make use of the Excel chi-square function CHISQ.TEST. Regardless of the application, the user must do the following before creating an Excel worksheet that will perform the test.

1. Select a sample from the population or populations and record the data
2. Summarize the data to show observed frequencies in a tabular format

Excel's PivotTable can be used to summarize the data in step 2 above. Since this procedure was previously presented in Appendix 2.2, we shall not describe it in this appendix. Rather we will begin the Excel chi-square test procedure with the understanding that the user has already determined the observed frequencies for the study.

WEB file
AutoLoyalty

Let us demonstrate the Excel chi-square test by considering the automobile loyalty example presented in Section 12.1. Using the data in the web file AutoLoyalty and the Excel PivotTable procedure, we obtained the observed frequencies shown in the Excel worksheet of Figure 12.5. The user must next insert Excel formulas in the worksheet to compute the expected frequencies. Using equation (12.1), the Excel formulas for expected frequencies are as shown in the background worksheet of Figure 12.5.

The last step is to insert the Excel function CHISQ.TEST. The format of this function is as follows:

=CHISQ.TEST(Observed Frequency Cells, Expected Frequency Cells)

In Figure 12.5, the observed frequency cells are B7 to D8, written B7:D8 and the expected frequency cells are B16 to D17, written B16:D17. The function =CHISQ.TEST(B7:D8,B16:D17) is shown in cell E20 of the background worksheet. This function does all the chi-square test computations and returns the p-value for the test.

FIGURE 12.5 EXCEL WORKSHEET FOR THE AUTOMOBILE LOYALITY STUDY

	A	B	C	D	E	F
1	**Chi Square Test**					
2						
3	**Observed Frequencies**					
4						
5			**Populations**			
6	**Likely Purchase**	Chevrolet Impala	Ford Fusion	Honda Accord	Total	
7	Yes	69	120	123	312	
8	No	56	80	52	188	
9	Total	125	200	175	500	
10						
11						
12	**Expected Frequencies**					
13						
14			**Populations**			
15	**Likely Purchase**	Chevrolet Impala	Ford Fusion	Honda Accord	Total	
16	Yes	78	124.8	109.2	312	
17	No	47	75.2	65.8	188	
18	Total	125	200	175	500	
19						
20				***p*-value**	0.0193	
21						

WEB file

ChiSquare

	A	B	C	D	E	F
1	**Chi Square Test**					
2						
3	**Observed Frequencies**					
4						
5			**Populations**			
6	**Likely Purchase**	Chevrolet Impala	Ford Fusion	Honda Accord	Total	
7	Yes	69	120	123	=SUM(B7:D7)	
8	No	56	80	52	=SUM(B8:D8)	
9	Total	=SUM(B7:B8)	=SUM(C7:C8)	=SUM(D7:D8)	=SUM(E7:E8)	
10						
11						
12	**Expected Frequencies**					
13						
14			**Populations**			
15	**Likely Purchase**	Chevrolet Impala	Ford Fusion	Honda Accord	Total	
16	Yes	=E7*B9/E9	=E7*C9/E9	=E7*D9/E9	=SUM(B16:D16)	
17	No	=E8*B9/E9	=E8*C9/E9	=E8*D9/E9	=SUM(B17:D17)	
18	Total	=SUM(B16:B17)	=SUM(C16:C17)	=SUM(D16:D17)	=SUM(E16:E17)	
19						
20				***p*-value**	=CHISQ.TEST(B7:D8,B16:D17)	
21						

The Excel worksheet shown in Figure 12.5 is available in the webfile ChiSquare.

The test of independence summarizes the observed frequencies in a tabular format very similar to the one shown in Figure 12.5. The formulas to compute expected frequencies are also very similar to the formulas shown in the background worksheet. For the goodness of fit test, the user provides the observed frequencies in a column rather than a table. The user must also provide the associated expected frequencies in another column. Lastly, the CHISQ.TEST function is used to obtain the p-value as described above.

Appendix 12.3 Chi-Square Tests Using StatTools

Test the Equality of Population Proportions and Test of Independence

The StatTools procedure is identical for both of these applications. In each case, the user must do the following before creating an Excel worksheet that can be used to perform the test.

1. Select a sample from the population or populations and record the data
2. Summarize the data to show the observed frequencies in a tabular format

We will begin the StatTools chi-square test procedure with the understanding that the user has already determined the observed frequencies for the study.

WEB file

AutoLoyalty

Let us demonstrate the StatTools chi-square test by considering the automobile loyalty example presented in Section 12.1. Using the data in the web file AutoLoyalty and the Excel PivotTable procedure, we obtained the observed frequencies shown in the Excel worksheet of Figure 12.5. Note that the observed frequencies including the row and column heading are located from cell A6 to cell D8. This is all the information needed to conduct the chi-square test with StatTools. The steps are as follows.

Step 1. Select **Statistical Inference**
Step 2. Select **Chi-Square Independence Test**
Step 3. When the Chi-Square Test for Independence dialog box appears:
Enter A6:D8 in the **Contingency Table Range** box
Click **Table Includes Row and Column Headers**
Click **OK**

A test of independence application would begin with a tabular summary of the observed frequencies for the two categorical variables. The three steps shown above will provide the test of independence results.

Goodness of Fit Test for a Normal Probability Distribution

WEB file

Chemline

StatTools provides a routine for the chi-square goodness of fit test when the population is assumed to have a normal probability distribution. Let us consider the sample of 50 Chemline employee aptitude test scores presented in Section 12.3. The assumption is that the population of test scores has a normal probability distribution. Open the Chemline file and begin by using the Data Set Manager to create a StatTools data set using the procedure described in the appendix in Chapter 1. The remaining steps are as follows.

Step 1. Select **Normality Tests**
Step 2. Select **Chi-square Test**
Step 3. When the Chi-Square Normality Text box appears:
Check the **Name** Score
Click **OK**

The StatTools add-in simplifies the steps required when doing chi-square tests with Excel.

StatTools automatically establishes the data ranges that determine the k categories for the data. Since StatTools may establish the test score categories differently than we did in Section 12.3, the p-values can differ slightly. However, the test conclusion remains the same.

CHAPTER 13

Experimental Design and Analysis of Variance

CONTENTS

UNIVERSITY OF GLAMORGAN PRIFYSGOL MORGANNWG
Learning Resources Centre

STATISTICS *in* PRACTICE

BURKE MARKETING SERVICES, INC.*
CINCINNATI, OHIO

Burke Marketing Services, Inc., is one of the most experienced market research firms in the industry. Burke writes more proposals, on more projects, every day than any other market research company in the world. Supported by state-of-the-art technology, Burke offers a wide variety of research capabilities, providing answers to nearly any marketing question.

In one study, a firm retained Burke to evaluate potential new versions of a children's dry cereal. To maintain confidentiality, we refer to the cereal manufacturer as the Anon Company. The four key factors that Anon's product developers thought would enhance the taste of the cereal were the following:

1. Ratio of wheat to corn in the cereal flake
2. Type of sweetener: sugar, honey, or artificial
3. Presence or absence of flavor bits with a fruit taste
4. Short or long cooking time

Burke designed an experiment to determine what effects these four factors had on cereal taste. For example, one test cereal was made with a specified ratio of wheat to corn, sugar as the sweetener, flavor bits, and a short cooking time; another test cereal was made with a different ratio of wheat to corn and the other three factors the same, and so on. Groups of children then taste-tested the cereals and stated what they thought about the taste of each.

Burke uses taste tests to provide valuable statistical information on what customers want from a product. © Mircea Foto/Shutterstock.com.

Analysis of variance was the statistical method used to study the data obtained from the taste tests. The results of the analysis showed the following:

- The flake composition and sweetener type were highly influential in taste evaluation.
- The flavor bits actually detracted from the taste of the cereal.
- The cooking time had no effect on the taste.

This information helped Anon identify the factors that would lead to the best-tasting cereal.

The experimental design employed by Burke and the subsequent analysis of variance were helpful in making a product design recommendation. In this chapter, we will see how such procedures are carried out.

*The authors are indebted to Dr. Ronald Tatham of Burke Marketing Services for providing this Statistics in Practice.

In Chapter 1 we stated that statistical studies can be classified as either experimental or observational. In an experimental statistical study, an experiment is conducted to generate the data. An experiment begins with identifying a variable of interest. Then one or more other variables, thought to be related, are identified and controlled, and data are collected about how those variables influence the variable of interest.

In an observational study, data are usually obtained through sample surveys and not a controlled experiment. Good design principles are still employed, but the rigorous controls associated with an experimental statistical study are often not possible. For instance, in a study of the relationship between smoking and lung cancer the researcher cannot assign a smoking habit to subjects. The researcher is restricted to simply observing the effects of smoking on people who already smoke and the effects of not smoking on people who do not already smoke.

Sir Ronald Aylmer Fisher (1890–1962) invented the branch of statistics known as experimental design. In addition to being accomplished in statistics, he was a noted scientist in the field of genetics.

In this chapter we introduce three types of experimental designs: a completely randomized design, a randomized block design, and a factorial experiment. For each design we show how a statistical procedure called analysis of variance (ANOVA) can be used to analyze the data available. ANOVA can also be used to analyze the data obtained through an observation a study. For instance, we will see that the ANOVA procedure used for a completely randomized experimental design also works for testing the equality of three or more population means when data are obtained through an observational study. In the following chapters we will see that ANOVA plays a key role in analyzing the results of regression studies involving both experimental and observational data.

In the first section, we introduce the basic principles of an experimental study and show how they are employed in a completely randomized design. In the second section, we then show how ANOVA can be used to analyze the data from a completely randomized experimental design. In later sections we discuss multiple comparison procedures and two other widely used experimental designs, the randomized block design and the factorial experiment.

13.1 An Introduction to Experimental Design and Analysis of Variance

Cause-and-effect relationships can be difficult to establish in observational studies; such relationships are easier to establish in experimental studies.

As an example of an experimental statistical study, let us consider the problem facing Chemitech, Inc. Chemitech developed a new filtration system for municipal water supplies. The components for the new filtration system will be purchased from several suppliers, and Chemitech will assemble the components at its plant in Columbia, South Carolina. The industrial engineering group is responsible for determining the best assembly method for the new filtration system. After considering a variety of possible approaches, the group narrows the alternatives to three: method A, method B, and method C. These methods differ in the sequence of steps used to assemble the system. Managers at Chemitech want to determine which assembly method can produce the greatest number of filtration systems per week.

In the Chemitech experiment, assembly method is the independent variable or **factor**. Because three assembly methods correspond to this factor, we say that three treatments are associated with this experiment; each **treatment** corresponds to one of the three assembly methods. The Chemitech problem is an example of a **single-factor experiment**; it involves one categorical factor (method of assembly). More complex experiments may consist of multiple factors; some factors may be categorical and others may be quantitative.

The three assembly methods or treatments define the three populations of interest for the Chemitech experiment. One population is all Chemitech employees who use assembly method A, another is those who use method B, and the third is those who use method C. Note that for each population the dependent or **response variable** is the number of filtration systems assembled per week, and the primary statistical objective of the experiment is to determine whether the mean number of units produced per week is the same for all three populations (methods).

Randomization is the process of assigning the treatments to the experimental units at random. Prior to the work of Sir R. A. Fisher, treatments were assigned on a systematic or subjective basis.

Suppose a random sample of three employees is selected from all assembly workers at the Chemitech production facility. In experimental design terminology, the three randomly selected workers are the **experimental units**. The experimental design that we will use for the Chemitech problem is called a **completely randomized design**. This type of design requires that each of the three assembly methods or treatments be assigned randomly to one of the experimental units or workers. For example, method A might be randomly assigned to the second worker, method B to the first worker, and method C to the third worker. The concept of *randomization,* as illustrated in this example, is an important principle of all experimental designs.

FIGURE 13.1 COMPLETELY RANDOMIZED DESIGN FOR EVALUATING THE CHEMITECH ASSEMBLY METHOD EXPERIMENT

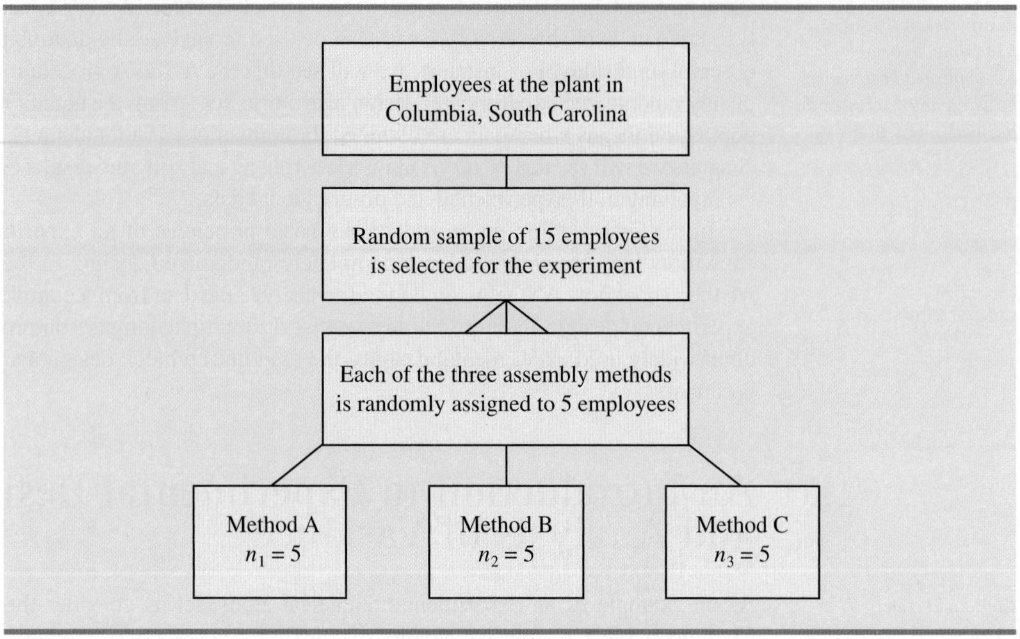

Note that this experiment would result in only one measurement or number of units assembled for each treatment. To obtain additional data for each assembly method, we must repeat or replicate the basic experimental process. Suppose, for example, that instead of selecting just three workers at random we selected 15 workers and then randomly assigned each of the three treatments to 5 of the workers. Because each method of assembly is assigned to 5 workers, we say that five replicates have been obtained. The process of *replication* is another important principle of experimental design. Figure 13.1 shows the completely randomized design for the Chemitech experiment.

Data Collection

Once we are satisfied with the experimental design, we proceed by collecting and analyzing the data. In the Chemitech case, the employees would be instructed in how to perform the assembly method assigned to them and then would begin assembling the new filtration systems using that method. After this assignment and training, the number of units assembled by each employee during one week is as shown in Table 13.1. The sample means, sample variances, and sample standard deviations for each assembly method are also provided. Thus, the sample mean number of units produced using method A is 62; the sample mean using method B is 66; and the sample mean using method C is 52. From these data, method B appears to result in higher production rates than either of the other methods.

The real issue is whether the three sample means observed are different enough for us to conclude that the means of the populations corresponding to the three methods of assembly are different. To write this question in statistical terms, we introduce the following notation.

$$\mu_1 = \text{mean number of units produced per week using method A}$$
$$\mu_2 = \text{mean number of units produced per week using method B}$$
$$\mu_3 = \text{mean number of units produced per week using method C}$$

TABLE 13.1 NUMBER OF UNITS PRODUCED BY 15 WORKERS

WEB file

Chemitech

	Method		
	A	**B**	**C**
	58	58	48
	64	69	57
	55	71	59
	66	64	47
	67	68	49
Sample mean	62	66	52
Sample variance	27.5	26.5	31.0
Sample standard deviation	5.244	5.148	5.568

Although we will never know the actual values of μ_1, μ_2, and μ_3, we want to use the sample means to test the following hypotheses.

If H_0 is rejected, we cannot conclude that all population means are different. Rejecting H_0 means that at least two population means have different values.

$$H_0\colon \mu_1 = \mu_2 = \mu_3$$
$$H_a\colon \text{Not all population means are equal}$$

As we will demonstrate shortly, analysis of variance (ANOVA) is the statistical procedure used to determine whether the observed differences in the three sample means are large enough to reject H_0.

Assumptions for Analysis of Variance

Three assumptions are required to use analysis of variance.

If the sample sizes are equal, analysis of variance is not sensitive to departures from the assumption of normally distributed populations.

1. **For each population, the response variable is normally distributed.** Implication: In the Chemitech experiment the number of units produced per week (response variable) must be normally distributed for each assembly method.
2. **The variance of the response variable, denoted σ^2, is the same for all of the populations.** Implication: In the Chemitech experiment, the variance of the number of units produced per week must be the same for each assembly method.
3. **The observations must be independent.** Implication: In the Chemitech experiment, the number of units produced per week for each employee must be independent of the number of units produced per week for any other employee.

Analysis of Variance: A Conceptual Overview

If the means for the three populations are equal, we would expect the three sample means to be close together. In fact, the closer the three sample means are to one another, the weaker the evidence we have for the conclusion that the population means differ. Alternatively, the more the sample means differ, the stronger the evidence we have for the conclusion that the population means differ. In other words, if the variability among the sample means is "small," it supports H_0; if the variability among the sample means is "large," it supports H_a.

If the null hypothesis, $H_0\colon \mu_1 = \mu_2 = \mu_3$, is true, we can use the variability among the sample means to develop an estimate of σ^2. First, note that if the assumptions for analysis

FIGURE 13.2 SAMPLING DISTRIBUTION OF $\bar{x}$ GIVEN H_0 IS TRUE

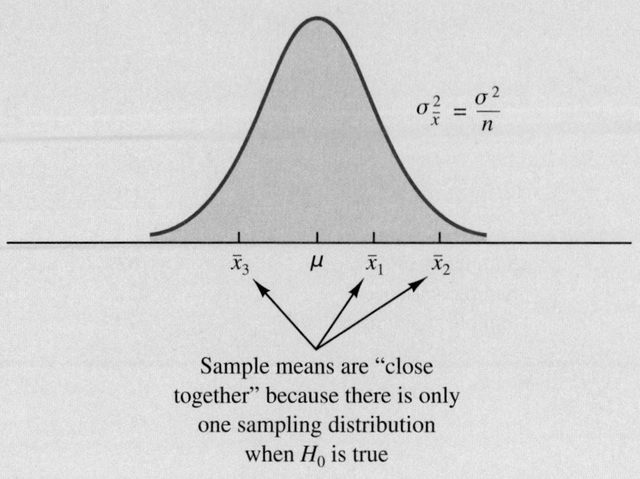

of variance are satisfied and the null hypothesis is true, each sample will have come from the same normal distribution with mean μ and variance σ^2. Recall from Chapter 7 that the sampling distribution of the sample mean $\bar{x}$ for a simple random sample of size n from a normal population will be normally distributed with mean μ and variance σ^2/n. Figure 13.2 illustrates such a sampling distribution.

Thus, if the null hypothesis is true, we can think of each of the three sample means, $\bar{x}_1 = 62$, $\bar{x}_2 = 66$, and $\bar{x}_3 = 52$ from Table 13.1, as values drawn at random from the sampling distribution shown in Figure 13.2. In this case, the mean and variance of the three $\bar{x}$ values can be used to estimate the mean and variance of the sampling distribution. When the sample sizes are equal, as in the Chemitech experiment, the best estimate of the mean of the sampling distribution of $\bar{x}$ is the mean or average of the sample means. In the Chemitech experiment, an estimate of the mean of the sampling distribution of $\bar{x}$ is $(62 + 66 + 52)/3 = 60$. We refer to this estimate as the *overall sample mean.* An estimate of the variance of the sampling distribution of $\bar{x}$, $\sigma^2_{\bar{x}}$, is provided by the variance of the three sample means.

$$s^2_{\bar{x}} = \frac{(62 - 60)^2 + (66 - 60)^2 + (52 - 60)^2}{3 - 1} = \frac{104}{2} = 52$$

Because $\sigma^2_{\bar{x}} = \sigma^2/n$, solving for σ^2 gives

$$\sigma^2 = n\sigma^2_{\bar{x}}$$

Hence,

$$\text{Estimate of } \sigma^2 = n\,(\text{Estimate of } \sigma^2_{\bar{x}}) = ns^2_{\bar{x}} = 5(52) = 260$$

The result, $ns^2_{\bar{x}} = 260$, is referred to as the *between-treatments* estimate of σ^2.

The between-treatments estimate of σ^2 is based on the assumption that the null hypothesis is true. In this case, each sample comes from the same population, and there is only

FIGURE 13.3 SAMPLING DISTRIBUTIONS OF $\bar{x}$ GIVEN H_0 IS FALSE

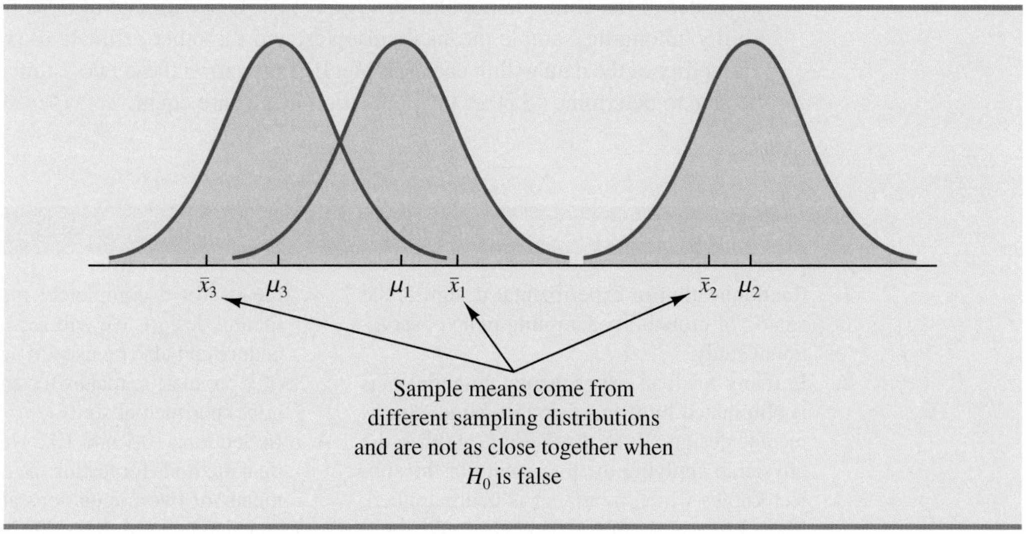
$\bar{x}_3$
μ_3
μ_1
$\bar{x}_1$
$\bar{x}_2$
μ_2
Sample means come from different sampling distributions and are not as close together when H_0 is false

one sampling distribution of $\bar{x}$. To illustrate what happens when H_0 is false, suppose the population means all differ. Note that because the three samples are from normal populations with different means, they will result in three different sampling distributions. Figure 13.3 shows that in this case, the sample means are not as close together as they were when H_0 was true. Thus, $s_{\bar{x}}^2$ will be larger, causing the between-treatments estimate of σ^2 to be larger. In general, when the population means are not equal, the between-treatments estimate will overestimate the population variance σ^2.

The variation within each of the samples also has an effect on the conclusion we reach in analysis of variance. When a simple random sample is selected from each population, each of the sample variances provides an unbiased estimate of σ^2. Hence, we can combine or pool the individual estimates of σ^2 into one overall estimate. The estimate of σ^2 obtained in this way is called the *pooled* or *within-treatments* estimate of σ^2. Because each sample variance provides an estimate of σ^2 based only on the variation within each sample, the within-treatments estimate of σ^2 is not affected by whether the population means are equal. When the sample sizes are equal, the within-treatments estimate of σ^2 can be obtained by computing the average of the individual sample variances. For the Chemitech experiment we obtain

$$\text{Within-treatments estimate of } \sigma^2 = \frac{27.5 + 26.5 + 31.0}{3} = \frac{85}{3} = 28.33$$

In the Chemitech experiment, the between-treatments estimate of σ^2 (260) is much larger than the within-treatments estimate of σ^2 (28.33). In fact, the ratio of these two estimates is 260/28.33 = 9.18. Recall, however, that the between-treatments approach provides a good estimate of σ^2 only if the null hypothesis is true; if the null hypothesis is false, the between-treatments approach overestimates σ^2. The within-treatments approach provides a good estimate of σ^2 in either case. Thus, if the null hypothesis is true, the two estimates will be similar and their ratio will be close to 1. If the null hypothesis is false, the between-treatments estimate will be larger than the within-treatments estimate, and their ratio will be large. In the next section we will show how large this ratio must be to reject H_0.

In summary, the logic behind ANOVA is based on the development of two independent estimates of the common population variance σ^2. One estimate of σ^2 is based on the variability among the sample means themselves, and the other estimate of σ^2 is based on the variability of the data within each sample. By comparing these two estimates of σ^2, we will be able to determine whether the population means are equal.

NOTES AND COMMENTS

1. Randomization in experimental design is the analog of probability sampling in an observational study.
2. In many medical experiments, potential bias is eliminated by using a double-blind experimental design. With this design, neither the physician applying the treatment nor the subject knows which treatment is being applied. Many other types of experiments could benefit from this type of design.
3. In this section we provided a conceptual overview of how analysis of variance can be used to test for the equality of k population means for a completely randomized experimental design. We will see that the same procedure can also be used to test for the equality of k population means for an observational or nonexperimental study.
4. In Sections 10.1 and 10.2 we presented statistical methods for testing the hypothesis that the means of two populations are equal. ANOVA can also be used to test the hypothesis that the means of two populations are equal. In practice, however, analysis of variance is usually not used except when dealing with three or more population means.

13.2 Analysis of Variance and the Completely Randomized Design

In this section we show how analysis of variance can be used to test for the equality of k population means for a completely randomized design. The general form of the hypotheses tested is

$$H_0: \mu_1 = \mu_2 = \cdots = \mu_k$$
$$H_a: \text{Not all population means are equal}$$

where

$$\mu_j = \text{mean of the } j\text{th population}$$

We assume that a simple random sample of size n_j has been selected from each of the k populations or treatments. For the resulting sample data, let

$$\begin{aligned} x_{ij} &= \text{value of observation } i \text{ for treatment } j \\ n_j &= \text{number of observations for treatment } j \\ \bar{x}_j &= \text{sample mean for treatment } j \\ s_j^2 &= \text{sample variance for treatment } j \\ s_j &= \text{sample standard deviation for treatment } j \end{aligned}$$

The formulas for the sample mean and sample variance for treatment j are as follow.

$$\bar{x}_j = \frac{\sum_{i=1}^{n_j} x_{ij}}{n_j} \tag{13.1}$$

$$s_j^2 = \frac{\sum_{i=1}^{n_j} (x_{ij} - \bar{x}_j)^2}{n_j - 1} \tag{13.2}$$

The overall sample mean, denoted $\bar{\bar{x}}$, is the sum of all the observations divided by the total number of observations. That is,

$$\bar{\bar{x}} = \frac{\sum_{j=1}^{k}\sum_{i=1}^{n_j} x_{ij}}{n_T} \tag{13.3}$$

where

$$n_T = n_1 + n_2 + \cdots + n_k \tag{13.4}$$

If the size of each sample is n, $n_T = kn$; in this case equation (13.3) reduces to

$$\bar{\bar{x}} = \frac{\sum_{j=1}^{k}\sum_{i=1}^{n_j} x_{ij}}{kn} = \frac{\sum_{j=1}^{k}\sum_{i=1}^{n_j} x_{ij}/n}{k} = \frac{\sum_{j=1}^{k} \bar{x}_j}{k} \tag{13.5}$$

In other words, whenever the sample sizes are the same, the overall sample mean is just the average of the k sample means.

Because each sample in the Chemitech experiment consists of $n = 5$ observations, the overall sample mean can be computed by using equation (13.5). For the data in Table 13.1 we obtained the following result.

$$\bar{\bar{x}} = \frac{62 + 66 + 52}{3} = 60$$

If the null hypothesis is true ($\mu_1 = \mu_2 = \mu_3 = \mu$), the overall sample mean of 60 is the best estimate of the population mean μ.

Between-Treatments Estimate of Population Variance

In the preceding section, we introduced the concept of a between-treatments estimate of σ^2 and showed how to compute it when the sample sizes were equal. This estimate of σ^2 is called the *mean square due to treatments* and is denoted MSTR. The general formula for computing MSTR is

$$\text{MSTR} = \frac{\sum_{j=1}^{k} n_j(\bar{x}_j - \bar{\bar{x}})^2}{k - 1} \tag{13.6}$$

The numerator in equation (13.6) is called the *sum of squares due to treatments* and is denoted SSTR. The denominator, $k - 1$, represents the degrees of freedom associated with SSTR. Hence, the mean square due to treatments can be computed using the following formula.

MEAN SQUARE DUE TO TREATMENTS

$$\text{MSTR} = \frac{\text{SSTR}}{k - 1} \tag{13.7}$$

where

$$\text{SSTR} = \sum_{j=1}^{k} n_j(\bar{x}_j - \bar{\bar{x}})^2 \tag{13.8}$$

If H_0 is true, MSTR provides an unbiased estimate of σ^2. However, if the means of the k populations are not equal, MSTR is not an unbiased estimate of σ^2; in fact, in that case, MSTR should overestimate σ^2.

For the Chemitech data in Table 13.1, we obtain the following results.

$$\text{SSTR} = \sum_{j=1}^{k} n_j(\bar{x}_j - \bar{\bar{x}})^2 = 5(62 - 60)^2 + 5(66 - 60)^2 + 5(52 - 60)^2 = 520$$

$$\text{MSTR} = \frac{\text{SSTR}}{k - 1} = \frac{520}{2} = 260$$

Within-Treatments Estimate of Population Variance

Earlier, we introduced the concept of a within-treatments estimate of σ^2 and showed how to compute it when the sample sizes were equal. This estimate of σ^2 is called the *mean square due to error* and is denoted MSE. The general formula for computing MSE is

$$\text{MSE} = \frac{\sum_{j=1}^{k}(n_j - 1)s_j^2}{n_T - k} \tag{13.9}$$

The numerator in equation (13.9) is called the *sum of squares due to error* and is denoted SSE. The denominator of MSE is referred to as the degrees of freedom associated with SSE. Hence, the formula for MSE can also be stated as follows.

MEAN SQUARE DUE TO ERROR

$$\text{MSE} = \frac{\text{SSE}}{n_T - k} \tag{13.10}$$

where

$$\text{SSE} = \sum_{j=1}^{k}(n_j - 1)s_j^2 \tag{13.11}$$

Note that MSE is based on the variation within each of the treatments; it is not influenced by whether the null hypothesis is true. Thus, MSE always provides an unbiased estimate of σ^2.

For the Chemitech data in Table 13.1 we obtain the following results.

$$\text{SSE} = \sum_{j=1}^{k}(n_j - 1)s_j^2 = (5-1)27.5 + (5-1)26.5 + (5-1)31 = 340$$

$$\text{MSE} = \frac{\text{SSE}}{n_T - k} = \frac{340}{15-3} = \frac{340}{12} = 28.33$$

Comparing the Variance Estimates: The F Test

An introduction to the F distribution and the use of the F distribution table were presented in Section 11.2.

If the null hypothesis is true, MSTR and MSE provide two independent, unbiased estimates of σ^2. Based on the material covered in Chapter 11 we know that for normal populations, the sampling distribution of the ratio of two independent estimates of σ^2 follows an F distribution. Hence, if the null hypothesis is true and the ANOVA assumptions are valid, the sampling distribution of MSTR/MSE is an F distribution with numerator degrees of freedom equal to $k - 1$ and denominator degrees of freedom equal to $n_T - k$. In other words, if the null hypothesis is true, the value of MSTR/MSE should appear to have been selected from this F distribution.

However, if the null hypothesis is false, the value of MSTR/MSE will be inflated because MSTR overestimates σ^2. Hence, we will reject H_0 if the resulting value of MSTR/MSE appears to be too large to have been selected from an F distribution with $k - 1$ numerator degrees of freedom and $n_T - k$ denominator degrees of freedom. Because the decision to reject H_0 is based on the value of MSTR/MSE, the test statistic used to test for the equality of k population means is as follows.

TEST STATISTIC FOR THE EQUALITY OF k POPULATION MEANS

$$F = \frac{\text{MSTR}}{\text{MSE}} \quad \textbf{(13.12)}$$

The test statistic follows an F distribution with $k - 1$ degrees of freedom in the numerator and $n_T - k$ degrees of freedom in the denominator.

Let us return to the Chemitech experiment and use a level of significance $\alpha = .05$ to conduct the hypothesis test. The value of the test statistic is

$$F = \frac{\text{MSTR}}{\text{MSE}} = \frac{260}{28.33} = 9.18$$

The numerator degrees of freedom is $k - 1 = 3 - 1 = 2$ and the denominator degrees of freedom is $n_T - k = 15 - 3 = 12$. Because we will only reject the null hypothesis for large values of the test statistic, the p-value is the upper tail area of the F distribution to the right of the test statistic $F = 9.18$. Figure 13.4 shows the sampling distribution of F = MSTR/MSE, the value of the test statistic, and the upper tail area that is the p-value for the hypothesis test.

From Table 4 of Appendix B we find the following areas in the upper tail of an F distribution with 2 numerator degrees of freedom and 12 denominator degrees of freedom.

Area in Upper Tail	**.10**	**.05**	**.025**	**.01**
F Value ($df_1 = 2, df_2 = 12$)	2.81	3.89	5.10	6.93

$F = 9.18$

FIGURE 13.4 COMPUTATION OF p-VALUE USING THE SAMPLING DISTRIBUTION OF MSTR/MSE

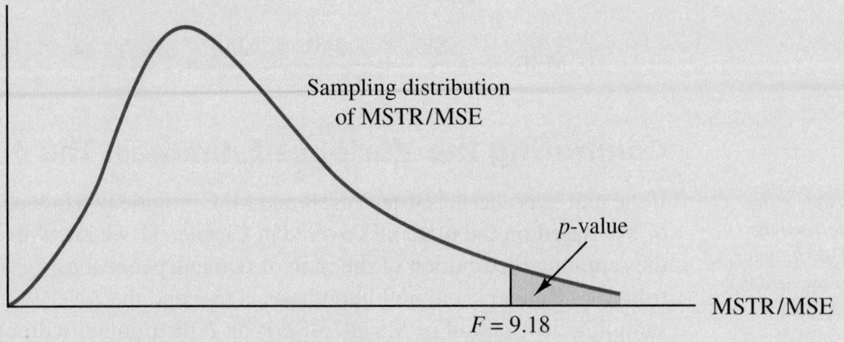

Appendix F shows how to compute p-values using Minitab or Excel.

Because $F = 9.18$ is greater than 6.93, the area in the upper tail at $F = 9.18$ is less than .01. Thus, the p-value is less than .01. Minitab or Excel can be used to show that the exact p-value is .004. With p-value $\leq \alpha = .05$, H_0 is rejected. The test provides sufficient evidence to conclude that the means of the three populations are not equal. In other words, analysis of variance supports the conclusion that the population mean number of units produced per week for the three assembly methods are not equal.

As with other hypothesis testing procedures, the critical value approach may also be used. With $\alpha = .05$, the critical F value occurs with an area of .05 in the upper tail of an F distribution with 2 and 12 degrees of freedom. From the F distribution table, we find $F_{.05} = 3.89$. Hence, the appropriate upper tail rejection rule for the Chemitech experiment is

$$\text{Reject } H_0 \text{ if } F \geq 3.89$$

With $F = 9.18$, we reject H_0 and conclude that the means of the three populations are not equal. A summary of the overall procedure for testing for the equality of k population means follows.

TEST FOR THE EQUALITY OF k POPULATION MEANS

$$H_0\colon \mu_1 = \mu_2 = \cdots = \mu_k$$
$$H_a\colon \text{Not all population means are equal}$$

TEST STATISTIC

$$F = \frac{\text{MSTR}}{\text{MSE}}$$

REJECTION RULE

p-value approach: Reject H_0 if p-value $\leq \alpha$

Critical value approach: Reject H_0 if $F \geq F_\alpha$

where the value of F_α is based on an F distribution with $k - 1$ numerator degrees of freedom and $n_T - k$ denominator degrees of freedom.

TABLE 13.2 ANOVA TABLE FOR A COMPLETELY RANDOMIZED DESIGN

Source of Variation	Sum of Squares	Degrees of Freedom	Mean Square	F	p-value
Treatments	SSTR	$k - 1$	$\text{MSTR} = \dfrac{\text{SSTR}}{k - 1}$	$\dfrac{\text{MSTR}}{\text{MSE}}$	
Error	SSE	$n_T - k$	$\text{MSE} = \dfrac{\text{SSE}}{n_T - k}$		
Total	SST	$n_T - 1$			

ANOVA Table

The results of the preceding calculations can be displayed conveniently in a table referred to as the analysis of variance or **ANOVA table**. The general form of the ANOVA table for a completely randomized design is shown in Table 13.2; Table 13.3 is the corresponding ANOVA table for the Chemitech experiment. The sum of squares associated with the source of variation referred to as "Total" is called the total sum of squares (SST). Note that the results for the Chemitech experiment suggest that SST = SSTR + SSE, and that the degrees of freedom associated with this total sum of squares is the sum of the degrees of freedom associated with the sum of squares due to treatments and the sum of squares due to error.

We point out that SST divided by its degrees of freedom $n_T - 1$ is nothing more than the overall sample variance that would be obtained if we treated the entire set of 15 observations as one data set. With the entire data set as one sample, the formula for computing the total sum of squares, SST, is

$$\text{SST} = \sum_{j=1}^{k}\sum_{i=1}^{n_j}(x_{ij} - \bar{\bar{x}})^2 \tag{13.13}$$

It can be shown that the results we observed for the analysis of variance table for the Chemitech experiment also apply to other problems. That is,

$$\text{SST} = \text{SSTR} + \text{SSE} \tag{13.14}$$

Analysis of variance can be thought of as a statistical procedure for partitioning the total sum of squares into separate components.

In other words, SST can be partitioned into two sums of squares: the sum of squares due to treatments and the sum of squares due to error. Note also that the degrees of freedom corresponding to SST, $n_T - 1$, can be partitioned into the degrees of freedom corresponding to SSTR, $k - 1$, and the degrees of freedom corresponding to SSE, $n_T - k$. The analysis of variance can be viewed as the process of **partitioning** the total sum of squares and the degrees of freedom into their corresponding sources: treatments and error. Dividing the sum of squares by the appropriate degrees of freedom provides the variance estimates, the F value, and the p-value used to test the hypothesis of equal population means.

TABLE 13.3 ANALYSIS OF VARIANCE TABLE FOR THE CHEMITECH EXPERIMENT

Source of Variation	Sum of Squares	Degrees of Freedom	Mean Square	F	p-value
Treatments	520	2	260.00	9.18	.004
Error	340	12	28.33		
Total	860	14			

FIGURE 13.5 MINITAB OUTPUT FOR THE CHEMITECH EXPERIMENT ANALYSIS OF VARIANCE

```
Source    DF       SS       MS      F      P
Factor     2    520.0    260.0   9.18  0.004
Error     12    340.0     28.3
Total     14    860.0

S = 5.323    R-Sq = 60.47%    R-Sq(adj) = 53.88%

                                   Individual 95% CIs For Mean Based on
                                   Pooled StDev
 Level   N     Mean   StDev   ---+---------+---------+---------+------
 A       5   62.000   5.244                  (-------*-------)
 B       5   66.000   5.148                         (------*-------)
 C       5   52.000   5.568   (------*-------)
                              ---+---------+---------+---------+------
Pooled StDev = 5.323            49.0      56.0      63.0      70.0
```

Computer Results for Analysis of Variance

Using statistical computer packages, analysis of variance computations with large sample sizes or a large number of populations can be performed easily. Appendixes 13.1–13.3 show the steps required to use Minitab, Excel, and StatTools to perform the analysis of variance computations. In Figure 13.5 we show output for the Chemitech experiment obtained using Minitab. The first part of the computer output contains the familiar ANOVA table format. Comparing Figure 13.5 with Table 13.3, we see that the same information is available, although some of the headings are slightly different. The heading Source is used for the source of variation column, Factor identifies the treatments row, and the sum of squares and degrees of freedom columns are interchanged.

Note that following the ANOVA table the computer output contains the respective sample sizes, the sample means, and the standard deviations. In addition, Minitab provides a figure that shows individual 95% confidence interval estimates of each population mean. In developing these confidence interval estimates, Minitab uses MSE as the estimate of σ^2. Thus, the square root of MSE provides the best estimate of the population standard deviation σ. This estimate of σ on the computer output is Pooled StDev; it is equal to 5.323. To provide an illustration of how these interval estimates are developed, we will compute a 95% confidence interval estimate of the population mean for method A.

From our study of interval estimation in Chapter 8, we know that the general form of an interval estimate of a population mean is

$$\bar{x} \pm t_{\alpha/2} \frac{s}{\sqrt{n}} \qquad \textbf{(13.15)}$$

where s is the estimate of the population standard deviation σ. Because the best estimate of σ is provided by the Pooled StDev, we use a value of 5.323 for s in expression (13.15). The degrees of freedom for the t value is 12, the degrees of freedom associated with the error sum of squares. Hence, with $t_{.025} = 2.179$ we obtain

$$62 \pm 2.179 \frac{5.323}{\sqrt{5}} = 62 \pm 5.19$$

Thus, the individual 95% confidence interval for method A goes from 62 − 5.19 = 56.81 to 62 + 5.19 = 67.19. Because the sample sizes are equal for the Chemitech experiment, the individual confidence intervals for methods B and C are also constructed by adding and subtracting 5.19 from each sample mean. Thus, in the figure provided by Minitab we see that the widths of the confidence intervals are the same.

Testing for the Equality of *k* Population Means: An Observational Study

We have shown how analysis of variance can be used to test for the equality of *k* population means for a completely randomized experimental design. It is important to understand that ANOVA can also be used to test for the equality of three or more population means using data obtained from an observational study. As an example, let us consider the situation at National Computer Products, Inc. (NCP).

NCP manufactures printers and fax machines at plants located in Atlanta, Dallas, and Seattle. To measure how much employees at these plants know about quality management, a random sample of 6 employees was selected from each plant and the employees selected were given a quality awareness examination. The examination scores for these 18 employees are shown in Table 13.4. The sample means, sample variances, and sample standard deviations for each group are also provided. Managers want to use these data to test the hypothesis that the mean examination score is the same for all three plants.

We define population 1 as all employees at the Atlanta plant, population 2 as all employees at the Dallas plant, and population 3 as all employees at the Seattle plant. Let

$$\mu_1 = \text{mean examination score for population 1}$$
$$\mu_2 = \text{mean examination score for population 2}$$
$$\mu_3 = \text{mean examination score for population 3}$$

Although we will never know the actual values of μ_1, μ_2, and μ_3, we want to use the sample results to test the following hypotheses.

$$H_0\colon \mu_1 = \mu_2 = \mu_3$$
$$H_a\colon \text{Not all population means are equal}$$

Note that the hypothesis test for the NCP observational study is exactly the same as the hypothesis test for the Chemitech experiment. Indeed, the same analysis of variance

TABLE 13.4 EXAMINATION SCORES FOR 18 EMPLOYEES

WEB file NCP

	Plant 1 Atlanta	Plant 2 Dallas	Plant 3 Seattle
	85	71	59
	75	75	64
	82	73	62
	76	74	69
	71	69	75
	85	82	67
Sample mean	79	74	66
Sample variance	34	20	32
Sample standard deviation	5.83	4.47	5.66

Exercise 8 will ask you to analyze the NCP data using the analysis of variance procedure.

methodology we used to analyze the Chemitech experiment can also be used to analyze the data from the NCP observational study.

Even though the same ANOVA methodology is used for the analysis, it is worth noting how the NCP observational statistical study differs from the Chemitech experimental statistical study. The individuals who conducted the NCP study had no control over how the plants were assigned to individual employees. That is, the plants were already in operation and a particular employee worked at one of the three plants. All that NCP could do was to select a random sample of 6 employees from each plant and administer the quality awareness examination. To be classified as an experimental study, NCP would have had to be able to randomly select 18 employees and then assign the plants to each employee in a random fashion.

NOTES AND COMMENTS

1. The overall sample mean can also be computed as a weighted average of the k sample means.

$$\bar{\bar{x}} = \frac{n_1\bar{x}_1 + n_2\bar{x}_2 + \cdots + n_k\bar{x}_k}{n_T}$$

In problems where the sample means are provided, this formula is simpler than equation (13.3) for computing the overall mean.

2. If each sample consists of n observations, equation (13.6) can be written as

$$\text{MSTR} = \frac{n\sum_{j=1}^{k}(\bar{x}_j - \bar{\bar{x}})^2}{k-1} = n\left[\frac{\sum_{j=1}^{k}(\bar{x}_j - \bar{\bar{x}})^2}{k-1}\right] = ns_{\bar{x}}^2$$

Note that this result is the same as presented in Section 13.1 when we introduced the concept of the between-treatments estimate of σ^2. Equation (13.6) is simply a generalization of this result to the unequal sample-size case.

3. If each sample has n observations, $n_T = kn$; thus, $n_T - k = k(n - 1)$, and equation (13.9) can be rewritten as

$$\text{MSE} = \frac{\sum_{j=1}^{k}(n-1)s_j^2}{k(n-1)} = \frac{(n-1)\sum_{j=1}^{k}s_j^2}{k(n-1)} = \frac{\sum_{j=1}^{k}s_j^2}{k}$$

In other words, if the sample sizes are the same, MSE is the average of the k sample variances. Note that it is the same result we used in Section 13.1 when we introduced the concept of the within-treatments estimate of σ^2.

Exercises

Methods

SELF test

1. The following data are from a completely randomized design.

	Treatment		
	A	B	C
	162	142	126
	142	156	122
	165	124	138
	145	142	140
	148	136	150
	174	152	128
Sample mean	156	142	134
Sample variance	164.4	131.2	110.4

a. Compute the sum of squares between treatments.
b. Compute the mean square between treatments.

c. Compute the sum of squares due to error.
d. Compute the mean square due to error.
e. Set up the ANOVA table for this problem.
f. At the $\alpha = .05$ level of significance, test whether the means for the three treatments are equal.

2. In a completely randomized design, seven experimental units were used for each of the five levels of the factor. Complete the following ANOVA table.

Source of Variation	Sum of Squares	Degrees of Freedom	Mean Square	F	p-value
Treatments	300				
Error					
Total	460				

3. Refer to exercise 2.
   a. What hypotheses are implied in this problem?
   b. At the $\alpha = .05$ level of significance, can we reject the null hypothesis in part (a)? Explain.

4. In an experiment designed to test the output levels of three different treatments, the following results were obtained: SST = 400, SSTR = 150, $n_T = 19$. Set up the ANOVA table and test for any significant difference between the mean output levels of the three treatments. Use $\alpha = .05$.

5. In a completely randomized design, 12 experimental units were used for the first treatment, 15 for the second treatment, and 20 for the third treatment. Complete the following analysis of variance. At a .05 level of significance, is there a significant difference between the treatments?

Source of Variation	Sum of Squares	Degrees of Freedom	Mean Square	F	p-value
Treatments	1200				
Error					
Total	1800				

6. Develop the analysis of variance computations for the following completely randomized design. At $\alpha = .05$, is there a significant difference between the treatment means?

WEB file

Exer6

	Treatment		
	A	B	C
	136	107	92
	120	114	82
	113	125	85
	107	104	101
	131	107	89
	114	109	117
	129	97	110
	102	114	120
		104	98
		89	106
$\bar{x}_j$	119	107	100
s_j^2	146.86	96.44	173.78

Applications

7. Three different methods for assembling a product were proposed by an industrial engineer. To investigate the number of units assembled correctly with each method, 30 employees were randomly selected and randomly assigned to the three proposed methods in such a way that each method was used by 10 workers. The number of units assembled correctly was recorded, and the analysis of variance procedure was applied to the resulting data set. The following results were obtained: SST = 10,800; SSTR = 4560.
   a. Set up the ANOVA table for this problem.
   b. Use $\alpha = .05$ to test for any significant difference in the means for the three assembly methods.
8. Refer to the NCP data in Table 13.4. Set up the ANOVA table and test for any significant difference in the mean examination score for the three plants. Use $\alpha = .05$.
9. To study the effect of temperature on yield in a chemical process, five batches were produced at each of three temperature levels. The results follow. Construct an analysis of variance table. Use a .05 level of significance to test whether the temperature level has an effect on the mean yield of the process.

Temperature		
50° C	60° C	70° C
34	30	23
24	31	28
36	34	28
39	23	30
32	27	31

10. Auditors must make judgments about various aspects of an audit on the basis of their own direct experience, indirect experience, or a combination of the two. In a study, auditors were asked to make judgments about the frequency of errors to be found in an audit. The judgments by the auditors were then compared to the actual results. Suppose the following data were obtained from a similar study; lower scores indicate better judgments.

WEB file
AudJudg

Direct	Indirect	Combination
17.0	16.6	25.2
18.5	22.2	24.0
15.8	20.5	21.5
18.2	18.3	26.8
20.2	24.2	27.5
16.0	19.8	25.8
13.3	21.2	24.2

Use $\alpha = .05$ to test to see whether the basis for the judgment affects the quality of the judgment. What is your conclusion?

11. Four different paints are advertised as having the same drying time. To check the manufacturer's claims, five samples were tested for each of the paints. The time in minutes until the paint was dry enough for a second coat to be applied was recorded. The following data were obtained.

WEB file

Paint

Paint 1	Paint 2	Paint 3	Paint 4
128	144	133	150
137	133	143	142
135	142	137	135
124	146	136	140
141	130	131	153

At the $\alpha = .05$ level of significance, test to see whether the mean drying time is the same for each type of paint.

12. The *Consumer Reports* Restaurant Customer Satisfaction Survey is based upon 148,599 visits to full-service restaurant chains (*Consumer Reports* website). One of the variables in the study is meal price, the average amount paid per person for dinner and drinks, minus the tip. Suppose a reporter for the *Sun Coast Times* thought that it would be of interest to her readers to conduct a similar study for restaurants located on the Grand Strand section in Myrtle Beach, South Carolina. The reporter selected a sample of 8 seafood restaurants, 8 Italian restaurants, and 8 steakhouses. The following data show the meal prices ($) obtained for the 24 restaurants sampled. Use $\alpha = .05$ to test whether there is a significant difference among the mean meal price for the three types of restaurants.

WEB file

GrandStrand

Italian	Seafood	Steakhouse
$12	$16	$24
13	18	19
15	17	23
17	26	25
18	23	21
20	15	22
17	19	27
24	18	31

13.3 Multiple Comparison Procedures

When we use analysis of variance to test whether the means of k populations are equal, rejection of the null hypothesis allows us to conclude only that the population means are *not all equal.* In some cases we will want to go a step further and determine where the differences among means occur. The purpose of this section is to show how **multiple comparison procedures** can be used to conduct statistical comparisons between pairs of population means.

Fisher's LSD

Suppose that analysis of variance provides statistical evidence to reject the null hypothesis of equal population means. In this case, Fisher's least significant difference (LSD) procedure can be used to determine where the differences occur. To illustrate the use of Fisher's LSD procedure in making pairwise comparisons of population means, recall the Chemitech experiment introduced in Section 13.1. Using analysis of variance, we concluded that the mean number of units produced per week are not the same for the three assembly methods. In this case, the follow-up question is: We believe the assembly methods differ, but where do the differences occur? That is, do the means of populations 1 and 2 differ? Or those of populations 1 and 3? Or those of populations 2 and 3? The following table summarizes Fisher's LSD procedure for comparing pairs of population means.

FISHER'S LSD PROCEDURE

$$H_0: \mu_i = \mu_j$$
$$H_a: \mu_i \neq \mu_j$$

TEST STATISTIC

$$t = \frac{\bar{x}_i - \bar{x}_j}{\sqrt{\text{MSE}\left(\frac{1}{n_i} + \frac{1}{n_j}\right)}} \tag{13.16}$$

REJECTION RULE

p-value approach: Reject H_0 if *p*-value $\leq \alpha$
Critical value approach: Reject H_0 if $t \leq -t_{\alpha/2}$ or $t \geq t_{\alpha/2}$

where the value of $t_{\alpha/2}$ is based on a t distribution with $n_T - k$ degrees of freedom.

Let us now apply this procedure to determine whether there is a significant difference between the means of population 1 (method A) and population 2 (method B) at the $\alpha = .05$ level of significance. Table 13.1 showed that the sample mean is 62 for method A and 66 for method B. Table 13.3 showed that the value of MSE is 28.33; it is the estimate of σ^2 and is based on 12 degrees of freedom. For the Chemitech data the value of the test statistic is

$$t = \frac{62 - 66}{\sqrt{28.33\left(\frac{1}{5} + \frac{1}{5}\right)}} = -1.19$$

Because we have a two-tailed test, the *p*-value is two times the area under the curve for the t distribution to the left of $t = -1.19$. Using Table 2 in Appendix B, the t distribution table for 12 degrees of freedom provides the following information.

Area in Upper Tail	**.20**	**.10**	**.05**	**.025**	**.01**	**.005**
***t* Value (12 *df*)**	.873	1.356	1.782	2.179	2.681	3.055

$t = 1.19$

Appendix F shows how to compute p-values using Excel or Minitab.

The t distribution table only contains positive t values. Because the t distribution is symmetric, however, we can find the area under the curve to the right of $t = 1.19$ and double it to find the *p*-value corresponding to $t = -1.19$. We see that $t = 1.19$ is between .20 and .10. Doubling these amounts, we see that the *p*-value must be between .40 and .20. Excel or Minitab can be used to show that the exact *p*-value is .2571. Because the *p*-value is greater than $\alpha = .05$, we cannot reject the null hypothesis. Hence, we cannot conclude that the population mean number of units produced per week for method A is different from the population mean for method B.

Many practitioners find it easier to determine how large the difference between the sample means must be to reject H_0. In this case the test statistic is $\bar{x}_i - \bar{x}_j$, and the test is conducted by the following procedure.

FISHER'S LSD PROCEDURE BASED ON THE TEST STATISTIC $\bar{x}_i - \bar{x}_j$

$$H_0: \mu_i = \mu_j$$
$$H_a: \mu_i \neq \mu_j$$

TEST STATISTIC

$$\bar{x}_i - \bar{x}_j$$

REJECTION RULE AT A LEVEL OF SIGNIFICANCE α

$$\text{Reject } H_0 \text{ if } |\bar{x}_i - \bar{x}_j| \geq \text{LSD}$$

where

$$\text{LSD} = t_{\alpha/2}\sqrt{\text{MSE}\left(\frac{1}{n_i} + \frac{1}{n_j}\right)} \tag{13.17}$$

For the Chemitech experiment the value of LSD is

$$\text{LSD} = 2.179\sqrt{28.33\left(\frac{1}{5} + \frac{1}{5}\right)} = 7.34$$

Note that when the sample sizes are equal, only one value for LSD is computed. In such cases we can simply compare the magnitude of the difference between any two sample means with the value of LSD. For example, the difference between the sample means for population 1 (method A) and population 3 (method C) is $62 - 52 = 10$. This difference is greater than LSD $= 7.34$, which means we can reject the null hypothesis that the population mean number of units produced per week for method A is equal to the population mean for method C. Similarly, with the difference between the sample means for populations 2 and 3 of $66 - 52 = 14 > 7.34$, we can also reject the hypothesis that the population mean for method B is equal to the population mean for method C. In effect, our conclusion is that methods A and B both differ from method C.

Fisher's LSD can also be used to develop a confidence interval estimate of the difference between the means of two populations. The general procedure follows.

CONFIDENCE INTERVAL ESTIMATE OF THE DIFFERENCE BETWEEN TWO POPULATION MEANS USING FISHER'S LSD PROCEDURE

$$\bar{x}_i - \bar{x}_j \pm \text{LSD} \tag{13.18}$$

where

$$\text{LSD} = t_{\alpha/2}\sqrt{\text{MSE}\left(\frac{1}{n_i} + \frac{1}{n_j}\right)} \tag{13.19}$$

and $t_{\alpha/2}$ is based on a t distribution with $n_T - k$ degrees of freedom.

If the confidence interval in expression (13.18) includes the value zero, we cannot reject the hypothesis that the two population means are equal. However, if the confidence interval does not include the value zero, we conclude that there is a difference between the population means. For the Chemitech experiment, recall that LSD = 7.34 (corresponding to $t_{.025} = 2.179$). Thus, a 95% confidence interval estimate of the difference between the means of populations 1 and 2 is $62 - 66 \pm 7.34 = -4 \pm 7.34 = -11.34$ to 3.34; because this interval includes zero, we cannot reject the hypothesis that the two population means are equal.

Type I Error Rates

We began the discussion of Fisher's LSD procedure with the premise that analysis of variance gave us statistical evidence to reject the null hypothesis of equal population means. We showed how Fisher's LSD procedure can be used in such cases to determine where the differences occur. Technically, it is referred to as a *protected* or *restricted* LSD test because it is employed only if we first find a significant F value by using analysis of variance. To see why this distinction is important in multiple comparison tests, we need to explain the difference between a *comparisonwise* Type I error rate and an *experimentwise* Type I error rate.

In the Chemitech experiment we used Fisher's LSD procedure to make three pairwise comparisons.

Test 1	**Test 2**	**Test 3**
$H_0\colon \mu_1 = \mu_2$	$H_0\colon \mu_1 = \mu_3$	$H_0\colon \mu_2 = \mu_3$
$H_a\colon \mu_1 \neq \mu_2$	$H_a\colon \mu_1 \neq \mu_3$	$H_a\colon \mu_2 \neq \mu_3$

In each case, we used a level of significance of $\alpha = .05$. Therefore, for each test, if the null hypothesis is true, the probability that we will make a Type I error is $\alpha = .05$; hence, the probability that we will not make a Type I error on each test is $1 - .05 = .95$. In discussing multiple comparison procedures we refer to this probability of a Type I error ($\alpha = .05$) as the **comparisonwise Type I error rate**; comparisonwise Type I error rates indicate the level of significance associated with a single pairwise comparison.

Let us now consider a slightly different question. What is the probability that in making three pairwise comparisons, we will commit a Type I error on at least one of the three tests? To answer this question, note that the probability that we will not make a Type I error on any of the three tests is $(.95)(.95)(.95) = .8574$.[1] Therefore, the probability of making at least one Type I error is $1 - .8574 = .1426$. Thus, when we use Fisher's LSD procedure to make all three pairwise comparisons, the Type I error rate associated with this approach is not .05, but actually .1426; we refer to this error rate as the *overall* or **experimentwise Type I error rate**. To avoid confusion, we denote the experimentwise Type I error rate as α_{EW}.

The experimentwise Type I error rate gets larger for problems with more populations. For example, a problem with five populations has 10 possible pairwise comparisons. If we tested all possible pairwise comparisons by using Fisher's LSD with a comparisonwise error rate of $\alpha = .05$, the experimentwise Type I error rate would be $1 - (1 - .05)^{10} = .40$. In such cases, practitioners look to alternatives that provide better control over the experimentwise error rate.

One alternative for controlling the overall experimentwise error rate, referred to as the Bonferroni adjustment, involves using a smaller comparisonwise error rate for each test. For example, if we want to test C pairwise comparisons and want the maximum probability of

[1]The assumption is that the three tests are independent, and hence the joint probability of the three events can be obtained by simply multiplying the individual probabilities. In fact, the three tests are not independent because MSE is used in each test; therefore, the error involved is even greater than that shown.

making a Type I error for the overall experiment to be α_{EW}, we simply use a comparisonwise error rate equal to α_{EW}/C. In the Chemitech experiment, if we want to use Fisher's LSD procedure to test all three pairwise comparisons with a maximum experimentwise error rate of $\alpha_{EW} = .05$, we set the comparisonwise error rate to be $\alpha = .05/3 = .017$. For a problem with five populations and 10 possible pairwise comparisons, the Bonferroni adjustment would suggest a comparisonwise error rate of .05/10 = .005. Recall from our discussion of hypothesis testing in Chapter 9 that for a fixed sample size, any decrease in the probability of making a Type I error will result in an increase in the probability of making a Type II error, which corresponds to accepting the hypothesis that the two population means are equal when in fact they are not equal. As a result, many practitioners are reluctant to perform individual tests with a low comparisonwise Type I error rate because of the increased risk of making a Type II error.

Several other procedures, such as Tukey's procedure and Duncan's multiple range test, have been developed to help in such situations. However, there is considerable controversy in the statistical community as to which procedure is "best." The truth is that no one procedure is best for all types of problems.

Exercises

Methods

SELF test

13. The following data are from a completely randomized design.

	Treatment A	Treatment B	Treatment C
	32	44	33
	30	43	36
	30	44	35
	26	46	36
	32	48	40
Sample mean	30	45	36
Sample variance	6.00	4.00	6.50

a. At the $\alpha = .05$ level of significance, can we reject the null hypothesis that the means of the three treatments are equal?

b. Use Fisher's LSD procedure to test whether there is a significant difference between the means for treatments A and B, treatments A and C, and treatments B and C. Use $\alpha = .05$.

c. Use Fisher's LSD procedure to develop a 95% confidence interval estimate of the difference between the means of treatments A and B.

14. The following data are from a completely randomized design. In the following calculations, use $\alpha = .05$.

	Treatment 1	Treatment 2	Treatment 3
	63	82	69
	47	72	54
	54	88	61
	40	66	48
$\bar{x}_j$	51	77	58
s_j^2	96.67	97.34	81.99

a. Use analysis of variance to test for a significant difference among the means of the three treatments.
b. Use Fisher's LSD procedure to determine which means are different.

Applications

SELF test

15. To test whether the mean time needed to mix a batch of material is the same for machines produced by three manufacturers, the Jacobs Chemical Company obtained the following data on the time (in minutes) needed to mix the material.

Manufacturer		
1	2	3
20	28	20
26	26	19
24	31	23
22	27	22

a. Use these data to test whether the population mean times for mixing a batch of material differ for the three manufacturers. Use $\alpha = .05$.
b. At the $\alpha = .05$ level of significance, use Fisher's LSD procedure to test for the equality of the means for manufacturers 1 and 3. What conclusion can you draw after carrying out this test?

SELF test

16. Refer to exercise 15. Use Fisher's LSD procedure to develop a 95% confidence interval estimate of the difference between the means for manufacturer 1 and manufacturer 2.
17. The following data are from an experiment designed to investigate the perception of corporate ethical values among individuals specializing in marketing (higher scores indicate higher ethical values).

Marketing Managers	Marketing Research	Advertising
6	5	6
5	5	7
4	4	6
5	4	5
6	5	6
4	4	6

a. Use $\alpha = .05$ to test for significant differences in perception among the three groups.
b. At the $\alpha = .05$ level of significance, we can conclude that there are differences in the perceptions for marketing managers, marketing research specialists, and advertising specialists. Use the procedures in this section to determine where the differences occur. Use $\alpha = .05$.

18. To test for any significant difference in the number of hours between breakdowns for four machines, the following data were obtained.

Machine 1	Machine 2	Machine 3	Machine 4
6.4	8.7	11.1	9.9
7.8	7.4	10.3	12.8
5.3	9.4	9.7	12.1
7.4	10.1	10.3	10.8
8.4	9.2	9.2	11.3
7.3	9.8	8.8	11.5

a. At the $\alpha = .05$ level of significance, what is the difference, if any, in the population mean times among the four machines?
b. Use Fisher's LSD procedure to test for the equality of the means for machines 2 and 4. Use a .05 level of significance.

19. Refer to exercise 18. Use the Bonferroni adjustment to test for a significant difference between all pairs of means. Assume that a maximum overall experimentwise error rate of .05 is desired.

20. The International League of Triple-A minor league baseball consists of 14 teams organized into three divisions: North, South, and West. The following data show the average attendance for the 14 teams in the International League (The Biz of Baseball website, January 2009). Also shown are the teams' records; W denotes the number of games won, L denotes the number of games lost, and PCT is the proportion of games played that were won.

WEB file

Triple-A

Team Name	Division	W	L	PCT	Attendance
Buffalo Bisons	North	66	77	.462	8812
Lehigh Valley IronPigs	North	55	89	.382	8479
Pawtucket Red Sox	North	85	58	.594	9097
Rochester Red Wings	North	74	70	.514	6913
Scranton-Wilkes Barre Yankees	North	88	56	.611	7147
Syracuse Chiefs	North	69	73	.486	5765
Charlotte Knights	South	63	78	.447	4526
Durham Bulls	South	74	70	.514	6995
Norfolk Tides	South	64	78	.451	6286
Richmond Braves	South	63	78	.447	4455
Columbus Clippers	West	69	73	.486	7795
Indianapolis Indians	West	68	76	.472	8538
Louisville Bats	West	88	56	.611	9152
Toledo Mud Hens	West	75	69	.521	8234

a. Use $\alpha = .05$ to test for any difference in the mean attendance for the three divisions.
b. Use Fisher's LSD procedure to determine where the differences occur. Use $\alpha = .05$.

13.4 Randomized Block Design

Thus far we have considered the completely randomized experimental design. Recall that to test for a difference among treatment means, we computed an F value by using the ratio

$$F = \frac{\text{MSTR}}{\text{MSE}} \tag{13.20}$$

A completely randomized design is useful when the experimental units are homogeneous. If the experimental units are heterogeneous, **blocking** *is often used to form homogeneous groups.*

A problem can arise whenever differences due to extraneous factors (ones not considered in the experiment) cause the MSE term in this ratio to become large. In such cases, the F value in equation (13.20) can become small, signaling no difference among treatment means when in fact such a difference exists.

In this section we present an experimental design known as a **randomized block design**. Its purpose is to control some of the extraneous sources of variation by removing such variation from the MSE term. This design tends to provide a better estimate of the true error variance and leads to a more powerful hypothesis test in terms of the ability to detect differences among treatment means. To illustrate, let us consider a stress study for air traffic controllers.

Air Traffic Controller Stress Test

A study measuring the fatigue and stress of air traffic controllers resulted in proposals for modification and redesign of the controller's workstation. After consideration of several designs for the workstation, three specific alternatives are selected as having the best potential for reducing controller stress. The key question is: To what extent do the three alternatives differ in terms of their effect on controller stress? To answer this question, we need to design an experiment that will provide measurements of air traffic controller stress under each alternative.

Experimental studies in business often involve experimental units that are highly heterogeneous; as a result, randomized block designs are often employed.

In a completely randomized design, a random sample of controllers would be assigned to each workstation alternative. However, controllers are believed to differ substantially in their ability to handle stressful situations. What is high stress to one controller might be only moderate or even low stress to another. Hence, when considering the within-group source of variation (MSE), we must realize that this variation includes both random error and error due to individual controller differences. In fact, managers expected controller variability to be a major contributor to the MSE term.

Blocking in experimental design is similar to stratification in sampling.

One way to separate the effect of the individual differences is to use a randomized block design. Such a design will identify the variability stemming from individual controller differences and remove it from the MSE term. The randomized block design calls for a single sample of controllers. Each controller in the sample is tested with each of the three workstation alternatives. In experimental design terminology, the workstation is the *factor of interest* and the controllers are the *blocks.* The three treatments or populations associated with the workstation factor correspond to the three workstation alternatives. For simplicity, we refer to the workstation alternatives as system A, system B, and system C.

The *randomized* aspect of the randomized block design is the random order in which the treatments (systems) are assigned to the controllers. If every controller were to test the three systems in the same order, any observed difference in systems might be due to the order of the test rather than to true differences in the systems.

To provide the necessary data, the three workstation alternatives were installed at the Cleveland Control Center in Oberlin, Ohio. Six controllers were selected at random and assigned to operate each of the systems. A follow-up interview and a medical examination of each controller participating in the study provided a measure of the stress for each controller on each system. The data are reported in Table 13.5.

Table 13.6 is a summary of the stress data collected. In this table we include column totals (treatments) and row totals (blocks) as well as some sample means that will be helpful in

TABLE 13.5 A RANDOMIZED BLOCK DESIGN FOR THE AIR TRAFFIC CONTROLLER STRESS TEST

WEB file AirTraffic

		Treatments		
		System A	**System B**	**System C**
Blocks	**Controller 1**	15	15	18
	Controller 2	14	14	14
	Controller 3	10	11	15
	Controller 4	13	12	17
	Controller 5	16	13	16
	Controller 6	13	13	13

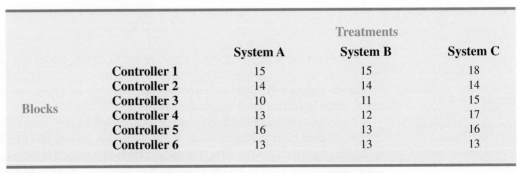

TABLE 13.6 SUMMARY OF STRESS DATA FOR THE AIR TRAFFIC CONTROLLER STRESS TEST

		Treatments			Row or Block Totals	Block Means
		System A	**System B**	**System C**		
Blocks	**Controller 1**	15	15	18	48	$\bar{x}_{1\cdot} = 48/3 = 16.0$
	Controller 2	14	14	14	42	$\bar{x}_{2\cdot} = 42/3 = 14.0$
	Controller 3	10	11	15	36	$\bar{x}_{3\cdot} = 36/3 = 12.0$
	Controller 4	13	12	17	42	$\bar{x}_{4\cdot} = 42/3 = 14.0$
	Controller 5	16	13	16	45	$\bar{x}_{5\cdot} = 45/3 = 15.0$
	Controller 6	13	13	13	39	$\bar{x}_{6\cdot} = 39/3 = 13.0$
Column or Treatment Totals		81	78	93	252	$\bar{\bar{x}} = \frac{252}{18} = 14.0$
Treatment Means		$\bar{x}_{\cdot 1} = \frac{81}{6} = 13.5$	$\bar{x}_{\cdot 2} = \frac{78}{6} = 13.0$	$\bar{x}_{\cdot 3} = \frac{93}{6} = 15.5$		

making the sum of squares computations for the ANOVA procedure. Because lower stress values are viewed as better, the sample data seem to favor system B with its mean stress rating of 13. However, the usual question remains: Do the sample results justify the conclusion that the population mean stress levels for the three systems differ? That is, are the differences statistically significant? An analysis of variance computation similar to the one performed for the completely randomized design can be used to answer this statistical question.

ANOVA Procedure

The ANOVA procedure for the randomized block design requires us to partition the sum of squares total (SST) into three groups: sum of squares due to treatments (SSTR), sum of squares due to blocks (SSBL), and sum of squares due to error (SSE). The formula for this partitioning follows.

$$\text{SST} = \text{SSTR} + \text{SSBL} + \text{SSE} \tag{13.21}$$

This sum of squares partition is summarized in the ANOVA table for the randomized block design as shown in Table 13.7. The notation used in the table is

$$k = \text{the number of treatments}$$
$$b = \text{the number of blocks}$$
$$n_T = \text{the total sample size } (n_T = kb)$$

Note that the ANOVA table also shows how the $n_T - 1$ total degrees of freedom are partitioned such that $k - 1$ degrees of freedom go to treatments, $b - 1$ go to blocks, and $(k - 1)(b - 1)$ go to the error term. The mean square column shows the sum of squares divided by the degrees of freedom, and $F = \text{MSTR/MSE}$ is the F ratio used to test for a significant difference among the treatment means. The primary contribution of the randomized block design is that, by including blocks, we remove the individual controller differences from the MSE term and obtain a more powerful test for the stress differences in the three workstation alternatives.

TABLE 13.7 ANOVA TABLE FOR THE RANDOMIZED BLOCK DESIGN WITH k TREATMENTS AND b BLOCKS

Source of Variation	Sum of Squares	Degrees of Freedom	Mean Square	F	p-value
Treatments	SSTR	$k - 1$	$\text{MSTR} = \dfrac{\text{SSTR}}{k-1}$	$\dfrac{\text{MSTR}}{\text{MSE}}$	
Blocks	SSBL	$b - 1$	$\text{MSBL} = \dfrac{\text{SSBL}}{b-1}$		
Error	SSE	$(k-1)(b-1)$	$\text{MSE} = \dfrac{\text{SSE}}{(k-1)(b-1)}$		
Total	SST	$n_T - 1$			

Computations and Conclusions

To compute the F statistic needed to test for a difference among treatment means with a randomized block design, we need to compute MSTR and MSE. To calculate these two mean squares, we must first compute SSTR and SSE; in doing so, we will also compute SSBL and SST. To simplify the presentation, we perform the calculations in four steps. In addition to k, b, and n_T as previously defined, the following notation is used.

x_{ij} = value of the observation corresponding to treatment j in block i
$\bar{x}_{\cdot j}$ = sample mean of the jth treatment
$\bar{x}_{i\cdot}$ = sample mean for the ith block
$\bar{\bar{x}}$ = overall sample mean

Step 1. Compute the total sum of squares (SST).

$$\text{SST} = \sum_{i=1}^{b}\sum_{j=1}^{k}(x_{ij} - \bar{\bar{x}})^2 \quad \textbf{(13.22)}$$

Step 2. Compute the sum of squares due to treatments (SSTR).

$$\text{SSTR} = b\sum_{j=1}^{k}(\bar{x}_{\cdot j} - \bar{\bar{x}})^2 \quad \textbf{(13.23)}$$

Step 3. Compute the sum of squares due to blocks (SSBL).

$$\text{SSBL} = k\sum_{i=1}^{b}(\bar{x}_{i\cdot} - \bar{\bar{x}})^2 \quad \textbf{(13.24)}$$

Step 4. Compute the sum of squares due to error (SSE).

$$\text{SSE} = \text{SST} - \text{SSTR} - \text{SSBL} \quad \textbf{(13.25)}$$

For the air traffic controller data in Table 13.6, these steps lead to the following sums of squares.

Step 1. $\text{SST} = (15 - 14)^2 + (15 - 14)^2 + (18 - 14)^2 + \cdots + (13 - 14)^2 = 70$
Step 2. $\text{SSTR} = 6[(13.5 - 14)^2 + (13.0 - 14)^2 + (15.5 - 14)^2] = 21$
Step 3. $\text{SSBL} = 3[(16 - 14)^2 + (14 - 14)^2 + (12 - 14)^2 + (14 - 14)^2 + (15 - 14)^2 + (13 - 14)^2] = 30$
Step 4. $\text{SSE} = 70 - 21 - 30 = 19$

TABLE 13.8 ANOVA TABLE FOR THE AIR TRAFFIC CONTROLLER STRESS TEST

Source of Variation	Sum of Squares	Degrees of Freedom	Mean Square	F	p-value
Treatments	21	2	10.5	10.5/1.9 = 5.53	.024
Blocks	30	5	6.0		
Error	19	10	1.9		
Total	70	17			

These sums of squares divided by their degrees of freedom provide the corresponding mean square values shown in Table 13.8.

Let us use a level of significance $\alpha = .05$ to conduct the hypothesis test. The value of the test statistic is

$$F = \frac{\text{MSTR}}{\text{MSE}} = \frac{10.5}{1.9} = 5.53$$

The numerator degrees of freedom is $k - 1 = 3 - 1 = 2$ and the denominator degrees of freedom is $(k - 1)(b - 1) = (3 - 1)(6 - 1) = 10$. Because we will only reject the null hypothesis for large values of the test statistic, the p-value is the area under the F distribution to the right of $F = 5.53$. From Table 4 of Appendix B we find that with the degrees of freedom 2 and 10, $F = 5.53$ is between $F_{.025} = 5.46$ and $F_{.01} = 7.56$. As a result, the area in the upper tail, or the p-value, is between .01 and .025. Alternatively, we can use Excel or Minitab to show that the exact p-value for $F = 5.53$ is .024. With p-value $\leq \alpha = .05$, we reject the null hypothesis H_0: $\mu_1 = \mu_2 = \mu_3$ and conclude that the population mean stress levels differ for the three workstation alternatives.

Some general comments can be made about the randomized block design. The experimental design described in this section is a *complete* block design; the word "complete" indicates that each block is subjected to all k treatments. That is, all controllers (blocks) were tested with all three systems (treatments). Experimental designs in which some but not all treatments are applied to each block are referred to as *incomplete* block designs. A discussion of incomplete block designs is beyond the scope of this text.

Because each controller in the air traffic controller stress test was required to use all three systems, this approach guarantees a complete block design. In some cases, however, blocking is carried out with "similar" experimental units in each block. For example, assume that in a pretest of air traffic controllers, the population of controllers was divided into groups ranging from extremely high-stress individuals to extremely low-stress individuals. The blocking could still be accomplished by having three controllers from each of the stress classifications participate in the study. Each block would then consist of three controllers in the same stress group. The randomized aspect of the block design would be the random assignment of the three controllers in each block to the three systems.

Finally, note that the ANOVA table shown in Table 13.7 provides an F value to test for treatment effects but *not* for blocks. The reason is that the experiment was designed to test a single factor—workstation design. The blocking based on individual stress differences was conducted to remove such variation from the MSE term. However, the study was not designed to test specifically for individual differences in stress.

Some analysts compute F = MSB/MSE and use that statistic to test for significance of the blocks. Then they use the result as a guide to whether the same type of blocking would be desired in future experiments. However, if individual stress difference is to be a factor in the study, a different experimental design should be used. A test of significance on blocks should not be performed as a basis for a conclusion about a second factor.

NOTES AND COMMENTS

The error degrees of freedom are less for a randomized block design than for a completely randomized design because $b - 1$ degrees of freedom are lost for the b blocks. If n is small, the potential effects due to blocks can be masked because of the loss of error degrees of freedom; for large n, the effects are minimized.

Exercises

Methods

SELF test

21. Consider the experimental results for the following randomized block design. Make the calculations necessary to set up the analysis of variance table.

		Treatments		
		A	**B**	**C**
Blocks	**1**	10	9	8
	2	12	6	5
	3	18	15	14
	4	20	18	18
	5	8	7	8

Use $\alpha = .05$ to test for any significant differences.

22. The following data were obtained for a randomized block design involving five treatments and three blocks: SST = 430, SSTR = 310, SSBL = 85. Set up the ANOVA table and test for any significant differences. Use $\alpha = .05$.

23. An experiment has been conducted for four treatments with eight blocks. Complete the following analysis of variance table.

Source of Variation	Sum of Squares	Degrees of Freedom	Mean Square	F
Treatments	900			
Blocks	400			
Error				
Total	1800			

Use $\alpha = .05$ to test for any significant differences.

Applications

24. An automobile dealer conducted a test to determine if the time in minutes needed to complete a minor engine tune-up depends on whether a computerized engine analyzer or an electronic analyzer is used. Because tune-up time varies among compact, intermediate, and full-sized cars, the three types of cars were used as blocks in the experiment. The data obtained follow.

		Analyzer	
		Computerized	Electronic
Car	Compact	50	42
	Intermediate	55	44
	Full-sized	63	46

Use $\alpha = .05$ to test for any significant differences.

25. The price drivers pay for gasoline often varies a great deal across regions throughout the United States. The following data show the price per gallon for regular gasoline for a random sample of gasoline service stations for three major brands of gasoline (Shell, BP, and Marathon) located in 11 metropolitan areas across the upper Midwest region (OhioGasPrices.com website, March 18, 2012).

WEB file

MidwestGas

Metropolitan Area	Shell	BP	Marathon
Akron, Ohio	3.77	3.83	3.78
Cincinnati, Ohio	3.72	3.83	3.87
Cleveland, Ohio	3.87	3.85	3.89
Columbus, Ohio	3.76	3.77	3.79
Ft. Wayne, Indiana	3.83	3.84	3.87
Indianapolis, Indiana	3.85	3.84	3.87
Lansing, Michigan	3.93	4.04	3.99
Lexington, Kentucky	3.79	3.78	3.79
Louisville, Kentucky	3.78	3.84	3.79
Muncie, Indiana	3.81	3.84	3.86
Toledo, Ohio	3.69	3.83	3.86

Use $\alpha = .05$ to test for any significant difference in the mean price of gasoline for the three brands.

26. The Scholastic Aptitude Test (SAT) contains three parts: critical reading, mathematics, and writing. Each part is scored on an 800-point scale. Information on test scores for the 2009 version of the SAT is available at the College Board website. A sample of SAT scores for six students follows.

WEB file

SATScores

Student	Critical Reading	Mathematics	Writing
1	526	534	530
2	594	590	586
3	465	464	445
4	561	566	553
5	436	478	430
6	430	458	420

a. Using a .05 level of significance, do students perform differently on the three portions of the SAT?
b. Which portion of the test seems to give the students the most trouble? Explain.

27. A study reported in the *Journal of the American Medical Association* investigated the cardiac demands of heavy snow shoveling. Ten healthy men underwent exercise testing with a treadmill and a cycle ergometer modified for arm cranking. The men then cleared two tracts of heavy, wet snow by using a lightweight plastic snow shovel and an electric snow thrower. Each subject's heart rate, blood pressure, oxygen uptake, and perceived exertion during snow removal were compared with the values obtained during treadmill

and arm-crank ergometer testing. Suppose the following table gives the heart rates in beats per minute for each of the 10 subjects.

WEB file

SnowShoveling

Subject	Treadmill	Arm-Crank Ergometer	Snow Shovel	Snow Thrower
1	177	205	180	98
2	151	177	164	120
3	184	166	167	111
4	161	152	173	122
5	192	142	179	151
6	193	172	205	158
7	164	191	156	117
8	207	170	160	123
9	177	181	175	127
10	174	154	191	109

At the .05 level of significance, test for any significant differences.

13.5 Factorial Experiment

The experimental designs we have considered thus far enable us to draw statistical conclusions about one factor. However, in some experiments we want to draw conclusions about more than one variable or factor. A **factorial experiment** is an experimental design that allows simultaneous conclusions about two or more factors. The term *factorial* is used because the experimental conditions include all possible combinations of the factors. For example, for a levels of factor A and b levels of factor B, the experiment will involve collecting data on ab treatment combinations. In this section we will show the analysis for a two-factor factorial experiment. The basic approach can be extended to experiments involving more than two factors.

As an illustration of a two-factor factorial experiment, we will consider a study involving the Graduate Management Admissions Test (GMAT), a standardized test used by graduate schools of business to evaluate an applicant's ability to pursue a graduate program in that field. Scores on the GMAT range from 200 to 800, with higher scores implying higher aptitude.

In an attempt to improve students' performance on the GMAT, a major Texas university is considering offering the following three GMAT preparation programs.

1. A three-hour review session covering the types of questions generally asked on the GMAT.
2. A one-day program covering relevant exam material, along with the taking and grading of a sample exam.
3. An intensive 10-week course involving the identification of each student's weaknesses and the setting up of individualized programs for improvement.

Hence, one factor in this study is the GMAT preparation program, which has three treatments: three-hour review, one-day program, and 10-week course. Before selecting the preparation program to adopt, further study will be conducted to determine how the proposed programs affect GMAT scores.

The GMAT is usually taken by students from three colleges: the College of Business, the College of Engineering, and the College of Arts and Sciences. Therefore, a second factor of interest in the experiment is whether a student's undergraduate college affects the GMAT score. This second factor, undergraduate college, also has three treatments: business, engineering, and arts and sciences. The factorial design for this experiment with three treatments corresponding to factor A, the preparation program, and three treatments corresponding to

TABLE 13.9 NINE TREATMENT COMBINATIONS FOR THE TWO-FACTOR GMAT EXPERIMENT

		Factor B: College		
		Business	Engineering	Arts and Sciences
Factor A:	Three-hour review	1	2	3
Preparation	One-day program	4	5	6
Program	10-week course	7	8	9

factor B, the undergraduate college, will have a total of $3 \times 3 = 9$ treatment combinations. These treatment combinations or experimental conditions are summarized in Table 13.9.

Assume that a sample of two students will be selected corresponding to each of the nine treatment combinations shown in Table 13.9: Two business students will take the three-hour review, two will take the one-day program, and two will take the 10-week course. In addition, two engineering students and two arts and sciences students will take each of the three preparation programs. In experimental design terminology, the sample size of two for each treatment combination indicates that we have two **replications**. Additional replications and a larger sample size could easily be used, but we elect to minimize the computational aspects for this illustration.

This experimental design requires that 6 students who plan to attend graduate school be randomly selected from *each* of the three undergraduate colleges. Then 2 students from each college should be assigned randomly to each preparation program, resulting in a total of 18 students being used in the study.

Let us assume that the randomly selected students participated in the preparation programs and then took the GMAT. The scores obtained are reported in Table 13.10.

The analysis of variance computations with the data in Table 13.10 will provide answers to the following questions.

- **Main effect (factor A):** Do the preparation programs differ in terms of effect on GMAT scores?
- **Main effect (factor B):** Do the undergraduate colleges differ in terms of effect on GMAT scores?
- **Interaction effect (factors A and B):** Do students in some colleges do better on one type of preparation program whereas others do better on a different type of preparation program?

The term **interaction** refers to a new effect that we can now study because we used a factorial experiment. If the interaction effect has a significant impact on the GMAT scores,

TABLE 13.10 GMAT SCORES FOR THE TWO-FACTOR EXPERIMENT

WEB file
GMATStudy

		Factor B: College		
		Business	Engineering	Arts and Sciences
Factor A: Preparation Program	Three-hour review	500	540	480
		580	460	400
	One-day program	460	560	420
		540	620	480
	10-week course	560	600	480
		600	580	410

TABLE 13.11 ANOVA TABLE FOR THE TWO-FACTOR FACTORIAL EXPERIMENT WITH r REPLICATIONS

Source of Variation	Sum of Squares	Degrees of Freedom	Mean Square	F	p-value
Factor A	SSA	$a - 1$	$\text{MSA} = \frac{\text{SSA}}{a - 1}$	$\frac{\text{MSA}}{\text{MSE}}$	
Factor B	SSB	$b - 1$	$\text{MSB} = \frac{\text{SSB}}{b - 1}$	$\frac{\text{MSB}}{\text{MSE}}$	
Interaction	SSAB	$(a - 1)(b - 1)$	$\text{MSAB} = \frac{\text{SSAB}}{(a - 1)(b - 1)}$	$\frac{\text{MSAB}}{\text{MSE}}$	
Error	SSE	$ab(r - 1)$	$\text{MSE} = \frac{\text{SSE}}{ab(r - 1)}$		
Total	SST	$n_T - 1$			

we can conclude that the effect of the type of preparation program depends on the undergraduate college.

ANOVA Procedure

The ANOVA procedure for the two-factor factorial experiment requires us to partition the sum of squares total (SST) into four groups: sum of squares for factor A (SSA), sum of squares for factor B (SSB), sum of squares for interaction (SSAB), and sum of squares due to error (SSE). The formula for this partitioning follows.

$$\text{SST} = \text{SSA} + \text{SSB} + \text{SSAB} + \text{SSE} \tag{13.26}$$

The partitioning of the sum of squares and degrees of freedom is summarized in Table 13.11. The following notation is used.

a = number of levels of factor A
b = number of levels of factor B
r = number of replications
n_T = total number of observations taken in the experiment; $n_T = abr$

Computations and Conclusions

To compute the F statistics needed to test for the significance of factor A, factor B, and interaction, we need to compute MSA, MSB, MSAB, and MSE. To calculate these four mean squares, we must first compute SSA, SSB, SSAB, and SSE; in doing so we will also compute SST. To simplify the presentation, we perform the calculations in five steps. In addition to a, b, r, and n_T as previously defined, the following notation is used.

x_{ijk} = observation corresponding to the kth replicate taken from treatment i of factor A and treatment j of factor B
$\bar{x}_{i\cdot}$ = sample mean for the observations in treatment i (factor A)
$\bar{x}_{\cdot j}$ = sample mean for the observations in treatment j (factor B)
$\bar{x}_{ij}$ = sample mean for the observations corresponding to the combination of treatment i (factor A) and treatment j (factor B)
$\bar{\bar{x}}$ = overall sample mean of all n_T observations

Step 1. Compute the total sum of squares.

$$\text{SST} = \sum_{i=1}^{a}\sum_{j=1}^{b}\sum_{k=1}^{r}(x_{ijk} - \bar{\bar{x}})^2 \tag{13.27}$$

Step 2. Compute the sum of squares for factor A.

$$\text{SSA} = br\sum_{i=1}^{a}(\bar{x}_{i\cdot} - \bar{\bar{x}})^2 \tag{13.28}$$

Step 3. Compute the sum of squares for factor B.

$$\text{SSB} = ar\sum_{j=1}^{b}(\bar{x}_{\cdot j} - \bar{\bar{x}})^2 \tag{13.29}$$

Step 4. Compute the sum of squares for interaction.

$$\text{SSAB} = r\sum_{i=1}^{a}\sum_{j=1}^{b}(\bar{x}_{ij} - \bar{x}_{i\cdot} - \bar{x}_{\cdot j} + \bar{\bar{x}})^2 \tag{13.30}$$

Step 5. Compute the sum of squares due to error.

$$\text{SSE} = \text{SST} - \text{SSA} - \text{SSB} - \text{SSAB} \tag{13.31}$$

Table 13.12 reports the data collected in the experiment and the various sums that will help us with the sum of squares computations. Using equations (13.27) through (13.31), we calculate the following sums of squares for the GMAT two-factor factorial experiment.

Step 1. $\text{SST} = (500 - 515)^2 + (580 - 515)^2 + (540 - 515)^2 + \cdots + (410 - 515)^2 = 82{,}450$

Step 2. $\text{SSA} = (3)(2)[(493.33 - 515)^2 + (513.33 - 515)^2 + (538.33 - 515)^2] = 6100$

Step 3. $\text{SSB} = (3)(2)[(540 - 515)^2 + (560 - 515)^2 + (445 - 515)^2] = 45{,}300$

Step 4. $\text{SSAB} = 2[(540 - 493.33 - 540 + 515)^2 + (500 - 493.33 - 560 + 515)^2 + \cdots + (445 - 538.33 - 445 + 515)^2] = 11{,}200$

Step 5. $\text{SSE} = 82{,}450 - 6100 - 45{,}300 - 11{,}200 = 19{,}850$

These sums of squares divided by their corresponding degrees of freedom provide the appropriate mean square values for testing the two main effects (preparation program and undergraduate college) and the interaction effect.

Because of the computational effort involved in any modest- to large-size factorial experiment, the computer usually plays an important role in performing the analysis of variance computations shown above and in the calculation of the p-values used to make the hypothesis testing decisions. Figure 13.6 shows the Minitab output for the analysis of variance for the GMAT two-factor factorial experiment. Let us use the Minitab output and a level of significance $\alpha = .05$ to conduct the hypothesis tests for the two-factor GMAT study. The p-value used to test for significant differences among the three preparation programs (factor A) is .299. Because the p-value $= .299$ is greater than $\alpha = .05$, there is no significant difference in the mean GMAT test scores for the three preparation programs. However, for the undergraduate college effect, the p-value $= .005$ is less than $\alpha = .05$; thus, there is a significant difference in the mean GMAT test scores among the three undergraduate colleges.

TABLE 13.12 GMAT SUMMARY DATA FOR THE TWO-FACTOR EXPERIMENT

		Factor B: College				
	Treatment combination totals (→ 1080)	**Business**	**Engineering**	**Arts and Sciences**	**Row Totals**	**Factor A Means**
Factor A: Preparation Program	**Three-hour review**	500 580 1080 $\bar{x}_{11} = \frac{1080}{2} = 540$	540 460 1000 $\bar{x}_{12} = \frac{1000}{2} = 500$	480 400 880 $\bar{x}_{13} = \frac{880}{2} = 440$	2960	$\bar{x}_{1\cdot} = \frac{2960}{6} = 493.33$
	One-day program	460 540 1000 $\bar{x}_{21} = \frac{1000}{2} = 500$	560 620 1180 $\bar{x}_{22} = \frac{1180}{2} = 590$	420 480 900 $\bar{x}_{23} = \frac{900}{2} = 450$	3080	$\bar{x}_{2\cdot} = \frac{3080}{6} = 513.33$
	10-week course	560 600 1160 $\bar{x}_{31} = \frac{1160}{2} = 580$	600 580 1180 $\bar{x}_{32} = \frac{1180}{2} = 590$	480 410 890 $\bar{x}_{33} = \frac{890}{2} = 445$	3230	$\bar{x}_{3\cdot} = \frac{3230}{6} = 538.33$
Column Totals		3240	3360	2670	9270	← Overall total
Factor B Means		$\bar{x}_{\cdot 1} = \frac{3240}{6} = 540$	$\bar{x}_{\cdot 2} = \frac{3360}{6} = 560$	$\bar{x}_{\cdot 3} = \frac{2670}{6} = 445$	$\bar{\bar{x}} = \frac{9270}{18} = 515$	

FIGURE 13.6 MINITAB OUTPUT FOR THE GMAT TWO-FACTOR DESIGN

SOURCE	DF	SS	MS	F	P
Factor A	2	6100	3050	1.38	0.299
Factor B	2	45300	22650	10.27	0.005
Interaction	4	11200	2800	1.27	0.350
Error	9	19850	2206		
Total	17	82450			

Finally, because the p-value of .350 for the interaction effect is greater than $\alpha = .05$, there is no significant interaction effect. Therefore, the study provides no reason to believe that the three preparation programs differ in their ability to prepare students from the different colleges for the GMAT.

Undergraduate college was found to be a significant factor. Checking the calculations in Table 13.12, we see that the sample means are: business students $\bar{x}_{\cdot 1} = 540$, engineering students $\bar{x}_{\cdot 2} = 560$, and arts and sciences students $\bar{x}_{\cdot 3} = 445$. Tests on individual treatment means can be conducted, yet after reviewing the three sample means, we would anticipate no difference in preparation for business and engineering graduates. However, the arts and sciences students appear to be significantly less prepared for the GMAT than students in the other colleges. Perhaps this observation will lead the university to consider other options for assisting these students in preparing for the Graduate Management Admission Test.

Exercises

Methods

SELF test

28. A factorial experiment involving two levels of factor A and three levels of factor B resulted in the following data.

		Factor B		
		Level 1	Level 2	Level 3
Factor A	Level 1	135 165	90 66	75 93
	Level 2	125 95	127 105	120 136

Test for any significant main effects and any interaction. Use $\alpha = .05$.

29. The calculations for a factorial experiment involving four levels of factor A, three levels of factor B, and three replications resulted in the following data: SST = 280, SSA = 26, SSB = 23, SSAB = 175. Set up the ANOVA table and test for any significant main effects and any interaction effect. Use $\alpha = .05$.

Applications

30. A mail-order catalog firm designed a factorial experiment to test the effect of the size of a magazine advertisement and the advertisement design on the number of catalog requests received (data in thousands). Three advertising designs and two different size advertisements were considered. The data obtained follow. Use the ANOVA procedure for

factorial designs to test for any significant effects due to type of design, size of advertisement, or interaction. Use $\alpha = .05$.

		Size of Advertisement	
		Small	**Large**
Design	**A**	8	12
		12	8
	B	22	26
		14	30
	C	10	18
		18	14

31. An amusement park studied methods for decreasing the waiting time (minutes) for rides by loading and unloading riders more efficiently. Two alternative loading/unloading methods have been proposed. To account for potential differences due to the type of ride and the possible interaction between the method of loading and unloading and the type of ride, a factorial experiment was designed. Use the following data to test for any significant effect due to the loading and unloading method, the type of ride, and interaction. Use $\alpha = .05$.

	Type of Ride		
	Roller Coaster	**Screaming Demon**	**Log Flume**
Method 1	41	52	50
	43	44	46
Method 2	49	50	48
	51	46	44

32. As part of a study designed to compare hybrid and similarly equipped conventional vehicles, *Consumer Reports* tested a variety of classes of hybrid and all-gas model cars and sport utility vehicles (SUVs). The following data show the miles-per-gallon rating *Consumer Reports* obtained for two hybrid small cars, two hybrid midsize cars, two hybrid small SUVs, and two hybrid midsize SUVs; also shown are the miles per gallon obtained for eight similarly equipped conventional models (*Consumer Reports,* October 2008).

WEB file

HybridTest

Make/Model	Class	Type	MPG
Honda Civic	Small Car	Hybrid	37
Honda Civic	Small Car	Conventional	28
Toyota Prius	Small Car	Hybrid	44
Toyota Corolla	Small Car	Conventional	32
Chevrolet Malibu	Midsize Car	Hybrid	27
Chevrolet Malibu	Midsize Car	Conventional	23
Nissan Altima	Midsize Car	Hybrid	32
Nissan Altima	Midsize Car	Conventional	25
Ford Escape	Small SUV	Hybrid	27
Ford Escape	Small SUV	Conventional	21
Saturn Vue	Small SUV	Hybrid	28
Saturn Vue	Small SUV	Conventional	22
Lexus RX	Midsize SUV	Hybrid	23
Lexus RX	Midsize SUV	Conventional	19
Toyota Highlander	Midsize SUV	Hybrid	24
Toyota Highlander	Midsize SUV	Conventional	18

At the $\alpha = .05$ level of significance, test for significant effects due to class, type, and interaction.

33. A study reported in *The Accounting Review* examined the separate and joint effects of two levels of time pressure (low and moderate) and three levels of knowledge (naive, declarative, and procedural) on key word selection behavior in tax research. Subjects were given a tax case containing a set of facts, a tax issue, and a key word index consisting of 1336 key words. They were asked to select the key words they believed would refer them to a tax authority relevant to resolving the tax case. Prior to the experiment, a group of tax experts determined that the text contained 19 relevant key words. Subjects in the naive group had little or no declarative or procedural knowledge, subjects in the declarative group had significant declarative knowledge but little or no procedural knowledge, and subjects in the procedural group had significant declarative knowledge and procedural knowledge. Declarative knowledge consists of knowledge of both the applicable tax rules and the technical terms used to describe such rules. Procedural knowledge is knowledge of the rules that guide the tax researcher's search for relevant key words. Subjects in the low time pressure situation were told they had 25 minutes to complete the problem, an amount of time which should be "more than adequate" to complete the case; subjects in the moderate time pressure situation were told they would have "only" 11 minutes to complete the case. Suppose 25 subjects were selected for each of the six treatment combinations and the sample means for each treatment combination are as follows (standard deviations are in parentheses).

		Knowledge		
		Naive	**Declarative**	**Procedural**
Time Pressure	**Low**	1.13 (1.12)	1.56 (1.33)	2.00 (1.54)
	Moderate	0.48 (0.80)	1.68 (1.36)	2.86 (1.80)

Use the ANOVA procedure to test for any significant differences due to time pressure, knowledge, and interaction. Use a .05 level of significance. Assume that the total sum of squares for this experiment is 327.50.

Summary

In this chapter we showed how analysis of variance can be used to test for differences among means of several populations or treatments. We introduced the completely randomized design, the randomized block design, and the two-factor factorial experiment. The completely randomized design and the randomized block design are used to draw conclusions about differences in the means of a single factor. The primary purpose of blocking in the randomized block design is to remove extraneous sources of variation from the error term. Such blocking provides a better estimate of the true error variance and a better test to determine whether the population or treatment means of the factor differ significantly.

We showed that the basis for the statistical tests used in analysis of variance and experimental design is the development of two independent estimates of the population variance σ^2. In the single-factor case, one estimator is based on the variation between the treatments; this estimator provides an unbiased estimate of σ^2 only if the means $\mu_1, \mu_2, \ldots, \mu_k$ are all equal. A second estimator of σ^2 is based on the variation of the observations within each sample; this estimator will always provide an unbiased estimate of σ^2. By computing the ratio of these two estimators (the F statistic) we developed a rejection rule for determining whether to reject the null hypothesis that the population or treatment means are equal. In all the experimental designs considered, the partitioning of the sum of squares and degrees of freedom into their various sources enabled us to compute the appropriate values for the analysis of variance calculations and tests. We also showed how Fisher's LSD procedure and the Bonferroni adjustment can be used to perform pairwise comparisons to determine which means are different.

Glossary

Factor Another word for the independent variable of interest.
Treatments Different levels of a factor.
Single-factor experiment An experiment involving only one factor with k populations or treatments.
Response variable Another word for the dependent variable of interest.
Experimental units The objects of interest in the experiment.
Completely randomized design An experimental design in which the treatments are randomly assigned to the experimental units.
ANOVA table A table used to summarize the analysis of variance computations and results. It contains columns showing the source of variation, the sum of squares, the degrees of freedom, the mean square, and the F value(s).
Partitioning The process of allocating the total sum of squares and degrees of freedom to the various components.
Multiple comparison procedures Statistical procedures that can be used to conduct statistical comparisons between pairs of population means.
Comparisonwise Type I error rate The probability of a Type I error associated with a single pairwise comparison.
Experimentwise Type I error rate The probability of making a Type I error on at least one of several pairwise comparisons.
Blocking The process of using the same or similar experimental units for all treatments. The purpose of blocking is to remove a source of variation from the error term and hence provide a more powerful test for a difference in population or treatment means.
Randomized block design An experimental design employing blocking.
Factorial experiment An experimental design that allows simultaneous conclusions about two or more factors.
Replications The number of times each experimental condition is repeated in an experiment.
Interaction The effect produced when the levels of one factor interact with the levels of another factor in influencing the response variable.

Key Formulas

Completely Randomized Design

Sample Mean for Treatment *j*

$$\bar{x}_j = \frac{\sum_{i=1}^{n_j} x_{ij}}{n_j} \tag{13.1}$$

Sample Variance for Treatment *j*

$$s_j^2 = \frac{\sum_{i=1}^{n_j} (x_{ij} - \bar{x}_j)^2}{n_j - 1} \tag{13.2}$$

Overall Sample Mean

$$\bar{\bar{x}} = \frac{\sum_{j=1}^{k}\sum_{i=1}^{n_j} x_{ij}}{n_T} \qquad \textbf{(13.3)}$$

$$n_T = n_1 + n_2 + \cdots + n_k \qquad \textbf{(13.4)}$$

Mean Square Due to Treatments

$$\text{MSTR} = \frac{\text{SSTR}}{k - 1} \qquad \textbf{(13.7)}$$

Sum of Squares Due to Treatments

$$\text{SSTR} = \sum_{j=1}^{k} n_j(\bar{x}_j - \bar{\bar{x}})^2 \qquad \textbf{(13.8)}$$

Mean Square Due to Error

$$\text{MSE} = \frac{\text{SSE}}{n_T - k} \qquad \textbf{(13.10)}$$

Sum of Squares Due to Error

$$\text{SSE} = \sum_{j=1}^{k}(n_j - 1)s_j^2 \qquad \textbf{(13.11)}$$

Test Statistic for the Equality of k Population Means

$$F = \frac{\text{MSTR}}{\text{MSE}} \qquad \textbf{(13.12)}$$

Total Sum of Squares

$$\text{SST} = \sum_{j=1}^{k}\sum_{i=1}^{n_j}(x_{ij} - \bar{\bar{x}})^2 \qquad \textbf{(13.13)}$$

Partitioning of Sum of Squares

$$\text{SST} = \text{SSTR} + \text{SSE} \qquad \textbf{(13.14)}$$

Multiple Comparison Procedures

Test Statistic for Fisher's LSD Procedure

$$t = \frac{\bar{x}_i - \bar{x}_j}{\sqrt{\text{MSE}\left(\frac{1}{n_i} + \frac{1}{n_j}\right)}} \qquad \textbf{(13.16)}$$

Fisher's LSD

$$\text{LSD} = t_{\alpha/2}\sqrt{\text{MSE}\left(\frac{1}{n_i} + \frac{1}{n_j}\right)} \qquad \textbf{(13.17)}$$

Randomized Block Design

Total Sum of Squares

$$\text{SST} = \sum_{i=1}^{b}\sum_{j=1}^{k}(x_{ij} - \bar{\bar{x}})^2 \tag{13.22}$$

Sum of Squares Due to Treatments

$$\text{SSTR} = b\sum_{j=1}^{k}(\bar{x}_{\cdot j} - \bar{\bar{x}})^2 \tag{13.23}$$

Sum of Squares Due to Blocks

$$\text{SSBL} = k\sum_{i=1}^{b}(\bar{x}_{i\cdot} - \bar{\bar{x}})^2 \tag{13.24}$$

Sum of Squares Due to Error

$$\text{SSE} = \text{SST} - \text{SSTR} - \text{SSBL} \tag{13.25}$$

Factorial Experiment

Total Sum of Squares

$$\text{SST} = \sum_{i=1}^{a}\sum_{j=1}^{b}\sum_{k=1}^{r}(x_{ijk} - \bar{\bar{x}})^2 \tag{13.27}$$

Sum of Squares for Factor A

$$\text{SSA} = br\sum_{i=1}^{a}(\bar{x}_{i\cdot} - \bar{\bar{x}})^2 \tag{13.28}$$

Sum of Squares for Factor B

$$\text{SSB} = ar\sum_{j=1}^{b}(\bar{x}_{\cdot j} - \bar{\bar{x}})^2 \tag{13.29}$$

Sum of Squares for Interaction

$$\text{SSAB} = r\sum_{i=1}^{a}\sum_{j=1}^{b}(\bar{x}_{ij} - \bar{x}_{i\cdot} - \bar{x}_{\cdot j} + \bar{\bar{x}})^2 \tag{13.30}$$

Sum of Squares for Error

$$\text{SSE} = \text{SST} - \text{SSA} - \text{SSB} - \text{SSAB} \tag{13.31}$$

Supplementary Exercises

34. In a completely randomized experimental design, three brands of paper towels were tested for their ability to absorb water. Equal-size towels were used, with four sections of towels tested per brand. The absorbency rating data follow. At a .05 level of significance, does there appear to be a difference in the ability of the brands to absorb water?

	Brand	
x	*y*	*z*
91	99	83
100	96	88
88	94	89
89	99	76

35. A study reported in the *Journal of Small Business Management* concluded that self-employed individuals do not experience higher job satisfaction than individuals who are not self-employed. In this study, job satisfaction is measured using 18 items, each of which is rated using a Likert-type scale with 1–5 response options ranging from strong agreement to strong disagreement. A higher score on this scale indicates a higher degree of job satisfaction. The sum of the ratings for the 18 items, ranging from 18–90, is used as the measure of job satisfaction. Suppose that this approach was used to measure the job satisfaction for lawyers, physical therapists, cabinetmakers, and systems analysts. The results obtained for a sample of 10 individuals from each profession follow.

WEB file
SatisJob

Lawyer	Physical Therapist	Cabinetmaker	Systems Analyst
44	55	54	44
42	78	65	73
74	80	79	71
42	86	69	60
53	60	79	64
50	59	64	66
45	62	59	41
48	52	78	55
64	55	84	76
38	50	60	62

At the $\alpha = .05$ level of significance, test for any difference in the job satisfaction among the four professions.

36. The U.S. Environmental Protection Agency (EPA) monitors levels of pollutants in the air for cities across the country. Ozone pollution levels are measured using a 500-point scale; lower scores indicate little health risk, and higher scores indicate greater health risk. The following data show the peak levels of ozone pollution in four cities (Birmingham, Alabama; Memphis, Tennessee; Little Rock, Arkansas; and Jackson, Mississippi) for 10 dates in 2012 (U.S. EPA website, March 20, 2012).

WEB file
OzoneLevels

	City			
Data	Birmingham AL	Memphis TN	Little Rock AR	Jackson MS
Jan 9	18	20	18	14
Jan17	23	31	22	30
Jan 18	19	25	22	21
Jan 31	29	36	28	35
Feb 1	27	31	28	24
Feb 6	26	31	31	25
Feb 14	31	24	19	25
Feb 17	31	31	28	28
Feb 20	33	35	35	34
Feb 29	20	42	42	21

Use $\alpha = .05$ to test for any significant difference in the mean peak ozone levels among the four cities.

37. The U.S. Census Bureau computes quarterly vacancy and homeownership rates by state and metropolitan statistical area. Each metropolitan statistical area (MSA) has at least one urbanized area of 50,000 or more inhabitants. The following data are the rental vacancy rates (%) for MSAs in four geographic regions of the United States for the first quarter of 2008 (U.S. Census Bureau website, January 2009).

WEB file RentalVacancy

Midwest	Northeast	South	West
16.2	2.7	16.6	7.9
10.1	11.5	8.5	6.6
8.6	6.6	12.1	6.9
12.3	7.9	9.8	5.6
10.0	5.3	9.3	4.3
16.9	10.7	9.1	15.2
16.9	8.6	5.6	5.7
5.4	5.5	9.4	4.0
18.1	12.7	11.6	12.3
11.9	8.3	15.6	3.6
11.0	6.7	18.3	11.0
9.6	14.2	13.4	12.1
7.6	1.7	6.5	8.7
12.9	3.6	11.4	5.0
12.2	11.5	13.1	4.7
13.6	16.3	4.4	3.3
		8.2	3.4
		24.0	5.5
		12.2	
		22.6	
		12.0	
		14.5	
		12.6	
		9.5	
		10.1	

Use $\alpha = .05$ to test whether there the mean vacancy rate is the same for each geographic region.

38. Three different assembly methods have been proposed for a new product. A completely randomized experimental design was chosen to determine which assembly method results in the greatest number of parts produced per hour, and 30 workers were randomly selected and assigned to use one of the proposed methods. The number of units produced by each worker follows.

WEB file Assembly

Method		
A	B	C
97	93	99
73	100	94
93	93	87
100	55	66
73	77	59
91	91	75
100	85	84
86	73	72
92	90	88
95	83	86

Use these data and test to see whether the mean number of parts produced is the same with each method. Use $\alpha = .05$.

39. In a study conducted to investigate browsing activity by shoppers, each shopper was initially classified as a nonbrowser, light browser, or heavy browser. For each shopper, the study obtained a measure to determine how comfortable the shopper was in a store. Higher scores indicated greater comfort. Suppose the following data were collected.

WEB file
Browsing

Nonbrowser	Light Browser	Heavy Browser
4	5	5
5	6	7
6	5	5
3	4	7
3	7	4
4	4	6
5	6	5
4	5	7

a. Use $\alpha = .05$ to test for differences among comfort levels for the three types of browsers.
b. Use Fisher's LSD procedure to compare the comfort levels of nonbrowsers and light browsers. Use $\alpha = .05$. What is your conclusion?

40. A research firm tests the miles-per-gallon characteristics of three brands of gasoline. Because of different gasoline performance characteristics in different brands of automobiles, five brands of automobiles are selected and treated as blocks in the experiment; that is, each brand of automobile is tested with each type of gasoline. The results of the experiment (in miles per gallon) follow.

		Gasoline Brands		
		I	II	III
Automobiles	A	18	21	20
	B	24	26	27
	C	30	29	34
	D	22	25	24
	E	20	23	24

a. At $\alpha = .05$, is there a significant difference in the mean miles-per-gallon characteristics of the three brands of gasoline?
b. Analyze the experimental data using the ANOVA procedure for completely randomized designs. Compare your findings with those obtained in part (a). What is the advantage of attempting to remove the block effect?

41. The compact car market in the United States is extremely competitive. Sales for six of the top models for six months in 2011 follow (*Motor Trend,* November 2, 2011).

WEB file
CompactCars

Month	Chevy Cruze	Ford Focus	Hyundai Elantra	Honda Civic	Toyota Corolla	VW Jetta
May	22,711	22,303	20,006	18,341	16,985	16,671
June	24,896	21,385	19,992	18,872	17,485	17,105
July	24,468	17,577	15,713	15,181	14,889	14,006
August	21,897	16,420	15,054	14,500	14,093	12,083
September	18,097	16,147	15,023	14,386	13,724	10,309
October	16,244	16,173	14,295	13,058	13,000	12,386

At the .05 level of significance, test for any significant difference in the mean number of cars sold per month for the six models.

42. Major League Baseball franchises rely on attendance for a large share of their total revenue, and weekend games are particularly important. The following table shows the attendance for the Houston Astros for games played during seven weekend series for the first three months (April, May, and June) of the 2011 season (ESPN website, January 12, 2012).

WEB file

HoustonAstros

Opponent	Friday	Saturday	Sunday
Florida Marlins	41,042	25,421	22,299
San Diego Padres	23,755	28,100	22,899
Milwaukee Brewers	25,734	26,514	23,908
New York Mets	28,791	31,140	28,406
Arizona Diamondbacks	21,834	31,405	21,882
Atlanta Braves	29,252	32,117	23,765
Tampa Bay Rays	26,682	27,208	23,965

At the .05 level of significance, test whether the mean attendance is the same for these three days. The Houston Astros are considering running a special promotion to increase attendance during one game of each weekend series during the second half of the season. Do these data suggest a particular day on which the Astros should schedule these promotions?

43. A factorial experiment was designed to test for any significant differences in the time needed to perform English to foreign language translations with two computerized language translators. Because the type of language translated was also considered a significant factor, translations were made with both systems for three different languages: Spanish, French, and German. Use the following data for translation time in hours.

	Language		
	Spanish	French	German
System 1	8	10	12
	12	14	16
System 2	6	14	16
	10	16	22

Test for any significant differences due to language translator, type of language, and interaction. Use $\alpha = .05$.

44. A manufacturing company designed a factorial experiment to determine whether the number of defective parts produced by two machines differed and if the number of defective parts produced also depended on whether the raw material needed by each machine was loaded manually or by an automatic feed system. The following data give the numbers of defective parts produced. Use $\alpha = .05$ to test for any significant effect due to machine, loading system, and interaction.

	Loading System	
	Manual	Automatic
Machine 1	30	30
	34	26
Machine 2	20	24
	22	28

Case Problem 1 Wentworth Medical Center

As part of a long-term study of individuals 65 years of age or older, sociologists and physicians at the Wentworth Medical Center in upstate New York investigated the relationship between geographic location and depression. A sample of 60 individuals, all in reasonably good health, was selected; 20 individuals were residents of Florida, 20 were residents of New York, and 20 were residents of North Carolina. Each of the individuals sampled was given a standardized test to measure depression. The data collected follow; higher test scores indicate higher levels of depression. These data are contained in the file Medical1.

A second part of the study considered the relationship between geographic location and depression for individuals 65 years of age or older who had a chronic health condition such as arthritis, hypertension, and/or heart ailment. A sample of 60 individuals with such conditions was identified. Again, 20 were residents of Florida, 20 were residents of New York, and 20 were residents of North Carolina. The levels of depression recorded for this study follow. These data are contained in the file named Medical2.

WEB file Medical1

WEB file Medical2

Data from Medical1			Data from Medical2		
Florida	New York	North Carolina	Florida	New York	North Carolina
3	8	10	13	14	10
7	11	7	12	9	12
7	9	3	17	15	15
3	7	5	17	12	18
8	8	11	20	16	12
8	7	8	21	24	14
8	8	4	16	18	17
5	4	3	14	14	8
5	13	7	13	15	14
2	10	8	17	17	16
6	6	8	12	20	18
2	8	7	9	11	17
6	12	3	12	23	19
6	8	9	15	19	15
9	6	8	16	17	13
7	8	12	15	14	14
5	5	6	13	9	11
4	7	3	10	14	12
7	7	8	11	13	13
3	8	11	17	11	11

Managerial Report

1. Use descriptive statistics to summarize the data from the two studies. What are your preliminary observations about the depression scores?
2. Use analysis of variance on both data sets. State the hypotheses being tested in each case. What are your conclusions?
3. Use inferences about individual treatment means where appropriate. What are your conclusions?

Case Problem 2 Compensation for Sales Professionals

Suppose that a local chapter of sales professionals in the greater San Francisco area conducted a survey of its membership to study the relationship, if any, between the years of experience and salary for individuals employed in inside and outside sales positions. On the survey, respondents were asked to specify one of three levels of years of experience: low (1–10 years), medium (11–20 years), and high (21 or more years). A portion of the data obtained follow. The complete data set, consisting of 120 observations, is contained in the file named SalesSalary.

WEB file
SalesSalary

Observation	Salary $	Position	Experience
1	53,938	Inside	Medium
2	52,694	Inside	Medium
3	70,515	Outside	Low
4	52,031	Inside	Medium
5	62,283	Outside	Low
6	57,718	Inside	Low
7	79,081	Outside	High
8	48,621	Inside	Low
9	72,835	Outside	High
10	54,768	Inside	Medium
.	.	.	.
.	.	.	.
.	.	.	.
115	58,080	Inside	High
116	78,702	Outside	Medium
117	83,131	Outside	Medium
118	57,788	Inside	High
119	53,070	Inside	Medium
120	60,259	Outside	Low

Managerial Report

1. Use descriptive statistics to summarize the data.
2. Develop a 95% confidence interval estimate of the mean annual salary for all salespersons, regardless of years of experience and type of position.
3. Develop a 95% confidence interval estimate of the mean salary for inside salespersons.
4. Develop a 95% confidence interval estimate of the mean salary for outside salespersons.
5. Use analysis of variance to test for any significant differences due to position. Use a .05 level of significance, and for now, ignore the effect of years of experience.
6. Use analysis of variance to test for any significant differences due to years of experience. Use a .05 level of significance, and for now, ignore the effect of position.
7. At the .05 level of significance test for any significant differences due to position, years of experience, and interaction.

Appendix 13.1 Analysis of Variance with Minitab

Completely Randomized Design

In Section 13.2 we showed how analysis of variance could be used to test for the equality of k population means using data from a completely randomized design. To illustrate how Minitab can be used for this type of experimental design, we show how to test whether

WEB file
Chemitech

the mean number of units produced per week is the same for each assembly method in the Chemitech experiment introduced in Section 13.1. The sample data are entered into the first three columns of a Minitab worksheet; column 1 is labeled A, column 2 is labeled B, and column 3 is labeled C. The following steps produce the Minitab output in Figure 13.5.

Step 1. Select the **Stat** menu
Step 2. Choose **ANOVA**
Step 3. Choose **One-way (Unstacked)**
Step 4. When the One-way Analysis of Variance dialog box appears:
Enter C1-C3 in the **Responses (in separate columns)** box
Click **OK**

Randomized Block Design

WEB file
AirTraffic

In Section 13.4 we showed how analysis of variance could be used to test for the equality of k population means using the data from a randomized block design. To illustrate how Minitab can be used for this type of experimental design, we show how to test whether the mean stress levels for air traffic controllers are the same for three workstations using the data in Table 13.5. The blocks (controllers), treatments (system), and stress level scores shown in Table 13.5 are entered into columns C1, C2, and C3 of a Minitab worksheet, respectively. The following steps produce the Minitab output corresponding to the ANOVA table shown in Table 13.8.

The treatments are entered in the Row factor box and the blocks are entered in the Column factor box.

Step 1. Select the **Stat** menu
Step 2. Choose **ANOVA**
Step 3. Choose **Two-way**
Step 4. When the Two-way Analysis of Variance dialog box appears:
Enter C3 in the **Response** box
Enter C2 in the **Row factor** box
Enter C1 in the **Column factor** box
Select **Fit Additive Model**
Click **OK**

Factorial Experiment

WEB file
GMATStudy

In Section 13.5 we showed how analysis of variance could be used to test for the equality of k population means using data from a factorial experiment. To illustrate how Minitab can be used for this type of experimental design, we show how to analyze the data for the two-factor GMAT experiment introduced in that section. The GMAT scores shown in Table 13.10 are entered into column 1 of a Minitab worksheet; column 1 is labeled Score, column 2 is labeled Program, and column 3 is labeled College. The following steps produce the Minitab output corresponding to the ANOVA table shown in Figure 13.6.

Step 1. Select the **Stat** menu
Step 2. Choose **ANOVA**
Step 3. Choose **Two-way**
Step 4. When the Two-way Analysis of Variance dialog box appears:
Enter C1 in the **Response** box
Enter C2 in the **Row factor** box
Enter C3 in the **Column factor** box
Click **OK**

Appendix 13.2 Analysis of Variance with Excel

Completely Randomized Design

WEB file
Chemitech

In Section 13.2 we showed how analysis of variance could be used to test for the equality of k population means using data from a completely randomized design. To illustrate how Excel can be used to test for the equality of k population means for this type of experimental design, we show how to test whether the mean number of units produced per week is the same for each assembly method in the Chemitech experiment introduced in Section 13.1. The sample data are entered into worksheet rows 2 to 6 of columns A, B, and C as shown in Figure 13.7. The following steps are used to obtain the output shown in cells A8:G22; the ANOVA portion of this output corresponds to the ANOVA table shown in Table 13.3.

Step 1. Click the **Data** tab on the Ribbon
Step 2. In the **Analysis** group, click **Data Analysis**
Step 3. Choose **Anova: Single Factor** from the list of Analysis Tools
Step 4. When the Anova: Single Factor dialog box appears:
Enter A1:C6 in **Input Range** box
Select **Columns**
Select **Labels in First Row**
Select **Output Range** and enter A8 in the box
Click **OK**

FIGURE 13.7 EXCEL SOLUTION FOR THE CHEMITECH EXPERIMENT

A1 fx Method A

	A	B	C	D	E	F	G	H
1	**Method A**	**Method B**	**Method C**					
2	58	58	48					
3	64	69	57					
4	55	71	59					
5	66	64	47					
6	67	68	49					
7								
8	Anova: Single Factor							
9								
10	SUMMARY							
11	*Groups*	*Count*	*Sum*	*Average*	*Variance*			
12	Method A	5	310	62	27.5			
13	Method B	5	330	66	26.5			
14	Method C	5	260	52	31			
15								
16								
17	ANOVA							
18	*Source of Variation*	*SS*	*df*	*MS*	*F*	*P-value*	*F crit*	
19	Between Groups	520	2	260	9.1765	0.0038	3.8853	
20	Within Groups	340	12	28.3333				
21								
22	Total	860	14					
23								

FIGURE 13.8 EXCEL SOLUTION FOR THE AIR TRAFFIC CONTROLLER STRESS TEST

A1 f_x Controller

	A	B	C	D	E	F	G	H
1	**Controller**	**System A**	**System B**	**System C**				
2	1	15	15	18				
3	2	14	14	14				
4	3	10	11	15				
5	4	13	12	17				
6	5	16	13	16				
7	6	13	13	13				
8								
9	Anova: Two-Factor Without Replication							
10								
11	*SUMMARY*	*Count*	*Sum*	*Average*	*Variance*			
12	1	3	48	16	3			
13	2	3	42	14	0			
14	3	3	36	12	7			
15	4	3	42	14	7			
16	5	3	45	15	3			
17	6	3	39	13	0			
18								
19	System A	6	81	13.5	4.3			
20	System B	6	78	13	2			
21	System C	6	93	15.5	3.5			
22								
23								
24	ANOVA							
25	*Source of Variation*	*SS*	*df*	*MS*	*F*	*P-value*	*F crit*	
26	Rows	30	5	6	3.1579	0.0574	3.3258	
27	Columns	21	2	10.5	5.5263	0.0242	4.1028	
28	Error	19	10	1.9				
29								
30	Total	70	17					
31								

Randomized Block Design

WEB file AirTraffic

In Section 13.4 we showed how analysis of variance could be used to test for the equality of k population means using data from a randomized block design. To illustrate how Excel can be used for this type of experimental design, we show how to test whether the mean stress levels for air traffic controllers are the same for three workstations. The stress level scores shown in Table 13.5 are entered into worksheet rows 2 to 7 of columns B, C, and D as shown in Figure 13.8. The cells in rows 2 to 7 of column A contain the number of each controller (1, 2, 3, 4, 5, 6). The following steps produce the Excel output shown in cells A9:G30. The ANOVA portion of this output corresponds to the ANOVA table shown in Table 13.8.

Step 1. Click the **Data** tab on the Ribbon
Step 2. In the **Analysis** group, click **Data Analysis**
Step 3. Choose **Anova: Two-Factor Without Replication** from the list of Analysis Tools
Step 4. When the Anova: Two-Factor Without Replication dialog box appears:
Enter A1:D7 in **Input Range** box
Select **Labels**
Select **Output Range** and enter A9 in the box
Click **OK**

Factorial Experiment

WEB file

GMATStudy

In Section 13.5 we showed how analysis of variance could be used to test for the equality of k population means using data from a factorial experiment. To illustrate how Excel can be used for this type of experimental design, we show how to analyze the data for the two-factor GMAT experiment introduced in that section. The GMAT scores shown in Table 13.10 are entered into worksheet rows 2 to 7 of columns B, C, and D as shown in Figure 13.9. The following steps are used to obtain the output shown in cells A9:G44; the ANOVA portion of this output corresponds to the Minitab output in Figure 13.6.

Step 1. Click the **Data** tab on the Ribbon
Step 2. In the **Analysis** group, click **Data Analysis**
Step 3. Choose **Anova: Two-Factor With Replication** from the list of Analysis Tools

FIGURE 13.9 EXCEL SOLUTION FOR THE TWO-FACTOR GMAT EXPERIMENT

A1 | *fx*

	A	B	C	D	E	F	G	H
1		**Business**	**Engineering**	**Arts and Sciences**				
2	**3-hour review**	500	540	480				
3		580	460	400				
4	**1-day program**	460	560	420				
5		540	620	480				
6	**10-week course**	560	600	480				
7		600	580	410				
8								
9	Anova: Two-Factor With Replication							
10								
11	SUMMARY	Business	Engineering	Arts and Sciences	Total			
12	*3-hour review*							
13	Count	2	2	2	6			
14	Sum	1080	1000	880	2960			
15	Average	540	500	440	493.3333			
16	Variance	3200	3200	3200	3946.667			
17								
18	*1-day program*							
19	Count	2	2	2	6			
20	Sum	1000	1180	900	3080			
21	Average	500	590	450	513.3333			
22	Variance	3200	1800	1800	5386.667			
23								
24	*10-week course*							
25	Count	2	2	2	6			
26	Sum	1160	1180	890	3230			
27	Average	580	590	445	538.3333			
28	Variance	800	200	2450	5936.667			
29								
30	*Total*							
31	Count	6	6	6				
32	Sum	3240	3360	2670				
33	Average	540	560	445				
34	Variance	2720	3200	1510				
35								
36								
37	ANOVA							
38	*Source of Variation*	*SS*	*df*	*MS*	*F*	*P-value*	*F crit*	
39	Sample	6100	2	3050	1.3829	0.2994	4.2565	
40	Columns	45300	2	22650	10.2695	0.0048	4.2565	
41	Interaction	11200	4	2800	1.2695	0.3503	3.6331	
42	Within	19850	9	2205.5556				
43								
44	Total	82450	17					
45								

Step 4. When the Anova: Two-Factor With Replication dialog box appears:
Enter A1:D7 in **Input Range** box
Enter 2 in **Rows per sample** box
Select **Output Range** and enter A9 in the box
Click **OK**

Appendix 13.3 Analysis of a Completely Randomized Design Using StatTools

In this appendix we show how StatTools can be used to test for the equality of k population means for a completely randomized design. We use the Chemitech data in Table 13.1 to illustrate. Begin by using the Data Set Manager to create a StatTools data set for these data using the procedure described in the appendix in Chapter 1. The following steps can be used to test for the equality of the three population means.

WEB file
Chemitech

Step 1. Click the **StatTools** tab on the Ribbon
Step 2. In the **Analyses** group, click **Statistical Inference**
Step 3. Choose the **One-Way ANOVA** option
Step 4. When the StatTools-One-Way ANOVA dialog box appears:
In the **Variables** section:
Click the **Format button** and select **Unstacked**
Select **Method A**
Select **Method B**
Select **Method C**
Select 95% in the **Confidence Level** box
Click **OK**

Note that in step 4 we selected the Unstacked option after clicking the Format button. The Unstacked option means that the data for the three treatments appear in separate columns of the worksheet. In a stacked format, only two columns would be used. For example, the data could have been organized as follows:

	A	B	C
1	**Method**	**Units Produced**	
2	Method A	58	
3	Method A	64	
4	Method A	55	
5	Method A	66	
6	Method A	67	
7	Method B	58	
8	Method B	69	
9	Method B	71	
10	Method B	64	
11	Method B	68	
12	Method C	48	
13	Method C	57	
14	Method C	59	
15	Method C	47	
16	Method C	49	
17			

Data are frequently recorded in a stacked format. For stacked data, simply select the Stacked option after clicking the Format button.

CHAPTER 14

Simple Linear Regression

CONTENTS

STATISTICS *in* PRACTICE

ALLIANCE DATA SYSTEMS*
DALLAS, TEXAS

Alliance Data Systems analysts discuss use of a regression model to predict sales for a direct marketing campaign. © Courtesy of Alliance Data Systems.

Alliance Data Systems (ADS) provides transaction processing, credit services, and marketing services for clients in the rapidly growing customer relationship management (CRM) industry. ADS clients are concentrated in four industries: retail, petroleum/convenience stores, utilities, and transportation. In 1983, Alliance began offering end-to-end credit processing services to the retail, petroleum, and casual dining industries; today they employ more than 6500 employees who provide services to clients around the world. Operating more than 140,000 point-of-sale terminals in the United States alone, ADS processes in excess of 2.5 billion transactions annually. The company ranks second in the United States in private label credit services by representing 49 private label programs with nearly 72 million cardholders. In 2001, ADS made an initial public offering and is now listed on the New York Stock Exchange.

As one of its marketing services, ADS designs direct mail campaigns and promotions. With its database containing information on the spending habits of more than 100 million consumers, ADS can target those consumers most likely to benefit from a direct mail promotion. The Analytical Development Group uses regression analysis to build models that measure and predict the responsiveness of consumers to direct market campaigns. Some regression models predict the probability of purchase for individuals receiving a promotion, and others predict the amount spent by those consumers making a purchase.

For one particular campaign, a retail store chain wanted to attract new customers. To predict the effect of the campaign, ADS analysts selected a sample from the consumer database, sent the sampled individuals promotional materials, and then collected transaction data on the consumers' response. Sample data were collected on the amount of purchase made by the consumers responding to the campaign, as well as a variety of consumer-specific variables thought to be useful in predicting sales. The consumer-specific variable that contributed most to predicting the amount purchased was the total amount of credit purchases at related stores over the past 39 months. ADS analysts developed an estimated regression equation relating the amount of purchase to the amount spent at related stores:

$$\hat{y} = 26.7 + 0.00205x$$

where

$\hat{y}$ = amount of purchase
x = amount spent at related stores

Using this equation, we could predict that someone spending \$10,000 over the past 39 months at related stores would spend \$47.20 when responding to the direct mail promotion. In this chapter, you will learn how to develop this type of estimated regression equation.

The final model developed by ADS analysts also included several other variables that increased the predictive power of the preceding equation. Some of these variables included the absence/presence of a bank credit card, estimated income, and the average amount spent per trip at a selected store. In the following chapter, we will learn how such additional variables can be incorporated into a multiple regression model.

*The authors are indebted to Philip Clemance, Director of Analytical Development at Alliance Data Systems, for providing this Statistics in Practice.

Managerial decisions often are based on the relationship between two or more variables. For example, after considering the relationship between advertising expenditures and sales, a marketing manager might attempt to predict sales for a given level of advertising expenditures. In another case, a public utility might use the relationship between the daily high temperature and the demand for electricity to predict electricity usage on the basis of next month's anticipated daily high temperatures. Sometimes a manager will rely on intuition to judge how two variables are related. However, if data can be obtained, a statistical procedure called *regression analysis* can be used to develop an equation showing how the variables are related.

The statistical methods used in studying the relationship between two variables were first employed by Sir Francis Galton (1822–1911). Galton was interested in studying the relationship between a father's height and the son's height. Galton's disciple, Karl Pearson (1857–1936), analyzed the relationship between the father's height and the son's height for 1078 pairs of subjects.

In regression terminology, the variable being predicted is called the **dependent variable**. The variable or variables being used to predict the value of the dependent variable are called the **independent variables**. For example, in analyzing the effect of advertising expenditures on sales, a marketing manager's desire to predict sales would suggest making sales the dependent variable. Advertising expenditure would be the independent variable used to help predict sales. In statistical notation, y denotes the dependent variable and x denotes the independent variable.

In this chapter we consider the simplest type of regression analysis involving one independent variable and one dependent variable in which the relationship between the variables is approximated by a straight line. It is called **simple linear regression**. Regression analysis involving two or more independent variables is called multiple regression analysis; multiple regression and cases involving curvilinear relationships are covered in Chapters 15 and 16.

14.1 Simple Linear Regression Model

Armand's Pizza Parlors is a chain of Italian-food restaurants located in a five-state area. Armand's most successful locations are near college campuses. The managers believe that quarterly sales for these restaurants (denoted by y) are related positively to the size of the student population (denoted by x); that is, restaurants near campuses with a large student population tend to generate more sales than those located near campuses with a small student population. Using regression analysis, we can develop an equation showing how the dependent variable y is related to the independent variable x.

Regression Model and Regression Equation

In the Armand's Pizza Parlors example, the population consists of all the Armand's restaurants. For every restaurant in the population, there is a value of x (student population) and a corresponding value of y (quarterly sales). The equation that describes how y is related to x and an error term is called the **regression model**. The regression model used in simple linear regression follows.

SIMPLE LINEAR REGRESSION MODEL

$$y = \beta_0 + \beta_1 x + \epsilon \tag{14.1}$$

β_0 and β_1 are referred to as the parameters of the model, and ϵ (the Greek letter epsilon) is a random variable referred to as the error term. The error term accounts for the variability in y that cannot be explained by the linear relationship between x and y.

The population of all Armand's restaurants can also be viewed as a collection of subpopulations, one for each distinct value of x. For example, one subpopulation consists of all Armand's restaurants located near college campuses with 8000 students; another subpopulation consists of all Armand's restaurants located near college campuses with 9000 students; and so on. Each subpopulation has a corresponding distribution of y values. Thus, a distribution of y values is associated with restaurants located near campuses with 8000 students; a distribution of y values is associated with restaurants located near campuses with 9000 students; and so on. Each distribution of y values has its own mean or expected value. The equation that describes how the expected value of y, denoted $E(y)$, is related to x is called the **regression equation**. The regression equation for simple linear regression follows.

SIMPLE LINEAR REGRESSION EQUATION

$$E(y) = \beta_0 + \beta_1 x \qquad \textbf{(14.2)}$$

The graph of the simple linear regression equation is a straight line; β_0 is the y-intercept of the regression line, β_1 is the slope, and $E(y)$ is the mean or expected value of y for a given value of x.

Examples of possible regression lines are shown in Figure 14.1. The regression line in Panel A shows that the mean value of y is related positively to x, with larger values of $E(y)$ associated with larger values of x. The regression line in Panel B shows the mean value of y is related negatively to x, with smaller values of $E(y)$ associated with larger values of x. The regression line in Panel C shows the case in which the mean value of y is not related to x; that is, the mean value of y is the same for every value of x.

Estimated Regression Equation

If the values of the population parameters β_0 and β_1 were known, we could use equation (14.2) to compute the mean value of y for a given value of x. In practice, the parameter values are not known and must be estimated using sample data. Sample statistics (denoted b_0 and b_1) are computed as estimates of the population parameters β_0 and β_1. Substituting the values of the sample statistics b_0 and b_1 for β_0 and β_1 in the regression equation, we obtain the

FIGURE 14.1 POSSIBLE REGRESSION LINES IN SIMPLE LINEAR REGRESSION

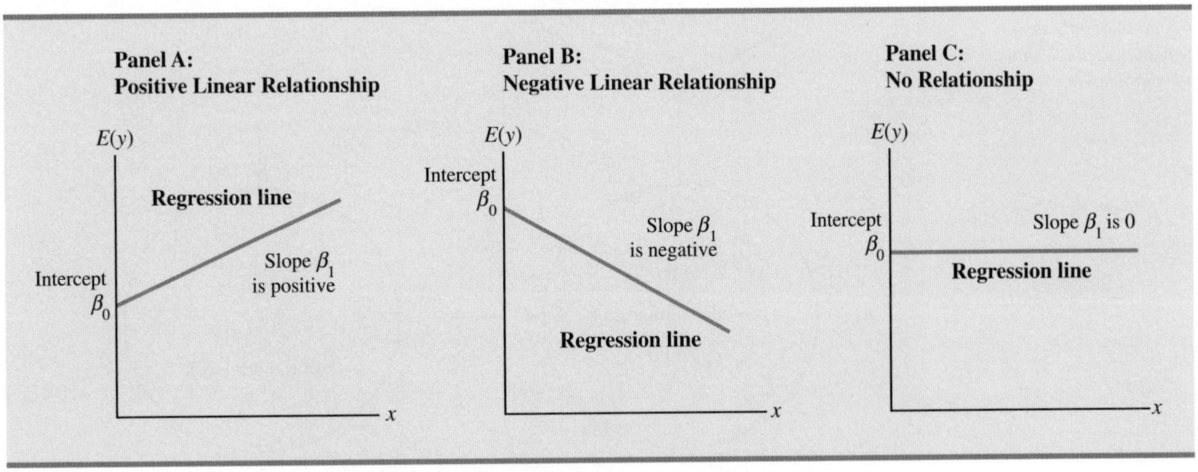

estimated regression equation. The estimated regression equation for simple linear regression follows.

ESTIMATED SIMPLE LINEAR REGRESSION EQUATION

$$\hat{y} = b_0 + b_1x \tag{14.3}$$

Figure 14.2 provides a summary of the estimation process for simple linear regression.

The graph of the estimated simple linear regression equation is called the *estimated regression line*; b_0 is the y-intercept and b_1 is the slope. In the next section, we show how the least squares method can be used to compute the values of b_0 and b_1 in the estimated regression equation.

The value of $\hat{y}$ provides both a point estimate of $E(y)$ for a given value of x and a prediction of an individual value of y for a given value of x.

In general, $\hat{y}$ is the point estimator of $E(y)$, the mean value of y for a given value of x. Thus, to estimate the mean or expected value of quarterly sales for all restaurants located near campuses with 10,000 students, Armand's would substitute the value of 10,000 for x in equation (14.3). In some cases, however, Armand's may be more interested in predicting sales for one particular restaurant. For example, suppose Armand's would like to predict quarterly sales for the restaurant they are considering building near Talbot College, a school with 10,000 students. As it turns out, the best predictor of y for a given value of x is also provided by $\hat{y}$. Thus, to predict quarterly sales for the restaurant located near Talbot College, Armand's would also substitute the value of 10,000 for x in equation (14.3).

The estimation of β_0 and β_1 is a statistical process much like the estimation of μ discussed in Chapter 7. β_0 and β_1 are the unknown parameters of interest, and b_0 and b_1 are the sample statistics used to estimate the parameters.

FIGURE 14.2 THE ESTIMATION PROCESS IN SIMPLE LINEAR REGRESSION

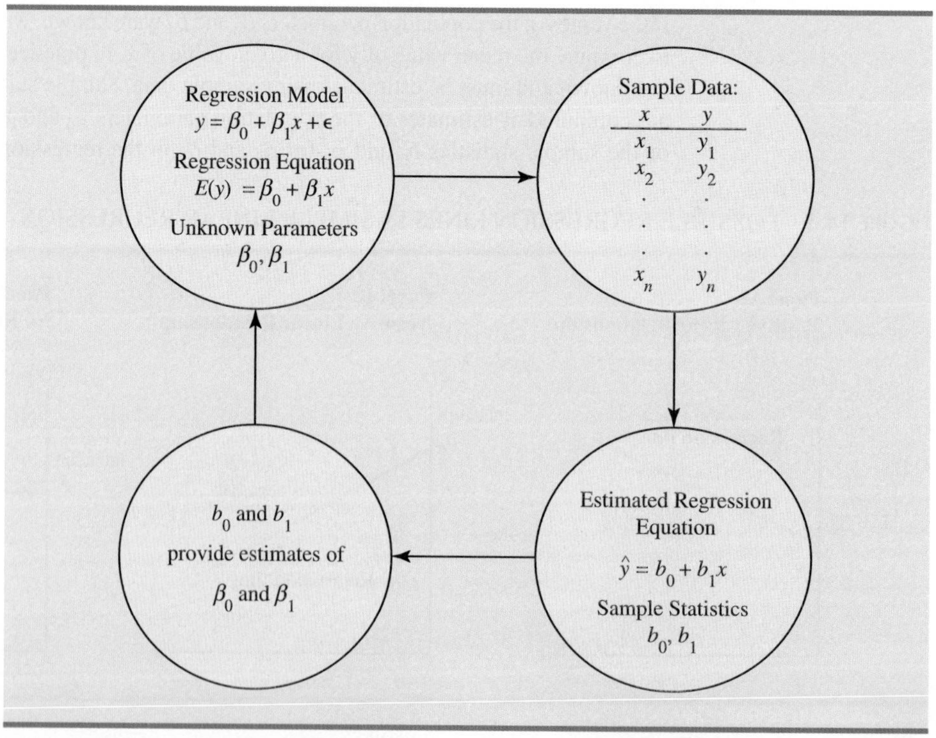

NOTES AND COMMENTS

1. Regression analysis cannot be interpreted as a procedure for establishing a cause-and-effect relationship between variables. It can only indicate how or to what extent variables are associated with each other. Any conclusions about cause and effect must be based upon the judgment of those individuals most knowledgeable about the application.
2. The regression equation in simple linear regression is $E(y) = \beta_0 + \beta_1 x$. More advanced texts in regression analysis often write the regression equation as $E(y|x) = \beta_0 + \beta_1 x$ to emphasize that the regression equation provides the mean value of y for a given value of x.

14.2 Least Squares Method

In simple linear regression, each observation consists of two values: one for the independent variable and one for the dependent variable.

The **least squares method** is a procedure for using sample data to find the estimated regression equation. To illustrate the least squares method, suppose data were collected from a sample of 10 Armand's Pizza Parlor restaurants located near college campuses. For the ith observation or restaurant in the sample, x_i is the size of the student population (in thousands) and y_i is the quarterly sales (in thousands of dollars). The values of x_i and y_i for the 10 restaurants in the sample are summarized in Table 14.1. We see that restaurant 1, with $x_1 = 2$ and $y_1 = 58$, is near a campus with 2000 students and has quarterly sales of \$58,000. Restaurant 2, with $x_2 = 6$ and $y_2 = 105$, is near a campus with 6000 students and has quarterly sales of \$105,000. The largest sales value is for restaurant 10, which is near a campus with 26,000 students and has quarterly sales of \$202,000.

Figure 14.3 is a scatter diagram of the data in Table 14.1. Student population is shown on the horizontal axis and quarterly sales is shown on the vertical axis. **Scatter diagrams** for regression analysis are constructed with the independent variable x on the horizontal axis and the dependent variable y on the vertical axis. The scatter diagram enables us to observe the data graphically and to draw preliminary conclusions about the possible relationship between the variables.

What preliminary conclusions can be drawn from Figure 14.3? Quarterly sales appear to be higher at campuses with larger student populations. In addition, for these data the relationship between the size of the student population and quarterly sales appears to be approximated by a straight line; indeed, a positive linear relationship is indicated between x

TABLE 14.1 STUDENT POPULATION AND QUARTERLY SALES DATA FOR 10 ARMAND'S PIZZA PARLORS

WEB file Armand's

Restaurant i	Student Population (1000s) x_i	Quarterly Sales ($1000s) y_i
1	2	58
2	6	105
3	8	88
4	8	118
5	12	117
6	16	137
7	20	157
8	20	169
9	22	149
10	26	202

FIGURE 14.3 SCATTER DIAGRAM OF STUDENT POPULATION AND QUARTERLY SALES FOR ARMAND'S PIZZA PARLORS

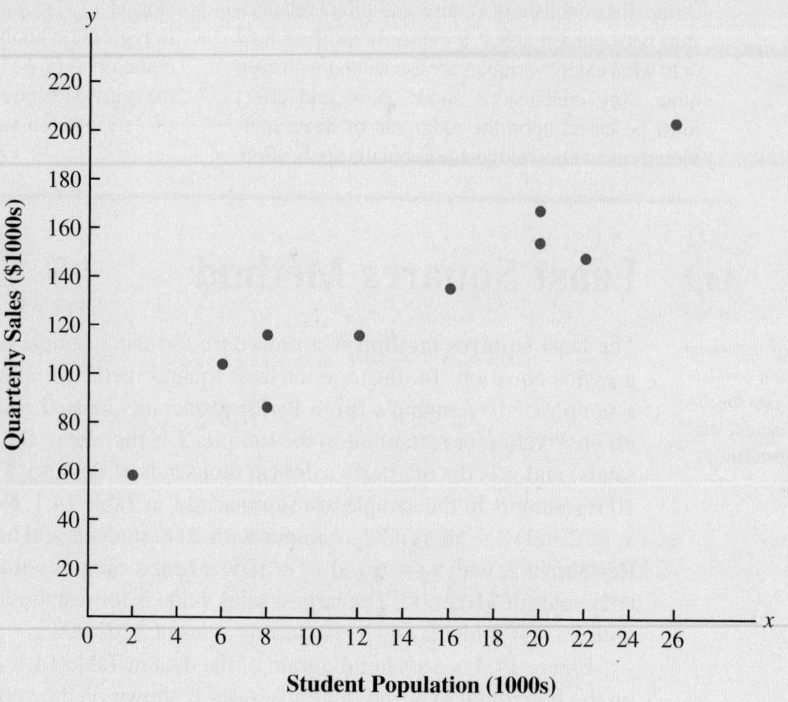

and y. We therefore choose the simple linear regression model to represent the relationship between quarterly sales and student population. Given that choice, our next task is to use the sample data in Table 14.1 to determine the values of b_0 and b_1 in the estimated simple linear regression equation. For the ith restaurant, the estimated regression equation provides

$$\hat{y}_i = b_0 + b_1 x_i \tag{14.4}$$

where

$\hat{y}_i$ = predicted value of quarterly sales ($1000s) for the ith restaurant
b_0 = the y-intercept of the estimated regression line
b_1 = the slope of the estimated regression line
x_i = size of the student population (1000s) for the ith restaurant

With y_i denoting the observed (actual) sales for restaurant i and $\hat{y}_i$ in equation (14.4) representing the predicted value of sales for restaurant i, every restaurant in the sample will have an observed value of sales y_i and a predicted value of sales $\hat{y}_i$. For the estimated regression line to provide a good fit to the data, we want the differences between the observed sales values and the predicted sales values to be small.

The least squares method uses the sample data to provide the values of b_0 and b_1 that minimize the *sum of the squares of the deviations* between the observed values of the dependent variable y_i and the predicted values of the dependent variable $\hat{y}_i$. The criterion for the least squares method is given by expression (14.5).

Carl Friedrich Gauss (1777–1855) proposed the least squares method.

LEAST SQUARES CRITERION

$$\min \Sigma(y_i - \hat{y}_i)^2 \quad \textbf{(14.5)}$$

where

y_i = observed value of the dependent variable for the ith observation
$\hat{y}_i$ = predicted value of the dependent variable for the ith observation

Differential calculus can be used to show (see Appendix 14.1) that the values of b_0 and b_1 that minimize expression (14.5) can be found by using equations (14.6) and (14.7).

In computing b_1 with a calculator, carry as many significant digits as possible in the intermediate calculations. We recommend carrying at least four significant digits.

SLOPE AND y-INTERCEPT FOR THE ESTIMATED REGRESSION EQUATION[1]

$$b_1 = \frac{\Sigma(x_i - \bar{x})(y_i - \bar{y})}{\Sigma(x_i - \bar{x})^2} \quad \textbf{(14.6)}$$

$$b_0 = \bar{y} - b_1\bar{x} \quad \textbf{(14.7)}$$

where

x_i = value of the independent variable for the ith observation
y_i = value of the dependent variable for the ith observation
$\bar{x}$ = mean value for the independent variable
$\bar{y}$ = mean value for the dependent variable
n = total number of observations

Some of the calculations necessary to develop the least squares estimated regression equation for Armand's Pizza Parlors are shown in Table 14.2. With the sample of 10 restaurants, we have $n = 10$ observations. Because equations (14.6) and (14.7) require $\bar{x}$ and $\bar{y}$ we begin the calculations by computing $\bar{x}$ and $\bar{y}$.

$$\bar{x} = \frac{\Sigma x_i}{n} = \frac{140}{10} = 14$$

$$\bar{y} = \frac{\Sigma y_i}{n} = \frac{1300}{10} = 130$$

Using equations (14.6) and (14.7) and the information in Table 14.2, we can compute the slope and intercept of the estimated regression equation for Armand's Pizza Parlors. The calculation of the slope (b_1) proceeds as follows.

[1] An alternate formula for b_1 is

$$b_1 = \frac{\Sigma x_i y_i - (\Sigma x_i \Sigma y_i)/n}{\Sigma x_i^2 - (\Sigma x_i)^2/n}$$

This form of equation (14.6) is often recommended when using a calculator to compute b_1.

TABLE 14.2 CALCULATIONS FOR THE LEAST SQUARES ESTIMATED REGRESSION EQUATION FOR ARMAND'S PIZZA PARLORS

Restaurant i	x_i	y_i	$x_i - \bar{x}$	$y_i - \bar{y}$	$(x_i - \bar{x})(y_i - \bar{y})$	$(x_i - \bar{x})^2$
1	2	58	−12	−72	864	144
2	6	105	−8	−25	200	64
3	8	88	−6	−42	252	36
4	8	118	−6	−12	72	36
5	12	117	−2	−13	26	4
6	16	137	2	7	14	4
7	20	157	6	27	162	36
8	20	169	6	39	234	36
9	22	149	8	19	152	64
10	26	202	12	72	864	144
Totals	140	1300			2840	568
	Σx_i	Σy_i			$\Sigma(x_i - \bar{x})(y_i - \bar{y})$	$\Sigma(x_i - \bar{x})^2$

$$b_1 = \frac{\Sigma(x_i - \bar{x})(y_i - \bar{y})}{\Sigma(x_i - \bar{x})^2} = \frac{2840}{568} = 5$$

The calculation of the y intercept (b_0) follows.

$$b_0 = \bar{y} - b_1\bar{x} = 130 - 5(14) = 60$$

Thus, the estimated regression equation is

$$\hat{y} = 60 + 5x$$

Figure 14.4 shows the graph of this equation on the scatter diagram.

The slope of the estimated regression equation ($b_1 = 5$) is positive, implying that as student population increases, sales increase. In fact, we can conclude (based on sales measured in \$1000s and student population in 1000s) that an increase in the student population of 1000 is associated with an increase of \$5000 in expected sales; that is, quarterly sales are expected to increase by \$5 per student.

Using the estimated regression equation to make predictions outside the range of the values of the independent variable should be done with caution because outside that range we cannot be sure that the same relationship is valid.

If we believe the least squares estimated regression equation adequately describes the relationship between x and y, it would seem reasonable to use the estimated regression equation to predict the value of y for a given value of x. For example, if we wanted to predict quarterly sales for a restaurant to be located near a campus with 16,000 students, we would compute

$$\hat{y} = 60 + 5(16) = 140$$

Hence, we would predict quarterly sales of \$140,000 for this restaurant. In the following sections we will discuss methods for assessing the appropriateness of using the estimated regression equation for estimation and prediction.

FIGURE 14.4 GRAPH OF THE ESTIMATED REGRESSION EQUATION FOR ARMAND'S PIZZA PARLORS: $\hat{y} = 60 + 5x$

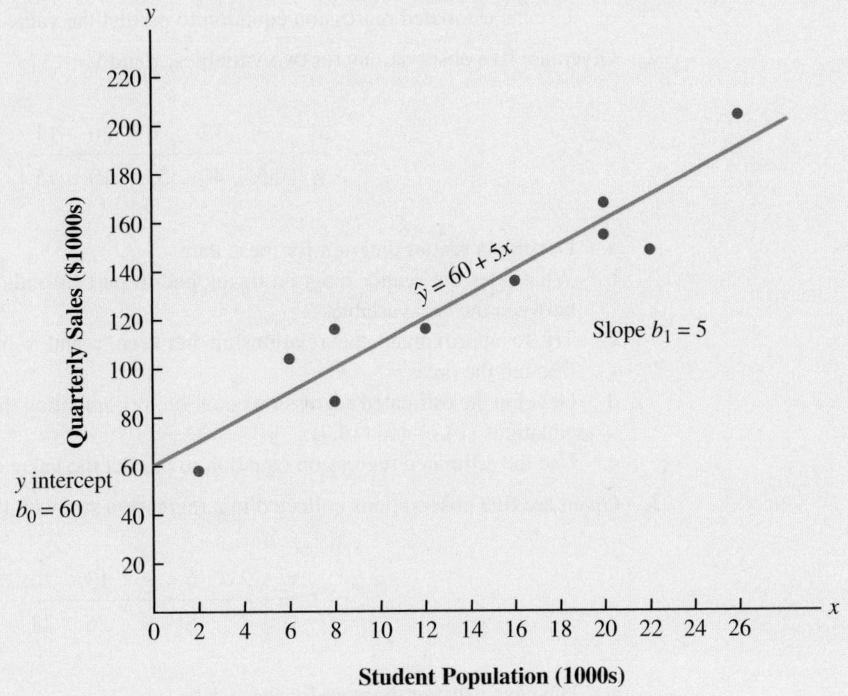

NOTES AND COMMENTS

The least squares method provides an estimated regression equation that minimizes the sum of squared deviations between the observed values of the dependent variable y_i and the predicted values of the dependent variable $\hat{y}_i$. This least squares criterion is used to choose the equation that provides the best fit. If some other criterion were used, such as minimizing the sum of the absolute deviations between y_i and $\hat{y}_i$, a different equation would be obtained. In practice, the least squares method is the most widely used.

Exercises

Methods

SELF test

1. Given are five observations for two variables, x and y.

x_i	1	2	3	4	5
y_i	3	7	5	11	14

   a. Develop a scatter diagram for these data.
   b. What does the scatter diagram developed in part (a) indicate about the relationship between the two variables?

c. Try to approximate the relationship between x and y by drawing a straight line through the data.
d. Develop the estimated regression equation by computing the values of b_0 and b_1 using equations (14.6) and (14.7).
e. Use the estimated regression equation to predict the value of y when $x = 4$.

2. Given are five observations for two variables, x and y.

x_i	3	12	6	20	14
y_i	55	40	55	10	15

a. Develop a scatter diagram for these data.
b. What does the scatter diagram developed in part (a) indicate about the relationship between the two variables?
c. Try to approximate the relationship between x and y by drawing a straight line through the data.
d. Develop the estimated regression equation by computing the values of b_0 and b_1 using equations (14.6) and (14.7).
e. Use the estimated regression equation to predict the value of y when $x = 10$.

3. Given are five observations collected in a regression study on two variables.

x_i	2	6	9	13	20
y_i	7	18	9	26	23

a. Develop a scatter diagram for these data.
b. Develop the estimated regression equation for these data.
c. Use the estimated regression equation to predict the value of y when $x = 6$.

Applications

SELF test

4. The following data give the percentage of women working in five companies in the retail and trade industry. The percentage of management jobs held by women in each company is also shown.

% Working	67	45	73	54	61
% Management	49	21	65	47	33

a. Develop a scatter diagram for these data with the percentage of women working in the company as the independent variable.
b. What does the scatter diagram developed in part (a) indicate about the relationship between the two variables?
c. Try to approximate the relationship between the percentage of women working in the company and the percentage of management jobs held by women in that company.
d. Develop the estimated regression equation by computing the values of b_0 and b_1.
e. Predict the percentage of management jobs held by women in a company that has 60% women employees.

5. Elliptical trainers are becoming one of the more popular exercise machines. Their smooth and steady low-impact motion makes them a preferred choice for individuals with knee and ankle problems. But selecting the right trainer can be a difficult process. Price and

quality are two important factors in any purchase decision. Are higher prices generally associated with higher quality elliptical trainers? *Consumer Reports* conducted extensive tests to develop an overall rating based on ease of use, ergonomics, construction, and exercise range. The following data show the price and rating for eight elliptical trainers tested (*Consumer Reports,* February 2008).

WEB file

Ellipticals

Brand and Model	Price ($)	Rating
Precor 5.31	3700	87
Keys Fitness CG2	2500	84
Octane Fitness Q37e	2800	82
LifeFitness X1 Basic	1900	74
NordicTrack AudioStrider 990	1000	73
Schwinn 430	800	69
Vision Fitness X6100	1700	68
ProForm XP 520 Razor	600	55

a. Develop a scatter diagram with price as the independent variable.
b. An exercise equipment store that sells primarily higher priced equipment has a sign over the display area that says "Quality: You Get What You Pay For." Based upon your analysis of the data for ellipical trainers, do you think this sign fairly reflects the price-quality relationship for elliptical trainers?
c. Use the least squares method to develop the estimated regression equation.
d. Use the estimated regression equation to predict the rating for an ellipitical trainer with a price of $1500.

6. The National Football League (NFL) records a variety of performance data for individuals and teams. To investigate the importance of passing on the percentage of games won by a team, the following data show the average number of passing yards per attempt (Yds/Att) and the percentage of games won (WinPct) for a random sample of 10 NFL teams for the 2011 season (NFL website, February 12, 2012).

WEB file

NFLPassing

Team	Yds/Att	WinPct
Arizona Cardinals	6.5	50
Atlanta Falcons	7.1	63
Carolina Panthers	7.4	38
Chicago Bears	6.4	50
Dallas Cowboys	7.4	50
New England Patriots	8.3	81
Philadelphia Eagles	7.4	50
Seattle Seahawks	6.1	44
St. Louis Rams	5.2	13
Tampa Bay Buccaneers	6.2	25

a. Develop a scatter diagram with the number of passing yards per attempt on the horizontal axis and the percentage of games won on the vertical axis.
b. What does the scatter diagram developed in part (a) indicate about the relationship between the two variables?
c. Develop the estimated regression equation that could be used to predict the percentage of games won given the average number of passing yards per attempt.
d. Provide an interpretation for the slope of the estimated regression equation.

e. For the 2011 season, the average number of passing yards per attempt for the Kansas City Chiefs was 6.2. Use the estimated regression equation developed in part (c) to predict the percentage of games won by the Kansas City Chiefs. (*Note:* For the 2011 season the Kansas City Chiefs record was 7 wins and 9 losses.) Compare your prediction to the actual percentage of games won by the Kansas City Chiefs.

7. A sales manager collected the following data on annual sales for new customer accounts and the number of years of experience for a sample of 10 salespersons.

WEB file
Sales

Salesperson	Years of Experience	Annual Sales ($1000s)
1	1	80
2	3	97
3	4	92
4	4	102
5	6	103
6	8	111
7	10	119
8	10	123
9	11	117
10	13	136

a. Develop a scatter diagram for these data with years of experience as the independent variable.
b. Develop an estimated regression equation that can be used to predict annual sales given the years of experience.
c. Use the estimated regression equation to predict annual sales for a salesperson with 9 years of experience.

8. The American Association of Individual Investors (AAII) On-Line Discount Broker Survey polls members on their experiences with discount brokers. As part of the survey, members were asked to rate the quality of the speed of execution with their broker as well as provide an overall satisfaction rating for electronic trades. Possible responses (scores) were no opinion (0), unsatisfied (1), somewhat satisfied (2), satisfied (3), and very satisfied (4). For each broker summary scores were computed by calculating a weighted average of the scores provided by each respondent. A portion of the survey results follow (AAII website, February 7, 2012).

WEB file
BrokerRatings

Brokerage	Speed	Satisfaction
Scottrade, Inc.	3.4	3.5
Charles Schwab	3.3	3.4
Fidelity Brokerage Services	3.4	3.9
TD Ameritrade	3.6	3.7
E*Trade Financial	3.2	2.9
Vanguard Brokerage Services	3.8	2.8
USAA Brokerage Services	3.8	3.6
Thinkorswim	2.6	2.6
Wells Fargo Investments	2.7	2.3
Interactive Brokers	4.0	4.0
Zecco.com	2.5	2.5

a. Develop a scatter diagram for these data with the speed of execution as the independent variable.

b. What does the scatter diagram developed in part (a) indicate about the relationship between the two variables?
c. Develop the least squares estimated regression equation.
d. Provide an interpretation for the slope of the estimated regression equation.
e. Suppose Zecco.com developed new software to increase their speed of execution rating. If the new software is able to increase their speed of execution rating from the current value of 2.5 to the average speed of execution rating for the other 10 brokerage firms that were surveyed, what value would you predict for the overall satisfaction rating?

9. Using a global-positioning-system (GPS)-based navigator for your car, you enter a destination and the system will plot a route, give spoken turn-by-turn directions, and show your progress along the route. Today, even budget units include features previously available only on more expensive models. *Consumer Reports* conducted extensive tests of GPS-based navigators and developed an overall rating based on factors such as ease of use, driver information, display, and battery life. The following data show the price and rating for a sample of 20 GPS units with a 4.3-inch screen that *Consumer Reports* tested (*Consumer Reports* website, April 17, 2012).

WEB file
GPS

Brand and Model	Price ($)	Rating
Garmin Nuvi 3490LMT	400	82
Garmin Nuvi 3450	330	80
Garmin Nuvi 3790T	350	77
Garmin Nuvi 3790LMT	400	77
Garmin Nuvi 3750	250	74
Garmin Nuvi 2475LT	230	74
Garmin Nuvi 2455LT	160	73
Garmin Nuvi 2370LT	270	71
Garmin Nuvi 2360LT	250	71
Garmin Nuvi 2360LMT	220	71
Garmin Nuvi 755T	260	70
Motorola Motonab TN565t	200	68
Motorola Motonab TN555	200	67
Garmin Nuvi 1350T	150	65
Garmin Nuvi 1350LMT	180	65
Garmin Nuvi 2300	160	65
Garmin Nuvi 1350	130	64
Tom Tom VIA 1435T	200	62
Garmin Nuvi 1300	140	62
Garmin Nuvi 1300LM	180	62

a. Develop a scatter diagram with price as the independent variable.
b. What does the scatter diagram developed in part (a) indicate about the relationship between the two variables?
c. Use the least squares method to develop the estimated regression equation.
d. Predict the rating for a GPS system with a 4.3-inch screen that has a price of $200.

10. On March 31, 2009, Ford Motor Company's shares were trading at a 26-year low of $2.63. Ford's board of directors gave the CEO a grant of options and restricted shares with an estimated value of $16 million. On April 26, 2011, the price of a share of Ford had increased to $15.58, and the CEO's grant was worth $202.8 million, a gain in value of $186.8 million. The following table shows the share price in 2009 and 2011 for 10 companies, the stock-option and share grants to the CEOs in late 2008 and 2009, and the value of the options and grants in 2011. Also shown are the percentage increases in the stock price and the percentage gains in the options values (*The Wall Street Journal,* April 27, 2011).

WEB file
CEOGrants

Company	Stock Price 2009 ($)	Stock Price 2011 ($)	% Increase in Stock Price	Options and Grants Value 2009 ($ millions)	Options and Grants Value 2011 ($ millions)	% Gain in Options Value
Ford Motor	2.63	15.58	492	16.0	202.8	1168
Abercrombie & Fitch	23.80	70.47	196	46.2	196.1	324
Nabors Industries	9.99	32.06	221	37.2	132.2	255
Starbucks	9.99	32.06	221	12.4	75.9	512
Salesforce.com	32.73	137.61	320	7.8	67.0	759
Starwood Hotels	12.70	60.28	375	5.8	57.1	884
Caterpillar	27.96	111.94	300	4.0	47.5	1088
Oracle	18.07	34.97	94	61.9	97.5	58
Capital One	12.24	54.61	346	6.0	40.6	577
Dow Chemical	8.43	39.97	374	5.0	38.8	676

a. Develop a scatter diagram for these data with the percentage increase in the stock price as the independent variable.
b. What does the scatter diagram developed in part (a) indicate about the relationship between the two variables?
c. Develop the least squares estimated regression equation.
d. Provide an interpretation for the slope of the estimated regression equation.
e. Do the rewards for the CEO appear to be based on performance increases as measured by the stock price?

11. Sporty cars are designed to provide better handling, acceleration, and a more responsive driving experience than a typical sedan. But, even within this select group of cars, performance as well as price can vary. *Consumer Reports* provided road-test scores and prices for the following 12 sporty cars (Consumer Reports website, October 2008). Prices are in thousands of dollars and road-test scores are based on a 0–100 rating scale, with higher values indicating better performance.

WEB file
SportyCars

Car	Price ($1000s)	Road-Test Score
Chevrolet Cobalt SS	24.5	78
Dodge Caliber SRT4	24.9	56
Ford Mustang GT (V8)	29.0	73
Honda Civic Si	21.7	78
Mazda RX-8	31.3	86
Mini Cooper S	26.4	74
Mitsubishi Lancer Evolution GSR	38.1	83
Nissan Sentra SE-R Spec V	23.3	66
Suburu Impreza WRX	25.2	81
Suburu Impreza WRX Sti	37.6	89
Volkswagen GTI	24.0	83
Volkswagen R32	33.6	83

a. Develop a scatter diagram with price as the independent variable.
b. What does the scatter diagram developed in part (a) indicate about the relationship between the two variables?
c. Use the least squares method to develop the estimated regression equation.
d. Provide an interpretation for the slope of the estimated regression equation.
e. Another sporty car that *Consumer Reports* tested is the BMW 135i; the price for this car was $36,700. Predict the road-test score for the BMW 135i using the estimated regression equation developed in part (c).

12. Concur Technologies, Inc., is a large expense-management company located in Redmond, Washington. *The Wall Street Journal* asked Concur to examine the data from 8.3 million expense reports to provide insights regarding business travel expenses. Their analysis of the data showed that New York was the most expensive city, with an average daily hotel room rate of $198 and an average amount spent on entertainment, including group meals and tickets for shows, sports, and other events, of $172. In comparison, the U.S. averages for these two categories were $89 for the room rate and $99 for entertainment. The following table shows the average daily hotel room rate and the amount spent on entertainment for a random sample of 9 of the 25 most visited U.S. cities (*The Wall Street Journal,* August 18, 2011).

WEB file

BusinessTravel

City	Room Rate ($)	Entertainment ($)
Boston	148	161
Denver	96	105
Nashville	91	101
New Orleans	110	142
Phoenix	90	100
San Diego	102	120
San Francisco	136	167
San Jose	90	140
Tampa	82	98

a. Develop a scatter diagram for these data with the room rate as the independent variable.
b. What does the scatter diagram developed in part (a) indicate about the relationship between the two variables?
c. Develop the least squares estimated regression equation.
d. Provide an interpretation for the slope of the estimated regression equation.
e. The average room rate in Chicago is $128, considerably higher than the U.S. average. Predict the entertainment expense per day for Chicago.

13. To the Internal Revenue Service, the reasonableness of total itemized deductions depends on the taxpayer's adjusted gross income. Large deductions, which include charity and medical deductions, are more reasonable for taxpayers with large adjusted gross incomes. If a taxpayer claims larger than average itemized deductions for a given level of income, the chances of an IRS audit are increased. Data (in thousands of dollars) on adjusted gross income and the average or reasonable amount of itemized deductions follow.

Adjusted Gross Income ($1000s)	Reasonable Amount of Itemized Deductions ($1000s)
22	9.6
27	9.6
32	10.1
48	11.1
65	13.5
85	17.7
120	25.5

a. Develop a scatter diagram for these data with adjusted gross income as the independent variable.
b. Use the least squares method to develop the estimated regression equation.
c. Predict the reasonable level of total itemized deductions for a taxpayer with an adjusted gross income of $52,500. If this taxpayer claimed itemized deductions of $20,400, would the IRS agent's request for an audit appear justified? Explain.

14. *PCWorld* rated four component characteristics for 10 ultraportable laptop computers: features, performance, design, and price. Each characteristic was rated using a 0–100 point scale. An overall rating, referred to as the *PCW World* Rating, was then developed for each laptop. The following table shows the features rating and the *PCW World* Rating for the 10 laptop computers (*PC World* website, February 5, 2009).

WEB file Laptop

Model	Features Rating	*PCW World* Rating
Thinkpad X200	87	83
VGN-Z598U	85	82
U6V	80	81
Elitebook 2530P	75	78
X360	80	78
Thinkpad X300	76	78
Ideapad U110	81	77
Micro Express JFT2500	73	75
Toughbook W7	79	73
HP Voodoo Envy133	68	72

a. Develop a scatter diagram with the features rating as the independent variable.
b. What does the scatter diagram developed in part (a) indicate about the relationship between the two variables?
c. Use the least squares method to develop the estimated regression equation.
d. Predict the *PCW World* Rating for a new laptop computer that has a features rating of 70.

14.3 Coefficient of Determination

For the Armand's Pizza Parlors example, we developed the estimated regression equation $\hat{y} = 60 + 5x$ to approximate the linear relationship between the size of the student population x and quarterly sales y. A question now is: How well does the estimated regression equation fit the data? In this section, we show that the **coefficient of determination** provides a measure of the goodness of fit for the estimated regression equation.

For the ith observation, the difference between the observed value of the dependent variable, y_i, and the predicted value of the dependent variable, $\hat{y}_i$, is called the ***i*th residual**. The ith residual represents the error in using $\hat{y}_i$ to estimate y_i. Thus, for the ith observation, the residual is $y_i - \hat{y}_i$. The sum of squares of these residuals or errors is the quantity that is minimized by the least squares method. This quantity, also known as the *sum of squares due to error,* is denoted by SSE.

SUM OF SQUARES DUE TO ERROR

$$\text{SSE} = \Sigma(y_i - \hat{y}_i)^2 \tag{14.8}$$

The value of SSE is a measure of the error in using the estimated regression equation to predict the values of the dependent variable in the sample.

In Table 14.3 we show the calculations required to compute the sum of squares due to error for the Armand's Pizza Parlors example. For instance, for restaurant 1 the values of the independent and dependent variables are $x_1 = 2$ and $y_1 = 58$. Using the estimated

TABLE 14.3 CALCULATION OF SSE FOR ARMAND'S PIZZA PARLORS

Restaurant i	x_i = Student Population (1000s)	y_i = Quarterly Sales ($1000s)	Predicted Sales $\hat{y}_i = 60 + 5x_i$	Error $y_i - \hat{y}_i$	Squared Error $(y_i - \hat{y}_i)^2$
1	2	58	70	−12	144
2	6	105	90	15	225
3	8	88	100	−12	144
4	8	118	100	18	324
5	12	117	120	−3	9
6	16	137	140	−3	9
7	20	157	160	−3	9
8	20	169	160	9	81
9	22	149	170	−21	441
10	26	202	190	12	144
					SSE = 1530

regression equation, we find that the predicted value of quarterly sales for restaurant 1 is $\hat{y}_1 = 60 + 5(2) = 70$. Thus, the error in using $\hat{y}_1$ to predict y_1 for restaurant 1 is $y_1 - \hat{y}_1 = 58 - 70 = -12$. The squared error, $(-12)^2 = 144$, is shown in the last column of Table 14.3. After computing and squaring the residuals for each restaurant in the sample, we sum them to obtain SSE = 1530. Thus, SSE = 1530 measures the error in using the estimated regression equation $\hat{y} = 60 + 5x$ to predict sales.

Now suppose we are asked to develop an estimate of quarterly sales without knowledge of the size of the student population. Without knowledge of any related variables, we would use the sample mean as an estimate of quarterly sales at any given restaurant. Table 14.2 showed that for the sales data, $\Sigma y_i = 1300$. Hence, the mean value of quarterly sales for the sample of 10 Armand's restaurants is $\bar{y} = \Sigma y_i/n = 1300/10 = 130$. In Table 14.4 we show the sum of squared deviations obtained by using the sample mean $\bar{y} = 130$ to predict the value of quarterly sales for each restaurant in the sample. For the ith restaurant in the sample, the difference $y_i - \bar{y}$ provides a measure of the error involved in using $\bar{y}$ to predict sales. The corresponding sum of squares, called the *total sum of squares,* is denoted SST.

TABLE 14.4 COMPUTATION OF THE TOTAL SUM OF SQUARES FOR ARMAND'S PIZZA PARLORS

Restaurant i	x_i = Student Population (1000s)	y_i = Quarterly Sales ($1000s)	Deviation $y_i - \bar{y}$	Squared Deviation $(y_i - \bar{y})^2$
1	2	58	−72	5184
2	6	105	−25	625
3	8	88	−42	1764
4	8	118	−12	144
5	12	117	−13	169
6	16	137	7	49
7	20	157	27	729
8	20	169	39	1521
9	22	149	19	361
10	26	202	72	5184
				SST = 15,730

TOTAL SUM OF SQUARES

$$\text{SST} = \Sigma(y_i - \bar{y})^2 \tag{14.9}$$

With SST = 15,730 and SSE = 1530, the estimated regression line provides a much better fit to the data than the line $y = \bar{y}$.

The sum at the bottom of the last column in Table 14.4 is the total sum of squares for Armand's Pizza Parlors; it is SST = 15,730.

In Figure 14.5 we show the estimated regression line $\hat{y} = 60 + 5x$ and the line corresponding to $\bar{y} = 130$. Note that the points cluster more closely around the estimated regression line than they do about the line $\bar{y} = 130$. For example, for the 10th restaurant in the sample we see that the error is much larger when $\bar{y} = 130$ is used to predict y_{10} than when $\hat{y}_{10} = 60 + 5(26) = 190$ is used. We can think of SST as a measure of how well the observations cluster about the $\bar{y}$ line and SSE as a measure of how well the observations cluster about the $\hat{y}$ line.

To measure how much the $\hat{y}$ values on the estimated regression line deviate from $\bar{y}$, another sum of squares is computed. This sum of squares, called the *sum of squares due to regression,* is denoted SSR.

SUM OF SQUARES DUE TO REGRESSION

$$\text{SSR} = \Sigma(\hat{y}_i - \bar{y})^2 \tag{14.10}$$

FIGURE 14.5 DEVIATIONS ABOUT THE ESTIMATED REGRESSION LINE AND THE LINE $y = \bar{y}$ FOR ARMAND'S PIZZA PARLORS

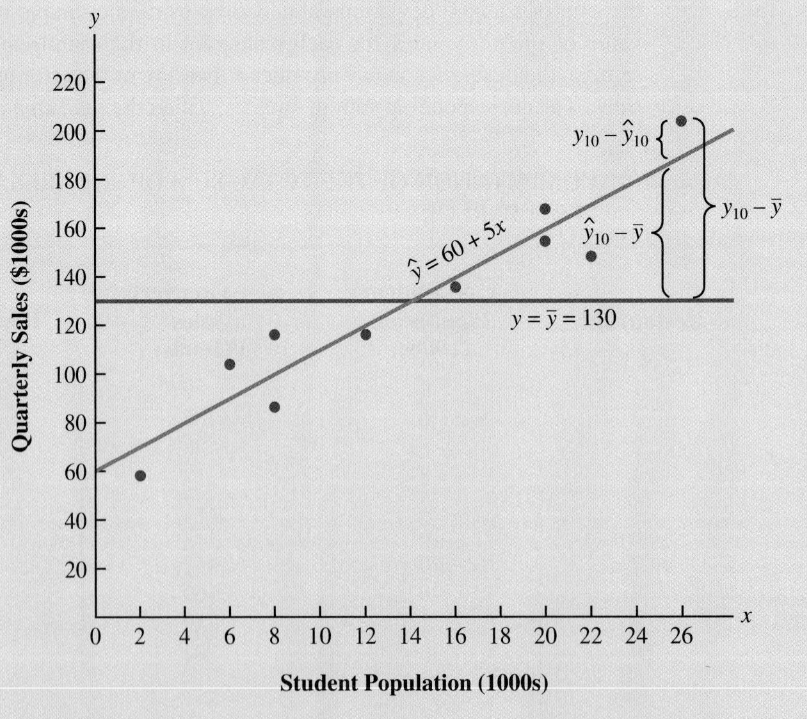

From the preceding discussion, we should expect that SST, SSR, and SSE are related. Indeed, the relationship among these three sums of squares provides one of the most important results in statistics.

SSR can be thought of as the explained portion of SST, and SSE can be thought of as the unexplained portion of SST.

RELATIONSHIP AMONG SST, SSR, AND SSE

$$\text{SST} = \text{SSR} + \text{SSE} \tag{14.11}$$

where

SST = total sum of squares
SSR = sum of squares due to regression
SSE = sum of squares due to error

Equation (14.11) shows that the total sum of squares can be partitioned into two components, the sum of squares due to regression and the sum of squares due to error. Hence, if the values of any two of these sum of squares are known, the third sum of squares can be computed easily. For instance, in the Armand's Pizza Parlors example, we already know that SSE = 1530 and SST = 15,730; therefore, solving for SSR in equation (14.11), we find that the sum of squares due to regression is

$$\text{SSR} = \text{SST} - \text{SSE} = 15{,}730 - 1530 = 14{,}200$$

Now let us see how the three sums of squares, SST, SSR, and SSE, can be used to provide a measure of the goodness of fit for the estimated regression equation. The estimated regression equation would provide a perfect fit if every value of the dependent variable y_i happened to lie on the estimated regression line. In this case, $y_i - \hat{y}_i$ would be zero for each observation, resulting in SSE = 0. Because SST = SSR + SSE, we see that for a perfect fit SSR must equal SST, and the ratio (SSR/SST) must equal one. Poorer fits will result in larger values for SSE. Solving for SSE in equation (14.11), we see that SSE = SST − SSR. Hence, the largest value for SSE (and hence the poorest fit) occurs when SSR = 0 and SSE = SST.

The ratio SSR/SST, which will take values between zero and one, is used to evaluate the goodness of fit for the estimated regression equation. This ratio is called the *coefficient of determination* and is denoted by r^2.

COEFFICIENT OF DETERMINATION

$$r^2 = \frac{\text{SSR}}{\text{SST}} \tag{14.12}$$

For the Armand's Pizza Parlors example, the value of the coefficient of determination is

$$r^2 = \frac{\text{SSR}}{\text{SST}} = \frac{14{,}200}{15{,}730} = .9027$$

When we express the coefficient of determination as a percentage, r^2 can be interpreted as the percentage of the total sum of squares that can be explained by using the estimated regression equation. For Armand's Pizza Parlors, we can conclude that 90.27% of the total sum of squares can be explained by using the estimated regression equation $\hat{y} = 60 + 5x$ to predict quarterly sales. In other words, 90.27% of the variability in sales can be explained by the linear relationship between the size of the student population and sales. We should be pleased to find such a good fit for the estimated regression equation.

Correlation Coefficient

In Chapter 3 we introduced the **correlation coefficient** as a descriptive measure of the strength of linear association between two variables, x and y. Values of the correlation coefficient are always between -1 and $+1$. A value of $+1$ indicates that the two variables x and y are perfectly related in a positive linear sense. That is, all data points are on a straight line that has a positive slope. A value of -1 indicates that x and y are perfectly related in a negative linear sense, with all data points on a straight line that has a negative slope. Values of the correlation coefficient close to zero indicate that x and y are not linearly related.

In Section 3.5 we presented the equation for computing the sample correlation coefficient. If a regression analysis has already been performed and the coefficient of determination r^2 computed, the sample correlation coefficient can be computed as follows.

SAMPLE CORRELATION COEFFICIENT

$$\begin{aligned} r_{xy} &= (\text{sign of } b_1)\sqrt{\text{Coefficient of determination}} \\ &= (\text{sign of } b_1)\sqrt{r^2} \end{aligned} \quad \textbf{(14.13)}$$

where

b_1 = the slope of the estimated regression equation $\hat{y} = b_0 + b_1x$

The sign for the sample correlation coefficient is positive if the estimated regression equation has a positive slope ($b_1 > 0$) and negative if the estimated regression equation has a negative slope ($b_1 < 0$).

For the Armand's Pizza Parlor example, the value of the coefficient of determination corresponding to the estimated regression equation $\hat{y} = 60 + 5x$ is .9027. Because the slope of the estimated regression equation is positive, equation (14.13) shows that the sample correlation coefficient is $+\sqrt{.9027} = +.9501$. With a sample correlation coefficient of $r_{xy} = +.9501$, we would conclude that a strong positive linear association exists between x and y.

In the case of a linear relationship between two variables, both the coefficient of determination and the sample correlation coefficient provide measures of the strength of the relationship. The coefficient of determination provides a measure between zero and one, whereas the sample correlation coefficient provides a measure between -1 and $+1$. Although the sample correlation coefficient is restricted to a linear relationship between two variables, the coefficient of determination can be used for nonlinear relationships and for relationships that have two or more independent variables. Thus, the coefficient of determination provides a wider range of applicability.

NOTES AND COMMENTS

1. In developing the least squares estimated regression equation and computing the coefficient of determination, we made no probabilistic assumptions about the error term ϵ, and no statistical tests for significance of the relationship between x and y were conducted. Larger values of r^2 imply that the least squares line provides a better fit to the data; that is, the observations are more closely grouped about the least squares line. But, using only r^2, we can draw no conclusion about whether the relationship between x and y is statistically significant. Such a conclusion must be based on considerations that involve the sample size and the properties of the appropriate sampling distributions of the least squares estimators.
2. As a practical matter, for typical data found in the social sciences, values of r^2 as low as .25 are often considered useful. For data in the physical and life sciences, r^2 values of .60 or greater are often found; in fact, in some cases, r^2 values greater than .90 can be found. In business applications, r^2 values vary greatly, depending on the unique characteristics of each application.

Exercises

Methods

SELF test

15. The data from exercise 1 follow.

x_i	1	2	3	4	5
y_i	3	7	5	11	14

The estimated regression equation for these data is $\hat{y} = .20 + 2.60x$.

a. Compute SSE, SST, and SSR using equations (14.8), (14.9), and (14.10).
b. Compute the coefficient of determination r^2. Comment on the goodness of fit.
c. Compute the sample correlation coefficient.

16. The data from exercise 2 follow.

x_i	3	12	6	20	14
y_i	55	40	55	10	15

The estimated regression equation for these data is $\hat{y} = 68 - 3x$.

a. Compute SSE, SST, and SSR.
b. Compute the coefficient of determination r^2. Comment on the goodness of fit.
c. Compute the sample correlation coefficient.

17. The data from exercise 3 follow.

x_i	2	6	9	13	20
y_i	7	18	9	26	23

The estimated regression equation for these data is $\hat{y} = 7.6 + .9x$. What percentage of the total sum of squares can be accounted for by the estimated regression equation? What is the value of the sample correlation coefficient?

Applications

SELF test

18. The following data show the brand, price (\$), and the overall score for six stereo headphones that were tested by *Consumer Reports* (*Consumer Reports* website, March 5, 2012). The overall score is based on sound quality and effectiveness of ambient noise reduction. Scores range from 0 (lowest) to 100 (highest). The estimated regression equation for these data is $\hat{y} = 23.194 + .318x$, where x = price (\$) and y = overall score.

Brand	Price ($)	Score
Bose	180	76
Skullcandy	150	71
Koss	95	61
Phillips/O'Neill	70	56
Denon	70	40
JVC	35	26

a. Compute SST, SSR, and SSE.
b. Compute the coefficient of determination r^2. Comment on the goodness of fit.
c. What is the value of the sample correlation coefficient?

19. In exercise 7 a sales manager collected the following data on x = annual sales and y = years of experience. The estimated regression equation for these data is $\hat{y} = 80 + 4x$.

WEB file
Sales

Salesperson	Years of Experience	Annual Sales ($1000s)
1	1	80
2	3	97
3	4	92
4	4	102
5	6	103
6	8	111
7	10	119
8	10	123
9	11	117
10	13	136

a. Compute SST, SSR, and SSE.
b. Compute the coefficient of determination r^2. Comment on the goodness of fit.
c. What is the value of the sample correlation coefficient?

20. *Bicycling,* the world's leading cycling magazine, reviews hundreds of bicycles throughout the year. Their "Road-Race" category contains reviews of bikes used by riders primarily interested in racing. One of the most important factors in selecting a bike for racing is the weight of the bike. The following data show the weight (pounds) and price ($) for 10 racing bikes reviewed by the magazine (*Bicycling* website, March 8, 2012).

WEB file
RacingBicycles

Brand	Weight	Price ($)
FELT F5	17.8	2100
PINARELLO Paris	16.1	6250
ORBEA Orca GDR	14.9	8370
EDDY MERCKX EMX-7	15.9	6200
BH RC1 Ultegra	17.2	4000
BH Ultralight 386	13.1	8600
CERVELO S5 Team	16.2	6000
GIANT TCR Advanced 2	17.1	2580
WILIER TRIESTINA Gran Turismo	17.6	3400
SPECIALIZED S-Works Amira SL4	14.1	8000

a. Use the data to develop an estimated regression equation that could be used to estimate the price for a bike given the weight.
b. Compute r^2. Did the estimated regression equation provide a good fit?
c. Predict the price for a bike that weighs 15 pounds.

21. An important application of regression analysis in accounting is in the estimation of cost. By collecting data on volume and cost and using the least squares method to develop an estimated regression equation relating volume and cost, an accountant can estimate the cost associated with a particular manufacturing volume. Consider the following sample of production volumes and total cost data for a manufacturing operation.

Production Volume (units)	Total Cost ($)
400	4000
450	5000
550	5400
600	5900
700	6400
750	7000

a. Use these data to develop an estimated regression equation that could be used to predict the total cost for a given production volume.
b. What is the variable cost per unit produced?
c. Compute the coefficient of determination. What percentage of the variation in total cost can be explained by production volume?
d. The company's production schedule shows 500 units must be produced next month. Predict the total cost for this operation?

22. Refer to exercise 5 where the following data were used to investigate whether higher prices are generally associated with higher ratings for elliptical trainers (*Consumer Reports,* February 2008).

WEB file
Ellipticals

Brand and Model	Price ($)	Rating
Precor 5.31	3700	87
Keys Fitness CG2	2500	84
Octane Fitness Q37e	2800	82
LifeFitness X1 Basic	1900	74
NordicTrack AudioStrider 990	1000	73
Schwinn 430	800	69
Vision Fitness X6100	1700	68
ProForm XP 520 Razor	600	55

With x = price ($) and y = rating, the estimated regression equation is $\hat{y} = 58.158 + .008449x$. For these data, SSE = 173.88.

a. Compute the coefficient of determination r^2.
b. Did the estimated regression equation provide a good fit? Explain.
c. What is the value of the sample correlation coefficient? Does it reflect a strong or weak relationship between price and rating?

14.4 Model Assumptions

In conducting a regression analysis, we begin by making an assumption about the appropriate model for the relationship between the dependent and independent variable(s). For the case of simple linear regression, the assumed regression model is

$$y = \beta_0 + \beta_1 x + \epsilon$$

Then the least squares method is used to develop values for b_0 and b_1, the estimates of the model parameters β_0 and β_1, respectively. The resulting estimated regression equation is

$$\hat{y} = b_0 + b_1 x$$

We saw that the value of the coefficient of determination (r^2) is a measure of the goodness of fit of the estimated regression equation. However, even with a large value of r^2, the estimated regression equation should not be used until further analysis of the appropriateness of the assumed model has been conducted. An important step in determining whether the assumed model is appropriate involves testing for the significance of the relationship. The tests of significance in regression analysis are based on the following assumptions about the error term ϵ.

ASSUMPTIONS ABOUT THE ERROR TERM ϵ IN THE REGRESSION MODEL

$$y = \beta_0 + \beta_1 x + \epsilon$$

1. The error term ϵ is a random variable with a mean or expected value of zero; that is, $E(\epsilon) = 0$.
   Implication: β_0 and β_1 are constants, therefore $E(\beta_0) = \beta_0$ and $E(\beta_1) = \beta_1$; thus, for a given value of x, the expected value of y is

   $$E(y) = \beta_0 + \beta_1 x \qquad \textbf{(14.14)}$$

   As we indicated previously, equation (14.14) is referred to as the regression equation.
2. The variance of ϵ, denoted by σ^2, is the same for all values of x.
   Implication: The variance of y about the regression line equals σ^2 and is the same for all values of x.
3. The values of ϵ are independent.
   Implication: The value of ϵ for a particular value of x is not related to the value of ϵ for any other value of x; thus, the value of y for a particular value of x is not related to the value of y for any other value of x.
4. The error term ϵ is a normally distributed random variable for all values of x.
   Implication: Because y is a linear function of ϵ, y is also a normally distributed random variable for all values of x.

Figure 14.6 illustrates the model assumptions and their implications; note that in this graphical interpretation, the value of $E(y)$ changes according to the specific value of x considered. However, regardless of the x value, the probability distribution of ϵ and hence the probability distributions of y are normally distributed, each with the same variance. The specific value of the error ϵ at any particular point depends on whether the actual value of y is greater than or less than $E(y)$.

FIGURE 14.6 ASSUMPTIONS FOR THE REGRESSION MODEL

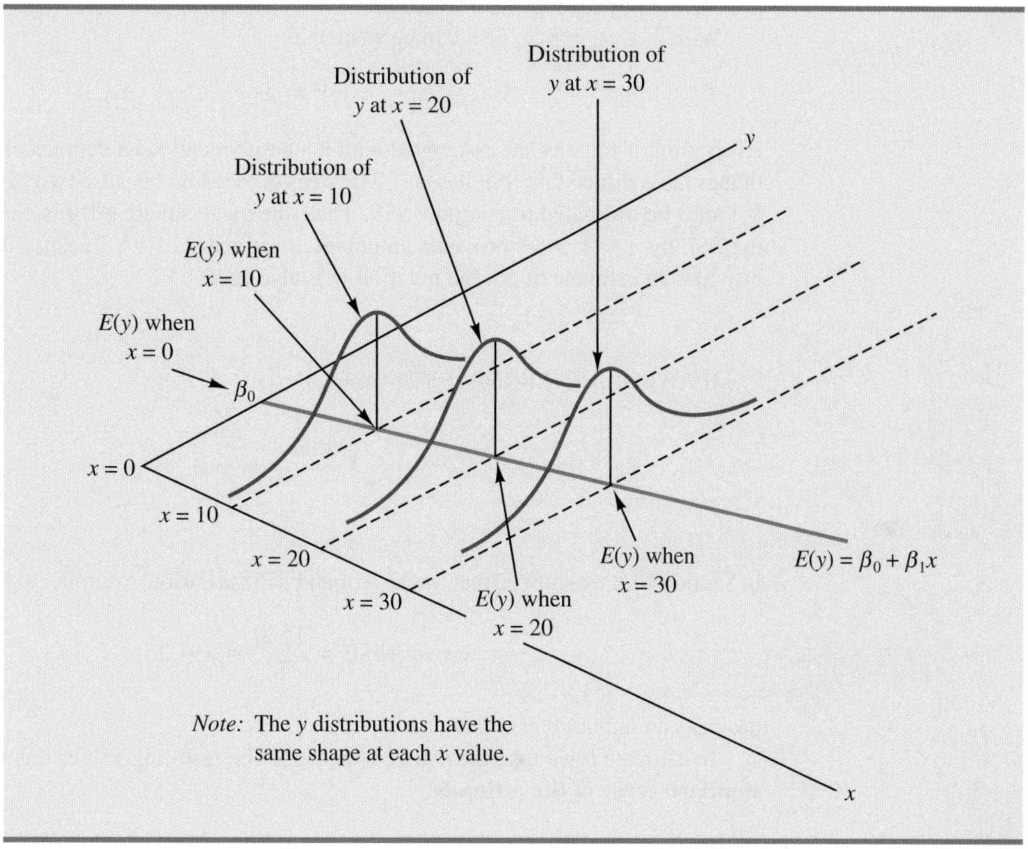

At this point, we must keep in mind that we are also making an assumption or hypothesis about the form of the relationship between x and y. That is, we assume that a straight line represented by $\beta_0 + \beta_1 x$ is the basis for the relationship between the variables. We must not lose sight of the fact that some other model, for instance $y = \beta_0 + \beta_1 x^2 + \epsilon$, may turn out to be a better model for the underlying relationship.

14.5 Testing for Significance

In a simple linear regression equation, the mean or expected value of y is a linear function of x: $E(y) = \beta_0 + \beta_1 x$. If the value of β_1 is zero, $E(y) = \beta_0 + (0)x = \beta_0$. In this case, the mean value of y does not depend on the value of x and hence we would conclude that x and y are not linearly related. Alternatively, if the value of β_1 is not equal to zero, we would conclude that the two variables are related. Thus, to test for a significant regression relationship, we must conduct a hypothesis test to determine whether the value of β_1 is zero. Two tests are commonly used. Both require an estimate of σ^2, the variance of ϵ in the regression model.

Estimate of σ^2

From the regression model and its assumptions we can conclude that σ^2, the variance of ϵ, also represents the variance of the y values about the regression line. Recall that the deviations of the y values about the estimated regression line are called residuals. Thus, SSE, the sum of squared residuals, is a measure of the variability of the actual observations about the estimated regression line. The **mean square error** (MSE) provides the estimate of σ^2; it is SSE divided by its degrees of freedom.

With $\hat{y}_i = b_0 + b_1x_i$, SSE can be written as

$$\text{SSE} = \Sigma(y_i - \hat{y}_i)^2 = \Sigma(y_i - b_0 - b_1x_i)^2$$

Every sum of squares has associated with it a number called its degrees of freedom. Statisticians have shown that SSE has $n - 2$ degrees of freedom because two parameters (β_0 and β_1) must be estimated to compute SSE. Thus, the mean square error is computed by dividing SSE by $n - 2$. MSE provides an unbiased estimator of σ^2. Because the value of MSE provides an estimate of σ^2, the notation s^2 is also used.

MEAN SQUARE ERROR (ESTIMATE OF σ^2)

$$s^2 = \text{MSE} = \frac{\text{SSE}}{n - 2} \tag{14.15}$$

In Section 14.3 we showed that for the Armand's Pizza Parlors example, SSE = 1530; hence,

$$s^2 = \text{MSE} = \frac{1530}{8} = 191.25$$

provides an unbiased estimate of σ^2.

To estimate σ we take the square root of s^2. The resulting value, s, is referred to as the **standard error of the estimate**.

STANDARD ERROR OF THE ESTIMATE

$$s = \sqrt{\text{MSE}} = \sqrt{\frac{\text{SSE}}{n - 2}} \tag{14.16}$$

For the Armand's Pizza Parlors example, $s = \sqrt{\text{MSE}} = \sqrt{191.25} = 13.829$. In the following discussion, we use the standard error of the estimate in the tests for a significant relationship between x and y.

t Test

The simple linear regression model is $y = \beta_0 + \beta_1x + \epsilon$. If x and y are linearly related, we must have $\beta_1 \neq 0$. The purpose of the t test is to see whether we can conclude that $\beta_1 \neq 0$. We will use the sample data to test the following hypotheses about the parameter β_1.

$$H_0\colon \beta_1 = 0$$
$$H_a\colon \beta_1 \neq 0$$

If H_0 is rejected, we will conclude that $\beta_1 \neq 0$ and that a statistically significant relationship exists between the two variables. However, if H_0 cannot be rejected, we will have insufficient evidence to conclude that a significant relationship exists. The properties of the sampling distribution of b_1, the least squares estimator of β_1, provide the basis for the hypothesis test.

First, let us consider what would happen if we used a different random sample for the same regression study. For example, suppose that Armand's Pizza Parlors used the sales records of a different sample of 10 restaurants. A regression analysis of this new sample might result in an estimated regression equation similar to our previous estimated regression equation $\hat{y} = 60 + 5x$. However, it is doubtful that we would obtain exactly the same equation (with an intercept of exactly 60 and a slope of exactly 5). Indeed, b_0 and b_1, the least squares estimators, are sample statistics with their own sampling distributions. The properties of the sampling distribution of b_1 follow.

SAMPLING DISTRIBUTION OF b_1

Expected Value

$$E(b_1) = \beta_1$$

Standard Deviation

$$\sigma_{b_1} = \frac{\sigma}{\sqrt{\Sigma(x_i - \bar{x})^2}} \tag{14.17}$$

Distribution Form

Normal

Note that the expected value of b_1 is equal to β_1, so b_1 is an unbiased estimator of β_1.

Because we do not know the value of σ, we develop an estimate of σ_{b_1}, denoted s_{b_1}, by estimating σ with s in equation (14.17). Thus, we obtain the following estimate of σ_{b_1}.

The standard deviation of b_1 is also referred to as the standard error of b_1. Thus, s_{b_1} provides an estimate of the standard error of b_1.

ESTIMATED STANDARD DEVIATION OF b_1

$$s_{b_1} = \frac{s}{\sqrt{\Sigma(x_i - \bar{x})^2}} \tag{14.18}$$

For Armand's Pizza Parlors, $s = 13.829$. Hence, using $\Sigma(x_i - \bar{x})^2 = 568$ as shown in Table 14.2, we have

$$s_{b_1} = \frac{13.829}{\sqrt{568}} = .5803$$

as the estimated standard deviation of b_1.

The t test for a significant relationship is based on the fact that the test statistic

$$\frac{b_1 - \beta_1}{s_{b_1}}$$

follows a t distribution with $n - 2$ degrees of freedom. If the null hypothesis is true, then $\beta_1 = 0$ and $t = b_1/s_{b_1}$.

Let us conduct this test of significance for Armand's Pizza Parlors at the $\alpha = .01$ level of significance. The test statistic is

$$t = \frac{b_1}{s_{b_1}} = \frac{5}{.5803} = 8.62$$

Appendixes 14.3 and 14.4 show how Minitab and Excel can be used to compute the p-value.

The t distribution table (Table 2 of Appendix D) shows that with $n - 2 = 10 - 2 = 8$ degrees of freedom, $t = 3.355$ provides an area of .005 in the upper tail. Thus, the area in the upper tail of the t distribution corresponding to the test statistic $t = 8.62$ must be less than .005. Because this test is a two-tailed test, we double this value to conclude that the p-value associated with $t = 8.62$ must be less than $2(.005) = .01$. Excel or Minitab show the p-value $= .000$. Because the p-value is less than $\alpha = .01$, we reject H_0 and conclude that β_1 is not equal to zero. This evidence is sufficient to conclude that a significant relationship exists between student population and quarterly sales. A summary of the t test for significance in simple linear regression follows.

t TEST FOR SIGNIFICANCE IN SIMPLE LINEAR REGRESSION

$$H_0\colon \beta_1 = 0$$
$$H_a\colon \beta_1 \neq 0$$

TEST STATISTIC

$$t = \frac{b_1}{s_{b_1}} \tag{14.19}$$

REJECTION RULE

p-value approach: Reject H_0 if p-value $\leq \alpha$

Critical value approach: Reject H_0 if $t \leq -t_{\alpha/2}$ or if $t \geq t_{\alpha/2}$

where $t_{\alpha/2}$ is based on a t distribution with $n - 2$ degrees of freedom.

Confidence Interval for β_1

The form of a confidence interval for β_1 is as follows:

$$b_1 \pm t_{\alpha/2} s_{b_1}$$

The point estimator is b_1 and the margin of error is $t_{\alpha/2} s_{b_1}$. The confidence coefficient associated with this interval is $1 - \alpha$, and $t_{\alpha/2}$ is the t value providing an area of $\alpha/2$ in the upper tail of a t distribution with $n - 2$ degrees of freedom. For example, suppose that we wanted to develop a 99% confidence interval estimate of β_1 for Armand's Pizza Parlors. From Table 2 of Appendix B we find that the t value corresponding to $\alpha = .01$ and $n - 2 = 10 - 2 = 8$ degrees of freedom is $t_{.005} = 3.355$. Thus, the 99% confidence interval estimate of β_1 is

$$b_1 \pm t_{\alpha/2} s_{b_1} = 5 \pm 3.355(.5803) = 5 \pm 1.95$$

or 3.05 to 6.95.

In using the t test for significance, the hypotheses tested were

$$H_0\colon \beta_1 = 0$$
$$H_a\colon \beta_1 \neq 0$$

At the $\alpha = .01$ level of significance, we can use the 99% confidence interval as an alternative for drawing the hypothesis testing conclusion for the Armand's data. Because 0, the hypothesized value of β_1, is not included in the confidence interval (3.05 to 6.95), we can reject

H_0 and conclude that a significant statistical relationship exists between the size of the student population and quarterly sales. In general, a confidence interval can be used to test any two-sided hypothesis about β_1. If the hypothesized value of β_1 is contained in the confidence interval, do not reject H_0. Otherwise, reject H_0.

F Test

An F test, based on the F probability distribution, can also be used to test for significance in regression. With only one independent variable, the F test will provide the same conclusion as the t test; that is, if the t test indicates $\beta_1 \neq 0$ and hence a significant relationship, the F test will also indicate a significant relationship. But with more than one independent variable, only the F test can be used to test for an overall significant relationship.

The logic behind the use of the F test for determining whether the regression relationship is statistically significant is based on the development of two independent estimates of σ^2. We explained how MSE provides an estimate of σ^2. If the null hypothesis H_0: $\beta_1 = 0$ is true, the sum of squares due to regression, SSR, divided by its degrees of freedom provides another independent estimate of σ^2. This estimate is called the *mean square due to regression,* or simply the *mean square regression,* and is denoted MSR. In general,

$$\text{MSR} = \frac{\text{SSR}}{\text{Regression degrees of freedom}}$$

For the models we consider in this text, the regression degrees of freedom is always equal to the number of independent variables in the model:

$$\text{MSR} = \frac{\text{SSR}}{\text{Number of independent variables}} \tag{14.20}$$

Because we consider only regression models with one independent variable in this chapter, we have MSR $=$ SSR/1 $=$ SSR. Hence, for Armand's Pizza Parlors, MSR $=$ SSR $=$ 14,200.

If the null hypothesis (H_0: $\beta_1 = 0$) is true, MSR and MSE are two independent estimates of σ^2 and the sampling distribution of MSR/MSE follows an F distribution with numerator degrees of freedom equal to one and denominator degrees of freedom equal to $n - 2$. Therefore, when $\beta_1 = 0$, the value of MSR/MSE should be close to one. However, if the null hypothesis is false ($\beta_1 \neq 0$), MSR will overestimate σ^2 and the value of MSR/MSE will be inflated; thus, large values of MSR/MSE lead to the rejection of H_0 and the conclusion that the relationship between x and y is statistically significant.

Let us conduct the F test for the Armand's Pizza Parlors example. The test statistic is

$$F = \frac{\text{MSR}}{\text{MSE}} = \frac{14{,}200}{191.25} = 74.25$$

The F test and the t test provide identical results for simple linear regression.

The F distribution table (Table 4 of Appendix B) shows that with one degree of freedom in the numerator and $n - 2 = 10 - 2 = 8$ degrees of freedom in the denominator, $F = 11.26$ provides an area of .01 in the upper tail. Thus, the area in the upper tail of the F distribution corresponding to the test statistic $F = 74.25$ must be less than .01. Thus, we conclude that the p-value must be less than .01. Excel or Minitab show the p-value $=$ .000. Because the p-value is less than $\alpha = .01$, we reject H_0 and conclude that a significant relationship exists between the size of the student population and quarterly sales. A summary of the F test for significance in simple linear regression follows.

If H_0 is false, MSE *still provides an unbiased estimate of σ^2 and* MSR *overestimates σ^2. If H_0 is true, both* MSE *and* MSR *provide unbiased estimates of σ^2; in this case the value of* MSR/MSE *should be close to 1.*

F TEST FOR SIGNIFICANCE IN SIMPLE LINEAR REGRESSION

$$H_0: \beta_1 = 0$$
$$H_a: \beta_1 \neq 0$$

TEST STATISTIC

$$F = \frac{\text{MSR}}{\text{MSE}} \tag{14.21}$$

REJECTION RULE

p-value approach: Reject H_0 if *p*-value $\leq \alpha$
Critical value approach: Reject H_0 if $F \geq F_\alpha$

where F_α is based on an *F* distribution with 1 degree of freedom in the numerator and $n - 2$ degrees of freedom in the denominator.

In Chapter 13 we covered analysis of variance (ANOVA) and showed how an **ANOVA table** could be used to provide a convenient summary of the computational aspects of analysis of variance. A similar ANOVA table can be used to summarize the results of the *F* test for significance in regression. Table 14.5 is the general form of the ANOVA table for simple linear regression. Table 14.6 is the ANOVA table with the *F* test computations performed for Armand's Pizza Parlors. Regression, Error, and Total are the labels for the three sources of variation, with SSR, SSE, and SST appearing as the corresponding sum of squares in

In every analysis of variance table the total sum of squares is the sum of the regression sum of squares and the error sum of squares; in addition, the total degrees of freedom is the sum of the regression degrees of freedom and the error degrees of freedom.

TABLE 14.5 GENERAL FORM OF THE ANOVA TABLE FOR SIMPLE LINEAR REGRESSION

Source of Variation	Sum of Squares	Degrees of Freedom	Mean Square	*F*	*p*-value
Regression	SSR	1	$\text{MSR} = \frac{\text{SSR}}{1}$	$F = \frac{\text{MSR}}{\text{MSE}}$	
Error	SSE	$n - 2$	$\text{MSE} = \frac{\text{SSE}}{n - 2}$		
Total	SST	$n - 1$			

TABLE 14.6 ANOVA TABLE FOR THE ARMAND'S PIZZA PARLORS PROBLEM

Source of Variation	Sum of Squares	Degrees of Freedom	Mean Square	*F*	*p*-value
Regression	14,200	1	$\frac{14{,}200}{1} = 14{,}200$	$\frac{14{,}200}{191.25} = 74.25$	.000
Error	1,530	8	$\frac{1530}{8} = 191.25$		
Total	15,730	9			

column 2. The degrees of freedom, 1 for SSR, $n - 2$ for SSE, and $n - 1$ for SST, are shown in column 3. Column 4 contains the values of MSR and MSE, column 5 contains the value of $F = \text{MSR}/\text{MSE}$, and column 6 contains the p-value corresponding to the F value in column 5. Almost all computer printouts of regression analysis include an ANOVA table summary of the F test for significance.

Some Cautions About the Interpretation of Significance Tests

Regression analysis, which can be used to identify how variables are associated with one another, cannot be used as evidence of a cause-and-effect relationship.

Rejecting the null hypothesis $H_0: \beta_1 = 0$ and concluding that the relationship between x and y is significant does not enable us to conclude that a cause-and-effect relationship is present between x and y. Concluding a cause-and-effect relationship is warranted only if the analyst can provide some type of theoretical justification that the relationship is in fact causal. In the Armand's Pizza Parlors example, we can conclude that there is a significant relationship between the size of the student population x and quarterly sales y; moreover, the estimated regression equation $\hat{y} = 60 + 5x$ provides the least squares estimate of the relationship. We cannot, however, conclude that changes in student population x *cause* changes in quarterly sales y just because we identified a statistically significant relationship. The appropriateness of such a cause-and-effect conclusion is left to supporting theoretical justification and to good judgment on the part of the analyst. Armand's managers felt that increases in the student population were a likely cause of increased quarterly sales. Thus, the result of the significance test enabled them to conclude that a cause-and-effect relationship was present.

In addition, just because we are able to reject $H_0: \beta_1 = 0$ and demonstrate statistical significance does not enable us to conclude that the relationship between x and y is linear. We can state only that x and y are related and that a linear relationship explains a significant portion of the variability in y over the range of values for x observed in the sample. Figure 14.7 illustrates this situation. The test for significance calls for the rejection of the null hypothesis $H_0: \beta_1 = 0$ and leads to the conclusion that x and y are significantly related, but the figure shows that the actual relationship between x and y is not linear. Although the

FIGURE 14.7 EXAMPLE OF A LINEAR APPROXIMATION OF A NONLINEAR RELATIONSHIP

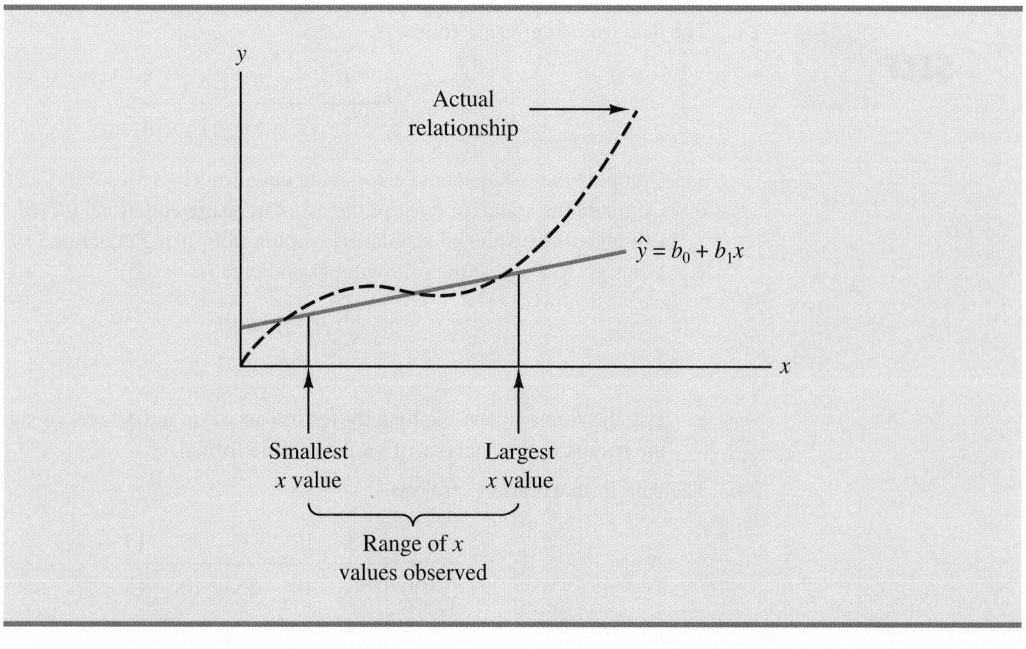

linear approximation provided by $\hat{y} = b_0 + b_1x$ is good over the range of x values observed in the sample, it becomes poor for x values outside that range.

Given a significant relationship, we should feel confident in using the estimated regression equation for predictions corresponding to x values within the range of the x values observed in the sample. For Armand's Pizza Parlors, this range corresponds to values of x between 2 and 26. Unless other reasons indicate that the model is valid beyond this range, predictions outside the range of the independent variable should be made with caution. For Armand's Pizza Parlors, because the regression relationship has been found significant at the .01 level, we should feel confident using it to predict sales for restaurants where the associated student population is between 2000 and 26,000.

NOTES AND COMMENTS

1. The assumptions made about the error term (Section 14.4) are what allow the tests of statistical significance in this section. The properties of the sampling distribution of b_1 and the subsequent t and F tests follow directly from these assumptions.
2. Do not confuse statistical significance with practical significance. With very large sample sizes, statistically significant results can be obtained for small values of b_1; in such cases, one must exercise care in concluding that the relationship has practical significance.
3. A test of significance for a linear relationship between x and y can also be performed by using the sample correlation coefficient r_{xy}. With ρ_{xy} denoting the population correlation coefficient, the hypotheses are as follows.

$$H_0: \rho_{xy} = 0$$
$$H_a: \rho_{xy} \neq 0$$

A significant relationship can be concluded if H_0 is rejected. The details of this test are provided in Appendix 14.2. However, the t and F tests presented previously in this section give the same result as the test for significance using the correlation coefficient. Conducting a test for significance using the correlation coefficient therefore is not necessary if a t or F test has already been conducted.

Exercises

Methods

SELF test

23. The data from exercise 1 follow.

x_i	1	2	3	4	5
y_i	3	7	5	11	14

a. Compute the mean square error using equation (14.15).
b. Compute the standard error of the estimate using equation (14.16).
c. Compute the estimated standard deviation of b_1 using equation (14.18).
d. Use the t test to test the following hypotheses ($\alpha = .05$):

$$H_0: \beta_1 = 0$$
$$H_a: \beta_1 \neq 0$$

e. Use the F test to test the hypotheses in part (d) at a .05 level of significance. Present the results in the analysis of variance table format.

24. The data from exercise 2 follow.

x_i	3	12	6	20	14
y_i	55	40	55	10	15

a. Compute the mean square error using equation (14.15).
b. Compute the standard error of the estimate using equation (14.16).
c. Compute the estimated standard deviation of b_1 using equation (14.18).
d. Use the t test to test the following hypotheses ($\alpha = .05$):

$$H_0: \beta_1 = 0$$
$$H_a: \beta_1 \neq 0$$

e. Use the F test to test the hypotheses in part (d) at a .05 level of significance. Present the results in the analysis of variance table format.

25. The data from exercise 3 follow.

x_i	2	6	9	13	20
y_i	7	18	9	26	23

a. What is the value of the standard error of the estimate?
b. Test for a significant relationship by using the t test. Use $\alpha = .05$.
c. Use the F test to test for a significant relationship. Use $\alpha = .05$. What is your conclusion?

Applications

SELF test

26. In exercise 18 the data on price ($) and the overall score for six stereo headphones tested by *Consumer Reports* were as follows (*Consumer Reports* website, March 5, 2012).

Brand	Price ($)	Score
Bose	180	76
Skullcandy	150	71
Koss	95	61
Phillips/O'Neill	70	56
Denon	70	40
JVC	35	26

a. Does the t test indicate a significant relationship between price and the overall score? What is your conclusion? Use $\alpha = .05$.
b. Test for a significant relationship using the F test. What is your conclusion? Use $\alpha = .05$.
c. Show the ANOVA table for these data.

27. The number of megapixels in a digital camera is one of the most important factors in determining picture quality. But, do digital cameras with more megapixels cost more? The following data show the number of megapixels and the price ($) for 10 digital cameras (*Consumer Reports,* March 2009).

WEB file
DigitalCameras

Brand and Model	Megapixels	Price ($)
Canon PowerShot SD1100 IS	8	180
Casio Exilim Card EX-510	10	200
Sony Cyber-shot DSC-T70	7	230
Pentax Optio M50	8	120
Canon PowerShot G10	15	470
Canon PowerShot A590 IS	8	140
Canon PowerShot E1	10	180
Fujifilm FinePix F00FD	12	310
Sony Cyber-shot DSC-W170	10	250
Canon PowerShot A470	7	110

a. Use these data to develop an estimated regression equation that can be used to predict the price of a digital camera given the number of megapixels.
b. At the .05 level of significance, are the number of megapixels and the price related? Explain.
c. Would you feel comfortable using the estimated regression equation developed in part (a) to predict the price of a digital camera given the number of megapixels? Explain.
d. The Canon Power Shot S95 digital camera has 10 megapixels. Predict the price of this camera using the estimated regression equation developed in part (a).

WEB file BrokerRatings

28. In exercise 8 ratings data on x = the quality of the speed of execution and y = overall satisfaction with electronic trades provided the estimated regression equation $\hat{y} = .2046 + .9077x$. At the .05 level of significance, test whether speed of execution and overall satisfaction are related. Show the ANOVA table. What is your conclusion?

29. Refer to exercise 21, where data on production volume and cost were used to develop an estimated regression equation relating production volume and cost for a particular manufacturing operation. Use $\alpha = .05$ to test whether the production volume is significantly related to the total cost. Show the ANOVA table. What is your conclusion?

30. Refer to excercise 5 where the following data were used to investigate whether higher prices are generally associated with higher ratings for elliptical trainers (*Consumer Reports,* February 2008).

WEB file Ellipticals

Brand and Model	Price ($)	Rating
Precor 5.31	3700	87
Keys Fitness CG2	2500	84
Octane Fitness Q37e	2800	82
LifeFitness X1 Basic	1900	74
NordicTrack AudioStrider 990	1000	73
Schwinn 430	800	69
Vision Fitness X6100	1700	68
ProForm XP 520 Razor	600	55

With x = price ($) and y = rating, the estimated regression equation is $\hat{y} = 58.158 + .008449x$. For these data, SSE = 173.88 and SST = 756. Does the evidence indicate a significant relationship between price and rating?

WEB file RacingBicycles

31. In exercise 20, data on x = weight (pounds) and y = price ($) for 10 road-racing bikes provided the estimated regression equation $\hat{y} = 28{,}574 - 1439x$. (*Bicycling* website, March 8, 2012). For these data SSE = 7,102,922.54 and SST = 52,120,800. Use the F test to determine whether the weight for a bike and the price are related at the .05 level of significance.

14.6 Using the Estimated Regression Equation for Estimation and Prediction

When using the simple linear regression model, we are making an assumption about the relationship between x and y. We then use the least squares method to obtain the estimated simple linear regression equation. If a significant relationship exists between x and y and the coefficient of determination shows that the fit is good, the estimated regression equation should be useful for estimation and prediction.

For the Armand's Pizza Parlors example, the estimated regression equation is $\hat{y} = 60 + 5x$. At the end of Section 14.1 we stated that $\hat{y}$ can be used as a *point estimator* of $E(y)$, the mean or expected value of y for a given value of x, and as a predictor of an individual value of y. For example, suppose Armand's managers want to estimate the mean quarterly sales for *all* restaurants located near college campuses with 10,000 students. Using the estimated regression equation $\hat{y} = 60 + 5x$, we see that for $x = 10$ (10,000 students), $\hat{y} = 60 + 5(10) = 110$. Thus, a *point estimate* of the mean quarterly sales for all restaurant locations near campuses with 10,000 students is $110,000. In this case we are using $\hat{y}$ as the point estimator of the mean value of y when $x = 10$.

We can also use the estimated regression equation to *predict* an individual value of y for a given value of x. For example, to predict quarterly sales for a new restaurant Armand's is considering building near Talbot College, a campus with 10,000 students, we would compute $\hat{y} = 60 + 5(10) = 110$. Hence, we would predict quarterly sales of $110,000 for such a new restaurant. In this case, we are using $\hat{y}$ as the *predictor* of y for a new observation when $x = 10$.

When we are using the estimated regression equation to estimate the mean value of y or to predict an individual value of y, it is clear that the estimate or prediction depends on the given value of x. For this reason, as we discuss in more depth the issues concerning estimation and prediction, the following notation will help clarify matters.

$$\begin{aligned}
x^* &= \text{the given value of the independent variable } x \\
y^* &= \text{the random variable denoting the possible values of the dependent} \\
&\quad\ \text{variable } y \text{ when } x = x^* \\
E(y^*) &= \text{the mean or expected value of the dependent variable } y \text{ when } x = x^* \\
\hat{y}^* &= b_0 + b_1x^* = \text{the point estimator of } E(y^*) \text{ and the predictor of an} \\
&\quad\ \text{individual value of } y^* \text{ when } x = x^*
\end{aligned}$$

To illustrate the use of this notation, suppose we want to estimate the mean value of quarterly sales for *all* Armand's restaurants located near a campus with 10,000 students. For this case, $x^* = 10$ and $E(y^*)$ denotes the unknown mean value of quarterly sales for all restaurants where $x^* = 10$. Thus, the point estimate of $E(y^*)$ is provided by $\hat{y}^* = 60 + 5(10) = 110$, or $110,000. But, using this notation, $\hat{y}^* = 110$ is also the predictor of quarterly sales for the new restaurant located near Talbot College, a school with 10,000 students.

Interval Estimation

Confidence intervals and prediction intervals show the precision of the regression results. Narrower intervals provide a higher degree of precision.

Point estimators and predictors do not provide any information about the precision associated with the estimate and/or prediction. For that we must develop confidence intervals and prediction intervals. A **confidence interval** is an interval estimate of the *mean value of* y for a given value of x. A **prediction interval** is used whenever we want to *predict an individual value of* y for a new observation corresponding to a given value of x. Although the predictor of y for a given value of x is the same as the point estimator of the mean value of y for a given value of x, the interval estimates we obtain for the two cases are different. As we will show, the margin of error is larger for a prediction interval. We begin by showing how to develop an interval estimate of the mean value of y.

Confidence Interval for the Mean Value of y

In general, we cannot expect $\hat{y}^*$ to equal $E(y^*)$ exactly. If we want to make an inference about how close $\hat{y}^*$ is to the true mean value $E(y^*)$, we will have to estimate the variance of $\hat{y}^*$. The formula for estimating the variance of $\hat{y}^*$, denoted by $s^2_{\hat{y}^*}$, is

$$s^2_{\hat{y}^*} = s^2\left[\frac{1}{n} + \frac{(x^* - \bar{x})^2}{\Sigma(x_i - \bar{x})^2}\right] \tag{14.22}$$

The estimate of the standard deviation of $\hat{y}^*$ is given by the square root of equation (14.22).

$$s_{\hat{y}^*} = s\sqrt{\frac{1}{n} + \frac{(x^* - \bar{x})^2}{\Sigma(x_i - \bar{x})^2}} \tag{14.23}$$

The computational results for Armand's Pizza Parlors in Section 14.5 provided $s = 13.829$. With $x^* = 10$, $\bar{x} = 14$, and $\Sigma(x_i - \bar{x})^2 = 568$, we can use equation (14.23) to obtain

$$\begin{aligned} s_{\hat{y}^*} &= 13.829\sqrt{\frac{1}{10} + \frac{(10 - 14)^2}{568}} \\ &= 13.829\sqrt{.1282} = 4.95 \end{aligned}$$

The general expression for a confidence interval follows.

The margin of error associated with this confidence interval is $t_{\alpha/2}s_{\hat{y}^*}$.

CONFIDENCE INTERVAL FOR $E(y^*)$

$$\hat{y}^* \pm t_{\alpha/2}s_{\hat{y}^*} \tag{14.24}$$

where the confidence coefficient is $1 - \alpha$ and $t_{\alpha/2}$ is based on the t distribution with $n - 2$ degrees of freedom.

Using expression (14.24) to develop a 95% confidence interval of the mean quarterly sales for all Armand's restaurants located near campuses with 10,000 students, we need the value of t for $\alpha/2 = .025$ and $n - 2 = 10 - 2 = 8$ degrees of freedom. Using Table 2 of Appendix B, we have $t_{.025} = 2.306$. Thus, with $\hat{y}^* = 110$ and a margin of error of $t_{\alpha/2}s_{\hat{y}^*} = 2.306(4.95) = 11.415$, the 95% confidence interval estimate is

$$110 \pm 11.415$$

In dollars, the 95% confidence interval for the mean quarterly sales of all restaurants near campuses with 10,000 students is \$110,000 ± \$11,415. Therefore, the 95% confidence interval for the mean quarterly sales when the student population is 10,000 is \$98,585 to \$121,415.

Note that the estimated standard deviation of $\hat{y}^*$ given by equation (14.23) is smallest when $x^* - \bar{x} = 0$. In this case the estimated standard deviation of $\hat{y}^*$ becomes

$$s_{\hat{y}^*} = s\sqrt{\frac{1}{n} + \frac{(\bar{x} - \bar{x})^2}{\Sigma(x_i - \bar{x})^2}} = s\sqrt{\frac{1}{n}}$$

This result implies that we can make the best or most precise estimate of the mean value of y whenever $x^* = \bar{x}$. In fact, the further x^* is from $\bar{x}$, the larger $x^* - \bar{x}$ becomes. As a result, the confidence interval for the mean value of y will become wider as x^* deviates more from $\bar{x}$. This pattern is shown graphically in Figure 14.8.

FIGURE 14.8 CONFIDENCE INTERVALS FOR THE MEAN SALES y AT GIVEN VALUES OF STUDENT POPULATION x

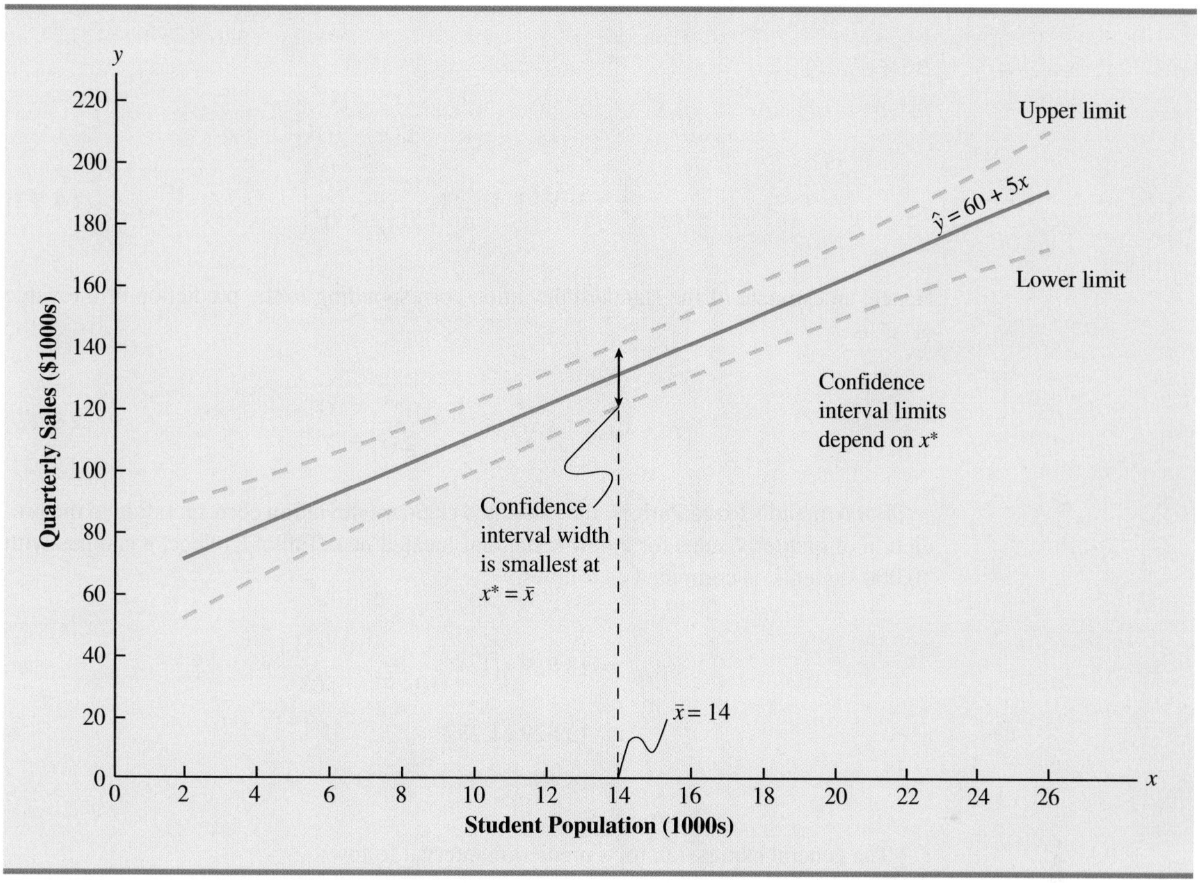

Prediction Interval for an Individual Value of y

Instead of estimating the mean value of quarterly sales for all Armand's restaurants located near campuses with 10,000 students, suppose we want to predict quarterly sales for a new restaurant Armand's is considering building near Talbot College, a campus with 10,000 students. As noted previously, the predictor of y^*, the value of y corresponding to the given x^*, is $\hat{y}^* = b_0 + b_1x^*$. For the new restaurant located near Talbot College, $x^* = 10$ and the prediction of quarterly sales is $\hat{y}^* = 60 + 5(10) = 110$, or \$110,000. Note that the prediction of quarterly sales for the new Armand's restaurant near Talbot College is the same as the point estimate of the mean sales for all Armand's restaurants located near campuses with 10,000 students.

To develop a prediction interval, let us first determine the variance associated with using $\hat{y}^*$ as a predictor of y when $x = x^*$. This variance is made up of the sum of the following two components.

1. The variance of the y^* values about the mean $E(y^*)$, an estimate of which is given by s^2
2. The variance associated with using $\hat{y}^*$ to estimate $E(y^*)$, an estimate of which is given by $s^2_{\hat{y}^*}$

The formula for estimating the variance corresponding to the prediction of the value of y when $x = x^*$, denoted s^2_{pred}, is

$$\begin{aligned} s^2_{\text{pred}} &= s^2 + s^2_{\hat{y}^*} \\ &= s^2 + s^2\left[\frac{1}{n} + \frac{(x^* - \bar{x})^2}{\Sigma(x_i - \bar{x})^2}\right] \\ &= s^2\left[1 + \frac{1}{n} + \frac{(x^* - \bar{x})^2}{\Sigma(x_i - \bar{x})^2}\right] \end{aligned} \tag{14.25}$$

Hence, an estimate of the standard deviation corresponding to the prediction of the value of y^* is

$$s_{\text{pred}} = s\sqrt{1 + \frac{1}{n} + \frac{(x^* - \bar{x})^2}{\Sigma(x_i - \bar{x})^2}} \tag{14.26}$$

For Armand's Pizza Parlors, the estimated standard deviation corresponding to the prediction of quarterly sales for a new restaurant located near Talbot College, a campus with 10,000 students, is computed as follows.

$$\begin{aligned} s_{\text{pred}} &= 13.829\sqrt{1 + \frac{1}{10} + \frac{(10 - 14)^2}{568}} \\ &= 13.829\sqrt{1.282} \\ &= 14.69 \end{aligned}$$

The general expression for a prediction interval follows.

PREDICTION INTERVAL FOR y^*

$$\hat{y}^* \pm t_{\alpha/2}s_{\text{pred}} \tag{14.27}$$

where the confidence coefficient is $1 - \alpha$ and $t_{\alpha/2}$ is based on a t distribution with $n - 2$ degrees of freedom.

The margin of error associated with this prediction interval is $t_{\alpha/2}s_{\text{pred}}$.

The 95% prediction interval for quarterly sales for the new Armand's restaurant located near Talbot College can be found using $t_{\alpha/2} = t_{.025} = 2.306$ and $s_{\text{pred}} = 14.69$. Thus, with $\hat{y}^* = 110$ and a margin of error of $t_{.025}s_{\text{pred}} = 2.306(14.69) = 33.875$, the 95% prediction interval is

$$110 \pm 33.875$$

In dollars, this prediction interval is \$110,000 ± \$33,875 or \$76,125 to \$143,875. Note that the prediction interval for the new restaurant located near Talbot College, a campus with 10,000 students, is wider than the confidence interval for the mean quarterly sales of all restaurants located near campuses with 10,000 students. The difference reflects the fact that we are able to estimate the mean value of y more precisely than we can predict an individual value of y.

FIGURE 14.9 CONFIDENCE AND PREDICTION INTERVALS FOR SALES y AT GIVEN VALUES OF STUDENT POPULATION x

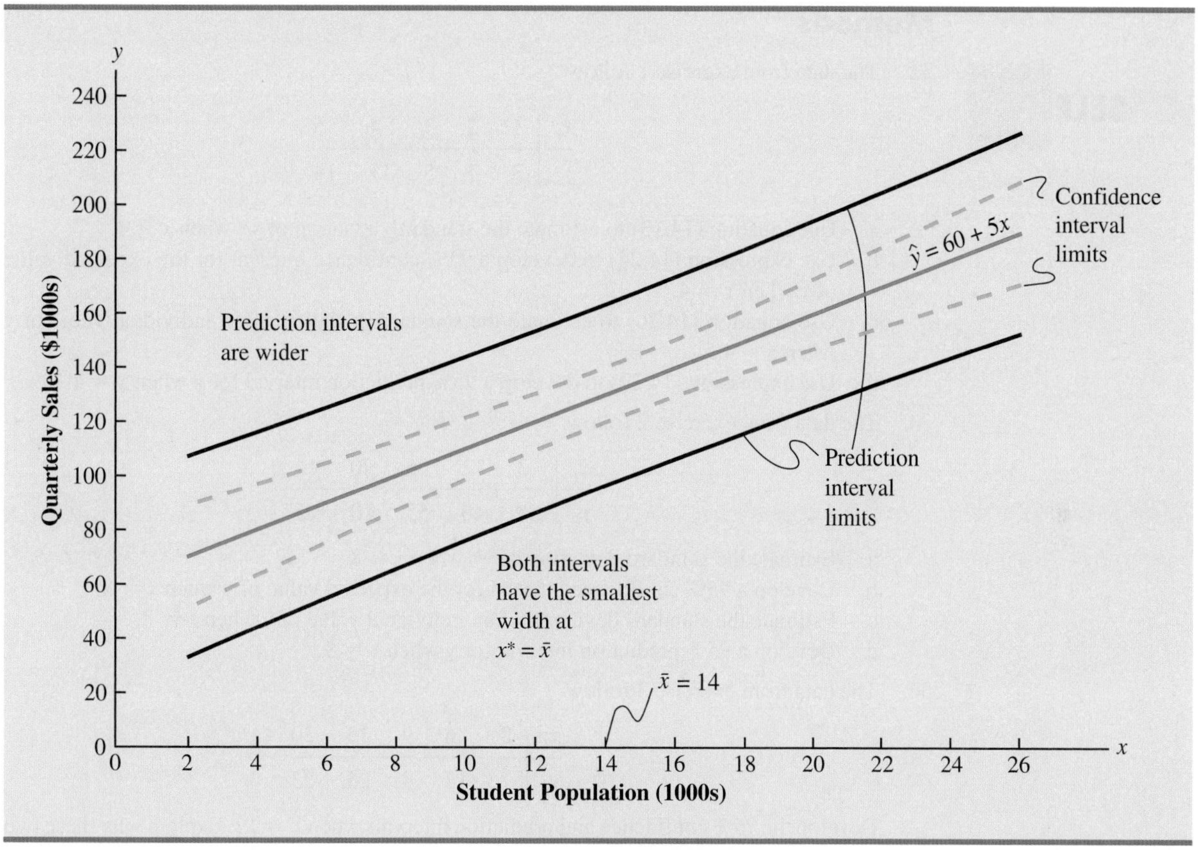

In general, the lines for the confidence interval limits and the prediction interval limits both have curvature.

Confidence intervals and prediction intervals are both more precise when the value of the independent variable x^* is closer to $\bar{x}$. The general shapes of confidence intervals and the wider prediction intervals are shown together in Figure 14.9.

NOTES AND COMMENTS

A prediction interval is used to predict the value of the dependent variable y for a *new observation*. As an illustration, we showed how to develop a prediction interval of quarterly sales for a new restaurant that Armand's is considering building near Talbot College, a campus with 10,000 students. The fact that the value of $x = 10$ is not one of the values of student population for the Armand's sample data in Table 14.1 is not meant to imply that prediction intervals cannot be developed for values of x in the sample data. But, for the ten restaurants that make up the data in Table 14.1, developing a prediction interval for quarterly sales for *one of these restaurants* does not make any sense because we already know the value of quarterly sales for each of these restaurants. In other words, a prediction interval only has meaning for something new, in this case a new observation corresponding to a particular value of x that may or may not equal one of the values of x in the sample.

Exercises

Methods

SELF test

32. The data from exercise 1 follow.

x_i	1	2	3	4	5
y_i	3	7	5	11	14

a. Use equation (14.23) to estimate the standard deviation of $\hat{y}^*$ when $x = 4$.
b. Use expression (14.24) to develop a 95% confidence interval for the expected value of y when $x = 4$.
c. Use equation (14.26) to estimate the standard deviation of an individual value of y when $x = 4$.
d. Use expression (14.27) to develop a 95% prediction interval for y when $x = 4$.

33. The data from exercise 2 follow.

x_i	3	12	6	20	14
y_i	55	40	55	10	15

a. Estimate the standard deviation of $\hat{y}^*$ when $x = 8$.
b. Develop a 95% confidence interval for the expected value of y when $x = 8$.
c. Estimate the standard deviation of an individual value of y when $x = 8$.
d. Develop a 95% prediction interval for y when $x = 8$.

34. The data from exercise 3 follow.

x_i	2	6	9	13	20
y_i	7	18	9	26	23

Develop the 95% confidence and prediction intervals when $x = 12$. Explain why these two intervals are different.

Applications

SELF test

35. The following data are the monthly salaries y and the grade point averages x for students who obtained a bachelor's degree in business administration.

GPA	Monthly Salary ($)
2.6	3600
3.4	3900
3.6	4300
3.2	3800
3.5	4200
2.9	3900

The estimated regression equation for these data is $\hat{y} = 2090.5 + 581.1x$ and MSE = 21,284.

a. Develop a point estimate of the starting salary for a student with a GPA of 3.0.
b. Develop a 95% confidence interval for the mean starting salary for all students with a 3.0 GPA.
c. Develop a 95% prediction interval for Ryan Dailey, a student with a GPA of 3.0.
d. Discuss the differences in your answers to parts (b) and (c).

WEB file

Sales

36. In exercise 7, the data on y = annual sales ($1000s) for new customer accounts and x = number of years of experience for a sample of 10 salespersons provided the estimated regression equation $\hat{y} = 80 + 4x$. For these data $\bar{x} = 7$, $\Sigma(x_i - \bar{x})^2 = 142$, and $s = 4.6098$.
   a. Develop a 95% confidence interval for the mean annual sales for all salespersons with nine years of experience.
   b. The company is considering hiring Tom Smart, a salesperson with nine years of experience. Develop a 95% prediction interval of annual sales for Tom Smart.
   c. Discuss the differences in your answers to parts (a) and (b).
37. In exercise 13, data were given on the adjusted gross income x and the amount of itemized deductions taken by taxpayers. Data were reported in thousands of dollars. With the estimated regression equation $\hat{y} = 4.68 + .16x$, the point estimate of a reasonable level of total itemized deductions for a taxpayer with an adjusted gross income of $52,500 is $13,080.
   a. Develop a 95% confidence interval for the mean amount of total itemized deductions for all taxpayers with an adjusted gross income of $52,500.
   b. Develop a 95% prediction interval estimate for the amount of total itemized deductions for a particular taxpayer with an adjusted gross income of $52,500.
   c. If the particular taxpayer referred to in part (b) claimed total itemized deductions of $20,400, would the IRS agent's request for an audit appear to be justified?
   d. Use your answer to part (b) to give the IRS agent a guideline as to the amount of total itemized deductions a taxpayer with an adjusted gross income of $52,500 should claim before an audit is recommended.
38. Refer to Exercise 21, where data on the production volume x and total cost y for a particular manufacturing operation were used to develop the estimated regression equation $\hat{y} = 1246.67 + 7.6x$.
   a. The company's production schedule shows that 500 units must be produced next month. What is the point estimate of the total cost for next month?
   b. Develop a 99% prediction interval for the total cost for next month.
   c. If an accounting cost report at the end of next month shows that the actual production cost during the month was $6000, should managers be concerned about incurring such a high total cost for the month? Discuss.
39. Almost all U.S. light-rail systems use electric cars that run on tracks built at street level. The Federal Transit Administration claims light-rail is one of the safest modes of travel, with an accident rate of .99 accidents per million passenger miles as compared to 2.29 for buses. The following data show the miles of track and the weekday ridership in thousands of passengers for six light-rail systems (*USA Today,* January 7, 2003).

City	Miles of Track	Ridership (1000s)
Cleveland	15	15
Denver	17	35
Portland	38	81
Sacramento	21	31
San Diego	47	75
San Jose	31	30
St. Louis	34	42

   a. Use these data to develop an estimated regression equation that could be used to predict the ridership given the miles of track.
   b. Did the estimated regression equation provide a good fit? Explain.
   c. Develop a 95% confidence interval for the mean weekday ridership for all light-rail systems with 30 miles of track.

d. Suppose that Charlotte is considering construction of a light-rail system with 30 miles of track. Develop a 95% prediction interval for the weekday ridership for the Charlotte system. Do you think that the prediction interval you developed would be of value to Charlotte planners in anticipating the number of weekday riders for their new light-rail system? Explain.

14.7 Computer Solution

Performing the regression analysis computations without the help of a computer can be quite time consuming. In this section we discuss how the computational burden can be minimized by using a computer software package such as Minitab.

We entered Armand's student population and sales data into a Minitab worksheet. The independent variable was named Pop and the dependent variable was named Sales to assist with interpretation of the computer output. Using Minitab, we obtained the printout for Armand's Pizza Parlors shown in Figure 14.10.[2] The interpretation of this printout follows.

1. Minitab prints the estimated regression equation as Sales = 60.0 + 5.00 Pop.
2. A table is printed that shows the values of the coefficients b_0 and b_1, the standard deviation of each coefficient, the t value obtained by dividing each coefficient value by its standard deviation, and the p-value associated with the t test. Because the p-value is zero (to three decimal places), the sample results indicate that the null hypothesis (H_0: $\beta_1 = 0$) should be rejected. Alternatively, we could compare 8.62 (located in the t-ratio column) to the appropriate critical value. This procedure for the t test was described in Section 14.5.

FIGURE 14.10 MINITAB OUTPUT FOR THE ARMAND'S PIZZA PARLORS PROBLEM

```
The regression equation is
Sales = 60.0 + 5.00 Pop        <-- Estimated regression equation

Predictor      Coef   SE Coef      T      p
Constant     60.000     9.226   6.50  0.000
Pop          5.0000    0.5803   8.62  0.000

S = 13.8293     R-Sq = 90.3%     R-Sq(adj) = 89.1%

Analysis of Variance

SOURCE          DF      SS      MS      F      p
Regression       1   14200   14200  74.25  0.000   } <-- ANOVA table
Residual Error   8    1530     191
Total            9   15730

Predicted Values for New Observations

New Obs     Fit   SE Fit        95% C.I.          95% P.I.    <-- Interval estimates
      1  110.00     4.95   (98.58, 121.42)   (76.13, 143.87)
```

[2]The Minitab steps necessary to generate the output are given in Appendix 14.3.

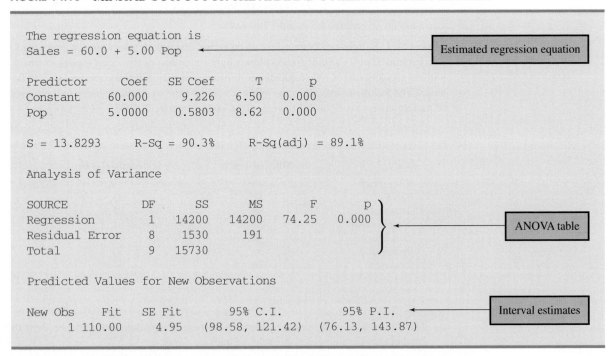

3. Minitab prints the standard error of the estimate, $s = 13.8293$, as well as information about the goodness of fit. Note that "R-sq = 90.3%" is the coefficient of determination expressed as a percentage. The value "R-Sq(adj) = 89.1%" is discussed in Chapter 15.
4. The ANOVA table is printed below the heading Analysis of Variance. Minitab uses the label Residual Error for the error source of variation. Note that DF is an abbreviation for degrees of freedom and that MSR is given as 14,200 and MSE as 191. The ratio of these two values provides the F value of 74.25 and the corresponding p-value of 0.000. Because the p-value is zero (to three decimal places), the relationship between Sales and Pop is judged statistically significant.
5. The 95% confidence interval estimate of the expected sales and the 95% prediction interval estimate of sales for an individual restaurant located near a campus with 10,000 students are printed below the ANOVA table. The confidence interval is (98.58, 121.42) and the prediction interval is (76.13, 143.87) as we showed in Section 14.6.

Exercises

Applications

SELF test

40. The commercial division of a real estate firm is conducting a regression analysis of the relationship between x, annual gross rents (in thousands of dollars), and y, selling price (in thousands of dollars) for apartment buildings. Data were collected on several properties recently sold and the following computer output was obtained.

```
The regression equation is
Y = 20.0 + 7.21 X

Predictor       Coef     SE Coef      T
Constant      20.000      3.2213   6.21
X              7.210      1.3626   5.29

Analysis of Variance

SOURCE              DF        SS
Regression           1   41587.3
Residual Error       7
Total                8   51984.1
```

a. How many apartment buildings were in the sample?
b. Write the estimated regression equation.
c. What is the value of s_{b_1}?
d. Use the F statistic to test the significance of the relationship at a .05 level of significance.
e. Predict the selling price of an apartment building with gross annual rents of \$50,000.

41. Following is a portion of the computer output for a regression analysis relating y = maintenance expense (dollars per month) to x = usage (hours per week) of a particular brand of computer terminal.

```
The regression equation is
Y = 6.1092 + .8951 X

Predictor        Coef    SE Coef
Constant       6.1092     0.9361
X              0.8951     0.1490

Analysis of Variance

SOURCE            DF        SS        MS
Regression         1   1575.76   1575.76
Residual Error     8    349.14     43.64
Total              9   1924.90
```

a. Write the estimated regression equation.
b. Use a t test to determine whether monthly maintenance expense is related to usage at the .05 level of significance.
c. Use the estimated regression equation to predict monthly maintenance expense for any terminal that is used 25 hours per week.

42. A regression model relating x, number of salespersons at a branch office, to y, annual sales at the office (in thousands of dollars) provided the following computer output from a regression analysis of the data.

```
The regression equation is
Y = 80.0 + 50.00 X

Predictor        Coef   SE Coef      T
Constant         80.0    11.333   7.06
X                50.0     5.482   9.12

Analysis of Variance

SOURCE            DF       SS       MS
Regression         1   6828.6   6828.6
Residual Error    28   2298.8     82.1
Total             29   9127.4
```

a. Write the estimated regression equation.
b. How many branch offices were involved in the study?
c. Compute the F statistic and test the significance of the relationship at a .05 level of significance.
d. Predict the annual sales at the Memphis branch office. This branch employs 12 salespersons.

43. Out-of-state tuition and fees at the top graduate schools of business can be very expensive, but the starting salary and bonus paid to graduates from many of these schools can be substantial. The following data show the out-of-state tuition and fees (rounded to the nearest $1000) and the average starting salary and bonus paid to recent graduates (rounded to the nearest $1000) for a sample of 20 graduate schools of business (*U.S. News & World Report 2009 Edition America's Best Graduate Schools*).

WEB file

BusinessSchools

School	Tuition & Fees ($1000s)	Salary & Bonus ($1000s)
Arizona State University	28	98
Babson College	35	94
Cornell University	44	119
Georgetown University	40	109
Georgia Institute of Technology	30	88
Indiana University—Bloomington	35	105
Michigan State University	26	99
Northwestern University	44	123
Ohio State University	35	97
Purdue University—West Lafayette	33	96
Rice University	36	102
Stanford University	46	135
University of California—Davis	35	89
University of Florida	23	71
University of Iowa	25	78
University of Minnesota—Twin Cities	37	100
University of Notre Dame	36	95
University of Rochester	38	99
University of Washington	30	94
University of Wisconsin—Madison	27	93

a. Develop a scatter diagram with salary and bonus as the dependent variable.
b. Does there appear to be any relationship between these variables? Explain.
c. Develop an estimated regression equation that can be used to predict the starting salary and bonus paid to graduates given the cost of out-of-state tuition and fees at the school.
d. Test for a significant relationship at the .05 level of significance. What is your conclusion?
e. Did the estimated regression equation provide a good fit? Explain.
f. Suppose that we randomly select a recent graduate of the University of Virginia graduate school of business. The school has an out-of-state tuition and fees of $43,000. Predict the starting salary and bonus for this graduate.

44. Automobile racing, high-performance driving schools, and driver education programs run by automobile clubs continue to grow in popularity. All these activities require the participant to wear a helmet that is certified by the Snell Memorial Foundation, a not-for-profit organization dedicated to research, education, testing, and development of helmet safety standards. Snell "SA" (Sports Application) rated professional helmets are designed for auto racing and provide extreme impact resistance and high fire protection. One of the key factors in selecting a helmet is weight, since lower weight helmets tend to place less stress on the neck. The following data show the weight and price for 18 SA helmets (SoloRacer website, April 20, 2008).

WEB file

RaceHelmets

Helmet	Weight (oz)	Price ($)
Pyrotect Pro Airflow	64	248
Pyrotect Pro Airflow Graphics	64	278
RCi Full Face	64	200
RaceQuip RidgeLine	64	200
HJC AR-10	58	300
HJC Si-12	47	700

(*continued*)

Helmet	Weight (oz)	Price ($)
HJC HX-10	49	900
Impact Racing Super Sport	59	340
Zamp FSA-1	66	199
Zamp RZ-2	58	299
Zamp RZ-2 Ferrari	58	299
Zamp RZ-3 Sport	52	479
Zamp RZ-3 Sport Painted	52	479
Bell M2	63	369
Bell M4	62	369
Bell M4 Pro	54	559
G Force Pro Force 1	63	250
G Force Pro Force 1 Grafx	63	280

a. Develop a scatter diagram with weight as the independent variable.
b. Does there appear to be any relationship between these two variables?
c. Develop the estimated regression equation that could be used to predict the price given the weight.
d. Test for the significance of the relationship at the .05 level of significance.
e. Did the estimated regression equation provide a good fit? Explain.

14.8 Residual Analysis: Validating Model Assumptions

Residual analysis *is the primary tool for determining whether the assumed regression model is appropriate.*

As we noted previously, the *residual* for observation *i* is the difference between the observed value of the dependent variable (y_i) and the predicted value of the dependent variable ($\hat{y}_i$).

RESIDUAL FOR OBSERVATION *i*

$$y_i - \hat{y}_i \tag{14.28}$$

where

y_i is the observed value of the dependent variable
$\hat{y}_i$ is the predicted value of the dependent variable

In other words, the *i*th residual is the error resulting from using the estimated regression equation to predict the value of the dependent variable. The residuals for the Armand's Pizza Parlors example are computed in Table 14.7. The observed values of the dependent variable are in the second column and the predicted values of the dependent variable, obtained using the estimated regression equation $\hat{y} = 60 + 5x$, are in the third column. An analysis of the corresponding residuals in the fourth column will help determine whether the assumptions made about the regression model are appropriate.

Let us now review the regression assumptions for the Armand's Pizza Parlors example. A simple linear regression model was assumed.

$$y = \beta_0 + \beta_1 x + \epsilon \tag{14.29}$$

TABLE 14.7 RESIDUALS FOR ARMAND'S PIZZA PARLORS

Student Population x_i	Sales y_i	Predicted Sales $\hat{y}_i = 60 + 5x_i$	Residuals $y_i - \hat{y}_i$
2	58	70	−12
6	105	90	15
8	88	100	−12
8	118	100	18
12	117	120	−3
16	137	140	−3
20	157	160	−3
20	169	160	9
22	149	170	−21
26	202	190	12

This model indicates that we assumed quarterly sales (y) to be a linear function of the size of the student population (x) plus an error term ϵ. In Section 14.4 we made the following assumptions about the error term ϵ.

1. $E(\epsilon) = 0$.
2. The variance of ϵ, denoted by σ^2, is the same for all values of x.
3. The values of ϵ are independent.
4. The error term ϵ has a normal distribution.

These assumptions provide the theoretical basis for the t test and the F test used to determine whether the relationship between x and y is significant, and for the confidence and prediction interval estimates presented in Section 14.6. If the assumptions about the error term ϵ appear questionable, the hypothesis tests about the significance of the regression relationship and the interval estimation results may not be valid.

The residuals provide the best information about ϵ; hence an analysis of the residuals is an important step in determining whether the assumptions for ϵ are appropriate. Much of residual analysis is based on an examination of graphical plots. In this section, we discuss the following residual plots.

1. A plot of the residuals against values of the independent variable x
2. A plot of residuals against the predicted values of the dependent variable y
3. A standardized residual plot
4. A normal probability plot

Residual Plot Against x

A **residual plot** against the independent variable x is a graph in which the values of the independent variable are represented by the horizontal axis and the corresponding residual values are represented by the vertical axis. A point is plotted for each residual. The first coordinate for each point is given by the value of x_i and the second coordinate is given by the corresponding value of the residual $y_i - \hat{y}_i$. For a residual plot against x with the Armand's Pizza Parlors data from Table 14.7, the coordinates of the first point are (2, −12), corresponding to $x_1 = 2$ and $y_1 - \hat{y}_1 = -12$; the coordinates of the second point are (6, 15), corresponding to $x_2 = 6$ and $y_2 - \hat{y}_2 = 15$; and so on. Figure 14.11 shows the resulting residual plot.

Before interpreting the results for this residual plot, let us consider some general patterns that might be observed in any residual plot. Three examples appear in Figure 14.12. If the assumption that the variance of ϵ is the same for all values of x and the assumed regression model is an adequate representation of the relationship between the variables, the

FIGURE 14.11 PLOT OF THE RESIDUALS AGAINST THE INDEPENDENT VARIABLE x FOR ARMAND'S PIZZA PARLORS

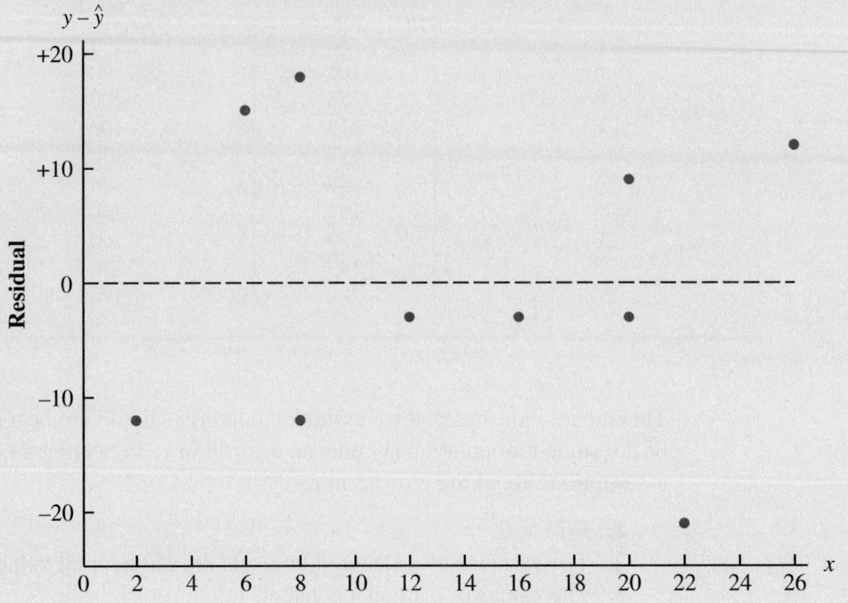

residual plot should give an overall impression of a horizontal band of points such as the one in Panel A of Figure 14.12. However, if the variance of ϵ is not the same for all values of x—for example, if variability about the regression line is greater for larger values of x—a pattern such as the one in Panel B of Figure 14.12 could be observed. In this case, the assumption of a constant variance of ϵ is violated. Another possible residual plot is shown in Panel C. In this case, we would conclude that the assumed regression model is not an adequate representation of the relationship between the variables. A curvilinear regression model or multiple regression model should be considered.

Now let us return to the residual plot for Armand's Pizza Parlors shown in Figure 14.11. The residuals appear to approximate the horizontal pattern in Panel A of Figure 14.12. Hence, we conclude that the residual plot does not provide evidence that the assumptions made for Armand's regression model should be challenged. At this point, we are confident in the conclusion that Armand's simple linear regression model is valid.

Experience and good judgment are always factors in the effective interpretation of residual plots. Seldom does a residual plot conform precisely to one of the patterns in Figure 14.12. Yet analysts who frequently conduct regression studies and frequently review residual plots become adept at understanding the differences between patterns that are reasonable and patterns that indicate the assumptions of the model should be questioned. A residual plot provides one technique to assess the validity of the assumptions for a regression model.

Residual Plot Against $\hat{y}$

Another residual plot represents the predicted value of the dependent variable $\hat{y}$ on the horizontal axis and the residual values on the vertical axis. A point is plotted for each residual. The first coordinate for each point is given by $\hat{y}_i$ and the second coordinate is given by the

FIGURE 14.12 RESIDUAL PLOTS FROM THREE REGRESSION STUDIES

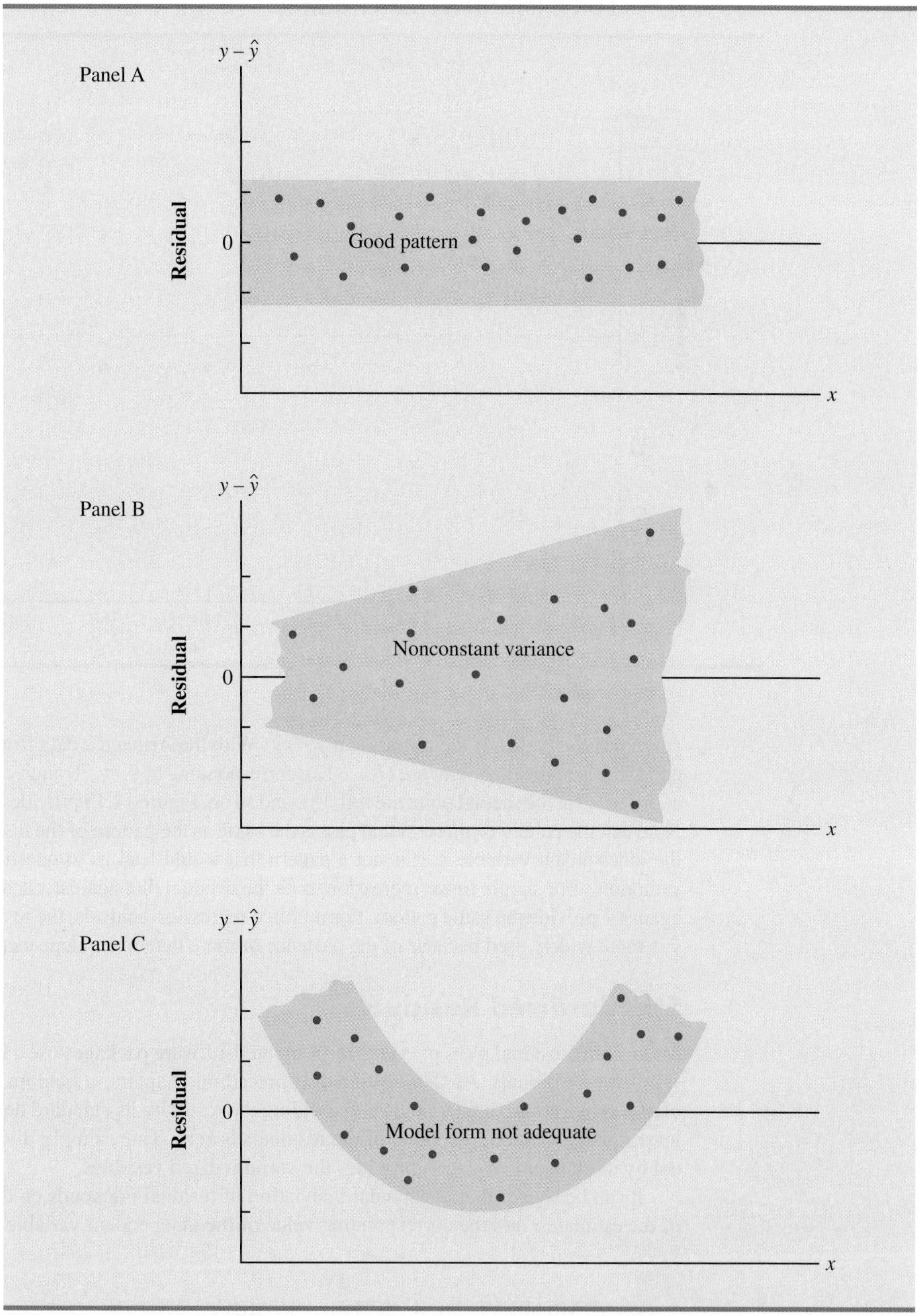

FIGURE 14.13 PLOT OF THE RESIDUALS AGAINST THE PREDICTED VALUES y FOR ARMAND'S PIZZA PARLORS

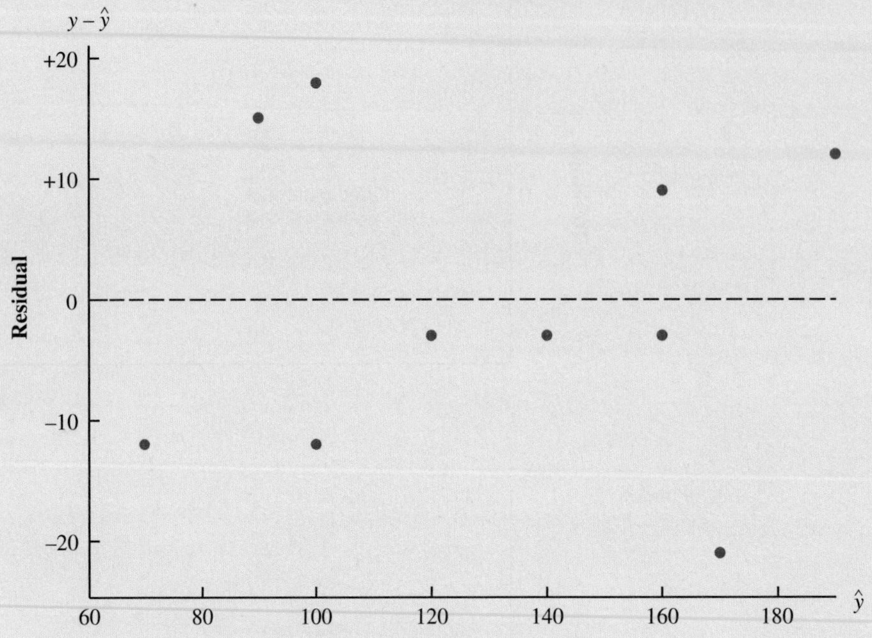

corresponding value of the *i*th residual $y_i - \hat{y}_i$. With the Armand's data from Table 14.7, the coordinates of the first point are (70, −12), corresponding to $\hat{y}_1 = 70$ and $y_1 - \hat{y}_1 = -12$; the coordinates of the second point are (90, 15); and so on. Figure 14.13 provides the residual plot. Note that the pattern of this residual plot is the same as the pattern of the residual plot against the independent variable *x*. It is not a pattern that would lead us to question the model assumptions. For simple linear regression, both the residual plot against *x* and the residual plot against $\hat{y}$ provide the same pattern. For multiple regression analysis, the residual plot against $\hat{y}$ is more widely used because of the presence of more than one independent variable.

Standardized Residuals

Many of the residual plots provided by computer software packages use a standardized version of the residuals. As demonstrated in preceding chapters, a random variable is standardized by subtracting its mean and dividing the result by its standard deviation. With the least squares method, the mean of the residuals is zero. Thus, simply dividing each residual by its standard deviation provides the **standardized residual**.

It can be shown that the standard deviation of residual *i* depends on the standard error of the estimate *s* and the corresponding value of the independent variable x_i.

STANDARD DEVIATION OF THE *i*th RESIDUAL[3]

$$s_{y_i - \hat{y}_i} = s\sqrt{1 - h_i} \tag{14.30}$$

[3]This equation actually provides an estimate of the standard deviation of the *i*th residual, because *s* is used instead of σ.

where

$$s_{y_i - \hat{y}_i} = \text{the standard deviation of residual } i$$
$$s = \text{the standard error of the estimate}$$
$$h_i = \frac{1}{n} + \frac{(x_i - \bar{x})^2}{\Sigma(x_i - \bar{x})^2} \quad \textbf{(14.31)}$$

Note that equation (14.30) shows that the standard deviation of the ith residual depends on x_i because of the presence of h_i in the formula.[4] Once the standard deviation of each residual is calculated, we can compute the standardized residual by dividing each residual by its corresponding standard deviation.

STANDARDIZED RESIDUAL FOR OBSERVATION i

$$\frac{y_i - \hat{y}_i}{s_{y_i - \hat{y}_i}} \quad \textbf{(14.32)}$$

Table 14.8 shows the calculation of the standardized residuals for Armand's Pizza Parlors. Recall that previous calculations showed $s = 13.829$. Figure 14.14 is the plot of the standardized residuals against the independent variable x.

Small departures from normality do not have a great effect on the statistical tests used in regression analysis.

The standardized residual plot can provide insight about the assumption that the error term ϵ has a normal distribution. If this assumption is satisfied, the distribution of the standardized residuals should appear to come from a standard normal probability distribution.[5] Thus, when looking at a standardized residual plot, we should expect to see approximately 95% of the standardized residuals between -2 and $+2$. We see in Figure 14.14 that for the

TABLE 14.8 COMPUTATION OF STANDARDIZED RESIDUALS FOR ARMAND'S PIZZA PARLORS

Restaurant i	x_i	$x_i - \bar{x}$	$(x_i - \bar{x})^2$	$\frac{(x_i - \bar{x})^2}{\Sigma(x_i - \bar{x})^2}$	h_i	$s_{y_i - \hat{y}_i}$	$y_i - \hat{y}_i$	Standardized Residual
1	2	−12	144	.2535	.3535	11.1193	−12	−1.0792
2	6	−8	64	.1127	.2127	12.2709	15	1.2224
3	8	−6	36	.0634	.1634	12.6493	−12	−.9487
4	8	−6	36	.0634	.1634	12.6493	18	1.4230
5	12	−2	4	.0070	.1070	13.0682	−3	−.2296
6	16	2	4	.0070	.1070	13.0682	−3	−.2296
7	20	6	36	.0634	.1634	12.6493	−3	−.2372
8	20	6	36	.0634	.1634	12.6493	9	.7115
9	22	8	64	.1127	.2127	12.2709	−21	−1.7114
10	26	12	144	.2535	.3535	11.1193	12	1.0792
		Total	568					

Note: The values of the residuals were computed in Table 14.7.

[4] h_i is referred to as the *leverage* of observation i. Leverage will be discussed further when we consider influential observations in Section 14.9.

[5] Because s is used instead of σ in equation (14.30), the probability distribution of the standardized residuals is not technically normal. However, in most regression studies, the sample size is large enough that a normal approximation is very good.

FIGURE 14.14 PLOT OF THE STANDARDIZED RESIDUALS AGAINST THE INDEPENDENT VARIABLE x FOR ARMAND'S PIZZA PARLORS

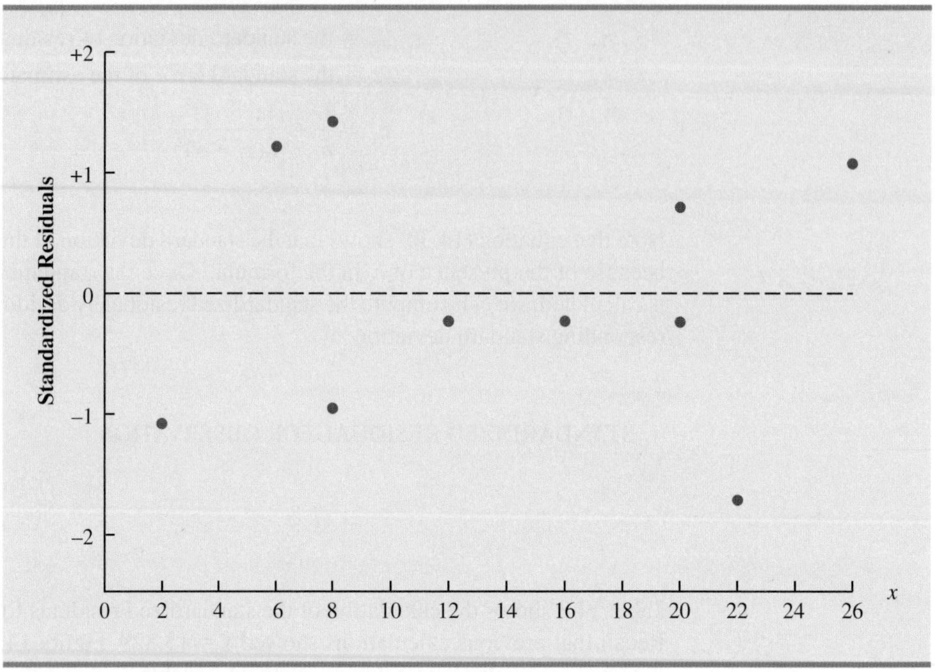

Armand's example all standardized residuals are between −2 and +2. Therefore, on the basis of the standardized residuals, this plot gives us no reason to question the assumption that ϵ has a normal distribution.

Because of the effort required to compute the estimated values of $\hat{y}$, the residuals, and the standardized residuals, most statistical packages provide these values as optional regression output. Hence, residual plots can be easily obtained. For large problems computer packages are the only practical means for developing the residual plots discussed in this section.

TABLE 14.9 NORMAL SCORES FOR $n = 10$

Order Statistic	Normal Score
1	−1.55
2	−1.00
3	−.65
4	−.37
5	−.12
6	.12
7	.37
8	.65
9	1.00
10	1.55

Normal Probability Plot

Another approach for determining the validity of the assumption that the error term has a normal distribution is the **normal probability plot**. To show how a normal probability plot is developed, we introduce the concept of *normal scores*.

Suppose 10 values are selected randomly from a normal probability distribution with a mean of zero and a standard deviation of one, and that the sampling process is repeated over and over with the values in each sample of 10 ordered from smallest to largest. For now, let us consider only the smallest value in each sample. The random variable representing the smallest value obtained in repeated sampling is called the first-order statistic.

Statisticians show that for samples of size 10 from a standard normal probability distribution, the expected value of the first-order statistic is −1.55. This expected value is called a normal score. For the case with a sample of size $n = 10$, there are 10 order statistics and 10 normal scores (see Table 14.9). In general, a data set consisting of n observations will have n order statistics and hence n normal scores.

Let us now show how the 10 normal scores can be used to determine whether the standardized residuals for Armand's Pizza Parlors appear to come from a standard normal probability distribution. We begin by ordering the 10 standardized residuals from Table 14.8. The 10 normal scores and the ordered standardized residuals are shown together in Table 14.10. If the normality assumption is satisfied, the smallest standardized residual should be close to the smallest normal score, the next smallest standardized residual should be close to the next smallest normal

TABLE 14.10 NORMAL SCORES AND ORDERED STANDARDIZED RESIDUALS FOR ARMAND'S PIZZA PARLORS

Normal Scores	Ordered Standardized Residuals
−1.55	−1.7114
−1.00	−1.0792
−.65	−.9487
−.37	−.2372
−.12	−.2296
.12	−.2296
.37	.7115
.65	1.0792
1.00	1.2224
1.55	1.4230

FIGURE 14.15 NORMAL PROBABILITY PLOT FOR ARMAND'S PIZZA PARLORS

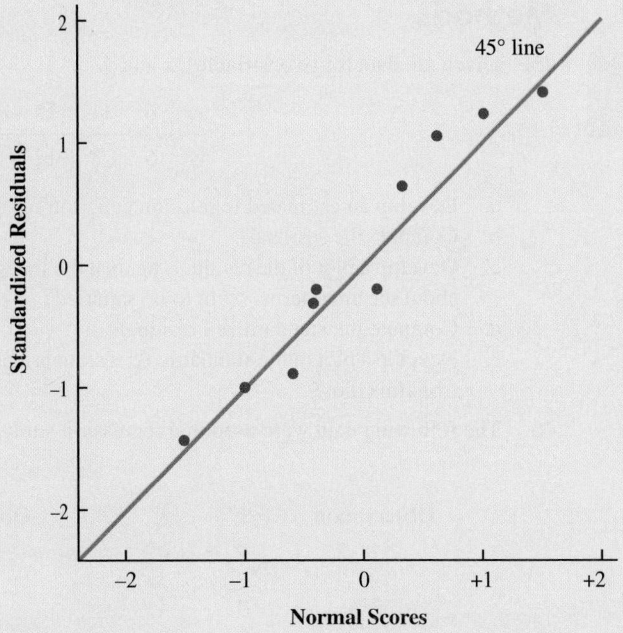

score, and so on. If we were to develop a plot with the normal scores on the horizontal axis and the corresponding standardized residuals on the vertical axis, the plotted points should cluster closely around a 45-degree line passing through the origin if the standardized residuals are approximately normally distributed. Such a plot is referred to as a *normal probability plot.*

Figure 14.15 is the normal probability plot for the Armand's Pizza Parlors example. Judgment is used to determine whether the pattern observed deviates from the line enough to conclude that the standardized residuals are not from a standard normal probability distribution. In Figure 14.15, we see that the points are grouped closely about the line. We therefore conclude that the assumption of the error term having a normal probability distribution is reasonable. In general, the more closely the points are clustered about the 45-degree line, the stronger the evidence supporting the normality assumption. Any substantial curvature in the normal probability plot is evidence that the residuals have not come from a normal distribution. Normal scores and the associated normal probability plot can be obtained easily from statistical packages such as Minitab.

NOTES AND COMMENTS

1. We use residual and normal probability plots to validate the assumptions of a regression model. If our review indicates that one or more assumptions are questionable, a different regression model or a transformation of the data should be considered. The appropriate corrective action when the assumptions are violated must be based on good judgment; recommendations from an experienced statistician can be valuable.
2. Analysis of residuals is the primary method statisticians use to verify that the assumptions associated with a regression model are valid. Even if no violations are found, it does not necessarily follow that the model will yield good predictions. However, if additional statistical tests support the conclusion of significance and the coefficient of determination is large, we should be able to develop good estimates and predictions using the estimated regression equation.

Exercises

Methods

SELF test

45. Given are data for two variables, x and y.

x_i	6	11	15	18	20
y_i	6	8	12	20	30

a. Develop an estimated regression equation for these data.
b. Compute the residuals.
c. Develop a plot of the residuals against the independent variable x. Do the assumptions about the error terms seem to be satisfied?
d. Compute the standardized residuals.
e. Develop a plot of the standardized residuals against $\hat{y}$. What conclusions can you draw from this plot?

46. The following data were used in a regression study.

Observation	x_i	y_i	Observation	x_i	y_i
1	2	4	6	7	6
2	3	5	7	7	9
3	4	4	8	8	5
4	5	6	9	9	11
5	7	4			

a. Develop an estimated regression equation for these data.
b. Construct a plot of the residuals. Do the assumptions about the error term seem to be satisfied?

Applications

SELF test

47. Data on advertising expenditures and revenue (in thousands of dollars) for the Four Seasons Restaurant follow.

Advertising Expenditures	Revenue
1	19
2	32
4	44
6	40
10	52
14	53
20	54

a. Let x equal advertising expenditures and y equal revenue. Use the method of least squares to develop a straight line approximation of the relationship between the two variables.
b. Test whether revenue and advertising expenditures are related at a .05 level of significance.
c. Prepare a residual plot of $y - \hat{y}$ versus $\hat{y}$. Use the result from part (a) to obtain the values of $\hat{y}$.
d. What conclusions can you draw from residual analysis? Should this model be used, or should we look for a better one?

48. Refer to exercise 7, where an estimated regression equation relating years of experience and annual sales was developed.
    a. Compute the residuals and construct a residual plot for this problem.
    b. Do the assumptions about the error terms seem reasonable in light of the residual plot?
49. Recent family home sales in San Antonio provided the following data (San Antonio Realty Watch website, November 2008).

WEB file

HomePrices

Square Footage	Price ($)
1580	142,500
1572	145,000
1352	115,000
2224	155,900
1556	95,000
1435	128,000
1438	100,000
1089	55,000
1941	142,000
1698	115,000
1539	115,000
1364	105,000
1979	155,000
2183	132,000
2096	140,000
1400	85,000
2372	145,000
1752	155,000
1386	80,000
1163	100,000

    a. Develop the estimated regression equation that can be used to predict the sales prices given the square footage.
    b. Construct a residual plot of the standardized residuals against the independent variable.
    c. Do the assumptions about the error term and model form seem reasonable in light of the residual plot?

14.9 Residual Analysis: Outliers and Influential Observations

In Section 14.8 we showed how residual analysis could be used to determine when violations of assumptions about the regression model occur. In this section, we discuss how residual analysis can be used to identify observations that can be classified as outliers or as being especially influential in determining the estimated regression equation. Some steps that should be taken when such observations occur are discussed.

Detecting Outliers

Figure 14.16 is a scatter diagram for a data set that contains an **outlier**, a data point (observation) that does not fit the trend shown by the remaining data. Outliers represent observations that are suspect and warrant careful examination. They may represent erroneous data; if so, the data should be corrected. They may signal a violation of model assumptions; if so, another model should be considered. Finally, they may simply be unusual values that occurred by chance. In this case, they should be retained.

FIGURE 14.16 DATA SET WITH AN OUTLIER

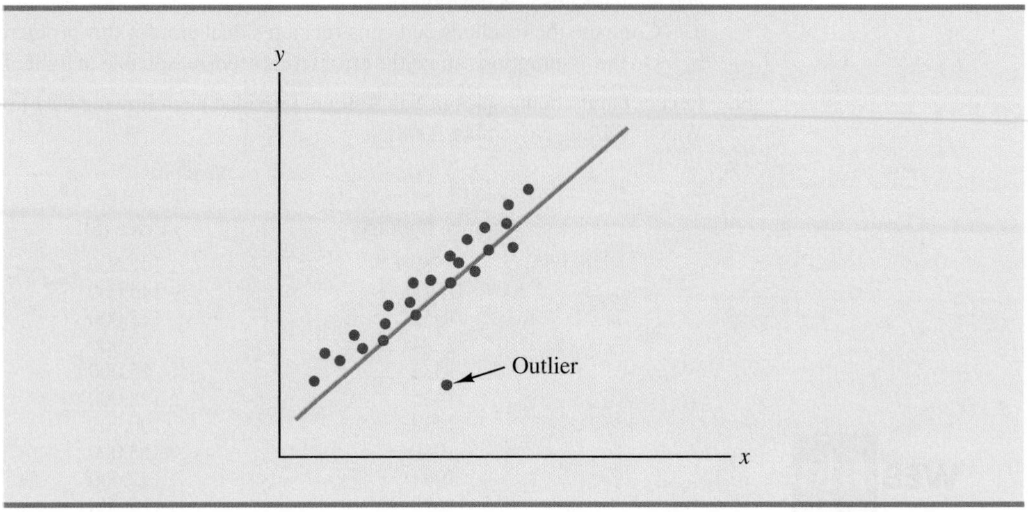

TABLE 14.11

DATA SET ILLUSTRATING THE EFFECT OF AN OUTLIER

x_i	y_i
1	45
1	55
2	50
3	75
3	40
3	45
4	30
4	35
5	25
6	15

To illustrate the process of detecting outliers, consider the data set in Table 14.11; Figure 14.17 is a scatter diagram. Except for observation 4 ($x_4 = 3$, $y_4 = 75$), a pattern suggesting a negative linear relationship is apparent. Indeed, given the pattern of the rest of the data, we would expect y_4 to be much smaller and hence would identify the corresponding observation as an outlier. For the case of simple linear regression, one can often detect outliers by simply examining the scatter diagram.

The standardized residuals can also be used to identify outliers. If an observation deviates greatly from the pattern of the rest of the data (e.g., the outlier in Figure 14.16), the corresponding standardized residual will be large in absolute value. Many computer packages automatically identify observations with standardized residuals that are large in absolute value. In Figure 14.18 we show the Minitab output from a regression analysis of the data in Table 14.11. The next to last line of the output shows that the standardized residual for observation 4 is 2.67. Minitab provides a list of each observation with a standardized residual

FIGURE 14.17 SCATTER DIAGRAM FOR OUTLIER DATA SET

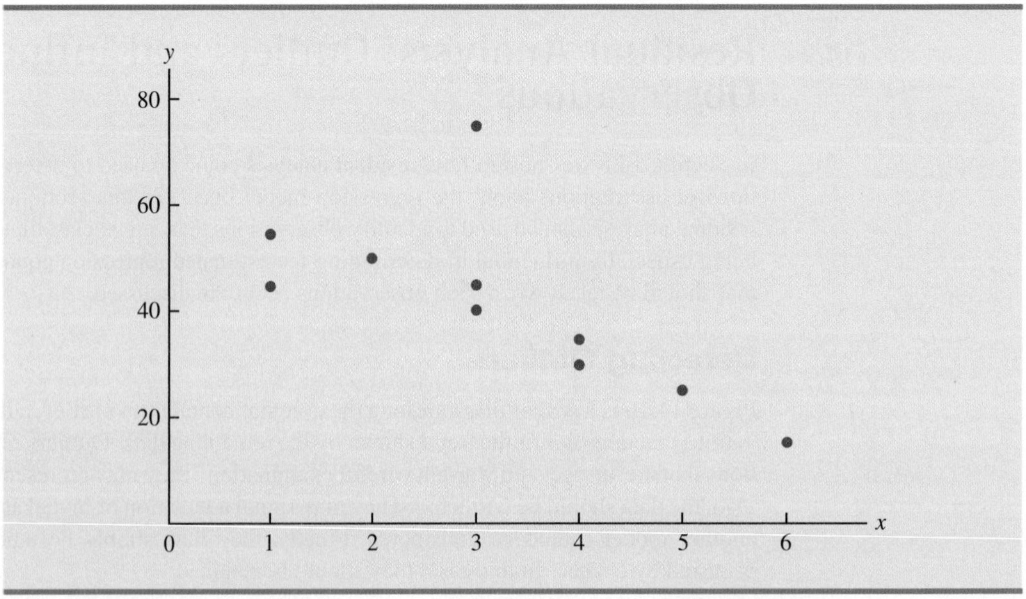

FIGURE 14.18 MINITAB OUTPUT FOR REGRESSION ANALYSIS OF THE OUTLIER DATA SET

```
The regression equation is
y = 65.0 - 7.33 x

Predictor      Coef  SE Coef      T      p
Constant     64.958    9.258   7.02  0.000
X            -7.331    2.608  -2.81  0.023

S = 12.6704   R-Sq = 49.7%   R-Sq(adj) = 43.4%

Analysis of Variance

SOURCE           DF      SS      MS     F      p
Regression        1  1268.2  1268.2  7.90  0.023
Residual Error    8  1284.3   160.5
Total             9  2552.5

Unusual Observations
Obs     x      y    Fit  SE Fit  Residual  St Resid
  4  3.00  75.00  42.97    4.04     32.03     2.67R

R denotes an observation with a large standardized residual.
```

of less than −2 or greater than +2 in the Unusual Observation section of the output; in such cases, the observation is printed on a separate line with an R next to the standardized residual, as shown in Figure 14.18. With normally distributed errors, standardized residuals should be outside these limits approximately 5% of the time.

In deciding how to handle an outlier, we should first check to see whether it is a valid observation. Perhaps an error was made in initially recording the data or in entering the data into the computer file. For example, suppose that in checking the data for the outlier in Table 14.17, we find an error; the correct value for observation 4 is $x_4 = 3$, $y_4 = 30$. Figure 14.19 is the Minitab output obtained after correction of the value of y_4. We see that

FIGURE 14.19 MINITAB OUTPUT FOR THE REVISED OUTLIER DATA SET

```
The regression equation is
Y = 59.2 - 6.95 X

Predictor      Coef  SE Coef      T      p
Constant     59.237    3.835  15.45  0.000
X            -6.949    1.080  -6.43  0.000

S = 5.24808   R-Sq = 83.8%   R-Sq(adj) = 81.8%

Analysis of Variance

SOURCE           DF      SS      MS      F      p
Regression        1  1139.7  1139.7  41.38  0.000
Residual Error    8   220.3    27.5
Total             9  1360.0
```

using the incorrect data value substantially affected the goodness of fit. With the correct data, the value of R-sq increased from 49.7% to 83.8% and the value of b_0 decreased from 64.958 to 59.237. The slope of the line changed from −7.331 to −6.949. The identification of the outlier enabled us to correct the data error and improve the regression results.

Detecting Influential Observations

Sometimes one or more observations exert a strong influence on the results obtained. Figure 14.20 shows an example of an **influential observation** in simple linear regression. The estimated regression line has a negative slope. However, if the influential observation were dropped from the data set, the slope of the estimated regression line would change from negative to positive and the *y*-intercept would be smaller. Clearly, this one observation is much more influential in determining the estimated regression line than any of the others; dropping one of the other observations from the data set would have little effect on the estimated regression equation.

Influential observations can be identified from a scatter diagram when only one independent variable is present. An influential observation may be an outlier (an observation with a y value that deviates substantially from the trend), it may correspond to an x value far away from its mean (e.g., see Figure 14.20), or it may be caused by a combination of the two (a somewhat off-trend y value and a somewhat extreme x value).

Because influential observations may have such a dramatic effect on the estimated regression equation, they must be examined carefully. We should first check to make sure that no error was made in collecting or recording the data. If an error occurred, it can be corrected and a new estimated regression equation can be developed. If the observation is valid, we might consider ourselves fortunate to have it. Such a point, if valid, can contribute to a better understanding of the appropriate model and can lead to a better estimated regression equation. The presence of the influential observation in Figure 14.20, if valid, would suggest trying to obtain data on intermediate values of x to understand better the relationship between x and y.

Observations with extreme values for the independent variables are called **high leverage points**. The influential observation in Figure 14.20 is a point with high leverage. The leverage of an observation is determined by how far the values of the independent variables are from their mean values. For the single-independent-variable case, the leverage of the *i*th observation, denoted h_i, can be computed by using equation (14.33).

FIGURE 14.20 DATA SET WITH AN INFLUENTIAL OBSERVATION

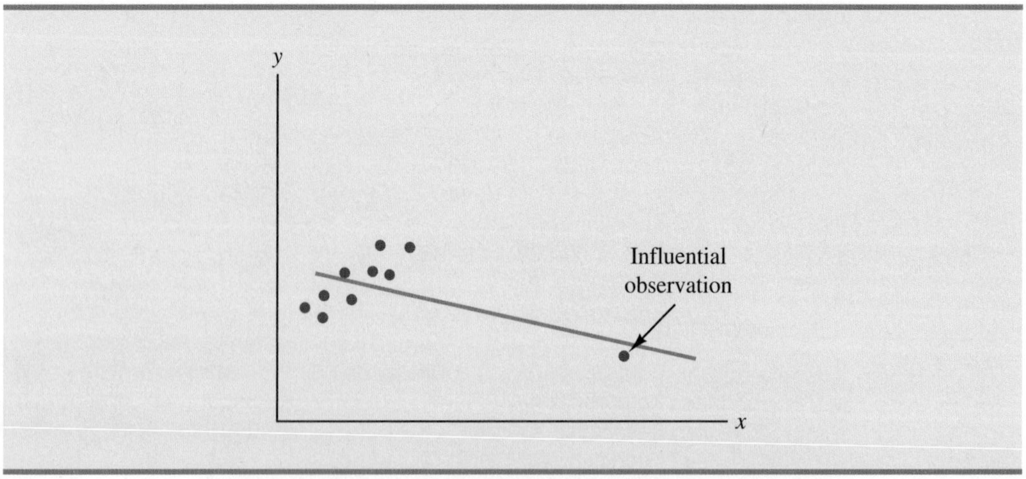

TABLE 14.12

DATA SET WITH A HIGH LEVERAGE OBSERVATION

x_i	y_i
10	125
10	130
15	120
20	115
20	120
25	110
70	100

LEVERAGE OF OBSERVATION *i*

$$h_i = \frac{1}{n} + \frac{(x_i - \bar{x})^2}{\Sigma(x_i - \bar{x})^2} \tag{14.33}$$

From the formula, it is clear that the farther x_i is from its mean $\bar{x}$, the higher the leverage of observation *i*.

Many statistical packages automatically identify observations with high leverage as part of the standard regression output. As an illustration of how the Minitab statistical package identifies points with high leverage, let us consider the data set in Table 14.12.

From Figure 14.21, a scatter diagram for the data set in Table 14.12, it is clear that observation 7 ($x = 70$, $y = 100$) is an observation with an extreme value of *x*. Hence, we would expect it to be identified as a point with high leverage. For this observation, the leverage is computed by using equation (14.33) as follows.

$$h_7 = \frac{1}{n} + \frac{(x_7 - \bar{x})^2}{\Sigma(x_i - \bar{x})^2} = \frac{1}{7} + \frac{(70 - 24.286)^2}{2621.43} = .94$$

For the case of simple linear regression, Minitab identifies observations as having high leverage if $h_i > 6/n$ or .99, whichever is smaller. For the data set in Table 14.12, $6/n = 6/7 = .86$. Because $h_7 = .94 > .86$, Minitab will identify observation 7 as an observation whose *x* value gives it large influence. Figure 14.22 shows the Minitab output for a regression analysis of this data set. Observation 7 ($x = 70$, $y = 100$) is identified as having large influence; it is printed on a separate line at the bottom, with an X in the right margin.

Computer software packages are essential for performing the computations to identify influential observations. Minitab's selection rule is discussed here.

Influential observations that are caused by an interaction of large residuals and high leverage can be difficult to detect. Diagnostic procedures are available that take both into account in determining when an observation is influential. One such measure, called Cook's *D* statistic, will be discussed in Chapter 15.

FIGURE 14.21 SCATTER DIAGRAM FOR THE DATA SET WITH A HIGH LEVERAGE OBSERVATION

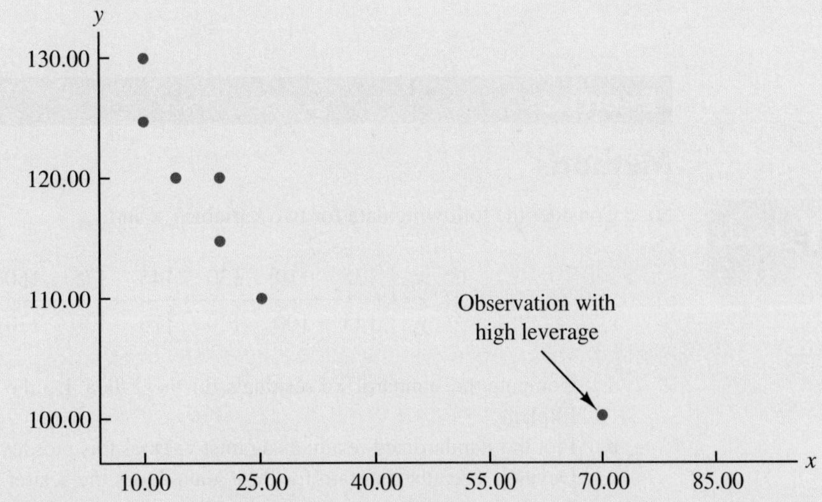

FIGURE 14.22 MINITAB OUTPUT FOR THE DATA SET WITH A HIGH LEVERAGE OBSERVATION

```
The regression equation is
y = 127 - 0.425 x

Predictor       Coef   SE Coef      T      p
Constant     127.466     2.961  43.04  0.000
X           -0.42507   0.09537  -4.46  0.007

S = 4.88282   R-sq = 79.9%   R-sq(adj) = 75.9%

Analysis of Variance

SOURCE          DF      SS      MS      F      p
Regression       1  473.65  473.65  19.87  0.007
Residual Error   5  119.21   23.84
Total            6  592.86

Unusual Observations
Obs     x       y     Fit  SE Fit  Residual  St Resid
  7  70.0  100.00   97.71    4.73      2.29    1.91 X

X denotes an observation whose X value gives it large influence.
```

NOTES AND COMMENTS

Once an observation is identified as potentially influential because of a large residual or high leverage, its impact on the estimated regression equation should be evaluated. More advanced texts discuss diagnostics for doing so. However, if one is not familiar with the more advanced material, a simple procedure is to run the regression analysis with and without the observation. This approach will reveal the influence of the observation on the results.

Exercises

Methods

SELF test

50. Consider the following data for two variables, x and y.

x_i	135	110	130	145	175	160	120
y_i	145	100	120	120	130	130	110

a. Compute the standardized residuals for these data. Do the data include any outliers? Explain.
b. Plot the standardized residuals against $\hat{y}$. Does this plot reveal any outliers?
c. Develop a scatter diagram for these data. Does the scatter diagram indicate any outliers in the data? In general, what implications does this finding have for simple linear regression?

51. Consider the following data for two variables, x and y.

x_i	4	5	7	8	10	12	12	22
y_i	12	14	16	15	18	20	24	19

a. Compute the standardized residuals for these data. Do the data include any outliers? Explain.
b. Compute the leverage values for these data. Do there appear to be any influential observations in these data? Explain.
c. Develop a scatter diagram for these data. Does the scatter diagram indicate any influential observations? Explain.

Applications

52. Charity Navigator is America's leading independent charity evaluator. The following data show the total expenses ($), the percentage of the total budget spent on administrative expenses, the percentage spent on fundraising, and the percentage spent on program expenses for 10 supersized charities (Charity Navigator website, April 12, 2012). Administrative expenses include overhead, administrative staff and associated costs, and organizational meetings. Fundraising expenses are what a charity spends to raise money, and program expenses are what the charity spends on the programs and services it exists to deliver. The sum of the three percentages does not add to 100% because of rounding.

WEB file

Charities

Charity	Total Expenses ($)	Administrative Expenses (%)	Fundraising Expenses (%)	Program Expenses (%)
American Red Cross	3,354,177,445	3.9	3.8	92.1
World Vision	1,205,887,020	4.0	7.5	88.3
Smithsonian Institution	1,080,995,083	23.5	2.6	73.7
Food For The Poor	1,050,829,851	.7	2.4	96.8
American Cancer Society	1,003,781,897	6.1	22.2	71.6
Volunteers of America	929,158,968	8.6	1.9	89.4
Dana-Farber Cancer Institute	877,321,613	13.1	1.6	85.2
AmeriCares	854,604,824	.4	.7	98.8
ALSAC—St. Jude Children's Research Hospital	829,662,076	9.6	16.9	73.4
City of Hope	736,176,619	13.7	3.0	83.1

a. Develop a scatter diagram with fundraising expenses (%) on the horizontal axis and program expenses (%) on the vertical axis. Looking at the data, do there appear to be any outliers and/or influential observations?
b. Develop an estimated regression equation that could be used to predict program expenses (%) given fundraising expenses (%).
c. Does the value for the slope of the estimated regression equation make sense in the context of this problem situation?
d. Use residual analysis to determine whether any outliers and/or influential observations are present. Briefly summarize your findings and conclusions.

53. Many countries, especially those in Europe, have significant gold holdings. But, many of these countries also have massive debts. The following data show the total value of gold holdings in billions of U.S. dollars and the debt as a percentage of the gross domestic product for nine countries (WordPress and Trading Economics websites, February 24, 2012).

WEB file

GoldHoldings

Country	Gold Value ($ billions)	Debt (% of GDP)
China	63	17.7
France	146	81.7
Germany	203	83.2
Indonesia	33	69.2
Italy	147	119.0
Netherlands	36	63.7
Russia	50	9.9
Switzerland	62	55.0
United States	487	93.2

a. Develop a scatter diagram for the total value of a country's gold holdings ($ billions) as the independent variable.
b. What does the scatter diagram developed in part (a) indicate about the relationship between the two variables? Do there appear to be any outliers and/or influential observations? Explain.
c. Using the entire data set, develop the estimated regression equation that can be used to predict the debt of a country given the total value of its gold holdings.
d. Use residual analysis to determine whether any outliers or influential observations are present.
e. Suppose that after looking at the scatter diagram in part (a) that you were able to visually identify what appears to be an influential observation. Drop this observation from the data set and fit an estimated regression equation to the remaining data. Compare the estimated slope for the new estimated regression equation to the estimated slope obtained in part (c). Does this approach confirm the conclusion you reached in part (d)? Explain.

54. The following data show the annual revenue ($ millions) and the estimated team value ($ millions) for the 32 teams in the National Football League (*Forbes* website, February 2009).

WEB file

NFLValues

Team	Revenue ($ millions)	Value ($ millions)
Arizona Cardinals	203	914
Atlanta Falcons	203	872
Baltimore Ravens	226	1062
Buffalo Bills	206	885
Carolina Panthers	221	1040
Chicago Bears	226	1064
Cincinnati Bengals	205	941
Cleveland Browns	220	1035
Dallas Cowboys	269	1612
Denver Broncos	226	1061
Detroit Lions	204	917
Green Bay Packers	218	1023
Houston Texans	239	1125
Indianapolis Colts	203	1076
Jacksonville Jaguars	204	876
Kansas City Chiefs	214	1016
Miami Dolphins	232	1044
Minnesota Vikings	195	839
New England Patriots	282	1324
New Orleans Saints	213	937

Team	Revenue ($ millions)	Value ($ millions)
New York Giants	214	1178
New York Jets	213	1170
Oakland Raiders	205	861
Philadelphia Eagles	237	1116
Pittsburgh Steelers	216	1015
San Diego Chargers	207	888
San Francisco 49ers	201	865
Seattle Seahawks	215	1010
St. Louis Rams	206	929
Tampa Bay Buccaneers	224	1053
Tennessee Titans	216	994
Washington Redskins	327	1538

a. Develop a scatter diagram with Revenue on the horizontal axis and Value on the vertical axis. Looking at the scatter diagram, does it appear that there are any outliers and/or influential observations in the data?
b. Develop the estimated regression equation that can be used to predict team value given the value of annual revenue.
c. Use residual analysis to determine whether any outliers and/or influential observations are present. Briefly summarize your findings and conclusions.

Summary

In this chapter we showed how regression analysis can be used to determine how a dependent variable y is related to an independent variable x. In simple linear regression, the regression model is $y = \beta_0 + \beta_1 x + \epsilon$. The simple linear regression equation $E(y) = \beta_0 + \beta_1 x$ describes how the mean or expected value of y is related to x. We used sample data and the least squares method to develop the estimated regression equation $\hat{y} = b_0 + b_1 x$. In effect, b_0 and b_1 are the sample statistics used to estimate the unknown model parameters β_0 and β_1.

The coefficient of determination was presented as a measure of the goodness of fit for the estimated regression equation; it can be interpreted as the proportion of the variation in the dependent variable y that can be explained by the estimated regression equation. We reviewed correlation as a descriptive measure of the strength of a linear relationship between two variables.

The assumptions about the regression model and its associated error term ϵ were discussed, and t and F tests, based on those assumptions, were presented as a means for determining whether the relationship between two variables is statistically significant. We showed how to use the estimated regression equation to develop confidence interval estimates of the mean value of y and prediction interval estimates of individual values of y.

The chapter concluded with a section on the computer solution of regression problems and two sections on the use of residual analysis to validate the model assumptions and to identify outliers and influential observations.

Glossary

Dependent variable The variable that is being predicted or explained. It is denoted by y.
Independent variable The variable that is doing the predicting or explaining. It is denoted by x.
Simple linear regression Regression analysis involving one independent variable and one dependent variable in which the relationship between the variables is approximated by a straight line.

Regression model The equation that describes how y is related to x and an error term; in simple linear regression, the regression model is $y = \beta_0 + \beta_1 x + \epsilon$.
Regression equation The equation that describes how the mean or expected value of the dependent variable is related to the independent variable; in simple linear regression, $E(y) = \beta_0 + \beta_1 x$.
Estimated regression equation The estimate of the regression equation developed from sample data by using the least squares method. For simple linear regression, the estimated regression equation is $\hat{y} = b_0 + b_1 x$.
Least squares method A procedure used to develop the estimated regression equation. The objective is to minimize $\Sigma(y_i - \hat{y}_i)^2$.
Scatter diagram A graph of bivariate data in which the independent variable is on the horizontal axis and the dependent variable is on the vertical axis.
Coefficient of determination A measure of the goodness of fit of the estimated regression equation. It can be interpreted as the proportion of the variability in the dependent variable y that is explained by the estimated regression equation.
***i*th residual** The difference between the observed value of the dependent variable and the value predicted using the estimated regression equation; for the ith observation the ith residual is $y_i - \hat{y}_i$.
Correlation coefficient A measure of the strength of the linear relationship between two variables (previously discussed in Chapter 3).
Mean square error The unbiased estimate of the variance of the error term σ^2. It is denoted by MSE or s^2.
Standard error of the estimate The square root of the mean square error, denoted by s. It is the estimate of σ, the standard deviation of the error term ϵ.
ANOVA table The analysis of variance table used to summarize the computations associated with the F test for significance.
Confidence interval The interval estimate of the mean value of y for a given value of x.
Prediction interval The interval estimate of an individual value of y for a given value of x.
Residual analysis The analysis of the residuals used to determine whether the assumptions made about the regression model appear to be valid. Residual analysis is also used to identify outliers and influential observations.
Residual plot Graphical representation of the residuals that can be used to determine whether the assumptions made about the regression model appear to be valid.
Standardized residual The value obtained by dividing a residual by its standard deviation.
Normal probability plot A graph of the standardized residuals plotted against values of the normal scores. This plot helps determine whether the assumption that the error term has a normal probability distribution appears to be valid.
Outlier A data point or observation that does not fit the trend shown by the remaining data.
Influential observation An observation that has a strong influence or effect on the regression results.
High leverage points Observations with extreme values for the independent variables.

Key Formulas

Simple Linear Regression Model

$$y = \beta_0 + \beta_1 x + \epsilon \tag{14.1}$$

Simple Linear Regression Equation

$$E(y) = \beta_0 + \beta_1 x \tag{14.2}$$

Estimated Simple Linear Regression Equation

$$\hat{y} = b_0 + b_1x \tag{14.3}$$

Least Squares Criterion

$$\min \Sigma(y_i - \hat{y}_i)^2 \tag{14.5}$$

Slope and *y*-Intercept for the Estimated Regression Equation

$$b_1 = \frac{\Sigma(x_i - \bar{x})(y_i - \bar{y})}{\Sigma(x_i - \bar{x})^2} \tag{14.6}$$

$$b_0 = \bar{y} - b_1\bar{x} \tag{14.7}$$

Sum of Squares Due to Error

$$\text{SSE} = \Sigma(y_i - \hat{y}_i)^2 \tag{14.8}$$

Total Sum of Squares

$$\text{SST} = \Sigma(y_i - \bar{y})^2 \tag{14.9}$$

Sum of Squares Due to Regression

$$\text{SSR} = \Sigma(\hat{y}_i - \bar{y})^2 \tag{14.10}$$

Relationship Among SST, SSR, and SSE

$$\text{SST} = \text{SSR} + \text{SSE} \tag{14.11}$$

Coefficient of Determination

$$r^2 = \frac{\text{SSR}}{\text{SST}} \tag{14.12}$$

Sample Correlation Coefficient

$$\begin{aligned} r_{xy} &= (\text{sign of } b_1)\sqrt{\text{Coefficient of determination}} \\ &= (\text{sign of } b_1)\sqrt{r^2} \end{aligned} \tag{14.13}$$

Mean Square Error (Estimate of σ^2)

$$s^2 = \text{MSE} = \frac{\text{SSE}}{n - 2} \tag{14.15}$$

Standard Error of the Estimate

$$s = \sqrt{\text{MSE}} = \sqrt{\frac{\text{SSE}}{n - 2}} \tag{14.16}$$

Standard Deviation of b_1

$$\sigma_{b_1} = \frac{\sigma}{\sqrt{\Sigma(x_i - \bar{x})^2}} \tag{14.17}$$

Estimated Standard Deviation of b_1

$$s_{b_1} = \frac{s}{\sqrt{\Sigma(x_i - \bar{x})^2}} \tag{14.18}$$

t Test Statistic

$$t = \frac{b_1}{s_{b_1}} \tag{14.19}$$

Mean Square Regression

$$\text{MSR} = \frac{\text{SSR}}{\text{Number of independent variables}} \tag{14.20}$$

F Test Statistic

$$F = \frac{\text{MSR}}{\text{MSE}} \tag{14.21}$$

Estimated Standard Deviation of $\hat{y}^*$

$$s_{\hat{y}^*} = s\sqrt{\frac{1}{n} + \frac{(x^* - \bar{x})^2}{\Sigma(x_i - \bar{x})^2}} \tag{14.23}$$

Confidence Interval for $E(y^*)$

$$\hat{y}^* \pm t_{\alpha/2}s_{\hat{y}^*} \tag{14.24}$$

Estimated Standard Deviation of an Individual Value

$$s_{\text{pred}} = s\sqrt{1 + \frac{1}{n} + \frac{(x^* - \bar{x})^2}{\Sigma(x_i - \bar{x})^2}} \tag{14.26}$$

Prediction Interval for y^*

$$\hat{y}^* \pm t_{\alpha/2}s_{\text{pred}} \tag{14.27}$$

Residual for Observation i

$$y_i - \hat{y}_i \tag{14.28}$$

Standard Deviation of the ith Residual

$$s_{y_i - \hat{y}_i} = s\sqrt{1 - h_i} \tag{14.30}$$

Standardized Residual for Observation i

$$\frac{y_i - \hat{y}_i}{s_{y_i - \hat{y}_i}} \tag{14.32}$$

Leverage of Observation i

$$h_i = \frac{1}{n} + \frac{(x_i - \bar{x})^2}{\Sigma(x_i - \bar{x})^2} \tag{14.33}$$

Supplementary Exercises

55. Does a high value of r^2 imply that two variables are causally related? Explain.
56. In your own words, explain the difference between an interval estimate of the mean value of y for a given x and an interval estimate for an individual value of y for a given x.

57. What is the purpose of testing whether $\beta_1 = 0$? If we reject $\beta_1 = 0$, does it imply a good fit?

58. The Dow Jones Industrial Average (DJIA) and the Standard & Poor's 500 (S&P 500) indexes are used as measures of overall movement in the stock market. The DJIA is based on the price movements of 30 large companies; the S&P 500 is an index composed of 500 stocks. Some say the S&P 500 is a better measure of stock market performance because it is broader based. The closing price for the DJIA and the S&P 500 for 15 weeks, beginning with January 6, 2012, follow (*Barron*'s website, April 17, 2012).

WEB file

DJIAS&P500

Date	DJIA	S&P
January 6	12,360	1278
January 13	12,422	1289
January 20	12,720	1315
January 27	12,660	1316
February 3	12,862	1345
February 10	12,801	1343
February 17	12,950	1362
February 24	12,983	1366
March 2	12,978	1370
March 9	12,922	1371
March 16	13,233	1404
March 23	13,081	1397
March 30	13,212	1408
April 5	13,060	1398
April 13	12,850	1370

a. Develop a scatter diagram with DJIA as the independent variable.
b. Develop the estimated regression equation.
c. Test for a significant relationship. Use $\alpha = .05$.
d. Did the estimated regression equation provide a good fit? Explain.
e. Suppose that the closing price for the DJIA is 13,500. Predict the closing price for the S&P 500.
f. Should we be concerned that the DJIA value of 13,500 used to predict the S&P 500 value in part (e) is beyond the range of the data used to develop the estimated regression equation?

59. The following data show Morningstar's Fair Value estimate and the Share Price for 28 companies. Fair Value is an estimate of a company's value per share that takes into account estimates of the company's growth, profitability, riskiness, and other factors over the next five years (*Morningstar Stocks 500,* 2008 edition).

WEB file

Stocks500

Company	Fair Value ($)	Share Price ($)
Air Products and Chemicals	80	98.63
Allied Waste Industries	17	11.02
America Mobile	83	61.39
AT&T	35	41.56
Bank of America	70	41.26
Barclays PLC	68	40.37
Citigroup	53	29.44
Costco Wholesale Corp.	75	69.76
Covidien, Ltd.	58	44.29
Darden Restaurants	52	27.71
Dun & Bradstreet	87	88.63

(*continued*)

Company	Fair Value ($)	Share Price ($)
Equifax	42	36.36
Gannett Co.	38	39.00
Genuine Parts	48	46.30
GlaxoSmithKline PLC	57	50.39
Iron Mountain	33	37.02
ITT Corporation	83	66.04
Johnson & Johnson	80	66.70
Las Vegas Sands	98	103.05
Macrovision	23	18.33
Marriott International	39	34.18
Nalco Holding Company	29	24.18
National Interstate	25	33.10
Portugal Telecom	15	13.02
Qualcomm	48	39.35
Royal Dutch Shell Ltd.	87	84.20
SanDisk	60	33.17
Time Warner	42	27.60

a. Develop the estimated regression equation that could be used to estimate the share price given the fair value.
b. At the .05 level of significance, is there a significant relationship between the two variables?
c. Use the estimated regression equation to predict the share price for a company that has a fair value of $50.
d. Do you believe the estimated regression equation would provide a good prediction of the share price? Use r^2 to support your answer.

60. One of the biggest changes in higher education in recent years has been the growth of online universities. The Online Education Database is an independent organization whose mission is to build a comprehensive list of the top accredited online colleges. The following table shows the retention rate (%) and the graduation rate (%) for 29 online colleges (Online Education Database website, January 2009).

WEB file
OnlineEdu

College	Retention Rate (%)	Graduation Rate (%)
Western International University	7	25
South University	51	25
University of Phoenix	4	28
American InterContinental University	29	32
Franklin University	33	33
Devry University	47	33
Tiffin University	63	34
Post University	45	36
Peirce College	60	36
Everest University	62	36
Upper Iowa University	67	36
Dickinson State University	65	37
Western Governors University	78	37
Kaplan University	75	38
Salem International University	54	39
Ashford University	45	41

College	Retention Rate (%)	Graduation Rate (%)
ITT Technical Institute	38	44
Berkeley College	51	45
Grand Canyon University	69	46
Nova Southeastern University	60	47
Westwood College	37	48
Everglades University	63	50
Liberty University	73	51
LeTourneau University	78	52
Rasmussen College	48	53
Keiser University	95	55
Herzing College	68	56
National University	100	57
Florida National College	100	61

a. Develop a scatter diagram with retention rate as the independent variable. What does the scatter diagram indicate about the relationship between the two variables?
b. Develop the estimated regression equation.
c. Test for a significant relationship. Use $\alpha = .05$.
d. Did the estimated regression equation provide a good fit?
e. Suppose you were the president of South University. After reviewing the results, would you have any concerns about the performance of your university as compared to other online universities?
f. Suppose you were the president of the University of Phoenix. After reviewing the results, would you have any concerns about the performance of your university as compared to other online universities?

61. Jensen Tire & Auto is in the process of deciding whether to purchase a maintenance contract for its new computer wheel alignment and balancing machine. Managers feel that maintenance expense should be related to usage, and they collected the following information on weekly usage (hours) and annual maintenance expense (in hundreds of dollars).

WEB file

Jensen

Weekly Usage (hours)	Annual Maintenance Expense
13	17.0
10	22.0
20	30.0
28	37.0
32	47.0
17	30.5
24	32.5
31	39.0
40	51.5
38	40.0

a. Develop the estimated regression equation that relates annual maintenance expense to weekly usage.
b. Test the significance of the relationship in part (a) at a .05 level of significance.
c. Jensen expects to use the new machine 30 hours per week. Develop a 95% prediction interval for the company's annual maintenance expense.
d. If the maintenance contract costs $3000 per year, would you recommend purchasing it? Why or why not?

62. In a manufacturing process the assembly line speed (feet per minute) was thought to affect the number of defective parts found during the inspection process. To test this theory, managers devised a situation in which the same batch of parts was inspected visually at a variety of line speeds. They collected the following data.

Line Speed	Number of Defective Parts Found
20	21
20	19
40	15
30	16
60	14
40	17

a. Develop the estimated regression equation that relates line speed to the number of defective parts found.
b. At a .05 level of significance, determine whether line speed and number of defective parts found are related.
c. Did the estimated regression equation provide a good fit to the data?
d. Develop a 95% confidence interval to predict the mean number of defective parts for a line speed of 50 feet per minute.

63. A sociologist was hired by a large city hospital to investigate the relationship between the number of unauthorized days that employees are absent per year and the distance (miles) between home and work for the employees. A sample of 10 employees was chosen, and the following data were collected.

WEB file

Absent

Distance to Work (miles)	Number of Days Absent
1	8
3	5
4	8
6	7
8	6
10	3
12	5
14	2
14	4
18	2

a. Develop a scatter diagram for these data. Does a linear relationship appear reasonable? Explain.
b. Develop the least squares estimated regression equation.
c. Is there a significant relationship between the two variables? Use $\alpha = .05$.
d. Did the estimated regression equation provide a good fit? Explain.
e. Use the estimated regression equation developed in part (b) to develop a 95% confidence interval for the expected number of days absent for employees living 5 miles from the company.

64. The regional transit authority for a major metropolitan area wants to determine whether there is any relationship between the age of a bus and the annual maintenance cost. A sample of 10 buses resulted in the following data.

WEB file

AgeCost

Age of Bus (years)	Maintenance Cost ($)
1	350
2	370
2	480
2	520
2	590
3	550
4	750
4	800
5	790
5	950

a. Develop the least squares estimated regression equation.
b. Test to see whether the two variables are significantly related with $\alpha = .05$.
c. Did the least squares line provide a good fit to the observed data? Explain.
d. Develop a 95% prediction interval for the maintenance cost for a specific bus that is 4 years old.

65. A marketing professor at Givens College is interested in the relationship between hours spent studying and total points earned in a course. Data collected on 10 students who took the course last quarter follow.

WEB file

HoursPts

Hours Spent Studying	Total Points Earned
45	40
30	35
90	75
60	65
105	90
65	50
90	90
80	80
55	45
75	65

a. Develop an estimated regression equation showing how total points earned is related to hours spent studying.
b. Test the significance of the model with $\alpha = .05$.
c. Predict the total points earned by Mark Sweeney. He spent 95 hours studying.
d. Develop a 95% prediction interval for the total points earned by Mark Sweeney.

66. Reuters reported the market beta for Xerox was 1.22 (Reuters website, January 30, 2009). Market betas for individual stocks are determined by simple linear regression. For each stock, the dependent variable is its quarterly percentage return (capital appreciation plus dividends) minus the percentage return that could be obtained from a risk-free investment (the Treasury Bill rate is used as the risk-free rate). The independent variable is the quarterly percentage return (capital appreciation plus dividends) for the stock market (S&P 500) minus the percentage return from a risk-free investment. An estimated regression equation is developed with quarterly data; the market beta for the stock is the slope of the estimated regression equation (b_1). The value of the market beta is often interpreted as a measure of the risk associated with the stock. Market betas greater than 1 indicate that the stock is more volatile than the market average; market betas less than 1 indicate that the

stock is less volatile than the market average. Suppose that the following figures are the differences between the percentage return and the risk-free return for 10 quarters for the S&P 500 and Horizon Technology.

WEB file
MktBeta

S&P 500	Horizon
1.2	−.7
−2.5	−2.0
−3.0	−5.5
2.0	4.7
5.0	1.8
1.2	4.1
3.0	2.6
−1.0	2.0
.5	−1.3
2.5	5.5

a. Develop an estimated regression equation that can be used to predict the market beta for Horizon Technology. What is Horizon Technology's market beta?
b. Test for a significant relationship at the .05 level of significance.
c. Did the estimated regression equation provide a good fit? Explain.
d. Use the market betas of Xerox and Horizon Technology to compare the risk associated with the two stocks.

67. The Transactional Records Access Clearinghouse at Syracuse University reported data showing the odds of an Internal Revenue Service audit. The following table shows the average adjusted gross income reported and the percent of the returns that were audited for 20 selected IRS districts.

WEB file
IRSAudit

District	Adjusted Gross Income ($)	Percent Audited
Los Angeles	36,664	1.3
Sacramento	38,845	1.1
Atlanta	34,886	1.1
Boise	32,512	1.1
Dallas	34,531	1.0
Providence	35,995	1.0
San Jose	37,799	0.9
Cheyenne	33,876	0.9
Fargo	30,513	0.9
New Orleans	30,174	0.9
Oklahoma City	30,060	0.8
Houston	37,153	0.8
Portland	34,918	0.7
Phoenix	33,291	0.7
Augusta	31,504	0.7
Albuquerque	29,199	0.6
Greensboro	33,072	0.6
Columbia	30,859	0.5
Nashville	32,566	0.5
Buffalo	34,296	0.5

a. Develop the estimated regression equation that could be used to predict the percent audited given the average adjusted gross income reported.
b. At the .05 level of significance, determine whether the adjusted gross income and the percent audited are related.

c. Did the estimated regression equation provide a good fit? Explain.
d. Use the estimated regression equation developed in part (a) to calculate a 95% confidence interval for the expected percent audited for districts with an average adjusted gross income of $35,000.

68. The Toyota Camry is one of the best-selling cars in North America. The cost of a previously owned Camry depends upon many factors, including the model year, mileage, and condition. To investigate the relationship between the car's mileage and the sales price for a 2007 model year Camry, the following data show the mileage and sale price for 19 sales (PriceHub website, February 24, 2012).

WEB file
Camry

Miles (1000s)	Price ($1000s)
22	16.2
29	16.0
36	13.8
47	11.5
63	12.5
77	12.9
73	11.2
87	13.0
92	11.8
101	10.8
110	8.3
28	12.5
59	11.1
68	15.0
68	12.2
91	13.0
42	15.6
65	12.7
110	8.3

a. Develop a scatter diagram with the car mileage on the horizontal axis and the price on the vertical axis.
b. What does the scatter diagram developed in part (a) indicate about the relationship between the two variables?
c. Develop the estimated regression equation that could be used to predict the price ($1000s) given the miles (1000s).
d. Test for a significant relationship at the .05 level of significance.
e. Did the estimated regression equation provide a good fit? Explain.
f. Provide an interpretation for the slope of the estimated regression equation.
g. Suppose that you are considering purchasing a previously owned 2007 Camry that has been driven 60,000 miles. Using the estimated regression equation developed in part (c), predict the price for this car. Is this the price you would offer the seller?

Case Problem 1 Measuring Stock Market Risk

One measure of the risk or volatility of an individual stock is the standard deviation of the total return (capital appreciation plus dividends) over several periods of time. Although the standard deviation is easy to compute, it does not take into account the extent to which the price of a given stock varies as a function of a standard market index, such as the S&P 500. As a result, many financial analysts prefer to use another measure of risk referred to as *beta*.

Betas for individual stocks are determined by simple linear regression. The dependent variable is the total return for the stock and the independent variable is the total return for

WEB file

Beta

the stock market.* For this case problem we will use the S&P 500 index as the measure of the total return for the stock market, and an estimated regression equation will be developed using monthly data. The beta for the stock is the slope of the estimated regression equation (b_1). The data contained in the file named Beta provides the total return (capital appreciation plus dividends) over 36 months for eight widely traded common stocks and the S&P 500.

The value of beta for the stock market will always be 1; thus, stocks that tend to rise and fall with the stock market will also have a beta close to 1. Betas greater than 1 indicate that the stock is more volatile than the market, and betas less than 1 indicate that the stock is less volatile than the market. For instance, if a stock has a beta of 1.4, it is 40% *more* volatile than the market, and if a stock has a beta of .4, it is 60% *less* volatile than the market.

Managerial Report

You have been assigned to analyze the risk characteristics of these stocks. Prepare a report that includes but is not limited to the following items.

a. Compute descriptive statistics for each stock and the S&P 500. Comment on your results. Which stocks are the most volatile?

b. Compute the value of beta for each stock. Which of these stocks would you expect to perform best in an up market? Which would you expect to hold their value best in a down market?

c. Comment on how much of the return for the individual stocks is explained by the market.

Case Problem 2 U.S. Department of Transportation

As part of a study on transportation safety, the U.S. Department of Transportation collected data on the number of fatal accidents per 1000 licenses and the percentage of licensed drivers under the age of 21 in a sample of 42 cities. Data collected over a one-year period follow. These data are contained in the file named Safety.

WEB file

Safety

Percent Under 21	Fatal Accidents per 1000 Licenses	Percent Under 21	Fatal Accidents per 1000 Licenses
13	2.962	17	4.100
12	0.708	8	2.190
8	0.885	16	3.623
12	1.652	15	2.623
11	2.091	9	0.835
17	2.627	8	0.820
18	3.830	14	2.890
8	0.368	8	1.267
13	1.142	15	3.224
8	0.645	10	1.014
9	1.028	10	0.493

*Various sources use different approaches for computing betas. For instance, some sources subtract the return that could be obtained from a risk-free investment (e.g., T-bills) from the dependent variable and the independent variable before computing the estimated regression equation. Some also use different indexes for the total return of the stock market; for instance, Value Line computes betas using the New York Stock Exchange composite index.

Percent Under 21	Fatal Accidents per 1000 Licenses	Percent Under 21	Fatal Accidents per 1000 Licenses
16	2.801	14	1.443
12	1.405	18	3.614
9	1.433	10	1.926
10	0.039	14	1.643
9	0.338	16	2.943
11	1.849	12	1.913
12	2.246	15	2.814
14	2.855	13	2.634
14	2.352	9	0.926
11	1.294	17	3.256

Managerial Report

1. Develop numerical and graphical summaries of the data.
2. Use regression analysis to investigate the relationship between the number of fatal accidents and the percentage of drivers under the age of 21. Discuss your findings.
3. What conclusion and recommendations can you derive from your analysis?

Case Problem 3 Selecting a Point-and-Shoot Digital Camera

Consumer Reports tested 166 different point-and-shoot digital cameras. Based upon factors such as the number of megapixels, weight (oz.), image quality, and ease of use, they developed an overall score for each camera tested. The overall score ranges from 0 to 100, with higher scores indicating better overall test results. Selecting a camera with many options can be a difficult process, and price is certainly a key issue for most consumers. By spending more, will a consumer really get a superior camera? And, do cameras that have more megapixels, a factor often considered to be a good measure of picture quality, cost more than cameras with fewer megapixels? Table 14.13 shows the brand, average retail price ($), number of megapixels, weight (oz.), and the overall score for 13 Canon and 15 Nikon subcompact cameras tested by *Consumer Reports* (*Consumer Reports* website, February 7, 2012).

Managerial Report

1. Develop numerical summaries of the data.
2. Using overall score as the dependent variable, develop three scatter diagrams, one using price as the independent variable, one using the number of megapixels as the independent variable, and one using weight as the independent variable. Which of the three independent variables appears to be the best predictor of overall score?
3. Using simple linear regression, develop an estimated regression equation that could be used to predict the overall score given the price of the camera. For this estimated regression equation, perform an analysis of the residuals and discuss your findings and conclusions.
4. Analyze the data using only the observations for the Canon cameras. Discuss the appropriateness of using simple linear regression and make any recommendations regarding the prediction of overall score using just the price of the camera.

TABLE 14.13 DATA FOR 28 POINT-AND-SHOOT DIGITAL CAMERAS

WEB file

Cameras

Observation	Brand	Price ($)	Megapixels	Weight (oz.)	Score
1	Canon	330	10	7	66
2	Canon	200	12	5	66
3	Canon	300	12	7	65
4	Canon	200	10	6	62
5	Canon	180	12	5	62
6	Canon	200	12	7	61
7	Canon	200	14	5	60
8	Canon	130	10	7	60
9	Canon	130	12	5	59
10	Canon	110	16	5	55
11	Canon	90	14	5	52
12	Canon	100	10	6	51
13	Canon	90	12	7	46
14	Nikon	270	16	5	65
15	Nikon	300	16	7	63
16	Nikon	200	14	6	61
17	Nikon	400	14	7	59
18	Nikon	120	14	5	57
19	Nikon	170	16	6	56
20	Nikon	150	12	5	56
21	Nikon	230	14	6	55
22	Nikon	180	12	6	53
23	Nikon	130	12	6	53
24	Nikon	80	12	7	52
25	Nikon	80	14	7	50
26	Nikon	100	12	4	46
27	Nikon	110	12	5	45
28	Nikon	130	14	4	42

Case Problem 4 Finding the Best Car Value

When trying to decide what car to buy, real value is not necessarily determined by how much you spend on the initial purchase. Instead, cars that are reliable and don't cost much to own often represent the best values. But, no matter how reliable or inexpensive a car may cost to own, it must also perform well.

To measure value, *Consumer Reports* developed a statistic referred to as a value score. The value score is based upon five-year owner costs, overall road-test scores, and predicted reliability ratings. Five-year owner costs are based on the expenses incurred in the first five years of ownership, including depreciation, fuel, maintenance and repairs, and so on. Using a national average of 12,000 miles per year, an average cost per mile driven is used as the measure of five-year owner costs. Road-test scores are the results of more than 50 tests and evaluations and are based upon a 100-point scale, with higher scores indicating better performance, comfort, convenience, and fuel economy. The highest road-test score obtained in the tests conducted by *Consumer Reports* was a 99 for a Lexus LS 460L. Predicted-reliability ratings (1 = Poor, 2 = Fair, 3 = Good, 4 = Very Good, and 5 = Excellent) are based on data from *Consumer Reports'* Annual Auto Survey.

A car with a value score of 1.0 is considered to be "average-value." A car with a value score of 2.0 is considered to be twice as good a value as a car with a value score of 1.0; a car with a value score of 0.5 is considered half as good as average; and so on. The data for 20 family sedans, including the price ($) of each car tested, follow.

WEB file

FamilySedans

Car	Price ($)	Cost/Mile	Road-Test Score	Predicted Reliability	Value Score
Nissan Altima 2.5 S (4-cyl.)	23,970	0.59	91	4	1.75
Kia Optima LX (2.4)	21,885	0.58	81	4	1.73
Subaru Legacy 2.5i Premium	23,830	0.59	83	4	1.73
Ford Fusion Hybrid	32,360	0.63	84	5	1.70
Honda Accord LX-P (4-cyl.)	23,730	0.56	80	4	1.62
Mazda6 i Sport (4-cyl.)	22,035	0.58	73	4	1.60
Hyundai Sonata GLS (2.4)	21,800	0.56	89	3	1.58
Ford Fusion SE (4-cyl.)	23,625	0.57	76	4	1.55
Chevrolet Malibu LT (4-cyl.)	24,115	0.57	74	3	1.48
Kia Optima SX (2.0T)	29,050	0.72	84	4	1.43
Ford Fusion SEL (V6)	28,400	0.67	80	4	1.42
Nissan Altima 3.5 SR (V6)	30,335	0.69	93	4	1.42
Hyundai Sonata Limited (2.0T)	28,090	0.66	89	3	1.39
Honda Accord EX-L (V6)	28,695	0.67	90	3	1.36
Mazda6 s Grand Touring (V6)	30,790	0.74	81	4	1.34
Ford Fusion SEL (V6, AWD)	30,055	0.71	75	4	1.32
Subaru Legacy 3.6R Limited	30,094	0.71	88	3	1.29
Chevrolet Malibu LTZ (V6)	28,045	0.67	83	3	1.20
Chrysler 200 Limited (V6)	27,825	0.70	52	5	1.20
Chevrolet Impala LT (3.6)	28,995	0.67	63	3	1.05

Managerial Report

1. Develop numerical summaries of the data.
2. Use regression analysis to develop an estimated regression equation that could be used to predict the value score given the price of the car.
3. Use regression analysis to develop an estimated regression equation that could be used to predict the value score given the five-year owner costs (cost/mile).
4. Use regression analysis to develop an estimated regression equation that could be used to predict the value score given the road-test score.
5. Use regression analysis to develop an estimated regression equation that could be used to predict the value score given the predicted-reliability.
6. What conclusions can you derive from your analysis?

Appendix 14.1 Calculus-Based Derivation of Least Squares Formulas

As mentioned in the chapter, the least squares method is a procedure for determining the values of b_0 and b_1 that minimize the sum of squared residuals. The sum of squared residuals is given by

$$\Sigma(y_i - \hat{y}_i)^2$$

Substituting $\hat{y}_i = b_0 + b_1 x_i$, we get

$$\Sigma(y_i - b_0 - b_1 x_i)^2 \tag{14.34}$$

as the expression that must be minimized.

To minimize expression (14.34), we must take the partial derivatives with respect to b_0 and b_1, set them equal to zero, and solve. Doing so, we get

$$\frac{\partial \Sigma(y_i - b_0 - b_1 x_i)^2}{\partial b_0} = -2\Sigma(y_i - b_0 - b_1 x_i) = 0 \tag{14.35}$$

$$\frac{\partial \Sigma(y_i - b_0 - b_1 x_i)^2}{\partial b_1} = -2\Sigma x_i(y_i - b_0 - b_1 x_i) = 0 \tag{14.36}$$

Dividing equation (14.35) by two and summing each term individually yields

$$-\Sigma y_i + \Sigma b_0 + \Sigma b_1 x_i = 0$$

Bringing Σy_i to the other side of the equal sign and noting that $\Sigma b_0 = nb_0$, we obtain

$$nb_0 + (\Sigma x_i)b_1 = \Sigma y_i \tag{14.37}$$

Similar algebraic simplification applied to equation (14.36) yields

$$(\Sigma x_i)b_0 + (\Sigma x_i^2)b_1 = \Sigma x_i y_i \tag{14.38}$$

Equations (14.37) and (14.38) are known as the *normal equations.* Solving equation (14.37) for b_0 yields

$$b_0 = \frac{\Sigma y_i}{n} - b_1 \frac{\Sigma x_i}{n} \tag{14.39}$$

Using equation (14.39) to substitute for b_0 in equation (14.38) provides

$$\frac{\Sigma x_i \Sigma y_i}{n} - \frac{(\Sigma x_i)^2}{n} b_1 + (\Sigma x_i^2)b_1 = \Sigma x_i y_i \tag{14.40}$$

By rearranging the terms in equation (14.40), we obtain

$$b_1 = \frac{\Sigma x_i y_i - (\Sigma x_i \Sigma y_i)/n}{\Sigma x_i^2 - (\Sigma x_i)^2/n} = \frac{\Sigma(x_i - \bar{x})(y_i - \bar{y})}{\Sigma(x_i - \bar{x})^2} \tag{14.41}$$

Because $\bar{y} = \Sigma y_i/n$ and $\bar{x} = \Sigma x_i/n$, we can rewrite equation (14.39) as

$$b_0 = \bar{y} - b_1 \bar{x} \tag{14.42}$$

Equations (14.41) and (14.42) are the formulas (14.6) and (14.7) we used in the chapter to compute the coefficients in the estimated regression equation.

Appendix 14.2 A Test for Significance Using Correlation

Using the sample correlation coefficient r_{xy}, we can determine whether the linear relationship between x and y is significant by testing the following hypotheses about the population correlation coefficient ρ_{xy}.

$$H_0: \rho_{xy} = 0$$
$$H_a: \rho_{xy} \neq 0$$

If H_0 is rejected, we can conclude that the population correlation coefficient is not equal to zero and that the linear relationship between the two variables is significant. This test for significance follows.

A TEST FOR SIGNIFICANCE USING CORRELATION

$$H_0: \rho_{xy} = 0$$
$$H_a: \rho_{xy} \neq 0$$

TEST STATISTIC

$$t = r_{xy}\sqrt{\frac{n-2}{1-r_{xy}^2}} \tag{14.43}$$

REJECTION RULE

p-value approach: Reject H_0 if p-value $\leq \alpha$
Critical value approach: Reject H_0 if $t \leq -t_{\alpha/2}$ or if $t \geq t_{\alpha/2}$

where $t_{\alpha/2}$ is based on a t distribution with $n - 2$ degrees of freedom.

In Section 14.3, we found that the sample with $n = 10$ provided the sample correlation coefficient for student population and quarterly sales of $r_{xy} = .9501$. The test statistic is

$$t = r_{xy}\sqrt{\frac{n-2}{1-r_{xy}^2}} = .9501\sqrt{\frac{10-2}{1-(.9501)^2}} = 8.61$$

The t distribution table shows that with $n - 2 = 10 - 2 = 8$ degrees of freedom, $t = 3.355$ provides an area of .005 in the upper tail. Thus, the area in the upper tail of the t distribution corresponding to the test statistic $t = 8.61$ must be less than .005. Because this test is a two-tailed test, we double this value to conclude that the p-value associated with $t = 8.61$ must be less than $2(.005) = .01$. Excel or Minitab show the p-value $= .000$. Because the p-value is less than $\alpha = .01$, we reject H_0 and conclude that ρ_{xy} is not equal to zero. This evidence is sufficient to conclude that a significant linear relationship exists between student population and quarterly sales.

Note that except for rounding, the test statistic t and the conclusion of a significant relationship are identical to the results obtained in Section 14.5 for the t test conducted using Armand's estimated regression equation $\hat{y} = 60 + 5x$. Performing regression analysis provides the conclusion of a significant relationship between x and y and in addition provides the equation showing how the variables are related. Most analysts therefore use modern computer packages to perform regression analysis and find that using correlation as a test of significance is unnecessary.

Appendix 14.3 Regression Analysis with Minitab

WEB file
Armand's

In Section 14.7 we discussed the computer solution of regression problems by showing Minitab's output for the Armand's Pizza Parlors problem. In this appendix, we describe the steps required to generate the Minitab computer solution. First, the data must be entered in a Minitab worksheet. Student population data are entered in column C1 and quarterly sales data are entered in column C2. The variable names Pop and Sales are entered as the column headings on the worksheet. In subsequent steps, we refer to the data by using the variable names Pop and Sales or the column indicators C1 and C2. The following steps describe how to use Minitab to produce the regression results shown in Figure 14.10.

Step 1. Select the **Stat** menu
Step 2. Select the **Regression** menu
Step 3. Choose **Regression**
Step 4. When the Regression dialog box appears:
  Enter Sales in the **Response** box
  Enter Pop in the **Predictors** box
  Click the **Options** button
 When the Regression-Options dialog box appears:
  Enter 10 in the **Prediction intervals for new observations** box
  Click **OK**
 When the Regression dialog box reappears:
  Click **OK**

The Minitab regression dialog box provides additional capabilities that can be obtained by selecting the desired options. For instance, to obtain a residual plot that shows the predicted value of the dependent variable $\hat{y}$ on the horizontal axis and the standardized residual values on the vertical axis, step 4 would be as follows:

Step 4. When the Regression dialog box appears:
  Enter Sales in the **Response** box
  Enter Pop in the **Predictors** box
  Click the **Graphs** button
 When the Regression-Graphs dialog box appears:
  Select **Standardized** under Residuals for Plots
  Select **Residuals versus fits** under Residual Plots
  Click **OK**
 When the Regression dialog box reappears:
  Click **OK**

Appendix 14.4 Regression Analysis with Excel

WEB file
Armand's

In this appendix we will illustrate how Excel's Regression tool can be used to perform the regression analysis computations for the Armand's Pizza Parlors problem. Refer to Figure 14.23 as we describe the steps involved. The labels Restaurant, Population, and Sales are entered into cells A1:C1 of the worksheet. To identify each of the 10 observations, we entered the numbers 1 through 10 into cells A2:A11. The sample data are entered into cells B2:C11. The following steps describe how to use Excel to produce the regression results.

Step 1. Click the **Data** tab on the Ribbon
Step 2. In the **Analysis** group, click **Data Analysis**
Step 3. Choose **Regression** from the list of Analysis Tools
Step 4. Click **OK**

FIGURE 14.23 EXCEL SOLUTION TO THE ARMAND'S PIZZA PARLORS PROBLEM

A1 fx Restaurant

	A	B	C	D	E	F	G	H	I	J
1	**Restaurant**	**Population**	**Sales**							
2	1	2	58							
3	2	6	105							
4	3	8	88							
5	4	8	118							
6	5	12	117							
7	6	16	137							
8	7	20	157							
9	8	20	169							
10	9	22	149							
11	10	26	202							
12										
13	SUMMARY OUTPUT									
14										
15	*Regression Statistics*									
16	Multiple R	0.9501								
17	R Square	0.9027								
18	Adjusted R Square	0.8906								
19	Standard Error	13.8293								
20	Observations	10								
21										
22	ANOVA									
23		*df*	*SS*	*MS*	*F*	*Significance F*				
24	Regression	1	14200	14200	74.2484	2.55E-05				
25	Residual	8	1530	191.25						
26	Total	9	15730							
27										
28		*Coefficients*	*Standard Error*	*t Stat*	*P-value*	*Lower 95%*	*Upper 95%*	*Lower 99.0%*	*Upper 99.0%*	
29	Intercept	60	9.2260	6.5033	0.0002	38.7247	81.2753	29.0431	90.9569	
30	Population	5	0.5803	8.6167	2.55E-05	3.6619	6.3381	3.0530	6.9470	
31										

Step 5. When the Regression dialog box appears:

Enter C1:C11 in the **Input Y Range** box
Enter B1:B11 in the **Input X Range** box
Select **Labels**
Select **Confidence Level**
Enter 99 in the **Confidence Level** box
Select **Output Range**
Enter A13 in the **Output Range** box
(Any upper left-hand corner cell indicating where the output is to begin may be entered here.)
Click **OK**

The first section of the output, entitled *Regression Statistics,* contains summary statistics such as the coefficient of determination (R Square). The second section of the output, titled ANOVA, contains the analysis of variance table. The last section of the output, which is not titled, contains the estimated regression coefficients and related information. We will begin our discussion of the interpretation of the regression output with the information contained in cells A28:I30.

Interpretation of Estimated Regression Equation Output

The y intercept of the estimated regression line, $b_0 = 60$, is shown in cell B29, and the slope of the estimated regression line, $b_1 = 5$, is shown in cell B30. The label Intercept in cell A29 and the label Population in cell A30 are used to identify these two values.

In Section 14.5 we showed that the estimated standard deviation of b_1 is $s_{b_1} = .5803$. Note that the value in cell C30 is .5803. The label Standard Error in cell C28 is Excel's way of indicating that the value in cell C30 is the standard error, or standard deviation, of b_1. Recall that the t test for a significant relationship required the computation of the t statistic, $t = b_1/s_{b_1}$. For the Armand's data, the value of t that we computed was $t = 5/.5803 = 8.62$. The label in cell D28, *t Stat,* reminds us that cell D30 contains the value of the t test statistic.

The value in cell E30 is the p-value associated with the t test for significance. Excel has displayed the p-value in cell E30 using scientific notation. To obtain the decimal value, we move the decimal point 5 places to the left, obtaining a value of .0000255. Because the p-value $= .0000255 < \alpha = .01$, we can reject H_0 and conclude that we have a significant relationship between student population and quarterly sales.

The information in cells F28:I30 can be used to develop confidence interval estimates of the y intercept and slope of the estimated regression equation. Excel always provides the lower and upper limits for a 95% confidence interval. Recall that in step 4 we selected Confidence Level and entered 99 in the Confidence Level box. As a result, Excel's Regression tool also provides the lower and upper limits for a 99% confidence interval. The value in cell H30 is the lower limit for the 99% confidence interval estimate of β_1 and the value in cell I30 is the upper limit. Thus, after rounding, the 99% confidence interval estimate of β_1 is 3.05 to 6.95. The values in cells F30 and G30 provide the lower and upper limits for the 95% confidence interval. Thus, the 95% confidence interval is 3.66 to 6.34.

Interpretation of ANOVA Output

The information in cells A22:F26 is a summary of the analysis of variance computations. The three sources of variation are labeled Regression, Residual, and Total. The label *df* in cell B23 stands for degrees of freedom, the label *SS* in cell C23 stands for sum of squares, and the label *MS* in cell D23 stands for mean square.

In Section 14.5 we stated that the mean square error, obtained by dividing the error or residual sum of squares by its degrees of freedom, provides an estimate of σ^2. The value in cell D25, 191.25, is the mean square error for the Armand's regression output. In Section 14.5 we showed that an F test could also be used to test for significance in regression. The value in cell F24, .0000255, is the p-value associated with the F test for significance. Because the p-value $= .0000255 < \alpha = .01$, we can reject H_0 and conclude that we have a significant relationship between student population and quarterly sales. The label Excel uses to identify the p-value for the F test for significance, shown in cell F23, is *Significance F.*

The label Significance F may be more meaningful if you think of the value in cell F24 as the observed level of significance for the F test.

Interpretation of Regression Statistics Output

The coefficient of determination, .9027, appears in cell B17; the corresponding label, R Square, is shown in cell A17. The square root of the coefficient of determination provides the sample correlation coefficient of .9501 shown in cell B16. Note that Excel uses the label Multiple R (cell A16) to identify this value. In cell A19, the label Standard Error is used to identify the value of the standard error of the estimate shown in cell B19. Thus, the standard error of the estimate is 13.8293. We caution the reader to keep in mind that in the Excel output, the label Standard Error appears in two different places. In the Regression Statistics section of the output, the label Standard Error refers to the estimate of σ. In the Estimated Regression Equation section of the output, the label *Standard Error* refers to s_{b_1}, the standard deviation of the sampling distribution of b_1.

Appendix 14.5 Regression Analysis Using StatTools

WEB file
Armand's

In this appendix we show how StatTools can be used to perform the regression analysis computations for the Armand's Pizza Parlors problem. Begin by using the Data Set Manager to create a StatTools data set for these data using the procedure described in the appendix in Chapter 1. The following steps describe how StatTools can be used to provide the regression results.

Step 1. Click the **StatTools** tab on the Ribbon
Step 2. In the **Analyses** group, click **Regression and Classification**
Step 3. Choose the **Regression** option
Step 4. When the StatTools-Regression dialog box appears:
  Select **Multiple** in the **Regression Type** box
  In the **Variables** section,
    Click the **Format button** and select **Unstacked**
    In the column labeled **I** select **Population**
    In the column labeled **D** select **Sales**
  Click **OK**

The regression analysis output will appear.

Note that in step 4 we selected Multiple in the Regression Type box. In StatTools, the Multiple option is used for both simple linear regression and multiple regression. The StatTools-Regression dialog box contains a number of more advanced options for developing prediction interval estimates and producing residual plots. The StatTools Help facility provides information on using all of these options.

CHAPTER 15

Multiple Regression

CONTENTS

STATISTICS *in* PRACTICE

dunnhumby*
LONDON, ENGLAND

Founded in 1989 by the husband-and-wife team of Clive Humby (a mathematician) and Edwina Dunn (a marketer), dunnhumby combines proven natural abilities with big ideas to find clues and patterns as to what customers are buying and why. The company turns these insights into actionable strategies that create dramatic growth and sustainable loyalty, ultimately improving brand value and the customer experience.

Employing more than 950 people in Europe, Asia, and the Americas, dunnhumby serves a prestigious list of companies, including Kroger, Tesco, Coca-Cola, General Mills, Kimberly-Clark, PepsiCo, Procter & Gamble, and Home Depot. dunnhumbyUSA is a joint venture between the Kroger Company and dunnhumby and has offices in New York, Chicago, Atlanta, Minneapolis, Cincinnati, and Portland.

The company's research begins with data collected about a client's customers. Data come from customer reward or discount card purchase records, electronic point-of-sale transactions, and traditional market research. Analysis of the data often translates billions of data points into detailed insights about the behavior, preferences, and lifestyles of the customers. Such insights allow for more effective merchandising programs to be activated, including strategy recommendations on pricing, promotion, advertising, and product assortment decisions.

dunnhumby uses logistic regression to predict customer shopping behavior. © Micro 10x/ Shutterstock.com.

Researchers have used a multiple regression technique referred to as logistic regression to help in their analysis of customer-based data. Using logistic regression, an estimated multiple regression equation of the following form is developed.

$$\hat{y} = b_0 + b_1x_1 + b_2x_2 + b_3x_3 + \cdots + b_px_p$$

The dependent variable $\hat{y}$ is a prediction of the probability that a customer belongs to a particular customer group. The independent variables $x_1, x_2, x_3, \ldots, x_p$ are measures of the customer's actual shopping behavior and may include the specific items purchased, number of items purchased, amount purchased, day of the week, hour of the day, and so on. The analysis helps identify the independent variables that are most relevant in predicting the customer's group and provides a better understanding of the customer population, enabling further analysis with far greater confidence. The focus of the analysis is on understanding the customer to the point of developing merchandising, marketing, and direct marketing programs that will maximize the relevancy and service to the customer group.

In this chapter, we will introduce multiple regression and show how the concepts of simple linear regression introduced in Chapter 14 can be extended to the multiple regression case. In addition, we will show how computer software packages are used for multiple regression. In the final section of the chapter we introduce logistic regression using an example that illustrates how the technique is used in a marketing research application.

*The authors are indebted to Paul Hunter, Senior Vice President of Solutions for dunnhumby for providing this Statistics in Practice.

In Chapter 14 we presented simple linear regression and demonstrated its use in developing an estimated regression equation that describes the relationship between two variables. Recall that the variable being predicted or explained is called the dependent variable and the variable being used to predict or explain the dependent variable is called the independent variable. In this chapter we continue our study of regression analysis by considering situations involving two or more independent variables. This subject area, called **multiple regression analysis**, enables us to consider more factors and thus obtain better predictions than are possible with simple linear regression.

15.1 Multiple Regression Model

Multiple regression analysis is the study of how a dependent variable y is related to two or more independent variables. In the general case, we will use p to denote the number of independent variables.

Regression Model and Regression Equation

The concepts of a regression model and a regression equation introduced in the preceding chapter are applicable in the multiple regression case. The equation that describes how the dependent variable y is related to the independent variables $x_1, x_2, \ldots, x_p$ and an error term is called the **multiple regression model**. We begin with the assumption that the multiple regression model takes the following form.

MULTIPLE REGRESSION MODEL

$$y = \beta_0 + \beta_1 x_1 + \beta_2 x_2 + \cdots + \beta_p x_p + \epsilon \qquad \textbf{(15.1)}$$

In the multiple regression model, $\beta_0, \beta_1, \beta_2, \ldots, \beta_p$ are the parameters and the error term ϵ (the Greek letter epsilon) is a random variable. A close examination of this model reveals that y is a linear function of $x_1, x_2, \ldots, x_p$ (the $\beta_0 + \beta_1 x_1 + \beta_2 x_2 + \cdots + \beta_p x_p$ part) plus the error term ϵ. The error term accounts for the variability in y that cannot be explained by the linear effect of the p independent variables.

In Section 15.4 we will discuss the assumptions for the multiple regression model and ϵ. One of the assumptions is that the mean or expected value of ϵ is zero. A consequence of this assumption is that the mean or expected value of y, denoted $E(y)$, is equal to $\beta_0 + \beta_1 x_1 + \beta_2 x_2 + \cdots + \beta_p x_p$. The equation that describes how the mean value of y is related to $x_1, x_2, \ldots, x_p$ is called the **multiple regression equation**.

MULTIPLE REGRESSION EQUATION

$$E(y) = \beta_0 + \beta_1 x_1 + \beta_2 x_2 + \cdots + \beta_p x_p \qquad \textbf{(15.2)}$$

Estimated Multiple Regression Equation

If the values of $\beta_0, \beta_1, \beta_2, \ldots, \beta_p$ were known, equation (15.2) could be used to compute the mean value of y at given values of $x_1, x_2, \ldots, x_p$. Unfortunately, these parameter values will not, in general, be known and must be estimated from sample data. A simple random sample is used to compute sample statistics $b_0, b_1, b_2, \ldots, b_p$ that are used as the point

FIGURE 15.1 THE ESTIMATION PROCESS FOR MULTIPLE REGRESSION

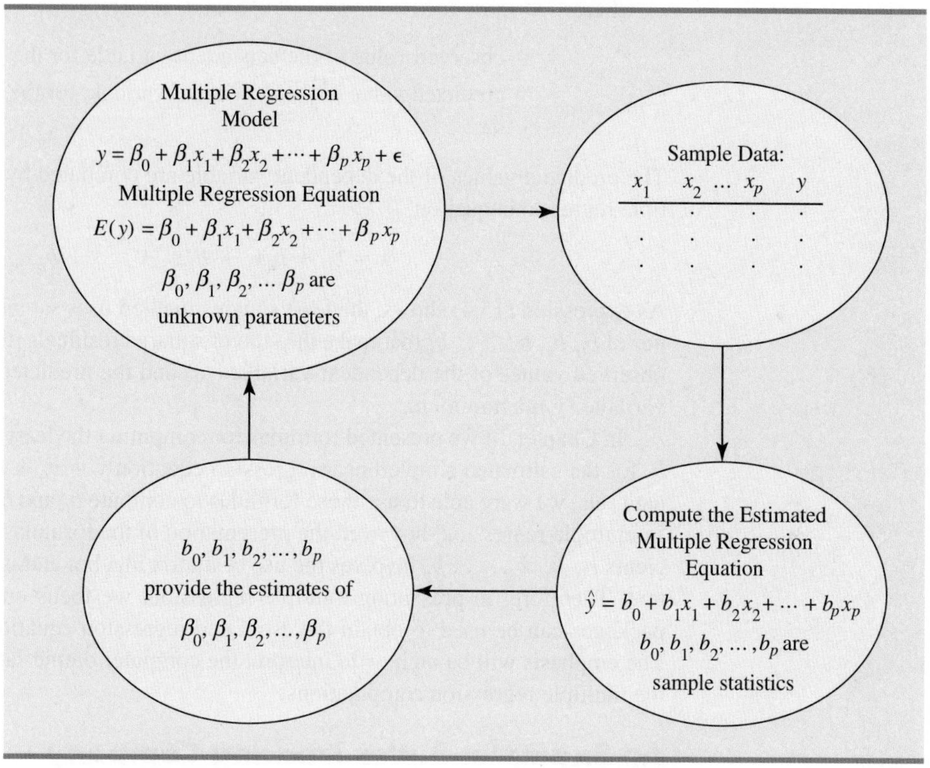

In simple linear regression, b_0 and b_1 were the sample statistics used to estimate the parameters β_0 and β_1. Multiple regression parallels this statistical inference process, with b_0, b_1, b_2, . . . , b_p denoting the sample statistics used to estimate the parameters β_0, β_1, β_2, . . . , β_p.

estimators of the parameters $\beta_0, \beta_1, \beta_2, \ldots, \beta_p$. These sample statistics provide the following **estimated multiple regression equation**.

ESTIMATED MULTIPLE REGRESSION EQUATION

$$\hat{y} = b_0 + b_1x_1 + b_2x_2 + \cdots + b_px_p \tag{15.3}$$

where

$b_0, b_1, b_2, \ldots, b_p$ are the estimates of $\beta_0, \beta_1, \beta_2, \ldots, \beta_p$

$\hat{y}$ = predicted value of the dependent variable

The estimation process for multiple regression is shown in Figure 15.1.

15.2 Least Squares Method

In Chapter 14, we used the **least squares method** to develop the estimated regression equation that best approximated the straight-line relationship between the dependent and independent variables. This same approach is used to develop the estimated multiple regression equation. The least squares criterion is restated as follows.

LEAST SQUARES CRITERION

$$\min \Sigma(y_i - \hat{y}_i)^2 \tag{15.4}$$

where

y_i = observed value of the dependent variable for the ith observation
$\hat{y}_i$ = predicted value of the dependent variable for the ith observation

The predicted values of the dependent variable are computed by using the estimated multiple regression equation,

$$\hat{y} = b_0 + b_1x_1 + b_2x_2 + \cdots + b_px_p$$

As expression (15.4) shows, the least squares method uses sample data to provide the values of $b_0, b_1, b_2, \ldots, b_p$ that make the sum of squared residuals [the deviations between the observed values of the dependent variable (y_i) and the predicted values of the dependent variable ($\hat{y}_i$)] a minimum.

In Chapter 14 we presented formulas for computing the least squares estimators b_0 and b_1 for the estimated simple linear regression equation $\hat{y} = b_0 + b_1x$. With relatively small data sets, we were able to use those formulas to compute b_0 and b_1 by manual calculations. In multiple regression, however, the presentation of the formulas for the regression coefficients $b_0, b_1, b_2, \ldots, b_p$ involves the use of matrix algebra and is beyond the scope of this text. Therefore, in presenting multiple regression, we focus on how computer software packages can be used to obtain the estimated regression equation and other information. The emphasis will be on how to interpret the computer output rather than on how to make the multiple regression computations.

An Example: Butler Trucking Company

As an illustration of multiple regression analysis, we will consider a problem faced by the Butler Trucking Company, an independent trucking company in southern California. A major portion of Butler's business involves deliveries throughout its local area. To develop better work schedules, the managers want to predict the total daily travel time for their drivers.

Initially the managers believed that the total daily travel time would be closely related to the number of miles traveled in making the daily deliveries. A simple random sample of 10 driving assignments provided the data shown in Table 15.1 and the scatter diagram shown in Figure 15.2. After reviewing this scatter diagram, the managers hypothesized that the simple linear regression model $y = \beta_0 + \beta_1x_1 + \epsilon$ could be used to describe the relationship between the total travel time (y) and the number of miles traveled (x_1). To estimate

TABLE 15.1 PRELIMINARY DATA FOR BUTLER TRUCKING

WEB file

Butler

Driving Assignment	x_1 = Miles Traveled	y = Travel Time (hours)
1	100	9.3
2	50	4.8
3	100	8.9
4	100	6.5
5	50	4.2
6	80	6.2
7	75	7.4
8	65	6.0
9	90	7.6
10	90	6.1

FIGURE 15.2 SCATTER DIAGRAM OF PRELIMINARY DATA FOR BUTLER TRUCKING

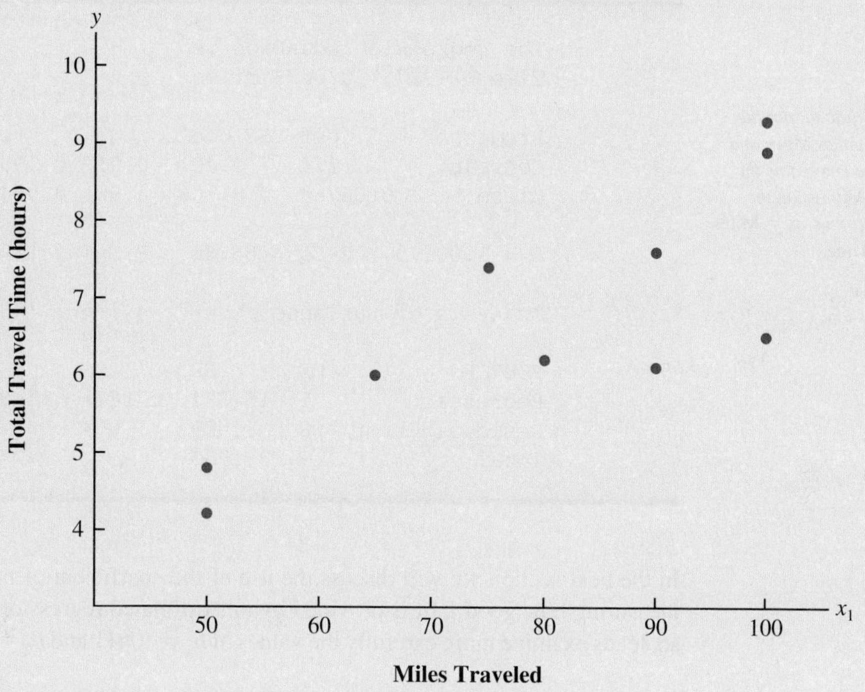

the parameters β_0 and β_1, the least squares method was used to develop the estimated regression equation.

$$\hat{y} = b_0 + b_1x_1 \tag{15.5}$$

In Figure 15.3, we show the Minitab computer output from applying simple linear regression to the data in Table 15.1. The estimated regression equation is

$$\hat{y} = 1.27 + .0678x_1$$

At the .05 level of significance, the F value of 15.81 and its corresponding p-value of .004 indicate that the relationship is significant; that is, we can reject H_0: $\beta_1 = 0$ because the p-value is less than $\alpha = .05$. Note that the same conclusion is obtained from the t value of 3.98 and its associated p-value of .004. Thus, we can conclude that the relationship between the total travel time and the number of miles traveled is significant; longer travel times are associated with more miles traveled. With a coefficient of determination (expressed as a percentage) of R-Sq = 66.4%, we see that 66.4% of the variability in travel time can be explained by the linear effect of the number of miles traveled. This finding is fairly good, but the managers might want to consider adding a second independent variable to explain some of the remaining variability in the dependent variable.

The Minitab steps necessary to generate the output shown in Figure 15.4 are given in Appendix 15.1.

In attempting to identify another independent variable, the managers felt that the number of deliveries could also contribute to the total travel time. The Butler Trucking data, with the number of deliveries added, are shown in Table 15.2. The Minitab computer solution with both miles traveled (x_1) and number of deliveries (x_2) as independent variables is shown in Figure 15.4. The estimated regression equation is

$$\hat{y} = -.869 + .0611x_1 + .923x_2 \tag{15.6}$$

FIGURE 15.3 MINITAB OUTPUT FOR BUTLER TRUCKING WITH ONE INDEPENDENT VARIABLE

```
The regression equation is
Time = 1.27 + 0.0678 Miles

Predictor      Coef   SE Coef     T      p
Constant      1.274     1.401  0.91  0.390
Miles       0.06783   0.01706  3.98  0.004

S = 1.00179   R-Sq = 66.4%   R-Sq(adj) = 62.2%

Analysis of Variance

SOURCE           DF      SS      MS      F      p
Regression        1  15.871  15.871  15.81  0.004
Residual Error    8   8.029   1.004
Total             9  23.900
```

In the Minitab output the variable names Miles *and* Time *were entered as the column headings on the worksheet; thus,* x_1 = Miles *and* y = Time.

In the next section we will discuss the use of the coefficient of multiple determination in measuring how good a fit is provided by this estimated regression equation. Before doing so, let us examine more carefully the values of $b_1 = .0611$ and $b_2 = .923$ in equation (15.6).

Note on Interpretation of Coefficients

One observation can be made at this point about the relationship between the estimated regression equation with only the miles traveled as an independent variable and the equation that includes the number of deliveries as a second independent variable. The value of b_1 is not the same in both cases. In simple linear regression, we interpret b_1 as an estimate of the change in y for a one-unit change in the independent variable. In multiple regression analysis, this interpretation must be modified somewhat. That is, in multiple regression analysis, we interpret each regression coefficient as follows: b_i represents an estimate of the change in y corresponding to a one-unit change in x_i when all other independent variables are held constant. In the Butler Trucking example involving two independent variables, $b_1 = .0611$. Thus,

TABLE 15.2 DATA FOR BUTLER TRUCKING WITH MILES TRAVELED (x_1) AND NUMBER OF DELIVERIES (x_2) AS THE INDEPENDENT VARIABLES

WEB file Butler

Driving Assignment	x_1 = Miles Traveled	x_2 = Number of Deliveries	y = Travel Time (hours)
1	100	4	9.3
2	50	3	4.8
3	100	4	8.9
4	100	2	6.5
5	50	2	4.2
6	80	2	6.2
7	75	3	7.4
8	65	4	6.0
9	90	3	7.6
10	90	2	6.1

FIGURE 15.4 MINITAB OUTPUT FOR BUTLER TRUCKING WITH TWO INDEPENDENT VARIABLES

```
The regression equation is
Time = - 0.869 + 0.0611 Miles + 0.923 Deliveries

Predictor          Coef    SE Coef      T      p
Constant        -0.8687     0.9515  -0.91  0.392
Miles          0.061135   0.009888   6.18  0.000
Deliveries       0.9234     0.2211   4.18  0.004

S = 0.573142   R-Sq = 90.4%   R-Sq(adj) = 87.6%

Analysis of Variance

SOURCE           DF      SS      MS      F      p
Regression        2  21.601  10.800  32.88  0.000
Residual Error    7   2.299   0.328
Total             9  23.900
```

In the Minitab output the variable names Miles, Deliveries, *and* Time *were entered as the column headings on the worksheet; thus,* x_1 = Miles, x_2 = Deliveries, *and* y = Time.

.0611 hours is an estimate of the expected increase in travel time corresponding to an increase of one mile in the distance traveled when the number of deliveries is held constant. Similarly, because $b_2 = .923$, an estimate of the expected increase in travel time corresponding to an increase of one delivery when the number of miles traveled is held constant is .923 hours.

Exercises

Note to student: The exercises involving data in this and subsequent sections were designed to be solved using a computer software package.

Methods

1. The estimated regression equation for a model involving two independent variables and 10 observations follows.

$$\hat{y} = 29.1270 + .5906x_1 + .4980x_2$$

   a. Interpret b_1 and b_2 in this estimated regression equation.
   b. Predict y when $x_1 = 180$ and $x_2 = 310$.

SELF test

2. Consider the following data for a dependent variable y and two independent variables, x_1 and x_2.

WEB file

Exer2

x_1	x_2	y
30	12	94
47	10	108
25	17	112
51	16	178
40	5	94
51	19	175
74	7	170
36	12	117
59	13	142
76	16	211

a. Develop an estimated regression equation relating y to x_1. Predict y if $x_1 = 45$.
b. Develop an estimated regression equation relating y to x_2. Predict y if $x_2 = 15$.
c. Develop an estimated regression equation relating y to x_1 and x_2. Predict y if $x_1 = 45$ and $x_2 = 15$.

3. In a regression analysis involving 30 observations, the following estimated regression equation was obtained.

$$\hat{y} = 17.6 + 3.8x_1 - 2.3x_2 + 7.6x_3 + 2.7x_4$$

a. Interpret b_1, b_2, b_3, and b_4 in this estimated regression equation.
b. Predict y when $x_1 = 10$, $x_2 = 5$, $x_3 = 1$, and $x_4 = 2$.

Applications

4. A shoe store developed the following estimated regression equation relating sales to inventory investment and advertising expenditures.

$$\hat{y} = 25 + 10x_1 + 8x_2$$

where

$$x_1 = \text{inventory investment (\$1000s)}$$
$$x_2 = \text{advertising expenditures (\$1000s)}$$
$$y = \text{sales (\$1000s)}$$

a. Predict the sales resulting from a \$15,000 investment in inventory and an advertising budget of \$10,000.
b. Interpret b_1 and b_2 in this estimated regression equation.

SELF test

5. The owner of Showtime Movie Theaters, Inc., would like to predict weekly gross revenue as a function of advertising expenditures. Historical data for a sample of eight weeks follow.

WEB file Showtime

Weekly Gross Revenue (\$1000s)	Television Advertising (\$1000s)	Newspaper Advertising (\$1000s)
96	5.0	1.5
90	2.0	2.0
95	4.0	1.5
92	2.5	2.5
95	3.0	3.3
94	3.5	2.3
94	2.5	4.2
94	3.0	2.5

a. Develop an estimated regression equation with the amount of television advertising as the independent variable.
b. Develop an estimated regression equation with both television advertising and newspaper advertising as the independent variables.
c. Is the estimated regression equation coefficient for television advertising expenditures the same in part (a) and in part (b)? Interpret the coefficient in each case.
d. Predict weekly gross revenue for a week when \$3500 is spent on television advertising and \$1800 is spent on newspaper advertising?

6. The National Football League (NFL) records a variety of performance data for individuals and teams. To investigate the importance of passing on the percentage of games won by a team, the following data show the conference (Conf), average number of passing yards per

attempt (Yds/Att), the number of interceptions thrown per attempt (Int/Att), and the percentage of games won (Win%) for a random sample of 16 NFL teams for the 2011 season (NFL website, February 12, 2012).

WEB file
NFLPassing

Team	Conf	Yds/Att	Int/Att	Win%
Arizona Cardinals	NFC	6.5	.042	50.0
Atlanta Falcons	NFC	7.1	.022	62.5
Carolina Panthers	NFC	7.4	.033	37.5
Cincinnati Bengals	AFC	6.2	.026	56.3
Detroit Lions	NFC	7.2	.024	62.5
Green Bay Packers	NFC	8.9	.014	93.8
Houstan Texans	AFC	7.5	.019	62.5
Indianapolis Colts	AFC	5.6	.026	12.5
Jacksonville Jaguars	AFC	4.6	.032	31.3
Minnesota Vikings	NFC	5.8	.033	18.8
New England Patriots	AFC	8.3	.020	81.3
New Orleans Saints	NFC	8.1	.021	81.3
Oakland Raiders	AFC	7.6	.044	50.0
San Francisco 49ers	NFC	6.5	.011	81.3
Tennessee Titans	AFC	6.7	.024	56.3
Washington Redskins	NFC	6.4	.041	31.3

a. Develop the estimated regression equation that could be used to predict the percentage of games won given the average number of passing yards per attempt.
b. Develop the estimated regression equation that could be used to predict the percentage of games won given the number of interceptions thrown per attempt.
c. Develop the estimated regression equation that could be used to predict the percentage of games won given the average number of passing yards per attempt and the number of interceptions thrown per attempt.
d. The average number of passing yards per attempt for the Kansas City Chiefs was 6.2 and the number of interceptions thrown per attempt was .036. Use the estimated regression equation developed in part (c) to predict the percentage of games won by the Kansas City Chiefs. (*Note:* For the 2011 season the Kansas City Chiefs' record was 7 wins and 9 losses.) Compare your prediction to the actual percentage of games won by the Kansas City Chiefs.

7. *PC World* rated four component characteristics for 10 ultraportable laptop computers: features; performance; design; and price. Each characteristic was rated using a 0–100 point scale. An overall rating, referred to as the *PCW World* Rating, was then developed for each laptop. The following table shows the performance rating, features rating, and the *PCW World* Rating for the 10 laptop computers (*PC World* website, February 5, 2009).

WEB file
Laptop

Model	Performance	Features	PCW Rating
Thinkpad X200	77	87	83
VGN-Z598U	97	85	82
U6V	83	80	81
Elitebook 2530P	77	75	78
X360	64	80	78
Thinkpad X300	56	76	78
Ideapad U110	55	81	77
Micro Express JFT2500	76	73	75
Toughbook W7	46	79	73
HP Voodoo Envy133	54	68	72

a. Determine the estimated regression equation that can be used to predict the *PCW World* Rating using the performance rating as the independent variable.
b. Determine the estimated regression equation that can be used to predict the *PCW World* Rating using both the performance rating and the features rating.
c. Predict the *PCW World* Rating for a laptop computer that has a performance rating of 80 and a features rating of 70.

8. The *Condé Nast Traveler* Gold List for 2012 provided ratings for the top 20 small cruise ships (*Condé Nast Traveler* website, March 1, 2012). The data shown below are the scores each ship received based upon the results from *Condé Nast Traveler*'s annual Readers' Choice Survey. Each score represents the percentage of respondents who rated a ship as excellent or very good on several criteria, including Shore Excursions and Food/Dining. An overall score was also reported and used to rank the ships. The highest ranked ship, the *Seabourn Odyssey,* has an overall score of 94.4, the highest component of which is 97.8 for Food/Dining.

WEB file
Ships

Ship	Overall	Shore Excursions	Food/Dining
Seabourn Odyssey	94.4	90.9	97.8
Seabourn Pride	93.0	84.2	96.7
National Geographic Endeavor	92.9	100.0	88.5
Seabourn Sojourn	91.3	94.8	97.1
Paul Gauguin	90.5	87.9	91.2
Seabourn Legend	90.3	82.1	98.8
Seabourn Spirit	90.2	86.3	92.0
Silver Explorer	89.9	92.6	88.9
Silver Spirit	89.4	85.9	90.8
Seven Seas Navigator	89.2	83.3	90.5
Silver Whisperer	89.2	82.0	88.6
National Geographic Explorer	89.1	93.1	89.7
Silver Cloud	88.7	78.3	91.3
Celebrity Xpedition	87.2	91.7	73.6
Silver Shadow	87.2	75.0	89.7
Silver Wind	86.6	78.1	91.6
SeaDream II	86.2	77.4	90.9
Wind Star	86.1	76.5	91.5
Wind Surf	86.1	72.3	89.3
Wind Spirit	85.2	77.4	91.9

a. Determine an estimated regression equation that can be used to predict the overall score given the score for Shore Excursions.
b. Consider the addition of the independent variable Food/Dining. Develop the estimated regression equation that can be used to predict the overall score given the scores for Shore Excursions and Food/Dining.
c. Predict the overall score for a cruise ship with a Shore Excursions score of 80 and a Food/Dining Score of 90.

9. Waterskiing and wakeboarding are two popular water-sports. Finding a model that best suits your intended needs, whether it is waterskiing, wakeboading, or general boating, can be a difficult task. *WaterSki* magazine did extensive testing for 88 boats and provided a wide variety of information to help consumers select the best boat. A portion of the data they reported for 20 boats with a length of between 20 and 22 feet follows (*WaterSki,* January/February 2006). Beam is the maximum width of the boat in inches, HP is the horsepower of the boat's engine, and TopSpeed is the top speed in miles per hour (mph).

WEB file
Boats

Make and Model	Beam	HP	TopSpeed
Calabria Cal Air Pro V-2	100.0	330	45.3
Correct Craft Air Nautique 210	91.0	330	47.3
Correct Craft Air Nautique SV-211	93.0	375	46.9
Correct Craft Ski Nautique 206 Limited	91.0	330	46.7
Gekko GTR 22	96.0	375	50.1
Gekko GTS 20	83.0	375	52.2
Malibu Response LXi	93.5	340	47.2
Malibu Sunsettter LXi	98.0	400	46.0
Malibu Sunsetter 21 XTi	98.0	340	44.0
Malibu Sunscape 21 LSV	98.0	400	47.5
Malibu Wakesetter 21 XTi	98.0	340	44.9
Malibu Wakesetter VLX	98.0	400	47.3
Malibu vRide	93.5	340	44.5
Malibu Ride XTi	93.5	320	44.5
Mastercraft ProStar 209	96.0	350	42.5
Mastercraft X-1	90.0	310	45.8
Mastercraft X-2	94.0	310	42.8
Mastercraft X-9	96.0	350	43.2
MB Sports 190 Plus	92.0	330	45.3
Svfara SVONE	91.0	330	47.7

a. Using these data, develop an estimated regression equation relating the top speed with the boat's beam and horsepower rating.

b. The Svfara SV609 has a beam of 85 inches and an engine with a 330 horsepower rating. Use the estimated regression equation developed in part (a) to predict the top speed for the Svfara SV609.

10. Major League Baseball (MLB) consists of teams that play in the American League and the National League. MLB collects a wide variety of team and player statistics. Some of the statistics often used to evaluate pitching performance are as follows:

ERA: The average number of earned runs given up by the pitcher per nine innings. An earned run is any run that the opponent scores off a particular pitcher except for runs scored as a result of errors.

SO/IP: The average number of strikeouts per inning pitched.

HR/IP: The average number of home runs per inning pitched.

R/IP: The number of runs given up per inning pitched.

The following data show values for these statistics for a random sample of 20 pitchers from the American League for the 2011 season (MLB website, March 1, 2012).

WEB file
MLBPitching

Player	Team	W	L	ERA	SO/IP	HR/IP	R/IP
Verlander, J	DET	24	5	2.40	1.00	.10	.29
Beckett, J	BOS	13	7	2.89	.91	.11	.34
Wilson, C	TEX	16	7	2.94	.92	.07	.40
Sabathia, C	NYY	19	8	3.00	.97	.07	.37
Haren, D	LAA	16	10	3.17	.81	.08	.38
McCarthy, B	OAK	9	9	3.32	.72	.06	.43
Santana, E	LAA	11	12	3.38	.78	.11	.42

(continued)

Player	Team	W	L	ERA	SO/IP	HR/IP	R/IP
Lester, J	BOS	15	9	3.47	.95	.10	.40
Hernandez, F	SEA	14	14	3.47	.95	.08	.42
Buehrle, M	CWS	13	9	3.59	.53	.10	.45
Pineda, M	SEA	9	10	3.74	1.01	.11	.44
Colon, B	NYY	8	10	4.00	.82	.13	.52
Tomlin, J	CLE	12	7	4.25	.54	.15	.48
Pavano, C	MIN	9	13	4.30	.46	.10	.55
Danks, J	CWS	8	12	4.33	.79	.11	.52
Guthrie, J	BAL	9	17	4.33	.63	.13	.54
Lewis, C	TEX	14	10	4.40	.84	.17	.51
Scherzer, M	DET	15	9	4.43	.89	.15	.52
Davis, W	TB	11	10	4.45	.57	.13	.52
Porcello, R	DET	14	9	4.75	.57	.10	.57

a. Develop an estimated regression equation that can be used to predict the average number of runs given up per inning given the average number of strikeouts per inning pitched.
b. Develop an estimated regression equation that can be used to predict the average number of runs given up per inning given the average number of home runs per inning pitched.
c. Develop an estimated regression equation that can be used to predict the average number of runs given up per inning given the average number of strikeouts per inning pitched and the average number of home runs per inning pitched.
d. A. J. Burnett, a pitcher for the New York Yankees, had an average number of strikeouts per inning pitched of .91 and an average number of home runs per inning of .16. Use the estimated regression equation developed in part (c) to predict the average number of runs given up per inning for A. J. Burnett. (*Note:* The actual value for R/IP was .6.)
e. Suppose a suggestion was made to also use the earned run average as another independent variable in part (c). What do you think of this suggestion?

15.3 Multiple Coefficient of Determination

In simple linear regression we showed that the total sum of squares can be partitioned into two components: the sum of squares due to regression and the sum of squares due to error. The same procedure applies to the sum of squares in multiple regression.

RELATIONSHIP AMONG SST, SSR, AND SSE

$$\text{SST} = \text{SSR} + \text{SSE} \tag{15.7}$$

where

$$\text{SST} = \text{total sum of squares} = \Sigma(y_i - \bar{y})^2$$
$$\text{SSR} = \text{sum of squares due to regression} = \Sigma(\hat{y}_i - \bar{y})^2$$
$$\text{SSE} = \text{sum of squares due to error} = \Sigma(y_i - \hat{y}_i)^2$$

Because of the computational difficulty in computing the three sums of squares, we rely on computer packages to determine those values. The analysis of variance part of the Minitab output in Figure 15.4 shows the three values for the Butler Trucking problem with two independent variables: SST = 23.900, SSR = 21.601, and SSE = 2.299. With only one independent variable (number of miles traveled), the Minitab output in Figure 15.3 shows that SST = 23.900, SSR = 15.871, and SSE = 8.029. The value of SST is the same in both cases because it does not depend on $\hat{y}$, but SSR increases and SSE decreases when a second independent variable (number of deliveries) is added. The implication is that the estimated multiple regression equation provides a better fit for the observed data.

In Chapter 14, we used the coefficient of determination, r^2 = SSR/SST, to measure the goodness of fit for the estimated regression equation. The same concept applies to multiple regression. The term **multiple coefficient of determination** indicates that we are measuring the goodness of fit for the estimated multiple regression equation. The multiple coefficient of determination, denoted R^2, is computed as follows.

MULTIPLE COEFFICIENT OF DETERMINATION

$$R^2 = \frac{\text{SSR}}{\text{SST}} \tag{15.8}$$

The multiple coefficient of determination can be interpreted as the proportion of the variability in the dependent variable that can be explained by the estimated multiple regression equation. Hence, when multiplied by 100, it can be interpreted as the percentage of the variability in y that can be explained by the estimated regression equation.

In the two-independent-variable Butler Trucking example, with SSR = 21.601 and SST = 23.900, we have

$$R^2 = \frac{21.601}{23.900} = .904$$

Therefore, 90.4% of the variability in travel time y is explained by the estimated multiple regression equation with miles traveled and number of deliveries as the independent variables. In Figure 15.4, we see that the multiple coefficient of determination (expressed as a percentage) is also provided by the Minitab output; it is denoted by R-Sq = 90.4%.

Adding independent variables causes the prediction errors to become smaller, thus reducing the sum of squares due to error, SSE. *Because* SSR = SST − SSE, *when* SSE *becomes smaller,* SSR *becomes larger, causing* R^2 = SSR/SST *to increase.*

Figure 15.3 shows that the R-Sq value for the estimated regression equation with only one independent variable, number of miles traveled (x_1), is 66.4%. Thus, the percentage of the variability in travel times that is explained by the estimated regression equation increases from 66.4% to 90.4% when number of deliveries is added as a second independent variable. In general, R^2 always increases as independent variables are added to the model.

Many analysts prefer adjusting R^2 for the number of independent variables to avoid overestimating the impact of adding an independent variable on the amount of variability explained by the estimated regression equation. With n denoting the number of observations and p denoting the number of independent variables, the **adjusted multiple coefficient of determination** is computed as follows.

If a variable is added to the model, R^2 becomes larger even if the variable added is not statistically significant. The adjusted multiple coefficient of determination compensates for the number of independent variables in the model.

ADJUSTED MULTIPLE COEFFICIENT OF DETERMINATION

$$R_a^2 = 1 - (1 - R^2)\frac{n - 1}{n - p - 1} \quad (15.9)$$

For the Butler Trucking example with $n = 10$ and $p = 2$, we have

$$R_a^2 = 1 - (1 - .904)\frac{10 - 1}{10 - 2 - 1} = .88$$

Thus, after adjusting for the two independent variables, we have an adjusted multiple coefficient of determination of .88. This value (expressed as a percentage) is provided by the Minitab output in Figure 15.4 as R-Sq(adj) = 87.6%; the value we calculated differs because we used a rounded value of R^2 in the calculation.

NOTES AND COMMENTS

If the value of R^2 is small and the model contains a large number of independent variables, the adjusted coefficient of determination can take a negative value; in such cases, Minitab sets the adjusted coefficient of determination to zero.

Exercises

Methods

11. In exercise 1, the following estimated regression equation based on 10 observations was presented.

$$\hat{y} = 29.1270 + .5906x_1 + .4980x_2$$

The values of SST and SSR are 6724.125 and 6216.375, respectively.

a. Find SSE.
b. Compute R^2.
c. Compute R_a^2.
d. Comment on the goodness of fit.

SELF test

12. In exercise 2, 10 observations were provided for a dependent variable y and two independent variables x_1 and x_2; for these data SST = 15,182.9, and SSR = 14,052.2.

a. Compute R^2.
b. Compute R_a^2.
c. Does the estimated regression equation explain a large amount of the variability in the data? Explain.

13. In exercise 3, the following estimated regression equation based on 30 observations was presented.

$$\hat{y} = 17.6 + 3.8x_1 - 2.3x_2 + 7.6x_3 + 2.7x_4$$

The values of SST and SSR are 1805 and 1760, respectively.

a. Compute R^2.
b. Compute R_a^2.
c. Comment on the goodness of fit.

Applications

14. In exercise 4, the following estimated regression equation relating sales to inventory investment and advertising expenditures was given.

$$\hat{y} = 25 + 10x_1 + 8x_2$$

The data used to develop the model came from a survey of 10 stores; for those data, SST = 16,000 and SSR = 12,000.
a. For the estimated regression equation given, compute R^2.
b. Compute R_a^2.
c. Does the model appear to explain a large amount of variability in the data? Explain.

SELF test

15. In exercise 5, the owner of Showtime Movie Theaters, Inc., used multiple regression analysis to predict gross revenue (y) as a function of television advertising (x_1) and newspaper advertising (x_2). The estimated regression equation was

WEB file Showtime

$$\hat{y} = 83.2 + 2.29x_1 + 1.30x_2$$

The computer solution provided SST = 25.5 and SSR = 23.435.
a. Compute and interpret R^2 and R_a^2.
b. When television advertising was the only independent variable, $R^2 = .653$ and $R_a^2 = .595$. Do you prefer the multiple regression results? Explain.

WEB file NFLPassing

16. In exercise 6, data were given on the average number of passing yards per attempt (Yds/Att), the number of interceptions thrown per attempt (Int/Att), and the percentage of games won (Win%) for a random sample of 16 National Football League (NFL) teams for the 2011 season (NFL website, February 12, 2012).
a. Did the estimated regression equation that uses only the average number of passing yards per attempt as the independent variable to predict the percentage of games won provide a good fit?
b. Discuss the benefit of using both the average number of passing yards per attempt and the number of interceptions thrown per attempt to predict the percentage of games won.

WEB file Boats

17. In exercise 9, an estimated regression equation was developed relating the top speed for a boat to the boat's beam and horsepower rating.
a. Compute and interpret and R^2 and R_a^2.
b. Does the estimated regression equation provide a good fit to the data? Explain.

WEB file MLBPitching

18. Refer to exercise 10, where Major League Baseball (MLB) pitching statistics were reported for a random sample of 20 pitchers from the American League for the 2011 season (MLB website, March 1, 2012).
a. In part (c) of exercise 10, an estimated regression equation was developed relating the average number of runs given up per inning pitched given the average number of strikeouts per inning pitched and the average number of home runs per inning pitched. What are the values of R^2 and R_a^2?
b. Does the estimated regression equation provide a good fit to the data? Explain.
c. Suppose the earned run average (ERA) is used as the dependent variable in part (c) instead of the average number of runs given up per inning pitched. Does the estimated regression equation that uses the ERA provide a good fit to the data? Explain.

15.4 Model Assumptions

In Section 15.1 we introduced the following multiple regression model.

MULTIPLE REGRESSION MODEL

$$y = \beta_0 + \beta_1x_1 + \beta_2x_2 + \cdots + \beta_px_p + \epsilon \quad \textbf{(15.10)}$$

The assumptions about the error term ϵ in the multiple regression model parallel those for the simple linear regression model.

ASSUMPTIONS ABOUT THE ERROR TERM ϵ IN THE MULTIPLE REGRESSION MODEL $y = \beta_0 + \beta_1x_1 + \cdots + \beta_px_p + \epsilon$

1. The error term ϵ is a random variable with mean or expected value of zero; that is, $E(\epsilon) = 0$.
   Implication: For given values of $x_1, x_2, \ldots, x_p$, the expected, or average, value of y is given by

   $$E(y) = \beta_0 + \beta_1x_1 + \beta_2x_2 + \cdots + \beta_px_p \quad \textbf{(15.11)}$$

   Equation (15.11) is the multiple regression equation we introduced in Section 15.1. In this equation, $E(y)$ represents the average of all possible values of y that might occur for the given values of $x_1, x_2, \ldots, x_p$.
2. The variance of ϵ is denoted by σ^2 and is the same for all values of the independent variables $x_1, x_2, \ldots, x_p$.
   Implication: The variance of y about the regression line equals σ^2 and is the same for all values of $x_1, x_2, \ldots, x_p$.
3. The values of ϵ are independent.
   Implication: The value of ϵ for a particular set of values for the independent variables is not related to the value of ϵ for any other set of values.
4. The error term ϵ is a normally distributed random variable reflecting the deviation between the y value and the expected value of y given by $\beta_0 + \beta_1x_1 + \beta_2x_2 + \cdots + \beta_px_p$.
   Implication: Because $\beta_0, \beta_1, \ldots, \beta_p$ are constants for the given values of $x_1, x_2, \ldots, x_p$, the dependent variable y is also a normally distributed random variable.

To obtain more insight about the form of the relationship given by equation (15.11), consider the following two-independent-variable multiple regression equation.

$$E(y) = \beta_0 + \beta_1x_1 + \beta_2x_2$$

The graph of this equation is a plane in three-dimensional space. Figure 15.5 provides an example of such a graph. Note that the value of ϵ shown is the difference between the actual y value and the expected value of y, $E(y)$, when $x_1 = x_1^*$ and $x_2 = x_2^*$.

FIGURE 15.5 GRAPH OF THE REGRESSION EQUATION FOR MULTIPLE REGRESSION ANALYSIS WITH TWO INDEPENDENT VARIABLES

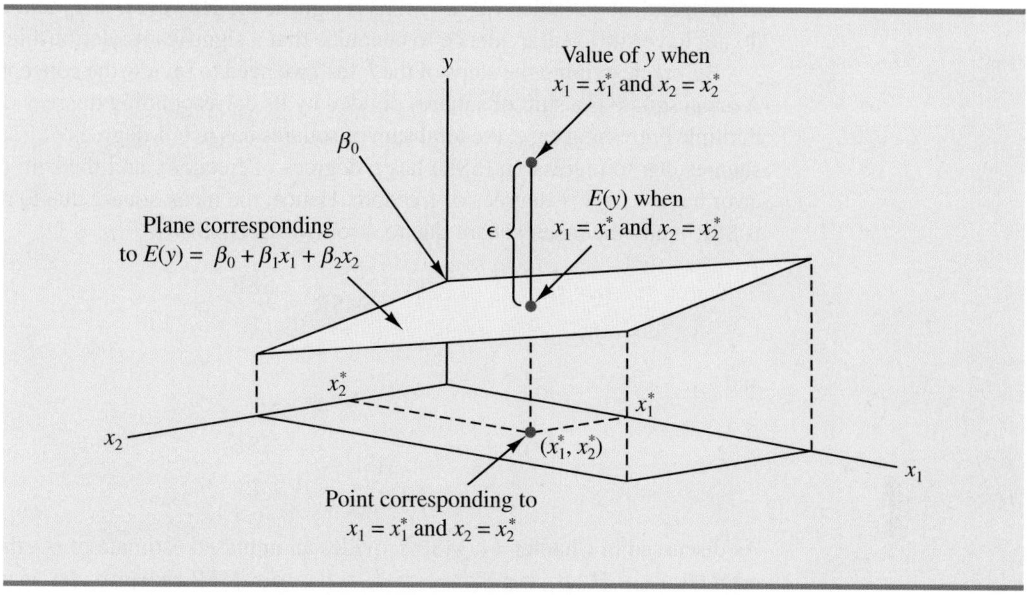

In regression analysis, the term *response variable* is often used in place of the term *dependent variable.* Furthermore, since the multiple regression equation generates a plane or surface, its graph is called a *response surface.*

15.5 Testing for Significance

In this section we show how to conduct significance tests for a multiple regression relationship. The significance tests we used in simple linear regression were a t test and an F test. In simple linear regression, both tests provide the same conclusion; that is, if the null hypothesis is rejected, we conclude that $\beta_1 \neq 0$. In multiple regression, the t test and the F test have different purposes.

1. The F test is used to determine whether a significant relationship exists between the dependent variable and the set of all the independent variables; we will refer to the F test as the test for *overall significance.*
2. If the F test shows an overall significance, the t test is used to determine whether each of the individual independent variables is significant. A separate t test is conducted for each of the independent variables in the model; we refer to each of these t tests as a test for *individual significance.*

In the material that follows, we will explain the F test and the t test and apply each to the Butler Trucking Company example.

F Test

The multiple regression model as defined in Section 15.4 is

$$y = \beta_0 + \beta_1 x_1 + \beta_2 x_2 + \cdots + \beta_p x_p + \epsilon$$

The hypotheses for the F test involve the parameters of the multiple regression model.

$$H_0\text{: } \beta_1 = \beta_2 = \cdots = \beta_p = 0$$
$$H_a\text{: One or more of the parameters is not equal to zero}$$

If H_0 is rejected, the test gives us sufficient statistical evidence to conclude that one or more of the parameters is not equal to zero and that the overall relationship between y and the set of independent variables $x_1, x_2, \ldots, x_p$ is significant. However, if H_0 cannot be rejected, we do not have sufficient evidence to conclude that a significant relationship is present.

Before describing the steps of the F test, we need to review the concept of *mean square.* A mean square is a sum of squares divided by its corresponding degrees of freedom. In the multiple regression case, the total sum of squares has $n - 1$ degrees of freedom, the sum of squares due to regression (SSR) has p degrees of freedom, and the sum of squares due to error has $n - p - 1$ degrees of freedom. Hence, the mean square due to regression (MSR) is SSR/p and the mean square due to error (MSE) is SSE/$(n - p - 1)$.

$$\text{MSR} = \frac{\text{SSR}}{p} \tag{15.12}$$

and

$$\text{MSE} = \frac{\text{SSE}}{n - p - 1} \tag{15.13}$$

As discussed in Chapter 14, MSE provides an unbiased estimate of σ^2, the variance of the error term ϵ. If $H_0: \beta_1 = \beta_2 = \cdots = \beta_p = 0$ is true, MSR also provides an unbiased estimate of σ^2, and the value of MSR/MSE should be close to 1. However, if H_0 is false, MSR overestimates σ^2 and the value of MSR/MSE becomes larger. To determine how large the value of MSR/MSE must be to reject H_0, we make use of the fact that if H_0 is true and the assumptions about the multiple regression model are valid, the sampling distribution of MSR/MSE is an F distribution with p degrees of freedom in the numerator and $n - p - 1$ in the denominator. A summary of the F test for significance in multiple regression follows.

F TEST FOR OVERALL SIGNIFICANCE

$H_0: \beta_1 = \beta_2 = \cdots = \beta_p = 0$

H_a: One or more of the parameters is not equal to zero

TEST STATISTIC

$$F = \frac{\text{MSR}}{\text{MSE}} \tag{15.14}$$

REJECTION RULE

p-value approach:	Reject H_0 if p-value $\leq \alpha$
Critical value approach:	Reject H_0 if $F \geq F_\alpha$

where F_α is based on an F distribution with p degrees of freedom in the numerator and $n - p - 1$ degrees of freedom in the denominator.

Let us apply the F test to the Butler Trucking Company multiple regression problem. With two independent variables, the hypotheses are written as follows.

$H_0: \beta_1 = \beta_2 = 0$

H_a: β_1 and/or β_2 is not equal to zero

FIGURE 15.6 MINITAB OUTPUT FOR BUTLER TRUCKING WITH TWO INDEPENDENT VARIABLES, MILES TRAVELED (x_1) AND NUMBER OF DELIVERIES (x_2)

```
The regression equation is
Time = - 0.869 + 0.0611 Miles + 0.923 Deliveries

Predictor         Coef    SE Coef      T      p
Constant       -0.8687     0.9515  -0.91  0.392
Miles         0.061135   0.009888   6.18  0.000
Deliveries      0.9234     0.2211   4.18  0.004

S = 0.573142   R-Sq = 90.4%   R-Sq(adj) = 87.6%

Analysis of Variance

SOURCE            DF      SS      MS      F      p
Regression         2  21.601  10.800  32.88  0.000
Residual Error     7   2.299   0.328
Total              9  23.900
```

Figure 15.6 is the Minitab output for the multiple regression model with miles traveled (x_1) and number of deliveries (x_2) as the two independent variables. In the analysis of variance part of the output, we see that MSR = 10.8 and MSE = .328. Using equation (15.14), we obtain the test statistic.

$$F = \frac{10.8}{.328} = 32.9$$

Note that the F value on the Minitab output is $F = 32.88$; the value we calculated differs because we used rounded values for MSR and MSE in the calculation. Using $\alpha = .01$, the p-value $= 0.000$ in the last column of the analysis of variance table (Figure 15.6) indicates that we can reject H_0: $\beta_1 = \beta_2 = 0$ because the p-value is less than $\alpha = .01$. Alternatively, Table 4 of Appendix B shows that with two degrees of freedom in the numerator and seven degrees of freedom in the denominator, $F_{.01} = 9.55$. With $32.9 > 9.55$, we reject H_0: $\beta_1 = \beta_2 = 0$ and conclude that a significant relationship is present between travel time y and the two independent variables, miles traveled and number of deliveries.

As noted previously, the mean square error provides an unbiased estimate of σ^2, the variance of the error term ϵ. Referring to Figure 15.6, we see that the estimate of σ^2 is MSE = .328. The square root of MSE is the estimate of the standard deviation of the error term. As defined in Section 14.5, this standard deviation is called the standard error of the estimate and is denoted s. Hence, we have $s = \sqrt{\text{MSE}} = \sqrt{.328} = .573$. Note that the value of the standard error of the estimate appears in the Minitab output in Figure 15.6.

Table 15.3 is the general analysis of variance (ANOVA) table that provides the F test results for a multiple regression model. The value of the F test statistic appears in the last column and can be compared to F_α with p degrees of freedom in the numerator and $n - p - 1$ degrees of freedom in the denominator to make the hypothesis test conclusion. By reviewing the Minitab output for Butler Trucking Company in Figure 15.6, we see that Minitab's analysis of variance table contains this information. Moreover, Minitab also provides the p-value corresponding to the F test statistic.

TABLE 15.3 ANOVA TABLE FOR A MULTIPLE REGRESSION MODEL WITH p INDEPENDENT VARIABLES

Source	Sum of Squares	Degrees of Freedom	Mean Square	F
Regression	SSR	p	$\text{MSR} = \dfrac{\text{SSR}}{p}$	$F = \dfrac{\text{MSR}}{\text{MSE}}$
Error	SSE	$n - p - 1$	$\text{MSE} = \dfrac{\text{SSE}}{n - p - 1}$	
Total	SST	$n - 1$		

t Test

If the F test shows that the multiple regression relationship is significant, a t test can be conducted to determine the significance of each of the individual parameters. The t test for individual significance follows.

t TEST FOR INDIVIDUAL SIGNIFICANCE

For any parameter β_i

$$H_0\colon \beta_i = 0$$
$$H_a\colon \beta_i \neq 0$$

TEST STATISTIC

$$t = \frac{b_i}{s_{b_i}} \tag{15.15}$$

REJECTION RULE

p-value approach: Reject H_0 if p-value $\leq \alpha$

Critical value approach: Reject H_0 if $t \leq -t_{\alpha/2}$ or if $t \geq t_{\alpha/2}$

where $t_{\alpha/2}$ is based on a t distribution with $n - p - 1$ degrees of freedom.

In the test statistic, s_{b_i} is the estimate of the standard deviation of b_i. The value of s_{b_i} will be provided by the computer software package.

Let us conduct the t test for the Butler Trucking regression problem. Refer to the section of Figure 15.6 that shows the Minitab output for the t-ratio calculations. Values of b_1, b_2, s_{b_1}, and s_{b_2} are as follows.

$$b_1 = .061135 \quad s_{b_1} = .009888$$
$$b_2 = .9234 \quad s_{b_2} = .2211$$

Using equation (15.15), we obtain the test statistic for the hypotheses involving parameters β_1 and β_2.

$$t = .061135/.009888 = 6.18$$
$$t = .9234/.2211 = 4.18$$

Note that both of these *t*-ratio values and the corresponding *p*-values are provided by the Minitab output in Figure 15.6. Using $\alpha = .01$, the *p*-values of .000 and .004 on the Minitab output indicate that we can reject H_0: $\beta_1 = 0$ and H_0: $\beta_2 = 0$. Hence, both parameters are statistically significant. Alternatively, Table 2 of Appendix B shows that with $n - p - 1 = 10 - 2 - 1 = 7$ degrees of freedom, $t_{.005} = 3.499$. With $6.18 > 3.499$, we reject H_0: $\beta_1 = 0$. Similarly, with $4.18 > 3.499$, we reject H_0: $\beta_2 = 0$.

Multicollinearity

We use the term *independent variable* in regression analysis to refer to any variable being used to predict or explain the value of the dependent variable. The term does not mean, however, that the independent variables themselves are independent in any statistical sense. On the contrary, most independent variables in a multiple regression problem are correlated to some degree with one another. For example, in the Butler Trucking example involving the two independent variables x_1 (miles traveled) and x_2 (number of deliveries), we could treat the miles traveled as the dependent variable and the number of deliveries as the independent variable to determine whether those two variables are themselves related. We could then compute the sample correlation coefficient $r_{x_1x_2}$ to determine the extent to which the variables are related. Doing so yields $r_{x_1x_2} = .16$. Thus, we find some degree of linear association between the two independent variables. In multiple regression analysis, **multicollinearity** refers to the correlation among the independent variables.

To provide a better perspective of the potential problems of multicollinearity, let us consider a modification of the Butler Trucking example. Instead of x_2 being the number of deliveries, let x_2 denote the number of gallons of gasoline consumed. Clearly, x_1 (the miles traveled) and x_2 are related; that is, we know that the number of gallons of gasoline used depends on the number of miles traveled. Hence, we would conclude logically that x_1 and x_2 are highly correlated independent variables.

Assume that we obtain the equation $\hat{y} = b_0 + b_1x_1 + b_2x_2$ and find that the *F* test shows the relationship to be significant. Then suppose we conduct a *t* test on β_1 to determine whether $\beta_1 \neq 0$, and we cannot reject H_0: $\beta_1 = 0$. Does this result mean that travel time is not related to miles traveled? Not necessarily. What it probably means is that with x_2 already in the model, x_1 does not make a significant contribution to determining the value of *y*. This interpretation makes sense in our example; if we know the amount of gasoline consumed, we do not gain much additional information useful in predicting *y* by knowing the miles traveled. Similarly, a *t* test might lead us to conclude $\beta_2 = 0$ on the grounds that, with x_1 in the model, knowledge of the amount of gasoline consumed does not add much.

A sample correlation coefficient greater than +.7 or less than −.7 for two independent variables is a rule of thumb warning of potential problems with multicollinearity.

To summarize, in *t* tests for the significance of individual parameters, the difficulty caused by multicollinearity is that it is possible to conclude that none of the individual parameters are significantly different from zero when an *F* test on the overall multiple regression equation indicates a significant relationship. This problem is avoided when there is little correlation among the independent variables.

Statisticians have developed several tests for determining whether multicollinearity is high enough to cause problems. According to the rule of thumb test, multicollinearity is a potential problem if the absolute value of the sample correlation coefficient exceeds .7 for any two of the independent variables. The other types of tests are more advanced and beyond the scope of this text.

When the independent variables are highly correlated, it is not possible to determine the separate effect of any particular independent variable on the dependent variable.

If possible, every attempt should be made to avoid including independent variables that are highly correlated. In practice, however, strict adherence to this policy is rarely possible. When decision makers have reason to believe substantial multicollinearity is present, they must realize that separating the effects of the individual independent variables on the dependent variable is difficult.

NOTES AND COMMENTS

Ordinarily, multicollinearity does not affect the way in which we perform our regression analysis or interpret the output from a study. However, when multicollinearity is severe—that is, when two or more of the independent variables are highly correlated with one another—we can have difficulty interpreting the results of t tests on the individual parameters. In addition to the type of problem illustrated in this section, severe cases of multicollinearity have been shown to result in least squares estimates that have the wrong sign. That is, in simulated studies where researchers created the underlying regression model and then applied the least squares technique to develop estimates of β_0, β_1, β_2, and so on, it has been shown that under conditions of high multicollinearity the least squares estimates can have a sign opposite that of the parameter being estimated. For example, β_2 might actually be $+10$ and b_2, its estimate, might turn out to be -2. Thus, little faith can be placed in the individual coefficients if multicollinearity is present to a high degree.

Exercises

Methods

SELF test

19. In exercise 1, the following estimated regression equation based on 10 observations was presented.

$$\hat{y} = 29.1270 + .5906x_1 + .4980x_2$$

Here SST = 6724.125, SSR = 6216.375, $s_{b_1} = .0813$, and $s_{b_2} = .0567$.

a. Compute MSR and MSE.
b. Compute F and perform the appropriate F test. Use $\alpha = .05$.
c. Perform a t test for the significance of β_1. Use $\alpha = .05$.
d. Perform a t test for the significance of β_2. Use $\alpha = .05$.

20. Refer to the data presented in exercise 2. The estimated regression equation for these data is

$$\hat{y} = -18.37 + 2.01x_1 + 4.74x_2$$

Here SST = 15,182.9, SSR = 14,052.2, $s_{b_1} = .2471$, and $s_{b_2} = .9484$.

a. Test for a significant relationship among x_1, x_2, and y. Use $\alpha = .05$.
b. Is β_1 significant? Use $\alpha = .05$.
c. Is β_2 significant? Use $\alpha = .05$.

21. The following estimated regression equation was developed for a model involving two independent variables.

$$\hat{y} = 40.7 + 8.63x_1 + 2.71x_2$$

After x_2 was dropped from the model, the least squares method was used to obtain an estimated regression equation involving only x_1 as an independent variable.

$$\hat{y} = 42.0 + 9.01x_1$$

a. Give an interpretation of the coefficient of x_1 in both models.
b. Could multicollinearity explain why the coefficient of x_1 differs in the two models? If so, how?

Applications

22. In exercise 4, the following estimated regression equation relating sales to inventory investment and advertising expenditures was given.

$$\hat{y} = 25 + 10x_1 + 8x_2$$

The data used to develop the model came from a survey of 10 stores; for these data SST = 16,000 and SSR = 12,000.

a. Compute SSE, MSE, and MSR.
b. Use an F test and a .05 level of significance to determine whether there is a relationship among the variables.

SELF test

23. Refer to exercise 5.

a. Use $\alpha = .01$ to test the hypotheses

$$H_0: \beta_1 = \beta_2 = 0$$
$$H_a: \beta_1 \text{ and/or } \beta_2 \text{ is not equal to zero}$$

for the model $y = \beta_0 + \beta_1 x_1 + \beta_2 x_2 + \epsilon$, where

$$x_1 = \text{television advertising (\$1000s)}$$
$$x_2 = \text{newspaper advertising (\$1000s)}$$

b. Use $\alpha = .05$ to test the significance of β_1. Should x_1 be dropped from the model?
c. Use $\alpha = .05$ to test the significance of β_2. Should x_2 be dropped from the model?

24. *The Wall Street Journal* conducted a study of basketball spending at top colleges. A portion of the data showing the revenue ($ millions), percentage of wins, and the coach's salary ($ millions) for 39 of the country's top basketball programs follows (*The Wall Street Journal,* March 11–12, 2006).

WEB file

Basketball

School	Revenue	%Wins	Salary
Alabama	6.5	61	1.00
Arizona	16.6	63	.70
Arkansas	11.1	72	.80
Boston College	3.4	80	.53
⋮	⋮	⋮	⋮
Washington	5.0	83	.89
West Virginia	4.9	67	.70
Wichita State	3.1	75	.41
Wisconsin	12.0	66	.70

a. Develop the estimated regression equation that can be used to predict the coach's salary given the revenue generated by the program and the percentage of wins.
b. Use the F test to determine the overall significance of the relationship. What is your conclusion at the .05 level of significance?
c. Use the t test to determine the significance of each independent variable. What is your conclusion at the .05 level of significance?

25. The *Condé Nast Traveler* Gold List for 2012 provided ratings for the top 20 small cruise ships (*Condé Nast Traveler* website, March 1, 2012). The data shown below are the scores each ship received based upon the results from *Condé Nast Traveler's* annual Readers' Choice Survey. Each score represents the percentage of respondents who rated a ship as excellent or very good on several criteria, including Itineraries/Schedule, Shore Excursions, and Food/Dining. An overall score was also reported and used to rank the ships. The highest ranked ship, the *Seabourn Odyssey,* has an overall score of 94.4, the highest component of which is 97.8 for Food/Dining.

Ship	Overall	Itineraries/ Schedule	Shore Excursions	Food/ Dining
Seabourn Odyssey	94.4	94.6	90.9	97.8
Seabourn Pride	93.0	96.7	84.2	96.7
National Geographic Endeavor	92.9	100.0	100.0	88.5
Seabourn Sojourn	91.3	88.6	94.8	97.1
Paul Gauguin	90.5	95.1	87.9	91.2
Seabourn Legend	90.3	92.5	82.1	98.8
Seabourn Spirit	90.2	96.0	86.3	92.0
Silver Explorer	89.9	92.6	92.6	88.9
Silver Spirit	89.4	94.7	85.9	90.8
Seven Seas Navigator	89.2	90.6	83.3	90.5
Silver Whisperer	89.2	90.9	82.0	88.6
National Geographic Explorer	89.1	93.1	93.1	89.7
Silver Cloud	88.7	92.6	78.3	91.3
Celebrity Xpedition	87.2	93.1	91.7	73.6
Silver Shadow	87.2	91.0	75.0	89.7
Silver Wind	86.6	94.4	78.1	91.6
SeaDream II	86.2	95.5	77.4	90.9
Wind Star	86.1	94.9	76.5	91.5
Wind Surf	86.1	92.1	72.3	89.3
Wind Spirit	85.2	93.5	77.4	91.9

WEB file CruiseShips

a. Determine the estimated regression equation that can be used to predict the overall score given the scores for Itineraries/Schedule, Shore Excursions, and Food/Dining.
b. Use the *F* test to determine the overall significance of the relationship. What is your conclusion at the .05 level of significance?
c. Use the *t* test to determine the significance of each independent variable. What is your conclusion at the .05 level of significance?
d. Remove any independent variable that is not significant from the estimated regression equation. What is your recommended estimated regression equation?

WEB file MLBPitching

26. In exercise 10, data showing the values of several pitching statistics for a random sample of 20 pitchers from the American League of Major League Baseball were provided. In part (c) of this exercise an estimated regression equation was developed to predict the average number of runs given up per inning pitched (R/IP) given the average number of strikeouts per inning pitched (SO/IP) and the average number of home runs per inning pitched (HR/IP).
a. Use the *F* test to determine the overall significance of the relationship. What is your conclusion at the .05 level of significance?
b. Use the *t* test to determine the significance of each independent variable. What is your conclusion at the .05 level of significance?

15.6 Using the Estimated Regression Equation for Estimation and Prediction

The procedures for estimating the mean value of y and predicting an individual value of y in multiple regression are similar to those in regression analysis involving one independent variable. First, recall that in Chapter 14 we showed that the point estimate of the expected value of y for a given value of x was the same as the point estimate of an individual value of y. In both cases, we used $\hat{y} = b_0 + b_1x$ as the point estimate.

In multiple regression we use the same procedure. That is, we substitute the given values of $x_1, x_2, \ldots, x_p$ into the estimated regression equation and use the corresponding value of $\hat{y}$ as the point estimate. Suppose that for the Butler Trucking example we want to use the

TABLE 15.4 THE 95% CONFIDENCE AND PREDICTION INTERVALS FOR BUTLER TRUCKING

Value of	Value of	Confidence Interval		Prediction Interval	
x_1	x_2	Lower Limit	Upper Limit	Lower Limit	Upper Limit
50	2	3.146	4.924	2.414	5.656
50	3	4.127	5.789	3.368	6.548
50	4	4.815	6.948	4.157	7.607
100	2	6.258	7.926	5.500	8.683
100	3	7.385	8.645	6.520	9.510
100	4	8.135	9.742	7.362	10.515

estimated regression equation involving x_1 (miles traveled) and x_2 (number of deliveries) to develop two interval estimates:

1. A *confidence interval* of the mean travel time for all trucks that travel 100 miles and make two deliveries
2. A *prediction interval* of the travel time for *one specific* truck that travels 100 miles and makes two deliveries

Using the estimated regression equation $\hat{y} = -.869 + .0611x_1 + .923x_2$ with $x_1 = 100$ and $x_2 = 2$, we obtain the following value of $\hat{y}$.

$$\hat{y} = -.869 + .0611(100) + .923(2) = 7.09$$

Hence, the point estimate of travel time in both cases is approximately seven hours.

To develop interval estimates for the mean value of y and for an individual value of y, we use a procedure similar to that for regression analysis involving one independent variable. The formulas required are beyond the scope of the text, but computer packages for multiple regression analysis will often provide confidence intervals once the values of x_1, $x_2, \ldots, x_p$ are specified by the user. In Table 15.4 we show the 95% confidence and prediction intervals for the Butler Trucking example for selected values of x_1 and x_2; these values were obtained using Minitab. Note that the interval estimate for an individual value of y is wider than the interval estimate for the expected value of y. This difference simply reflects the fact that for given values of x_1 and x_2 we can estimate the mean travel time for all trucks with more precision than we can predict the travel time for one specific truck.

Exercises

Methods

27. In exercise 1, the following estimated regression equation based on 10 observations was presented.

$$\hat{y} = 29.1270 + .5906x_1 + .4980x_2$$

a. Develop a point estimate of the mean value of y when $x_1 = 180$ and $x_2 = 310$.
b. Develop a point estimate for an individual value of y when $x_1 = 180$ and $x_2 = 310$.

SELF test

28. Refer to the data in exercise 2. The estimated regression equation for those data is

$$\hat{y} = -18.4 + 2.01x_1 + 4.74x_2$$

a. Develop a 95% confidence interval for the mean value of y when $x_1 = 45$ and $x_2 = 15$.
b. Develop a 95% prediction interval for y when $x_1 = 45$ and $x_2 = 15$.

Applications

SELF test

29. In exercise 5, the owner of Showtime Movie Theaters, Inc., used multiple regression analysis to predict gross revenue (y) as a function of television advertising (x_1) and newspaper advertising (x_2). The estimated regression equation was

$$\hat{y} = 83.2 + 2.29x_1 + 1.30x_2$$

   a. What is the gross revenue expected for a week when \$3500 is spent on television advertising ($x_1 = 3.5$) and \$1800 is spent on newspaper advertising ($x_2 = 1.8$)?
   b. Provide a 95% confidence interval for the mean revenue of all weeks with the expenditures listed in part (a).
   c. Provide a 95% prediction interval for next week's revenue, assuming that the advertising expenditures will be allocated as in part (a).

WEB file
Boats

30. In exercise 9 an estimated regression equation was developed relating the top speed for a boat to the boat's beam and horsepower rating.
   a. Develop a 95% confidence interval for the mean top speed of a boat with a beam of 85 inches and an engine with a 330 horsepower rating.
   b. The Svfara SV609 has a beam of 85 inches and an engine with a 330 horsepower rating. Develop a 95% prediction interval for the mean top speed for the Svfara SV609.

31. The American Association of Individual Investors (AAII) On-Line Discount Broker Survey polls members on their experiences with electronic trades handled by discount brokers. As part of the survey, members were asked to rate their satisfaction with the trade price and the speed of execution, as well as provide an overall satisfaction rating. Possible responses (scores) were no opinion (0), unsatisfied (1), somewhat satisfied (2), satisfied (3), and very satisfied (4). For each broker, summary scores were computed by computing a weighted average of the scores provided by each respondent. A portion of the survey results follows (AAII website, February 7, 2012).

WEB file
Broker

Brokerage	Trade Price	Speed of Execution	Satisfaction Electronic Trades
Scottrade, Inc.	3.4	3.4	3.5
Charles Schwab	3.2	3.3	3.4
Fidelity Brokerage Services	3.1	3.4	3.9
TD Ameritrade	2.9	3.6	3.7
E*Trade Financial	2.9	3.2	2.9
(Not listed)	2.5	3.2	2.7
Vanguard Brokerage Services	2.6	3.8	2.8
USAA Brokerage Services	2.4	3.8	3.6
Thinkorswim	2.6	2.6	2.6
Wells Fargo Investments	2.3	2.7	2.3
Interactive Brokers	3.7	4.0	4.0
Zecco.com	2.5	2.5	2.5
Firstrade Securities	3.0	3.0	4.0
Banc of America Investment Services	4.0	1.0	2.0

   a. Develop an estimated regression equation using trade price and speed of execution to predict overall satisfaction with the broker.
   b. Finger Lakes Investments has developed a new electronic trading system and would like to predict overall customer satisfaction assuming they can provide satisfactory levels of service levels (3) for both trade price and speed of execution. Use the estimated repression equation developed in part (a) to predict overall satisfaction level for Finger Lakes Investments if they can achieve these performance levels.

c. Develop a 95% confidence interval estimate of the overall satisfaction of electronic trades for all brokers that provide satisfactory levels of service for both trade price and speed of execution.
d. Develop a 95% prediction interval of overall satisfaction for Finger Lakes Investments assuming they achieve service levels of 3 for both trade price and speed of execution.

15.7 Categorical Independent Variables

The independent variables may be categorical or quantitative.

Thus far, the examples we have considered involved quantitative independent variables such as student population, distance traveled, and number of deliveries. In many situations, however, we must work with **categorical independent variables** such as gender (male, female), method of payment (cash, credit card, check), and so on. The purpose of this section is to show how categorical variables are handled in regression analysis. To illustrate the use and interpretation of a categorical independent variable, we will consider a problem facing the managers of Johnson Filtration, Inc.

An Example: Johnson Filtration, Inc.

Johnson Filtration, Inc., provides maintenance service for water-filtration systems throughout southern Florida. Customers contact Johnson with requests for maintenance service on their water-filtration systems. To estimate the service time and the service cost, Johnson's managers want to predict the repair time necessary for each maintenance request. Hence, repair time in hours is the dependent variable. Repair time is believed to be related to two factors, the number of months since the last maintenance service and the type of repair problem (mechanical or electrical). Data for a sample of 10 service calls are reported in Table 15.5.

Let y denote the repair time in hours and x_1 denote the number of months since the last maintenance service. The regression model that uses only x_1 to predict y is

$$y = \beta_0 + \beta_1 x_1 + \epsilon$$

Using Minitab to develop the estimated regression equation, we obtained the output shown in Figure 15.7. The estimated regression equation is

$$\hat{y} = 2.15 + .304x_1 \quad \textbf{(15.16)}$$

At the .05 level of significance, the p-value of .016 for the t (or F) test indicates that the number of months since the last service is significantly related to repair time. R-sq = 53.4% indicates that x_1 alone explains 53.4% of the variability in repair time.

TABLE 15.5 DATA FOR THE JOHNSON FILTRATION EXAMPLE

Service Call	Months Since Last Service	Type of Repair	Repair Time in Hours
1	2	electrical	2.9
2	6	mechanical	3.0
3	8	electrical	4.8
4	3	mechanical	1.8
5	2	electrical	2.9
6	7	electrical	4.9
7	9	mechanical	4.2
8	8	mechanical	4.8
9	4	electrical	4.4
10	6	electrical	4.5

FIGURE 15.7 MINITAB OUTPUT FOR JOHNSON FILTRATION WITH MONTHS SINCE LAST SERVICE (x_1) AS THE INDEPENDENT VARIABLE

```
The regression equation is
Time = 2.15 + 0.304 Months

Predictor      Coef   SE Coef      T      p
Constant     2.1473    0.6050   3.55  0.008
Months       0.3041    0.1004   3.03  0.016

S = 0.781022   R-Sq = 53.4%   R-Sq(adj) = 47.6%

Analysis of Variance

SOURCE           DF       SS      MS     F      p
Regression        1   5.5960  5.5960  9.17  0.016
Residual Error    8   4.8800  0.6100
Total             9  10.4760
```

In the Minitab output the variable names Months *and* Time *were entered as the column headings on the worksheet; thus,* x_1 = Months *and* y = Time.

To incorporate the type of repair into the regression model, we define the following variable.

$$x_2 = \begin{cases} 0 \text{ if the type of repair is mechanical} \\ 1 \text{ if the type of repair is electrical} \end{cases}$$

In regression analysis x_2 is called a **dummy** or *indicator* **variable**. Using this dummy variable, we can write the multiple regression model as

$$y = \beta_0 + \beta_1 x_1 + \beta_2 x_2 + \epsilon$$

Table 15.6 is the revised data set that includes the values of the dummy variable. Using Minitab and the data in Table 15.6, we can develop estimates of the model parameters. The Minitab output in Figure 15.8 shows that the estimated multiple regression equation is

$$\hat{y} = .93 + .388x_1 + 1.26x_2 \tag{15.17}$$

TABLE 15.6 DATA FOR THE JOHNSON FILTRATION EXAMPLE WITH TYPE OF REPAIR INDICATED BY A DUMMY VARIABLE ($x_2 = 0$ FOR MECHANICAL; $x_2 = 1$ FOR ELECTRICAL)

WEB file

Johnson

Customer	Months Since Last Service (x_1)	Type of Repair (x_2)	Repair Time in Hours (y)
1	2	1	2.9
2	6	0	3.0
3	8	1	4.8
4	3	0	1.8
5	2	1	2.9
6	7	1	4.9
7	9	0	4.2
8	8	0	4.8
9	4	1	4.4
10	6	1	4.5

FIGURE 15.8 MINITAB OUTPUT FOR JOHNSON FILTRATION WITH MONTHS SINCE LAST SERVICE (x_1) AND TYPE OF REPAIR (x_2) AS THE INDEPENDENT VARIABLES

In the Minitab output the variable names Months, Type, *and* Time *were entered as the column headings on the worksheet; thus,* x_1 = Months, x_2 = Type, *and* y = Time.

```
The regression equation is
Time = 0.930 + 0.388 Months + 1.26 Type

Predictor       Coef   SE Coef     T      p
Constant      0.9305    0.4670  1.99  0.087
Months       0.38762   0.06257  6.20  0.000
Type          1.2627    0.3141  4.02  0.005

S = 0.459048    R-Sq = 85.9%    R-Sq(adj) = 81.9%

Analysis of Variance

SOURCE          DF        SS       MS      F      p
Regression       2    9.0009   4.5005  21.36  0.001
Residual Error   7    1.4751   0.2107
Total            9   10.4760
```

At the .05 level of significance, the p-value of .001 associated with the F test ($F = 21.36$) indicates that the regression relationship is significant. The t test part of the printout in Figure 15.8 shows that both months since last service (p-value $= .000$) and type of repair (p-value $= .005$) are statistically significant. In addition, R-Sq = 85.9% and R-Sq (adj) = 81.9% indicate that the estimated regression equation does a good job of explaining the variability in repair times. Thus, equation (15.17) should prove helpful in predicting the repair time necessary for the various service calls.

Interpreting the Parameters

The multiple regression equation for the Johnson Filtration example is

$$E(y) = \beta_0 + \beta_1 x_1 + \beta_2 x_2 \tag{15.18}$$

To understand how to interpret the parameters β_0, β_1, and β_2 when a categorical variable is present, consider the case when $x_2 = 0$ (mechanical repair). Using $E(y \mid \text{mechanical})$ to denote the mean or expected value of repair time *given* a mechanical repair, we have

$$E(y \mid \text{mechanical}) = \beta_0 + \beta_1 x_1 + \beta_2(0) = \beta_0 + \beta_1 x_1 \tag{15.19}$$

Similarly, for an electrical repair ($x_2 = 1$), we have

$$\begin{aligned} E(y \mid \text{electrical}) &= \beta_0 + \beta_1 x_1 + \beta_2(1) = \beta_0 + \beta_1 x_1 + \beta_2 \\ &= (\beta_0 + \beta_2) + \beta_1 x_1 \end{aligned} \tag{15.20}$$

Comparing equations (15.19) and (15.20), we see that the mean repair time is a linear function of x_1 for both mechanical and electrical repairs. The slope of both equations is β_1, but the y-intercept differs. The y-intercept is β_0 in equation (15.19) for mechanical repairs and $(\beta_0 + \beta_2)$ in equation (15.20) for electrical repairs. The interpretation of β_2 is that it indicates the difference between the mean repair time for an electrical repair and the mean repair time for a mechanical repair.

If β_2 is positive, the mean repair time for an electrical repair will be greater than that for a mechanical repair; if β_2 is negative, the mean repair time for an electrical repair will be less than that for a mechanical repair. Finally, if $\beta_2 = 0$, there is no difference in the mean repair time between electrical and mechanical repairs and the type of repair is not related to the repair time.

Using the estimated multiple regression equation $\hat{y} = .93 + .388x_1 + 1.26x_2$, we see that .93 is the estimate of β_0 and 1.26 is the estimate of β_2. Thus, when $x_2 = 0$ (mechanical repair)

$$\hat{y} = .93 + .388x_1 \tag{15.21}$$

and when $x_2 = 1$ (electrical repair)

$$\begin{aligned}\hat{y} &= .93 + .388x_1 + 1.26(1) \\ &= 2.19 + .388x_1\end{aligned} \tag{15.22}$$

In effect, the use of a dummy variable for type of repair provides two estimated regression equations that can be used to predict the repair time, one corresponding to mechanical repairs and one corresponding to electrical repairs. In addition, with $b_2 = 1.26$, we learn that, on average, electrical repairs require 1.26 hours longer than mechanical repairs.

Figure 15.9 is the plot of the Johnson data from Table 15.6. Repair time in hours (y) is represented by the vertical axis and months since last service (x_1) is represented by the horizontal axis. A data point for a mechanical repair is indicated by an M and a data point for an electrical repair is indicated by an E. Equations (15.21) and (15.22) are plotted on the graph to show graphically the two equations that can be used to predict the repair time, one corresponding to mechanical repairs and one corresponding to electrical repairs.

FIGURE 15.9 SCATTER DIAGRAM FOR THE JOHNSON FILTRATION REPAIR DATA FROM TABLE 15.6

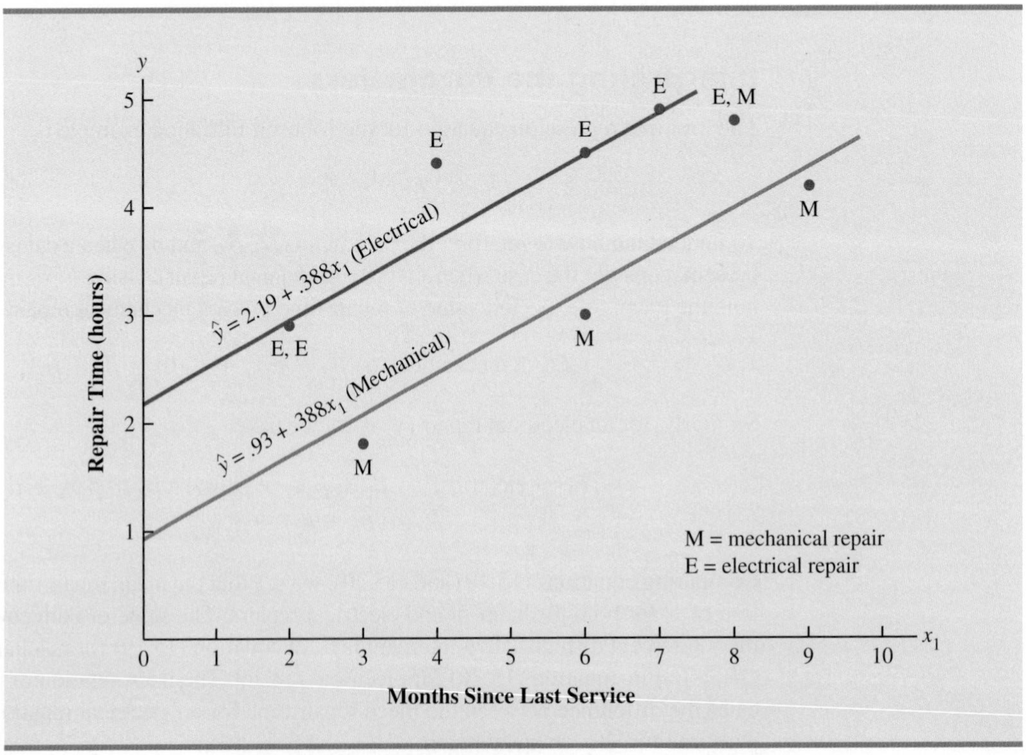

More Complex Categorical Variables

A categorical variable with k levels must be modeled using k − 1 dummy variables. Care must be taken in defining and interpreting the dummy variables.

Because the categorical variable for the Johnson Filtration example had two levels (mechanical and electrical), defining a dummy variable with zero indicating a mechanical repair and one indicating an electrical repair was easy. However, when a categorical variable has more than two levels, care must be taken in both defining and interpreting the dummy variables. As we will show, if a categorical variable has k levels, $k - 1$ dummy variables are required, with each dummy variable being coded as 0 or 1.

For example, suppose a manufacturer of copy machines organized the sales territories for a particular state into three regions: A, B, and C. The managers want to use regression analysis to help predict the number of copiers sold per week. With the number of units sold as the dependent variable, they are considering several independent variables (the number of sales personnel, advertising expenditures, and so on). Suppose the managers believe sales region is also an important factor in predicting the number of copiers sold. Because sales region is a categorical variable with three levels, A, B and C, we will need $3 - 1 = 2$ dummy variables to represent the sales region. Each variable can be coded 0 or 1 as follows.

$$x_1 = \begin{cases} 1 \text{ if sales region B} \\ 0 \text{ otherwise} \end{cases}$$
$$x_2 = \begin{cases} 1 \text{ if sales region C} \\ 0 \text{ otherwise} \end{cases}$$

With this definition, we have the following values of x_1 and x_2.

Region	x_1	x_2
A	0	0
B	1	0
C	0	1

Observations corresponding to region A would be coded $x_1 = 0, x_2 = 0$; observations corresponding to region B would be coded $x_1 = 1, x_2 = 0$; and observations corresponding to region C would be coded $x_1 = 0, x_2 = 1$.

The regression equation relating the expected value of the number of units sold, $E(y)$, to the dummy variables would be written as

$$E(y) = \beta_0 + \beta_1 x_1 + \beta_2 x_2$$

To help us interpret the parameters β_0, β_1, and β_2, consider the following three variations of the regression equation.

$$E(y \mid \text{region A}) = \beta_0 + \beta_1(0) + \beta_2(0) = \beta_0$$
$$E(y \mid \text{region B}) = \beta_0 + \beta_1(1) + \beta_2(0) = \beta_0 + \beta_1$$
$$E(y \mid \text{region C}) = \beta_0 + \beta_1(0) + \beta_2(1) = \beta_0 + \beta_2$$

Thus, β_0 is the mean or expected value of sales for region A; β_1 is the difference between the mean number of units sold in region B and the mean number of units sold in region A; and β_2 is the difference between the mean number of units sold in region C and the mean number of units sold in region A.

Two dummy variables were required because sales region is a categorical variable with three levels. But the assignment of $x_1 = 0, x_2 = 0$ to indicate region A, $x_1 = 1, x_2 = 0$ to indicate region B, and $x_1 = 0, x_2 = 1$ to indicate region C was arbitrary. For example, we could have chosen $x_1 = 1, x_2 = 0$ to indicate region A, $x_1 = 0, x_2 = 0$ to indicate region B, and $x_1 = 0, x_2 = 1$ to indicate region C. In that case, β_1 would have been interpreted as the mean difference between regions A and B and β_2 as the mean difference between regions C and B.

The important point to remember is that when a categorical variable has k levels, $k - 1$ dummy variables are required in the multiple regression analysis. Thus, if the sales region example had a fourth region, labeled D, three dummy variables would be necessary. For example, the three dummy variables can be coded as follows.

$$x_1 = \begin{cases} 1 \text{ if sales region B} \\ 0 \text{ otherwise} \end{cases} \qquad x_2 = \begin{cases} 1 \text{ if sales region C} \\ 0 \text{ otherwise} \end{cases} \qquad x_3 = \begin{cases} 1 \text{ if sales region D} \\ 0 \text{ otherwise} \end{cases}$$

Exercises

Methods

SELF test

32. Consider a regression study involving a dependent variable y, a quantitative independent variable x_1, and a categorical independent variable with two levels (level 1 and level 2).
    a. Write a multiple regression equation relating x_1 and the categorical variable to y.
    b. What is the expected value of y corresponding to level 1 of the categorical variable?
    c. What is the expected value of y corresponding to level 2 of the categorical variable?
    d. Interpret the parameters in your regression equation.

33. Consider a regression study involving a dependent variable y, a quantitative independent variable x_1, and a categorical independent variable with three possible levels (level 1, level 2, and level 3).
    a. How many dummy variables are required to represent the categorical variable?
    b. Write a multiple regression equation relating x_1 and the categorical variable to y.
    c. Interpret the parameters in your regression equation.

Applications

SELF test

34. Management proposed the following regression model to predict sales at a fast-food outlet.

$$y = \beta_0 + \beta_1 x_1 + \beta_2 x_2 + \beta_3 x_3 + \epsilon$$

where

$$\begin{aligned} x_1 &= \text{number of competitors within one mile} \\ x_2 &= \text{population within one mile (1000s)} \\ x_3 &= \begin{cases} 1 \text{ if drive-up window present} \\ 0 \text{ otherwise} \end{cases} \\ y &= \text{sales (\$1000s)} \end{aligned}$$

The following estimated regression equation was developed after 20 outlets were surveyed.

$$\hat{y} = 10.1 - 4.2x_1 + 6.8x_2 + 15.3x_3$$

a. What is the expected amount of sales attributable to the drive-up window?
b. Predict sales for a store with two competitors, a population of 8000 within one mile, and no drive-up window.
c. Predict sales for a store with one competitor, a population of 3000 within one mile, and a drive-up window.

35. Refer to the Johnson Filtration problem introduced in this section. Suppose that in addition to information on the number of months since the machine was serviced and whether a mechanical or an electrical repair was necessary, the managers obtained a list showing which repairperson performed the service. The revised data follow.

WEB file
Repair

Repair Time in Hours	Months Since Last Service	Type of Repair	Repairperson
2.9	2	Electrical	Dave Newton
3.0	6	Mechanical	Dave Newton
4.8	8	Electrical	Bob Jones
1.8	3	Mechanical	Dave Newton
2.9	2	Electrical	Dave Newton
4.9	7	Electrical	Bob Jones
4.2	9	Mechanical	Bob Jones
4.8	8	Mechanical	Bob Jones
4.4	4	Electrical	Bob Jones
4.5	6	Electrical	Dave Newton

a. Ignore for now the months since the last maintenance service (x_1) and the repairperson who performed the service. Develop the estimated simple linear regression equation to predict the repair time (y) given the type of repair (x_2). Recall that $x_2 = 0$ if the type of repair is mechanical and 1 if the type of repair is electrical.
b. Does the equation that you developed in part (a) provide a good fit for the observed data? Explain.
c. Ignore for now the months since the last maintenance service and the type of repair associated with the machine. Develop the estimated simple linear regression equation to predict the repair time given the repairperson who performed the service. Let $x_3 = 0$ if Bob Jones performed the service and $x_3 = 1$ if Dave Newton performed the service.
d. Does the equation that you developed in part (c) provide a good fit for the observed data? Explain.

36. This problem is an extension of the situation described in exercise 35.
a. Develop the estimated regression equation to predict the repair time given the number of months since the last maintenance service, the type of repair, and the repairperson who performed the service.
b. At the .05 level of significance, test whether the estimated regression equation developed in part (a) represents a significant relationship between the independent variables and the dependent variable.
c. Is the addition of the independent variable x_3, the repairperson who performed the service, statistically significant? Use $\alpha = .05$. What explanation can you give for the results observed?

37. The *Consumer Reports* Restaurant Customer Satisfaction Survey is based upon 148,599 visits to full-service restaurant chains (*Consumer Reports* website, February 11, 2009). Assume the following data are representative of the results reported. The variable type indicates whether the restaurant is an Italian restaurant or a seafood/steakhouse. Price indicates the average amount paid per person for dinner and drinks, minus the tip. Score reflects diners' overall satisfaction, with higher values indicating greater overall satisfaction. A score of 80 can be interpreted as very satisfied.

WEB file
RestaurantRatings

Restaurant	Type	Price ($)	Score
Bertucci's	Italian	16	77
Black Angus Steakhouse	Seafood/Steakhouse	24	79
Bonefish Grill	Seafood/Steakhouse	26	85

(*continued*)

Restaurant	Type	Price ($)	Score
Bravo! Cucina Italiana	Italian	18	84
Buca di Beppo	Italian	17	81
Bugaboo Creek Steak House	Seafood/Steakhouse	18	77
Carrabba's Italian Grill	Italian	23	86
Charlie Brown's Steakhouse	Seafood/Steakhouse	17	75
Il Fornaio	Italian	28	83
Joe's Crab Shack	Seafood/Steakhouse	15	71
Johnny Carino's Italian	Italian	17	81
Lone Star Steakhouse & Saloon	Seafood/Steakhouse	17	76
LongHorn Steakhouse	Seafood/Steakhouse	19	81
Maggiano's Little Italy	Italian	22	83
McGrath's Fish House	Seafood/Steakhouse	16	81
Olive Garden	Italian	19	81
Outback Steakhouse	Seafood/Steakhouse	20	80
Red Lobster	Seafood/Steakhouse	18	78
Romano's Macaroni Grill	Italian	18	82
The Old Spaghetti Factory	Italian	12	79
Uno Chicago Grill	Italian	16	76

a. Develop the estimated regression equation to show how overall customer satisfaction is related to the independent variable average meal price.
b. At the .05 level of significance, test whether the estimated regression equation developed in part (a) indicates a significant relationship between overall customer satisfaction and average meal price.
c. Develop a dummy variable that will account for the type of restaurant (Italian or seafood/steakhouse).
d. Develop the estimated regression equation to show how overall customer satisfaction is related to the average meal price and the type of restaurant.
e. Is type of restaurant a significant factor in overall customer satisfaction?
f. Predict the *Consumer Reports* customer satisfaction score for a seafood/steakhouse that has an average meal price of $20. How much would the predicted score have changed for an Italian restaurant?

38. A 10-year study conducted by the American Heart Association provided data on how age, blood pressure, and smoking relate to the risk of strokes. Assume that the following data are from a portion of this study. Risk is interpreted as the probability (times 100) that the patient will have a stroke over the next 10-year period. For the smoking variable, define a dummy variable with 1 indicating a smoker and 0 indicating a nonsmoker.

WEB file
Stroke

Risk	Age	Pressure	Smoker
12	57	152	No
24	67	163	No
13	58	155	No
56	86	177	Yes
28	59	196	No
51	76	189	Yes
18	56	155	Yes
31	78	120	No
37	80	135	Yes
15	78	98	No
22	71	152	No
36	70	173	Yes

Risk	Age	Pressure	Smoker
15	67	135	Yes
48	77	209	Yes
15	60	199	No
36	82	119	Yes
8	66	166	No
34	80	125	Yes
3	62	117	No
37	59	207	Yes

a. Develop an estimated regression equation that relates risk of a stroke to the person's age, blood pressure, and whether the person is a smoker.
b. Is smoking a significant factor in the risk of a stroke? Explain. Use $\alpha = .05$.
c. What is the probability of a stroke over the next 10 years for Art Speen, a 68-year-old smoker who has blood pressure of 175? What action might the physician recommend for this patient?

15.8 Residual Analysis

In Chapter 14 we pointed out that standardized residuals are frequently used in residual plots and in the identification of outliers. The general formula for the standardized residual for observation i follows.

STANDARDIZED RESIDUAL FOR OBSERVATION i

$$\frac{y_i - \hat{y}_i}{s_{y_i - \hat{y}_i}} \tag{15.23}$$

where

$s_{y_i - \hat{y}_i}$ = the standard deviation of residual i

The general formula for the standard deviation of residual i is defined as follows.

STANDARD DEVIATION OF RESIDUAL i

$$s_{y_i - \hat{y}_i} = s\sqrt{1 - h_i} \tag{15.24}$$

where

s = standard error of the estimate
h_i = leverage of observation i

As we stated in Chapter 14, the **leverage** of an observation is determined by how far the values of the independent variables are from their means. The computation of h_i, $s_{y_i - \hat{y}_i}$, and hence the standardized residual for observation i in multiple regression analysis is too complex to be

TABLE 15.7 RESIDUALS AND STANDARDIZED RESIDUALS FOR THE BUTLER TRUCKING REGRESSION ANALYSIS

Miles Traveled (x_1)	Deliveries (x_2)	Travel Time (y)	Predicted Value ($\hat{y}$)	Residual ($y - \hat{y}$)	Standardized Residual
100	4	9.3	8.93846	.361541	.78344
50	3	4.8	4.95830	−.158304	−.34962
100	4	8.9	8.93846	−.038460	−.08334
100	2	6.5	7.09161	−.591609	−1.30929
50	2	4.2	4.03488	.165121	.38167
80	2	6.2	5.86892	.331083	.65431
75	3	7.4	6.48667	.913331	1.68917
65	4	6.0	6.79875	−.798749	−1.77372
90	3	7.6	7.40369	.196311	.36703
90	2	6.1	6.48026	−.380263	−.77639

done by hand. However, the standardized residuals can be easily obtained as part of the output from statistical software packages. Table 15.7 lists the predicted values, the residuals, and the standardized residuals for the Butler Trucking example presented previously in this chapter; we obtained these values by using the Minitab statistical software package. The predicted values in the table are based on the estimated regression equation $\hat{y} = -.869 + .0611x_1 + .923x_2$.

The standardized residuals and the predicted values of y from Table 15.7 are used in Figure 15.10, the standardized residual plot for the Butler Trucking multiple regression example. This standardized residual plot does not indicate any unusual abnormalities. Also, all the standardized residuals are between -2 and $+2$; hence, we have no reason to question the assumption that the error term ϵ is normally distributed. We conclude that the model assumptions are reasonable.

FIGURE 15.10 STANDARDIZED RESIDUAL PLOT FOR BUTLER TRUCKING

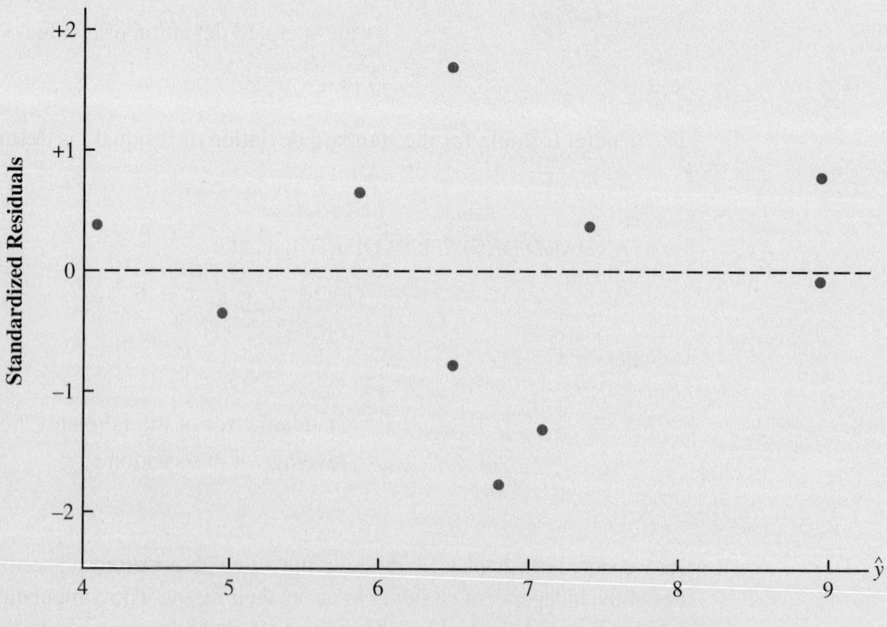

A normal probability plot also can be used to determine whether the distribution of ϵ appears to be normal. The procedure and interpretation for a normal probability plot were discussed in Section 14.8. The same procedure is appropriate for multiple regression. Again, we would use a statistical software package to perform the computations and provide the normal probability plot.

Detecting Outliers

An **outlier** is an observation that is unusual in comparison with the other data; in other words, an outlier does not fit the pattern of the other data. In Chapter 14 we showed an example of an outlier and discussed how standardized residuals can be used to detect outliers. Minitab classifies an observation as an outlier if the value of its standardized residual is less than -2 or greater than $+2$. Applying this rule to the standardized residuals for the Butler Trucking example (see Table 15.7), we do not detect any outliers in the data set.

In general, the presence of one or more outliers in a data set tends to increase s, the standard error of the estimate, and hence increase $s_{y_i - \hat{y}_i}$, the standard deviation of residual i. Because $s_{y_i - \hat{y}_i}$ appears in the denominator of the formula for the standardized residual (15.23), the size of the standardized residual will decrease as s increases. As a result, even though a residual may be unusually large, the large denominator in expression (15.23) may cause the standardized residual rule to fail to identify the observation as being an outlier. We can circumvent this difficulty by using a form of the standardized residuals called **studentized deleted residuals**.

Studentized Deleted Residuals and Outliers

Suppose the ith observation is deleted from the data set and a new estimated regression equation is developed with the remaining $n - 1$ observations. Let $s_{(i)}$ denote the standard error of the estimate based on the data set with the ith observation deleted. If we compute the standard deviation of residual i using $s_{(i)}$ instead of s, and then compute the standardized residual for observation i using the revised $s_{y_i - \hat{y}_i}$ value, the resulting standardized residual is called a studentized deleted residual. If the ith observation is an outlier, $s_{(i)}$ will be less than s. The absolute value of the ith studentized deleted residual therefore will be larger than the absolute value of the standardized residual. In this sense, studentized deleted residuals may detect outliers that standardized residuals do not detect.

Many statistical software packages provide an option for obtaining studentized deleted residuals. Using Minitab, we obtained the studentized deleted residuals for the Butler Trucking example; the results are reported in Table 15.8. The t distribution can be used to

TABLE 15.8 STUDENTIZED DELETED RESIDUALS FOR BUTLER TRUCKING

Miles Traveled (x_1)	Deliveries (x_2)	Travel Time (y)	Standardized Residual	Studentized Deleted Residual
100	4	9.3	.78344	.75939
50	3	4.8	−.34962	−.32654
100	4	8.9	−.08334	−.07720
100	2	6.5	−1.30929	−1.39494
50	2	4.2	.38167	.35709
80	2	6.2	.65431	.62519
75	3	7.4	1.68917	2.03187
65	4	6.0	−1.77372	−2.21314
90	3	7.6	.36703	.34312
90	2	6.1	−.77639	−.75190

TABLE 15.9 LEVERAGE AND COOK'S DISTANCE MEASURES FOR BUTLER TRUCKING

Miles Traveled (x_1)	Deliveries (x_2)	Travel Time (y)	Leverage (h_i)	Cook's D (D_i)
100	4	9.3	.351704	.110994
50	3	4.8	.375863	.024536
100	4	8.9	.351704	.001256
100	2	6.5	.378451	.347923
50	2	4.2	.430220	.036663
80	2	6.2	.220557	.040381
75	3	7.4	.110009	.117562
65	4	6.0	.382657	.650029
90	3	7.6	.129098	.006656
90	2	6.1	.269737	.074217

determine whether the studentized deleted residuals indicate the presence of outliers. Recall that p denotes the number of independent variables and n denotes the number of observations. Hence, if we delete the ith observation, the number of observations in the reduced data set is $n - 1$; in this case the error sum of squares has $(n - 1) - p - 1$ degrees of freedom. For the Butler Trucking example with $n = 10$ and $p = 2$, the degrees of freedom for the error sum of squares with the ith observation deleted is $9 - 2 - 1 = 6$. At a .05 level of significance, the t distribution (Table 2 of Appendix B) shows that with six degrees of freedom, $t_{.025} = 2.447$. If the value of the ith studentized deleted residual is less than -2.447 or greater than $+2.447$, we can conclude that the ith observation is an outlier. The studentized deleted residuals in Table 15.8 do not exceed those limits; therefore, we conclude that outliers are not present in the data set.

Influential Observations

In Section 14.9 we discussed how the leverage of an observation can be used to identify observations for which the value of the independent variable may have a strong influence on the regression results. As we indicated in the discussion of standardized residuals, the leverage of an observation, denoted h_i, measures how far the values of the independent variables are from their mean values. The leverage values are easily obtained as part of the output from statistical software packages. Minitab computes the leverage values and uses the rule of thumb $h_i > 3(p + 1)/n$ to identify **influential observations**. For the Butler Trucking example with $p = 2$ independent variables and $n = 10$ observations, the critical value for leverage is $3(2 + 1)/10 = .9$. The leverage values for the Butler Trucking example obtained by using Minitab are reported in Table 15.9. Because h_i does not exceed .9, we do not detect influential observations in the data set.

TABLE 15.10 DATA SET ILLUSTRATING POTENTIAL PROBLEM USING THE LEVERAGE CRITERION

x_i	y_i	Leverage h_i
1	18	.204170
1	21	.204170
2	22	.164205
3	21	.138141
4	23	.125977
4	24	.125977
5	26	.127715
15	39	.909644

Using Cook's Distance Measure to Identify Influential Observations

A problem that can arise in using leverage to identify influential observations is that an observation can be identified as having high leverage and not necessarily be influential in terms of the resulting estimated regression equation. For example, Table 15.10 is a data set consisting of eight observations and their corresponding leverage values (obtained by using Minitab). Because the leverage for the eighth observation is $.91 > .75$ (the critical leverage value), this observation is identified as influential. Before reaching any final conclusions, however, let us consider the situation from a different perspective.

FIGURE 15.11 SCATTER DIAGRAM FOR THE DATA SET IN TABLE 15.10

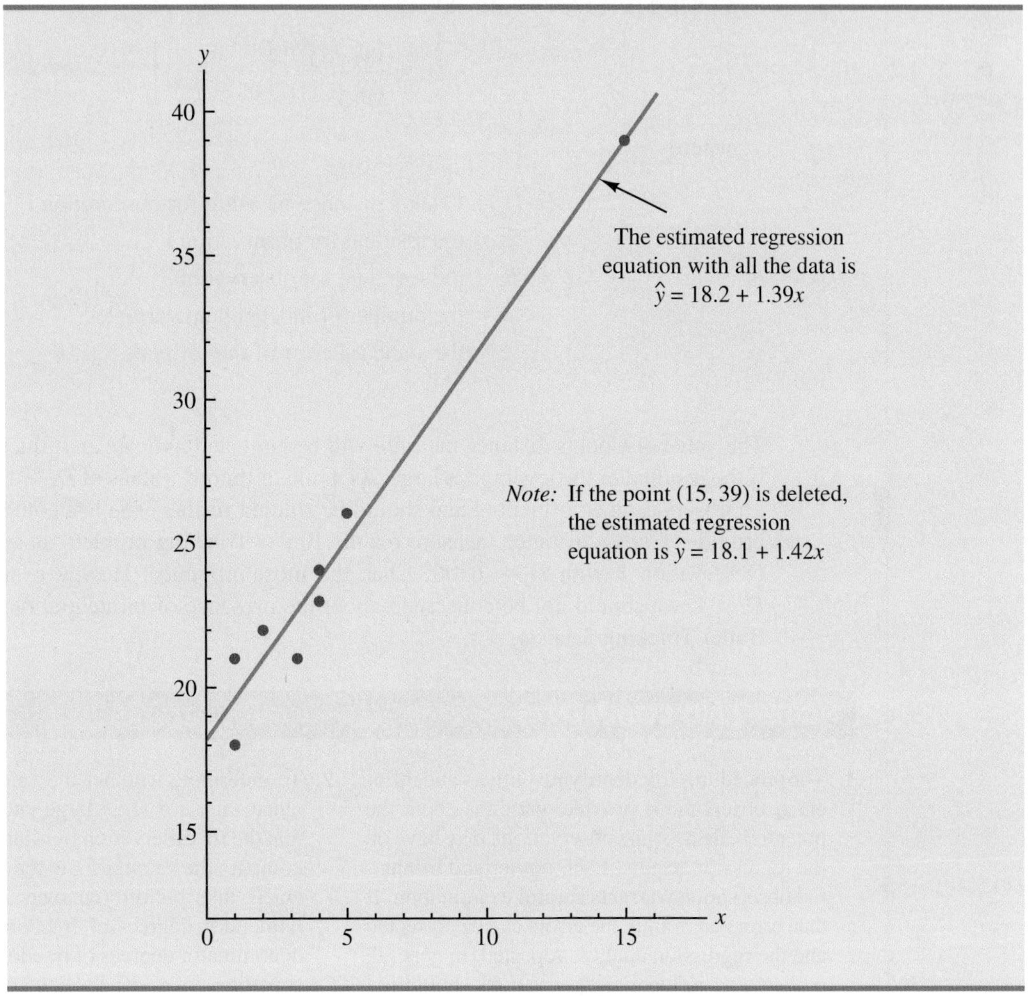

Figure 15.11 shows the scatter diagram corresponding to the data set in Table 15.10. We used Minitab to develop the following estimated regression equation for these data.

$$\hat{y} = 18.2 + 1.39x$$

The straight line in Figure 15.11 is the graph of this equation. Now, let us delete the observation $x = 15$, $y = 39$ from the data set and fit a new estimated regression equation to the remaining seven observations; the new estimated regression equation is

$$\hat{y} = 18.1 + 1.42x$$

We note that the y-intercept and slope of the new estimated regression equation are very close to the values obtained using all the data. Although the leverage criterion identified the eighth observation as influential, this observation clearly had little influence on the results obtained. Thus, in some situations using only leverage to identify influential observations can lead to wrong conclusions.

Cook's distance measure uses both the leverage of observation i, h_i, and the residual for observation i, $(y_i - \hat{y}_i)$, to determine whether the observation is influential.

COOK'S DISTANCE MEASURE

$$D_i = \frac{(y_i - \hat{y}_i)^2}{(p + 1)s^2}\left[\frac{h_i}{(1 - h_i)^2}\right] \tag{15.25}$$

where

D_i = Cook's distance measure for observation i
$y_i - \hat{y}_i$ = the residual for observation i
h_i = the leverage for observation i
p = the number of independent variables
s = the standard error of the estimate

The value of Cook's distance measure will be large and indicate an influential observation if the residual or the leverage is large. As a rule of thumb, values of $D_i > 1$ indicate that the ith observation is influential and should be studied further. The last column of Table 15.9 provides Cook's distance measure for the Butler Trucking problem as given by Minitab. Observation 8 with $D_i = .650029$ has the most influence. However, applying the rule $D_i > 1$, we should not be concerned about the presence of influential observations in the Butler Trucking data set.

NOTES AND COMMENTS

1. The procedures for identifying outliers and influential observations provide warnings about the potential effects some observations may have on the regression results. Each outlier and influential observation warrants careful examination. If data errors are found, the errors can be corrected and the regression analysis repeated. In general, outliers and influential observations should not be removed from the data set unless clear evidence shows that they are not based on elements of the population being studied and should not have been included in the original data set.
2. To determine whether the value of Cook's distance measure D_i is large enough to conclude that the ith observation is influential, we can also compare the value of D_i to the 50th percentile of an F distribution (denoted $F_{.50}$) with $p + 1$ numerator degrees of freedom and $n - p - 1$ denominator degrees of freedom. F tables corresponding to a .50 level of significance must be available to carry out the test. The rule of thumb we provided ($D_i > 1$) is based on the fact that the table value is close to one for a wide variety of cases.

Exercises

Methods

SELF test

39. Data for two variables, x and y, follow.

x_i	1	2	3	4	5
y_i	3	7	5	11	14

a. Develop the estimated regression equation for these data.
b. Plot the standardized residuals versus $\hat{y}$. Do there appear to be any outliers in these data? Explain.
c. Compute the studentized deleted residuals for these data. At the .05 level of significance, can any of these observations be classified as an outlier? Explain.

40. Data for two variables, x and y, follow.

x_i	22	24	26	28	40
y_i	12	21	31	35	70

a. Develop the estimated regression equation for these data.
b. Compute the studentized deleted residuals for these data. At the .05 level of significance, can any of these observations be classified as an outlier? Explain.
c. Compute the leverage values for these data. Do there appear to be any influential observations in these data? Explain.
d. Compute Cook's distance measure for these data. Are any observations influential? Explain.

Applications

SELF test

41. Exercise 5 gave the following data on weekly gross revenue, television advertising, and newspaper advertising for Showtime Movie Theaters.

WEB file
Showtime

Weekly Gross Revenue ($1000s)	Television Advertising ($1000s)	Newspaper Advertising ($1000s)
96	5.0	1.5
90	2.0	2.0
95	4.0	1.5
92	2.5	2.5
95	3.0	3.3
94	3.5	2.3
94	2.5	4.2
94	3.0	2.5

a. Find an estimated regression equation relating weekly gross revenue to television and newspaper advertising.
b. Plot the standardized residuals against $\hat{y}$. Does the residual plot support the assumptions about ϵ? Explain.
c. Check for any outliers in these data. What are your conclusions?
d. Are there any influential observations? Explain.

42. The following data show the curb weight, horsepower, and ¼-mile speed for 16 popular sports and GT cars. Suppose that the price of each sports and GT car is also available. The complete data set is as follows:

WEB file
Auto2

Sports & GT Car	Price ($1000s)	Curb Weight (lb.)	Horsepower	Speed at ¼ Mile (mph)
Acura Integra Type R	25.035	2577	195	90.7
Acura NSX-T	93.758	3066	290	108.0
BMW Z3 2.8	40.900	2844	189	93.2
Chevrolet Camaro Z28	24.865	3439	305	103.2
Chevrolet Corvette Convertible	50.144	3246	345	102.1
Dodge Viper RT/10	69.742	3319	450	116.2
Ford Mustang GT	23.200	3227	225	91.7
Honda Prelude Type SH	26.382	3042	195	89.7
Mercedes-Benz CLK320	44.988	3240	215	93.0
Mercedes-Benz SLK230	42.762	3025	185	92.3
Mitsubishi 3000GT VR-4	47.518	3737	320	99.0

(continued)

Sports & GT Car	Price ($1000s)	Curb Weight (lb.)	Horsepower	Speed at ¼ Mile (mph)
Nissan 240SX SE	25.066	2862	155	84.6
Pontiac Firebird Trans Am	27.770	3455	305	103.2
Porsche Boxster	45.560	2822	201	93.2
Toyota Supra Turbo	40.989	3505	320	105.0
Volvo C70	41.120	3285	236	97.0

a. Find the estimated regression equation that uses price and horsepower to predict ¼-mile speed.
b. Plot the standardized residuals against $\hat{y}$. Does the residual plot support the assumption about ϵ? Explain.
c. Check for any outliers. What are your conclusions?
d. Are there any influential observations? Explain.

WEB file

LPGA

43. The Ladies Professional Golfers Association (LPGA) maintains statistics on performance and earnings for members of the LPGA Tour. Year-end performance statistics for the 30 players who had the highest total earnings in LPGA Tour events for 2005 appear in the file named LPGA (LPGA website, 2006). Earnings ($1000s) is the total earnings in thousands of dollars; Scoring Avg. is the average score for all events; Greens in Reg. is the percentage of time a player is able to hit the green in regulation; and Putting Avg. is the average number of putts taken on greens hit in regulation. A green is considered hit in regulation if any part of the ball is touching the putting surface and the difference between the value of par for the hole and the number of strokes taken to hit the green is at least 2.
a. Develop an estimated regression equation that can be used to predict the average score for all events given the percentage of time a player is able to hit the green in regulation and the average number of putts taken on greens hit in regulation.
b. Plot the standardized residuals against $\hat{y}$. Does the residual plot support the assumption about ϵ? Explain.
c. Check for any outliers. What are your conclusions?
d. Are there any influential observations? Explain.

15.9 Logistic Regression

In many regression applications the dependent variable may only assume two discrete values. For instance, a bank might like to develop an estimated regression equation for predicting whether a person will be approved for a credit card. The dependent variable can be coded as $y = 1$ if the bank approves the request for a credit card and $y = 0$ if the bank rejects the request for a credit card. Using logistic regression we can estimate the probability that the bank will approve the request for a credit card given a particular set of values for the chosen independent variables.

Let us consider an application of logistic regression involving a direct mail promotion being used by Simmons Stores. Simmons owns and operates a national chain of women's apparel stores. Five thousand copies of an expensive four-color sales catalog have been printed, and each catalog includes a coupon that provides a $50 discount on purchases of $200 or more. The catalogs are expensive and Simmons would like to send them to only those customers who have the highest probability of using the coupon.

Management thinks that annual spending at Simmons Stores and whether a customer has a Simmons credit card are two variables that might be helpful in predicting whether a customer who receives the catalog will use the coupon. Simmons conducted a pilot study using a random sample of 50 Simmons credit card customers and 50 other customers who do not have a Simmons credit card. Simmons sent the catalog to each of the 100 customers selected. At the end of a test period, Simmons noted whether the customer used the coupon. The sample data for the first 10 catalog recipients are shown in Table 15.11. The amount each customer spent last year at Simmons is shown in thousands of dollars and the credit card information has been coded as 1 if the customer has a Simmons credit card and 0 if not. In the Coupon column, a 1 is recorded if the sampled customer used the coupon and 0 if not.

We might think of building a multiple regression model using the data in Table 15.11 to help Simmons estimate whether a catalog recipient will use the coupon. We would use Annual Spending ($1000) and Simmons Card as independent variables and Coupon as the dependent variable. Because the dependent variable may only assume the values of 0 or 1, however, the ordinary multiple regression model is not applicable. This example shows the type of situation for which logistic regression was developed. Let us see how logistic regression can be used to help Simmons estimate which type of customer is most likely to take advantage of their promotion.

Logistic Regression Equation

In many ways logistic regression is like ordinary regression. It requires a dependent variable, y, and one or more independent variables. In multiple regression analysis, the mean or expected value of y is referred to as the multiple regression equation.

$$E(y) = \beta_0 + \beta_1 x_1 + \beta_2 x_2 + \cdots + \beta_p x_p \tag{15.26}$$

In logistic regression, statistical theory as well as practice has shown that the relationship between $E(y)$ and $x_1, x_2, \ldots, x_p$ is better described by the following nonlinear equation.

LOGISTIC REGRESSION EQUATION

$$E(y) = \frac{e^{\beta_0 + \beta_1 x_1 + \beta_2 x_2 + \cdots + \beta_p x_p}}{1 + e^{\beta_0 + \beta_1 x_1 + \beta_2 x_2 + \cdots + \beta_p x_p}} \tag{15.27}$$

If the two values of the dependent variable y are coded as 0 or 1, the value of $E(y)$ in equation (15.27) provides the *probability* that $y = 1$ given a particular set of values for the

TABLE 15.11 PARTIAL SAMPLE DATA FOR THE SIMMONS STORES EXAMPLE

WEB file
Simmons

Customer	Annual Spending ($1000)	Simmons Card	Coupon
1	2.291	1	0
2	3.215	1	0
3	2.135	1	0
4	3.924	0	0
5	2.528	1	0
6	2.473	0	1
7	2.384	0	0
8	7.076	0	0
9	1.182	1	1
10	3.345	0	0

independent variables $x_1, x_2, \ldots, x_p$. Because of the interpretation of $E(y)$ as a probability, the **logistic regression equation** is often written as follows.

INTERPRETATION OF $E(y)$ AS A PROBABILITY IN LOGISTIC REGRESSION

$$E(y) = P(y = 1 | x_1, x_2, \ldots, x_p) \tag{15.28}$$

To provide a better understanding of the characteristics of the logistic regression equation, suppose the model involves only one independent variable x and the values of the model parameters are $\beta_0 = -7$ and $\beta_1 = 3$. The logistic regression equation corresponding to these parameter values is

$$E(y) = P(y = 1 | x) = \frac{e^{\beta_0 + \beta_1 x}}{1 + e^{\beta_0 + \beta_1 x}} = \frac{e^{-7+3x}}{1 + e^{-7+3x}} \tag{15.29}$$

Figure 15.12 shows a graph of equation (15.29). Note that the graph is S-shaped. The value of $E(y)$ ranges from 0 to 1, with the value of $E(y)$ gradually approaching 1 as the value of x becomes larger and the value of $E(y)$ approaching 0 as the value of x becomes smaller. Note also that the values of $E(y)$, representing probability, increase fairly rapidly as x increases from 2 to 3. The fact that the values of $E(y)$ range from 0 to 1 and that the curve is S-shaped makes equation (15.29) ideally suited to model the probability the dependent variable is equal to 1.

Estimating the Logistic Regression Equation

In simple linear and multiple regression the least squares method is used to compute $b_0, b_1, \ldots, b_p$ as estimates of the model parameters ($\beta_0, \beta_1, \ldots, \beta_p$). The nonlinear form of the logistic regression equation makes the method of computing estimates more complex and beyond the scope of this text. We will use computer software to provide the estimates. The **estimated logistic regression equation** is

FIGURE 15.12 LOGISTIC REGRESSION EQUATION FOR $\beta_0 = -7$ AND $\beta_1 = 3$

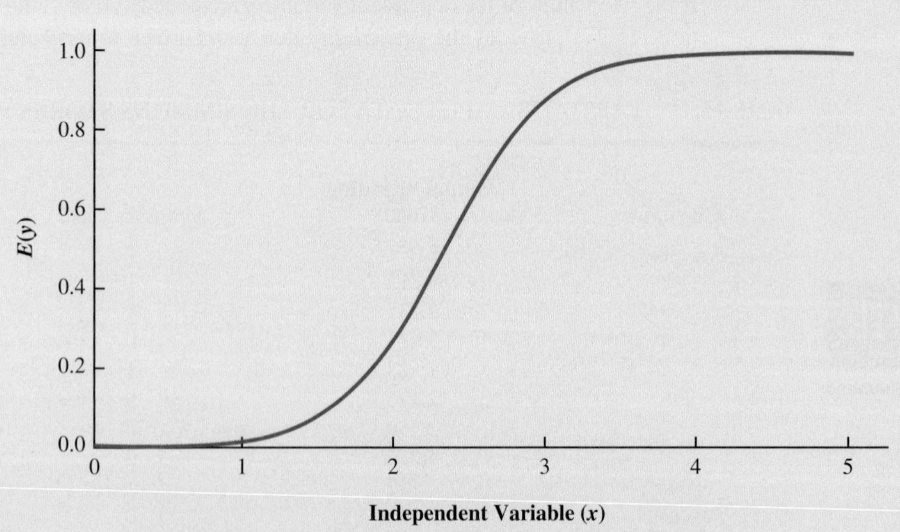

ESTIMATED LOGISTIC REGRESSION EQUATION

$$\hat{y} = \text{estimate of } P(y = 1|x_1, x_2, \ldots, x_p) = \frac{e^{b_0 + b_1x_1 + b_2x_2 + \cdots + b_px_p}}{1 + e^{b_0 + b_1x_1 + b_2x_2 + \cdots + b_px_p}} \quad \textbf{(15.30)}$$

Here, $\hat{y}$ provides an estimate of the probability that $y = 1$ given a particular set of values for the independent variables.

Let us now return to the Simmons Stores example. The variables in the study are defined as follows:

$$y = \begin{cases} 0 \text{ if the customer did not use the coupon} \\ 1 \text{ if the customer used the coupon} \end{cases}$$

$$x_1 = \text{annual spending at Simmons Stores (\$1000s)}$$

$$x_2 = \begin{cases} 0 \text{ if the customer does not have a Simmons credit card} \\ 1 \text{ if the customer has a Simmons credit card} \end{cases}$$

Thus, we choose a logistic regression equation with two independent variables.

$$E(y) = \frac{e^{\beta_0 + \beta_1x_1 + \beta_2x_2}}{1 + e^{\beta_0 + \beta_1x_1 + \beta_2x_2}} \quad \textbf{(15.31)}$$

In Appendix 15.3 we show how Minitab is used to generate the output in Figure 15.13.

Using the sample data (see Table 15.11), Minitab's binary logistic regression procedure was used to compute estimates of the model parameters β_0, β_1, and β_2. A portion of the output obtained is shown in Figure 15.13. We see that $b_0 = -2.14637$, $b_1 = .341643$, and $b_2 = 1.09873$. Thus, the estimated logistic regression equation is

$$\hat{y} = \frac{e^{b_0 + b_1x_1 + b_2x_2}}{1 + e^{b_0 + b_1x_1 + b_2x_2}} = \frac{e^{-2.14637 + .341643x_1 + 1.09873x_2}}{1 + e^{-2.14637 + .341643x_1 + 1.09873x_2}} \quad \textbf{(15.32)}$$

We can now use equation (15.32) to estimate the probability of using the coupon for a particular type of customer. For example, to estimate the probability of using the coupon for customers who spend \$2000 annually and do not have a Simmons credit card, we substitute $x_1 = 2$ and $x_2 = 0$ into equation (15.32).

FIGURE 15.13 PARTIAL LOGISTIC REGRESSION OUTPUT FOR THE SIMMONS STORES EXAMPLE

In the Minitab output, x_1 = Spending and x_2 = Card.

```
Logistic Regression Table
                                                  Odds     95%   CI
Predictor        Coef    SE Coef      Z       p  Ratio   Lower  Upper
Constant     -2.14637   0.577245  -3.72   0.000
Spending     0.341643   0.128672   2.66   0.008   1.41    1.09   1.81
Card          1.09873   0.444696   2.47   0.013   3.00    1.25   7.17

Log-Likelihood = -60.487
Test that all slopes are zero: G = 13.628, DF = 2, P-Value = 0.001
```

$$\hat{y} = \frac{e^{-2.14637+.341643(2)+1.09873(0)}}{1 + e^{-2.14637+.341643(2)+1.09873(0)}} = \frac{e^{-1.4631}}{1 + e^{-1.4631}} = \frac{.2315}{1.2315} = .1880$$

Thus, an estimate of the probability of using the coupon for this particular group of customers is approximately 0.19. Similarly, to estimate the probability of using the coupon for customers who spent \$2000 last year and have a Simmons credit card, we substitute $x_1 = 2$ and $x_2 = 1$ into equation (15.32).

$$\hat{y} = \frac{e^{-2.14637+.341643(2)+1.09873(1)}}{1 + e^{-2.14637+.341643(2)+1.09873(1)}} = \frac{e^{-.3644}}{1 + e^{-.3644}} = \frac{.6946}{1.6946} = .4099$$

Thus, for this group of customers, the probability of using the coupon is approximately 0.41. It appears that the probability of using the coupon is much higher for customers with a Simmons credit card. Before reaching any conclusions, however, we need to assess the statistical significance of our model.

Testing for Significance

Testing for significance in logistic regression is similar to testing for significance in multiple regression. First we conduct a test for overall significance. For the Simmons Stores example, the hypotheses for the test of overall significance follow:

$$H_0\colon \beta_1 = \beta_2 = 0$$
$$H_a\colon \text{One or both of the parameters is not equal to zero}$$

The test for overall significance is based upon the value of a G test statistic. If the null hypothesis is true, the sampling distribution of G follows a chi-square distribution with degrees of freedom equal to the number of independent variables in the model. Although the computation of G is beyond the scope of the book, the value of G and its corresponding p-value are provided as part of Minitab's binary logistic regression output. Referring to the last line in Figure 15.13, we see that the value of G is 13.628, its degrees of freedom are 2, and its p-value is .001. Thus, at any level of significance $\alpha \geq .001$, we would reject the null hypothesis and conclude that the overall model is significant.

If the G test shows an overall significance, a z test can be used to determine whether each of the individual independent variables is making a significant contribution to the overall model. For the independent variables x_i, the hypotheses are

$$H_0\colon \beta_i = 0$$
$$H_a\colon \beta_i \neq 0$$

If the null hypothesis is true, the value of the estimated coefficient divided by its standard error follows a standard normal probability distribution. The column labeled Z in the Minitab output contains the values of $z_i = b_i/s_{b_i}$ for each of the estimated coefficients and the column labeled p contains the corresponding p-values. Suppose we use $\alpha = .05$ to test for the significance of the independent variables in the Simmons model. For the independent variable x_1 the z value is 2.66 and the corresponding p-value is .008. Thus, at the .05 level of significance we can reject $H_0\colon \beta_1 = 0$. In a similar fashion we can also reject $H_0\colon \beta_2 = 0$ because the p-value corresponding to $z = 2.47$ is .013. Hence, at the .05 level of significance, both independent variables are statistically significant.

Managerial Use

We described how to develop the estimated logistic regression equation and how to test it for significance. Let us now use it to make a decision recommendation concerning the Simmons Stores catalog promotion. For Simmons Stores, we already computed $P(y = 1|x_1 = 2, x_2 = 1) = .4099$ and $P(y = 1|x_1 = 2, x_2 = 0) = .1880$. These probabilities indicate that for customers with annual spending of \$2000 the presence of a Simmons credit card increases the probability of using the coupon. In Table 15.12 we show estimated probabilities for values of annual spending ranging from \$1000 to \$7000 for both customers who have a Simmons credit card and customers who do not have a Simmons credit card. How can Simmons use this information to better target customers for the new promotion? Suppose Simmons wants to send the promotional catalog only to customers who have a 0.40 or higher probability of using the coupon. Using the estimated probabilities in Table 15.12, Simmons promotion strategy would be:

Customers who have a Simmons credit card: Send the catalog to every customer who spent \$2000 or more last year.

Customers who do not have a Simmons credit card: Send the catalog to every customer who spent \$6000 or more last year.

Looking at the estimated probabilities further, we see that the probability of using the coupon for customers who do not have a Simmons credit card but spend \$5000 annually is .3922. Thus, Simmons may want to consider revising this strategy by including those customers who do not have a credit card, as long as they spent \$5000 or more last year.

Interpreting the Logistic Regression Equation

Interpreting a regression equation involves relating the independent variables to the business question that the equation was developed to answer. With logistic regression, it is difficult to interpret the relation between the independent variables and the probability that $y = 1$ directly because the logistic regression equation is nonlinear. However, statisticians have shown that the relationship can be interpreted indirectly using a concept called the odds ratio.

The **odds in favor of an event occurring** is defined as the probability the event will occur divided by the probability the event will not occur. In logistic regression the event of interest is always $y = 1$. Given a particular set of values for the independent variables, the odds in favor of $y = 1$ can be calculated as follows:

$$\text{odds} = \frac{P(y = 1|x_1, x_2, \ldots, x_p)}{P(y = 0|x_1, x_2, \ldots, x_p)} = \frac{P(y = 1|x_1, x_2, \ldots, x_p)}{1 - P(y = 1|x_1, x_2, \ldots, x_p)} \quad \textbf{(15.33)}$$

The **odds ratio** measures the impact on the odds of a one-unit increase in only one of the independent variables. The odds ratio is the odds that $y = 1$ given that one of the

TABLE 15.12 ESTIMATED PROBABILITIES FOR SIMMONS STORES

		Annual Spending						
		\$1000	**\$2000**	**\$3000**	**\$4000**	**\$5000**	**\$6000**	**\$7000**
Credit Card	**Yes**	.3305	.4099	.4943	.5791	.6594	.7315	.7931
	No	.1413	.1880	.2457	.3144	.3922	.4759	.5610

independent variables has been increased by one unit (odds_1) divided by the odds that $y = 1$ given no change in the values for the independent variables (odds_0).

ODDS RATIO

$$\text{Odds Ratio} = \frac{\text{odds}_1}{\text{odds}_0} \tag{15.34}$$

For example, suppose we want to compare the odds of using the coupon for customers who spend \$2000 annually and have a Simmons credit card ($x_1 = 2$ and $x_2 = 1$) to the odds of using the coupon for customers who spend \$2000 annually and do not have a Simmons credit card ($x_1 = 2$ and $x_2 = 0$). We are interested in interpreting the effect of a one-unit increase in the independent variable x_2. In this case

$$\text{odds}_1 = \frac{P(y = 1|x_1 = 2, x_2 = 1)}{1 - P(y = 1|x_1 = 2, x_2 = 1)}$$

and

$$\text{odds}_0 = \frac{P(y = 1|x_1 = 2, x_2 = 0)}{1 - P(y = 1|x_1 = 2, x_2 = 0)}$$

Previously we showed that an estimate of the probability that $y = 1$ given $x_1 = 2$ and $x_2 = 1$ is .4099, and an estimate of the probability that $y = 1$ given $x_1 = 2$ and $x_2 = 0$ is .1880. Thus,

$$\text{estimate of odds}_1 = \frac{.4099}{1 - .4099} = .6946$$

and

$$\text{estimate of odds}_0 = \frac{.1880}{1 - .1880} = .2315$$

The estimated odds ratio is

$$\text{Estimated odds ratio} = \frac{.6946}{.2315} = 3.00$$

Thus, we can conclude that the estimated odds in favor of using the coupon for customers who spent \$2000 last year and have a Simmons credit card are 3 times greater than the estimated odds in favor of using the coupon for customers who spent \$2000 last year and do not have a Simmons credit card.

The odds ratio for each independent variable is computed while holding all the other independent variables constant. But it does not matter what constant values are used for the other independent variables. For instance, if we computed the odds ratio for the Simmons credit card variable (x_2) using \$3000, instead of \$2000, as the value for the annual spending variable (x_1), we would still obtain the same value for the estimated odds ratio (3.00). Thus, we can conclude that the estimated odds of using the coupon for customers who have a Simmons credit card are 3 times greater than the estimated odds of using the coupon for customers who do not have a Simmons credit card.

The odds ratio is standard output for logistic regression software packages. Refer to the Minitab output in Figure 15.13. The column with the heading Odds Ratio contains the estimated odds ratios for each of the independent variables. The estimated odds ratio for x_1 is 1.41 and the estimated odds ratio for x_2 is 3.00. We already showed how to interpret the estimated odds ratio for the binary independent variable x_2. Let us now consider the interpretation of the estimated odds ratio for the continuous independent variable x_1.

The value of 1.41 in the Odds Ratio column of the Minitab output tells us that the estimated odds in favor of using the coupon for customers who spent \$3000 last year is 1.41 times greater than the estimated odds in favor of using the coupon for customers who spent \$2000 last year. Moreover, this interpretation is true for any one-unit change in x_1. For instance, the estimated odds in favor of using the coupon for someone who spent \$5000 last year is 1.41 times greater than the odds in favor of using the coupon for a customer who spent \$4000 last year. But suppose we are interested in the change in the odds for an increase of more than one unit for an independent variable. Note that x_1 can range from 1 to 7. The odds ratio given by the Minitab output does not answer this question. To answer this question we must explore the relationship between the odds ratio and the regression coefficients.

A unique relationship exists between the odds ratio for a variable and its corresponding regression coefficient. For each independent variable in a logistic regression equation it can be shown that

$$\text{Odds ratio} = e^{\beta_i}$$

To illustrate this relationship, consider the independent variable x_1 in the Simmons example. The estimated odds ratio for x_1 is

$$\text{Estimated odds ratio} = e^{b_1} = e^{.341643} = 1.41$$

Similarly, the estimated odds ratio for x_2 is

$$\text{Estimated odds ratio} = e^{b_2} = e^{1.09873} = 3.00$$

This relationship between the odds ratio and the coefficients of the independent variables makes it easy to compute estimated odds ratios once we develop estimates of the model parameters. Moreover, it also provides us with the ability to investigate changes in the odds ratio of more than or less than one unit for a continuous independent variable.

The odds ratio for an independent variable represents the change in the odds for a one-unit change in the independent variable holding all the other independent variables constant. Suppose that we want to consider the effect of a change of more than one unit, say c units. For instance, suppose in the Simmons example that we want to compare the odds of using the coupon for customers who spend \$5000 annually ($x_1 = 5$) to the odds of using the coupon for customers who spend \$2000 annually ($x_1 = 2$). In this case $c = 5 - 2 = 3$ and the corresponding estimated odds ratio is

$$e^{cb_1} = e^{3(.341643)} = e^{1.0249} = 2.79$$

This result indicates that the estimated odds of using the coupon for customers who spend \$5000 annually is 2.79 times greater than the estimated odds of using the coupon for customers who spend \$2000 annually. In other words, the estimated odds ratio for an increase of \$3000 in annual spending is 2.79.

In general, the odds ratio enables us to compare the odds for two different events. If the value of the odds ratio is 1, the odds for both events are the same. Thus, if the independent variable we are considering (such as Simmons credit card status) has a positive impact on the probability of the event occurring, the corresponding odds ratio will be greater than 1. Most logistic regression software packages provide a confidence interval for the odds ratio. The Minitab output in Figure 15.13 provides a 95% confidence interval for each of the odds ratios. For example, the point estimate of the odds ratio for x_1 is 1.41 and the 95% confidence interval is 1.09 to 1.81. Because the confidence interval does not contain the value of 1, we can conclude that x_1, has a significant effect on the estimated odds ratio. Similarly, the 95% confidence interval for the odds ratio for x_2 is 1.25 to 7.17. Because this interval does not contain the value of 1, we can also conclude that x_2 has a significant effect on the odds ratio.

Logit Transformation

An interesting relationship can be observed between the odds in favor of $y = 1$ and the exponent for e in the logistic regression equation. It can be shown that

$$\ln(\text{odds}) = \beta_0 + \beta_1 x_1 + \beta_2 x_2 + \cdots + \beta_p x_p$$

This equation shows that the natural logarithm of the odds in favor of $y = 1$ is a linear function of the independent variables. This linear function is called the **logit**. We will use the notation $g(x_1, x_2, \ldots, x_p)$ to denote the logit.

LOGIT

$$g(x_1, x_2, \ldots, x_p) = \beta_0 + \beta_1 x_1 + \beta_2 x_2 + \cdots + \beta_p x_p \quad \textbf{(15.35)}$$

Substituting $g(x_1, x_2, \ldots, x_p)$ for $\beta_1 + \beta_1 x_1 + \beta_2 x_2 + \cdots + \beta_p x_p$ in equation (15.27), we can write the logistic regression equation as

$$E(y) = \frac{e^{g(x_1, x_2, \ldots, x_p)}}{1 + e^{g(x_1, x_2, \ldots, x_p)}} \quad \textbf{(15.36)}$$

Once we estimate the parameters in the logistic regression equation, we can compute an estimate of the logit. Using $\hat{g}(x_1, x_2, \ldots, x_p)$ to denote the **estimated logit**, we obtain

ESTIMATED LOGIT

$$\hat{g}(x_1, x_2, \ldots, x_p) = b_0 + b_1 x_1 + b_2 x_2 + \cdots + b_p x_p \quad \textbf{(15.37)}$$

Thus, in terms of the estimated logit, the estimated regression equation is

$$\hat{y} = \frac{e^{b_0 + b_1 x_1 + b_2 x_2 + \cdots + b_p x_p}}{1 + e^{b_0 + b_1 x_1 + b_2 x_2 + \cdots + b_p x_p}} = \frac{e^{\hat{g}(x_1, x_2, \ldots, x_p)}}{1 + e^{\hat{g}(x_1, x_2, \ldots, x_p)}} \quad \textbf{(15.38)}$$

For the Simmons Stores example, the estimated logit is

$$\hat{g}(x_1, x_2) = -2.14637 + .341643x_1 + 1.09873x_2$$

and the estimated regression equation is

$$\hat{y} = \frac{e^{\hat{g}(x_1, x_2)}}{1 + e^{\hat{g}(x_1, x_2)}} = \frac{e^{-2.14637 + .341643x_1 + 1.09873x_2}}{1 + e^{-2.14637 + .341643x_1 + 1.09873x_2}}$$

Thus, because of the unique relationship between the estimated logit and the estimated logistic regression equation, we can compute the estimated probabilities for Simmons Stores by dividing $e^{\hat{g}(x_1, x_2)}$ by $1 + e^{\hat{g}(x_1, x_2)}$.

NOTES AND COMMENTS

1. Because of the unique relationship between the estimated coefficients in the model and the corresponding odds ratios, the overall test for significance based upon the G statistic is also a test of overall significance for the odds ratios. In addition, the z test for the individual significance of a model parameter also provides a statistical test of significance for the corresponding odds ratio.

2. In simple and multiple regression, the coefficient of determination is used to measure the goodness of fit. In logistic regression, no single measure provides a similar interpretation. A discussion of goodness of fit is beyond the scope of our introductory treatment of logistic regression.

Exercises

Applications

WEB file Simmons

44. Refer to the Simmons Stores example introduced in this section. The dependent variable is coded as $y = 1$ if the customer used the coupon and 0 if not. Suppose that the only information available to help predict whether the customer will use the coupon is the customer's credit card status, coded as $x = 1$ if the customer has a Simmons credit card and $x = 0$ if not.
    a. Write the logistic regression equation relating x to y.
    b. What is the interpretation of $E(y)$ when $x = 0$?
    c. For the Simmons data in Table 15.11, use Minitab to compute the estimated logit.
    d. Use the estimated logit computed in part (c) to estimate the probability of using the coupon for customers who do not have a Simmons credit card and to estimate the probability of using the coupon for customers who have a Simmons credit card.
    e. What is the estimated odds ratio? What is its interpretation?

45. In Table 15.12 we provided estimates of the probability of using the coupon in the Simmons Stores catalog promotion. A different value is obtained for each combination of values for the independent variables.
    a. Compute the odds in favor of using the coupon for a customer with annual spending of \$4000 who does not have a Simmons credit card ($x_1 = 4, x_2 = 0$).
    b. Use the information in Table 15.12 and part (a) to compute the odds ratio for the Simmons credit card variable $x_2 = 0$, holding annual spending constant at $x_1 = 4$.
    c. In the text, the odds ratio for the credit card variable was computed using the information in the \$2000 column of Table 15.12. Did you get the same value for the odds ratio in part (b)?

46. Community Bank would like to increase the number of customers who use payroll direct deposit. Management is considering a new sales campaign that will require each branch manager to call each customer who does not currently use payroll direct deposit. As an incentive to sign up for payroll direct deposit, each customer contacted will be offered free checking for two years. Because of the time and cost associated with the new campaign, management would like to focus their efforts on customers who have the highest probability of signing up for payroll direct deposit. Management believes that the average monthly balance in a customer's checking account may be a useful predictor of whether the customer will sign up for direct payroll deposit. To investigate the relationship between these two variables, Community Bank tried the new campaign using a sample of 50 checking account customers who do not currently use payroll direct deposit. The sample data show the average monthly checking account balance (in hundreds of dollars) and whether the customer contacted signed up for payroll direct deposit (coded 1 if the customer signed up for payroll direct deposit and 0 if not). The data are contained in the data set named Bank; a portion of the data follows.

WEB file
Bank

Customer	x = Monthly Balance	y = Direct Deposit
1	1.22	0
2	1.56	0
3	2.10	0
4	2.25	0
5	2.89	0
6	3.55	0
7	3.56	0
8	3.65	1
.	.	.
.	.	.
.	.	.
48	18.45	1
49	24.98	0
50	26.05	1

a. Write the logistic regression equation relating x to y.
b. For the Community Bank data, use Minitab to compute the estimated logistic regression equation.
c. Conduct a test of significance using the G test statistic. Use $\alpha = .05$.
d. Estimate the probability that customers with an average monthly balance of $1000 will sign up for direct payroll deposit.
e. Suppose Community Bank only wants to contact customers who have a .50 or higher probability of signing up for direct payroll deposit. What is the average monthly balance required to achieve this level of probability?
f. What is the estimated odds ratio? What is its interpretation?

47. Over the past few years the percentage of students who leave Lakeland College at the end of the first year has increased. Last year Lakeland started a voluntary one-week orientation program to help first-year students adjust to campus life. If Lakeland is able to show that the orientation program has a positive effect on retention, they will consider making the program a requirement for all first-year students. Lakeland's administration also suspects that students with lower GPAs have a higher probability of leaving Lakeland at the end of the first year. In order to investigate the relation of these variables to retention, Lakeland selected a random sample of 100 students from last year's entering class. The data are contained in the data set named Lakeland; a portion of the data follows.

WEB file
Lakeland

Student	GPA	Program	Return
1	3.78	1	1
2	2.38	0	1
3	1.30	0	0
4	2.19	1	0
5	3.22	1	1
6	2.68	1	1
.	.	.	.
.	.	.	.
.	.	.	.
98	2.57	1	1
99	1.70	1	1
100	3.85	1	1

The dependent variable was coded as $y = 1$ if the student returned to Lakeland for the sophomore year and $y = 0$ if not. The two independent variables are:

$$x_1 = \text{GPA at the end of the first semester}$$

$$x_2 = \begin{cases} 0 \text{ if the student did not attend the orientation program} \\ 1 \text{ if the student attended the orientation program} \end{cases}$$

a. Write the logistic regression equation relating x_1 and x_2 to y.
b. What is the interpretation of $E(y)$ when $x_2 = 0$?
c. Use both independent variables and Minitab to compute the estimated logit.
d. Conduct a test for overall significance using $\alpha = .05$.
e. Use $\alpha = .05$ to determine whether each of the independent variables is significant.
f. Use the estimated logit computed in part (c) to estimate the probability that students with a 2.5 grade point average who did not attend the orientation program will return to Lakeland for their sophomore year. What is the estimated probability for students with a 2.5 grade point average who attended the orientation program?
g. What is the estimated odds ratio for the orientation program? Interpret it.
h. Would you recommend making the orientation program a required activity? Why or why not?

WEB file
TireRatings

48. The Tire Rack maintains an independent consumer survey to help drivers help each other by sharing their long-term tire experiences. The data contained in the file named TireRatings show survey results for 68 all-season tires (Tire Rack website, March 21, 2012). Performance traits are rated using the following 10-point scale.

Superior		Excellent		Good		Fair		Unacceptable	
10	9	8	7	6	5	4	3	2	1

The values for the variable labeled Wet are the average of the ratings for each tire's wet traction performance and the values for the variable labeled Noise are the average of the ratings for the noise level generated by each tire. Respondents were also asked whether they would buy the tire again using the following 10-point scale:

Definitely		Probably		Possibly		Probably Not		Definitely Not	
10	9	8	7	6	5	4	3	2	1

The values for the variable labeled Buy Again are the average of the buy-again responses. For the purposes of this exercise, we created the following binary dependent variable:

$$\text{Purchase} = \begin{cases} 1 \text{ if the value of the Buy-Again variable is 7 or greater} \\ 0 \text{ if the value of the Buy-Again variable is less than 7} \end{cases}$$

Thus, if Purchase = 1, the respondent would probably or definitely buy the tire again.

a. Write the logistic regression equation relating x_1 = Wet performance rating and x_2 = Noise performance rating to y = Purchase.
b. Use Minitab to compute the estimated logit.
c. Use the estimated logit to compute an estimate of the probability that a customer will probably or definitely purchase a particular tire again with a Wet performance rating of 8 and a Noise performance rating of 8.
d. Suppose that the Wet and Noise performance ratings were 7. How does that affect the probability that a customer will probably or definitely purchase a particular tire again with these performance ratings?
e. If you were the CEO of a tire company, what do the results for parts (c) and (d) tell you?

Summary

In this chapter, we introduced multiple regression analysis as an extension of simple linear regression analysis presented in Chapter 14. Multiple regression analysis enables us to understand how a dependent variable is related to two or more independent variables. The mulitple regression equation $E(y) = \beta_0 + \beta_1 x_1 + \beta_2 x_2 + \cdots + \beta_p x_p$ shows that the mean or expected value of the dependent variable y, denoted $E(y)$, is related to the values of the independent variables $x_1, x_2, \ldots, x_p$. Sample data and the least squares method are used to develop the estimated multiple regression equation $\hat{y} = b_0 + b_1 x_1 + b_2 x_2 + \cdots + b_p x_p$. In effect $b_0, b_1, b_2, \ldots, b_p$ are sample statistics used to estimate the unknown model parameters $\beta_0, \beta_1, \beta_2, \ldots, \beta_p$. Computer printouts were used throughout the chapter to emphasize the fact that statistical software packages are the only realistic means of performing the numerous computations required in multiple regression analysis.

The multiple coefficient of determination was presented as a measure of the goodness of fit of the estimated regression equation. It determines the proportion of the variation of y that can be explained by the estimated regression equation. The adjusted multiple coefficient of determination is a similar measure of goodness of fit that adjusts for the number of independent variables and thus avoids overestimating the impact of adding more independent variables.

An F test and a t test were presented as ways to determine statistically whether the relationship among the variables is significant. The F test is used to determine whether there is a significant overall relationship between the dependent variable and the set of all independent variables. The t test is used to determine whether there is a significant relationship between the dependent variable and an individual independent variable given the other independent variables in the regression model. Correlation among the independent variables, known as multicollinearity, was discussed.

The section on categorical independent variables showed how dummy variables can be used to incorporate categorical data into multiple regression analysis. The section on residual analysis showed how residual analysis can be used to validate the model assumptions, detect outliers, and identify influential observations. Standardized residuals, leverage, studentized deleted residuals, and Cook's distance measure were discussed. The chapter concluded with a section on how logistic regression can be used to model situations in which the dependent variable may only assume two values.

Glossary

Multiple regression analysis Regression analysis involving two or more independent variables.

Multiple regression model The mathematical equation that describes how the dependent variable y is related to the independent variables $x_1, x_2, \ldots, x_p$ and an error term ϵ.

Multiple regression equation The mathematical equation relating the expected value or mean value of the dependent variable to the values of the independent variables; that is, $E(y) = \beta_0 + \beta_1 x_1 + \beta_2 x_2 + \cdots + \beta_p x_p$.

Estimated multiple regression equation The estimate of the multiple regression equation based on sample data and the least squares method; it is $\hat{y} = b_0 + b_1 x_1 + b_2 x_2 + \cdots + b_p x_p$.

Least squares method The method used to develop the estimated regression equation. It minimizes the sum of squared residuals (the deviations between the observed values of the dependent variable, y_i, and the predicted values of the dependent variable, $\hat{y}_i$).

Multiple coefficient of determination A measure of the goodness of fit of the estimated multiple regression equation. It can be interpreted as the proportion of the variability in the dependent variable that is explained by the estimated regression equation.

Adjusted multiple coefficient of determination A measure of the goodness of fit of the estimated multiple regression equation that adjusts for the number of independent variables in the model and thus avoids overestimating the impact of adding more independent variables.
Multicollinearity The term used to describe the correlation among the independent variables.
Categorical independent variable An independent variable with categorical data.
Dummy variable A variable used to model the effect of categorical independent variables. A dummy variable may take only the value zero or one.
Leverage A measure of how far the values of the independent variables are from their mean values.
Outlier An observation that does not fit the pattern of the other data.
Studentized deleted residuals Standardized residuals that are based on a revised standard error of the estimate obtained by deleting observation i from the data set and then performing the regression analysis and computations.
Influential observation An observation that has a strong influence on the regression results.
Cook's distance measure A measure of the influence of an observation based on both the leverage of observation i and the residual for observation i.
Logistic regression equation The mathematical equation relating $E(y)$, the probability that $y = 1$, to the values of the independent variables; that is, $E(y) = P(y = 1|x_1, x_2, \ldots, x_p) = \dfrac{e^{\beta_0+\beta_1x_1+\beta_2x_2+\cdots+\beta_px_p}}{1 + e^{\beta_0+\beta_1x_1+\beta_2x_2+\cdots+\beta_px_p}}$.
Estimated logistic regression equation The estimate of the logistic regression equation based on sample data; that is, $\hat{y} = \text{estimate of } P(y = 1|x_1, x_2, \ldots, x_p) = \dfrac{e^{b_0+b_1x_1+b_2x_2+\cdots+b_px_p}}{1 + e^{b_0+b_1x_1+b_2x_2+\cdots+b_px_p}}$.
Odds in favor of an event occurring The probability the event will occur divided by the probability the event will not occur.
Odds ratio The odds that $y = 1$ given that one of the independent variables increased by one unit (odds_1) divided by the odds that $y = 1$ given no change in the values for the independent variables (odds_0); that is, Odds ratio $= \text{odds}_1/\text{odds}_0$.
Logit The natural logarithm of the odds in favor of $y = 1$; that is, $g(x_1, x_2, \ldots, x_p) = \beta_0 + \beta_1x_1 + \beta_2x_2 + \cdots + \beta_px_p$.
Estimated logit An estimate of the logit based on sample data; that is, $\hat{g}(x_1, x_2, \ldots, x_p) = b_0 + b_1x_1 + b_2x_2 + \cdots + b_px_p$.

Key Formulas

Multiple Regression Model

$$y = \beta_0 + \beta_1x_1 + \beta_2x_2 + \cdots + \beta_px_p + \epsilon \qquad \textbf{(15.1)}$$

Multiple Regression Equation

$$E(y) = \beta_0 + \beta_1x_1 + \beta_2x_2 + \cdots + \beta_px_p \qquad \textbf{(15.2)}$$

Estimated Multiple Regression Equation

$$\hat{y} = b_0 + b_1x_1 + b_2x_2 + \cdots + b_px_p \qquad \textbf{(15.3)}$$

Least Squares Criterion

$$\min \Sigma(y_i - \hat{y}_i)^2 \qquad \textbf{(15.4)}$$

Relationship Among SST, SSR, and SSE

$$SST = SSR + SSE \tag{15.7}$$

Multiple Coefficient of Determination

$$R^2 = \frac{SSR}{SST} \tag{15.8}$$

Adjusted Multiple Coefficient of Determination

$$R_a^2 = 1 - (1 - R^2)\frac{n - 1}{n - p - 1} \tag{15.9}$$

Mean Square Due to Regression

$$MSR = \frac{SSR}{p} \tag{15.12}$$

Mean Square Due to Error

$$MSE = \frac{SSE}{n - p - 1} \tag{15.13}$$

***F* Test Statistic**

$$F = \frac{MSR}{MSE} \tag{15.14}$$

***t* Test Statistic**

$$t = \frac{b_i}{s_{b_i}} \tag{15.15}$$

Standardized Residual for Observation *i*

$$\frac{y_i - \hat{y}_i}{s_{y_i - \hat{y}_i}} \tag{15.23}$$

Standard Deviation of Residual *i*

$$s_{y_i - \hat{y}_i} = s\sqrt{1 - h_i} \tag{15.24}$$

Cook's Distance Measure

$$D_i = \frac{(y_i - \hat{y}_i)^2}{(p + 1)s^2}\left[\frac{h_i}{(1 - h_i)^2}\right] \tag{15.25}$$

Logistic Regression Equation

$$E(y) = \frac{e^{\beta_0+\beta_1x_1+\beta_2x_2+\cdots+\beta_px_p}}{1 + e^{\beta_0+\beta_1x_1+\beta_2x_2+\cdots+\beta_px_p}} \tag{15.27}$$

Estimated Logistic Regression Equation

$$\hat{y} = \text{estimate of } P(y = 1|x_1, x_2, \ldots, x_p) = \frac{e^{b_0+b_1x_1+b_2x_2+\cdots+b_px_p}}{1 + e^{b_0+b_1x_1+b_2x_2+\cdots+b_px_p}} \quad \textbf{(15.30)}$$

Odds Ratio

$$\text{Odds ratio} = \frac{\text{odds}_1}{\text{odds}_0} \quad \textbf{(15.34)}$$

Logit

$$g(x_1, x_2, \ldots, x_p) = \beta_0 + \beta_1x_1 + \beta_2x_2 + \cdots + \beta_px_p \quad \textbf{(15.35)}$$

Estimated Logit

$$\hat{g}(x_1, x_2, \ldots, x_p) = b_0 + b_1x_1 + b_2x_2 + \cdots + b_px_p \quad \textbf{(15.37)}$$

Supplementary Exercises

49. The admissions officer for Clearwater College developed the following estimated regression equation relating the final college GPA to the student's SAT mathematics score and high-school GPA.

$$\hat{y} = -1.41 + .0235x_1 + .00486x_2$$

where

$$x_1 = \text{high-school grade point average}$$
$$x_2 = \text{SAT mathematics score}$$
$$y = \text{final college grade point average}$$

a. Interpret the coefficients in this estimated regression equation.
b. Predict the final college GPA for a student who has a high-school average of 84 and a score of 540 on the SAT mathematics test.

50. The personnel director for Electronics Associates developed the following estimated regression equation relating an employee's score on a job satisfaction test to his or her length of service and wage rate.

$$\hat{y} = 14.4 - 8.69x_1 + 13.5x_2$$

where

$$x_1 = \text{length of service (years)}$$
$$x_2 = \text{wage rate (dollars)}$$
$$y = \text{job satisfaction test score (higher scores indicate greater job satisfaction)}$$

a. Interpret the coefficients in this estimated regression equation.
b. Predict the job satisfaction test score for an employee who has four years of service and makes \$6.50 per hour.

51. A partial computer output from a regression analysis follows.

```
The regression equation is
Y = 8.103 + 7.602 X1 + 3.111 X2

Predictor          Coef       SE Coef        T
Constant         _______       2.667       _____
X1               _______       2.105       _____
X2               _______       0.613       _____

S = 3.335     R-Sq = 92.3%     R-Sq(adj) = _____%

Analysis of Variance

SOURCE            DF        SS        MS        F
Regression       _____     1612      ______   _____
Residual Error    12      ______     ______
Total            _____    ______
```

a. Compute the missing entries in this output.
b. Use the F test and $\alpha = .05$ to see whether a significant relationship is present.
c. Use the t test and $\alpha = .05$ to test H_0: $\beta_1 = 0$ and H_0: $\beta_2 = 0$.
d. Compute R_a^2.

52. Recall that in exercise 49, the admissions officer for Clearwater College developed the following estimated regression equation relating final college GPA to the student's SAT mathematics score and high-school GPA.

$$\hat{y} = -1.41 + .0235x_1 + .00486x_2$$

where

x_1 = high-school grade point average
x_2 = SAT mathematics score
y = final college grade point average

A portion of the Minitab computer output follows.

```
The regression equation is
Y = -1.41 + .0235 X1 + .00486 X2

Predictor          Coef         SE Coef        T
Constant         -1.4053        0.4848       _____
X1               0.023467       0.008666     _____
X2               _______        0.001077     _____

S = 0.1298     R-Sq = ______     R-Sq(adj) = ______

Analysis of Variance

SOURCE            DF        SS          MS        F
Regression       _____     1.76209     _____    _____
Residual Error   _____     _______     _____
Total              9       1.88000
```

a. Complete the missing entries in this output.
b. Use the F test and a .05 level of significance to see whether a significant relationship is present.
c. Use the t test and $\alpha = .05$ to test $H_0: \beta_1 = 0$ and $H_0: \beta_2 = 0$.
d. Did the estimated regression equation provide a good fit to the data? Explain.

53. Recall that in exercise 50 the personnel director for Electronics Associates developed the following estimated regression equation relating an employee's score on a job satisfaction test to length of service and wage rate.

$$\hat{y} = 14.4 - 8.69x_1 + 13.5x_2$$

where

x_1 = length of service (years)
x_2 = wage rate (dollars)
y = job satisfaction test score (higher scores indicate greater job satisfaction)

A portion of the Minitab computer output follows.

```
The regression equation is
Y = 14.4 - 8.69 X1 + 13.52 X2

Predictor          Coef      SE Coef        T
Constant         14.448        8.191     1.76
X1               ______        1.555    _____
X2               13.517        2.085    _____

S = 3.773      R-Sq = _____%   R-Sq(adj) = _____%

Analysis of Variance

SOURCE            DF        SS        MS        F
Regression         2     _____     _____    _____
Residual Error  ____     71.17     _____
Total              7     720.0
```

a. Complete the missing entries in this output.
b. Compute F and test using $\alpha = .05$ to see whether a significant relationship is present.
c. Did the estimated regression equation provide a good fit to the data? Explain.
d. Use the t test and $\alpha = .05$ to test $H_0: \beta_1 = 0$ and $H_0: \beta_2 = 0$.

54. The Tire Rack, America's leading online distributor of tires and wheels, conducts extensive testing to provide customers with products that are right for their vehicle, driving style, and driving conditions. In addition, the Tire Rack maintains an independent consumer survey to help drivers help each other by sharing their long-term tire experiences. The following data show survey ratings (1 to 10 scale with 10 the highest rating) for 18 maximum performance summer tires (Tire Rack website, February 3, 2009). The variable Steering rates the tire's steering responsiveness, Tread Wear rates quickness of wear based on the driver's expectations, and Buy Again rates the driver's overall tire satisfaction and desire to purchase the same tire again.

WEB file
TireRack

Tire	Steering	Tread Wear	Buy Again
Goodyear Assurance TripleTred	8.9	8.5	8.1
Michelin HydroEdge	8.9	9.0	8.3
Michelin Harmony	8.3	8.8	8.2
Dunlop SP 60	8.2	8.5	7.9
Goodyear Assurance ComforTred	7.9	7.7	7.1
Yokohama Y372	8.4	8.2	8.9
Yokohama Aegis LS4	7.9	7.0	7.1
Kumho Power Star 758	7.9	7.9	8.3
Goodyear Assurance	7.6	5.8	4.5
Hankook H406	7.8	6.8	6.2
Michelin Energy LX4	7.4	5.7	4.8
Michelin MX4	7.0	6.5	5.3
Michelin Symmetry	6.9	5.7	4.2
Kumho 722	7.2	6.6	5.0
Dunlop SP 40 A/S	6.2	4.2	3.4
Bridgestone Insignia SE200	5.7	5.5	3.6
Goodyear Integrity	5.7	5.4	2.9
Dunlop SP20 FE	5.7	5.0	3.3

a. Develop an estimated regression equation that can be used to predict the Buy Again rating given based on the Steering rating. At the .05 level of significance, test for a significant relationship.
b. Did the estimated regression equation developed in part (a) provide a good fit to the data? Explain.
c. Develop an estimated regression equation that can be used to predict the Buy Again rating given the Steering rating and the Tread Wear rating.
d. Is the addition of the Tread Wear independent variable significant? Use $\alpha = .05$.

WEB file
2012FuelEcon

55. The Department of Energy and the U.S. Environmental Protection Agency's *2012 Fuel Economy Guide* provides fuel efficiency data for 2012 model year cars and trucks (Department of Energy website, April 16, 2012). The file named 2012FuelEcon provides a portion of the data for 309 cars. The column labeled Manufacturer shows the name of the company that manufactured the car; the column labeled Displacement shows the engine's displacement in liters; the column labeled Fuel shows the required or recommended type of fuel (regular or premium gasoline); the column labeled Drive identifies the type of drive (F for front wheel, R for rear wheel, and A for all wheel); and the column labeled Hwy MPG shows the fuel efficiency rating for highway driving in terms of miles per gallon.
a. Develop an estimated regression equation that can be used to predict the fuel efficiency for highway driving given the engine's displacement. Test for significance using $\alpha = .05$.
b. Consider the addition of the dummy variable FuelPremium, where the value of FuelPremium is 1 if the required or recommended type of fuel is premium gasoline and 0 if the type of fuel is regular gasoline. Develop the estimated regression equation that can be used to predict the fuel efficiency for highway driving given the engine's displacement and the dummy variable FuelPremium.
c. Use $\alpha = .05$ to determine whether the dummy variable added in part (b) is significant.
d. Consider the addition of the dummy variables FrontWheel and RearWheel. The value of FrontWheel is 1 if the car has front wheel drive and 0 otherwise; the value of RearWheel is 1 if the car has rear wheel drive and 0 otherwise. Thus, for a car that has all-wheel drive, the value of FrontWheel and the value of RearWheel is 0. Develop the estimated regression equation that can be used to predict the fuel efficiency for

highway driving given the engine's displacement, the dummy variable FuelPremium, and the dummy variables FrontWheel and RearWheel.

e. For the estimated regression equation developed in part (d), test for overall significance and individual significance using $\alpha = .05$.

56. A portion of a data set containing information for 45 mutual funds that are part of the *Morningstar Funds 500* for 2008 follows. The complete data set is available in the file named MutualFunds. The data set includes the following five variables:

Fund Type: The type of fund, labeled DE (Domestic Equity), IE (International Equity), and FI (Fixed Income).

Net Asset Value ($): The closing price per share on December 31, 2007.

5-Year Average Return (%): The average annual return for the fund over the past five years.

Expense Ratio (%): The percentage of assets deducted each fiscal year for fund expenses.

Morningstar Rank: The risk adjusted star rating for each fund; Morningstar ranks go from a low of 1-Star to a high of 5-Stars.

WEB file
MutualFunds

Fund Name	Fund Type	Net Asset Value ($)	5-Year Average Return (%)	Expense Ratio (%)	Morningstar Rank
Amer Cent Inc & Growth Inv	DE	28.88	12.39	.67	2-Star
American Century Intl. Disc	IE	14.37	30.53	1.41	3-Star
American Century Tax-Free Bond	FI	10.73	3.34	.49	4-Star
American Century Ultra	DE	24.94	10.88	.99	3-Star
Ariel	DE	46.39	11.32	1.03	2-Star
Artisan Intl Val	IE	25.52	24.95	1.23	3-Star
Artisan Small Cap	DE	16.92	15.67	1.18	3-Star
Baron Asset	DE	50.67	16.77	1.31	5-Star
Brandywine	DE	36.58	18.14	1.08	4-Star
⋮	⋮	⋮	⋮	⋮	⋮

a. Develop an estimated regression equation that can be used to predict the 5-year average return given the type of fund. At the .05 level of significance, test for a significant relationship.

b. Did the estimated regression equation developed in part (a) provide a good fit to the data? Explain.

c. Develop the estimated regression equation that can be used to predict the 5-year average return given the type of fund, the net asset value, and the expense ratio. At the .05 level of significance, test for a significant relationship. Do you think any variables should be deleted from the estimated regression equation? Explain.

d. Morningstar Rank is a categorical variable. Because the data set contains only funds with four ranks (2-Star through 5-Star), use the following dummy variables: 3StarRank = 1 for a 3-Star fund, 0 otherwise; 4StarRank = 1 for a 4-Star fund, 0 otherwise; and 5StarRank = 1 for a 5-Star fund, 0 otherwise. Develop an estimated regression equation that can be used to predict the 5-year average return given the type of fund, the expense ratio, and the Morningstar Rank. Using $\alpha = .05$, remove any independent variables that are not significant.

e. Use the estimated regression equation developed in part (d) to predict the 5-year average return for a domestic equity fund with an expense ratio of 1.05% and a 3-Star Morningstar Rank.

57. *Fortune* magazine publishes an annual list of the 100 best companies to work for. The data in the file named FortuneBest shows a portion of the data for a random sample of 30 of the companies that made the top 100 list for 2012 (*Fortune,* February 6, 2012). The column labeled Rank shows the rank of the company in the Fortune 100 list; the column labeled Size indicates whether the company is a small, midsize, or large company; the column labeled Salaried ($1000s) shows the average annual salary for salaried employees rounded to the nearest $1000; and the column labeled Hourly ($1000s) shows the average annual salary for hourly employees rounded to the nearest $1000. *Fortune* defines large companies as having more than 10,000 employees, midsize companies as having between 2500 and 10,000 employees, and small companies as having fewer than 2500 employees.

WEB file FortuneBest

Rank	Company	Size	Salaried ($1000s)	Hourly ($1000s)
4	Wegmans Food Markets	Large	56	29
6	NetApp	Midsize	143	76
7	Camden Property Trust	Small	71	37
8	Recreational Equipment (REI)	Large	103	28
10	Quicken Loans	Midsize	78	54
11	Zappos.com	Midsize	48	25
12	Mercedes-Benz USA	Small	118	50
20	USAA	Large	96	47
22	The Container Store	Midsize	71	45
25	Ultimate Software	Small	166	56
37	Plante Moran	Small	73	45
42	Baptist Health South Florida	Large	126	80
50	World Wide Technology	Small	129	31
53	Methodist Hospital	Large	100	83
58	Perkins Coie	Small	189	63
60	American Express	Large	114	35
64	TDIndustries	Small	93	47
66	QuikTrip	Large	69	44
72	EOG Resources	Small	189	81
75	FactSet Research Systems	Small	103	51
80	Stryker	Large	71	43
81	SRC	Small	84	33
84	Booz Allen Hamilton	Large	105	77
91	CarMax	Large	57	34
93	GoDaddy.com	Midsize	105	71
94	KPMG	Large	79	59
95	Navy Federal Credit Union	Midsize	77	39
97	Schweitzer Engineering Labs	Small	99	28
99	Darden Restaurants	Large	57	24
100	Intercontinental Hotels Group	Large	63	26

a. Use these data to develop an estimated regression equation that could be used to predict the average annual salary for salaried employees given the average annual salary for hourly employees.

b. Use $\alpha = .05$ to test for overall significance.

c. To incorporate the effect of size, a categorical variable with three levels, we used two dummy variables: Size-Midsize and Size-Small. The value of Size-Midsize = 1 if the company is a midsize company and 0 otherwise. And, the value of Size-Small = 1 if the company is a small company and 0 otherwise. Develop an estimated regression equation that could be used to predict the average annual salary for salaried employees given the average annual salary for hourly employees and the size of the company.

d. For the estimated regression equation developed in part (c), use the t test to determine the significance of the independent variables. Use $\alpha = .05$.

e. Based upon your findings in part (d), develop an estimated regression equation that can be used to predict the average annual salary for salaried employees given the average annual salary for hourly employees and the size of the company.

Case Problem 1 Consumer Research, Inc.

Consumer Research, Inc., is an independent agency that conducts research on consumer attitudes and behaviors for a variety of firms. In one study, a client asked for an investigation of consumer characteristics that can be used to predict the amount charged by credit card users. Data were collected on annual income, household size, and annual credit card charges for a sample of 50 consumers. The following data are contained in the file named Consumer.

WEB file
Consumer

Income ($1000s)	Household Size	Amount Charged ($)	Income ($1000s)	Household Size	Amount Charged ($)
54	3	4016	54	6	5573
30	2	3159	30	1	2583
32	4	5100	48	2	3866
50	5	4742	34	5	3586
31	2	1864	67	4	5037
55	2	4070	50	2	3605
37	1	2731	67	5	5345
40	2	3348	55	6	5370
66	4	4764	52	2	3890
51	3	4110	62	3	4705
25	3	4208	64	2	4157
48	4	4219	22	3	3579
27	1	2477	29	4	3890
33	2	2514	39	2	2972
65	3	4214	35	1	3121
63	4	4965	39	4	4183
42	6	4412	54	3	3730
21	2	2448	23	6	4127
44	1	2995	27	2	2921
37	5	4171	26	7	4603
62	6	5678	61	2	4273
21	3	3623	30	2	3067
55	7	5301	22	4	3074
42	2	3020	46	5	4820
41	7	4828	66	4	5149

Managerial Report

1. Use methods of descriptive statistics to summarize the data. Comment on the findings.
2. Develop estimated regression equations, first using annual income as the independent variable and then using household size as the independent variable. Which variable is the better predictor of annual credit card charges? Discuss your findings.
3. Develop an estimated regression equation with annual income and household size as the independent variables. Discuss your findings.
4. What is the predicted annual credit card charge for a three-person household with an annual income of $40,000?
5. Discuss the need for other independent variables that could be added to the model. What additional variables might be helpful?

Case Problem 2 Predicting Winnings for NASCAR Drivers

Matt Kenseth won the 2012 Daytona 500, the most important race of the NASCAR season. His win was no surprise because for the 2011 season he finished fourth in the point standings with 2330 points, behind Tony Stewart (2403 points), Carl Edwards (2403 points), and Kevin Harvick (2345 points). In 2011 he earned $6,183,580 by winning three Poles (fastest driver in qualifying), winning three races, finishing in the top five 12 times, and finishing in the top ten 20 times. NASCAR's point system in 2011 allocated 43 points to the driver who finished first, 42 points to the driver who finished second, and so on down to 1 point for the driver who finished in the 43rd position. In addition any driver who led a lap received 1 bonus point, the driver who led the most laps received an additional bonus point, and the race winner was awarded 3 bonus points. But, the maximum number of points a driver could earn in any race was 48. Table 15.13 shows data for the 2011 season for the top 35 drivers (NASCAR website, February 28, 2011).

TABLE 15.13 NASCAR RESULTS FOR THE 2011 SEASON

WEB file NASCAR

Driver	Points	Poles	Wins	Top 5	Top 10	Winnings ($)
Tony Stewart	2403	1	5	9	19	6,529,870
Carl Edwards	2403	3	1	19	26	8,485,990
Kevin Harvick	2345	0	4	9	19	6,197,140
Matt Kenseth	2330	3	3	12	20	6,183,580
Brad Keselowski	2319	1	3	10	14	5,087,740
Jimmie Johnson	2304	0	2	14	21	6,296,360
Dale Earnhardt Jr.	2290	1	0	4	12	4,163,690
Jeff Gordon	2287	1	3	13	18	5,912,830
Denny Hamlin	2284	0	1	5	14	5,401,190
Ryan Newman	2284	3	1	9	17	5,303,020
Kurt Busch	2262	3	2	8	16	5,936,470
Kyle Busch	2246	1	4	14	18	6,161,020
Clint Bowyer	1047	0	1	4	16	5,633,950
Kasey Kahne	1041	2	1	8	15	4,775,160
A. J. Allmendinger	1013	0	0	1	10	4,825,560
Greg Biffle	997	3	0	3	10	4,318,050
Paul Menard	947	0	1	4	8	3,853,690
Martin Truex Jr.	937	1	0	3	12	3,955,560
Marcos Ambrose	936	0	1	5	12	4,750,390
Jeff Burton	935	0	0	2	5	3,807,780
Juan Montoya	932	2	0	2	8	5,020,780
Mark Martin	930	2	0	2	10	3,830,910
David Ragan	906	2	1	4	8	4,203,660
Joey Logano	902	2	0	4	6	3,856,010
Brian Vickers	846	0	0	3	7	4,301,880
Regan Smith	820	0	1	2	5	4,579,860
Jamie McMurray	795	1	0	2	4	4,794,770
David Reutimann	757	1	0	1	3	4,374,770
Bobby Labonte	670	0	0	1	2	4,505,650
David Gilliland	572	0	0	1	2	3,878,390
Casey Mears	541	0	0	0	0	2,838,320
Dave Blaney	508	0	0	1	1	3,229,210
Andy Lally	398	0	0	0	0	2,868,220
Robby Gordon	268	0	0	0	0	2,271,890
J. J. Yeley	192	0	0	0	0	2,559,500

Managerial Report

1. Suppose you wanted to predict Winnings ($) using only the number of poles won (Poles), the number of wins (Wins), the number of top five finishes (Top 5), or the number of top ten finishes (Top 10). Which of these four variables provides the best single predictor of winnings?
2. Develop an estimated regression equation that can be used to predict Winnings ($) given the number of poles won (Poles), the number of wins (Wins), the number of top five finishes (Top 5), and the number of top ten (Top 10) finishes. Test for individual significance and discuss your findings and conclusions.
3. Create two new independent variables: Top 2–5 and Top 6–10. Top 2–5 represents the number of times the driver finished between second and fifth place and Top 6–10 represents the number of times the driver finished between sixth and tenth place. Develop an estimated regression equation that can be used to predict Winnings ($) using Poles, Wins, Top 2–5, and Top 6–10. Test for individual significance and discuss your findings and conclusions.
4. Based upon the results of your analysis, what estimated regression equation would you recommend using to predict Winnings ($)? Provide an interpretation of the estimated regression coefficients for this equation.

Case Problem 3 Finding the Best Car Value

When trying to decide what car to buy, real value is not necessarily determined by how much you spend on the initial purchase. Instead, cars that are reliable and don't cost much to own often represent the best values. But no matter how reliable or inexpensive a car may cost to own, it must also perform well.

To measure value, *Consumer Reports* developed a statistic referred to as a value score. The value score is based upon five-year owner costs, overall road-test scores, and predicted-reliability ratings. Five-year owner costs are based upon the expenses incurred in the first five years of ownership, including depreciation, fuel, maintenance and repairs, and so on. Using a national average of 12,000 miles per year, an average cost per mile driven is used as the measure of five-year owner costs. Road-test scores are the results of more than 50 tests and evaluations and are based on a 100-point scale, with higher scores indicating better performance, comfort, convenience, and fuel economy. The highest road-test score obtained in the tests conducted by *Consumer Reports* was a 99 for a Lexus LS 460L. Predicted-reliability ratings (1 = Poor, 2 = Fair, 3 = Good, 4 = Very Good, and 5 = Excellent) are based upon data from *Consumer Reports*' Annual Auto Survey.

WEB file
CarValues

A car with a value score of 1.0 is considered to be an "average-value" car. A car with a value score of 2.0 is considered to be twice as good a value as a car with a value score of 1.0; a car with a value score of 0.5 is considered half as good as average; and so on. The data for three sizes of cars (13 small sedans, 20 family sedans, and 21 upscale sedans), including the price ($) of each car tested, are contained in the file named CarValues (*Consumer Reports* website, April 18, 2012). To incorporate the effect of size of car, a categorical variable with three values (small sedan, family sedan, and upscale sedan), use the following dummy variables:

$$\text{Family-Sedan} = \begin{cases} 1 \text{ if the car is a Family Sedan} \\ 0 \text{ otherwise} \end{cases}$$

$$\text{Upscale-Sedan} = \begin{cases} 1 \text{ if the car is an Upscale Sedan} \\ 0 \text{ otherwise} \end{cases}$$

Managerial Report

1. Treating Cost/Mile as the dependent variable, develop an estimated regression with Family-Sedan and Upscale-Sedan as the independent variables. Discuss your findings.
2. Treating Value Score as the dependent variable, develop an estimated regression equation using Cost/Mile, Road-Test Score, Predicted Reliability, Family-Sedan, and Upscale-Sedan as the independent variables.
3. Delete any independent variables that are not significant from the estimated regression equation developed in part 2 using a .05 level of significance. After deleting any independent variables that are not significant, develop a new estimated regression equation.
4. Suppose someone claims that "smaller cars provide better values than larger cars." For the data in this case, the Small Sedans represent the smallest type of car and the Upscale Sedans represent the largest type of car. Does your analysis support this claim?
5. Use regression analysis to develop an estimated regression equation that could be used to predict the value score given the value of the Road-Test Score.
6. Use regression analysis to develop an estimated regression equation that could be used to predict the value score given the Predicted Reliability.
7. What conclusions can you derive from your analysis?

Appendix 15.1 Multiple Regression with Minitab

WEB file

Butler

In Section 15.2 we discussed the computer solution of multiple regression problems by showing Minitab's output for the Butler Trucking Company problem. In this appendix we describe the steps required to generate the Minitab computer solution. First, the data must be entered in a Minitab worksheet. The miles traveled are entered in column C1, the number of deliveries are entered in column C2, and the travel times (hours) are entered in column C3. The variable names Miles, Deliveries, and Time were entered as the column headings on the worksheet. In subsequent steps, we refer to the data by using the variable names Miles, Deliveries, and Time or the column indicators C1, C2, and C3. The following steps describe how to use Minitab to produce the regression results shown in Figure 15.4.

Step 1. Select the **Stat** menu
Step 2. Select the **Regression** menu
Step 3. Choose **Regression**
Step 4. When the **Regression** dialog box appears:
Enter Time in the **Response** box
Enter Miles and Deliveries in the **Predictors** box
Click **OK**

Appendix 15.2 Multiple Regression with Excel

WEB file

Butler

In Section 15.2 we discussed the computer solution of multiple regression problems by showing Minitab's output for the Butler Trucking Company problem. In this appendix we describe how to use Excel's Regression tool to develop the estimated multiple regression equation for the Butler Trucking problem. Refer to Figure 15.14 as we describe the tasks involved. First, the labels Assignment, Miles, Deliveries, and Time are entered into cells A1:D1 of the worksheet, and the sample data into cells B2:D11. The numbers 1–10 in cells A2:A11 identify each observation.

FIGURE 15.14 EXCEL OUTPUT FOR BUTLER TRUCKING WITH TWO INDEPENDENT VARIABLES

	A	B	C	D	E	F	G	H	I	J
1	Assignment	Miles	Deliveries	Time						
2	1	100	4	9.3						
3	2	50	3	4.8						
4	3	100	4	8.9						
5	4	100	2	6.5						
6	5	50	2	4.2						
7	6	80	2	6.2						
8	7	75	3	7.4						
9	8	65	4	6						
10	9	90	3	7.6						
11	10	90	2	6.1						
12										
13	SUMMARY OUTPUT									
14										
15	*Regression Statistics*									
16	Multiple R	0.9507								
17	R Square	0.9038								
18	Adjusted R Square	0.8763								
19	Standard Error	0.5731								
20	Observations	10								
21										
22	ANOVA									
23		*df*	*SS*	*MS*	*F*	*Significance F*				
24	Regression	2	21.6006	10.8003	32.8784	0.0003				
25	Residual	7	2.2994	0.3285						
26	Total	9	23.9							
27										
28		*Coefficients*	*Standard Error*	*t Stat*	*P-value*	*Lower 95%*	*Upper 95%*	*Lower 99.0%*	*Upper 99.0%*	
29	Intercept	-0.8687	0.9515	-0.9129	0.3916	-3.1188	1.3813	-4.1986	2.4612	
30	Miles	0.0611	0.0099	6.1824	0.0005	0.0378	0.0845	0.0265	0.0957	
31	Deliveries	0.9234	0.2211	4.1763	0.0042	0.4006	1.4463	0.1496	1.6972	
32										

The following steps describe how to use the Regression tool for the multiple regression analysis.

Step 1. Click the **Data** tab on the Ribbon
Step 2. In the **Analysis** group, click **Data Analysis**
Step 3. Choose **Regression** from the list of Analysis Tools
Step 4. When the Regression dialog box appears:
Enter D1:D11 in the **Input Y Range** box
Enter B1:C11 in the **Input X Range** box
Select **Labels**
Select **Confidence Level**
Enter 99 in the **Confidence Level** box
Select **Output Range**
Enter A13 in the **Output Range** box (to identify the upper left corner of the section of the worksheet where the output will appear)
Click **OK**

In the Excel output shown in Figure 15.14 the label for the independent variable x_1 is Miles (see cell A30), and the label for the independent variable x_2 is Deliveries (see cell A31). The estimated regression equation is

$$\hat{y} = -.8687 + .0611x_1 + .9234x_2$$

Note that using Excel's Regression tool for multiple regression is almost the same as using it for simple linear regression. The major difference is that in the multiple regression case a larger range of cells is required in order to identify the independent variables.

Appendix 15.3 Logistic Regression with Minitab

WEB file Simmons

Minitab calls logistic regression with a dependent variable that can only assume the values 0 and 1 Binary Logistic Regression. In this appendix we describe the steps required to use Minitab's Binary Logistic Regression procedure to generate the computer output for the Simmons Stores problem shown in Figure 15.13. First, the data must be entered in a Minitab worksheet. The amounts customers spent last year at Simmons (in thousands of dollars) are entered into column C2, the credit card data (1 if a Simmons card; 0 otherwise) are entered into column C3, and the coupon use data (1 if the customer used the coupon; 0 otherwise) are entered in column C4. The variable names Spending, Card, and Coupon are entered as the column headings on the worksheet. In subsequent steps, we refer to the data by using the variable names Spending, Card, and Coupon or the column indicators C2, C3, and C4. The following steps will generate the logistic regression output.

Step 1. Select the **Stat** menu
Step 2. Select the **Regression** menu
Step 3. Choose **Binary Logistic Regression**
Step 4. When the **Binary Logistic Regression** dialog box appears:
  Enter Coupon in the **Response** box
  Enter Spending and Card in the **Model** box
  Click **OK**

The information in Figure 15.13 will now appear as a portion of the output.

Appendix 15.4 Multiple Regression Analysis Using StatTools

WEB file Butler

In this appendix we show how StatTools can be used to perform the regression analysis computations for the Butler Trucking problem. Begin by using the Data Set Manager to create a StatTools data set for these data using the procedure described in the appendix in Chapter 1. The following steps describe how StatTools can be used to provide the regression results.

Step 1. Click the **StatTools** tab on the Ribbon
Step 2. In the **Analyses** group, click **Regression and Classification**
Step 3. Choose the **Regression** option
Step 4. When the StatTools-Regression dialog box appears:
  Select **Multiple** in the **Regression Type** box
  In the **Variables** section:
    Click the **Format** button and select **Unstacked**
    In the column labeled **I** select **Miles**
    In the column labeled **I** select **Deliveries**
    In the column labeled **D** select **Time**
  Click **OK**

The regression analysis output will appear.

The StatTools-Regression dialog box contains a number of more advanced options for developing prediction interval estimates and producing residual plots. The StatTools Help facility provides information on using all of these options.

CHAPTER 16

Regression Analysis: Model Building

CONTENTS

STATISTICS *in* PRACTICE

MONSANTO COMPANY*
ST. LOUIS, MISSOURI

Monsanto Company traces its roots to one entrepreneur's investment of $500 and a dusty warehouse on the Mississippi riverfront, where in 1901 John F. Queeney began manufacturing saccharin. Today, Monsanto is one of the nation's largest chemical companies, producing more than a thousand products ranging from industrial chemicals to synthetic playing surfaces used in modern sports stadiums. Monsanto is a worldwide corporation with manufacturing facilities, laboratories, technical centers, and marketing operations in 65 countries.

Monsanto's Nutrition Chemical Division manufactures and markets a methionine supplement used in poultry, swine, and cattle feed products. Because poultry growers work with high volumes and low profit margins, cost-effective poultry feed products with the best possible nutrition value are needed. Optimal feed composition will result in rapid growth and high final body weight for a given level of feed intake. The chemical industry works closely with poultry growers to optimize poultry feed products. Ultimately, success depends on keeping the cost of poultry low in comparison with the cost of beef and other meat products.

Monsanto used regression analysis to model the relationship between body weight y and the amount of methionine x added to the poultry feed. Initially, the following simple linear estimated regression equation was developed.

$$\hat{y} = .21 + .42x$$

This estimated regression equation proved statistically significant; however, the analysis of the residuals indicated that a curvilinear relationship would be a better model of the relationship between body weight and methionine.

Monsanto researchers used regression analysis to develop an optimal feed composition for poultry growers. © Krugloff/Shutterstock.com.

Further research conducted by Monsanto showed that although small amounts of methionine tended to increase body weight, at some point body weight leveled off and additional amounts of the methionine were of little or no benefit. In fact, when the amount of methionine increased beyond nutritional requirements, body weight tended to decline. The following estimated multiple regression equation was used to model the curvilinear relationship between body weight and methionine.

$$\hat{y} = -1.89 + 1.32x - .506x^2$$

Use of the regression results enabled Monsanto to determine the optimal level of methionine to be used in poultry feed products.

In this chapter we will extend the discussion of regression analysis by showing how curvilinear models such as the one used by Monsanto can be developed. In addition, we will describe a variety of tools that help determine which independent variables lead to the best estimated regression equation.

*The authors are indebted to James R. Ryland and Robert M. Schisla, Senior Research Specialists, Monsanto Nutrition Chemical Division, for providing this Statistics in Practice.

Model building is the process of developing an estimated regression equation that describes the relationship between a dependent variable and one or more independent variables. The major issues in model building are finding the proper functional form of the relationship and selecting the independent variables to be included in the model. In Section 16.1 we establish the framework for model building by introducing the concept of a general linear model. Section 16.2, which provides the foundation for the more sophisticated computer-based procedures, introduces a general approach for determining when to add or delete independent variables. In Section 16.3 we consider a larger regression problem involving eight independent variables and 25 observations; this problem is used to illustrate the variable selection procedures presented in Section 16.4, including stepwise regression, the forward selection procedure, the backward elimination procedure, and best-subsets regression. In Section 16.5 we show how multiple regression analysis can provide another approach to solving experimental design problems, and in Section 16.6 we show how the Durbin-Watson test can be used to detect serial or autocorrelation.

16.1 General Linear Model

Suppose we collected data for one dependent variable y and k independent variables $x_1, x_2, \ldots, x_k$. Our objective is to use these data to develop an estimated regression equation that provides the best relationship between the dependent and independent variables. As a general framework for developing more complex relationships among the independent variables, we introduce the concept of a **general linear model** involving p independent variables.

If you can write a regression model in the form of equation (16.1), the standard multiple regression procedures described in Chapter 15 are applicable.

GENERAL LINEAR MODEL

$$y = \beta_0 + \beta_1 z_1 + \beta_2 z_2 + \cdots + \beta_p z_p + \epsilon \quad (16.1)$$

In equation (16.1), each of the independent variables z_j (where $j = 1, 2, \ldots, p$) is a function of $x_1, x_2, \ldots, x_k$ (the variables for which data are collected). In some cases, each z_j may be a function of only one x variable. The simplest case is when we collect data for just one variable x_1 and want to predict y by using a straight-line relationship. In this case $z_1 = x_1$ and equation (16.1) becomes

$$y = \beta_0 + \beta_1 x_1 + \epsilon \quad (16.2)$$

Equation (16.2) is the simple linear regression model introduced in Chapter 14 with the exception that the independent variable is labeled x_1 instead of x. In the statistical modeling literature, this model is called a *simple first-order model with one predictor variable.*

TABLE 16.1
DATA FOR THE REYNOLDS EXAMPLE

Months Employed	Scales Sold
41	275
106	296
76	317
104	376
22	162
12	150
85	367
111	308
40	189
51	235
9	83
12	112
6	67
56	325
19	189

WEB file
Reynolds

Modeling Curvilinear Relationships

More complex types of relationships can be modeled with equation (16.1). To illustrate, let us consider the problem facing Reynolds, Inc., a manufacturer of industrial scales and laboratory equipment. Managers at Reynolds want to investigate the relationship between length of employment of their salespeople and the number of electronic laboratory scales sold. Table 16.1 gives the number of scales sold by 15 randomly selected salespeople for the most recent sales period and the number of months each salesperson has been employed by the firm. Figure 16.1 is the scatter diagram for these data. The scatter diagram indicates a possible curvilinear relationship between the length of time employed and the number of units sold. Before considering how to develop a curvilinear relationship for Reynolds, let us consider the Minitab output in Figure 16.2 corresponding to a simple first-order model; the estimated regression is

$$\text{Sales} = 111 + 2.38 \text{ Months}$$

where

$$\text{Sales} = \text{number of electronic laboratory scales sold}$$
$$\text{Months} = \text{the number of months the salesperson has been employed}$$

FIGURE 16.1 SCATTER DIAGRAM FOR THE REYNOLDS EXAMPLE

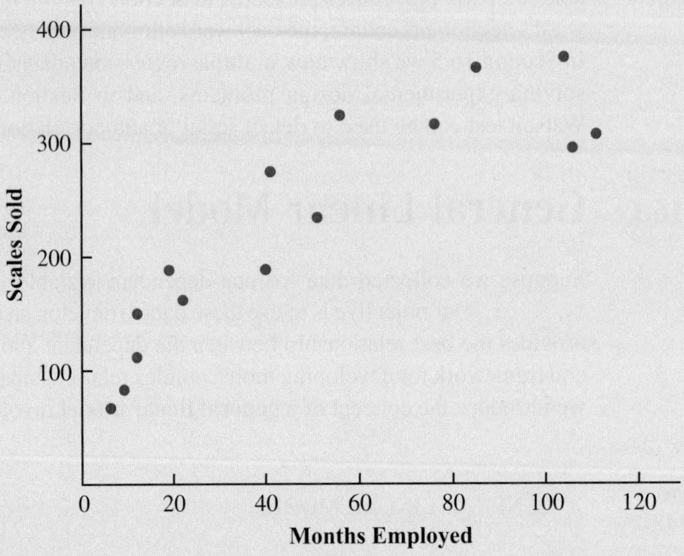

Figure 16.3 is the corresponding standardized residual plot. Although the computer output shows that the relationship is significant (p-value = .000) and that a linear relationship explains a high percentage of the variability in sales (R-Sq = 78.1%), the standardized residual plot suggests that a curvilinear relationship is needed.

To account for the curvilinear relationship, we set $z_1 = x_1$ and $z_2 = x_1^2$ in equation (16.1) to obtain the model

$$y = \beta_0 + \beta_1 x_1 + \beta_2 x_1^2 + \epsilon \tag{16.3}$$

This model is called a *second-order model with one predictor variable.* To develop an estimated regression equation corresponding to this second-order model, the statistical

FIGURE 16.2 MINITAB OUTPUT FOR THE REYNOLDS EXAMPLE: FIRST-ORDER MODEL

```
The regression equation is
Sales = 111 + 2.38 Months

Predictor      Coef   SE Coef     T      p
Constant     111.23     21.63  5.14  0.000
Months       2.3768    0.3489  6.81  0.000

S = 49.5158   R-Sq = 78.1%   R-Sq(adj) = 76.4%

Analysis of Variance

SOURCE          DF      SS      MS      F      p
Regression       1  113783  113783  46.41  0.000
Residual Error  13   31874    2452
Total           14  145657
```

FIGURE 16.3 STANDARDIZED RESIDUAL PLOT FOR THE REYNOLDS EXAMPLE: FIRST-ORDER MODEL

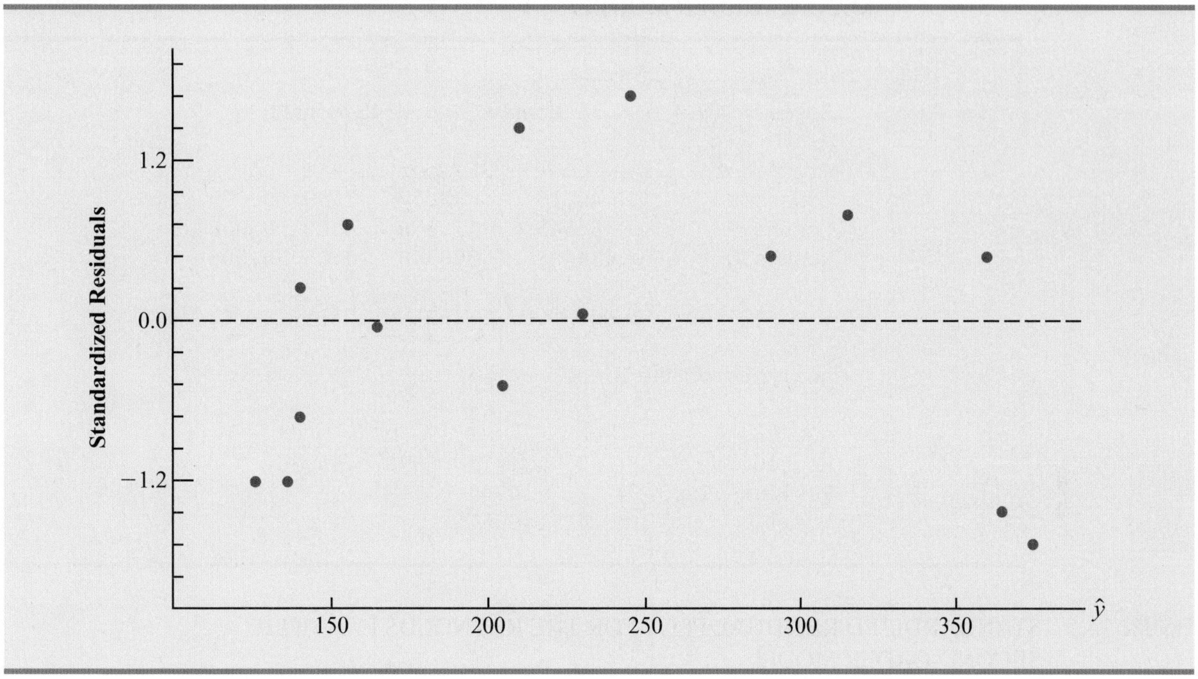

software package we are using needs the original data in Table 16.1, as well as that data corresponding to adding a second independent variable that is the square of the number of months the employee has been with the firm. In Figure 16.4 we show the Minitab output corresponding to the second-order model; the estimated regression equation is

The data for the MonthsSq *independent variable is obtained by squaring the values of* Months.

$$\text{Sales} = 45.3 + 6.34 \text{ Months} - .0345 \text{ MonthsSq}$$

where

MonthsSq = the square of the number of months the salesperson has been employed

Figure 16.5 is the corresponding standardized residual plot. It shows that the previous curvilinear pattern has been removed. At the .05 level of significance, the computer output shows that the overall model is significant (p-value for the F test is .000); note also that the p-value corresponding to the t-ratio for MonthsSq (p-value = .002) is less than .05, and hence we can conclude that adding MonthsSq to the model involving Months is significant. With R-Sq(adj) = 88.6%, we should be pleased with the fit provided by this estimated regression equation. More important, however, is seeing how easy it is to handle curvilinear relationships in regression analysis.

Clearly, many types of relationships can be modeled by using equation (16.1). The regression techniques with which we have been working are definitely not limited to linear, or straight-line, relationships. In multiple regression analysis the word *linear* in the term "general linear model" refers only to the fact that $\beta_0, \beta_1, \ldots, \beta_p$ all have exponents of 1; it does not imply that the relationship between y and the x_i's is linear. Indeed, in this section we have seen one example of how equation (16.1) can be used to model a curvilinear relationship.

FIGURE 16.4 MINITAB OUTPUT FOR THE REYNOLDS EXAMPLE: SECOND-ORDER MODEL

```
The regression equation is
Sales = 45.3 + 6.34 Months - 0.0345 MonthsSq

Predictor          Coef    SE Coef      T      p
Constant          45.35      22.77   1.99  0.070
Months            6.345      1.058   6.00  0.000
MonthsSq      -0.034486   0.008948  -3.85  0.002

S = 34.4528   R-Sq = 90.2%   R-Sq(adj) = 88.6%

Analysis of Variance

SOURCE           DF      SS     MS      F      p
Regression        2  131413  65707  55.36  0.000
Residual Error   12   14244   1187
Total            14  145657
```

FIGURE 16.5 STANDARDIZED RESIDUAL PLOT FOR THE REYNOLDS EXAMPLE: SECOND-ORDER MODEL

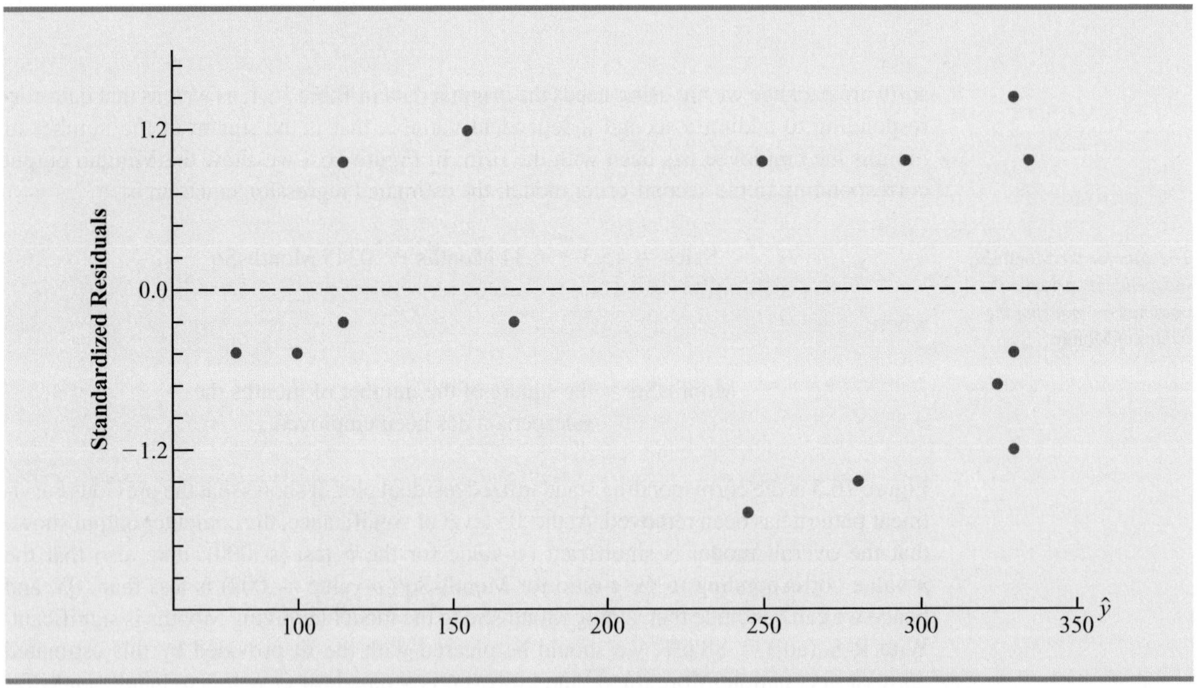

Interaction

If the original data set consists of observations for y and two independent variables x_1 and x_2, we can develop a second-order model with two predictor variables by setting $z_1 = x_1$, $z_2 = x_2$, $z_3 = x_1^2$, $z_4 = x_2^2$, and $z_5 = x_1x_2$ in the general linear model of equation (16.1). The model obtained is

$$y = \beta_0 + \beta_1 x_1 + \beta_2 x_2 + \beta_3 x_1^2 + \beta_4 x_2^2 + \beta_5 x_1 x_2 + \epsilon \qquad (16.4)$$

WEB file

Tyler

TABLE 16.2 DATA FOR THE TYLER PERSONAL CARE EXAMPLE

Price	Advertising Expenditure ($1000s)	Sales (1000s)	Price	Advertising Expenditure ($1000s)	Sales (1000s)
$2.00	50	478	$2.00	100	810
$2.50	50	373	$2.50	100	653
$3.00	50	335	$3.00	100	345
$2.00	50	473	$2.00	100	832
$2.50	50	358	$2.50	100	641
$3.00	50	329	$3.00	100	372
$2.00	50	456	$2.00	100	800
$2.50	50	360	$2.50	100	620
$3.00	50	322	$3.00	100	390
$2.00	50	437	$2.00	100	790
$2.50	50	365	$2.50	100	670
$3.00	50	342	$3.00	100	393

In this second-order model, the variable $z_5 = x_1x_2$ is added to account for the potential effects of the two variables acting together. This type of effect is called **interaction**.

To provide an illustration of interaction and what it means, let us review the regression study conducted by Tyler Personal Care for one of its new shampoo products. Two factors believed to have the most influence on sales are unit selling price and advertising expenditure. To investigate the effects of these two variables on sales, prices of $2.00, $2.50, and $3.00 were paired with advertising expenditures of $50,000 and $100,000 in 24 test markets. The unit sales (in thousands) that were observed are reported in Table 16.2.

Table 16.3 is a summary of these data. Note that the sample mean sales corresponding to a price of $2.00 and an advertising expenditure of $50,000 is 461,000, and the sample mean sales corresponding to a price of $2.00 and an advertising expenditure of $100,000 is 808,000. Hence, with price held constant at $2.00, the difference in the sample mean sales between advertising expenditures of $50,000 and $100,000 is 808,000 − 461,000 = 347,000 units. When the price of the product is $2.50, the difference in the sample mean sales is 646,000 − 364,000 = 282,000 units. Finally, when the price is $3.00, the difference in the sample mean sales is 375,000 − 332,000 = 43,000 units. Clearly, the difference in the sample mean sales between advertising expenditures of $50,000 and $100,000 depends on the price of the product. In other words, at higher selling prices, the effect of increased advertising expenditure diminishes. These observations provide evidence of interaction between the price and advertising expenditure variables.

TABLE 16.3 SAMPLE MEAN UNIT SALES (1000s) FOR THE TYLER PERSONAL CARE EXAMPLE

		Price		
		$2.00	**$2.50**	**$3.00**
Advertising Expenditure	**$50,000**	461	364	332
	$100,000	808	646	375

Mean sales of 808,000 units when price = $2.00 and advertising expenditure = $100,000

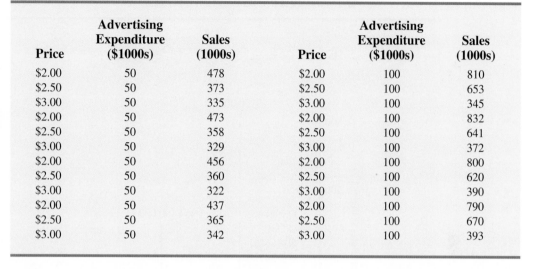

FIGURE 16.6 SAMPLE MEAN UNIT SALES (1000s) AS A FUNCTION OF SELLING PRICE AND ADVERTISING EXPENDITURE

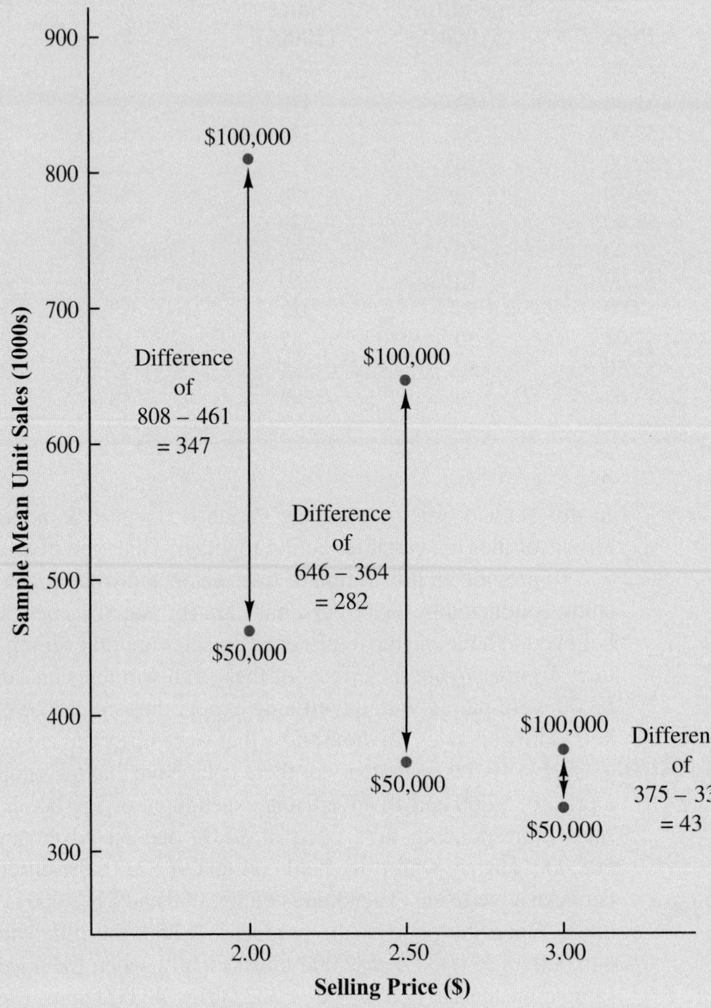

To provide another perspective of interaction, Figure 16.6 shows the sample mean sales for the six price-advertising expenditure combinations. This graph also shows that the effect of advertising expenditure on the sample mean sales depends on the price of the product; we again see the effect of interaction. When interaction between two variables is present, we cannot study the effect of one variable on the response y independently of the other variable. In other words, meaningful conclusions can be developed only if we consider the joint effect that both variables have on the response.

To account for the effect of interaction, we will use the following regression model.

$$y = \beta_0 + \beta_1 x_1 + \beta_2 x_2 + \beta_3 x_1 x_2 + \epsilon \qquad (16.5)$$

where

y = unit sales (1000s)
x_1 = price ($)
x_2 = advertising expenditure ($1000s)

Note that equation (16.5) reflects Tyler's belief that the number of units sold depends linearly on selling price and advertising expenditure (accounted for by the $\beta_1 x_1$ and $\beta_2 x_2$ terms), and that there is interaction between the two variables (accounted for by the $\beta_3 x_1 x_2$ term).

To develop an estimated regression equation, a general linear model involving three independent variables (z_1, z_2, and z_3) was used.

$$y = \beta_0 + \beta_1 z_1 + \beta_2 z_2 + \beta_3 z_3 + \epsilon \quad (16.6)$$

where

$$z_1 = x_1$$
$$z_2 = x_2$$
$$z_3 = x_1 x_2$$

Figure 16.7 is the Minitab output corresponding to the interaction model for the Tyler Personal Care example. The resulting estimated regression equation is

$$\text{Sales} = -276 + 175 \text{ Price} + 19.7 \text{ AdvExp} - 6.08 \text{ PriceAdv}$$

where

The data for the PriceAdv *independent variable is obtained by multiplying each value of* Price *times the corresponding value of* AdvExp.

$$\begin{aligned} \text{Sales} &= \text{unit sales (1000s)} \\ \text{Price} &= \text{price of the product (\$)} \\ \text{AdvExp} &= \text{advertising expenditure (\$1000s)} \\ \text{PriceAdv} &= \text{interaction term (Price times AdvExp)} \end{aligned}$$

Because the model is significant (p-value for the F test is .000) and the p-value corresponding to the t test for PriceAdv is .000, we conclude that interaction is significant given the linear effect of the price of the product and the advertising expenditure. Thus, the regression results show that the effect of advertising expenditure on sales depends on the price.

FIGURE 16.7 MINITAB OUTPUT FOR THE TYLER PERSONAL CARE EXAMPLE

```
The regression equation is
Sales = - 276 + 175 Price + 19.7 AdvExpen - 6.08 PriceAdv

Predictor       Coef  SE Coef       T      p
Constant      -275.8    112.8   -2.44  0.024
Price         175.00    44.55    3.93  0.001
Adver         19.680    1.427   13.79  0.000
PriceAdv     -6.0800   0.5635  -10.79  0.000

S = 28.1739   R-Sq = 97.8%   R-Sq(adj) = 97.5%

Analysis of Variance

SOURCE          DF      SS      MS       F      p
Regression       3  709316  236439  297.87  0.000
Residual Error  20   15875     794
Total           23  725191
```

TABLE 16.4 MILES-PER-GALLON RATINGS AND WEIGHTS FOR 12 AUTOMOBILES

Weight	Miles per Gallon
2289	28.7
2113	29.2
2180	34.2
2448	27.9
2026	33.3
2702	26.4
2657	23.9
2106	30.5
3226	18.1
3213	19.5
3607	14.3
2888	20.9

Transformations Involving the Dependent Variable

In showing how the general linear model can be used to model a variety of possible relationships between the independent variables and the dependent variable, we have focused attention on transformations involving one or more of the independent variables. Often it is worthwhile to consider transformations involving the dependent variable y. As an illustration of when we might want to transform the dependent variable, consider the data in Table 16.4, which shows the miles-per-gallon ratings and weights for 12 automobiles. The scatter diagram in Figure 16.8 indicates a negative linear relationship between these two variables. Therefore, we use a simple first-order model to relate the two variables. The Minitab output is shown in Figure 16.9; the resulting estimated regression equation is

$$\text{MPG} = 56.1 - 0.0116\ \text{Weight}$$

where

$$\text{MPG} = \text{miles-per-gallon rating}$$
$$\text{Weight} = \text{weight of the car in pounds}$$

The model is significant (p-value for the F test is .000) and the fit is very good (R-sq = 93.5%). However, we note in Figure 16.9 that observation 3 is identified as having a large standardized residual.

WEB file

MPG

Figure 16.10 is the standardized residual plot corresponding to the first-order model. The pattern we observe does not look like the horizontal band we should expect to find if the assumptions about the error term are valid. Instead, the variability in the residuals appears to increase as the value of $\hat{y}$ increases. In other words, we see the wedge-shaped pattern referred to in Chapters 14 and 15 as being indicative of a nonconstant variance. We are not justified in reaching any conclusions about the statistical significance of the resulting estimated regression equation when the underlying assumptions for the tests of significance do not appear to be satisfied.

FIGURE 16.8 SCATTER DIAGRAM FOR THE MILES-PER-GALLON EXAMPLE

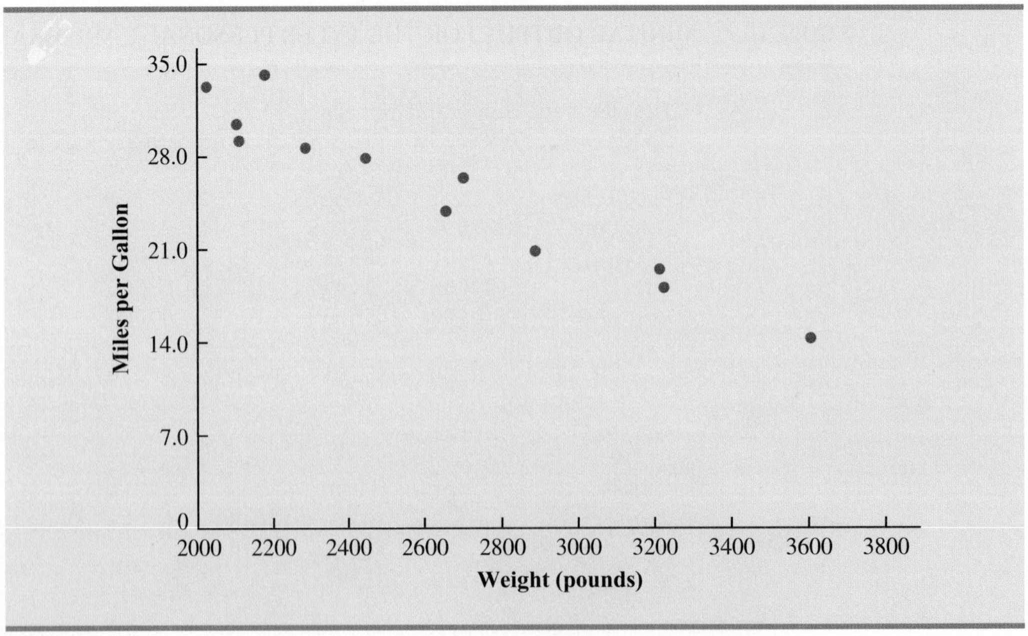

FIGURE 16.9 MINITAB OUTPUT FOR THE MILES-PER-GALLON EXAMPLE

```
The regression equation is
MPG = 56.1 - 0.0116 Weight

Predictor        Coef    SE Coef       T      p
Constant       56.096      2.582   21.72  0.000
Weight     -0.0116436  0.0009677  -12.03  0.000

S = 1.67053   R-Sq = 93.5%   R-Sq(adj) = 92.9%

Analysis of Variance

SOURCE          DF      SS      MS       F      p
Regression       1  403.98  403.98  144.76  0.000
Residual Error  10   27.91    2.79
Total           11  431.88

Unusual Observations
Obs  Weight     MPG     Fit  SE Fit  Residual  St Resid
  3    2180  34.200  30.713   0.644     3.487     2.26R

R denotes an observation with a large standardized residual.
```

FIGURE 16.10 STANDARDIZED RESIDUAL PLOT FOR THE MILES-PER-GALLON EXAMPLE

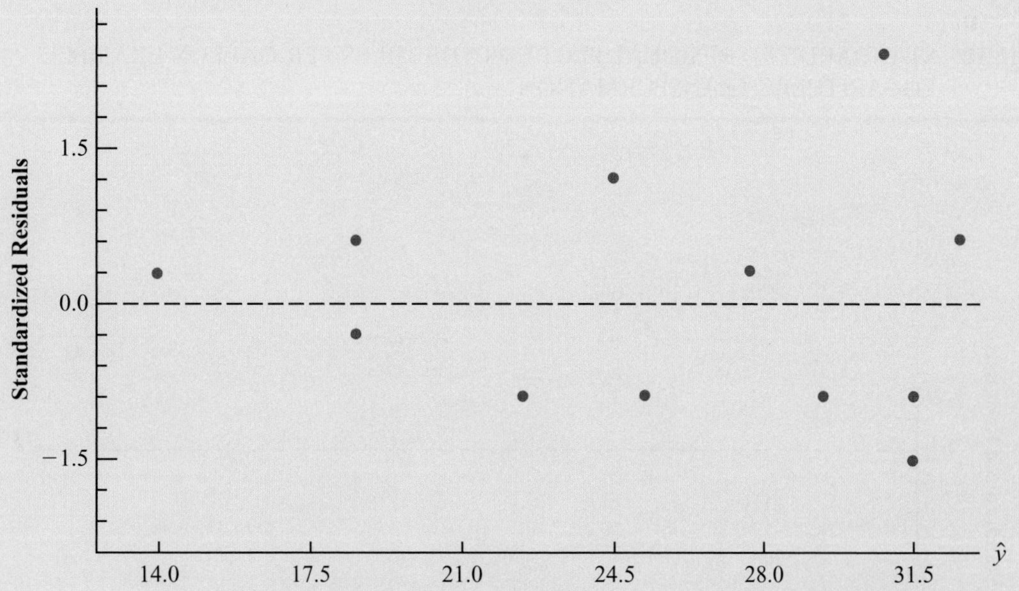

Often the problem of nonconstant variance can be corrected by transforming the dependent variable to a different scale. For instance, if we work with the logarithm of the dependent variable instead of the original dependent variable, the effect will be to compress the values of the dependent variable and thus diminish the effects of nonconstant variance.

Most statistical packages provide the ability to apply logarithmic transformations using either the base 10 (common logarithm) or the base $e = 2.71828\ldots$ (natural logarithm). We applied a natural logarithmic transformation to the miles-per-gallon data and developed the estimated regression equation relating weight to the natural logarithm of miles-per-gallon. The regression results obtained by using the natural logarithm of miles-per-gallon as the dependent variable, labeled LogeMPG in the output, are shown in Figure 16.11; Figure 16.12 is the corresponding standardized residual plot.

FIGURE 16.11 MINITAB OUTPUT FOR THE MILES-PER-GALLON EXAMPLE: LOGARITHMIC TRANSFORMATION

```
The regression equation is
LogeMPG = 4.52 -0.000501 Weight

Predictor            Coef     SE Coef       T      p
Constant          4.52423     0.09932   45.55  0.000
Weight        -0.00050110  0.00003722  -13.46  0.000

S = 0.0642547   R-Sq = 94.8%   R-Sq(adj) = 94.2%

Analysis of Variance

SOURCE          DF       SS        MS       F      p
Regression       1  0.74822   0.74822  181.22  0.000
Residual Error  10  0.04129   0.00413
Total           11  0.78950
```

FIGURE 16.12 STANDARDIZED RESIDUAL PLOT FOR THE MILES-PER-GALLON EXAMPLE: LOGARITHMIC TRANSFORMATION

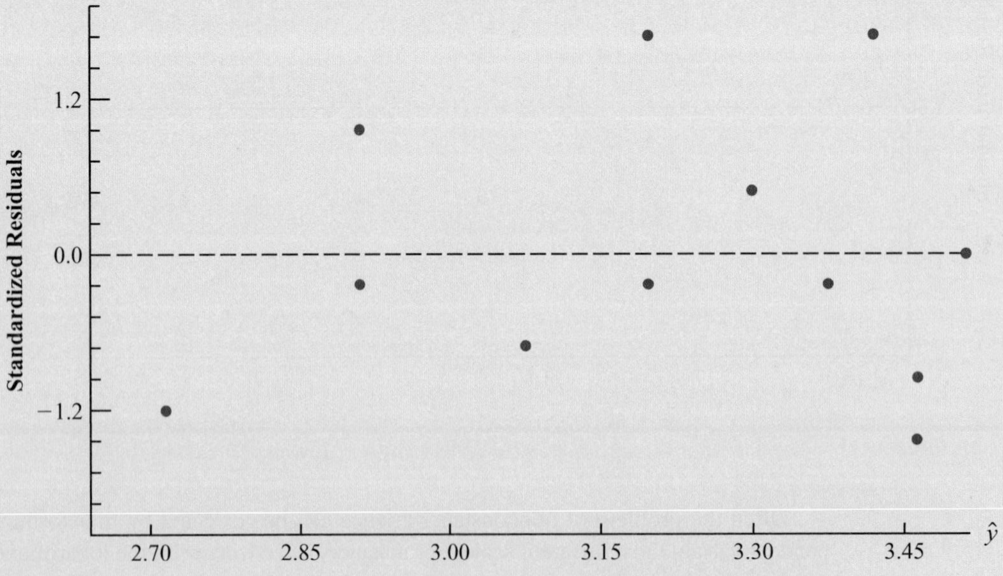

Looking at the residual plot in Figure 16.12, we see that the wedge-shaped pattern has now disappeared. Moreover, none of the observations are identified as having a large standardized residual. The model with the logarithm of miles per gallon as the dependent variable is statistically significant and provides an excellent fit to the observed data. Hence, we would recommend using the estimated regression equation

$$\text{LogeMPG} = 4.52 - .000501 \text{ Weight}$$

To predict the miles-per-gallon rating for an automobile that weighs 2500 pounds, we first develop an estimate of the logarithm of the miles-per-gallon rating.

$$\text{LogeMPG} = 4.52 - .000501(2500) = 3.2675$$

The miles-per-gallon estimate is obtained by finding the number whose natural logarithm is 3.2675. Using a calculator with an exponential function, or raising e to the power 3.2675, we obtain 26.2 miles per gallon.

Another approach to problems of nonconstant variance is to use $1/y$ as the dependent variable instead of y. This type of transformation is called a *reciprocal transformation.* For instance, if the dependent variable is measured in miles per gallon, the reciprocal transformation would result in a new dependent variable whose units would be 1/(miles per gallon) or gallons per mile. In general, there is no way to determine whether a logarithmic transformation or a reciprocal transformation will perform best without actually trying each of them.

Nonlinear Models That Are Intrinsically Linear

Models in which the parameters ($\beta_0, \beta_1, \ldots, \beta_p$) have exponents other than 1 are called nonlinear models. However, for the case of the exponential model, we can perform a transformation of variables that will enable us to perform regression analysis with equation (16.1), the general linear model. The exponential model involves the following regression equation.

$$E(y) = \beta_0 \beta_1^x \quad \textbf{(16.7)}$$

This regression equation is appropriate when the dependent variable y increases or decreases by a constant percentage, instead of by a fixed amount, as x increases.

As an example, suppose sales for a product y are related to advertising expenditure x (in thousands of dollars) according to the following regression equation.

$$E(y) = 500(1.2)^x$$

Thus, for $x = 1$, $E(y) = 500(1.2)^1 = 600$; for $x = 2$, $E(y) = 500(1.2)^2 = 720$; and for $x = 3$, $E(y) = 500(1.2)^3 = 864$. Note that $E(y)$ is not increasing by a constant amount in this case, but by a constant percentage; the percentage increase is 20%.

We can transform this nonlinear regression equation to a linear regression equation by taking the logarithm of both sides of equation (16.7).

$$\log E(y) = \log \beta_0 + x \log \beta_1 \quad \textbf{(16.8)}$$

Now if we let $y' = \log E(y)$, $\beta_0' = \log \beta_0$, and $\beta_1' = \log \beta_1$, we can rewrite equation (16.8) as

$$y' = \beta_0' + \beta_1' x$$

It is clear that the formulas for simple linear regression can now be used to develop estimates of β_0' and β_1'. Denoting the estimates as b_0' and b_1' leads to the following estimated regression equation.

$$\hat{y}' = b_0' + b_1' x \quad \textbf{(16.9)}$$

To obtain predictions of the original dependent variable y given a value of x, we would first substitute the value of x into equation (16.9) and compute $\hat{y}'$. The antilog of $\hat{y}'$ would be the prediction of y, or the expected value of y.

Many nonlinear models cannot be transformed into an equivalent linear model. However, such models have had limited use in business and economic applications. Furthermore, the mathematical background needed for study of such models is beyond the scope of this text.

Exercises

Methods

SELF test

1. Consider the following data for two variables, x and y.

x	22	24	26	30	35	40
y	12	21	33	35	40	36

   a. Develop an estimated regression equation for the data of the form $\hat{y} = b_0 + b_1x$.
   b. Use the results from part (a) to test for a significant relationship between x and y. Use $\alpha = .05$.
   c. Develop a scatter diagram for the data. Does the scatter diagram suggest an estimated regression equation of the form $\hat{y} = b_0 + b_1x + b_2x^2$? Explain.
   d. Develop an estimated regression equation for the data of the form $\hat{y} = b_0 + b_1x + b_2x^2$.
   e. Refer to part (d). Is the relationship between x, x^2, and y significant? Use $\alpha = .05$.
   f. Predict the value of y when $x = 25$.

2. Consider the following data for two variables, x and y.

x	9	32	18	15	26
y	10	20	21	16	22

   a. Develop an estimated regression equation for the data of the form $\hat{y} = b_0 + b_1x$. Comment on the adequacy of this equation for predicting y.
   b. Develop an estimated regression equation for the data of the form $\hat{y} = b_0 + b_1x + b_2x^2$. Comment on the adequacy of this equation for predicting y.
   c. Predict the value of y when $x = 20$.

3. Consider the following data for two variables, x and y.

x	2	3	4	5	7	7	7	8	9
y	4	5	4	6	4	6	9	5	11

   a. Does there appear to be a linear relationship between x and y? Explain.
   b. Develop the estimated regression equation relating x and y.
   c. Plot the standardized residuals versus $\hat{y}$ for the estimated regression equation developed in part (b). Do the model assumptions appear to be satisfied? Explain.
   d. Perform a logarithmic transformation on the dependent variable y. Develop an estimated regression equation using the transformed dependent variable. Do the model assumptions appear to be satisfied by using the transformed dependent variable? Does a reciprocal transformation work better in this case? Explain.

Applications

4. A highway department is studying the relationship between traffic flow and speed. The following model has been hypothesized.

$$y = \beta_0 + \beta_1 x + \epsilon$$

where

y = traffic flow in vehicles per hour
x = vehicle speed in miles per hour

The following data were collected during rush hour for six highways leading out of the city.

Traffic Flow (y)	Vehicle Speed (x)
1256	35
1329	40
1226	30
1335	45
1349	50
1124	25

a. Develop an estimated regression equation for the data.
b. Use $\alpha = .01$ to test for a significant relationship.

SELF test

5. In working further with the problem of exercise 4, statisticians suggested the use of the following curvilinear estimated regression equation.

$$\hat{y} = b_0 + b_1 x + b_2 x^2$$

a. Use the data of exercise 4 to estimate the parameters of this estimated regression equation.
b. Use $\alpha = .01$ to test for a significant relationship.
c. Predict the traffic flow in vehicles per hour at a speed of 38 miles per hour.

6. A study of emergency service facilities investigated the relationship between the number of facilities and the average distance traveled to provide the emergency service. The following table gives the data collected.

Number of Facilities	Average Distance (miles)
9	1.66
11	1.12
16	.83
21	.62
27	.51
30	.47

a. Develop a scatter diagram for these data, treating average distance traveled as the dependent variable.
b. Does a simple linear regression model appear to be appropriate? Explain.
c. Develop an estimated regression equation for the data that you believe will best explain the relationship between these two variables.

7. In 2011, home prices and mortgage rates fell so far that in a number of cities the monthly cost of owning a home was less expensive than renting. The following data show the average asking rent and the monthly mortgage on the median-priced home

(including taxes and insurance) for 10 cities where the average monthly mortgage payment was less than the average asking rent (*The Wall Street Journal,* November 26–27, 2011).

WEB file
RentMortgage

City	Rent ($)	Mortgage ($)
Atlanta	840	539
Chicago	1062	1002
Detroit	823	626
Jacksonville, Fla.	779	711
Las Vegas	796	655
Miami	1071	977
Minneapolis	953	776
Orlando, Fla.	851	695
Phoenix	762	651
St. Louis	723	654

a. Develop a scatter diagram for these data, treating the average asking rent as the independent variable. Does a simple linear regression model appear to be appropriate?
b. Use a simple linear regression model to develop an estimated regression equation to predict the monthly mortgage on the median-priced home given the average asking rent. Construct a standardized residual plot. Based upon the standardized residual plot, does a simple linear regression model appear to be appropriate?
c. Using a second-order model, develop an estimated regression equation to predict the monthly mortgage on the median-priced home given the average asking rent.
d. Do you prefer the estimated regression equation developed in part (a) or part (c)? Explain.

8. Corvette, Ferrari, and Jaguar produced a variety of classic cars that continue to increase in value. The following data, based upon the Martin Rating System for Collectible Cars, show the rarity rating (1–20) and the high price ($1000) for 15 classic cars (*BusinessWeek* website, February 2006).

WEB file
ClassicCars

Year	Make	Model	Rating	Price ($1000)
1984	Chevrolet	Corvette	18	1600.0
1956	Chevrolet	Corvette 265/225-hp	19	4000.0
1963	Chevrolet	Corvette coupe (340-bhp 4-speed)	18	1000.0
1978	Chevrolet	Corvette coupe Silver Anniversary	19	1300.0
1960–1963	Ferrari	250 GTE 2+2	16	350.0
1962–1964	Ferrari	250 GTL Lusso	19	2650.0
1962	Ferrari	250 GTO	18	375.0
1967–1968	Ferrari	275 GTB/4 NART Spyder	17	450.0
1968–1973	Ferrari	365 GTB/4 Daytona	17	140.0
1962–1967	Jaguar	E-type OTS	15	77.5
1969–1971	Jaguar	E-type Series II OTS	14	62.0
1971–1974	Jaguar	E-type Series III OTS	16	125.0
1951–1954	Jaguar	XK 120 roadster (steel)	17	400.0
1950–1953	Jaguar	XK C-type	16	250.0
1956–1957	Jaguar	XKSS	13	70.0

a. Develop a scatter diagram of the data using the rarity rating as the independent variable and price as the independent variable. Does a simple linear regression model appear to be appropriate?
b. Develop an estimated multiple regression equation with x = rarity rating and x^2 as the two independent variables.
c. Consider the nonlinear relationship shown by equation (16.7). Use logarithms to develop an estimated regression equation for this model.
d. Do you prefer the estimated regression equation developed in part (b) or part (c)? Explain.

WEB file

MetroAreas

9. *Kiplinger's Personal Finance Magazine* rated 359 U.S. metropolitan areas to determine the best cities to live, work, and play. The data contained in the data set named MetroAreas show the data from the Kiplinger study for the 50 metropolitan areas with a population of 1,000,000 or more (Kiplinger's website, March 2, 2009). The data set includes the following variables: Population, Income, Cost of Living Index, and Creative (%). Population is the size of the population in 1000s; Income is the median household income in $1000s; Cost of Living Index is based on 100 being the national average; and Creative (%) is the percentage of the workforce in creative fields such as science, engineering, architecture, education, art, and entertainment. Workers in creative fields are generally considered an important factor in the vitality and livability of a city and a key to future economic prosperity.
   a. Develop a scatter diagram for these data with median household income as the independent variable and the percentage of the workforce in creative fields as the dependent variable. Does a simple linear regression model appear to be appropriate?
   b. Develop a scatter diagram for these data with the cost of living index as the independent variable and the percentage of the workforce in creative fields as the dependent variable. Does a simple linear regression model appear to be appropriate?
   c. Use the data provided to develop the best estimated multiple regression equation for estimating the percentage of the workforce in creative fields.
   d. The Tucson, Arizona, metropolitan area has a population of 946,362, a median household income of $42,984, and cost of living index of 99. Develop a prediction of the percentage of the workforce in creative fields for Tucson. Are there any factors that should be considered before using this predicted value?

16.2 Determining When to Add or Delete Variables

In this section we will show how an F test can be used to determine whether it is advantageous to add one or more independent variables to a multiple regression model. This test is based on a determination of the amount of reduction in the error sum of squares resulting from adding one or more independent variables to the model. We will first illustrate how the test can be used in the context of the Butler Trucking example.

In Chapter 15, the Butler Trucking example was introduced to illustrate the use of multiple regression analysis. Recall that the managers wanted to develop an estimated regression equation to predict total daily travel time for trucks using two independent variables: miles traveled and number of deliveries. With miles traveled x_1 as the only independent variable, the least squares procedure provided the following estimated regression equation.

$$\hat{y} = 1.27 + .0678x_1$$

In Chapter 15 we showed that the error sum of squares for this model was SSE = 8.029. When x_2, the number of deliveries, was added as a second independent variable, we obtained the following estimated regression equation.

$$\hat{y} = -.869 + .0611x_1 + .923x_2$$

The error sum of squares for this model was SSE = 2.299. Clearly, adding x_2 resulted in a reduction of SSE. The question we want to answer is: Does adding the variable x_2 lead to a *significant* reduction in SSE?

We use the notation SSE(x_1) to denote the error sum of squares when x_1 is the only independent variable in the model, SSE(x_1, x_2) to denote the error sum of squares when x_1 and x_2 are both in the model, and so on. Hence, the reduction in SSE resulting from adding x_2 to the model involving just x_1 is

$$\text{SSE}(x_1) - \text{SSE}(x_1, x_2) = 8.029 - 2.299 = 5.730$$

An F test is conducted to determine whether this reduction is significant.

The numerator of the F statistic is the reduction in SSE divided by the number of independent variables added to the original model. Here only one variable, x_2, has been added; thus, the numerator of the F statistic is

$$\frac{\text{SSE}(x_1) - \text{SSE}(x_1, x_2)}{1} = 5.730$$

The result is a measure of the reduction in SSE per independent variable added to the model. The denominator of the F statistic is the mean square error for the model that includes all of the independent variables. For Butler Trucking this corresponds to the model containing both x_1 and x_2; thus, $p = 2$ and

$$\text{MSE} = \frac{\text{SSE}(x_1, x_2)}{n - p - 1} = \frac{2.299}{7} = .3284$$

The following F statistic provides the basis for testing whether the addition of x_2 is statistically significant.

$$F = \frac{\dfrac{\text{SSE}(x_1) - \text{SSE}(x_1, x_2)}{1}}{\dfrac{\text{SSE}(x_1, x_2)}{n - p - 1}} \tag{16.10}$$

The numerator degrees of freedom for this F test is equal to the number of variables added to the model, and the denominator degrees of freedom is equal to $n - p - 1$.

For the Butler Trucking problem, we obtain

$$F = \frac{\dfrac{5.730}{1}}{\dfrac{2.299}{7}} = \frac{5.730}{.3284} = 17.45$$

Refer to Table 4 of Appendix B. We find that for a level of significance of $\alpha = .05$, $F_{.05} = 5.59$. Because $F = 17.45 > F_{.05} = 5.59$, we can reject the null hypothesis that x_2 is not statistically significant; in other words, adding x_2 to the model involving only x_1 results in a significant reduction in the error sum of squares.

When we want to test for the significance of adding only one more independent variable to a model, the result found with the F test just described could also be obtained by using the t test for the significance of an individual parameter (described in Section 15.4). Indeed, the F statistic we just computed is the square of the t statistic used to test the significance of an individual parameter.

Because the t test is equivalent to the F test when only one independent variable is being added to the model, we can now further clarify the proper use of the t test for testing the significance of an individual parameter. If an individual parameter is not significant, the corresponding variable can be dropped from the model. However, if the t test shows that two or more parameters are not significant, no more than one independent variable can ever be dropped from a model on the basis of a t test; if one variable is dropped, a second variable that was not significant initially might become significant.

We now turn to a consideration of whether the addition of more than one independent variable—as a set—results in a significant reduction in the error sum of squares.

General Case

Consider the following multiple regression model involving q independent variables, where $q < p$.

$$y = \beta_0 + \beta_1 x_1 + \beta_2 x_2 + \cdots + \beta_q x_q + \epsilon \tag{16.11}$$

If we add variables $x_{q+1}, x_{q+2}, \ldots, x_p$ to this model, we obtain a model involving p independent variables.

$$\begin{aligned} y = \beta_0 &+ \beta_1 x_1 + \beta_2 x_2 + \cdots + \beta_q x_q \\ &+ \beta_{q+1} x_{q+1} + \beta_{q+2} x_{q+2} + \cdots + \beta_p x_p + \epsilon \end{aligned} \tag{16.12}$$

To test whether the addition of $x_{q+1}, x_{q+2}, \ldots, x_p$ is statistically significant, the null and alternative hypotheses can be stated as follows.

H_0: $\beta_{q+1} = \beta_{q+2} = \cdots = \beta_p = 0$
H_a: One or more of the parameters is not equal to zero

The following F statistic provides the basis for testing whether the additional independent variables are statistically significant.

$$F = \frac{\dfrac{\text{SSE}(x_1, x_2, \ldots, x_q) - \text{SSE}(x_1, x_2, \ldots, x_q, x_{q+1}, \ldots, x_p)}{p - q}}{\dfrac{\text{SSE}(x_1, x_2, \ldots, x_q, x_{q+1}, \ldots, x_p)}{n - p - 1}} \tag{16.13}$$

This computed F value is then compared with F_α, the table value with $p - q$ numerator degrees of freedom and $n - p - 1$ denominator degrees of freedom. If $F > F_\alpha$, we reject H_0 and conclude that the set of additional independent variables is statistically significant. Note that for the special case where $q = 1$ and $p = 2$, equation (16.13) reduces to equation (16.10).

Many computer packages, such as Minitab, provide extra sums of squares corresponding to the order in which each independent variable enters the model; in such cases, the computation of the F test for determining whether to add or delete a set of variables is simplified.

Many students find equation (16.13) somewhat complex. To provide a simpler description of this F ratio, we can refer to the model with the smaller number of independent variables as the reduced model and the model with the larger number of independent variables as the full model. If we let SSE(reduced) denote the error sum of squares for the reduced model and SSE(full) denote the error sum of squares for the full model, we can write the numerator of (16.13) as

$$\frac{\text{SSE(reduced)} - \text{SSE(full)}}{\text{number of extra terms}} \tag{16.14}$$

Note that "number of extra terms" denotes the difference between the number of independent variables in the full model and the number of independent variables in the reduced model. The denominator of equation (16.13) is the error sum of squares for the full model divided by the corresponding degrees of freedom; in other words, the denominator is the mean square error for the full model. Denoting the mean square error for the full model as MSE(full) enables us to write it as

$$F = \frac{\dfrac{\text{SSE(reduced)} - \text{SSE(full)}}{\text{number of extra terms}}}{\text{MSE(full)}} \tag{16.15}$$

To illustrate the use of this F statistic, suppose we have a regression problem involving 30 observations. One model with the independent variables x_1, x_2, and x_3 has an error sum of squares of 150 and a second model with the independent variables x_1, x_2, x_3, x_4, and x_5 has an error sum of squares of 100. Did the addition of the two independent variables x_4 and x_5 result in a significant reduction in the error sum of squares?

First, note that the degrees of freedom for SST is $30 - 1 = 29$ and that the degrees of freedom for the regression sum of squares for the full model is five (the number of independent variables in the full model). Thus, the degrees of freedom for the error sum of squares for the full model is $29 - 5 = 24$, and hence MSE(full) $= 100/24 = 4.17$. Therefore the F statistic is

$$F = \frac{\dfrac{150 - 100}{2}}{4.17} = 6.00$$

This computed F value is compared with the table F value with two numerator and 24 denominator degrees of freedom. At the .05 level of significance, Table 4 of Appendix B shows $F_{.05} = 3.40$. Because $F = 6.00$ is greater than 3.40, we conclude that the addition of variables x_4 and x_5 is statistically significant.

Use of p-Values

The p-value criterion can also be used to determine whether it is advantageous to add one or more independent variables to a multiple regression model. In the preceding example, we showed how to perform an F test to determine if the addition of two independent variables, x_4 and x_5, to a model with three independent variables, x_1, x_2, and x_3, was statistically significant. For this example, the computed F statistic was 6.00 and we concluded (by comparing $F = 6.00$ to the critical value $F_{.05} = 3.40$) that the addition of variables x_4 and x_5 was significant. Using Minitab or Excel, the p-value associated with $F = 6.00$ (2 numerator and 24 denominator degrees of freedom) is .008. With a p-value $= .008 < \alpha = .05$, we also conclude that the addition of the two independent variables is statistically significant. It is difficult to determine the p-value directly from tables of the F distribution, but computer software packages, such as Minitab or Excel, make computing the p-value easy.

NOTES AND COMMENTS

Computation of the F statistic can also be based on the difference in the regression sums of squares. To show this form of the F statistic, we first note that

$$\text{SSE(reduced)} = \text{SST} - \text{SSR(reduced)}$$
$$\text{SSE(full)} = \text{SST} - \text{SSR(full)}$$

Hence

$$\begin{aligned}\text{SSE(reduced)} - \text{SSE(full)} &= [\text{SST} - \text{SSR(reduced)}] - [\text{SST} - \text{SSR(full)}] \\ &= \text{SSR(full)} - \text{SSR(reduced)}\end{aligned}$$

Thus,

$$F = \frac{\dfrac{\text{SSR(full)} - \text{SSR(reduced)}}{\text{number of extra terms}}}{\text{MSE(full)}}$$

Exercises

Methods

10. In a regression analysis involving 27 observations, the following estimated regression equation was developed.

$$\hat{y} = 25.2 + 5.5x_1$$

For this estimated regression equation SST = 1550 and SSE = 520.

a. At $\alpha = .05$, test whether x_1 is significant.

Suppose that variables x_2 and x_3 are added to the model and the following regression equation is obtained.

$$\hat{y} = 16.3 + 2.3x_1 + 12.1x_2 - 5.8x_3$$

For this estimated regression equation SST = 1550 and SSE = 100.

b. Use an F test and a .05 level of significance to determine whether x_2 and x_3 contribute significantly to the model.

SELF test

11. In a regression analysis involving 30 observations, the following estimated regression equation was obtained.

$$\hat{y} = 17.6 + 3.8x_1 - 2.3x_2 + 7.6x_3 + 2.7x_4$$

For this estimated regression equation SST = 1805 and SSR = 1760.

a. At $\alpha = .05$, test the significance of the relationship among the variables.

Suppose variables x_1 and x_4 are dropped from the model and the following estimated regression equation is obtained.

$$\hat{y} = 11.1 - 3.6x_2 + 8.1x_3$$

For this model SST = 1805 and SSR = 1705.

b. Compute SSE(x_1, x_2, x_3, x_4).

c. Compute SSE(x_2, x_3).

d. Use an F test and a .05 level of significance to determine whether x_1 and x_4 contribute significantly to the model.

Applications

WEB file
LPGATour

12. The Ladies Professional Golfers Association (LPGA) maintains statistics on performance and earnings for members of the LPGA Tour. Year-end performance statistics for the 30 players who had the highest total earnings in LPGA Tour events for 2005 appear in the file named LPGATour (LPGA Tour website, 2006). Earnings ($1000) is the total earnings in thousands of dollars; Scoring Avg. is the average score for all events; Greens in Reg. is the percentage of time a player is able to hit the green in regulation; Putting Avg. is the average number of putts taken on greens hit in regulation; and Sand Saves is the percentage of time a player is able to get "up and down" once in a greenside sand bunker. A green is considered hit in regulation if any part of the ball is touching the putting surface and the difference between the value of par for the hole and the number of strokes taken to hit the green is at least 2.

a. Develop an estimated regression equation that can be used to predict the average score for all events given the average number of putts taken on greens hit in regulation.

b. Develop an estimated regression equation that can be used to predict the average score for all events given the percentage of time a player is able to hit the green in regulation, the average number of putts taken on greens hit in regulation, and the percentage of time a player is able to get "up and down" once in a greenside sand bunker.

c. At the .05 level of significance, test whether the two independent variables added in part (b), the percentage of time a player is able to hit the green in regulation and the percentage of time a player is able to get "up and down" once in a greenside sand bunker, contribute significantly to the estimated regression equation developed in part (a). Explain.

WEB file LPGATour

13. Refer to exercise 12.
   a. Develop an estimated regression equation that can be used to predict the total earnings for all events given the average number of putts taken on greens hit in regulation.
   b. Develop an estimated regression equation that can be used to predict the total earnings for all events given the percentage of time a player is able to hit the green in regulation, the average number of putts taken on greens hit in regulation, and the percentage of time a player is able to get "up and down" once in a greenside sand bunker.
   c. At the .05 level of significance, test whether the two independent variables added in part (b), the percentage of time a player is able to hit the green in regulation and the percentage of time a player is able to get "up and down" once in a greenside sand bunker, contribute significantly to the estimated regression equation developed in part (a). Explain.
   d. In general, lower scores should lead to higher earnings. To investigate this option to predicting total earnings, develop an estimated regression equation that can be used to predict total earnings for all events given the average score for all events. Would you prefer to use this equation to predict total earnings or the estimated regression equation developed in part (b)? Explain.

14. A 10-year study conducted by the American Heart Association provided data on how age, blood pressure, and smoking relate to the risk of strokes. Data from a portion of this study follow. Risk is interpreted as the probability (times 100) that a person will have a stroke over the next 10-year period. For the smoker variable, 1 indicates a smoker and 0 indicates a nonsmoker.

WEB file Stroke

Risk	Age	Blood Pressure	Smoker
12	57	152	0
24	67	163	0
13	58	155	0
56	86	177	1
28	59	196	0
51	76	189	1
18	56	155	1
31	78	120	0
37	80	135	1
15	78	98	0
22	71	152	0
36	70	173	1
15	67	135	1
48	77	209	1
15	60	199	0
36	82	119	1
8	66	166	0
34	80	125	1
3	62	117	0
37	59	207	1

   a. Develop an estimated regression equation that can be used to predict the risk of stroke given the age and blood-pressure level.
   b. Consider adding two independent variables to the model developed in part (a), one for the interaction between age and blood-pressure level and the other for whether the

person is a smoker. Develop an estimated regression equation using these four independent variables.

c. At a .05 level of significance, test to see whether the addition of the interaction term and the smoker variable contribute significantly to the estimated regression equation developed in part (a).

15. In baseball, an earned run is any run that the opposing team scores off the pitcher except for runs scored as a result of errors. The earned run average (ERA), the statistic most often used to compare the performance of pitchers, is computed as follows:

$$\text{ERA} = \left(\frac{\text{earned runs given up}}{\text{innings pitched}}\right)9$$

WEB file
MLBPitching

Note that the average number of earned runs per inning pitched is multiplied by nine, the number of innings in a regulation game. Thus, ERA represents the average number of runs the pitcher gives up per nine innings. For instance, in 2008, Roy Halladay, a pitcher for the Toronto Blue Jays, pitched 246 innings and gave up 76 earned runs; his ERA was $(76/246)9 = 2.78$. To investigate the relationship between ERA and other measures of pitching performance, data for 50 Major League Baseball pitchers for the 2008 season appear in the data set named MLBPitching (MLB website, February 2009). Descriptions for variables which appear on the data set follow:

W	Number of games won
L	Number of games lost
WPCT	Percentage of games won
H/9	Average number of hits given up per nine innings
HR/9	Average number of home runs given up per nine innings
BB/9	Average number of bases on balls given up per nine innings

a. Develop an estimated regression equation that can be used to predict the earned run average given the average number hits given up per nine innings.

b. Develop an estimated regression equation that can be used to predict the earned run average given the average number hits given up per nine innings, the average number of home runs given up per nine innings, and the average number of bases on balls given up per nine innings.

c. At the .05 level of significance, test whether the two independent variables added in part (b), the average number of home runs given up per nine innings and the average number of bases on ball given up per nine innings, contribute significantly to the estimated regression equation developed in part (a).

16.3 Analysis of a Larger Problem

In introducing multiple regression analysis, we used the Butler Trucking example extensively. The small size of this problem was an advantage in exploring introductory concepts but would make it difficult to illustrate some of the variable selection issues involved in model building. To provide an illustration of the variable selection procedures discussed in the next section, we introduce a data set consisting of 25 observations on eight independent variables. Permission to use these data was provided by Dr. David W. Cravens of the Department of Marketing at Texas Christian University. Consequently, we refer to the data set as the Cravens data.[1]

The Cravens data are for a company that sells products in several sales territories, each of which is assigned to a single sales representative. A regression analysis was conducted

[1]For details see David W. Cravens, Robert B. Woodruff, and Joe C. Stamper, "An Analytical Approach for Evaluating Sales Territory Performance," *Journal of Marketing*, 36 (January 1972): 31–37. Copyright © 1972 American Marketing Association.

TABLE 16.5 CRAVENS DATA

WEB file
Cravens

Sales	Time	Poten	AdvExp	Share	Change	Accounts	Work	Rating
3,669.88	43.10	74,065.1	4,582.9	2.51	.34	74.86	15.05	4.9
3,473.95	108.13	58,117.3	5,539.8	5.51	.15	107.32	19.97	5.1
2,295.10	13.82	21,118.5	2,950.4	10.91	−.72	96.75	17.34	2.9
4,675.56	186.18	68,521.3	2,243.1	8.27	.17	195.12	13.40	3.4
6,125.96	161.79	57,805.1	7,747.1	9.15	.50	180.44	17.64	4.6
2,134.94	8.94	37,806.9	402.4	5.51	.15	104.88	16.22	4.5
5,031.66	365.04	50,935.3	3,140.6	8.54	.55	256.10	18.80	4.6
3,367.45	220.32	35,602.1	2,086.2	7.07	−.49	126.83	19.86	2.3
6,519.45	127.64	46,176.8	8,846.2	12.54	1.24	203.25	17.42	4.9
4,876.37	105.69	42,053.2	5,673.1	8.85	.31	119.51	21.41	2.8
2,468.27	57.72	36,829.7	2,761.8	5.38	.37	116.26	16.32	3.1
2,533.31	23.58	33,612.7	1,991.8	5.43	−.65	142.28	14.51	4.2
2,408.11	13.82	21,412.8	1,971.5	8.48	.64	89.43	19.35	4.3
2,337.38	13.82	20,416.9	1,737.4	7.80	1.01	84.55	20.02	4.2
4,586.95	86.99	36,272.0	10,694.2	10.34	.11	119.51	15.26	5.5
2,729.24	165.85	23,093.3	8,618.6	5.15	.04	80.49	15.87	3.6
3,289.40	116.26	26,878.6	7,747.9	6.64	.68	136.58	7.81	3.4
2,800.78	42.28	39,572.0	4,565.8	5.45	.66	78.86	16.00	4.2
3,264.20	52.84	51,866.1	6,022.7	6.31	−.10	136.58	17.44	3.6
3,453.62	165.04	58,749.8	3,721.1	6.35	−.03	138.21	17.98	3.1
1,741.45	10.57	23,990.8	861.0	7.37	−1.63	75.61	20.99	1.6
2,035.75	13.82	25,694.9	3,571.5	8.39	−.43	102.44	21.66	3.4
1,578.00	8.13	23,736.3	2,845.5	5.15	.04	76.42	21.46	2.7
4,167.44	58.44	34,314.3	5,060.1	12.88	.22	136.58	24.78	2.8
2,799.97	21.14	22,809.5	3,552.0	9.14	−.74	88.62	24.96	3.9

to determine whether a variety of predictor (independent) variables could explain sales in each territory. A random sample of 25 sales territories resulted in the data in Table 16.5; the variable definitions are given in Table 16.6.

As a preliminary step, let us consider the sample correlation coefficients between each pair of variables. Figure 16.13 is the correlation matrix obtained using Minitab. Note that the sample correlation coefficient between Sales and Time is .623, between Sales and Poten is .598, and so on.

TABLE 16.6 VARIABLE DEFINITIONS FOR THE CRAVENS DATA

Variable	Definition
Sales	Total sales credited to the sales representative
Time	Length of time employed in months
Poten	Market potential; total industry sales in units for the sales territory*
AdvExp	Advertising expenditure in the sales territory
Share	Market share; weighted average for the past four years
Change	Change in the market share over the previous four years
Accounts	Number of accounts assigned to the sales representative*
Work	Workload; a weighted index based on annual purchases and concentrations of accounts
Rating	Sales representative overall rating on eight performance dimensions; an aggregate rating on a 1–7 scale

*These data were coded to preserve confidentiality.

FIGURE 16.13 SAMPLE CORRELATION COEFFICIENTS FOR THE CRAVENS DATA

	Sales	Time	Poten	AdvExp	Share	Change	Accounts	Work
Time	0.623							
Poten	0.598	0.454						
AdvExp	0.596	0.249	0.174					
Share	0.484	0.106	-0.21	0.264				
Change	0.489	0.251	0.268	0.377	0.085			
Accounts	0.754	0.758	0.479	0.200	0.403	0.327		
Work	-0.117	-0.179	-0.259	-0.272	0.349	-0.288	-0.199	
Rating	0.402	0.101	0.359	0.411	-0.024	0.549	0.229	-0.277

Looking at the sample correlation coefficients between the independent variables, we see that the correlation between Time and Accounts is .758; hence, if Accounts were used as an independent variable, Time would not add much more explanatory power to the model. Recall the rule-of-thumb test from the discussion of multicollinearity in Section 15.4: Multicollinearity can cause problems if the absolute value of the sample correlation coefficient exceeds .7 for any two of the independent variables. If possible, then, we should avoid including both Time and Accounts in the same regression model. The sample correlation coefficient of .549 between Change and Rating is also high and may warrant further consideration.

Looking at the sample correlation coefficients between Sales and each of the independent variables can give us a quick indication of which independent variables are, by themselves, good predictors. We see that the single best predictor of Sales is Accounts, because it has the highest sample correlation coefficient (.754). Recall that for the case of one independent variable, the square of the sample correlation coefficient is the coefficient of determination. Thus, Accounts can explain $(.754)^2(100)$, or 56.85%, of the variability in Sales. The next most important independent variables are Time, Poten, and AdvExp, each with a sample correlation coefficient of approximately .6.

Although there are potential multicollinearity problems, let us consider developing an estimated regression equation using all eight independent variables. The Minitab computer package provided the results in Figure 16.14. The eight-variable multiple regression model has an R-Sq (adj) value of 88.3%. Note, however, that the p-values for the t tests of individual parameters show that only Poten, AdvExp, and Share are significant at the $\alpha = .05$ level, given the effect of all the other variables. Hence, we might be inclined to investigate the results that would be obtained if we used just those three variables. Figure 16.15 shows the Minitab results obtained for the estimated regression equation with those three variables. We see that the estimated regression equation has an R-Sq (adj) value of 82.7%, which, although not quite as good as that for theeight-independent-variable estimated regression equation, is high.

How can we find an estimated regression equation that will do the best job given the data available? One approach is to compute all possible regressions. That is, we could develop 8 one-variable estimated regression equations (each of which corresponds to one of the independent variables), 28 two-variable estimated regression equations (the number of combinations of eight variables taken two at a time), and so on. In all, for the Cravens data, 255 different estimated regression equations involving one or more independent variables would have to be fitted to the data.

With the excellent computer packages available today, it is possible to compute all possible regressions. But doing so involves a great amount of computation and requires the model builder to review a large volume of computer output, much of which is associated with obviously poor models. Statisticians prefer a more systematic approach to selecting the subset of independent variables that provide the best estimated regression equation. In the next section, we introduce some of the more popular approaches.

FIGURE 16.14 MINITAB OUTPUT FOR THE MODEL INVOLVING ALL EIGHT INDEPENDENT VARIABLES

```
The regression equation is
Sales = - 1508 + 2.01 Time + 0.0372 Poten + 0.151 AdvExp + 199 Share
        + 291 Change + 5.55 Accounts + 19.8 Work + 8 Rating

Predictor       Coef    SE Coef      T      p
Constant     -1507.8      778.6  -1.94  0.071
Time           2.010      1.931   1.04  0.313
Poten       0.037206   0.008202   4.54  0.000
AdvExp       0.15094    0.04711   3.21  0.006
Share         199.08      67.03   2.97  0.009
Change         290.9      186.8   1.56  0.139
Accounts       5.550      4.775   1.16  0.262
Work           19.79      33.68   0.59  0.565
Rating           8.2      128.5   0.06  0.950

S = 449.015   R-Sq = 92.2%   R-Sq(adj) = 88.3%

Analysis of Variance

SOURCE           DF        SS       MS      F      p
Regression        8  38153712  4769214  23.66  0.000
Residual Error   16   3225837   201615
Total            24  41379549
```

FIGURE 16.15 MINITAB OUTPUT FOR THE MODEL INVOLVING Poten, AdvExp, AND Share

```
The regression equation is
Sales = - 1604 + 0.0543 Poten + 0.167 AdvExp + 283 Share

Predictor       Coef    SE Coef      T      p
Constant     -1603.6      505.6  -3.17  0.005
Poten       0.054286   0.007474   7.26  0.000
AdvExp       0.16748    0.04427   3.78  0.001
Share         282.75      48.76   5.80  0.000

S = 545.515   R-Sq = 84.9%   R-Sq(adj) = 82.7%

Analysis of Variance

SOURCE           DF        SS        MS      F      p
Regression        3  35130228  11710076  39.35  0.000
Residual Error   21   6249321    297587
Total            24  41379549
```

16.4 Variable Selection Procedures

Variable selection procedures are particularly useful in the early stages of building a model, but they cannot substitute for experience and judgment on the part of the analyst.

In this section we discuss four **variable selection procedures**: stepwise regression, forward selection, backward elimination, and best-subsets regression. Given a data set with several possible independent variables, we can use these procedures to identify which independent variables provide the best model. The first three procedures are iterative; at each step of the procedure a single independent variable is added or deleted and the new model is evaluated. The process continues until a stopping criterion indicates that the procedure cannot find a better model. The last procedure (best subsets) is not a one-variable-at-a-time procedure; it evaluates regression models involving different subsets of the independent variables.

In the stepwise regression, forward selection, and backward elimination procedures, the criterion for selecting an independent variable to add or delete from the model at each step is based on the F statistic introduced in Section 16.2. Suppose, for instance, that we are considering adding x_2 to a model involving x_1 or deleting x_2 from a model involving x_1 and x_2. To test whether the addition or deletion of x_2 is statistically significant, the null and alternative hypotheses can be stated as follows:

$$H_0: \beta_2 = 0$$
$$H_a: \beta_2 \neq 0$$

In Section 16.2 (see equation (16.10)) we showed that

$$F = \frac{\dfrac{\text{SSE}(x_1) - \text{SSE}(x_1, x_2)}{1}}{\dfrac{\text{SSE}(x_1, x_2)}{n - p - 1}}$$

can be used as a criterion for determining whether the presence of x_2 in the model causes a significant reduction in the error sum of squares. The p-value corresponding to this F statistic is the criterion used to determine whether an independent variable should be added or deleted from the regression model. The usual rejection rule applies: Reject H_0 if p-value $\leq \alpha$.

Stepwise Regression

The stepwise regression procedure begins each step by determining whether any of the variables *already in the model* should be removed. It does so by first computing an F statistic and a corresponding p-value for each independent variable in the model. The level of significance α for determining whether an independent variable should be removed from the model is referred to in Minitab as *Alpha to remove.* If the p-value for any independent variable is greater than *Alpha to remove,* the independent variable with the largest p-value is removed from the model and the stepwise regression procedure begins a new step.

If no independent variable can be removed from the model, the procedure attempts to enter another independent variable into the model. It does so by first computing an F statistic and corresponding p-value for each independent variable that is not in the model. The level of significance α for determining whether an independent variable should be entered into the model is referred to in Minitab as *Alpha to enter.* The independent variable with the smallest p-value is entered into the model provided its p-value is less than or equal to *Alpha to enter.* The procedure continues in this manner until no independent variables can be deleted from or added to the model.

Figure 16.16 shows the results obtained by using the Minitab stepwise regression procedure for the Cravens data using values of .05 for *Alpha to remove* and .05 for *Alpha to enter.*

FIGURE 16.16 MINITAB STEPWISE REGRESSION OUTPUT FOR THE CRAVENS DATA

```
Alpha-to-Enter: 0.05      Alpha-to-Remove: 0.05

Response is Sales on 8 predictors, with N = 25

     Step          1        2         3          4
Constant      709.32    50.29   -327.24   -1441.93

Accounts        21.7     19.0      15.6        9.2
T-Value         5.50     6.41      5.19       3.22
P-Value        0.000    0.000     0.000      0.004

AdvExp                  0.227     0.216      0.175
T-Value                  4.50      4.77       4.74
P-Value                 0.000     0.000      0.000

Poten                            0.0219     0.0382
T-Value                            2.53       4.79
P-Value                           0.019      0.000

Share                                          190
T-Value                                       3.82
P-Value                                      0.001

S                881      650       583        454
R-Sq           56.85    77.51     82.77      90.04
R-Sq(adj)      54.97    75.47     80.31      88.05
Mallows Cp      67.6     27.2      18.4        5.4
```

The stepwise procedure terminated after four steps. The estimated regression equation identified by the Minitab stepwise regression procedure is

$$\hat{y} = -1441.93 + 9.2 \text{ Accounts} + .175 \text{ AdvExp} + .0382 \text{ Poten} + 190 \text{ Share}$$

Because the stepwise procedure does not consider every possible subset for a given number of independent variables, it will not necessarily select the estimated regression equation with the highest R-sq value.

Note also in Figure 16.16 that $s = \sqrt{\text{MSE}}$ has been reduced from 881 with the best one-variable model (using Accounts) to 454 after four steps. The value of R-sq has been increased from 56.85% to 90.04%, and the recommended estimated regression equation has an R-Sq(adj) value of 88.05%.

In summary, at each step of the stepwise regression procedure the first consideration is to see whether any independent variable can be removed from the current model. If none of the independent variables can be removed from the model, the procedure checks to see whether any of the independent variables that are not currently in the model can be entered. Because of the nature of the stepwise regression procedure, an independent variable can enter the model at one step, be removed at a subsequent step, and then enter the model at a later step. The procedure stops when no independent variables can be removed from or entered into the model.

Forward Selection

The forward selection procedure starts with no independent variables. It adds variables one at a time using the same procedure as stepwise regression for determining whether an independent variable should be entered into the model. However, the forward selection procedure does not permit a variable to be removed from the model once it has been entered. The procedure stops if the *p*-value for each of the independent variables not in the model is greater than *Alpha to enter.*

The estimated regression equation obtained using Minitab's forward selection procedure is

$$\hat{y} = -1441.93 + 9.2 \text{ Accounts} + .175 \text{ AdvExp} + .0382 \text{ Poten} + 190 \text{ Share}$$

Thus, for the Cravens data, the forward selection procedure (using .05 for *Alpha to enter*) leads to the same estimated regression equation as the stepwise procedure.

Backward Elimination

The backward elimination procedure begins with a model that includes all the independent variables. It then deletes one independent variable at a time using the same procedure as stepwise regression. However, the backward elimination procedure does not permit an independent variable to be reentered once it has been removed. The procedure stops when none of the independent variables in the model have a *p*-value greater than *Alpha to remove.*

The estimated regression equation obtained using Minitab's backward elimination procedure for the Cravens data (using .05 for *Alpha to remove*) is

$$\hat{y} = -1312 + 3.8 \text{ Time} + .0444 \text{ Poten} + .152 \text{ AdvExp} + 259 \text{ Share}$$

Comparing the estimated regression equation identified using the backward elimination procedure to the estimated regression equation identified using the forward selection procedure, we see that three independent variables—AdvExp, Poten, and Share—are common to both. However, the backward elimination procedure has included Time instead of Accounts.

Forward selection and backward elimination may lead to different models.

Forward selection and backward elimination are the two extremes of model building; the forward selection procedure starts with no independent variables in the model and adds independent variables one at a time, whereas the backward elimination procedure starts with all independent variables in the model and deletes variables one at a time. The two procedures may lead to the same estimated regression equation. It is possible, however, for them to lead to two different estimated regression equations, as we saw with the Cravens data. Deciding which estimated regression equation to use remains a topic for discussion. Ultimately, the analyst's judgment must be applied. The best-subsets model building procedure we discuss next provides additional model-building information to be considered before a final decision is made.

Best-Subsets Regression

Stepwise regression, forward selection, and backward elimination are approaches to choosing the regression model by adding or deleting independent variables one at a time. None of them guarantees that the best model for a given number of variables will be found. Hence, these one-variable-at-a-time methods are properly viewed as heuristics for selecting a good regression model.

The complete best-subsets output also includes values for the Mallows Cp statistic. More advanced texts discuss the use of this statistic.

Some software packages use a procedure called best-subsets regression that enables the user to find, given a specified number of independent variables, the best regression model. Minitab has such a procedure. Figure 16.17 is a portion of the computer output obtained by using the best-subsets procedure for the Cravens data set.

This output identifies the two best one-variable estimated regression equations, the two best two-variable equations, the two best three-variable equations, and so on. The criterion used in determining which estimated regression equations are best for any number of

FIGURE 16.17 PORTION OF MINITAB BEST-SUBSETS REGRESSION OUTPUT

Vars	R-Sq	R-Sq(adj)	S	Time	Poten	AdvExp	Share	Change	Accounts	WorK	RatIng
1	56.8	55.0	881.09						X		
1	38.8	36.1	1049.3	X							
2	77.5	75.5	650.39			X			X		
2	74.6	72.3	691.11		X			X			
3	84.9	82.7	545.52		X	X	X				
3	82.8	80.3	582.64		X	X			X		
4	90.0	88.1	453.84		X	X	X		X		
4	89.6	87.5	463.93	X	X	X	X				
5	91.5	89.3	430.21	X	X	X	X	X			
5	91.2	88.9	436.75		X	X	X	X	X		
6	92.0	89.4	427.99	X	X	X	X	X	X		
6	91.6	88.9	438.20		X	X	X	X	X	X	
7	92.2	89.0	435.66	X	X	X	X	X	X	X	
7	92.0	88.8	440.29	X	X	X	X	X	X		X
8	92.2	88.3	449.02	X	X	X	X	X	X	X	X

predictors is the value of the coefficient of determination (R-Sq). For instance, Accounts, with an R-Sq = 56.8%, provides the best estimated regression equation using only one independent variable; AdvExp and Accounts, with an R-Sq = 77.5%, provides the best estimated regression equation using two independent variables; and Poten, AdvExp, and Share, with an R-Sq = 84.9%, provides the best estimated regression equation with three independent variables. For the Cravens data, the adjusted coefficient of determination (R-Sq (adj) = 89.4%) is largest for the model with six independent variables: Time, Poten, AdvExp, Share, Change, and Accounts. However, the best model with four independent variables (Poten, AdvExp, Share, Accounts) has an adjusted coefficient of determination almost as high (R-Sq (adj) = 88.1%). All other things being equal, a simpler model with fewer variables is usually preferred.

Making the Final Choice

The analysis performed on the Cravens data to this point is good preparation for choosing a final model, but more analysis should be conducted before the final choice. As we noted in Chapters 14 and 15, a careful analysis of the residuals should be done. We want the residual plot for the chosen model to resemble approximately a horizontal band. Let us assume the residuals are not a problem and that we want to use the results of the best-subsets procedure to help choose the model.

The best-subsets procedure shows us that the best four-variable model contains the independent variables Poten, AdvExp, Share, and Accounts. This result also happens to be the four-variable model identified with the stepwise regression procedure. Table 16.7 is helpful in making the final choice. It shows several possible models consisting of some or all of these four independent variables.

TABLE 16.7 SELECTED MODELS INVOLVING Accounts, AdvExp, Poten, AND Share

Model	Independent Variables	R-Sq (adj)
1	Accounts	55.0
2	AdvExp, Accounts	75.5
3	Poten, Share	72.3
4	Poten, AdvExp, Accounts	80.3
5	Poten, AdvExp, Share	82.7
6	Poten, AdvExp, Share, Accounts	88.1

From Table 16.7, we see that the model with just AdvExp and Accounts is good. The adjusted coefficient of determination is R-Sq (adj) = 75.5%, and the model with all four variables provides only a 12.6-percentage-point improvement. The simpler two-variable model might be preferred, for instance, if it is difficult to measure market potential (Poten). However, if the data are readily available and highly accurate predictions of sales are needed, the model builder would clearly prefer the model with all four variables.

NOTES AND COMMENTS

1. The stepwise procedure requires that *Alpha to remove* be greater than or equal to *Alpha to enter*. This requirement prevents the same variable from being removed and then reentered at the same step.

2. Functions of the independent variables can be used to create new independent variables for use with any of the procedures in this section. For instance, if we wanted x_1x_2 in the model to account for interaction, we would use the data for x_1 and x_2 to create the data for $z = x_1x_2$.

3. None of the procedures that add or delete variables one at a time can be guaranteed to identify the best regression model. But they are excellent approaches to finding good models—especially when little multicollinearity is present.

Exercises

Applications

16. A study provided data on variables that may be related to the number of weeks a manufacturing worker has been jobless. The dependent variable in the study (Weeks) was defined as the number of weeks a worker has been jobless due to a layoff. The following independent variables were used in the study.

WEB file
Layoffs

Age	The age of the worker
Educ	The number of years of education
Married	A dummy variable; 1 if married, 0 otherwise
Head	A dummy variable; 1 if the head of household, 0 otherwise
Tenure	The number of years on the previous job
Manager	A dummy variable; 1 if management occupation, 0 otherwise
Sales	A dummy variable; 1 if sales occupation, 0 otherwise

The data are available in the file named Layoffs.

a. Develop the best one-variable estimated regression equation.

b. Use the stepwise procedure to develop the best estimated regression equation. Use values of .05 for *Alpha to enter* and *Alpha to remove*.

c. Use the forward selection procedure to develop the best estimated regression equation. Use a value of .05 for *Alpha to enter*.

d. Use the backward elimination procedure to develop the best estimated regression equation. Use a value of .05 for *Alpha to remove*.

e. Use the best-subsets regression procedure to develop the best estimated regression equation.

WEB file LPGATour2

17. The Ladies Professional Golfers Association (LPGA) maintains statistics on performance and earnings for members of the LPGA Tour. Year-end performance statistics for the 30 players who had the highest total earnings in LPGA Tour events for 2005 appear in the file named LPGATour2 (LPGA Tour website, 2006). Earnings ($1000) is the total earnings in thousands of dollars; Scoring Avg. is the average score for all events; Drive Average is the average length of a players drive in yards; Greens in Reg. is the percentage of time a player is able to hit the green in regulation; Putting Avg. is the average number of putts taken on greens hit in regulation; and Sand Saves is the percentage of time a player is able to get "up and down" once in a greenside sand bunker. A green is considered hit in regulation if any part of the ball is touching the putting surface and the difference between the value of par for the hole and the number of strokes taken to hit the green is at least 2. Let DriveGreens denote a new independent variable that represents the interaction between the average length of a player's drive and the percentage of time a player is able to hit the green in regulation. Use the methods in this section to develop the best estimated multiple regression equation for predicting a player's average score for all events.

18. Jeff Sagarin has been providing sports ratings for *USA Today* since 1985. In baseball his predicted RPG (runs/game) statistic takes into account the entire player's offensive statistics, and is claimed to be the best measure of a player's true offensive value. The following data show the RPG and a variety of offensive statistics for the 2005 Major League Baseball (MLB) season for 20 members of the New York Yankees (*USA Today* website, March 3, 2006). The labels on columns are defined as follows: RPG, predicted runs per game statistic; H, hits; 2B, doubles; 3B, triples; HR, home runs; RBI, runs batted in; BB, bases on balls (walks); SO, strikeouts; SB, stolen bases; CS, caught stealing; OBP, on-base percentage; SLG, slugging percentage; and AVG, batting average.

WEB file Yankees

Player	RPG	H	2B	3B	HR	RBI	BB	SO	SB	CS	OBP	SLG	AVG
D Jeter	6.51	202	25	5	19	70	77	117	14	5	.389	.450	.309
H Matsui	6.32	192	45	3	23	116	63	78	2	2	.367	.496	.305
A Rodriguez	9.06	194	29	1	48	130	91	139	21	6	.421	.610	.321
G Sheffield	6.93	170	27	0	34	123	78	76	10	2	.379	.512	.291
R Cano	5.01	155	34	4	14	62	16	68	1	3	.320	.458	.297
B Williams	4.14	121	19	1	12	64	53	75	1	2	.321	.367	.249
J Posada	5.36	124	23	0	19	71	66	94	1	0	.352	.430	.262
J Giambi	9.11	113	14	0	32	87	108	109	0	0	.440	.535	.271
T Womack	2.91	82	8	1	0	15	12	49	27	5	.276	.280	.249
T Martinez	5.08	73	9	0	17	49	38	54	2	0	.328	.439	.241
M Bellhorn	4.07	63	20	0	8	30	52	112	3	0	.324	.357	.210
R Sierra	3.27	39	12	0	4	29	9	41	0	0	.265	.371	.229
J Flaherty	1.83	21	5	0	2	11	6	26	0	0	.206	.252	.165
B Crosby	3.48	27	0	1	1	6	4	14	4	1	.304	.327	.276
M Lawton	5.15	6	0	0	2	4	7	8	1	0	.263	.250	.125
R Sanchez	3.36	12	1	0	0	2	2	3	0	1	.326	.302	.279
A Phillips	2.13	6	4	0	1	4	1	13	0	0	.171	.325	.150
M Cabrera	1.19	4	0	0	0	0	0	2	0	0	.211	.211	.211
R Johnson	3.44	4	2	0	0	0	1	4	0	0	.300	.333	.222
F Escalona	5.31	4	1	0	0	2	1	4	0	0	.375	.357	.286

Let the dependent variable be the RPG statistic.

a. Develop the best one-variable estimated regression equation.
b. Use the methods in this section to develop the best estimated multiple regression equation for predicting a player's RPG.

WEB file

Stroke

19. Refer to exercise 14. Using age, blood pressure, whether a person is a smoker, and any interaction involving those variables, develop an estimated regression equation that can be used to predict risk. Briefly describe the process you used to develop an estimated regression equation for these data.

16.5 Multiple Regression Approach to Experimental Design

In Section 15.7 we discussed the use of dummy variables in multiple regression analysis. In this section we show how the use of dummy variables in a multiple regression equation can provide another approach to solving experimental design problems. We will demonstrate the multiple regression approach to experimental design by applying it to the Chemitech, Inc., completely randomized design introduced in Chapter 13.

Recall that Chemitech developed a new filtration system for municipal water supplies. The components for the new filtration system will be purchased from several suppliers, and Chemitech will assemble the components at its plant in Columbia, South Carolina. Three different assembly methods, referred to as methods A, B, and C, have been proposed. Managers at Chemitech want to determine which assembly method can produce the greatest number of filtration systems per week.

A random sample of 15 employees was selected, and each of the three assembly methods was randomly assigned to 5 employees. The number of units assembled by each employee is shown in Table 16.8. The sample mean number of units produced with each of the three assembly methods is as follows:

Assembly Method	Mean Number Produced
A	62
B	66
C	52

Although method B appears to result in higher production rates than either of the other methods, the issue is whether the three sample means observed are different enough for us to conclude that the means of the populations corresponding to the three methods of assembly are different.

We begin the regression approach to this problem by defining dummy variables that will be used to indicate which assembly method was used. Because the Chemitech problem has

TABLE 16.8 NUMBER OF UNITS PRODUCED BY 15 WORKERS

	Method	
A	**B**	**C**
58	58	48
64	69	57
55	71	59
66	64	47
67	68	49

TABLE 16.9 DUMMY VARIABLES FOR THE CHEMITECH EXPERIMENT

A	B	
1	0	Observation is associated with assembly method A
0	1	Observation is associated with assembly method B
0	0	Observation is associated with assembly method C

three assembly methods or treatments, we need two dummy variables. In general, if the factor being investigated involves k distinct levels or treatments, we need to define $k - 1$ dummy variables. For the Chemitech experiment we define dummy variables A and B as shown in Table 16.9.

We can use the dummy variables to relate the number of units produced per week, y, to the method of assembly the employee uses.

$$\begin{aligned} E(y) &= \text{Expected value of the number of units produced per week} \\ &= \beta_0 + \beta_1\text{A} + \beta_2\text{B} \end{aligned}$$

Thus, if we are interested in the expected value of the number of units assembled per week for an employee who uses method C, our procedure for assigning numerical values to the dummy variables would result in setting A = B = 0. The multiple regression equation then reduces to

$$E(y) = \beta_0 + \beta_1(0) + \beta_2(0) = \beta_0$$

We can interpret β_0 as the expected value of the number of units assembled per week for an employee who uses method C. In other words, β_0 is the mean number of units assembled per week using method C.

Next let us consider the forms of the multiple regression equation for each of the other methods. For method A the values of the dummy variables are A = 1 and B = 0, and

$$E(y) = \beta_0 + \beta_1(1) + \beta_2(0) = \beta_0 + \beta_1$$

For method B we set A = 0 and B = 1, and

$$E(y) = \beta_0 + \beta_1(0) + \beta_2(1) = \beta_0 + \beta_2$$

We see that $\beta_0 + \beta_1$ represents the mean number of units assembled per week using method A, and $\beta_0 + \beta_2$ represents the mean number of units assembled per week using method B.

We now want to estimate the coefficients β_0, β_1, and β_2 and hence develop an estimate of the mean number of units assembled per week for each method. Table 16.10 shows the sample data, consisting of 15 observations of A, B, and y. Figure 16.18 shows the corresponding Minitab multiple regression output. We see that the estimates of β_0, β_1, and β_2 are $b_0 = 52$, $b_1 = 10$, and $b_2 = 14$. Thus, the best estimate of the mean number of units assembled per week for each assembly method is as follows:

Assembly Method	Prediction of $E(y)$
A	$b_0 + b_1 = 52 + 10 = 62$
B	$b_0 = 52 + 14 = 66$
C	$b_0 = 52$

TABLE 16.10 INPUT DATA FOR THE CHEMITECH COMPLETELY RANDOMIZED DESIGN

WEB file

Chemitech

A	B	*y*
1	0	58
1	0	64
1	0	55
1	0	66
1	0	67
0	1	58
0	1	69
0	1	71
0	1	64
0	1	68
0	0	48
0	0	57
0	0	59
0	0	47
0	0	49

Note that the estimate of the mean number of units produced with each of the three assembly methods obtained from the regression analysis is the same as the sample mean shown previously.

Now let us see how we can use the output from the multiple regression analysis to perform the ANOVA test on the difference among the means for the three plants. First, we observe that if the means do not differ

$$E(y) \text{ for method A} - E(y) \text{ for method C} = 0$$
$$E(y) \text{ for method B} - E(y) \text{ for method C} = 0$$

FIGURE 16.18 MULTIPLE REGRESSION OUTPUT FOR THE CHEMITECH COMPLETELY RANDOMIZED DESIGN

```
The regression equation is
y = 52.0 + 10.0 A + 14.0 B

Predictor      Coef  SE Coef      T      P
Constant     52.000    2.380  21.84  0.000
A            10.000    3.367   2.97  0.012
B            14.000    3.367   4.16  0.001

S = 5.32291   R-Sq = 60.5%   R-Sq(adj) = 53.9%

Analysis of Variance

SOURCE          DF       SS      MS     F      P
Regression       2   520.00  260.00  9.18  0.004
Residual Error  12   340.00   28.33
Total           14   860.00
```

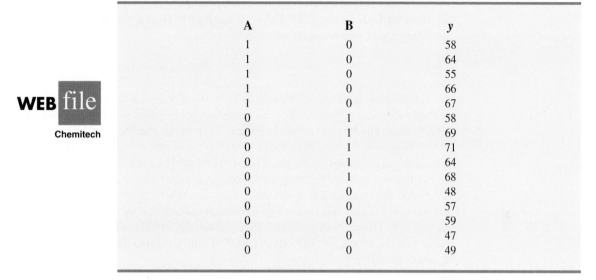

Because β_0 equals $E(y)$ for method C and $\beta_0 + \beta_1$ equals $E(y)$ for method A, the first difference is equal to $(\beta_0 + \beta_1) - \beta_0 = \beta_1$. Moreover, because $\beta_0 + \beta_2$ equals $E(y)$ for method B, the second difference is equal to $(\beta_0 + \beta_2) - \beta_0 = \beta_2$. We would conclude that the three methods do not differ if $\beta_1 = 0$ and $\beta_2 = 0$. Hence, the null hypothesis for a test for difference of means can be stated as

$$H_0 : \beta_1 = \beta_2 = 0$$

Suppose the level of significance is $\alpha = .05$. Recall that to test this type of null hypothesis about the significance of the regression relationship we use the F test for overall significance. The Minitab output in Figure 16.18 shows that the p-value corresponding to $F = 9.18$ is .004. Because the p-value $= .004 < \alpha = .05$, we reject $H_0 : \beta_1 = \beta_2 = 0$ and conclude that the means for the three assembly methods are not the same. Because the F test shows that the multiple regression relationship is significant, a t test can be conducted to determine the significance of the individual parameters, β_1 and β_2. Using $\alpha = .05$, the p-values of .012 and .001 on the Minitab output indicate that we can reject $H_0 : \beta_1 = 0$ and $H_0 : \beta_2 = 0$. Hence, both parameters are statistically significant. Thus, we can also conclude that the means for methods A and C are different and that the means for methods B and C are different.

Exercises

Methods

SELF test

20. Consider a completely randomized design involving four treatments: A, B, C, and D. Write a multiple regression equation that can be used to analyze these data. Define all variables.
21. Write a multiple regression equation that can be used to analyze the data for a randomized block design involving three treatments and two blocks. Define all variables.
22. Write a multiple regression equation that can be used to analyze the data for a two-factorial design with two levels for factor A and three levels for factor B. Define all variables.

Applications

SELF test

23. The Jacobs Chemical Company wants to estimate the mean time (minutes) required to mix a batch of material on machines produced by three different manufacturers. To limit the cost of testing, four batches of material were mixed on machines produced by each of the three manufacturers. The times needed to mix the material follow.

Manufacturer 1	Manufacturer 2	Manufacturer 3
20	28	20
26	26	19
24	31	23
22	27	22

a. Write a multiple regression equation that can be used to analyze the data.
b. What are the best estimates of the coefficients in your regression equation?

c. In terms of the regression equation coefficients, what hypotheses must we test to see whether the mean time to mix a batch of material is the same for all three manufacturers?
d. For an $\alpha = .05$ level of significance, what conclusion should be drawn?

24. Four different paints are advertised as having the same drying time. To check the manufacturers' claims, five samples were tested for each of the paints. The time in minutes until the paint was dry enough for a second coat to be applied was recorded for each sample. The data obtained follow.

Paint 1	Paint 2	Paint 3	Paint 4
128	144	133	150
137	133	143	142
135	142	137	135
124	146	136	140
141	130	131	153

a. Use $\alpha = .05$ to test for any significant differences in mean drying time among the paints.
b. What is your estimate of the mean drying time for paint 2? How is it obtained from the computer output?

25. An automobile dealer conducted a test to determine whether the time needed to complete a minor engine tune-up depends on whether a computerized engine analyzer or an electronic analyzer is used. Because tune-up time varies among compact, intermediate, and full-sized cars, the three types of cars were used as blocks in the experiment. The data (time in minutes) obtained follow.

		Car		
		Compact	**Intermediate**	**Full Size**
Analyzer	**Computerized**	50	55	63
	Electronic	42	44	46

Use $\alpha = .05$ to test for any significant differences.

26. A mail-order catalog firm designed a factorial experiment to test the effect of the size of a magazine advertisement and the advertisement design on the number (in thousands) of catalog requests received. Three advertising designs and two sizes of advertisements were considered. The following data were obtained. Test for any significant effects due to type of design, size of advertisement, or interaction. Use $\alpha = .05$.

		Size of Advertisement	
		Small	**Large**
Design	**A**	8	12
		12	8
	B	22	26
		14	30
	C	10	18
		18	14

16.6 Autocorrelation and the Durbin-Watson Test

Often, the data used for regression studies in business and economics are collected over time. It is not uncommon for the value of y at time t, denoted by y_t, to be related to the value of y at previous time periods. In such cases, we say **autocorrelation** (also called **serial correlation**) is present in the data. If the value of y in time period t is related to its value in time period $t - 1$, first-order autocorrelation is present. If the value of y in time period t is related to the value of y in time period $t - 2$, second-order autocorrelation is present, and so on.

One of the assumptions of the regression model is the error terms are independent. However, when autocorrelation is present, this assumption is violated. In the case of first-order autocorrelation, the error at time t, denoted ϵ_t, will be related to the error at time period $t - 1$, denoted ϵ_{t-1}. Two cases of first-order autocorrelation are illustrated in Figure 16.19. Panel A is the case of positive autocorrelation; panel B is the case of negative autocorrelation. With positive autocorrelation we expect a positive residual in one period to be followed by a positive residual in the next period, a negative residual in one period to be followed by a negative residual in the next period, and so on. With negative autocorrelation, we expect a positive residual in one period to be followed by a negative residual in the next period, then a positive residual, and so on.

When autocorrelation is present, serious errors can be made in performing tests of statistical significance based upon the assumed regression model. It is therefore important to be able to detect autocorrelation and take corrective action. We will show how the Durbin-Watson statistic can be used to detect first-order autocorrelation.

Suppose the values of ϵ are not independent but are related in the following manner:

$$\epsilon_t = \rho\epsilon_{t-1} + z_t \tag{16.16}$$

where ρ is a parameter with an absolute value less than one and z_t is a normally and independently distributed random variable with a mean of zero and a variance of σ^2. From equation (16.16) we see that if $\rho = 0$, the error terms are not related, and each has a mean of zero and a variance of σ^2. In this case, there is no autocorrelation and the regression assumptions

FIGURE 16.19 TWO DATA SETS WITH FIRST-ORDER AUTOCORRELATION

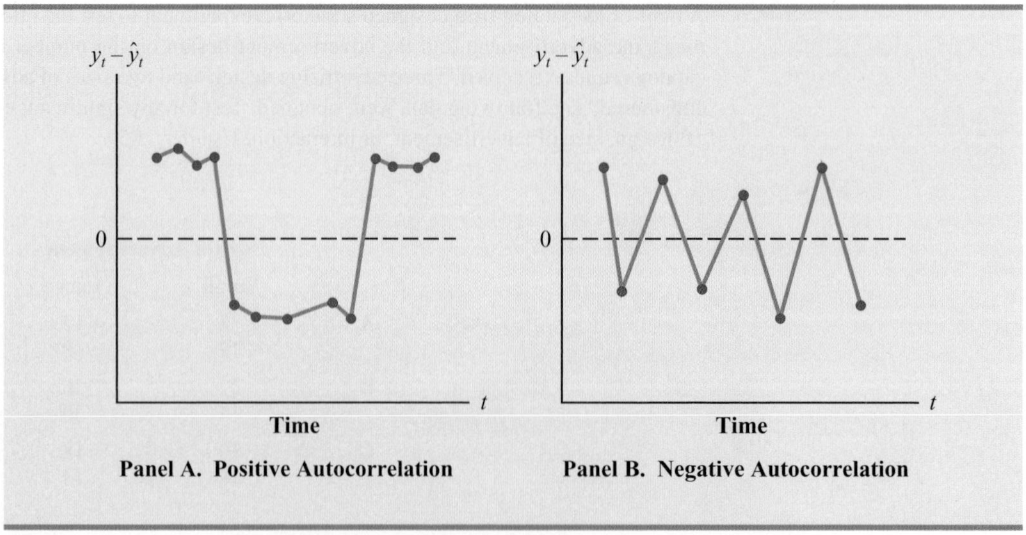

are satisfied. If $\rho > 0$, we have positive autocorrelation; if $\rho < 0$, we have negative autocorrelation. In either of these cases, the regression assumptions about the error term are violated.

The **Durbin-Watson test** for autocorrelation uses the residuals to determine whether $\rho = 0$. To simplify the notation for the Durbin-Watson statistic, we denote the ith residual by $e_i = y_i - \hat{y}_i$. The Durbin-Watson test statistic is computed as follows.

DURBIN-WATSON TEST STATISTIC

$$d = \frac{\sum_{t=2}^{n}(e_t - e_{t-1})^2}{\sum_{t=1}^{n} e_t^2} \quad (16.17)$$

If successive values of the residuals are close together (positive autocorrelation), the value of the Durbin-Watson test statistic will be small. If successive values of the residuals are far apart (negative autocorrelation), the value of the Durbin-Watson statistic will be large.

The Durbin-Watson test statistic ranges in value from zero to four, with a value of two indicating no autocorrelation is present. Durbin and Watson developed tables that can be used to determine when their test statistic indicates the presence of autocorrelation. Table 16.11 shows lower and upper bounds (d_L and d_U) for hypothesis tests using $\alpha = .05$; n denotes the number of observations. The null hypothesis to be tested is always that there is no autocorrelation.

$$H_0: \rho = 0$$

The alternative hypothesis to test for positive autocorrelation is

$$H_a: \rho > 0$$

TABLE 16.11 CRITICAL VALUES FOR THE DURBIN-WATSON TEST FOR AUTOCORRELATION

Note: Entries in the table are the critical values for a one-tailed Durbin-Watson test for autocorrelation. For a two-tailed test, the level of significance is doubled.

Significance Points of d_L and d_U: $\alpha = .05$
Number of Independent Variables

	1		2		3		4		5	
n*	d_L	d_U	d_L	d_U	d_L	d_U	d_L	d_U	d_L	d_U
15	1.08	1.36	.95	1.54	.82	1.75	.69	1.97	.56	2.21
20	1.20	1.41	1.10	1.54	1.00	1.68	.90	1.83	.79	1.99
25	1.29	1.45	1.21	1.55	1.12	1.66	1.04	1.77	.95	1.89
30	1.35	1.49	1.28	1.57	1.21	1.65	1.14	1.74	1.07	1.83
40	1.44	1.54	1.39	1.60	1.34	1.66	1.29	1.72	1.23	1.79
50	1.50	1.59	1.46	1.63	1.42	1.67	1.38	1.72	1.34	1.77
70	1.58	1.64	1.55	1.67	1.52	1.70	1.49	1.74	1.46	1.77
100	1.65	1.69	1.63	1.72	1.61	1.74	1.59	1.76	1.57	1.78

*Interpolate linearly for intermediate n values.

FIGURE 16.20 HYPOTHESIS TEST FOR AUTOCORRELATION USING THE DURBIN-WATSON TEST

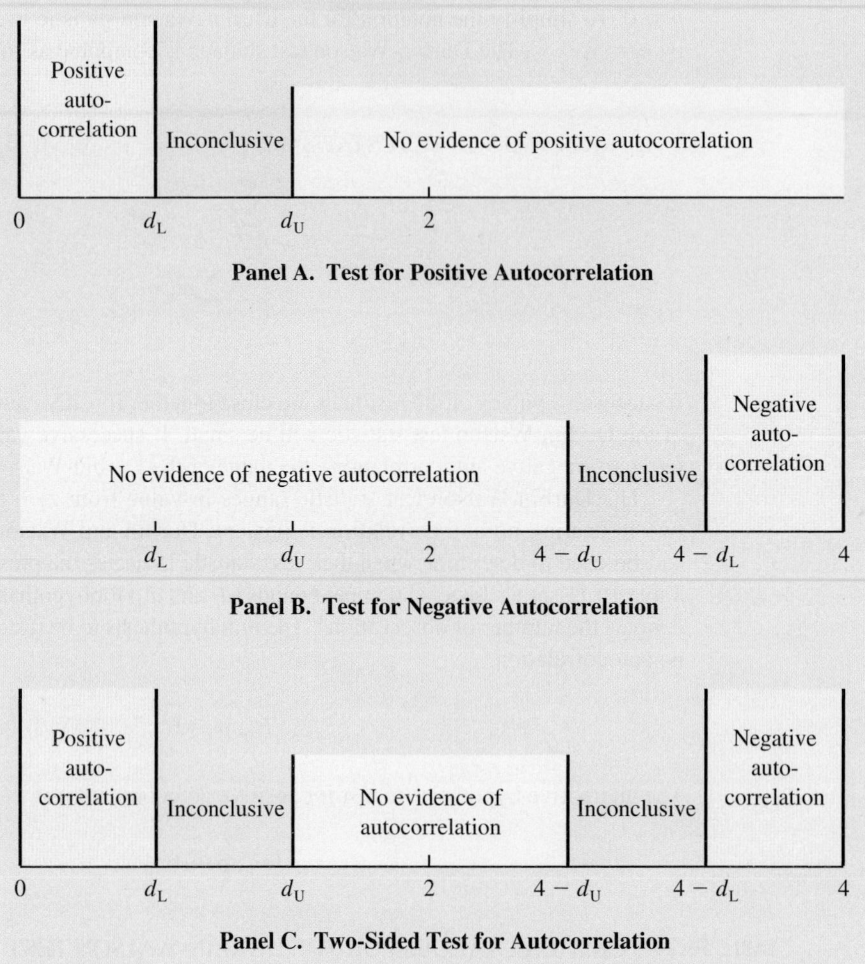

The alternative hypothesis to test for negative autocorrelation is

$$H_a: \rho < 0$$

A two-sided test is also possible. In this case the alternative hypothesis is

$$H_a: \rho \neq 0$$

Figure 16.20 shows how the values of d_L and d_U in Table 16.11 are used to test for autocorrelation. Panel A illustrates the test for positive autocorrelation. If $d < d_L$, we conclude that positive autocorrelation is present. If $d_L \leq d \leq d_U$, we say the test is inconclusive. If $d > d_U$, we conclude that there is no evidence of positive autocorrelation.

Panel B illustrates the test for negative autocorrelation. If $d > 4 - d_L$, we conclude that negative autocorrelation is present. If $4 - d_U \leq d \leq 4 - d_L$, we say the test is inconclusive. If $d < 4 - d_U$, we conclude that there is no evidence of negative autocorrelation.

Panel C illustrates the two-sided test. If $d < d_L$ or $d > 4 - d_L$, we reject H_0 and conclude that autocorrelation is present. If $d_L \leq d \leq d_U$ or $4 - d_U \leq d \leq 4 - d_L$, we say the test is inconclusive. If $d_U < d < 4 - d_U$, we conclude that there is no evidence of autocorrelation.

If significant autocorrelation is identified, we should investigate whether we omitted one or more key independent variables that have time-ordered effects on the dependent variable. If no such variables can be identified, including an independent variable that measures the time of the observation (for instance, the value of this variable could be one for the first observation, two for the second observation, and so on) will sometimes eliminate or reduce the autocorrelation. When these attempts to reduce or remove autocorrelation do not work, transformations on the dependent or independent variables can prove helpful; a discussion of such transformations can be found in more advanced texts on regression analysis.

Note that the Durbin-Watson tables list the smallest sample size as 15. The reason is that the test is generally inconclusive for smaller sample sizes; in fact, many statisticians believe the sample size should be at least 50 for the test to produce worthwhile results.

Exercises

Applications

27. The following data show the daily closing prices (in dollars per share) for a stock.

WEB file

ClosingPrice

Date	Price ($)
Nov. 3	82.87
Nov. 4	83.00
Nov. 7	83.61
Nov. 8	83.15
Nov. 9	82.84
Nov. 10	83.99
Nov. 11	84.55
Nov. 14	84.36
Nov. 15	85.53
Nov. 16	86.54
Nov. 17	86.89
Nov. 18	87.77
Nov. 21	87.29
Nov. 22	87.99
Nov. 23	88.80
Nov. 25	88.80
Nov. 28	89.11
Nov. 29	89.10
Nov. 30	88.90
Dec. 1	89.21

a. Define the independent variable Period, where Period = 1 corresponds to the data for November 3, Period = 2 corresponds to the data for November 4, and so on. Develop the estimated regression equation that can be used to predict the closing price given the value of Period.
b. At the .05 level of significance, test for any positive autocorrelation in the data.

28. Refer to the Cravens data set in Table 16.5. In Section 16.3 we showed that the estimated regression equation involving Accounts, AdvExp, Poten, and Share had an adjusted coefficient

of determination of 88.1%. Use the .05 level of significance and apply the Durbin-Watson test to determine whether positive autocorrelation is present.

Summary

In this chapter we discussed several concepts used by model builders to help identify the best estimated regression equation. First, we introduced the concept of a general linear model to show how the methods discussed in Chapters 14 and 15 could be extended to handle curvilinear relationships and interaction effects. Then we discussed how transformations involving the dependent variable could be used to account for problems such as nonconstant variance in the error term.

In many applications of regression analysis, a large number of independent variables are considered. We presented a general approach based on an F statistic for adding or deleting variables from a regression model. We then introduced a larger problem involving 25 observations and eight independent variables. We saw that one issue encountered in solving larger problems is finding the best subset of the independent variables. To help in that task, we discussed several variable selection procedures: stepwise regression, forward selection, backward elimination, and best-subsets regression.

In Section 16.5, we extended the discussion of how multiple regression models could be developed to provide another approach for solving analysis of variance and experimental design problems. The chapter concluded with an application of residual analysis to show the Durbin-Watson test for autocorrelation.

Glossary

General linear model A model of the form $y = \beta_0 + \beta_1 z_1 + \beta_2 z_2 + \cdots + \beta_p z_p + \epsilon$, where each of the independent variables z_j $(j = 1, 2, \ldots, p)$ is a function of $x_1, x_2, \ldots, x_k$, the variables for which data have been collected.

Interaction The effect of two independent variables acting together.

Variable selection procedures Methods for selecting a subset of the independent variables for a regression model.

Autocorrelation Correlation in the errors that arises when the error terms at successive points in time are related.

Serial correlation Same as autocorrelation.

Durbin-Watson test A test to determine whether first-order autocorrelation is present.

Key Formulas

General Linear Model

$$y = \beta_0 + \beta_1 z_1 + \beta_2 z_2 + \cdots + \beta_p z_p + \epsilon \quad \textbf{(16.1)}$$

F Test Statistic for Adding or Deleting $p - q$ Variables

$$F = \frac{\dfrac{\text{SSE}(x_1, x_2, \ldots, x_q) - \text{SSE}(x_1, x_2, \ldots, x_q, x_{q+1}, \ldots, x_p)}{p - q}}{\dfrac{\text{SSE}(x_1, x_2, \ldots, x_q, x_{q+1}, \ldots, x_p)}{n - p - 1}} \quad \textbf{(16.13)}$$

First-Order Autocorrelation

$$\epsilon_t = \rho\epsilon_{t-1} + z_t \tag{16.16}$$

Durbin-Watson Test Statistic

$$d = \frac{\sum_{t=2}^{n}(e_t - e_{t-1})^2}{\sum_{t=1}^{n} e_t^2} \tag{16.17}$$

Supplementary Exercises

WEB file CorporateBonds

29. A sample containing years to maturity and yield (%) for 40 corporate bonds is contained in the data file named CorporateBonds (*Barron's*, April 2, 2012).
    a. Develop a scatter diagram of the data using x = years to maturity as the independent variable. Does a simple linear regression model appear to be appropriate?
    b. Develop an estimated regression equation with x = years to maturity and x^2 as the independent variables.
    c. As an alternative to fitting a second-order model, fit a model using the natural logarithm of price as the independent variable; that is, $\hat{y} = b_0 + b_1\ln(x)$. Does the estimated regression using the natural logarithm of x provide a better fit than the estimated regression developed in part (b)? Explain.
30. *Consumer Reports* tested 19 different brands and models of road, fitness, and comfort bikes. Road bikes are designed for long road trips; fitness bikes are designed for regular workouts or daily commutes; and comfort bikes are designed for leisure rides on typically flat roads. The following data show the type, weight (lb.), and price ($) for the 19 bicycles tested (*Consumer Reports* website, February 2009).

WEB file Bikes

Brand and Model	Type	Weight	Price($)
Klein RÃªve v	Road	20	1800
Giant OCR Composite 3	Road	22	1800
Giant OCR 1	Road	22	1000
Specialized Roubaix	Road	21	1300
Trek Pilot 2.1	Road	21	1320
Cannondale Synapse 4	Road	21	1050
LeMond Poprad	Road	22	1350
Raleigh Cadent 1.0	Road	24	650
Giant FCR3	Fitness	23	630
Schwinn Super Sport GS	Fitness	23	700
Fuji Absolute 2.0	Fitness	24	700
Jamis Coda Comp	Fitness	26	830
Cannondale Road Warrior 400	Fitness	25	700
Schwinn Sierra GS	Comfort	31	340
Mongoose Switchback SX	Comfort	32	280
Giant Sedona DX	Comfort	32	360
Jamis Explorer 4.0	Comfort	35	600
Diamondback Wildwood Deluxe	Comfort	34	350
Specialized Crossroads Sport	Comfort	31	330

a. Develop a scatter diagram with weight as the independent variable and price as the dependent variable. Does a simple linear regression model appear to be appropriate?
b. Develop an estimated multiple regression equation with x = weight and x^2 as the two independent variables.

c. Use the following dummy variables to develop an estimated regression equation that can be used to predict the price given the type of bike: Type_Fitness = 1 if the bike is a fitness bike, 0 otherwise; and Type_Comfort = 1 if the bike is a comfort bike; 0 otherwise. Compare the results obtained to the results obtained in part (b).

d. To account for possible interaction between the type of bike and the weight of the bike, develop a new estimated regression equation that can be used to predict the price of the bike given the type, the weight of the bike, and any interaction between weight and each of the dummy variables defined in part (c). What estimated regression equation appears to be the best predictor of price? Explain.

31. A study investigated the relationship between audit delay (Delay), the length of time from a company's fiscal year-end to the date of the auditor's report, and variables that describe the client and the auditor. Some of the independent variables that were included in this study follow.

Industry	A dummy variable coded 1 if the firm was an industrial company or 0 if the firm was a bank, savings and loan, or insurance company.
Public	A dummy variable coded 1 if the company was traded on an organized exchange or over the counter; otherwise coded 0.
Quality	A measure of overall quality of internal controls, as judged by the auditor, on a five-point scale ranging from "virtually none" (1) to "excellent" (5).
Finished	A measure ranging from 1 to 4, as judged by the auditor, where 1 indicates "all work performed subsequent to year-end" and 4 indicates "most work performed prior to year-end."

A sample of 40 companies provided the following data.

WEB file

Audit

Delay	Industry	Public	Quality	Finished
62	0	0	3	1
45	0	1	3	3
54	0	0	2	2
71	0	1	1	2
91	0	0	1	1
62	0	0	4	4
61	0	0	3	2
69	0	1	5	2
80	0	0	1	1
52	0	0	5	3
47	0	0	3	2
65	0	1	2	3
60	0	0	1	3
81	1	0	1	2
73	1	0	2	2
89	1	0	2	1
71	1	0	5	4
76	1	0	2	2
68	1	0	1	2
68	1	0	5	2
86	1	0	2	2
76	1	1	3	1
67	1	0	2	3
57	1	0	4	2
55	1	1	3	2
54	1	0	5	2
69	1	0	3	3
82	1	0	5	1
94	1	0	1	1

Delay	Industry	Public	Quality	Finished
74	1	1	5	2
75	1	1	4	3
69	1	0	2	2
71	1	0	4	4
79	1	0	5	2
80	1	0	1	4
91	1	0	4	1
92	1	0	1	4
46	1	1	4	3
72	1	0	5	2
85	1	0	5	1

a. Develop the estimated regression equation using all of the independent variables.
b. Did the estimated regression equation developed in part (a) provide a good fit? Explain.
c. Develop a scatter diagram showing Delay as a function of Finished. What does this scatter diagram indicate about the relationship between Delay and Finished?
d. On the basis of your observations about the relationship between Delay and Finished, develop an alternative estimated regression equation to the one developed in (a) to explain as much of the variability in Delay as possible.

32. Refer to the data in exercise 31. Consider a model in which only Industry is used to predict Delay. At a .01 level of significance, test for any positive autocorrelation in the data.

33. Refer to the data in exercise 31.
 a. Develop an estimated regression equation that can be used to predict Delay by using Industry and Quality.
 b. Plot the residuals obtained from the estimated regression equation developed in part (a) as a function of the order in which the data are presented. Does any autocorrelation appear to be present in the data? Explain.
 c. At the .05 level of significance, test for any positive autocorrelation in the data.

34. A study was conducted to investigate browsing activity by shoppers. Shoppers were classified as nonbrowsers, light browsers, and heavy browsers. For each shopper in the study, a measure was obtained to determine how comfortable the shopper was in the store. Higher scores indicated greater comfort. Assume that the following data are from this study. Use a .05 level of significance to test for differences in comfort levels among the three types of browsers.

WEB file
Browsing

Nonbrowser	Light Browser	Heavy Browser
4	5	5
5	6	7
6	5	5
3	4	7
3	7	4
4	4	6
5	6	5
4	5	7

WEB file
CarMileage

35. The Department of Energy and the U.S. Environmental Protection Agency's *2012 Fuel Economy Guide* provides fuel efficiency data for 2012 model year cars and trucks (Department of Energy website, April 16, 2012). The file named CarMileage provides a portion of the data for 316 cars. The column labeled Size identifies the size of the car (Compact, Midsize, and Large) and the column labeled Hwy MPG shows the fuel efficiency rating for highway driving in terms of miles per gallon. Use $\alpha = .05$ and test for any significant difference in the mean fuel efficiency rating for highway driving among the three sizes of cars.

Case Problem 1 Analysis of PGA Tour Statistics

The Professional Golfers Association (PGA) maintains data on performance and earnings for members of the PGA Tour. Based on total earnings in PGA Tour events, the top 125 players are exempt for the following season. Making the top 125 money list is important because a player who is "exempt" has qualified to be a full-time member of the PGA Tour for the following season.

WEB file PGATour

Scoring average is generally considered the most important statistic in terms of success on the PGA Tour. To investigate the relationship between scoring average and variables such as driving distance, driving accuracy, greens in regulation, sand saves, and average putts per round, year-end performance data for the 125 players who had the highest total earnings in PGA Tour events for 2008 are contained in the file named PGATour (PGA Tour website, 2009). Each row of the data set corresponds to a PGA Tour player, and the data have been sorted based upon total earnings. Descriptions for the variables in the data set follow.

Money	Total earnings in PGA Tour events.
Scoring Average	The average number of strokes per completed round.
DrDist (Driving Distance)	DrDist is the average number of yards per measured drive. On the PGA Tour driving distance is measured on two holes per round. Care is taken to select two holes which face in opposite directions to counteract the effect of wind. Drives are measured to the point at which they come to rest regardless of whether they are in the fairway or not.
DrAccu (Driving Accuracy)	The percentage of time a tee shot comes to rest in the fairway (regardless of club). Driving accuracy is measured on every hole, excluding par 3's.
GIR (Greens in Regulation)	The percentage of time a player was able to hit the green in regulation. A green is considered hit in regulation if any portion of the ball is touching the putting surface after the GIR stroke has been taken. The GIR stroke is determined by subtracting 2 from par (1st stroke on a par 3, 2nd on a par 4, 3rd on a par 5). In other words, a green is considered hit in regulation if the player has reached the putting surface in par minus two strokes.
Sand Saves	The percentage of time a player was able to get "up and down" once in a greenside sand bunker (regardless of score). "Up and down" indicates it took the player 2 shots or less to put the ball in the hole from a greenside sand bunker.
PPR (Putts per Round)	The average number of putts per round.
Scrambling	The percentage of time a player missed the green in regulation but still made par or better.
Bounce Back	The percentage of time a player is over par on a hole and then under par on the following hole. In other words, it is the percentage of holes with a bogey or worse followed on the next hole with a birdie or better.

Managerial Report

Suppose that you have been hired by the commissioner of the PGA Tour to analyze the data for a presentation to be made at the annual PGA Tour meeting. The commissioner has asked whether it would be possible to use these data to determine the performance measures that are the best predictors of a player's average score. Use the methods presented in this and previous chapters to analyze the data. Prepare a report for the PGA Tour commissioner that summarizes your analysis, including key statistical results, conclusions, and recommendations. Include any appropriate technical material in an appendix.

Case Problem 2 Rating Wines from the Piedmont Region of Italy

WEB file
WineRatings

Wine Spectator magazine contains articles and reviews on every aspect of the wine industry, including ratings of wine from around the world. In a recent issue they reviewed and scored 475 wines from the Piedmont region of Italy using a 100-point scale (*Wine Spectator,* April 30, 2011). The following table shows how the *Wine Spectator* score each wine received is used to rate each wine as being classic, outstanding, very good, good, mediocre, or not recommended.

Score	Rating
95–100	Classic: a great wine
90–94	Outstanding: a wine of superior character and style
85–89	Very good: a wine with special qualities
80–84	Good: a solid, well-made wine
75–79	Mediocre: a drinkable wine that may have minor flaws
below 75	Not Recommended

A key question for most consumers is whether paying more for a bottle of wine will result in a better wine. To investigate this question for wines from the Piedmont region we selected a random sample of 100 of the 475 wines that *Wine Spectator* reviewed. The data, contained in the file named WineRatings, shows the price ($), the *Wine Spectator* score, and the rating for each wine.

Managerial Report

1. Develop a table that shows the number of wines that were classified as classic, outstanding, very good, good, mediocre, and not recommended and the average price. Does there appear to be any relationship between the price of the wine and the *Wine Spectator* rating? Are there any other aspects of your initial summary of the data that stand out?
2. Develop a scatter diagram with price on the horizontal axis and the *Wine Spectator* score on the vertical axis. Does the relationship between price and score appear to be linear?
3. Using linear regression, develop an estimated regression equation that can be used to predict the score given the price of the wine.
4. Using a second-order model, develop an estimated regression equation that can be used to predict the score given the price of the wine.
5. Compare the results from fitting a linear model and fitting a second-order model.
6. As an alternative to fitting a second-order model, fit a model using the natural logarithm of price as the independent variable. Compare the results with the second-order model.
7. Based upon your analysis, would you say that spending more for a bottle of wine will provide a better wine?
8. Suppose that you want to spend a maximum of $30 for a bottle of wine. In this case, will spending closer to your upper limit for price result in a better wine than a much lower price?

Appendix 16.1 Variable Selection Procedures with Minitab

WEB file
Cravens

In Section 16.4 we discussed the use of variable selection procedures in solving multiple regression problems. In Figure 16.16 we showed the Minitab stepwise regression output for the Cravens data, and in Figure 16.17 we showed the Minitab best-subsets output. In this appendix we describe the steps required to generate the output in both of these figures, as well as the steps required to use the forward selection and backward elimination procedures. First, the data in Table 16.5 must be entered in a Minitab worksheet. The values of Sales, Time, Poten, AdvExp, Share, Change, Accounts, Work, and Rating are entered into columns C1–C9 of a Minitab worksheet.

Using Minitab's Stepwise Procedure

The following steps can be used to produce the Minitab stepwise regression output for the Cravens data.

Step 1. Select the **Stat** menu
Step 2. Select the **Regression** menu
Step 3. Choose **Stepwise**
Step 4. When the **Stepwise Regression** dialog box appears:
  Enter Sales in the **Response** box
  Enter Time, Poten, AdvExp, Share, Change, Accounts, Work, and Rating in the **Predictors** box
  Select the **Methods** button
Step 5. When the **Stepwise-Methods** dialog box appears:
  Select **Stepwise (forward and backward)**
  Enter .05 in the **Alpha to enter** box
  Enter .05 in the **Alpha to remove** box
  Click **OK**
Step 6. When the **Stepwise Regression** dialog box reappears:
  Click **OK**

Using Minitab's Forward Selection Procedure

To use Minitab's forward selection procedure, we simply modify step 5 in Minitab's stepwise regression procedure as shown here:

Step 5. When the **Stepwise-Methods** dialog box appears:
  Select **Forward selection**
  Enter .05 in the **Alpha to enter** box
  Click **OK**

Using Minitab's Backward Elimination Procedure

To use Minitab's backward elimination procedure, we simply modify step 5 in Minitab's stepwise regression procedure as shown here:

Step 5. When the **Stepwise-Methods** dialog box appears:
  Select **Backward elimination**
  Enter .05 in the **Alpha to remove** box
  Click **OK**

Using Minitab's Best-Subsets Procedure

The following steps can be used to produce the Minitab best-subsets regression output for the Craven data.

Step 1. Select the **Stat** menu
Step 2. Select the **Regression** menu
Step 3. Choose **Best Subsets**
Step 4. When the **Best Subsets Regression** dialog box appears:
Enter Sales in the **Response** box
Enter Time, Poten, AdvExp, Share, Change, Accounts, Work, and Rating in the **Predictors** box
Click **OK**

Appendix 16.2 Variable Selection Procedures Using StatTools

WEB file
Cravens

In this appendix we show how StatTools can be used to perform three variable selection procedures: stepwise regression, forward selection, and backward elimination. First, we show how StatTools can provide the stepwise regression output for the Cravens problem.

Begin by using the Data Set Manager to create a StatTools data set for these data using the procedure described in the appendix in Chapter 1. The following steps describe how StatTools can be used to provide the stepwise regression results.

Step 1. Click the **StatTools** tab on the Ribbon
Step 2. In the **Analyses** group, click **Regression and Classification**
Step 3. Choose the **Regression** option
Step 4. When the StatTools-Regression dialog box appears:
Select **Stepwise** in the **Regression Type** box
In the **Variables** section:
Click the **Format** button and select **Unstacked**
In the column labeled **D** select **Sales**
In the column labeled **I** select **Time, Poten, AdvExp, Share, Change, Accounts, Work,** and **Rating**
In the **Parameters** section:
Select **Use p-Values**
Enter .05 in the **p-Value to Enter** box
Enter .05 in the **p-Value to Leave** box
In the **Advance Options** section, select **Include Detailed Step Information**
Click **OK**

The stepwise regression output for the Cravens problem will appear.

The StatTools-Regression dialog box contains a number of more advanced options for developing prediction interval estimates and producing residual plots. The StatTools Help facility provides information on using all these options. StatTools can also be used to perform the forward selection and backward elimination procedures. The steps required are very similar to the steps for the stepwise procedure. The major difference is that in step 4 you would select either Forward or Backward in the Regression Type box. If you choose Forward, you would enter a value in the p-Value to Enter box and if you choose Backward you would enter a value the p-Value to Leave box.

CHAPTER 17

Time Series Analysis and Forecasting

CONTENTS

STATISTICS *in* PRACTICE

NEVADA OCCUPATIONAL HEALTH CLINIC*
SPARKS, NEVADA

Nevada Occupational Health Clinic is a privately owned medical clinic in Sparks, Nevada. The clinic specializes in industrial medicine. Operating at the same site for more than 20 years, the clinic had been in a rapid growth phase. Monthly billings increased from \$57,000 to more than \$300,000 in 26 months, when the main clinic building burned to the ground.

The clinic's insurance policy covered physical property and equipment as well as loss of income due to the interruption of regular business operations. Settling the property insurance claim was a relatively straightforward matter of determining the value of the physical property and equipment lost during the fire. However, determining the value of the income lost during the seven months that it took to rebuild the clinic was a complicated matter involving negotiations between the business owners and the insurance company. No preestablished rules could help calculate "what would have happened" to the clinic's billings if the fire had not occurred. To estimate the lost income, the clinic used a forecasting method to project the growth in business that would have been realized during the seven-month lost-business period. The actual history of billings prior to the fire provided the basis for a forecasting model with linear trend and seasonal components as discussed in this chapter. This forecasting model enabled the clinic to establish an accurate estimate of the loss, which was eventually accepted by the insurance company.

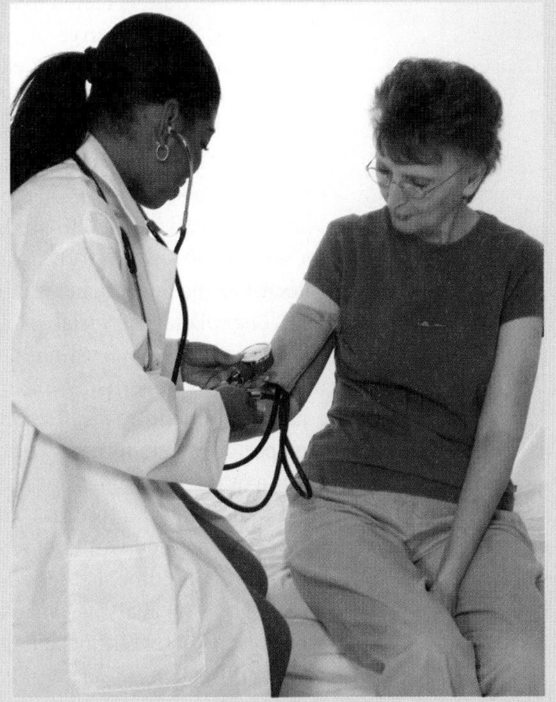

A physician checks a patient's blood pressure at the Nevada Occupational Health Clinic. © Bob Pardue–Medical Lifestyle/Alamy.

*The authors are indebted to Bard Betz, Director of Operations, and Curtis Brauer, Executive Administrative Assistant, Nevada Occupational Health Clinic, for providing this Statistics in Practice.

A forecast is simply a prediction of what will happen in the future. Managers must learn to accept that regardless of the technique used, they will not be able to develop perfect forecasts.

The purpose of this chapter is to provide an introduction to time series analysis and forecasting. Suppose we are asked to provide quarterly forecasts of sales for one of our company's products over the coming one-year period. Production schedules, raw material purchasing, inventory policies, and sales quotas will all be affected by the quarterly forecasts we provide. Consequently, poor forecasts may result in poor planning and increased costs for the company. How should we go about providing the quarterly sales forecasts? Good judgment, intuition, and an awareness of the state of the economy may give us a rough idea or "feeling" of what is likely to happen in the future, but converting that feeling into a number that can be used as next year's sales forecast is difficult.

Forecasting methods can be classified as qualitative or quantitative. Qualitative methods generally involve the use of expert judgment to develop forecasts. Such methods are appropriate when historical data on the variable being forecast are either not applicable or unavailable. Quantitative forecasting methods can be used when (1) past information about the variable being forecast is available, (2) the information can be quantified, and (3) it is reasonable to assume that the pattern of the past will continue into the future. In such cases, a forecast can be developed using a time series method or a causal method. We will focus exclusively on quantitative forecasting methods in this chapter.

If the historical data are restricted to past values of the variable to be forecast, the forecasting procedure is called a *time series method* and the historical data are referred to as a time series. The objective of time series analysis is to discover a pattern in the historical data or time series and then extrapolate the pattern into the future; the forecast is based solely on past values of the variable and/or on past forecast errors.

Causal forecasting methods are based on the assumption that the variable we are forecasting has a cause-effect relationship with one or more other variables. In the discussion of regression analysis in Chapters 14, 15, and 16, we showed how one or more independent variables could be used to predict the value of a single dependent variable. Looking at regression analysis as a forecasting tool, we can view the time series value that we want to forecast as the dependent variable. Hence, if we can identify a good set of related independent, or explanatory, variables, we may be able to develop an estimated regression equation for predicting or forecasting the time series. For instance, the sales for many products are influenced by advertising expenditures, so regression analysis may be used to develop an equation showing how sales and advertising are related. Once the advertising budget for the next period is determined, we could substitute this value into the equation to develop a prediction or forecast of the sales volume for that period. Note that if a time series method were used to develop the forecast, advertising expenditures would not be considered; that is, a time series method would base the forecast solely on past sales.

By treating time as the independent variable and the time series as a dependent variable, regression analysis can also be used as a time series method. To help differentiate the application of regression analysis in these two cases, we use the terms *cross-sectional regression* and *time series regression*. Thus, time series regression refers to the use of regression analysis when the independent variable is time. Because our focus in this chapter is on time series methods, we leave the discussion of the application of regression analysis as a causal forecasting method to more advanced texts on forecasting.

17.1 Time Series Patterns

WEB file Gasoline

A **time series** is a sequence of observations on a variable measured at successive points in time or over successive periods of time. The measurements may be taken every hour, day, week, month, or year, or at any other regular interval.[1] The pattern of the data is an important factor in understanding how the time series has behaved in the past. If such behavior can be expected to continue in the future, we can use the past pattern to guide us in selecting an appropriate forecasting method.

TABLE 17.1

GASOLINE SALES TIME SERIES

Week	Sales (1000s of gallons)
1	17
2	21
3	19
4	23
5	18
6	16
7	20
8	18
9	22
10	20
11	15
12	22

To identify the underlying pattern in the data, a useful first step is to construct a **time series plot**. A time series plot is a graphical presentation of the relationship between time and the time series variable; time is on the horizontal axis and the time series values are shown on the vertical axis. Let us review some of the common types of data patterns that can be identified when examining a time series plot.

Horizontal Pattern

A **horizontal pattern** exists when the data fluctuate around a constant mean. To illustrate a time series with a horizontal pattern, consider the 12 weeks of data in Table 17.1. These data

[1]We limit our discussion to time series in which the values of the series are recorded at equal intervals. Cases in which the observations are made at unequal intervals are beyond the scope of this text.

FIGURE 17.1 GASOLINE SALES TIME SERIES PLOT

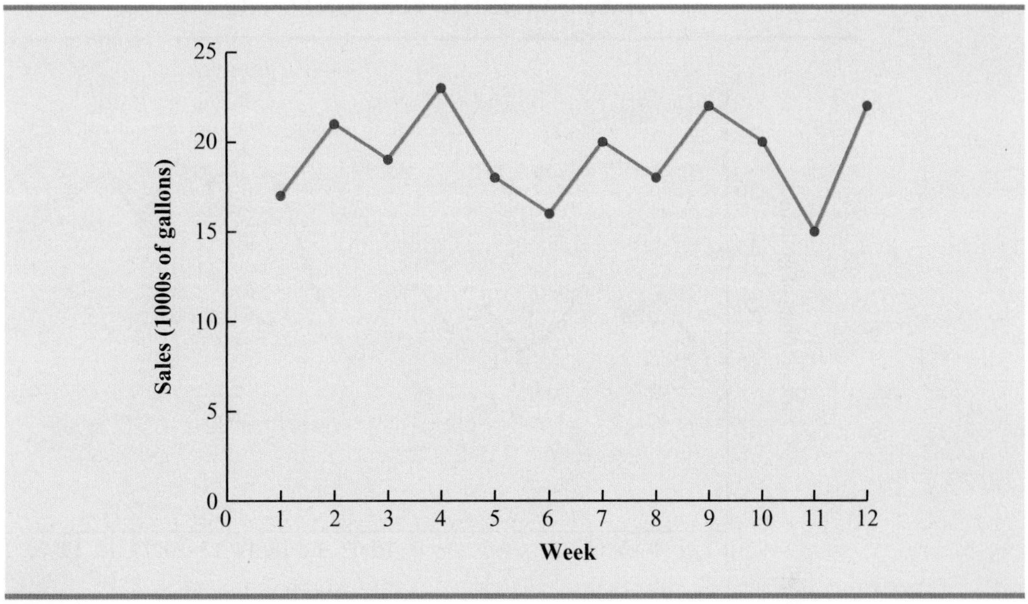

WEB file

GasolineRevised

TABLE 17.2

GASOLINE SALES TIME SERIES AFTER OBTAINING THE CONTRACT WITH THE VERMONT STATE POLICE

Week	Sales (1000s of gallons)
1	17
2	21
3	19
4	23
5	18
6	16
7	20
8	18
9	22
10	20
11	15
12	22
13	31
14	34
15	31
16	33
17	28
18	32
19	30
20	29
21	34
22	33

show the number of gallons of gasoline sold by a gasoline distributor in Bennington, Vermont, over the past 12 weeks. The average value or mean for this time series is 19.25 or 19,250 gallons per week. Figure 17.1 shows a time series plot for these data. Note how the data fluctuate around the sample mean of 19,250 gallons. Although random variability is present, we would say that these data follow a horizontal pattern.

The term **stationary time series**[2] is used to denote a time series whose statistical properties are independent of time. In particular this means that

1. The process generating the data has a constant mean.
2. The variability of the time series is constant over time.

A time series plot for a stationary time series will always exhibit a horizontal pattern. But simply observing a horizontal pattern is not sufficient evidence to conclude that the time series is stationary. More advanced texts on forecasting discuss procedures for determining if a time series is stationary and provide methods for transforming a time series that is not stationary into a stationary series.

Changes in business conditions can often result in a time series that has a horizontal pattern shifting to a new level. For instance, suppose the gasoline distributor signs a contract with the Vermont State Police to provide gasoline for state police cars located in southern Vermont. With this new contract, the distributor expects to see a major increase in weekly sales starting in week 13. Table 17.2 shows the number of gallons of gasoline sold for the original time series and for the 10 weeks after signing the new contract. Figure 17.2 shows the corresponding time series plot. Note the increased level of the time series beginning in week 13. This change in the level of the time series makes it more difficult to choose an appropriate forecasting method. Selecting a forecasting method that adapts well to changes in the level of a time series is an important consideration in many practical applications.

---

[2]For a formal definition of stationary see G. E. P., Box, G. M. Jenkins, and G. C. Reinsell, *Time Series Analysis: Forecasting and Control*, 3rd ed. Englewood Cliffs, NJ: Prentice Hall, 1994, p. 23.

FIGURE 17.2 GASOLINE SALES TIME SERIES PLOT AFTER OBTAINING THE CONTRACT WITH THE VERMONT STATE POLICE

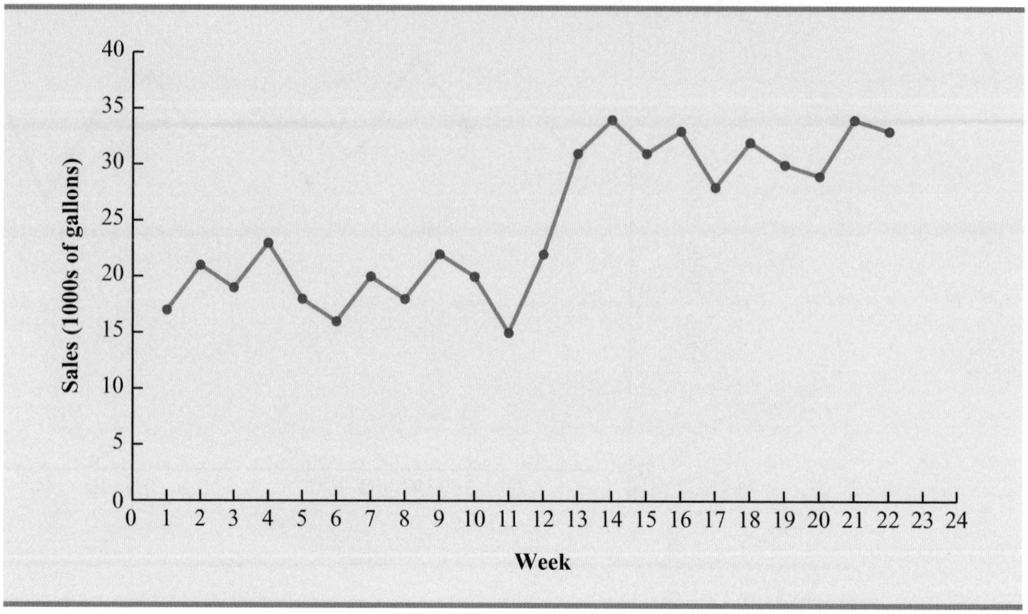

Trend Pattern

WEB file

Bicycle

Although time series data generally exhibit random fluctuations, a time series may also show gradual shifts or movements to relatively higher or lower values over a longer period of time. If a time series plot exhibits this type of behavior, we say that a **trend pattern** exists. A trend is usually the result of long-term factors such as population increases or decreases, changing demographic characteristics of the population, technology, and/or consumer preferences.

To illustrate a time series with a trend pattern, consider the time series of bicycle sales for a particular manufacturer over the past 10 years, as shown in Table 17.3 and Figure 17.3. Note that 21,600 bicycles were sold in year one, 22,900 were sold in year two, and so on. In year 10, the most recent year, 31,400 bicycles were sold. Visual inspection of the time series plot shows some up and down movement over the past 10 years, but the time series also seems to have a systematically increasing or upward trend.

TABLE 17.3

BICYCLE SALES TIME SERIES

Year	Sales (1000s)
1	21.6
2	22.9
3	25.5
4	21.9
5	23.9
6	27.5
7	31.5
8	29.7
9	28.6
10	31.4

The trend for the bicycle sales time series appears to be linear and increasing over time, but sometimes a trend can be described better by other types of patterns. For instance, the data in Table 17.4 and the corresponding time series plot in Figure 17.4 show the sales for a cholesterol drug since the company won FDA approval for it 10 years ago. The time series increases in a nonlinear fashion; that is, the rate of change of revenue does not increase by a constant amount from one year to the next. In fact, the revenue appears to be growing in an exponential fashion. Exponential relationships such as this are appropriate when the percentage change from one period to the next is relatively constant.

Seasonal Pattern

The trend of a time series can be identified by analyzing multiyear movements in historical data. Seasonal patterns are recognized by seeing the same repeating patterns over successive periods of time. For example, a manufacturer of swimming pools expects low sales activity in the fall and winter months, with peak sales in the spring and summer months. Manufacturers of snow removal equipment and heavy clothing, however, expect just the

FIGURE 17.3 BICYCLE SALES TIME SERIES PLOT

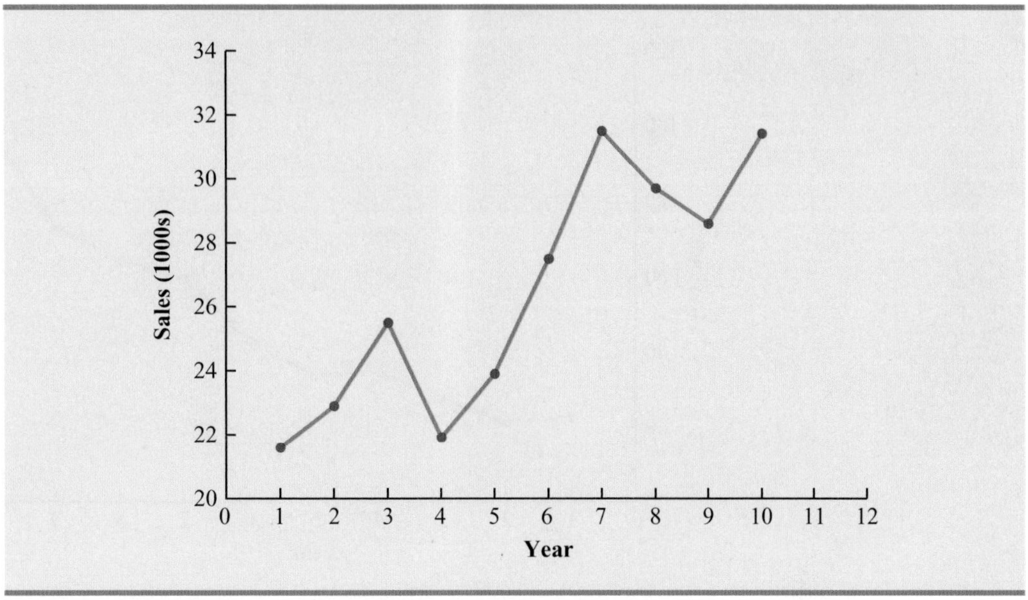

WEB file

Cholesterol

TABLE 17.4

CHOLESTEROL REVENUE TIME SERIES ($MILLIONS)

Year	Revenue
1	23.1
2	21.3
3	27.4
4	34.6
5	33.8
6	43.2
7	59.5
8	64.4
9	74.2
10	99.3

opposite yearly pattern. Not surprisingly, the pattern for a time series plot that exhibits a repeating pattern over a one-year period due to seasonal influences is called a **seasonal pattern**. While we generally think of seasonal movement in a time series as occurring within one year, time series data can also exhibit seasonal patterns of less than one year in duration. For example, daily traffic volume shows within-the-day "seasonal" behavior, with peak levels occurring during rush hours, moderate flow during the rest of the day and early evening, and light flow from midnight to early morning.

As an example of a seasonal pattern, consider the number of umbrellas sold at a clothing store over the past five years. Table 17.5 shows the time series and Figure 17.5 shows the corresponding time series plot. The time series plot does not indicate any long-term trend in sales. In fact, unless you look carefully at the data, you might conclude that the data follow a horizontal pattern. But closer inspection of the time series plot reveals a regular pattern in the data. That is, the first and third quarters have moderate sales, the second quarter has the highest sales, and the fourth quarter tends to have the lowest sales volume. Thus, we would conclude that a quarterly seasonal pattern is present.

Trend and Seasonal Pattern

Some time series include a combination of a trend and seasonal pattern. For instance, the data in Table 17.6 and the corresponding time series plot in Figure 17.6 show television set sales for a particular manufacturer over the past four years. Clearly, an increasing trend is present. But, Figure 17.6 also indicates that sales are lowest in the second quarter of each year and increase in quarters 3 and 4. Thus, we conclude that a seasonal pattern also exists for television set sales. In such cases we need to use a forecasting method that has the capability to deal with both trend and seasonality.

Cyclical Pattern

A **cyclical pattern** exists if the time series plot shows an alternating sequence of points below and above the trend line lasting more than one year. Many economic time series exhibit

FIGURE 17.4 CHOLESTEROL REVENUE TIMES SERIES PLOT ($MILLIONS)

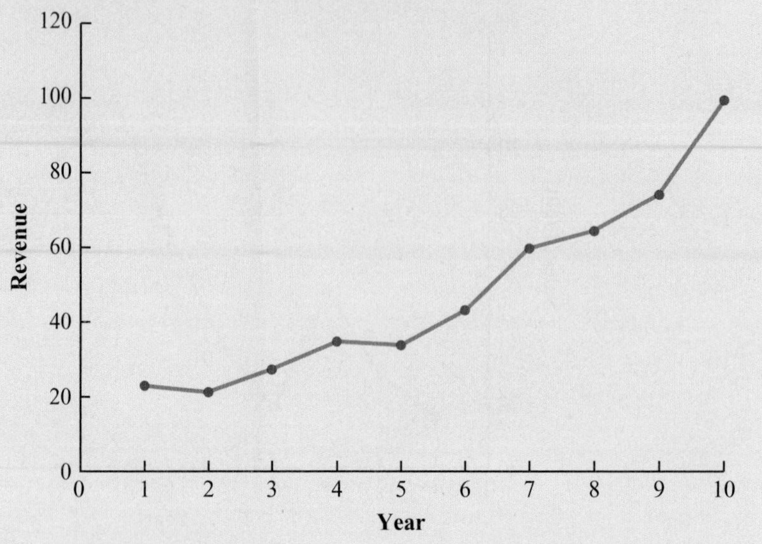

TABLE 17.5 UMBRELLA SALES TIME SERIES

Year	Quarter	Sales
1	1	125
	2	153
	3	106
	4	88
2	1	118
	2	161
	3	133
	4	102
3	1	138
	2	144
	3	113
	4	80
4	1	109
	2	137
	3	125
	4	109
5	1	130
	2	165
	3	128
	4	96

WEB file
Umbrella

cyclical behavior with regular runs of observations below and above the trend line. Often, the cyclical component of a time series is due to multiyear business cycles. For example, periods of moderate inflation followed by periods of rapid inflation can lead to time series that alternate below and above a generally increasing trend line (e.g., a time series for housing costs). Business cycles are extremely difficult, if not impossible, to forecast. As a

FIGURE 17.5 UMBRELLA SALES TIME SERIES PLOT

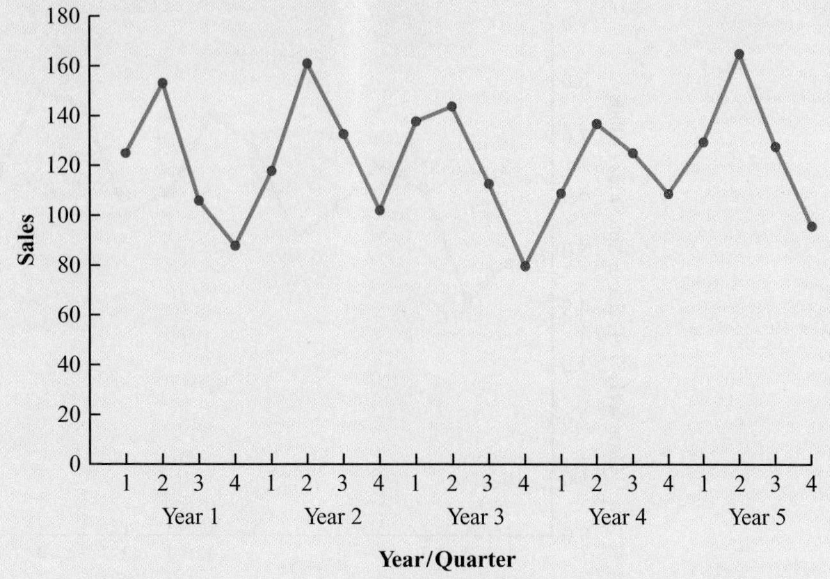

TABLE 17.6 QUARTERLY TELEVISION SET SALES TIME SERIES

WEB file
TVSales

Year	Quarter	Sales (1000s)
1	1	4.8
	2	4.1
	3	6.0
	4	6.5
2	1	5.8
	2	5.2
	3	6.8
	4	7.4
3	1	6.0
	2	5.6
	3	7.5
	4	7.8
4	1	6.3
	2	5.9
	3	8.0
	4	8.4

result, cyclical effects are often combined with long-term trend effects and referred to as trend-cycle effects. In this chapter we do not deal with cyclical effects that may be present in the time series.

Selecting a Forecasting Method

The underlying pattern in the time series is an important factor in selecting a forecasting method. Thus, a time series plot should be one of the first things developed when trying to determine which forecasting method to use. If we see a horizontal pattern, then we need to

FIGURE 17.6 QUARTERLY TELEVISION SET SALES TIME SERIES PLOT

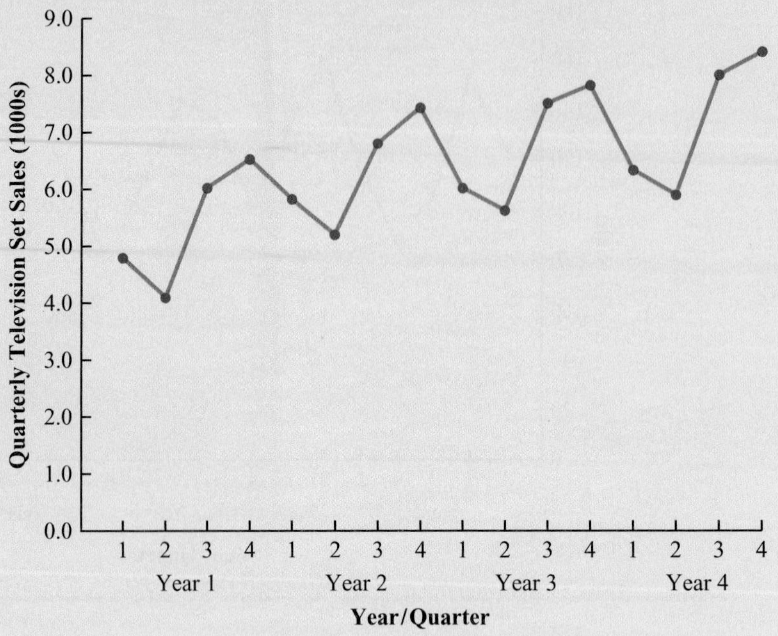

select a method appropriate for this type of pattern. Similarly, if we observe a trend in the data, then we need to use a forecasting method that has the capability to handle trend effectively. The next two sections illustrate methods that can be used in situations where the underlying pattern is horizontal; in other words, no trend or seasonal effects are present. We then consider methods appropriate when trend and/or seasonality are present in the data.

17.2 Forecast Accuracy

In this section we begin by developing forecasts for the gasoline time series shown in Table 17.1 using the simplest of all the forecasting methods: an approach that uses the most recent week's sales volume as the forecast for the next week. For instance, the distributor sold 17 thousand gallons of gasoline in week 1; this value is used as the forecast for week 2. Next, we use 21, the actual value of sales in week 2, as the forecast for week 3, and so on. The forecasts obtained for the historical data using this method are shown in Table 17.7 in the column labeled Forecast. Because of its simplicity, this method is often referred to as a *naive* forecasting method.

How accurate are the forecasts obtained using this *naive* forecasting method? To answer this question we will introduce several measures of forecast accuracy. These measures are used to determine how well a particular forecasting method is able to reproduce the time series data that are already available. By selecting the method that has the best accuracy for the data already known, we hope to increase the likelihood that we will obtain better forecasts for future time periods.

The key concept associated with measuring forecast accuracy is **forecast error**, defined as

$$\text{Forecast Error} = \text{Actual Value} - \text{Forecast}$$

TABLE 17.7 COMPUTING FORECASTS AND MEASURES OF FORECAST ACCURACY USING THE MOST RECENT VALUE AS THE FORECAST FOR THE NEXT PERIOD

Week	Time Series Value	Forecast	Forecast Error	Absolute Value of Forecast Error	Squared Forecast Error	Percentage Error	Absolute Value of Percentage Error
1	17						
2	21	17	4	4	16	19.05	19.05
3	19	21	−2	2	4	−10.53	10.53
4	23	19	4	4	16	17.39	17.39
5	18	23	−5	5	25	−27.78	27.78
6	16	18	−2	2	4	−12.50	12.50
7	20	16	4	4	16	20.00	20.00
8	18	20	−2	2	4	−11.11	11.11
9	22	18	4	4	16	18.18	18.18
10	20	22	−2	2	4	−10.00	10.00
11	15	20	−5	5	25	−33.33	33.33
12	22	15	7	7	49	31.82	31.82
		Totals	5	41	179	1.19	211.69

For instance, because the distributor actually sold 21 thousand gallons of gasoline in week 2 and the forecast, using the sales volume in week 1, was 17 thousand gallons, the forecast error in week 2 is

$$\text{Forecast Error in week 2} = 21 - 17 = 4$$

The fact that the forecast error is positive indicates that in week 2 the forecasting method underestimated the actual value of sales. Next, we use 21, the actual value of sales in week 2, as the forecast for week 3. Since the actual value of sales in week 3 is 19, the forecast error for week 3 is $19 - 21 = -2$. In this case, the negative forecast error indicates that in week 3 the forecast overestimated the actual value. Thus, the forecast error may be positive or negative, depending on whether the forecast is too low or too high. A complete summary of the forecast errors for this naive forecasting method is shown in Table 17.7 in the column labeled Forecast Error.

In regression analysis, a residual is defined as the difference between the observed value of the dependent variable and the estimated value. The forecast errors are analogous to the residuals in regression analysis.

A simple measure of forecast accuracy is the mean or average of the forecast errors. Table 17.7 shows that the sum of the forecast errors for the gasoline sales time series is 5; thus, the mean or average forecast error is $5/11 = .45$. Note that although the gasoline time series consists of 12 values, to compute the mean error we divided the sum of the forecast errors by 11 because there are only 11 forecast errors. Because the mean forecast error is positive, the method is underforecasting; in other words, the observed values tend to be greater than the forecasted values. Because positive and negative forecast errors tend to offset one another, the mean error is likely to be small; thus, the mean error is not a very useful measure of forecast accuracy.

The **mean absolute error**, denoted MAE, is a measure of forecast accuracy that avoids the problem of positive and negative forecast errors offsetting one another. As you might expect given its name, MAE is the average of the absolute values of the forecast errors. Table 17.7 shows that the sum of the absolute values of the forecast errors is 41; thus,

$$\text{MAE} = \text{average of the absolute value of forecast errors} = \frac{41}{11} = 3.73$$

In regression analysis the mean square error (MSE) is the residual sum of squares divided by its degrees of freedom. In forecasting, MSE is the average of the sum of squared forecast errors.

Another measure that avoids the problem of positive and negative forecast errors offsetting each other is obtained by computing the average of the squared forecast errors. This measure of forecast accuracy, referred to as the **mean squared error**, is denoted MSE. From Table 17.7, the sum of the squared errors is 179; hence,

$$\text{MSE} = \text{average of the sum of squared forecast errors} = \frac{179}{11} = 16.27$$

The size of MAE and MSE depends upon the scale of the data. As a result, it is difficult to make comparisons for different time intervals, such as comparing a method of forecasting monthly gasoline sales to a method of forecasting weekly sales, or to make comparisons across different time series. To make comparisons like these we need to work with relative or percentage error measures. The **mean absolute percentage error**, denoted MAPE, is such a measure. To compute MAPE we must first compute the percentage error for each forecast. For example, the percentage error corresponding to the forecast of 17 in week 2 is computed by dividing the forecast error in week 2 by the actual value in week 2 and multiplying the result by 100. For week 2 the percentage error is computed as follows:

$$\text{Percentage error for week 2} = \frac{4}{21}(100) = 19.05\%$$

Thus, the forecast error for week 2 is 19.05% of the observed value in week 2. A complete summary of the percentage errors is shown in Table 17.7 in the column labeled Percentage Error. In the next column, we show the absolute value of the percentage error.

Table 17.7 shows that the sum of the absolute values of the percentage errors is 211.69; thus,

$$\text{MAPE} = \text{average of the absolute value of percentage forecast errors} = \frac{211.69}{11} = 19.24\%$$

Summarizing, using the naive (most recent observation) forecasting method, we obtained the following measures of forecast accuracy:

$$\text{MAE} = 3.73$$
$$\text{MSE} = 16.27$$
$$\text{MAPE} = 19.24\%$$

These measures of forecast accuracy simply measure how well the forecasting method is able to forecast historical values of the time series. Now, suppose we want to forecast sales for a future time period, such as week 13. In this case the forecast for week 13 is 22, the actual value of the time series in week 12. Is this an accurate estimate of sales for week 13? Unfortunately, there is no way to address the issue of accuracy associated with forecasts for future time periods. But, if we select a forecasting method that works well for the historical data, and we think that the historical pattern will continue into the future, we should obtain results that will ultimately be shown to be good.

Before closing this section, let's consider another method for forecasting the gasoline sales time series in Table 17.1. Suppose we use the average of all the historical data available as the forecast for the next period. We begin by developing a forecast for week 2. Since there is only one historical value available prior to week 2, the forecast for week 2 is just the time series value in week 1; thus, the forecast for week 2 is 17 thousand gallons of gasoline. To compute the forecast for week 3, we take the average of the sales values in weeks 1 and 2. Thus,

TABLE 17.8 COMPUTING FORECASTS AND MEASURES OF FORECAST ACCURACY USING THE AVERAGE OF ALL THE HISTORICAL DATA AS THE FORECAST FOR THE NEXT PERIOD

Week	Time Series Value	Forecast	Forecast Error	Absolute Value of Forecast Error	Squared Forecast Error	Percentage Error	Absolute Value of Percentage Error
1	17						
2	21	17.00	4.00	4.00	16.00	19.05	19.05
3	19	19.00	0.00	0.00	0.00	0.00	0.00
4	23	19.00	4.00	4.00	16.00	17.39	17.39
5	18	20.00	−2.00	2.00	4.00	−11.11	11.11
6	16	19.60	−3.60	3.60	12.96	−22.50	22.50
7	20	19.00	1.00	1.00	1.00	5.00	5.00
8	18	19.14	−1.14	1.14	1.31	−6.35	6.35
9	22	19.00	3.00	3.00	9.00	13.64	13.64
10	20	19.33	0.67	0.67	0.44	3.33	3.33
11	15	19.40	−4.40	4.40	19.36	−29.33	29.33
12	22	19.00	3.00	3.00	9.00	13.64	13.64
		Totals	4.53	26.81	89.07	2.76	141.34

$$\text{Forecast for week 3} = \frac{17 + 21}{2} = 19$$

Similarly, the forecast for week 4 is

$$\text{Forecast for week 4} = \frac{17 + 21 + 19}{3} = 19$$

The forecasts obtained using this method for the gasoline time series are shown in Table 17.8 in the column labeled Forecast. Using the results shown in Table 17.8, we obtained the following values of MAE, MSE, and MAPE:

$$\text{MAE} = \frac{26.81}{11} = 2.44$$

$$\text{MSE} = \frac{89.07}{11} = 8.10$$

$$\text{MAPE} = \frac{141.34}{11} = 12.85\%$$

We can now compare the accuracy of the two forecasting methods we have considered in this section by comparing the values of MAE, MSE, and MAPE for each method.

	Naive Method	Average of Past Values
MAE	3.73	2.44
MSE	16.27	8.10
MAPE	19.24%	12.85%

For every measure, the average of past values provides more accurate forecasts than using the most recent observation as the forecast for the next period. In general, if the underlying time series is stationary, the average of all the historical data will always provide the best results.

But suppose that the underlying time series is not stationary. In Section 17.1 we mentioned that changes in business conditions can often result in a time series that has a horizontal pattern shifting to a new level. We discussed a situation in which the gasoline distributor signed a contract with the Vermont State Police to provide gasoline for state police cars located in southern Vermont. Table 17.2 shows the number of gallons of gasoline sold for the original time series and the 10 weeks after signing the new contract, and Figure 17.2 shows the corresponding time series plot. Note the change in level in week 13 for the resulting time series. When a shift to a new level like this occurs, it takes a long time for the forecasting method that uses the average of all the historical data to adjust to the new level of the time series. But, in this case, the simple naive method adjusts very rapidly to the change in level because it uses the most recent observation available as the forecast.

Measures of forecast accuracy are important factors in comparing different forecasting methods, but we have to be careful not to rely upon them too heavily. Good judgment and knowledge about business conditions that might affect the forecast also have to be carefully considered when selecting a method. And historical forecast accuracy is not the only consideration, especially if the time series is likely to change in the future.

In the next section we will introduce more sophisticated methods for developing forecasts for a time series that exhibits a horizontal pattern. Using the measures of forecast accuracy developed here, we will be able to determine if such methods provide more accurate forecasts than we obtained using the simple approaches illustrated in this section. The methods that we will introduce also have the advantage of adapting well in situations where the time series changes to a new level. The ability of a forecasting method to adapt quickly to changes in level is an important consideration, especially in short-term forecasting situations.

Exercises

Methods

SELF test

1. Consider the following time series data.

Week	1	2	3	4	5	6
Value	18	13	16	11	17	14

Using the naive method (most recent value) as the forecast for the next week, compute the following measures of forecast accuracy.

a. Mean absolute error.
b. Mean squared error.
c. Mean absolute percentage error.
d. What is the forecast for week 7?

SELF test

2. Refer to the time series data in exercise 1. Using the average of all the historical data as a forecast for the next period, compute the following measures of forecast accuracy.

a. Mean absolute error.
b. Mean squared error.
c. Mean absolute percentage error.
d. What is the forecast for week 7?

SELF test

3. Exercises 1 and 2 used different forecasting methods. Which method appears to provide the more accurate forecasts for the historical data? Explain.
4. Consider the following time series data.

Month	1	2	3	4	5	6	7
Value	24	13	20	12	19	23	15

a. Compute MSE using the most recent value as the forecast for the next period. What is the forecast for month 8?
b. Compute MSE using the average of all the data available as the forecast for the next period. What is the forecast for month 8?
c. Which method appears to provide the better forecast?

17.3 Moving Averages and Exponential Smoothing

In this section we discuss three forecasting methods that are appropriate for a time series with a horizontal pattern: moving averages, weighted moving averages, and exponential smoothing. These methods also adapt well to changes in the level of a horizontal pattern such as we saw with the extended gasoline sales time series (Table 17.2 and Figure 17.2). However, without modification they are not appropriate when significant trend, cyclical, or seasonal effects are present. Because the objective of each of these methods is to "smooth out" the random fluctuations in the time series, they are referred to as smoothing methods. These methods are easy to use and generally provide a high level of accuracy for short-range forecasts, such as a forecast for the next time period.

Moving Averages

The **moving averages** method uses the average of the most recent k data values in the time series as the forecast for the next period. Mathematically, a moving average forecast of order k is as follows:

MOVING AVERAGE FORECAST OF ORDER k

$$F_{t+1} = \frac{\sum(\text{most recent } k \text{ data values})}{k} = \frac{Y_t + Y_{t-1} + \ldots + Y_{t-k+1}}{k} \quad (17.1)$$

where

F_{t+1} = forecast of the times series for period $t + 1$
Y_t = actual value of the time series in period t

The term *moving* is used because every time a new observation becomes available for the time series, it replaces the oldest observation in the equation and a new average is computed. As a result, the average will change, or move, as new observations become available.

To illustrate the moving averages method, let us return to the gasoline sales data in Table 17.1 and Figure 17.1. The time series plot in Figure 17.1 indicates that the gasoline sales time series has a horizontal pattern. Thus, the smoothing methods of this section are applicable.

To use moving averages to forecast a time series, we must first select the order, or number of time series values, to be included in the moving average. If only the most recent values of the time series are considered relevant, a small value of k is preferred. If more past values are considered relevant, then a larger value of k is better. As mentioned earlier, a time series with a horizontal pattern can shift to a new level over time. A moving average will adapt to the new level of the series and resume providing good forecasts in k periods. Thus, a smaller value of k will track shifts in a time series more quickly. But larger values of k will be more effective in smoothing out the random fluctuations over time. So managerial judgment based on an understanding of the behavior of a time series is helpful in choosing a good value for k.

To illustrate how moving averages can be used to forecast gasoline sales, we will use a three-week moving average ($k = 3$). We begin by computing the forecast of sales in week 4 using the average of the time series values in weeks 1–3.

$$F_4 = \text{average of weeks 1–3} = \frac{17 + 21 + 19}{3} = 19$$

Thus, the moving average forecast of sales in week 4 is 19 or 19,000 gallons of gasoline. Because the actual value observed in week 4 is 23, the forecast error in week 4 is $23 - 19 = 4$.

Next, we compute the forecast of sales in week 5 by averaging the time series values in weeks 2–4.

$$F_5 = \text{average of weeks 2–4} = \frac{21 + 19 + 23}{3} = 21$$

Hence, the forecast of sales in week 5 is 21 and the error associated with this forecast is $18 - 21 = -3$. A complete summary of the three-week moving average forecasts for the gasoline sales time series is provided in Table 17.9. Figure 17.7 shows the original time series plot and the three-week moving average forecasts. Note how the graph of the moving average forecasts has tended to smooth out the random fluctuations in the time series.

TABLE 17.9 SUMMARY OF THREE-WEEK MOVING AVERAGE CALCULATIONS

Week	Time Series Value	Forecast	Forecast Error	Absolute Value of Forecast Error	Squared Forecast Error	Percentage Error	Absolute Value of Percentage Error
1	17						
2	21						
3	19						
4	23	19	4	4	16	17.39	17.39
5	18	21	−3	3	9	−16.67	16.67
6	16	20	−4	4	16	−25.00	25.00
7	20	19	1	1	1	5.00	5.00
8	18	18	0	0	0	0.00	0.00
9	22	18	4	4	16	18.18	18.18
10	20	20	0	0	0	0.00	0.00
11	15	20	−5	5	25	−33.33	33.33
12	22	19	3	3	9	13.64	13.64
		Totals	0	24	92	−20.79	129.21

FIGURE 17.7 GASOLINE SALES TIME SERIES PLOT AND THREE-WEEK MOVING AVERAGE FORECASTS

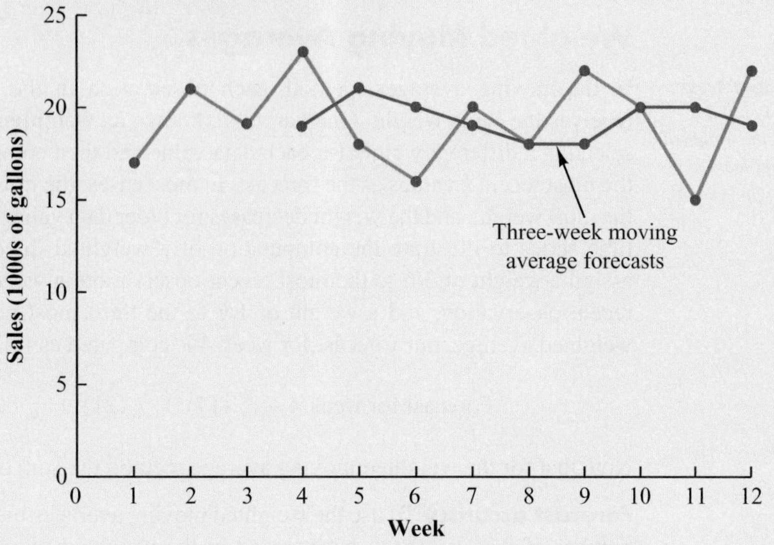

To forecast sales in week 13, the next time period in the future, we simply compute the average of the time series values in weeks 10, 11, and 12.

$$F_{13} = \text{average of weeks 10–12} = \frac{20 + 15 + 22}{3} = 19$$

Thus, the forecast for week 13 is 19 or 19,000 gallons of gasoline.

Forecast accuracy In Section 17.2 we discussed three measures of forecast accuracy: MAE, MSE, and MAPE. Using the three-week moving average calculations in Table 17.9, the values for these three measures of forecast accuracy are

$$\text{MAE} = \frac{24}{9} = 2.67$$

$$\text{MSE} = \frac{92}{9} = 10.22$$

$$\text{MAPE} = \frac{129.21}{9} = 14.36\%$$

In situations where you need to compare forecasting methods for different time periods, such as comparing a forecast of weekly sales to a forecast of monthly sales, relative measures such as MAPE are preferred.

In Section 17.2 we also showed that using the most recent observation as the forecast for the next week (a moving average of order $k = 1$) resulted in values of MAE = 3.73, MSE = 16.27, and MAPE = 19.24%. Thus, in each case the three-week moving average approach provided more accurate forecasts than simply using the most recent observation as the forecast.

To determine if a moving average with a different order k can provide more accurate forecasts, we recommend using trial and error to determine the value of k that minimizes MSE. For the gasoline sales time series, it can be shown that the minimum value of MSE corresponds to a moving average of order $k = 6$ with MSE = 6.79. If we are willing to assume that the order of the moving average that is best for the historical data will also be best for future values of the time series, the most accurate moving average forecasts of gasoline sales can be obtained using a moving average of order $k = 6$.

Weighted Moving Averages

A moving average forecast of order $k = 3$ is just a special case of the weighted moving averages method in which each weight is equal to 1/3.

In the moving averages method, each observation in the moving average calculation receives the same weight. One variation, known as **weighted moving averages**, involves selecting a different weight for each data value and then computing a weighted average of the most recent k values as the forecast. In most cases, the most recent observation receives the most weight, and the weight decreases for older data values. Let us use the gasoline sales time series to illustrate the computation of a weighted three-week moving average. We assign a weight of 3/6 to the most recent observation, a weight of 2/6 to the second most recent observation, and a weight of 1/6 to the third most recent observation. Using this weighted average, our forecast for week 4 is computed as follows:

$$\text{Forecast for week 4} = \tfrac{1}{6}(17) + \tfrac{2}{6}(21) + \tfrac{3}{6}(19) = 19.33$$

Note that for the weighted moving average method the sum of the weights is equal to 1.

Forecast accuracy To use the weighted moving averages method, we must first select the number of data values to be included in the weighted moving average and then choose weights for each of the data values. In general, if we believe that the recent past is a better predictor of the future than the distant past, larger weights should be given to the more recent observations. However, when the time series is highly variable, selecting approximately equal weights for the data values may be best. The only requirement in selecting the weights is that their sum must equal 1. To determine whether one particular combination of number of data values and weights provides a more accurate forecast than another combination, we recommend using MSE as the measure of forecast accuracy. That is, if we assume that the combination that is best for the past will also be best for the future, we would use the combination of number of data values and weights that minimizes MSE for the historical time series to forecast the next value in the time series.

Exponential Smoothing

There are a number of exponential smoothing procedures. The method presented here is often referred to as single exponential smoothing. In the next section we show how an exponential smoothing method that uses two smoothing constants can be used to forecast a time series with a linear trend.

Exponential smoothing also uses a weighted average of past time series values as a forecast; it is a special case of the weighted moving averages method in which we select only one weight—the weight for the most recent observation. The weights for the other data values are computed automatically and become smaller as the observations move farther into the past. The exponential smoothing equation follows.

EXPONENTIAL SMOOTHING FORECAST

$$F_{t+1} = \alpha Y_t + (1 - \alpha)F_t \tag{17.2}$$

where

F_{t+1} = forecast of the time series for period $t + 1$
Y_t = actual value of the time series in period t
F_t = forecast of the time series for period t
α = smoothing constant ($0 \leq \alpha \leq 1$)

Equation (17.2) shows that the forecast for period $t + 1$ is a weighted average of the actual value in period t and the forecast for period t. The weight given to the actual value in period t is the **smoothing constant** α and the weight given to the forecast in period t is $1 - \alpha$. It turns out that the exponential smoothing forecast for any period is actually a weighted average of *all the previous actual values* of the time series. Let us illustrate by working with a time series involving only three periods of data: Y_1, Y_2, and Y_3.

To initiate the calculations, we let F_1 equal the actual value of the time series in period 1; that is, $F_1 = Y_1$. Hence, the forecast for period 2 is

$$\begin{aligned} F_2 &= \alpha Y_1 + (1 - \alpha)F_1 \\ &= \alpha Y_1 + (1 - \alpha)Y_1 \\ &= Y_1 \end{aligned}$$

We see that the exponential smoothing forecast for period 2 is equal to the actual value of the time series in period 1.

The forecast for period 3 is

$$F_3 = \alpha Y_2 + (1 - \alpha)F_2 = \alpha Y_2 + (1 - \alpha)Y_1$$

Finally, substituting this expression for F_3 in the expression for F_4, we obtain

$$\begin{aligned} F_4 &= \alpha Y_3 + (1 - \alpha)F_3 \\ &= \alpha Y_3 + (1 - \alpha)[\alpha Y_2 + (1 - \alpha)Y_1] \\ &= \alpha Y_3 + \alpha(1 - \alpha)Y_2 + (1 - \alpha)^2 Y_1 \end{aligned}$$

The term exponential smoothing comes from the exponential nature of the weighting scheme for the historical values.

We now see that F_4 is a weighted average of the first three time series values. The sum of the coefficients, or weights, for Y_1, Y_2, and Y_3 equals 1. A similar argument can be made to show that, in general, any forecast F_{t+1} is a weighted average of all the previous time series values.

Despite the fact that exponential smoothing provides a forecast that is a weighted average of all past observations, all past data do not need to be saved to compute the forecast for the next period. In fact, equation (17.2) shows that once the value for the smoothing constant α is selected, only two pieces of information are needed to compute the forecast: Y_t, the actual value of the time series in period t, and F_t, the forecast for period t.

To illustrate the exponential smoothing approach, let us again consider the gasoline sales time series in Table 17.1 and Figure 17.1. As indicated previously, to start the calculations we set the exponential smoothing forecast for period 2 equal to the actual value of the time series in period 1. Thus, with $Y_1 = 17$, we set $F_2 = 17$ to initiate the computations. Referring to the time series data in Table 17.1, we find an actual time series value in period 2 of $Y_2 = 21$. Thus, period 2 has a forecast error of $21 - 17 = 4$.

Continuing with the exponential smoothing computations using a smoothing constant of $\alpha = .2$, we obtain the following forecast for period 3:

$$F_3 = .2Y_2 + .8F_2 = .2(21) + .8(17) = 17.8$$

Once the actual time series value in period 3, $Y_3 = 19$, is known, we can generate a forecast for period 4 as follows:

$$F_4 = .2Y_3 + .8F_3 = .2(19) + .8(17.8) = 18.04$$

Continuing the exponential smoothing calculations, we obtain the weekly forecast values shown in Table 17.10. Note that we have not shown an exponential smoothing forecast

TABLE 17.10 SUMMARY OF THE EXPONENTIAL SMOOTHING FORECASTS AND FORECAST ERRORS FOR THE GASOLINE SALES TIME SERIES WITH SMOOTHING CONSTANT $\alpha = .2$

Week	Time Series Value	Forecast	Forecast Error	Squared Forecast Error
1	17			
2	21	17.00	4.00	16.00
3	19	17.80	1.20	1.44
4	23	18.04	4.96	24.60
5	18	19.03	−1.03	1.06
6	16	18.83	−2.83	8.01
7	20	18.26	1.74	3.03
8	18	18.61	−0.61	0.37
9	22	18.49	3.51	12.32
10	20	19.19	0.81	0.66
11	15	19.35	−4.35	18.92
12	22	18.48	3.52	12.39
		Totals	10.92	98.80

or a forecast error for week 1 because no forecast was made. For week 12, we have $Y_{12} = 22$ and $F_{12} = 18.48$. We can we use this information to generate a forecast for week 13.

$$F_{13} = .2Y_{12} + .8F_{12} = .2(22) + .8(18.48) = 19.18$$

Thus, the exponential smoothing forecast of the amount sold in week 13 is 19.18, or 19,180 gallons of gasoline. With this forecast, the firm can make plans and decisions accordingly.

Figure 17.8 shows the time series plot of the actual and forecast time series values. Note in particular how the forecasts "smooth out" the irregular or random fluctuations in the time series.

Forecast accuracy In the preceding exponential smoothing calculations, we used a smoothing constant of $\alpha = .2$. Although any value of α between 0 and 1 is acceptable, some values will yield better forecasts than others. Insight into choosing a good value for α can be obtained by rewriting the basic exponential smoothing model as follows:

$$\begin{aligned} F_{t+1} &= \alpha Y_t + (1 - \alpha)F_t \\ F_{t+1} &= \alpha Y_t + F_t - \alpha F_t \\ F_{t+1} &= F_t + \alpha(Y_t - F_t) \end{aligned} \quad \textbf{(17.3)}$$

Thus, the new forecast F_{t+1} is equal to the previous forecast F_t plus an adjustment, which is the smoothing constant α times the most recent forecast error, $Y_t - F_t$. That is, the forecast in period $t + 1$ is obtained by adjusting the forecast in period t by a fraction of the forecast error. If the time series contains substantial random variability, a small value of the smoothing constant is preferred. The reason for this choice is that if much of the forecast error is due to random variability, we do not want to overreact and adjust the forecasts too quickly. For a time series with relatively little random variability, forecast errors are more likely to represent a change in the level of the series. Thus, larger values of the smoothing

FIGURE 17.8 ACTUAL AND FORECAST GASOLINE SALES TIME SERIES WITH SMOOTHING CONSTANT $\alpha = .2$

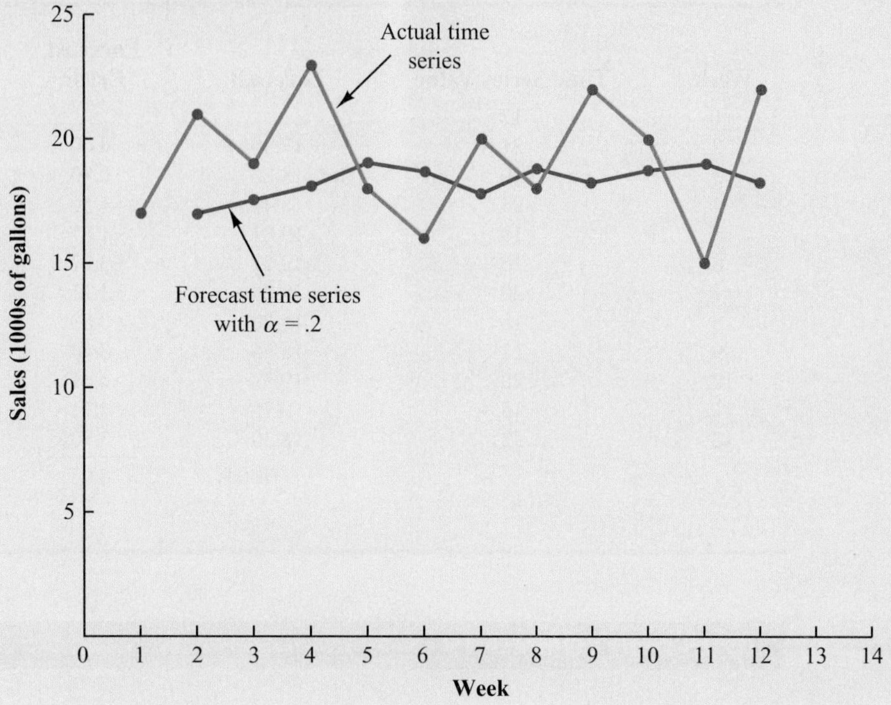

constant provide the advantage of quickly adjusting the forecasts; this allows the forecasts to react more quickly to changing conditions.

The criterion we will use to determine a desirable value for the smoothing constant α is the same as the criterion we proposed for determining the order or number of periods of data to include in the moving averages calculation. That is, we choose the value of α that minimizes the MSE. A summary of the MSE calculations for the exponential smoothing forecast of gasoline sales with $\alpha = .2$ is shown in Table 17.10. Note that there is one less squared error term than the number of time periods because we had no past values with which to make a forecast for period 1. The value of the sum of squared forecast errors is 98.80; hence MSE $= 98.80/11 = 8.98$. Would a different value of α provide better results in terms of a lower MSE value? Perhaps the most straightforward way to answer this question is simply to try another value for α. We will then compare its mean squared error with the MSE value of 8.98 obtained by using a smoothing constant of $\alpha = .2$.

The exponential smoothing results with $\alpha = .3$ are shown in Table 17.11. The value of the sum of squared forecast errors is 102.83; hence MSE $= 102.83/11 = 9.35$. With MSE $=$ 9.35, we see that, for the current data set, a smoothing constant of $\alpha = .3$ results in less forecast accuracy than a smoothing constant of $\alpha = .2$. Thus, we would be inclined to prefer the original smoothing constant of $\alpha = .2$. Using a trial-and-error calculation with other values of α, we can find a "good" value for the smoothing constant. This value can be used in the exponential smoothing model to provide forecasts for the future. At a later date, after new time series observations are obtained, we analyze the newly collected time series data to determine whether the smoothing constant should be revised to provide better forecasting results.

TABLE 17.11 SUMMARY OF THE EXPONENTIAL SMOOTHING FORECASTS AND FORECAST ERRORS FOR THE GASOLINE SALES TIME SERIES WITH SMOOTHING CONSTANT $\alpha = .3$

Week	Time Series Value	Forecast	Forecast Error	Squared Forecast Error
1	17			
2	21	17.00	4.00	16.00
3	19	18.20	0.80	0.64
4	23	18.44	4.56	20.79
5	18	19.81	−1.81	3.28
6	16	19.27	−3.27	10.69
7	20	18.29	1.71	2.92
8	18	18.80	−0.80	0.64
9	22	18.56	3.44	11.83
10	20	19.59	0.41	0.17
11	15	19.71	−4.71	22.18
12	22	18.30	3.70	13.69
		Totals	8.03	102.83

NOTES AND COMMENTS

1. Spreadsheet packages are an effective aid in choosing a good value of α for exponential smoothing. With the time series data and the forecasting formulas in a spreadsheet, you can experiment with different values of α and choose the value that provides the smallest forecast error using one or more of the measures of forecast accuracy (MAE, MSE, or MAPE).
2. We presented the moving average and exponential smoothing methods in the context of a stationary time series. These methods can also be used to forecast a nonstationary time series which shifts in level but exhibits no trend or seasonality. Moving averages with small values of k adapt more quickly than moving averages with larger values of k. Exponential smoothing models with smoothing constants closer to one adapt more quickly than models with smaller values of the smoothing constant.

Exercises

Methods

SELF test

5. Consider the following time series data.

Week	1	2	3	4	5	6
Value	18	13	16	11	17	14

   a. Construct a time series plot. What type of pattern exists in the data?
   b. Develop the three-week moving average forecasts for this time series. Compute MSE and a forecast for week 7.
   c. Use $\alpha = .2$ to compute the exponential smoothing forecasts for the time series. Compute MSE and a forecast for week 7.

d. Compare the three-week moving average approach with the exponential smoothing approach using $\alpha = .2$. Which appears to provide more accurate forecasts based on MSE? Explain.

e. Use a smoothing constant of $\alpha = .4$ to compute the exponential smoothing forecasts. Does a smoothing constant of .2 or .4 appear to provide more accurate forecasts based on MSE? Explain.

6. Consider the following time series data.

Month	1	2	3	4	5	6	7
Value	24	13	20	12	19	23	15

Construct a time series plot. What type of pattern exists in the data?

a. Develop the three-week moving average forecasts for this time series. Compute MSE and a forecast for week 8.

b. Use $\alpha = .2$ to compute the exponential smoothing forecasts for the time series. Compute MSE and a forecast for week 8.

c. Compare the three-week moving average approach with the exponential smoothing approach using $\alpha = .2$. Which appears to provide more accurate forecasts based on MSE?

d. Use a smoothing constant of $\alpha = .4$ to compute the exponential smoothing forecasts. Does a smoothing constant of .2 or .4 appear to provide more accurate forecasts based on MSE? Explain.

WEB file
Gasoline

7. Refer to the gasoline sales time series data in Table 17.1.

a. Compute four-week and five-week moving averages for the time series.

b. Compute the MSE for the four-week and five-week moving average forecasts.

c. What appears to be the best number of weeks of past data (three, four, or five) to use in the moving average computation? Recall that MSE for the three-week moving average is 10.22.

WEB file
Gasoline

8. Refer again to the gasoline sales time series data in Table 17.1.

a. Using a weight of 1/2 for the most recent observation, 1/3 for the second most recent observation, and 1/6 for third most recent observation, compute a three-week weighted moving average for the time series.

b. Compute the MSE for the weighted moving average in part (a). Do you prefer this weighted moving average to the unweighted moving average? Remember that the MSE for the unweighted moving average is 10.22.

c. Suppose you are allowed to choose any weights as long as they sum to 1. Could you always find a set of weights that would make the MSE at least as small for a weighted moving average than for an unweighted moving average? Why or why not?

WEB file
Gasoline

9. With the gasoline time series data from Table 17.1, show the exponential smoothing forecasts using $\alpha = .1$.

a. Applying the MSE measure of forecast accuracy, would you prefer a smoothing constant of $\alpha = .1$ or $\alpha = .2$ for the gasoline sales time series?

b. Are the results the same if you apply MAE as the measure of accuracy?

c. What are the results if MAPE is used?

10. With a smoothing constant of $\alpha = .2$, equation (17.2) shows that the forecast for week 13 of the gasoline sales data from Table 17.1 is given by $F_{13} = .2Y_{12} + .8F_{12}$. However, the forecast for week 12 is given by $F_{12} = .2Y_{11} + .8F_{11}$. Thus, we could combine these two results to show that the forecast for week 13 can be written

$$F_{13} = .2Y_{12} + .8(.2Y_{11} + .8F_{11}) = .2Y_{12} + .16Y_{11} + .64Y_{11} + .64F_{11}$$

a. Making use of the fact that $F_{11} = .2Y_{10} + .8F_{10}$ (and similarly for F_{10} and F_9), continue to expand the expression for F_{13} until it is written in terms of the past data values Y_{12}, Y_{11}, Y_{10}, Y_9, Y_8, and the forecast for period 8.

b. Refer to the coefficients or weights for the past values Y_{12}, Y_{11}, Y_{10}, Y_9, Y_8. What observation can you make about how exponential smoothing weights past data values in arriving at new forecasts? Compare this weighting pattern with the weighting pattern of the moving averages method.

Applications

11. For the Hawkins Company, the monthly percentages of all shipments received on time over the past 12 months are 80, 82, 84, 83, 83, 84, 85, 84, 82, 83, 84, and 83.
    a. Construct a time series plot. What type of pattern exists in the data?
    b. Compare the three-month moving average approach with the exponential smoothing approach for $\alpha = .2$. Which provides more accurate forecasts using MSE as the measure of forecast accuracy?
    c. What is the forecast for next month?
12. Corporate triple-A bond interest rates for 12 consecutive months follow.

    9.5 9.3 9.4 9.6 9.8 9.7 9.8 10.5 9.9 9.7 9.6 9.6

    a. Construct a time series plot. What type of pattern exists in the data?
    b. Develop three-month and four-month moving averages for this time series. Does the three-month or four-month moving average provide more accurate forecasts based on MSE? Explain.
    c. What is the moving average forecast for the next month?

SELF test

13. The values of Alabama building contracts (in $ millions) for a 12-month period follow.

    240 350 230 260 280 320 220 310 240 310 240 230

    a. Construct a time series plot. What type of pattern exists in the data?
    b. Compare the three-month moving average approach with the exponential smoothing forecast using $\alpha = .2$. Which provides more accurate forecasts based on MSE?
    c. What is the forecast for the next month?
14. The following time series shows the sales of a particular product over the past 12 months.

WEB file

ProductSales

Month	Sales	Month	Sales
1	105	7	145
2	135	8	140
3	120	9	100
4	105	10	80
5	90	11	100
6	120	12	110

    a. Construct a time series plot. What type of pattern exists in the data?
    b. Use $\alpha = .3$ to compute the exponential smoothing forecasts for the time series.
    c. Use a smoothing constant of $\alpha = .5$ to compute the exponential smoothing forecasts. Does a smoothing constant of .3 or .5 appear to provide more accurate forecasts based on MSE?
15. Ten weeks of data on the Commodity Futures Index are 7.35, 7.40, 7.55, 7.56, 7.60, 7.52, 7.52, 7.70, 7.62, and 7.55.
    a. Construct a time series plot. What type of pattern exists in the data?
    b. Compute the exponential smoothing forecasts for $\alpha = .2$.
    c. Compute the exponential smoothing forecasts for $\alpha = .3$.
    d. Which exponential smoothing constant provides more accurate forecasts based on MSE? Forecast week 11.

16. The U.S. Census Bureau tracks the median price for new home sales by month of year. The median prices for April for the years 1990 to 2011 follow (U.S. Census Bureau website, April 16, 2012).

WEB file HomePrices

Year	Price ($1000s)	Year	Price ($1000s)
1990	130.0	2001	175.2
1991	121.0	2002	187.1
1992	120.0	2003	189.5
1993	127.0	2004	222.3
1994	129.0	2005	236.3
1995	134.0	2006	257.0
1996	140.0	2007	242.5
1997	150.0	2008	246.4
1998	148.0	2009	219.2
1999	159.9	2010	208.3
2000	162.6	2011	224.7

a. Construct a time series plot. Comment on any pattern you observe. Discuss some of the factors that may have resulted in this time series plot.
b. Given the time series plot in part (a), do you think the forecasting methods developed in this section are appropriate for this time series? Explain.
c. To forecast a value for April 2012, how much of the past data would you use? Explain.

17.4 Trend Projection

WEB file Bicycle

We present three forecasting methods in this section that are appropriate for time series exhibiting a trend pattern. First, we show how simple linear regression can be used to forecast a time series with a linear trend. We then illustrate how to develop forecasts using Holt's linear exponential smoothing, an extension of single exponential smoothing that uses two smoothing constants: one to account for the level of the time series and a second to account for the linear trend in the data. Finally, we show how the curve-fitting capability of regression analysis can also be used to forecast time series with a curvilinear or nonlinear trend.

Linear Trend Regression

TABLE 17.12 BICYCLE SALES TIME SERIES

Year	Sales (1000s)
1	21.6
2	22.9
3	25.5
4	21.9
5	23.9
6	27.5
7	31.5
8	29.7
9	28.6
10	31.4

In Section 17.1 we used the bicycle sales time series in Table 17.3 and Figure 17.3 to illustrate a time series with a trend pattern. Let us now use this time series to illustrate how simple linear regression can be used to forecast a time series with a linear trend. The data for the bicycle time series are repeated in Table 17.12 and Figure 17.9.

Although the time series plot in Figure 17.9 shows some up and down movement over the past 10 years, we might agree that the linear trend line shown in Figure 17.10 provides a reasonable approximation of the long-run movement in the series. We can use the methods of simple linear regression (see Chapter 14) to develop such a linear trend line for the bicycle sales time series.

In Chapter 14, the estimated regression equation describing a straight-line relationship between an independent variable x and a dependent variable y is written as

$$\hat{y} = b_0 + b_1x$$

FIGURE 17.9 BICYCLE SALES TIME SERIES PLOT

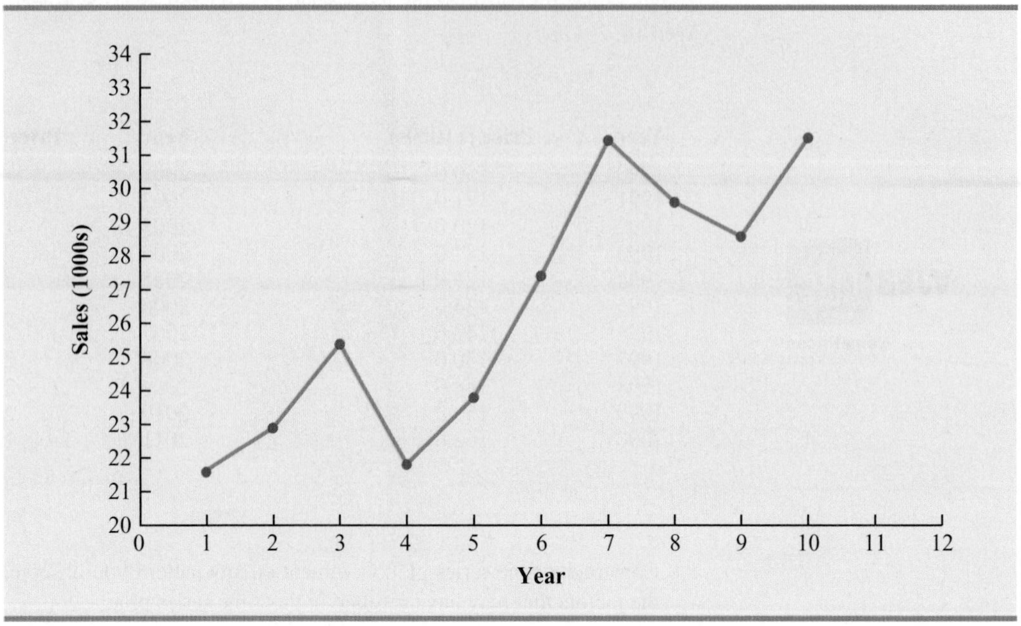

FIGURE 17.10 TREND REPRESENTED BY A LINEAR FUNCTION FOR THE BICYCLE SALES TIME SERIES

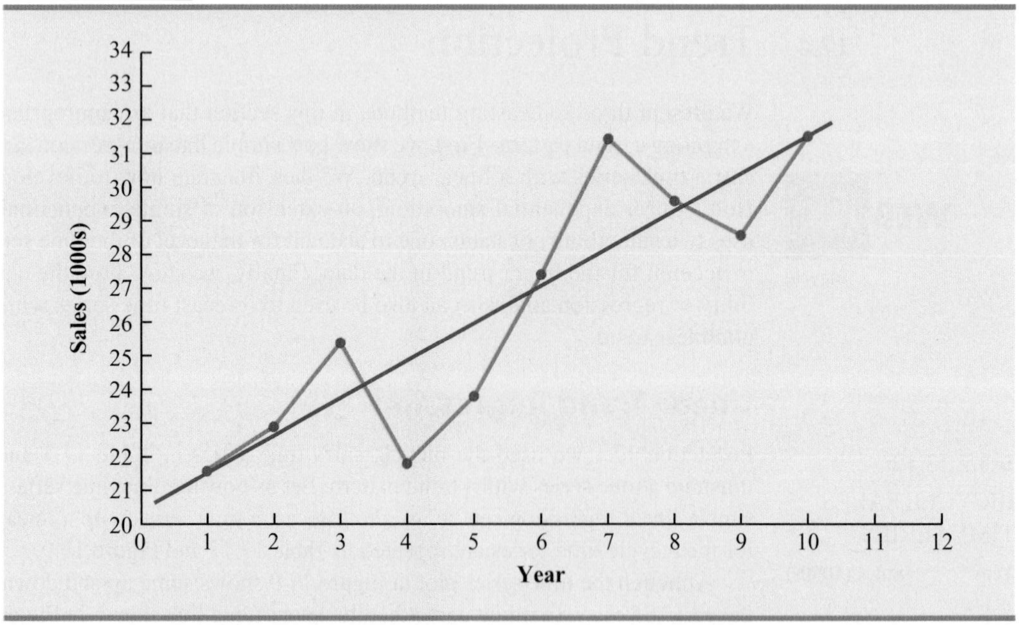

where $\hat{y}$ is the estimated or predicted value of y. To emphasize the fact that in forecasting the independent variable is time, we will replace x with t and $\hat{y}$ with T_t to emphasize that we are estimating the trend for a time series. Thus, for estimating the linear trend in a time series we will use the following estimated regression equation.

LINEAR TREND EQUATION

$$T_t = b_0 + b_1 t \tag{17.4}$$

where

T_t = linear trend forecast in period t
b_0 = intercept of the linear trend line
b_1 = slope of the linear trend line
t = time period

In equation (17.4), the time variable begins at $t = 1$ corresponding to the first time series observation (year 1 for the bicycle sales time series) and continues until $t = n$ corresponding to the most recent time series observation (year 10 for the bicycle sales time series). Thus, for the bicycle sales time series $t = 1$ corresponds to the oldest time series value and $t = 10$ corresponds to the most recent year.

Formulas for computing the estimated regression coefficients (b_1 and b_0) in equation (17.4) follow.

COMPUTING THE SLOPE AND INTERCEPT FOR A LINEAR TREND*

$$b_1 = \frac{\sum_{t=1}^{n}(t - \bar{t})(Y_t - \bar{Y})}{\sum_{t=1}^{n}(t - \bar{t})^2} \tag{17.5}$$

$$b_0 = \bar{Y} - b_1\bar{t} \tag{17.6}$$

where

Y_t = value of the time series in period t
n = number of time periods (number of observations)
$\bar{Y}$ = average value of the time series
$\bar{t}$ = average value of t

*An alternate formula for b_1 is

$$b_1 = \frac{\sum_{t=1}^{n} tY_t - \left(\sum_{t=1}^{n} t \sum_{t=1}^{n} Y_t\right)/n}{\sum_{t=1}^{n} t^2 - \left(\sum_{t=1}^{n} t\right)^2/n}$$

This form of equation (17.5) is often recommended when using a calculator to compute b_1.

To compute the linear trend equation for the bicycle sales time series, we begin the calculations by computing $\bar{t}$ and $\bar{Y}$ using the information in Table 17.12.

$$\bar{t} = \frac{\sum_{t=1}^{n} t}{n} = \frac{55}{10} = 5.5$$

$$\bar{Y} = \frac{\sum_{t=1}^{n} Y_t}{n} = \frac{264.5}{10} = 26.45$$

TABLE 17.13 SUMMARY OF LINEAR TREND CALCULATIONS FOR THE BICYCLE SALES TIME SERIES

	t	Y_t	$t - \bar{t}$	$Y_t - \bar{Y}$	$(t - \bar{t})(Y_t - \bar{Y})$	$(t - \bar{t})^2$
	1	21.6	−4.5	−4.85	21.825	20.25
	2	22.9	−3.5	−3.55	12.425	12.25
	3	25.5	−2.5	−0.95	2.375	6.25
	4	21.9	−1.5	−4.55	6.825	2.25
	5	23.9	−0.5	−2.55	1.275	0.25
	6	27.5	0.5	1.05	0.525	0.25
	7	31.5	1.5	5.05	7.575	2.25
	8	29.7	2.5	3.25	8.125	6.25
	9	28.6	3.5	2.15	7.525	12.25
	10	31.4	4.5	4.95	22.275	20.25
Totals	55	264.5			90.750	82.50

Using these values, and the information in Table 17.13, we can compute the slope and intercept of the trend line for the bicycle sales time series.

$$b_1 = \frac{\sum_{t=1}^{n}(t - \bar{t})(Y_t - \bar{Y})}{\sum_{t=1}^{n}(t - \bar{t})^2} = \frac{90.75}{82.5} = 1.1$$

$$b_0 = \bar{Y} - b_1\bar{t} = 26.45 - 1.1(5.5) = 20.4$$

Therefore, the linear trend equation is

$$T_t = 20.4 + 1.1t$$

The slope of 1.1 indicates that over the past 10 years the firm experienced an average growth in sales of about 1100 units per year. If we assume that the past 10-year trend in sales is a good indicator of the future, this trend equation can be used to develop forecasts for future time periods. For example, substituting $t = 11$ into the equation yields next year's trend projection or forecast, T_{11}.

$$T_{11} = 20.4 + 1.1(11) = 32.5$$

Thus, using trend projection, we would forecast sales of 32,500 bicycles next year.

To compute the accuracy associated with the trend projection forecasting method, we will use the MSE. Table 17.14 shows the computation of the sum of squared errors for the bicycle sales time series. Thus, for the bicycle sales time series,

$$\text{MSE} = \frac{\sum_{t=1}^{n}(Y_t - F_t)^2}{n} = \frac{30.7}{10} = 3.07$$

Because linear trend regression in forecasting uses the same regression analysis procedure introduced in Chapter 14, we can use the standard regression analysis procedures in Minitab or Excel to perform the calculations. Figure 17.11 shows the computer output for the bicycle sales time series obtained using Minitab's regression analysis module.

TABLE 17.14 SUMMARY OF THE LINEAR TREND FORECASTS AND FORECAST ERRORS FOR THE BICYCLE SALES TIME SERIES

Year	Sales (1000s) Y_t	Forecast T_t	Forecast Error	Squared Forecast Error
1	21.6	21.5	0.1	0.01
2	22.9	22.6	0.3	0.09
3	25.5	23.7	1.8	3.24
4	21.9	24.8	−2.9	8.41
5	23.9	25.9	−2.0	4.00
6	27.5	27.0	0.5	0.25
7	31.5	28.1	3.4	11.56
8	29.7	29.2	0.5	0.25
9	28.6	30.3	−1.7	2.89
10	31.4	31.4	0.0	0.00
			Total	30.70

In Figure 17.11 the value of MSE in the ANOVA table is

$$\text{MSE} = \frac{\text{Sum of Squares Due to Error}}{\text{Degrees of Freedom}} = \frac{30.7}{8} = 3.837$$

MSD in Minitab's Trend Analysis output is the mean squared deviation, the average of the squared forecast errors.

This value of MSE differs from the value of MSE that we computed previously because the sum of squared errors is divided by 8 instead of 10; thus, MSE in the regression output is not the average of the squared forecast errors. Most forecasting packages, however, compute MSE by taking the average of the squared errors. Thus, when using time series packages to develop a trend equation, the value of MSE that is reported may differ slightly from the value you would obtain using a general regression approach. For instance, in Figure 17.12, we show the graphical portion of the computer output obtained using Minitab's Trend Analysis time series procedure. Note that MSD = 3.07 is the average of the squared forecast errors.

FIGURE 17.11 MINITAB REGRESSION OUTPUT FOR THE BICYCLE SALES TIME SERIES

```
The regression equation is
Y = 20.4 + 1.10 t

Predictor     Coef   SE Coef       T      p
Constant    20.400     1.338   15.24  0.000
t           1.1000    0.2157    5.10  0.001

S = 1.95895   R-sq = 76.5%   R-sq(adj) = 73.5%

Analysis of Variance

SOURCE          DF        SS      MS      F      p
Regression       1    99.825  99.825  26.01  0.001
Residual Error   8    30.700   3.837
Total            9   130.525
```

FIGURE 17.12 MINITAB TIME SERIES LINEAR TREND ANALYSIS OUTPUT FOR THE BICYCLE SALES TIME SERIES

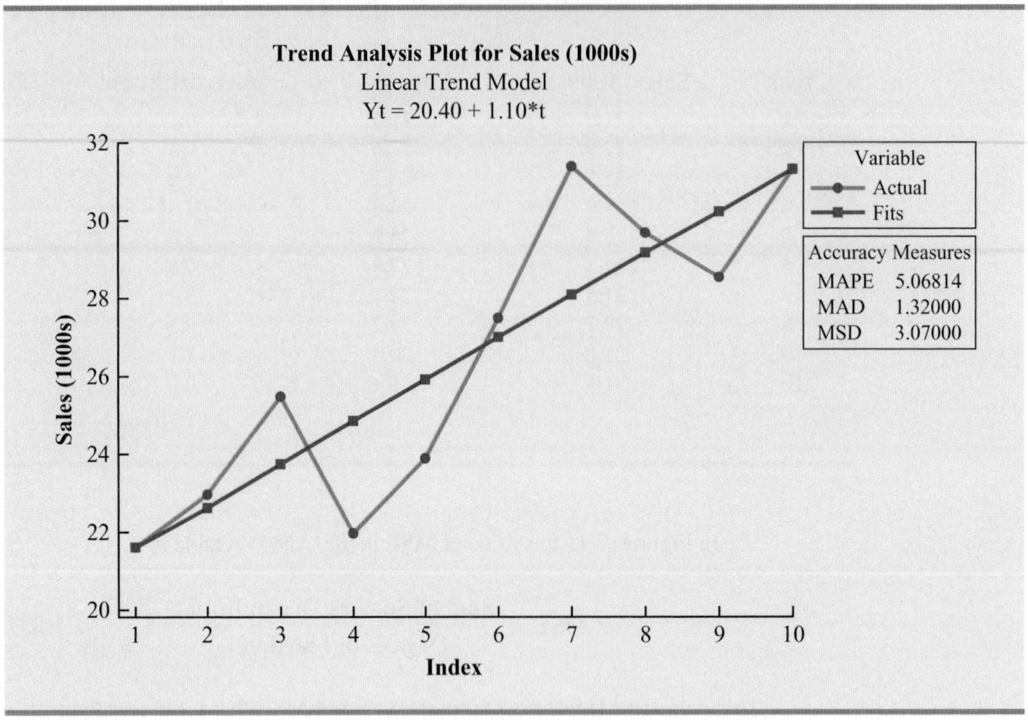

Holt's Linear Exponential Smoothing

Holt's linear exponential smoothing is often called double exponential smoothing.

Charles Holt developed a version of exponential smoothing that can be used to forecast a time series with a linear trend. Recall that the exponential smoothing procedure discussed in Section 17.3 uses the smoothing constant α to "smooth out" the randomness or irregular fluctuations in a time series; and, forecasts for time period $t + 1$ are obtained using the equation

$$F_{t+1} = \alpha Y_t + (1 - \alpha)F_t$$

Forecasts for Holt's **linear exponential smoothing** method are obtained using two smoothing constants, α and β, and three equations.

EQUATIONS FOR HOLT'S LINEAR EXPONENTIAL SMOOTHING

$$L_t = \alpha Y_t + (1 - \alpha)(L_{t-1} + b_{t-1}) \quad \textbf{(17.7)}$$
$$b_t = \beta(L_t - L_{t-1}) + (1 - \beta)\, b_{t-1} \quad \textbf{(17.8)}$$
$$F_{t+k} = L_t + b_t\, k \quad \textbf{(17.9)}$$

where

L_t = estimate of the level of the time series in period t

b_t = estimate of the slope of the time series in period t

α = smoothing constant for the level of the time series

β = smoothing constant for the slope of the time series
F_{t+k} = forecast for k periods ahead
k = the number of periods ahead to be forecast

Let us apply Holt's method to the bicycle sales time series in Table 17.12 using $\alpha = .1$ and $\beta = .2$. To get the method started, we need values for L_1, the estimate of the level of the time series in year 1, and b_1, the estimate of the slope of the time series in year 1. A commonly used approach is to set $L_1 = Y_1$ and $b_1 = Y_2 - Y_1$. Using this startup procedure, we obtain

$$L_1 = Y_1 = 21.6$$
$$b_1 = Y_2 - Y_1 = 22.9 - 21.6 = 1.3$$

Using equation (17.9) with $k = 1$, the forecast of sales in year 2 is $F_2 = L_1 + b_1 = 21.6 + 1.3(1) = 22.9$. Then we move on using equations (17.7) to (17.9) to compute estimates of the level and trend for year 2 as well as a forecast for year 3.

First we use equation (17.7) and the smoothing constant $\alpha = .1$ to compute an estimate of the level of the time series in year 2.

$$L_2 = .1(22.9) + .9(21.6 + 1.3) = 22.9$$

Note that 21.6 + 1.3 is the forecast of sales for year 2. Thus, the estimate of the level of the time series in year 2 obtained using equation (17.7) is simply a weighted average of the observed value in year 2 (using a weight of $\alpha = .1$) and the forecast for year 2 (using a weight of $1 - \alpha = 1 - .1 = .9$). In general, large values of α place more weight on the observed value (Y_t) whereas smaller values place more weight on the forecasted value ($L_{t-1} + b_{t-1}$).

Next we use equation (17.8) and the smoothing constant $\beta = .2$ to compute an estimate of the slope of the time series in year 2.

$$b_2 = .2(22.9 - 21.6) + (1 - .2)(1.3) = 1.3$$

The estimate the slope of the time series in year 2 is a weighted average of the difference in the estimated level of the time series between year 2 and year 1 (using a weight of $\beta = .2$) and the estimate of the slope in year 1(using a weight of $1 - \beta = 1 - .2 = .8$). In general, higher values of β place more weight on the difference between the estimated levels, whereas smaller values place more weight on the estimate of the slope from the last period.

Using the estimates of L_2 and b_2 just obtained, the forecast of sales for year 3 is computed using equation (17.9):

$$F_3 = L_2 + b_2 = 22.9 + 1.3(1) = 24.2$$

The other calculations are made in a similar manner and are shown in Table 17.15. The sum of the squared forecast errors is 39.678; hence MSE = 39.678/9 = 4.41.

Will different values for the smoothing constants α and β provide more accurate forecasts? To answer this question we would have to try different combinations of α and β to determine if a combination can be found that will provide a value of MSE lower than 4.41, the value we obtained using smoothing constants $\alpha = .1$ and $\beta = .2$. Searching for good values of α and β can be done by trial and error or using more advanced statistical software packages that have an option for selecting the optimal set of smoothing constants.

TABLE 17.15 SUMMARY CALCULATIONS FOR HOLT'S LINEAR EXPONENTIAL SMOOTHING FOR THE BICYCLE SALES TIME SERIES USING $\alpha = .1$ AND $\beta = .2$

Year	Sales (1000s) Y_t	Estimated Level L_t	Estimated Trend b_t	Forecast F_t	Forecast Error	Squared Forecast Error
1	21.6	21.600	1.300			
2	22.9	22.900	1.300	22.900	0.000	0.000
3	25.5	24.330	1.326	24.200	1.300	1.690
4	21.9	25.280	1.251	25.656	−3.756	14.108
5	23.9	26.268	1.198	26.531	−2.631	6.924
6	27.5	27.470	1.199	27.466	0.034	0.001
7	31.5	28.952	1.256	28.669	2.831	8.016
8	29.7	30.157	1.245	30.207	−0.507	0.257
9	28.6	31.122	1.189	31.402	−2.802	7.851
10	31.4	32.220	1.171	32.311	−0.911	0.830
					Total	39.678

Note that the estimate of the level of the time series in year 10 is $L_1 = 32.220$ and the estimate of the slope in year 10 is $b_1 = 1.171$. If we assume that the past 10-year trend in sales is a good indicator of the future, equation (17.9) can be used to develop forecasts for future time periods. For example, substituting $t = 11$ into equation (17.9) yields next year's trend projection or forecast, F_{11}.

$$F_{11} = L_{10} + b_{10}(1) = 32.220 + 1.171 = 33.391$$

Thus, using Holt's linear exponential smoothing we would forecast sales of 33,391 bicycles next year.

Nonlinear Trend Regression

WEB file

Cholesterol

The use of a linear function to model trend is common. However, as we discussed previously, sometimes time series have a curvilinear or nonlinear trend. As an example, consider the annual revenue in millions of dollars for a cholesterol drug for the first 10 years of sales. Table 17.16 shows the time series and Figure 17.13 shows the corresponding time series plot. For instance, revenue in year 1 was \$23.1 million; revenue in year 2 was \$21.3 million; and so on. The time series plot indicates an overall increasing or upward trend. But, unlike the bicycle sales time series, a linear trend does not appear to be appropriate. Instead, a curvilinear function appears to be needed to model the long-term trend.

TABLE 17.16 CHOLESTEROL REVENUE TIME SERIES (\$MILLIONS)

Year (t)	Revenue (\$millions)
1	23.1
2	21.3
3	27.4
4	34.6
5	33.8
6	43.2
7	59.5
8	64.4
9	74.2
10	99.3

Quadratic trend equation A variety of nonlinear functions can be used to develop an estimate of the trend for the cholesterol time series. For instance, consider the following quadratic trend equation:

$$T_t = b_0 + b_1 t + b_2 t^2 \tag{17.10}$$

For the cholesterol time series, $t = 1$ corresponds to year 1, $t = 2$ corresponds to year 2, and so on.

The general linear model discussed in Section 16.1 can be used to compute the values of b_0, b_1, and b_2. There are two independent variables, year and year squared, and the dependent variable is the sales revenue in millions of dollars. Thus, the first observation is 1,

FIGURE 17.13 CHOLESTEROL REVENUE TIMES SERIES PLOT ($MILLIONS)

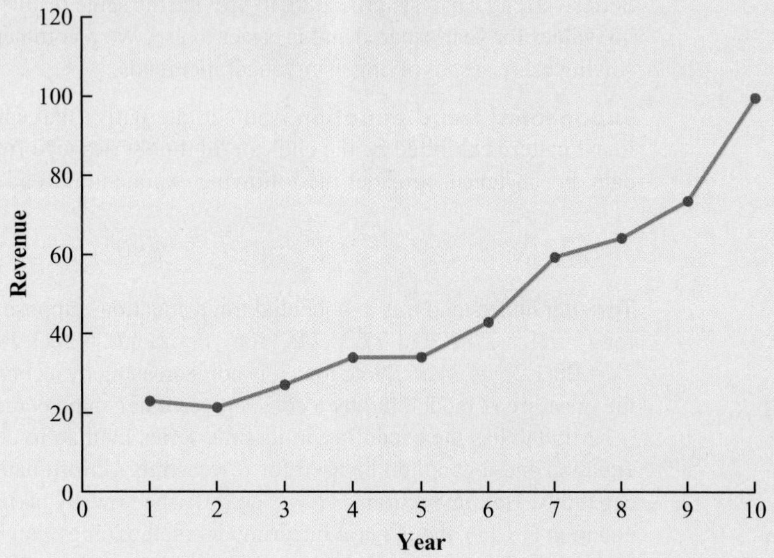

1, 23.1; the second observation is 2, 4, 21.3; the third observation is 3, 9, 27.4; and so on. Figure 17.14 shows the Minitab multiple regression output for the quadratic trend model; the estimated regression equation is

$$\text{Revenue (\$millions)} = 24.2 - 2.11\ \text{Year} + 0.922\ \text{YearSq}$$

where

$$\text{Year} = 1, 2, 3, \ldots, 10$$
$$\text{YearSq} = 1, 4, 9, \ldots, 100$$

FIGURE 17.14 MINITAB QUADRATIC TREND REGRESSION OUTPUT FOR THE BICYCLE SALES TIME SERIES

```
The regression equation is
Revenue = 24.2 - 2.11 Year + 0.922 YearSq

Predictor      Coef   SE Coef       T      p
Constant     24.182     4.676    5.17  0.001
Year         -2.106     1.953   -1.08  0.317
YearSq       0.9216    0.1730    5.33  0.001

S = 3.97578   R-Sq = 98.1%   R-Sq(adj) = 97.6%

Analysis of Variance

SOURCE          DF      SS      MS       F      p
Regression       2  5770.1  2885.1  182.52  0.000
Residual Error   7   110.6    15.8
Total            9  5880.8
```

Using the standard multiple regression procedure requires us to compute the values for year squared as a second independent variable. Alternatively, we can use Minitab's Time Series—Trend Analysis procedure to provide the same results. It does not require developing values for year squared and is easier to use. We recommend using this approach when solving exercises involving using quadratic trends.

Exponential trend equation Another alternative that can be used to model the nonlinear pattern exhibited by the cholesterol time series is to fit an exponential model to the data. For instance, consider the following exponential trend equation:

$$T_t = b_0(b_1)^t \tag{17.11}$$

To better understand this exponential trend equation, suppose $b_0 = 20$ and $b_1 = 1.2$. Then, for $t = 1$, $T_1 = 20(1.2)^1 = 24$; for $t = 2$, $T_2 = 20(1.2)^2 = 28.8$; and for $t = 3$, $T_3 = 20(1.2)^3 = 34.56$. Note that T_t is not increasing by a constant amount as in the case of the linear trend model, but by a constant percentage; the percentage increase is 20%.

Minitab has the capability in its time series module to compute an exponential trend equation and it can then be used for forecasting. Unfortunately, Excel does not have this capability. But, in Section 16.1, we do describe how, by taking logarithms of the terms in equation (17.11), the general linear model methodology can be used to compute an exponential trend equation.

Minitab's time series module is quite easy to use to develop an exponential trend equation. There is no need to deal with logarithms and use regression analysis to compute the exponential trend equation. In Figure 17.15, we show the graphical portion of the computer output obtained using Minitab's Trend Analysis time series procedure to fit an exponential trend equation.

FIGURE 17.15 MINITAB TIME SERIES EXPONENTIAL GROWTH TREND ANALYSIS OUTPUT FOR THE CHOLESTEROL SALES TIME SERIES

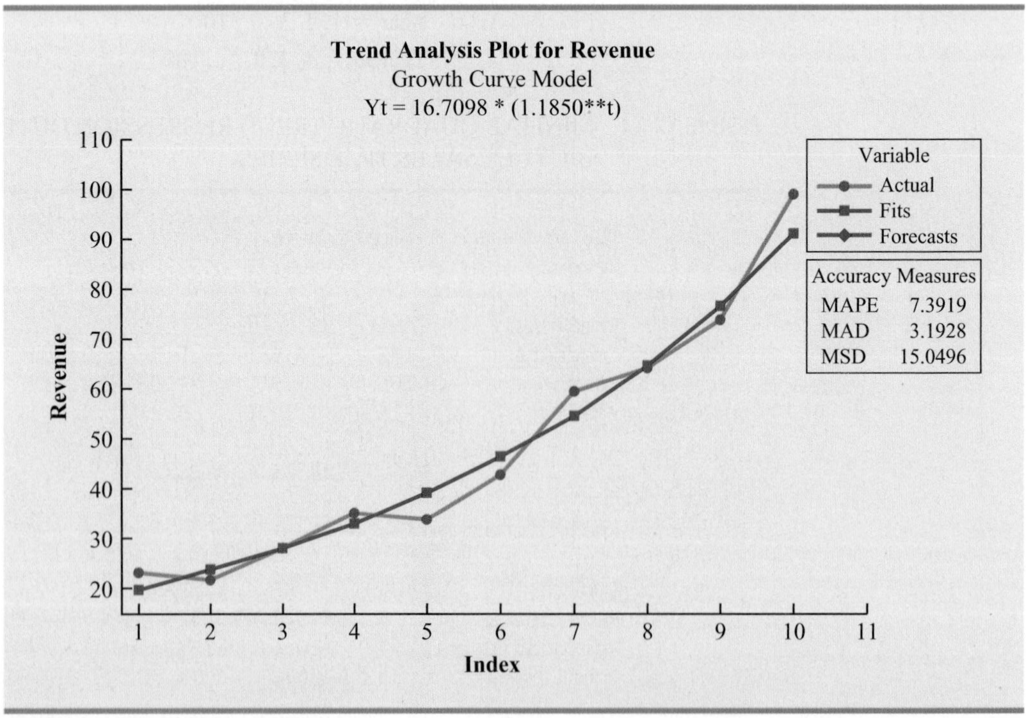

NOTES AND COMMENTS

Linear trend regression is based upon finding the estimated regression equation that minimizes the sum of squared forecast errors and therefore MSE. So, we would expect linear trend regression to outperform Holt's linear exponential smoothing in terms of MSE. For example, for the bicycle sales time series, the value of MSE using linear trend regression is 3.07 as compared to a value of 3.97 using Holt's linear exponential smoothing. Linear trend regression also provides a more accurate forecast using the MAE measure of forecast accuracy; for the bicycle sales time series, linear trend regression results in a value of MAE of 1.32 versus a value of 1.67 using Holt's linear method. However, based on MAPE, Holt's linear exponential smoothing (MAPE = 5.07%) outperforms linear trend regression (6.42%). Hence, for the bicycle sales time series, deciding which method provides the more accurate forecasts depends upon which measure of forecast accuracy is used.

Exercises

Methods

SELF test

17. Consider the following time series data.

t	1	2	3	4	5
Y_t	6	11	9	14	15

   a. Construct a time series plot. What type of pattern exists in the data?
   b. Develop the linear trend equation for this time series.
   c. What is the forecast for $t = 6$?

18. Refer to the time series in exercise 17. Use Holt's linear exponential smoothing method with $\alpha = .3$ and $\beta = .5$ to develop a forecast for $t = 6$.

19. Consider the following time series.

t	1	2	3	4	5	6	7
Y_t	120	110	100	96	94	92	88

   a. Construct a time series plot. What type of pattern exists in the data?
   b. Develop the linear trend equation for this time series.
   c. What is the forecast for $t = 8$?

20. Consider the following time series.

t	1	2	3	4	5	6	7
Y_t	82	60	44	35	30	29	35

   a. Construct a time series plot. What type of pattern exists in the data?
   b. Using Minitab or Excel, develop the quadratic trend equation for the time series.
   c. What is the forecast for $t = 8$?

SELF test

21. The general fund budget for the state of Kentucky for 1988 (Period 1) to 2011 (Period 24) follows (*Northern Kentucky Enquirer*, January 18, 2012).

WEB file
KYBudget

Year	Period	Budget ($billions)	Year	Period	Budget ($billions)
1988	1	3.03	2000	13	6.48
1989	2	3.29	2001	14	6.65
1990	3	3.56	2002	15	6.56
1991	4	4.31	2003	16	6.78
1992	5	4.36	2004	17	6.98
1993	6	4.51	2005	18	7.65
1994	7	4.65	2006	19	8.38
1995	8	5.15	2007	20	8.57
1996	9	5.34	2008	21	8.66
1997	10	5.66	2009	22	8.43
1998	11	6.01	2010	23	8.23
1999	12	6.20	2011	24	8.76

a. Construct a time series plot. What type of pattern exists in the data?
b. Develop a linear trend equation for this time series.
c. What is the forecast for 2012?

22. The Seneca Children's Fund (SCF) is a local charity that runs a summer camp for disadvantaged children. The fund's board of directors has been working very hard in recent years to decrease the amount of overhead expenses, a major factor in how charities are rated by independent agencies. The following data show the percentage of the money SCF has raised that was spent on administrative and fund-raising expenses for 2006–2012.

Year	Period (t)	Expense (%)
2006	1	13.9
2007	2	12.2
2008	3	10.5
2009	4	10.4
2010	5	11.5
2011	6	10.0
2012	7	8.5

a. Construct a time series plot. What type of pattern exists in the data?
b. Develop the linear trend equation for this time series.
c. Forecast the percentage of administrative expenses for 2013.
d. If SCF can maintain their current trend in reducing administrative expenses, how long will it take them to achieve a level of 5% or less?

23. The president of a small manufacturing firm is concerned about the continual increase in manufacturing costs over the past several years. The following figures provide a time series of the cost per unit for the firm's leading product over the past eight years.

Year	Cost/Unit ($)	Year	Cost/Unit ($)
1	20.00	5	26.60
2	24.50	6	30.00
3	28.20	7	31.00
4	27.50	8	36.00

a. Construct a time series plot. What type of pattern exists in the data?
b. Develop the linear trend equation for this time series.
c. What is the average cost increase that the firm has been realizing per year?
d. Compute an estimate of the cost/unit for next year.

24. FRED® (Federal Reserve Economic Data), a database of more than 3000 U.S. economic time series, contains historical data on foreign exchange rates. The following data show the foreign exchange rate for the United States and China (Federal Reserve Bank of St.Louis website). The units for Rate are the number of Chinese yuan to one U.S. dollar.

WEB file
ExchangeRate

Year	Month	Rate
2007	October	7.5019
2007	November	7.4210
2007	December	7.3682
2008	January	7.2405
2008	February	7.1644
2008	March	7.0722
2008	April	6.9997
2008	May	6.9725
2008	June	6.8993
2008	July	6.8355

a. Construct a time series plot. Does a linear trend appear to be present?
b. Using Minitab or Excel, develop the linear trend equation for this time series.
c. Use the trend equation to forecast the exchange rate for August 2008.
d. Would you feel comfortable using the trend equation to forecast the exchange rate for December 2008?

25. Automobile unit sales at B. J. Scott Motors, Inc., provided the following 10-year time series.

Year	Sales	Year	Sales
1	400	6	260
2	390	7	300
3	320	8	320
4	340	9	340
5	270	10	370

a. Construct a time series plot. Comment on the appropriateness of a linear trend.
b. Using Minitab or Excel, develop a quadratic trend equation that can be used to forecast sales.
c. Using the trend equation developed in part (b), forecast sales in year 11.
d. Suggest an alternative to using a quadratic trend equation to forecast sales. Explain.

26. Giovanni Food Products produces and sells frozen pizzas to public schools throughout the eastern United States. Using a very aggressive marketing strategy they have been able to increase their annual revenue by approximately \$10 million over the past 10 years. But increased competition has slowed their growth rate in the past few years. The annual revenue, in millions of dollars, for the previous 10 years is shown.

WEB file
Pasta

Year	Revenue
1	8.53
2	10.84
3	12.98
4	14.11
5	16.31
6	17.21
7	18.37
8	18.45
9	18.40
10	18.43

a. Construct a time series plot. Comment on the appropriateness of a linear trend.
b. Using Minitab or Excel, develop a quadratic trend equation that can be used to forecast revenue.
c. Using the trend equation developed in part (b), forecast revenue in year 11.

27. The number of users of Facebook from 2004 through 2011 follows (Facebook website, April 16, 2012).

WEB file Facebook

Year	Period	Users (Millions)
2004	1	1
2005	2	6
2006	3	12
2007	4	58
2008	5	145
2009	6	360
2010	7	608
2011	8	845

a. Construct a time series plot. What type of pattern exists?
b. Using Minitab or Excel, develop a quadratic trend equation.

17.5 Seasonality and Trend

In this section we show how to develop forecasts for a time series that has a seasonal pattern. To the extent that seasonality exists, we need to incorporate it into our forecasting models to ensure accurate forecasts. We begin by considering a seasonal time series with no trend and then discuss how to model seasonality with trend.

WEB file Umbrella

Seasonality Without Trend

As an example, consider the number of umbrellas sold at a clothing store over the past five years. Table 17.17 shows the time series and Figure 17.16 shows the corresponding time series plot. The time series plot does not indicate any long-term trend in sales. In fact, unless you look carefully at the data, you might conclude that the data follow a horizontal pattern and that single exponential smoothing could be used to forecast sales. But closer inspection of the time series plot reveals a pattern in the data. That is, the first and third quarters have moderate sales, the second quarter has the highest sales, and the fourth quarter tends to be the lowest quarter in terms of sales volume. Thus, we would conclude that a quarterly seasonal pattern is present.

TABLE 17.17
UMBRELLA SALES TIME SERIES

Year	Quarter	Sales
1	1	125
	2	153
	3	106
	4	88
2	1	118
	2	161
	3	133
	4	102
3	1	138
	2	144
	3	113
	4	80
4	1	109
	2	137
	3	125
	4	109
5	1	130
	2	165
	3	128
	4	96

In Chapter 15 we showed how dummy variables can be used to deal with categorical independent variables in a multiple regression model. We can use the same approach to model a time series with a seasonal pattern by treating the season as a categorical variable. Recall that when a categorical variable has k levels, $k - 1$ dummy variables are required. So, if there are four seasons, we need three dummy variables. For instance, in the umbrella sales time series season is a categorical variable with four levels: quarter 1, quarter 2, quarter 3, and quarter 4. Thus, to model the seasonal effects in the umbrella time series we need $4 - 1 = 3$ dummy variables. The three dummy variables can be coded as follows:

$$\text{Qtr1} = \begin{cases} 1 \text{ if Quarter 1} \\ 0 \text{ otherwise} \end{cases} \quad \text{Qtr2} = \begin{cases} 1 \text{ if Quarter 2} \\ 0 \text{ otherwise} \end{cases} \quad \text{Qtr3} = \begin{cases} 1 \text{ if Quarter 3} \\ 0 \text{ otherwise} \end{cases}$$

FIGURE 17.16 UMBRELLA SALES TIME SERIES PLOT

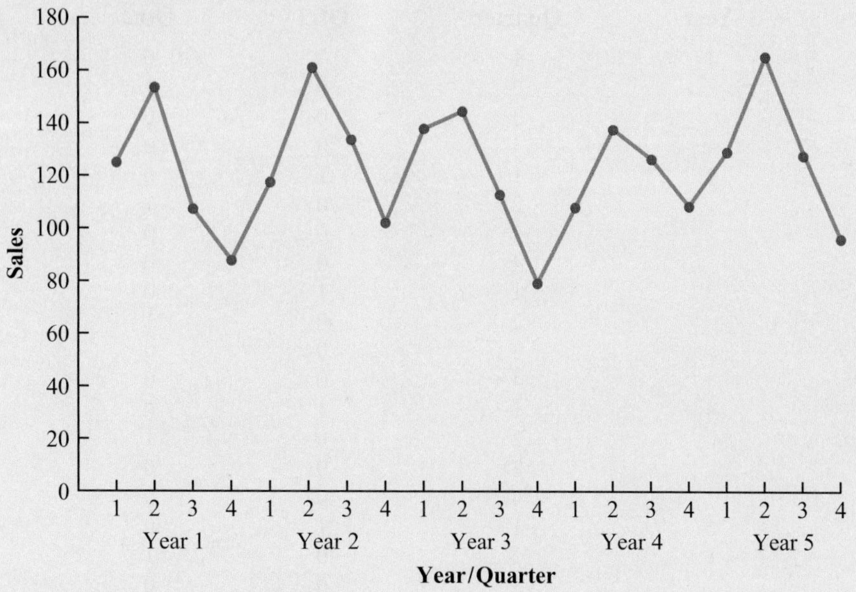

Using $\hat{Y}$ to denote the estimated or forecasted value of sales, the general form of the estimated regression equation relating the number of umbrellas sold to the quarter the sales take place follows:

$$\hat{Y} = b_0 + b_1 \text{Qtr1} + b_2 \text{Qtr2} + b_3 \text{Qtr3}$$

Table 17.18 is the umbrella sales time series with the coded values of the dummy variables shown. Using the data in Table 17.18 and Minitab's regression procedure, we obtained the computer output shown in Figure 17.17. The estimated multiple regression equation obtained is

$$\text{Sales} = 95.0 + 29.0 \text{ Qtr1} + 57.0 \text{ Qtr2} + 26.0 \text{ Qtr3}$$

We can use this equation to forecast quarterly sales for next year.

$$\text{Quarter 1: Sales} = 95.0 + 29.0(1) + 57.0(0) + 26.0(0) = 124$$
$$\text{Quarter 2: Sales} = 95.0 + 29.0(0) + 57.0(1) + 26.0(0) = 152$$
$$\text{Quarter 3: Sales} = 95.0 + 29.0(0) + 57.0(0) + 26.0(1) = 121$$
$$\text{Quarter 4: Sales} = 95.0 + 29.0(0) + 57.0(1) + 26.0(0) = 95$$

It is interesting to note that we could have obtained the quarterly forecasts for next year simply by computing the average number of umbrellas sold in each quarter, as shown in the following table.

Year	Quarter 1	Quarter 2	Quarter 3	Quarter 4
1	125	153	106	88
2	118	161	133	102
3	138	144	113	80
4	109	137	125	109
5	130	165	128	96
Average	124	152	121	95

TABLE 17.18 UMBRELLA SALES TIME SERIES WITH DUMMY VARIABLES

Year	Quarter	Qtr1	Qtr2	Qtr3	Sales
1	1	1	0	0	125
	2	0	1	0	153
	3	0	0	1	106
	4	0	0	0	88
2	1	1	0	0	118
	2	0	1	0	161
	3	0	0	1	133
	4	0	0	0	102
3	1	1	0	0	138
	2	0	1	0	144
	3	0	0	1	113
	4	0	0	0	80
4	1	1	0	0	109
	2	0	1	0	137
	3	0	0	1	125
	4	0	0	0	109
5	1	1	0	0	130
	2	0	1	0	165
	3	0	0	1	128
	4	0	0	0	96

Nonetheless, the regression output shown in Figure 17.17 provides additional information that can be used to assess the accuracy of the forecast and determine the significance of the results. And, for more complex types of problem situations, such as dealing with a time series that has both trend and seasonal effects, this simple averaging approach will not work.

Seasonality and Trend

WEB file TVSales

Let us now extend the regression approach to include situations where the time series contains both a seasonal effect and a linear trend by showing how to forecast the quarterly television set sales time series introduced in Section 17.1. The data for the television set time series are shown in Table 17.19. The time series plot in Figure 17.18 indicates that sales are lowest in the second quarter of each year and increase in quarters 3 and 4. Thus, we conclude that a seasonal pattern exists for television set sales. But the time series also has an upward linear trend that will need to be accounted for in order to develop accurate forecasts of quarterly sales. This is easily handled by combining the dummy variable approach for

TABLE 17.19 TELEVISION SET SALES TIME SERIES

Year	Quarter	Sales (1000s)
1	1	4.8
	2	4.1
	3	6.0
	4	6.5
2	1	5.8
	2	5.2
	3	6.8
	4	7.4
3	1	6.0
	2	5.6
	3	7.5
	4	7.8
4	1	6.3
	2	5.9
	3	8.0
	4	8.4

FIGURE 17.17 MINITAB REGRESSION OUTPUT FOR THE UMBRELLA SALES TIME SERIES

```
The regression equation is
Sales = 95.0 + 29.0 Qtr1 + 57.0 Qtr2 + 26.0 Qtr3

Predictor    Coef  SE Coef      T      P
Constant   95.000    5.065  18.76  0.000
Qtr1       29.000    7.162   4.05  0.001
Qtr2       57.000    7.162   7.96  0.000
Qtr3       26.000    7.162   3.63  0.002
```

FIGURE 17.18 TELEVISION SET SALES TIME SERIES PLOT

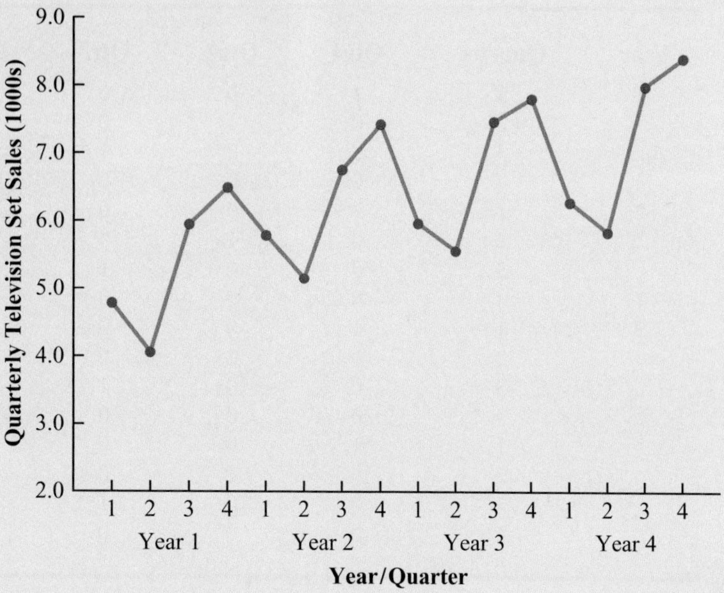

seasonality with the time series regression approach we discussed in Section 17.3 for handling linear trend.

The general form of the estimated multiple regression equation for modeling both the quarterly seasonal effects and the linear trend in the television set time series is as follows:

$$\hat{Y}_t = b_0 + b_1 \text{ Qtr1} + b_2 \text{ Qtr2} + b_3 \text{ Qtr3} + b_4 t$$

where

$\hat{Y}_t$ = estimate or forecast of sales in period t

Qtr1 = 1 if time period t corresponds to the first quarter of the year; 0 otherwise

Qtr2 = 1 if time period t corresponds to the second quarter of the year; 0 otherwise

Qtr3 = 1 if time period t corresponds to the third quarter of the year; 0 otherwise

t = time period

Table 17.20 is the revised television set sales time series that includes the coded values of the dummy variables and the time period t. Using the data in Table 17.20, and Minitab's regression procedure, we obtained the computer output shown in Figure 17.19. The estimated multiple regression equation is

$$\text{Sales} = 6.07 - 1.36 \text{ Qtr1} - 2.03 \text{ Qtr2} - .304 \text{ Qtr3} + .146t \qquad (17.12)$$

We can now use equation (17.12) to forecast quarterly sales for next year. Next year is year 5 for the television set sales time series; that is, time periods 17, 18, 19, and 20.

Forecast for Time Period 17 (Quarter 1 in Year 5)

$$\text{Sales} = 6.07 - 1.36(1) - 2.03(0) - .304(0) + .146(17) = 7.19$$

Forecast for Time Period 18 (Quarter 2 in Year 5)

$$\text{Sales} = 6.07 - 1.36(0) - 2.03(1) - .304(0) + .146(18) = 6.67$$

TABLE 17.20 TELEVISION SET SALES TIME SERIES WITH DUMMY VARIABLES AND TIME PERIOD

Year	Quarter	Qtr1	Qtr2	Qtr3	Period	Sales (1000s)
1	1	1	0	0	1	4.8
	2	0	1	0	2	4.1
	3	0	0	1	3	6.0
	4	0	0	0	4	6.5
2	1	1	0	0	5	5.8
	2	0	1	0	6	5.2
	3	0	0	1	7	6.8
	4	0	0	0	8	7.4
3	1	1	0	0	9	6.0
	2	0	1	0	10	5.6
	3	0	0	1	11	7.5
	4	0	0	0	12	7.8
4	1	1	0	0	13	6.3
	2	0	1	0	14	5.9
	3	0	0	1	15	8.0
	4	0	0	0	16	8.4

Forecast for Time Period 19 (Quarter 3 in Year 5)

$$\text{Sales} = 6.07 - 1.36(0) - 2.03(0) - .304(1) + .146(19) = 8.54$$

Forecast for Time Period 20 (Quarter 4 in Year 5)

$$\text{Sales} = 6.07 - 1.36(0) - 2.03(0) - .304(0) + .146(20) = 8.99$$

Thus, accounting for the seasonal effects and the linear trend in television set sales, the estimates of quarterly sales in year 5 are 7190, 6670, 8540, and 8990.

The dummy variables in the estimated multiple regression equation actually provide four estimated multiple regression equations, one for each quarter. For instance, if time period t corresponds to quarter 1, the estimate of quarterly sales is

$$\text{Quarter 1: Sales} = 6.07 - 1.36(1) - 2.03(0) - .304(0) + .146t = 4.71 + .146t$$

FIGURE 17.19 MINITAB REGRESSION OUTPUT FOR THE UMBRELLA SALES TIME SERIES

```
The regression equation is
Sales (1000s) = 6.07 - 1.36 Qtr1 - 2.03 Qtr2 - 0.304
                Qtr3 + 0.146 Period

Predictor      Coef  SE Coef       T      P
Constant     6.0688   0.1625   37.35  0.000
Qtr1        -1.3631   0.1575   -8.66  0.000
Qtr2        -2.0337   0.1551  -13.11  0.000
Qtr3        -0.3044   0.1537   -1.98  0.073
Period      0.14562  0.01211   12.02  0.000
```

Similarly, if time period t corresponds to quarters 2, 3, and 4, the estimates of quarterly sales are

$$\text{Quarter 2: Sales} = 6.07 - 1.36(0) - 2.03(1) - .304(0) + .146t = 4.04 + .146t$$
$$\text{Quarter 3: Sales} = 6.07 - 1.36(0) - 2.03(0) - .304(1) + .146t = 5.77 + .146t$$
$$\text{Quarter 4: Sales} = 6.07 - 1.36(0) - 2.03(0) - .304(0) + .146t = 6.07 + .146t$$

The slope of the trend line for each quarterly forecast equation is .146, indicating a growth in sales of about 146 sets per quarter. The only difference in the four equations is that they have different intercepts. For instance, the intercept for the quarter 1 equation is 4.71 and the intercept for the quarter 4 equation is 6.07. Thus, sales in quarter 1 are $4.71 - 6.07 = -1.36$ or 1360 sets less than in quarter 4. In other words, the estimated regression coefficient for Qtr1 in equation (17.12) provides an estimate of the difference in sales between quarter 1 and quarter 4. Similar interpretations can be provided for −2.03, the estimated regression coefficient for dummy variable Qtr2, and −.304, the estimated regression coefficient for dummy variable Qtr3.

Models Based on Monthly Data

Whenever a categorical variable such as season has k levels, k − 1 dummy variables are required.

In the preceding television set sales example, we showed how dummy variables can be used to account for the quarterly seasonal effects in the time series. Because there were 4 levels for the categorical variable season, 3 dummy variables were required. However, many businesses use monthly rather than quarterly forecasts. For monthly data, season is a categorical variable with 12 levels and thus $12 - 1 = 11$ dummy variables are required. For example, the 11 dummy variables could be coded as follows:

$$\text{Month1} = \begin{cases} 1 \text{ if January} \\ 0 \text{ otherwise} \end{cases}$$
$$\text{Month2} = \begin{cases} 1 \text{ if February} \\ 0 \text{ otherwise} \end{cases}$$

.

.

.

$$\text{Month11} = \begin{cases} 1 \text{ if November} \\ 0 \text{ otherwise} \end{cases}$$

Other than this change, the multiple regression approach for handling seasonality remains the same.

Exercises

Methods

SELF test

28. Consider the following time series.

Quarter	Year 1	Year 2	Year 3
1	71	68	62
2	49	41	51
3	58	60	53
4	78	81	72

a. Construct a time series plot. What type of pattern exists in the data?
b. Use the following dummy variables to develop an estimated regression equation to account for seasonal effects in the data: Qtr1 = 1 if Quarter 1, 0 otherwise; Qtr2 = 1 if Quarter 2, 0 otherwise; Qtr3 = 1 if Quarter 3, 0 otherwise.
c. Compute the quarterly forecasts for next year.

29. Consider the following time series data.

Quarter	Year 1	Year 2	Year 3
1	4	6	7
2	2	3	6
3	3	5	6
4	5	7	8

a. Construct a time series plot. What type of pattern exists in the data?
b. Use the following dummy variables to develop an estimated regression equation to account for any seasonal and linear trend effects in the data: Qtr1 = 1 if Quarter 1, 0 otherwise; Qtr2 = 1 if Quarter 2, 0 otherwise; Qtr3 = 1 if Quarter 3, 0 otherwise.
c. Compute the quarterly forecasts for next year.

Applications

30. The quarterly sales data (number of copies sold) for a college textbook over the past three years follow.

WEB file

Textbooks

Quarter	Year 1	Year 2	Year 3
1	1690	1800	1850
2	940	900	1100
3	2625	2900	2930
4	2500	2360	2615

a. Construct a time series plot. What type of pattern exists in the data?
b. Use the following dummy variables to develop an estimated regression equation to account for any seasonal effects in the data: Qtr1 = 1 if Quarter 1, 0 otherwise; Qtr2 = 1 if Quarter 2, 0 otherwise; Qtr3 = 1 if Quarter 3, 0 otherwise.
c. Compute the quarterly forecasts for next year.
d. Let $t = 1$ to refer to the observation in quarter 1 of year 1; $t = 2$ to refer to the observation in quarter 2 of year 1; . . . and $t = 12$ to refer to the observation in quarter 4 of year 3. Using the dummy variables defined in part (b) and t, develop an estimated regression equation to account for seasonal effects and any linear trend in the time series. Based upon the seasonal effects in the data and linear trend, compute the quarterly forecasts for next year.

31. Air pollution control specialists in southern California monitor the amount of ozone, carbon dioxide, and nitrogen dioxide in the air on an hourly basis. The hourly time series data exhibit seasonality, with the levels of pollutants showing patterns that vary over the hours in the day. On July 15, 16, and 17, the following levels of nitrogen dioxide were observed for the 12 hours from 6:00 A.M. to 6:00 P.M.

WEB file

Pollution

July 15:	25	28	35	50	60	60	40	35	30	25	25	20
July 16:	28	30	35	48	60	65	50	40	35	25	20	20
July 17:	35	42	45	70	72	75	60	45	40	25	25	25

a. Construct a time series plot. What type of pattern exists in the data?
b. Use the following dummy variables to develop an estimated regression equation to account for the seasonal effects in the data.

Hour1 = 1 if the reading was made between 6:00 A.M. and 7:00 A.M.; 0 otherwise
Hour2 = 1 if if the reading was made between 7:00 A.M. and 8:00 A.M.; 0 otherwise
.
.
.
Hour11 = 1 if the reading was made between 4:00 P.M. and 5:00 P.M., 0 otherwise.

Note that when the values of the 11 dummy variables are equal to 0, the observation corresponds to the 5:00 P.M. to 6:00 P.M. hour.

c. Using the estimated regression equation developed in part (a), compute estimates of the levels of nitrogen dioxide for July 18.
d. Let $t = 1$ to refer to the observation in hour 1 on July 15; $t = 2$ to refer to the observation in hour 2 of July 15; . . . and $t = 36$ to refer to the observation in hour 12 of July 17. Using the dummy variables defined in part (b) and t, develop an estimated regression equation to account for seasonal effects and any linear trend in the time series. Based upon the seasonal effects in the data and linear trend, compute estimates of the levels of nitrogen dioxide for July 18.

32. South Shore Construction builds permanent docks and seawalls along the southern shore of Long Island, New York. Although the firm has been in business only five years, revenue has increased from $308,000 in the first year of operation to $1,084,000 in the most recent year. The following data show the quarterly sales revenue in thousands of dollars.

WEB file

SouthShore

Quarter	Year 1	Year 2	Year 3	Year 4	Year 5
1	20	37	75	92	176
2	100	136	155	202	282
3	175	245	326	384	445
4	13	26	48	82	181

a. Construct a time series plot. What type of pattern exists in the data?
b. Use the following dummy variables to develop an estimated regression equation to account for seasonal effects in the data. Qtr1 = 1 if Quarter 1, 0 otherwise; Qtr2 = 1 if Quarter 2, 0 otherwise; Qtr3 = 1 if Quarter 3, 0 otherwise. Based only on the seasonal effects in the data, compute estimates of quarterly sales for year 6.
c. Let Period = 1 to refer to the observation in quarter 1 of year 1; Period = 2 to refer to the observation in quarter 2 of year 1; . . . and Period = 20 to refer to the observation in quarter 4 of year 5. Using the dummy variables defined in part (b) and Period, develop an estimated regression equation to account for seasonal effects and any linear trend in the time series. Based upon the seasonal effects in the data and linear trend, compute estimates of quarterly sales for year 6.

33. Electric power consumption is measured in kilowatt-hours (kWh). The local utility company offers an interrupt program whereby commercial customers that participate receive favorable rates but must agree to cut back consumption if the utility requests them to do so. Timko Products has agreed to cut back consumption from noon to 8:00 P.M. on Thursday. To determine Timko's savings, the utility must estimate Timko's normal power usage for this period of time. Data on Timko's electric power consumption for the previous 72 hours are shown below.

WEB file

Power

Time Period	Monday	Tuesday	Wednesday	Thursday
12–4 A.M.	—	19,281	31,209	27,330
4–8 A.M	—	33,195	37,014	32,715
8–12 noon	—	99,516	119,968	152,465
12–4 P.M.	124,299	123,666	156,033	
4–8 P.M.	113,545	111,717	128,889	
8–12 midnight	41,300	48,112	73,923	

a. Construct a time series plot. What type of pattern exists in the data?
b. Use the following dummy variables to develop an estimated regression equation to account for any seasonal effects in the data.

Time1 = 1 for time period 12–4 A.M.; 0 otherwise
Time2 = 1 for time period 4–8 A.M; 0 otherwise
Time3 = 1 for time period 8–12 noon; 0 otherwise
Time4 = 1 for time period 12–4 P.M; 0 otherwise
Time5 = 1 for time period 4–8 P.M; 0 otherwise

c. Use the estimated regression equation developed in part (b) to estimate Timko's normal usage over the period of interrupted service.
d. Let Period = 1 to refer to the observation for Monday in the time period 12–4 P.M.; Period = 2 to refer to the observation for Monday in the time period 4–8 P.M; . . . and Period = 18 to refer to the observation for Thursday in the time period 8–12 noon. Using the dummy variables defined in part (b) and Period, develop an estimated regression equation to account for seasonal effects and any linear trend in the time series.
e. Using the estimated regression equation developed in part (d), estimate Timko's normal usage over the period of interrupted service.

34. Three years of monthly lawn-maintenance expenses ($) for a six-unit apartment house in southern Florida follow.

WEB file

AptExp

Month	Year 1	Year 2	Year 3
January	170	180	195
February	180	205	210
March	205	215	230
April	230	245	280
May	240	265	290
June	315	330	390
July	360	400	420
August	290	335	330
September	240	260	290
October	240	270	295
November	230	255	280
December	195	220	250

a. Construct a time series plot. What type of pattern exists in the data?
b. Develop an estimated regression equation that can be used to account for any seasonal and linear trend effects in the data. Use the following dummy variables to account for the seasonal effects in the data: Jan = 1 if January, 0 otherwise; Feb = 1 if February, 0 otherwise; Mar = 1 if March, 0 otherwise; . . . Nov = 1 if November, 0 otherwise. Note that using this coding method, when all the 11 dummy variables are 0, the observation corresponds to an expense in December.
c. Compute the monthly forecasts for next year based upon both trend and seasonal effects.

17.6 Time Series Decomposition

In this section we turn our attention to what is called **time series decomposition**. Time series decomposition can be used to separate or decompose a time series into seasonal, trend, and irregular components. While this method can be used for forecasting, its primary applicability is to get a better understanding of the time series. Many business and economic time series are maintained and published by government agencies such as the Census Bureau and the Bureau of Labor Statistics. These agencies use time series decomposition to create deseasonalized time series.

Understanding what is really going on with a time series often depends upon the use of deseasonalized data. For instance, we might be interested in learning whether electrical power consumption is increasing in our area. Suppose we learn that electric power consumption in September is down 3% from the previous month. Care must be exercised in using such information, because whenever a seasonal influence is present, such comparisons may be misleading if the data have not been deseasonalized. The fact that electric power consumption is down 3% from August to September might be only the seasonal effect associated with a decrease in the use of air conditioning and not because of a long-term decline in the use of electric power. Indeed, after adjusting for the seasonal effect, we might even find that the use of electric power increased. Many other time series, such as unemployment statistics, home sales, and retail sales, are subject to strong seasonal influences. It is important to deseasonalize such data before making a judgment about any long-term trend.

Time series decomposition methods assume that Y_t, the actual time series value at period t, is a function of three components: a trend component; a seasonal component; and an irregular or error component. How these three components are combined to generate the observed values of the time series depends upon whether we assume the relationship is best described by an additive or a multiplicative model.

An **additive decomposition model** takes the following form:

$$Y_t = \text{Trend}_t + \text{Seasonal}_t + \text{Irregular}_t \tag{17.13}$$

where

$$\begin{aligned} \text{Trend}_t &= \text{trend value at time period } t \\ \text{Seasonal}_t &= \text{seasonal value at time period } t \\ \text{Irregular}_t &= \text{irregular value at time period } t \end{aligned}$$

The irregular component corresponds to the error term ε in the simple linear regression model we discussed in Chapter 14.

In an additive model the values for the three components are simply added together to obtain the actual time series value Y_t. The irregular or error component accounts for the variability in the time series that cannot be explained by the trend and seasonal components.

An additive model is appropriate in situations where the seasonal fluctuations do not depend upon the level of the time series. The regression model for incorporating seasonal and trend effects in Section 17.5 is an additive model. If the sizes of the seasonal fluctuations in earlier time periods are about the same as the sizes of the seasonal fluctuations in later time periods, an additive model is appropriate. However, if the seasonal fluctuations change over time, growing larger as the sales volume increases because of a long-term linear trend, then a multiplicative model should be used. Many business and economic time series follow this pattern.

A **multiplicative decomposition model** takes the following form:

$$Y_t = \text{Trend}_t \times \text{Seasonal}_t \times \text{Irregular}_t \tag{17.14}$$

where

$$\text{Trend}_t = \text{trend value at time period } t$$
$$\text{Seasonal}_t = \text{seasonal index at time period } t$$
$$\text{Irregular}_t = \text{irregular index at time period } t$$

The Census Bureau uses a multiplicative model in conjunction with its methodology for deseasonalizing time series.

In this model, the trend and seasonal and irregular components are multiplied to give the value of the time series. Trend is measured in units of the item being forecast. However, the seasonal and irregular components are measured in relative terms, with values above 1.00 indicating effects above the trend and values below 1.00 indicating effects below the trend.

Because this is the method most often used in practice, we will restrict our discussion of time series decomposition to showing how to develop estimates of the trend and seasonal components for a multiplicative model. As an illustration we will work with the quarterly television set sales time series introduced in Section 17.5; the quarterly sales data are shown in Table 17.19 and the corresponding time series plot is presented in Figure 17.18. After demonstrating how to decompose a time series using the multiplicative model, we will show how the seasonal indices and trend component can be recombined to develop a forecast.

Calculating the Seasonal Indexes

Figure 17.18 indicates that sales are lowest in the second quarter of each year and increase in quarters 3 and 4. Thus, we conclude that a seasonal pattern exists for the television set sales time series. The computational procedure used to identify each quarter's seasonal influence begins by computing a moving average to remove the combined seasonal and irregular effects from the data, leaving us with a time series that contains only trend and any remaining random variation not removed by the moving average calculations.

Because we are working with a quarterly series, we will use four data values in each moving average. The moving average calculation for the first four quarters of the television set sales data is

$$\text{First moving average} = \frac{4.8 + 4.1 + 6.0 + 6.5}{4} = \frac{21.4}{4} = 5.35$$

Note that the moving average calculation for the first four quarters yields the average quarterly sales over year 1 of the time series. Continuing the moving average calculations, we next add the 5.8 value for the first quarter of year 2 and drop the 4.8 for the first quarter of year 1. Thus, the second moving average is

$$\text{Second moving average} = \frac{4.1 + 6.0 + 6.5 + 5.8}{4} = \frac{22.4}{4} = 5.60$$

Similarly, the third moving average calculation is $(6.0 + 6.5 + 5.8 + 5.2)/4 = 5.875$.

Before we proceed with the moving average calculations for the entire time series, let us return to the first moving average calculation, which resulted in a value of 5.35. The 5.35 value is the average quarterly sales volume for year 1. As we look back at the calculation of the 5.35 value, associating 5.35 with the "middle" of the moving average group makes sense. Note, however, that with four quarters in the moving average, there is no middle period. The 5.35 value really corresponds to period 2.5, the last half of quarter 2 and the first half of quarter 3. Similarly, if we go to the next moving average value of 5.60, the middle period corresponds to period 3.5, the last half of quarter 3 and the first half of quarter 4.

The two moving average values we computed do not correspond directly to the original quarters of the time series. We can resolve this difficulty by computing the average of the two moving averages. Since the center of the first moving average is period 2.5 (half a period or quarter early) and the center of the second moving average is period 3.5 (half a period or quarter late), the average of the two moving averages is centered at quarter 3, exactly where it should be. This moving average is referred to as a *centered moving average*. Thus, the centered moving average for period 3 is (5.35 + 5.60)/2 = 5.475. Similarly, the centered moving average value for period 4 is (5.60 + 5.875)/2 = 5.738. Table 17.21 shows a complete summary of the moving average and centered moving average calculations for the television set sales data.

What do the centered moving averages in Table 17.21 tell us about this time series? Figure 17.20 shows a time series plot of the actual time series values and the centered moving average values. Note particularly how the centered moving average values tend to "smooth out" both the seasonal and irregular fluctuations in the time series. The centered moving averages represent the trend in the data and any random variation that was not removed by using moving averages to smooth the data.

Previously we showed that the multiplicative decomposition model is

$$Y_t = \text{Trend}_t \times \text{Seasonal}_t \times \text{Irregular}_t$$

TABLE 17.21 CENTERED MOVING AVERAGE CALCULATIONS FOR THE TELEVISION SET SALES TIME SERIES

Year	Quarter	Sales (1000s)	Four-Quarter Moving Average	Centered Moving Average
1	1	4.8		
1	2	4.1		
			5.350	
1	3	6.0		5.475
			5.600	
1	4	6.5		5.738
			5.875	
2	1	5.8		5.975
			6.075	
2	2	5.2		6.188
			6.300	
2	3	6.8		6.325
			6.350	
2	4	7.4		6.400
			6.450	
3	1	6.0		6.538
			6.625	
3	2	5.6		6.675
			6.725	
3	3	7.5		6.763
			6.800	
3	4	7.8		6.838
			6.875	
4	1	6.3		6.938
			7.000	
4	2	5.9		7.075
			7.150	
4	3	8.0		
4	4	8.4		

FIGURE 17.20 QUARTERLY TELEVISION SET SALES TIME SERIES AND CENTERED MOVING AVERAGE

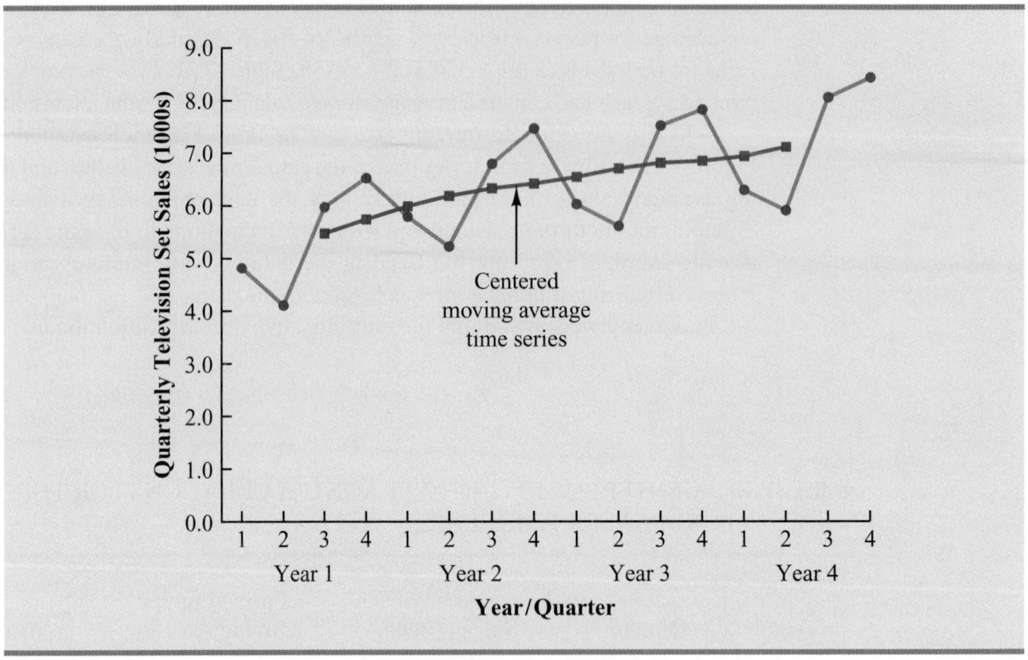

By dividing each side of this equation by the trend component T_t, we can identify the combined seasonal-irregular effect in the time series.

The seasonal-irregular values are often referred to as the de-trended values of the time series.

$$\frac{Y_t}{\text{Trend}_t} = \frac{\text{Trend}_t \times \text{Seasonal}_t \times \text{Irregular}_t}{\text{Trend}_t} = \text{Seasonal}_t \times \text{Irregular}_t$$

For example, the third quarter of year 1 shows a trend value of 5.475 (the centered moving average). So 6.0/5.475 = 1.096 is the combined seasonal-irregular value. Table 17.22 summarizes the seasonal-irregular values for the entire time series.

Consider the seasonal-irregular values for the third quarter: 1.096, 1.075, and 1.109. Seasonal-irregular values greater than 1.00 indicate effects above the trend estimate and values below 1.00 indicate effects below the trend estimate. Thus, the three seasonal-irregular values for quarter 3 show an above-average effect in the third quarter. Since the year-to-year fluctuations in the seasonal-irregular values are primarily due to random error, we can average the computed values to eliminate the irregular influence and obtain an estimate of the third-quarter seasonal influence.

$$\text{Seasonal effect of quarter 3} = \frac{1.096 + 1.075 + 1.109}{3} = 1.09$$

We refer to 1.09 as the *seasonal index* for the third quarter. Table 17.23 summarizes the calculations involved in computing the seasonal indexes for the television set sales time series. The seasonal indexes for the four quarters are .93, .84, 1.09, and 1.14.

Interpretation of the seasonal indexes in Table 17.23 provides some insight about the seasonal component in television set sales. The best sales quarter is the fourth quarter, with sales averaging 14% above the trend estimate. The worst, or slowest, sales quarter is the second quarter; its seasonal index of .84 shows that the sales average is 16% below the trend estimate. The seasonal component corresponds clearly to the intuitive expectation that television viewing interest and thus television purchase patterns tend to peak in the

TABLE 17.22 SEASONAL IRREGULAR VALUES FOR THE TELEVISION SET SALES TIME SERIES

Year	Quarter	Sales (1000s)	Centered Moving Average	Seasonal-Irregular Value
1	1	4.8		
1	2	4.1		
1	3	6.0	5.475	1.096
1	4	6.5	5.738	1.133
2	1	5.8	5.975	0.971
2	2	5.2	6.188	0.840
2	3	6.8	6.325	1.075
2	4	7.4	6.400	1.156
3	1	6.0	6.538	0.918
3	2	5.6	6.675	0.839
3	3	7.5	6.763	1.109
3	4	7.8	6.838	1.141
4	1	6.3	6.938	0.908
4	2	5.9	7.075	0.834
4	3	8.0		
4	4	8.4		

TABLE 17.23 SEASONAL INDEX CALCULATIONS FOR THE TELEVISION SET SALES TIME SERIES

Quarter	Seasonal-Irregular Values			Seasonal Index
1	0.971	0.918	0.908	0.93
2	0.840	0.839	0.834	0.84
3	1.096	1.075	1.109	1.09
4	1.133	1.156	1.141	1.14

fourth quarter because of the coming winter season and reduction in outdoor activities. The low second-quarter sales reflect the reduced interest in television viewing due to the spring and presummer activities of potential customers.

One final adjustment is sometimes necessary in obtaining the seasonal indexes. Because the multiplicative model requires that the average seasonal index equal 1.00, the sum of the four seasonal indexes in Table 17.23 must equal 4.00. In other words, the seasonal effects must even out over the year. The average of the seasonal indexes in our example is equal to 1.00, and hence this type of adjustment is not necessary. In other cases, a slight adjustment may be necessary. To make the adjustment, multiply each seasonal index by the number of seasons divided by the sum of the unadjusted seasonal indexes. For instance, for quarterly data, multiply each seasonal index by 4/(sum of the unadjusted seasonal indexes). Some of the exercises will require this adjustment to obtain the appropriate seasonal indexes.

Deseasonalizing the Time Series

A time series that has had the seasonal effects removed is referred to as a **deseasonalized time series**, and the process of using the seasonal indexes to remove the seasonal effects from a time series is referred to as deseasonalizing the time series. Using a multiplicative

TABLE 17.24 DESEASONALIZED VALUES FOR THE TELEVISION SET SALES TIME SERIES

Year	Quarter	Time Period	Sales (1000s)	Seasonal Index	Deseasonalized Sales
1	1	1	4.8	0.93	5.16
	2	2	4.1	0.84	4.88
	3	3	6.0	1.09	5.50
	4	4	6.5	1.14	5.70
2	1	5	5.8	0.93	6.24
	2	6	5.2	0.84	6.19
	3	7	6.8	1.09	6.24
	4	8	7.4	1.14	6.49
3	1	9	6.0	0.93	6.45
	2	10	5.6	0.84	6.67
	3	11	7.5	1.09	6.88
	4	12	7.8	1.14	6.84
4	1	13	6.3	0.93	6.77
	2	14	5.9	0.84	7.02
	3	15	8.0	1.09	7.34
	4	16	8.4	1.14	7.37

decomposition model, we deseasonalize a time series by dividing each observation by its corresponding seasonal index. The multiplicative decomposition model is

Economic time series adjusted for seasonal variations are often reported in publications such as the Survey of Current Business, The Wall Street Journal, *and* BusinessWeek.

$$Y_t = \text{Trend}_t \times \text{Seasonal}_t \times \text{Irregular}_t$$

So, when we divide each time series observation (Y_t) by its corresponding seasonal index, the resulting data show only trend and random variability (the irregular component). The deseasonalized time series for television set sales is summarized in Table 17.24. A graph of the deseasonalized time series is shown in Figure 17.21.

FIGURE 17.21 DESEASONALIZED TELEVISION SET SALES TIME SERIES

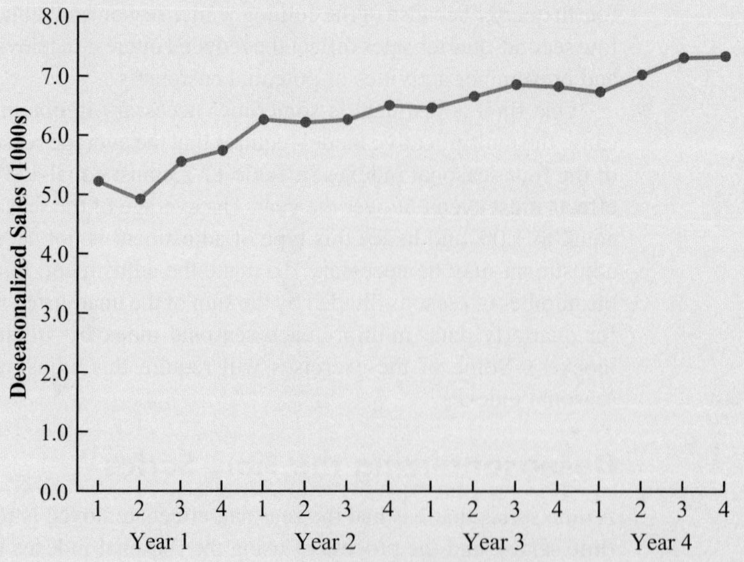

Using the Deseasonalized Time Series to Identify Trend

The graph of the deseasonalized television set sales time series shown in Figure 17.21 appears to have an upward linear trend. To identify this trend, we will fit a linear trend equation to the deseasonalized time series using the same method shown in Section 17.4. The only difference is that we will be fitting a trend line to the deseasonalized data instead of the original data.

Recall that for a linear trend the estimated regression equation can be written as

$$T_t = b_0 + b_1 t$$

where

$$\begin{aligned} T_t &= \text{linear trend forecast in period } t \\ b_0 &= \text{intercept of the linear trend line} \\ b_1 &= \text{slope of the trend line} \\ t &= \text{time period} \end{aligned}$$

In Section 17.4 we provided formulas for computing the values of b_0 and b_1. To fit a linear trend line to the deseasonalized data in Table 17.24, the only change is that the deseasonalized time series values are used instead of the observed values Y_t in computing b_0 and b_1.

Figure 17.22 shows the computer output obtained using Minitab's regression analysis procedure to estimate the trend line for the deseasonalized television set time series. The estimated linear trend equation is

$$\text{Deseasonalized Sales} = 5.10 + 0.148t$$

The slope of 0.148 indicates that over the past 16 quarters, the firm averaged a deseasonalized growth in sales of about 148 sets per quarter. If we assume that the past 16-quarter trend in sales data is a reasonably good indicator of the future, this equation can be used to develop a trend projection for future quarters. For example, substituting $t = 17$ into the equation yields next quarter's deseasonalized trend projection, T_{17}.

$$T_{17} = 5.10 + 0.148(17) = 7.616$$

FIGURE 17.22 MINITAB REGRESSION OUTPUT FOR THE DESEASONALIZED TELEVISION SET SALES TIME SERIES

```
The regression equation is
Deseasonalized Sales = 5.10 + 0.148 Period

Predictor      Coef  SE Coef      T      P
Constant     5.1050   0.1133  45.07  0.000
Period      0.14760  0.01171  12.60  0.000

S = 0.215985   R-Sq = 91.9%   R-Sq(adj) = 91.3%

Analysis of Variance

Source          DF      SS      MS       F      P
Regression       1  7.4068  7.4068  158.78  0.000
Residual Error  14  0.6531  0.0466
Total           15  8.0599
```

TABLE 17.25 QUARTERLY FORECASTS FOR THE TELEVISION SET SALES TIME SERIES

Year	Quarter	Deseasonalized Trend Forecast	Seasonal Index	Quarterly Forecast
5	1	7616	0.93	(7616)(0.93) = 7083
	2	7764	0.84	(7764)(0.84) = 6522
	3	7912	1.09	(7912)(1.09) = 8624
	4	8060	1.14	(8060)(1.14) = 9188

Thus, using the deseasonalized data, the linear trend forecast for next quarter (period 17) is 7616 television sets. Similarly, the deseasonalized trend forecasts for the next three quarters (periods 18, 19, and 20) are 7764, 7912, and 8060 television sets, respectively.

Seasonal Adjustments

The final step in developing the forecast when both trend and seasonal components are present is to use the seasonal indexes to adjust the deseasonalized trend projections. Returning to the television set sales example, we have a deseasonalized trend projection for the next four quarters. Now we must adjust the forecast for the seasonal effect. The seasonal index for the first quarter of year 5 ($t = 17$) is 0.93, so we obtain the quarterly forecast by multiplying the deseasonalized forecast based on trend ($T_{17} = 7616$) by the seasonal index (0.93). Thus, the forecast for the next quarter is 7616(0.93) = 7083. Table 17.25 shows the quarterly forecast for quarters 17 through 20. The high-volume fourth quarter has a 9188-unit forecast, and the low-volume second quarter has a 6522-unit forecast.

Models Based on Monthly Data

In the preceding television set sales example, we used quarterly data to illustrate the computation of seasonal indexes. However, many businesses use monthly rather than quarterly forecasts. In such cases, the procedures introduced in this section can be applied with minor modifications. First, a 12-month moving average replaces the four-quarter moving average; second, 12 monthly seasonal indexes, rather than four quarterly seasonal indexes, must be computed. Other than these changes, the computational and forecasting procedures are identical.

Cyclical Component

Mathematically, the multiplicative model of equation (17.14) can be expanded to include a cyclical component.

$$Y_t = \text{Trend}_t \times \text{Cyclical}_t \times \text{Seasonal}_t \times \text{Irregular}_t \quad (17.15)$$

The cyclical component, like the seasonal component, is expressed as a percentage of trend. As mentioned in Section 17.1, this component is attributable to multiyear cycles in the time series. It is analogous to the seasonal component, but over a longer period of time. However, because of the length of time involved, obtaining enough relevant data to estimate the cyclical component is often difficult. Another difficulty is that cycles usually vary in length. Because it is so difficult to identify and/or separate cyclical effects from long-term trend effects, in practice these effects are often combined and referred to as a combined trend-cycle component. We leave further discussion of the cyclical component to specialized texts on forecasting methods.

NOTES AND COMMENTS

1. There are a number of different approaches to computing the seasonal indexes. In this section each seasonal index was computed by averaging the corresponding seasonal-irregular values. Another approach, and the one used by Minitab, is to use the median of the seasonal-irregular values as the seasonal index.
2. Calendar adjustments are often made before deseasonalizing a time series. For example, if a time series consists of monthly sales values, the value for February sales may be less than for another month simply because there are fewer days in February. To account for this factor, we would first divide each month's sales value by the number of days in the month to obtain a daily average. Since the average number of days in a month is approximately 365/12 = 30.4167, we then multiply the daily averages by 30.4167 to obtain adjusted monthly values. For the examples and exercises in this chapter, you can assume that any required calendar adjustments have already been made.

Exercises

Methods

SELF test

35. Consider the following time series data.

Quarter	Year 1	Year 2	Year 3
1	4	6	7
2	2	3	6
3	3	5	6
4	5	7	8

   a. Construct a time series plot. What type of pattern exists in the data?
   b. Show the four-quarter and centered moving average values for this time series.
   c. Compute seasonal indexes and adjusted seasonal indexes for the four quarters.

36. Refer to exercise 35.
   a. Deseasonalize the time series using the adjusted seasonal indexes computed in part (c) of exercise 35.
   b. Using Minitab or Excel, compute the linear trend regression equation for the deseasonalized data.
   c. Compute the deseasonalized quarterly trend forecast for Year 4.
   d. Use the seasonal indexes to adjust the deseasonalized trend forecasts computed in part (c).

Applications

37. The quarterly sales data (number of copies sold) for a college textbook over the past three years follow.

WEB file
TextSales

Quarter	Year 1	Year 2	Year 3
1	1690	1800	1850
2	940	900	1100
3	2625	2900	2930
4	2500	2360	2615

a. Construct a time series plot. What type of pattern exists in the data?
b. Show the four-quarter and centered moving average values for this time series.
c. Compute the seasonal and adjusted seasonal indexes for the four quarters.
d. When does the publisher have the largest seasonal index? Does this result appear reasonable? Explain.
e. Deseasonalize the time series.
f. Compute the linear trend equation for the deseasonalized data and forecast sales using the linear trend equation.
g. Adjust the linear trend forecasts using the adjusted seasonal indexes computed in part (c).

38. Three years of monthly lawn-maintenance expenses ($) for a six-unit apartment house in southern Florida follow.

WEB file
AptExp

Month	Year 1	Year 2	Year 3
January	170	180	195
February	180	205	210
March	205	215	230
April	230	245	280
May	240	265	290
June	315	330	390
July	360	400	420
August	290	335	330
September	240	260	290
October	240	270	295
November	230	255	280
December	195	220	250

a. Construct a time series plot. What type of pattern exists in the data?
b. Identify the monthly seasonal indexes for the three years of lawn-maintenance expenses for the apartment house in southern Florida as given here. Use a 12-month moving average calculation.
c. Deseasonalize the time series.
d. Compute the linear trend equation for the deseasonalized data.
e. Compute the deseasonalized trend forecasts and then adjust the trend forecasts using the seasonal indexes to provide a forecast for monthly expenses in year 4.

39. Air pollution control specialists in southern California monitor the amount of ozone, carbon dioxide, and nitrogen dioxide in the air on an hourly basis. The hourly time series data exhibit seasonality, with the levels of pollutants showing patterns over the hours in the day. On July 15, 16, and 17, the following levels of nitrogen dioxide were observed in the downtown area for the 12 hours from 6:00 A.M. to 6:00 P.M.

WEB file
Pollution

July 15:	25	28	35	50	60	60	40	35	30	25	25	20
July 16:	28	30	35	48	60	65	50	40	35	25	20	20
July 17:	35	42	45	70	72	75	60	45	40	25	25	25

a. Construct a time series plot. What type of pattern exists in the data?
b. Identify the hourly seasonal indexes for the 12 readings each day.
c. Deseasonalize the time series.
d. Using Minitab or Excel, compute the linear trend equation for the deseasonalized data.
e. Compute the deseasonalized trend forecasts for the 12 hours for July 18 and then adjust the trend forecasts using the seasonal indexes computed in part (b).

40. Electric power consumption is measured in kilowatt-hours (kWh). The local utility company offers an interrupt program whereby commercial customers that participate receive favorable rates but must agree to cut back consumption if the utility requests them to do so. Timko

Products cut back consumption at 12:00 noon Thursday. To assess the savings, the utility must estimate Timko's usage without the interrupt. The period of interrupted service was from noon to 8:00 P.M. Data on electric power consumption for the previous 72 hours are available.

WEB file

Power

Time Period	Monday	Tuesday	Wednesday	Thursday
12–4 A.M.	—	19,281	31,209	27,330
4–8 A.M	—	33,195	37,014	32,715
8–12 noon	—	99,516	119,968	152,465
12–4 P.M.	124,299	123,666	156,033	
4–8 P.M.	113,545	111,717	128,889	
8–12 midnight	41,300	48,112	73,923	

a. Is there a seasonal effect over the 24-hour period?
b. Compute seasonal indexes for the six 4-hour periods.
c. Use trend adjusted for seasonal indexes to estimate Timko's normal usage over the period of interrupted service.

Summary

This chapter provided an introduction to the basic methods of time series analysis and forecasting. First, we showed that the underlying pattern in the time series can often be identified by constructing a time series plot. Several types of data patterns can be distinguished, including a horizontal pattern, a trend pattern, and a seasonal pattern. The forecasting methods we have discussed are based on which of these patterns are present in the time series.

For a time series with a horizontal pattern, we showed how moving averages and exponential smoothing can be used to develop a forecast. The moving averages method consists of computing an average of past data values and then using that average as the forecast for the next period. In the exponential smoothing method, a weighted average of past time series values is used to compute a forecast. These methods also adapt well when a horizontal pattern shifts to a different level and resumes a horizontal pattern.

An important factor in determining what forecasting method to use involves the accuracy of the method. We discussed three measures of forecast accuracy: mean absolute error (MAE), mean squared error (MSE), and mean absolute percentage error (MAPE). Each of these measures is designed to determine how well a particular forecasting method is able to reproduce the time series data that are already available. By selecting a method that has the best accuracy for the data already known, we hope to increase the likelihood that we will obtain better forecasts for future time periods.

For time series that have only a long-term linear trend, we showed how simple time series regression can be used to make trend projections. We also discussed how an extension of single exponential smoothing, referred to as Holt's linear exponential smoothing, can be used to forecast a time series with a linear trend. For a time series with a curvilinear or nonlinear trend, we showed how multiple regression can be used to fit a quadratic trend equation or an exponential trend equation to the data.

For a time series with a seasonal pattern, we showed how the use of dummy variables in a multiple regression model can be used to develop an estimated regression equation with seasonal effects. We then extended the regression approach to include situations where the time series contains both a seasonal and a linear trend effect by showing how to combine the dummy variable approach for handling seasonality with the time series regression approach for handling linear trend.

In the last section of the chapter we showed how time series decomposition can be used to separate or decompose a time series into seasonal and trend components and then to deseasonalize the time series. We showed how to compute seasonal indexes for a multiplicative model, how to use the seasonal indexes to deseasonalize the time series, and how to use regression analysis on the deseasonalized data to estimate the trend component. The final step in developing a forecast when both trend and seasonal components are present is to use the seasonal indexes to adjust the trend projections.

Glossary

Time series A sequence of observations on a variable measured at successive points in time or over successive periods of time.
Time series plot A graphical presentation of the relationship between time and the time series variable. Time is shown on the horizontal axis and the time series values are shown on the verical axis.
Horizontal pattern A horizontal pattern exists when the data fluctuate around a constant mean.
Stationary time series A time series whose statistical properties are indepepndent of time. For a stationary time series the process generating the data has a constant mean and the variability of the time series is constant over time.
Trend pattern A trend pattern exists if the time series plot shows gradual shifts or movements to relatively higher or lower values over a longer period of time.
Seasonal pattern A seasonal pattern exists if the time series plot exhibits a repeating pattern over successive periods. The successive periods are often one-year intervals, which is where the name seasonal pattern comes from.
Cyclical pattern A cyclical pattern exists if the time series plot shows an alternating sequence of points below and above the trend line lasting more than one year.
Forecast error The difference between the actual time series value and the forecast.
Mean absolute error (MAE) The average of the absolute values of the forecast errors.
Mean squared error (MSE) The average of the sum of squared forecast errors.
Mean absolute percentage error (MAPE) The average of the absolute values of the percentage forecast errors.
Moving averages A forecasting method that uses the average of the most recent k data values in the time series as the forecast for the next period.
Weighted moving averages A forecasting method that involves selecting a different weight for the most recent k data values values in the time series and then computing a weighted average of the values. The sum of the weights must equal one.
Exponential smoothing A forecasting method that uses a weighted average of past time series values as the forecast; it is a special case of the weighted moving averages method in which we select only one weight—the weight for the most recent observation.
Smoothing constant A parameter of the exponential smoothing model that provides the weight given to the most recent time series value in the calculation of the forecast value.
Linear exponential smoothing An extension of single exponential smoothing that uses two smoothing constants to enable forecasts to be developed for a time series with a linear trend.
Time series decompostition A time series method that is used to separate or decompose a time series into seasonal and trend components.
Additive decomposition model In an additive decomposition model the actual time series value at time period t is obtained by adding the values of a trend component, a seasonal component, and an irregular component.
Multiplicative decomposition model In a multiplicative decomposition model the actual time series value at time period t is obtained by multiplying the values of a trend component, a seasonal component, and an irregular component.
Deseasonalized time series A time series from which the effect of season has been removed by dividing each original time series observation by the corresponding seasonal index.

Key Formulas

Moving Average Forecast of Order k

$$F_{t+1} = \frac{\sum(\text{most recent } k \text{ data values})}{k} \tag{17.1}$$

Exponential Smoothing Forecast

$$F_{t+1} = \alpha Y_t + (1 - \alpha) F_t \tag{17.2}$$

Linear Trend Equation

$$T_t = b_0 + b_1 t \tag{17.4}$$

where

$$b_1 = \frac{\sum_{t=1}^{n}(t - \bar{t})(Y_t - \bar{Y})}{\sum_{t=1}^{n}(t - \bar{t})^2} \tag{17.5}$$

$$b_0 = \bar{Y} - b_1\bar{t} \tag{17.6}$$

Holt's Linear Exponential Smoothing

$$L_t = \alpha Y_t + (1 - \alpha)(L_{t-1} + b_{t-1}) \tag{17.7}$$

$$b_t = \beta(L_t - L_{t-1}) + (1 - \beta)\, b_{t-1} \tag{17.8}$$

$$F_{t+k} = L_t + b_t k \tag{17.9}$$

Quadratic Trend Equation

$$T_t = b_0 + b_1 t + b_2 t^2 \tag{17.10}$$

Exponential Trend Equation

$$T_t = b_0(b_1)^t \tag{17.11}$$

Additive Decomposition Model

$$Y_t = \text{Trend}_t + \text{Seasonal}_t + \text{Irregular}_t \tag{17.13}$$

Multiplicative Decomposition Model

$$Y_t = \text{Trend}_t \times \text{Seasonal}_t \times \text{Irregular}_t \tag{17.14}$$

Supplementary Exercises

41. The weekly demand (in cases) for a particular brand of automatic dishwasher detergent for a chain of grocery stores located in Columbus, Ohio, follows.

WEB file
Dishwasher

Week	Demand	Week	Demand
1	22	6	24
2	18	7	20
3	23	8	19
4	21	9	18
5	17	10	21

a. Construct a time series plot. What type of pattern exists in the data?
b. Use a three-week moving average to develop a forecast for week 11.
c. Use exponential smoothing with a smoothing constant of $\alpha = .2$ to develop a forecast for week 11.
d. Which of the two methods do you prefer? Why?

42. The following table reports the percentage of stocks in a portfolio for nine quarters from 2010 to 2012.

WEB file
Portfolio

Quarter	Stock %
1st—2010	29.8
2nd—2010	31.0
3rd—2010	29.9
4th—2010	30.1
1st—2011	32.2
2nd—2011	31.5
3rd—2011	32.0
4th—2011	31.9
1st—2012	30.0

a. Construct a time series plot. What type of pattern exists in the data?
b. Use exponential smoothing to forecast this time series. Consider smoothing constants of $\alpha = .2$, .3, and .4. What value of the smoothing constant provides the most accurate forecasts?
c. What is the forecast of the percentage of stocks in a typical portfolio for the second quarter of 2009?

43. United Dairies, Inc., supplies milk to several independent grocers throughout Dade County, Florida. Managers at United Dairies want to develop a forecast of the number of half-gallons of milk sold per week. Sales data for the past 12 weeks follow.

WEB file
UDFMilk

Week	Sales	Week	Sales
1	2750	7	3300
2	3100	8	3100
3	3250	9	2950
4	2800	10	3000
5	2900	11	3200
6	3050	12	3150

a. Construct a time series plot. What type of pattern exists in the data?
b. Use exponential smoothing withf $\alpha = .4$ to develop a forecast of demand for week 13.

44. To avoid a monthly service fee in an interest-bearing checking account, customers must maintain a minimum average daily balance. Bankrate's 2008 survey of 249 banks and thrifts in the top 25 metropolitan areas showed that you need to maintain an average balance of $3,462 to avoid a monthly service fee. With an average fee of $11.97 and an average interest rate of only 0.24 percent, customers with interest-bearing checking accounts

are not getting much value for basically providing the bank with a line of credit equal to the average monthly balance required to avoid the monthly service fee (Bankrate website, October 27, 2008). The following table shows the minimum average balance required to avoid paying a monthly service fee from 2001–2008.

Year	Balance ($)
2001	2435
2002	2593
2003	2258
2004	2087
2005	2294
2006	2660
2007	3317
2008	3462

a. Construct a time series plot. What type of pattern exists in the data?
b. Using Minitab or Excel, develop a linear trend equation for the time series. Compute an estimate of the average balance required to avoid a monthly service fee for 2009.
c. Using Minitab or Excel, develop a quadratic trend equation for the time series. Compute an estimate of the average balance required to avoid a monthly service fee for 2009.
d. Using MSE, which approach provides the most accurate forecasts for the historical data?
e. For these data would you recommend that the forecast for 2009 be developed using the linear trend equation or the quadratic trend equation? Explain.

45. The Garden Avenue Seven sells CDs of its musical performances. The following table reports sales (in units) for the past 18 months. The group's manager wants an accurate method for forecasting future sales.

WEB file
CDSales

Month	Sales	Month	Sales	Month	Sales
1	293	7	381	13	549
2	283	8	431	14	544
3	322	9	424	15	601
4	355	10	433	16	587
5	346	11	470	17	644
6	379	12	481	18	660

a. Construct a time series plot. What type of pattern exists in the data?
b. Use exponential smoothing with $\alpha = .3, .4$, and $.5$. Which value of α provides the most accurate forecasts?
c. Use trend projection to provide a forecast. What is the value of MSE?
d. Which method of forecasting would you recommend to the manager? Why?

46. The Mayfair Department Store in Davenport, Iowa, is trying to determine the amount of sales lost while it was shut down during July and August because of damage caused by the Mississippi River flood. Sales data for January through June follow.

Month	Sales ($1000s)	Month	Sales ($1000s)
January	185.72	April	210.36
February	167.84	May	255.57
March	205.11	June	261.19

a. Use exponential smoothing, with $\alpha = .4$, to develop a forecast for July and August. (*Hint*: Use the forecast for July as the actual sales in July in developing the August forecast.) Comment on the use of exponential smoothing for forecasts more than one period into the future.
b. Use trend projection to forecast sales for July and August.
c. Mayfair's insurance company proposed a settlement based on lost sales of $240,000 in July and August. Is this amount fair? If not, what amount would you recommend as a counteroffer?

47. Canton Supplies, Inc., is a service firm that employs approximately 100 individuals. Managers of Canton Supplies are concerned about meeting monthly cash obligations and want to develop a forecast of monthly cash requirements. Because of a recent change in operating policy, only the past seven months of data that follow are considered to be relevant.

Month	1	2	3	4	5	6	7
Cash Required ($1000s)	205	212	218	224	230	240	246

a. Construct a time series plot. What type of pattern exists in the data?
b. Use Holt's linear exponential smoothing with $\alpha = .6$ and $\beta = .4$ to forecast cash requirements for each of the next two months.
c. Using Minitab or Excel, develop a linear trend equation to forecast cash requirements for each of the next two months.
d. Would you recommend using Holt's linear exponential smoothing with $\alpha = .6$ and $\beta = .4$ to forecast cash requirements for each of the next two months or the linear trend equation? Explain.

48. The Costello Music Company has been in business for five years. During that time, sales of pianos increased from 12 units in the first year to 76 units in the most recent year. Fred Costello, the firm's owner, wants to develop a forecast of piano sales for the coming year. The historical data follow.

Year	1	2	3	4	5
Sales	12	28	34	50	76

a. Construct a time series plot. What type of pattern exists in the data?
b. Develop the linear trend equation for the time series. What is the average increase in sales that the firm has been realizing per year?
c. Forecast sales for years 6 and 7.

49. Consider the Costello Music Company problem in exercise 48. The quarterly sales data follow.

WEB file
PianoSales

Year	Quarter 1	Quarter 2	Quarter 3	Quarter 4	Total Yearly Sales
1	4	2	1	5	12
2	6	4	4	14	28
3	10	3	5	16	34
4	12	9	7	22	50
5	18	10	13	35	76

a. Use the following dummy variables to develop an estimated regression equation to account for any seasonal and linear trend effects in the data: Qtr1 = 1 if Quarter 1, 0 otherwise; Qtr2 = 1 if Quarter 2, 0 otherwise; and Qtr3 = 1 if Quarter 3, 0 otherwise.
b. Compute the quarterly forecasts for next year.

50. Refer to the Costello Music Company problem in exercise 49.
a. Using time series decomposition, compute the seasonal indexes for the four quarters.
b. When does Costello Music experience the largest seasonal effect? Does this result appear reasonable? Explain.

51. Refer to the Costello Music Company time series in exercise 49.
    a. Deseasonalize the data and use the deseasonalized time series to identify the trend.
    b. Use the results of part (a) to develop a quarterly forecast for next year based on trend.
    c. Use the seasonal indexes developed in exercise 50 to adjust the forecasts developed in part (b) to account for the effect of season.

52. Hudson Marine has been an authorized dealer for C&D marine radios for the past seven years. The following table reports the number of radios sold each year.

Year	1	2	3	4	5	6	7
Number Sold	35	50	75	90	105	110	130

    a. Construct a time series plot. Does a linear trend appear to be present?
    b. Using Minitab or Excel, develop a linear trend equation for this time series.
    c. Use the linear trend equation developed in part (b) to develop a forecast for annual sales in year 8.

53. Refer to the Hudson Marine problem in exercise 52. Suppose the quarterly sales values for the seven years of historical data are as follows.

WEB file

HudsonMarine

Year	Quarter 1	Quarter 2	Quarter 3	Quarter 4	Total Yearly Sales
1	6	15	10	4	35
2	10	18	15	7	50
3	14	26	23	12	75
4	19	28	25	18	90
5	22	34	28	21	105
6	24	36	30	20	110
7	28	40	35	27	130

    a. Use the following dummy variables to develop an estimated regression equation to account for any season and linear trend effects in the data: Qtr1 = 1 if Quarter 1, 0 otherwise; Qtr2 = 1 if Quarter 2, 0 otherwise; and Qtr3 = 1 if Quarter 3, 0 otherwise.
    b. Compute the quarterly forecasts for next year.

54. Refer to the Hudson Marine problem in exercise 53.
    a. Compute the centered moving average values for this time series.
    b. Construct a time series plot that also shows the centered moving average and original time series on the same graph. Discuss the differences between the original time series plot and the centered moving average time series.
    c. Compute the seasonal indexes for the four quarters.
    d. When does Hudson Marine experience the largest seasonal effect? Does this result seem reasonable? Explain.

55. Refer to the Hudson Marine data in exercise 53.
    a. Deseasonalize the data and use the deseasonalized time series to identify the trend.
    b. Use the results of part (a) to develop a quarterly forecast for next year based on trend.
    c. Use the seasonal indexes developed in exercise 54 to adjust the forecasts developed in part (b) to account for the effect of season.

Case Problem 1 Forecasting Food and Beverage Sales

The Vintage Restaurant, on Captiva Island near Fort Myers, Florida, is owned and operated by Karen Payne. The restaurant just completed its third year of operation. Since opening her restaurant, Karen has sought to establish a reputation for the Vintage as a high-quality dining establishment that specializes in fresh seafood. Through the efforts of Karen and her staff, her restaurant has become one of the best and fastest growing restaurants on the island.

TABLE 17.26 FOOD AND BEVERAGE SALES FOR THE VINTAGE RESTAURANT ($1000s)

WEB file

Vintage

Month	First Year	Second Year	Third Year
January	242	263	282
February	235	238	255
March	232	247	265
April	178	193	205
May	184	193	210
June	140	149	160
July	145	157	166
August	152	161	174
September	110	122	126
October	130	130	148
November	152	167	173
December	206	230	235

To better plan for future growth of the restaurant, Karen needs to develop a system that will enable her to forecast food and beverage sales by month for up to one year in advance. Table 17.26 shows the value of food and beverage sales ($1000s) for the first three years of operation.

Managerial Report

Perform an analysis of the sales data for the Vintage Restaurant. Prepare a report for Karen that summarizes your findings, forecasts, and recommendations. Include the following:

1. A time series plot. Comment on the underlying pattern in the time series.
2. An analysis of the seasonality of the data. Indicate the seasonal indexes for each month, and comment on the high and low seasonal sales months. Do the seasonal indexes make intuitive sense? Discuss.
3. Deseasonalize the time series. Does there appear to be any trend in the deseasonalized time series?
4. Using the time series decomposition method, forecast sales for January through December of the fourth year.
5. Using the dummy variable regression approach, forecast sales for January through December of the fourth year.
6. Provide summary tables of your calculations and any graphs in the appendix of your report.

Assume that January sales for the fourth year turn out to be $295,000. What was your forecast error? If this error is large, Karen may be puzzled about the difference between your forecast and the actual sales value. What can you do to resolve her uncertainty in the forecasting procedure?

Case Problem 2 Forecasting Lost Sales

The Carlson Department Store suffered heavy damage when a hurricane struck on August 31. The store was closed for four months (September through December), and Carlson is now involved in a dispute with its insurance company about the amount of lost sales during the time the store was closed. Two key issues must be resolved: (1) the amount of sales Carlson would have made if the hurricane had not struck and (2) whether Carlson is entitled to any compensation for excess sales due to increased business activity after the storm. More than

WEB file

CarlsonSales

TABLE 17.27 SALES FOR CARLSON DEPARTMENT STORE ($MILLIONS)

Month	Year 1	Year 2	Year 3	Year 4	Year 5
January		1.45	2.31	2.31	2.56
February		1.80	1.89	1.99	2.28
March		2.03	2.02	2.42	2.69
April		1.99	2.23	2.45	2.48
May		2.32	2.39	2.57	2.73
June		2.20	2.14	2.42	2.37
July		2.13	2.27	2.40	2.31
August		2.43	2.21	2.50	2.23
September	1.71	1.90	1.89	2.09	
October	1.90	2.13	2.29	2.54	
November	2.74	2.56	2.83	2.97	
December	4.20	4.16	4.04	4.35	

$8 billion in federal disaster relief and insurance money came into the county, resulting in increased sales at department stores and numerous other businesses.

Table 17.27 gives Carlson's sales data for the 48 months preceding the storm. Table 17.28 reports total sales for the 48 months preceding the storm for all department stores in the county, as well as the total sales in the county for the four months the Carlson Department Store was closed. Carlson's managers asked you to analyze these data and develop estimates of the lost sales at the Carlson Department Store for the months of September through December. They also asked you to determine whether a case can be made for excess storm-related sales during the same period. If such a case can be made, Carlson is entitled to compensation for excess sales it would have earned in addition to ordinary sales.

Managerial Report

Prepare a report for the managers of the Carlson Department Store that summarizes your findings, forecasts, and recommendations. Include the following:

1. An estimate of sales for Carlson Department Store had there been no hurricane.
2. An estimate of countywide department store sales had there been no hurricane.
3. An estimate of lost sales for the Carlson Department Store for September through December.

WEB file

CountySales

TABLE 17.28 DEPARTMENT STORE SALES FOR THE COUNTY ($MILLIONS)

Month	Year 1	Year 2	Year 3	Year 4	Year 5
January		46.80	46.80	43.80	48.00
February		48.00	48.60	45.60	51.60
March		60.00	59.40	57.60	57.60
April		57.60	58.20	53.40	58.20
May		61.80	60.60	56.40	60.00
June		58.20	55.20	52.80	57.00
July		56.40	51.00	54.00	57.60
August		63.00	58.80	60.60	61.80
September	55.80	57.60	49.80	47.40	69.00
October	56.40	53.40	54.60	54.60	75.00
November	71.40	71.40	65.40	67.80	85.20
December	117.60	114.00	102.00	100.20	121.80

In addition, use the countywide actual department stores sales for September through December and the estimate in part (2) to make a case for or against excess storm-related sales.

Appendix 17.1 Forecasting with Minitab

In this appendix we show how Minitab can be used to develop forecasts using the following forecasting methods: moving averages, exponential smoothing, trend projection, Holt's linear exponential smoothing, and time series decomposition.

Moving Averages

WEB file
Gasoline

To show how Minitab can be used to develop forecasts using the moving averages method, we will develop a forecast for the gasoline sales time series in Table 17.1 and Figure 17.1. The sales data for the 12 weeks are entered into column 2 of the worksheet. The following steps can be used to produce a three-week moving average forecast for week 13.

Step 1. Select the **Stat** menu
Step 2. Choose **Time Series**
Step 3. Choose **Moving Average**
Step 4. When the Moving Average dialog box appears:
Enter C2 in the **Variable** box
Enter 3 in the **MA length** box
Select **Generate forecasts**
Enter 1 in the **Number of forecasts** box
Enter 12 in the **Starting from origin** box
Click **OK**

Measures of forecast accuracy and the forecast for week 13 are shown in the session window. The mean absolute error is labeled MAD and the mean squared error is labeled MSD in the Minitab output.

Exponential Smoothing

WEB file
Gasoline

To show how Minitab can be used to develop an exponential smoothing forecast, we will again develop a forecast of sales in week 13 for the gasoline sales time series in Table 17.1 and Figure 17.1. The sales data for the 12 weeks are entered into column 2 of the worksheet. The following steps can be used to produce a forecast for week 13 using a smoothing constant of $\alpha = .2$.

Step 1. Select the **Stat** menu
Step 2. Choose **Time Series**
Step 3. Choose **Single Exp Smoothing**
Step 4. When the Single Exponential Smoothing dialog box appears:
Enter C2 in the **Variable** box
Select the **Use** option for the Weight to Use in Smoothing
Enter 0.2 in the Use box
Select **Generate forecasts**
Enter 1 in the **Number of forecasts** box
Enter 12 in the **Starting from origin** box
Select **Options**
Step 5. When the Single Exponential Smoothing-Options dialog box appears:
Enter 1 in the **Use average of first K observations** box
Click **OK**
Step 6. When the Single Exponential Smoothing dialog box appears:
Click **OK**

Measures of forecast accuracy and the exponential smoothing forecast for week 13 are shown in the session window. The mean absolute error is labeled MAD and the mean squared error is labeled MSD in the Minitab output.*

Trend Projection

WEB file
Bicycle

To show how Minitab can be used for trend projection, we develop a forecast for the bicycle sales time series in Table 17.3 and Figure 17.3. The year numbers are entered into column 1 and the sales data are entered into column 2 of the worksheet. The following steps can be used to produce a forecast for year 11 using trend projection.

Step 1. Select the **Stat** menu
Step 2. Choose **Time Series**
Step 3. Choose **Trend Analysis**
Step 4. When the Trend Analysis dialog box appears:
Enter C2 in the **Variable** box
Choose **Linear** for the Model Type
Select **Generate forecasts**
Enter 1 in the **Number of forecasts** box
Enter 10 in the **Starting from origin** box
Click **OK**

The equation for linear trend, measures of forecast accuracy, and the forecast for the next year are shown in the session window. The mean absolute error is labeled MAD and the mean square error is labeled MSD in the Minitab output. To generate forecasts for a quadratic or exponential trend select **Quadratic** of **Exponential growth** instead of **Linear** in step 4.

Holt's Linear Exponential Smoothing

WEB file
Bicycle

To show how Minitab can be used to develop forecasts using Holt's linear exponential smoothing method, we develop a forecast for the bicycle sales time series in Table 17.3 and Figure 17.3. In Minitab, Holt's linear exponential smoothing method is referred to as Double Exponential Smoothing. The year numbers are entered into column 1 and the sales data are entered into column 2 of the worksheet. The following steps can be used to forecast sales in year 11 using Holt's linear exponential smoothing with $\alpha = .1$ and $\beta = .2$.

Step 1. Select the **Stat** menu
Step 2. Choose **Time Series**
Step 3. Choose **Double Exp Smoothing**
Step 4. When the Double Exponential Smoothing dialog box appears:
Enter C2 in the **Variable** box
Select the **Use** option for the Weights to Use in Smoothing
Enter .1 in the **level** box
Enter .2 in the **trend** box
Select **Generate forecasts**
Enter 1 in the **Number of forecasts** box
Enter 10 in the **Starting from origin** box
Click **OK**

*The value of MSD computed by Minitab is not the same as the value of MSE that we computed in Section 17.3. Minitab uses a forecast of 17 for week 1 and computes MSD using all 12 weeks of data. In Section 17.3 we compute MSE using only the data for weeks 2 through 12, because we had no past values with which to make a forecast for week 1.

Measures of forecast accuracy and Holt's linear exponential smoothing forecast for year 11 are shown in the session window. The mean absolute error is labeled MAD and the mean square error is labeled MSD in the Minitab output.

Time Series Decomposition

WEB file

TVSales

To show how Minitab can be used to forecast a time series with trend and seasonality using time series decomposition, we develop a forecast for the television set sales time series in Table 17.6 and Figure 17.6. In Minitab, the user has the option of either a multiplicative or additive decomposition model. We will illustrate how to use the multiplicative approach as described in section 17.6. The year numbers are entered into column 1, the quarter values are entered into column 2, and the sales data are entered into column 3 of the worksheet. The following steps can be used to produce a forecast for the next quarter.

Step 1. Select the **Stat** menu
Step 2. Choose **Time Series**
Step 3. Choose **Decomposition**
Step 4. When the Decomposition dialog box appears:
Enter C3 in the **Variable** box
Enter 4 in the **Season Length** box
Select **Multiplicative** for Method Type
Select **Trend plus Seasonal** for Model Components
Select **Generate forecasts**
Enter 1 in the **Number of forecasts** box
Enter 16 in the **Starting from origin** box
Click **OK**

The seasonal indexes,[†] measures of forecast accuracy, and the forecast for the next quarter are shown in the session window. The mean absolute error is labeled MAD and the mean square error is labeled MSD in the Minitab output.

Appendix 17.2 Forecasting with Excel

In this appendix we show how Excel can be used to develop forecasts using three forecasting methods: moving averages, exponential smoothing, and trend projection.

Moving Averages

WEB file

Gasoline

To show how Excel can be used to develop forecasts using the moving averages method, we will develop a forecast for the gasoline sales time series in Table 17.1 and Figure 17.1. The sales data for the 12 weeks are entered into worksheet rows 2 through 13 of column B. The following steps can be used to produce a three-week moving average.

Step 1. Click the **Data** tab on the Ribbon
Step 2. In the **Analysis** group, click **Data Analysis**
Step 3. Choose **Moving Average** from the list of Analysis Tools
Step 4. When the Moving Average dialog box appears:
Enter B2:B13 in the **Input Range** box
Enter 3 in the **Interval** box
Enter C2 in the **Output Range** box
Click **OK**

[†]The results differ slightly from the results shown in Table 17.12 because Minitab computes the seasonal indexes using the median of the seasonal-irregular values.

The three-week moving averages will appear in column C of the worksheet. The forecast for week 4 appears next to the sales value for week 3, and so on. Forecasts for periods of other length can be computed easily by entering a different value in the **Interval** box.

Exponential Smoothing

WEB file

Gasoline

To show how Excel can be used for exponential smoothing, we again develop a forecast for the gasoline sales time series in Table 17.1 and Figure 17.1. The sales data for the 12 weeks are entered into worksheet rows 2 through 13 of column B. The following steps can be used to produce a forecast using a smoothing constant of $\alpha = .2$.

Step 1. Click the **Data** tab on the Ribbon
Step 2. In the **Analysis** group, click **Data Analysis**
Step 3. Choose **Exponential Smoothing** from the list of Analysis Tools
Step 4. When the Exponential Smoothing dialog box appears:
  Enter B2:B13 in the **Input Range** box
  Enter .8 in the **Damping factor** box
  Enter C2 in the **Output Range** box
  Click **OK**

The exponential smoothing forecasts will appear in column C of the worksheet. Note that the value we entered in the Damping factor box is $1 - \alpha$; forecasts for other smoothing constants can be computed easily by entering a different value for $1 - \alpha$ in the Damping factor box.

Trend Projection

WEB file

Bicycle

To show how Excel can be used for trend projection, we develop a forecast for the bicycle sales time series in Table 17.3 and Figure 17.3. The data, with appropriate labels in row 1, are entered into worksheet rows 1 through 11 of columns A and B. The following steps can be used to produce a forecast for year 11 by trend projection.

Step 1. Select an empty cell in the worksheet
Step 2. Select the **Formulas** tab on the Ribbon
Step 3. In the **Function Library** group, click **Insert** Function
Step 4. When the Insert Function dialog box appears:
  Choose **Statistical** in the Or select a category box
  Choose **Forecast** in the Select a function box
Step 5. When the Forecast Arguments dialog box appears:
  Enter 11 in the **x** box
  Enter B2:B11 in the **Known y's** box
  Enter A2:A11 in the **Known x's** box
  Click **OK**

The forecast for year 11, in this case 32.5, will appear in the cell selected in step 1.

Appendix 17.3 Forecasting Using StatTools

In this appendix we show how StatTools can be used to develop forecasts using three forecasting methods: moving averages, exponential smoothing, and Holt's linear exponential smoothing.

Moving Averages

WEB file

Gasoline

To show how StatTools can be used to develop forecasts using the moving averages method we will develop a forecast for the gasoline sales time series in Table 17.1 and Figure 17.1. Begin by using the Data Set Manager to create a StatTools data set for these data using the procedure described in the appendix in Chapter 1. The following steps will generate a three-week moving average forecast for week 13.

Step 1. Click the **StatTools** tab on the Ribbon
Step 2. In the **Analyses Group**, click **Time Series and Forecasting**
Step 3. Choose the **Forecast** option
Step 4. When the StatTools-Forecast dialog box appears:
In the **Variables** section, select **Sales**
Select the **Forecast Settings** tab
In the **Method** section, select **Moving Average**
In the Parameters section, enter 3 in the **Span** box
Select the **Time Scale** tab
In the **Seasonal Period** section, select **None**
In the **Label Style** section, select **Integer**
Click **OK**

The following output will appear: three measures of forecast accuracy; time series plots showing the original data, the forecasts, and the forecast errors; and a table showing the forecasts and forecast errors. Note that StatTools uses the term "Mean Abs Err" to identify the value of MAE; "Root Mean Sq Err" to identify the square root of the value of MSE; and "Mean Abs Per% Err" to identify the value of MAPE.

Exponential Smoothing

WEB file
Gasoline

To show how StatTools can be used to develop an exponential smoothing forecast, we will again develop a forecast of sales in week 13 for the gasoline time series shown in Table 17.1 and Figure 17.1. Begin by using the Data Set Manager to create a StatTools data set for these data using the procedure described in the appendix in Chapter 1. The following steps will produce a forecast using a smoothing constant of $\alpha = .2$.

Step 1. Click the **StatTools** tab on the Ribbon
Step 2. In the **Analyses Group**, click **Time Series and Forecasting**
Step 3. Choose the **Forecast** option
Step 4. When the StatTools-Forecast dialog box appears:
In the **Variables** section, select **Sales**
Select the **Forecast Settings** tab
In the **Method** section, select **Exponential Smoothing (Simple)**
Remove the check mark in the **Optimize Parameters** box
In the Parameters section, enter .2 in the **Level (a)** box
Select the Time Scale tab
In the **Seasonal Period** section, select **None**
In the **Label Style** section, select **Integer**
Click **OK**

The following output will appear: three measures of forecast accuracy; time series plots showing the original data, the forecasts, and the forecast errors; and a table showing the forecasts and forecast errors. Note that StatTools uses the term "Mean Abs Err" to identify the value of MAE; "Root Mean Sq Err" to identify the square root of the value of MSE; and "Mean Abs Per% Err" to identify the value of MAPE.

Holt's Linear Exponential Smoothing

WEB file
Bicycle

To show how StatTools can be used for trend projection, we develop a forecast for the bicycle sales time series in Table 17.3 and Figure 17.3 using Holt's linear exponential smoothing. Begin by using the Data Set Manager to create a StatTools data set for these data using the procedure described in the appendix in Chapter 1. The following steps will produce a forecast using smoothing constants of $\alpha = .1$ and $\beta = .2$.

Step 1. Click the **StatTools** tab on the Ribbon
Step 2. In the **Analyses Group**, click **Time Series and Forecasting**
Step 3. Choose the **Forecast** option
Step 4. When the StatTools-Forecast dialog box appears:
In the **Variables** section, select **Sales**
Select the **Forecast Settings** tab
In the **Method** section, select **Exponential Smoothing (Holt's)**
Remove the check mark in the **Optimize Parameters** box
In the Parameters section, enter .1 in the **Level (a)** box
In the Parameters section, enter .2 in the **Trend (b)** box
Select the Time Scale tab
In the **Seasonal Period** section, select **None**
In the **Label Style** section, select **Integer**
Click **OK**

The following output will appear: three measures of forecast accuracy; time series plots showing the original data, the forecasts, and the forecast errors; and a table showing the forecasts and forecast errors. Note that StatTools uses the term "Mean Abs Err" to identify the value of MAE; "Root Mean Sq Err" to identify the square root of the of MSE; and "Mean Abs Per% Err" to identify the value of MAPE. The StatTools output differs slightly from the results shown in Section 17.4 because StatTools uses a different approach to compute the estimate of the slope in period 1. With larger data sets the choice of startup values is not critical.

CHAPTER 18

Nonparametric Methods

CONTENTS

STATISTICS *in* PRACTICE

WEST SHELL REALTORS*
CINCINNATI, OHIO

West Shell Realtors was founded in 1958 with one office and a sales staff of three people. In 1964, the company began a long-term expansion program, with new offices added almost yearly. Over the years, West Shell grew to become one of the largest realtors in Greater Cincinnati, with offices in southwest Ohio, southeast Indiana, and northern Kentucky.

Statistical analysis helps real estate firms such as West Shell monitor sales performance. Monthly reports are generated for each of West Shell's offices as well as for the total company. Statistical summaries of total sales dollars, number of units sold, and median selling price per unit are essential in keeping both office managers and the company's top management informed of progress and trouble spots in the organization.

In addition to monthly summaries of ongoing operations, the company uses statistical considerations to guide corporate plans and strategies. West Shell has implemented a strategy of planned expansion. Each time an expansion plan calls for the establishment of a new sales office, the company must address the question of office location. Selling prices of homes, turnover rates, and forecast sales volumes are the types of data used in evaluating and comparing alternative locations.

In one instance, West Shell identified two suburbs, Clifton and Roselawn, as prime candidates for a new office. A variety of factors were considered in comparing the two areas, including selling prices of homes. West Shell employed nonparametric statistical methods to help identify any differences in sales patterns for the two areas.

Samples of 25 sales in the Clifton area and 18 sales in the Roselawn area were taken, and the Mann-Whitney-Wilcoxon rank-sum test was chosen as an appropriate statistical test of the difference in the pattern of selling prices. At the .05 level of significance, the Mann-Whitney-Wilcoxon test did not allow rejection of the null hypothesis that the two populations of selling prices were identical. Thus, West Shell was able to focus on criteria other than selling prices of homes in the site selection process.

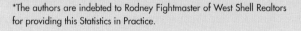

West Shell uses statistical analysis of home sales to remain competitive. © Bloomberg/Getty Images.

In this chapter we will show how nonparametric statistical tests such as the Mann-Whitney-Wilcoxon test are applied. We will also discuss the proper interpretation of such tests.

*The authors are indebted to Rodney Fightmaster of West Shell Realtors for providing this Statistics in Practice.

The statistical methods for inference presented previously in the text are generally known as **parametric methods**. These methods begin with an assumption about the probability distribution of the population which is often that the population has a normal distribution. Based upon this assumption, statisticians are able to derive the sampling distribution that can be used to make inferences about one or more parameters of the population, such as the population mean μ or the population standard deviation σ. For example, in Chapter 9 we presented a method for making an inference about a population mean that was based on an assumption that the population had a normal probability distribution with unknown parameters μ and σ. Using the sample standard deviation s to estimate the population standard deviation σ, the test statistic for making an inference about the population mean was shown to have a t distribution. As a result, the t distribution was used to compute confidence intervals and conduct hypothesis tests about the mean of a normally distributed population.

In this chapter we present **nonparametric methods** which can be used to make inferences about a population without requiring an assumption about the specific form of the population's probability distribution. For this reason, these nonparametric methods are also called **distribution-free methods**.

Most of the statistical methods referred to as parametric methods require quantitative data, while nonparametric methods allow inferences based on either categorical or quantitative data. However, the computations used in the nonparametric methods are generally done with categorical data. Whenever the data are quantitative, we will transform the data into categorical data in order to conduct the nonparametric test. In the first section of the chapter, we show how the binomial distribution uses two categories of data to make an inference about a population median. In the next three sections, we show how rank-ordered data are used in nonparametric tests about two or more populations. In the final section, we use rank-ordered data to compute the rank correlation for two variables.

18.1 Sign Test

The **sign test** is a versatile nonparametric method for hypothesis testing that uses the binomial distribution with $p = .50$ as the sampling distribution. It does not require an assumption about the distribution of the population. In this section we present two applications of the sign test: one involving a hypothesis test about a population median and one involving a matched-sample test about the difference between two populations.

Hypothesis Test About a Population Median

In Chapter 9, we described how to conduct hypothesis tests about a population mean. In this section, we show how the sign test can be used to conduct a hypothesis test about a population median. If we consider a population where no data value is exactly equal to the median, the median is the measure of central tendency that divides the population so that 50% of the values are greater than the median and 50% of the values are less than the median. Whenever a population distribution is skewed, the median is often preferred over the mean as the best measure of central location for the population. The sign test provides a nonparametric procedure for testing a hypothesis about the value of a population median.

In order to demonstrate the sign test, we consider the weekly sales of Cape May Potato Chips by the Lawler Grocery Store chain. Lawler's management made the decision to carry the new potato chip product based on the manufacturer's estimate that the median sales should be \$450 per week on a per store basis. After carrying the product for three-months, Lawler's management requested the following hypothesis test about the population median weekly sales.

$$H_0\text{: Median} = 450$$
$$H_a\text{: Median} \neq 450$$

Data showing one-week sales at 10 randomly selected Lawler's stores are provided in Table 18.1.

TABLE 18.1 ONE-WEEK SALES OF CAPE MAY POTATO CHIPS AT 10 LAWLER GROCERY STORES

Store Number	Weekly Sales ($)	Store Number	Weekly Sales ($)
56	485	63	474
19	562	39	662
36	415	84	380
128	860	102	515
12	426	44	721

Observations equal to the hypothesized value are discarded and the analysis proceeds with the observations having either a plus sign or a minus sign.

In conducting the sign test, we compare each sample observation to the hypothesized value of the population median. If the observation is greater than the hypothesized value, we record a plus sign "+." If the observation is less than the hypothesized value, we record a minus sign "−." If an observation is exactly equal to the hypothesized value, the observation is eliminated from the sample and the analysis proceeds with the smaller sample size, using only the observations where a plus sign or a minus sign has been recorded. It is the conversion of the sample data to either a plus sign or a minus sign that gives the nonparametric method its name: the sign test.

Consider the sample data in Table 18.1. The first observation, 485, is greater than the hypothesized median 450; a plus sign is recorded. The second observation, 562, is greater than the hypothesized median 450; a plus sign is recorded. Continuing with the 10 observations in the sample provides the plus and minus signs as shown in Table 18.2. Note that there are 7 plus signs and 3 minus signs.

The assigning of the plus signs and minus signs has made the situation a binomial distribution application. The sample size $n = 10$ is the number of trials. There are two outcomes possible per trial, a plus sign or a minus sign, and the trials are independent. Let p denote the probability of a plus sign. If the population median is 450, p would equal .50 as there should be 50% plus signs and 50% minus signs in the population. Thus, in terms of the binomial probability p, the sign test hypotheses about the population median

$$H_0\text{: Median} = 450$$
$$H_a\text{: Median} \neq 450$$

are converted to the following hypotheses about the binomial probability p.

$$H_0\text{: } p = .50$$
$$H_a\text{: } p \neq .50$$

TABLE 18.2 LAWLER SAMPLE DATA FOR THE SIGN TEST ABOUT THE POPULATION MEDIAN WEEKLY SALES

Store Number	Weekly Sales ($)	Sign	Store Number	Weekly Sales ($)	Sign
56	485	+	63	474	+
19	562	+	39	662	+
36	415	−	84	380	−
128	860	+	102	515	+
12	426	−	44	721	+

TABLE 18.3

BINOMIAL PROBABILITIES WITH $n = 10$ AND $p = .50$

Number of Plus Signs	Probability
0	.0010
1	.0098
2	.0439
3	.1172
4	.2051
5	.2461
6	.2051
7	.1172
8	.0439
9	.0098
10	.0010

Binomial probabilities are provided in Table 5 of Appendix B when the sample size is less than or equal to 20. Excel or Minitab can be used to provide binomial probabilities for any sample size.

If H_0 cannot be rejected, we cannot conclude that p is different from .50 and thus we cannot conclude that the population median is different from 450. However, if H_0 is rejected, we can conclude that p is not equal to .50 and thus the population median is not equal to 450.

With $n = 10$ stores or trials and $p = .50$, we used Table 5 in Appendix B to obtain the binomial probabilities for the number of plus signs under the assumption H_0 is true. These probabilities are shown in Table 18.3. Figure 18.1 shows a graphical representation of this binomial distribution.

Let us proceed to show how the binomial distribution can be used to test the hypothesis about the population median. We will use a .10 level of significance for the test. Since the observed number of plus signs for the sample data, 7, is in the upper tail of the binomial distribution, we begin by computing the probability of obtaining 7 or more plus signs. This probability is the probability of 7, 8, 9, or 10 plus signs. Adding these probabilities shown in Table 18.3, we have $.1172 + .0439 + .0098 + .0010 = .1719$. Since we are using a two-tailed hypothesis test, this upper tail probability is doubled to obtain the p-value $= 2(.1719) = .3438$. With p-value $> \alpha$, we cannot reject H_0. In terms of the binomial probability p, we cannot reject H_0: $p = .50$, and thus we cannot reject the hypothesis that the population median is \$450.

In this example, the hypothesis test about the population median was formulated as a two-tailed test. However, one-tailed sign tests about a population median are also possible. For example, we could have formulated the hypotheses as an upper tail test so that the null and alternative hypotheses would be written as follows:

$$H_0\text{: Median} \leq 450$$
$$H_a\text{: Median} > 450$$

The corresponding p-value is equal to the binomial probability that the number of plus signs is greater than or equal to 7 found in the sample. This one-tailed p-value would have been $.1172 + .0439 + .0098 + .0010 = .1719$. If the example were converted to a lower tail test, the p-value would have been the probability of obtaining 7 or fewer plus signs.

FIGURE 18.1 BINOMIAL SAMPLING DISTRIBUTION FOR THE NUMBER OF PLUS SIGNS WHEN $n = 10$ AND $p = .50$

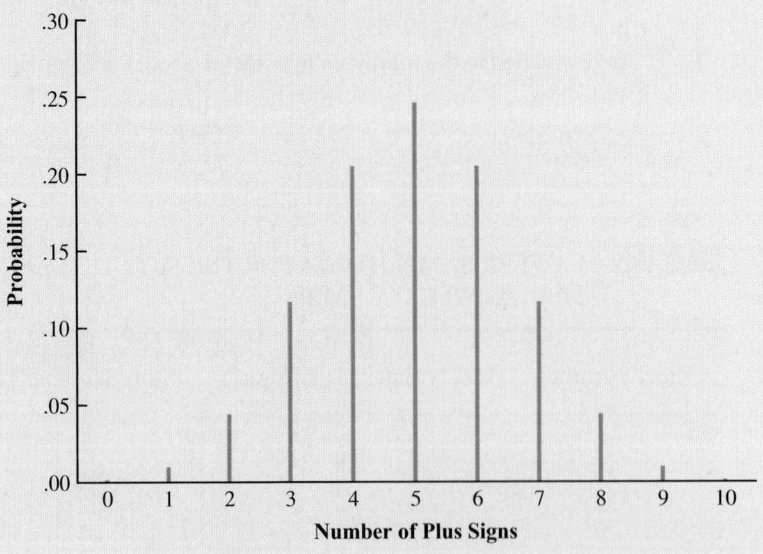

The application we have just described makes use of the binomial distribution with $p = .50$. The binomial probabilities provided in Table 5 of Appendix B can be used to compute the p-value when the sample size is 20 or less. With larger sample sizes, we rely on the normal distribution approximation of the binomial distribution to compute the p-value; this makes the computations quicker and easier. A large sample application of the sign test is illustrated in the following example.

One year ago the median price of a new home was $236,000. However, a current downturn in the economy has real estate firms using sample data on recent home sales to determine if the population median price of a new home is less today than it was a year ago. The hypothesis test about the population median price of a new home is as follows:

$$H_0\text{: Median} \geq 236{,}000$$
$$H_a\text{: Median} < 236{,}000$$

We will use a .05 level of significance to conduct this test.

WEB file
HomeSales

A random sample of 61 recent new home sales found 22 homes sold for more than $236,000, 38 homes sold for less than $236,000, and one home sold for $236,000. After deleting the home that sold for the hypothesized median price of $236,000, the sign test continues with 22 plus signs, 38 minus signs, and a sample of 60 homes.

The null hypothesis that the population median is greater than or equal to $236,000 is expressed by the binomial distribution hypothesis H_0: $p \geq .50$. If H_0 were true as an equality, we would expect $.50(60) = 30$ homes to have a plus sign. The sample result showing 22 plus signs is in the lower tail of the binomial distribution. Thus, the p-value is the probability of 22 or fewer plus signs when $p = .50$. While it is possible to compute the exact binomial probabilities for 0, 1, 2, . . . to 22 and sum these probabilities, we will use the normal distribution approximation of the binomial distribution to make this computation easier. For this approximation, the mean and standard deviation of the normal distribution are as follows.

NORMAL APPROXIMATION OF THE SAMPLING DISTRIBUTION OF THE NUMBER OF PLUS SIGNS WHEN H_0: $p = .50$

$$\text{Mean: } \mu = .50n \tag{18.1}$$
$$\text{Standard deviation: } \sigma = \sqrt{.25n} \tag{18.2}$$

Distribution form: Approximately normal for $n > 20$

Using equations (18.1) and (18.2) with $n = 60$ homes and $p = .50$, the sampling distribution of the number of plus signs can be approximated by a normal distribution with

$$\mu = .50n = .50(60) = 30$$
$$\sigma = \sqrt{.25n} = \sqrt{.25(60)} = 3.873$$

Let us now use the normal distribution to approximate the binomial probability of 22 or fewer plus signs. Before we proceed, remember that the binomial probability distribution is discrete and the normal probability distribution is continuous. To account for this, the binomial probability of 22 is computed by the normal probability interval 21.5 to 22.5. The .5 added to and subtracted from 22 is called the continuity correction factor. Thus, to compute

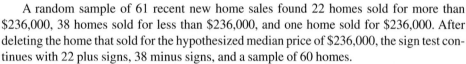

FIGURE 18.2 NORMAL DISTRIBUTION APPROXIMATION OF THE p-VALUE FOR THE SIGN TEST ABOUT THE MEDIAN PRICE OF NEW HOMES

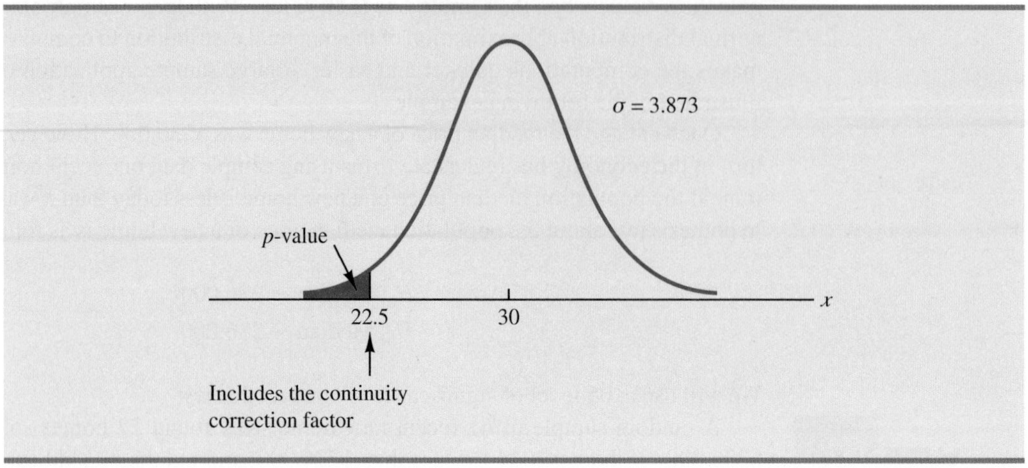

the p-value for 22 or fewer plus signs we use the normal distribution with $\mu = 30$ and $\sigma = 3.873$ to compute the probability that the normal random variable, x, has a value less than or equal to 22.5. A graph of this p-value is shown in Figure 18.2.

Using this normal distribution, we compute the p-value as follows:

$$p\text{-value} = P(x \le 22.5) = P\left(z \le \frac{22.5 - 30}{3.873}\right) = P(z \le -1.94)$$

Using the table of areas for a normal probability distribution, we see that the cumulative probability for $z = -1.94$ provides the p-value $= .0262$. With $.0262 < .05$, we reject the null hypothesis and conclude that the median price of a new home is less than the \$236,000 median price a year ago.

NOTES AND COMMENTS

1. The examples used to illustrate a hypothesis test about a population median involved weekly sales data and home price data. The probability distributions for these types of variables are usually not symmetrical and are most often skewed to the right. In such cases, the population median rather than the population mean becomes the preferred measure of central location. In general, when the population is not symmetrical, the nonparametric sign test for the population median is often the more appropriate statistical test.
2. The binomial sampling distribution for the sign test can be used to compute a confidence interval estimate of the population median. However, the computations are rather complex and would rarely be done by hand. Statistical packages such as Minitab can be used to obtain a confidence interval for a population median. The Minitab procedure to do this is described in Appendix 18.1. Using the price of homes example in this section, Minitab provides the 95% confidence interval for the median price of a new home as \$183,000 to \$231,000.

Hypothesis Test with Matched Samples

In Chapter 10, we introduced a matched-sample experimental design where each of n experimental units provided a pair of observations, one from population 1 and one from population 2. Using quantitative data and assuming that the differences between the pairs of matched observations were normally distributed, the t distribution was used to make an inference about the difference between the means of the two populations.

In the following example we will use the nonparametric sign test to analyze matched-sample data. Unlike the t distribution procedure, which required quantitative data and the assumption that the differences were normally distributed, the sign test enables us to analyze categorical as well as quantitative data and requires no assumption about the distribution of the differences. This type of matched-sample design occurs in market research when a sample of n potential customers is asked to compare two brands of a product such as coffee, soft drinks, or detergents. Without obtaining a quantitative measure of each individual's preference for the brands, each individual is asked to state a brand preference. Consider the following example.

Sun Coast Farms produces an orange juice product called Citrus Valley. The primary competition for Citrus Valley comes from the producer of an orange juice known as Tropical Orange. In a consumer preference comparison of the two brands, 14 individuals were given unmarked samples of the two orange juice products. The brand each individual tasted first was selected randomly. If the individual selected Citrus Valley as the more preferred, a plus sign was recorded. If the individual selected Tropical Orange as the more preferred, a minus sign was recorded. If the individual was unable to express a difference in preference for the two products, no sign was recorded. The data for the 14 individuals in the study are shown in Table 18.4.

Deleting the two individuals who could not express a preference for either brand, the data have been converted to a sign test with 2 plus signs and 10 minus signs for the $n = 12$ individuals who could express a preference for one of the two brands. Letting p indicate the proportion of the population of customers who prefer Citrus Valley orange juice, we want to test the hypotheses that there is no difference between the preferences for the two brands as follows:

$$H_0: p = .50$$
$$H_a: p \neq .50$$

If H_0 cannot be rejected, we cannot conclude that there is a difference in preference for the two brands. However, if H_0 can be rejected, we can conclude that the consumer preferences differ for the two brands. We will use a .05 level of significance for this hypothesis test.

We will conduct the sign test exactly as we did earlier in this section. The sampling distribution for the number of plus signs is a binomial distribution with $p = .50$ and $n = 12$.

TABLE 18.4 PREFERENCE DATA FOR THE SUN COAST FARMS TASTE TEST

Individual	Preference	Sign	Individual	Preference	Sign
1	Tropical Orange	−	8	Tropical Orange	−
2	Tropical Orange	−	9	Tropical Orange	−
3	Citrus Valley	+	10	No Preference	
4	Tropical Orange	−	11	Tropical Orange	−
5	Tropical Orange	−	12	Citrus Valley	+
6	No Preference		13	Tropical Orange	−
7	Tropical Orange	−	14	Tropical Orange	−

TABLE 18.5

BINOMIAL PROBABILITIES WITH $n = 12$ AND $p = .50$

Number of Plus Signs	Probability
0	.0002
1	.0029
2	.0161
3	.0537
4	.1208
5	.1934
6	.2256
7	.1934
8	.1208
9	.0537
10	.0161
11	.0029
12	.0002

Using Table 5 in Appendix B we obtain the binomial probabilities for the number of plus signs, as shown in Table 18.5. Under the assumption H_0 is true, we would expect $.50n = .50(12) = 6$ plus signs. With only two plus signs in the sample, the results are in the lower tail of the binomial distribution. To compute the p-value for this two-tailed test, we first compute the probability of 2 or fewer plus signs and then double this value. Using the binomial probabilities of 0, 1, and 2 shown in Table 18.5, the p-value is $2(.0002 + .0029 + .0161) = .0384$. With $.0384 < .05$, we reject H_0. The taste test provides evidence that consumer preference differs significantly for the two brands of orange juice. We would advise Sun Coast Farms of this result and conclude that the competitor's Tropical Orange product is the more preferred. Sun Coast Farms can then pursue a strategy to address this issue.

Similar to other uses of the sign test, one-tailed tests may be used depending upon the application. Also, as the sample size becomes large, the normal distribution approximation of the binomial distribution will ease the computations as shown earlier in this section. While the Sun Coast Farms sign test for matched samples used categorical preference data, the sign test for matched samples can be used with quantitative data as well. This would be particularly helpful if the paired differences are not normally distributed and are skewed. In this case a positive difference is assigned a plus sign, a negative difference is assigned a negative sign, and a zero difference is removed from the sample. The sign test computations proceed as before.

Exercises

Methods

SELF test

1. The following hypothesis test is to be conducted.

$$H_0\text{: Median} \leq 150$$
$$H_a\text{: Median} > 150$$

A sample of 30 provided 22 observations greater than 150, 3 observations equal to 150, and 5 observations less than 150. Use $\alpha = .01$. What is your conclusion?

SELF test

2. Ten individuals participated in a taste test involving two brands of a product. Sample results show 7 preferred brand A, 2 preferred brand B, and 1 was unable to state a preference. With $\alpha = .05$, test for a significant difference in the preferences for the two brands. What is your conclusion?

Applications

3. The median number of part-time employees at fast-food restaurants in a particular city was known to be 18 last year. City officials think the use of part-time employees may be increasing. A sample of nine fast-food restaurants showed that seven restaurants were employing more than 18 part-time employees, one restaurant was employing exactly 18 part-time employees, and one restaurant was employing fewer than 18 part-time employees. Can it be concluded that the median number of part-time employees has increased? Test using $\alpha = .05$.

4. Net assets for the 50 largest stock mutual funds show a median of \$15 billion (*The Wall Street Journal,* March 2, 2009). A sample of 10 of the 50 largest bond mutual funds follows.

Bond Fund	Net Assets	Bond Fund	Net Assets
Fidelity Intl Bond	6.1	T Rowe Price New Income	6.9
Franklin CA TF	11.7	Vanguard GNMA	15.0
American Funds	22.4	Oppenheimer Intl Bond	6.6
Vanguard Short Term	9.6	Dodge & Cox Income	14.5
PIMCO: Real Return	4.9	iShares: TIPS Bond	9.6

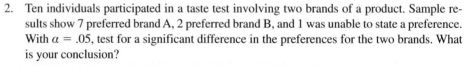

Using the median, can it be concluded that bond mutual funds are smaller and have fewer net assets than stock mutual funds? Use $\alpha = .05$.

a. What are the hypotheses for this test?
b. What is the p-value? What is your conclusion?

5. The median annual income of subscribers to *Shutterbug* magazine is \$75,000 (Home Theater website, August 18, 2008). A sample of 300 subscribers to *Popular Photography & Imaging* magazine found 165 subscribers with an annual income over \$75,000 and 135 with an annual income under \$75,000. Can you conclude that the median annual income of *Popular Photography & Imaging* subscribers differs from the median annual income of *Shutterbug* subscribers? Use $\alpha = .05$.

WEB file
ChicagoIncome

6. The median annual income for families living in the United States is \$56,200 (*The New York Times Almanac,* 2008). Annual incomes in thousands of dollars for a sample of 50 families living in Chicago, Illinois, are shown. Use the sample data to see if it can be concluded that the families living in Chicago have a median annual income greater than \$56,200. Use $\alpha = .05$. What is your conclusion?

66.3	60.2	49.9	75.4	73.7
65.7	61.1	123.8	57.3	48.5
74.0	146.3	92.2	43.7	86.9
59.7	64.2	56.2	48.9	109.6
39.8	60.9	79.7	42.3	52.6
60.9	43.5	61.7	54.7	95.2
70.4	43.8	57.8	83.5	56.5
51.3	42.9	87.5	43.6	67.2
48.7	79.1	61.9	53.4	56.2
57.0	49.6	109.5	42.1	74.6

7. Are stock splits beneficial to stockholders? SNL Financial studied stock splits in the banking industry over an 18-month period. For a sample of 20 stock splits, 14 led to an increase in investment value, 4 led to a decrease in investment value, and 2 resulted in no change. Conduct a sign test to determine if it can be concluded that stock splits are beneficial for holders of bank stocks.

a. What are the null and alternative hypotheses?
b. Using $\alpha = .05$, what is your conclusion?

8. A Pew Research Center survey asked adults if their ideal place to live would have a faster pace of life or a slower pace of life (*USA Today,* February 13, 2009). A preliminary sample of 16 respondents showed 4 preferred a faster pace of life, 11 preferred a slower place of life, and 1 said it did not matter.

a. Are these data sufficient to conclude there is a difference between the preferences for a faster pace of life or a slower pace of life? Use $\alpha = .05$. What is your conclusion?
b. Considering the entire sample of 16 respondents, what is the percentage who would like a faster pace of life? What is the percentage who would like a slower pace of life? What recommendation do you have for the study?

9. A poll taken during the recession in 2008 asked 600 adults a series of questions about the state of the economy and their children's future. One question was, "Do you expect your children to have a better life than you have had, a worse life, or a life about the same as yours?" The responses showed 242 better, 310 worse, and 48 about the same. Use the sign test and $\alpha = .05$ to determine whether there is a difference between the number of adults who feel their children will have a better life compared to a worse life. What is your conclusion?

10. Nielsen Media Research identified *American Idol* and *Dancing with the Stars* as the two top-rated prime-time television shows (*USA Today*, April 14, 2008). In a local television preference survey, 750 individuals were asked to indicate their favorite prime-time television show: Three hundred thirty selected *American Idol*, 270 selected *Dancing with the Stars,* and 150 selected another television show. Use a .05 level of significance to test the hypothesis that there is no difference in the preference for the *American Idol* and *Dancing with the Stars* television shows. What is your conclusion?

11. Competition in the personal computer market is intense. A sample of 450 purchases showed 202 Brand A computers, 175 Brand B computers, and 73 other computers. Use a .05 level of significance to test the null hypothesis that Brand A and Brand B have the same share of the personal computer market. What is your conclusion?

18.2 Wilcoxon Signed-Rank Test

In Chapter 10, we introduced a matched-sample experimental design where each of n experimental units provided a pair of observations, one from population 1 and one from population 2. The parametric test for this experiment requires quantitative data and the assumption that the differences between the paired observations are normally distributed. The t distribution can then be used to make an inference about the difference between the means of the two populations.

If the population of differences is skewed, the sign test for matched samples presented in Section 18.1 is recommended.

The **Wilcoxon signed-rank test** is a nonparametric procedure for analyzing data from a matched-sample experiment. The test uses quantitative data but does not require the assumption that the differences between the paired observations are normally distributed. It only requires the assumption that the differences between the paired observations have a symmetric distribution. This occurs whenever the shapes of the two populations are the same and the focus is on determining if there is a difference between the medians of the two populations. Let us demonstrate the Wilcoxon signed-rank test with the following example.

Consider a manufacturing firm that is attempting to determine whether two production methods differ in terms of task completion time. Using a matched-samples experimental design, 11 randomly selected workers completed the production task two times, once using method A and once using method B. The production method that the worker used first was randomly selected. The completion times for the two methods and the differences between the completion times are shown in Table 18.6. A positive difference indicates that method

TABLE 18.6 PRODUCTION TASK COMPLETION TIMES (MINUTES)

	Method		
Worker	A	B	Difference
1	10.2	9.5	.7
2	9.6	9.8	−.2
3	9.2	8.8	.4
4	10.6	10.1	.5
5	9.9	10.3	−.4
6	10.2	9.3	.9
7	10.6	10.5	.1
8	10.0	10.0	.0
9	11.2	10.6	.6
10	10.7	10.2	.5
11	10.6	9.8	.8

A required more time; a negative difference indicates that method B required more time. Do the data indicate that the two production methods differ significantly in terms of completion times? If we assume that the differences have a symmetric distribution but not necessarily a normal distribution, the Wilcoxon signed-rank test applies.

The examples in this section take the point of view that the two populations have the same shape and if they do differ, it is only in location. This enables the hypotheses for the Wilcoxon signed-rank test to be stated in terms of the population medians.

In particular, we will use the Wilcoxon signed-rank test for the difference between the median completion times for the two production methods. The hypotheses are as follows:

$$H_0\text{: Median for method A} - \text{Median for method B} = 0$$
$$H_a\text{: Median for method A} - \text{Median for method B} \neq 0$$

If H_0 cannot be rejected, we will not be able to conclude that the median completion times are different. However, if H_0 is rejected, we will conclude that the median completion times are different. We will use a .05 level of significance for the test.

Differences of 0 are discarded and the analysis continues with the smaller sample size involving the nonzero differences.

The first step in the Wilcoxon signed-rank test is to discard the difference of zero for worker 8 and then compute the absolute value of the differences for the remaining 10 workers as shown in column 3 of Table 18.7. Next we rank these absolute differences from lowest to highest as shown in column 4. The smallest absolute difference of .1 for worker 7 is assigned the rank of 1. The second smallest absolute difference of .2 for worker 2 is assigned the rank of 2. This ranking of absolute differences continues with the largest absolute difference of .9 for worker 6 being assigned the rank of 10. The tied absolute differences of .4 for workers 3 and 5 are assigned the average rank of 3.5. Similarly, the tied absolute differences of .5 for workers 4 and 10 are assigned the average rank of 5.5.

Ties among absolute differences are assigned the average of their ranks.

Once the ranks of the absolute differences have been determined, each rank is given the *sign* of the original difference for the worker. The negative signed ranks are placed in column 5 and the positive signed ranks are placed in column 6 (see Table 18.7). For example, the difference for worker 1 was a positive .7 (see column 2) and the rank of the absolute difference was 8 (see column 4). Thus, the rank for worker 1 is shown as a positive signed rank in column 6. The difference for worker 2 was a negative .2 and the rank of the absolute difference was 2. Thus, the rank for worker 2 is shown as a negative signed rank of −2 in column 5. Continuing this process generates the negative and positive signed ranks as shown in Table 18.7.

TABLE 18.7 RANKING THE ABSOLUTE DIFFERENCES AND THE SIGNED RANKS FOR THE PRODUCTION TASK COMPLETION TIMES

Worker	Difference	Absolute Difference	Rank	Signed Ranks Negative	Signed Ranks Positive
1	.7	.7	8		8
2	−.2	.2	2	−2	
3	.4	.4	3.5		3.5
4	.5	.5	5.5		5.5
5	−.4	.4	3.5	−3.5	
6	.9	.9	10		10
7	.1	.1	1		1
8	.0				
9	.6	.6	7		7
10	.5	.5	5.5		5.5
11	.8	.8	9		9
				Sum of Positive Signed Ranks	$T^+ = 49.5$

Let T^+ denote the sum of the positive signed ranks, which is $T^+ = 49.5$. To conduct the Wilcoxon signed-rank test, we will use T^+ as the test statistic. If the medians of the two populations are equal and the number of matched pairs is 10 or more, the sampling distribution of T^+ can be approximated by a normal distribution as follows.

SAMPLING DISTRIBUTION OF T^+ FOR THE WILCOXON SIGNED-RANK TEST

$$\text{Mean: } \mu_{T^+} = \frac{n(n+1)}{4} \qquad \textbf{(18.3)}$$

$$\text{Standard deviation: } \sigma_{T^+} = \sqrt{\frac{n(n+1)(2n+1)}{24}} \qquad \textbf{(18.4)}$$

Distribution Form: Approximately normal for $n \geq 10$

After discarding the observation of a zero difference for worker 8, the analysis continues with the $n = 10$ matched pairs. Using equations (18.3) and (18.4), we have

$$\mu_{T^+} = \frac{n(n+1)}{4} = \frac{10(10+1)}{4} = 27.5$$

$$\sigma_{T^+} = \sqrt{\frac{n(n+1)(2n+1)}{24}} = \sqrt{\frac{10(10+1)(20+1)}{24}} = \sqrt{\frac{2310}{24}} = 9.8107$$

Figure 18.3 shows the sampling distribution of the T^+ test statistic.

Let us compute the two-tailed p-value for the hypothesis that the median completion times for the two production methods are equal. Since the test statistic $T^+ = 49.5$ is in the upper tail of the sampling distribution, we begin by computing the upper tail probability $P(T^+ \geq 49.5)$. Since the sum of the positive ranks T^+ is discrete and the normal distribution is continuous, we will obtain the best approximation by including the continuity correction factor. Thus, the discrete probability of $T^+ = 49.5$ is approximated by the normal probability interval, 49 to 50, and the probability that $T^+ \geq 49.5$ is approximated by:

$$P(T^+ \geq 49.5) = P\left(z \geq \frac{49 - 27.5}{9.8107}\right) = P(z \geq 2.19)$$

FIGURE 18.3 SAMPLING DISTRIBUTION OF T^+ FOR THE PRODUCTION TASK COMPLETION TIME EXAMPLE

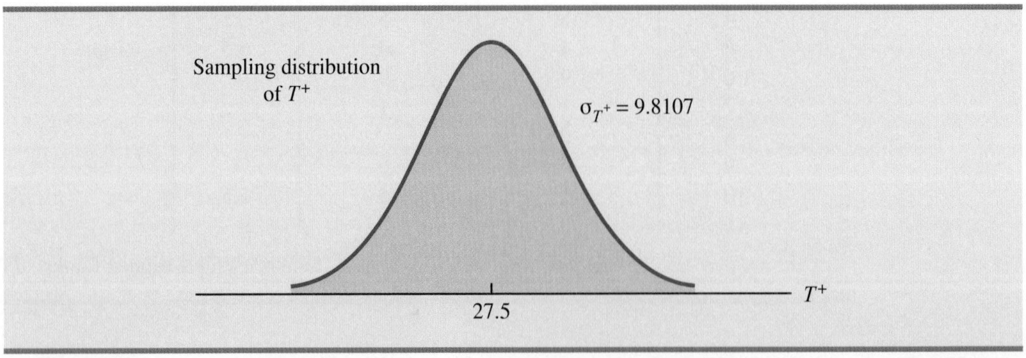

Using the standard normal distribution table and $z = 2.19$, we see that the two-tailed p-value $= 2(1-.9857) = .0286$. With the p-value $\leq .05$, we reject H_0 and conclude that the median completion times for the two production methods are not equal. With T^+ being in the upper tail of the sampling distribution, we see that method A led to the longer completion times. We would expect management to conclude that method B is the faster or better production method.

One-tailed Wilcoxon signed-rank tests are possible. For example, if initially we had been looking for statistical evidence to conclude method A had the larger median completion time and method B has the smaller median completion time, we would have formulated an upper tail hypothesis test as follows:

$$H_0\text{: Median for method A} - \text{Median for method B} \leq 0$$
$$H_a\text{: Median for method A} - \text{Median for method B} > 0$$

Rejecting H_0 would provide the conclusion that method A has the greater median completion time and method B has the smaller median completion time. Lower tail hypothesis tests are also possible.

The Wilcoxon signed-rank test can be used to test the hypothesis about the median of a symmetric population. If the population is skewed, the sign test presented in Section 18.1 is preferred.

As a final note, in Section 18.1 we showed how the sign test could be used for both a hypothesis test about a population median and a hypothesis test with matched samples. In this section, we have demonstrated the use of the Wilcoxon signed-rank test for a hypothesis test with matched samples. However, the Wilcoxon signed-rank test can also be used for a nonparametric test about a population median. This test makes no assumption about the population distribution other than that it is symmetric. If this assumption is appropriate, the Wilcoxon signed-rank test is the preferred nonparametric test for a population median. However, if the population is skewed, the sign test presented in Section 18.1 is preferred. With the Wilcoxon signed-rank test, the differences between the observations and the hypothesized value of the population median are used instead of the differences between the matched-pair observations. Otherwise the calculations are exactly as shown in this section. Exercise 17 will ask you to use the Wilcoxon signed-rank test to conduct a hypothesis test about the median of a symmetric population.

NOTES AND COMMENTS

1. The Wilcoxon signed-rank test for a population median is based on the assumption that the population is symmetric. With this assumption, the population median is equal to the population mean. Thus, the Wilcoxon signed-rank test can also be used as a test about the mean of a symmetric population.
2. The Wilcoxon signed-rank procedure can also be used to compute a confidence interval for the median of a symmetric population. However, the computations are rather complex and would rarely be done by hand. Statistical packages such as Minitab can be used to obtain this confidence interval.

Exercises

Applications

In the following exercises involving paired differences, consider that it is reasonable to assume the populations being compared have approximately the same shape and that the distribution of paired differences is approximately symmetric.

SELF test

12. Two fuel additives are tested to determine their effect on miles per gallon for passenger cars. Test results for 12 cars follow; each car was tested with both fuel additives. Use $\alpha = .05$ and the Wilcoxon signed-rank test to see whether there is a significant difference between the median miles per gallon for the additives.

WEB file Additive

	Additive			Additive	
Car	1	2	Car	1	2
1	20.12	18.05	7	16.16	17.20
2	23.56	21.77	8	18.55	14.98
3	22.03	22.57	9	21.87	20.03
4	19.15	17.06	10	24.23	21.15
5	21.23	21.22	11	23.21	22.78
6	24.77	23.80	12	25.02	23.70

SELF test

13. A sample of 10 men was used in a study to test the effects of a relaxant on the time required to fall asleep. Data for 10 subjects showing the number of minutes required to fall asleep with and without the relaxant follow. Use a .05 level of significance to determine whether the relaxant reduces the median time required to fall asleep. What is your conclusion?

WEB file Relaxant

	Relaxant			Relaxant	
Subject	No	Yes	Subject	No	Yes
1	15	10	6	7	5
2	12	10	7	8	10
3	22	12	8	10	7
4	8	11	9	14	11
5	10	9	10	9	6

14. Percents of on-time arrivals for flights in 2006 and 2007 were collected for 11 randomly selected airports. Data for these airports follow (Research and Innovative Technology Administration website, August 29, 2008). Use $\alpha = .05$ to test the hypothesis that there is no difference between the median percent of on-time arrivals for the two years. What is your conclusion?

WEB file OnTime

	Percent On Time	
Airport	2006	2007
Boston Logan	71.78	69.69
Chicago O'Hare	68.23	65.88
Chicago Midway	77.98	78.40
Denver	78.71	75.78
Fort Lauderdale	77.59	73.45
Houston	77.67	78.68
Los Angeles	76.67	76.38
Miami	76.29	70.98
New York (JFK)	69.39	62.84
Orlando	79.91	76.49
Washington (Dulles)	75.55	72.42

15. A test was conducted for two overnight mail delivery services. Two samples of identical deliveries were set up so that both delivery services were notified of the need for a delivery at the same time. The hours required to make each delivery follow. Do the data shown suggest a difference in the median delivery times for the two services? Use a .05 level of significance for the test.

WEB file Overnight

	Service	
Delivery	1	2
1	24.5	28.0
2	26.0	25.5
3	28.0	32.0
4	21.0	20.0
5	18.0	19.5
6	36.0	28.0
7	25.0	29.0
8	21.0	22.0
9	24.0	23.5
10	26.0	29.5
11	31.0	30.0

16. The PGA Players Championship was held at the Sedgefield Country Club in Greensboro, North Carolina, August 11–17, 2008. Shown here are first-round and second-round scores for a random sample of 11 golfers. Use $\alpha = .05$ to determine whether the first- and second-round median scores for golfers in the Players Championship differed significantly. What is your conclusion?

WEB file GolfScores

Golfer	1st Round	2nd Round
Marvin Laird	63	74
Jimmy Walker	70	73
Kevin Chappell	72	70
Kevin Duke	65	71
Andrew Buckle	70	74
Paul Claxton	69	73
Larry Mize	72	71
Chris Riley	68	70
Bubba Watson	70	68
Carlos Franco	71	71
Richard Johnson	72	69

17. The Scholastic Aptitude Test (SAT) consists of three parts: critical reading, mathematics, and writing. Each part of the test is scored on a 200- to 800-point scale with a median of approximately 500 (*The World Almanac,* 2009). Scores for each part of the test can be assumed to be symmetric. Use the following data to test the hypothesis that the population median score for the students taking the writing portion of the SAT is 500. Using $\alpha = .05$, what is your conclusion?

WEB file WritingScore

635	701	439	447	464
502	405	453	471	476
447	590	337	387	514

18.3 Mann-Whitney-Wilcoxon Test

In Chapter 10, we introduced a procedure for conducting a hypothesis test about the difference between the means of two populations using two independent samples, one from population 1 and one from population 2. This parametric test required quantitative data and the assumption that both populations had a normal distribution. In the case where the population standard deviations σ_1 and σ_2 were unknown, the sample standard deviations s_1 and

FIGURE 18.4 TWO POPULATIONS ARE NOT IDENTICAL WITH POPULATION 1 TENDING TO PROVIDE THE SMALLER VALUES

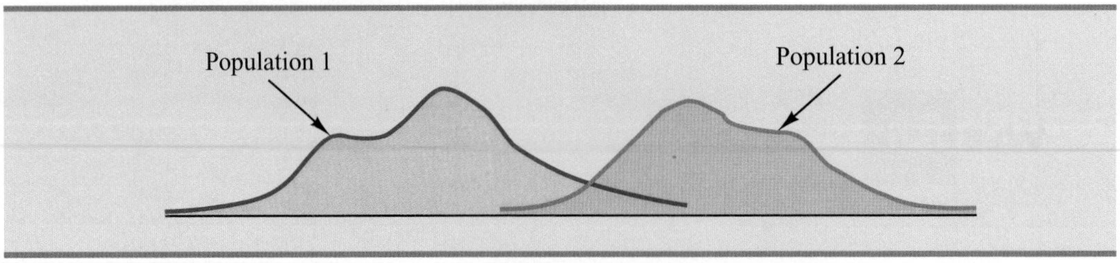

s_2 provided estimates of σ_1 and σ_2 and the t distribution was used to make an inference about the difference between the means of the two populations.

In this section we present a nonparametric test for the difference between two populations based on two independent samples. Advantages of this nonparametric procedure are that it can be used with either ordinal data[1] or quantitative data and it does not require the assumption that the populations have a normal distribution. Versions of the test were developed jointly by Mann and Whitney and also by Wilcoxon. As a result, the test has been referred to as the *Mann-Whitney test* and the *Wilcoxon rank-sum test*. The tests are equivalent and both versions provide the same conclusion. In this section, we will refer to this nonparametric test as the **Mann-Whitney-Wilcoxon (MWW) test**.

We begin the MWW test by stating the most general form of the null and alternative hypotheses as follows:

$$H_0\text{: The two populations are identical}$$
$$H_a\text{: The two populations are not identical}$$

The alternative hypothesis that the two populations are not identical requires some clarification. If H_0 is rejected, we are using the test to conclude that the populations are not identical and that population 1 tends to provide either smaller or larger values than population 2. A situation where population 1 tends to provide smaller values than population 2 is shown in Figure 18.4. Note that it is not necessary that all values from population 1 be less than all values from population 2. However, the figure correctly shows, the conclusion that H_a is true; the two populations are not identical and population 1 tends to provide smaller values than population 2. In a two-tailed test, we consider the alternative hypothesis that either population may provide the smaller or larger values. One-tailed versions of the test can be formulated with the alternative hypothesis that population 1 provides either the smaller or the larger values compared to population 2.

We will first illustrate the MWW test using small samples with rank-ordered data. This will give you an understanding of how the rank-sum statistic is computed and how it is used to determine whether to reject the null hypothesis that the two populations are identical. Later in the section, we will introduce a large-sample approximation based on the normal distribution that will simplify the calculations required by the MWW test.

Let us consider the on-the-job performance ratings for employees at a Showtime Cinemas 20-screen multiplex movie theater. During an employee performance review, the theater manager rated all 35 employees from best (rating 1) to worst (rating 35) in the theater's annual report. Knowing that the part-time employees were primarily college and high school students, the district manager asked if there was evidence of a significant difference in performance for college students compared to high school students. In terms of the

[1]Ordinal data are categorical data that can be rank ordered. This scale of measurement was discussed more fully in Section 1.2 of Chapter 1.

TABLE 18.8 PERFORMANCE RATINGS FOR A SAMPLE OF COLLEGE STUDENTS AND A SAMPLE OF HIGH SCHOOL STUDENTS WORKING AT SHOWTIME CINEMAS

College Student	Manager's Performance Rating	High School Student	Manager's Performance Rating
1	15	1	18
2	3	2	20
3	23	3	32
4	8	4	9
		5	25

population of college students and the population of high school students who could be considered for employment at the theater, the hypotheses were stated as follows:

H_0: College and high school student populations are identical in terms of performance

H_a: College and high school student populations are not identical in terms of performance

We will use a .05 level of significance for this test.

We begin by selecting a random sample of four college students and a random sample of five high school students working at Showtime Cinemas. The theater manager's overall performance rating based on all 35 employees was recorded for each of these employees, as shown in Table 18.8. The first college student selected was rated 15th in the manager's annual performance report, the second college student selected was rated 3rd in the manager's annual performance report, and so on.

The data in this example show how the MWW test can be used with ordinal (rank-ordered) data. Exercise 25 provides another application that uses this type of data.

The next step in the MWW procedure is to rank the *combined* samples from low to high. Since there is a total of 9 students, we rank the performance rating data in Table 18.8 from 1 to 9. The lowest value of 3 for college student 2 receives a rank of 1 and the second lowest value of 8 for college student 4 receives a rank of 2. The highest value of 32 for high school student 3 receives a rank of 9. The combined-sample ranks for all 9 students are shown in Table 18.9.

Next we sum the ranks for each sample as shown in Table 18.9. The MWW procedure may use the sum of the ranks for either sample. However, in our application of the MWW test we will follow the common practice of using the first sample which is the sample of four college students. The sum of ranks for the first sample will be the test statistic W for the MWW test. This sum, as shown in Table 18.9, is $W = 4 + 1 + 7 + 2 = 14$.

TABLE 18.9 RANKS FOR THE NINE STUDENTS IN THE SHOWTIME CINEMAS COMBINED SAMPLES

College Student	Manager's Performance Rating	Rank	High School Student	Manager's Performance Rating	Rank
1	15	4	1	18	5
2	3	1	2	20	6
3	23	7	3	32	9
4	8	2	4	9	3
	Sum of Ranks	14	5	25	8
				Sum of Ranks	31

Let us consider why the sum of the ranks will help us select between the two hypotheses: H_0: The two populations are identical and H_a: The two populations are not identical. Letting C denote a college student and H denote a high school student, suppose the ranks of the nine students had the following order with the four college students having the four lowest ranks.

Rank	1	2	3	4	5	6	7	8	9
Student	C	C	C	C	H	H	H	H	H

Notice that this permutation or ordering separates the two samples, with the college students all having a lower rank than the high school students. This is a strong indication that the two populations are not identical. The sum of ranks for the college students in this case is $W = 1 + 2 + 3 + 4 = 10$.

Now consider a ranking where the four college students have the four highest ranks.

Rank	1	2	3	4	5	6	7	8	9
Student	H	H	H	H	H	C	C	C	C

Notice that this permutation or ordering separates the two samples again, but this time the college students all have a higher rank than the high school students. This is another strong indication that the two populations are not identical. The sum of ranks for the college students in this case is $W = 6 + 7 + 8 + 9 = 30$. Thus, we see that the sum of the ranks for the college students must be between 10 and 30. Values of W near 10 imply that college students have lower ranks than the high school students, whereas values of W near 30 imply that college students have higher ranks than the high school students. Either of these extremes would signal the two populations are not identical. However, if the two populations are identical, we would expect a mix in the ordering of the C's and H's so that the sum of ranks W is closer to the average of the two extremes, or nearer to $(10 + 30)/2 = 20$.

Making the assumption that the two populations are identical, we used a computer program to compute all possible orderings for the nine students. For each ordering, we computed the sum of the ranks for the college students. This provided the probability distribution showing the exact sampling distribution of W in Figure 18.5. The exact probabilities associated with

FIGURE 18.5 EXACT SAMPLING DISTRIBUTION OF THE SUM OF THE RANKS FOR THE SAMPLE OF COLLEGE STUDENTS

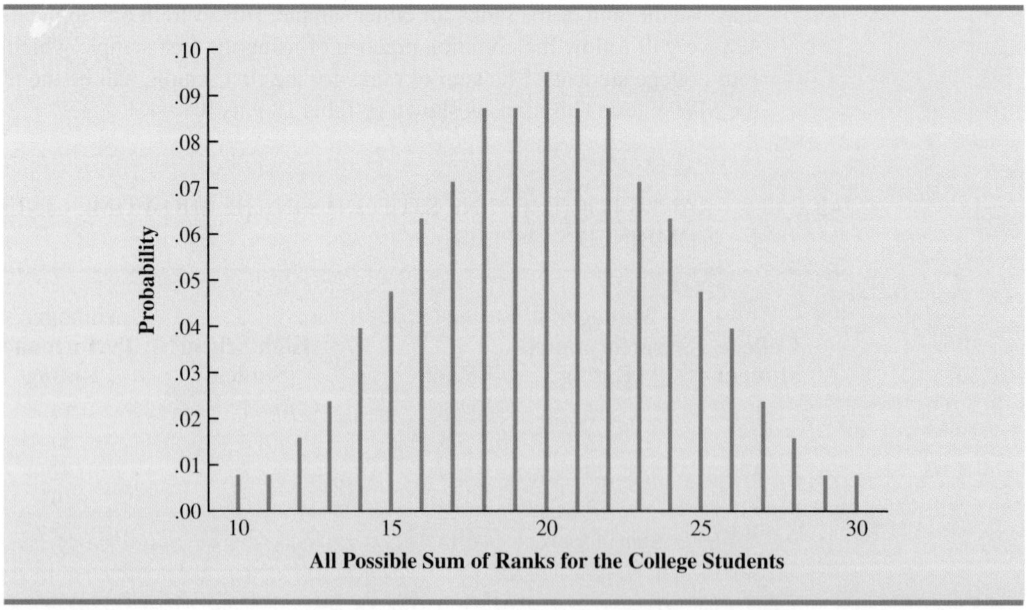

TABLE 18.10 PROBABILITIES FOR THE EXACT SAMPLING DISTRIBUTION OF THE SUM OF THE RANKS FOR THE SAMPLE OF COLLEGE STUDENTS

W	Probability	W	Probability
10	0.0079	20	0.0952
11	0.0079	21	0.0873
12	0.0159	22	0.0873
13	0.0238	23	0.0714
14	0.0397	24	0.0635
15	0.0476	25	0.0476
16	0.0635	26	0.0397
17	0.0714	27	0.0238
18	0.0873	28	0.0159
19	0.0873	29	0.0079
		30	0.0079

the values of W are summarized in Table 18.10. While we will not ask you to generate this exact sampling distribution, we will use it to test the hypothesis that the two populations of students are identical.

Let us use the sampling distribution of W in Figure 18.5 to compute the p-value for the test just as we have done using other sampling distributions. Table 18.9 shows that the sum of ranks for the four college student is $W = 14$. Because this value of W is in the lower tail of the sampling distribution, we begin by computing the lower tail probability $P(W \leq 14)$. Thus, we have

$$\begin{aligned} P(W \leq 14) &= P(10) + P(11) + P(12) + P(13) + P(14) \\ &= .0079 + .0079 + .0159 + .0238 + .0397 = .0952 \end{aligned}$$

The two-tailed p-value $= 2(.0952) = .1904$. With $\alpha = .05$ as the level of significance and p-value $> .05$, the MWW test conclusion is that we cannot reject the null hypothesis that the populations of college and high school students are identical. While the sample of four college students and the sample of five high school students did not provide statistical evidence to conclude there is a difference between the two populations, this is an ideal time to suggest withholding judgment. Further study with larger samples should be considered before drawing a final conclusion.

Most applications of the MWW test involve larger sample sizes than shown in this first example. For such applications, a large sample approximation of the sampling distribution of W based on the normal distribution is employed. In fact, note that the sampling distribution of W in Figure 18.5 shows a normal distribution is a pretty good approximation for sample sizes as small as four and five. We will use the same combined-sample ranking procedure that we used in the previous example but will use the normal distribution approximation rather than the exact sampling distribution of W to compute the p-value and draw the conclusion.

We illustrate the use of the normal distribution approximation for the MWW test by considering the situation at Third National Bank. The bank manager is monitoring the balances maintained in checking accounts at two branch banks and is wondering if the populations of account balances at the two branch banks are identical. Two independent samples of checking accounts are taken with sample sizes $n_1 = 12$ at branch 1 and $n_2 = 10$ at branch 2. The data are shown in Table 18.11.

Doing the ranking of the combined samples by hand will take some time. Computer routines can be used to do this ranking quickly and efficiently.

As before, the first step in the MWW test is to rank the *combined* data from the lowest to highest values. Using the combined 22 observations in Table 18.11, we find the

TABLE 18.11 ACCOUNT BALANCES FOR TWO BRANCHES OF THIRD NATIONAL BANK

Branch 1		Branch 2	
Account	Balance ($)	Account	Balance ($)
1	1095	1	885
2	955	2	850
3	1200	3	915
4	1195	4	950
5	925	5	800
6	950	6	750
7	805	7	865
8	945	8	1000
9	875	9	1050
10	1055	10	935
11	1025		
12	975		

smallest value of $750 (Branch 2 Account 6) and assign it a rank of 1. The second smallest value of $800 (Branch 2 Account 5) is assigned a rank of 2. The third smallest value of $805 (Branch 1 Account 7) is assigned a rank of 3, and so on. In ranking the combined data, we may find that two or more values are the same. In that case, the tied values are assigned the *average* rank of their positions in the combined data set. For example, the balance of $950 occurs for both Branch 1 Account 6 and Branch 2 Account 4. In the combined data set, the two values of $950 are in positions 12 and 13 when the combined data are ranked from low to high. As a result, these two accounts are assigned the average rank $(12 + 13)/2 = 12.5$. Table 18.12 shows the assigned ranks for the combined samples.

TABLE 18.12 ASSIGNED RANKS FOR THE COMBINED ACCOUNT BALANCE SAMPLES

Branch	Account	Balance	Rank
2	6	750	1
2	5	800	2
1	7	805	3
2	2	850	4
2	7	865	5
1	9	875	6
2	1	885	7
2	3	915	8
1	5	925	9
2	10	935	10
1	8	945	11
1	6	950	12.5
2	4	950	12.5
1	2	955	14
1	12	975	15
2	8	1000	16
1	11	1025	17
2	9	1050	18
1	10	1055	19
1	1	1095	20
1	4	1195	21
1	3	1200	22

TABLE 18.13 COMBINED RANKING OF THE DATA IN THE TWO SAMPLES FROM THIRD NATIONAL BANK

Branch 1			Branch 2		
Account	**Balance ($)**	**Rank**	**Account**	**Balance ($)**	**Rank**
1	1095	20	1	885	7
2	955	14	2	850	4
3	1200	22	3	915	8
4	1195	21	4	950	12.5
5	925	9	5	800	2
6	950	12.5	6	750	1
7	805	3	7	865	5
8	945	11	8	1000	16
9	875	6	9	1050	18
10	1055	19	10	935	10
11	1025	17		Sum of Ranks	83.5
12	975	15			
	Sum of Ranks	169.5			

We now return to the two separate samples and show the ranks from Table 18.12 for each account balance. These results are provided in Table 18.13. The next step is to sum the ranks for each sample: 169.5 for sample 1 and 83.5 for sample 2 are shown. As stated previously, we will always follow the procedure of using the sum of the ranks for sample 1 as the test statistic W. Thus, we have $W = 169.5$. When both samples sizes are 7 or more, a normal approximation of the sampling distribution of W can be used. Under the assumption that the null hypothesis is true and the populations are identical, the sampling distribution of the test statistic W is as follows.

SAMPLING DISTRIBUTION OF *W* WITH IDENTICAL POPULATIONS

$$\text{Mean: } \mu_W = \tfrac{1}{2}n_1(n_1 + n_2 + 1) \tag{18.5}$$

$$\text{Standard deviation: } \sigma_W = \sqrt{\tfrac{1}{12}n_1n_2(n_1 + n_2 + 1)} \tag{18.6}$$

Distribution form: Approximately normal provided $n_1 \geq 7$ and $n_2 \geq 7$

Given the sample sizes $n_1 = 12$ and $n_2 = 10$, equations (18.5) and (18.6) provide the following mean and standard deviation for the sampling distribution:

$$\mu_W = \tfrac{1}{2}n_1(n_1 + n_2 + 1) = \tfrac{1}{2}(12)(12 + 10 + 1) = 138$$

$$\sigma_W = \sqrt{\tfrac{1}{12}n_1n_2(n_1 + n_2 + 1)} = \sqrt{\tfrac{1}{12}(12)(10)(12 + 10 + 1)} = 15.1658$$

Figure 18.6 shows the normal distribution used for the sampling distribution of W.

Let us proceed with the MWW test and use a .05 level of significance to draw a conclusion. Since the test statistic W is discrete and the normal distribution is continuous, we will again use the continuity correction factor for the normal distribution approximation.

FIGURE 18.6 SAMPLING DISTRIBUTION OF W FOR THE THIRD NATIONAL BANK EXAMPLE

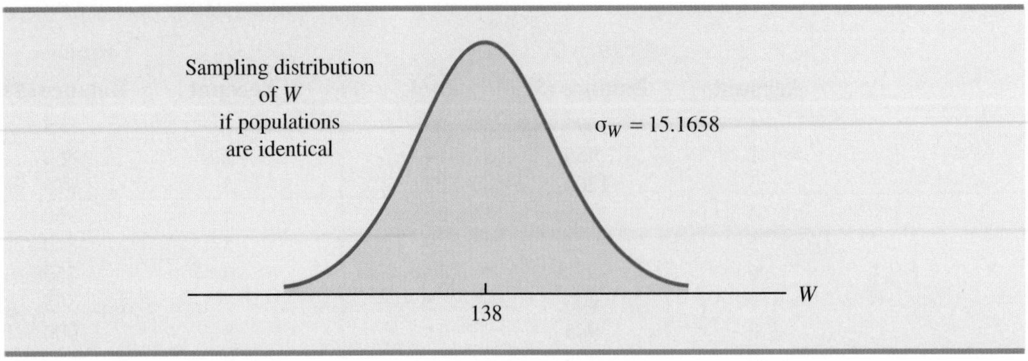

With $W = 169.5$ in the upper tail of the sampling distribution, we have the following p-value calculation:

$$P(W \geq 169.5) = P\left(z \geq \frac{169 - 138}{15.1658}\right) = P(z \geq 2.04)$$

Using the standard normal random variable and $z = 2.04$, the two-tailed p-value $= 2(1-.9793) = .0414$. With p-value $\leq .05$, reject H_0 and conclude that the two populations of account balances are not identical. The upper tail value for test statistic W indicates that the population of account balances at branch 1 tends to be larger.

If the assumption can be made that the two populations have the same shape, the MWW test becomes a test about the difference between the medians of the two populations.

As a final comment, some applications of the MWW test make it appropriate to assume that the two populations have identical shapes and if the populations differ, it is only by a shift in the location of the distributions. If the two populations have the same shape, the hypothesis test may be stated in terms of the difference between the two population medians. Any difference between the medians can be interpreted as the shift in location of one population compared to the other. In this case, the three forms of the MWW test about the medians of the two populations are as follows:

Two-Tailed Test	Lower Tail Test	Upper Tail Test
H_0: Median$_1$ − Median$_2$ = 0	H_0: Median$_1$ − Median$_2$ ≥ 0	H_0: Median$_1$ − Median$_2$ ≤ 0
H_a: Median$_1$ − Median$_2$ ≠ 0	H_a: Median$_1$ − Median$_2$ < 0	H_a: Median$_1$ − Median$_2$ > 0

NOTES AND COMMENTS

The Minitab procedure for the MWW test is described in Appendix 18.1. Minitab makes the assumption that the two populations have the same shape. As a result, Minitab describes the test results in terms of a test about the medians of the two populations. If you do not feel comfortable making the "same shape" assumption, Minitab results are still applicable. However, you need to interpret the results as a test of the null hypothesis that the two populations are identical.

Exercises

Applications

SELF test

18. Two fuel additives are being tested to determine their effect on gasoline mileage. Seven cars were tested with additive 1 and nine cars were tested with additive 2. The following data show the miles per gallon obtained with the two additives. Use $\alpha = .05$ and the MWW test to see whether there is a significant difference between gasoline mileage for the two additives.

Additive 1	Additive 2
17.3	18.7
18.4	17.8
19.1	21.3
16.7	21.0
18.2	22.1
18.6	18.7
17.5	19.8
	20.7
	20.2

SELF test

19. Samples of starting annual salaries for individuals entering the public accounting and financial planning professions follow. Annual salaries are shown in thousands of dollars.

WEB file

AcctPlanners

Public Accountant	Financial Planner
50.2	49.0
58.8	49.2
56.3	53.1
58.2	55.9
54.2	51.9
55.0	53.6
50.9	49.7
59.5	53.9
57.0	51.8
51.9	48.9

a. Use a .05 level of significance and test the hypothesis that there is no difference between the starting annual salaries of public accountants and financial planners. What is your conclusion?
b. What are the sample median annual salaries for the two professions?

20. The gap between the earnings of men and women with equal education is narrowing but has not closed. Sample data for seven men and seven women with bachelor's degrees are as follows. Data are shown in thousands of dollars.

Men	Women
35.6	49.5
80.5	40.4
50.2	32.9
67.2	45.5
43.2	30.8
54.9	52.5
60.3	29.8

a. What is the median salary for men? For women?
b. Use $\alpha = .05$ and conduct the hypothesis test for identical population distributions. What is your conclusion?

21. Unisys maintains a hurricane database which provides information on hurricanes in the Atlantic/Caribbean/Gulf of Mexico and the Eastern Pacific Ocean. Maximum wind speeds in knots for a sample of hurricanes over the past four hurricane seasons are shown (Unisys weather website, April 2009).

WEB file
Hurricanes

Atlantic/Carribean/Gulf of Mexico		Eastern Pacific Ocean	
Storm	**Max Wind Speed**	**Storm**	**Max Wind Speed**
Frances	125	Darby	105
Jeanne	110	Frank	75
Lisa	65	Isis	65
Emily	135	Hilary	90
Ophelia	80	Max	70
Rita	150	Bud	110
Wilma	150	Daniel	130
Ernesto	65	Sergio	95
Florence	80	Cosme	65
Helene	105	Flossie	120
Dean	145	Henriette	75
Karen	60	Ivo	70

Use $\alpha = .05$ and test to determine whether the distribution of hurricane wind speeds is the same for these two regions. What is your conclusion?

22. Each year *Bloomberg Businessweek* publishes statistics on the world's 1000 largest companies. A company's price/earnings (P/E) ratio is the company's current stock price divided by the latest 12 months' earnings per share. The following table shows the P/E ratios for a sample of 10 Japanese companies and 12 U.S. companies. Is the difference between the P/E ratios for the two countries significant? Use the MWW test and $\alpha = .01$ to support your conclusion.

WEB file
JapanUS

Japan		United States	
Company	**P/E Ratio**	**Company**	**P/E Ratio**
Sumitomo Corp.	153	Gannet	19
Kinden	21	Motorola	24
Heiwa	18	Schlumberger	24
NCR Japan	125	Oracle Systems	43
Suzuki Motor	31	Gap	22
Fuji Bank	213	Winn-Dixie	14
Sumintomo Chemical	64	Ingersoll-Rand	21
Seibu Railway	666	American Electric	14
Shiseido	33	Hercules	21
Toho Gas	68	Times Mirror	38
		WellPoint Health	15
		Northern States Power	14

23. Police records show the following numbers of daily crime reports for a sample of days during the winter months and a sample of days during the summer months. Use a .05 level of significance to determine whether there is a significant difference between the winter and summer months in terms of the number of crime reports. What is your conclusion?

WEB file
PoliceRecords

Winter	Summer
18	28
20	18
15	24
16	32
21	18
20	29
12	23
16	38
19	28
20	18

24. A certain brand of microwave oven was priced at 10 stores in Dallas and 13 stores in San Antonio. The data follow. Use a .05 level of significance and test whether prices for the microwave oven are the same in the two cities.

WEB file
Microwave

Dallas	San Antonio
445	460
489	451
405	435
485	479
439	475
449	445
436	429
420	434
430	410
405	422
	425
	459
	430

25. The National Football League (NFL) holds its annual draft of the nation's best college football players in April each year. Prior to the draft, various sporting news services project the players who will be drafted along with the order in which each will be selected. The better players are selected earlier in the draft. For the 2009 draft, the colleges from the Southeastern Conference (SEC) and the Atlantic Coast Conference (ACC) were projected to have the most players who would be selected during the first round (SportProjection website, March 15, 2009). The player's college and the projected draft position for seven players from each conference are as follows.

Southeastern Conference		Atlantic Coast Conference	
Player's College	**Projected Draft Position**	**Player's College**	**Projected Draft Position**
Georgia	1	Georgia Tech	3
Alabama	2	Wake Forest	6
Vanderbilt	14	Virginia	8
Florida	18	Wake Forest	23
Mississippi	20	Florida State	25
Mississippi	24	Maryland	26
Auburn	27	Virginia	29

Using the projected draft position as an indicator of preference the NFL teams have for the two conferences, use the MWW test to determine if there is any difference between the NFL preferences for players from these two conferences? Use $\alpha = .05$. What is the p-value? What is your conclusion?

18.4 Kruskal-Wallis Test

In this section we extend the nonparametric procedures to hypothesis tests involving three or more populations. We considered a parametric test for this situation in Chapter 13 when we used quantitative data and assumed that the populations had normal distributions with the same standard deviations. Based on an independent random sample from each population, we used the F distribution to test for differences among the population means.

The nonparametric **Kruskal-Wallis test** is based on the analysis of independent random samples from each of k populations. This procedure can be used with either ordinal data or quantitative data and does not require the assumption that the populations have normal distributions. The general form of the null and alternative hypotheses is as follows:

$$H_0\text{: All populations are identical}$$
$$H_a\text{: Not all populations are identical}$$

If H_0 is rejected, we will conclude that there is a difference among the populations with one or more populations tending to provide smaller or larger values compared to the other populations. We will demonstrate the Kruskal-Wallis test using the following example.

Williams Manufacturing Company hires employees for its management staff from three different colleges. Recently, the company's personnel director began reviewing the annual performance reports for the management staff in an attempt to determine whether there are differences in the performance ratings among the managers who graduated from the three colleges. Performance rating data are available for independent samples of seven managers who graduated from college A, six managers who graduated from college B, and seven managers who graduated from college C. These data are summarized in Table 18.14. The performance rating shown for each manger is recorded on a scale from 0 to 100, with 100 being the highest possible rating. Suppose we want to test whether the three populations of managers are identical in terms of performance ratings. We will use a .05 level of significance for the test.

TABLE 18.14 PERFORMANCE EVALUATION RATINGS FOR 20 WILLIAMS EMPLOYEES

College A	College B	College C
25	60	50
70	20	70
60	30	60
85	15	80
95	40	90
90	35	70
80		75

The first step in the Kruskal-Wallis procedure is to rank the combined samples from lowest to highest values. Using all 20 observations in Table 18.14, the lowest rating of 15 for the 4th manager in the college B sample receives a rank of 1. The highest rating of 95 for the 5th manager in the college A sample receives a rank of 20. The performance rating data and their assigned ranks are shown in Table 18.15. Note that we assigned the average ranks to tied performance ratings of 60, 70, 80, and 90. Table 18.15 also shows the sum of ranks for each of the three samples.

TABLE 18.15 COMBINED RANKINGS FOR THE THREE SAMPLES

College A	Rank	College B	Rank	College C	Rank
25	3	60	9	50	7
70	12	20	2	70	12
60	9	30	4	60	9
85	17	15	1	80	15.5
95	20	40	6	90	18.5
90	18.5	35	5	70	12
80	15.5	Sum of Ranks	27	75	14
Sum of Ranks	95			Sum of Ranks	88

The Kruskal-Wallis test statistic uses the sum of the ranks for the three samples and is computed as follows.

KRUSKAL-WALLIS TEST STATISTIC

$$H = \left[\frac{12}{n_T(n_T + 1)}\sum_{i=1}^{k}\frac{R_i^2}{n_i}\right] - 3(n_T + 1) \quad \textbf{(18.7)}$$

where

k = the number of populations

n_i = the number of observations in sample i

$n_T = \sum_{i=1}^{k} n_i$ = the total number of observations in all samples

R_i = the sum of the ranks for sample i

Kruskal and Wallis were able to show that, under the null hypothesis assumption of identical populations, the sampling distribution of H can be approximated by a chi-square distribution with $(k - 1)$ degrees of freedom. This approximation is generally acceptable if the sample sizes for each of the k populations are all greater than or equal to five. The null hypothesis of identical populations will be rejected if the test statistic H is large. As a result, the Kruskal-Wallis test is always expressed as an upper tail test. The computation of the test statistic for the sample data in Table 18.15 is as follows.

The sample sizes are

$$n_1 = 7 \qquad n_2 = 6 \qquad n_3 = 7$$

and

$$n_T = \sum_{i=1}^{3} n_i = 7 + 6 + 7 = 20$$

Using the sum of ranks for each sample, the value of the Kruskal-Wallis test statistic is as follows:

$$H = \left[\frac{12}{n_T(n_T + 1)}\sum_{i=1}^{k}\frac{R_i^2}{n_i}\right] - 3(n_T + 1) = \frac{12}{20(21)}\left[\frac{(95)^2}{7} + \frac{(27)^2}{6} + \frac{(88)^2}{7}\right] - 3(20 + 1) = 8.92$$

We can now use the chi-square distribution table (Table 3 of Appendix B) to determine the p-value for the test. Using $k - 1 = 3 - 1 = 2$ degrees of freedom, we find $\chi^2 = 7.378$ has an area of .025 in the upper tail of the chi-square distribution and $\chi^2 = 9.21$ has an area of .01 in the upper tail of the chi-square distribution. With $H = 8.92$ between 7.378 and 9.21, we can conclude that the area in the upper tail of the chi-square distribution is between .025 and .01. Because this is an upper tail test, we conclude that the p-value is between .025 and .01. Using Minitab or Excel will show the exact p-value for $\chi^2 = 8.92$ is .0116. Because p-value $\leq \alpha = .05$, we reject H_0 and conclude that the three populations are not all the same. The three populations of performance ratings are not identical and differ significantly depending upon the college. Because the sum of the ranks is relatively low for the sample of managers who graduated from college B, it would be reasonable for the company to either reduce its recruiting from college B, or at least evaluate the college B graduates more thoroughly before making a hiring decision.

If the assumption can be made that the populations all have the same shape, the Kruskal-Wallis test becomes a test about the medians of the k populations.

As a final comment, we note that in some applications of the Kruskal-Wallis test it may be appropriate to make the assumption that the populations have identical shapes and if they differ, it is only by a shift in location for one or more of the populations. If the k populations are assumed to have the same shape, the hypothesis test can be stated in terms of the population medians. In this case, the hypotheses for the Kruskal-Wallis test would be written as follows:

$$H_0\text{: Median}_1 = \text{Median}_2 = \cdots = \text{Median}_k$$
$$H_a\text{: Not all Medians are equal}$$

NOTES AND COMMENTS

1. The example in this section used quantitative data on employee performance ratings to conduct the Kruskal-Wallis test. This test could also have been used if the data were the ordinal rankings of the 20 employees in terms of performance. In this case, the test would use the ordinal data directly. The step of converting the quantitative data into rank-ordered data would not be necessary. Exercise 30 illustrates this situation.
2. The Minitab procedure for the Kruskal-Wallis test is described in Appendix 18.1. Minitab makes the assumption that the populations all have the same shape. As a result, Minitab describes the Kruskal-Wallis test as a test of differences among the population medians. If you do not feel comfortable making the "same shape" assumption, you can still use Minitab. However, you will need to interpret the results as a test of the null hypothesis that all populations are identical.

Exercises

Applications

SELF test

26. A sample of 15 consumers provided the following product ratings for three different products. Five consumers were randomly assigned to test and rate each product. Use the Kruskal-Wallis test and $\alpha = .05$ to determine whether there is a significant difference among the ratings for the products.

Product		
A	B	C
50	80	60
62	95	45
75	98	30
48	87	58
65	90	57

WEB file
TestPrepare

27. Three admission test preparation programs are being evaluated. The scores obtained by a sample of 20 people who used the programs provided the following data. Use the Kruskal-Wallis test to determine whether there is a significant difference among the three test preparation programs. Use $\alpha = .05$.

	Program	
A	**B**	**C**
540	450	600
400	540	630
490	400	580
530	410	490
490	480	590
610	370	620
	550	570

SELF test

28. Forty-minute workouts of one of the following activities three days a week will lead to a loss of weight. The following sample data show the number of calories burned during 40-minute workouts for three different activities. Do these data indicate differences in the amount of calories burned for the three activities? Use a .05 level of significance. What is your conclusion?

Swimming	Tennis	Cycling
408	415	385
380	485	250
425	450	295
400	420	402
427	530	268

WEB file
CruiseShips

29. *Condé Nast Traveler* magazine conducts an annual survey of its readers in order to rate the top 80 cruise ships in the world (*Condé Nast Traveler,* February 2008). With 100 the highest possible rating, the overall ratings for a sample of ships from the Holland America, Princess, and Royal Caribbean cruise lines are shown. Use the Kruskal-Wallis test with $\alpha = .05$ to determine whether the overall ratings among the three cruise lines differ significantly. What is your conclusion?

Holland America		Princess		Royal Caribbean	
Ship	**Rating**	**Ship**	**Rating**	**Ship**	**Rating**
Amsterdam	84.5	Coral	85.1	Adventure	84.8
Maasdam	81.4	Dawn	79.0	Jewel	81.8
Ooterdam	84.0	Island	83.9	Mariner	84.0
Volendam	78.5	Princess	81.1	Navigator	85.9
Westerdam	80.9	Star	83.7	Serenade	87.4

30. A large corporation sends many of its first-level managers to an off-site supervisory skills training course. Four different management development centers offer this course. The director of human resources would like to know whether there is a difference among the quality of training provided at the four centers. An independent random sample of five employees was chosen from each training center. The employees were then ranked 1 to 20 in terms of supervisory skills. A rank of 1 was assigned to the employee with the best supervisory skills. The ranks are shown. Use $\alpha = .05$ and test whether there is a significant difference among the quality of training provided by the four programs.

Course			
A	B	C	D
3	2	19	20
14	7	16	4
10	1	9	15
12	5	18	6
13	11	17	8

31. The better-selling candies are often high in calories. Assume that the following data show the calorie content from samples of M&M's, Kit Kat, and Milky Way II. Test for significant differences among the calorie content of these three candies. At a .05 level of significance, what is your conclusion?

M&M's	Kit Kat	Milky Way II
230	225	200
210	205	208
240	245	202
250	235	190
230	220	180

18.5 Rank Correlation

The Pearson product moment correlation coefficient introduced in Chapter 3 is a measure of the linear association between two variables using quantitative data. In this section, we provide a correlation measure of association between two variables when ordinal or rank-ordered data are available. The **Spearman rank-correlation coefficient** has been developed for this purpose.

SPEARMAN RANK-CORRELATION COEFFICIENT

$$r_s = 1 - \frac{6 \sum_{i=1}^{n} d_i^2}{n(n^2 + 1)} \tag{18.8}$$

where

n = the number of observations in the sample
x_i = the rank of observation i with respect to the first variable
y_i = the rank of observation i with respect to the second variable
$d_i = x_i - y_i$

Let us illustrate the use of the Spearman rank-correlation coefficient. A company wants to determine whether individuals who had a greater potential at the time of employment turn out to have higher sales records. To investigate, the personnel director reviewed the original job interview reports, academic records, and letters of recommendation for 10 current members of the sales force. After the review, the director ranked the 10 individuals

TABLE 18.16 SALES POTENTIAL AND ACTUAL TWO-YEAR SALES DATA

Salesperson	Ranking of Potential	Two-Year Sales (units)	Ranking According to Two-Year Sales
A	2	400	1
B	4	360	3
C	7	300	5
D	1	295	6
E	6	280	7
F	3	350	4
G	10	200	10
H	9	260	8
I	8	220	9
J	5	385	2

WEB file

PotentialActual

in terms of their potential for success at the time of employment and assigned the individual who had the most potential the rank of 1. Data were then collected on the actual sales for each individual during their first two years of employment. On the basis of the actual sales records, a second ranking of the 10 individuals based on sales performance was obtained. Table 18.16 provides the ranks based on potential as well as the ranks based on the actual performance.

Let us compute the Spearman rank-correlation coefficient for the data in Table 18.16. The computations are summarized in Table 18.17. We first compute the difference between the two ranks for each salesperson, d_i, as shown in column 4. The sum of the d_i^2 in column 5 is 44. This value and the sample size $n = 10$ are used to compute the rank-correlation coefficient $r_s = .733$ shown in Table 18.17.

The Spearman rank-correlation coefficient ranges from -1.0 to $+1.0$ and its interpretation is similar to the Pearson product moment correlation coefficient for quantitative data.

TABLE 18.17 COMPUTATION OF THE SPEARMAN RANK-CORRELATION COEFFICIENT FOR SALES POTENTIAL AND SALES PERFORMANCE

Salesperson	x_i = Ranking of Potential	y_i = Ranking of Sales Performance	$d_i = x_i - y_i$	d_i^2
A	2	1	1	1
B	4	3	1	1
C	7	5	2	4
D	1	6	−5	25
E	6	7	−1	1
F	3	4	−1	1
G	10	10	0	0
H	9	8	1	1
I	8	9	−1	1
J	5	2	3	9
				$\Sigma d_i^2 = 44$

$$r_s = 1 - \frac{6\sum d_i^2}{n(n^2+1)} = 1 - \frac{6(44)}{10(100-1)} = .733$$

A rank-correlation coefficient near +1.0 indicates a strong positive association between the ranks for the two variables, while a rank-correlation coefficient near −1.0 indicates a strong negative association between the ranks for the two variables. A rank-correlation coefficient of 0 indicates no association between the ranks for the two variables. In the example, r_s = .733 indicates a positive correlation between the ranks based on potential and the ranks based on sales performance. Individuals who ranked higher in potential at the time of employment tended to rank higher in two-year sales performance.

At this point, we may want to use the sample rank correlation r_s to make an inference about the population rank correlation coefficient ρ_s. To do this, we test the following hypotheses:

$$H_0: \rho_s = 0$$
$$H_a: \rho_s \neq 0$$

Under the assumption that the null hypothesis is true and the population rank-correlation coefficient is 0, the following sampling distribution of r_s can be used to conduct the test.

SAMPLING DISTRIBUTION OF r_s

$$\text{Mean: } \mu_{r_s} = 0 \qquad \textbf{(18.9)}$$

$$\text{Standard deviation: } \sigma_{r_s} = \sqrt{\frac{1}{n-1}} \qquad \textbf{(18.10)}$$

Distribution form: Approximately normal provided $n \geq 10$

The sample rank-correlation coefficient for sales potential and sales performance is $r_s = .733$. Using equation (18.9), we have $\mu_{r_s} = 0$, and using equation (18.10), we have $\sigma_{r_s} = \sqrt{1/(10-1)} = .333$. With the sampling distribution of r_s approximated by a normal distribution, the standard normal random variable z becomes the test statistic with

$$z = \frac{r_s - \mu_{r_s}}{\sigma_{r_s}} = \frac{.733 - 0}{.333} = 2.20$$

Using the standard normal probability table and $z = 2.20$, we find the two-tailed p-value = $2(1 - .9861) = .0278$. With a .05 level of significance, p-value $\leq \alpha$. Thus, we reject the null hypothesis that the population rank-correlation coefficient is zero. The test result shows that there is a significant rank correlation between potential at the time of employment and actual sales performance.

NOTES AND COMMENTS

The Spearman rank-correlation coefficient provides the same value that is obtained by using the Pearson product moment correlation coefficient procedure with the rank-ordered data. In Appendixes 18.1 and 18.2, we show how Minitab and Excel correlation tools for the Pearson product moment correlation coefficient can be used to compute the Spearman rank-correlation coefficient.

Exercises

Methods

SELF test

32. Consider the following set of rankings for a sample of 10 elements.

Element	x	y	Element	x	y
1	10	8	6	2	7
2	6	4	7	8	6
3	7	10	8	5	3
4	3	2	9	1	1
5	4	5	10	9	9

a. Compute the Spearman rank-correlation coefficient for the data.
b. Use $\alpha = .05$ and test for significant rank correlation. What is your conclusion?

33. Consider the following two sets of rankings for six items.

Case One			Case Two		
Item	First Ranking	Second Ranking	Item	First Ranking	Second Ranking
A	1	1	A	1	6
B	2	2	B	2	5
C	3	3	C	3	4
D	4	4	D	4	3
E	5	5	E	5	2
F	6	6	F	6	1

Note that in the first case the rankings are identical, whereas in the second case the rankings are exactly opposite. What value should you expect for the Spearman rank-correlation coefficient for each of these cases? Explain. Calculate the rank-correlation coefficient for each case.

Applications

SELF test

34. The following data show the rankings of 11 states based on expenditure per student (ranked 1 highest to 11 lowest) and student-teacher ratio (ranked 1 lowest to 11 highest).

WEB file
Student

State	Expenditure per Student	Student-Teacher Ratio
Arizona	9	10
Colorado	5	8
Florida	4	6
Idaho	2	11
Iowa	6	4
Louisiana	11	3
Massachusetts	1	1
Nebraska	7	2
North Dakota	8	7
South Dakota	10	5
Washington	3	9

a. What is the rank correlation between expenditure per student and student-teacher ratio. Discuss.
b. At the $\alpha = .05$ level, does there appear to be a relationship between expenditure per student and student-teacher ratio?

35. A national study by Harris Interactive, Inc., evaluated the top technology companies and their reputations. The following shows how 10 technology companies ranked in reputation and how the companies ranked in percentage of respondents who said they would purchase the company's stock. A positive rank correlation is anticipated because it seems reasonable to expect that a company with a higher reputation would have the more desirable stock to purchase.

WEB file

Techs

Company	Reputation	Stock Purchase
Microsoft	1	3
Intel	2	4
Dell	3	1
Lucent	4	2
Texas Instruments	5	9
Cisco Systems	6	5
Hewlett-Packard	7	10
IBM	8	6
Motorola	9	7
Yahoo	10	8

a. Compute the rank correlation between reputation and stock purchase.
b. Test for a significant positive rank correlation. What is the p-value?
c. At $\alpha = .05$, what is your conclusion?

36. The rankings of a sample of professional golfers in both driving distance and putting are shown. What is the rank correlation between driving distance and putting for these golfers? Test for significance of the correlation coefficient at the .10 level of significance.

WEB file

ProGolfers

Golfer	Driving Distance	Putting
Fred Couples	1	5
David Duval	5	6
Ernie Els	4	10
Nick Faldo	9	2
Tom Lehman	6	7
Justin Leonard	10	3
Davis Love III	2	8
Phil Mickelson	3	9
Greg Norman	7	4
Mark O'Meara	8	1

37. A student organization surveyed both current students and recent graduates to obtain information on the quality of teaching at a particular university. An analysis of the responses provided the following teaching-ability rankings. Do the rankings given by the current students agree with the rankings given by the recent graduates? Use $\alpha = .10$ and test for a significant rank correlation.

WEB file

Professors

Professor	Current Students	Recent Graduates
1	4	6
2	6	8
3	8	5
4	3	1
5	1	2
6	2	3
7	5	7
8	10	9
9	7	4
10	9	10

Summary

In this chapter we have presented statistical procedures that are classified as nonparametric methods. Because these methods can be applied to categorical data as well as quantitative data and because they do not require an assumption about the distribution of the population, they expand the number of situations that can be subjected to statistical analysis.

The sign test is a nonparametric procedure for testing a hypothesis about a population median or for testing a hypothesis with matched samples. The data must be summarized in two categories, one denoted by a plus sign and one denoted by a minus sign. The Wilcoxon signed-rank test analyzes matched samples from two populations when quantitative data are available. No assumption is required other than the distribution of the paired differences is symmetric. The Wilcoxon signed-rank test is used to determine if the median of the population of paired differences is zero. This test can also be used to make inferences about the median of a symmetric population.

The Mann-Whitney-Wilcoxon test is a nonparametric procedure for the difference between two populations based on two independent samples. It is an alternative to the parametric t test for the difference between the means of the two populations. The combined ranks for the data from the two samples are obtained and the test statistic for the MWW test is the sum of ranks for the first sample. In most applications, the samples sizes are large enough to use a normal approximation with the continuity correction factor in conducting the hypothesis test. If no assumption is made about the populations, the MWW procedure tests whether the two populations are identical. If the assumption can be made that the two populations have the same shape, the test provides an inference about the difference between the medians of the two populations.

The Kruskal-Wallis test extends the MWW test to three or more populations. It is an alternative to the parametric analysis of variance test for the differences among the means of three or more normally distributed populations. The Kruskal-Wallis test does not require any assumption about the distribution of the populations and uses the null hypothesis that the k populations are identical. If the assumption can be made that the populations have the same shape, the test provides an inference about differences among the medians of the k populations. In the last section of the chapter we introduced the Spearman rank-correlation coefficient as a measure of association between two variables based on rank-ordered data.

Glossary

Parametric methods Statistical methods that begin with an assumption about the probability distribution of the population which is often that the population has a normal distribution. A sampling distribution for the test statistic can then be derived and used to make an inference about one or more parameters of the population such as the population mean μ or the population standard deviation σ.

Nonparametric methods Statistical methods that require no assumption about the form of the probability distribution of the population and are often referred to as distribution-free methods. Several of the methods can be applied with categorical as well as quantitative data.

Distribution-free methods Statistical methods that make no assumption about the probability distribution of the population.

Sign test A nonparametric test for a hypothesis about a population median or for identifying differences between two populations based on matched samples. The data are summarized in two categories, denoted by a plus sign or a minus sign, and the binomial distribution with $p = .50$ provides the sampling distribution for the test statistic.

Wilcoxon signed-rank test A nonparametric test for the difference between the medians of two populations based on matched samples. The procedure uses quantitative data and is based on the assumption that the distribution of differences is symmetric. The paired-difference data are used to make an inference about the medians of the two populations. This test can also be used to make inferences about the median of a symmetric population.

Mann-Whitney-Wilcoxon (MWW) test A nonparametric test for the difference between two populations based on an independent sample from each population. The null hypothesis is that the two populations are identical. If the assumption can be made that the populations have the same shape, this test provides an inference about the difference between the medians of the two populations.

Kruskal-Wallis test A nonparametric test for the differences among three or more populations based on the analysis of an independent sample from each population. The null hypothesis is that the populations are identical. If the assumption can be made that the populations have the same shape, this test provides an inference about the differences among the medians of the populations.

Spearman rank-correlation coefficient A correlation measure of the association between two variables based on rank-ordered data.

Key Formulas

Sign Test: Normal Approximation

$$\text{Mean: } \mu = .50n \tag{18.1}$$

$$\text{Standard Deviation: } \sigma = \sqrt{.25n} \tag{18.2}$$

Wilcoxon Signed-Rank Test: Normal Approximation

$$\text{Mean: } \mu_{T^+} = \frac{n(n+1)}{4} \tag{18.3}$$

$$\text{Standard deviation: } \sigma_{T^+} = \sqrt{\frac{n(n+1)(2n+1)}{24}} \tag{18.4}$$

Mann-Whitney-Wilcoxon Test: Normal Approximation

$$\text{Mean: } \mu_W = \tfrac{1}{2}n_1(n_1 + n_2 + 1) \tag{18.5}$$

$$\text{Standard deviation: } \sigma_W = \sqrt{\tfrac{1}{12}\, n_1 n_2 (n_1 + n_2 + 1)} \tag{18.6}$$

Kruskal-Wallis Test Statistic

$$H = \left[\frac{12}{n_T(n_T+1)} \sum_{i=1}^{k} \frac{R_i^2}{n_i}\right] - 3(n_T + 1) \tag{18.7}$$

Spearman Rank-Correlation Coefficient

$$r_s = 1 - \frac{6\sum_{i=1}^{n} d_i^2}{n(n^2+1)} \tag{18.8}$$

Supplementary Exercises

38. A survey asked the following question: Do you favor or oppose providing tax-funded vouchers or tax deductions to parents who send their children to private schools? Of the 2010 individuals surveyed, 905 favored the proposal, 1045 opposed the proposal, and 60 offered no opinion. Do the data indicate a significant difference in the preferences for the financial support of parents who send their children to private schools? Use a .05 level of significance.

39. Due to a recent decline in the housing market, the national median sales price for single-family homes is \$180,000 (The National Association of Realtors, January 2009). Assume that the following data were obtained from samples of recent sales of single-family homes in St. Louis and Denver.

Metropolitan Area	Less Than \$180,000	Equal to \$180,000	Greater Than \$180,000
St. Louis	32	2	18
Denver	13	1	27

a. Is the median sales price in St. Louis significantly lower than the national median of \$180,000? Use a statistical test with $\alpha = .05$ to support your conclusion.
b. Is the median sales price in Denver significantly higher than the national median of \$180,000? Use a statistical test with $\alpha = .05$ to support your conclusion.

40. Twelve homemakers were asked to estimate the retail selling price of two models of refrigerators. Their estimates of selling price are shown in the following table. Use these data and test at the .05 level of significance to determine whether there is a difference between the two models in terms of homemakers' perceptions of selling price.

WEB file
Refrigerators

Homemaker	Model 1	Model 2	Homemaker	Model 1	Model 2
1	\$850	\$1100	7	\$900	\$1090
2	960	920	8	890	1120
3	940	890	9	1100	1200
4	900	1050	10	700	890
5	790	1120	11	810	900
6	820	1000	12	920	900

41. A study was designed to evaluate the weight-gain potential of a new poultry feed. A sample of 12 chickens was used in a six-week study. The weight of each chicken was recorded before and after the six-week test period. The differences between the before and after weights of the 12 chickens are as follows: 1.5, 1.2, −.2, .0, .5, .7, .8, 1.0, .0, .6, .2, −.01. A positive difference indicates a weight gain and a negative difference indicates a weight loss. Use a .05 level of significance to determine if the new feed provides a significant weight gain for the chickens.

42. The following data are product weights for the same items produced on two different production lines. Test for a difference between the product weights for the two lines. Use $\alpha = .05$.

WEB file
ProductWeights

Line 1	Line 2
13.6	13.7
13.8	14.1
14.0	14.2
13.9	14.0
13.4	14.6
13.2	13.5
13.3	14.4
13.6	14.8
12.9	14.5
14.4	14.3
	15.0
	14.9

43. A client wants to determine whether there is a significant difference in the time required to complete a program evaluation with the three different methods that are in common use. The times (in hours) required for each of 18 evaluators to conduct a program evaluation follow. Use $\alpha = .05$ and test to see whether there is a significant difference in the time required by the three methods.

WEB file
Methods

Method 1	Method 2	Method 3
68	62	58
74	73	67
65	75	69
76	68	57
77	72	59
72	70	62

44. A sample of 20 engineers employed with a company for three years has been rank ordered with respect to managerial potential. Some of the engineers attended the company's management-development course, others attended an off-site management-development program at a local university, and the remainder did not attend any program. Use the following rankings and $\alpha = .025$ to test for a significant difference in the managerial potential of the three groups.

WEB file
Programs

No Program	Company Program	Off-Site Program
16	12	7
9	20	1
10	17	4
15	19	2
11	6	3
13	18	8
	14	5

45. Course evaluation ratings for four college instructors are shown in the following table. Use $\alpha = .05$ and test for a significant difference among the rating for these instructors. What is your conclusion?

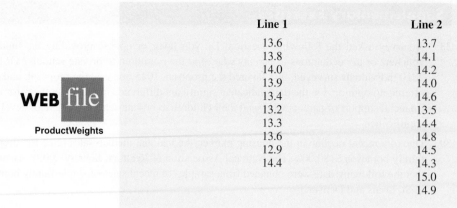

WEB file Evaluations

Instructor			
Black	**Jennings**	**Swanson**	**Wilson**
88	87	88	80
80	78	76	85
79	82	68	56
68	85	82	71
96	99	85	89
69	99	82	87
	85	84	
	94	83	
		81	

46. A sample of 15 students received the following rankings on midterm and final examinations in a statistics course. Compute the Spearman rank-correlation coefficient for the data and test for a significant correlation with $\alpha = .10$. What is the p-value and what is your conclusion?

WEB file Exams

Rank		Rank		Rank	
Midterm	**Final**	**Midterm**	**Final**	**Midterm**	**Final**
1	4	6	2	11	14
2	7	7	5	12	15
3	1	8	12	13	11
4	3	9	6	14	10
5	8	10	9	15	13

47. Nielsen Research provides weekly ratings of nationally broadcast television programs. The ratings of the 84 prime-time programs broadcast by the four major televisions networks (ABC, CBS, FOX, and NBC) for the week of April 14–20, 2008, are provided in the file named NielsenResearch. The ratings range from 1 to 103. Shown are the ratings for 12 shows in the file (days and times for shows that aired multiple episodes are shown). Do these data suggest that the overall ratings for the four networks differ significantly? Use the Kruskal-Wallis test with a .10 level of significance. What is the p-value and what is your conclusion?

WEB file NielsenResearch

Program	Network	Rating
20/20	ABC	60
30 Rock	NBC	44
48 Hours Mystery (Sat. 10:00 P.M)	CBS	51
48 Hours Mystery (Sat. 9:00 P.M)	CBS	78
48 Hours Mystery (Tues. 10:00 P.M)	CBS	63
60 Minutes	CBS	13
According to Jim (Tues. 8:00 P.M)	ABC	89
According to Jim (Tues. 8:30 P.M)	ABC	91
American Dad (Sun. 7:30 P.M)	FOX	100
American Dad (Sun. 9:30 P.M)	FOX	65
American Idol (Tues. 8:00 P.M)	FOX	1
American Idol (Wed. 9:00 P.M)	FOX	2

Appendix 18.1 Nonparametric Methods with Minitab

Minitab can be used for all the nonparametric methods introduced in this chapter.

Sign Test for a Hypothesis Test About a Population Median

WEB file HomeSales

We illustrate a hypothesis test about a population median using the sales price data for new homes in Section 18.1. The prices appear in column C1 of the Minitab worksheet named HomeSales. The following steps can be used to test the hypotheses H_0: Median $\geq$ \$236,000 versus H_a: Median $<$ \$236,000.

Step 1. Select the **Stat** menu
Step 2. Choose **Nonparametrics**
Step 3. Choose **1-Sample Sign**
Step 4. When the 1-Sample Sign dialogue box appears:
Enter C1 in the **Variables** box
Select **Test Median**
Enter the hypothesized value 236000 in the **Test Median** box
Select **less than** from the **Alternative** menu
Click **OK**

Minitab provides the *p*-value as well as a point estimate of the population median.

This Minitab procedure can also be used to obtain an interval estimate of the population median. In step 4, select **Confidence interval** instead of **Test median**, enter the **Confidence level**, and click **OK**. For sample sizes greater than 50, Minitab uses a normal approximation to the binomial sampling distribution with the continuity correction factor for both the hypothesis test and the confidence interval calculations.

Sign Test for a Hypothesis Test with Matched Samples

WEB file SunCoast

In order to use Minitab's sign test procedure for a hypothesis test with matched samples, we will use a numerical code for a plus sign, a minus sign, and the no preference data. We will use the Sun Coast Farms hypothesis test in Section 18.1 to illustrate this procedure. The data file SunCoast shows that column C1 contains numbers identifying each of the 14 individuals participating in the taste test and that column C2 uses a $+1$ for each plus sign, a -1 for each minus sign and a 0 for each no preference. If the null hypothesis of no preference is true, the median of the population of $+1$'s, -1's, and 0's will be zero. Thus, we follow the steps for testing a population median with the median hypothesized to be zero. For the Sun Coast Farms hypothesis test we will use a two-tailed test as follows.

Step 1. Select the **Stat** menu
Step 2. Choose **Nonparametrics**
Step 3. Choose **1-Sample Sign**
Step 4. When the 1-Sample Sign dialogue box appears:
Enter C2 in the **Variables** box
Select **Test Median**
Enter the hypothesized value 0 in the **Test Median** box
Select not equal from the **Alternative** menu
Click **OK**

Wilcoxon Signed-Rank Test with Matched Samples

WEB file MatchedSample

The following steps can be used to test hypotheses about the difference between two population medians using matched-sample data. We will use the production task completion time data in Section 18.2 to illustrate. The data file MatchedSample provides the production times for method A in column C1, the production times for method B in column C2, and the differences in column C3. The following steps can be used to test the hypotheses H_0: Median $= 0$ and H_a: Median $\neq 0$ for the population of differences.

Step 1. Select the **Stat** menu
Step 2. Choose **Nonparametrics**
Step 3. Choose **1-Sample Wilcoxon**
Step 4. When the 1-Sample Wilcoxon dialogue box appears:
Enter C3 in the **Variables** box
Select **Test Median**
Enter the hypothesized value 0 in the **Test Median** box
Select **not equal** from the **Alternative** menu
Click **OK**

Note that the Minitab procedure uses the paired difference data in column C3. Although the data file shows the times for each production method in columns C1 and C2, these data were not used to obtain the Minitab output.

The same procedure can also be used to test a hypothesis about the median of a *symmetric* population. Enter the actual data in any column of the worksheet and follow the preceding steps. Enter the hypothesized value of the population median in the **Test Median** box and select the desired alternative hypothesis in the **Alternative** box. Click **OK** to obtain the results. For this test, you do not have to enter the difference data. The Minitab routine will make the calculations automatically. But remember, this test is valid for only the median of a symmetric population.

Mann-Whitney-Wilcoxon Test

WEB file ThirdNational

The following steps can be used to test a hypothesis that two populations are identical using two independent samples, one from each population. We will use the Third National Bank example in Section 18.3 to illustrate. The data file ThirdNational provides the 12 account balances for branch 1 in column C1 and the ten account balances for branch 2 in column C2. The following steps will implement the Minitab procedure for testing H_0: The two populations are identical versus H_a: The two populations are not identical.

Step 1. Select the **Stat** menu
Step 2. Choose **Nonparametrics**
Step 3. Choose **Mann-Whitney**
Step 4. When the Mann-Whitney dialogue box appears:
Enter C1 in the **First sample** box
Enter C2 in the **Second sample** box
Select **not equal** from the **Alternative** menu
Click **OK**

Minitab will report the value of the test statistic and the corresponding *p*-value for the test. Since Minitab automatically assumes the two populations have the same shape, the output describes the results in terms of the difference between the medians of the two populations. Note that the output also provides both a point estimate and a confidence interval estimate of the difference between the medians. With the Greek letter η (eta) sometimes used to denote a population median, the Minitab output uses ETA1 and ETA2 as abbreviations for the two population medians.

Kruskal-Wallis Test

WEB file Williams

The following steps can be used to test a hypothesis that three or more populations are identical using independent samples, one from each population. We will use the Williams Manufacturing Company data in Section 18.4 to illustrate. The data file Williams provides the college the employee attended (A, B, or C) in column C1 and the annual performance rating in column C2. Minitab's terminology is to refer to the college as the factor and the performance rating as the response. The following steps will implement the Minitab procedure for testing H_0: All populations are identical versus H_a: Not all populations are identical. If the assumption is made that the populations have the same shape, the hypotheses can be stated in terms of the population medians.

Step 1. Select the **Stat** menu
Step 2. Choose **Nonparametrics**
Step 3. Choose **Kruskal-Wallis**
Step 4. When the Kruskal-Wallis dialogue box appears:
Enter C2 in the **Response** box
Enter C1 in the **Factor** box
Click **OK**

Spearman Rank Correlation

WEB file
PotentialActual

The Spearman rank-correlation coefficient is the same as the Pearson correlation coefficient computed for the ordinal, or rank-ordered, data. So we can compute the Spearman rank-correlation coefficient for the rank-ordered data by using Minitab's procedure to compute the Pearson correlation coefficient. We will use the sales potential and actual two-year sales data in Section 18.5 to illustrate. The data file PotentialActual provides the ranking of potential of each employee in column C2 and the ranking of the actual two-year sales of each employee in column C3. The following Minitab steps can be used to calculate Spearman's rank correlation for the two variables.

Step 1. Select the **Stat** menu
Step 2. Choose **Basic Statistics**
Step 3. Choose **Correlation**
Step 4. When the Correlation dialogue box appears:
Enter C2 C3 in the **Variables** box
Uncheck **Display p-values**
Click **OK**

The Minitab output provides a value of .733 for the Pearson correlation coefficient. Since the data were rank-ordered data, this is also the Spearman rank-correlation coefficient. However, the *p*-value for the Pearson correlation coefficient is not appropriate for rank-ordered data and should not be interpreted as the *p*-value for the Spearman rank-correlation coefficient.

Appendix 18.2 Nonparametric Methods with Excel

Excel does not have nonparametric procedures in its Data Analysis package. But we will show how Excel's BINOM.DIST function can be used to conduct a sign test and how a Data Analysis procedure can be used to compute a rank-correlation coefficient. The StatTools Excel add-in can be used for the Wilcoxon signed-rank test and the Mann-Whitney-Wilcoxon test (see Appendix 18.3).

Sign Test

The sign test uses a binomial sampling distribution with $p = .50$ to conduct a hypothesis test about a population median or a hypothesis test with matched samples. Excel's BINOM.DIST function can be used to compute exact binomial probabilities for these tests.

Since the BINOM.DIST probabilities are exact, there is no need to use the normal distribution approximation calculation when using Excel for the sign test.

Let x = the number of plus signs
n = the sample size for the observations with a plus sign or a minus sign

The BINOM.DIST function can be used as follows:

$$\text{Lower tail probability} = \text{BINOM.DIST}(x, n, .50, \text{True})$$
$$\text{Upper tail probability} = 1 - \text{BINOM.DIST}(x - 1, n, .50, \text{True})$$

You can see from the lower tail probability expression, the BINOM.DIST function provides the cumulative binomial probability of *less than or equal to* x. The .50 in the function is the value of $p = .50$ and the term True is used to obtain the cumulative binomial probability. The upper tail probability is 1 − (the cumulative probability) as shown. Note that since the binomial distribution is discrete, $(x - 1)$ is used in the upper tail probability calculation. For example, the upper tail probability $P(x \geq 7) = 1 - P(x \leq 6)$.

Using Excel for the Lawler Grocery Store hypothesis test about a population median, we have 7 plus signs and 3 minus signs for the sample of 10 stores. The number of plus signs was in the upper tail with $P(x \geq 7)$ given by the function

$$= 1 - \text{BINOM.DIST}(x - 1, n, .50, \text{True}) = 1 - \text{BINOM.DIST}(6, 10, .50, \text{True}) = .1719$$

Since this is a two-tailed hypothesis test, we have p-value = 2(.1719) = .3438.

In Section 18.1 we also considered the lower tail test about the population median price of a new home:

$$H_0\text{: Median} \geq 236{,}000$$
$$H_a\text{: Median} < 236{,}000$$

After deleting the home that sold for exactly $236,000, the sample provided 22 plus signs and 38 minus signs for a sample of 60 homes. Since this is a lower tail test, the p-value is given by the lower tail probability $P(x \leq 22)$, which is as follows:

$$=\text{BINOM.DIST}(x, n, .50, \text{True}) = \text{BINOM.DIST}(22, 60, .50, \text{True}) = .0259$$

By using the BINOM.DIST function, we have the capability of computing the exact p-value for any application of the sign test.

Spearman Rank Correlation

WEB file
PotentialActual

Excel does not have a specific procedure for computing the Spearman rank-correlation coefficient. However, this correlation coefficient is the same as the Pearson correlation coefficient provided you are using rank-ordered data. As a result, we can compute the Spearman rank-correlation coefficient by applying Excel's Pearson correlation coefficient procedure to the rank-ordered data. We illustrate using the data on sales potential and actual two-year sales from Section 18.5. The data file PotentialActual provides the ranking of the 10 individuals in terms of potential in column B and the ranking of the 10 individuals in terms of actual two-year sales in column C. The following steps provide the Spearman rank-correlation coefficient.

Step 1. Click the **Data** tab on the Ribbon
Step 2. In the **Analysis** group, click **Data Analysis**
Step 3. Choose **Correlation** from the list of **Analysis Tools**

Step 4. When the Correlation Dialog box appears:
Enter B1:C11 in the **Input Range** box
Select **Grouped by Columns**
Select **Labels in First Row**
Select **Output Range**
Enter D1 in the **Output Range** box
Click **OK**

The Spearman rank-correlation coefficient will appear in cell E3.

Appendix 18.3 Nonparametric Methods with StatTools

In this appendix we show how to use StatTools for the Wilcoxon Signed-Rank test and the Mann-Whitney-Wilcoxon test.

Wilcoxon Signed-Rank Test with Matched Samples

WEB file
MatchedSample

The following steps can be used to test hypotheses about the difference between two population medians based on matched samples. We will use the production task completion time data in Section 18.2 to illustrate. The data file MatchedSample provides the production times for method A in column A, the production times for method B in column B, and the differences between the two methods in column C. Begin by using the Data Set Manager to create a StatTools data set using the procedure described in the appendix to Chapter 1. The following steps can then be used to test the hypotheses H_0: Median = 0 and H_a: Median $\neq$ 0 for the population of differences.

Step 1. Click the **StatTools** tab on the Ribbon
Step 2. In the **Analyses Group**, select **Nonparametric Tests**
Step 3. Choose **Wilcoxon Signed-Rank Test**
Step 4. When the Wilcoxon Signed-Rank Test dialog box appears:
Select **One-Sample Analysis** in the **Analysis Type** box
Check the **Difference** variable
Enter 0 in the **Null Hypothesis Value** box
Select **Not Equal to Null Value** in the **Alternative Hypothesis Type** box
Click **OK**

The same procedure can also be used to test a hypothesis about the median of a symmetric population. Enter the data in any column of the worksheet. Then follow the preceding steps. Enter the hypothesized value of the population median in the **Null Hypothesis Value** box and select the desired alternative hypothesis in the **Alternative Hypothesis Type** box. Click **OK** to obtain the results. For this test, you do not have to enter the difference data because the StatTools routine will make the calculations automatically. But remember, this test is valid only for the median of a symmetric population.

Mann-Whitney-Wilcoxon Test

WEB file
ThirdNational

The following steps can be used to test a hypothesis that two populations are identical using two independent samples, one from each population. We will use the Third National Bank example in Section 18.3 to illustrate. The data file ThirdNational provides the 12 account balances for branch 1 in column A and the 10 account balances for branch 2 in column B. Begin by using the Data Set Manager to create a StatTools data set using the procedure described in the appendix to Chapter 1. The following steps can then be used for testing the hypotheses H_0: The two populations are identical and H_a: The two populations are not identical.

Step 1. Click the **StatTools** tab on the Ribbon
Step 2. In the **Analyses Group**, select **Nonparametric Tests**
Step 3. Choose **Mann-Whitney Test**
Step 4. When the Mann-Whitney Test dialog box appears:
Select **General Version** in the **Analysis Type** box
Check the **Branch 1** variable
Check the **Branch 2** variable
Select **Either distribution smaller (Two-Tailed Test)** in the **Alternative Hypothesis Type** box
Click **OK**
Step 5. When the StatTools dialog box appears:
Click **OK**
When the Choose Variable Ordering dialog box appears:
Click **OK**

If you want to make the assumption that the two populations have the same shape, select **Median Version** in the **Analysis Type** box. The test results will be the same, with the output showing the hypotheses are about the difference between the two population medians.

CHAPTER 19

Statistical Methods for Quality Control

CONTENTS

STATISTICS *in* PRACTICE

DOW CHEMICAL COMPANY*
FREEPORT, TEXAS

In 1940 the Dow Chemical Company purchased 800 acres of Texas land on the Gulf Coast to build a magnesium production facility. That original site has expanded to cover more than 5000 acres and holds one of the largest petrochemical complexes in the world. Among the products from Dow Texas Operations are magnesium, styrene, plastics, adhesives, solvent, glycol, and chlorine. Some products are made solely for use in other processes, but many end up as essential ingredients in products such as pharmaceuticals, toothpastes, dog food, water hoses, ice chests, milk cartons, garbage bags, shampoos, and furniture.

Dow's Texas Operations produce more than 30% of the world's magnesium, an extremely lightweight metal used in products ranging from tennis racquets to suitcases to "mag" wheels. The Magnesium Department was the first group in Texas Operations to train its technical people and managers in the use of statistical quality control. Some of the earliest successful applications of statistical quality control were in chemical processing.

In one application involving the operation of a drier, samples of the output were taken at periodic intervals; the average value for each sample was computed and recorded on a chart called an $\bar{x}$ chart. Such a chart enabled Dow analysts to monitor trends in the output that might indicate the process was not operating correctly. In one instance, analysts began to observe values for the sample mean that were not indicative of a process operating within its design limits. On further examination of the control chart and the operation itself, the analysts found that the variation could be traced to problems involving one operator. The $\bar{x}$ chart recorded after retraining the operator showed a significant improvement in the process quality.

Statistical quality control has enabled Dow Chemical Company to improve its processing methods and output. © Bill Pugliano/Getty Images.

Dow achieves quality improvements everywhere it applies statistical quality control. Documented savings of several hundred thousand dollars per year are realized, and new applications are continually being discovered.

In this chapter we will show how an $\bar{x}$ chart such as the one used by Dow can be developed. Such charts are a part of statistical quality control known as statistical process control. We will also discuss methods of quality control for situations in which a decision to accept or reject a group of items is based on a sample.

*The authors are indebted to Clifford B. Wilson, Magnesium Technical Manager, The Dow Chemical Company, for providing this Statistics in Practice.

ASQ's Vision: "By making quality a global priority, an organizational imperative, and a personal ethic, the American Society for Quality becomes the community for everyone who seeks quality concepts, technology, and tools to improve themselves and their world" (ASQ website).

The American Society for Quality (ASQ) defines quality as "the totality of features and characteristics of a product or service that bears on its ability to satisfy given needs." In other words, quality measures how well a product or service meets customer needs. Organizations recognize that to be competitive in today's global economy, they must strive for a high level of quality. As a result, they place increased emphasis on methods for monitoring and maintaining quality.

Today, the customer-driven focus that is fundamental to high-performing organizations has changed the scope that quality issues encompass, from simply eliminating defects on a production line to developing broad-based corporate quality strategies. Broadening the scope of quality naturally leads to the concept of **total quality (TQ)**.

> Total Quality (TQ) is a people-focused management system that aims at continual increase in customer satisfaction at continually lower real cost. TQ is a total system approach (not a

separate area or work program) and an integral part of high-level strategy; it works horizontally across function and departments, involves all employees, top to bottom, and extends backward and forward to include the supply chain and the customer chain. TQ stresses learning and adaptation to continual change as keys to organization success.[1]

Regardless of how it is implemented in different organizations, total quality is based on three fundamental principles: a focus on customers and stakeholders; participation and teamwork throughout the organization; and a focus on continuous improvement and learning. In the first section of the chapter we provide a brief introduction to three quality management frameworks: the Malcolm Baldrige Quality Award, ISO 9000 standards, and the Six Sigma philosophy. In the last two sections we introduce two statistical tools that can be used to monitor quality: statistical process control and acceptance sampling.

19.1 Philosophies and Frameworks

After World War II, Dr. W. Edwards Deming became a consultant to Japanese industry; he is credited with being the person who convinced top managers in Japan to use the methods of statistical quality control.

In the early twentieth century, quality control practices were limited to inspecting finished products and removing defective items. But this all changed as the result of the pioneering efforts of a young engineer named Walter A. Shewhart. After completing his doctorate in physics from the University of California in 1917, Dr. Shewhart joined the Western Electric Company, working in the inspection engineering department. In 1924 Dr. Shewhart prepared a memorandum that included a set of principles that are the basis for what is known today as process control. And his memo also contained a diagram that would be recognized as a statistical control chart. Continuing his work in quality at Bell Telephone Laboratories until his retirement in 1956, he brought together the disciplines of statistics, engineering, and economics and in doing so changed the course of industrial history. Dr. Shewhart is recognized as the father of statistical quality control and was the first honorary member of the ASQ.

Two other individuals who have had great influence on quality are Dr. W. Edwards Deming, a student of Dr. Shewhart, and Joseph Juran. These men helped educate the Japanese in quality management shortly after World War II. Although quality is everybody's job, Deming stressed that the focus on quality must be led by managers. He developed a list of 14 points that he believed represent the key responsibilities of managers. For instance, Deming stated that managers must cease dependence on mass inspection; must end the practice of awarding business solely on the basis of price; must seek continual improvement in all production processes and service; must foster a team-oriented environment; and must eliminate goals, slogans, and work standards that prescribe numerical quotas. Perhaps most important, managers must create a work environment in which a commitment to quality and productivity is maintained at all times.

Juran proposed a simple definition of quality: *fitness for use.* Juran's approach to quality focused on three quality processes: quality planning, quality control, and quality improvement. In contrast to Deming's philosophy, which required a major cultural change in the organization, Juran's programs were designed to improve quality by working within the current organizational system. Nonetheless, the two philosophies are similar in that they both focus on the need for top management to be involved and stress the need for continuous improvement, the importance of training, and the use of quality control techniques.

Many other individuals played significant roles in the quality movement, including Philip B. Crosby, A. V. Feigenbaum, Karou Ishikawa, and Genichi Taguchi. More specialized texts dealing exclusively with quality provide details of the contributions of each of

[1] J. R. Evans and W. M. Lindsay, *Managing for Quality and Performance Excellence*, 8th ed. (Cincinnati, OH: South-Western, 2011), p. 11.

these individuals. The contributions of all individuals involved in the quality movement helped define a set of best practices and led to numerous awards and certification programs. The two most significant programs are the U.S. Malcolm Baldrige National Quality Award and the international ISO 9000 certification process. In recent years, use of Six Sigma—a methodology for improving organizational performance based on rigorous data collection and statistical analysis—has also increased.

Malcolm Baldrige National Quality Award

The U.S. Commerce Department's National Institute of Standards and Technology (NIST) manages the Baldrige National Quality Program. More information can be obtained at the NIST website.

The Malcolm Baldrige National Quality Award is given by the president of the United States to organizations that apply and are judged to be outstanding in seven areas: leadership; strategic planning; customer and market focus; measurement, analysis, and knowledge management; human resource focus; process management; and business results. Congress established the award program in 1987 to recognize U.S. organizations for their achievements in quality and performance and to raise awareness about the importance of quality as a competitive edge. The award is named for Malcolm Baldrige, who served as secretary of commerce from 1981 until his death in 1987.

2004 was the final year for the Baldrige Stock Study because of the increase in the number of recipients that are either nonprofit or privately held businesses.

Since the presentation of the first awards in 1988, the Baldrige National Quality Program (BNQP) has grown in stature and impact. Approximately 2 million copies of the criteria have been distributed since 1988, and wide-scale reproduction by organizations and electronic access add to that number significantly. For eight years in a row, a hypothetical stock index, made up of publicly traded U.S. companies that had received the Baldrige Award, outperformed the Standard & Poor's 500. In one year, the "Baldrige Index" outperformed the S&P 500 by 4.4 to 1. Bob Barnett, executive vice president of Motorola, Inc., said, "We applied for the Award, not with the idea of winning, but with the goal of receiving the evaluation of the Baldrige Examiners. That evaluation was comprehensive, professional, and insightful . . . making it perhaps the most cost-effective, value-added business consultation available anywhere in the world today."

ISO 9000

ISO 9000 standards are revised periodically to improve the quality of the standard.

ISO 9000 is a series of five international standards published in 1987 by the International Organization for Standardization (ISO), Geneva, Switzerland. Companies can use the standards to help determine what is needed to maintain an efficient quality conformance system. For example, the standards describe the need for an effective quality system, for ensuring that measuring and testing equipment is calibrated regularly, and for maintaining an adequate record-keeping system. ISO 9000 registration determines whether a company complies with its own quality system. Overall, ISO 9000 registration covers less than 10% of the Baldrige Award criteria.

Six Sigma

In the late 1980s Motorola recognized the need to improve the quality of its products and services; their goal was to achieve a level of quality so good that for every million opportunities no more than 3.4 defects will occur. This level of quality is referred to as the six sigma level of quality, and the methodology created to reach this quality goal is referred to as **Six Sigma**.

An organization may undertake two kinds of Six Sigma projects:

- DMAIC (Define, Measure, Analyze, Improve, and Control) to help redesign existing processes
- DFSS (Design for Six Sigma) to design new products, processes, or services

In helping to redesign existing processes and design new processes, Six Sigma places a heavy emphasis on statistical analysis and careful measurement. Today, Six Sigma is a major tool in helping organizations achieve Baldrige levels of business performance and process quality. Many Baldrige examiners view Six Sigma as the ideal approach for implementing Baldrige improvement programs.

Six Sigma limits and defects per million opportunities In Six Sigma terminology, a *defect* is any mistake or error that is passed on to the customer. The Six Sigma process defines quality performance as defects per million opportunities (dpmo). As we indicated previously, Six Sigma represents a quality level of at most 3.4 dpmo. To illustrate how this quality level is measured, let us consider the situation at KJW Packaging.

KJW operates a production line where boxes of cereal are filled. The filling process has a mean of $\mu = 16.05$ ounces and a standard deviation of $\sigma = .10$ ounces. It is reasonable to assume the filling weights are normally distributed. The distribution of filling weights is shown in Figure 19.1. Suppose management considers 15.45 to 16.65 ounces to be acceptable quality limits for the filling process. Thus, any box of cereal that contains less than 15.45 or more than 16.65 ounces is considered to be a defect. Using Excel, it can be shown that 99.9999998% of the boxes filled will have between $16.05 - 6(.10) = 15.45$ ounces and $16.05 + 6(.10) = 16.65$ ounces. In other words, only .0000002% of the boxes filled will contain less than 15.45 ounces or more than 16.65 ounces. Thus, the likelihood of obtaining a defective box of cereal from the filling process appears to be extremely unlikely, because on average only two boxes in 10 million will be defective.

Using Excel, NORM.S.DIST (6,TRUE)−NORM.S.DIST (−6,TRUE) = .999999998.

Motorola's early work on Six Sigma convinced them that a process mean can shift on average by as much as 1.5 standard deviations. For instance, suppose that the process mean for KJW increases by 1.5 standard deviations or $1.5(.10) = .15$ ounces. With such a shift, the normal distribution of filling weights would now be centered at $\mu = 16.05 + .15 = 16.20$ ounces. With a process mean of $\mu = 16.05$ ounces, the probability of obtaining a box of cereal with more than 16.65 ounces is extremely small. But how does this probability

Using Excel, 1−NORM.S.DIST (4.5,TRUE) = .0000034.

FIGURE 19.1 NORMAL DISTRIBUTION OF CEREAL BOX FILLING WEIGHTS WITH A PROCESS MEAN $\mu = 16.05$

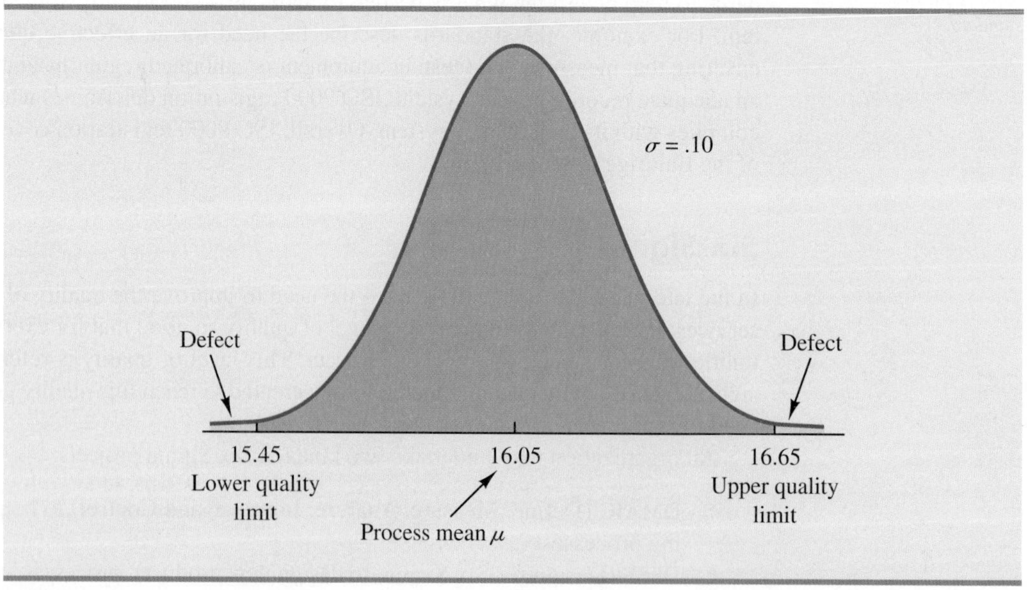

FIGURE 19.2 NORMAL DISTRIBUTION OF CEREAL BOX FILLING WEIGHTS WITH A PROCESS MEAN $\mu = 16.20$

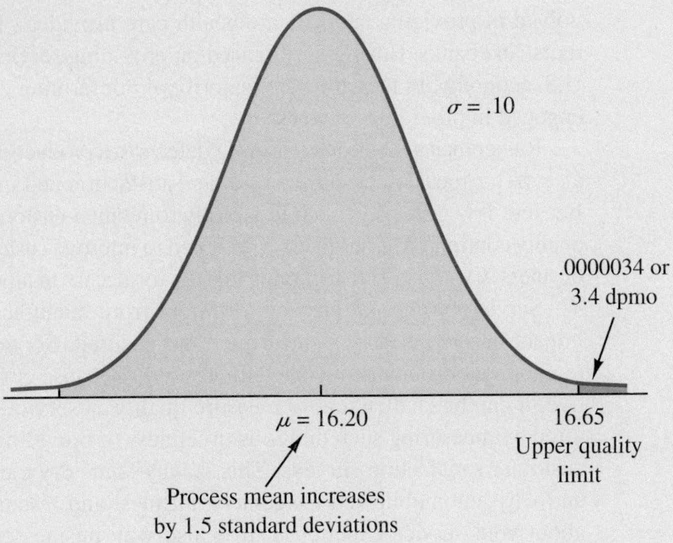

change if the mean of the process shifts up to $\mu = 16.20$ ounces? Figure 19.2 shows that for this case, the upper quality limit of 16.65 is 4.5 standard deviations to the right of the new mean $\mu = 16.20$ ounces. Using this mean and Excel, we find that the probability of obtaining a box with more than 16.65 ounces is .0000034. Thus, if the process mean shifts up by 1.5 standard deviations, approximately 1,000,000(.0000034) = 3.4 boxes of cereal will exceed the upper limit of 16.65 ounces. In Six Sigma terminology, the quality level of the process is said to be 3.4 defects per million opportunities. If management of KJW considers 15.45 to 16.65 ounces to be acceptable quality limits for the filling process, the KJW filling process would be considered a Six Sigma process. Thus, if the process mean stays within 1.5 standard deviations of its target value $\mu = 16.05$ ounces, a maximum of only 3.4 defects per million boxes filled can be expected.

Organizations that want to achieve and maintain a Six Sigma level of quality must emphasize methods for monitoring and maintaining quality. *Quality assurance* refers to the entire system of policies, procedures, and guidelines established by an organization to achieve and maintain quality. Quality assurance consists of two principal functions: quality engineering and quality control. The object of *quality engineering* is to include quality in the design of products and processes and to identify quality problems prior to production. **Quality control** consists of a series of inspections and measurements used to determine whether quality standards are being met. If quality standards are not being met, corrective or preventive action can be taken to achieve and maintain conformance. In the next two sections we present two statistical methods used in quality control. The first method, *statistical process control,* uses graphical displays known as control charts to monitor a process; the goal is to determine whether the process can be continued or whether corrective action should be taken to achieve a desired quality level. The second method, *acceptance sampling,* is used in situations where a decision to accept or reject a group of items must be based on the quality found in a sample.

Quality in the Service Sector

While its roots are in manufacturing, quality control is also very important for businesses that focus primarily on providing services. Examples of businesses that are primarily involved in providing services are health care providers, law firms, hotels, airlines, restaurants, and banks. Businesses focused on providing services are a very important part of the U.S. economy. In fact, the vast majority of nonfarming employees in the United States are engaged in providing services.

Rather than a focus on measuring defects in a production process, quality efforts in the service sector focus on ensuring customer satisfaction and improving the customer experience. Because it is generally much less costly to retain a customer than it is to acquire a new one, quality control processes that are designed to improve customer service are critical to a service business. Customer satisfaction is the key to success in any service-oriented business.

Service businesses are very different from manufacturing businesses and this has an impact on how quality is measured and ensured. Services provided are often intangible (e.g., advice from a residence hall adviser). Because customer satisfaction is very subjective, it can be challenging to measure quality in services. However, quality can be monitored by measuring such things as timeliness of providing service as well as by conducting customer satisfaction surveys. This is why some dry cleaners guarantee one-hour service and why automobile service centers, airlines, and restaurants ask you to fill out a survey about your service experience. It is also why businesses use customer loyalty cards. By tracking your buying behavior, they can better understand the wants and needs of their customers and consequently provide better service.

19.2 Statistical Process Control

In this section we consider quality control procedures for a production process whereby goods are manufactured continuously. On the basis of sampling and inspection of production output, a decision will be made to either continue the production process or adjust it to bring the items or goods being produced up to acceptable quality standards.

Continuous improvement is one of the most important concepts of the total quality management movement. The most important use of a control chart is in improving the process.

Despite high standards of quality in manufacturing and production operations, machine tools will invariably wear out, vibrations will throw machine settings out of adjustment, purchased materials will be defective, and human operators will make mistakes. Any or all of these factors can result in poor quality output. Fortunately, procedures are available for monitoring production output so that poor quality can be detected early and the production process can be adjusted or corrected.

If the variation in the quality of the production output is due to **assignable causes** such as tools wearing out, incorrect machine settings, poor quality raw materials, or operator error, the process should be adjusted or corrected as soon as possible. Alternatively, if the variation is due to what are called **common causes**—that is, randomly occurring variations in materials, temperature, humidity, and so on, which the manufacturer cannot possibly control—the process does not need to be adjusted. The main objective of statistical process control is to determine whether variations in output are due to assignable causes or common causes.

Process control procedures are closely related to hypothesis testing procedures discussed earlier in this text. Control charts provide an ongoing test of the hypothesis that the process is in control.

Whenever assignable causes are detected, we conclude that the process is *out of control.* In that case, corrective action will be taken to bring the process back to an acceptable level of quality. However, if the variation in the output of a production process is due only to common causes, we conclude that the process is *in statistical control,* or simply *in control;* in such cases, no changes or adjustments are necessary.

The statistical procedures for process control are based on the hypothesis testing methodology presented in Chapter 9. The null hypothesis H_0 is formulated in terms of the production process being in control. The alternative hypothesis H_a is formulated in terms of

TABLE 19.1 THE OUTCOMES OF STATISTICAL PROCESS CONTROL

		State of Production Process	
		H_0 True Process in Control	H_0 False Process Out of Control
Decision	**Continue Process**	Correct decision	Type II error (allowing an out-of-control process to continue)
	Adjust Process	Type I error (adjusting an in-control process)	Correct decision

the production process being out of control. Table 19.1 shows that correct decisions to continue an in-control process and adjust an out-of-control process are possible. However, as with other hypothesis testing procedures, both a Type I error (adjusting an in-control process) and a Type II error (allowing an out-of-control process to continue) are also possible.

Control Charts

A **control chart** provides a basis for deciding whether the variation in the output is due to common causes (in control) or assignable causes (out of control). Whenever an out-of-control situation is detected, adjustments or other corrective action will be taken to bring the process back into control.

Control charts based on data that can be measured on a continuous scale are called variables control charts. The $\bar{x}$ chart is a variables control chart.

Control charts can be classified by the type of data they contain. An **$\bar{x}$ chart** is used if the quality of the output of the process is measured in terms of a variable such as length, weight, temperature, and so on. In that case, the decision to continue or to adjust the production process will be based on the mean value found in a sample of the output. To introduce some of the concepts common to all control charts, let us consider some specific features of an $\bar{x}$ chart.

Figure 19.3 shows the general structure of an $\bar{x}$ chart. The center line of the chart corresponds to the mean of the process when the process is *in control.* The vertical line identifies the scale of measurement for the variable of interest. Each time a sample is taken from the production process, a value of the sample mean $\bar{x}$ is computed and a data point showing the value of $\bar{x}$ is plotted on the control chart.

FIGURE 19.3 $\bar{x}$ CHART STRUCTURE

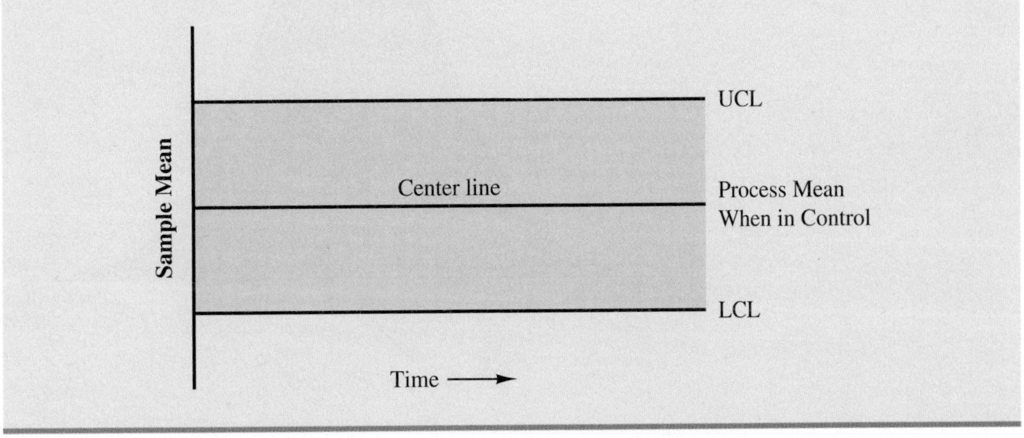

The two lines labeled UCL and LCL are important in determining whether the process is in control or out of control. The lines are called the *upper control limit* and the *lower control limit,* respectively. They are chosen so that when the process is in control, there will be a high probability that the value of $\bar{x}$ will be between the two control limits. Values outside the control limits provide strong statistical evidence that the process is out of control and corrective action should be taken.

Over time, more and more data points will be added to the control chart. The order of the data points will be from left to right as the process is sampled. In essence, every time a point is plotted on the control chart, we are carrying out a hypothesis test to determine whether the process is in control.

In addition to the $\bar{x}$ chart, other control charts can be used to monitor the range of the measurements in the sample (***R* chart**), the proportion defective in the sample (***p* chart**), and the number of defective items in the sample (***np* chart**). In each case, the control chart has a LCL, a center line, and an UCL similar to the $\bar{x}$ chart in Figure 19.3. The major difference among the charts is what the vertical axis measures; for instance, in a p chart the measurement scale denotes the proportion of defective items in the sample instead of the sample mean. In the following discussion, we will illustrate the construction and use of the $\bar{x}$ chart, R chart, p chart, and np chart.

$\bar{x}$ Chart: Process Mean and Standard Deviation Known

To illustrate the construction of an $\bar{x}$ chart, let us reconsider the situation at KJW Packaging. Recall that KJW operates a production line where cartons of cereal are filled. When the process is operating correctly—and hence the system is in control—the mean filling weight is $\mu = 16.05$ ounces, and the process standard deviation is $\sigma = .10$ ounces. In addition, the filling weights are assumed to be normally distributed. This distribution is shown in Figure 19.4.

The sampling distribution of $\bar{x}$, as presented in Chapter 7, can be used to determine the variation that can be expected in $\bar{x}$ values for a process that is in control. Let us first briefly review the properties of the sampling distribution of $\bar{x}$. First, recall that the expected value or mean of $\bar{x}$ is equal to μ, the mean filling weight when the production line is in control.

FIGURE 19.4 NORMAL DISTRIBUTION OF CEREAL CARTON FILLING WEIGHTS

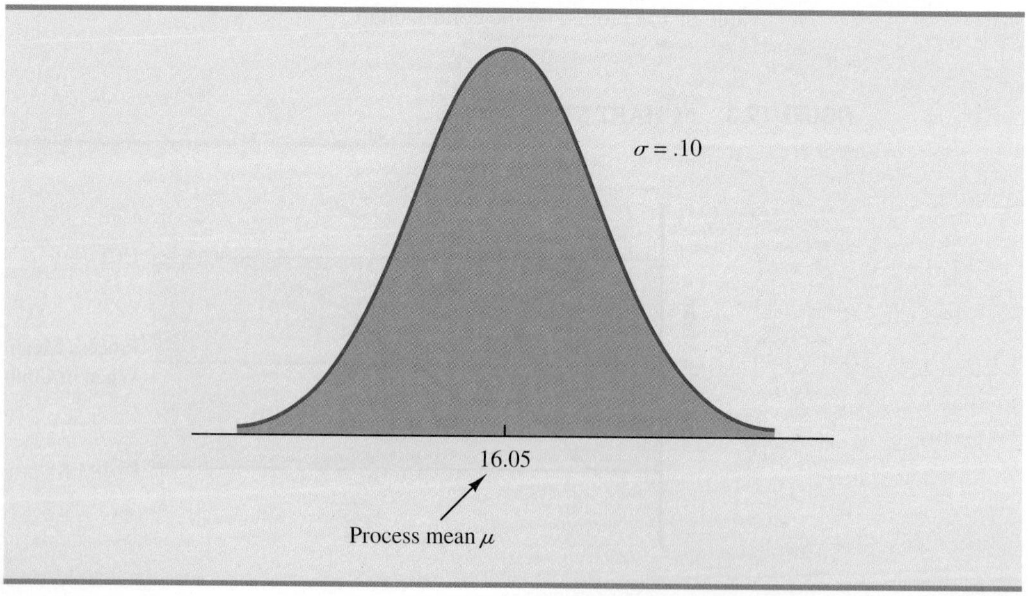

FIGURE 19.5 SAMPLING DISTRIBUTION OF $\bar{x}$ FOR A SAMPLE OF n FILLING WEIGHTS

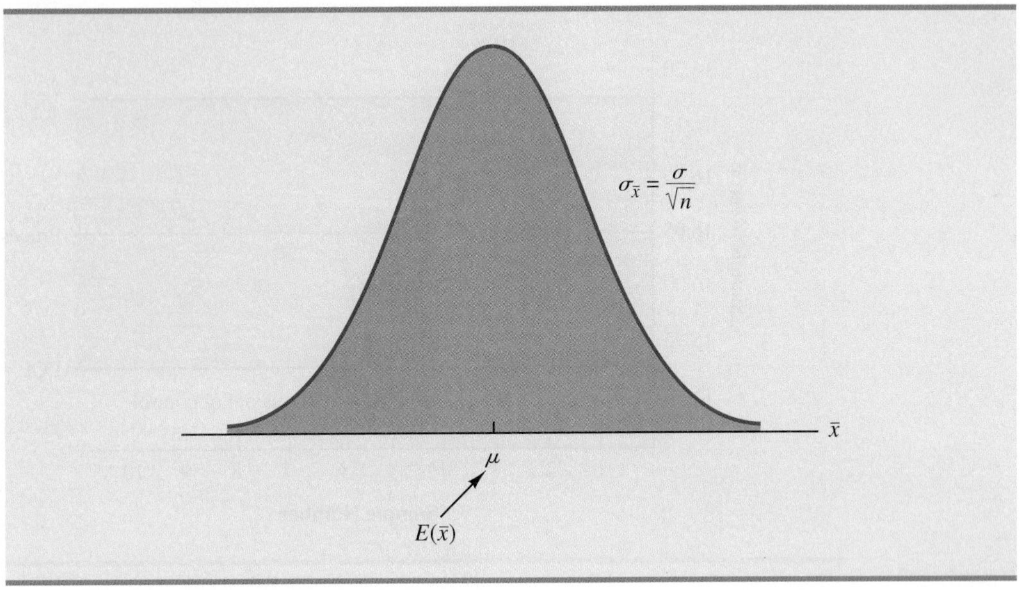

For samples of size n, the equation for the standard deviation of $\bar{x}$, called the standard error of the mean, is

$$\sigma_{\bar{x}} = \frac{\sigma}{\sqrt{n}} \tag{19.1}$$

In addition, because the filling weights are normally distributed, the sampling distribution of $\bar{x}$ is normally distributed for any sample size. Thus, the sampling distribution of $\bar{x}$ is a normal distribution with mean μ and standard deviation $\sigma_{\bar{x}}$. This distribution is shown in Figure 19.5.

The sampling distribution of $\bar{x}$ is used to determine what values of $\bar{x}$ are reasonable if the process is in control. The general practice in quality control is to define as reasonable any value of $\bar{x}$ that is within 3 standard deviations, or standard errors, above or below the mean value. Recall from the study of the normal probability distribution that approximately 99.7% of the values of a normally distributed random variable are within ± 3 standard deviations of its mean value. Thus, if a value of $\bar{x}$ is within the interval $\mu - 3\sigma_{\bar{x}}$ to $\mu + 3\sigma_{\bar{x}}$, we will assume that the process is in control. In summary, then, the control limits for an $\bar{x}$ chart are as follows.

CONTROL LIMITS FOR AN $\bar{x}$ CHART: PROCESS MEAN AND STANDARD DEVIATION KNOWN

$$\text{UCL} = \mu + 3\sigma_{\bar{x}} \tag{19.2}$$
$$\text{LCL} = \mu - 3\sigma_{\bar{x}} \tag{19.3}$$

Reconsider the KJW Packaging example with the process distribution of filling weights shown in Figure 19.4 and the sampling distribution of $\bar{x}$ shown in Figure 19.5. Assume that a quality control inspector periodically samples six cartons and uses the sample mean filling weight to determine whether the process is in control or out of control. Using equation (19.1), we find that the standard error of the mean is $\sigma_{\bar{x}} = \sigma/\sqrt{n} = .10/\sqrt{6} = .04$. Thus, with the process mean at 16.05, the control limits are UCL $= 16.05 + 3(.04) = 16.17$ and

FIGURE 19.6 THE $\bar{x}$ CHART FOR THE CEREAL CARTON FILLING PROCESS

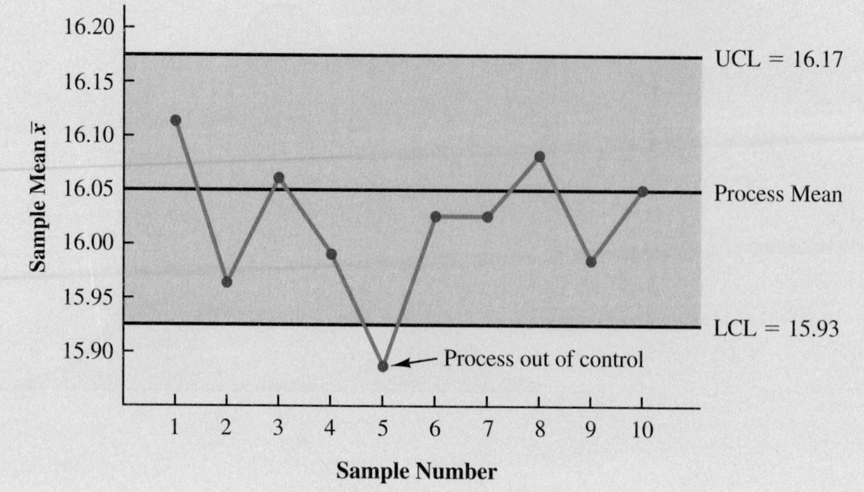

LCL = 16.05 − 3(.04) = 15.93. Figure 19.6 is the control chart with the results of 10 samples taken over a 10-hour period. For ease of reading, the sample numbers 1 through 10 are listed below the chart.

Note that the mean for the fifth sample in Figure 19.6 shows there is strong evidence that the process is out of control. The fifth sample mean is below the LCL, indicating that assignable causes of output variation are present and that underfilling is occurring. As a result, corrective action was taken at this point to bring the process back into control. The fact that the remaining points on the $\bar{x}$ chart are within the upper and lower control limits indicates that the corrective action was successful.

$\bar{x}$ Chart: Process Mean and Standard Deviation Unknown

In the KJW Packaging example, we showed how an $\bar{x}$ chart can be developed when the mean and standard deviation of the process are known. In most situations, the process mean and standard deviation must be estimated by using samples that are selected from the process when it is in control. For instance, KJW might select a random sample of five boxes each morning and five boxes each afternoon for 10 days of in-control operation. For each subgroup, or sample, the mean and standard deviation of the sample are computed. The overall averages of both the sample means and the sample standard deviations are used to construct control charts for both the process mean and the process standard deviation.

It is important to maintain control over both the mean and the variability of a process.

In practice, it is more common to monitor the variability of the process by using the range instead of the standard deviation because the range is easier to compute. The range can be used to provide good estimates of the process standard deviation; thus it can be used to construct upper and lower control limits for the $\bar{x}$ chart with little computational effort. To illustrate, let us consider the problem facing Jensen Computer Supplies, Inc.

Jensen Computer Supplies (JCS) manufactures 3.5-inch-diameter solid-state drives; they just finished adjusting their production process so that it is operating in control. Suppose random samples of five drives were selected during the first hour of operation, five drives were selected during the second hour of operation, and so on, until 20 samples were obtained. Table 19.2 provides the diameter of each drive sampled as well as the mean $\bar{x}_j$ and range R_j for each of the samples.

TABLE 19.2 DATA FOR THE JENSEN COMPUTER SUPPLIES PROBLEM

WEB file Jensen

Sample Number	Observations					Sample Mean $\bar{x}_j$	Sample Range R_j
1	3.5056	3.5086	3.5144	3.5009	3.5030	3.5065	.0135
2	3.4882	3.5085	3.4884	3.5250	3.5031	3.5026	.0368
3	3.4897	3.4898	3.4995	3.5130	3.4969	3.4978	.0233
4	3.5153	3.5120	3.4989	3.4900	3.4837	3.5000	.0316
5	3.5059	3.5113	3.5011	3.4773	3.4801	3.4951	.0340
6	3.4977	3.4961	3.5050	3.5014	3.5060	3.5012	.0099
7	3.4910	3.4913	3.4976	3.4831	3.5044	3.4935	.0213
8	3.4991	3.4853	3.4830	3.5083	3.5094	3.4970	.0264
9	3.5099	3.5162	3.5228	3.4958	3.5004	3.5090	.0270
10	3.4880	3.5015	3.5094	3.5102	3.5146	3.5047	.0266
11	3.4881	3.4887	3.5141	3.5175	3.4863	3.4989	.0312
12	3.5043	3.4867	3.4946	3.5018	3.4784	3.4932	.0259
13	3.5043	3.4769	3.4944	3.5014	3.4904	3.4935	.0274
14	3.5004	3.5030	3.5082	3.5045	3.5234	3.5079	.0230
15	3.4846	3.4938	3.5065	3.5089	3.5011	3.4990	.0243
16	3.5145	3.4832	3.5188	3.4935	3.4989	3.5018	.0356
17	3.5004	3.5042	3.4954	3.5020	3.4889	3.4982	.0153
18	3.4959	3.4823	3.4964	3.5082	3.4871	3.4940	.0259
19	3.4878	3.4864	3.4960	3.5070	3.4984	3.4951	.0206
20	3.4969	3.5144	3.5053	3.4985	3.4885	3.5007	.0259

The estimate of the process mean μ is given by the overall sample mean.

OVERALL SAMPLE MEAN

$$\bar{\bar{x}} = \frac{\bar{x}_1 + \bar{x}_2 + \cdots + \bar{x}_k}{k} \tag{19.4}$$

where

$\bar{x}_j$ = mean of the jth sample $j = 1, 2, \ldots, k$
k = number of samples

For the JCS data in Table 19.2, the overall sample mean is $\bar{\bar{x}} = 3.4995$. This value will be the center line for the $\bar{x}$ chart. The range of each sample, denoted R_j, is simply the difference between the largest and smallest values in each sample. The average range for k samples is computed as follows.

AVERAGE RANGE

$$\bar{R} = \frac{R_1 + R_2 + \cdots + R_k}{k} \tag{19.5}$$

where

R_j = range of the jth sample, $j = 1, 2, \ldots, k$
k = number of samples

For the JCS data in Table 19.2, the average range is $\bar{R} = .0253$.

In the preceding section we showed that the upper and lower control limits for the $\bar{x}$ chart are

$$\bar{\bar{x}} \pm 3\frac{\sigma}{\sqrt{n}} \quad \textbf{(19.6)}$$

The overall sample mean $\bar{\bar{x}}$ is used to estimate μ and the sample ranges are used to develop an estimate of σ.

Hence, to construct the control limits for the $\bar{x}$ chart, we need to estimate μ and σ, the mean and standard deviation of the process. An estimate of μ is given by $\bar{\bar{x}}$. An estimate of σ can be developed by using the range data.

It can be shown that an estimator of the process standard deviation σ is the average range divided by d_2, a constant that depends on the sample size n. That is,

$$\text{Estimator of } \sigma = \frac{\bar{R}}{d_2} \quad \textbf{(19.7)}$$

The *American Society for Testing and Materials Manual on Presentation of Data and Control Chart Analysis* provides values for d_2 as shown in Table 19.3. For instance, when $n = 5$, $d_2 = 2.326$, and the estimate of σ is the average range divided by 2.326.

TABLE 19.3 FACTORS FOR $\bar{x}$ AND R CONTROL CHARTS

Observations in Sample, n	d_2	A_2	d_3	D_3	D_4
2	1.128	1.880	0.853	0	3.267
3	1.693	1.023	0.888	0	2.574
4	2.059	0.729	0.880	0	2.282
5	2.326	0.577	0.864	0	2.114
6	2.534	0.483	0.848	0	2.004
7	2.704	0.419	0.833	0.076	1.924
8	2.847	0.373	0.820	0.136	1.864
9	2.970	0.337	0.808	0.184	1.816
10	3.078	0.308	0.797	0.223	1.777
11	3.173	0.285	0.787	0.256	1.744
12	3.258	0.266	0.778	0.283	1.717
13	3.336	0.249	0.770	0.307	1.693
14	3.407	0.235	0.763	0.328	1.672
15	3.472	0.223	0.756	0.347	1.653
16	3.532	0.212	0.750	0.363	1.637
17	3.588	0.203	0.744	0.378	1.622
18	3.640	0.194	0.739	0.391	1.608
19	3.689	0.187	0.734	0.403	1.597
20	3.735	0.180	0.729	0.415	1.585
21	3.778	0.173	0.724	0.425	1.575
22	3.819	0.167	0.720	0.434	1.566
23	3.858	0.162	0.716	0.443	1.557
24	3.895	0.157	0.712	0.451	1.548
25	3.931	0.153	0.708	0.459	1.541

Source: Reprinted with permission from Table 27 of ASTM STP 15D, *ASTM Manual on Presentation of Data and Control Chart Analysis,* Copyright ASTM International, 100 Barr Harbor Drive, West Conshohocken, PA 19428.

FIGURE 19.7 $\bar{x}$ CHART FOR THE JENSEN COMPUTER SUPPLIES PROBLEM

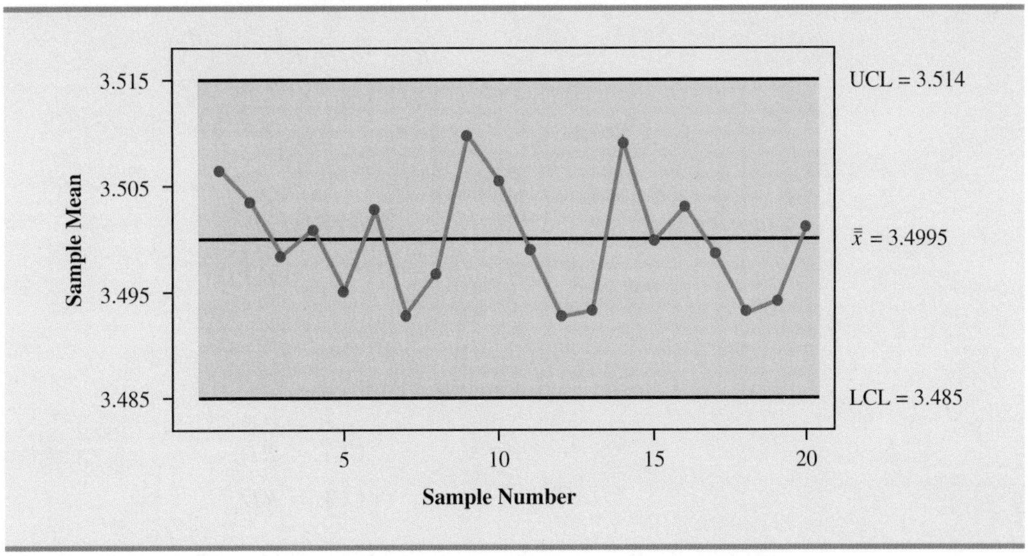

If we substitute $\bar{R}/d_2$ for σ in expression (19.6), we can write the control limits for the $\bar{x}$ chart as

$$\bar{\bar{x}} \pm 3\frac{\bar{R}/d_2}{\sqrt{n}} = \bar{\bar{x}} \pm \frac{3}{d_2\sqrt{n}}\bar{R} = \bar{\bar{x}} \pm A_2\bar{R} \tag{19.8}$$

Note that $A_2 = 3/(d_2\sqrt{n})$ is a constant that depends only on the sample size. Values for A_2 are provided in Table 19.3. For $n = 5$, $A_2 = .577$; thus, the control limits for the $\bar{x}$ chart are

$$3.4995 \pm (.577)(.0253) = 3.4995 \pm .0146$$

Hence, UCL = 3.514 and LCL = 3.485.

In the chapter appendix, we show how to construct $\bar{x}$ and R charts using StatTools.

Figure 19.7 shows the $\bar{x}$ chart for the Jensen Computer Supplies problem. We used the data in Table 19.2 and StatTools' X/R Charts routine to construct the chart. The center line is shown at the overall sample mean $\bar{\bar{x}} = 3.4995$. The upper control limit (UCL) is 3.514 and the lower control (LCL) is 3.485. The $\bar{x}$ chart shows the 20 sample means plotted over time. Because all 20 sample means are within the control limits, we confirm that the process mean was in control during the sampling period.

R Chart

Let us now consider a range chart (*R* chart) that can be used to control the variability of a process. To develop the *R* chart, we need to think of the range of a sample as a random variable with its own mean and standard deviation. The average range $\bar{R}$ provides an estimate of the mean of this random variable. Moreover, it can be shown that an estimate of the standard deviation of the range is

$$\hat{\sigma}_R = d_3\frac{\bar{R}}{d_2} \tag{19.9}$$

where d_2 and d_3 are constants that depend on the sample size; values of d_2 and d_3 are provided in Table 19.3. Thus, the UCL for the *R* chart is given by

$$\bar{R} + 3\hat{\sigma}_R = \bar{R}\left(1 + 3\frac{d_3}{d_2}\right) \tag{19.10}$$

and the LCL is

$$\bar{R} - 3\hat{\sigma}_R = \bar{R}\left(1 - 3\frac{d_3}{d_2}\right) \tag{19.11}$$

If we let

$$D_4 = 1 + 3\frac{d_3}{d_2} \tag{19.12}$$

$$D_3 = 1 - 3\frac{d_3}{d_2} \tag{19.13}$$

we can write the control limits for the R chart as

$$\text{UCL} = \bar{R}D_4 \tag{19.14}$$

$$\text{LCL} = \bar{R}D_3 \tag{19.15}$$

Values for D_3 and D_4 are also provided in Table 19.3. Note that for $n = 5$, $D_3 = 0$ and $D_4 = 2.114$. Thus, with $\bar{R} = .0253$, the control limits are

$$\text{UCL} = .0253(2.114) = .053$$
$$\text{LCL} = .0253(0) = 0$$

If the R chart indicates that the process is out of control, the $\bar{x}$ chart should not be interpreted until the R chart indicates the process variability is in control.

Figure 19.8 shows the R chart for the Jensen Computer Supplies problem. We used the data in Table 19.2 and StatTools' X/R Charts routine to construct the chart. The center line is shown at the overall mean of the 20 sample ranges, $\bar{R} = .0253$. The UCL is .053 and the LCL is .000. The R chart shows the 20 sample ranges plotted over time. Because all 20 sample ranges are within the control limits, we confirm that the process variability was in control during the sampling period.

FIGURE 19.8 *R* CHART FOR THE JENSEN COMPUTER SUPPLIES PROBLEM

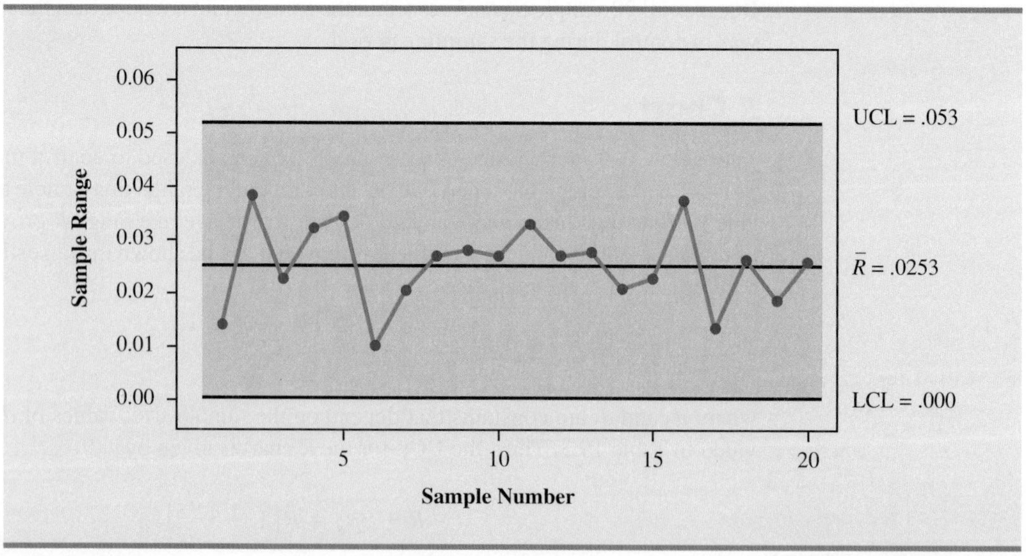

p Chart

Control charts that are based on data indicating the presence of a defect or a number of defects are called attributes control charts. A p chart is an attributes control chart.

Let us consider the case in which the output quality is measured by either nondefective or defective items. The decision to continue or to adjust the production process will be based on $\bar{p}$, the proportion of defective items found in a sample. The control chart used for proportion-defective data is called a p chart.

To illustrate the construction of a p chart, consider the use of automated mail-sorting machines in a post office. These automated machines scan the zip codes on letters and divert each letter to its proper carrier route. Even when a machine is operating properly, some letters are diverted to incorrect routes. Assume that when a machine is operating correctly, or in a state of control, 3% of the letters are incorrectly diverted. Thus p, the proportion of letters incorrectly diverted when the process is in control, is .03.

The sampling distribution of $\bar{p}$, as presented in Chapter 7, can be used to determine the variation that can be expected in $\bar{p}$ values for a process that is in control. Recall that the expected value or mean of $\bar{p}$ is p, the proportion defective when the process is in control. With samples of size n, the formula for the standard deviation of $\bar{p}$, called the standard error of the proportion, is

$$\sigma_{\bar{p}} = \sqrt{\frac{p(1-p)}{n}} \qquad \textbf{(19.16)}$$

We also learned in Chapter 7 that the sampling distribution of $\bar{p}$ can be approximated by a normal distribution whenever the sample size is large. With $\bar{p}$, the sample size can be considered large whenever the following two conditions are satisfied.

$$np \geq 5$$
$$n(1-p) \geq 5$$

In summary, whenever the sample size is large, the sampling distribution of $\bar{p}$ can be approximated by a normal distribution with mean p and standard deviation $\sigma_{\bar{p}}$. This distribution is shown in Figure 19.9.

FIGURE 19.9 SAMPLING DISTRIBUTION OF $\bar{p}$

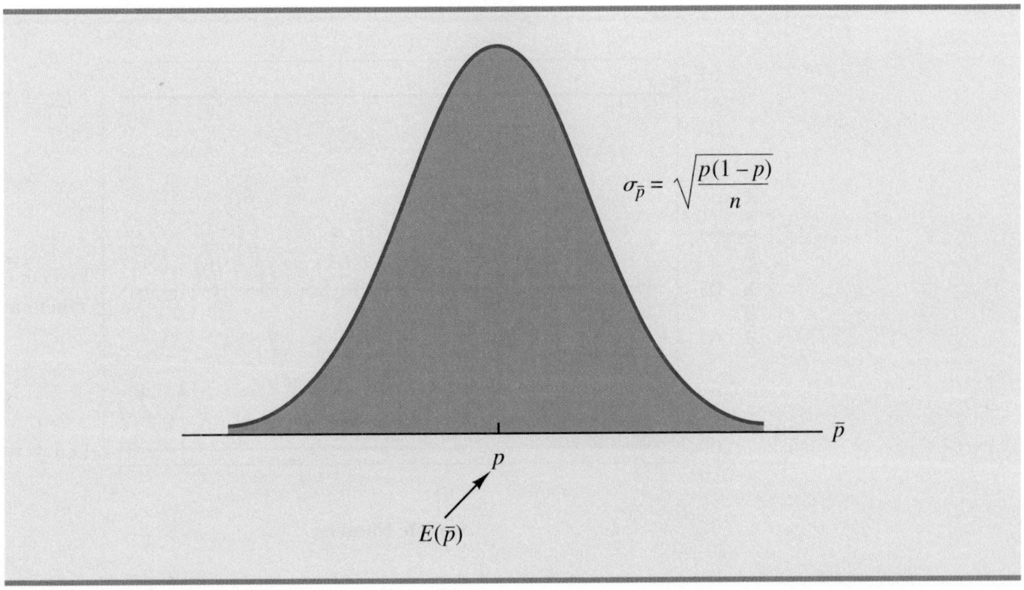

To establish control limits for a p chart, we follow the same procedure we used to establish control limits for an chart. That is, the limits for the control chart are set at 3 standard deviations, or standard errors, above and below the proportion defective when the process is in control. Thus, we have the following control limits.

CONTROL LIMITS FOR A p CHART

$$\text{UCL} = p + 3\sigma_{\bar{p}} \tag{19.17}$$

$$\text{LCL} = p - 3\sigma_{\bar{p}} \tag{19.18}$$

With $p = .03$ and samples of size $n = 200$, equation (19.16) shows that the standard error is

$$\sigma_{\bar{p}} = \sqrt{\frac{.03(1 - .03)}{200}} = .0121$$

Hence, the control limits are UCL $= .03 + 3(.0121) = .0663$ and LCL $= .03 - 3(.0121) = -.0063$. Whenever equation (19.18) provides a negative value for LCL, LCL is set equal to zero in the control chart.

Figure 19.10 is the p chart for the mail-sorting process. The points plotted show the sample proportion defective found in samples of letters taken from the process. All points are within the control limits, providing no evidence to conclude that the sorting process is out of control.

If the proportion of defective items for a process that is in control is not known, that value is first estimated by using sample data. Suppose, for example, that k different samples, each of size n, are selected from a process that is in control. The fraction or proportion of defective items in each sample is then determined. Treating all the data collected as one large sample, we can compute the proportion of defective items for all the data; that value can then be used to estimate p, the proportion of defective items observed when the process is in control. Note that this estimate of p also enables us to estimate the standard error of the proportion; upper and lower control limits can then be established.

FIGURE 19.10 p CHART FOR THE PROPORTION DEFECTIVE IN A MAIL-SORTING PROCESS

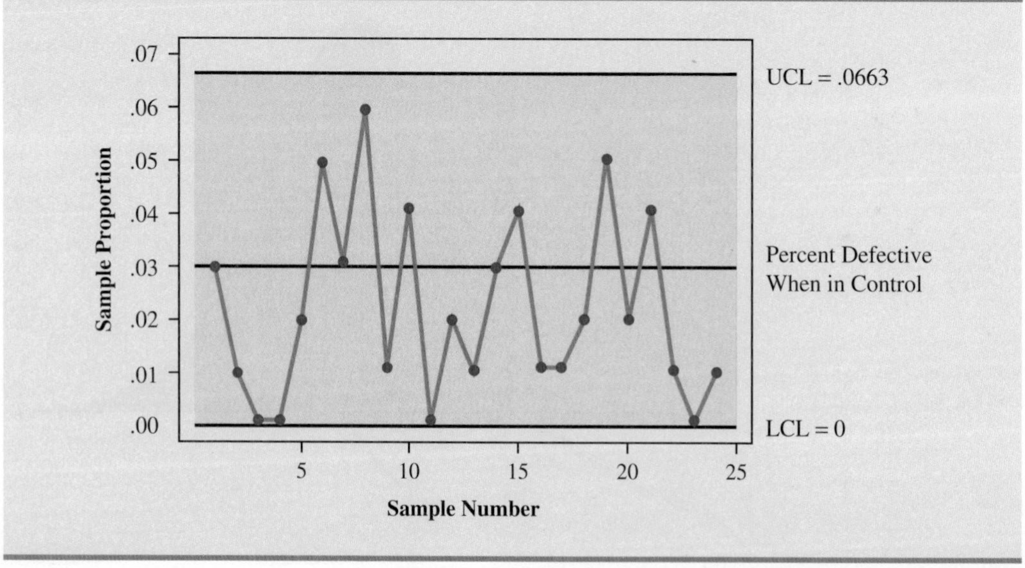

np Chart

An np chart is a control chart developed for the number of defective items in a sample. In this case, n is the sample size and p is the probability of observing a defective item when the process is in control. Whenever the sample size is large, that is, when $np \geq 5$ and $n(1 - p) \geq 5$, the distribution of the number of defective items observed in a sample of size n can be approximated by a normal distribution with mean np and standard deviation $\sqrt{np(1 - p)}$. Thus, for the mail-sorting example, with $n = 200$ and $p = .03$, the number of defective items observed in a sample of 200 letters can be approximated by a normal distribution with a mean of $200(.03) = 6$ and a standard deviation of $\sqrt{200(.03)(.97)} = 2.4125$.

The control limits for an np chart are set at 3 standard deviations above and below the expected number of defective items observed when the process is in control. Thus, we have the following control limits.

CONTROL LIMITS FOR AN np CHART

$$UCL = np + 3\sqrt{np(1 - p)} \tag{19.19}$$

$$LCL = np - 3\sqrt{np(1 - p)} \tag{19.20}$$

For the mail-sorting process example, with $p = .03$ and $n = 200$, the control limits are $UCL = 6 + 3(2.4125) = 13.2375$ and $LCL = 6 - 3(2.4125) = -1.2375$. When LCL is negative, LCL is set equal to zero in the control chart. Hence, if the number of letters diverted to incorrect routes is greater than 13, the process is concluded to be out of control.

The information provided by an np chart is equivalent to the information provided by the p chart; the only difference is that the np chart is a plot of the number of defective items observed, whereas the p chart is a plot of the proportion of defective items observed. Thus, if we were to conclude that a particular process is out of control on the basis of a p chart, the process would also be concluded to be out of control on the basis of an np chart.

Interpretation of Control Charts

The location and pattern of points in a control chart enable us to determine, with a small probability of error, whether a process is in statistical control. A primary indication that a process may be out of control is a data point outside the control limits, such as point 5 in Figure 19.6. Finding such a point is statistical evidence that the process is out of control; in such cases, corrective action should be taken as soon as possible.

In addition to points outside the control limits, certain patterns of the points within the control limits can be warning signals of quality control problems. For example, assume that all the data points are within the control limits but that a large number of points are on one side of the center line. This pattern may indicate that an equipment problem, a change in materials, or some other assignable cause of a shift in quality has occurred. Careful investigation of the production process should be undertaken to determine whether quality has changed.

Even if all points are within the upper and lower control limits, a process may not be in control. Trends in the sample data points or unusually long runs above or below the center line may also indicate out-of-control conditions.

Another pattern to watch for in control charts is a gradual shift, or trend, over time. For example, as tools wear out, the dimensions of machined parts will gradually deviate from their designed levels. Gradual changes in temperature or humidity, general equipment deterioration, dirt buildup, or operator fatigue may also result in a trend pattern in control charts. Six or seven points in a row that indicate either an increasing or decreasing trend should be cause for concern, even if the data points are all within the control limits. When such a pattern occurs, the process should be reviewed for possible changes or shifts in quality. Corrective action to bring the process back into control may be necessary.

NOTES AND COMMENTS

1. Because the control limits for the $\bar{x}$ chart depend on the value of the average range, these limits will not have much meaning unless the process variability is in control. In practice, the R chart is usually constructed before the $\bar{x}$ chart; if the R chart indicates that the process variability is in control, then the $\bar{x}$ chart is constructed. The StatTools X/R Charts option provides the $\bar{x}$ chart and the R chart simultaneously. The steps of this procedure are described in the chapter appendix.
2. An *np* chart is used to monitor a process in terms of the number of defects. The Motorola Six Sigma Quality Level sets a goal of producing no more than 3.4 defects per million operations; this goal implies $p = .0000034$.

Exercises

Methods

1. A process that is in control has a mean of $\mu = 12.5$ and a standard deviation of $\sigma = .8$.
   a. Construct the $\bar{x}$ control chart for this process if samples of size 4 are to be used.
   b. Repeat part (a) for samples of size 8 and 16.
   c. What happens to the limits of the control chart as the sample size is increased? Discuss why this is reasonable.
2. Twenty-five samples, each of size 5, were selected from a process that was in control. The sum of all the data collected was 677.5 pounds.
   a. What is an estimate of the process mean (in terms of pounds per unit) when the process is in control?
   b. Develop the $\bar{x}$ control chart for this process if samples of size 5 will be used. Assume that the process standard deviation is .5 when the process is in control, and that the mean of the process is the estimate developed in part (a).
3. Twenty-five samples of 100 items each were inspected when a process was considered to be operating satisfactorily. In the 25 samples, a total of 135 items were found to be defective.
   a. What is an estimate of the proportion defective when the process is in control?
   b. What is the standard error of the proportion if samples of size 100 will be used for statistical process control?
   c. Compute the upper and lower control limits for the control chart.

SELF test

4. A process sampled 20 times with a sample of size 8 resulted in $\bar{\bar{x}} = 28.5$ and $\bar{R} = 1.6$. Compute the upper and lower control limits for the $\bar{x}$ and R charts for this process.

Applications

5. Temperature is used to measure the output of a production process. When the process is in control, the mean of the process is $\mu = 128.5$ and the standard deviation is $\sigma = .4$.
   a. Construct the $\bar{x}$ chart for this process if samples of size 6 are to be used.
   b. Is the process in control for a sample providing the following data?

128.8	128.2	129.1	128.7	128.4	129.2

   c. Is the process in control for a sample providing the following data?

129.3	128.7	128.6	129.2	129.5	129.0

6. A quality control process monitors the weight per carton of laundry detergent. Control limits are set at UCL = 20.12 ounces and LCL = 19.90 ounces. Samples of size 5 are used

for the sampling and inspection process. What are the process mean and process standard deviation for the manufacturing operation?

7. The Goodman Tire and Rubber Company periodically tests its tires for tread wear under simulated road conditions. To study and control the manufacturing process, 20 samples, each containing three radial tires, were chosen from different shifts over several days of operation, with the following results. Assuming that these data were collected when the manufacturing process was believed to be operating in control, develop the R and $\bar{x}$ charts.

WEB file

Tires

Sample	Tread Wear*		
1	31	42	28
2	26	18	35
3	25	30	34
4	17	25	21
5	38	29	35
6	41	42	36
7	21	17	29
8	32	26	28
9	41	34	33
10	29	17	30
11	26	31	40
12	23	19	25
13	17	24	32
14	43	35	17
15	18	25	29
16	30	42	31
17	28	36	32
18	40	29	31
19	18	29	28
20	22	34	26

*Hundredths of an inch

8. Over several weeks of normal, or in-control, operation, 20 samples of 150 packages each of synthetic-gut tennis strings were tested for breaking strength. A total of 141 packages of the 3000 tested failed to conform to the manufacturer's specifications.
   a. What is an estimate of the process proportion defective when the system is in control?
   b. Compute the upper and lower control limits for a p chart.
   c. With the results of part (b), what conclusion should be made about the process if tests on a new sample of 150 packages find 12 defective? Do there appear to be assignable causes in this situation?
   d. Compute the upper and lower control limits for an np chart.
   e. Answer part (c) using the results of part (d).
   f. Which control chart would be preferred in this situation? Explain.

9. An airline operates a call center to handle customer questions and complaints. The airline monitors a sample of calls to help ensure that the service being provided is of high quality. Ten random samples of 100 calls each were monitored under normal conditions. The center can be thought of as being in control when these 10 samples were taken. The number of calls in each sample not resulting in a satisfactory resolution for the customer is as follows:

4	5	3	2	3	3	4	6	4	7

   a. What is an estimate of the proportion of calls not resulting in a satisfactory outcome for the customer when the center is in control?

b. Construct the upper and lower limits for a p chart for the manufacturing process, assuming each sample has 100 calls.
c. With the results of part (b), what conclusion should be made if a sample of 100 has 12 calls not resulting in a satisfactory resolution for the customer?
d. Compute the upper and lower limits for the np chart.
e. With the results of part (d), what conclusion should be made if a sample of 100 calls has 12 not resulting in a satisfactory conclusion for the customer?

19.3 Acceptance Sampling

In acceptance sampling, the items of interest can be incoming shipments of raw materials or purchased parts as well as finished goods from final assembly. Suppose we want to decide whether to accept or reject a group of items on the basis of specified quality characteristics. In quality control terminology, the group of items is a **lot**, and **acceptance sampling** is a statistical method that enables us to base the accept-reject decision on the inspection of a sample of items from the lot.

The general steps of acceptance sampling are shown in Figure 19.11. After a lot is received, a sample of items is selected for inspection. The results of the inspection are compared to specified quality characteristics. If the quality characteristics are satisfied, the lot is accepted and sent to production or shipped to customers. If the lot is rejected, managers must decide on its disposition. In some cases, the decision may be to keep the lot and remove the unacceptable or nonconforming items. In other cases, the lot may be returned to the supplier

FIGURE 19.11 ACCEPTANCE SAMPLING PROCEDURE

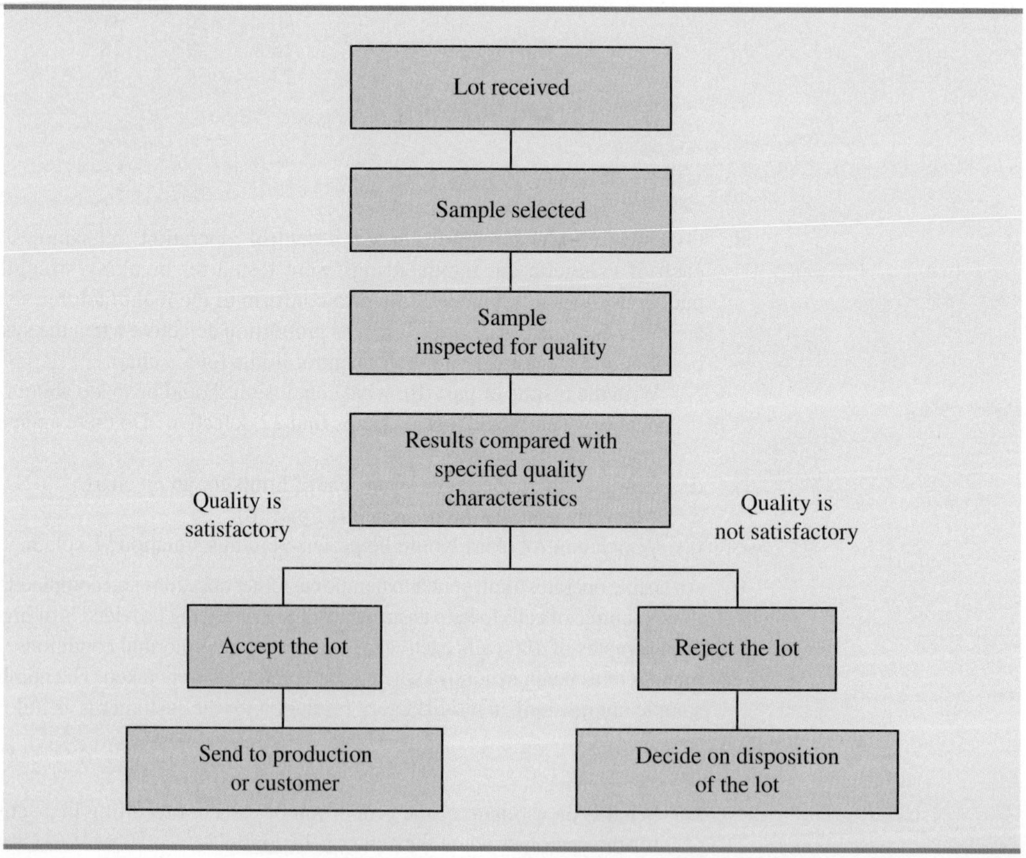

TABLE 19.4 THE OUTCOMES OF ACCEPTANCE SAMPLING

		State of the Lot	
		H_0 True Good-Quality Lot	H_0 False Poor-Quality Lot
Decision	Accept the Lot	Correct decision	Type II error (accepting a poor-quality lot)
	Reject the Lot	Type I error (rejecting a good-quality lot)	Correct decision

at the supplier's expense; the extra work and cost placed on the supplier can motivate the supplier to provide high-quality lots. Finally, if the rejected lot consists of finished goods, the goods must be scrapped or reworked to meet acceptable quality standards.

Acceptance sampling has the following advantages over 100% inspection:
1. Usually less expensive
2. Less product damage due to less handling and testing
3. Fewer inspectors required
4. The only approach possible if destructive testing must be used

The statistical procedure of acceptance sampling is based on the hypothesis testing methodology presented in Chapter 9. The null and alternative hypotheses are stated as follows.

$$H_0\text{: Good-quality lot}$$
$$H_a\text{: Poor-quality lot}$$

Table 19.4 shows the results of the hypothesis testing procedure. Note that correct decisions correspond to accepting a good-quality lot and rejecting a poor-quality lot. However, as with other hypothesis testing procedures, we need to be aware of the possibilities of making a Type I error (rejecting a good-quality lot) or a Type II error (accepting a poor-quality lot).

The probability of a Type I error creates a risk for the producer of the lot and is known as the **producer's risk**. For example, a producer's risk of .05 indicates a 5% chance that a good-quality lot will be erroneously rejected. The probability of a Type II error, on the other hand, creates a risk for the consumer of the lot and is known as the **consumer's risk**. For example, a consumer's risk of .10 means a 10% chance that a poor-quality lot will be erroneously accepted and thus used in production or shipped to the customer. Specific values for the producer's risk and the consumer's risk can be controlled by the person designing the acceptance sampling procedure. To illustrate how to assign risk values, let us consider the problem faced by KALI, Inc.

KALI, Inc.: An Example of Acceptance Sampling

KALI, Inc., manufactures home appliances that are marketed under a variety of trade names. However, KALI does not manufacture every component used in its products. Several components are purchased directly from suppliers. For example, one of the components that KALI purchases for use in home air conditioners is an overload protector, a device that turns off the compressor if it overheats. The compressor can be seriously damaged if the overload protector does not function properly, and therefore KALI is concerned about the quality of the overload protectors. One way to ensure quality would be to test every component received through an approach known as 100% inspection. However, to determine proper functioning of an overload protector, the device must be subjected to time-consuming and expensive tests, and KALI cannot justify testing every overload protector it receives.

Instead, KALI uses an acceptance sampling plan to monitor the quality of the overload protectors. The acceptance sampling plan requires that KALI's quality control inspectors select and test a sample of overload protectors from each shipment. If very few defective units are found in the sample, the lot is probably of good quality and should be accepted.

However, if a large number of defective units are found in the sample, the lot is probably of poor quality and should be rejected.

An acceptance sampling plan consists of a sample size n and an acceptance criterion c. The **acceptance criterion** is the maximum number of defective items that can be found in the sample and still indicate an acceptable lot. For example, for the KALI problem let us assume that a sample of 15 items will be selected from each incoming shipment or lot. Furthermore, assume that the manager of quality control states that the lot can be accepted only if no defective items are found. In this case, the acceptance sampling plan established by the quality control manager is $n = 15$ and $c = 0$.

This acceptance sampling plan is easy for the quality control inspector to implement. The inspector simply selects a sample of 15 items, performs the tests, and reaches a conclusion based on the following decision rule.

- *Accept the lot* if zero defective items are found.
- *Reject the lot* if one or more defective items are found.

Before implementing this acceptance sampling plan, the quality control manager wants to evaluate the risks or errors possible under the plan. The plan will be implemented only if both the producer's risk (Type I error) and the consumer's risk (Type II error) are controlled at reasonable levels.

Computing the Probability of Accepting a Lot

The key to analyzing both the producer's risk and the consumer's risk is a "what-if" type of analysis. That is, we will assume that a lot has some known percentage of defective items and compute the probability of accepting the lot for a given sampling plan. By varying the assumed percentage of defective items, we can examine the effect of the sampling plan on both types of risks.

Let us begin by assuming that a large shipment of overload protectors has been received and that 5% of the overload protectors in the shipment are defective. For a shipment or lot with 5% of the items defective, what is the probability that the $n = 15$, $c = 0$ sampling plan will lead us to accept the lot? Because each overload protector tested will be either defective or nondefective and because the lot size is large, the number of defective items in a sample of 15 has a *binomial distribution.* The binomial probability function, which was presented in Chapter 5, follows.

BINOMIAL PROBABILITY FUNCTION FOR ACCEPTANCE SAMPLING

$$f(x) = \frac{n!}{x!(n - x)!} p^x(1 - p)^{(n-x)} \tag{19.21}$$

where

$$\begin{aligned} n &= \text{the sample size} \\ p &= \text{the proportion of defective items in the lot} \\ x &= \text{the number of defective items in the sample} \\ f(x) &= \text{the probability of } x \text{ defective items in the sample} \end{aligned}$$

For the KALI acceptance sampling plan, $n = 15$; thus, for a lot with 5% defective ($p = .05$), we have

$$f(x) = \frac{15!}{x!(15 - x)!} (.05)^x(1 - .05)^{(15-x)} \tag{19.22}$$

TABLE 19.5 PROBABILITY OF ACCEPTING THE LOT FOR THE KALI PROBLEM WITH $n = 15$ AND $c = 0$

Percent Defective in the Lot	Probability of Accepting the Lot
1	.8601
2	.7386
3	.6333
4	.5421
5	.4633
10	.2059
15	.0874
20	.0352
25	.0134

Using equation (19.22), $f(0)$ will provide the probability that zero overload protectors will be defective and the lot will be accepted. In using equation (19.22), recall that $0! = 1$. Thus, the probability computation for $f(0)$ is

$$f(0) = \frac{15!}{0!(15 - 0)!}(.05)^0(1 - .05)^{(15-0)}$$

$$= \frac{15!}{0!(15)!}(.05)^0(.95)^{15} = (.95)^{15} = .4633$$

We now know that the $n = 15$, $c = 0$ sampling plan has a .4633 probability of accepting a lot with 5% defective items. Hence, there must be a corresponding $1 - .4633 = .5367$ probability of rejecting a lot with 5% defective items.

Excel's BINOM.DIST function can be used to simplify making these binomial probability calculations. Using this function, we can determine that if the lot contains 10% defective items, there is a .2059 probability that the $n = 15$, $c = 0$ sampling plan will indicate an acceptable lot. The probability that the $n = 15$, $c = 0$ sampling plan will lead to the acceptance of lots with 1%, 2%, 3%, . . . defective items is summarized in Table 19.5.

Using the probabilities in Table 19.5, a graph of the probability of accepting the lot versus the percent defective in the lot can be drawn as shown in Figure 19.12. This graph, or curve, is called the **operating characteristic (OC) curve** for the $n = 15$, $c = 0$ acceptance sampling plan.

Perhaps we should consider other sampling plans, ones with different sample sizes n or different acceptance criteria c. First consider the case in which the sample size remains $n = 15$ but the acceptance criterion increases from $c = 0$ to $c = 1$. That is, we will now accept the lot if zero or one defective component is found in the sample. For a lot with 5% defective items ($p = .05$), the binomial probability function in equation (19.21), or Excel's BINOM.DIST function, can be used to compute $f(0) = .4633$ and $f(1) = .3658$. Thus, there is a $.4633 + .3658 = .8291$ probability that the $n = 15$, $c = 1$ plan will lead to the acceptance of a lot with 5% defective items.

Continuing these calculations, we obtain Figure 19.13, which shows the operating characteristic curves for four alternative acceptance sampling plans for the KALI problem. Samples of size 15 and 20 are considered. Note that regardless of the proportion defective in the lot, the $n = 15$, $c = 1$ sampling plan provides the highest probabilities of accepting the lot. The $n = 20$, $c = 0$ sampling plan provides the lowest probabilities of accepting the lot; however, that plan also provides the highest probabilities of rejecting the lot.

FIGURE 19.12 OPERATING CHARACTERISTIC CURVE FOR THE $n = 15$, $c = 0$ ACCEPTANCE SAMPLING PLAN

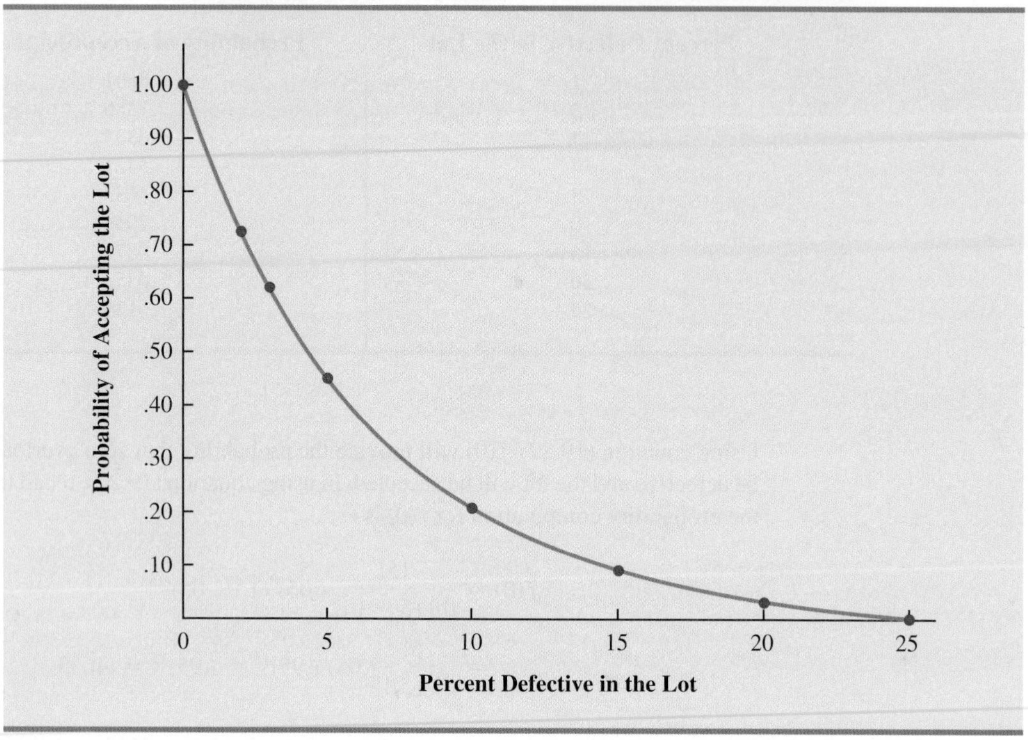

FIGURE 19.13 OPERATING CHARACTERISTIC CURVES FOR FOUR ACCEPTANCE SAMPLING PLANS

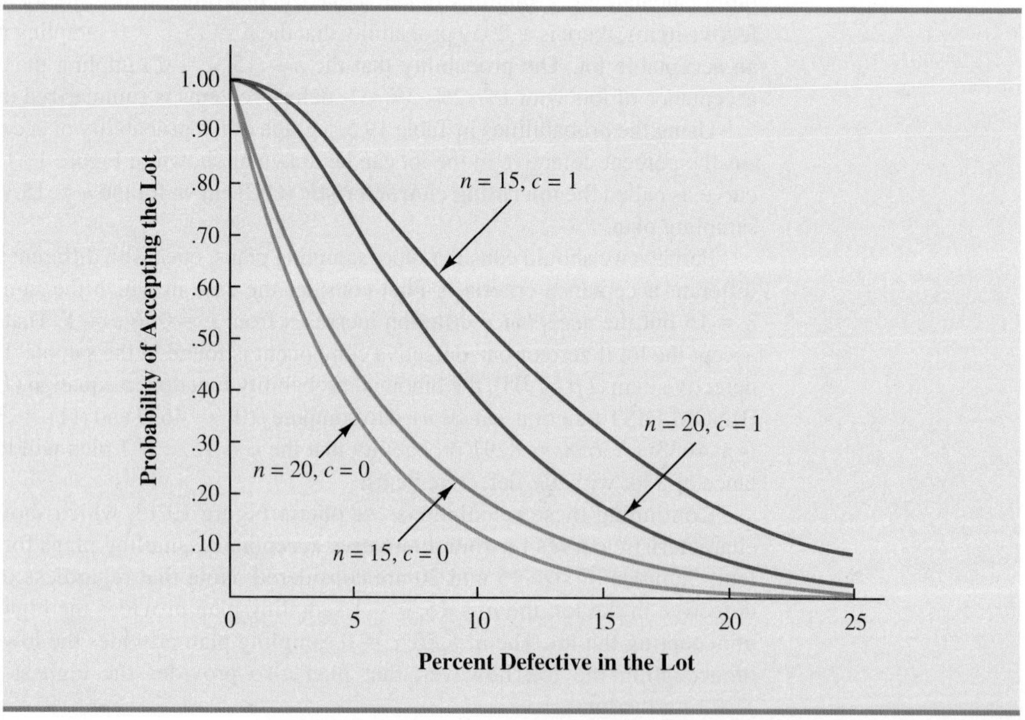

Selecting an Acceptance Sampling Plan

Now that we know how to use the binomial distribution to compute the probability of accepting a lot with a given proportion defective, we are ready to select the values of n and c that determine the desired acceptance sampling plan for the application being studied. To develop this plan, managers must specify two values for the fraction defective in the lot. One value, denoted p_0, will be used to control for the producer's risk, and the other value, denoted p_1, will be used to control for the consumer's risk.

We will use the following notation.

α = the producer's risk; the probability of rejecting a lot with p_0 defective items

β = the consumer's risk; the probability of accepting a lot with p_1 defective items

Suppose that for the KALI problem, the managers specify that $p_0 = .03$ and $p_1 = .15$. From the OC curve for $n = 15$, $c = 0$ in Figure 19.14, we see that $p_0 = .03$ provides a producer's risk of approximately $1 - .63 = .37$, and $p_1 = .15$ provides a consumer's risk of approximately .09. Thus, if the managers are willing to tolerate both a .37 probability of rejecting a lot with 3% defective items (producer's risk) and a .09 probability of accepting a lot with 15% defective items (consumer's risk), the $n = 15$, $c = 0$ acceptance sampling plan would be acceptable.

Suppose, however, that the managers request a producer's risk of $\alpha = .10$ and a consumer's risk of $\beta = .19$. We see that now the $n = 15$, $c = 0$ sampling plan has a better-than-desired consumer's risk but an unacceptably large producer's risk. The fact that $\alpha = .37$ indicates that 37% of the lots will be erroneously rejected when only 3% of the items in them are defective. The producer's risk is too high, and a different acceptance sampling plan should be considered.

FIGURE 19.14 OPERATING CHARACTERISTIC CURVE FOR $n = 15$, $c = 0$ WITH $p_0 = .03$ AND $p_1 = .15$

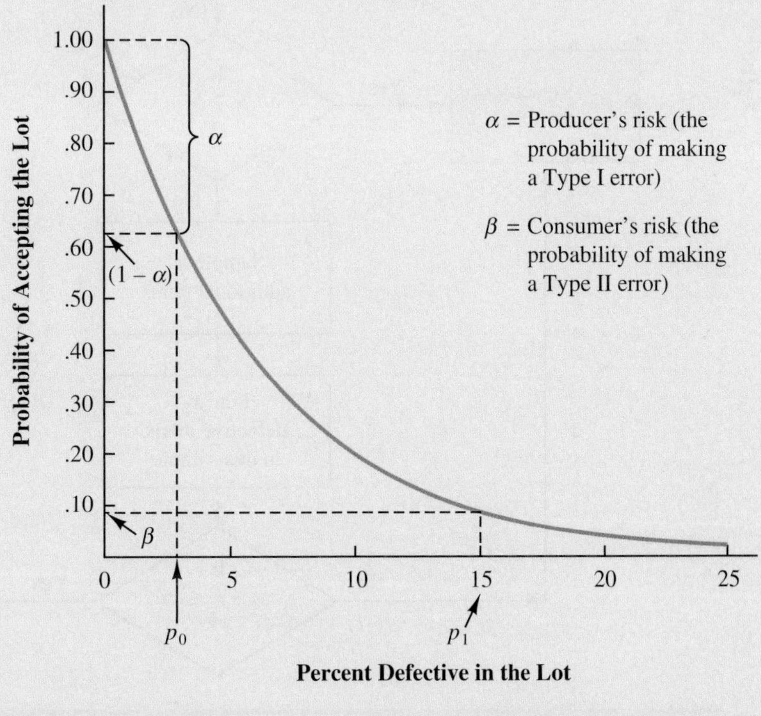

Exercise 13 at the end of this section will ask you to compute the producer's risk and the consumer's risk for the n = 20, c = 1 sampling plan.

Using $p_0 = .03$, $\alpha = .10$, $p_1 = .15$, and $\beta = .20$, Figure 19.13 shows that the acceptance sampling plan with $n = 20$ and $c = 1$ comes closest to meeting both the producer's and the consumer's risk requirements.

As shown in this section, several computations and several operating characteristic curves may need to be considered to determine the sampling plan with the desired producer's and consumer's risk. Fortunately, tables of sampling plans are published. For example, the American Military Standard Table, MIL-STD-105D, provides information helpful in designing acceptance sampling plans. More advanced texts on quality control, such as those listed in the bibliography, describe the use of such tables. The advanced texts also discuss the role of sampling costs in determining the optimal sampling plan.

FIGURE 19.15 A TWO-STAGE ACCEPTANCE SAMPLING PLAN

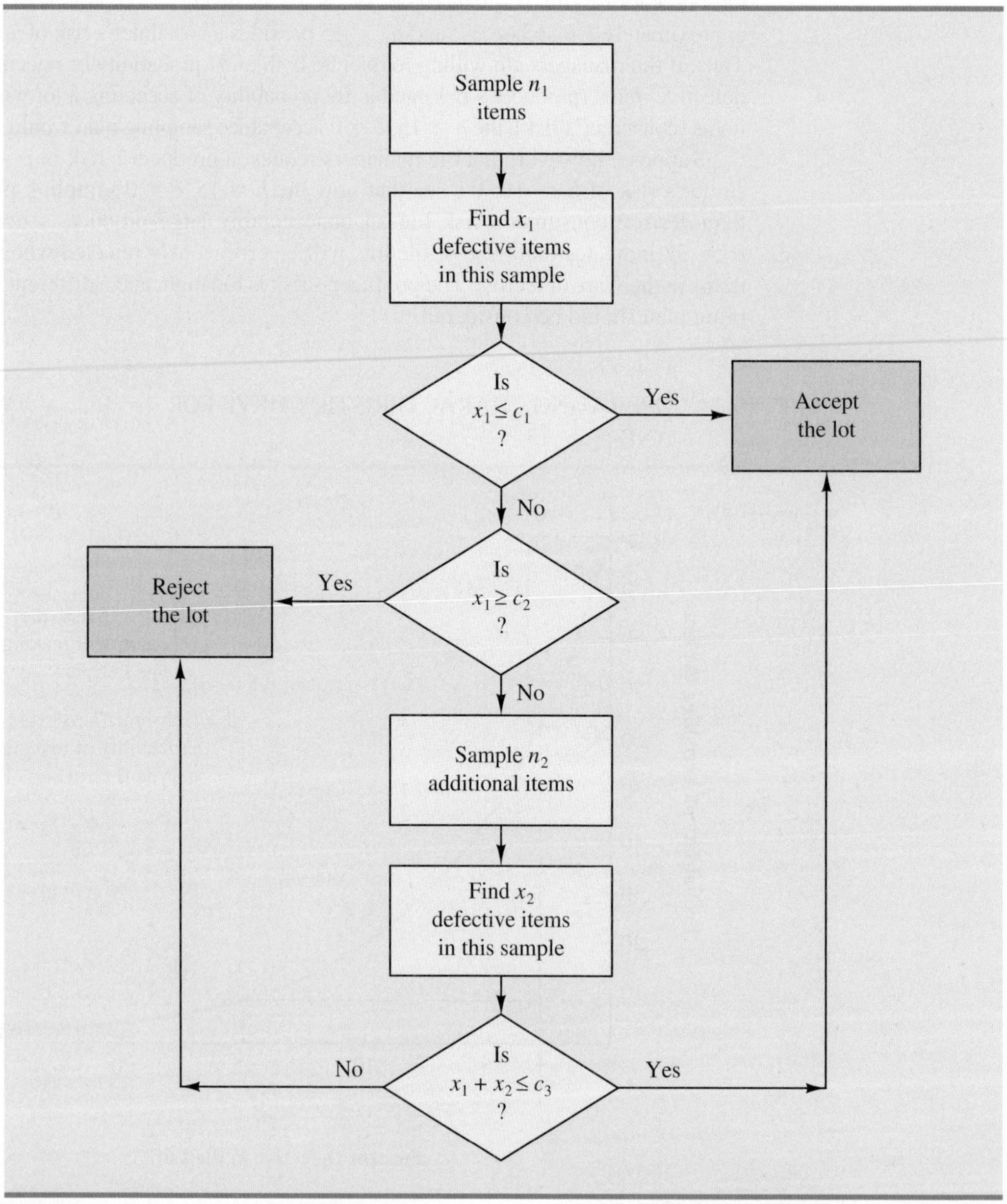

Multiple Sampling Plans

The acceptance sampling procedure we presented for the KALI problem is a *single-sample* plan. It is called a single-sample plan because only one sample or sampling stage is used. After the number of defective components in the sample is determined, a decision must be made to accept or reject the lot. An alternative to the single-sample plan is a **multiple sampling plan**, in which two or more stages of sampling are used. At each stage a decision is made among three possibilities: stop sampling and accept the lot, stop sampling and reject the lot, or continue sampling. Although more complex, multiple sampling plans often result in a smaller total sample size than single-sample plans with the same α and β probabilities.

The logic of a two-stage, or double-sample, plan is shown in Figure 19.15. Initially a sample of n_1 items is selected. If the number of defective components x_1 is less than or equal to c_1, accept the lot. If x_1 is greater than or equal to c_2, reject the lot. If x_1 is between c_1 and c_2 ($c_1 < x_1 < c_2$), select a second sample of n_2 items. Determine the combined, or total, number of defective components from the first sample (x_1) and the second sample (x_2). If $x_1 + x_2 \leq c_3$, accept the lot; otherwise reject the lot. The development of the double-sample plan is more difficult because the sample sizes n_1 and n_2 and the acceptance numbers c_1, c_2, and c_3 must meet both the producer's and consumer's risks desired.

NOTES AND COMMENTS

1. The use of the binomial distribution for acceptance sampling is based on the assumption of large lots. If the lot size is small, the hypergeometric distribution is appropriate.
2. In the MIL-STD-105D sampling tables, p_0 is called the acceptable quality level (AQL). In some sampling tables, p_1 is called the lot tolerance percent defective (LTPD) or the rejectable quality level (RQL). Many of the published sampling plans also use quality indexes such as the indifference quality level (IQL) and the average outgoing quality limit (AOQL). The more advanced texts listed in the bibliography provide a complete discussion of these other indexes.
3. In this section we provided an introduction to *attributes sampling plans.* In these plans each item sampled is classified as nondefective or defective. In *variables sampling plans,* a sample is taken and a measurement of the quality characteristic is taken. For example, for gold jewelry a measurement of quality may be the amount of gold it contains. A simple statistic such as the average amount of gold in the sample jewelry is computed and compared with an allowable value to determine whether to accept or reject the lot.

Exercises

Methods

SELF test

10. For an acceptance sampling plan with $n = 25$ and $c = 0$, find the probability of accepting a lot that has a defect rate of 2%. What is the probability of accepting the lot if the defect rate is 6%?
11. Consider an acceptance sampling plan with $n = 20$ and $c = 0$. Compute the producer's risk for each of the following cases.
    a. The lot has a defect rate of 2%.
    b. The lot has a defect rate of 6%.

12. Repeat exercise 11 for the acceptance sampling plan with $n = 20$ and $c = 1$. What happens to the producer's risk as the acceptance number c is increased? Explain.

Applications

13. Refer to the KALI problem presented in this section. The quality control manager requested a producer's risk of .10 when p_0 was .03 and a consumer's risk of .20 when p_1 was .15. Consider the acceptance sampling plan based on a sample size of 20 and an acceptance number of 1. Answer the following questions.
    a. What is the producer's risk for the $n = 20$, $c = 1$ sampling plan?
    b. What is the consumer's risk for the $n = 20$, $c = 1$ sampling plan?
    c. Does the $n = 20$, $c = 1$ sampling plan satisfy the risks requested by the quality control manager? Discuss.
14. To inspect incoming shipments of raw materials, a manufacturer is considering samples of sizes 10, 15, and 19. Use the binomial probabilities from Table 5 of Appendix B to select a sampling plan that provides a producer's risk of $\alpha = .03$ when p_0 is .05 and a consumer's risk of $\beta = .12$ when p_1 is .30.
15. A domestic manufacturer of watches purchases quartz crystals from a Swiss firm. The crystals are shipped in lots of 1000. The acceptance sampling procedure uses 20 randomly selected crystals.
    a. Construct operating characteristic curves for acceptance numbers of 0, 1, and 2.
    b. If p_0 is .01 and $p_1 = .08$, what are the producer's and consumer's risks for each sampling plan in part (a)?

Summary

In this chapter we discussed how statistical methods can be used to assist in the control of quality. We first presented the $\bar{x}$, R, p, and np control charts as graphical aids in monitoring process quality. Control limits are established for each chart; samples are selected periodically, and the data points plotted on the control chart. Data points outside the control limits indicate that the process is out of control and that corrective action should be taken. Patterns of data points within the control limits can also indicate potential quality control problems and suggest that corrective action may be warranted.

We also considered the technique known as acceptance sampling. With this procedure, a sample is selected and inspected. The number of defective items in the sample provides the basis for accepting or rejecting the lot. The sample size and the acceptance criterion can be adjusted to control both the producer's risk (Type I error) and the consumer's risk (Type II error).

Glossary

Total quality (TQ) A total system approach to improving customer satisfaction and lowering real cost through a strategy of continuous improvement and learning.

Six Sigma A methodology that uses measurement and statistical analysis to achieve a level of quality so good that for every million opportunities no more than 3.4 defects will occur.

Quality control A series of inspections and measurements that determine whether quality standards are being met.
Assignable causes Variations in process outputs that are due to factors such as machine tools wearing out, incorrect machine settings, poor-quality raw materials, operator error, and so on. Corrective action should be taken when assignable causes of output variation are detected.
Common causes Normal or natural variations in process outputs that are due purely to chance. No corrective action is necessary when output variations are due to common causes.
Control chart A graphical tool used to help determine whether a process is in control or out of control.
$\bar{x}$ chart A control chart used when the quality of the output of a process is measured in terms of the mean value of a variable such as a length, weight, temperature, and so on.
***R* chart** A control chart used when the quality of the output of a process is measured in terms of the range of a variable.
***p* chart** A control chart used when the quality of the output of a process is measured in terms of the proportion defective.
***np* chart** A control chart used to monitor the quality of the output of a process in terms of the number of defective items.
Lot A group of items such as incoming shipments of raw materials or purchased parts as well as finished goods from final assembly.
Acceptance sampling A statistical method in which the number of defective items found in a sample is used to determine whether a lot should be accepted or rejected.
Producer's risk The risk of rejecting a good-quality lot; a Type I error.
Consumer's risk The risk of accepting a poor-quality lot; a Type II error.
Acceptance criterion The maximum number of defective items that can be found in the sample and still indicate an acceptable lot.
Operating characteristic (OC) curve A graph showing the probability of accepting the lot as a function of the percentage defective in the lot. This curve can be used to help determine whether a particular acceptance sampling plan meets both the producer's and the consumer's risk requirements.
Multiple sampling plan A form of acceptance sampling in which more than one sample or stage is used. On the basis of the number of defective items found in a sample, a decision will be made to accept the lot, reject the lot, or continue sampling.

Key Formulas

Standard Error of the Mean

$$\sigma_{\bar{x}} = \frac{\sigma}{\sqrt{n}} \tag{19.1}$$

Control Limits for an $\bar{x}$ Chart: Process Mean and Standard Deviation Known

$$\text{UCL} = \mu + 3\sigma_{\bar{x}} \tag{19.2}$$

$$\text{LCL} = \mu - 3\sigma_{\bar{x}} \tag{19.3}$$

Overall Sample Mean

$$\bar{\bar{x}} = \frac{\bar{x}_1 + \bar{x}_2 + \cdots + \bar{x}_k}{k} \tag{19.4}$$

Average Range

$$\bar{R} = \frac{R_1 + R_2 + \cdots + R_k}{k} \tag{19.5}$$

Control Limits for an $\bar{x}$ Chart: Process Mean and Standard Deviation Unknown

$$\bar{\bar{x}} \pm A_2\bar{R} \tag{19.8}$$

Control Limits for an R Chart

$$\text{UCL} = \bar{R}D_4 \tag{19.14}$$
$$\text{LCL} = \bar{R}D_3 \tag{19.15}$$

Standard Error of the Proportion

$$\sigma_{\bar{p}} = \sqrt{\frac{p(1-p)}{n}} \tag{19.16}$$

Control Limits for a p Chart

$$\text{UCL} = p + 3\sigma_{\bar{p}} \tag{19.17}$$
$$\text{LCL} = p - 3\sigma_{\bar{p}} \tag{19.18}$$

Control Limits for an np Chart

$$\text{UCL} = np + 3\sqrt{np(1-p)} \tag{19.19}$$
$$\text{LCL} = np - 3\sqrt{np(1-p)} \tag{19.20}$$

Binomial Probability Function for Acceptance Sampling

$$f(x) = \frac{n!}{x!(n-x)!}p^x(1-p)^{(n-x)} \tag{19.21}$$

Supplementary Exercises

16. Samples of size 5 provided the following 20 sample means for a production process that is believed to be in control.

95.72	95.24	95.18
95.44	95.46	95.32
95.40	95.44	95.08
95.50	95.80	95.22
95.56	95.22	95.04
95.72	94.82	95.46
95.60	95.78	

a. Based on these data, what is an estimate of the mean when the process is in control?
b. Assume that the process standard deviation is $\sigma = .50$. Develop the $\bar{x}$ control chart for this production process. Assume that the mean of the process is the estimate developed in part (a).
c. Do any of the 20 sample means indicate that the process was out of control?

17. Product filling weights are normally distributed with a mean of 350 grams and a standard deviation of 15 grams.
    a. Develop the control limits for the $\bar{x}$ chart for samples of size 10, 20, and 30.
    b. What happens to the control limits as the sample size is increased?
    c. What happens when a Type I error is made?
    d. What happens when a Type II error is made?
    e. What is the probability of a Type I error for samples of size 10, 20, and 30?
    f. What is the advantage of increasing the sample size for control chart purposes? What error probability is reduced as the sample size is increased?

18. Twenty-five samples of size 5 resulted in $\bar{\bar{x}} = 5.42$ and $\bar{R} = 2.0$. Compute control limits for the $\bar{x}$ and R charts, and estimate the standard deviation of the process.

19. The following are quality control data for a manufacturing process at Kensport Chemical Company. The data show the temperature in degrees centigrade at five points in time during a manufacturing cycle. The company is interested in using control charts to monitor the temperature of its manufacturing process. Construct the $\bar{x}$ chart and R chart. What conclusions can be made about the quality of the process?

Sample	$\bar{x}$	R	Sample	$\bar{x}$	R
1	95.72	1.0	11	95.80	.6
2	95.24	.9	12	95.22	.2
3	95.18	.8	13	95.56	1.3
4	95.44	.4	14	95.22	.5
5	95.46	.5	15	95.04	.8
6	95.32	1.1	16	95.72	1.1
7	95.40	.9	17	94.82	.6
8	95.44	.3	18	95.46	.5
9	95.08	.2	19	95.60	.4
10	95.50	.6	20	95.74	.6

20. The following were collected for the Master Blend Coffee production process. The data show the filling weights based on samples of 3-pound cans of coffee. Use these data to construct the $\bar{x}$ and R charts. What conclusions can be made about the quality of the production process?

WEB file

Coffee

	Observations				
Sample	1	2	3	4	5
1	3.05	3.08	3.07	3.11	3.11
2	3.13	3.07	3.05	3.10	3.10
3	3.06	3.04	3.12	3.11	3.10
4	3.09	3.08	3.09	3.09	3.07
5	3.10	3.06	3.06	3.07	3.08
6	3.08	3.10	3.13	3.03	3.06
7	3.06	3.06	3.08	3.10	3.08
8	3.11	3.08	3.07	3.07	3.07
9	3.09	3.09	3.08	3.07	3.09
10	3.06	3.11	3.07	3.09	3.07

21. An insurance company samples claim forms for errors created by its employees as well as the amount of time it takes to process a claim.
    a. When the process is in control, the proportion of claims with an error is .033. A p chart has LCL = 0 and UCL = .068. Plot the following seven sample results: .035, .062,

.055, .049, .058, .066, and .055. Comment on whether there might be concern about the quality of the process.

b. An $\bar{x}$ chart for the mean processing time has LCL = 22.2 and UCL = 24.5. The mean is $\mu = 23.35$ when the claim process is in control. Plot the following seven sample results: 22.4, 22.6, 22.65, 23.2, 23.4, 23.85, and 24.1. Comment on whether there might be concern about the quality of the process.

22. Managers of 1200 different retail outlets make twice-a-month restocking orders from a central warehouse. Past experience shows that 4% of the orders result in one or more errors such as wrong item shipped, wrong quantity shipped, and item requested but not shipped. Random samples of 200 orders are selected monthly and checked for accuracy.
    a. Construct a control chart for this situation.
    b. Six months of data show the following numbers of orders with one or more errors: 10, 15, 6, 13, 8, and 17. Plot the data on the control chart. What does your plot indicate about the order process?

23. An $n = 10$, $c = 2$ acceptance sampling plan is being considered; assume that $p_0 = .05$ and $p_1 = .20$.
    a. Compute both producer's and consumer's risk for this acceptance sampling plan.
    b. Would the producer, the consumer, or both be unhappy with the proposed sampling plan?
    c. What change in the sampling plan, if any, would you recommend?

24. An acceptance sampling plan with $n = 15$ and $c = 1$ has been designed with a producer's risk of .075.
    a. Was the value of p_0 .01, .02, .03, .04, or .05? What does this value mean?
    b. What is the consumer's risk associated with this plan if p_1 is .25?

25. A manufacturer produces lots of a canned food product. Let p denote the proportion of the lots that do not meet the product quality specifications. An $n = 25$, $c = 0$ acceptance sampling plan will be used.
    a. Compute points on the operating characteristic curve when $p = .01, .03, .10,$ and $.20$.
    b. Plot the operating characteristic curve.
    c. What is the probability that the acceptance sampling plan will reject a lot containing .01 defective?

Appendix 19.1 Control Charts with Minitab

WEB file
Jensen

In this appendix we describe the steps required to generate Minitab control charts using the Jensen Computer Supplies data shown in Table 19.2. The sample number appears in column C1. The first observation is in column C2, the second observation is in column C3, and so on. The following steps describe how to use Minitab to produce both the $\bar{x}$ chart and R chart simultaneously.

Step 1. Select the **Stat** menu
Step 2. Choose **Control Charts**
Step 3. Choose **Variables Charts for Subgroups**
Step 4. Choose **Xbar-R**
Step 5. When the Xbar-R Chart dialog box appears:
  Select **Observations for a subgroup are in one row of columns**
  In the box below, enter C2-C6
  Select **Xbar-R Options**

Step 6. When the Xbar-R-Options dialog box appears:
Select the **Tests** tab
Select **Perform selected tests for special causes**
Choose **1 point > K standard deviations from center line***
Enter 3 in the **K** box
Click **OK**

Step 7. When the Xbar-R Chart dialog box appears:
Click **OK**

The $\bar{x}$ chart and the R chart will be shown together on the Minitab output. The choices available under step 3 of the preceding Minitab procedure provide access to a variety of control chart options. For example, the $\bar{x}$ and the R chart can be selected separately. Additional options include the p chart, the np chart, and others.

Appendix 19.2 Control Charts Using StatTools

WEB file
Jensen

In this appendix we show how StatTools can be used to construct an $\bar{x}$ chart and an R chart for the Jensen Computer Supplies data shown in Table 19.2. Figure 19.16 is an Excel worksheet containing the Jensen data with the StatTools - Xbar and R Control Charts dialog box in the foreground. Begin by using the Data Set Manager to create a

FIGURE 19.16 EXCEL DATA WORKSHEET FOR JENSEN COMPUTER SUPPLIES

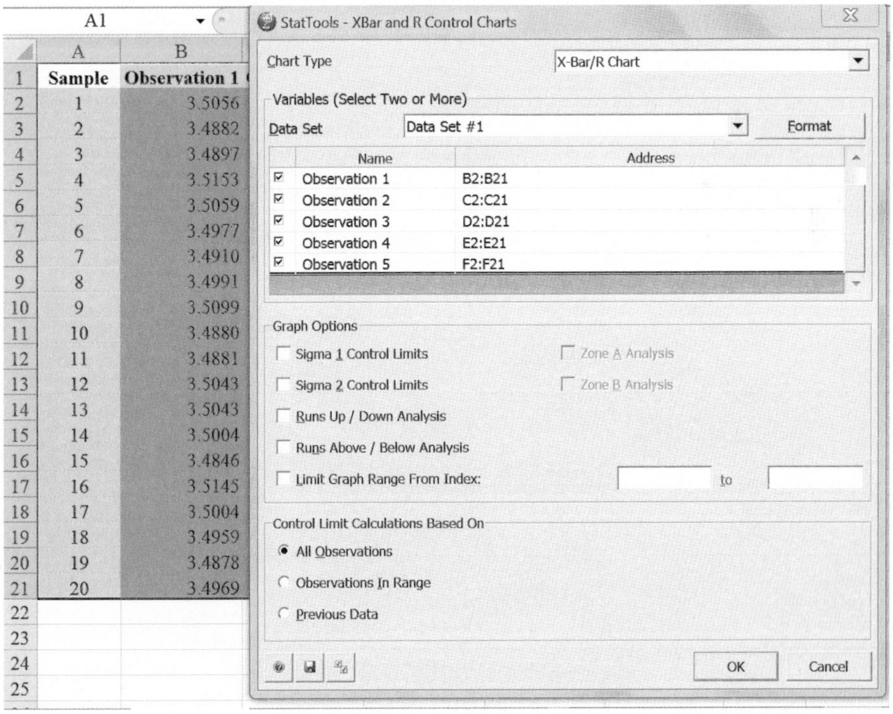

*Minitab provides several additional tests for detecting special causes of variation and out-of-control conditions. The user may select several of these tests simultaneously.

StatTools data set for these data using the procedure described in the appendix in Chapter 1. The following steps describe how StatTools can be used to provide both control charts.

Step 1. Click the **StatTools** tab on the Ribbon
Step 2. In the **Analyses** group, click **Quality Control**
Step 3. Choose the **X/R Charts** option
Step 4. When the StatTools - Xbar and R Control Charts dialog box appears (Figure 19.16);
Select **X-Bar/R Chart** in the **Chart Type** box
In the **Variables** section, select **Observation 1, Observation 2, Observation 3, Observation 4,** and **Observation 5**
Click **OK**

An $\bar{x}$ chart similar to the one in Figure 19.7 will appear. It will be followed by an R chart similar to the one in Figure 19.8.

CHAPTER 20

Index Numbers

CONTENTS

STATISTICS *in* PRACTICE

U.S. DEPARTMENT OF LABOR,
BUREAU OF LABOR STATISTICS
WASHINGTON, D.C.

The U.S. Department of Labor, through its Bureau of Labor Statistics, compiles and distributes indexes and statistics that are indicators of business and economic activity in the United States. For instance, the Bureau compiles and publishes the Consumer Price Index, the Producer Price Index, and statistics on average hours and earnings of various groups of workers. Perhaps the most widely quoted index produced by the Bureau of Labor Statistics is the Consumer Price Index. It is often used as a measure of inflation.

In March 2009, the Bureau of Labor Statistics reported that the Consumer Price Index (CPI) increased by .5% in February. The February level of 212.2 was .3% higher than in February 2008. On a seasonally adjusted basis, the CPI increased .4% in February after rising .3% in January. The 8.3% increase in the gasoline price index seemed to cause most of the increase. The food index actually declined .1%. Some economists considered the CPI increase good news because it reduced the likelihood of a deflationary period.

The Bureau of Labor Statistics, one day earlier, had reported that the Producer Price Index (PPI) increased by .1% in February, seasonally adjusted. The increase followed a .8% increase in January and a 1.9% decline in December. The PPI measures price changes in wholesale markets and is often seen as a leading indicator of changes in the Consumer Price Index. The slower rate of increase in February was heavily influenced by the declining rate of increase in energy goods. The energy goods index rose by 1.3% in February after rising by 3.7% in Janurary.

In this chapter we will see how various indexes, such as the Consumer and Producer Price Indexes, are computed and how they should be interpreted.

Gasoline prices are a component of the Consumer Price Index. © Randy Green/Alamy Limited.

Each month the U.S. government publishes a variety of indexes designed to help individuals understand current business and economic conditions. Perhaps the most widely known and cited of these indexes is the Consumer Price Index (CPI). As its name implies, the CPI is an indicator of what is happening to prices consumers pay for items purchased. Specifically, the CPI measures changes in price over a period of time. With a given starting point or *base period* and its associated index of 100, the CPI can be used to compare current period consumer prices with those in the base period. For example, a CPI of 125 reflects the condition that consumer prices as a whole are running approximately 25% above the base period prices for the same items. Although relatively few individuals know exactly what this number means, they do know enough about the CPI to understand that an increase means higher prices.

Even though the CPI is perhaps the best-known index, many other governmental and private-sector indexes are available to help us measure and understand how economic conditions in one period compare with economic conditions in other periods. The purpose of this chapter is to describe the most widely used types of indexes. We will begin by constructing some simple index numbers to gain a better understanding of how indexes are computed.

20.1 Price Relatives

The simplest form of a price index shows how the current price per unit for a given item compares to a base period price per unit for the same item. For example, Table 20.1 reports the cost of one gallon of regular gasoline for the years 1990 through 2011. To facilitate comparisons with other years, the actual cost-per-gallon figure can be converted to a **price relative**, which expresses the unit price in each period as a percentage of the unit price in a base period.

$$\text{Price relative in period } t = \frac{\text{Price in period } t}{\text{Base period price}}(100) \qquad \textbf{(20.1)}$$

TABLE 20.1 REGULAR GASOLINE (ALL FORMULATIONS) COST

Year	Price per Gallon ($)
1990	1.30
1991	1.10
1992	1.09
1993	1.07
1994	1.08
1995	1.11
1996	1.22
1997	1.20
1998	1.03
1999	1.14
2000	1.48
2001	1.42
2002	1.34
2003	1.56
2004	1.85
2005	2.27
2006	2.57
2007	2.80
2008	3.25
2009	2.35
2010	2.78
2011	3.52

Source: U.S. Energy Information Administration.

For the gasoline prices in Table 20.1 and with 1990 as the base year, the price relatives for one gallon of regular gasoline in the years 1990 through 2011 can be calculated. These price relatives are listed in Table 20.2. Note how easily the price in any one year can be compared with the price in the base year by knowing the price relative. For example, the price relative of 85.4 in 1995 shows that the price of gasoline in 1995 was 14.6% below the 1990 base-year price. Similarly, the 2002 price relative of 103.1 shows a 3.1% increase in the gasoline price in 2002 from the 1990 base-year price. And the 2011 price relative of 270.8 shows a 170.8% increase in the price of regular gesoline from the 1990 base-year price. Price relatives, such as the ones for regular gasoline, are extremely helpful in terms of understanding and interpreting changing economic and business conditions over time.

20.2 Aggregate Price Indexes

TABLE 20.2 PRICE RELATIVES FOR ONE GALLON OF REGULAR GASOLINE

Year	Price Relative (Base 1990)
1990	(1.30/1.30)100 = 100.0
1991	(1.10/1.30)100 = 84.6
1992	(1.09/1.30)100 = 83.8
1993	(1.07/1.30)100 = 82.3
1994	(1.08/1.30)100 = 83.1
1995	(1.11/1.30)100 = 85.4
1996	(1.22/1.30)100 = 93.8
1997	(1.20/1.30)100 = 92.3
1998	(1.03/1.30)100 = 79.2
1999	(1.14/1.30)100 = 87.7
2000	(1.48/1.30)100 = 113.8
2001	(1.42/1.30)100 = 109.2
2002	(1.34/1.30)100 = 103.1
2003	(1.56/1.30)100 = 120.0
2004	(1.85/1.30)100 = 142.3
2005	(2.27/1.30)100 = 174.6
2006	(2.57/1.30)100 = 197.7
2007	(2.80/1.30)100 = 215.4
2008	(3.25/1.30)100 = 250.0
2009	(2.35/1.30)100 = 180.8
2010	(2.78/1.30)100 = 213.8
2011	(3.52/1.30)100 = 270.8

Although price relatives can be used to identify price changes over time for individual items, we are often more interested in the general price change for a group of items taken as a whole. For example, if we want an index that measures the change in the overall cost of living over time, we will want the index to be based on the price changes for a variety of items, including food, housing, clothing, transportation, medical care, and so on. An **aggregate price index** is developed for the specific purpose of measuring the combined change of a group of items.

Consider the development of an aggregate price index for a group of items categorized as normal automotive operating expenses. For illustration, we limit the items included in the group to gasoline, oil, tire, and insurance expenses.

Table 20.3 gives the data for the four components of our automotive operating expense index for the years 1990 and 2011. With 1990 as the base period, an aggregate price index for the four components will give us a measure of the change in normal automotive operating expenses over the 1990–2011 period.

An unweighted aggregate index can be developed by simply summing the unit prices in the year of interest (e.g., 2011) and dividing that sum by the sum of the unit prices in the base year (1990). Let

$$P_{it} = \text{unit price for item } i \text{ in period } t$$
$$P_{i0} = \text{unit price for item } i \text{ in the base period}$$

TABLE 20.3 DATA FOR AUTOMOTIVE OPERATING EXPENSE INDEX

Item	Unit Price ($) 1990	2011
Gallon of gasoline	1.30	3.52
Quart of oil	2.10	6.25
Tire	130.00	145.00
Insurance policy	820.00	1040.00

An unweighted aggregate price index in period t, denoted by I_t, is given by

$$I_t = \frac{\Sigma P_{it}}{\Sigma P_{i0}}(100) \tag{20.2}$$

where the sums are for all items in the group.

An unweighted aggregate index for normal automotive operating expenses in 2011 ($t = 2011$) is given by

$$I_{2011} = \frac{3.52 + 6.25 + 145.00 + 1040.00}{1.30 + 2.10 + 130.00 + 820.00}(100)$$

$$= \frac{1194.77}{953.4}(100) = 125.3$$

From the unweighted aggregate price index, we might conclude that the price of normal automotive operating expenses has only increased 25.3% over the period from 1990 to 2011. But note that the unweighted aggregate approach to establishing a composite price index for automotive expenses is heavily influenced by the items with large per-unit prices. Consequently, items with relatively low unit prices such as gasoline and oil are dominated by the high unit-price items such as tires and insurance. The unweighted aggregate index for automotive operating expenses is too heavily influenced by price changes in tires and insurance.

If quantity of usage is the same for each item, an unweighted index gives the same value as a weighted index. In practice, however, quantities of usage are rarely the same.

Because of the sensitivity of an unweighted index to one or more high-priced items, this form of aggregate index is not widely used. A weighted aggregate price index provides a better comparison when usage quantities differ.

The philosophy behind the **weighted aggregate price index** is that each item in the group should be weighted according to its importance. In most cases, the quantity of usage is the best measure of importance. Hence, one must obtain a measure of the quantity of usage for the various items in the group. Table 20.4 gives annual usage information for each item of automotive operating expense based on the typical operation of a midsize automobile for approximately 15,000 miles per year. The quantity weights listed show the expected annual usage for this type of driving situation.

TABLE 20.4 ANNUAL USAGE INFORMATION FOR AUTOMOTIVE OPERATING EXPENSE INDEX

Item	Quantity Weights*
Gallons of gasoline	1000
Quarts of oil	15
Tires	2
Insurance policy	1

*Based on 15,000 miles per year. Tire usage is based on a 30,000-mile tire life.

Let Q_i = quantity of usage for item i. The weighted aggregate price index in period t is given by

$$I_t = \frac{\Sigma P_{it}Q_i}{\Sigma P_{i0}Q_i}(100) \tag{20.3}$$

where the sums are for all items in the group. Applied to our automotive operating expenses, the weighted aggregate price index is based on dividing total operating costs in 2011 by total operating costs in 1990.

Let $t = 2011$, and use the quantity weights in Table 20.4. We obtain the following weighted aggregate price index for automotive operating expenses in 2011.

$$I_{2011} = \frac{3.52(1000) + 6.25(15) + 145.00(2) + 1040.00(1)}{1.30(1000) + 2.10(15) + 130.00(2) + 820.00(1)}(100)$$

$$= \frac{4943.75}{2411.5} = 205.0$$

From this weighted aggregate price index, we would conclude that the price of automotive operating expenses has increased 105% over the period from 1990 through 2011.

Clearly, compared with the unweighted aggregate index, the weighted index provides a more accurate indication of the price change for automotive operating expenses over the 1990–2011 period. Taking the quantity of usage of gasoline into account helps to offset the smaller percentage increase in insurance costs. The weighted index shows a larger increase in automotive operating expenses than the unweighted index. In general, the weighted aggregate index with quantities of usage as weights is the preferred method for establishing a price index for a group of items.

In the weighted aggregate price index formula (20.3), note that the quantity term Q_i does not have a second subscript to indicate the time period. The reason is that the quantities Q_i are considered fixed and do not vary with time as the prices do. The fixed weights or quantities are specified by the designer of the index at levels believed to be representative of typical usage. Once established, they are held constant or fixed for all periods of time the index is in use. Indexes for years other than 2011 require the gathering of new price data P_{it}, but the weighting quantities Q_i remain the same.

In a special case of the fixed-weight aggregate index, the quantities are determined from base-year usages. In this case we write $Q_i = Q_{i0}$, with the zero subscript indicating base-year quantity weights; formula (20.3) becomes

$$I_t = \frac{\Sigma P_{it} Q_{i0}}{\Sigma P_{i0} Q_{i0}}(100) \tag{20.4}$$

Whenever the fixed quantity weights are determined from base-year usage, the weighted aggregate index is given the name **Laspeyres index**.

Another option for determining quantity weights is to revise the quantities each period. A quantity Q_{it} is determined for each year that the index is computed. The weighted aggregate index in period t with these quantity weights is given by

$$I_t = \frac{\Sigma P_{it} Q_{it}}{\Sigma P_{i0} Q_{it}}(100) \tag{20.5}$$

Note that the same quantity weights are used for the base period (period 0) and for period t. However, the weights are based on usage in period t, not the base period. This weighted aggregate index is known as the **Paasche index**. It has the advantage of being based on current usage patterns. However, this method of computing a weighted aggregate index presents two disadvantages: The normal usage quantities Q_{it} must be redetermined each year, thus adding to the time and cost of data collection, and each year the index numbers for previous years must be recomputed to reflect the effect of the new quantity weights. Because of these disadvantages, the Laspeyres index is more widely used. The automotive operating expense index was computed with base-period quantities; hence, it is a Laspeyres index. Had usage figures for 2011 been used, it would be a Paasche index. Indeed, because of more fuel efficient cars, gasoline usage decreased and a Paasche index differs from a Laspeyres index.

Exercises

Methods

SELF test

1. The following table reports prices and usage quantities for two items in 2009 and 2011.

	Quantity		Unit Price ($)	
Item	**2009**	**2011**	**2009**	**2011**
A	1500	1800	7.50	7.75
B	2	1	630.00	1500.00

a. Compute price relatives for each item in 2011 using 2009 as the base period.
b. Compute an unweighted aggregate price index for the two items in 2011 using 2009 as the base period.
c. Compute a weighted aggregate price index for the two items using the Laspeyres method.
d. Compute a weighted aggregate price index for the two items using the Paasche method.

2. An item with a price relative of 132 cost $10.75 in 2011. Its base year was 1994.
a. What was the percentage increase or decrease in cost of the item over the 17-year period?
b. What did the item cost in 1994?

Applications

SELF test

3. A large manufacturer purchases an identical component from three independent suppliers that differ in unit price and quantity supplied. The relevant data for 2009 and 2011 are given here.

		Unit Price ($)	
Supplier	**Quantity (2007)**	**2009**	**2011**
A	150	5.45	6.00
B	200	5.60	5.95
C	120	5.50	6.20

a. Compute the price relatives for each of the component suppliers separately. Compare the price increases by the suppliers over the two-year period.
b. Compute an unweighted aggregate price index for the component part in 2011.
c. Compute a 2011 weighted aggregate price index for the component part. What is the interpretation of this index for the manufacturing firm?

4. R&B Beverages, Inc., provides a complete line of beer, wine, and soft drink products for distribution through retail outlets in central Iowa. Unit price data for 2008 and 2011 and quantities sold in cases for 2008 follow.

		Unit Price ($)	
Item	**2008 Quantity (cases)**	**2008**	**2011**
Beer	35,000	17.50	20.15
Wine	5,000	100.00	118.00
Soft drink	60,000	8.00	8.80

Compute a weighted aggregate index for the R&B Beverage sales in 2011, with 2008 as the base period.

5. Under the last-in, first-out (LIFO) inventory valuation method, a price index for inventory must be established for tax purposes. The quantity weights are based on year-ending inventory levels. Use the beginning-of-the-year price per unit as the base-period price and develop a weighted aggregate index for the total inventory value at the end of the year. What type of weighted aggregate price index must be developed for the LIFO inventory valuation?

		Unit Price ($)	
Product	**Ending Inventory**	**Beginning**	**Ending**
A	500	.15	.19
B	50	1.60	1.80
C	100	4.50	4.20
D	40	12.00	13.20

20.3 Computing an Aggregate Price Index from Price Relatives

In Section 20.1 we defined the concept of a price relative and showed how a price relative can be computed with knowledge of the current-period unit price and the base-period unit price. We now want to show how aggregate price indexes like the ones developed in Section 20.2 can be computed directly from information about the price relative of each item in the group. Because of the limited use of unweighted indexes, we restrict our attention to weighted aggregate price indexes. Let us return to the automotive operating expense index of the preceding section. The necessary information for the four items is given in Table 20.5.

One must be sure prices and quantities are in the same units. For example, if prices are per case, quantity must be the number of cases and not, for instance, the number of individual units.

Let w_i be the weight applied to the price relative for item i. The general expression for a weighted average of price relatives is given by

$$I_t = \frac{\sum \frac{P_{it}}{P_{i0}}(100)w_i}{\Sigma w_i} \tag{20.6}$$

The proper choice of weights in equation (20.6) will enable us to compute a weighted aggregate price index from the price relatives. The proper choice of weights is given by multiplying the base-period price by the quantity of usage.

$$w_i = P_{i0}Q_i \tag{20.7}$$

Substituting $w_i = P_{i0}Q_i$ into equation (20.6) provides the following expression for a weighted price relatives index.

$$I_t = \frac{\sum \frac{P_{it}}{P_{i0}}(100)(P_{i0}Q_i)}{\Sigma P_{i0}Q_i} \tag{20.8}$$

With the canceling of the P_{i0} terms in the numerator, an equivalent expression for the weighted price relatives index is

$$I_t = \frac{\Sigma P_{it}Q_i}{\Sigma P_{i0}Q_i}(100)$$

Thus, we see that the weighted price relatives index with $w_i = P_{i0}Q_i$ provides a price index identical to the weighted aggregate index presented in Section 20.2 by equation (20.3). Use

TABLE 20.5 PRICE RELATIVES FOR AUTOMOTIVE OPERATING EXPENSE INDEX

	Unit Price ($)			
Item	1990 (P_0)	2011 (P_t)	Price Relative $(P_t/P_0)100$	Annual Usage
Gallon of gasoline	1.30	3.52	270.8	1000
Quart of oil	2.10	6.25	297.6	15
Tire	130.00	145.00	111.5	2
Insurance policy	820.00	1040.00	126.8	1

TABLE 20.6 AUTOMOTIVE OPERATING EXPENSE INDEX (1990–2011) BASED ON WEIGHTED PRICE RELATIVES

Item	Price Relatives $(P_{it}/P_{i0})(100)$	Base Price ($) P_{i0}	Quantity Q_i	Weight $w_i = P_{i0}Q_i$	Weighted Price Relatives $(P_{it}/P_{i0})(100)w_i$
Gasoline	270.8	1.30	1000	1300.00	352,040.00
Oil	297.6	2.10	15	31.50	9,374.40
Tire	111.5	130.00	2	260.00	28,990.00
Insurance	126.8	820.00	1	820.00	103,976.00
			Totals	2411.50	494,380.40

$$I_{2011} = \frac{494,380.40}{2411.50} = 205$$

of base-period quantities (i.e., $Q_i = Q_{i0}$) in equation (20.7) leads to a Laspeyres index. Use of current-period quantities (i.e., $Q_i = Q_{it}$) in equation (20.7) leads to a Paasche index.

Let us return to the automotive operating expense data. We can use the price relatives in Table 20.5 and equation (20.6) to compute a weighted average of price relatives. The results obtained by using the weights specified by equation (20.7) are reported in Table 20.6. The index number 205 represents a 105% increase in automotive operating expenses, which is the same as the increase identified by the weighted aggregate index computation in Section 20.2.

Exercises

Methods

SELF test

6. Price relatives for three items, along with base-period prices and usage are shown in the following table. Compute a weighted aggregate price index for the current period.

		Base Period	
Item	Price Relative	Price	Usage
A	150	22.00	20
B	90	5.00	50
C	120	14.00	40

Applications

SELF test

7. The Mitchell Chemical Company produces a special industrial chemical that is a blend of three chemical ingredients. The beginning-year cost per pound, the ending-year cost per pound, and the blend proportions follow.

	Cost per Pound ($)		
Ingredient	Beginning	Ending	Quantity (pounds) per 100 Pounds of Product
A	2.50	3.95	25
B	8.75	9.90	15
C	.99	.95	60

a. Compute the price relatives for the three ingredients.
b. Compute a weighted average of the price relatives to develop a one-year cost index for raw materials used in the product. What is your interpretation of this index value?

8. An investment portfolio consists of four stocks. The purchase price, current price, and number of shares are reported in the following table.

Stock	Purchase Price/Share ($)	Current Price/Share ($)	Number of Shares
Holiday Trans	15.50	17.00	500
NY Electric	18.50	20.25	200
KY Gas	26.75	26.00	500
PQ Soaps	42.25	45.50	300

Construct a weighted average of price relatives as an index of the performance of the portfolio to date. Interpret this price index.

9. Compute the price relatives for the R&B Beverages products in exercise 4. Use a weighted average of price relatives to show that this method provides the same index as the weighted aggregate method.

20.4 Some Important Price Indexes

We identified the procedures used to compute price indexes for single items or groups of items. Now let us consider some price indexes that are important measures of business and economic conditions. Specifically, we consider the Consumer Price Index, the Producer Price Index, and the Dow Jones averages.

Consumer Price Index

The CPI includes charges for services (e.g., doctor and dentist bills) and all taxes directly associated with the purchase and use of an item.

The **Consumer Price Index (CPI)**, published monthly by the U.S. Bureau of Labor Statistics, is the primary measure of the cost of living in the United States. The group of items used to develop the index consists of a *market basket* of 400 items including food, housing, clothing, transportation, and medical items. The CPI is a weighted aggregate price index with fixed weights.[1] The weight applied to each item in the market basket derives from a usage survey of urban families throughout the United States.

The February 2012 CPI, computed with a 1982–1984 base index of 100, was 227.7. This figure means that the cost of purchasing the market basket of goods and services increased 127.7% since the base period 1982–1984. The 50-year time series of the CPI from 1960–2010 is shown in Figure 20.1. Note how the CPI measure reflects the sharp inflationary behavior of the economy in the late 1970s and early 1980s.

Producer Price Index

The PPI is designed as a measure of price changes for domestic goods; imports are not included.

The **Producer Price Index (PPI)**, also published monthly by the U.S. Bureau of Labor Statistics, measures the monthly changes in prices in primary markets in the United States. The PPI is based on prices for the first transaction of each product in nonretail markets. All

[1]The Bureau of Labor Statistics actually publishes two Consumer Price Indexes: one for all urban consumers (CPI-U) and a revised Consumer Price Index for urban wage earners and clerical workers (CPI-W). The CPI-U is the one most widely quoted, and it is published regularly in *The Wall Street Journal*.

FIGURE 20.1 CONSUMER PRICE INDEX, 1960–2010 (BASE 1982–1984 = 100)

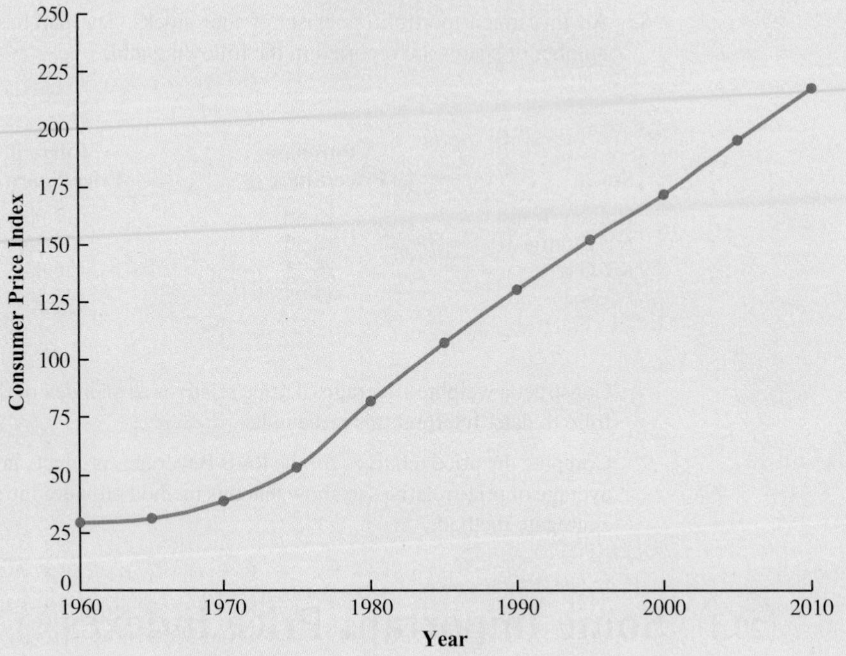

commodities sold in commercial transactions in these markets are represented. The survey covers raw, manufactured, and processed goods at each level of processing and includes the output of industries classified as manufacturing, agriculture, forestry, fishing, mining, gas and electricity, and public utilities. One of the common uses of this index is as a leading indicator of the future trend of consumer prices and the cost of living. An increase in the PPI reflects producer price increases that will eventually be passed on to the consumer through higher retail prices.

Weights for the various items in the PPI are based on the value of shipments. The weighted average of price relatives is calculated by the Laspeyres method. The February 2012 PPI, computed with a 1982 base index of 100, was 201.6.

Dow Jones Averages

Charles Henry Dow published his first stock average on July 3, 1884, in the Customer's Afternoon Letter. *Eleven stocks, nine of which were railroad issues, were included in the first index. An average comparable to the DJIA was first published on October 1, 1928.*

The **Dow Jones averages** are indexes designed to show price trends and movements associated with common stocks. The best known of the Dow Jones indexes is the Dow Jones Industrial Average (DJIA), which is based on common stock prices of 30 large companies. It is the sum of these stock prices divided by a number, which is revised from time to time to adjust for stock splits and switching of companies in the index. Unlike the other price indexes that we studied, it is not expressed as a percentage of base-year prices. The specific firms used in February 2012 to compute the DJIA are listed in Table 20.7.

Other Dow Jones averages are computed for 20 transportation stocks and for 15 utility stocks. The Dow Jones averages are computed and published daily in *The Wall Street Journal* and other financial publications.

TABLE 20.7 THE 30 COMPANIES USED IN THE DOW JONES INDUSTRIAL AVERAGE (FEBRUARY 2012)

3m	Disney	Kraft Foods
Alcoa	DuPont	McDonald's
American Express	ExxonMobil	Merck
AT&T	General Electric	Microsoft
Bank of America	Hewlett-Packard	Pfizer
Boeing	Home Depot	Procter & Gamble
Caterpillar	IBM	Travelers
Chevron Corp.	Intel	United Technologies
Cisco Systems	Johnson & Johnson	Verizon
Coca-Cola	J. P. Morgan Chase	Wal-Mart Stores

20.5 Deflating a Series by Price Indexes

Time series are deflated to remove the effects of inflation.

Many business and economic series reported over time, such as company sales, industry sales, and inventories, are measured in dollar amounts. These time series often show an increasing growth pattern over time, which is generally interpreted as indicating an increase in the physical volume associated with the activities. For example, a total dollar amount of inventory up by 10% might be interpreted to mean that the physical inventory is 10% larger. Such interpretations can be misleading if a time series is measured in terms of dollars, and the total dollar amount is a combination of both price and quantity changes. Hence, in periods when price changes are significant, the changes in the dollar amounts may not be indicative of quantity changes unless we are able to adjust the time series to eliminate the price change effect.

For example, from 1976 to 1980, the total amount of spending in the construction industry increased approximately 75%. That figure suggests excellent growth in construction activity. However, construction prices were increasing just as fast as—or sometimes even faster than—the 75% rate. In fact, while total construction spending was increasing, construction activity was staying relatively constant or, as in the case of new housing starts, decreasing. To interpret construction activity correctly for the 1976–1980 period, we must adjust the total spending series by a price index to remove the price increase effect. Whenever we remove the price increase effect from a time series, we say we are *deflating the series.*

In relation to personal income and wages, we often hear discussions about issues such as "real wages" or the "purchasing power" of wages. These concepts are based on the notion of deflating an hourly wage index. For example, Figure 20.2 shows the pattern of hourly wages of electricians for the period 2007–2011. We see a trend of wage increases from $23.12 per hour to $25.44 per hour. Should electricians be pleased with this growth in hourly wages? The answer depends on what happened to the purchasing power of their wages. If we can compare the purchasing power of the $23.12 hourly wage in 2007 with the purchasing power of the $25.44 hourly wage in 2011, we will be better able to judge the relative improvement in wages.

Table 20.8 reports both the hourly wage rate and the CPI (computed with a 1982–1984 base index of 100) for the period 2007–2011. With these data, we will show how the CPI can be used to deflate the index of hourly wages. The deflated series is found by dividing

FIGURE 20.2 ACTUAL AVERAGE HOURLY WAGES OF ELECTRICIANS

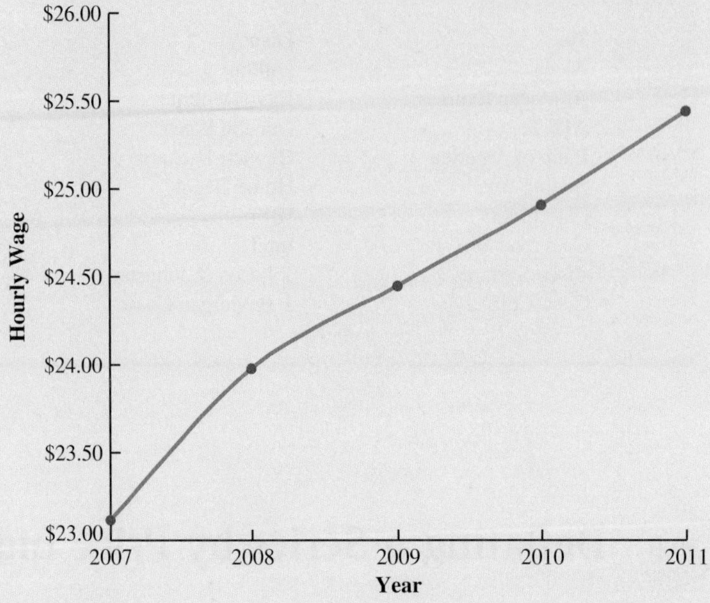

the hourly wage rate in each year by the corresponding value of the CPI and multiplying by 100. The deflated hourly wage index for electricians is given in Table 20.9; Figure 20.3 is a graph showing the deflated, or real, wages.

Real wages are a better measure of purchasing power than actual wages. Indeed, many union contracts call for wages to be adjusted in accordance with changes in the cost of living.

What does the deflated series of wages tell us about the real wages or purchasing power of electricians during the 2007–2011 period? In terms of base period dollars (1982–1984 = 100), the hourly wage rate remained relatively flat over the period. After removing the inflationary effect we see that the purchasing power of the workers only increased by $.16 over the four-year period. This effect is seen in Figure 20.3. Thus, the advantage of using price indexes to deflate a series is that they give us a clearer picture of the real dollar changes that are occurring.

This process of deflating a series measured over time has an important application in the computation of the gross domestic product (GDP). The GDP is the total value of all

TABLE 20.8 HOURLY WAGES OF ELECTRICIANS AND CONSUMER PRICE INDEX, 2007–2011

Year	Hourly Wage ($)	CPI
2007	23.12	207.3
2008	23.98	215.3
2009	24.45	214.5
2010	24.91	218.1
2011	25.44	224.9

Source: Bureau of Labor Statistics. CPI is computed with a 1982–1984 base index of 100.

TABLE 20.9 DEFLATED SERIES OF HOURLY WAGES FOR ELECTRICIANS, 2007–2011

Year	Deflated Hourly Wage
2007	(\$23.12/207.3)(100) = \$11.15
2008	(\$23.98/215.3)(100) = \$11.14
2009	(\$24.45/214.5)(100) = \$11.40
2010	(\$24.91/218.1)(100) = \$11.42
2011	(\$25.44/224.9)(100) = \$11.31

FIGURE 20.3 REAL HOURLY WAGES OF ELECTRICIANS, 2007–2011

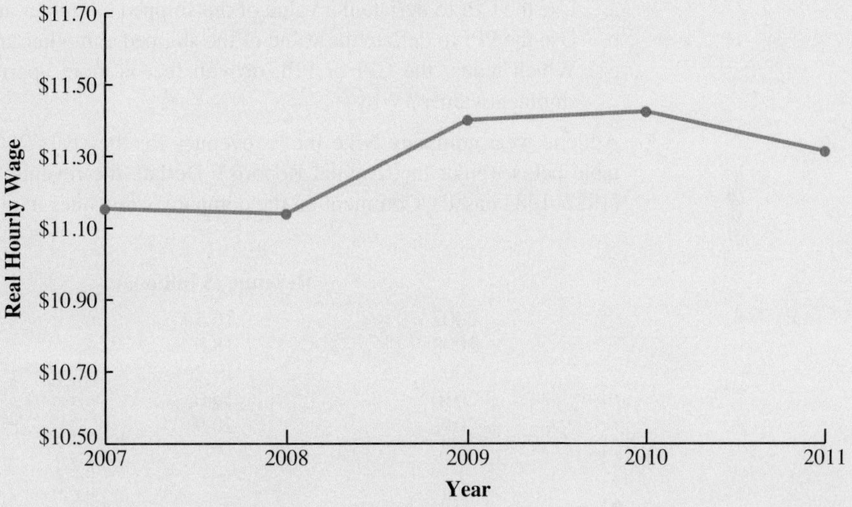

goods and services produced in a given country. Obviously, over time the GDP will show gains that are in part due to price increases if the GDP is not deflated by a price index. Therefore, to adjust the total value of goods and services to reflect actual changes in the volume of goods and services produced and sold, the GDP must be computed with a price index deflator. The process is similar to that discussed in the real wages computation.

Exercises

Applications

SELF test

10. Registered nurses in 2007 made an average hourly wage of \$30.04. In 2011, their hourly wage had risen to \$33.23. Given that the CPI for 2007 was 207.3 and the 2011 CPI was 224.9, answer the following.
    a. Give the real wages for registered nurses for 2007 and 2011 by deflating the hourly wage rates.
    b. What is the percentage change in the actual hourly wage for registered nurses from 2007 to 2011?
    c. For registered nurses, what was the percentage change in real wages from 2007 to 2011?

11. The average hourly wage rate for construction laborers in 2001 was $13.36. In 2011 construction laborers made $16.43 per hour. The CPI for 2001 was 177.1 and for 2011, 224.9. Calculate the percentage increase or decrease in real hourly wages from 2001 to 2011.
12. Shipments of product from manufacturer to the retailer are tracked by the U.S. Census Bureau. The value of shipments for computer and electronic products in September 2009, 2010, and 2011 are shown in the table below, along with the CPI and PPI for each of these months.

	Manufacturer Shipments ($ billions)	CPI	PPI
2009	$29.1	216.0	173.4
2010	$33.3	218.4	180.2
2011	$32.9	226.9	192.5

a. Use the CPI to deflate the value of the shipped computer and electronics products.
b. Use the PPI to deflate the value of the shipped computer and electronics products.
c. Which index, the CPI or PPI, do you feel is more appropriate for deflating these shipment values? Why?

13. Athletic wear company Nike Inc.'s revenues for the years 2007–2011 are shown in the table below (Nike Inc. Annual Reports). Deflate the revenue dollars based on the CPI (1982–1984 based). Comment on the company's revenues in terms of deflated dollars.

	Revenue ($ billions)	CPI
2007	16.3	207.3
2008	18.6	215.3
2009	19.2	214.5
2010	19.0	218.1
2011	20.9	224.9

20.6 Price Indexes: Other Considerations

In the preceding sections we described several methods used to compute price indexes, discussed the use of some important indexes, and presented a procedure for using price indexes to deflate a time series. Several other issues must be considered to enhance our understanding of how price indexes are constructed and how they are used. Some are discussed in this section.

Selection of Items

The primary purpose of a price index is to measure the price change over time for a specified class of items, products, and so on. Whenever the class of items is very large, the index cannot be based on all items in the class. Rather, a sample of representative items must be used. By collecting price and quantity information for the sampled items, we hope to obtain a good idea of the price behavior of all items that the index is representing. For example, in the Consumer Price Index the total number of items that might be considered in the population of normal purchase items for a consumer could be 2000 or more. However, the index is based on the price-quantity characteristics of just 400 items. The selection of the specific items in the index is not a trivial task. Surveys of user purchase patterns as well as good judgment go into the selection process. A simple random sample is not used to select the 400 items.

After the initial selection process, the group of items in the index must be periodically reviewed and revised whenever purchase patterns change. Thus, the issue of which items to include in an index must be resolved before an index can be developed and again before it is revised.

Selection of a Base Period

Most indexes are established with a base-period value of 100 at some specific time. All future values of the index are then related to the base-period value. What base period is appropriate for an index is not an easy question to answer. It must be based on the judgment of the developer of the index.

Many of the indexes established by the U.S. government as of 2011 use a 1982 base period. As a general guideline, the base period should not be too far from the current period. For example, a Consumer Price Index with a 1945 base period would be difficult for most individuals to understand because of unfamiliarity with conditions in 1945. The base period for most indexes therefore is adjusted periodically to a more recent period of time. The CPI base period was changed from 1967 to the 1982–1984 average in 1988. The PPI currently uses 1982 as its base period (i.e., 1982 = 100).

Quality Changes

The purpose of a price index is to measure changes in prices over time. Ideally, price data are collected for the same set of items at several times, and then the index is computed. A basic assumption is that the prices are identified for the same items each period. A problem is encountered when a product changes in quality from one period to the next. For example, a manufacturer may alter the quality of a product by using less expensive materials, fewer features, and so on, from year to year. The price may go up in following years, but the price is for a lower-quality product. Consequently, the price may actually go up more than is represented by the list price for the item. It is difficult, if not impossible, to adjust an index for decreases in the quality of an item.

A substantial quality improvement may also cause an increase in the price of a product. A portion of the price related to the quality improvement should be excluded from the index computation. However, adjusting an index for a price increase that is related to higher quality of an item is extremely difficult, if not impossible.

Although common practice is to ignore minor quality changes in developing a price index, major quality changes must be addressed because they can alter the product description from period to period. If a product description is changed, the index must be modified to account for it; in some cases, the product might be deleted from the index.

In some situations, however, a substantial improvement in quality is followed by a decrease in the price. This less typical situation has been the case with personal computers during the 1990s and early 2000s.

20.7 Quantity Indexes

In addition to the price indexes described in the preceding sections, other types of indexes are useful. In particular, one other application of index numbers is to measure changes in quantity levels over time. This type of index is called a **quantity index**.

Recall that in the development of the weighted aggregate price index in Section 20.2, to compute an index number for period t we needed data on unit prices at a base period (P_0) and period t (P_t). Equation (20.3) provided the weighted aggregate price index as

$$I_t = \frac{\Sigma P_{it} Q_i}{\Sigma P_{i0} Q_i}(100)$$

The numerator, $\Sigma P_{it}Q_i$, represents the total value of fixed quantities of the index items in period t. The denominator, $\Sigma P_{i0}Q_i$, represents the total value of the same fixed quantities of the index items in year 0.

Computation of a weighted aggregate quantity index is similar to that of a weighted aggregate price index. Quantities for each item are measured in the base period and period t, with Q_{i0} and Q_{it}, respectively, representing those quantities for item i. The quantities are then weighted by a fixed price, the value added, or some other factor. The "value added" to a product is the sales value minus the cost of purchased inputs. The formula for computing a weighted aggregate quantity index for period t is

$$I_t = \frac{\Sigma Q_{it}w_i}{\Sigma Q_{i0}w_i}(100) \tag{20.9}$$

In some quantity indexes the weight for item i is taken to be the base-period price (P_{i0}), in which case the weighted aggregate quantity index is

$$I_t = \frac{\Sigma Q_{it}P_{i0}}{\Sigma Q_{i0}P_{i0}}(100) \tag{20.10}$$

Quantity indexes can also be computed on the basis of weighted quantity relatives. One formula for this version of a quantity index follows.

$$I_t = \frac{\sum \frac{Q_{it}}{Q_{i0}}(Q_{i0}P_i)}{\Sigma Q_{i0}P_i}(100) \tag{20.11}$$

This formula is the quantity version of the weighted price relatives formula developed in Section 20.3 as in equation (20.8).

The **Index of Industrial Production**, developed by the Federal Reserve Board, is probably the best-known quantity index. It is reported monthly and the base period is 2002. The index is designed to measure changes in volume of production levels for a variety of manufacturing classifications in addition to mining and utilities. In February 2012 the index was 96.2.

Exercises

Methods

SELF test

14. Data on quantities of three items sold in 1997 and 2011 are given here along with the sales prices of the items in 1997. Compute a weighted aggregate quantity index for 2011.

	Quantity Sold		
Item	**1997**	**2011**	**Price/Unit 1997 ($)**
A	350	300	18.00
B	220	400	4.90
C	730	850	15.00

Applications

SELF test

15. A trucking firm handles four commodities for a particular distributor. Total shipments for the commodities in 1996 and 2011, as well as the 1996 prices, are reported in the following table.

Commodity	Shipments 1996	Shipments 2011	Price/Shipment 1996
A	120	95	$1200
B	86	75	$1800
C	35	50	$2000
D	60	70	$1500

Develop a weighted aggregate quantity index with a 1996 base. Comment on the growth or decline in quantities over the 1996–2011 period.

16. An automobile dealer reports the 1994 and 2011 sales for three models in the following table. Compute quantity relatives and use them to develop a weighted aggregate quantity index for 2011 using the two years of data.

Model	Sales 1994	Sales 2011	Mean Price per Sale (1994)
Sedan	200	170	$15,200
Sport	100	80	$17,000
Wagon	75	60	$16,800

Summary

Price and quantity indexes are important measures of changes in price and quantity levels within the business and economic environment. Price relatives are simply the ratio of the current unit price of an item to a base-period unit price multiplied by 100, with a value of 100 indicating no difference in the current and base-period prices. Aggregate price indexes are created as a composite measure of the overall change in prices for a given group of items or products. Usually the items in an aggregate price index are weighted by their quantity of usage. A weighted aggregate price index can also be computed by weighting the price relatives by the usage quantities for the items in the index.

The Consumer Price Index and the Producer Price Index are both widely quoted indexes with 1982–1984 and 1982, respectively, as base years. The Dow Jones Industrial Average is another widely quoted price index. It is a weighted sum of the prices of 30 common stocks of large companies. Unlike many other indexes, it is not stated as a percentage of some base-period value.

Often price indexes are used to deflate some other economic series reported over time. We saw how the CPI could be used to deflate hourly wages to obtain an index of real wages. Selection of the items to be included in the index, selection of a base period for the index, and adjustment for changes in quality are important additional considerations in the development of an index number. Quantity indexes were briefly discussed, and the Index of Industrial Production was mentioned as an important quantity index.

Glossary

Price relative A price index for a given item that is computed by dividing a current unit price by a base-period unit price and multiplying the result by 100.

Aggregate price index A composite price index based on the prices of a group of items.

Weighted aggregate price index A composite price index in which the prices of the items in the composite are weighted by their relative importance.

Laspeyres index A weighted aggregate price index in which the weight for each item is its base-period quantity.
Paasche index A weighted aggregate price index in which the weight for each item is its current-period quantity.
Consumer Price Index (CPI) A monthly price index that uses the price changes in a market basket of consumer goods and services to measure the changes in consumer prices over time.
Producer Price Index (PPI) A monthly price index designed to measure changes in prices of goods sold in primary markets (i.e., first purchase of a commodity in nonretail markets).
Dow Jones averages Aggregate price indexes designed to show price trends and movements associated with common stocks.
Quantity index An index designed to measure changes in quantities over time.
Index of Industrial Production A quantity index designed to measure changes in the physical volume or production levels of industrial goods over time.

Key Formulas

Price Relative in Period *t*

$$\frac{\text{Price in period } t}{\text{Base period price}}(100) \tag{20.1}$$

Unweighted Aggregate Price Index in Period *t*

$$I_t = \frac{\Sigma P_{it}}{\Sigma P_{i0}}(100) \tag{20.2}$$

Weighted Aggregate Price Index in Period *t*

$$I_t = \frac{\Sigma P_{it}Q_i}{\Sigma P_{i0}Q_i}(100) \tag{20.3}$$

Weighted Average of Price Relatives

$$I_t = \frac{\Sigma \frac{P_{it}}{P_{i0}}(100)w_i}{\Sigma w_i} \tag{20.6}$$

Weighting Factor for Equation (20.6)

$$w_i = P_{i0}Q_i \tag{20.7}$$

Weighted Aggregate Quantity Index

$$I_t = \frac{\Sigma Q_{it}w_i}{\Sigma Q_{i0}w_i}(100) \tag{20.9}$$

Supplementary Exercises

17. Many factors influence the retail price of computers. The table below shows the average retail price for a computer, including both desktops and laptops, during the month of November for each of the indicated years (*Wall Street Journal,* December 13, 2010).

Year	Average Price ($)
2007	795
2008	705
2009	580
2010	615

a. Use 2007 as the base year and develop a price index for the retail price of a computer over this four-year period.
b. Use 2008 as the base year and develop a price index for the retail price of a computer over this four-year period.

18. Nickerson Manufacturing Company has the following data on quantities shipped and unit costs for each of its four products:

Products	Base-Period Quantities (2005)	Mean Shipping Cost per Unit ($) 2005	2011
A	2000	10.50	15.90
B	5000	16.25	32.00
C	6500	12.20	17.40
D	2500	20.00	35.50

a. Compute the price relative for each product.
b. Compute a weighted aggregate price index that reflects the shipping cost change over the four-year period.

19. Use the price data in exercise 18 to compute a Paasche index for the shipping cost if 2011 quantities are 4000, 3000, 7500, and 3000 for each of the four products.

20. Boran Stockbrokers, Inc., selects four stocks for the purpose of developing its own index of stock market behavior. Prices per share for a 2009 base period, January 2011, and March 2011 follow. Base-year quantities are set on the basis of historical volumes for the four stocks.

			Price per Share ($)		
Stock	Industry	2009 Quantity	2009 Base	January 2011	March 2011
A	Oil	100	31.50	22.75	22.50
B	Computer	150	65.00	49.00	47.50
C	Steel	75	40.00	32.00	29.50
D	Real Estate	50	18.00	6.50	3.75

Use the 2009 base period to compute the Boran index for January 2011 and March 2011. Comment on what the index tells you about what is happening in the stock market.

21. Compute the price relatives for the four stocks making up the Boran index in exercise 20. Use the weighted aggregates of price relatives to compute the January 2011 and March 2011 Boran indexes.

22. Suppose on average a male shaver in 2001 bought one razor handle and used 17 razor blades in a year and that the price relatives for 2001 to 2011 are as appears in the following table on next page. Develop a Male Shaver Expense Index based on weighted price relatives for 2011.

Item	2001 per Capita Use	Base Price ($)	2001–2011 Price Relatives
Razor handle	1	7.46	126.9
Blades	17	1.90	153.7

23. Seafood price and quantity data are reported by the U.S. Census Bureau. Data for the years 2000 and 2009 follow.

	2000 Qty. (000 lbs.)	2000 Price ($/pound)	2009 Price ($/pound)
Halibut	75,190	1.91	2.33
Lobster	83,180	3.62	3.09
Tuna	50,779	1.87	1.97

a. Compute a price relative for each type of seafood.
b. Compute a weighted aggregate price index for the U.S. domestic seafood catch. Comment on the change in seafood prices over the nine-year period.

24. Actuaries are analysts who specialize in the mathematics of risk. Actuaries often work for insurance companies and are responsible for setting premiums for insurance policies. Below are the median salaries for actuaries for the years 2008–2011. The CPI for each of the years is also given (U.S. Census Bureau). Use the CPI to deflate the salary data to constant dollars. Comment on the salary when viewed in constant dollars.

Year	Annual Median Salary ($1000)	CPI
2008	84,810	215.3
2009	87,210	214.5
2010	87,650	218.1
2011	91,060	224.9

25. The closing price of Apple stock, adjusted for splits and dividends, for the first trading day of December 2007–2011 are given below. The CPI for December of each year is also given (Yahoo Finance and the U.S. Census Bureau). Deflate the stock price series and comment on the financial performance of Apple stock.

Year	Price ($)	CPI
2007	198.08	210.0
2008	85.35	210.2
2009	210.73	215.9
2010	322.56	219.2
2011	405.00	225.7

26. A major manufacturing company reports the quantity and product value information for 2007 and 2011 in the table that follows. Compute a weighted aggregate quantity index for the data. Comment on what this quantity index means.

	Quantities		
Product	2007	2011	Values ($)
A	800	1200	30.00
B	600	500	20.00
C	200	500	25.00

APPENDIXES

Appendix A: References and Bibliography

General

Freedman, D., R. Pisani, and R. Purves. *Statistics*, 4th ed. W. W. Norton, 2007.

Hogg, R. V., and E. A. Tanis. *Probability and Statistical Inference,* 8th ed. Prentice Hall, 2009.

McKean, J. W., R. V. Hogg, and A. T. Craig. *Introduction to Mathematical Statistics,* 7th ed. Prentice Hall, 2012.

Miller, I., and M. Miller. *John E. Freund's Mathematical Statistics,* 7th ed. Pearson, 2003.

Moore, D. S., G. P. McCabe, and B. Craig. *Introduction to the Practice of Statistics,* 7th ed. Freeman, 2010.

Wackerly, D. D., W. Mendenhall, and R. L. Scheaffer. *Mathematical Statistics with Applications,* 7th ed. Cengage Learning, 2007.

Experimental Design

Cochran, W. G., and G. M. Cox. *Experimental Designs,* 2nd ed. Wiley, 1992.

Hicks, C. R., and K. V. Turner. *Fundamental Concepts in the Design of Experiments,* 5th ed. Oxford University Press, 1999.

Montgomery, D. C. *Design and Analysis of Experiments,* 8th ed. Wiley, 2012.

Winer, B. J., K. M. Michels, and D. R. Brown. *Statistical Principles in Experimental Design,* 3rd ed. McGraw-Hill, 1991.

Wu, C. F. Jeff, and M. Hamada. *Experiments: Planning, Analysis, and Optimization,* 2nd ed. Wiley, 2009.

Time Series and Forecasting

Bowerman, B. L., and R. T. O'Connell. *Forecasting and Time Series: An Applied Approach,* 3rd ed. Brooks/Cole, 2000.

Box, G. E. P., G. M. Jenkins, and G. C. Reinsel. *Time Series Analysis: Forecasting and Control,* 4th ed. Wiley, 2008.

Makridakis, S. G., S. C. Wheelwright, and R. J. Hyndman. *Forecasting Methods and Applications,* 3rd ed. Wiley, 1997.

Wilson, J. H., B. Keating, and John Galt Solutions, Inc. *Business Forecasting with Accompanying Excel-Based ForecastX™,* 5th ed. McGraw-Hill/Irwin, 2007.

Index Numbers

U.S. Department of Commerce. *Survey of Current Business.*

U.S. Department of Labor, Bureau of Labor Statistics. *CPI Detailed Report.*

U.S. Department of Labor. *Producer Price Indexes.*

Nonparametric Methods

Conover, W. J. *Practical Nonparametric Statistics,* 3rd ed. Wiley, 1999.

Gibbons, J. D., and S. Chakraborti. *Nonparametric Statistical Inference,* 5th ed. CRC Press, 2010.

Higgins, J. J. *Introduction to Modern Nonparametric Statistics*. Thomson-Brooks/Cole, 2004.

Hollander, M., and D. A. Wolfe. *Non-Parametric Statistical Methods,* 2nd ed. Wiley, 1999.

Probability

Hogg, R. V., and E. A. Tanis. *Probability and Statistical Inference,* 8th ed. Prentice Hall, 2009.

Ross, S. M. *Introduction to Probability Models,* 10th ed. Academic Press, 2009.

Wackerly, D. D., W. Mendenhall, and R. L. Scheaffer. *Mathematical Statistics with Applications,* 7th ed. Cengage Learning, 2007.

Quality Control

DeFeo, J. A., and J. M. Juran, *Juran's Quality Handbook,* 6th ed. McGraw-Hill, 2010.

Evans, J. R., and W. M. Lindsay. *The Management and Control of Quality,* 6th ed. South-Western, 2006.

Montgomery, D. C. *Introduction to Statistical Quality Control,* 6th ed. Wiley, 2008.

Regression Analysis

Chatterjee, S., and A. S. Hadi. *Regression Analysis by Example,* 4th ed. Wiley, 2006.

Draper, N. R., and H. Smith. *Applied Regression Analysis,* 3rd ed. Wiley, 1998.

Graybill, F. A., and H. K. Iyer. *Regression Analysis: Concepts and Applications.* Wadsworth, 1994.

Hosmer, D. W., and S. Lemeshow. *Applied Logistic Regression,* 2nd ed. Wiley, 2000.

Kleinbaum, D. G., L. L. Kupper, and K. E. Muller. *Applied Regression Analysis and Multivariate Methods,* 4th ed. Cengage Learning, 2007.

Neter, J., W. Wasserman, M. H. Kutner, and C. Nashtsheim. *Applied Linear Statistical Models,* 5th ed. McGraw-Hill, 2004.

Mendenhall, M., T. Sincich., and T. R. Dye. *A Second Course in Statistics: Regression Analysis,* 7th ed. Prentice Hall, 2011.

Decision Analysis

Clemen, R. T., and T. Reilly. *Making Hard Decisions with Decision Tools.* Cengage Learning, 2004.

Goodwin, P., and G. Wright. *Decision Analysis for Management Judgment,* 4th ed. Wiley, 2010.

Pratt, J. W., H. Raiffa, and R. Schlaifer. *Introduction to Statistical Decision Theory.* MIT Press, 1995.

Sampling

Cochran, W. G. *Sampling Techniques,* 3rd ed. Wiley, 1977.

Hansen, M. H., W. N. Hurwitz, W. G. Madow, and M. N. Hanson. *Sample Survey Methods and Theory.* Wiley, 1993.

Kish, L. *Survey Sampling.* Wiley, 2008.

Levy, P. S., and S. Lemeshow. *Sampling of Populations: Methods and Applications,* 4th ed. Wiley, 2009.

Scheaffer, R. L., W. Mendenhall, and L. Ott. *Elementary Survey Sampling,* 7th ed. Duxbury Press, 2011.

Data Visualization

Cleveland, W. S. *Visualizing Data.* Hobart Press, 1993.

Cleveland, W. S. *The Elements of Graphing Data,* 2nd ed. Hobart Press, 1994.

Few, S. *Show Me the Numbers: Designing Tables and Graphs to Enlighten.* Analytics Press, 2004.

Few, S. *Information Dashboard Design: The Effective Visual Communication of Data.* O'Reilly Media, 2006.

Few, S. *Now You See It: Simple Visualization Techniques for Quantitative Analysis.* Analytics Press, 2009.

Fry, B. *Visualizing Data: Exploring and Explaining Data with the Processing Environment.* O'Reilly Media, 2008.

Robbins, N. B. *Creating More Effective Graphs.* Wiley, 2004.

Telea, A. C. *Data Visualization Principles and Practice.* A.K. Peters Ltd., 2008.

Tufte, E. R. *Envisioning Information.* Graphics Press, 1990.

Tufte, E. R. *The Visual Display of Quantitative Information,* 2nd ed. Graphics Press, 1990.

Tufte, E. R. *Visual Explanations: Images and Quantities, Evidence and Narrative.* Graphics Press, 1997.

Tufte, E. R. *Visual and Statistical Thinking: Displays of Evidence for Making Decisions.* Graphics Press, 1997.

Tufte, E. R. *Beautiful Evidence.* Graphics Press, 2006.

Wong, D. M. *The Wall Street Journal Guide to Information Graphics.* W. W. Norton & Company, 2010.

Young, F. W., P. M. Valero-Mora, and M. Friendly. *Visual Statistics: Seeing Data with Dynamic Interactive Graphics.* Wiley, 2006.

Appendix B: Tables

TABLE 1 CUMULATIVE PROBABILITIES FOR THE STANDARD NORMAL DISTRIBUTION

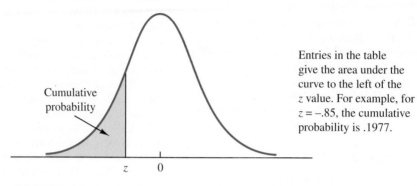

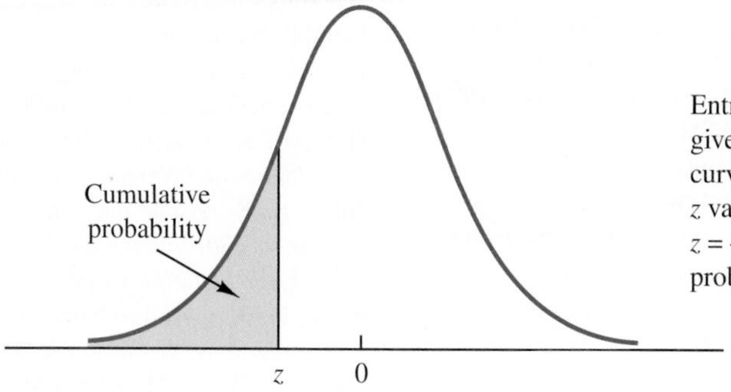

Entries in the table give the area under the curve to the left of the z value. For example, for $z = -.85$, the cumulative probability is .1977.

z	.00	.01	.02	.03	.04	.05	.06	.07	.08	.09
−3.0	.0013	.0013	.0013	.0012	.0012	.0011	.0011	.0011	.0010	.0010
−2.9	.0019	.0018	.0018	.0017	.0016	.0016	.0015	.0015	.0014	.0014
−2.8	.0026	.0025	.0024	.0023	.0023	.0022	.0021	.0021	.0020	.0019
−2.7	.0035	.0034	.0033	.0032	.0031	.0030	.0029	.0028	.0027	.0026
−2.6	.0047	.0045	.0044	.0043	.0041	.0040	.0039	.0038	.0037	.0036
−2.5	.0062	.0060	.0059	.0057	.0055	.0054	.0052	.0051	.0049	.0048
−2.4	.0082	.0080	.0078	.0075	.0073	.0071	.0069	.0068	.0066	.0064
−2.3	.0107	.0104	.0102	.0099	.0096	.0094	.0091	.0089	.0087	.0084
−2.2	.0139	.0136	.0132	.0129	.0125	.0122	.0119	.0116	.0113	.0110
−2.1	.0179	.0174	.0170	.0166	.0162	.0158	.0154	.0150	.0146	.0143
−2.0	.0228	.0222	.0217	.0212	.0207	.0202	.0197	.0192	.0188	.0183
−1.9	.0287	.0281	.0274	.0268	.0262	.0256	.0250	.0244	.0239	.0233
−1.8	.0359	.0351	.0344	.0336	.0329	.0322	.0314	.0307	.0301	.0294
−1.7	.0446	.0436	.0427	.0418	.0409	.0401	.0392	.0384	.0375	.0367
−1.6	.0548	.0537	.0526	.0516	.0505	.0495	.0485	.0475	.0465	.0455
−1.5	.0668	.0655	.0643	.0630	.0618	.0606	.0594	.0582	.0571	.0559
−1.4	.0808	.0793	.0778	.0764	.0749	.0735	.0721	.0708	.0694	.0681
−1.3	.0968	.0951	.0934	.0918	.0901	.0885	.0869	.0853	.0838	.0823
−1.2	.1151	.1131	.1112	.1093	.1075	.1056	.1038	.1020	.1003	.0985
−1.1	.1357	.1335	.1314	.1292	.1271	.1251	.1230	.1210	.1190	.1170
−1.0	.1587	.1562	.1539	.1515	.1492	.1469	.1446	.1423	.1401	.1379
−.9	.1841	.1814	.1788	.1762	.1736	.1711	.1685	.1660	.1635	.1611
−.8	.2119	.2090	.2061	.2033	.2005	.1977	.1949	.1922	.1894	.1867
−.7	.2420	.2389	.2358	.2327	.2296	.2266	.2236	.2206	.2177	.2148
−.6	.2743	.2709	.2676	.2643	.2611	.2578	.2546	.2514	.2483	.2451
−.5	.3085	.3050	.3015	.2981	.2946	.2912	.2877	.2843	.2810	.2776
−.4	.3446	.3409	.3372	.3336	.3300	.3264	.3228	.3192	.3156	.3121
−.3	.3821	.3783	.3745	.3707	.3669	.3632	.3594	.3557	.3520	.3483
−.2	.4207	.4168	.4129	.4090	.4052	.4013	.3974	.3936	.3897	.3859
−.1	.4602	.4562	.4522	.4483	.4443	.4404	.4364	.4325	.4286	.4247
−.0	.5000	.4960	.4920	.4880	.4840	.4801	.4761	.4721	.4681	.4641

TABLE 1 CUMULATIVE PROBABILITIES FOR THE STANDARD NORMAL DISTRIBUTION (*Continued*)

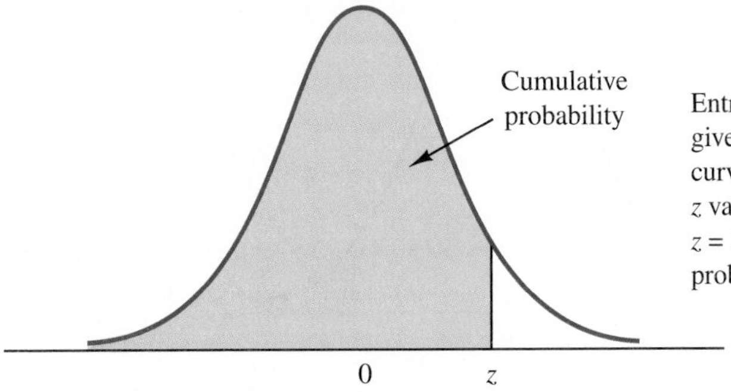

Entries in the table give the area under the curve to the left of the z value. For example, for $z = 1.25$, the cumulative probability is .8944.

z	.00	.01	.02	.03	.04	.05	.06	.07	.08	.09
.0	.5000	.5040	.5080	.5120	.5160	.5199	.5239	.5279	.5319	.5359
.1	.5398	.5438	.5478	.5517	.5557	.5596	.5636	.5675	.5714	.5753
.2	.5793	.5832	.5871	.5910	.5948	.5987	.6026	.6064	.6103	.6141
.3	.6179	.6217	.6255	.6293	.6331	.6368	.6406	.6443	.6480	.6517
.4	.6554	.6591	.6628	.6664	.6700	.6736	.6772	.6808	.6844	.6879
.5	.6915	.6950	.6985	.7019	.7054	.7088	.7123	.7157	.7190	.7224
.6	.7257	.7291	.7324	.7357	.7389	.7422	.7454	.7486	.7517	.7549
.7	.7580	.7611	.7642	.7673	.7704	.7734	.7764	.7794	.7823	.7852
.8	.7881	.7910	.7939	.7967	.7995	.8023	.8051	.8078	.8106	.8133
.9	.8159	.8186	.8212	.8238	.8264	.8289	.8315	.8340	.8365	.8389
1.0	.8413	.8438	.8461	.8485	.8508	.8531	.8554	.8577	.8599	.8621
1.1	.8643	.8665	.8686	.8708	.8729	.8749	.8770	.8790	.8810	.8830
1.2	.8849	.8869	.8888	.8907	.8925	.8944	.8962	.8980	.8997	.9015
1.3	.9032	.9049	.9066	.9082	.9099	.9115	.9131	.9147	.9162	.9177
1.4	.9192	.9207	.9222	.9236	.9251	.9265	.9279	.9292	.9306	.9319
1.5	.9332	.9345	.9357	.9370	.9382	.9394	.9406	.9418	.9429	.9441
1.6	.9452	.9463	.9474	.9484	.9495	.9505	.9515	.9525	.9535	.9545
1.7	.9554	.9564	.9573	.9582	.9591	.9599	.9608	.9616	.9625	.9633
1.8	.9641	.9649	.9656	.9664	.9671	.9678	.9686	.9693	.9699	.9706
1.9	.9713	.9719	.9726	.9732	.9738	.9744	.9750	.9756	.9761	.9767
2.0	.9772	.9778	.9783	.9788	.9793	.9798	.9803	.9808	.9812	.9817
2.1	.9821	.9826	.9830	.9834	.9838	.9842	.9846	.9850	.9854	.9857
2.2	.9861	.9864	.9868	.9871	.9875	.9878	.9881	.9884	.9887	.9890
2.3	.9893	.9896	.9898	.9901	.9904	.9906	.9909	.9911	.9913	.9916
2.4	.9918	.9920	.9922	.9925	.9927	.9929	.9931	.9932	.9934	.9936
2.5	.9938	.9940	.9941	.9943	.9945	.9946	.9948	.9949	.9951	.9952
2.6	.9953	.9955	.9956	.9957	.9959	.9960	.9961	.9962	.9963	.9964
2.7	.9965	.9966	.9967	.9968	.9969	.9970	.9971	.9972	.9973	.9974
2.8	.9974	.9975	.9976	.9977	.9977	.9978	.9979	.9979	.9980	.9981
2.9	.9981	.9982	.9982	.9983	.9984	.9984	.9985	.9985	.9986	.9986
3.0	.9987	.9987	.9987	.9988	.9988	.9989	.9989	.9989	.9990	.9990

TABLE 2 *t* DISTRIBUTION

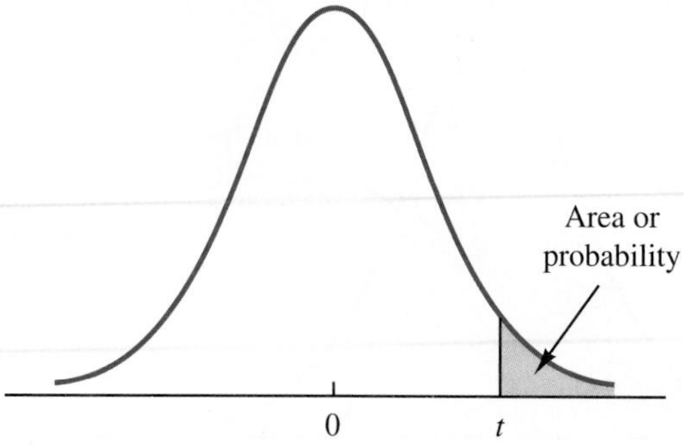

Entries in the table give t values for an area or probability in the upper tail of the t distribution. For example, with 10 degrees of freedom and a .05 area in the upper tail, $t_{.05} = 1.812$.

Degrees of Freedom	Area in Upper Tail					
	.20	**.10**	**.05**	**.025**	**.01**	**.005**
1	1.376	3.078	6.314	12.706	31.821	63.656
2	1.061	1.886	2.920	4.303	6.965	9.925
3	.978	1.638	2.353	3.182	4.541	5.841
4	.941	1.533	2.132	2.776	3.747	4.604
5	.920	1.476	2.015	2.571	3.365	4.032
6	.906	1.440	1.943	2.447	3.143	3.707
7	.896	1.415	1.895	2.365	2.998	3.499
8	.889	1.397	1.860	2.306	2.896	3.355
9	.883	1.383	1.833	2.262	2.821	3.250
10	.879	1.372	1.812	2.228	2.764	3.169
11	.876	1.363	1.796	2.201	2.718	3.106
12	.873	1.356	1.782	2.179	2.681	3.055
13	.870	1.350	1.771	2.160	2.650	3.012
14	.868	1.345	1.761	2.145	2.624	2.977
15	.866	1.341	1.753	2.131	2.602	2.947
16	.865	1.337	1.746	2.120	2.583	2.921
17	.863	1.333	1.740	2.110	2.567	2.898
18	.862	1.330	1.734	2.101	2.552	2.878
19	.861	1.328	1.729	2.093	2.539	2.861
20	.860	1.325	1.725	2.086	2.528	2.845
21	.859	1.323	1.721	2.080	2.518	2.831
22	.858	1.321	1.717	2.074	2.508	2.819
23	.858	1.319	1.714	2.069	2.500	2.807
24	.857	1.318	1.711	2.064	2.492	2.797
25	.856	1.316	1.708	2.060	2.485	2.787
26	.856	1.315	1.706	2.056	2.479	2.779
27	.855	1.314	1.703	2.052	2.473	2.771
28	.855	1.313	1.701	2.048	2.467	2.763
29	.854	1.311	1.699	2.045	2.462	2.756
30	.854	1.310	1.697	2.042	2.457	2.750
31	.853	1.309	1.696	2.040	2.453	2.744
32	.853	1.309	1.694	2.037	2.449	2.738
33	.853	1.308	1.692	2.035	2.445	2.733
34	.852	1.307	1.691	2.032	2.441	2.728

TABLE 2 *t* DISTRIBUTION (*Continued*)

Degrees of Freedom	Area in Upper Tail .20	.10	.05	.025	.01	.005
35	.852	1.306	1.690	2.030	2.438	2.724
36	.852	1.306	1.688	2.028	2.434	2.719
37	.851	1.305	1.687	2.026	2.431	2.715
38	.851	1.304	1.686	2.024	2.429	2.712
39	.851	1.304	1.685	2.023	2.426	2.708
40	.851	1.303	1.684	2.021	2.423	2.704
41	.850	1.303	1.683	2.020	2.421	2.701
42	.850	1.302	1.682	2.018	2.418	2.698
43	.850	1.302	1.681	2.017	2.416	2.695
44	.850	1.301	1.680	2.015	2.414	2.692
45	.850	1.301	1.679	2.014	2.412	2.690
46	.850	1.300	1.679	2.013	2.410	2.687
47	.849	1.300	1.678	2.012	2.408	2.685
48	.849	1.299	1.677	2.011	2.407	2.682
49	.849	1.299	1.677	2.010	2.405	2.680
50	.849	1.299	1.676	2.009	2.403	2.678
51	.849	1.298	1.675	2.008	2.402	2.676
52	.849	1.298	1.675	2.007	2.400	2.674
53	.848	1.298	1.674	2.006	2.399	2.672
54	.848	1.297	1.674	2.005	2.397	2.670
55	.848	1.297	1.673	2.004	2.396	2.668
56	.848	1.297	1.673	2.003	2.395	2.667
57	.848	1.297	1.672	2.002	2.394	2.665
58	.848	1.296	1.672	2.002	2.392	2.663
59	.848	1.296	1.671	2.001	2.391	2.662
60	.848	1.296	1.671	2.000	2.390	2.660
61	.848	1.296	1.670	2.000	2.389	2.659
62	.847	1.295	1.670	1.999	2.388	2.657
63	.847	1.295	1.669	1.998	2.387	2.656
64	.847	1.295	1.669	1.998	2.386	2.655
65	.847	1.295	1.669	1.997	2.385	2.654
66	.847	1.295	1.668	1.997	2.384	2.652
67	.847	1.294	1.668	1.996	2.383	2.651
68	.847	1.294	1.668	1.995	2.382	2.650
69	.847	1.294	1.667	1.995	2.382	2.649
70	.847	1.294	1.667	1.994	2.381	2.648
71	.847	1.294	1.667	1.994	2.380	2.647
72	.847	1.293	1.666	1.993	2.379	2.646
73	.847	1.293	1.666	1.993	2.379	2.645
74	.847	1.293	1.666	1.993	2.378	2.644
75	.846	1.293	1.665	1.992	2.377	2.643
76	.846	1.293	1.665	1.992	2.376	2.642
77	.846	1.293	1.665	1.991	2.376	2.641
78	.846	1.292	1.665	1.991	2.375	2.640
79	.846	1.292	1.664	1.990	2.374	2.639

TABLE 2 *t* DISTRIBUTION (*Continued*)

Degrees of Freedom	Area in Upper Tail .20	.10	.05	.025	.01	.005
80	.846	1.292	1.664	1.990	2.374	2.639
81	.846	1.292	1.664	1.990	2.373	2.638
82	.846	1.292	1.664	1.989	2.373	2.637
83	.846	1.292	1.663	1.989	2.372	2.636
84	.846	1.292	1.663	1.989	2.372	2.636
85	.846	1.292	1.663	1.988	2.371	2.635
86	.846	1.291	1.663	1.988	2.370	2.634
87	.846	1.291	1.663	1.988	2.370	2.634
88	.846	1.291	1.662	1.987	2.369	2.633
89	.846	1.291	1.662	1.987	2.369	2.632
90	.846	1.291	1.662	1.987	2.368	2.632
91	.846	1.291	1.662	1.986	2.368	2.631
92	.846	1.291	1.662	1.986	2.368	2.630
93	.846	1.291	1.661	1.986	2.367	2.630
94	.845	1.291	1.661	1.986	2.367	2.629
95	.845	1.291	1.661	1.985	2.366	2.629
96	.845	1.290	1.661	1.985	2.366	2.628
97	.845	1.290	1.661	1.985	2.365	2.627
98	.845	1.290	1.661	1.984	2.365	2.627
99	.845	1.290	1.660	1.984	2.364	2.626
100	.845	1.290	1.660	1.984	2.364	2.626
∞	.842	1.282	1.645	1.960	2.326	2.576

TABLE 3 CHI-SQUARE DISTRIBUTION

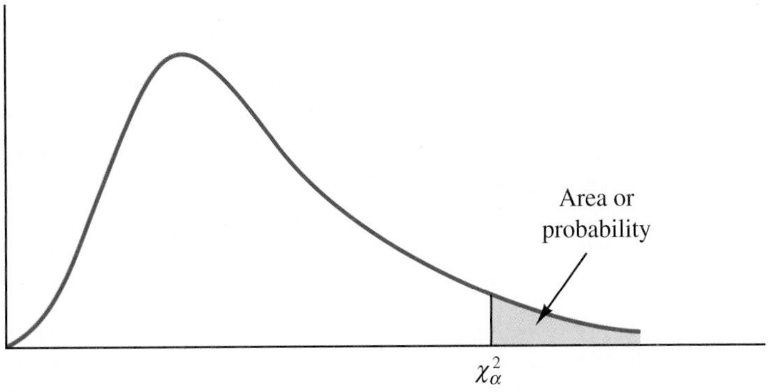

Entries in the table give χ^2_α values, where α is the area or probability in the upper tail of the chi-square distribution. For example, with 10 degrees of freedom and a .01 area in the upper tail, $\chi^2_{.01} = 23.209$.

Degrees of Freedom	Area in Upper Tail									
	.995	.99	.975	.95	.90	.10	.05	.025	.01	.005
1	.000	.000	.001	.004	.016	2.706	3.841	5.024	6.635	7.879
2	.010	.020	.051	.103	.211	4.605	5.991	7.378	9.210	10.597
3	.072	.115	.216	.352	.584	6.251	7.815	9.348	11.345	12.838
4	.207	.297	.484	.711	1.064	7.779	9.488	11.143	13.277	14.860
5	.412	.554	.831	1.145	1.610	9.236	11.070	12.832	15.086	16.750
6	.676	.872	1.237	1.635	2.204	10.645	12.592	14.449	16.812	18.548
7	.989	1.239	1.690	2.167	2.833	12.017	14.067	16.013	18.475	20.278
8	1.344	1.647	2.180	2.733	3.490	13.362	15.507	17.535	20.090	21.955
9	1.735	2.088	2.700	3.325	4.168	14.684	16.919	19.023	21.666	23.589
10	2.156	2.558	3.247	3.940	4.865	15.987	18.307	20.483	23.209	25.188
11	2.603	3.053	3.816	4.575	5.578	17.275	19.675	21.920	24.725	26.757
12	3.074	3.571	4.404	5.226	6.304	18.549	21.026	23.337	26.217	28.300
13	3.565	4.107	5.009	5.892	7.041	19.812	22.362	24.736	27.688	29.819
14	4.075	4.660	5.629	6.571	7.790	21.064	23.685	26.119	29.141	31.319
15	4.601	5.229	6.262	7.261	8.547	22.307	24.996	27.488	30.578	32.801
16	5.142	5.812	6.908	7.962	9.312	23.542	26.296	28.845	32.000	34.267
17	5.697	6.408	7.564	8.672	10.085	24.769	27.587	30.191	33.409	35.718
18	6.265	7.015	8.231	9.390	10.865	25.989	28.869	31.526	34.805	37.156
19	6.844	7.633	8.907	10.117	11.651	27.204	30.144	32.852	36.191	38.582
20	7.434	8.260	9.591	10.851	12.443	28.412	31.410	34.170	37.566	39.997
21	8.034	8.897	10.283	11.591	13.240	29.615	32.671	35.479	38.932	41.401
22	8.643	9.542	10.982	12.338	14.041	30.813	33.924	36.781	40.289	42.796
23	9.260	10.196	11.689	13.091	14.848	32.007	35.172	38.076	41.638	44.181
24	9.886	10.856	12.401	13.848	15.659	33.196	36.415	39.364	42.980	45.558
25	10.520	11.524	13.120	14.611	16.473	34.382	37.652	40.646	44.314	46.928
26	11.160	12.198	13.844	15.379	17.292	35.563	38.885	41.923	45.642	48.290
27	11.808	12.878	14.573	16.151	18.114	36.741	40.113	43.195	46.963	49.645
28	12.461	13.565	15.308	16.928	18.939	37.916	41.337	44.461	48.278	50.994
29	13.121	14.256	16.047	17.708	19.768	39.087	42.557	45.722	49.588	52.335

TABLE 3 CHI-SQUARE DISTRIBUTION (*Continued*)

Degrees of Freedom	Area in Upper Tail									
	.995	**.99**	**.975**	**.95**	**.90**	**.10**	**.05**	**.025**	**.01**	**.005**
30	13.787	14.953	16.791	18.493	20.599	40.256	43.773	46.979	50.892	53.672
35	17.192	18.509	20.569	22.465	24.797	46.059	49.802	53.203	57.342	60.275
40	20.707	22.164	24.433	26.509	29.051	51.805	55.758	59.342	63.691	66.766
45	24.311	25.901	28.366	30.612	33.350	57.505	61.656	65.410	69.957	73.166
50	27.991	29.707	32.357	34.764	37.689	63.167	67.505	71.420	76.154	79.490
55	31.735	33.571	36.398	38.958	42.060	68.796	73.311	77.380	82.292	85.749
60	35.534	37.485	40.482	43.188	46.459	74.397	79.082	83.298	88.379	91.952
65	39.383	41.444	44.603	47.450	50.883	79.973	84.821	89.177	94.422	98.105
70	43.275	45.442	48.758	51.739	55.329	85.527	90.531	95.023	100.425	104.215
75	47.206	49.475	52.942	56.054	59.795	91.061	96.217	100.839	106.393	110.285
80	51.172	53.540	57.153	60.391	64.278	96.578	101.879	106.629	112.329	116.321
85	55.170	57.634	61.389	64.749	68.777	102.079	107.522	112.393	118.236	122.324
90	59.196	61.754	65.647	69.126	73.291	107.565	113.145	118.136	124.116	128.299
95	63.250	65.898	69.925	73.520	77.818	113.038	118.752	123.858	129.973	134.247
100	67.328	70.065	74.222	77.929	82.358	118.498	124.342	129.561	135.807	140.170

TABLE 4 F DISTRIBUTION

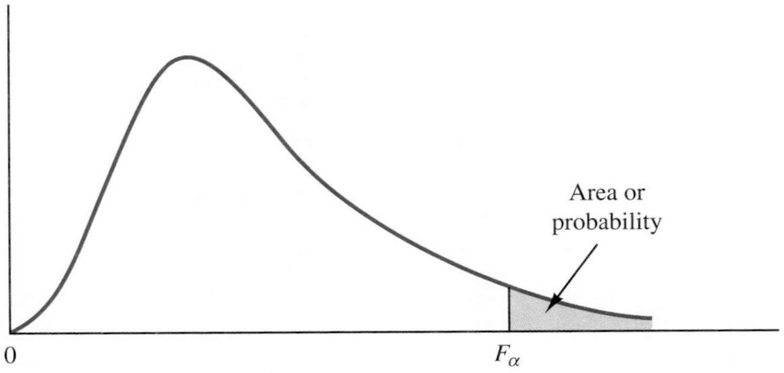

Entries in the table give F_α values, where α is the area or probability in the upper tail of the F distribution. For example, with 4 numerator degrees of freedom, 8 denominator degrees of freedom, and a .05 area in the upper tail, $F_{.05} = 3.84$.

Denominator Degrees of Freedom	Area in Upper Tail	Numerator Degrees of Freedom																	
		1	2	3	4	5	6	7	8	9	10	15	20	25	30	40	60	100	1000
1	.10	39.86	49.50	53.59	55.83	57.24	58.20	58.91	59.44	59.86	60.19	61.22	61.74	62.05	62.26	62.53	62.79	63.01	63.30
	.05	161.45	199.50	215.71	224.58	230.16	233.99	236.77	238.88	240.54	241.88	245.95	248.02	249.26	250.10	251.14	252.20	253.04	254.19
	.025	647.79	799.48	864.15	899.60	921.83	937.11	948.20	956.64	963.28	968.63	984.87	993.08	998.09	1001.40	1005.60	1009.79	1013.16	1017.76
	.01	4052.18	4999.34	5403.53	5624.26	5763.96	5858.95	5928.33	5980.95	6022.40	6055.93	6156.97	6208.66	6239.86	6260.35	6286.43	6312.97	6333.92	6362.80
2	.10	8.53	9.00	9.16	9.24	9.29	9.33	9.35	9.37	9.38	9.39	9.42	9.44	9.45	9.46	9.47	9.47	9.48	9.49
	.05	18.51	19.00	19.16	19.25	19.30	19.33	19.35	19.37	19.38	19.40	19.43	19.45	19.46	19.46	19.47	19.48	19.49	19.49
	.025	38.51	39.00	39.17	39.25	39.30	39.33	39.36	39.37	39.39	39.40	39.43	39.45	39.46	39.46	39.47	39.48	39.49	39.50
	.01	98.50	99.00	99.16	99.25	99.30	99.33	99.36	99.38	99.39	99.40	99.43	99.45	99.46	99.47	99.48	99.48	99.49	99.50
3	.10	5.54	5.46	5.39	5.34	5.31	5.28	5.27	5.25	5.24	5.23	5.20	5.18	5.17	5.17	5.16	5.15	5.14	5.13
	.05	10.13	9.55	9.28	9.12	9.01	8.94	8.89	8.85	8.81	8.79	8.70	8.66	8.63	8.62	8.59	8.57	8.55	8.53
	.025	17.44	16.04	15.44	15.10	14.88	14.73	14.62	14.54	14.47	14.42	14.25	14.17	14.12	14.08	14.04	13.99	13.96	13.91
	.01	34.12	30.82	29.46	28.71	28.24	27.91	27.67	27.49	27.34	27.23	26.87	26.69	26.58	26.50	26.41	26.32	26.24	26.14
4	.10	4.54	4.32	4.19	4.11	4.05	4.01	3.98	3.95	3.94	3.92	3.87	3.84	3.83	3.82	3.80	3.79	3.78	3.76
	.05	7.71	6.94	6.59	6.39	6.26	6.16	6.09	6.04	6.00	5.96	5.86	5.80	5.77	5.75	5.72	5.69	5.66	5.63
	.025	12.22	10.65	9.98	9.60	9.36	9.20	9.07	8.98	8.90	8.84	8.66	8.56	8.50	8.46	8.41	8.36	8.32	8.26
	.01	21.20	18.00	16.69	15.98	15.52	15.21	14.98	14.80	14.66	14.55	14.20	14.02	13.91	13.84	13.75	13.65	13.58	13.47
5	.10	4.06	3.78	3.62	3.52	3.45	3.40	3.37	3.34	3.32	3.30	3.324	3.21	3.19	3.17	3.16	3.14	3.13	3.11
	.05	6.61.	5.79	5.41	5.19	5.05	4.95	4.88	4.82	4.77	4.74	4.62	4.56	4.52	4.50	4.46	4.43	4.41	4.37
	.025	10.01	8.43	7.76	7.39	7.15	6.98	6.85	6.76	6.68	6.62	6.43	6.33	6.27	6.23	6.18	6.12	6.08	6.02
	.01	16.26	13.27	12.06	11.39	10.97	10.67	10.46	10.29	10.16	10.05	9.72	9.55	9.45	9.38	9.29	9.20	9.13	9.03

TABLE 4 *F* DISTRIBUTION (*Continued*)

Denominator Degrees of Freedom	Area in Upper Tail	Numerator Degrees of Freedom 1	2	3	4	5	6	7	8	9	10	15	20	25	30	40	60	100	1000
6	.10	3.78	3.46	3.29	3.18	3.11	3.05	3.01	2.98	2.96	2.94	2.87	2.84	2.81	2.80	2.78	2.76	2.75	2.72
	.05	5.99	5.14	4.76	4.53	4.39	4.28	4.21	4.15	4.10	4.06	3.94	3.87	3.83	3.81	3.77	3.74	3.71	3.67
	.025	8.81	7.26	6.60	6.23	5.99	5.82	5.70	5.60	5.52	5.46	5.27	5.17	5.11	5.07	5.01	4.96	4.92	4.86
	.01	13.75	10.92	9.78	9.15	8.75	8.47	8.26	8.10	7.98	7.87	7.56	7.40	7.30	7.23	7.14	7.06	6.99	6.89
7	.10	3.59	3.26	3.07	2.96	2.88	2.83	2.78	2.75	2.72	2.70	2.63	2.59	2.57	2.56	2.54	2.51	2.50	2.47
	.05	5.59	4.74	4.35	4.12	3.97	3.87	3.79	3.73	3.68	3.64	3.51	3.44	3.40	3.38	3.34	3.30	3.27	3.23
	.025	8.07	6.54	5.89	5.52	5.29	5.12	4.99	4.90	4.82	4.76	4.57	4.47	4.40	4.36	4.31	4.25	4.21	4.15
	.01	12.25	9.55	8.45	7.85	7.46	7.19	6.99	6.84	6.72	6.62	6.31	6.16	6.06	5.99	5.91	5.82	5.75	5.66
8	.10	3.46	3.11	2.92	2.81	2.73	2.67	2.62	2.59	2.56	2.54	2.46	2.42	2.40	2.38	2.36	2.34	2.32	2.30
	.05	5.32	4.46	4.07	3.84	3.69	3.58	3.50	3.44	3.39	3.35	3.22	3.15	3.11	3.08	3.04	3.01	2.97	2.93
	.025	7.57	6.06	5.42	5.05	4.82	4.65	4.53	4.43	4.36	4.30	4.10	4.00	3.94	3.89	3.84	3.78	3.74	3.68
	.01	11.26	8.65	7.59	7.01	6.63	6.37	6.18	6.03	5.91	5.81	5.52	5.36	5.26	5.20	5.12	5.03	4.96	4.87
9	.10	3.36	3.01	2.81	2.69	2.61	2.55	2.51	2.47	2.44	2.42	2.34	2.30	2.27	2.25	2.23	2.21	2.19	2.16
	.05	5.12	4.26	3.86	3.63	3.48	3.37	3.29	3.23	3.18	3.14	3.01	2.94	2.89	2.86	2.83	2.79	2.76	2.71
	.025	7.21	5.71	5.08	4.72	4.48	4.32	4.20	4.10	4.03	3.96	3.77	3.67	3.60	3.56	3.51	3.45	3.40	3.34
	.01	10.56	8.02	6.99	6.42	6.06	5.80	5.61	5.47	5.35	5.26	4.96	4.81	4.71	4.65	4.57	4.48	4.41	4.32
10	.10	3.29	2.92	2.73	2.61	2.52	2.46	2.41	2.38	2.35	2.32	2.24	2.20	2.17	2.16	2.13	2.11	2.09	2.06
	.05	4.96	4.10	3.71	3.48	3.33	3.22	3.14	3.07	3.02	2.98	2.85	2.77	2.73	2.70	2.66	2.62	2.59	2.54
	.025	6.94	5.46	4.83	4.47	4.24	4.07	3.95	3.85	3.78	3.72	3.52	3.42	3.35	3.31	3.26	3.20	3.15	3.09
	.01	10.04	7.56	6.55	5.99	5.64	5.39	5.20	5.06	4.94	4.85	4.56	4.41	4.31	4.25	4.17	4.08	4.01	3.92
11	.10	3.23	2.86	2.66	2.54	2.45	2.39	2.34	2.30	2.27	2.25	2.17	2.12	2.10	2.08	2.05	2.03	2.01	1.98
	.05	4.84	3.98	3.59	3.36	3.20	3.09	3.01	2.95	2.90	2.85	2.72	2.65	2.60	2.57	2.53	2.49	2.46	2.41
	.025	6.72	5.26	4.63	4.28	4.04	3.88	3.76	3.66	3.59	3.53	3.33	3.23	3.16	3.12	3.06	3.00	2.96	2.89
	.01	9.65	7.21	6.22	5.67	5.32	5.07	4.89	4.74	4.63	4.54	4.25	4.10	4.01	3.94	3.86	3.78	3.71	3.61
12	.10	3.18	2.81	2.61	2.48	2.39	2.33	2.28	2.24	2.21	2.19	2.10	2.06	2.03	2.01	1.99	1.96	1.94	1.91
	.05	4.75	3.89	3.49	3.26	3.11	3.00	2.91	2.85	2.80	2.75	2.62	2.54	2.50	2.47	2.43	2.38	2.35	2.30
	.025	6.55	5.10	4.47	4.12	3.89	3.73	3.61	3.51	3.44	3.37	3.18	3.07	3.01	2.96	2.91	2.85	2.80	2.73
	.01	9.33	6.93	5.95	5.41	5.06	4.82	4.64	4.50	4.39	4.30	4.01	3.86	3.76	3.70	3.62	3.54	3.47	3.37
13	.10	3.14	2.76	2.56	2.43	2.35	2.28	2.23	2.20	2.16	2.14	2.05	2.01	1.98	1.96	1.93	1.90	1.88	1.85
	.05	4.67	3.81	3.41	3.18	3.03	2.92	2.83	2.77	2.71	2.67	2.53	2.46	2.41	2.38	2.34	2.30	2.26	2.21
	.025	6.41	4.97	4.35	4.00	3.77	3.60	3.48	3.39	3.31	3.25	3.05	2.95	2.88	2.84	2.78	2.72	2.67	2.60
	.01	9.07	6.70	5.74	5.21	4.86	4.62	4.44	4.30	4.19	4.10	3.82	3.66	3.57	3.51	3.43	3.34	3.27	3.18
14	.10	3.10	2.73	2.52	2.39	2.31	2.24	2.19	2.15	2.12	2.10	2.01	1.96	1.93	1.99	1.89	1.86	1.83	1.80
	.05	4.60	3.74	3.34	3.11	2.96	2.85	2.76	2.70	2.65	2.60	2.46	2.39	2.34	2.31	2.27	2.22	2.19	2.14
	.025	6.30	4.86	4.24	3.89	3.66	3.50	3.38	3.29	3.21	3.15	2.95	2.84	2.78	2.73	2.67	2.61	2.56	2.50
	.01	8.86	6.51	5.56	5.04	4.69	4.46	4.28	4.14	4.03	3.94	3.66	3.51	3.41	3.35	3.27	3.18	3.11	3.02
15	.10	3.07	2.70	2.49	2.36	2.27	2.21	2.16	2.12	2.09	2.06	1.97	1.92	1.89	1.87	1.85	1.82	1.79	1.76
	.05	4.54	3.68	3.29	3.06	2.90	2.79	2.71	2.64	2.59	2.54	2.40	2.33	2.28	2.25	2.20	2.16	2.12	2.07
	.025	6.20	4.77	4.15	3.80	3.58	3.41	3.29	3.20	3.12	3.06	2.86	2.76	2.69	2.64	2.59	2.52	2.47	2.40
	.01	8.68	6.36	5.42	4.89	4.56	4.32	4.14	4.00	3.89	3.80	3.52	3.37	3.28	3.21	3.13	3.05	2.98	2.88

Denominator Degrees of Freedom	Area in Upper Tail	Numerator Degrees of Freedom																	
		1	2	3	4	5	6	7	8	9	10	15	20	25	30	40	60	100	1000
16	.10	3.05	2.67	2.46	2.33	2.24	2.18	2.13	2.09	2.06	2.03	1.94	1.89	1.86	1.84	1.81	1.78	1.76	1.72
	.05	4.49	3.63	3.24	3.01	2.85	2.74	2.66	2.59	2.54	2.49	2.35	2.28	2.23	2.19	2.15	2.11	2.07	2.02
	.025	6.12	4.69	4.08	3.73	3.50	3.34	3.22	3.12	3.05	2.99	2.79	2.68	2.61	2.57	2.51	2.45	2.40	2.32
	.01	8.53	6.23	5.29	4.77	4.44	4.20	4.03	3.89	3.78	3.69	3.41	3.26	3.16	3.10	3.02	2.93	2.86	2.76
17	.10	3.03	2.64	2.44	2.31	2.22	2.15	2.10	2.06	2.03	2.00	1.91	1.86	1.83	1.81	1.78	1.75	1.73	1.69
	.05	4.45	3.59	3.20	2.96	2.81	2.70	2.61	2.55	2.49	2.45	2.31	2.23	2.18	2.15	2.10	2.06	2.02	1.97
	.025	6.04	4.62	4.01	3.66	3.44	3.28	3.16	3.06	2.98	2.92	2.72	2.62	2.55	2.50	2.44	2.38	2.33	2.26
	.01	8.40	6.11	5.19	4.67	4.34	4.10	3.93	3.79	3.68	3.59	3.31	3.16	3.07	3.00	2.92	2.83	2.76	2.66
18	.10	3.01	2.62	2.42	2.29	2.20	2.13	2.08	2.04	2.00	1.98	1.89	1.84	1.80	1.78	1.75	1.72	1.70	1.66
	.05	4.41	3.55	3.16	2.93	2.77	2.66	2.58	2.51	2.46	2.41	2.27	2.19	2.14	2.11	2.06	2.02	1.98	1.92
	.025	5.98	4.56	3.95	3.61	3.38	3.22	3.10	3.01	2.93	2.87	2.67	2.56	2.49	2.44	2.38	2.32	2.27	2.20
	.01	8.29	6.01	5.09	4.58	4.25	4.01	3.84	3.71	3.60	3.51	3.23	3.08	2.98	2.92	2.84	2.75	2.68	2.58
19	.10	2.99	2.61	2.40	2.27	2.18	2.11	2.06	2.02	1.98	1.96	1.86	1.81	1.78	1.76	1.73	1.70	1.67	1.64
	.05	4.38	3.52	3.13	2.90	2.74	2.63	2.54	2.48	2.42	2.38	2.23	2.16	2.11	2.07	2.03	1.98	1.94	1.88
	.025	5.92	4.51	3.90	3.56	3.33	3.17	3.05	2.96	2.88	2.82	2.62	2.51	2.44	2.39	2.33	2.27	2.22	2.14
	.01	8.18	5.93	5.01	4.50	4.17	3.94	3.77	3.63	3.52	3.43	3.15	3.00	2.91	2.84	2.76	2.67	2.60	2.50
20	.10	2.97	2.59	2.38	2.25	2.16	2.09	2.04	2.00	1.96	1.94	1.84	1.79	1.76	1.74	1.71	1.68	1.65	1.61
	.05	4.35	3.49	3.10	2.87	2.71	2.60	2.51	2.45	2.39	2.35	2.20	2.12	2.07	2.04	1.99	1.95	1.91	1.85
	.025	5.87	4.46	3.86	3.51	3.29	3.13	3.01	2.91	2.84	2.77	2.57	2.46	2.40	2.35	2.29	2.22	2.17	2.09
	.01	8.10	5.85	4.94	4.43	4.10	3.87	3.70	3.56	3.46	3.37	3.09	2.94	2.84	2.78	2.69	2.61	2.54	2.43
21	.10	2.96	2.57	2.36	2.23	2.14	2.08	2.02	1.98	1.95	1.92	1.83	1.78	1.74	1.72	1.69	1.66	1.63	1.59
	.05	4.32	3.47	3.07	2.84	2.68	2.57	2.49	2.42	2.37	2.32	2.18	2.10	2.05	2.01	1.96	1.92	1.88	1.82
	.025	5.83	4.42	3.82	3.48	3.25	3.09	2.97	2.87	2.80	2.73	2.53	2.42	2.36	2.31	2.25	2.18	2.13	2.05
	.01	8.02	5.78	4.87	4.37	4.04	3.81	3.64	3.51	3.40	3.31	3.03	2.88	2.79	2.72	2.64	2.55	2.48	2.37
22	.10	2.95	2.56	2.35	2.22	2.13	2.06	2.01	1.97	1.93	1.90	1.81	1.76	1.73	1.70	1.67	1.64	1.61	1.57
	.05	4.30	3.44	3.05	2.82	2.66	2.55	2.46	2.40	2.34	2.30	2.15	2.07	2.02	1.98	1.94	1.89	1.85	1.79
	.025	5.79	4.38	3.78	3.44	3.22	3.05	2.93	2.84	2.76	2.70	2.50	2.39	2.32	2.27	2.21	2.14	2.09	2.01
	.01	7.95	5.72	4.82	4.31	3.99	3.76	3.59	3.45	3.35	3.26	2.98	2.83	2.73	2.67	2.58	2.50	2.42	2.32
23	.10	2.94	2.55	2.34	2.21	2.11	2.05	1.99	1.95	1.92	1.89	1.80	1.74	1.71	1.69	1.66	1.62	1.59	1.55
	.05	4.28	3.42	3.03	2.80	2.64	2.53	2.44	2.37	2.32	2.27	2.13	2.05	2.00	1.96	1.91	1.86	1.82	1.76
	.025	5.75	4.35	3.75	3.41	3.18	3.02	2.90	2.81	2.73	2.67	2.47	2.36	2.29	2.24	2.18	2.11	2.06	1.98
	.01	7.88	5.66	4.76	4.26	3.94	3.71	3.54	3.41	3.30	3.21	2.93	2.78	2.69	2.62	2.54	2.45	2.37	2.27
24	.10	2.93	2.54	2.33	2.19	2.10	2.04	1.98	1.94	1.91	1.88	1.78	1.73	1.70	1.67	1.64	1.61	1.58	1.54
	.05	4.26	3.40	3.01	2.78	2.62	2.51	2.42	2.36	2.30	2.25	2.11	2.03	1.97	1.94	1.89	1.84	1.80	1.74
	.025	5.72	4.32	3.72	3.38	3.15	2.99	2.87	2.78	2.70	2.64	2.44	2.33	2.26	2.21	2.15	2.08	2.02	1.94
	.01	7.82	5.61	4.72	4.22	3.90	3.67	3.50	3.36	3.26	3.17	2.89	2.74	2.64	2.58	2.49	2.40	2.33	2.22

TABLE 4 *F* DISTRIBUTION (*Continued*)

Denominator Degrees of Freedom	Area in Upper Tail	Numerator Degrees of Freedom																	
		1	2	3	4	5	6	7	8	9	10	15	20	25	30	40	60	100	1000
25	.10	2.92	2.53	2.32	2.18	2.09	2.02	1.97	1.93	1.89	1.87	1.77	1.72	1.68	1.66	1.63	1.59	1.56	1.52
	.05	4.24	3.39	2.99	2.76	2.60	2.49	2.40	2.34	2.28	2.24	2.09	2.01	1.96	1.92	1.87	1.82	1.78	1.72
	.025	5.69	4.29	3.69	3.35	3.13	2.97	2.85	2.75	2.68	2.61	2.41	2.30	2.23	2.18	2.12	2.05	2.00	1.91
	.01	7.77	5.57	4.68	4.18	3.85	3.63	3.46	3.32	3.22	3.13	2.85	2.70	2.60	2.54	2.45	2.36	2.29	2.18
26	.10	2.91	2.52	2.31	2.17	2.08	2.01	1.96	1.92	1.88	1.86	1.76	1.71	1.67	1.65	1.61	1.58	1.55	1.51
	.05	4.23	3.37	2.98	2.74	2.59	2.47	2.39	2.32	2.27	2.22	2.07	1.99	1.94	1.90	1.85	1.80	1.76	1.70
	.025	5.66	4.27	3.67	3.33	3.10	2.94	2.82	2.73	2.65	2.59	2.39	2.28	2.21	2.16	2.09	2.03	1.97	1.89
	.01	7.72	5.53	4.64	4.14	3.82	3.59	3.42	3.29	3.18	3.09	2.81	2.66	2.57	2.50	2.42	2.33	2.25	2.14
27	.10	2.90	2.51	2.30	2.17	2.07	2.00	1.95	1.91	1.87	1.85	1.75	1.70	1.66	1.64	1.60	1.57	1.54	1.50
	.05	4.21	3.35	2.96	2.73	2.57	2.46	2.37	2.31	2.25	2.20	2.06	1.97	1.92	1.88	1.84	1.79	1.74	1.68
	.025	5.63	4.24	3.65	3.31	3.08	2.92	2.80	2.71	2.63	2.57	2.36	2.25	2.18	2.13	2.07	2.00	1.94	1.86
	.01	7.68	5.49	4.60	4.11	3.78	3.56	3.39	3.26	3.15	3.06	2.78	2.63	2.54	2.47	2.38	2.29	2.22	2.11
28	.10	2.89	2.50	2.29	2.16	2.06	2.00	1.94	1.90	1.87	1.84	1.74	1.69	1.65	1.63	1.59	1.56	1.53	1.48
	.05	4.20	3.34	2.95	2.71	2.56	2.45	2.36	2.29	2.24	2.19	2.04	1.96	1.91	1.87	1.82	1.77	1.73	1.66
	.025	5.61	4.22	3.63	3.29	3.06	2.90	2.78	2.69	2.61	2.55	2.34	2.23	2.16	2.11	2.05	1.98	1.92	1.84
	.01	7.64	5.45	4.57	4.07	3.75	3.53	3.36	3.23	3.12	3.03	2.75	2.60	2.51	2.44	2.35	2.26	2.19	2.08
29	.10	2.89	2.50	2.28	2.15	2.06	1.99	1.93	1.89	1.86	1.83	1.73	1.68	1.64	1.62	1.58	1.55	1.52	1.47
	.05	4.18	3.33	2.93	2.70	2.55	2.43	2.35	2.28	2.22	2.18	2.03	1.94	1.89	1.85	1.81	1.75	1.71	1.65
	.025	5.59	4.20	3.61	3.27	3.04	2.88	2.76	2.67	2.59	2.53	2.32	2.21	2.14	2.09	2.03	1.96	1.90	1.82
	.01	7.60	5.42	4.54	4.04	3.73	3.50	3.33	3.20	3.09	3.00	2.73	2.57	2.48	2.41	2.33	2.23	2.16	2.05
30	.10	2.88	2.49	2.28	2.14	2.05	1.98	1.93	1.88	1.85	1.82	1.72	1.67	1.63	1.61	1.57	1.54	1.51	1.46
	.05	4.17	3.32	2.92	2.69	2.53	2.42	2.33	2.27	2.21	2.16	2.01	1.93	1.88	1.84	1.79	1.74	1.70	1.63
	.025	5.57	4.18	3.59	3.25	3.03	2.87	2.75	2.65	2.57	2.51	2.31	2.20	2.12	2.07	2.01	1.94	1.88	1.80
	.01	7.56	5.39	4.51	4.02	3.70	3.47	3.30	3.17	3.07	2.98	2.70	2.55	2.45	2.39	2.30	2.21	2.13	2.02
40	.10	2.84	2.44	2.23	2.09	2.00	1.93	1.87	1.83	1.79	1.76	1.66	1.61	1.57	1.54	1.51	1.47	1.43	1.38
	.05	4.08	3.23	2.84	2.61	2.45	2.34	2.25	2.18	2.12	2.08	1.92	1.84	1.78	1.74	1.69	1.64	1.59	1.52
	.025	5.42	4.05	3.46	3.13	2.90	2.74	2.62	2.53	2.45	2.39	2.18	2.07	1.99	1.94	1.88	1.80	1.74	1.65
	.01	7.31	5.18	4.31	3.83	3.51	3.29	3.12	2.99	2.89	2.80	2.52	2.37	2.27	2.20	2.11	2.02	1.94	1.82
60	.10	2.79	2.39	2.18	2.04	1.95	1.87	1.82	1.77	1.74	1.71	1.60	1.54	1.50	1.48	1.44	1.40	1.36	1.30
	.05	4.00	3.15	2.76	2.53	2.37	2.25	2.17	2.10	2.04	1.99	1.84	1.75	1.69	1.65	1.59	1.53	1.48	1.40
	.025	5.29	3.93	3.34	3.01	2.79	2.63	2.51	2.41	2.33	2.27	2.06	1.94	1.87	1.82	1.74	1.67	1.60	1.49
	.01	7.08	4.98	4.13	3.65	3.34	3.12	2.95	2.82	2.72	2.63	2.35	2.20	2.10	2.03	1.94	1.84	1.75	1.62
100	.10	2.76	2.36	2.14	2.00	1.91	1.83	1.78	1.73	1.69	1.66	1.56	1.49	1.45	1.42	1.38	1.34	1.29	1.22
	.05	3.94	3.09	2.70	2.46	2.31	2.19	2.10	2.03	1.97	1.93	1.77	1.68	1.62	1.57	1.52	1.45	1.39	1.30
	.025	5.18	3.83	3.25	2.92	2.70	2.54	2.42	2.32	2.24	2.18	1.97	1.85	1.77	1.71	1.64	1.56	1.48	1.36
	.01	6.90	4.82	3.98	3.51	3.21	2.99	2.82	2.69	2.59	2.50	2.22	2.07	1.97	1.89	1.80	1.69	1.60	1.45
1000	.10	2.71	2.31	2.09	1.95	1.85	1.78	1.72	1.68	1.64	1.61	1.49	1.43	1.38	1.35	1.30	1.25	1.20	1.08
	.05	3.85	3.00	2.61	2.38	2.22	2.11	2.02	1.95	1.89	1.84	1.68	1.58	1.52	1.47	1.41	1.33	1.26	1.11
	.025	5.04	3.70	3.13	2.80	2.58	2.42	2.30	2.20	2.13	2.06	1.85	1.72	1.64	1.58	1.50	1.41	1.32	1.13
	.01	6.66	4.63	3.80	3.34	3.04	2.82	2.66	2.53	2.43	2.34	2.06	1.90	1.79	1.72	1.61	1.50	1.38	1.16

TABLE 5 BINOMIAL PROBABILITIES

Entries in the table give the probability of x successes in n trials of a binomial experiment, where p is the probability of a success on one trial. For example, with six trials and $p = .05$, the probability of two successes is .0305.

		p								
n	x	.01	.02	.03	.04	.05	.06	.07	.08	.09
2	0	.9801	.9604	.9409	.9216	.9025	.8836	.8649	.8464	.8281
	1	.0198	.0392	.0582	.0768	.0950	.1128	.1302	.1472	.1638
	2	.0001	.0004	.0009	.0016	.0025	.0036	.0049	.0064	.0081
3	0	.9703	.9412	.9127	.8847	.8574	.8306	.8044	.7787	.7536
	1	.0294	.0576	.0847	.1106	.1354	.1590	.1816	.2031	.2236
	2	.0003	.0012	.0026	.0046	.0071	.0102	.0137	.0177	.0221
	3	.0000	.0000	.0000	.0001	.0001	.0002	.0003	.0005	.0007
4	0	.9606	.9224	.8853	.8493	.8145	.7807	.7481	.7164	.6857
	1	.0388	.0753	.1095	.1416	.1715	.1993	.2252	.2492	.2713
	2	.0006	.0023	.0051	.0088	.0135	.0191	.0254	.0325	.0402
	3	.0000	.0000	.0001	.0002	.0005	.0008	.0013	.0019	.0027
	4	.0000	.0000	.0000	.0000	.0000	.0000	.0000	.0000	.0001
5	0	.9510	.9039	.8587	.8154	.7738	.7339	.6957	.6591	.6240
	1	.0480	.0922	.1328	.1699	.2036	.2342	.2618	.2866	.3086
	2	.0010	.0038	.0082	.0142	.0214	.0299	.0394	.0498	.0610
	3	.0000	.0001	.0003	.0006	.0011	.0019	.0030	.0043	.0060
	4	.0000	.0000	.0000	.0000	.0000	.0001	.0001	.0002	.0003
	5	.0000	.0000	.0000	.0000	.0000	.0000	.0000	.0000	.0000
6	0	.9415	.8858	.8330	.7828	.7351	.6899	.6470	.6064	.5679
	1	.0571	.1085	.1546	.1957	.2321	.2642	.2922	.3164	.3370
	2	.0014	.0055	.0120	.0204	.0305	.0422	.0550	.0688	.0833
	3	.0000	.0002	.0005	.0011	.0021	.0036	.0055	.0080	.0110
	4	.0000	.0000	.0000	.0000	.0001	.0002	.0003	.0005	.0008
	5	.0000	.0000	.0000	.0000	.0000	.0000	.0000	.0000	.0000
	6	.0000	.0000	.0000	.0000	.0000	.0000	.0000	.0000	.0000
7	0	.9321	.8681	.8080	.7514	.6983	.6485	.6017	.5578	.5168
	1	.0659	.1240	.1749	.2192	.2573	.2897	.3170	.3396	.3578
	2	.0020	.0076	.0162	.0274	.0406	.0555	.0716	.0886	.1061
	3	.0000	.0003	.0008	.0019	.0036	.0059	.0090	.0128	.0175
	4	.0000	.0000	.0000	.0001	.0002	.0004	.0007	.0011	.0017
	5	.0000	.0000	.0000	.0000	.0000	.0000	.0000	.0001	.0001
	6	.0000	.0000	.0000	.0000	.0000	.0000	.0000	.0000	.0000
	7	.0000	.0000	.0000	.0000	.0000	.0000	.0000	.0000	.0000
8	0	.9227	.8508	.7837	.7214	.6634	.6096	.5596	.5132	.4703
	1	.0746	.1389	.1939	.2405	.2793	.3113	.3370	.3570	.3721
	2	.0026	.0099	.0210	.0351	.0515	.0695	.0888	.1087	.1288
	3	.0001	.0004	.0013	.0029	.0054	.0089	.0134	.0189	.0255
	4	.0000	.0000	.0001	.0002	.0004	.0007	.0013	.0021	.0031
	5	.0000	.0000	.0000	.0000	.0000	.0000	.0001	.0001	.0002
	6	.0000	.0000	.0000	.0000	.0000	.0000	.0000	.0000	.0000
	7	.0000	.0000	.0000	.0000	.0000	.0000	.0000	.0000	.0000
	8	.0000	.0000	.0000	.0000	.0000	.0000	.0000	.0000	.0000

TABLE 5 BINOMIAL PROBABILITIES (*Continued*)

		p								
n	*x*	.01	.02	.03	.04	.05	.06	.07	.08	.09
9	0	.9135	.8337	.7602	.6925	.6302	.5730	.5204	.4722	.4279
	1	.0830	.1531	.2116	.2597	.2985	.3292	.3525	.3695	.3809
	2	.0034	.0125	.0262	.0433	.0629	.0840	.1061	.1285	.1507
	3	.0001	.0006	.0019	.0042	.0077	.0125	.0186	.0261	.0348
	4	.0000	.0000	.0001	.0003	.0006	.0012	.0021	.0034	.0052
	5	.0000	.0000	.0000	.0000	.0000	.0001	.0002	.0003	.0005
	6	.0000	.0000	.0000	.0000	.0000	.0000	.0000	.0000	.0000
	7	.0000	.0000	.0000	.0000	.0000	.0000	.0000	.0000	.0000
	8	.0000	.0000	.0000	.0000	.0000	.0000	.0000	.0000	.0000
	9	.0000	.0000	.0000	.0000	.0000	.0000	.0000	.0000	.0000
10	0	.9044	.8171	.7374	.6648	.5987	.5386	.4840	.4344	.3894
	1	.0914	.1667	.2281	.2770	.3151	.3438	.3643	.3777	.3851
	2	.0042	.0153	.0317	.0519	.0746	.0988	.1234	.1478	.1714
	3	.0001	.0008	.0026	.0058	.0105	.0168	.0248	.0343	.0452
	4	.0000	.0000	.0001	.0004	.0010	.0019	.0033	.0052	.0078
	5	.0000	.0000	.0000	.0000	.0001	.0001	.0003	.0005	.0009
	6	.0000	.0000	.0000	.0000	.0000	.0000	.0000	.0000	.0001
	7	.0000	.0000	.0000	.0000	.0000	.0000	.0000	.0000	.0000
	8	.0000	.0000	.0000	.0000	.0000	.0000	.0000	.0000	.0000
	9	.0000	.0000	.0000	.0000	.0000	.0000	.0000	.0000	.0000
	10	.0000	.0000	.0000	.0000	.0000	.0000	.0000	.0000	.0000
12	0	.8864	.7847	.6938	.6127	.5404	.4759	.4186	.3677	.3225
	1	.1074	.1922	.2575	.3064	.3413	.3645	.3781	.3837	.3827
	2	.0060	.0216	.0438	.0702	.0988	.1280	.1565	.1835	.2082
	3	.0002	.0015	.0045	.0098	.0173	.0272	.0393	.0532	.0686
	4	.0000	.0001	.0003	.0009	.0021	.0039	.0067	.0104	.0153
	5	.0000	.0000	.0000	.0001	.0002	.0004	.0008	.0014	.0024
	6	.0000	.0000	.0000	.0000	.0000	.0000	.0001	.0001	.0003
	7	.0000	.0000	.0000	.0000	.0000	.0000	.0000	.0000	.0000
	8	.0000	.0000	.0000	.0000	.0000	.0000	.0000	.0000	.0000
	9	.0000	.0000	.0000	.0000	.0000	.0000	.0000	.0000	.0000
	10	.0000	.0000	.0000	.0000	.0000	.0000	.0000	.0000	.0000
	11	.0000	.0000	.0000	.0000	.0000	.0000	.0000	.0000	.0000
	12	.0000	.0000	.0000	.0000	.0000	.0000	.0000	.0000	.0000
15	0	.8601	.7386	.6333	.5421	.4633	.3953	.3367	.2863	.2430
	1	.1303	.2261	.2938	.3388	.3658	.3785	.3801	.3734	.3605
	2	.0092	.0323	.0636	.0988	.1348	.1691	.2003	.2273	.2496
	3	.0004	.0029	.0085	.0178	.0307	.0468	.0653	.0857	.1070
	4	.0000	.0002	.0008	.0022	.0049	.0090	.0148	.0223	.0317
	5	.0000	.0000	.0001	.0002	.0006	.0013	.0024	.0043	.0069
	6	.0000	.0000	.0000	.0000	.0000	.0001	.0003	.0006	.0011
	7	.0000	.0000	.0000	.0000	.0000	.0000	.0000	.0001	.0001
	8	.0000	.0000	.0000	.0000	.0000	.0000	.0000	.0000	.0000
	9	.0000	.0000	.0000	.0000	.0000	.0000	.0000	.0000	.0000
	10	.0000	.0000	.0000	.0000	.0000	.0000	.0000	.0000	.0000
	11	.0000	.0000	.0000	.0000	.0000	.0000	.0000	.0000	.0000
	12	.0000	.0000	.0000	.0000	.0000	.0000	.0000	.0000	.0000
	13	.0000	.0000	.0000	.0000	.0000	.0000	.0000	.0000	.0000
	14	.0000	.0000	.0000	.0000	.0000	.0000	.0000	.0000	.0000
	15	.0000	.0000	.0000	.0000	.0000	.0000	.0000	.0000	.0000

TABLE 5 BINOMIAL PROBABILITIES (*Continued*)

		p								
n	*x*	**.01**	**.02**	**.03**	**.04**	**.05**	**.06**	**.07**	**.08**	**.09**
18	0	.8345	.6951	.5780	.4796	.3972	.3283	.2708	.2229	.1831
	1	.1517	.2554	.3217	.3597	.3763	.3772	.3669	.3489	.3260
	2	.0130	.0443	.0846	.1274	.1683	.2047	.2348	.2579	.2741
	3	.0007	.0048	.0140	.0283	.0473	.0697	.0942	.1196	.1446
	4	.0000	.0004	.0016	.0044	.0093	.0167	.0266	.0390	.0536
	5	.0000	.0000	.0001	.0005	.0014	.0030	.0056	.0095	.0148
	6	.0000	.0000	.0000	.0000	.0002	.0004	.0009	.0018	.0032
	7	.0000	.0000	.0000	.0000	.0000	.0000	.0001	.0003	.0005
	8	.0000	.0000	.0000	.0000	.0000	.0000	.0000	.0000	.0001
	9	.0000	.0000	.0000	.0000	.0000	.0000	.0000	.0000	.0000
	10	.0000	.0000	.0000	.0000	.0000	.0000	.0000	.0000	.0000
	11	.0000	.0000	.0000	.0000	.0000	.0000	.0000	.0000	.0000
	12	.0000	.0000	.0000	.0000	.0000	.0000	.0000	.0000	.0000
	13	.0000	.0000	.0000	.0000	.0000	.0000	.0000	.0000	.0000
	14	.0000	.0000	.0000	.0000	.0000	.0000	.0000	.0000	.0000
	15	.0000	.0000	.0000	.0000	.0000	.0000	.0000	.0000	.0000
	16	.0000	.0000	.0000	.0000	.0000	.0000	.0000	.0000	.0000
	17	.0000	.0000	.0000	.0000	.0000	.0000	.0000	.0000	.0000
	18	.0000	.0000	.0000	.0000	.0000	.0000	.0000	.0000	.0000
20	0	.8179	.6676	.5438	.4420	.3585	.2901	.2342	.1887	.1516
	1	.1652	.2725	.3364	.3683	.3774	.3703	.3526	.3282	.3000
	2	.0159	.0528	.0988	.1458	.1887	.2246	.2521	.2711	.2818
	3	.0010	.0065	.0183	.0364	.0596	.0860	.1139	.1414	.1672
	4	.0000	.0006	.0024	.0065	.0133	.0233	.0364	.0523	.0703
	5	.0000	.0000	.0002	.0009	.0022	.0048	.0088	.0145	.0222
	6	.0000	.0000	.0000	.0001	.0003	.0008	.0017	.0032	.0055
	7	.0000	.0000	.0000	.0000	.0000	.0001	.0002	.0005	.0011
	8	.0000	.0000	.0000	.0000	.0000	.0000	.0000	.0001	.0002
	9	.0000	.0000	.0000	.0000	.0000	.0000	.0000	.0000	.0000
	10	.0000	.0000	.0000	.0000	.0000	.0000	.0000	.0000	.0000
	11	.0000	.0000	.0000	.0000	.0000	.0000	.0000	.0000	.0000
	12	.0000	.0000	.0000	.0000	.0000	.0000	.0000	.0000	.0000
	13	.0000	.0000	.0000	.0000	.0000	.0000	.0000	.0000	.0000
	14	.0000	.0000	.0000	.0000	.0000	.0000	.0000	.0000	.0000
	15	.0000	.0000	.0000	.0000	.0000	.0000	.0000	.0000	.0000
	16	.0000	.0000	.0000	.0000	.0000	.0000	.0000	.0000	.0000
	17	.0000	.0000	.0000	.0000	.0000	.0000	.0000	.0000	.0000
	18	.0000	.0000	.0000	.0000	.0000	.0000	.0000	.0000	.0000
	19	.0000	.0000	.0000	.0000	.0000	.0000	.0000	.0000	.0000
	20	.0000	.0000	.0000	.0000	.0000	.0000	.0000	.0000	.0000

TABLE 5 BINOMIAL PROBABILITIES (*Continued*)

		p								
n	*x*	.10	.15	.20	.25	.30	.35	.40	.45	.50
2	0	.8100	.7225	.6400	.5625	.4900	.4225	.3600	.3025	.2500
	1	.1800	.2550	.3200	.3750	.4200	.4550	.4800	.4950	.5000
	2	.0100	.0225	.0400	.0625	.0900	.1225	.1600	.2025	.2500
3	0	.7290	.6141	.5120	.4219	.3430	.2746	.2160	.1664	.1250
	1	.2430	.3251	.3840	.4219	.4410	.4436	.4320	.4084	.3750
	2	.0270	.0574	.0960	.1406	.1890	.2389	.2880	.3341	.3750
	3	.0010	.0034	.0080	.0156	.0270	.0429	.0640	.0911	.1250
4	0	.6561	.5220	.4096	.3164	.2401	.1785	.1296	.0915	.0625
	1	.2916	.3685	.4096	.4219	.4116	.3845	.3456	.2995	.2500
	2	.0486	.0975	.1536	.2109	.2646	.3105	.3456	.3675	.3750
	3	.0036	.0115	.0256	.0469	.0756	.1115	.1536	.2005	.2500
	4	.0001	.0005	.0016	.0039	.0081	.0150	.0256	.0410	.0625
5	0	.5905	.4437	.3277	.2373	.1681	.1160	.0778	.0503	.0312
	1	.3280	.3915	.4096	.3955	.3602	.3124	.2592	.2059	.1562
	2	.0729	.1382	.2048	.2637	.3087	.3364	.3456	.3369	.3125
	3	.0081	.0244	.0512	.0879	.1323	.1811	.2304	.2757	.3125
	4	.0004	.0022	.0064	.0146	.0284	.0488	.0768	.1128	.1562
	5	.0000	.0001	.0003	.0010	.0024	.0053	.0102	.0185	.0312
6	0	.5314	.3771	.2621	.1780	.1176	.0754	.0467	.0277	.0156
	1	.3543	.3993	.3932	.3560	.3025	.2437	.1866	.1359	.0938
	2	.0984	.1762	.2458	.2966	.3241	.3280	.3110	.2780	.2344
	3	.0146	.0415	.0819	.1318	.1852	.2355	.2765	.3032	.3125
	4	.0012	.0055	.0154	.0330	.0595	.0951	.1382	.1861	.2344
	5	.0001	.0004	.0015	.0044	.0102	.0205	.0369	.0609	.0938
	6	.0000	.0000	.0001	.0002	.0007	.0018	.0041	.0083	.0156
7	0	.4783	.3206	.2097	.1335	.0824	.0490	.0280	.0152	.0078
	1	.3720	.3960	.3670	.3115	.2471	.1848	.1306	.0872	.0547
	2	.1240	.2097	.2753	.3115	.3177	.2985	.2613	.2140	.1641
	3	.0230	.0617	.1147	.1730	.2269	.2679	.2903	.2918	.2734
	4	.0026	.0109	.0287	.0577	.0972	.1442	.1935	.2388	.2734
	5	.0002	.0012	.0043	.0115	.0250	.0466	.0774	.1172	.1641
	6	.0000	.0001	.0004	.0013	.0036	.0084	.0172	.0320	.0547
	7	.0000	.0000	.0000	.0001	.0002	.0006	.0016	.0037	.0078
8	0	.4305	.2725	.1678	.1001	.0576	.0319	.0168	.0084	.0039
	1	.3826	.3847	.3355	.2670	.1977	.1373	.0896	.0548	.0312
	2	.1488	.2376	.2936	.3115	.2965	.2587	.2090	.1569	.1094
	3	.0331	.0839	.1468	.2076	.2541	.2786	.2787	.2568	.2188
	4	.0046	.0185	.0459	.0865	.1361	.1875	.2322	.2627	.2734
	5	.0004	.0026	.0092	.0231	.0467	.0808	.1239	.1719	.2188
	6	.0000	.0002	.0011	.0038	.0100	.0217	.0413	.0703	.1094
	7	.0000	.0000	.0001	.0004	.0012	.0033	.0079	.0164	.0313
	8	.0000	.0000	.0000	.0000	.0001	.0002	.0007	.0017	.0039

TABLE 5 BINOMIAL PROBABILITIES (*Continued*)

		p								
n	*x*	.10	.15	.20	.25	.30	.35	.40	.45	.50
9	0	.3874	.2316	.1342	.0751	.0404	.0207	.0101	.0046	.0020
	1	.3874	.3679	.3020	.2253	.1556	.1004	.0605	.0339	.0176
	2	.1722	.2597	.3020	.3003	.2668	.2162	.1612	.1110	.0703
	3	.0446	.1069	.1762	.2336	.2668	.2716	.2508	.2119	.1641
	4	.0074	.0283	.0661	.1168	.1715	.2194	.2508	.2600	.2461
	5	.0008	.0050	.0165	.0389	.0735	.1181	.1672	.2128	.2461
	6	.0001	.0006	.0028	.0087	.0210	.0424	.0743	.1160	.1641
	7	.0000	.0000	.0003	.0012	.0039	.0098	.0212	.0407	.0703
	8	.0000	.0000	.0000	.0001	.0004	.0013	.0035	.0083	.0176
	9	.0000	.0000	.0000	.0000	.0000	.0001	.0003	.0008	.0020
10	0	.3487	.1969	.1074	.0563	.0282	.0135	.0060	.0025	.0010
	1	.3874	.3474	.2684	.1877	.1211	.0725	.0403	.0207	.0098
	2	.1937	.2759	.3020	.2816	.2335	.1757	.1209	.0763	.0439
	3	.0574	.1298	.2013	.2503	.2668	.2522	.2150	.1665	.1172
	4	.0112	.0401	.0881	.1460	.2001	.2377	.2508	.2384	.2051
	5	.0015	.0085	.0264	.0584	.1029	.1536	.2007	.2340	.2461
	6	.0001	.0012	.0055	.0162	.0368	.0689	.1115	.1596	.2051
	7	.0000	.0001	.0008	.0031	.0090	.0212	.0425	.0746	.1172
	8	.0000	.0000	.0001	.0004	.0014	.0043	.0106	.0229	.0439
	9	.0000	.0000	.0000	.0000	.0001	.0005	.0016	.0042	.0098
	10	.0000	.0000	.0000	.0000	.0000	.0000	.0001	.0003	.0010
12	0	.2824	.1422	.0687	.0317	.0138	.0057	.0022	.0008	.0002
	1	.3766	.3012	.2062	.1267	.0712	.0368	.0174	.0075	.0029
	2	.2301	.2924	.2835	.2323	.1678	.1088	.0639	.0339	.0161
	3	.0853	.1720	.2362	.2581	.2397	.1954	.1419	.0923	.0537
	4	.0213	.0683	.1329	.1936	.2311	.2367	.2128	.1700	.1208
	5	.0038	.0193	.0532	.1032	.1585	.2039	.2270	.2225	.1934
	6	.0005	.0040	.0155	.0401	.0792	.1281	.1766	.2124	.2256
	7	.0000	.0006	.0033	.0115	.0291	.0591	.1009	.1489	.1934
	8	.0000	.0001	.0005	.0024	.0078	.0199	.0420	.0762	.1208
	9	.0000	.0000	.0001	.0004	.0015	.0048	.0125	.0277	.0537
	10	.0000	.0000	.0000	.0000	.0002	.0008	.0025	.0068	.0161
	11	.0000	.0000	.0000	.0000	.0000	.0001	.0003	.0010	.0029
	12	.0000	.0000	.0000	.0000	.0000	.0000	.0000	.0001	.0002
15	0	.2059	.0874	.0352	.0134	.0047	.0016	.0005	.0001	.0000
	1	.3432	.2312	.1319	.0668	.0305	.0126	.0047	.0016	.0005
	2	.2669	.2856	.2309	.1559	.0916	.0476	.0219	.0090	.0032
	3	.1285	.2184	.2501	.2252	.1700	.1110	.0634	.0318	.0139
	4	.0428	.1156	.1876	.2252	.2186	.1792	.1268	.0780	.0417
	5	.0105	.0449	.1032	.1651	.2061	.2123	.1859	.1404	.0916
	6	.0019	.0132	.0430	.0917	.1472	.1906	.2066	.1914	.1527
	7	.0003	.0030	.0138	.0393	.0811	.1319	.1771	.2013	.1964
	8	.0000	.0005	.0035	.0131	.0348	.0710	.1181	.1647	.1964
	9	.0000	.0001	.0007	.0034	.0016	.0298	.0612	.1048	.1527
	10	.0000	.0000	.0001	.0007	.0030	.0096	.0245	.0515	.0916
	11	.0000	.0000	.0000	.0001	.0006	.0024	.0074	.0191	.0417
	12	.0000	.0000	.0000	.0000	.0001	.0004	.0016	.0052	.0139
	13	.0000	.0000	.0000	.0000	.0000	.0001	.0003	.0010	.0032
	14	.0000	.0000	.0000	.0000	.0000	.0000	.0000	.0001	.0005
	15	.0000	.0000	.0000	.0000	.0000	.0000	.0000	.0000	.0000

TABLE 5 BINOMIAL PROBABILITIES (*Continued*)

		p								
n	*x*	**.10**	**.15**	**.20**	**.25**	**.30**	**.35**	**.40**	**.45**	**.50**
18	0	.1501	.0536	.0180	.0056	.0016	.0004	.0001	.0000	.0000
	1	.3002	.1704	.0811	.0338	.0126	.0042	.0012	.0003	.0001
	2	.2835	.2556	.1723	.0958	.0458	.0190	.0069	.0022	.0006
	3	.1680	.2406	.2297	.1704	.1046	.0547	.0246	.0095	.0031
	4	.0700	.1592	.2153	.2130	.1681	.1104	.0614	.0291	.0117
	5	.0218	.0787	.1507	.1988	.2017	.1664	.1146	.0666	.0327
	6	.0052	.0301	.0816	.1436	.1873	.1941	.1655	.1181	.0708
	7	.0010	.0091	.0350	.0820	.1376	.1792	.1892	.1657	.1214
	8	.0002	.0022	.0120	.0376	.0811	.1327	.1734	.1864	.1669
	9	.0000	.0004	.0033	.0139	.0386	.0794	.1284	.1694	.1855
	10	.0000	.0001	.0008	.0042	.0149	.0385	.0771	.1248	.1669
	11	.0000	.0000	.0001	.0010	.0046	.0151	.0374	.0742	.1214
	12	.0000	.0000	.0000	.0002	.0012	.0047	.0145	.0354	.0708
	13	.0000	.0000	.0000	.0000	.0002	.0012	.0045	.0134	.0327
	14	.0000	.0000	.0000	.0000	.0000	.0002	.0011	.0039	.0117
	15	.0000	.0000	.0000	.0000	.0000	.0000	.0002	.0009	.0031
	16	.0000	.0000	.0000	.0000	.0000	.0000	.0000	.0001	.0006
	17	.0000	.0000	.0000	.0000	.0000	.0000	.0000	.0000	.0001
	18	.0000	.0000	.0000	.0000	.0000	.0000	.0000	.0000	.0000
20	0	.1216	.0388	.0115	.0032	.0008	.0002	.0000	.0000	.0000
	1	.2702	.1368	.0576	.0211	.0068	.0020	.0005	.0001	.0000
	2	.2852	.2293	.1369	.0669	.0278	.0100	.0031	.0008	.0002
	3	.1901	.2428	.2054	.1339	.0716	.0323	.0123	.0040	.0011
	4	.0898	.1821	.2182	.1897	.1304	.0738	.0350	.0139	.0046
	5	.0319	.1028	.1746	.2023	.1789	.1272	.0746	.0365	.0148
	6	.0089	.0454	.1091	.1686	.1916	.1712	.1244	.0746	.0370
	7	.0020	.0160	.0545	.1124	.1643	.1844	.1659	.1221	.0739
	8	.0004	.0046	.0222	.0609	.1144	.1614	.1797	.1623	.1201
	9	.0001	.0011	.0074	.0271	.0654	.1158	.1597	.1771	.1602
	10	.0000	.0002	.0020	.0099	.0308	.0686	.1171	.1593	.1762
	11	.0000	.0000	.0005	.0030	.0120	.0336	.0710	.1185	.1602
	12	.0000	.0000	.0001	.0008	.0039	.0136	.0355	.0727	.1201
	13	.0000	.0000	.0000	.0002	.0010	.0045	.0146	.0366	.0739
	14	.0000	.0000	.0000	.0000	.0002	.0012	.0049	.0150	.0370
	15	.0000	.0000	.0000	.0000	.0000	.0003	.0013	.0049	.0148
	16	.0000	.0000	.0000	.0000	.0000	.0000	.0003	.0013	.0046
	17	.0000	.0000	.0000	.0000	.0000	.0000	.0000	.0002	.0011
	18	.0000	.0000	.0000	.0000	.0000	.0000	.0000	.0000	.0002
	19	.0000	.0000	.0000	.0000	.0000	.0000	.0000	.0000	.0000
	20	.0000	.0000	.0000	.0000	.0000	.0000	.0000	.0000	.0000

TABLE 5 BINOMIAL PROBABILITIES (*Continued*)

		p								
n	*x*	.55	.60	.65	.70	.75	.80	.85	.90	.95
2	0	.2025	.1600	.1225	.0900	.0625	.0400	.0225	.0100	.0025
	1	.4950	.4800	.4550	.4200	.3750	.3200	.2550	.1800	.0950
	2	.3025	.3600	.4225	.4900	.5625	.6400	.7225	.8100	.9025
3	0	.0911	.0640	.0429	.0270	.0156	.0080	.0034	.0010	.0001
	1	.3341	.2880	.2389	.1890	.1406	.0960	.0574	.0270	.0071
	2	.4084	.4320	.4436	.4410	.4219	.3840	.3251	.2430	.1354
	3	.1664	.2160	.2746	.3430	.4219	.5120	.6141	.7290	.8574
4	0	.0410	.0256	.0150	.0081	.0039	.0016	.0005	.0001	.0000
	1	.2005	.1536	.1115	.0756	.0469	.0256	.0115	.0036	.0005
	2	.3675	.3456	.3105	.2646	.2109	.1536	.0975	.0486	.0135
	3	.2995	.3456	.3845	.4116	.4219	.4096	.3685	.2916	.1715
	4	.0915	.1296	.1785	.2401	.3164	.4096	.5220	.6561	.8145
5	0	.0185	.0102	.0053	.0024	.0010	.0003	.0001	.0000	.0000
	1	.1128	.0768	.0488	.0284	.0146	.0064	.0022	.0005	.0000
	2	.2757	.2304	.1811	.1323	.0879	.0512	.0244	.0081	.0011
	3	.3369	.3456	.3364	.3087	.2637	.2048	.1382	.0729	.0214
	4	.2059	.2592	.3124	.3601	.3955	.4096	.3915	.3281	.2036
	5	.0503	.0778	.1160	.1681	.2373	.3277	.4437	.5905	.7738
6	0	.0083	.0041	.0018	.0007	.0002	.0001	.0000	.0000	.0000
	1	.0609	.0369	.0205	.0102	.0044	.0015	.0004	.0001	.0000
	2	.1861	.1382	.0951	.0595	.0330	.0154	.0055	.0012	.0001
	3	.3032	.2765	.2355	.1852	.1318	.0819	.0415	.0146	.0021
	4	.2780	.3110	.3280	.3241	.2966	.2458	.1762	.0984	.0305
	5	.1359	.1866	.2437	.3025	.3560	.3932	.3993	.3543	.2321
	6	.0277	.0467	.0754	.1176	.1780	.2621	.3771	.5314	.7351
7	0	.0037	.0016	.0006	.0002	.0001	.0000	.0000	.0000	.0000
	1	.0320	.0172	.0084	.0036	.0013	.0004	.0001	.0000	.0000
	2	.1172	.0774	.0466	.0250	.0115	.0043	.0012	.0002	.0000
	3	.2388	.1935	.1442	.0972	.0577	.0287	.0109	.0026	.0002
	4	.2918	.2903	.2679	.2269	.1730	.1147	.0617	.0230	.0036
	5	.2140	.2613	.2985	.3177	.3115	.2753	.2097	.1240	.0406
	6	.0872	.1306	.1848	.2471	.3115	.3670	.3960	.3720	.2573
	7	.0152	.0280	.0490	.0824	.1335	.2097	.3206	.4783	.6983
8	0	.0017	.0007	.0002	.0001	.0000	.0000	.0000	.0000	.0000
	1	.0164	.0079	.0033	.0012	.0004	.0001	.0000	.0000	.0000
	2	.0703	.0413	.0217	.0100	.0038	.0011	.0002	.0000	.0000
	3	.1719	.1239	.0808	.0467	.0231	.0092	.0026	.0004	.0000
	4	.2627	.2322	.1875	.1361	.0865	.0459	.0185	.0046	.0004
	5	.2568	.2787	.2786	.2541	.2076	.1468	.0839	.0331	.0054
	6	.1569	.2090	.2587	.2965	.3115	.2936	.2376	.1488	.0515
	7	.0548	.0896	.1373	.1977	.2670	.3355	.3847	.3826	.2793
	8	.0084	.0168	.0319	.0576	.1001	.1678	.2725	.4305	.6634

TABLE 5 BINOMIAL PROBABILITIES (*Continued*)

		p								
n	*x*	**.55**	**.60**	**.65**	**.70**	**.75**	**.80**	**.85**	**.90**	**.95**
9	0	.0008	.0003	.0001	.0000	.0000	.0000	.0000	.0000	.0000
	1	.0083	.0035	.0013	.0004	.0001	.0000	.0000	.0000	.0000
	2	.0407	.0212	.0098	.0039	.0012	.0003	.0000	.0000	.0000
	3	.1160	.0743	.0424	.0210	.0087	.0028	.0006	.0001	.0000
	4	.2128	.1672	.1181	.0735	.0389	.0165	.0050	.0008	.0000
	5	.2600	.2508	.2194	.1715	.1168	.0661	.0283	.0074	.0006
	6	.2119	.2508	.2716	.2668	.2336	.1762	.1069	.0446	.0077
	7	.1110	.1612	.2162	.2668	.3003	.3020	.2597	.1722	.0629
	8	.0339	.0605	.1004	.1556	.2253	.3020	.3679	.3874	.2985
	9	.0046	.0101	.0207	.0404	.0751	.1342	.2316	.3874	.6302
10	0	.0003	.0001	.0000	.0000	.0000	.0000	.0000	.0000	.0000
	1	.0042	.0016	.0005	.0001	.0000	.0000	.0000	.0000	.0000
	2	.0229	.0106	.0043	.0014	.0004	.0001	.0000	.0000	.0000
	3	.0746	.0425	.0212	.0090	.0031	.0008	.0001	.0000	.0000
	4	.1596	.1115	.0689	.0368	.0162	.0055	.0012	.0001	.0000
	5	.2340	.2007	.1536	.1029	.0584	.0264	.0085	.0015	.0001
	6	.2384	.2508	.2377	.2001	.1460	.0881	.0401	.0112	.0010
	7	.1665	.2150	.2522	.2668	.2503	.2013	.1298	.0574	.0105
	8	.0763	.1209	.1757	.2335	.2816	.3020	.2759	.1937	.0746
	9	.0207	.0403	.0725	.1211	.1877	.2684	.3474	.3874	.3151
	10	.0025	.0060	.0135	.0282	.0563	.1074	.1969	.3487	.5987
12	0	.0001	.0000	.0000	.0000	.0000	.0000	.0000	.0000	.0000
	1	.0010	.0003	.0001	.0000	.0000	.0000	.0000	.0000	.0000
	2	.0068	.0025	.0008	.0002	.0000	.0000	.0000	.0000	.0000
	3	.0277	.0125	.0048	.0015	.0004	.0001	.0000	.0000	.0000
	4	.0762	.0420	.0199	.0078	.0024	.0005	.0001	.0000	.0000
	5	.1489	.1009	.0591	.0291	.0115	.0033	.0006	.0000	.0000
	6	.2124	.1766	.1281	.0792	.0401	.0155	.0040	.0005	.0000
	7	.2225	.2270	.2039	.1585	.1032	.0532	.0193	.0038	.0002
	8	.1700	.2128	.2367	.2311	.1936	.1329	.0683	.0213	.0021
	9	.0923	.1419	.1954	.2397	.2581	.2362	.1720	.0852	.0173
	10	.0339	.0639	.1088	.1678	.2323	.2835	.2924	.2301	.0988
	11	.0075	.0174	.0368	.0712	.1267	.2062	.3012	.3766	.3413
	12	.0008	.0022	.0057	.0138	.0317	.0687	.1422	.2824	.5404
15	0	.0000	.0000	.0000	.0000	.0000	.0000	.0000	.0000	.0000
	1	.0001	.0000	.0000	.0000	.0000	.0000	.0000	.0000	.0000
	2	.0010	.0003	.0001	.0000	.0000	.0000	.0000	.0000	.0000
	3	.0052	.0016	.0004	.0001	.0000	.0000	.0000	.0000	.0000
	4	.0191	.0074	.0024	.0006	.0001	.0000	.0000	.0000	.0000
	5	.0515	.0245	.0096	.0030	.0007	.0001	.0000	.0000	.0000
	6	.1048	.0612	.0298	.0116	.0034	.0007	.0001	.0000	.0000
	7	.1647	.1181	.0710	.0348	.0131	.0035	.0005	.0000	.0000
	8	.2013	.1771	.1319	.0811	.0393	.0138	.0030	.0003	.0000
	9	.1914	.2066	.1906	.1472	.0917	.0430	.0132	.0019	.0000
	10	.1404	.1859	.2123	.2061	.1651	.1032	.0449	.0105	.0006
	11	.0780	.1268	.1792	.2186	.2252	.1876	.1156	.0428	.0049

TABLE 5 BINOMIAL PROBABILITIES (*Continued*)

		p								
n	*x*	.55	.60	.65	.70	.75	.80	.85	.90	.95
	12	.0318	.0634	.1110	.1700	.2252	.2501	.2184	.1285	.0307
	13	.0090	.0219	.0476	.0916	.1559	.2309	.2856	.2669	.1348
	14	.0016	.0047	.0126	.0305	.0668	.1319	.2312	.3432	.3658
	15	.0001	.0005	.0016	.0047	.0134	.0352	.0874	.2059	.4633
18	0	.0000	.0000	.0000	.0000	.0000	.0000	.0000	.0000	.0000
	1	.0000	.0000	.0000	.0000	.0000	.0000	.0000	.0000	.0000
	2	.0001	.0000	.0000	.0000	.0000	.0000	.0000	.0000	.0000
	3	.0009	.0002	.0000	.0000	.0000	.0000	.0000	.0000	.0000
	4	.0039	.0011	.0002	.0000	.0000	.0000	.0000	.0000	.0000
	5	.0134	.0045	.0012	.0002	.0000	.0000	.0000	.0000	.0000
	6	.0354	.0145	.0047	.0012	.0002	.0000	.0000	.0000	.0000
	7	.0742	.0374	.0151	.0046	.0010	.0001	.0000	.0000	.0000
	8	.1248	.0771	.0385	.0149	.0042	.0008	.0001	.0000	.0000
	9	.1694	.1284	.0794	.0386	.0139	.0033	.0004	.0000	.0000
	10	.1864	.1734	.1327	.0811	.0376	.0120	.0022	.0002	.0000
	11	.1657	.1892	.1792	.1376	.0820	.0350	.0091	.0010	.0000
	12	.1181	.1655	.1941	.1873	.1436	.0816	.0301	.0052	.0002
	13	.0666	.1146	.1664	.2017	.1988	.1507	.0787	.0218	.0014
	14	.0291	.0614	.1104	.1681	.2130	.2153	.1592	.0700	.0093
	15	.0095	.0246	.0547	.1046	.1704	.2297	.2406	.1680	.0473
	16	.0022	.0069	.0190	.0458	.0958	.1723	.2556	.2835	.1683
	17	.0003	.0012	.0042	.0126	.0338	.0811	.1704	.3002	.3763
	18	.0000	.0001	.0004	.0016	.0056	.0180	.0536	.1501	.3972
20	0	.0000	.0000	.0000	.0000	.0000	.0000	.0000	.0000	.0000
	1	.0000	.0000	.0000	.0000	.0000	.0000	.0000	.0000	.0000
	2	.0000	.0000	.0000	.0000	.0000	.0000	.0000	.0000	.0000
	3	.0002	.0000	.0000	.0000	.0000	.0000	.0000	.0000	.0000
	4	.0013	.0003	.0000	.0000	.0000	.0000	.0000	.0000	.0000
	5	.0049	.0013	.0003	.0000	.0000	.0000	.0000	.0000	.0000
	6	.0150	.0049	.0012	.0002	.0000	.0000	.0000	.0000	.0000
	7	.0366	.0146	.0045	.0010	.0002	.0000	.0000	.0000	.0000
	8	.0727	.0355	.0136	.0039	.0008	.0001	.0000	.0000	.0000
	9	.1185	.0710	.0336	.0120	.0030	.0005	.0000	.0000	.0000
	10	.1593	.1171	.0686	.0308	.0099	.0020	.0002	.0000	.0000
	11	.1771	.1597	.1158	.0654	.0271	.0074	.0011	.0001	.0000
	12	.1623	.1797	.1614	.1144	.0609	.0222	.0046	.0004	.0000
	13	.1221	.1659	.1844	.1643	.1124	.0545	.0160	.0020	.0000
	14	.0746	.1244	.1712	.1916	.1686	.1091	.0454	.0089	.0003
	15	.0365	.0746	.1272	.1789	.2023	.1746	.1028	.0319	.0022
	16	.0139	.0350	.0738	.1304	.1897	.2182	.1821	.0898	.0133
	17	.0040	.0123	.0323	.0716	.1339	.2054	.2428	.1901	.0596
	18	.0008	.0031	.0100	.0278	.0669	.1369	.2293	.2852	.1887
	19	.0001	.0005	.0020	.0068	.0211	.0576	.1368	.2702	.3774
	20	.0000	.0000	.0002	.0008	.0032	.0115	.0388	.1216	.3585

TABLE 6 VALUES OF $e^{-\mu}$

μ	$e^{-\mu}$	μ	$e^{-\mu}$	μ	$e^{-\mu}$
.00	1.0000	2.00	.1353	4.00	.0183
.05	.9512	2.05	.1287	4.05	.0174
.10	.9048	2.10	.1225	4.10	.0166
.15	.8607	2.15	.1165	4.15	.0158
.20	.8187	2.20	.1108	4.20	.0150
.25	.7788	2.25	.1054	4.25	.0143
.30	.7408	2.30	.1003	4.30	.0136
.35	.7047	2.35	.0954	4.35	.0129
.40	.6703	2.40	.0907	4.40	.0123
.45	.6376	2.45	.0863	4.45	.0117
.50	.6065	2.50	.0821	4.50	.0111
.55	.5769	2.55	.0781	4.55	.0106
.60	.5488	2.60	.0743	4.60	.0101
.65	.5220	2.65	.0707	4.65	.0096
.70	.4966	2.70	.0672	4.70	.0091
.75	.4724	2.75	.0639	4.75	.0087
.80	.4493	2.80	.0608	4.80	.0082
.85	.4274	2.85	.0578	4.85	.0078
.90	.4066	2.90	.0550	4.90	.0074
.95	.3867	2.95	.0523	4.95	.0071
1.00	.3679	3.00	.0498	5.00	.0067
1.05	.3499	3.05	.0474	6.00	.0025
1.10	.3329	3.10	.0450	7.00	.0009
1.15	.3166	3.15	.0429	8.00	.000335
1.20	.3012	3.20	.0408	9.00	.000123
				10.00	.000045
1.25	.2865	3.25	.0388		
1.30	.2725	3.30	.0369		
1.35	.2592	3.35	.0351		
1.40	.2466	3.40	.0334		
1.45	.2346	3.45	.0317		
1.50	.2231	3.50	.0302		
1.55	.2122	3.55	.0287		
1.60	.2019	3.60	.0273		
1.65	.1920	3.65	.0260		
1.70	.1827	3.70	.0247		
1.75	.1738	3.75	.0235		
1.80	.1653	3.80	.0224		
1.85	.1572	3.85	.0213		
1.90	.1496	3.90	.0202		
1.95	.1423	3.95	.0193		

TABLE 7 POISSON PROBABILITIES

Entries in the table give the probability of x occurrences for a Poisson process with a mean μ. For example, when $\mu = 2.5$, the probability of four occurrences is .1336.

	μ									
x	**0.1**	**0.2**	**0.3**	**0.4**	**0.5**	**0.6**	**0.7**	**0.8**	**0.9**	**1.0**
0	.9048	.8187	.7408	.6703	.6065	.5488	.4966	.4493	.4066	.3679
1	.0905	.1637	.2222	.2681	.3033	.3293	.3476	.3595	.3659	.3679
2	.0045	.0164	.0333	.0536	.0758	.0988	.1217	.1438	.1647	.1839
3	.0002	.0011	.0033	.0072	.0126	.0198	.0284	.0383	.0494	.0613
4	.0000	.0001	.0002	.0007	.0016	.0030	.0050	.0077	.0111	.0153
5	.0000	.0000	.0000	.0001	.0002	.0004	.0007	.0012	.0020	.0031
6	.0000	.0000	.0000	.0000	.0000	.0000	.0001	.0002	.0003	.0005
7	.0000	.0000	.0000	.0000	.0000	.0000	.0000	.0000	.0000	.0001

	μ									
x	**1.1**	**1.2**	**1.3**	**1.4**	**1.5**	**1.6**	**1.7**	**1.8**	**1.9**	**2.0**
0	.3329	.3012	.2725	.2466	.2231	.2019	.1827	.1653	.1496	.1353
1	.3662	.3614	.3543	.3452	.3347	.3230	.3106	.2975	.2842	.2707
2	.2014	.2169	.2303	.2417	.2510	.2584	.2640	.2678	.2700	.2707
3	.0738	.0867	.0998	.1128	.1255	.1378	.1496	.1607	.1710	.1804
4	.0203	.0260	.0324	.0395	.0471	.0551	.0636	.0723	.0812	.0902
5	.0045	.0062	.0084	.0111	.0141	.0176	.0216	.0260	.0309	.0361
6	.0008	.0012	.0018	.0026	.0035	.0047	.0061	.0078	.0098	.0120
7	.0001	.0002	.0003	.0005	.0008	.0011	.0015	.0020	.0027	.0034
8	.0000	.0000	.0001	.0001	.0001	.0002	.0003	.0005	.0006	.0009
9	.0000	.0000	.0000	.0000	.0000	.0000	.0001	.0001	.0001	.0002

	μ									
x	**2.1**	**2.2**	**2.3**	**2.4**	**2.5**	**2.6**	**2.7**	**2.8**	**2.9**	**3.0**
0	.1225	.1108	.1003	.0907	.0821	.0743	.0672	.0608	.0550	.0498
1	.2572	.2438	.2306	.2177	.2052	.1931	.1815	.1703	.1596	.1494
2	.2700	.2681	.2652	.2613	.2565	.2510	.2450	.2384	.2314	.2240
3	.1890	.1966	.2033	.2090	.2138	.2176	.2205	.2225	.2237	.2240
4	.0992	.1082	.1169	.1254	.1336	.1414	.1488	.1557	.1622	.1680
5	.0417	.0476	.0538	.0602	.0668	.0735	.0804	.0872	.0940	.1008
6	.0146	.0174	.0206	.0241	.0278	.0319	.0362	.0407	.0455	.0504
7	.0044	.0055	.0068	.0083	.0099	.0118	.0139	.0163	.0188	.0216
8	.0011	.0015	.0019	.0025	.0031	.0038	.0047	.0057	.0068	.0081
9	.0003	.0004	.0005	.0007	.0009	.0011	.0014	.0018	.0022	.0027
10	.0001	.0001	.0001	.0002	.0002	.0003	.0004	.0005	.0006	.0008
11	.0000	.0000	.0000	.0000	.0000	.0001	.0001	.0001	.0002	.0002
12	.0000	.0000	.0000	.0000	.0000	.0000	.0000	.0000	.0000	.0001

TABLE 7 POISSON PROBABILITIES (*Continued*)

	μ									
x	3.1	3.2	3.3	3.4	3.5	3.6	3.7	3.8	3.9	4.0
0	.0450	.0408	.0369	.0344	.0302	.0273	.0247	.0224	.0202	.0183
1	.1397	.1304	.1217	.1135	.1057	.0984	.0915	.0850	.0789	.0733
2	.2165	.2087	.2008	.1929	.1850	.1771	.1692	.1615	.1539	.1465
3	.2237	.2226	.2209	.2186	.2158	.2125	.2087	.2046	.2001	.1954
4	.1734	.1781	.1823	.1858	.1888	.1912	.1931	.1944	.1951	.1954
5	.1075	.1140	.1203	.1264	.1322	.1377	.1429	.1477	.1522	.1563
6	.0555	.0608	.0662	.0716	.0771	.0826	.0881	.0936	.0989	.1042
7	.0246	.0278	.0312	.0348	.0385	.0425	.0466	.0508	.0551	.0595
8	.0095	.0111	.0129	.0148	.0169	.0191	.0215	.0241	.0269	.0298
9	.0033	.0040	.0047	.0056	.0066	.0076	.0089	.0102	.0116	.0132
10	.0010	.0013	.0016	.0019	.0023	.0028	.0033	.0039	.0045	.0053
11	.0003	.0004	.0005	.0006	.0007	.0009	.0011	.0013	.0016	.0019
12	.0001	.0001	.0001	.0002	.0002	.0003	.0003	.0004	.0005	.0006
13	.0000	.0000	.0000	.0000	.0001	.0001	.0001	.0001	.0002	.0002
14	.0000	.0000	.0000	.0000	.0000	.0000	.0000	.0000	.0000	.0001

	μ									
x	4.1	4.2	4.3	4.4	4.5	4.6	4.7	4.8	4.9	5.0
0	.0166	.0150	.0136	.0123	.0111	.0101	.0091	.0082	.0074	.0067
1	.0679	.0630	.0583	.0540	.0500	.0462	.0427	.0395	.0365	.0337
2	.1393	.1323	.1254	.1188	.1125	.1063	.1005	.0948	.0894	.0842
3	.1904	.1852	.1798	.1743	.1687	.1631	.1574	.1517	.1460	.1404
4	.1951	.1944	.1933	.1917	.1898	.1875	.1849	.1820	.1789	.1755
5	.1600	.1633	.1662	.1687	.1708	.1725	.1738	.1747	.1753	.1755
6	.1093	.1143	.1191	.1237	.1281	.1323	.1362	.1398	.1432	.1462
7	.0640	.0686	.0732	.0778	.0824	.0869	.0914	.0959	.1002	.1044
8	.0328	.0360	.0393	.0428	.0463	.0500	.0537	.0575	.0614	.0653
9	.0150	.0168	.0188	.0209	.0232	.0255	.0280	.0307	.0334	.0363
10	.0061	.0071	.0081	.0092	.0104	.0118	.0132	.0147	.0164	.0181
11	.0023	.0027	.0032	.0037	.0043	.0049	.0056	.0064	.0073	.0082
12	.0008	.0009	.0011	.0014	.0016	.0019	.0022	.0026	.0030	.0034
13	.0002	.0003	.0004	.0005	.0006	.0007	.0008	.0009	.0011	.0013
14	.0001	.0001	.0001	.0001	.0002	.0002	.0003	.0003	.0004	.0005
15	.0000	.0000	.0000	.0000	.0001	.0001	.0001	.0001	.0001	.0002

	μ									
x	5.1	5.2	5.3	5.4	5.5	5.6	5.7	5.8	5.9	6.0
0	.0061	.0055	.0050	.0045	.0041	.0037	.0033	.0030	.0027	.0025
1	.0311	.0287	.0265	.0244	.0225	.0207	.0191	.0176	.0162	.0149
2	.0793	.0746	.0701	.0659	.0618	.0580	.0544	.0509	.0477	.0446
3	.1348	.1293	.1239	.1185	.1133	.1082	.1033	.0985	.0938	.0892
4	.1719	.1681	.1641	.1600	.1558	.1515	.1472	.1428	.1383	.1339

TABLE 7 POISSON PROBABILITIES (*Continued*)

	μ									
x	**5.1**	**5.2**	**5.3**	**5.4**	**5.5**	**5.6**	**5.7**	**5.8**	**5.9**	**6.0**
5	.1753	.1748	.1740	.1728	.1714	.1697	.1678	.1656	.1632	.1606
6	.1490	.1515	.1537	.1555	.1571	.1587	.1594	.1601	.1605	.1606
7	.1086	.1125	.1163	.1200	.1234	.1267	.1298	.1326	.1353	.1377
8	.0692	.0731	.0771	.0810	.0849	.0887	.0925	.0962	.0998	.1033
9	.0392	.0423	.0454	.0486	.0519	.0552	.0586	.0620	.0654	.0688
10	.0200	.0220	.0241	.0262	.0285	.0309	.0334	.0359	.0386	.0413
11	.0093	.0104	.0116	.0129	.0143	.0157	.0173	.0190	.0207	.0225
12	.0039	.0045	.0051	.0058	.0065	.0073	.0082	.0092	.0102	.0113
13	.0015	.0018	.0021	.0024	.0028	.0032	.0036	.0041	.0046	.0052
14	.0006	.0007	.0008	.0009	.0011	.0013	.0015	.0017	.0019	.0022
15	.0002	.0002	.0003	.0003	.0004	.0005	.0006	.0007	.0008	.0009
16	.0001	.0001	.0001	.0001	.0001	.0002	.0002	.0002	.0003	.0003
17	.0000	.0000	.0000	.0000	.0000	.0001	.0001	.0001	.0001	.0001

	μ									
x	**6.1**	**6.2**	**6.3**	**6.4**	**6.5**	**6.6**	**6.7**	**6.8**	**6.9**	**7.0**
0	.0022	.0020	.0018	.0017	.0015	.0014	.0012	.0011	.0010	.0009
1	.0137	.0126	.0116	.0106	.0098	.0090	.0082	.0076	.0070	.0064
2	.0417	.0390	.0364	.0340	.0318	.0296	.0276	.0258	.0240	.0223
3	.0848	.0806	.0765	.0726	.0688	.0652	.0617	.0584	.0552	.0521
4	.1294	.1249	.1205	.1162	.1118	.1076	.1034	.0992	.0952	.0912
5	.1579	.1549	.1519	.1487	.1454	.1420	.1385	.1349	.1314	.1277
6	.1605	.1601	.1595	.1586	.1575	.1562	.1546	.1529	.1511	.1490
7	.1399	.1418	.1435	.1450	.1462	.1472	.1480	.1486	.1489	.1490
8	.1066	.1099	.1130	.1160	.1188	.1215	.1240	.1263	.1284	.1304
9	.0723	.0757	.0791	.0825	.0858	.0891	.0923	.0954	.0985	.1014
10	.0441	.0469	.0498	.0528	.0558	.0588	.0618	.0649	.0679	.0710
11	.0245	.0265	.0285	.0307	.0330	.0353	.0377	.0401	.0426	.0452
12	.0124	.0137	.0150	.0164	.0179	.0194	.0210	.0227	.0245	.0264
13	.0058	.0065	.0073	.0081	.0089	.0098	.0108	.0119	.0130	.0142
14	.0025	.0029	.0033	.0037	.0041	.0046	.0052	.0058	.0064	.0071
15	.0010	.0012	.0014	.0016	.0018	.0020	.0023	.0026	.0029	.0033
16	.0004	.0005	.0005	.0006	.0007	.0008	.0010	.0011	.0013	.0014
17	.0001	.0002	.0002	.0002	.0003	.0003	.0004	.0004	.0005	.0006
18	.0000	.0001	.0001	.0001	.0001	.0001	.0001	.0002	.0002	.0002
19	.0000	.0000	.0000	.0000	.0000	.0000	.0000	.0001	.0001	.0001

	μ									
x	**7.1**	**7.2**	**7.3**	**7.4**	**7.5**	**7.6**	**7.7**	**7.8**	**7.9**	**8.0**
0	.0008	.0007	.0007	.0006	.0006	.0005	.0005	.0004	.0004	.0003
1	.0059	.0054	.0049	.0045	.0041	.0038	.0035	.0032	.0029	.0027
2	.0208	.0194	.0180	.0167	.0156	.0145	.0134	.0125	.0116	.0107
3	.0492	.0464	.0438	.0413	.0389	.0366	.0345	.0324	.0305	.0286
4	.0874	.0836	.0799	.0764	.0729	.0696	.0663	.0632	.0602	.0573

TABLE 7 POISSON PROBABILITIES (*Continued*)

	μ									
x	**7.1**	**7.2**	**7.3**	**7.4**	**7.5**	**7.6**	**7.7**	**7.8**	**7.9**	**8.0**
5	.1241	.1204	.1167	.1130	.1094	.1057	.1021	.0986	.0951	.0916
6	.1468	.1445	.1420	.1394	.1367	.1339	.1311	.1282	.1252	.1221
7	.1489	.1486	.1481	.1474	.1465	.1454	.1442	.1428	.1413	.1396
8	.1321	.1337	.1351	.1363	.1373	.1382	.1388	.1392	.1395	.1396
9	.1042	.1070	.1096	.1121	.1144	.1167	.1187	.1207	.1224	.1241
10	.0740	.0770	.0800	.0829	.0858	.0887	.0914	.0941	.0967	.0993
11	.0478	.0504	.0531	.0558	.0585	.0613	.0640	.0667	.0695	.0722
12	.0283	.0303	.0323	.0344	.0366	.0388	.0411	.0434	.0457	.0481
13	.0154	.0168	.0181	.0196	.0211	.0227	.0243	.0260	.0278	.0296
14	.0078	.0086	.0095	.0104	.0113	.0123	.0134	.0145	.0157	.0169
15	.0037	.0041	.0046	.0051	.0057	.0062	.0069	.0075	.0083	.0090
16	.0016	.0019	.0021	.0024	.0026	.0030	.0033	.0037	.0041	.0045
17	.0007	.0008	.0009	.0010	.0012	.0013	.0015	.0017	.0019	.0021
18	.0003	.0003	.0004	.0004	.0005	.0006	.0006	.0007	.0008	.0009
19	.0001	.0001	.0001	.0002	.0002	.0002	.0003	.0003	.0003	.0004
20	.0000	.0000	.0001	.0001	.0001	.0001	.0001	.0001	.0001	.0002
21	.0000	.0000	.0000	.0000	.0000	.0000	.0000	.0000	.0001	.0001

	μ									
x	**8.1**	**8.2**	**8.3**	**8.4**	**8.5**	**8.6**	**8.7**	**8.8**	**8.9**	**9.0**
0	.0003	.0003	.0002	.0002	.0002	.0002	.0002	.0002	.0001	.0001
1	.0025	.0023	.0021	.0019	.0017	.0016	.0014	.0013	.0012	.0011
2	.0100	.0092	.0086	.0079	.0074	.0068	.0063	.0058	.0054	.0050
3	.0269	.0252	.0237	.0222	.0208	.0195	.0183	.0171	.0160	.0150
4	.0544	.0517	.0491	.0466	.0443	.0420	.0398	.0377	.0357	.0337
5	.0882	.0849	.0816	.0784	.0752	.0722	.0692	.0663	.0635	.0607
6	.1191	.1160	.1128	.1097	.1066	.1034	.1003	.0972	.0941	.0911
7	.1378	.1358	.1338	.1317	.1294	.1271	.1247	.1222	.1197	.1171
8	.1395	.1392	.1388	.1382	.1375	.1366	.1356	.1344	.1332	.1318
9	.1256	.1269	.1280	.1290	.1299	.1306	.1311	.1315	.1317	.1318
10	.1017	.1040	.1063	.1084	.1104	.1123	.1140	.1157	.1172	.1186
11	.0749	.0776	.0802	.0828	.0853	.0878	.0902	.0925	.0948	.0970
12	.0505	.0530	.0555	.0579	.0604	.0629	.0654	.0679	.0703	.0728
13	.0315	.0334	.0354	.0374	.0395	.0416	.0438	.0459	.0481	.0504
14	.0182	.0196	.0210	.0225	.0240	.0256	.0272	.0289	.0306	.0324
15	.0098	.0107	.0116	.0126	.0136	.0147	.0158	.0169	.0182	.1094
16	.0050	.0055	.0060	.0066	.0072	.0079	.0086	.0093	.0101	.0109
17	.0024	.0026	.0029	.0033	.0036	.0040	.0044	.0048	.0053	.0058
18	.0011	.0012	.0014	.0015	.0017	.0019	.0021	.0024	.0026	.0029
19	.0005	.0005	.0006	.0007	.0008	.0009	.0010	.0011	.0012	.0014
20	.0002	.0002	.0002	.0003	.0003	.0004	.0004	.0005	.0005	.0006
21	.0001	.0001	.0001	.0001	.0001	.0002	.0002	.0002	.0002	.0003
22	.0000	.0000	.0000	.0000	.0001	.0001	.0001	.0001	.0001	.0001

TABLE 7 POISSON PROBABILITIES (*Continued*)

x	μ 9.1	9.2	9.3	9.4	9.5	9.6	9.7	9.8	9.9	10
0	.0001	.0001	.0001	.0001	.0001	.0001	.0001	.0001	.0001	.0000
1	.0010	.0009	.0009	.0008	.0007	.0007	.0006	.0005	.0005	.0005
2	.0046	.0043	.0040	.0037	.0034	.0031	.0029	.0027	.0025	.0023
3	.0140	.0131	.0123	.0115	.0107	.0100	.0093	.0087	.0081	.0076
4	.0319	.0302	.0285	.0269	.0254	.0240	.0226	.0213	.0201	.0189
5	.0581	.0555	.0530	.0506	.0483	.0460	.0439	.0418	.0398	.0378
6	.0881	.0851	.0822	.0793	.0764	.0736	.0709	.0682	.0656	.0631
7	.1145	.1118	.1091	.1064	.1037	.1010	.0982	.0955	.0928	.0901
8	.1302	.1286	.1269	.1251	.1232	.1212	.1191	.1170	.1148	.1126
9	.1317	.1315	.1311	.1306	.1300	.1293	.1284	.1274	.1263	.1251
10	.1198	.1210	.1219	.1228	.1235	.1241	.1245	.1249	.1250	.1251
11	.0991	.1012	.1031	.1049	.1067	.1083	.1098	.1112	.1125	.1137
12	.0752	.0776	.0799	.0822	.0844	.0866	.0888	.0908	.0928	.0948
13	.0526	.0549	.0572	.0594	.0617	.0640	.0662	.0685	.0707	.0729
14	.0342	.0361	.0380	.0399	.0419	.0439	.0459	.0479	.0500	.0521
15	.0208	.0221	.0235	.0250	.0265	.0281	.0297	.0313	.0330	.0347
16	.0118	.0127	.0137	.0147	.0157	.0168	.0180	.0192	.0204	.0217
17	.0063	.0069	.0075	.0081	.0088	.0095	.0103	.0111	.0119	.0128
18	.0032	.0035	.0039	.0042	.0046	.0051	.0055	.0060	.0065	.0071
19	.0015	.0017	.0019	.0021	.0023	.0026	.0028	.0031	.0034	.0037
20	.0007	.0008	.0009	.0010	.0011	.0012	.0014	.0015	.0017	.0019
21	.0003	.0003	.0004	.0004	.0005	.0006	.0006	.0007	.0008	.0009
22	.0001	.0001	.0002	.0002	.0002	.0002	.0003	.0003	.0004	.0004
23	.0000	.0001	.0001	.0001	.0001	.0001	.0001	.0001	.0002	.0002
24	.0000	.0000	.0000	.0000	.0000	.0000	.0000	.0001	.0001	.0001

x	μ 11	12	13	14	15	16	17	18	19	20
0	.0000	.0000	.0000	.0000	.0000	.0000	.0000	.0000	.0000	.0000
1	.0002	.0001	.0000	.0000	.0000	.0000	.0000	.0000	.0000	.0000
2	.0010	.0004	.0002	.0001	.0000	.0000	.0000	.0000	.0000	.0000
3	.0037	.0018	.0008	.0004	.0002	.0001	.0000	.0000	.0000	.0000
4	.0102	.0053	.0027	.0013	.0006	.0003	.0001	.0001	.0000	.0000
5	.0224	.0127	.0070	.0037	.0019	.0010	.0005	.0002	.0001	.0001
6	.0411	.0255	.0152	.0087	.0048	.0026	.0014	.0007	.0004	.0002
7	.0646	.0437	.0281	.0174	.0104	.0060	.0034	.0018	.0010	.0005
8	.0888	.0655	.0457	.0304	.0194	.0120	.0072	.0042	.0024	.0013
9	.1085	.0874	.0661	.0473	.0324	.0213	.0135	.0083	.0050	.0029
10	.1194	.1048	.0859	.0663	.0486	.0341	.0230	.0150	.0095	.0058
11	.1194	.1144	.1015	.0844	.0663	.0496	.0355	.0245	.0164	.0106
12	.1094	.1144	.1099	.0984	.0829	.0661	.0504	.0368	.0259	.0176
13	.0926	.1056	.1099	.1060	.0956	.0814	.0658	.0509	.0378	.0271
14	.0728	.0905	.1021	.1060	.1024	.0930	.0800	.0655	.0514	.0387

TABLE 7 POISSON PROBABILITIES (*Continued*)

	μ									
x	**11**	**12**	**13**	**14**	**15**	**16**	**17**	**18**	**19**	**20**
15	.0534	.0724	.0885	.0989	.1024	.0992	.0906	.0786	.0650	.0516
16	.0367	.0543	.0719	.0866	.0960	.0992	.0963	.0884	.0772	.0646
17	.0237	.0383	.0550	.0713	.0847	.0934	.0963	.0936	.0863	.0760
18	.0145	.0256	.0397	.0554	.0706	.0830	.0909	.0936	.0911	.0844
19	.0084	.0161	.0272	.0409	.0557	.0699	.0814	.0887	.0911	.0888
20	.0046	.0097	.0177	.0286	.0418	.0559	.0692	.0798	.0866	.0888
21	.0024	.0055	.0109	.0191	.0299	.0426	.0560	.0684	.0783	.0846
22	.0012	.0030	.0065	.0121	.0204	.0310	.0433	.0560	.0676	.0769
23	.0006	.0016	.0037	.0074	.0133	.0216	.0320	.0438	.0559	.0669
24	.0003	.0008	.0020	.0043	.0083	.0144	.0226	.0328	.0442	.0557
25	.0001	.0004	.0010	.0024	.0050	.0092	.0154	.0237	.0336	.0446
26	.0000	.0002	.0005	.0013	.0029	.0057	.0101	.0164	.0246	.0343
27	.0000	.0001	.0002	.0007	.0016	.0034	.0063	.0109	.0173	.0254
28	.0000	.0000	.0001	.0003	.0009	.0019	.0038	.0070	.0117	.0181
29	.0000	.0000	.0001	.0002	.0004	.0011	.0023	.0044	.0077	.0125
30	.0000	.0000	.0000	.0001	.0002	.0006	.0013	.0026	.0049	.0083
31	.0000	.0000	.0000	.0000	.0001	.0003	.0007	.0015	.0030	.0054
32	.0000	.0000	.0000	.0000	.0001	.0001	.0004	.0009	.0018	.0034
33	.0000	.0000	.0000	.0000	.0000	.0001	.0002	.0005	.0010	.0020
34	.0000	.0000	.0000	.0000	.0000	.0000	.0001	.0002	.0006	.0012
35	.0000	.0000	.0000	.0000	.0000	.0000	.0000	.0001	.0003	.0007
36	.0000	.0000	.0000	.0000	.0000	.0000	.0000	.0001	.0002	.0004
37	.0000	.0000	.0000	.0000	.0000	.0000	.0000	.0000	.0001	.0002
38	.0000	.0000	.0000	.0000	.0000	.0000	.0000	.0000	.0000	.0001
39	.0000	.0000	.0000	.0000	.0000	.0000	.0000	.0000	.0000	.0001

Appendix C: Summation Notation

Summations

Definition

$$\sum_{i=1}^{n} x_i = x_1 + x_2 + \cdots + x_n \tag{C.1}$$

Example for $x_1 = 5, x_2 = 8, x_3 = 14$:

$$\begin{aligned}\sum_{i=1}^{3} x_i &= x_1 + x_2 + x_3 \\ &= 5 + 8 + 14 \\ &= 27\end{aligned}$$

Result 1

For a constant c:

$$\sum_{i=1}^{n} c = \underbrace{(c + c + \cdots + c)}_{n \text{ times}} = nc \tag{C.2}$$

Example for $c = 5, n = 10$:

$$\sum_{i=1}^{10} 5 = 10(5) = 50$$

Example for $c = \bar{x}$:

$$\sum_{i=1}^{n} \bar{x} = n\bar{x}$$

Result 2

$$\begin{aligned}\sum_{i=1}^{n} cx_i &= cx_1 + cx_2 + \cdots + cx_n \\ &= c(x_1 + x_2 + \cdots + x_n) = c\sum_{i=1}^{n} x_i\end{aligned} \tag{C.3}$$

Example for $x_1 = 5, x_2 = 8, x_3 = 14, c = 2$:

$$\sum_{i=1}^{3} 2x_i = 2\sum_{i=1}^{3} x_i = 2(27) = 54$$

Result 3

$$\sum_{i=1}^{n} (ax_i + by_i) = a\sum_{i=1}^{n} x_i + b\sum_{i=1}^{n} y_i \tag{C.4}$$

Example for $x_1 = 5, x_2 = 8, x_3 = 14, a = 2, y_1 = 7, y_2 = 3, y_3 = 8, b = 4$:

$$\begin{aligned}\sum_{i=1}^{3}(2x_i + 4y_i) &= 2\sum_{i=1}^{3}x_i + 4\sum_{i=1}^{3}y_i \\ &= 2(27) + 4(18) \\ &= 54 + 72 \\ &= 126\end{aligned}$$

Double Summations

Consider the following data involving the variable x_{ij}, where i is the subscript denoting the row position and j is the subscript denoting the column position:

		Column		
		1	**2**	**3**
Row	**1**	$x_{11} = 10$	$x_{12} = 8$	$x_{13} = 6$
	2	$x_{21} = 7$	$x_{22} = 4$	$x_{23} = 12$

Definition

$$\begin{aligned}\sum_{i=1}^{n}\sum_{j=1}^{m}x_{ij} = {} & (x_{11} + x_{12} + \cdots + x_{1m}) + (x_{21} + x_{22} + \cdots + x_{2m}) \\ & + (x_{31} + x_{32} + \cdots + x_{3m}) + \cdots + (x_{n1} + x_{n2} + \cdots + x_{nm})\end{aligned} \quad \textbf{(C.5)}$$

Example:

$$\begin{aligned}\sum_{i=1}^{2}\sum_{j=1}^{3}x_{ij} &= x_{11} + x_{12} + x_{13} + x_{21} + x_{22} + x_{23} \\ &= 10 + 8 + 6 + 7 + 4 + 12 \\ &= 47\end{aligned}$$

Definition

$$\sum_{i=1}^{n}x_{ij} = x_{1j} + x_{2j} + \cdots + x_{nj} \quad \textbf{(C.6)}$$

Example:

$$\begin{aligned}\sum_{i=1}^{2}x_{i2} &= x_{12} + x_{22} \\ &= 8 + 4 \\ &= 12\end{aligned}$$

Shorthand Notation

Sometimes when a summation is for all values of the subscript, we use the following shorthand notations:

$$\sum_{i=1}^{n}x_i = \sum x_i \quad \textbf{(C.7)}$$

$$\sum_{i=1}^{n}\sum_{j=1}^{m}x_{ij} = \sum\sum x_{ij} \quad \textbf{(C.8)}$$

$$\sum_{i=1}^{n}x_{ij} = \sum_{i}x_{ij} \quad \textbf{(C.9)}$$

Appendix D: Self-Test Solutions and Answers to Even-Numbered Exercises

Chapter 1

2. a. 10
b. 5
c. Categorical variables: Size and Fuel
Quantitiative variables: Cylinders, City MPG, and Highway MPG
d.

Variable	Measurement Scale
Size	Ordinal
Cylinders	Ratio
City MPG	Ratio
Highway MPG	Ratio
Fuel	Nominal

3. a. Average for city driving = 182/10 = 18.2 mpg
b. Average for highway driving = 261/10 = 26.1 mpg
On average, the miles per gallon for highway driving is 7.9 mpg greater than for city driving
c. 3 of 10 or 30% have four-cyclinder engines
d. 6 of 10 or 60% use regular fuel

4. a. 7
b. 5
c. Categorical variables: State, Campus Setting, and NCAA Division
d. Quantitiative variables: Endowment and Applicants Admitted

6. a. Quantitative
b. Categorical
c. Categorical
d. Quantitative
e. Categorical

8. a. 1015
b. Categorical
c. Percentages
d. .10(1015) = 101.5; 101 or 102 respondents

10. a. Categorical
b. Percentages
c. 15%
d. Support against

12. a. All visitors to Hawaii
b. Yes
c. First and fourth questions provide quantitative data
Second and third questions provide categorical data

13. a. Federal spending ($ trillions)
b. Quantitative
c. Time series
d. Federal spending has increased over time

14. a. Graph with time series line for each company
b. Hertz leader in 2007–2008; Avis increasing and now similar to Hertz; Dollar declining
c. A bar chart of cross-sectional data
Bar heights: Hertz 290, Dollar 108, Avis 270

18. a. 67%
b. 612
c. Categorical

20. a. 43% of managers were bullish or very bullish, and 21% of managers expected health care to be the leading industry over the next 12 months
b. The average 12-month return estimate is 11.2% for the population of investment managers
c. The sample average of 2.5 years is an estimate of how long the population of investment managers think it will take to resume sustainable growth

22. a. The population consists of all customers of the chain stores in Charlotte, North Carolina
b. Some of the ways the grocery store chain could use to collect the data are
- Customers entering or leaving the store could be surveyed
- A survey could be mailed to customers who have a shopper's club card
- Customers could be given a printed survey when they check out
- Customers could be given a coupon that asks them to complete a brief online survey; if they do, they will receive a 5% discount on their next shopping trip

24. a. Correct
b. Incorrect
c. Correct
d. Incorrect
e. Incorrect

Chapter 2

2. a. .20
b. 40
c/d.

Class	Frequency	Percent Frequency
A	44	22
B	36	18
C	80	40
D	40	20
Total	200	100

3. a. $360° \times 58/120 = 174°$

b. $360° \times 42/120 = 126°$

c.

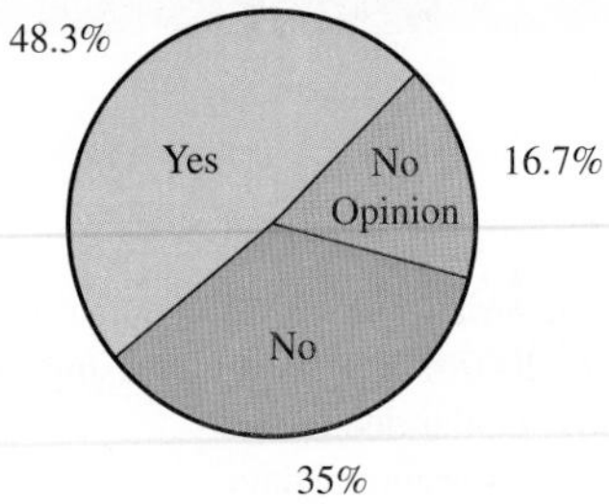

d.

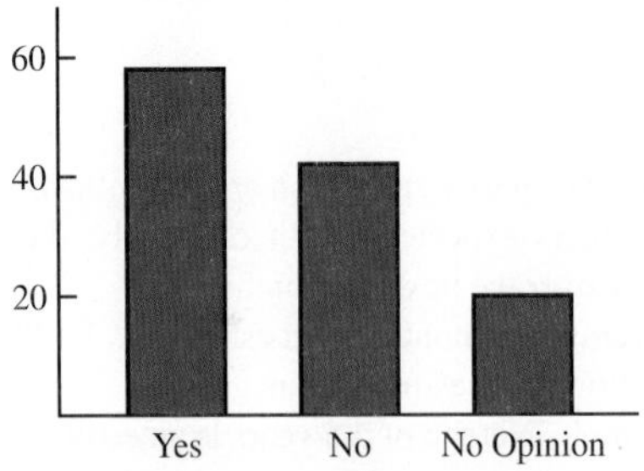

4. a. These data are categorical

b.

Show	Relative Frequency	Percent Frequency
Jep	10	20
JJ	8	16
OWS	7	14
THM	12	24
WoF	13	26
Total	50	100

c.

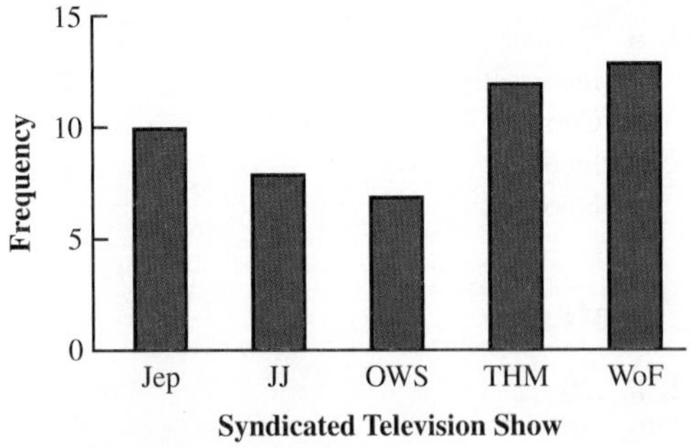

Syndicated Television Show

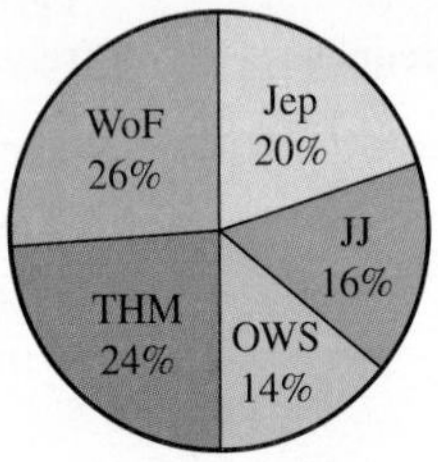

d. The largest viewing audience is for *Wheel of Fortune* and the second largest is for *Two and a Half Men*

6. a.

Network	Relative Frequency	Percent Frequency
ABC	6	24
CBS	9	36
FOX	1	4
NBC	9	36
Total	25	100

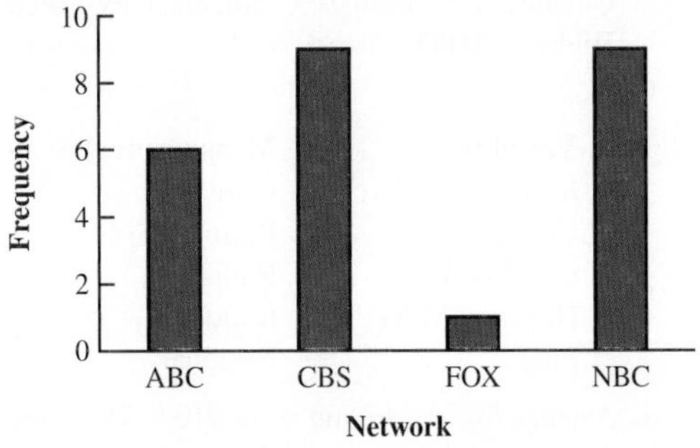

b. For these data, NBC and CBS tie for the number of top-rated shows; each has 9 (36%) of the top 25; ABC is third with 6 (24%) and the much younger FOX network has 1 (4%)

7. a.

Rating	Frequency	Percent Frequency
Excellent	20	40
Very Good	23	46
Good	4	8
Fair	1	2
Poor	2	4
Total	50	100

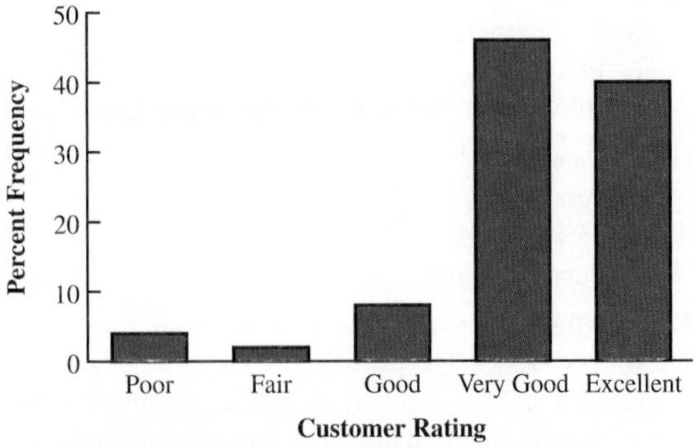

Management should be very pleased; 86% of the ratings are very good or excellent.

b. Review explanations from the three with Fair or Poor ratings to identify reasons for the low ratings

8. a.

Position	Frequency	Relative Frequency
P	17	.309
H	4	.073
1	5	.091
2	4	.073
3	2	.036
S	5	.091
L	6	.109
C	5	.091
R	7	.127
Totals	55	1.000

b. Pitcher
c. 3rd base
d. Right field
e. Infielders 16 to outfielders 18

10. a/b.

Rating	Frequency	Percent Frequency
Excellent	20	2
Good	101	10
Fair	528	52
Bad	244	24
Terrible	122	12
Total	1015	100

c.

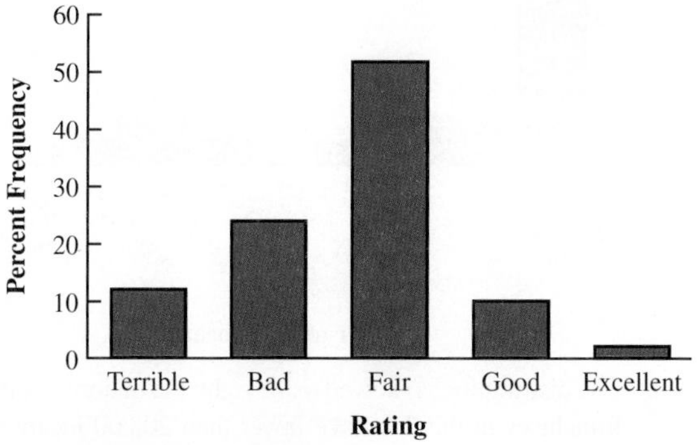

d. 36% a bad or a terrible job
12% a good or excellent job
e. 50% a bad or a terrible job
4% a good or excellent job
More pessimism in Spain

12.

Class	Cumulative Frequency	Cumulative Relative Frequency
≤19	10	.20
≤29	24	.48
≤39	41	.82
≤49	48	.96
≤59	50	1.00

14. b/c.

Class	Frequency	Percent Frequency
6.0–7.9	4	20
8.0–9.9	2	10
10.0–11.9	8	40
12.0–13.9	3	15
14.0–15.9	3	15
Totals	20	100

15. Leaf unit = .1

6	3
7	5 5 7
8	1 3 4 8
9	3 6
10	0 4 5
11	3

16. Leaf unit = 10

11	6
12	0 2
13	0 6 7
14	2 2 7
15	5
16	0 2 8
17	0 2 3

17. a/b.

Waiting Time	Frequency	Relative Frequency
0–4	4	.20
5–9	8	.40
10–14	5	.25
15–19	2	.10
20–24	1	.05
Totals	20	1.00

c/d.

Waiting Time	Cumulative Frequency	Cumulative Relative Frequency
≤4	4	.20
≤9	12	.60
≤14	17	.85
≤19	19	.95
≤24	20	1.00

e. 12/20 = .60

18. a.

Salary	Frequency
150–159	1
160–169	3
170–179	7
180–189	5
190–199	1
200–209	2
210–219	1
Total	20

b.

Salary	Percent Frequency
150–159	5
160–169	15
170–179	35
180–189	25
190–199	5
200–209	10
210–219	5
Total	100

c.

Salary	Cumulative Percent Frequency
Less than or equal to 159	5
Less than or equal to 169	20
Less than or equal to 179	55
Less than or equal to 189	80
Less than or equal to 199	85
Less than or equal to 209	95
Less than or equal to 219	100
Total	100

e. There is skewness to the right

f. 15%

20. a. Lowest = 12, Highest = 23

b.

Hours in Meetings per Week	Frequency	Percent Frequency
11–12	1	4
13–14	2	8
15–16	6	24
17–18	3	12
19–20	5	20
21–22	4	16
23–24	4	16
	25	100

c.

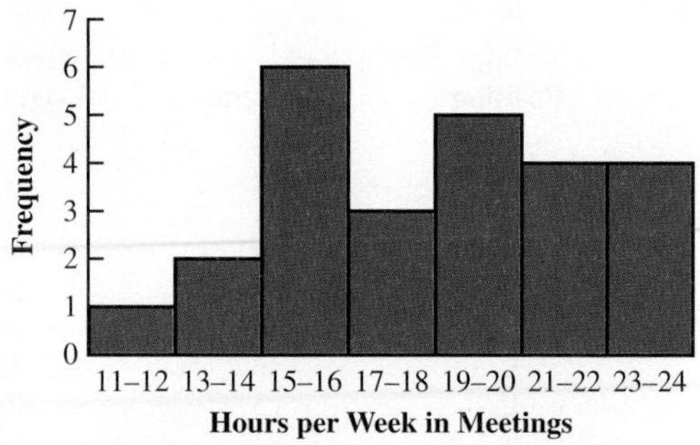

d. The distribution is slightly skewed to the left

22. a.

# U.S. Locations	Frequency	Percent Frequency
0–4999	10	50
5000–9999	3	15
10000–14999	2	10
15000–19999	1	5
20000–24999	0	0
25000–29999	1	5
30000–34999	2	10
35000–39999	1	5
Total:	20	100

b.

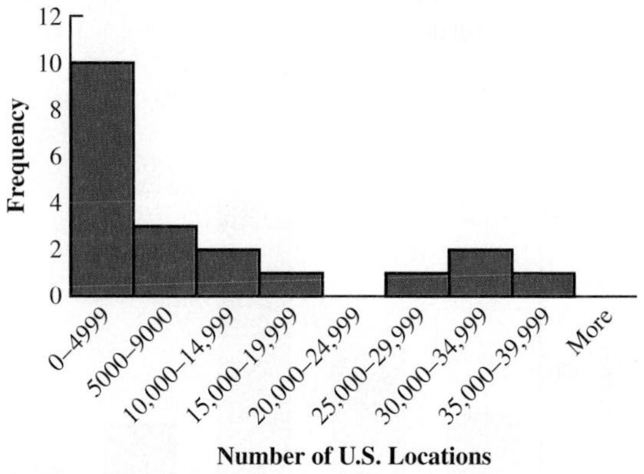

c. The distribution is skewed to the right; the majority of the franchises in this list have fewer than 20,000 locations (50% + 15% + 15% = 80%); McDonald's, Subway, and 7-Eleven have the highest number of locations

24. Median Pay

6	6 7 7
7	2 4 6 7 7 8 9
8	0 0 1 3 7
9	9
10	0 6
11	0
12	1

Top Pay

10	0 6 9
11	1 6 9
12	2 5 6
13	0 5 8 8
14	0 6
15	2 5 7
16	
17	
18	
19	
20	
21	4
22	1

The median pay for these careers is generally in the $70 and $80 thousands. The top pay is rather evenly distributed between $100 and $160 thousand.

26. a.

2	14
2	67
3	011123
3	5677
4	003333344
4	6679
5	00022
5	5679
6	14
6	6
7	2

b. 40–44 with 9
c. 43 with 5

27. a.

		y: 1	2	Total
x	A	5	0	5
	B	11	2	13
	C	2	10	12
	Total	18	12	30

b.

		y: 1	2	Total
x	A	100.0	0.0	100.0
	B	84.6	15.4	100.0
	C	16.7	83.3	100.0

c.

		y: 1	2
x	A	27.8	0.0
	B	61.1	16.7
	C	11.1	83.3
	Total	100.0	100.0

d. *A* values are always in $y = 1$
B values are most often in $y = 1$
C values are most often in $y = 2$

28. a.

		y: 20–39	40–59	60–79	80–100	Grand Total
x	10–29			1	4	5
	30–49	2		4		6
	50–69	1	3	1		5
	70–90	4				4
	Grand Total	7	3	6	4	20

b.

		y: 20–39	40–59	60–79	80–100	Grand Total
x	10–29			20.0	80.0	100
	30–49	33.3		66.7		100
	50–69	20.0	60.0	20.0		100
	70–90	100.0				100

c.

		y: 20–39	40–59	60–79	80–100
x	10–29	0.0	0.0	16.7	100.0
	30–49	28.6	0.0	66.7	0.0
	50–69	14.3	100.0	16.7	0.0
	70–90	57.1	0.0	0.0	0.0
	Grand Total	100	100	100	100

d. Higher values of *x* are associated with lower values of *y* and vice versa

30. a.

Education Level	Household Income ($1000s): Under 25	25.0–49.9	50.0–74.9	75.0–99.9	100 or more	Total
Not H.S. Graduate	32.10	18.71	9.13	5.26	2.20	13.51
H.S. Graduate	37.52	37.05	33.04	25.73	16.00	29.97
Some College	21.42	28.44	30.74	31.71	24.43	27.21
Bachelor's Degree	6.75	11.33	18.72	25.19	32.26	18.70
Beyond Bach. Deg.	2.21	4.48	8.37	12.11	25.11	10.61
Total	100.00	100.00	100.00	100.00	100.00	100.00

13.51% of the heads of households did not graduate from high school
b. 25.11%, 53.54%
c. Positive relationship between income and education level

32. a.

	5-Year Average Return						
Fund Type	**0–9.99**	**10–19.99**	**20–29.99**	**30–39.99**	**40–49.99**	**50–59.99**	**Total**
DE	1	25	1	0	0	0	27
FI	9	1	0	0	0	0	10
IE	0	2	3	2	0	1	8
Total	10	28	4	2	0	1	45

b.

5-Year Average Return	Frequency
0–9.99	10
10–19.99	28
20–29.99	4
30–39.99	2
40–49.99	0
50–59.99	1
Total	45

c.

Fund Type	Frequency
DE	27
FI	10
IE	8
Total	45

d. The margin of the crosstabulation shows these frequency distributions

e. Higher returns—International Equity funds
Lower returns—Fixed Income funds

34. b. Higher 5-year returns are associated with higher net asset values.

36. a.

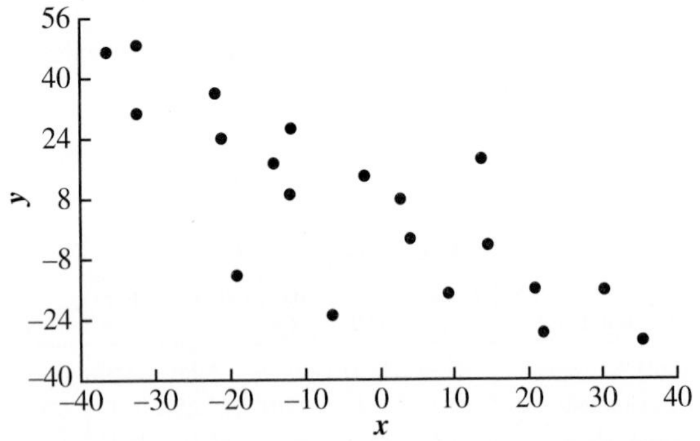

b. A negative relationship between x and y; y decreases as x increases

38. a.

		y		
		Yes	**No**	
x	**Low**	66.667	33.333	100
	Medium	30.000	70.000	100
	High	80.000	20.000	100

b.

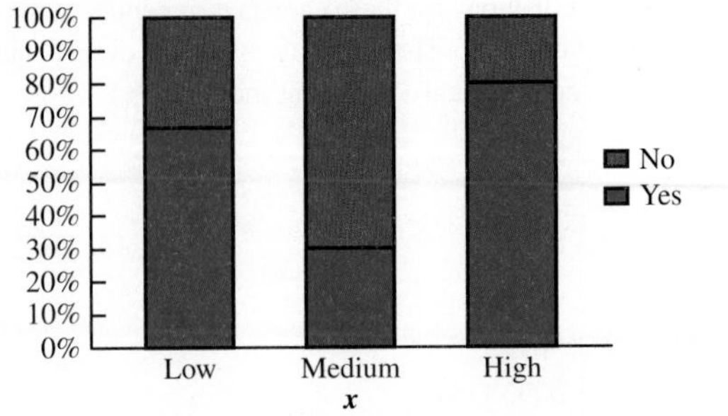

40. a.

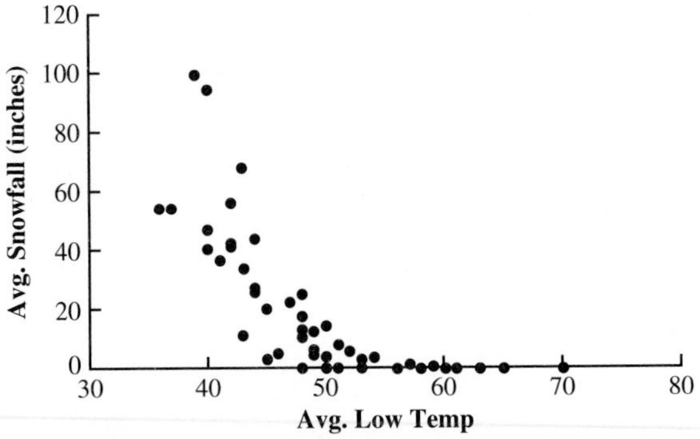

b. Colder average low temperature seems to lead to higher amounts of snowfall

c. Two cities have an average snowfall of nearly 100 inches of snowfall: Buffalo, New York and Rochester, New York; both are located near large lakes in New York

42. a.

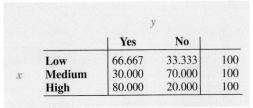

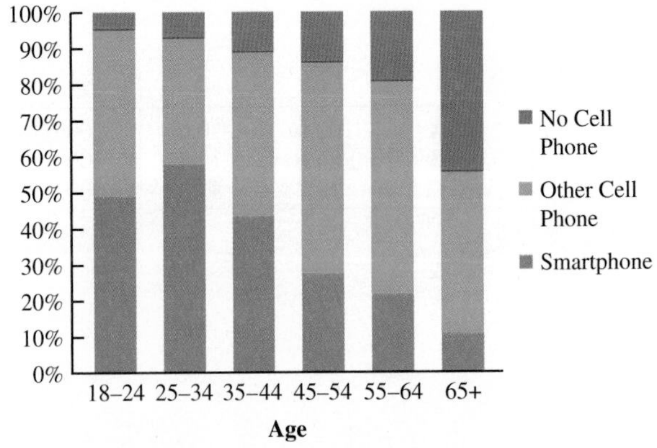

b. After an increase in age 25–34, smartphone ownership decreases as age increases; the percentage of people with no cell phone increases with age; there is less variation across age groups in the percentage who own other cell phones

c. Unless a newer device replaces the smartphone, we would expect smartphone ownership would become less sensitive to age; this would be true because current users will become older and because the device will become to be seen more as a necessity than a luxury

44. a.

SAT Score	Frequency
800–999	1
1000–1199	3
1200–1399	6
1400–1599	10
1600–1799	7
1800–1999	2
2000–2199	1
Total	30

b. Nearly symmetrical

c. 33% of the scores fall between 1400 and 1599
A score below 800 or above 2200 is unusual
The average is near or slightly above 1500

46. a.

Population in Millions	Frequency	Percent Frequency
0.0–2.4	15	30.0
2.5–4.9	13	26.0
5.0–7.4	10	20.0
7.5–9.9	5	10.0
10.0–12.4	1	2.0
12.5–14.9	2	4.0
15.0–17.4	0	0.0
17.5–19.9	2	4.0
20.0–22.4	0	0.0
22.5–24.9	0	0.0
25.0–27.4	1	2.0
27.5–29.9	0	0.0
30.0–32.4	0	0.0
32.5–34.9	0	0.0
35.0–37.4	1	2.0
37.5–39.9	0	0.0
More	0	0.0

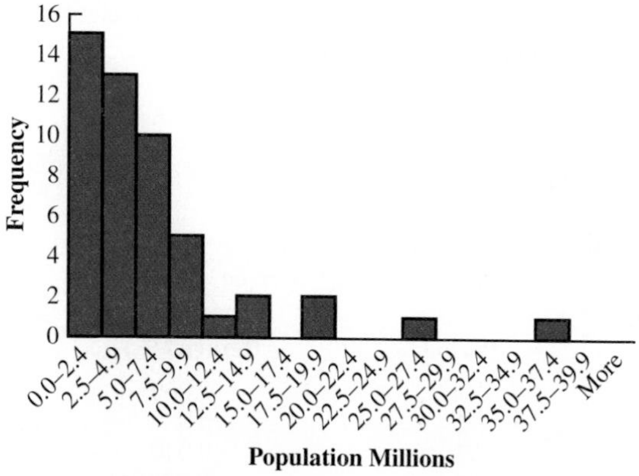

b. The distribution is skewed to the right

c. Fifteen states (30%) have a population less than 2.5 million; over half the states have population less than 5 million (28 states—56%); only seven states have a population greater than 10 million (California, Florida, Illinois, New York, Ohio, Pennsylvania, and Texas); the largest state is California (37.3 million) and the smallest states are Vermont and Wyoming (600 thousand)

48. a.

Industry	Frequency	Percent Frequency
Bank	26	13
Cable	44	22
Car	42	21
Cell	60	30
Collection	28	14
Total	200	100

b.

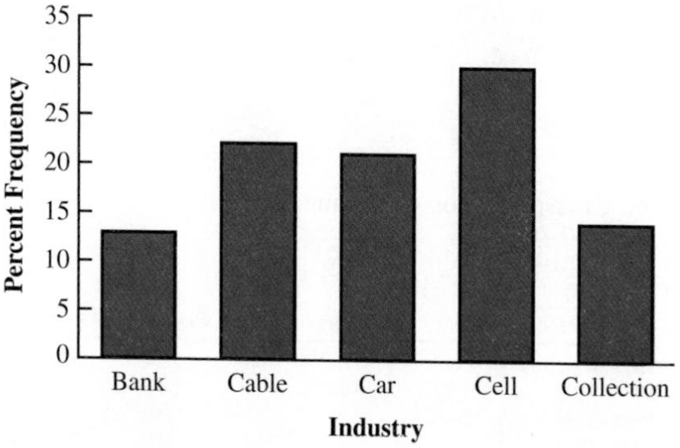

c. The cellular phone providers had the highest number of complaints

d. The percentage frequency distribution shows that the two financial industries (banks and collection agencies) had about the same number of complaints; new car dealers and cable and satellite television companies also had about the same number of complaints

50. a.

Level of Support	Percent Frequency
Strongly favor	30.10
Favor more than oppose	34.83
Oppose more than favor	21.13
Strongly oppose	13.94
Total	100.00

Overall favor higher tax = 30.10% + 34.83%
= 64.93%

b. 20.2, 19.5, 20.6, 20.7, 19.0
Roughly 20% per country

c. The crosstabulation with column percentages:

	Country				
Support	Great Britain	Italy	Spain	Germany	United States
Strongly favor	31.00	31.96	45.99	19.98	20.98
Favor more than oppose	34.04	39.04	32.01	36.99	32.06
Oppose more than favor	23.00	17.99	13.98	24.03	26.96
Strongly oppose	11.96	11.01	8.03	18.99	20.00
Total	100.00	100.00	100.00	100.00	100.00

Considering the percentage of respondents who favor the higher tax by either saying "strongly favor" or "favor more than oppose," 65.04%, 71.00%, 78.00%, 56.97%, and 53.04% for the five countries; all show more than 50% support, but all European countries show more support for the tax than the United States; Italy and Spain show the highest level of support

52. a. Row totals: 247; 54; 82; 121
Column totals: 149; 317; 17; 7; 14

b.

Year	Freq.	Fuel	Freq.
1973 or before	247	Elect.	149
1974–79	54	Nat. Gas	317
1980–86	82	Oil	17
1987–91	121	Propane	7
Total	504	Other	14
		Total	504

c. Crosstabulation of column percentages

Year Constructed	Fuel Type: Elect.	Nat. Gas	Oil	Propane	Other
1973 or before	26.9	57.7	70.5	71.4	50.0
1974–1979	16.1	8.2	11.8	28.6	0.0
1980–1986	24.8	12.0	5.9	0.0	42.9
1987–1991	32.2	22.1	11.8	0.0	7.1
Total	100.0	100.0	100.0	100.0	100.0

d. Crosstabulation of row percentages

Year Constructed	Fuel Type: Elect.	Nat. Gas	Oil	Propane	Other	Total
1973 or before	16.2	74.1	4.9	2.0	2.8	100.0
1974–1979	44.5	48.1	3.7	3.7	0.0	100.0
1980–1986	45.1	46.4	1.2	0.0	7.3	100.0
1987–1991	39.7	57.8	1.7	0.0	0.8	100.0

54. c. Older colleges and universities tend to have higher graduation rates.

56. a.

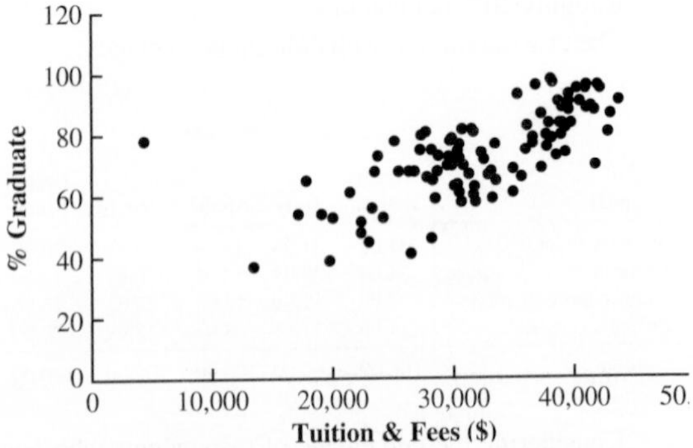

b. There appears to be a strong positive relationship between Tuition & Fees and % Graduation.

58. a.

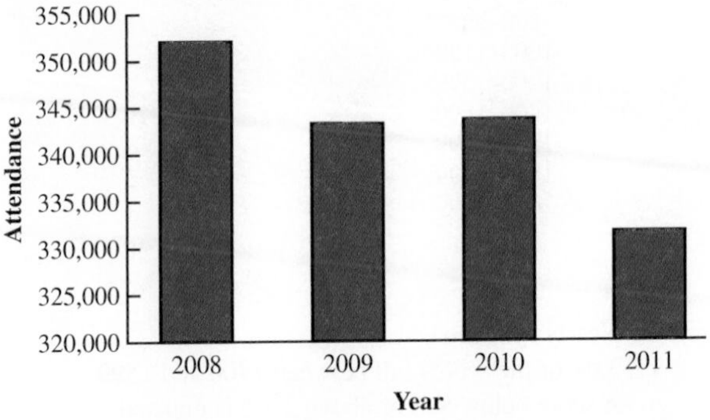

Zoo attendance appears to be dropping over time

b.

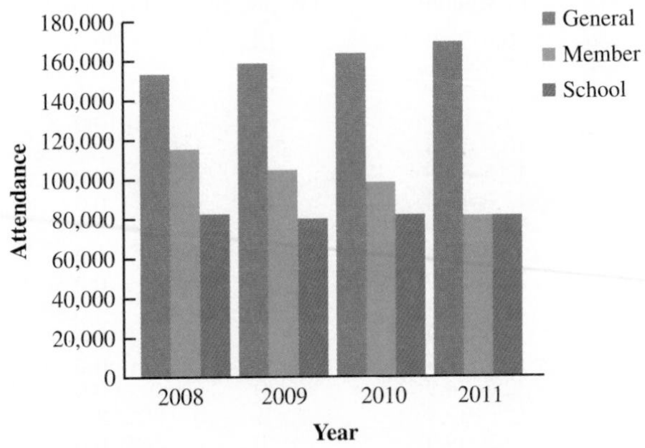

c. General attendance is increasing, but not enough to offset the decrease in member attendance; school membership appears fairly stable

Chapter 3

2. 16, 16.5

4.

Period	Return (%)
1	−0.060
2	−0.080
3	−0.040
4	0.020
5	0.054

The mean growth factor over the five periods is

$$\bar{x}_g = \sqrt[n]{(x_1)(x_2)\cdots(x_5)}$$
$$= \sqrt[5]{(0.940)(0.920)(0.960)(1.020)(1.054)}$$
$$= \sqrt[5]{(0.8925)} = 0.9775$$

So the mean growth rate $(0.9775 - 1)100\% = -2.25\%$

5. Arrange data in order: 15, 20, 25, 25, 27, 28, 30, 34

$i = \frac{20}{100}(8) = 1.6$; round up to position 2

20th percentile = 20

$i = \frac{25}{100}(8) = 2$; use positions 2 and 3

25th percentile $= \frac{20 + 25}{2} = 22.5$

$i = \frac{65}{100}(8) = 5.2$; round up to position 6

65th percentile = 28

$i = \frac{75}{100}(8) = 6$; use positions 6 and 7

75th percentile $= \frac{28 + 30}{2} = 29$

6. 59.73, 57, 53

8. a. 18.42
b. 6.32
c. 34.3%
d. Reductions of only .65 shots and .9% made shots per game
Yes, agree but not dramatically

10. a. $\bar{x} = \frac{\Sigma x_i}{n} = \frac{3200}{20} = 160$

Order the data from low 100 to high 360

Median: $i = \left(\frac{50}{100}\right)20 = 10$; use 10th and 11th positions

Median $= \left(\frac{130 + 140}{2}\right) = 135$

Mode = 120 (occur 3 times)

b. $i = \left(\frac{25}{100}\right)20 = 5$; use 5th and 6th positions

$Q_1 = \left(\frac{115 + 115}{2}\right) = 115$

$i = \left(\frac{75}{100}\right)20 = 15$; use 15th and 16th positions

$Q_3 = \left(\frac{180 + 195}{2}\right) = 187.5$

c. $i = \left(\frac{90}{100}\right)20 = 18$; use 18th and 19th positions

90th percentile $= \left(\frac{235 + 255}{2}\right) = 245$

90% of the tax returns cost $245 or less

12. a. The minimum number of viewers who watched a new episode is 13.3 million, and the maximum number is 16.5 million

b. The mean number of viewers who watched a new episode is 15.04 million or approximately 15.0 million; the median is also 15.0 million; the data are multimodal (13.6, 14.0, 16.1, and 16.2 million); in such cases the mode is usually not reported

c. The data are first arranged in ascending order. The index for the first quartile is $i = \left(\frac{25}{100}\right)21 = 5.25$, so the first quartile is the value of the 6th observation in the sorted data, or 14.1; the index for the third quartile is $i = \left(\frac{75}{100}\right)21 = 15.75$, so the third quartile is the value of the 16th observation in the sorted data, or 16.0

d. A graph showing the viewership data over the air dates follows: Period 1 corresponds to the first episode of the season, period 2 corresponds to the second episode, and so on

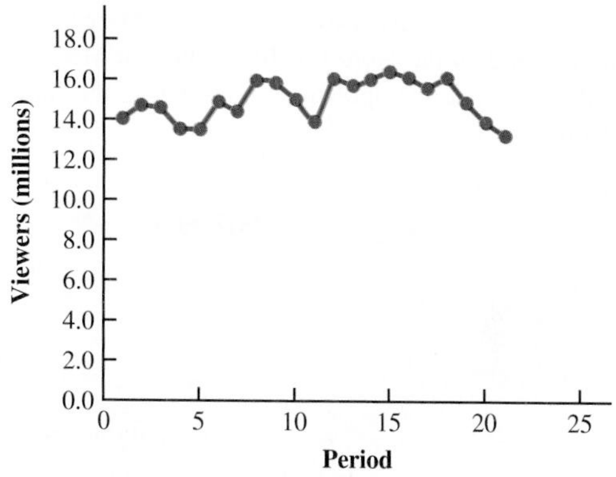

This graph shows that viewership of *The Big Bang Theory* has been relatively stable over the 2011–2012 television season

14. For March 2011

The index for the first quartile is $i = \left(\frac{25}{100}\right)50 = 12.50$, so the first quartile is the value of the 13th observation in the sorted data, or 6.8

The index for the median quartile is $i = \left(\frac{50}{100}\right)50 = 25.0$, so the median (or second quartile) is the average of the values of the 25th and 26th observations in the sorted data, or 8.0

The index for the third quartile is $i = \left(\frac{75}{100}\right)50 = 37.50$, so the third quartile is the value of the 38th observation in the sorted data, or 9.4

For March 2012

The minimum is 3.0

The index for the first quartile is $i = \left(\frac{25}{100}\right)50 = 12.50$, so the first quartile is the value of the 13th observation in the sorted data, or 6.8

The index for the median quartile is $i = \left(\frac{50}{100}\right)50 = 25.0$, so the median (or second quartile) is the average of the values of the 25th and 26th observations in the sorted data, or 7.35

The index for the third quartile is $i = \left(\frac{75}{100}\right)50 = 37.50$, so the third quartile is the value of the 38th observation in the sorted data, or 8.6

It may be easier to compare these results if we place them in a table

	March 2011	March 2012
First Quartile	6.8	6.8
Median	8.0	7.35
Third Quartile	9.4	8.6

The results show that in March 2012 approximately 25% of the states had an unemployment rate of 6.8% or less, the same as in March 2011; however, the median of 7.35% and the third quartile of 8.6% in March 2012 are both less than the corresponding values in March 2011, indicating that unemployment rates across the states are decreasing

16. a.

Grade x_i	Weight w_i
4 (A)	9
3 (B)	15
2 (C)	33
1 (D)	3
0 (F)	0
	60 credit hours

$$\bar{x} = \frac{\Sigma w_i x_i}{\Sigma w_i} = \frac{9(4) + 15(3) + 33(2) + 3(1)}{9 + 15 + 33 + 3}$$

$$= \frac{150}{60} = 2.5$$

b. Yes

18. 3.8, 3.7

20.

	Stivers		Trippi	
Year	End of Year Value ($)	Growth Factor	End of Year Value ($)	Growth Factor
2004	11,000	1.100	5,600	1.120
2005	12,000	1.091	6,300	1.125
2006	13,000	1.083	6,900	1.095
2007	14,000	1.077	7,600	1.101
2008	15,000	1.071	8,500	1.118
2009	16,000	1.067	9,200	1.082
2010	17,000	1.063	9,900	1.076
2011	18,000	1.059	10,600	1.071

For the Stivers mutual fund we have
$18000 = 10000[(x_1)(x_2) \cdots (x_8)]$, so $[(x_1)(x_2) \cdots (x_8)]$
$= 1.8$ and

$$\bar{x}_g = \sqrt[n]{(x_1)(x_2) \cdots (x_8)} = \sqrt[8]{1.80} = 1.07624$$

So the mean annual return for the Stivers mutual fund is $(1.07624 - 1)100 = 7.624\%$

For the Trippi mutual fund we have
$10600 = 5000\,[(x_1)(x_2) \cdots (x_8)]$, so $[(x_1)(x_2) \cdots (x_8)]$
$= 2.12$ and

$$\bar{x}_g = \sqrt[n]{(x_1)(x_2) \cdots (x_8)} = \sqrt[8]{2.12} = 1.09848$$

So the mean annual return for the Trippi mutual fund is $(1.09848 - 1)100 = 9.848\%$

While the Stivers mutual fund has generated a nice annual return of 7.6%, the annual return of 9.8% earned by the Trippi mutual fund is far superior

22. $25{,}000{,}000 = 10{,}000{,}000\,[(x_1)(x_2) \cdots (x_6)]$,
so $[(x_1)(x_2) \cdots (x_6)] = 2.50$

so $\bar{x}_g = \sqrt[n]{(x_1)(x_2) \cdots (x_6)} = \sqrt[6]{2.50} = 1.165$

So the mean annual growth rate is $(1.165 - 1)100 = 16.5\%$

24. 16, 4

25. Range $= 34 - 15 = 19$
Arrange data in order: 15, 20, 25, 25, 27, 28, 30, 34

$$i = \frac{25}{100}(8) = 2;\ Q_1 = \frac{20 + 25}{2} = 22.5$$

$$i = \frac{75}{100}(8) = 6;\ Q_3 = \frac{28 + 30}{2} = 29$$

$$\text{IQR} = Q_3 - Q_1 = 29 - 22.5 = 6.5$$

$$\bar{x} = \frac{\Sigma x_i}{n} = \frac{204}{8} = 25.5$$

x_i	$(x_i - \bar{x})$	$(x_i - \bar{x})^2$
27	1.5	2.25
25	−.5	.25
20	−5.5	30.25
15	−10.5	110.25
30	4.5	20.25
34	8.5	72.25
28	2.5	6.25
25	−.5	.25
		242.00

$$s^2 = \frac{\Sigma(x_i - \bar{x})^2}{n - 1} = \frac{242}{8 - 1} = 34.57$$

$$s = \sqrt{34.57} = 5.88$$

26. a. Range $= 190 - 168 = 22$

b. $\bar{x} = \dfrac{\Sigma x_i}{n} = \dfrac{1068}{6} = 178$

$$s^2 = \frac{\Sigma(x_i - \bar{x})^2}{n - 1}$$

$$= \frac{4^2 + (-10)^2 + 6^2 + 12^2 + (-8)^2 + (-4)^2}{6 - 1}$$

$$= \frac{376}{5} = 75.2$$

c. $s = \sqrt{75.2} = 8.67$

d. $\dfrac{s}{\bar{x}}(100) = \dfrac{8.67}{178}(100\%) = 4.87\%$

28. a. The mean serve speed is 180.95, the variance is 21.42, and the standard deviation is 4.63

b. Although the mean serve speed for the 20 Women's Singles serve speed leaders for the 2011 Wimbledon tournament is slightly higher, the difference is very small; furthermore, given the variation in the 20 Women's Singles serve speed leaders from the 2012 Australian Open and the 20 Women's Singles serve speed leaders from the 2011 Wimbledon tournament, the difference in the mean serve speeds is most likely due to random variation in the players' performances

30. *Dawson:* range = 2, $s = .67$
Clark: range = 8, $s = 2.58$

32. a. 1285, 433 Freshmen spend more
b. 1720, 352
c. 404, 131.5
d. 367.04, 96.96
e. Freshmen have more variability

34. *Quarter-milers:* $s = .0564$, Coef. of Var. = 5.8%
Milers: $s = .1295$, Coef. of Var. = 2.9%

36. .20, 1.50, 0, −.50, −2.20

37. a. $z = \frac{20 - 30}{5} = -2, z = \frac{40 - 30}{5} = 2 \quad 1 - \frac{1}{2^2} = .75$
At least 75%
b. $z = \frac{15 - 30}{5} = -3, z = \frac{45 - 30}{5} = 3 \quad 1 - \frac{1}{3^2} = .89$
At least 89%
c. $z = \frac{22 - 30}{5} = -1.6, z = \frac{38 - 30}{5} = 1.6 \quad 1 - \frac{1}{1.6^2} = .61$
At least 61%
d. $z = \frac{18 - 30}{5} = -2.4, z = \frac{42 - 30}{5} = 2.4 \quad 1 - \frac{1}{2.4^2} = .83$
At least 83%
e. $z = \frac{12 - 30}{5} = -3.6, z = \frac{48 - 30}{5} = 3.6 \quad 1 - \frac{1}{3.6^2} = .92$
At least 92%

38. a. 95%
b. Almost all
c. 68%

39. a. $z = 2$ standard deviations
$1 - \frac{1}{z^2} = 1 - \frac{1}{2^2} = \frac{3}{4}$; at least 75%
b. $z = 2.5$ standard deviations
$1 - \frac{1}{z^2} = 1 - \frac{1}{2.5^2} = .84$; at least 84%
c. $z = 2$ standard deviations
Empirical rule: 95%

40. a. 68%
b. 81.5%
c. 2.5%

42. a. −.67
b. 1.50
c. Neither is an outlier
d. Yes; $z = 8.25$

44. a. 76.5, 7
b. 16%, 2.5%
c. 12.2, 7.89; no outliers

46. 15, 22.5, 26, 29, 34

48. Arrange data in order: 5, 6, 8, 10, 10, 12, 15, 16, 18
$i = \frac{25}{100}(9) = 2.25$; round up to position 3
$Q_1 = 8$
Median (5th position) = 10
$i = \frac{75}{100}(9) = 6.75$; round up to position 7
$Q_3 = 15$
5-number summary: 5, 8, 10, 15, 18

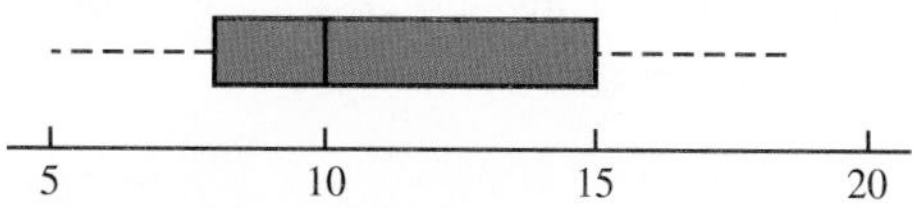

50. a. Men's 1st place 43.73 minutes faster
b. Medians: 109.64, 131.67
Men's median time 22.03 minutes faster
c. 65.30, 87.18, 109.64, 128.40, 148.70
109.03, 122.08, 131.67, 147.18, 189.28
d. Men's Limits: 25.35 to 190.23; no outliers
Women's Limits: 84.43 to 184.83; 2 outliers
e. Women runners show less variation

51. a. Arrange data in order low to high
$i = \frac{25}{100}(21) = 5.25$; round up to 6th position
$Q_1 = 1872$
Median (11th position) = 4019
$i = \frac{75}{100}(21) = 15.75$; round up to 16th position
$Q_3 = 8305$
5-number summary: 608, 1872, 4019, 8305, 14,138
b. IQR = $Q_3 - Q_1 = 8305 - 1872 = 6433$
Lower limit: 1872 − 1.5(6433) = −7777.5
Upper limit: 8305 + 1.5(6433) = 17,955
c. No; data are within limits
d. 41,138 > 27,604; 41,138 would be an outlier; data value would be reviewed and corrected
e.

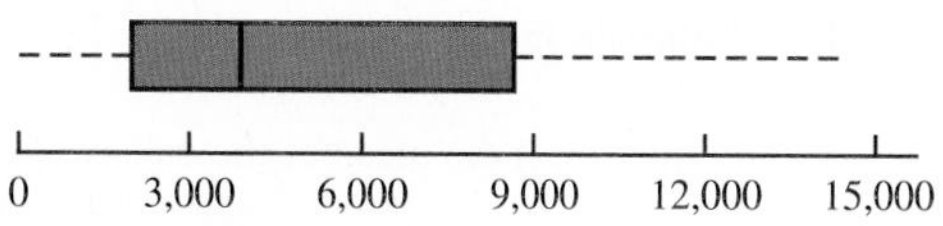

52. a. 73.5
b. 68, 71.5, 73.5, 74.5, 77
c. Limits: 67 and 79; no outliers
d. 66, 68, 71, 73, 75; 60.5 and 80.5
63, 65, 66, 67.6, 69; 61.25 and 71.25
75, 77, 78.5, 79.5, 81; 73.25 and 83.25
No outliers for any of the services
e. Verizon is highest rated
Sprint is lowest rated

54. a. 18.2, 15.35
b. 11.7, 23.5

c. 3.4, 11.7, 15.35, 23.5, 41.3
d. Yes; Alger Small Cap 41.3

55. b. There appears to be a negative linear relationship between x and y

c.

x_i	y_i	$x_i - \bar{x}$	$y_i - \bar{y}$	$(x_i - \bar{x})(y_i - \bar{y})$
4	50	−4	4	−16
6	50	−2	4	−8
11	40	3	−6	−18
3	60	−5	14	−70
16	30	8	−16	−128
40	230	0	0	−240

$\bar{x} = 8$; $\bar{y} = 46$

$$s_{xy} = \frac{\Sigma(x_i - \bar{x})(y_i - \bar{y})}{n - 1} = \frac{-240}{4} = -60$$

The sample covariance indicates a negative linear association between x and y

d. $r_{xy} = \dfrac{s_{xy}}{s_x s_y} = \dfrac{-60}{(5.43)(11.40)} = -.969$

The sample correlation coefficient of −.969 is indicative of a strong negative linear relationship

56. b. There appears to be a positive linear relationship between x and y
c. $s_{xy} = 26.5$
d. $r_{xy} = .693$

58. −.91; negative relationship

60. b. .910
c. Strong positive linear relationship; no

62. a. The mean is 2.95 and the median is 3.0
b. The index for the first quartile is $i = \left(\dfrac{25}{100}\right)20 = 5$, so the first quartile is the mean of the values of the 5th and 6th observations in the sorted data, or $\dfrac{1 + 1}{2} = 1$

The index for the third quartile is $i = \left(\dfrac{75}{100}\right)20 = 15$, so the third quartile is the mean of the values of the 15th and 16th observations in the sorted data, or $\dfrac{4 + 5}{2} = 4.5$

c. The range is 7 and the interquartile range is 4.5 – 1 = 3.5
d. The variance is 4.37 and standard deviation is 2.09
e. Because most people dine out relatively few times per week and a few families dine out very frequently, we would expect the data to be positively skewed; the skewness measure of 0.34 indicates the data are somewhat skewed to the right
f. The lower limit is −4.25 and the upper limit is 9.75; no values in the data are less than the lower limit or greater than the upper limit, so the Minitab boxplot indicates there are no outliers

Boxplot of Dining Out

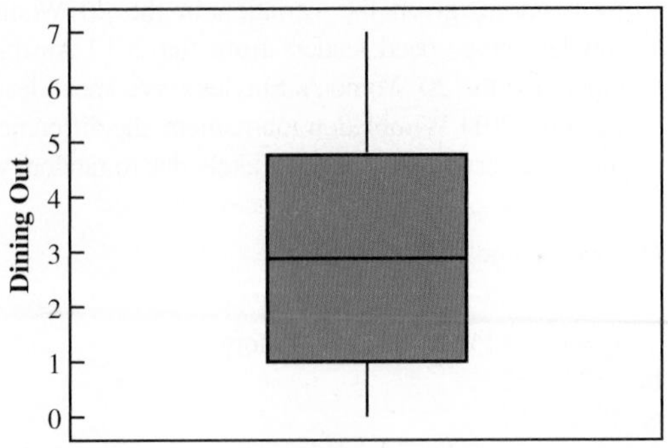

64. a. The mean and median patient wait times for offices with a wait-tracking system are 17.2 and 13.5, respectively; the mean and median patient wait times for offices without a wait-tracking system are 29.1 and 23.5, respectively
b. The variance and standard deviation of patient wait times for offices with a wait-tracking system are 86.2 and 9.3, respectively; the variance and standard deviation of patient wait times for offices without a wait-tracking system are 275.7 and 16.6, respectively
c. Offices with a wait-tracking system have substantially shorter patient wait times than offices without a wait-tracking system
d. $z = \dfrac{37 - 29.1}{16.6} = 0.48$
e. $z = \dfrac{37 - 17.2}{9.3} = 2.13$

As indicated by the positive z-scores, both patients had wait times that exceeded the means of their respective samples; even though the patients had the same wait time, the z-score for the sixth patient in the sample who visited an office with a wait-tracking system is much larger because that patient is part of a sample with a smaller mean and a smaller standard deviation
f. The z-scores for all patients follow

Without Wait-Tracking System	With Wait-Tracking System
−0.31	1.49
2.28	−0.67
−0.73	−0.34
−0.55	0.09
0.11	−0.56
0.90	2.13
−1.03	−0.88
−0.37	−0.45
−0.79	−0.56
0.48	−0.24

The z-scores do not indicate the existence of any outliers in either sample

66. a. \$670
b. \$456
c. $z = 3$; yes
d. Save time and prevent a penalty cost

68. a. 215.9
b. 55%
c. 175.0, 628.3
d. 48.8, 175.0, 215.9, 628.3, 2325.0
e. Yes, any price over 1308.25
f. 482.1; prefer median

70. a. 364 rooms
b. \$457
c. −.293; slight negative correlation
Higher cost per night tends to be associated with smaller hotels

72. a. .268, low or weak positive correlation
b. Very poor predictor; spring training is practice and does not count toward standings or playoffs

74. a. 60.68
b. $s^2 = 31.23$; $s = 5.59$

Chapter 4

2. $\binom{6}{3} = \frac{6!}{3!3!} = \frac{6 \cdot 5 \cdot 4 \cdot 3 \cdot 2 \cdot 1}{(3 \cdot 2 \cdot 1)(3 \cdot 2 \cdot 1)} = 20$

ABC	ACE	BCD	BEF
ABD	ACF	BCE	CDE
ABE	ADE	BCF	CDF
ABF	ADF	BDE	CEF
ACD	AEF	BDF	DEF

4. b. (H,H,H), (H,H,T), (H,T,H), (H,T,T), (T,H,H), (T,H,T), (T,T,H), (T,T,T)
c. $\frac{1}{8}$

6. $P(E_1) = .40$, $P(E_2) = .26$, $P(E_3) = .34$
The relative frequency method was used

8. a. 4: Commission Positive—Council Approves
Commission Positive—Council Disapproves
Commission Negative—Council Approves
Commission Negative—Council Disapproves

9. $\binom{50}{4} = \frac{50!}{4!46!} = \frac{50 \cdot 49 \cdot 48 \cdot 47}{4 \cdot 3 \cdot 2 \cdot 1} = 230{,}300$

10. a. Using the table, $P(\text{Debt}) = .94$
b. Five of the 8 institutions, $P(\text{over } 60\%) = 5/8 = .625$
c. Two of the 8 institutions, $P(\text{more than } \$30{,}000) = 2/8 = .25$
d. $P(\text{No debt}) = 1 - P(\text{debt}) = 1 - .72 = .28$
e. A weighted average with 72% having average debt of \$32,980 and 28% having no debt

$$\text{Average debt per graduate} = \frac{.72(\$32{,}980) + .28(\$0)}{.72 + .28} = \$23{,}746$$

12. a. 3,478,761
b. 1/3,478,761
c. 1/146,107,962

14. a. $\frac{1}{4}$
b. $\frac{1}{2}$
c. $\frac{3}{4}$

15. a. S = {ace of clubs, ace of diamonds, ace of hearts, ace of spades}
b. S = {2 of clubs, 3 of clubs, . . . , 10 of clubs, J of clubs, Q of clubs, K of clubs, A of clubs}
c. There are 12; jack, queen, or king in each of the four suits
d. For (a): 4/52 = 1/13 = .08
For (b): 13/52 = 1/4 = .25
For (c): 12/52 = .23

16. a. 36
c. $\frac{1}{6}$
d. $\frac{5}{18}$
e. No; $P(\text{odd}) = P(\text{even}) = \frac{1}{2}$
f. Classical

17. a. (4, 6), (4, 7), (4, 8)
b. .05 + .10 + .15 = .30
c. (2, 8), (3, 8), (4, 8)
d. .05 + .05 + .15 = .25
e. .15

18. a. .0222
b. .8226
c. .1048

20. a. .108
b. .096
c. .434

22. a. .40, .40, .60
b. .80, yes
c. $A^c = \{E_3, E_4, E_5\}$; $C^c = \{E_1, E_4\}$;
$P(A^c) = .60$; $P(C^c) = .40$
d. (E_1, E_2, E_5); .60
e. .80

23. a. $P(A) = P(E_1) + P(E_4) + P(E_6)$
$= .05 + .25 + .10 = .40$
$P(B) = P(E_2) + P(E_4) + P(E_7)$
$= .20 + .25 + .05 = .50$
$P(C) = P(E_2) + P(E_3) + P(E_5) + P(E_7)$
$= .20 + .20 + .15 + .05 = .60$
b. $A \cup B = \{E_1, E_2, E_4, E_6, E_7\}$;
$P(A \cup B) = P(E_1) + P(E_2) + P(E_4) + P(E_6) + P(E_7)$
$= .05 + .20 + .25 + .10 + .05$
$= .65$
c. $A \cap B = \{E_4\}$; $P(A \cap B) = P(E_4) = .25$
d. Yes, they are mutually exclusive
e. $B^c = \{E_1, E_3, E_5, E_6\}$;
$P(B^c) = P(E_1) + P(E_3) + P(E_5) + P(E_6)$
$= .05 + .20 + .15 + .10$
$= .50$

24. a. .05
b. .70

26. a. .64
b. .48
c. .36
d. .76

28. Let B = rented a car for business reasons
P = rented a car for personal reasons
a. $P(B \cup P) = P(B) + P(P) - P(B \cap P)$
$= .540 + .458 - .300$
$= .698$
b. $P(\text{Neither}) = 1 - .698 = .302$

30. a. $P(A \mid B) = \dfrac{P(A \cap B)}{P(B)} = \dfrac{.40}{.60} = .6667$
b. $P(B \mid A) = \dfrac{P(A \cap B)}{P(A)} = \dfrac{.40}{.50} = .80$
c. No, because $P(A \mid B) \neq P(A)$

32. a.

	Car	Light Truck	Total
U.S.	.1330	.2939	.4269
Non-U.S.	.3478	.2253	.5731
Total	.4808	.5192	1.0000

b. .4269, .5731 Non-U.S. higher
.4808, .5192 Light Truck slightly higher
c. .3115, .6885 Light Truck higher
d. .6909, .3931 Car higher
e. .5661, U.S. higher for Light Trucks

33. a.

		Undergraduate Major			
		Business	Engineering	Other	Totals
Intended Enrollment Status	Full-Time	0.270	0.151	0.192	0.613
	Part-Time	0.115	0.123	0.149	0.387
	Totals	0.385	0.274	0.341	1.000

b. $P(\text{Business}) = 0.385$, $P(\text{Engineering}) = 0.274$, and $P(\text{Other}) = 0.341$, so Business is the undergraduate major that produces the most potential MBA students
c. $P(\text{Engineering} \mid \text{Full-Time})$
$= \dfrac{P(\text{Engineering} \cap \text{Full-Time})}{P(\text{Full-Time})} = \dfrac{0.151}{0.613} = 0.246$
d. $P(\text{Full-Time} \mid \text{Business})$
$= \dfrac{P(\text{Full-Time} \cap \text{Business})}{P(\text{Business})} = \dfrac{0.270}{0.385} = 0.701$
e. For independence, we must have that $P(A)P(B) = P(A \cap B)$; from the joint probability table in part (a) of this problem, we have

$P(A) = 0.613$
$P(B) = 0.385$

So $P(A)\,P(B) = (0.387)(0.385) = 0.236$
But

$P(A \cap B) = (0.270)$

Because $P(A)P(B) \neq P(A \cap B)$, the events are not independent

34. a.

	On Time	Late	Total
Southwest	.3336	.0664	.40
US Airways	.2629	.0871	.35
JetBlue	.1753	.0747	.25
Total	.7718	.2282	1.00

b. Southwest (.40)
c. .7718
d. US Airways (.3817); Southwest (.2910)

36. a. We have that $P(\text{Make the Shot}) = .93$ for each foul shot, so the probability that Jamal Crawford will make two consecutive foul shots is $P(\text{Make the Shot})\,P(\text{Make the Shot}) = (.93)(.93) = .8649$
b. There are three unique ways that Jamal Crawford can make at least one shot—he can make the first shot and miss the second shot, miss the first shot and make the second shot, or make both shots; since the event "Miss the Shot" is the complement of the event "Make the Shot," $P(\text{Miss the Shot}) = 1 - P(\text{Make the Shot}) = 1 - .93 = .07$; thus

$P(\text{Make the Shot})\,P(\text{Miss the Shot}) = (.93)(.07) = .0651$
$P(\text{Miss the Shot})\,P(\text{Make the Shot}) = (.07)(.93) = .0651$
$P(\text{Make the Shot})\,P(\text{Make the Shot}) = (.93)(.93) = \underline{.8649}$
.9951

c. We can find this probability in two ways; we can calculate the probability directly

$P(\text{Miss the Shot})\,P(\text{Miss the Shot}) = (.07)(.07) = .0049$

Or we can recognize that the event "Miss both Shots" is the complement of the event "Make at Least One of the Two Shots," so

$P(\text{Miss the Shot})\,P(\text{Miss the Shot}) = 1 - .9951 = .0049$

d. For the Portland Trail Blazers' center, we have

$P(\text{Make the Shot}) = .58$ for each foul shot, so the probability that the Portland Trail Blazers' center will make two consecutive foul shots is $P(\text{Make the Shot})\,P(\text{Make the Shot}) = (.58)(.58) = .3364$

Again, there are three unique ways that the Portland Trail Blazers' center can make at least one shot—he can make the first shot and miss the second shot, miss the first shot and make the second shot, or make both shots; since the event "Miss the Shot" is the complement of the event "Make the Shot," $P(\text{Miss the Shot}) = 1 - P(\text{Make the Shot}) = 1 - .58 = .42$; thus

$P(\text{Make the Shot})\,P(\text{Miss the Shot}) = (.58)(.42) = .2436$
$P(\text{Miss the Shot})\,P(\text{Make the Shot}) = (.42)(.58) = .2436$
$P(\text{Make the Shot})\,P(\text{Make the Shot}) = (.58)(.58) = \underline{.3364}$
.8236

We can again find the probability the Portland Trail Blazers' center will miss both shots in two ways; we can calculate the probability directly

$P(\text{Miss the Shot})\,P(\text{Miss the Shot}) = (.42)(.42) = .1764$

Or we can recognize that the event "Miss both Shots" is the complement of the event "Make at Least One of the Two Shots," so

$P(\text{Miss the Shot})\, P(\text{Miss the Shot}) = 1 - .9951 = .1764$

Intentionally fouling the Portland Trail Blazers' center is a better strategy than intentionally fouling Jamal Crawford

38. a.

	Met Proficiency Standards		
Grade	**Yes**	**No**	**Total**
3	.11196	.05663	0.16858
4	.08271	.08205	0.16476
5	.08517	.07922	0.16439
6	.08588	.07777	0.16365
7	.09671	.07031	0.16702
8	.09618	.07542	0.17159
Total	.55861	.44139	1.00000

b. The column marginal probabilities are .55861 and .44139; the row marginal probabilities are .16858, .16476, and so on

c. .6641; .5020

d. .2004; .1481

39. a. Yes, because $P(A_1 \cap A_2) = 0$

b. $P(A_1 \cap B) = P(A_1)P(B \mid A_1) = .40(.20) = .08$

$P(A_2 \cap B) = P(A_2)P(B \mid A_2) = .60(.05) = .03$

c. $P(B) = P(A_1 \cap B) + P(A_2 \cap B) = .08 + .03 = .11$

d. $P(A_1 \mid B) = \dfrac{.08}{.11} = .7273$

$P(A_2 \mid B) = \dfrac{.03}{.11} = .2727$

40. a. .10, .20, .09

b. .51

c. .26, .51, .23

42. M = missed payment

D_1 = customer defaults

D_2 = customer does not default

$P(D_1) = .05, P(D_2) = .95, P(M \mid D_2) = .2, P(M \mid D_1) = 1$

a. $$P(D_1 \mid M) = \frac{P(D_1)P(M \mid D_1)}{P(D_1)P(M \mid D_1) + P(D_2)P(M \mid D_2)} = \frac{(.05)(1)}{(.05)(1) + (.95)(.2)} = \frac{.05}{.24} = .21$$

b. Yes, the probability of default is greater than .20

44. a. $P(A_1) = .095$

$P(A_2) = .905$

$P(W \mid A_1) = .60$

$P(W \mid A_2) = .49$

b.

Events	$P(A_i)$	$P(W \mid A_i)$	$P(A_i \cap W)$	$P(A_i \mid W)$
A_1	0.095	0.60	0.05700	0.1139
A_2	0.905	0.49	0.44345	0.8861
		$P(W) =$	0.50045	1.0000

$P(A_1 \mid W) = .1139$

c.

Events	$P(A_i)$	$P(M \mid A_i)$	$P(A_i \cap M)$	$P(A_i \mid M)$
A_1	0.095	0.40	0.03800	0.0761
A_2	0.905	0.51	0.46155	0.9239
		$P(M) =$	0.49995	1.0000

$P(A_1 \mid M) = .0761$

d. $P(W) = .50045$

$P(M) = .49965$

46. a. .60

b. .26

c. .40

d. .74

48. a. 0.5029

b. 0.5758

c. No, from part (a) we have $P(F) = 0.5029$ and from part (b) we have $P(A \mid F) = 0.5758$; since $P(F) \neq P(A \mid F)$, events A and F are not independent

50. a. .76

b. .24

52. b. .2022

c. .4618

d. .4005

54. a. .7766

b. $P(\text{OKAY} \mid 30\text{–}49)$

$$= \frac{P(\text{OKAY} \cap 30-49)}{P(30-49)} = \frac{0.0907}{0.3180} = 0.2852$$

c. $P(50+ \mid \text{NOT OKAY})$

$$= \frac{P(50+ \cap \text{NOT OKAY})}{P(\text{NOT OKAY})} = \frac{0.4008}{0.7766} = 0.5161$$

d. The attitude about this practice is not independent of the age of the respondent; we can show this in several ways: One example is to use the result from part (b); We have

$P(\text{OKAY} \mid 30\text{–}49) = 0.2852$

and

$P(\text{OKAY}) = 0.2234$

If the attitude about this practice were independent of the age of the respondent, we would expect these probabilities to be equal; since these probabilities are not equal, the data suggest the attitude about this practice is not independent of the age of the respondent

e. Respondents in the 50+ age category are far more likely to say this practice is NOT OKAY than are respondents in the 18–29 age category:

$P(\text{NOT OKAY} \mid 50+)$

$$= \frac{P(\text{NOT OKAY} \cap 50+)}{P(50+)} = \frac{0.4008}{0.4731} = 0.8472$$

$P(\text{NOT OKAY} \mid 18\text{–}29)$

$$= \frac{P(\text{NOT OKAY} \cap 18\text{–}29)}{P(18\text{–}29)} = \frac{0.1485}{0.2089} = 0.7109$$

56. a. .25
b. .125
c. .0125
d. .10
e. No

58. a.

	Young Adult	Older Adult	Total
Blogger	.0432	.0368	.08
Nonblogger	.2208	.6992	.92
Total	.2640	.7360	1.00

b. .2640
c. .0432
d. .1636

60. a. $P(\text{spam}|\textit{shipping!})$

$$= \frac{P(\text{spam})P(\textit{shipping!}|\text{spam})}{P(\text{spam})P(\textit{shipping!}|\text{spam}) + P(\text{ham})P(\textit{shipping!}|\text{ham})}$$

$$= \frac{(0.10)(0.051)}{(0.10)(0.051) + (0.90)(0.0015)} = 0.791$$

$P(\text{ham}|\textit{shipping!})$

$$= \frac{P(\text{ham})P(\textit{shipping!}|\text{ham})}{P(\text{ham})P(\textit{shipping!}|\text{ham}) + P(\text{spam})P(\textit{shipping!}|\text{spam})}$$

$$= \frac{(0.90)(0.0015)}{(0.90)(0.0015) + (0.10)(0.051)} = 0.209$$

If a message includes the word *shipping!*, the probability the message is spam is high (0.7910), and so the message should be flagged as spam

b. $P(\text{spam}|\textit{today!})$

$$= \frac{P(\text{spam})P(\textit{today!}|\text{spam})}{P(\text{spam})P(\textit{today!}|\text{spam}) + P(\text{ham})P(\textit{today!}|\text{ham})}$$

$$= \frac{(0.10)(0.045)}{(0.10)(0.045) + (0.90)(0.0022)} = 0.694$$

$P(\text{spam}|\textit{here!})$

$$= \frac{P(\text{spam})P(\textit{here!}|\text{spam})}{P(\text{spam})P(\textit{here!}|\text{spam}) + P(\text{ham})P(\textit{here!}|\text{ham})}$$

$$= \frac{(0.10)(0.034)}{(0.10)(0.034) + (0.90)(0.0022)} = 0.632$$

A message that includes the word *today!* is more likely to be spam because $P(\textit{today!}|\text{spam})$ is larger than $P(\textit{here!}|\text{spam})$; because *today!* occurs more often in unwanted messages (spam), it is easier to distinguish spam from ham in messages that include *today!*

c. $P(\text{spam}|\textit{available})$

$$= \frac{P(\text{spam})P(\textit{available}|\text{spam})}{P(\text{spam})P(\textit{available}|\text{spam}) + P(\text{ham})P(\textit{available}|\text{ham})}$$

$$= \frac{(0.10)(0.014)}{(0.10)(0.014) + (0.90)(0.0041)} = 0.275$$

$P(\text{spam}|\textit{fingertips!})$

$$= \frac{P(\text{spam})P(\textit{fingertips!}|\text{spam})}{P(\text{spam})P(\textit{fingertips!}|\text{spam}) + P(\text{ham})P(\textit{fingertips!}|\text{ham})}$$

$$= \frac{(0.10)(0.014)}{(0.10)(0.014) + (0.90)(0.0011)} = 0.586$$

A message that includes the word *fingertips!* is more likely to be spam; because $P(\textit{fingertips!}|\text{ham})$ is smaller than $P(\textit{available}|\text{ham})$; because *available* occurs more often in legitimate messages (ham), it is more difficult to distinguish spam from ham in messages that include *available*

d. It is easier to distinguish spam from ham when a word occurs more often in unwanted messages (spam) and/or less often in legitimate messages (ham)

Chapter 5

1. a. Head, Head (H, H)
Head, Tail (H, T)
Tail, Head (T, H)
Tail, Tail (T, T)

b. x = number of heads on two coin tosses

c.

Outcome	Values of x
(H, H)	2
(H, T)	1
(T, H)	1
(T, T)	0

d. Discrete; 0, 1, and 2

2. a. x = time in minutes to assemble product
b. Any positive value: $x > 0$
c. Continuous

3. Let Y = position is offered
N = position is not offered

a. $S = \{(Y, Y, Y), (Y, Y, N), (Y, N, Y), (Y, N, N), (N, Y, Y), (N, Y, N), (N, N, Y), (N, N, N)\}$

b. Let N = number of offers made; N is a discrete random variable

c.

Experimental Outcome	(Y, Y, Y)	(Y, Y, N)	(Y, N, Y)	(Y, N, N)	(N, Y, Y)	(N, Y, N)	(N, N, Y)	(N, N, N)
Value of N	3	2	2	1	2	1	1	0

4. $x = 0, 1, 2, \ldots, 9$

6. a. 0, 1, 2, . . . , 20; discrete
b. 0, 1, 2, . . . ; discrete
c. 0, 1, 2, . . . , 50; discrete
d. $0 \le x \le 8$; continuous
e. $x > 0$; continuous

7. a. $f(x) \geq 0$ for all values of x
$\Sigma f(x) = 1$; therefore, it is a valid probability distribution
b. Probability $x = 30$ is $f(30) = .25$
c. Probability $x \leq 25$ is $f(20) + f(25) = .20 + .15 = .35$
d. Probability $x > 30$ is $f(35) = .40$

8. a.

x	$f(x)$
1	3/20 = .15
2	5/20 = .25
3	8/20 = .40
4	4/20 = .20
Total	1.00

b.

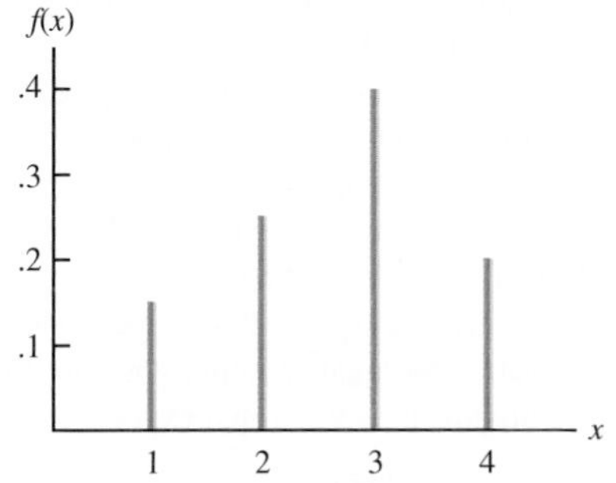

c. $f(x) \geq 0$ for $x = 1, 2, 3, 4$
$\Sigma f(x) = 1$

10. a.

x	1	2	3	4	5
$f(x)$	.05	.09	.03	.42	.41

b.

x	1	2	3	4	5
$f(x)$	.04	.10	.12	.46	.28

c. .83
d. .28
e. Senior executives are more satisfied

12. a. Yes
b. .15
c. .10

14. a. .05
b. .70
c. .40

16. a.

y	$f(y)$	$yf(y)$
2	.20	.4
4	.30	1.2
7	.40	2.8
8	.10	.8
Totals	1.00	5.2

$E(y) = \mu = 5.2$

b.

y	$y - \mu$	$(y - \mu)^2$	$f(y)$	$(y - \mu)^2 f(y)$
2	−3.20	10.24	.20	2.048
4	−1.20	1.44	.30	.432
7	1.80	3.24	.40	1.296
8	2.80	7.84	.10	.784
			Total	4.560

$Var(y) = 4.56$

$\sigma = \sqrt{4.56} = 2.14$

18. a/b.

x	$f(x)$	$xf(x)$	$x - \mu$	$(x - \mu)^2$	$(x - \mu)^2 f(x)$
0	0.04	0.00	−1.84	3.39	0.12
1	0.34	0.34	−0.84	0.71	0.24
2	0.41	0.82	0.16	0.02	0.01
3	0.18	0.53	1.16	1.34	0.24
4	0.04	0.15	2.16	4.66	0.17
Total	1.00	1.84			0.79
		↑ $E(x)$			↑ $Var(x)$

c/d.

y	$f(y)$	$yf(y)$	$y - \mu$	$(y - \mu)^2$	$y - \mu^2 f(y)$
0	0.00	0.00	−2.93	8.58	0.01
1	0.03	0.03	−1.93	3.72	0.12
2	0.23	0.45	−0.93	0.86	0.20
3	0.52	1.55	0.07	0.01	0.00
4	0.22	0.90	1.07	1.15	0.26
Total	1.00	2.93			0.59
		↑ $E(y)$			↑ $Var(y)$

e. The number of bedrooms in owner-occupied houses is greater than in renter-occupied houses; the expected number of bedrooms is $2.93 - 1.84 = 1.09$ greater, and the variability in the number of bedrooms is less for the owner-occupied houses

20. a. 430
b. −90; concern is to protect against the expense of a large loss

22. a. 445
b. \$1250 loss

24. a. Medium: 145; large: 140
b. Medium: 2725; large: 12,400

25. a.

$E(x) = .2(50) + .5(30) + .3(40) = 37$

$E(y) = .2(80) + .5(50) + .3(60) = 59$

$Var(x) = .2(50 - 37)^2 + .5(30 - 37)^2 + .3(40 - 37)^2 = 61$

$Var(y) = .2(80 - 59)^2 + .5(50 - 59)^2 + .3(60 - 59)^2 = 129$

b.

$x + y$	$f(x + y)$
130	.2
80	.5
100	.3

c.

$x + y$	$f(x + y)$	$(x + y)f(x + y)$	$x + y - E(x + y)$
130	.2	26	34
80	.5	40	−16
100	.3	30	4
		$E(x + y) = 96$	

$[x + y - E(x + y)]^2$	$[x + y - E(x + y)]^2 f(x + y)$
1156	231.2
256	128.0
16	4.8
	$Var(x + y) = 364$

d.

$\sigma_{xy} = [Var(x + y) - Var(x) - Var(y)]/2 = (364-61-129)/2=87$
$Var(x) = 61$ and $Var(y) = 129$ were computed in part (a), so

$$\sigma_x = \sqrt{61} = 7.8102 \qquad \sigma_y = \sqrt{129} = 11.3578$$

$$\rho_{xy} = \frac{\sigma_{xy}}{\sigma_x \sigma_y} = \frac{87}{(7.8102)(11.3578)} = .98$$

The random variables x and y are positively related; both the covariance and correlation coefficient are positive; indeed, they are very highly correlated; the correlation coefficient is almost equal to 1

e. $Var(x + y) = Var(x) + Var(y) + 2\sigma_{xy}$
$= 61 + 129 + 2(87) = 364$
$Var(x) + Var(y) = 61 + 129 = 190$

The variance of the sum of x and y is greater than the sum of the variances by two times the covariance: $2(87) = 174$; the reason it is positive is that, in this case, the variables are positively related; whenever two random variables are positively related, the variance of the sum of the random variables will be greater than the sum of the variances of the individual random variables

26. a. 5%, 1%, stock 1 is riskier
b. \$42.25, \$25.00
c. 5.825, 2.236
d. 6.875%, 3.329%
e. −.6, strong negative relationship

27. a. Dividing each of the frequencies in the table by the total number of restaurants provides the joint probability table below; the bivariate probability for each pair of quality and meal price is shown in the body of the table; this is the bivariate probability distribution; for instance, the probability of a rating of 2 on quality and a rating of 3 on meal price is given by $f(2, 3) = .18$; the marginal probability distribution for quality, x, is in the rightmost column; the marginal probability for meal price, y, is in the bottom row

	Meal Price y			
Quality x	**1**	**2**	**3**	**Total**
1	0.14	0.13	0.01	0.28
2	0.11	0.21	0.18	0.50
3	0.01	0.05	0.16	0.22
Total	0.26	0.39	0.35	1

b. $E(x) = 1(.28) + 2(.50) + 3(.22) = 1.94$
$Var(x) = .28(1 - 1.94)^2 + .50(2 - 1.94)^2 + .22(3 - 1.94)^2 = .4964$

c. $E(y) = 1(.26) + 2(.39) + 3(.35) = 2.09$
$Var(y) = .26(1 - 2.09)^2 + .39(2 - 2.09)^2 + .35(3 - 2.09)^2 = .6019$

d. $\sigma_{xy} = [Var(x + y) = Var(x) - Var(y)]/2$
$= [1.6691-.4964-.6019]/2 = .2854$
Since the covariance $\sigma_{xy} = .2854$ is positive, we can conclude that as the quality rating goes up, the meal price goes up; this is as we would expect

e. $\rho_{xy} = \dfrac{\sigma_{xy}}{\sigma_x \sigma_y} = \dfrac{.2854}{\sqrt{.4964}\sqrt{.6019}} = .5221$

With a correlation coefficient of .5221, we would call this a moderately positive relationship; it is not likely to find a low-cost restaurant that is also high quality, but, it is possible; there are 3 of them leading to $f(3,1) = .01$

28. a. $E(y) = \$45.30$, $Var(y) = 3.81$, $\sigma_y = \$1.95$
b. $E(x) = \$90.50$, $Var(x) = 24.75$, $\sigma_x = \$4.97$
c.

z	$f(z)$
128	.05
130	.20
133	.20
138	.25
140	.20
143	.10
	1.00

d. \$135.8, 21.26, \$4.61
e. No, $\sigma_{xy} = -3.65$
f. No, they are less

30. a. 333.486, −1.974
b. 9.055%, 19.89%
c. 9.425%, 11.63%
d. 7.238%, 4.94%
e. Aggresive: 50% Core Bonds, 50% REITS
Conservative: 80% Core Bonds, 20% REITS

31. a.

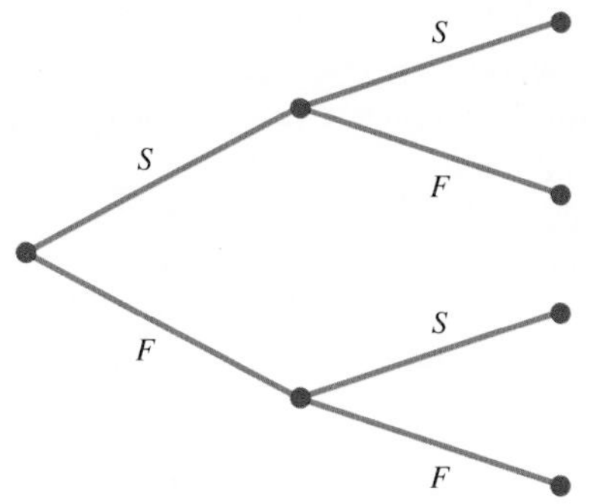

b. $f(1) = \binom{2}{1}(.4)^1(.6)^1 = \frac{2!}{1!1!}(.4)(.6) = .48$

c. $f(0) = \binom{2}{0}(.4)^0(.6)^2 = \frac{2!}{0!2!}(1)(.36) = .36$

d. $f(2) = \binom{2}{2}(.4)^2(.6)^0 = \frac{2!}{2!0!}(.16)(.1) = .16$

e. $P(x \geq 1) = f(1) + f(2) = .48 + .16 = .64$

f. $E(x) = np = 2(.4) = .8$
$Var(x) = np(1 - p) = 2(.4)(.6) = .48$
$\sigma = \sqrt{.48} = .6928$

32. a. .3487
b. .1937
c. .9298
d. .6513
e. 1
f. .9, .95

34. a. .2789
b. .4181
c. .0733

36. a. Probability of a defective part being produced must be .03 for each part selected; parts must be selected independently

b. Let D = defective
G = not defective

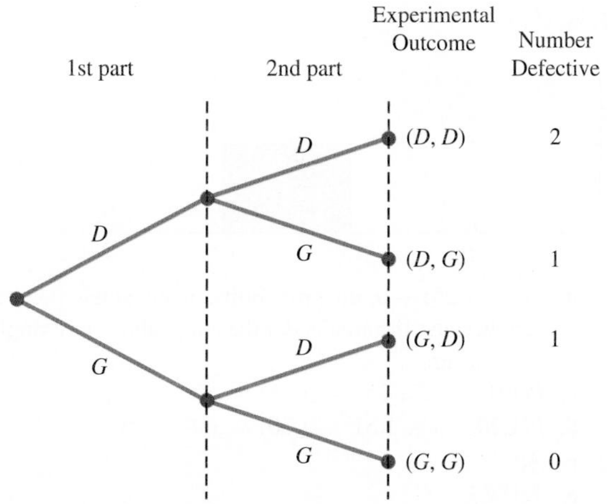

c. Two outcomes result in exactly one defect

d. $P(\text{no defects}) = (.97)(.97) = .9409$
$P(1 \text{ defect}) = 2(.03)(.97) = .0582$
$P(2 \text{ defects}) = (.03)(.03) = .0009$

38. a. .90
b. .99
c. .999
d. Yes

40. a. .2262
b. .8355

42. a. .1897
b. .9757
c. $f(12) = .0008$; yes
d. 5

44. a. $f(x) = \frac{3^x e^{-3}}{x!}$
b. .2241
c. .1494
d. .8008

45. a. $f(x) = \frac{2^x e^{-2}}{x!}$

b. $\mu = 6$ for 3 time periods

c. $f(x) = \frac{6^x e^{-6}}{x!}$

d. $f(2) = \frac{2^2 e^{-2}}{2!} = \frac{4(.1353)}{2} = .2706$

e. $f(6) = \frac{6^6 e^{-6}}{6!} = .1606$

f. $f(5) = \frac{4^5 e^{-4}}{5!} = .1563$

46. a. .1952
b. .1048
c. .0183
d. .0907

48. a. $f(0) = \frac{7^0 e^{-7}}{0!} = e^{-7} = .0009$

b. probability $= 1 - [f(0) + f(1)]$
$f(1) = \frac{7^1 e^{-7}}{1!} = 7e^{-7} = .0064$
probability $= 1 - [.0009 + .0064] = .9927$

c. $\mu = 3.5$
$f(0) = \frac{3.5^0 e^{-3.5}}{0!} = e^{-3.5} = .0302$
probability $= 1 - f(0) = 1 - .0302 = .9698$

d.
probability $= 1 - [f(0) + f(1) + f(2) + f(3) + f(4)]$
$= 1 - [.0009 + .0064 + .0223 + .0521 + .0912]$
$= .8271$

50. a. $\mu = 1.25$
b. .2865
c. .3581
d. .3554

52. a. $f(1) = \dfrac{\binom{3}{1}\binom{10-3}{4-1}}{\binom{10}{4}} = \dfrac{\left(\frac{3!}{1!2!}\right)\left(\frac{7!}{3!4!}\right)}{\frac{10!}{4!6!}}$

$= \dfrac{(3)(35)}{210} = .50$

b. $f(2) = \dfrac{\binom{3}{2}\binom{10-3}{2-2}}{\binom{10}{2}} = \dfrac{(3)(1)}{45} = .067$

c. $f(0) = \dfrac{\binom{3}{0}\binom{10-3}{2-0}}{\binom{10}{2}} = \dfrac{(1)(21)}{45} = .4667$

d. $f(2) = \dfrac{\binom{3}{2}\binom{10-3}{4-2}}{\binom{10}{4}} = \dfrac{(3)(21)}{210} = .30$

e. $x = 4$ is *greater than* $r = 3$; thus, $f(4) = 0$

54. a. .5250
b. .8167

56. $N = 60$, $n = 10$
a. $r = 20$, $x = 0$

$f(0) = \dfrac{\binom{20}{0}\binom{40}{10}}{\binom{60}{10}} = \dfrac{(1)\left(\frac{40!}{10!30!}\right)}{\frac{60!}{10!50!}}$

$= \left(\dfrac{40!}{10!30!}\right)\left(\dfrac{10!50!}{60!}\right)$

$= \dfrac{40 \cdot 39 \cdot 38 \cdot 37 \cdot 36 \cdot 35 \cdot 34 \cdot 33 \cdot 32 \cdot 31}{60 \cdot 59 \cdot 58 \cdot 57 \cdot 56 \cdot 55 \cdot 54 \cdot 53 \cdot 52 \cdot 51}$

$= .0112$

b. $r = 20$, $x = 1$

$f(1) = \dfrac{\binom{20}{1}\binom{40}{9}}{\binom{60}{10}} = 20\left(\dfrac{40!}{9!31!}\right)\left(\dfrac{10!50!}{60!}\right)$

$= .0725$

c. $1 - f(0) - f(1) = 1 - .0112 - .0725 = .9163$
d. Same as the probability one will be from Hawaii; .0725

58. a. .2917
b. .0083
c. .5250, .1750; 1 bank
d. .7083
e. .90, .49, .70

60. a.

x	1	2	3	4	5
$f(x)$	.24	.21	.10	.21	.24

b. 3.00, 2.34
c. Bonds: $E(x) = 1.36$, $Var(x) = .23$
Stocks: $E(x) = 4$, $Var(x) = 1$

62. a. .05, .40, 0 terminal only used when a purchase is made

b.

x	$f(x)$	$E(x) = 1.14$	$Var(x) = .3804$
0	.13		
1	.60		
2	.27		
	1.00		

c. .5, .45
d.

t	$f(t)$	$E(t) = 1.64$	$Var(t) = .5504$
1	.50		
2	.38		
3	.10		
4	.02		
	1.00		

e. $\sigma_{xy} = -.14$, $\rho_{xy} = -.3384$, negative relationship

64. a. .0596
b. .3585
c. 100
d. 95, 9.75

66. a. .9510
b. .0480
c. .0490

68. a. 240
b. 12.96
c. 12.96

70. .1912

72. a. .2240
b. .5767

74. a. .4667
b. .4667
c. .0667

Chapter 6

1. a.

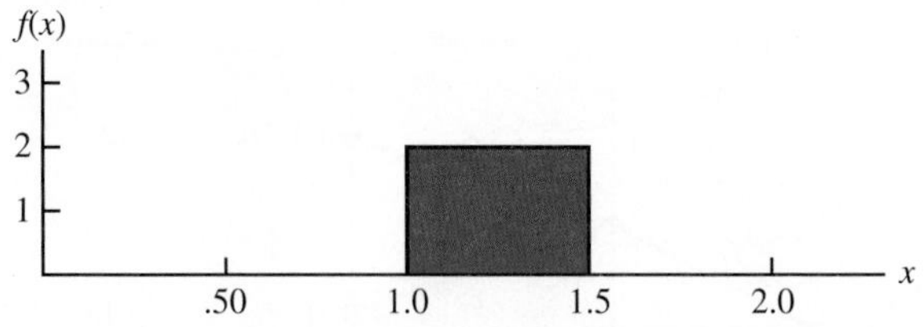

b. $P(x = 1.25) = 0$; the probability of any single point is zero because the area under the curve above any single point is zero
c. $P(1.0 \le x \le 1.25) = 2(.25) = .50$
d. $P(1.20 < x < 1.5) = 2(.30) = .60$

2. b. .50
c. .60
d. 15
e. 8.33

4. a.

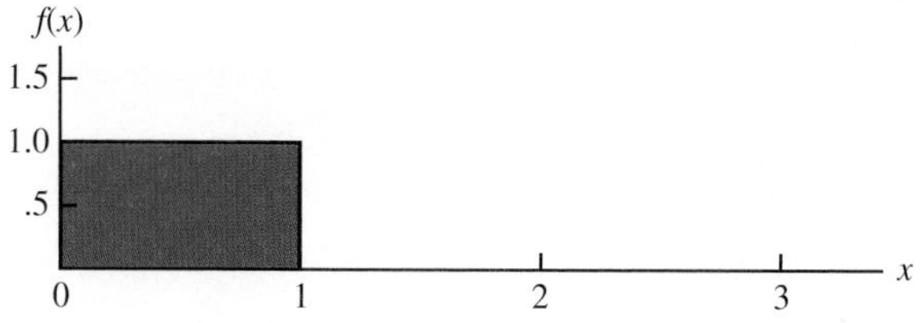

b. $P(.25 < x < .75) = 1(.50) = .50$
c. $P(x \leq .30) = 1(.30) = .30$
d. $P(x > .60) = 1(.40) = .40$

6. a. .125
b. .50
c. .25

10. a. .9332
b. .8413
c. .0919
d. .4938

12. a. .2967
b. .4418
c. .3300
d. .5910
e. .8849
f. .2389

13. a. $P(-1.98 \leq z \leq .49) = P(z \leq .49) - P(z < -1.98) = .6879 - .0239 = .6640$
b. $P(.52 \leq z \leq 1.22) = P(z \leq 1.22) - P(z < .52) = .8888 - .6985 = .1903$
c. $P(-1.75 \leq z \leq -1.04) = P(z \leq -1.04) - P(z < -1.75) = .1492 - .0401 = .1091$

14. a. $z = 1.96$
b. $z = 1.96$
c. $z = .61$
d. $z = 1.12$
e. $z = .44$
f. $z = .44$

15. a. The z value corresponding to a cumulative probability of .2119 is $z = -.80$
b. Compute $.9030/2 = .4515$; the cumulative probability of $.5000 + .4515 = .9515$ corresponds to $z = 1.66$
c. Compute $.2052/2 = .1026$; z corresponds to a cumulative probability of $.5000 + .1026 = .6026$, so $z = .26$
d. The z value corresponding to a cumulative probability of .9948 is $z = 2.56$
e. The area to the left of z is $1 - .6915 = .3085$, so $z = -.50$

16. a. $z = 2.33$
b. $z = 1.96$
c. $z = 1.645$
d. $z = 1.28$

18. $\mu = 14.4$ and $\sigma = 4.4$

a. At $x = 20$, $z = \dfrac{20 - 14.4}{4.4} = 1.27$

$P(z \leq 1.27) = .8980$
$P(x \geq 20) = 1 - .8980 = .1020$

b. At $x = 10$, $z = \dfrac{10 - 14.4}{4.4} = -1.00$

$P(z \leq -1.00) = .1587$
So, $P(x \leq 10) = .1587$

c. A z-value of 1.28 cuts off an area of approximately 10% in the upper tail

$x = 14.4 + 4.4(1.28) = 20.03$

A return of \$20.03% or higher will put a domestic stock fund in the top 10%

20. a. .1788
b. 69.15%
c. .0495

22. a. .6553
b. 13.05 hours
c. .9838

24. a. 200, 26.04
b. .2206
c. .1251
d. 242.84 million

26. a. $\mu = np = 100(.20) = 20$
$\sigma^2 = np(1 - p) = 100(.20)(.80) = 16$
$\sigma = \sqrt{16} = 4$
b. Yes, because $np = 20$ and $n(1 - p) = 80$
c. $P(23.5 \leq x \leq 24.5)$

$z = \dfrac{24.5 - 20}{4} = 1.13 \qquad P(z \leq 1.13) = .8708$

$z = \dfrac{23.5 - 20}{4} = .88 \qquad P(z \leq .88) = .8106$

$P(23.5 \leq x \leq 24.5) = P(.88 \leq z \leq 1.13) = .8708 - .8106 = .0602$

d. $P(17.5 \leq x \leq 22.5)$

$z = \dfrac{22.5 - 20}{4} = .63 \qquad P(z \leq .63) = .7357$

$z = \dfrac{17.5 - 20}{4} = -.63 \qquad P(z \leq -.63) = .2643$

$P(17.5 \leq x \leq 22.5) = P(-.63 \leq z \leq .63) = .7357 - .2643 = .4714$

e. $P(x \leq 15.5)$

$z = \dfrac{15.5 - 20}{4} = -1.13 \qquad P(z \leq -1.13) = .1292$

$P(x \leq 15.5) = P(z \leq -1.13) = .1292$

28. a. $\mu = np = 250(.20) = 50$
b. $\sigma^2 = np(1 - p) = 250(.20)(1 - 20) = 40$
$\sigma = \sqrt{40} = 6.3246$
$P(x < 40) = P(x \leq 39.5)$

$z = \dfrac{x - \mu}{\sigma} = \dfrac{39.5 - 50}{6.3246} = -1.66 \qquad \text{Area} = .0485$

$P(x \leq 39.5) = .0485$

c. $P(55 \leq x \leq 60) = P(54.5 \leq x \leq 60.5)$

$z = \dfrac{x - \mu}{\sigma} = \dfrac{54.5 - 50}{6.3246} = .71 \qquad \text{Area} = .7611$

$z = \frac{x - \mu}{\sigma} = \frac{60.5 - 50}{6.3246} = 1.66$ Area = .9515

$P(54.5 \le x \le 60.5) = .9515 - .7611 = .1904$

d. $P(x \ge 70) = P(x \ge 69.5)$

$z = \frac{x - \mu}{\sigma} = \frac{69.5 - 50}{6.3246} = 3.08$ Area = .9990

$P(x \ge 69.5) = 1 - .9990 = .0010$

30. a. 144
b. .1841
c. .9943

32. a. .5276
b. .3935
c. .4724
d. .1341

33. a. $P(x \le x_0) = 1 - e^{-x_0/3}$
b. $P(x \le 2) = 1 - e^{-2/3} = 1 - .5134 = .4866$
c. $P(x \ge 3) = 1 - P(x \le 3) = 1 - (1 - e^{-3/3})$
$= e^{-1} = .3679$
d. $P(x \le 5) = 1 - e^{-5/3} = 1 - .1889 = .8111$
e. $P(2 \le x \le 5) = P(x \le 5) - P(x \le 2)$
$= .8111 - .4866 = .3245$

34. a. $f(x) = \frac{1}{20}e^{-x/20}$
b. .5276
c. .3679
d. .5105

35. a.

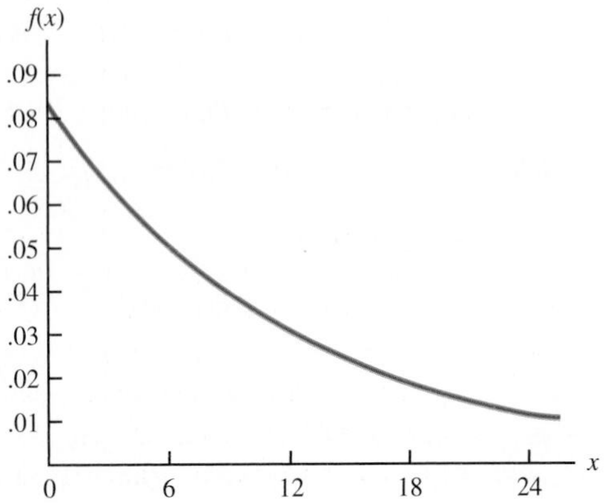

b. $P(x \le 12) = 1 - e^{-12/12} = 1 - .3679 = .6321$
c. $P(x \le 6) = 1 - e^{-6/12} = 1 - .6065 = .3935$
d. $P(x \ge 30) = 1 - P(x < 30)$
$= 1 - (1 - e^{-30/12})$
$= .0821$

36. a. .3935
b. .2386
c. .1353

38. a. $f(x) = 5.5e^{-5.5x}$
b. .2528
c. .6002

40. a. $16,312
b. 7.64%
c. $22,948

42. a. $\sigma = 25.5319$
b. .9401
c. 706 or more

44. a. .0228
b. $50

46. a. 38.3%
b. 3.59% better, 96.41% worse
c. 38.21%

48. $\mu = 19.23$ ounces

50. a. Lose $240
b. .1788
c. .3557
d. .0594

52. a. 1/7 minute
b. $7e^{-7x}$
c. .0009
d. .2466

54. a. 2 minutes
b. .2212
c. .3935
d. .0821

Chapter 7

1. a. AB, AC, AD, AE, BC, BD, BE, CD, CE, DE
b. With 10 samples, each has a 1/10 probability
c. E and C because 8 and 0 do not apply; 5 identifies E; 7 does not apply; 5 is skipped because E is already in the sample; 3 identifies C; 2 is not needed because the sample of size 2 is complete

2. 22, 147, 229, 289

3. 459, 147, 385, 113, 340, 401, 215, 2, 33, 348

4. a. Bell South, LSI Logic, General Electric
b. 120

6. 2782, 493, 825, 1807, 289

8. ExxonMobil, Chevron, Travelers, Microsoft, Pfizer, and Intel

10. a. finite; **b.** infinite; **c.** infinite; **d.** finite; **e.** infinite

11. a. $\bar{x} = \frac{\Sigma x_i}{n} = \frac{54}{6} = 9$

b. $s = \sqrt{\frac{\Sigma(x_i - \bar{x})^2}{n - 1}}$

$\Sigma(x_i - \bar{x})^2 = (-4)^2 + (-1)^2 + 1^2 + (-2)^2 + 1^2 + 5^2$
$= 48$

$s = \sqrt{\frac{48}{6 - 1}} = 3.1$

12. a. .50
b. .3667

13. a. $\bar{x} = \frac{\Sigma x_i}{n} = \frac{465}{5} = 93$

b.

	x_i	$(x_i - \bar{x})$	$(x_i - \bar{x})^2$
	94	+1	1
	100	+7	49
	85	−8	64
	94	+1	1
	92	−1	1
Totals	465	0	116

$$s = \sqrt{\frac{\Sigma(x_i - \bar{x})^2}{n-1}} = \sqrt{\frac{116}{4}} = 5.39$$

14. a. .45
b. .15
c. .45

16. a. All U.S. adults age 50 and older
b. .8216
c. 315
d. .8310
e. Target population is same as sampled population; if restricted to AARP members, inferences are questionable

18. a. 200
b. 5
c. Normal with $E(\bar{x}) = 200$ and $\sigma_{\bar{x}} = 5$
d. The probability distribution of $\bar{x}$

19. a. The sampling distribution is normal with
$E(\bar{x}) = \mu = 200$
$\sigma_{\bar{x}} = \sigma/\sqrt{n} = 50/\sqrt{100} = 5$
For ± 5, $195 \le \bar{x} \le 205$
Using the standard normal probability table:

$$\text{At } \bar{x} = 205,\ z = \frac{\bar{x} - \mu}{\sigma_{\bar{x}}} = \frac{5}{5} = 1$$

$P(z \le 1) = .8413$

$$\text{At } \bar{x} = 195,\ z = \frac{\bar{x} - \mu}{\sigma_{\bar{x}}} = \frac{-5}{5} = -1$$

$P(z < -1) = .1587$
$P(195 \le \bar{x} \le 205) = .8413 - .1587 = .6826$

b. For ± 10, $190 \le \bar{x} \le 210$
Using the standard normal probability table:

$$\text{At } \bar{x} = 210,\ z = \frac{\bar{x} - \mu}{\sigma_{\bar{x}}} = \frac{10}{5} = 2$$

$P(z \le 2) = .9772$

$$\text{At } \bar{x} = 190,\ z = \frac{\bar{x} - \mu}{\sigma_{\bar{x}}} = \frac{-10}{5} = -2$$

$P(z < -2) = .0228$
$P(190 \le \bar{x} \le 210) = .9722 - .0228 = .9544$

20. 3.54, 2.50, 2.04, 1.77
$\sigma_{\bar{x}}$ decreases as n increases

22. a. Normal with $E(\bar{x}) = 51{,}800$ and $\sigma_{\bar{x}} = 516.40$
b. $\sigma_{\bar{x}}$ decreases to 365.15
c. $\sigma_{\bar{x}}$ decreases as n increases

23. a.

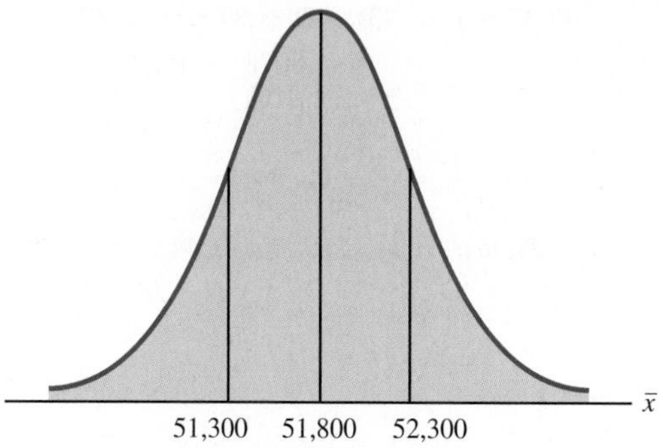

$$\sigma_{\bar{x}} = \frac{\sigma}{\sqrt{n}} = \frac{4000}{\sqrt{60}} = 516.40$$

$$\text{At } \bar{x} = 52{,}300,\ z = \frac{52{,}300 - 51{,}800}{516.40} = .97$$

$P(\bar{x} \le 52{,}300) = P(z \le .97) = .8340$

$$\text{At } \bar{x} = 51{,}300,\ z = \frac{51{,}300 - 51{,}800}{516.40} = -.97$$

$P(\bar{x} < 51{,}300) = P(z < -.97) = .1660$
$P(51{,}300 \le \bar{x} \le 52{,}300) = .8340 - .1660 = .6680$

b. $$\sigma_{\bar{x}} = \frac{\sigma}{\sqrt{n}} = \frac{4000}{\sqrt{120}} = 365.15$$

$$\text{At } \bar{x} = 52{,}300,\ z = \frac{52{,}300 - 51{,}800}{365.15} = 1.37$$

$P(\bar{x} \le 52{,}300) = P(z \le 1.37) = .9147$

$$\text{At } \bar{x} = 51{,}300,\ z = \frac{51{,}300 - 51{,}800}{365.15} = -1.37$$

$P(\bar{x} < 51{,}300) = P(z < -1.37) = .0853$
$P(51{,}300 \le \bar{x} \le 52{,}300) = .9147 - .0853 = .8294$

24. a. Normal with $E(\bar{x}) = 17.5$ and $\sigma_{\bar{x}} = .57$
b. .9198
c. .6212

26. a. .4246, .5284, .6922, .9586
b. Higher probability the sample mean will be close to population mean

28. a. Normal with $E(\bar{x}) = 95$ and $\sigma_{\bar{x}} = 2.56$
b. .7580
c. .8502
d. Part (c), larger sample size

30. a. $n/N = .01$; no
b. 1.29, 1.30; little difference
c. .8764

32. a. $E(\bar{p}) = .40$

$$\sigma_{\bar{p}} = \sqrt{\frac{p(1-p)}{n}} = \sqrt{\frac{(.40)(.60)}{200}} = .0346$$

Within $\pm .03$ means $.37 \le \bar{p} \le .43$

$$z = \frac{\bar{p} - p}{\sigma_{\bar{p}}} = \frac{.03}{.0346} = .87$$

$P(.37 \le \bar{p} \le .43) = P(-.87 \le z \le .87)$
$= .8078 - .1922$
$= .6156$

b. $z = \dfrac{\bar{p} - p}{\sigma_{\bar{p}}} = \dfrac{.05}{.0346} = 1.44$

$P(.35 \le \bar{p} \le .45) = P(-1.44 \le z \le 1.44)$
$= .9251 - .0749$
$= .8502$

34. a. .6156
b. .7814
c. .9488
d. .9942
e. Higher probability with larger n

35. a.

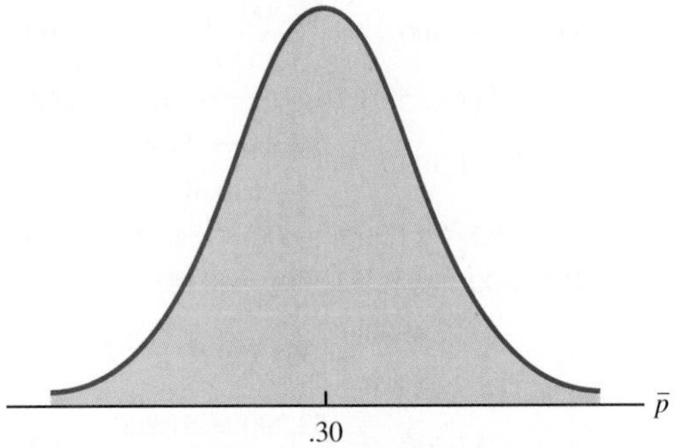

$\sigma_{\bar{p}} = \sqrt{\dfrac{p(1-p)}{n}} = \sqrt{\dfrac{.30(.70)}{100}} = .0458$

The normal distribution is appropriate because $np = 100(.30) = 30$ and $n(1 - p) = 100(.70) = 70$ are both greater than 5

b. $P(.20 \le \bar{p} \le .40) = ?$

$z = \dfrac{.40 - .30}{.0458} = 2.18$

$P(.20 \le \bar{p} \le .40) = P(-2.18 \le z \le 2.18)$
$= .9854 - .0146$
$= .9708$

c. $P(.25 \le \bar{p} \le .35) = ?$

$z = \dfrac{.35 - .30}{.0458} = 1.09$

$P(.25 \le \bar{p} \le .35) = P(-1.09 \le z \le 1.09)$
$= .8621 - .1379$
$= .7242$

36. a. Normal with $E(\bar{p}) = .55$ and $\sigma_{\bar{p}} = .0352$
b. .8444
c. Normal with $E(\bar{p}) = .45$ and $\sigma_{\bar{p}} = .0352$
d. .8444
e. No, standard error is the same in both parts
f. .9556, the probability is larger because the increased sample size reduces the standard error

38. a. Normal with $E(\bar{p}) = .42$ and $\sigma_{\bar{p}} = .0285$
b. .7062
c. .9198
d. Probabilities would increase

40. a. Normal with $E(\bar{p}) = .76$ and $\sigma_{\bar{p}} = .0214$
b. .8384
c. .9452

42. 122, 99, 25, 55, 115, 102, 61

44. a. Normal with $E(\bar{x}) = 406$ and $\sigma_{\bar{x}} = 10$
b. .8664
c. $z = -2.60$, .0047, Yes

46. a. 955
b. .50
c. .7062
d. .8230

48. a. 625
b. .7888

50. a. Normal with $E(\bar{p}) = .28$ and $\sigma_{\bar{p}} = .0290$
b. .8324
c. .5098

52. a. .8882
b. .0233

54. a. 48
b. Normal, $E(\bar{p}) = .25$, $\sigma_{\bar{p}} = .0625$
c. .2119

Chapter 8

2. Use $\bar{x} \pm z_{\alpha/2}(\sigma/\sqrt{n})$
a. $32 \pm 1.645(6/\sqrt{50})$
32 ± 1.4; 30.6 to 33.4
b. $32 \pm 1.96(6/\sqrt{50})$
32 ± 1.66; 30.34 to 33.66
c. $32 \pm 2.576(6/\sqrt{50})$
32 ± 2.19; 29.81 to 34.19

4. 54

5. a. With 99% confidence $z_{\alpha/2} = z_{.005} = 2.576$
Margin of Error $= 2.576\,\sigma/\sqrt{n} = 2.576\,(6/\sqrt{64}) = 1.93$
b. Confidence Interval: 21.52 ± 1.93 or 19.59 to 23.45

6. 8.1 to 8.9

8. a. Population is at least approximately normal
b. 3.41
c. 4.48

10. a. \$113,638 to \$124,672
b. \$112,581 to \$125,729
c. \$110,515 to \$127,795
d. Width increases as confidence level increases

12. a. 2.179
b. −1.676
c. 2.457
d. −1.708 and 1.708
e. −2.014 and 2.014

13. a. $\bar{x} = \dfrac{\Sigma x_i}{n} = \dfrac{80}{8} = 10$

b. $s = \sqrt{\dfrac{\Sigma(x_i - \bar{x})^2}{n - 1}} = \sqrt{\dfrac{84}{7}} = 3.464$

c. $t_{.025}\left(\frac{s}{\sqrt{n}}\right) = 2.365\left(\frac{3.46}{\sqrt{8}}\right) = 2.9$

d. $\bar{x} \pm t_{.025}\left(\frac{s}{\sqrt{n}}\right)$

10 ± 2.9 (7.1 to 12.9)

14. a. 21.5 to 23.5
b. 21.3 to 23.7
c. 20.9 to 24.1
d. A larger margin of error and a wider interval

15. $\bar{x} \pm t_{\alpha/2}(s/\sqrt{n})$

90% confidence: $df = 64$ and $t_{.05} = 1.669$

$19.5 \pm 1.669\left(\frac{5.2}{\sqrt{65}}\right)$

19.5 ± 1.08 or (18.42 to 20.58)

95% confidence: $df = 64$ and $t_{.025} = 1.998$

$19.5 \pm 1.998\left(\frac{5.2}{\sqrt{65}}\right)$

19.5 ± 1.29 or (18.21 to 20.79)

16. a. 9.7063, 7.9805
b. 7.1536 to 12.2590
c. 3.8854, 1.6194
d. 3.3674 to 4.4034

18. a. 22 weeks
b. 3.8020
c. 18.20 to 25.80
d. Larger n next time

20. $\bar{x} = 22$; 21.48 to 22.52

22. a. \$9,269 to \$12,541
b. 1523
c. 4,748,714, \$34 million

24. a. Planning value of $\sigma = \frac{\text{Range}}{4} = \frac{36}{4} = 9$

b. $n = \frac{z_{.025}^2\sigma^2}{E^2} = \frac{(1.96)^2(9)^2}{(3)^2} = 34.57$; use $n = 35$

c. $n = \frac{(1.96)^2(9)^2}{(2)^2} = 77.79$; use $n = 78$

25. a. Use $n = \frac{z_{\alpha/2}^2\sigma^2}{E^2}$

$n = \frac{(1.96)^2(6.84)^2}{(1.5)^2} = 79.88$; use $n = 80$

b. $n = \frac{(1.645)^2(6.84)^2}{(2)^2} = 31.65$; use $n = 32$

26. a. 25
b. 49
c. 97

28. a. 328
b. 465
c. 803
d. n gets larger; no to 99% confidence

30. 1537

31. a. $\bar{p} = \frac{100}{400} = .25$

b. $\sqrt{\frac{\bar{p}(1-\bar{p})}{n}} = \sqrt{\frac{.25(.75)}{400}} = .0217$

c. $\bar{p} \pm z_{.025}\sqrt{\frac{\bar{p}(1-\bar{p})}{n}}$

$.25 \pm 1.96(.0217)$

$.25 \pm .0424$; .2076 to .2924

32. a. .6733 to .7267
b. .6682 to .7318

34. 1068

35. a. $\bar{p} = \frac{1760}{2000} = .88$

b. Margin of error

$z_{.05}\sqrt{\frac{\bar{p}(1-\bar{p})}{n}} = 1.645\sqrt{\frac{.88(1-.88)}{2000}} = .0120$

c. Confidence interval

$.88 \pm .0120$ or .868 to .892

d. Margin of error

$z_{.05}\sqrt{\frac{\bar{p}(1-\bar{p})}{n}} = 1.96\sqrt{\frac{.88(1-.88)}{2000}} = .0142$

95% confidence interval

$.88 \pm .0142$ or .8658 to .8942

36. a. .23
b. .1716 to .2884

38. a. .1790
b. .0738, .5682 to .7158
c. 354

39. a. $n = \frac{z_{.025}^2 p^*(1-p^*)}{E^2} = \frac{(1.96)^2(.156)(1-.156)}{(.03)^2}$

$= 562$

b. $n = \frac{z_{.005}^2 p^*(1-p^*)}{E^2} = \frac{(2.576)^2(.156)(1-.156)}{(.03)^2}$

$= 970.77$; use 971

40. .0346 (.4854 to .5546)

42. a. .0442
b. 601, 1068, 2401, 9604

44. a. 4.00
b. \$29.77 to \$37.77

46. a. 122
b. \$1751 to \$1995
c. \$172, 316 million
d. Less than \$1873

48. a. 14 minutes
b. 13.38 to 14.62
c. 32 per day
d. Staff reduction

50. 37

52. 176

54. a. .5420
b. .0508
c. .4912 to .5928

56. a. .8273
b. .7957 to .8589

58. a. 1267
b. 1509

60. a. .3101
b. .2898 to .3304
c. 8219; no, this sample size is unnecessarily large

Chapter 9

2. a. H_0: $\mu \le 14$
H_a: $\mu > 14$
b. No evidence that the new plan increases sales
c. The research hypothesis $\mu > 14$ is supported; the new plan increases sales

4. a. H_0: $\mu \ge 220$
H_a: $\mu < 220$
b. Cannot conclude proposed method reduces cost.
c. Can conclude proposed method reduces cost.

5. a. Conclude that the population mean monthly cost of electricity in the Chicago neighborhood is greater than \$104 and hence higher than in the comparable neighborhood in Cincinnati
b. The Type I error is rejecting H_0 when it is true; this error occurs if the researcher concludes that the population mean monthly cost of electricity is greater than \$104 in the Chicago neighborhood when the population mean cost is actually less than or equal to \$104
c. The Type II error is accepting H_0 when it is false; this error occurs if the researcher concludes that the population mean monthly cost for the Chicago neighborhood is less than or equal to \$104 when it is not

6. a. H_0: $\mu \le 1$
H_a: $\mu > 1$
b. Claiming $\mu > 1$ when it is not true
c. Claiming $\mu \le 1$ when it is not true

8. a. H_0: $\mu \ge 220$
H_a: $\mu < 220$
b. Claiming $\mu < 220$ when it is not true
c. Claiming $\mu \ge 220$ when it is not true

10. a. $z = \dfrac{\bar{x} - \mu_0}{\sigma/\sqrt{n}} = \dfrac{26.4 - 25}{6/\sqrt{40}} = 1.48$
b. Using normal table with $z = 1.48$: p-value $= 1.0000 - .9306 = .0694$
c. p-value $> .01$, do not reject H_0
d. Reject H_0 if $z \ge 2.33$
$1.48 < 2.33$, do not reject H_0

11. a. $z = \dfrac{\bar{x} - \mu_0}{\sigma/\sqrt{n}} = \dfrac{14.15 - 15}{3/\sqrt{50}} = -2.00$
b. p-value $= 2(.0228) = .0456$
c. p-value $\le .05$, reject H_0
d. Reject H_0 if $z \le -1.96$ or $z \ge 1.96$
$-2.00 \le -1.96$, reject H_0

12. a. .1056; do not reject H_0
b. .0062; reject H_0
c. ≈ 0; reject H_0
d. .7967; do not reject H_0

14. a. .3844; do not reject H_0
b. .0074; reject H_0
c. .0836; do not reject H_0

15. a. H_0: $\mu \ge 1056$
H_a: $\mu < 1056$
b. $z = \dfrac{\bar{x} - \mu_0}{\sigma/\sqrt{n}} = \dfrac{910 - 1056}{1600/\sqrt{400}} = -1.83$
p-value $= .0336$
c. p-value $\le .05$, reject H_0; the mean refund of "last-minute" filers is less than \$1056
d. Reject H_0 if $z \le -1.645$
$-1.83 \le -1.645$; reject H_0

16. a. H_0: $\mu \le 3173$
H_a: $\mu > 3173$
b. .0207
c. Reject H_0, conclude mean credit card balance for undergraduate student has increased

18. a. H_0: $\mu = 4.1$
H_a: $\mu \ne 4.1$
b. -2.21, .0272
c. Reject H_0; return for Mid-Cap Growth Funds differs from that for U.S. Diversified Funds

20. a. H_0: $\mu \ge 32.79$
H_a: $\mu < 32.79$
b. -2.73
c. .0032
d. Reject H_0; conclude the mean monthly Internet bill is less in the southern state

22. a. H_0: $\mu = 8$
H_a: $\mu \ne 8$
b. .1706
c. Do not reject H_0; we cannot conclude the mean waiting time differs from 8 minutes
d. 7.83 to 8.97; yes

24. a. $t = \dfrac{\bar{x} - \mu_0}{s/\sqrt{n}} = \dfrac{17 - 18}{4.5/\sqrt{48}} = -1.54$
b. Degrees of freedom $= n - 1 = 47$
Area in lower tail is between .05 and .10
p-value (two-tail) is between .10 and .20
Exact p-value $= .1303$
c. p-value $> .05$; do not reject H_0
d. With $df = 47$, $t_{.025} = 2.012$
Reject H_0 if $t \le -2.012$ or $t \ge 2.012$
$t = -1.54$; do not reject H_0

26. a. Between .02 and .05; exact p-value $= .0397$; reject H_0
b. Between .01 and .02; exact p-value $= .0125$; reject H_0
c. Between .10 and .20; exact p-value $= .1285$; do not reject H_0

27. a. H_0: $\mu \geq 238$
H_a: $\mu < 238$

b. $t = \dfrac{\bar{x} - \mu_0}{s/\sqrt{n}} = \dfrac{231 - 238}{80/\sqrt{100}} = -.88$

Degrees of freedom $= n - 1 = 99$
p-value is between .10 and .20
Exact p-value = .1905

c. p-value > .05; do not reject H_0
Cannot conclude mean weekly benefit in Virginia is less than the national mean

d. $df = 99$, $t_{.05} = -1.66$
Reject H_0 if $t \leq -1.66$
$-.88 > -1.66$; do not reject H_0

28. a. H_0: $\mu \geq 9$
H_a: $\mu < 9$

b. Between .005 and .01
Exact p-value = .0072

c. Reject H_0; mean tenure of a CEO is less than 9 years

30. a. H_0: $\mu = 6.4$
H_a: $\mu \neq 6.4$

b. Between .10 and .20
Exact p-value = .1268

c. Do not reject H_0; cannot conclude that the group consensus is wrong

32. a. H_0: $\mu = 10{,}192$
H_a: $\mu \neq 10{,}192$

b. Between .02 and .05
Exact p-value = .0304

c. Reject H_0; mean price at dealership differs from national mean price

34. a. H_0: $\mu = 2$
H_a: $\mu \neq 2$

b. 2.2

c. .516

d. Between .20 and .40
Exact p-value = .2535

e. Do not reject H_0; no reason to change from 2 hours for cost estimating

36. a. $z = \dfrac{\bar{p} - p_0}{\sqrt{\dfrac{p_0(1 - p_0)}{n}}} = \dfrac{.68 - .75}{\sqrt{\dfrac{.75(1 - .75)}{300}}} = -2.80$

p-value = .0026
p-value ≤ .05; reject H_0

b. $z = \dfrac{.72 - .75}{\sqrt{\dfrac{.75(1 - .75)}{300}}} = -1.20$

p-value = .1151
p-value > .05; do not reject H_0

c. $z = \dfrac{.70 - .75}{\sqrt{\dfrac{.75(1 - .75)}{300}}} = -2.00$

p-value = .0228
p-value ≤ .05; reject H_0

d. $z = \dfrac{.77 - .75}{\sqrt{\dfrac{.75(1 - .75)}{300}}} = .80$

p-value = .7881
p-value > .05; do not reject H_0

38. a. H_0: $p = .64$
H_a: $p \neq .64$

b. $\bar{p} = 52/100 = .52$

$$z = \frac{\bar{p} - p_0}{\sqrt{\dfrac{p_0(1 - p_0)}{n}}} = \frac{.52 - .64}{\sqrt{\dfrac{.64(1 - .64)}{100}}} = -2.50$$

p-value = 2(.0062) = .0124

c. p-value ≤ .05; reject H_0
Proportion differs from the reported .64

d. Yes, because $\bar{p} = .52$ indicates that fewer believe the supermarket brand is as good as the name brand

40. a. 21

b. H_0: $p \geq .46$
H_a: $p < .46$
p-value ≈.0436

c. Yes, .0436

42. a. $\bar{p} = .15$

b. .0718 to .2282

c. The return rate for the Houston store is different than the national average

44. a. H_0: $p \leq .51$
H_a: $p > .51$

b. $\bar{p} = .58$, p-value = .0026

c. Reject H_0; people working the night shift get drowsy more often

46.

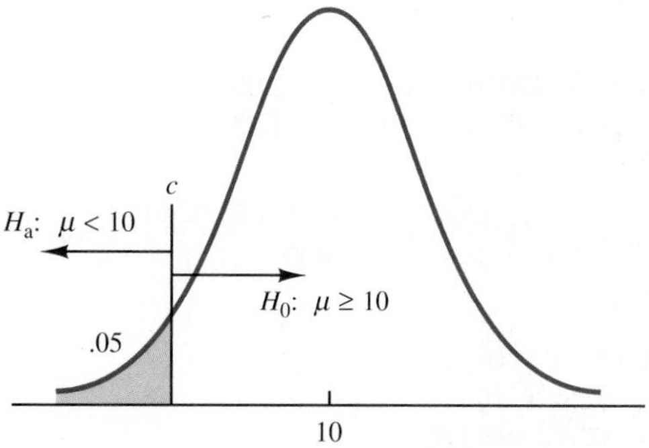

$$c = 10 - 1.645(5/\sqrt{120}) = 9.25$$

Reject H_0 if $\bar{x} \leq 9.25$

a. When $\mu = 9$,

$$z = \frac{9.25 - 9}{5/\sqrt{120}} = .55$$

$P(\text{Reject } H_0) = (1.0000 - .7088) = .2912$

b. Type II error

c. When $\mu = 8$,

$$z = \frac{9.25 - 8}{5/\sqrt{120}} = 2.74$$

$$\beta = (1.0000 - .9969) = .0031$$

48. a. Concluding $\mu \leq 15$ when it is not true
b. .2676
c. .0179

49. a. H_0: $\mu \geq 25$
H_a: $\mu < 25$
Reject H_0 if $z \leq -2.05$

$$z = \frac{\bar{x} - \mu_0}{\sigma/\sqrt{n}} = \frac{\bar{x} - 25}{3/\sqrt{30}} = -2.05$$

Solve for $\bar{x} = 23.88$
Decision Rule: Accept H_0 if $\bar{x} > 23.88$
Reject H_0 if $\bar{x} \leq 23.88$

b. For $\mu = 23$,

$$z = \frac{23.88 - 23}{3/\sqrt{30}} = 1.61$$

$$\beta = 1.0000 - .9463 = .0537$$

c. For $\mu = 24$,

$$z = \frac{23.88 - 24}{3/\sqrt{30}} = -.22$$

$$\beta = 1.0000 - .4129 = .5871$$

d. The Type II error cannot be made in this case; note that when $\mu = 25.5$, H_0 is true; the Type II error can only be made when H_0 is false

50. a. Concluding $\mu = 28$ when it is not true
b. .0853, .6179, .6179, .0853
c. .9147

52. .1151, .0015
Increasing n reduces β

54. $$n = \frac{(z_\alpha + z_\beta)^2\sigma^2}{(\mu_0 - \mu_a)^2} = \frac{(1.645 + 1.28)^2(5)^2}{(10 - 9)^2} = 214$$

56. 109

57. At $\mu_0 = 400$, $\alpha = .02$; $z_{.02} = 2.05$
At $\mu_a = 385$, $\beta = .10$; $z_{.10} = 1.28$
With $\sigma = 30$,

$$n = \frac{(z_\alpha + z_\beta)^2\sigma^2}{(\mu_0 - \mu_a)^2} = \frac{(2.05 + 1.28)^2(30)^2}{(400 - 385)^2} = 44.4 \text{ or } 45$$

58. 324

60. a. H_0: $\mu = 16$
H_a: $\mu \neq 16$
b. .0286; reject H_0
Readjust line
c. .2186; do not reject H_0
Continue operation
d. $z = 2.19$; reject H_0
$z = -1.23$; do not reject H_0
Yes, same conclusion

62. a. H_0: $\mu \leq 119{,}155$
H_a: $\mu > 119{,}155$
b. .0047
c. Reject H_0; mean annual income for theatergoers in Bay Area is higher

64. $t = -1.05$
p-value between .20 and .40
Exact p-value = .2999
Do not reject H_0; there is no evidence to conclude that the age at which women had their first child has changed

66. $t = 2.26$
p-value between .01 and .025
Exact p-value = .0155
Reject H_0; mean cost is greater than $125,000

68. a. H_0: $p \leq .80$
H_a: $p > .80$
Conclude that airline travelers feel security will be improved
b. Cannot reject H_0; mandatory use is not recommended

70. a. H_0: $p \leq .80$
H_a: $p > .80$
b. .84
c. .0418
d. Reject H_0; more than 80% of customers are satisfied with service of home agents

72. H_0: $p \geq .90$
H_a: $p < .90$
p-value = .0808
Do not reject H_0; claim of at least 90% cannot be rejected

74. a. H_0: $\mu \leq 72$
H_a: $\mu > 72$
b. .2912
c. .7939
d. 0, because H_0 is true

76. a. 45
b. .0192, .2358, .7291, .7291, .2358, .0192

Chapter 10

1. a. $\bar{x}_1 - \bar{x}_2 = 13.6 - 11.6 = 2$
b. $z_{\alpha/2} = z_{.05} = 1.645$

$$\bar{x}_1 - \bar{x}_2 \pm 1.645\sqrt{\frac{\sigma_1^2}{n_1} + \frac{\sigma_2^2}{n_2}}$$

$$2 \pm 1.645\sqrt{\frac{(2.2)^2}{50} + \frac{(3)^2}{35}}$$

$$2 \pm .98 \quad (1.02 \text{ to } 2.98)$$

c. $z_{\alpha/2} = z_{.05} = 1.96$

$$2 \pm 1.96\sqrt{\frac{(2.2)^2}{50} + \frac{(3)^2}{35}}$$

$$2 \pm 1.17 \quad (.83 \text{ to } 3.17)$$

2. a. $$z = \frac{(\bar{x}_1 - \bar{x}_2) - D_0}{\sqrt{\frac{\sigma_1^2}{n_1} + \frac{\sigma_2^2}{n_2}}} = \frac{(25.2 - 22.8) - 0}{\sqrt{\frac{(5.2)^2}{40} + \frac{(6)^2}{50}}} = 2.03$$

b. p-value = $1.0000 - .9788 = .0212$
c. p-value $\leq .05$; reject H_0

4. a. $\bar{x}_1 - \bar{x}_2 = 85.36 - 81.40 = 3.96$

b. $z_{.025}\sqrt{\frac{\sigma_1^2}{n_1} + \frac{\sigma_2^2}{n_2}} = 1.96\sqrt{\frac{(4.55)^2}{37} + \frac{(3.97)^2}{44}} = 1.88$

c. 3.96 ± 1.88 (2.08 to 5.84)

6. p-value $= .0351$
Reject H_0; mean price in Atlanta lower than mean price in Houston

8. a. Reject H_0; customer service has improved for Rite Aid
b. Do not reject H_0; the difference is not statistically significant
c. p-value $= .0336$; reject H_0; customer service has improved for Expedia
d. 1.80
e. The increase for J.C. Penney is not statistically significant

9. a. $\bar{x}_1 - \bar{x}_2 = 22.5 - 20.1 = 2.4$

b. $$df = \frac{\left(\frac{s_1^2}{n_1} + \frac{s_2^2}{n_2}\right)^2}{\frac{1}{n_1 - 1}\left(\frac{s_1^2}{n_1}\right)^2 + \frac{1}{n_2 - 1}\left(\frac{s_2^2}{n_2}\right)^2}$$

$$= \frac{\left(\frac{2.5^2}{20} + \frac{4.8^2}{30}\right)^2}{\frac{1}{19}\left(\frac{2.5^2}{20}\right)^2 + \frac{1}{29}\left(\frac{4.8^2}{30}\right)^2} = 45.8$$

c. $df = 45$, $t_{.025} = 2.014$

$$t_{.025}\sqrt{\frac{s_1^2}{n_1} + \frac{s_2^2}{n_2}} = 2.014\sqrt{\frac{2.5^2}{20} + \frac{4.8^2}{30}} = 2.1$$

d. 2.4 ± 2.1 (.3 to 4.5)

10. a. $$t = \frac{(\bar{x}_1 - \bar{x}_2) - 0}{\sqrt{\frac{s_1^2}{n_1} + \frac{s_2^2}{n_2}}} = \frac{(13.6 - 10.1) - 0}{\sqrt{\frac{5.2^2}{35} + \frac{8.5^2}{40}}} = 2.18$$

b. $$df = \frac{\left(\frac{s_1^2}{n_1} + \frac{s_2^2}{n_2}\right)^2}{\frac{1}{n_1 - 1}\left(\frac{s_1^2}{n_1}\right)^2 + \frac{1}{n_2 - 1}\left(\frac{s_2^2}{n_2}\right)^2}$$

$$= \frac{\left(\frac{5.2^2}{35} + \frac{8.5^2}{40}\right)^2}{\frac{1}{34}\left(\frac{5.2^2}{35}\right)^2 + \frac{1}{39}\left(\frac{8.5^2}{40}\right)^2} = 65.7$$

Use $df = 65$

c. $df = 65$, area in tail is between .01 and .025; two-tailed p-value is between .02 and .05
Exact p-value $= .0329$

d. p-value $\leq .05$; reject H_0

12. a. $\bar{x}_1 - \bar{x}_2 = 22.5 - 18.6 = 3.9$ miles

b. $$df = \frac{\left(\frac{s_1^2}{n_1} + \frac{s_2^2}{n_2}\right)^2}{\frac{1}{n_1 - 1}\left(\frac{s_1^2}{n_1}\right)^2 + \frac{1}{n_2 - 1}\left(\frac{s_2^2}{n_2}\right)^2}$$

$$= \frac{\left(\frac{8.4^2}{50} + \frac{7.4^2}{40}\right)^2}{\frac{1}{49}\left(\frac{8.4^2}{50}\right)^2 + \frac{1}{39}\left(\frac{7.4^2}{40}\right)^2} = 87.1$$

Use $df = 87$, $t_{.025} = 1.988$

$$3.9 \pm 1.988\sqrt{\frac{8.4^2}{50} + \frac{7.4^2}{40}}$$

3.9 ± 3.3 (.6 to 7.2)

14. a. $H_0: \mu_1 - \mu_2 \geq 0$
$H_a: \mu_1 - \mu_2 < 0$
b. -2.41
c. Using t table, p-value is between .005 and .01
Exact p-value $= .009$
d. Reject H_0; nursing salaries are lower in Tampa

16. a. $H_0: \mu_1 - \mu_2 \leq 0$
$H_a: \mu_1 - \mu_2 > 0$
b. 38
c. $t = 1.80$, $df = 25$
Using t table, p-value is between .025 and .05
Exact p-value $= .0420$
d. Reject H_0; conclude higher mean score if college grad

18. a. $H_0: \mu_1 - \mu_2 = 0$
$H_a: \mu_1 - \mu_2 \neq 0$
b. 50.6 and 52.8 minutes
c. p-value greater than .40
Do not reject H_0; cannot conclude population mean delay times differ

19. a. 1, 2, 0, 0, 2
b. $\bar{d} = \Sigma d_i/n = 5/5 = 1$

c. $$s_d = \sqrt{\frac{\Sigma(d_i - \bar{d})^2}{n - 1}} = \sqrt{\frac{4}{5 - 1}} = 1$$

d. $$t = \frac{\bar{d} - \mu}{s_d/\sqrt{n}} = \frac{1 - 0}{1/\sqrt{5}} = 2.24$$

$df = n - 1 = 4$
Using t table, p-value is between .025 and .05
Exact p-value $= .0443$
p-value $\leq .05$; reject H_0

20. a. 3, −1, 3, 5, 3, 0, 1
b. 2
c. 2.08
d. 2
e. .07 to 3.93

21. $H_0: \mu_d \leq 0$
$H_a: \mu_d > 0$
$\bar{d} = .625$
$s_d = 1.30$

$$t = \frac{\bar{d} - \mu_d}{s_d/\sqrt{n}} = \frac{.625 - 0}{1.30/\sqrt{8}} = 1.36$$

$df = n - 1 = 7$
Using t table, p-value is between .10 and .20
Exact p-value $= .1080$
p-value $> .05$; do not reject H_0; cannot conclude commercial improves mean potential to purchase

22. a. \$3.41
b. \$1.67 to \$5.15
Very nice increase

24. a. $H_0: \mu_d \leq 0$
$H_a: \mu_d > 0$
$\bar{d} = 23, t = 2.05$
p-value between .05 and .025
Reject H_0; conclude airfares have increased
b. \$487, \$464
c. 5% increase in airfares

26. a. $t = -1.42$
Using t table, p-value is between .10 and .20
Exact p-value = .1718
Do not reject H_0; no difference in mean scores
b. −1.05
c. 1.28; yes

28. a. $\bar{p}_1 - \bar{p}_2 = .48 - .36 = .12$

b. $\bar{p}_1 - \bar{p}_2 \pm z_{.05}\sqrt{\dfrac{\bar{p}_1(1-\bar{p}_1)}{n_1} + \dfrac{\bar{p}_2(1-\bar{p}_2)}{n_2}}$

$.12 \pm 1.645\sqrt{\dfrac{.48(1-.48)}{400} + \dfrac{.36(1-.36)}{300}}$

.12 ± .0614 (.0586 to .1814)

c. $.12 \pm 1.96\sqrt{\dfrac{.48(1-.48)}{400} + \dfrac{.36(1-.36)}{300}}$

.12 ± .0731 (.0469 to .1931)

29. a. $\bar{p} = \dfrac{n_1\bar{p}_1 + n_2\bar{p}_2}{n_1 + n_2} = \dfrac{200(.22) + 300(.16)}{200 + 300} = .1840$

$$z = \frac{\bar{p}_1 - \bar{p}_2}{\sqrt{\bar{p}(1-\bar{p})\left(\dfrac{1}{n_1} + \dfrac{1}{n_2}\right)}}$$

$$= \frac{.22 - .16}{\sqrt{.1840(1-.1840)\left(\dfrac{1}{200} + \dfrac{1}{300}\right)}} = 1.70$$

p-value = 1.0000 − .9554 = .0446
b. p-value ≤ .05; reject H_0

30. $\bar{p}_1 = 220/400 = .55$ $\bar{p}_2 = 192/400 = .48$

$\bar{p}_1 - \bar{p}_2 \pm z_{.025}\sqrt{\dfrac{\bar{p}_1(1-\bar{p}_1)}{n_1} + \dfrac{\bar{p}_2(1-\bar{p}_2)}{n_2}}$

$.55 - .48 \pm 1.96\sqrt{\dfrac{.55(1-.55)}{400} + \dfrac{.48(1-.48)}{400}}$

.07 ± .0691 (.0009 to .1391)

32. a. $H_0: p_w \leq p_m$
$H_a: p_w > p_m$
b. $\bar{p}_w = .3699$
c. $\bar{p}_m = .3400$
d. p-value = .1093
Do not reject H_0; cannot conclude women are more likely to ask directions

34. a. .64
b. .45
c. .19 ± .0813 (.1087 to .2713)

36. a. $H_0: p_1 - p_2 \leq 0$
$H_a: p_1 - p_2 > 0$
b. .84, .81
c. p-value = .0094
Reject H_0; conclude an increase
d. .005 to .055; yes due to increase

38. a. $H_0: \mu_1 - \mu_2 = 0$
$H_a: \mu_1 - \mu_2 \neq 0$
$z = 2.79$
p-value = .0052
Reject H_0; a significant difference between systems exists

40. a. $H_0: \mu_1 - \mu_2 \leq 0$
$H_a: \mu_1 - \mu_2 > 0$
b. $t = .60, df = 57$
Using t table, p-value is greater than .20
Exact p-value = .2754
Do not reject H_0; cannot conclude that funds with loads have a higher mean rate of return

42. a. A decline of \$2.45
b. 2.45 ± 2.15 (.30 to 4.60)
c. 8% decrease
d. \$23.93

44. a. p-value ≈ 0, reject H_0
b. .0468 to .1332

46. a. .35 and .47
b. .12 ± .1037 (.0163 to .2237)
c. Yes, we would expect occupancy rates to be higher

Chapter 11

2. $s^2 = 25$
a. With 19 degrees of freedom, $\chi^2_{.05} = 30.144$ and $\chi^2_{.95} = 10.117$

$$\frac{19(25)}{30.144} \leq \sigma^2 \leq \frac{19(25)}{10.117}$$

$$15.76 \leq \sigma^2 \leq 46.95$$

b. With 19 degrees of freedom, $\chi^2_{.025} = 32.852$ and $\chi^2_{.975} = 8.907$

$$\frac{19(25)}{32.852} \leq \sigma^2 \leq \frac{19(25)}{8.907}$$

$$14.46 \leq \sigma^2 \leq 53.33$$

c. $3.8 \leq \sigma \leq 7.3$

4. a. .22 to .71
b. .47 to .84

6. a. \$41
b. 23.52
c. 17.37 to 36.40

8. a. .4748
b. .6891

c. .2383 to 1.3687
.4882 to 1.1699

9. H_0: $\sigma^2 \leq .0004$
H_a: $\sigma^2 > .0004$

$$\chi^2 = \frac{(n-1)s^2}{\sigma_0^2} = \frac{(30-1)(.0005)}{.0004} = 36.25$$

From table with 29 degrees of freedom, p-value is greater than .10
p-value $> .05$; do not reject H_0
The product specification does not appear to be violated

10. a. 84
b. 118.71
c. 10.90
d. $\chi^2 = 11.54$; p-value $> .20$
Do not reject hypothesis $\sigma = 12$

12. a. .8106
b. $\chi^2 = 9.49$
p-value greater than .20
Do not reject H_0; cannot conclude the variance for the other magazine is different

14. a. $F = 2.4$
p-value between .025 and .05
Reject H_0
b. $F_{.05} = 2.2$; reject H_0

15. a. Larger sample variance is s_1^2

$$F = \frac{s_1^2}{s_2^2} = \frac{8.2}{4} = 2.05$$

Degrees of freedom: 20, 25
From table, area in tail is between .025 and .05
p-value for two-tailed test is between .05 and .10
p-value $> .05$; do not reject H_0
b. For a two-tailed test:
$F_{\alpha/2} = F_{.025} = 2.30$
Reject H_0 if $F \geq 2.30$
$2.05 < 2.30$; do not reject H_0

16. $F = 1.59$
p-value is between .05 and .025
Reject H_0; the Fidelity Fund has greater variance

17. a. Population 1 is 4-year-old automobiles
H_0: $\sigma_1^2 \leq \sigma_2^2$
H_a: $\sigma_1^2 > \sigma_2^2$

b. $$F = \frac{s_1^2}{s_2^2} = \frac{170^2}{100^2} = 2.89$$

Degrees of freedom: 25, 24
From tables, p-value is less than .01
p-value $\leq .01$; reject H_0
Conclude that 4-year-old automobiles have a larger variance in annual repair costs compared to 2-year-old automobiles, which is expected because older automobiles are more likely to have more expensive repairs that lead to greater variance in the annual repair costs

18. $F = 1.44$
p-value greater than .20
Do not reject H_0; the difference between the variances is not statistically significant

20. $F = 5.29$
p-value ≈ 0
Reject H_0; population variances are not equal for seniors and managers

22. a. $F = 4$
p-value less than .01
Reject H_0; greater variability in stopping distance on wet pavement

24. 10.72 to 24.68

26. a. $\chi^2 = 27.44$
p-value between .01 and .025
Reject H_0; variance exceeds maximum requirements
b. .00012 to .00042

28. $\chi^2 = 31.50$
p-value between .05 and .10
Reject H_0; conclude that population variance is greater than 1

30. a. $n = 15$
b. 6.25 to 11.13

32. $F = 1.39$
Do not reject H_0; cannot conclude the variances of grade point averages are different

34. $F = 2.08$
p-value between .05 and .10
Reject H_0; conclude the population variances are not equal

Chapter 12

1. H_0: $p_1 = p_2 = p_3$
H_a: Not all population proportions are equal
Expected frequencies (e_{ij}):

	1	2	3	Total
Yes	132.0	158.4	105.6	396
No	118.0	141.6	94.4	354
Total	250	300	200	750

Chi-square calculations $(f_{ij} - e_{ij})^2/e_{ij}$:

	1	2	3	Total
Yes	2.45	.45	.87	3.77
No	2.75	.50	.98	4.22
				$\chi^2 = 7.99$

$df = k - 1 = (3 - 1) = 2$
χ^2 table with $\chi^2 = 7.99$ shows p-value between .025 and .01
p-value $\leq .05$, reject H_0; not all population proportions are equal

2. a. $\bar{p}_1 = 150/250 = .60$
$\bar{p}_2 = 150/300 = .50$
$\bar{p}_3 = 96/200 = .48$
b. For 1 vs. 2

$$CV_{12} = \sqrt{\chi^2_{\alpha;\,k-1}}\sqrt{\frac{\bar{p}_1(1-\bar{p}_1)}{n_1} + \frac{\bar{p}_2(1-\bar{p}_2)}{n_2}}$$

$$= \sqrt{5.991}\sqrt{\frac{.60(1-.60)}{250} + \frac{.50(1-.50)}{300}} = .1037$$

p_i	p_j	Difference	n_i	n_j	Critical Value	Significant Diff > CV
.60	.50	.10	250	300	.1037	
.60	.48	.12	250	200	.1150	Yes
.50	.48	.02	300	200	.1117	

One comparison is significant, 1 vs. 3

4. a. H_0: $p_1 = p_2 = p_3$
H_a: Not all population proportions are equal
b. Expected frequencies (e_{ij}):

Component	A	B	C	Total
Defective	25	25	25	75
Good	475	475	475	1425
Total	500	500	500	1500

Chi-square calculations $(f_{ij} - e_{ij})^2/e_{ij}$:

Component	A	B	C	Total
Defective	4.00	1.00	9.00	14.00
Good	.21	.05	.47	0.74
				$\chi^2 = 14.74$

$df = k - 1 = (3 - 1) = 2$
χ^2 table, $\chi^2 = 14.74$, p-value is less than .01
p-value $\leq .05$, reject H_0; three suppliers do not provide equal proportions of defective components
c. $\bar{p}_1 = 15/500 = .03$
$\bar{p}_2 = 20/500 = .04$
$\bar{p}_3 = 40/500 = .08$
For supplier A vs. supplier B

$$CV_{ij} = \sqrt{\chi^2_{\alpha;\,k-1}}\sqrt{\frac{\bar{p}_i(1-\bar{p}_i)}{n_i} + \frac{\bar{p}_j(1-\bar{p}_j)}{n_j}}$$

$$= \sqrt{5.991}\sqrt{\frac{.03(1-.03)}{500} + \frac{.04(1-.04)}{500}} = .0284$$

Comparison	p_i	p_j	Difference	n_i	n_j	Critical Value	Significant Diff > CV
A vs. B	.03	.04	.01	500	500	.0284	
A vs. C	.03	.08	.05	500	500	.0351	Yes
B vs. C	.04	.08	.04	500	500	.0366	Yes

Supplier A and supplier B are both significantly different from supplier C

6. a. .14, .09
b. $\chi^2 = 3.41$, $df = 1$
p-value between .10 and .05
Reject H_0; conclude two offices do not have equal error rates
c. z provides options for one-tailed tests

8. $\chi^2 = 5.70$, $df = 4$
p-value greater than .10
Do no reject H_0; no evidence suppliers differ in quality

9. H_0: The column variable is independent of the row variable
H_a: The column variable is not independent of the row variable
Expected frequencies (e_{ij}):

	A	B	C	Total
P	28.5	39.9	45.6	114
Q	21.5	30.1	34.4	86
Total	50	70	80	200

Chi-square calculations $(f_{ij} - e_{ij})^2 / e_{ij}$:

	A	B	C	Total
P	2.54	.42	.42	3.38
Q	3.36	.56	.56	4.48
				$\chi^2 = 7.86$

$df = (2 - 1)(3 - 1) = 2$
Using the χ^2 table, p-value between .01 and .025
p-value $\leq .05$, reject H_0; conclude variables are not independent

10. $\chi^2 = 19.77$, $df = 4$
p-value less than .005
Reject H_0; conclude variables are not independent

11. a. H_0: Ticket purchased is independent of flight
H_a: Ticket purchased is not independent of flight
Expected frequencies:

$e_{11} = 35.59$ $e_{12} = 15.41$
$e_{21} = 150.73$ $e_{22} = 65.27$
$e_{31} = 455.68$ $e_{32} = 197.32$

Observed Frequency (f_i)	Expected Frequency (e_i)	Chi-square $(f_i - e_i)^2 / e_i$
29	35.59	1.22
22	15.41	2.82
95	150.73	20.61
121	65.27	47.59
518	455.68	8.52
135	197.32	19.68
920		$\chi^2 = 100.43$

$df = (r - 1)(c - 1) = (3 - 1)(2 - 1) = 2$
Using the χ^2 table, p-value is less than .005
p-value $\leq .05$, reject H_0; conclude ticket purchased is not independent of the type of flight

b. Column Percentages

	Type of Flight	
Type of Ticket	**Domestic**	**International**
First Class	4.5%	7.9%
Business Class	14.8%	43.5%
Economy Class	80.7%	48.6%

A higher percentage of first-class and business-class tickets are purchased for international flights

12. a. $\chi^2 = 9.44$, $df = 2$
p-value is less than .01
Reject H_0; plan not independent of type of company

b.

Employment Plan	Private	Public
Add Employees	.5139	.2963
No Change	.2639	.3148
Lay Off Employees	.2222	.3889

Employment opportunities better for private companies

14. a. $\chi^2 = 6.57$, $df = 6$
p-value greater than .10
Do not reject H_0; cannot reject assumption of independence

b. 29%, 46%, and 25%
Outstanding is most frequent owner rating

16. a. 6446

b. $\chi^2 = 425.4$, $df = 15$
p-value $= 0$
Reject H_0; attitude not independent of country

c. Italy (58%), Spain (32%)

18. $\chi^2 = 45.36$, $df = 4$
p-value less than .005
Reject H_0; conclude that the ratings of the hosts are not independent

19. a. Expected frequencies:
$e_1 = 200(.40) = 80$, $e_2 = 200(.40) = 80$
$e_3 = 200(.20) = 40$
Observed frequencies: $f_1 = 60, f_2 = 120, f_3 = 20$

$$\chi^2 = \frac{(60 - 80)^2}{80} + \frac{(120 - 80)^2}{80} + \frac{(20 - 40)^2}{40}$$

$$= \frac{400}{80} + \frac{1600}{80} + \frac{400}{40} = 35$$

$df = k - 1 = 3 - 1 = 2$
$\chi^2 = 35$ shows p-value less than .005
p-value $\leq .01$; reject H_0
Conclude proportions differ from .40, .40, and .20

b. Reject H_0 if $\chi^2 \geq 9.210$; reject H_0

20. With $n = 30$, use six classes, each probability $= .1667$
$\bar{x} = 22.8$, $s = 6.27$

z	Cutoff value of x
$-.98$	$22.8 - .98\,(6.27) = 16.66$
$-.43$	$22.8 - .43\,(6.27) = 20.11$
0	$22.8 + 0\,(6.27) = 22.80$
.43	$22.8 + .43\,(6.27) = 25.49$
.98	$22.8 + .98\,(6.27) = 28.94$

Interval	Observed Frequency	Expected Frequency	Difference
less than 16.66	3	5	−2
16.66–20.11	7	5	2
20.11–22.80	5	5	0
22.80–25.49	7	5	2
25.49–28.94	3	5	−2
28.94 and up	5	5	0

$$\chi^2 = \frac{(-2)^2}{5} + \frac{(2)^2}{5} + \frac{(0)^2}{5} + \frac{(2)^2}{5} + \frac{(-2)^2}{5} + \frac{(0)^2}{5} + \frac{16}{5}$$

$= 3.20$
$df = k - 2 - 1 = 6 - 2 - 1 = 3$
$\chi^2 = 3.20$ shows p-value greater than .10
p-value $> .05$; do not reject H_0; normal distribution cannot be rejected

21. H_0: The population proportions are .29, .28, .25, and .18
H_a: The proportions are not as shown above
Expected frequencies: $300(.29) = 87$, $300(.28) = 84$
$300(.25) = 75$, $300(.18) = 54$
Observed frequencies: $f_1 = 95, f_2 = 70, f_3 = 89, f_4 = 46$

$$\chi^2 = \frac{(95 - 87)^2}{87} + \frac{(70 - 84)^2}{84} + \frac{(89 - 75)^2}{75}$$

$$+ \frac{(46 - 54)^2}{54} = 6.87$$

$df = k - 1 = 4 - 1 = 3$
$\chi^2 = 6.87$ shows p-value between .05 and .10
p-value $> .05$; do not reject H_0; no significant change in viewing audience proportions

22. $\chi^2 = 5.85$, $df = 5$
p-value greater than .10
Do not reject H_0; cannot reject the hypothesis that percentages are as stated

24. a. $\chi^2 = 14.33$, $df = 6$
p-value between .05 and .025
Reject H_0; conclude proportion not the same for each day of the week

b. 15.17, 11.90, 12.62, 11.19, 13.10, 16.43, 19.05
Saturday and Friday have the highest percentage

26. $\chi^2 = 2.8$, $df = 3$
p-value greater than .10
Do not reject H_0; assumption of a normal distribution cannot be rejected

28. a. 8.8%, 11.7%, 9.0%, 8.5%
b. $\chi^2 = 2.48, df = 3$
p-value greater than .10
Do not reject H_0; cannot reject assumption that the population proportions are equal

30. a. $\chi^2 = 9.56, df = 2$
p-value is less than .01
Reject H_0; conclude preferred pace is not independent of gender
b. Women slower 75.17% to 67.65%
Men faster 26.47% to 16.55%

32. $\chi^2 = 6.17, df = 6$
p-value greater than .10
Do not reject H_0; assumption county and day of week are independent is not rejected

34. $\chi^2 = 2.00, df = 5$
p-value greater than .10
Do not reject H_0; assumption of normal distribution cannot be rejected

Chapter 13

1. a. $\bar{\bar{x}} = (156 + 142 + 134)/3 = 144$

$$\text{SSTR} = \sum_{j=1}^{k} n_j(\bar{x}_j - \bar{\bar{x}})^2$$
$$= 6(156 - 144)^2 + 6(142 - 144)^2 + 6(134 - 144)^2$$
$$= 1488$$

b. $\text{MSTR} = \dfrac{\text{SSTR}}{k-1} = \dfrac{1488}{2} = 744$

c. $s_1^2 = 164.4, \quad s_2^2 = 131.2, \quad s_3^2 = 110.4$

$$\text{SSE} = \sum_{j=1}^{k} (n_j - 1)s_j^2$$
$$= 5(164.4) + 5(131.2) + 5(110.4)$$
$$= 2030$$

d. $\text{MSE} = \dfrac{\text{SSE}}{n_T - k} = \dfrac{2030}{18 - 3} = 135.3$

e.

Source of Variation	Sum of Squares	Degrees of Freedom	Mean Square	F	p-value
Treatments	1488	2	744	5.50	.0162
Error	2030	15	135.3		
Total	3518	17			

f. $F = \dfrac{\text{MSTR}}{\text{MSE}} = \dfrac{744}{135.3} = 5.50$

From the F table (2 numerator degrees of freedom and 15 denominator), p-value is between .01 and .025
Using Excel or Minitab, the p-value corresponding to $F = 5.50$ is .0162
Because p-value $\le \alpha = .05$, we reject the hypothesis that the means for the three treatments are equal

2.

Source of Variation	Sum of Squares	Degrees of Freedom	Mean Square	F	p-value
Treatments	300	4	75	14.07	.0000
Error	160	30	5.33		
Total	460	34			

4.

Source of Variation	Sum of Squares	Degrees of Freedom	Mean Square	F	p-value
Treatments	150	2	75	4.80	.0233
Error	250	16	15.63		
Total	400	18			

Reject H_0 because p-value $\le \alpha = .05$

6. Because p-value $= .0082$ is less than $\alpha = .05$, we reject the null hypothesis that the means of the three treatments are equal

8. $\bar{\bar{x}} = (79 + 74 + 66)/3 = 73$

$$\text{SSTR} = \sum_{j=1}^{k} n_j(\bar{x}_j - \bar{\bar{x}})^2 = 6(79 - 73)^2 + 6(74 - 73)^2$$
$$+ 6(66 - 73)^2 = 516$$

$$\text{MSTR} = \frac{\text{SSTR}}{k-1} = \frac{516}{2} = 258$$

$$s_1^2 = 34 \qquad s_2^2 = 20 \qquad s_3^2 = 32$$

$$\text{SSE} = \sum_{j=1}^{k} (n_j - 1)s_j^2 = 5(34) + 5(20) + 5(32) = 430$$

$$\text{MSE} = \frac{\text{SSE}}{n_T - k} = \frac{430}{18 - 3} = 28.67$$

$$F = \frac{\text{MSTR}}{\text{MSE}} = \frac{258}{28.67} = 9.00$$

Source of Variation	Sum of Squares	Degrees of Freedom	Mean Square	F	p-value
Treatments	516	2	258	9.00	.003
Error	430	15	28.67		
Total	946	17			

Using F table (2 numerator degrees of freedom and 15 denominator), p-value is less than .01
Using Excel or Minitab, the p-value corresponding to $F = 9.00$ is .003
Because p-value $\le \alpha = .05$, we reject the null hypothesis that the means for the three plants are equal; in other words, analysis of variance supports the conclusion that the population mean examination scores at the three NCP plants are not equal

10. p-value $= .0000$
Because p-value $\le \alpha = .05$, we reject the null hypothesis that the means for the three groups are equal

12. p-value = .0038
Because p-value $\leq \alpha = .05$, we reject the null hypothesis that the mean meal prices are the same for the three types of restaurants

13. a. $\bar{\bar{x}} = (30 + 45 + 36)/3 = 37$

$$\text{SSTR} = \sum_{j=1}^{k} n_j(\bar{x}_j - \bar{\bar{x}})^2 = 5(30 - 37)^2 + 5(45 - 37)^2 + 5(36 - 37)^2 = 570$$

$$\text{MSTR} = \frac{\text{SSTR}}{k - 1} = \frac{570}{2} = 285$$

$$\text{SSE} = \sum_{j=1}^{k} (n_j - 1)s_j^2 = 4(6) + 4(4) + 4(6.5) = 66$$

$$\text{MSE} = \frac{\text{SSE}}{n_T - k} = \frac{66}{15 - 3} = 5.5$$

$$F = \frac{\text{MSTR}}{\text{MSE}} = \frac{285}{5.5} = 51.82$$

Using F table (2 numerator degrees of freedom and 12 denominator), p-value is less than .01
Using Excel or Minitab, the p-value corresponding to $F = 51.82$ is .0000
Because p-value $\leq \alpha = .05$, we reject the null hypothesis that the means of the three populations are equal

b. $$\text{LSD} = t_{\alpha/2}\sqrt{\text{MSE}\left(\frac{1}{n_i} + \frac{1}{n_j}\right)}$$

$$= t_{.025}\sqrt{5.5\left(\frac{1}{5} + \frac{1}{5}\right)}$$

$$= 2.179\sqrt{2.2} = 3.23$$

$|\bar{x}_1 - \bar{x}_2| = |30 - 45| = 15 > \text{LSD}$; significant difference
$|\bar{x}_1 - \bar{x}_3| = |30 - 36| = 6 > \text{LSD}$; significant difference
$|\bar{x}_2 - \bar{x}_3| = |45 - 36| = 9 > \text{LSD}$; significant difference

c. $$\bar{x}_1 - \bar{x}_2 \pm t_{\alpha/2}\sqrt{\text{MSE}\left(\frac{1}{n_1} + \frac{1}{n_2}\right)}$$

$$(30 - 45) \pm 2.179\sqrt{5.5\left(\frac{1}{5} + \frac{1}{5}\right)}$$

$$-15 \pm 3.23 = -18.23 \text{ to } -11.77$$

14. a. Significant; p-value = .0106
b. LSD = 15.34
1 and 2; significant
1 and 3; not significant
2 and 3; significant

15. a.

	Manufacturer 1	Manufacturer 2	Manufacturer 3
Sample Mean	23	28	21
Sample Variance	6.67	4.67	3.33

$\bar{\bar{x}} = (23 + 28 + 21)/3 = 24$

$$\text{SSTR} = \sum_{j=1}^{k} n_j(\bar{x}_j - \bar{\bar{x}})^2 = 4(23 - 24)^2 + 4(28 - 24)^2 + 4(21 - 24)^2 = 104$$

$$\text{MSTR} = \frac{\text{SSTR}}{k - 1} = \frac{104}{2} = 52$$

$$\text{SSE} = \sum_{j=1}^{k} (n_j - 1)s_j^2 = 3(6.67) + 3(4.67) + 3(3.33) = 44.01$$

$$\text{MSE} = \frac{\text{SSE}}{n_T - k} = \frac{44.01}{12 - 3} = 4.89$$

$$F = \frac{\text{MSTR}}{\text{MSE}} = \frac{52}{4.89} = 10.63$$

Using F table (2 numerator degrees of freedom and 9 denominator), p-value is less than .01
Using Excel or Minitab, the p-value corresponding to $F = 10.63$ is .0043
Because p-value $\leq \alpha = .05$, we reject the null hypothesis that the mean time needed to mix a batch of material is the same for each manufacturer

b. $$\text{LSD} = t_{\alpha/2}\sqrt{\text{MSE}\left(\frac{1}{n_1} + \frac{1}{n_3}\right)}$$

$$= t_{.025}\sqrt{4.89\left(\frac{1}{4} + \frac{1}{4}\right)}$$

$$= 2.262\sqrt{2.45} = 3.54$$

Since $|\bar{x}_1 - \bar{x}_3| = |23 - 21| = 2 < 3.54$, there does not appear to be any significant difference between the means for manufacturer 1 and manufacturer 3

16. $\bar{x}_1 - \bar{x}_2 \pm \text{LSD}$
$23 - 28 \pm 3.54$
$-5 \pm 3.54 = -8.54$ to -1.46

18. a. Significant; p-value = .0000
b. Significant; $2.3 > \text{LSD} = 1.19$

20. a. Significant; p-value = .011
b. Comparing North and South
$|7702 - 5566| = 2136 > \text{LSD} = 1620.76$
significant difference
Comparing North and West
$|7702 - 8430| = 728 > \text{LSD} = 1620.76$
no significant difference
Comparing South and West
$|5566 - 8430| = 2864 > \text{LSD} = 1775.45$
significant difference

21. ***Treatment Means***
$\bar{x}_{\cdot 1} = 13.6, \quad \bar{x}_{\cdot 2} = 11.0, \quad \bar{x}_{\cdot 3} = 10.6$

Block Means
$\bar{x}_{1\cdot} = 9, \bar{x}_{2\cdot} = 7.67, \bar{x}_{3\cdot} = 15.67, \bar{x}_{4\cdot} = 18.67, \bar{x}_{5\cdot} = 7.67$

Overall Mean
$\bar{\bar{x}} = 176/15 = 11.73$

Step 1

$$\text{SST} = \sum_i \sum_j (x_{ij} - \bar{\bar{x}})^2$$
$$= (10 - 11.73)^2 + (9 - 11.73)^2 + \cdots + (8 - 11.73)^2$$
$$= 354.93$$

Step 2

$$\text{SSTR} = b \sum_j (\bar{x}_{\cdot j} - \bar{\bar{x}})^2$$
$$= 5[(13.6 - 11.73)^2 + (11.0 - 11.73)^2 + (10.6 - 11.73)^2] = 26.53$$

Step 3

$$\text{SSBL} = k \sum_j (\bar{x}_{i\cdot} - \bar{\bar{x}})^2$$
$$= 3[(9 - 11.73)^2 + (7.67 - 11.73)^2 + (15.67 - 11.73)^2 + (18.67 - 11.73)^2 + (7.67 - 11.73)^2] = 312.32$$

Step 4

$$\text{SSE} = \text{SST} - \text{SSTR} - \text{SSBL}$$
$$= 354.93 - 26.53 - 312.32 = 16.08$$

Source of Variation	Sum of Squares	Degrees of Freedom	Mean Square	F	p-value
Treatments	26.53	2	13.27	6.60	.0203
Blocks	312.32	4	78.08		
Error	16.08	8	2.01		
Total	354.93	14			

From the F table (2 numerator degrees of freedom and 8 denominator), p-value is between .01 and .025

Actual p-value = .0203

Because p-value $\leq \alpha = .05$, we reject the null hypothesis that the means of the three treatments are equal

22.

Source of Variation	Sum of Squares	Degrees of Freedom	Mean Square	F	p-value
Treatments	310	4	77.5	17.69	.0005
Blocks	85	2	42.5		
Error	35	8	4.38		
Total	430	14			

Significant; p-value $\leq \alpha = .05$

24. p-value = .0453

Because p-value $\leq \alpha = .05$, we reject the null hypothesis that the mean tune-up times are the same for both analyzers

26. a. Significant; p-value = .0231

b. Writing section

28. *Step 1*

$$\text{SST} = \sum_i \sum_j \sum_k (x_{ijk} - \bar{\bar{x}})^2$$
$$= (135 - 111)^2 + (165 - 111)^2 + \cdots + (136 - 111)^2 = 9028$$

Step 2

$$\text{SSA} = br \sum_i (\bar{x}_{i\cdot} - \bar{\bar{x}})^2$$
$$= 3(2)[(104 - 111)^2 + (118 - 111)^2] = 588$$

Step 3

$$\text{SSB} = ar \sum_j (\bar{x}_{\cdot j} - \bar{\bar{x}})^2$$
$$= 2(2)[(130 - 111)^2 + (97 - 111)^2 + (106 - 111)^2]$$
$$= 2328$$

Step 4

$$\text{SSAB} = r \sum_i \sum_j (\bar{x}_{ij} - \bar{x}_{i\cdot} - \bar{x}_{\cdot j} + \bar{\bar{x}})^2$$
$$= 2[(150 - 104 - 130 + 111)^2 + (78 - 104 - 97 + 111)^2 + \cdots + (128 - 118 - 106 + 111)^2] = 4392$$

Step 5

$$\text{SSE} = \text{SST} - \text{SSA} - \text{SSB} - \text{SSAB}$$
$$= 9028 - 588 - 2328 - 4392 = 1720$$

Source of Variation	Sum of Squares	Degrees of Freedom	Mean Square	F	p-value
Factor A	588	1	588	2.05	.2022
Factor B	2328	2	1164	4.06	.0767
Interaction	4392	2	2196	7.66	.0223
Error	1720	6	286.67		
Total	9028	11			

Factor A: $F = 2.05$

Using F table (1 numerator degree of freedom and 6 denominator), p-value is greater than .10

Using Excel or Minitab, the p-value corresponding to $F = 2.05$ is .2022

Because p-value $> \alpha = .05$, Factor A is not significant

Factor B: $F = 4.06$

Using F table (2 numerator degrees of freedom and 6 denominator), p-value is between .05 and .10

Using Excel or Minitab, the p-value corresponding to $F = 4.06$ is .0767

Because p-value $> \alpha = .05$, Factor B is not significant

Interaction: $F = 7.66$

Using F table (2 numerator degrees of freedom and 6 denominator), p-value is between .01 and .025

Using Excel or Minitab, the p-value corresponding to $F = 7.66$ is .0223

Because p-value $\leq \alpha = .05$, interaction is significant

30. Design: p-value = .0104; significant
Size: p-value = .1340; not significant
Interaction: p-value = .2519; not significant

32. Class: p-value = .0002; significant
Type: p-value = .0006; significant
Interaction: p-value = .4229; not significant

34. Significant; p-value = .0134

36. Not significant; p-value = .088

38. Not significant; p-value = .2455

40. a. Significant; p-value = .0175

42. The blocks correspond to the 7 weekend series (Opponent) and the treatments correspond to the days for the series (Day)

The Minitab two-way output is

Source	DF	SS	MS	F	P
Day	2	101469683	50734841	2.69	0.109
Opponent	6	79329936	13221656	0.70	0.655
Error	12	226577122	18881427		
Total	20	407376741			

Because the p-value for Day (.109) is greater than $\alpha = .05$, there is no significant difference in the mean attendance per game for games played on Friday, Saturday, and Sunday; these data do not suggest a particular day on which the Astros should schedule these promotions

44. Type of machine (p-value = .0226) is significant; type of loading system (p-value = .7913) and interaction (p-value = .0671) are not significant

Chapter 14

1. a.

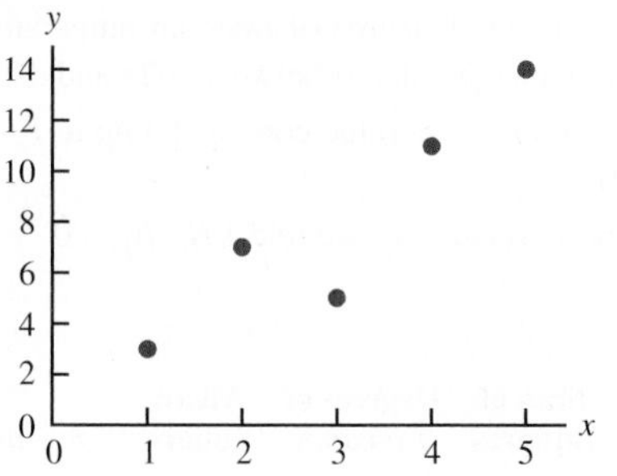

b. There appears to be a positive linear relationship between x and y

c. Many different straight lines can be drawn to provide a linear approximation of the relationship between x and y; in part (d) we will determine the equation of a straight line that "best" represents the relationship according to the least squares criterion

d. Summations needed to compute the slope and y-intercept:

$$\bar{x} = \frac{\Sigma x_i}{n} = \frac{15}{5} = 3, \quad \bar{y} = \frac{\Sigma y_i}{n} = \frac{40}{5} = 8,$$

$$\Sigma(x_i - \bar{x})(y_i - \bar{y}) = 26, \quad \Sigma(x_i - \bar{x})^2 = 10$$

$$b_1 = \frac{\Sigma(x_i - \bar{x})(y_i - \bar{y})}{\Sigma(x_i - \bar{x})^2} = \frac{26}{10} = 2.6$$

$$b_0 = \bar{y} - b_1\bar{x} = 8 - (2.6)(3) = 0.2$$

$$\hat{y} = 0.2 - 2.6x$$

e. $\hat{y} = .2 + 2.6x = .2 + 2.6(4) = 10.6$

2. b. There appears to be a negative linear relationship between x and y

d. $\hat{y} = 68 - 3x$

e. 38

4. a.

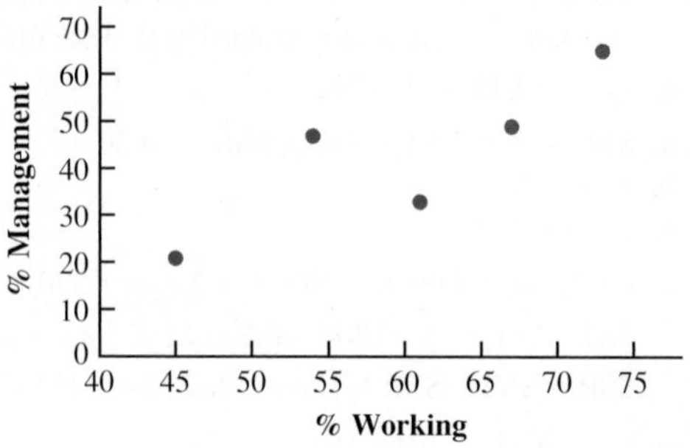

b. There appears to be a positive linear relationship between the percentage of women working in the five companies (x) and the percentage of management jobs held by women in that company (y)

c. Many different straight lines can be drawn to provide a linear approximation of the relationship between x and y; in part (d) we will determine the equation of a straight line that "best" represents the relationship according to the least squares criterion

d. $$\bar{x} = \frac{\Sigma x_i}{n} = \frac{300}{5} = 60 \quad \bar{y} = \frac{\Sigma y_i}{n} = \frac{215}{5} = 43$$

$$\Sigma(x_i - \bar{x})(y_i - \bar{y}) = 624 \quad \Sigma(x_i - \bar{x})^2 = 480$$

$$b_1 = \frac{\Sigma(x_i - \bar{x})(y_i - \bar{y})}{\Sigma(x_i - \bar{x})^2} = \frac{624}{480} = 1.3$$

$$b_0 = \bar{y} - b_1\bar{x} = 43 - 1.3(60) = -35$$

$$\hat{y} = -35 + 1.3x$$

e. $\hat{y} = -35 + 1.3x = -35 + 1.3(60) = 43\%$

6. c. $\hat{y} = -70.391 + 17.175x$

e. 43.8 or approximately 44%

8. c. $\hat{y} = .2046 + .9077x$

e. 3.29 or approximately 3.3

10. c. $\hat{y} = -167.81 + 2.7149x$

e. Yes

12. c. $\hat{y} = 17.49 + 1.0334x$

d. $150

14. c. $\hat{y} = 37.1217 + .51758x$

d. 73

15. a. $\hat{y}_i = .2 + 2.6x_i$ and $\bar{y} = 8$

x_i	y_i	$\hat{y}_i$	$y_i - \hat{y}_i$	$(y_i - \hat{y}_i)^2$	$y_i - \bar{y}$	$(y_i - \bar{y})^2$
1	3	2.8	.2	.04	−5	25
2	7	5.4	1.6	2.56	−1	1
3	5	8.0	−3.0	9.00	−3	9
4	11	10.6	.4	.16	3	9
5	14	13.2	.8	.64	6	36
				SSE = 12.40		SST = 80

SSR = SST − SSE = 80 − 12.4 = 67.6

b. $r^2 = \dfrac{\text{SSR}}{\text{SST}} = \dfrac{67.6}{80} = .845$

The least squares line provided a good fit; 84.5% of the variability in y has been explained by the least squares line

c. $r_{xy} = \sqrt{.845} = +.9192$

16. a. SSE = 230, SST = 1850, SSR = 1620

b. $r^2 = .876$

c. $r_{xy} = -.936$

18. a. $\bar{x} = \Sigma x_i/n = 600/6 = 100$ $\bar{y} = \Sigma y_i/n = 330/6 = 55$

$\text{SST} = \Sigma(y_i - \bar{y})^2 = 1800$ $\text{SSE} = \Sigma(y_i - \hat{y}_i)^2 = 287.624$

SSR = SST − SSR = 1800 − 287.624 = 1512.376

b. $r^2 = \dfrac{\text{SSR}}{\text{SST}} = \dfrac{1512.376}{1800} = .84$

c. $r = \sqrt{r^2} = \sqrt{.84} = .917$

20. a. $\hat{y} = 28{,}574 - 1439x$

b. $r^2 = .864$

c. \$6989

22. a. .77

b. Yes

c. $r_{xy} = +.88$, strong

23. a. $s^2 = \text{MSE} = \dfrac{\text{SSE}}{n-2} = \dfrac{12.4}{3} = 4.133$

b. $s = \sqrt{\text{MSE}} = \sqrt{4.133} = 2.033$

c. $\Sigma(x_i - \bar{x})^2 = 10$

$$s_{b_1} = \frac{s}{\sqrt{\Sigma(x_i - \bar{x})^2}} = \frac{2.033}{\sqrt{10}} = .643$$

d. $t = \dfrac{b_1 - \beta_1}{s_{b_1}} = \dfrac{2.6 - 0}{.643} = 4.044$

From the t table (3 degrees of freedom), area in tail is between .01 and .025

p-value is between .02 and .05

Using Excel or Minitab, the p-value corresponding to $t = 4.04$ is .0272

Because p-value $\leq \alpha$, we reject H_0: $\beta_1 = 0$

e. $\text{MSR} = \dfrac{\text{SSR}}{1} = 67.6$

$$F = \frac{\text{MSR}}{\text{MSE}} = \frac{67.6}{4.133} = 16.36$$

From the F table (1 numerator degree of freedom and 3 denominator), p-value is between .025 and .05

Using Excel or Minitab, the p-value corresponding to $F = 16.36$ is .0272

Because p-value $\leq \alpha$, we reject H_0: $\beta_1 = 0$

Source of Variation	Sum of Squares	Degrees of Freedom	Mean Square	F	p-value
Regression	67.6	1	67.6	16.36	.0272
Error	12.4	3	4.133		
Total	80	4			

24. a. 76.6667

b. 8.7560

c. .6526

d. Significant; p-value = .0193

e. Significant; p-value = .0193

26. a. In the statement of exercise 18, $\hat{y} = 23.194 + .318x$

In solving exercise 18, we found SSE = 287.624

$s^2 = \text{MSE} = \text{SSE}/(n-2) = 287.624/4 = 71.906$

$s = \sqrt{\text{MSE}} = \sqrt{71.906} = 8.4797$

$\Sigma(x - \bar{x})^2 = 14{,}950$

$$s_{b_1} = \frac{s}{\sqrt{\Sigma(x - \bar{x})^2}} = \frac{8.4797}{\sqrt{14{,}950}} = .0694$$

$$t = \frac{b_1}{s_{b_1}} = \frac{.318}{.0694} = 4.58$$

Using t table (4 degrees of freedom), area in tail is between .005 and .01

p-value is between .01 and .02

Using Excel, the p-value corresponding to $t = 4.58$ is .010

Because p-value $\leq \alpha$, we reject H_0: $\beta_1 = 0$; there is a significant relationship between price and overall score

b. In exercise 18 we found SSR = 1512.376

MSR = SSR/1 = 1512.376/1 = 1512.376

F = MSR/MSE = 1512.376/71.906 = 21.03

Using F table (1 degree of freedom numerator and 4 denominator), p-value is between .025 and .01

Using Excel, the p-value corresponding to $F = 11.74$ is .010

Because p-value $\leq \alpha$, we reject H_0: $\beta_1 = 0$

c.

Source of Variation	Sum of Squares	Degrees of Freedom	Mean Square	F	p-value
Regression	1512.376	1	1512.376	21.03	.010
Error	287.624	4	71.906		
Total	1800	5			

28. They are related; p-value = .000

30. Significant; p-value = .0042

32. a. $s = 2.033$

$\bar{x} = 3$, $\Sigma(x_i - \bar{x})^2 = 10$

$$s_{\hat{y}^*} = s\sqrt{\frac{1}{n} + \frac{(x^* - \bar{x})^2}{\Sigma(x_i - \bar{x})^2}}$$

$$= 2.033\sqrt{\frac{1}{5} + \frac{(4-3)^2}{10}} = 1.11$$

b. $\hat{y}^* = .2 + 2.6x^* = .2 + 2.6(4) = 10.6$

$\hat{y}^* \pm t_{\alpha/2} s_{\hat{y}^*}$

$10.6 \pm 3.182(1.11)$

10.6 ± 3.53, or 7.07 to 14.13

c. $s_{\text{pred}} = s\sqrt{1 + \frac{1}{n} + \frac{(x^* - \bar{x})^2}{\Sigma(x_i - \bar{x})^2}}$

$= 2.033\sqrt{1 + \frac{1}{5} + \frac{(4-3)^2}{10}} = 2.32$

d. $\hat{y}^* \pm t_{\alpha/2}s_{\text{pred}}$
$10.6 \pm 3.182(2.32)$
10.6 ± 7.38, or 3.22 to 17.98

34. Confidence interval: 8.65 to 21.15
Prediction interval: -4.50 to 41.30

35. a. $\hat{y}^* = 2090.5 + 581.1x^* = 2090.5 + 581.1(3) = 3833.8$

b. $s = \sqrt{\text{MSE}} = \sqrt{21{,}284} = 145.89$
$\bar{x} = 3.2$, $\Sigma(x_i - \bar{x})^2 = 0.74$

$s_{\hat{y}^*} = s\sqrt{\frac{1}{n} + \frac{(x^* - \bar{x})^2}{\Sigma(x_i - \bar{x})^2}}$

$= 145.89\sqrt{\frac{1}{6} + \frac{(3-3.2)^2}{0.74}} = 68.54$

$\hat{y}^* \pm t_{\alpha/2}s_{\hat{y}^*}$

$3833.8 \pm 2.776(68.54) = 3833.8 \pm 190.27$
or \$3643.53 to \$4024.07

c. $s_{\text{pred}} = s\sqrt{1 + \frac{1}{n} + \frac{(x^* - \bar{x})^2}{\Sigma(x_i - \bar{x})^2}}$

$= 145.89\sqrt{1 + \frac{1}{6} + \frac{(3-3.2)^2}{0.74}} = 161.19$

$\hat{y}^* \pm t_{\alpha/2}s_{\text{pred}}$

$3833.8 \pm 2.776(161.19) = 3833.8 \pm 447.46$
or \$3386.34 to \$4281.26

d. As expected, the prediction interval is much wider than the confidence interval. This is due to the fact that it is more difficult to predict the starting salary for one new student with a GPA of 3.0 than it is to estimate the mean for all students with a GPA of 3.0.

36. a. \$112,190 to \$119,810
b. \$104,710 to \$127,290

38. a. \$5046.67
b. \$3815.10 to \$6278.24
c. Not out of line

40. a. 9
b. $\hat{y} = 20.0 + 7.21x$
c. 1.3626
d. SSE = SST − SSR = 51,984.1 − 41,587.3 = 10,396.8
MSE = 10,396.8/7 = 1485.3

$F = \frac{\text{MSR}}{\text{MSE}} = \frac{41{,}587.3}{1485.3} = 28.0$

From the F table (1 numerator degree of freedom and 7 denominator), p-value is less than .01

Using Excel or Minitab, the p-value corresponding to $F = 28.0$ is .0011

Because p-value $\leq \alpha = .05$, we reject H_0: $\beta_1 = 0$

e. $\hat{y} = 20.0 + 7.21(50) = 380.5$, or \$380,500

42. a. $\hat{y} = 80.0 + 50.0x$
b. 30
c. Significant; p-value = .000
d. \$680,000

44. b. Yes
c. $\hat{y} = 2044.38 - 28.35$ weight
d. Significant; p-value = .000
e. .774; a good fit

45. a. $\bar{x} = \frac{\Sigma x_i}{n} = \frac{70}{5} = 14$, $\bar{y} = \frac{\Sigma y_i}{n} = \frac{76}{5} = 15.2$,

$\Sigma(x_i - \bar{x})(y_i - \bar{y}) = 200$, $\Sigma(x_i - \bar{x})^2 = 126$

$b_1 = \frac{\Sigma(x_i - \bar{x})(y_i - \bar{y})}{\Sigma(x_i - \bar{x})^2} = \frac{200}{126} = 1.5873$

$b_0 = \bar{y} - b_1\bar{x} = 15.2 - (1.5873)(14) = -7.0222$

$\hat{y} = -7.02 + 1.59x$

b.

x_i	y_i	$\hat{y}_i$	$y_i - \hat{y}_i$
6	6	2.52	3.48
11	8	10.47	−2.47
15	12	16.83	−4.83
18	20	21.60	−1.60
20	30	24.78	5.22

c.

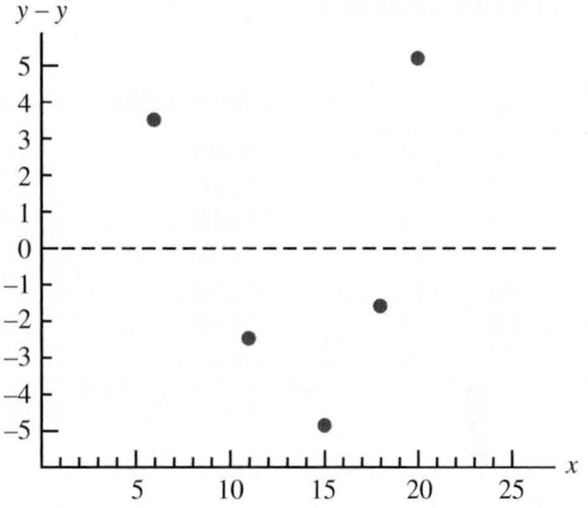

With only five observations, it is difficult to determine whether the assumptions are satisfied; however, the plot does suggest curvature in the residuals, which would indicate that the error term assumptions are not satisfied; the scatter diagram for these data also indicates that the underlying relationship between x and y may be curvilinear

d. $s^2 = 23.78$

$h_i = \frac{1}{n} + \frac{(x_i - \bar{x})^2}{\Sigma(x_i - \bar{x})^2} = \frac{1}{5} + \frac{(x_i - 14)^2}{126}$

x_i	h_i	$s_{y_i - \hat{y}_i}$	$y_i - \hat{y}_i$	Standardized Residuals
6	.7079	2.64	3.48	1.32
11	.2714	4.16	−2.47	−.59
15	.2079	4.34	−4.83	−1.11
18	.3270	4.00	−1.60	−.40
20	.4857	3.50	5.22	1.49

e. The plot of the standardized residuals against $\hat{y}$ has the same shape as the original residual plot; as stated in part (c), the curvature observed indicates that the assumptions regarding the error term may not be satisfied

46. a. $\hat{y} = 2.32 + .64x$

b. No; the variance appears to increase for larger values of x

47. a. Let x = advertising expenditures and y = revenue

$\hat{y} = 29.4 + 1.55x$

b. SST = 1002, SSE = 310.28, SSR = 691.72

$$\text{MSR} = \frac{\text{SSR}}{1} = 691.72$$

$$\text{MSE} = \frac{\text{SSE}}{n-2} = \frac{310.28}{5} = 62.0554$$

$$F = \frac{\text{MSR}}{\text{MSE}} = \frac{691.72}{62.0554} = 11.15$$

From the F table (1 numerator degree of freedom and 5 denominator), p-value is between .01 and .025

Using Excel or Minitab, p-value = .0206

Because p-value $\leq \alpha = .05$, we conclude that the two variables are related

c.

x_i	y_i	$\hat{y}_i = 29.40 + 1.55x_i$	$y_i - \hat{y}_i$
1	19	30.95	−11.95
2	32	32.50	−.50
4	44	35.60	8.40
6	40	38.70	1.30
10	52	44.90	7.10
14	53	51.10	1.90
20	54	60.40	−6.40

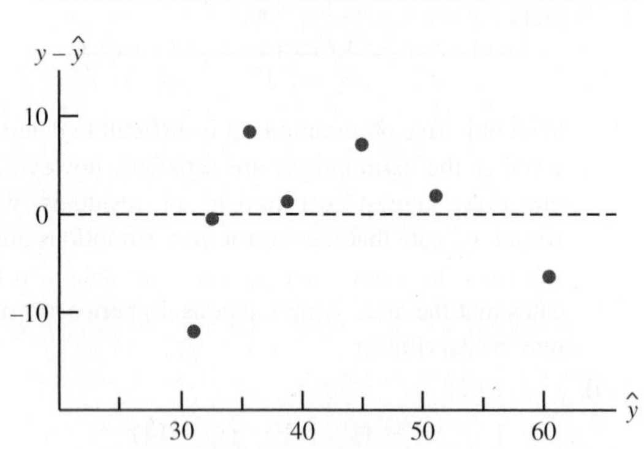

d. The residual plot leads us to question the assumption of a linear relationship between x and y; even though the relationship is significant at the $\alpha = .05$ level, it would be extremely dangerous to extrapolate beyond the range of the data

48. b. Yes

50. a. Using Minitab, we obtained the estimated regression equation $\hat{y} = 66.1 + .402x$; a portion of the Minitab output is shown in Figure D14.50; the fitted values and standardized residuals are shown:

x_i	y_i	$\hat{y}_i$	Standardized Residuals
135	145	120.41	2.11
110	100	110.35	−1.08
130	120	118.40	.14
145	120	124.43	−.38
175	130	136.50	−.78
160	130	130.47	−.04
120	110	114.38	−.41

b.

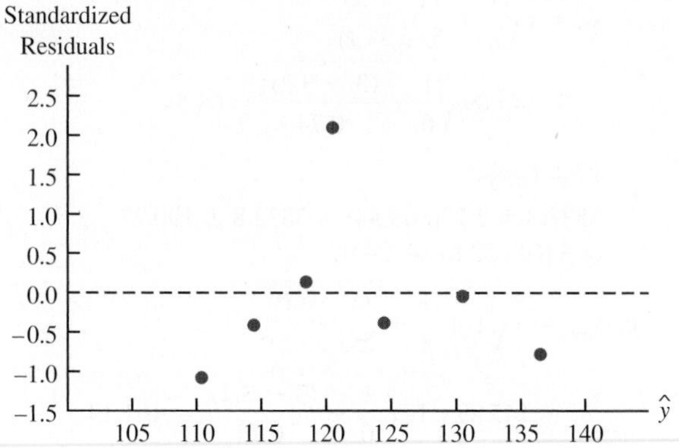

The standardized residual plot indicates that the observation $x = 135$, $y = 145$ may be an outlier; note that this observation has a standardized residual of 2.11

c. The scatter diagram is shown:

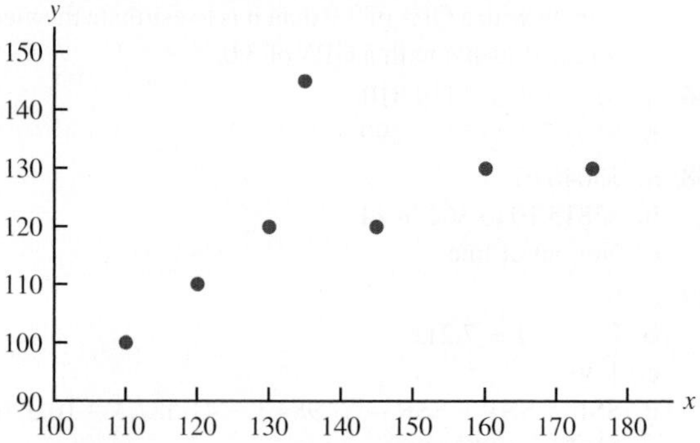

The scatter diagram also indicates that the observation $x = 135$, $y = 145$ may be an outlier; the implication is that for simple linear regression outliers can be identified by looking at the scatter diagram

52. b. $\hat{y} = 91.0 - 0.917x$

b. Smithsonian Institution: outlier

American Cancer Society: influential observation

54. b. Value = −252 + 5.83 Revenue

c. There are five unusual observations (9, 19, 21, 22, and 32).

58. b. $\hat{y} = -669 + .157$ DJIA
c. Significant; p-value = .001
d. $r^2 = .949$; excellent fit

60. b. GR(%) = 25.4 + .285 RR(%)
c. Significant; p-value = .000
d. No; $r^2 = .449$
e. Yes
f. Yes

62. a. $\hat{y} = 22.2 - .148x$
b. Significant relationship; p-value = .028
c. Good fit; $r^2 = .739$
d. 12.294 to 17.271

64. a. $\hat{y} = 220 + 132x$
b. Significant; p-value = .000
c. $r^2 = .873$; very good fit
d. \$559.50 to \$933.90

66. a. Market beta = .95
b. Significant; p-value = .029
c. $r^2 = .470$; not a good fit
d. Xerox has a higher risk

68. b. There appears to be a negative linear relationship between the two variables
c. $\hat{y} = 16.5 - .0588$ Miles
d. Significant; p-value = .000
e. $r^2 = .539$; reasonably good fit
g. approximately \$13,000; no

FIGURE D14.50

```
The regression equation is
Y = 66.1 + 0.402 X

Predictor        Coef     SE Coef       T        p
Constant        66.10       32.06     2.06    0.094
X              0.4023      0.2276     1.77    0.137

S = 12.62       R-sq = 38.5%      R-sq(adj) = 26.1%

Analysis of Variance

SOURCE              DF          SS        MS       F        p
Regression           1       497.2     497.2    3.12    0.137
Residual Error       5       795.7     159.1
Total                6      1292.9

Unusual Observations
Obs       X          Y        Fit    SE Fit    Residual    St Resid
  1     135     145.00     120.42      4.87       24.58       2.11R

 R denotes an observation with a large standardized residual
```

Chapter 15

2. a. The estimated regression equation is
$\hat{y} = 45.06 + 1.94x_1$
An estimate of y when $x_1 = 45$ is
$\hat{y} = 45.06 + 1.94(45) = 132.36$
b. The estimated regression equation is
$\hat{y} = 85.22 + 4.32x_2$
An estimate of y when $x_2 = 15$ is
$\hat{y} = 85.22 + 4.32(15) = 150.02$
c. The estimated regression equation is
$\hat{y} = -18.37 + 2.01x_1 + 4.74x_2$
An estimate of y when $x_1 = 45$ and $x_2 = 15$ is
$\hat{y} = -18.37 + 2.01(45) + 4.74(15) = 143.18$

4. a. \$255,000

5. a. The Minitab output is shown in Figure D15.5a
b. The Minitab output is shown in Figure D15.5b
c. It is 1.60 in part (a) and 2.29 in part (b); in part (a) the coefficient is an estimate of the change in revenue due to a one-unit change in television advertising expenditures; in part (b) it represents an estimate of the change in revenue due to a one-unit change in television advertising expenditures when the amount of newspaper advertising is held constant
d. Revenue = 83.2 + 2.29(3.5) + 1.30(1.8) = 93.56 or \$93,560

FIGURE D15.5a

```
The regression equation is
Revenue = 88.6 + 1.60 TVAdv

Predictor     Coef     SE Coef       T        p
Constant     88.638     1.582     56.02    0.000
TVAdv        1.6039     0.4778     3.36    0.015

S = 1.215       R-sq = 65.3%       R-sq(adj) = 59.5%

Analysis of Variance

SOURCE            DF       SS        MS        F        p
Regression         1    16.640    16.640    11.27    0.015
Residual Error     6     8.860     1.477
Total              7    25.500
```

FIGURE D15.5b

```
The regression equation is
Revenue = 83.2 + 2.29 TVAdv + 1.30 NewsAdv

Predictor     Coef     SE Coef       T        p
Constant     83.230     1.574     52.88    0.000
TVAdv        2.2902     0.3041     7.53    0.001
NewsAdv      1.3010     0.3207     4.06    0.010

S = 0.6426       R-sq = 91.9%       R-sq(adj) = 88.7%

Analysis of Variance

SOURCE            DF       SS        MS        F        p
Regression         2    23.435    11.718    28.38    0.002
Residual Error     5     2.065     0.413
Total              7    25.500
```

6. a. Win% = −58.8 + 16.4 Yds/Att
b. Win% = 97.5 − 1600 Int/Att
c. Win% = −5.8 + 12.9 Yds/Att − 1084 Int/Att
d. 35%

8. a. Overall = 69.3 + .235 Shore Excursions
b. Overall = 45.2 + .253 Shore Excursions + .248 Food/Dining
c. 87.76 or approximately 88.

10. a. R/IP = .676 − .284 SO/IP
b. R/IP = .308 + 1.35 HR/IP
c. R/IP = .537 − .248 SO/IP + 1.03 HR/IP
d. .48
e. Suggestion does not make sense

12. a. $R^2 = \dfrac{SSR}{SST} = \dfrac{14{,}052.2}{15{,}182.9} = .926$

b. $R_a^2 = 1 - (1 - R^2)\dfrac{n - 1}{n - p - 1}$

$= 1 - (1 - .926)\dfrac{10 - 1}{10 - 2 - 1} = .905$

c. Yes; after adjusting for the number of independent variables in the model, we see that 90.5% of the variability in y has been accounted for

14. a. .75
b. .68

15. a. $R^2 = \dfrac{SSR}{SST} = \dfrac{23.435}{25.5} = .919$

$R_a^2 = 1 - (1 - R^2)\dfrac{n - 1}{n - p - 1}$

$= 1 - (1 - .919)\dfrac{8 - 1}{8 - 2 - 1} = .887$

b. Multiple regression analysis is preferred because both R^2 and R_a^2 show an increased percentage of the variability of y explained when both independent variables are used

16. a. No, $R^2 = .577$
b. Better fit with multiple regression

18. a. $R^2 = .563$, $R_a^2 = .512$
b. The fit is not very good

19. a. $\text{MSR} = \frac{\text{SSR}}{p} = \frac{6216.375}{2} = 3108.188$

$\text{MSE} = \frac{\text{SSE}}{n - p - 1} = \frac{507.75}{10 - 2 - 1} = 72.536$

b. $F = \frac{\text{MSR}}{\text{MSE}} = \frac{3108.188}{72.536} = 42.85$

From the F table (2 numerator degrees of freedom and 7 denominator), p-value is less than .01

Using Excel or Minitab the p-value corresponding to $F = 42.85$ is .0001

Because p-value $\leq \alpha$, the overall model is significant

c. $t = \frac{b_1}{s_{b_1}} = \frac{.5906}{.0813} = 7.26$

p-value $= .0002$

Because p-value $\leq \alpha$, β_1 is significant

d. $t = \frac{b_2}{s_{b_2}} = \frac{.4980}{.0567} = 8.78$

p-value $= .0001$

Because p-value $\leq \alpha$, β_2 is significant

20. a. Significant; p-value $= .000$
b. Significant; p-value $= .000$
c. Significant; p-value $= .002$

22. a. $\text{SSE} = 4000$, $s^2 = 571.43$, $\text{MSR} = 6000$
b. Significant; p-value $= .008$

23. a. $F = 28.38$

p-value $= .002$

Because p-value $\leq \alpha$, there is a significant relationship

b. $t = 7.53$

p-value $= .001$

Because p-value $\leq \alpha$, β_1 is significant and x_1 should not be dropped from the model

c. $t = 4.06$

p-value $= .010$

Because p-value $\leq \alpha$, β_2 is significant and x_2 should not be dropped from the model

24. a. $\hat{y} = -.682 + .0498$ Revenue $+ .0147$ % Wins
b. Significant; p-value $= .001$
c. Revenue is significant; p-value $= .001$
%Wins is significant; p-value $= .025$

26. a. Significant; p-value $= .001$
b. All significant; p-values are all $< \alpha = .05$

28. a. Using Minitab, the 95% confidence interval is 132.16 to 154.16
b. Using Minitab, the 95% prediction interval is 111.13 at 175.18

29. a. See Minitab output in Figure D15.5b.

$\hat{y} = 83.23 + 2.29(3.5) + 1.30(1.8) = 93.555$ or \$93,555

b. Minitab results: 92.840 to 94.335, or \$92,840 to \$94,335
c. Minitab results: 91.774 to 95.401, or \$91,774 to \$95,401

30. a. 46.758 to 50.646
b. 44.815 to 52.589

32. a. $E(y) = \beta_0 + \beta_1 x_1 + \beta_2 x_2$

where $x_2 = \begin{cases} 0 \text{ if level 1} \\ 1 \text{ if level 2} \end{cases}$

b. $E(y) = \beta_0 + \beta_1 x_1 + \beta_2(0) = \beta_0 + \beta_1 x_1$
c. $E(y) = \beta_0 + \beta_1 x_1 + \beta_2(1) = \beta_0 + \beta_1 x_1 + \beta_2$
d. $\beta_2 = E(y \mid \text{level 2}) - E(y \mid \text{level 1})$
β_1 is the change in $E(y)$ for a 1-unit change in x_1 holding x_2 constant

34. a. \$15,300
b. $\hat{y} = 10.1 - 4.2(2) + 6.8(8) + 15.3(0) = 56.1$
Sales prediction: \$56,100
c. $\hat{y} = 10.1 - 4.2(1) + 6.8(3) + 15.3(1) = 41.6$
Sales prediction: \$41,600

36. a. $\hat{y} = 1.86 + 0.291$ Months $+ 1.10$ Type $- 0.609$ Person
b. Significant; p-value $= .002$
c. Person is not significant; p-value $= .167$

38. a. $\hat{y} = -91.8 + 1.08$ Age $+ .252$ Pressure $+ 8.74$ Smoker
b. Significant; p-value $= .01$
c. 95% prediction interval is 21.35 to 47.18 or a probability of .2135 to .4718; quit smoking and begin some type of treatment to reduce his blood pressure

39. a. The Minitab output is shown in Figure D15.39
b. Minitab provides the following values:

x_i	y_i	$\hat{y}_i$	Standardized Residual
1	3	2.8	.16
2	7	5.4	.94
3	5	8.0	−1.65
4	11	10.6	.24
5	14	13.2	.62

FIGURE D15.39

```
The regression equation is
Y = 0.20 + 2.60 X

Predictor       Coef    SE Coef       T        p
Constant       0.200      2.132    0.09    0.931
X             2.6000     0.6429    4.04    0.027

S = 2.033     R-sq = 84.5%      R-sq(adj) = 79.3%

Analysis of Variance
SOURCE            DF        SS        MS        F       p
Regression         1    67.600    67.600    16.35   0.027
Residual Error     3    12.400     4.133
Total              4    80.000
```

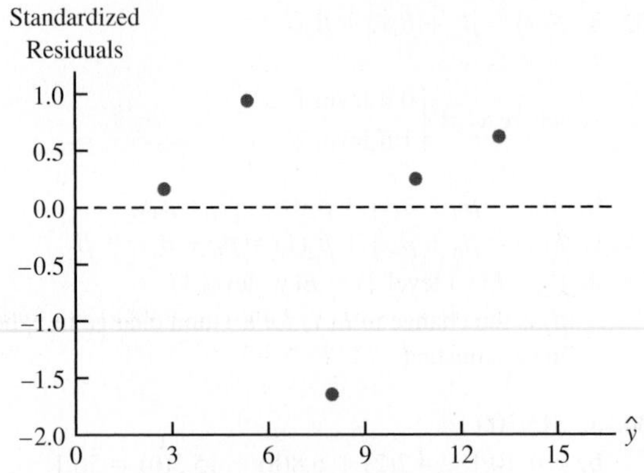

The point (3, 5) does not appear to follow the trend of the remaining data; however, the value of the standardized residual for this point, −1.65, is not large enough for us to conclude that (3, 5) is an outlier

c. Minitab provides the following values:

x_i	y_i	Studentized Deleted Residual
1	3	.13
2	7	.91
3	5	−4.42
4	11	.19
5	14	.54

$t_{.025} = 4.303$ ($n - p - 2 = 5 - 1 - 2 = 2$ degrees of freedom)

Because the studentized deleted residual for (3, 5) is $-4.42 < -4.303$, we conclude that the 3rd observation is an outlier

40. a. $\hat{y} = -53.3 + 3.11x$

b. −1.94, −.12, 1.79, .40, −1.90; no

c. .38, .28, .22, .20, .92; no

d. .60, .00, .26, .03, 11.09; yes, the fifth observation

41. a. The Minitab output appears in Figure D15.5b; the estimated regression equation is

Revenue = 83.2 + 2.29 TVAdv + 1.30 NewsAdv

b. Minitab provides the following values:

$\hat{y}_i$	Standardized Residual	$\hat{y}_i$	Standardized Residual
96.63	−1.62	94.39	1.10
90.41	−1.08	94.24	−.40
94.34	1.22	94.42	−1.12
92.21	−.37	93.35	1.08

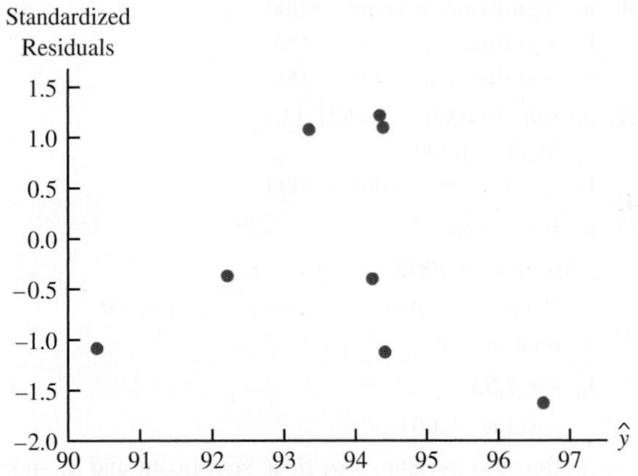

With relatively few observations, it is difficult to determine whether any of the assumptions regarding ϵ have been violated; for instance, an argument could be made that there does not appear to be any pattern in the plot; alternatively, an argument could be made that there is a curvilinear pattern in the plot

c. The values of the standardized residuals are greater than −2 and less than +2; thus, using this test, there are no outliers

As a further check for outliers, we used Minitab to compute the following studentized deleted residuals:

Observation	Studentized Deleted Residual	Observation	Studentized Deleted Residual
1	−2.11	5	1.13
2	−1.10	6	−.36
3	1.31	7	−1.16
4	−.33	8	1.10

$t_{.025} = 2.776$ ($n - p - 2 = 8 - 2 - 2 = 4$ degrees of freedom)

Because none of the studentized deleted residuals are less than −2.776 or greater than 2.776, we conclude that there are no outliers in the data

d. Minitab provides the following values:

Observation	h_i	D_i
1	.63	1.52
2	.65	.70
3	.30	.22
4	.23	.01
5	.26	.14
6	.14	.01
7	.66	.81
8	.13	.06

The critical leverage value is

$$\frac{3(p+1)}{n} = \frac{3(2+1)}{8} = 1.125$$

Because none of the values exceed 1.125, we conclude that there are no influential observations; however, using Cook's distance measure, we see that $D_1 > 1$ (rule of thumb critical value); thus, we conclude that the first observation is influential

Final conclusion: observation 1 is an influential observation

42. b. Unusual trend
c. No outliers
d. Observation 2 is an influential observation

44. a. $E(y) = \dfrac{e^{\beta_0+\beta_1 x}}{1 + e^{\beta_0+\beta_1 x}}$
b. Estimate of the probability that a customer who does not have a Simmons credit card will make a purchase
c. $\hat{g}(x) = -0.9445 + 1.0245x$
d. .28 for customers who do not have a Simmons credit card
.52 for customers who have a Simmons credit card
e. Estimated odds ratio = 2.79

46. a. $E(y) = \dfrac{e^{\beta_0+\beta_1 x}}{1 + e^{\beta_0+\beta_1 x}}$
b. $E(y) = \dfrac{e^{-2.6355+0.22018x}}{1 + e^{-2.6355+0.22018x}}$
c. Significant; p-value = .0002
d. .39
e. \$1200
f. Estimated odds ratio = 1.25

48. a. $E(y) = \dfrac{e^{\beta_0+\beta_1 x_1+\beta_2 x_2}}{1 + e^{\beta_0+\beta_1 x_1+\beta_2 x_2}}$
b. $\hat{g}(x) = -39.4982 + 3.37449$ Wet $+ 1.81628$ Noise
c. .88
d. Probability is .04

50. b. 67.39

52. a. $\hat{y} = -1.41 + .0235x_1 + .00486x_2$
b. Significant; p-value = .0001
c. Both significant
d. $R^2 = .937$; $R_a^2 = 9.19$; good fit

54. a. Buy Again = −7.522 + 1.8151 Steering
b. Yes
c. Buy Again = −5.388 + .6899 Steering + .9113 Treadwear
d. Significant; p-value = .001

56. a. $\hat{y} = 4.9090 + 10.4658$ FundDE $+ 21.6823$ FundIE
b. $R^2 = .6144$; reasonably good fit
c. $\hat{y} = 1.1899 + 6.8969$ FundDE $+ 17.6800$ FundIE $+ 0.0265$ Net Asset Value (\$) $+ 6.4564$ Expense Ratio (%)
Net Asset Value (\$) is not significant and can be deleted
d. $\hat{y} = -4.6074 + 8.1713$ FundDE $+ 19.5194$ FundIE $+ 5.5197$ Expense Ratio (%) $+ 5.9237$ 3StarRank $+ 8.2367$ 4StarRank $+ 6.6241$ 5StarRank
e. 15.28%

Chapter 16

1. a. The Minitab output is shown in Figure D16.1a
b. Because the p-value corresponding to $F = 6.85$ is $.059 > \alpha = .05$, the relationship is not significant
c.

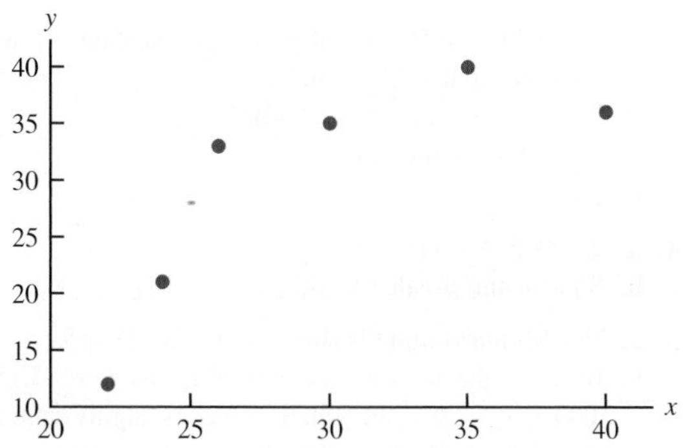

The scatter diagram suggests that a curvilinear relationship may be appropriate
d. The Minitab output is shown in Figure D16.1d
e. Because the p-value corresponding to $F = 25.68$ is $.013 < \alpha = .05$, the relationship is significant
f. $\hat{y} = -168.88 + 12.187(25) - .17704(25)^2 = 25.145$

FIGURE D16.1a

```
The regression equation is
Y = - 6.8 + 1.23 X

Predictor        Coef     SE Coef         T        p
Constant        -6.77       14.17     -0.48    0.658
X              1.2296      0.4697      2.62    0.059

S = 7.269      R-sq = 63.1%      R-sq(adj) = 53.9%

Analysis of Variance

SOURCE            DF        SS        MS       F        p
Regression         1    362.13    362.13    6.85    0.059
Residual Error     4    211.37     52.84
Total              5    573.50
```

FIGURE D16.1d

```
The regression equation is
Y = - 169 + 12.2 X - 0.177 XSQ

Predictor        Coef     SE Coef         T        p
Constant      -168.88       39.79     -4.74    0.024
X              12.187       2.663      4.58    0.020
XSQ          -0.17704     0.04290     -4.13    0.026

S = 3.248      R-sq = 94.5%      R-sq(adj) = 90.8%

Analysis of Variance

SOURCE            DF        SS        MS        F        p
Regression         2    541.85    270.92    25.68    0.013
Residual Error     3     31.65     10.55
Total              5    573.50
```

2. a. $\hat{y} = 9.32 + .424x$; p-value $= .117$ indicates a weak relationship between x and y

b. $\hat{y} = -8.10 + 2.41x - .0480x^2$

$R_a^2 = .932$; a good fit

c. 20.965

4. a. $\hat{y} = 943 + 8.71x$

b. Significant; p-value $= .005 < \alpha = .01$

5. a. The Minitab output is shown in Figure D16.5a

b. Because the p-value corresponding to $F = 73.15$ is $.003 < \alpha = .01$, the relationship is significant; we would reject H_0: $\beta_1 = \beta_2 = 0$

c. See Figure D16.5c

6. b. No, the relationship appears to be curvilinear

c. Several possible models; e.g.,

$\hat{y} = 2.90 - .185x + .00351x^2$

8. a. It appears that a simple linear regression model is not appropriate

b. Price = 33829 − 4571 Rating + 154 RatingSq

c. logPrice = −10.2 + 10.4 logRating

d. Part (c); higher percentage of variability is explained

10. a. Significant; p-value $= .000$

b. Significant; p-value $= .000$

11. a. SSE = 1805 − 1760 = 45

$$F = \frac{\text{MSR}}{\text{MSE}} = \left(\frac{1760/4}{45/25}\right) = 244.44$$

Because p-value $= .000$, the relationship is significant

b. $\text{SSE}(x_1, x_2, x_3, x_4) = 45$

c. $\text{SSE}(x_2, x_3) = 1805 - 1705 = 100$

d. $$F = \frac{(100 - 45)/2}{1.8} = 15.28$$

Because p-value $= .000$, x_1 and x_2 are significant

FIGURE D16.5a

```
The regression equation is
Y = 433 + 37.4 X -0.383 XSQ

Predictor        Coef    SE Coef       T        p
Constant        432.6      141.2    3.06    0.055
X              37.429      7.807    4.79    0.017
XSQ           -0.3829     0.1036   -3.70    0.034

S = 15.83      R-sq = 98.0%      R-sq(adj) = 96.7%

Analysis of Variance

SOURCE            DF       SS       MS        F        p
Regression         2    36643    18322    73.15    0.003
Residual Error     3      751      250
Total              5    37395
```

FIGURE D16.5c

```
    Fit    Stdev.Fit            95% C.I.               95% P.I.
1302.01       9.93     (1270.41, 1333.61)    (1242.55, 1361.47)
```

12. a. The Minitab output is shown in Figure D16.12a

b. The Minitab output is shown in Figure D16.12b

c. $F = \dfrac{[\text{SSE(reduced)} - \text{SSE(full)}]/(\#\text{ extra terms})}{\text{MSE(full)}}$

$= \dfrac{(7.2998 - 4.3240)/2}{.1663} = 8.95$

The p-value associated with $F = 8.95$ (2 numerator degrees of freedom and 26 denominator) is .001; with a p-value $< \alpha = .05$, the addition of the two independent variables is significant

14. a. $\hat{y} = -111 + 1.32 \text{ Age} + .296 \text{ Pressure}$

b. $\hat{y} = -123 + 1.51 \text{ Age} + .448 \text{ Pressure} + 8.87 \text{ Smoker} - .00276 \text{ AgePress}$

c. Significant; p-value $= .000$

16. a. Weeks $= -8.9 + 1.51$ Age

b. Weeks $= -.07 + 1.73$ Age $- 2.7$ Manager $- 15.1$ Head $- 17.4$ Sales

c. Same as part (b)

d. Same as part (b)

e. Weeks $= 13.1 + 1.64$ Age $- 9.76$ Married $- 19.4$ Head $- 29.0$ Manager $- 19.0$ Sales

FIGURE D16.12a

```
The regression equation is
Scoring Avg. = 46.3 + 14.1 Putting Avg.

Predictor          Coef    SE Coef      T       p
Constant         46.277      6.026   7.68   0.000
Putting Avg.     14.103      3.356   4.20   0.000

S = 0.510596    R-Sq = 38.7%    R-Sq(adj) = 36.5%

Analysis of Variance

SOURCE            DF        SS        MS       F        p
Regression         1    4.6036    4.6036   17.66   0.0000
Residual Error    28    7.2998    0.2607
Total             29   11.9035
```

FIGURE D16.12b

```
The regression equation is
Scoring Avg. = 59.0 - 10.3 Greens in Reg.
   + 11.4 Putting Avg - 1.81 Sand Saves

Predictor          Coef    SE Coef        T      p
Constant         59.022      5.774    10.22  0.000
Greens in Reg.  -10.281      2.877    -3.57  0.001
Putting Avg.     11.413      2.760     4.14  0.000
Sand Saves      -1.8130     0.9210    -1.97  0.060

S = 0.407808      R-Sq = 63.7%      R-Sq(adj) = 59.5%

Analysis of Variance

Source           DF        SS       MS       F      p
Regression        3    7.5795   2.5265   15.19  0.000
Residual Error   26    4.3240   0.1663
Total            29   11.9035
```

18. a. RPG = −4.05 + 27.6 OBP

b. A variety of models will provide a good fit; the five-variable model identified using Minitab's Stepwise Regression procedure with Alpha-to-Enter = .10 and Alpha-to-Remove = .10 follows:

RPG = −.0909 + 32.2 OBP + .109 HR − 21.5 AVG + .244 3B − .0223 BB

20.

x_1	x_2	x_3	Treatment
0	0	0	A
1	0	0	B
0	1	0	C
0	0	1	D

$$E(y) = \beta_0 + \beta_1 x_1 + \beta_2 x_2 + \beta_3 x_3$$

22. Factor A: $x_1 = 0$ if level 1 and 1 if level 2

Factor B:

x_2	x_3	Level
0	0	1
1	0	2
0	1	3

$$E(y) = \beta_0 + \beta_1 x_1 + \beta_2 x_2 + \beta_3 x_1 x_2 + \beta_4 x_1 x_3$$

23. a. The dummy variables are defined as follows:

D1	D2	Mfg.
0	0	1
1	0	2
0	1	3

$$E(y) = \beta_0 + \beta_1 \text{D1} + \beta_2 \text{D2}$$

b. The Minitab output is shown below:

```
The regression equation is
TIME = 23.0 + 5.00 D1 - 2.00 D2

Predictor      Coef   SE Coef      T      p
Constant     23.000     1.106  20.80  0.000
D1            5.000     1.563   3.20  0.011
D2           -2.000     1.563  -1.28  0.233

S = 2.211        R-Sq = 70.3%   R-Sq(adj) = 63.7%

Analysis of Variance

SOURCE          DF        SS       MS      F      p
Regression       2   104.000   52.000   1064  0.004
Residual Error   9    44.000    4.889
Total           11   148.000
```

c. H_0: $\beta_1 = \beta_2 = 0$

d. The p-value of .004 is less than $\alpha = .05$; therefore, we can reject H_0 and conclude that the mean time to mix a batch of material is not the same for each manufacturer

24. a. Not significant at the .05 level of significance; p-value = .093

b. 139

26. Overall significant; p-value = .029
Individually, none of the variables are significant at the .05 level of significance; a larger sample size would be helpful

28. d = 1.60; test is inconclusive

30. a.

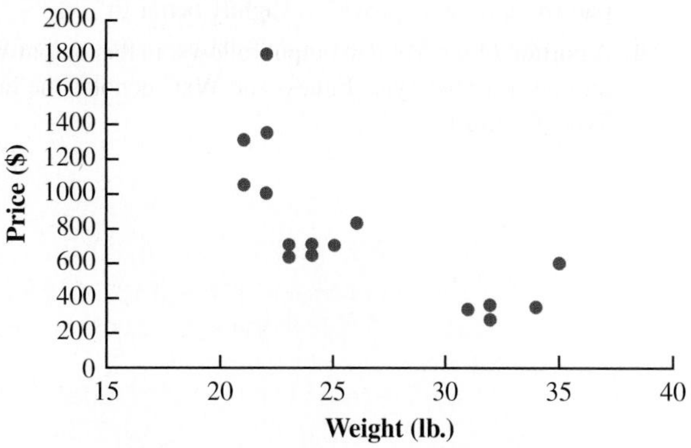

There appears to be a curvilinear relationship between weight and price

b. A portion of the Minitab output follows:

```
The regression equation is
Price = 11376 - 728 Weight + 12.0 WeightSq

Predictor      Coef    SE Coef      T      p
Constant      11376      2565    4.43  0.000
Weight       -728.3     193.7   -3.76  0.002
WeightSq     11.974     3.539    3.38  0.004

S = 242.804    R-Sq = 77.0%    R-Sq(adj) = 74.1%

Analysis of Variance

SOURCE          DF       SS       MS       F      p
Regression       2  3161747  1580874   26.82  0.000
Residual Error  16   943263    58954
Total           18  4105011
```

The results obtained support the conclusion that there is a curvilinear relationship between weight and price

c. A portion of the Minitab output follows:

```
The regression equation is
Price = 1284 - 572 Type_Fitness - 907 Type_Comfort

Predictor         Coef   SE Coef      T      p
Constant       1283.75     95.22  13.48  0.000
Type_Fitness    -571.8     153.5  -3.72  0.002
Type_Comfort    -907.1     145.5  -6.24  0.000

S = 269.328    R-Sq = 71.7%    R-Sq(adj) = 68.2%

Analysis of Variance

SOURCE          DF       SS       MS       F      p
Regression       2  2944410  1472205   20.30  0.000
Residual Error  16  1160601    72538
Total           18  4105011
```

Type of bike appears to be a significant factor in predicting price, but the estimated regression equation developed in part (b) appears to provide a slightly better fit

d. A portion of the Minitab output follows; in this output WxF denotes the interaction between the weight of the bike and the dummy variable Type_Fitness and WxC denotes the interaction between the weight of the bike and the dummy variable Type_Comfort

```
The regression equation is
Price = 5924 - 215 Weight - 6343 Type_Fitness - 7232
        Type_Comfort + 261 WxF + 266 WxC

Predictor          Coef    SE Coef       T       p
Constant           5924       1547    3.83   0.002
Weight          -214.56      71.42   -3.00   0.010
Type_Fitness      -6343       2596   -2.44   0.030
Type_Comfort      -7232       2518   -2.87   0.013
WxF               261.3      111.8    2.34   0.036
WxC              266.41      93.98    2.83   0.014

S = 224.438    R-Sq = 84.0%    R-Sq(adj) = 77.9%

Analysis of Variance

SOURCE           DF        SS       MS       F       p
Regression        5   3450170   690034   13.70   0.000
Residual Error   13    654841    50372
Total            18   4105011
```

By taking into account the type of bike, the weight, and the interaction between these two factors, this estimated regression equation provides an excellent fit

32. a. Delay = 63.0 + 11.1 Industry; no significant positive autocorrelation

34. Significant differences between comfort levels for the three types of browsers; p-value = .034

Chapter 17

1. The following table shows the calculations for parts (a), (b), and (c):

Week	Time Series Value	Forecast	Forecast Error	Absolute Value of Forecast Error	Squared Forecast Error	Percentage Error	Absolute Value of Percentage Error
1	18						
2	13	18	−5	5	25	−38.46	38.46
3	16	13	3	3	9	18.75	18.75
4	11	16	−5	5	25	−45.45	45.45
5	17	11	6	6	36	35.29	35.29
6	14	17	−3	3	9	−21.43	21.43
			Totals	22	104	−51.30	159.38

a. $\text{MAE} = \frac{22}{5} = 4.4$

b. $\text{MSE} = \frac{104}{5} = 20.8$

c. $\text{MAPE} = \frac{159.38}{5} = 31.88$

d. Forecast for week 7 is 14

2. The following table shows the calculations for parts (a), (b), and (c):

Week	Time Series Value	Forecast	Forecast Error	Absolute Value of Forecast Error	Squared Forecast Error	Percentage Error	Absolute Value of Percentage Error
1	18						
2	13	18.00	−5.00	5.00	25.00	−38.46	38.46
3	16	15.50	0.50	0.50	0.25	3.13	3.13
4	11	15.67	−4.67	4.67	21.81	−42.45	42.45
5	17	14.50	2.50	2.50	6.25	14.71	14.71
6	14	15.00	−1.00	1.00	1.00	−7.14	7.14
			Totals	13.67	54.31	−70.21	105.86

a. $\text{MAE} = \frac{13.67}{5} = 2.73$

b. $\text{MSE} = \frac{54.31}{5} = 10.86$

c. $\text{MAPE} = \frac{105.89}{5} = 21.18$

d. Forecast for week 7 is

$$\frac{18 + 13 + 16 + 11 + 17 + 14}{6} = 14.83$$

4. a. $\text{MSE} = \frac{363}{6} = 60.5$

Forecast for month 8 is 15

b. $\text{MSE} = \frac{216.72}{6} = 36.12$

Forecast for month 8 is 18

c. The average of all the previous values is better because MSE is smaller

5. a. The data appear to follow a horizontal pattern

b. Three-week moving average

Week	Time Series Value	Forecast	Forecast Error	Squared Forecast Error
1	18			
2	13			
3	16			
4	11	15.67	−4.67	21.78
5	17	13.33	3.67	13.44
6	14	14.67	−0.67	0.44
			Total	35.67

$$\text{MSE} = \frac{35.67}{3} = 11.89$$

$$\text{The forecast for week 7} = \frac{(11 + 17 + 14)}{3} = 14$$

c. Smoothing constant = .2

Week	Time Series Value	Forecast	Forecast Error	Squared Forecast Error
1	18			
2	13	18.00	−5.00	25.00
3	16	17.00	−1.00	1.00
4	11	16.80	−5.80	33.64
5	17	15.64	1.36	1.85
6	14	15.91	−1.91	3.66
			Total	65.15

$$\text{MSE} = \frac{65.15}{5} = 13.03$$

The forecast for week 7 is $.2(14) + (1 - .2)15.91 = 15.53$

d. The three-week moving average provides a better forecast since it has a smaller MSE

e. Smoothing constant = .4

Week	Time Series Value	Forecast	Forecast Error	Squared Forecast Error
1	18			
2	13	18.00	−5.00	25.00
3	16	16.00	0.00	0.00
4	11	16.00	−5.00	25.00
5	17	14.00	3.00	9.00
6	14	15.20	−1.20	1.44
			Total	60.44

$$\text{MSE} = \frac{60.44}{5} = 12.09$$

The exponential smoothing forecast using $\alpha = .4$ provides a better forecast than the exponential smoothing forecast using $\alpha = .2$ since it has a smaller MSE

6. a. The data appear to follow a horizontal pattern

b. $\text{MSE} = \frac{110}{4} = 27.5$

The forecast for week 8 is 19

c. $\text{MSE} = \frac{252.87}{6} = 42.15$

The forecast for week 7 is 19.12

d. The three-week moving average provides a better forecast since it has a smaller MSE

e. MSE = 39.79

The exponential smoothing forecast using $\alpha = .4$ provides a better forecast than the exponential smoothing forecast using $\alpha = .2$ since it has a smaller MSE

8. a.

Week	4	5	6	7	8	9	10	11	12
Forecast	19.33	21.33	19.83	17.83	18.33	18.33	20.33	20.33	17.83

b. MSE = 11.49

Prefer the unweighted moving average here; it has a smaller MSE

c. You could always find a weighted moving average at least as good as the unweighted one; actually the unweighted moving average is a special case of the weighted ones where the weights are equal

10. b. The more recent data receive the greater weight or importance in determining the forecast; the moving averages method weights the last *n* data values equally in determining the forecast

12. a. The data appear to follow a horizontal pattern

b. MSE(3-Month) = .12

MSE(4-Month) = .14

Use 3-Month moving averages

c. 9.63

13. a. The data appear to follow a horizontal pattern

b.

Month	Time-Series Value	3-Month Moving Average Forecast	(Error)^2	$\alpha = .2$ Forecast	(Error)^2
1	240				
2	350			240.00	12100.00
3	230			262.00	1024.00
4	260	273.33	177.69	255.60	19.36
5	280	280.00	0.00	256.48	553.19
6	320	256.67	4010.69	261.18	3459.79
7	220	286.67	4444.89	272.95	2803.70
8	310	273.33	1344.69	262.36	2269.57
9	240	283.33	1877.49	271.89	1016.97
10	310	256.67	2844.09	265.51	1979.36
11	240	286.67	2178.09	274.41	1184.05
12	230	263.33	1110.89	267.53	1408.50
		Totals	17,988.52		27,818.49

MSE (3-Month) = 17,988.52/9 = 1998.72

MSE ($\alpha = .2$) = 27,818.49/11 = 2528.95

Based on the preceding MSE values, the 3-Month moving averages appear better; however, exponential smoothing was penalized by including month 2, which was difficult for any method to forecast; using only the errors for months 4 to 12, the MSE for exponential smoothing is

$$\text{MSE}(\alpha = .2) = 14{,}694.49/9 = 1632.72$$

Thus, exponential smoothing was better considering months 4 to 12

c. Using exponential smoothing,

$$F_{13} = \alpha Y_{12} + (1 - \alpha)F_{12}$$
$$= .20(230) + .80(267.53) = 260$$

14. a. The data appear to follow a horizontal pattern

b. Values for months 2–12 are as follows:

105.00	114.00	115.80	112.56	105.79	110.05
120.54	126.38	118.46	106.92	104.85	

MSE = 510.29

c. Values for months 2–12 are as follows:

105.00	120.00	120.00	112.50	101.25	110.63
127.81	133.91	116.95	98.48	99.24	

MSE = 540.55

Conclusion: A smoothing constant of .3 is better than a smoothing constant of .5 since the MSE is less for 0.3

16. a.

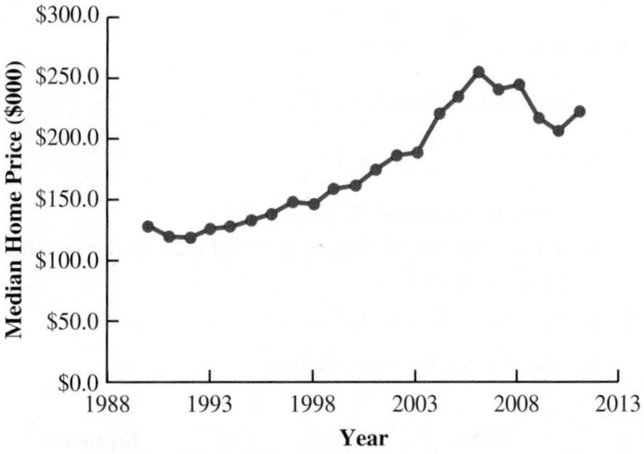

The time series plot exhibits a trend pattern; although the recession of 2008 led to a downturn in prices, the median price rose from 2010 to 2011

b. The methods discussed in this section are only applicable for a time series that has a horizontal pattern; because the time series plot exhibits a trend pattern, the methods discussed in this section are not appropriate

c. In 2003 the median price was $189,500, and in 2004 the median price was $222,300, so, it appears that the time series shifted to a new level in 2004; the time series plot using just the data for 2004 and later follows

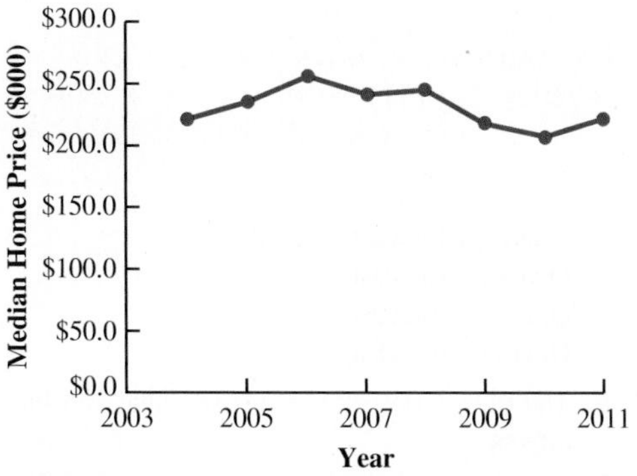

This time series plot exhibits a horizontal pattern; therefore, the methods discussed in this section are appropriate

17. **a.** The time series plot shows a linear trend

b. $\bar{t} = \frac{\sum_{t=1}^{n} t}{n} = \frac{15}{5} = 3 \quad \bar{Y} = \frac{\sum_{t=1}^{n} Y_t}{n} = \frac{55}{5} = 11$

$\Sigma(t - \bar{t})(Y_t - \bar{Y}) = 21 \quad \Sigma(t - \bar{t})^2 = 10$

$$b_1 = \frac{\sum_{t=1}^{n}(t - \bar{t})(Y_t - \bar{Y})}{\sum_{t=1}^{n}(t - \bar{t})^2} = \frac{21}{10} = 2.1$$

$b_0 = \bar{Y} - b_1\bar{t} = 11 - (2.1)(3) = 4.7$

$T_t = 4.7 + 2.1t$

c. $T_6 = 4.7 + 2.1(6) = 17.3$

18. Forecast for week 6 is 21.16

20. **a.** The time series plot exhibits a curvilinear trend
b. $T_t = 107.857 - 28.9881t + 2.65476t^2$
c. 45.86

21. **a.**

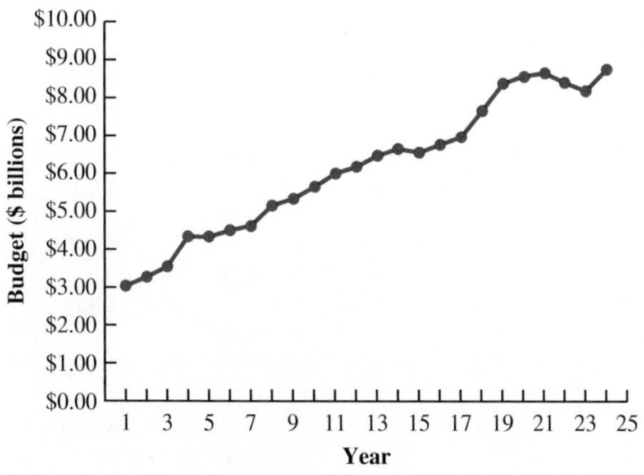

b. $\bar{t} = \frac{\sum_{t=1}^{n} t}{n} = \frac{300}{24} = 12.5 \quad \bar{Y} = \frac{\sum_{t=1}^{n} Y_t}{n} = \frac{148.2}{24} = 6.175$

$\Sigma(t - \bar{t})(Y_t - \bar{Y}) = 290.86 \quad \Sigma(t - \bar{t})^2 = 1150$

$$b_1 = \frac{\sum_{t=1}^{n}(t - \bar{t})(Y_t - \bar{Y})}{\sum_{t=1}^{n}(t - \bar{t})^2} = \frac{290.86}{1150} = .25292$$

$b_0 = \bar{Y} - b_1\bar{x} = 6.175 - (.25292)(12.5) = 3.0135$

c. $\hat{y} = 3.0135 + .25292(25) = 9.34$
Forecast for 2012 is \$9.34 billion

22. **a.**

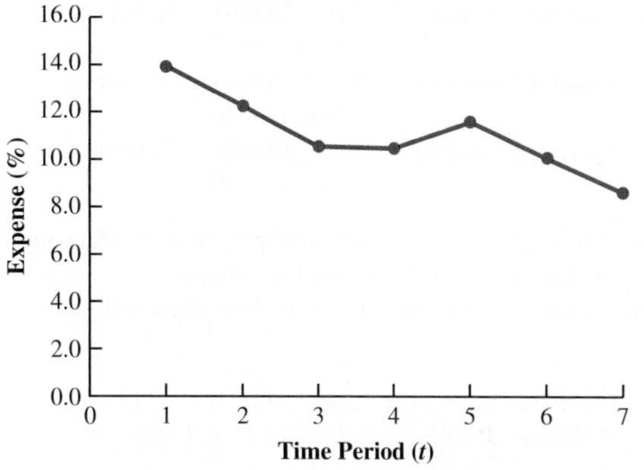

The time series plot shows a downward linear trend

b. $\bar{t} = \frac{\sum_{t=1}^{n} t}{n} = \frac{28}{7} = 4 \quad \bar{Y} = \frac{\sum_{t=1}^{n} Y_t}{n} = \frac{77}{7} = 11$

$\Sigma(t - \bar{t})(Y_t - \bar{Y}) = -19.6 \quad \Sigma(t - \bar{t})^2 = 28$

$$b_1 = \frac{\sum_{t=1}^{n}(t - \bar{t})(Y_t - \bar{Y})}{\sum_{t=1}^{n}(t - \bar{t})^2} = \frac{-19.6}{28} = -.7$$

$b_0 = \bar{Y} - b_1\bar{t} = 11 - (-.7)(4) = 13.8$

$T_t = 13.8 - .7t$

c. 2013 corresponds to time period $t = 8$, $T_8 = 13.8 - .7(8) = 8.2$
d. If SCF can continue to decrease the percentage of funds spent on administrative and fundraising by .7% per year, the forecast of expenses for 2018 is 4.70%

24. **a.** The time series plot shows a linear trend
b. $T_t = 7.5623 - .07541t$
c. 6.7328
d. Given the uncertainty in global market conditions, making a prediction for December using only time is not recommended

26. **a.** A linear trend is not appropriate
b. $T_t = 5.702 + 2.889t - 1618t^2$
c. 17.90

28. **a.** The time series plot shows a horizontal pattern, but there is a seasonal pattern in the data; for instance, in each year the lowest value occurs in quarter 2 and the highest value occurs in quarter 4

b. A portion of the Minitab regression output is shown;

```
The regression equation is
Value = 77.0 - 10.0 Qtr1 - 30.0
        Qtr2 - 20.0 Qtr3
```

c. The quarterly forecasts for next year are as follows:

Quarter 1 forecast = 77.0 − 10.0(1) − 30.0(0) − 20.0(0) = 67

Quarter 2 forecast = 77.0 − 10.0(0) − 30.0(1) − 20.0(0) = 47

Quarter 3 forecast = 77.0 − 10.0(0) − 30.0(0) − 20.0(1) = 57

Quarter 4 forecast = 77.0 − 10.0(0) − 30.0(0) − 20.0(0) = 77

30. a. There appears to be a seasonal pattern in the data and perhaps a moderate upward linear trend

b. A portion of the Minitab regression output follows:

```
The regression equation is
Value = 2492 - 712 Qtr1 - 1512
        Qtr2 + 327 Qtr3
```

c. The quarterly forecasts for next year are as follows:

Quarter 1 forecast is 1780
Quarter 2 forecast is 980
Quarter 3 forecast is 2819
Quarter 4 forecast is 2492

d. A portion of the Minitab regression output follows:

```
The regression equation is
Value = 2307 - 642 Qtr1 - 1465
        Qtr2 + 350 Qtr3 + 23.1 t
```

The quarterly forecasts for next year are as follows:

Quarter 1 forecast is 2058
Quarter 2 forecast is 1258
Quarter 3 forecast is 3096
Quarter 4 forecast is 2769

32. a. The time series plot shows both a linear trend and seasonal effects

b. A portion of the Minitab regression output follows:

```
The regression equation is
Revenue = 70.0 + 10.0 Qtr1 + 105
          Qtr2 + 245 Qtr3
```

Quarter 1 forecast is 80
Quarter 2 forecast is 175
Quarter 3 forecast is 315
Quarter 4 forecast is 70

c. A portion of the Minitab regression output follows

```
The regression equation is
Revenue = -70.1 + 45.0 Qtr1 + 128
          Qtr2 + 257 Qtr3 + 11.7 Period
```

Quarter 1 forecast = is 221
Quarter 2 forecast = is 315
Quarter 3 forecast = is 456
Quarter 4 forecast = is 211

34. a. The time series plot shows seasonal and linear trend effects

b. *Note*: Jan = 1 if January, 0 otherwise; Feb = 1 if February, 0 otherwise; and so on
A portion of the Minitab regression output follows:

```
The regression equation is
Expense = 175 - 18.4 Jan - 3.72 Feb +
          12.7 Mar + 45.7 Apr + 57.1
          May + 135 Jun + 181 Jul + 105
          Aug + 47.6 Sep + 50.6 Oct +
          35.3 Nov + 1.96 Period
```

c. *Note*: The next time period in the time series is Period = 37 (January of Year 4); the forecasts for January–December are 229; 246; 264; 299; 312; 392; 440; 366; 311; 316; 302; 269

35. a. The time series plot indicates a linear trend and a seasonal pattern

b.

Year	Quarter	Time Series Value	Four-Quarter Moving Average	Centered Moving Average
1	1	4		
	2	2		
			3.50	
	3	3		3.750
			4.00	
	4	5		4.125
			4.25	
2	1	6		4.500
			4.75	
	2	3		5.000
			5.25	
	3	5		5.375
			5.50	
	4	7		5.875
			6.25	
3	1	7		6.375
			6.50	
	2	6		6.625
			6.75	
	3	6		
	4	8		

c.

Year	Quarter	Time Series Value	Centered Moving Average	Seasonal-Irregular Component
1	1	4		
	2	2		
	3	3	3.750	0.800
	4	5	4.125	1.212
2	1	6	4.500	1.333
	2	3	5.000	0.600
	3	5	5.375	0.930
	4	7	5.875	1.191
3	1	7	6.375	1.098
	2	6	6.625	0.906
	3	6		
	4	8		

Quarter	Seasonal-Irregular Values		Seasonal Index	Adjusted Seasonal Index
1	1.333	1.098	1.216	1.205
2	0.600	0.906	0.752	0.746
3	0.800	0.930	0.865	0.857
4	1.212	1.191	1.201	1.191
		Total	4.036	

$$\text{Adjustment for seasonal index} = \frac{4.000}{4.036} = 0.991$$

36. a.

Year	Quarter	Deseasonalized Value
1	1	3.320
	2	2.681
	3	3.501
	4	4.198
2	1	4.979
	2	4.021
	3	5.834
	4	5.877
3	1	5.809
	2	8.043
	3	7.001
	4	6.717

b. Let Period = 1 denote the time series value in Year 1—Quarter 1; Period = 2 denote the time series value in Year 1—Quarter 2; and so on; a portion of the Minitab regression output treating Period as the independent variable and the Deseasonlized Values as the values of the dependent variable follows:

```
The regression equation is
Deseasonalized Value = 2.42 + 0.422
                       Period
```

c. The quarterly deseasonalized trend forecasts for Year 4 (Periods 13, 14, 15, and 16) are as follows:

Forecast for quarter 1 is 7.906
Forecast for quarter 2 is 8.328
Forecast for quarter 3 is 8.750
Forecast for quarter 4 is 9.172

d. Adjusting the quarterly deseasonalized trend forecasts provides the following quarterly estimates:

Forecast for quarter 1 is 9.527
Forecast for quarter 2 is 6.213
Forecast for quarter 3 is 7.499
Forecast for quarter 4 is 10.924

38. a. The time series plot shows a linear trend and seasonal effects

b. 0.71 0.78 0.83 0.97 1.02 1.30 1.50 1.23 0.98 0.99 0.93 0.79

c.

Month	Deseasonalized Expense
1	239.44
2	230.77
3	246.99
4	237.11
5	235.29
6	242.31
7	240.00
8	235.77
9	244.90
10	242.42
11	247.31
12	246.84
13	253.52
14	262.82
15	259.04
16	252.58
17	259.80
18	253.85
19	266.67
20	272.36
21	265.31
22	272.73
23	274.19
24	278.48
25	274.65
26	269.23
27	277.11
28	288.66
29	284.31
30	300.00
31	280.00
32	268.29
33	295.92
34	297.98
35	301.08
36	316.46

d. Let Period = 1 denote the time series value in January—Year 1; Period = 2 denote the time series value in February—Year 2; and so on; a portion of the Minitab regression output treating Period as the independent variable and the Deseasonlized Values as the values of the dependent variable follows:

```
The regression equation is
Deseasonalized Expense = 228 + 1.96
                         Period
```

e.

Month	Monthly Forecast
January	213.37
February	235.93
March	252.69
April	297.21
May	314.53
June	403.42
July	486.42
August	386.52
September	309.88
October	314.98
November	297.71
December	254.44

40. a. The time series plot indicates a seasonal effect; power consumption is lowest in the time period 12–4 A.M., steadily increases to the highest value in the 12–4 P.M. time period, and then decreases again. There may also be some linear trend in the data

b.

Time Period	Adjusted Seasonal Index
12–4 A.M.	0.3256
4–8 A.M.	0.4476
8–12 noon	1.3622
12–4 P.M.	1.6959
4–8 P.M.	1.4578
8–12 midnight	0.7109

c. The following Minitab output shows the results of fitting a linear trend equation to the deseasonalized time series:

```
The regression equation is
Deseasonalized Power = 63108 + 1854 t
```

Deseasonalized Power (t = 19) = 63,108 + 1854(19) = 98,334

Forecast for 12–4 P.M. = 1.6959(98,334) = 166,764.63 or approximately 166,765 kWh

Deseasonalized Power (t = 20) = 63,108 + 1854(20) = 100,188

Forecast for 4–8 P.M. = 1.4578(100,188) = 146,054.07 or approximately 146,054 kWh

Thus, the forecast of power consumption from noon to 8 P.M. is 166,765 + 146,054 = 312,819 kWh

42. a. The time series plot indicates a horizontal pattern

b.

$$\text{MSE}(\alpha = .2) = 1.40$$
$$\text{MSE}(\alpha = .3) = 1.27$$
$$\text{MSE}(\alpha = .4) = 1.23$$

A smoothing constant of $\alpha = .4$ provides the best forecast because it has a smaller MSE

c. 31.00

44. a. There appears to be an increasing trend in the data

b. A portion the Minitab regression output follows (*Note*: $t = 1$ corresponds to 2001, $t = 2$ corresponds to 2002, and so on)

```
The regression equation is
Balance($) = 1984 + 146 t
```

The forecast for 2009 ($t = 9$) is Balance($) = 1984 + 146(9) = $3298

c. A portion of the Minitab regression output follows (*Note*: $t = 1$ corresponds to 2001, $t = 2$ corresponds to 2002, and so on)

```
The regression equation is
Balance($) = 2924 - 419 t + 62.7 tsq
```

The forecast for 2009 ($t = 9$) is Balance ($) = 2924 − 419(9) + 62.7(9)^2 = $4232

d. The quadratic trend equation provides the best forecast accuracy for the historical data

e. Linear trend equation

46. a. The forecast for July is 236.97

Forecast for August, using forecast for July as the actual sales in July, is 236.97

Exponential smoothing provides the same forecast for every period in the future; this is why it is not usually recommended for long-term forecasting

b. Using Minitab's regression procedure we obtained the linear trend equation

$$T_t = 149.72 + 18.451t$$

Forecast for July is 278.88

Forecast for August is 297.33

c. The proposed settlement is not fair since it does not account for the upward trend in sales; based upon trend projection, the settlement should be based on forecasted lost sales of $278,880 in July and $297,330 in August

48. a. The time series plot shows a linear trend

b. $T_t = -5 + 15t$

The slope of 15 indicates that the average increase in sales is 15 pianos per year

c. 85, 100

50. a.

Quarter	Adjusted Seasonal Index
1	1.2717
2	0.6120
3	0.4978
4	1.6185

Note: Adjustment for seasonal index $= \frac{4}{3.8985} = 1.0260$

b. The largest effect is in quarter 4; this seems reasonable since retail sales are generally higher during October, November, and December

52. a. Yes, a linear trend pattern appears to be present

b. A portion of the Minitab regression output follows:

```
The regression equation is
Number Sold = 22.9 + 15.5 Year
```

c. Forecast in year 8 is or approximately 147 units

54. b. The centered moving average values smooth out the time series by removing seasonal effects and some of the random variability; the centered moving average time series shows the trend in the data

c.

Quarter	Adjusted Seasonal Index
1	0.899
2	1.362
3	1.118
4	0.621

d. Hudson Marine experiences the largest seasonal increase in quarter 2; since this quarter occurs prior to the peak summer boating season, this result seems reasonable, but the largest seasonal effect is the seasonal decrease in quarter 4; this is also reasonable because of decreased boating in the fall and winter

Chapter 18

1. $n = 27$ cases with a value different than 150
Normal approximation $\mu = .5n = .5(27) = 13.5$

$\sigma = \sqrt{.25\,n} = \sqrt{.25(27)} = 2.5981$

With the number of plus signs = 22 in the upper tail, use continuity correction factor as follows

$$P(x \geq 21.5) = P\left(z \geq \frac{21.5 - 13.5}{2.5981}\right) = P(z \geq 3.08)$$

p-value $= (1.0000 - .9990) = .0010$
p-value $\leq .01$; reject H_0; conclude population median > 150

2. Dropping the no preference, the binomial probabilities for $n = 9$ and $p = .50$ are as follows

x	Probability	x	Probability
0	0.0020	5	0.2461
1	0.0176	6	0.1641
2	0.0703	7	0.0703
3	0.1641	8	0.0176
4	0.2461	9	0.0020

Number of plus signs = 7
$P(x \geq 7) = P(7) + P(8) + P(9)$
$= .0703 + .0176 + .0020$
$= .0899$
Two-tailed p-value $= 2(.0899) = .1798$
p-value $> .05$, do not reject H_0; conclude no indication that a difference exists

4. a. H_0: Median ≥ 15
H_a: Median < 15
b. $n = 9$; number of plus signs = 1
p-value $= .0196$
Reject H_0; bond mutual funds have lower median

6. $n = 48$; $z = 1.88$
p-value $= .0301$
Reject H_0; conclude median > \$56.2 thousand

8. a. $n = 15$
p-value $= .1186$
Do not reject H_0; no significant difference for the pace
b. 25%, 68.8%; recommend larger sample

10. $n = 600$; $z = 2.41$
p-value $= .0160$
Reject H_0; significant difference, *American Idol* preferred

12. H_0: Median for Additive 1 − Median for Additive 2 = 0
H_a: Median for Additive 1 − Median for Additive 2 ≠ 0

Difference	Absolute Difference	Rank	Signed Ranks Negative	Signed Ranks Positive
2.07	2.07	9		9
1.79	1.79	7		7
−0.54	0.54	3	−3	
2.09	2.09	10		10
0.01	0.01	1		1
0.97	0.97	4		4
−1.04	1.04	5	−5	
3.57	3.57	12		12
1.84	1.84	8		8
3.08	3.08	11		11
0.43	0.43	2		2
1.32	1.32	6		6
			Sum of Positive Signed Ranks	$T^+ = 70$

$$\mu_{T^+} = \frac{n(n+1)}{4} = \frac{12(13)}{4} = 39$$

$$\sigma_{T^+} = \sqrt{\frac{n(n+1)(2n+1)}{24}} = \sqrt{\frac{12(13)(25)}{24}} = 12.7475$$

$$P(T^+ \geq 70) = P\left(z \geq \frac{69.5 - 39}{12.7475}\right) = P(z \geq 2.39)$$

p-value = 2(1.0000 − .9916) = .0168
p-value ≤ .05, reject H_0; conclude significant difference between additives

13. H_0: Median time without Relaxant 1 − Median time with Relaxant ≤ 0
H_a: Median time without Relaxant 1 − Median time with Relaxant > 0

Difference	Absolute Difference	Rank	Signed Ranks Negative	Signed Ranks Positive
5	5	9		9
2	2	3		3
10	10	10		10
−3	3	6.5	−6.5	
1	1	1		1
2	2	3		3
−2	2	3	−3	
3	3	6.5		6.5
3	3	6.5		6.5
3	3	6.5		6.5
			Sum of Positive Signed Ranks	T^+ = 45.5

$$\mu_{T^+} = \frac{n(n+1)}{4} = \frac{10(11)}{4} = 27.5$$

$$\sigma_{T^+} = \sqrt{\frac{n(n+1)(2n+1)}{24}} = \sqrt{\frac{10(11)(12)}{24}} = 9.8107$$

$$P(T^+ \geq 45.5) = P\left(z \geq \frac{45 - 27.5}{12.7475}\right) = P(z \geq 1.78)$$

p-value = (1.0000 − .9925) = .0375
p-value ≤ .05; reject H_0; conclude without the relaxant has a greater median time

14. $n = 11$; $T^+ = 61$; $z = 2.45$
p-value = .0142
Reject H_0; conclude significant difference; on-time % better in 2006

16. $n = 10$; $T^+ = 12.5$; $z = -1.48$
p-value = .1388
Do not reject H_0; conclude no difference between median scores

18. H_0: The two populations of additives are identical
H_a: The two populations of additives are not identical

Additive 1	Rank	Additive 2	Rank
17.3	2	18.7	8.5
18.4	6	17.8	4
19.1	10	21.3	15
16.7	1	21.0	14
18.2	5	22.1	16
18.6	7	18.7	8.5
17.5	3	19.8	11
		20.7	13
		20.2	12
W =	34		

$$\mu_W = \frac{1}{2} n_1(n_1 + n_2 + 1) = \frac{1}{2} 7(7 + 9 + 1) = 59.5$$

$$\sigma_W = \sqrt{\frac{1}{12} n_1 n_2 (n_1 + n_2 + 1)} = \sqrt{\frac{1}{12} 7(9)(7 + 9 + 1)}$$
$$= 9.4472$$

With $W = 34$ in lower tail, use the continuity correction

$$P(W \leq 34) = P\left(z \leq \frac{34.5 - 59.5}{9.4472}\right) = P(z \leq -2.65)$$

p-value = 2(.0040) = .0080
p-value < .05; reject H_0; conclude additives are not identical
Additive 2 tends to provide higher miles per gallon

19. a. H_0: The two populations of salaries are identical
H_a: The two populations of salaries are not identical

Public Accountant	Rank	Financial Planner	Rank
50.2	5	49.0	2
58.8	19	49.2	3
56.3	16	53.1	10
58.2	18	55.9	15
54.2	13	51.9	8.5
55.0	14	53.6	11
50.9	6	49.7	4
59.5	20	53.9	12
57.0	17	51.8	7
51.9	8.5	48.9	1
W =	136.5		

$$\mu_W = \frac{1}{2} n_1(n_1 + n_2 + 1) = \frac{1}{2} 10(10 + 10 + 1) = 105$$

$$\sigma_W = \sqrt{\frac{1}{12} n_1 n_2 (n_1 + n_2 + 1)} = \sqrt{\frac{1}{12} 10(10)(10 + 10 + 1)}$$
$$= 13.2288$$

With $W = 136.5$ in upper tail, use the continuity correction

$$P(W \geq 136.5) = P\left(z \geq \frac{136 - 105}{13.2288}\right) = P(z \geq 2.34)$$

p-value = 2(1.0000 − .9904) = .0192
p-value ≤ .05; reject H_0; conclude populations are not identical
Public accountants tend to have higher salaries

b. Public Accountant $\frac{(55.0 + 56.3)}{2}$ = \$55.65 thousand

Financial Planner $\frac{(51.8 + 51.9)}{2}$ = \$51.85 thousand

20. a. \$54,900, \$40,400
b. $W = 69$; $z = 2.04$
p-value = .0414
Reject H_0; conclude a difference between salaries; men higher

22. $W = 157$; $z = 2.74$
p-value = .0062

Reject H_0; conclude a difference between ratios; Japan tends to be higher

24. $W = 116$; $z = -.22$
p-value $= .8258$
Do not reject H_0; conclude no evidence prices differ

26. H_0: All populations of product ratings are identical
H_a: Not all populations of product ratings are identical

	A	B	C
	4	11	7
	8	14	2
	10	15	1
	3	12	6
	9	13	5
Sum of Ranks	34	65	21

$$H = \left[\frac{12}{15(16)}\left(\frac{34^2}{5} + \frac{65^2}{5} + \frac{21^2}{5}\right)\right] - 3(16) = 10.22$$

χ^2 table with $df = 2$, $\chi^2 = 10.22$; the p-value is between .005 and .01

p-value $\leq .01$; reject H_0; conclude the populations of ratings are not identical

28. H_0: All populations of calories burned are identical
H_a: Not all populations calories burned are identical

	Swimming	Tennis	Cycling
	8	9	5
	4	14	1
	11	13	3
	6	10	7
	12	15	2
Sum of Ranks	41	61	18

$$H = \left[\frac{12}{15(16)}\left(\frac{41^2}{5} + \frac{61^2}{5} + \frac{18^2}{5}\right)\right] - 3(16) = 9.26$$

χ^2 table with $df = 2$, $\chi^2 = 9.26$; the p-value is between .005 and .01

p-value $\leq .05$ reject H_0; conclude that the populations of calories burned are not identical

30. $H = 8.03$ with $df = 3$
p-value is between .025 and .05
Reject H_0; conclude a difference between quality of courses

32. a. $\Sigma d_i^2 = 52$

$$r_s = 1 - \frac{6\Sigma d_i^2}{n(n^2 - 1)} = 1 - \frac{6(52)}{10(99)} = .685$$

b. $$\sigma_{r_s} = \sqrt{\frac{1}{n-1}} = \sqrt{\frac{1}{9}} = .3333$$

$$z = \frac{r_s - 0}{\sigma_{r_s}} = \frac{.685}{.3333} = 2.05$$

p-value $= 2(1.0000 - .9798) = .0404$
p-value $\leq .05$, reject H_0; conclude significant positive rank correlation

34. $\Sigma d_i^2 = 250$

$$r_s = 1 - \frac{6\Sigma d_i^2}{n(n^2 - 1)} = 1 - \frac{6(250)}{11(120)} = -.136$$

$$\sigma_{r_s} = \sqrt{\frac{1}{n-1}} = \sqrt{\frac{1}{10}} = .3162$$

$$z = \frac{r_s - 0}{\sigma_{r_s}} = \frac{-.136}{.3162} = -.43$$

p-value $= 2(.3336) = .6672$
p-value $> .05$, do not reject H_0; we cannot conclude that there is a significant relationship

36. $r_s = -.709$, $z = -2.13$
p-value $= .0332$
Reject H_0; conclude a significant rank correlation

38. Number of plus signs $= 905$, $z = -3.15$
p-value less than .0020
Reject H_0; conclude a significant difference between the preferences

40. $n = 12$; $T^+ = 6$; $z = -2.55$
p-value $= .0108$
Reject H_0; conclude significant difference between prices

42. $W = 70$; $z = -2.93$
p-value $= .0034$
Reject H_0; conclude populations of weights are not identical

44. $H = 12.61$ with $df = 2$
p-value is less than .005
Reject H_0; conclude the populations of ratings are not identical

46. $r_s = .757$, $z = 2.83$
p-value $= .0046$
Reject H_0; conclude a significant positive rank correlation

Chapter 19

2. a. 5.42
b. UCL $= 6.09$, LCL $= 4.75$

4. *R chart:*
UCL $= \bar{R}D_4 = 1.6(1.864) = 2.98$
LCL $= \bar{R}D_3 = 1.6(.136) = .22$
$\bar{x}$ chart:
UCL $= \bar{\bar{x}} + A_2\bar{R} = 28.5 + .373(1.6) = 29.10$
LCL $= \bar{\bar{x}} - A_2\bar{R} = 28.5 - .373(1.6) = 27.90$

6. 20.01, .082

8. a. .0470
b. UCL $= .0989$, LCL $= -0.0049$ (use LCL $= 0$)
c. $\bar{p} = .08$; in control
d. UCL $= 14.826$, LCL $= -0.726$ (use LCL $= 0$)
Process is out of control if more than 14 defective
e. In control with 12 defective
f. *np* chart

10. $f(x) = \dfrac{n!}{x!(n-x)!} p^x(1-p)^{n-x}$

When $p = .02$, the probability of accepting the lot is

$$f(0) = \frac{25!}{0!(25-0)!}(.02)^0(1-.02)^{25} = .6035$$

When $p = .06$, the probability of accepting the lot is

$$f(0) = \frac{25!}{0!(25-0)!}(.06)^0(1-.06)^{25} = .2129$$

12. $p_0 = .02$; producer's risk = .0599
$p_0 = .06$; producer's risk = .3396
Producer's risk decreases as the acceptance number c is increased

14. $n = 20, c = 3$

16. a. 95.4
b. UCL = 96.07, LCL = 94.73
c. No

18.

	R Chart	$\bar{x}$ Chart
UCL	4.23	6.57
LCL	0	4.27

Estimate of standard deviation = .86

20.

	R Chart	$\bar{x}$ Chart
UCL	.1121	3.112
LCL	0	3.051

22. a. UCL = .0817, LCL = −.0017 (use LCL = 0)

24. a. .03
b. $\beta = .0802$

Chapter 20

1. a.

Item	Price Relative
A	103 = (7.75/7.50)(100)
B	238 = (1500/630)(100)

b. $I_{2011} = \dfrac{7.75 + 1500.00}{7.50 + 630.00}(100) = \dfrac{1507.75}{637.50}(100) = 237$

c. $I_{2011} = \dfrac{7.75(1500) + 1500.00(2)}{7.50(1500) + 630.00(2)}(100)$

$= \dfrac{14{,}625.00}{12{,}510.00}(100) = 117$

d. $I_{2011} = \dfrac{7.75(1800) + 1500.00(1)}{7.50(1800) + 630.00(1)}(100)$

$= \dfrac{15{,}450.00}{14{,}130.00}(100) = 109$

2. a. 32%
b. $8.14

3. a. Price relatives for A = (6.00/5.45)100 = 110
B = (5.95/5.60)100 = 106
C = (6.20/5.50)100 = 113

b. $I_{2011} = \dfrac{6.00 + 5.95 + 6.20}{5.45 + 5.60 + 5.50}(100) = 110$

c. $I_{2011} = \dfrac{6.00(150) + 5.95(200) + 6.20(120)}{5.45(150) + 5.60(200) + 5.50(120)}(100)$

$= 109$

9% increase over the two-year period

4. $I_{2011} = 114$

6.

Item	Price Relative	Base Period Price	Usage	Weight	Weighted Price Relative
A	150	22.00	20	440	66,000
B	90	5.00	50	250	22,500
C	120	14.00	40	560	67,200
			Totals	1250	155,700

$$I = \frac{155{,}700}{1250} = 125$$

7. a. Price relatives for A = (3.95/2.50)100 = 158
B = (9.90/8.75)100 = 113
C = (.95/.99)100 = 96

b.

Item	Price Relative	Base Price	Quantity	Weight $P_{i0}Q_i$	Weighted Price Relative
A	158	2.50	25	62.5	9,875
B	113	8.75	15	131.3	14,837
C	96	.99	60	59.4	5,702
			Totals	253.2	30,414

$$I = \frac{30{,}414}{253.2} = 120$$

Cost of raw materials is up 20% for the chemical

8. $I = 105$; portfolio is up 5%

10. a. Deflated 2007 wages: $\dfrac{\$30.04}{207.3}(100) = \$14.49$

Deflated 2011 wages: $\dfrac{\$33.23}{224.9}(100) = \$14.78$

b. $\dfrac{33.23}{30.04}(100) = 1.11$; the percentange increase in actual wages is 11%

c. $\dfrac{14.78}{14.49}(100) = 1.02$; the percentage increase in real wages is 2%

12. a. 2009: $\frac{\$29.1}{216.0}(100) = \$13.47$

2010: $\frac{\$33.3}{218.4}(100) = \$15.25$

2011: $\frac{\$32.9}{226.9}(100) = \$14.50$

b. 2009: $\frac{\$29.1}{173.4}(100) = \$16.78$

2010: $\frac{\$33.3}{180.2}(100) = \$18.48$

2011: $\frac{\$32.9}{192.5}(100) = \$17.09$

c. PPI is more appropriate than CPI because these figures reflect prices paid by retailers (rather than by consumers)

14. $I = \frac{300(18.00) + 400(4.90) + 850(15.00)}{350(18.00) + 220(4.90) + 730(15.00)}(100)$

$= \frac{20{,}110}{18{,}328}(100) = 110$

15. $I = \frac{95(1200) + 75(1800) + 50(2000) + 70(1500)}{120(1200) + 86(1800) + 35(2000) + 60(1500)}(100)$

$= 99$

Quantities are down slightly

16. $I = 83$

18. a. 151, 197, 143, 178

b. $I = 170$

20. $I_{\text{Jan}} = 73.5$, $I_{\text{Mar}} = 70.1$

22.

Item	Price Relatives $(P_{it}/P_{i0})100$	Base Price P_{i0}	Quantity Q_i	Weight $W_i = P_{i0}Q_i$	Weighted Price Relatives $(P_{it}/P_{i0})100\ W_i$
Handle	126.9	7.46	1	7.46	946.67
Blades	153.7	1.9	17	32.30	4964.51
			Totals	39.76	5911.18

$I_{2011} = (5911.18/39.76) = 148.7$

24. 2008 (84810/215.3)100 = \$39,392
2009 (87210/214.5)100 = \$40,657
2010 (87650/218.1)100 = \$40,188
2011 (91060/224.9)100 = \$40,489
(40489/39392) = 1.028

The median salary increased 2.8% over the period

26. $I = 143$; quantity is up 43%

Appendix E: Microsoft Excel 2010 and Tools for Statistical Analysis

Microsoft Excel 2010, part of the Microsoft Office 2010 system, is a spreadsheet program that can be used to organize and analyze data, perform complex calculations, and create a wide variety of graphical displays. We assume that readers are familiar with basic Excel operations such as selecting cells, entering formulas, copying, and so on. But we do not assume readers are familiar with Excel 2010 or the use of Excel for statistical analysis.

The purpose of this appendix is twofold. First, we provide an overview of Excel 2010 and discuss the basic operations needed to work with Excel 2010 workbooks and worksheets. Second, we provide an overview of the tools that are available for conducting statistical analysis with Excel. These include Excel functions and formulas which allow users to conduct their own analyses and add-ins that provide more comprehensive analysis tools.

Excel's Data Analysis add-in, included with the basic Excel system, is a valuable tool for conducting statistical analysis. In the last section of this appendix we provide instruction for installing the Data Analysis add-in. Other add-ins have been developed by outside suppliers to supplement the basic statistical capabilities provided by Excel. In the last section we also discuss StatTools, a commercially available add-in developed by Palisade Corporation.

Overview of Microsoft Excel 2010

A workbook is a file containing one or more worksheets.

When using Excel for statistical analysis, data is displayed in workbooks, each of which contains a series of worksheets that typically include the original data as well as any resulting analysis, including charts. Figure 1 shows the layout of a blank workbook created each time Excel is opened. The workbook is named Book1, and consists of three worksheets named Sheet1, Sheet2, and Sheet3. Excel highlights the worksheet currently displayed (Sheet1) by setting the name on the worksheet tab in bold. To select a different worksheet simply click on the corresponding tab. Note that cell A1 is initially selected.

The wide bar located across the top of the workbook is referred to as the Ribbon. Tabs, located at the top of the Ribbon, provide quick access to groups of related commands. There are nine tabs shown on the workbook in Figure 1: File; Home; Insert; Page Layout; Formulas; Data; Review; View; and Add-Ins. Each tab contains a series of groups of related commands. Note that the Home tab is selected when Excel is opened. Figure 2 displays the groups available when the Home tab is selected. Under the Home tab there are seven groups: Clipboard; Font; Alignment; Number; Styles; Cells; and Editing. Commands are arranged within each group. For example, to change selected text to boldface, click the Home tab and click the Bold button in the Font group.

Figure 3 illustrates the location of the Quick Access Toolbar and the Formula Bar. The Quick Access Toolbar allows you to quickly access workbook options. To add or remove features on the Quick Access Toolbar click the Customize Quick Access Toolbar button on the Quick Access Toolbar.

The Formula Bar (see Figure 3) contains a Name box, the Insert Function button *fx*, and a Formula box. In Figure 3, "A1" appears in the name box because cell A1 is selected. You can select any other cell in the worksheet by using the mouse to move the cursor to another cell and clicking or by typing the new cell location in the Name box. The Formula box is used to display the formula in the currently selected cell. For instance, if you had entered =A1+A2 into cell A3, whenever you select cell A3 the formula =A1+A2 will be

FIGURE 1 BLANK WORKBOOK CREATED WHEN EXCEL IS OPENED

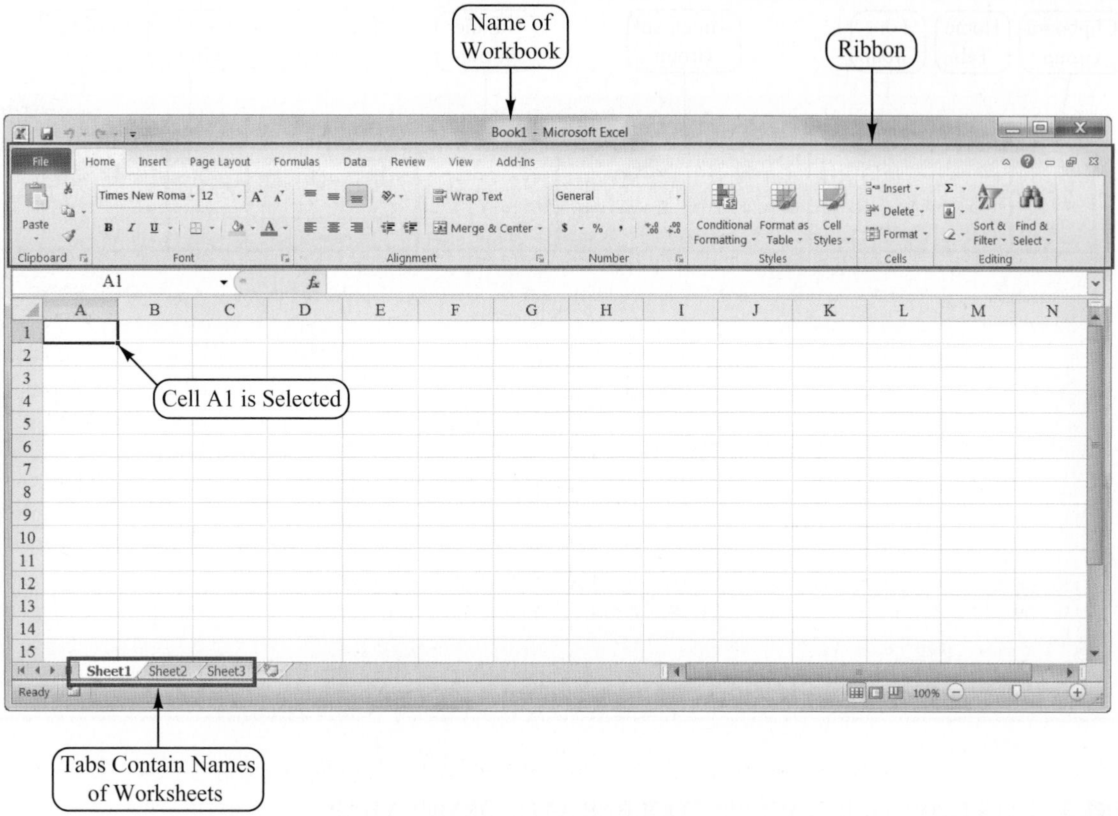

shown in the Formula box. This feature makes it very easy to see and edit a formula in a particular cell. The Insert Function button allows you to quickly access all of the functions available in Excel. Later we show how to find and use a particular function.

Basic Workbook Operations

Figure 4 illustrates the worksheet options that can be performed after right-clicking on a worksheet tab. For instance, to change the name of the current worksheet from "Sheet1" to "Data," right-click the worksheet tab named "Sheet1" and select the Rename option. The current worksheet name (Sheet1) will be highlighted. Then, simply type the new name (Data) and press the Enter key to rename the worksheet.

Suppose that you wanted to create a copy of "Sheet 1." After right-clicking the tab named "Sheet1," select the Move or Copy option. When the Move or Copy dialog box appears, select Create a Copy and click OK. The name of the copied worksheet will appear as "Sheet1 (2)." You can then rename it, if desired.

To add a worksheet to the workbook, right-click any worksheet tab and select the Insert option; when the Insert dialog box appears, select Worksheet and click OK. An additional blank worksheet titled "Sheet 4" will appear in the workbook. You can also insert a new worksheet by clicking the Insert Worksheet tab button that appears to the right of the last worksheet tab displayed. Worksheets can be deleted by right-clicking the worksheet tab and choosing Delete. After clicking Delete a window will appear warning you that any data appearing in the worksheet will be lost. Click Delete to confirm that you do want to

FIGURE 2 PORTION OF THE HOME TAB

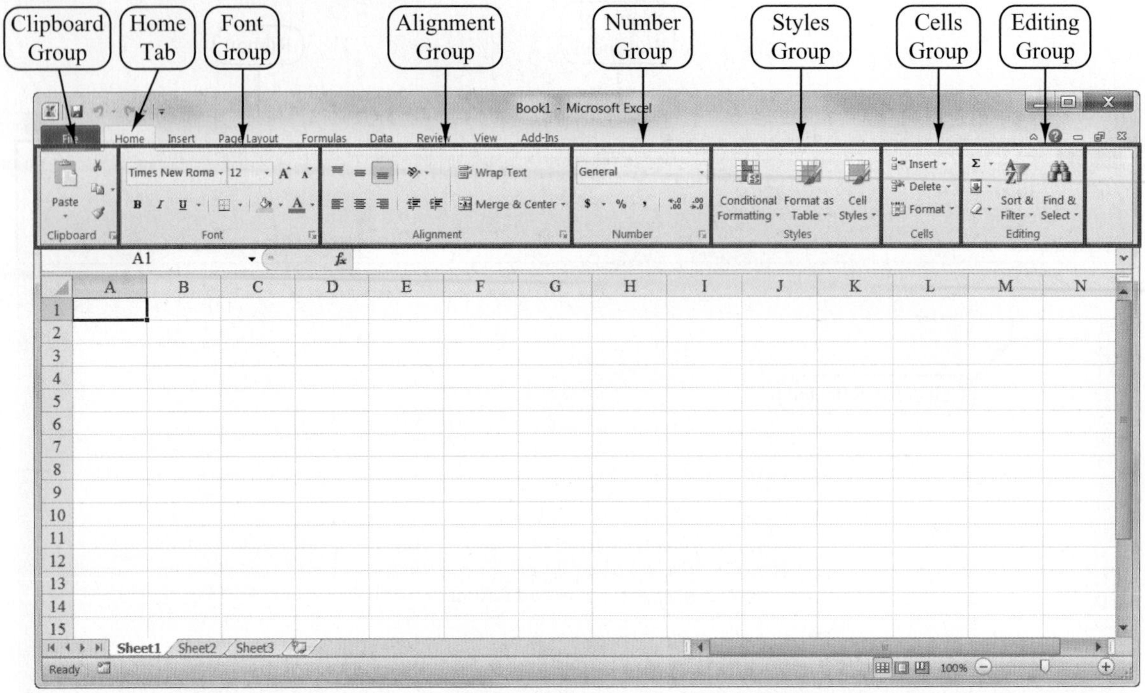

FIGURE 3 EXCEL 2010 QUICK ACCESS TOOLBAR AND FORMULA BAR

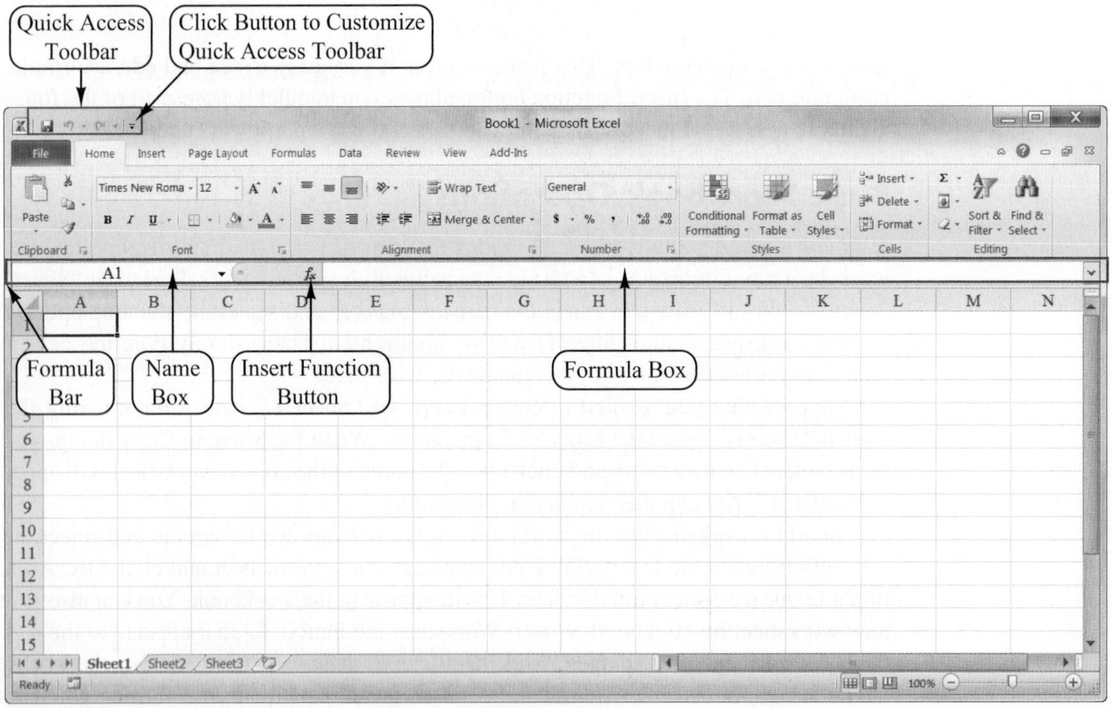

FIGURE 4 WORKSHEET OPTIONS OBTAINED AFTER RIGHT-CLICKING ON A WORKSHEET TAB

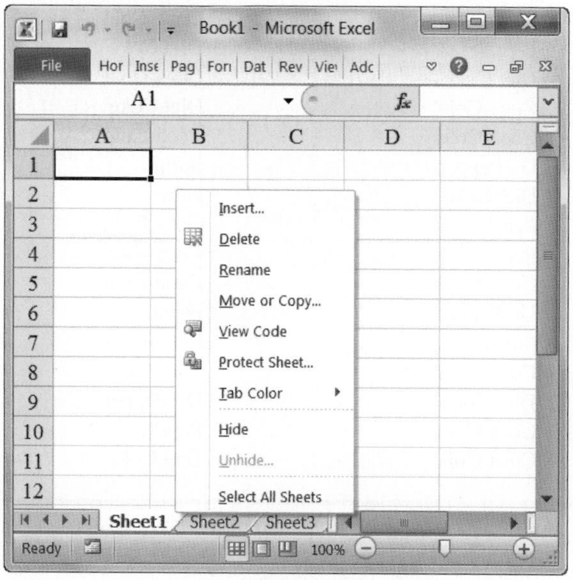

delete the worksheet. Worksheets can also be moved to other workbooks or a different position in the current workbook by using the Move or Copy option.

Creating, Saving, and Opening Files

Data can be entered into an Excel worksheet by manually entering the data into the worksheet or by opening another workbook that already contains the data. As an illustration of manually entering, saving, and opening a file we will use the example from Chapter 2 involving data for a sample of 50 soft drink purchases. The original data are shown in Table 1.

Suppose you just opened Excel and want to work with this data. A blank workbook containing three worksheets will be displayed. The soft drink data can now be entered by simply typing it into one of the worksheets. If Excel is currently running and no blank workbook is displayed, you can create a blank workbook using the following steps:

Step 1: Click the **File** tab
Step 2: Click **New** in the list of options
Step 3: When the Available Templates dialog box appears:
Double-click **Blank Workbook**

A new workbook containing three worksheets labeled Sheet1, Sheet2, and Sheet3 will appear.

Suppose we want to enter the data for the sample of 50 soft drink purchases into Sheet1 of the new workbook. First, we enter the label "Brand Purchased" into cell A1; then, we enter the data for the 50 soft drink purchases into cells A2:A51. As a reminder that this worksheet contains the data, we will change the name of the worksheet from "Sheet1" to "Data" using the procedure described previously. Figure 5 shows the data worksheet that we just developed.

TABLE 1 DATA FROM A SAMPLE OF 50 SOFT DRINK PURCHASES

Coca-Cola	Sprite	Pepsi
Diet Coke	Coca-Cola	Coca-Cola
Pepsi	Diet Coke	Coca-Cola
Diet Coke	Coca-Cola	Coca-Cola
Coca-Cola	Diet Coke	Pepsi
Coca-Cola	Coca-Cola	Dr. Pepper
Dr. Pepper	Sprite	Coca-Cola
Diet Coke	Pepsi	Diet Coke
Pepsi	Coca-Cola	Pepsi
Pepsi	Coca-Cola	Pepsi
Coca-Cola	Coca-Cola	Pepsi
Dr. Pepper	Pepsi	Pepsi
Sprite	Coca-Cola	Coca-Cola
Coca-Cola	Sprite	Dr. Pepper
Diet Coke	Dr. Pepper	Pepsi
Coca-Cola	Pepsi	Sprite
Coca-Cola	Diet Coke	

FIGURE 5 WORKSHEET CONTAINING THE SOFT DRINK DATA

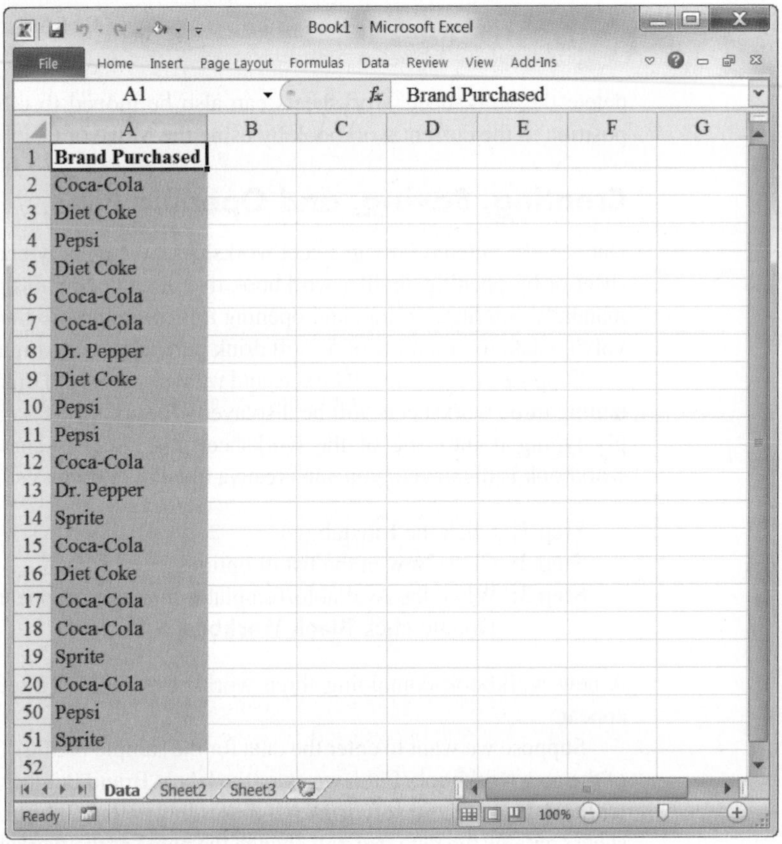

Note: Rows 21–49 are hidden.

Before doing any analysis with these data, we recommend that you first save the file; this will prevent you from having to reenter the data in case something happens that causes Excel to close. To save the file as an Excel 2010 workbook using the filename SoftDrink we perform the following steps:

Step 1: Click the **File** tab
Step 2: Click **Save** in the list of options
Step 3: When the Save As dialog box appears:
Select the location where you want to save the file
Type the filename **SoftDrink** in the **File name** box
Click **Save**

Excel's Save command is designed to save the file as an Excel 2010 workbook. As you work with the file to do statistical analysis you should follow the practice of periodically saving the file so you will not lose any statistical analysis you may have performed. Simply click the File tab and select Save in the list of options.

Keyboard shortcut: To save the file, press CTRL+S

Sometimes you may want to create a copy of an existing file. For instance, suppose you would like to save the soft drink data and any resulting statistical analysis in a new file named "SoftDrink Analysis." The following steps show how to create a copy of the SoftDrink workbook and analysis with the new filename, "SoftDrink Analysis."

Step 1: Click the **File** tab
Step 2: Click **Save As**
Step 3: When the Save As dialog box appears:
Select the location where you want to save the file
Type the filename **SoftDrink Analysis** in the **File name** box
Click **Save**

Once the workbook has been saved, you can continue to work with the data to perform whatever type of statistical analysis is appropriate. When you are finished working with the file simply click the close window button ✖ located at the top right-hand corner of the Ribbon. To access the SoftDrink Analysis file at another point in time you can open the file by performing the following steps:

Step 1: Click the **File** tab
Step 2: Click **Open**
Step 3: When the Open dialog box appears:
Select the location where you previously saved the file
Enter the filename **SoftDrink Analysis** in the **File name** box
Click **Open**

The procedures we showed for saving or opening a workbook begin by clicking File tab to access the Save and Open commands. Once you have used Excel for a while you will probably find it more convenient to add these commands to the Quick Access Toolbar.

Using Excel Functions

Excel 2010 provides a wealth of functions for data management and statistical analysis. If we know what function is needed, and how to use it, we can simply enter the function into the appropriate worksheet cell. However, if we are not sure what functions are available to accomplish a task or are not sure how to use a particular function, Excel can provide assistance. Many new functions for statistical analysis have been added with Excel 2010. To illustrate we will use the SoftDrink Analysis workbook created in the previous subsection.

FIGURE 6 INSERT FUNCTION DIALOG BOX

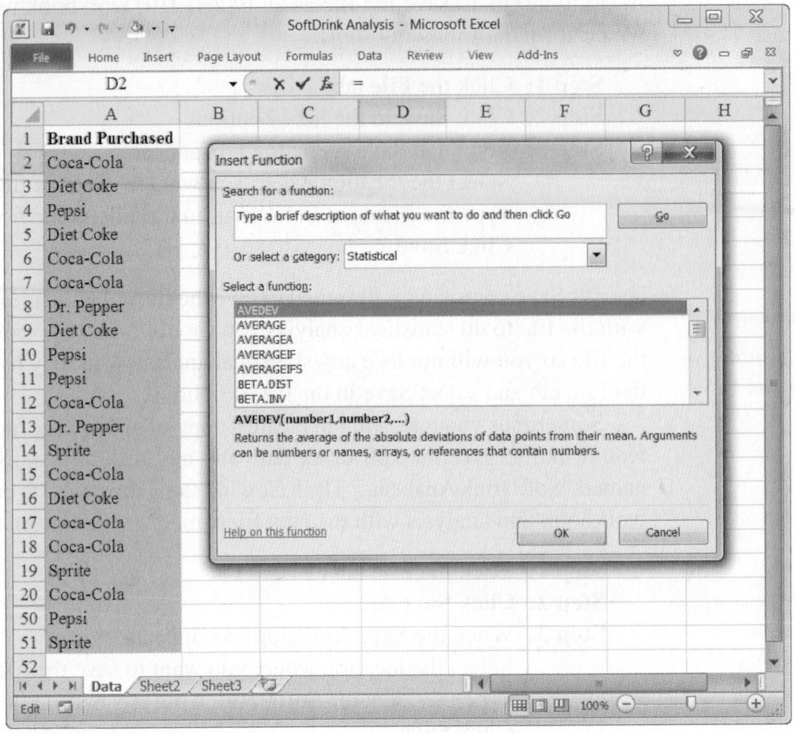

Note: Rows 21–49 are hidden.

Finding the Right Excel Function

To identify the functions available in Excel, select the cell where you want to insert the function; we have selected cell D2. Click the **Formulas** tab on the Ribbon and then click the **Insert Function** button in the **Function Library** group. Alternatively, click the *fx* button on the formula bar. Either approach provides the **Insert Function** dialog box shown in Figure 6.

The **Search for a function** box at the top of the Insert Function dialog box enables us to type a brief description of what we want to do. After doing so and clicking **Go**, Excel will search for and display, in the **Select a function** box, the functions that may accomplish our task. In many situations, however, we may want to browse through an entire category of functions to see what is available. For this task, the **Or select a category** box is helpful. It contains a drop-down list of several categories of functions provided by Excel. Figure 6 shows that we selected the **Statistical** category. As a result, Excel's statistical functions appear in alphabetic order in the Select a function box. We see the AVEDEV function listed first, followed by the AVERAGE function, and so on.

The AVEDEV function is highlighted in Figure 6, indicating it is the function currently selected. The proper syntax for the function and a brief description of the function appear below the Select a function box. We can scroll through the list in the Select a function box to display the syntax and a brief description for each of the statistical functions that are available. For instance, scrolling down farther, we select the COUNTIF function as shown in Figure 7. Note that COUNTIF is now highlighted, and that immediately below the Select a function box we see **COUNTIF(range,criteria)**, which indicates that the COUNTIF function contains two inputs, range and criteria. In addition, we see that the description of the COUNTIF function is "Counts the number of cells within a range that meet the given condition."

FIGURE 7 DESCRIPTION OF THE COUNTIF FUNCTION IN THE INSERT FUNCTION DIALOG BOX

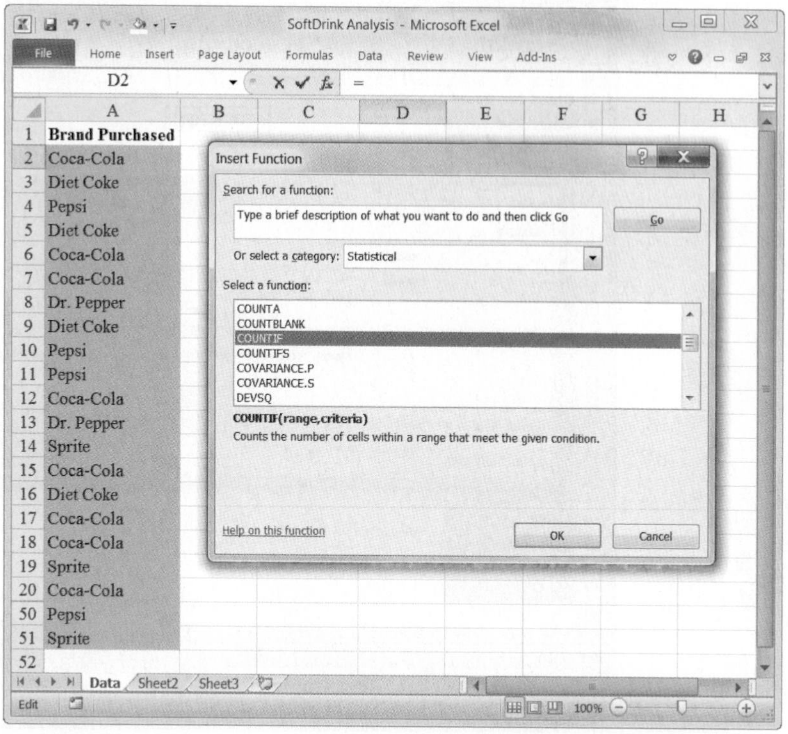

Note: Rows 21–49 are hidden.

If the function selected (highlighted) is the one we want to use, we click **OK**; the **Function Arguments** dialog box then appears. The Function Arguments dialog box for the COUNTIF function is shown in Figure 8. This dialog box assists in creating the appropriate arguments for the function selected. When finished entering the arguments, we click OK; Excel then inserts the function into a worksheet cell.

Inserting a Function into a Worksheet Cell

We will now show how to use the Insert Function and Function Arguments dialog boxes to select a function, develop its arguments, and insert the function into a worksheet cell.

Suppose we want to construct a frequency distribution for the soft drink purchase data in Table 1. Figure 9 displays an Excel worksheet containing the soft drink data and labels for the frequency distribution we would like to construct. We see that the frequency of Coca-Cola purchases will go into cell D2, the frequency of Diet Coke purchases will go into cell D3, and so on. Suppose we want to use the COUNTIF function to compute the frequencies and would like some assistance from Excel.

Step 1. Select cell D2
Step 2. Click *fx* on the formula bar
Step 3. When the Insert Function dialog box appears:
Select **Statistical** in the **Or select a category** box
Select **COUNTIF** in the **Select a function** box
Click **OK**

FIGURE 8 FUNCTION ARGUMENTS DIALOG BOX FOR THE COUNTIF FUNCTION

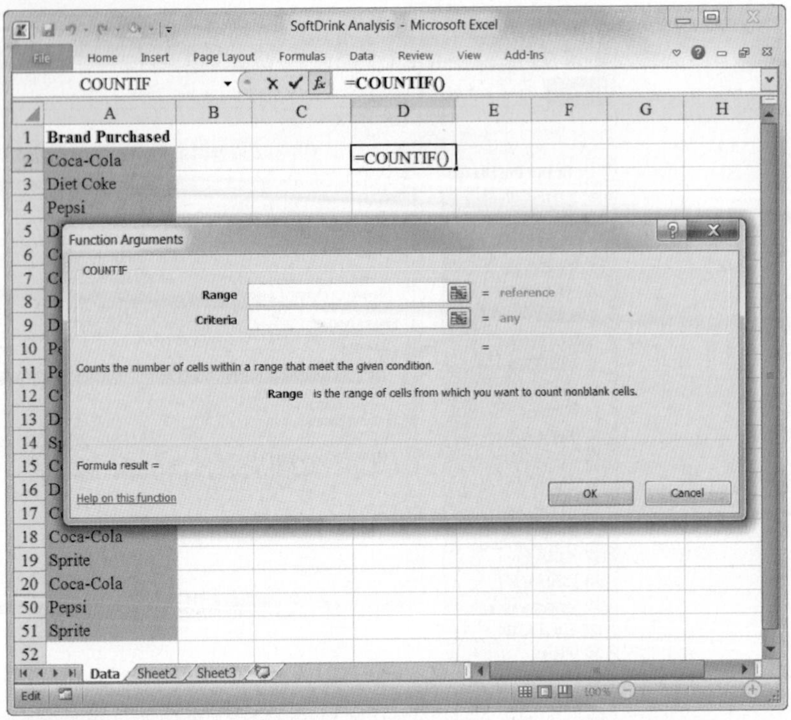

Note: Rows 21–49 are hidden.

FIGURE 9 EXCEL WORKSHEET WITH SOFT DRINK DATA AND LABELS FOR THE FREQUENCY DISTRIBUTION WE WOULD LIKE TO CONSTRUCT

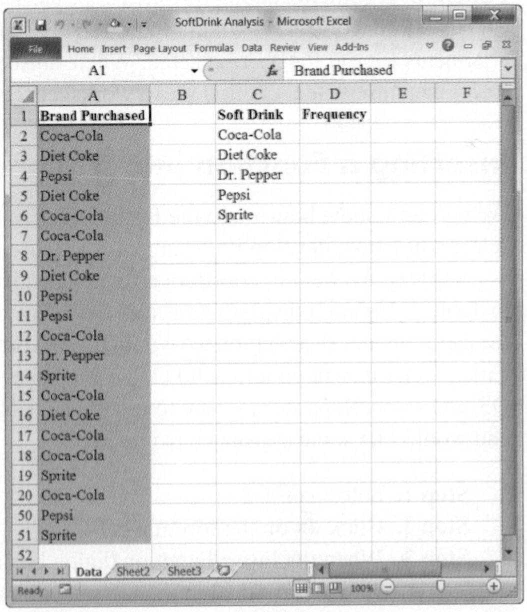

Note: Rows 21–49 are hidden.

FIGURE 10 COMPLETED FUNCTION ARGUMENTS DIALOG BOX FOR THE COUNTIF FUNCTION

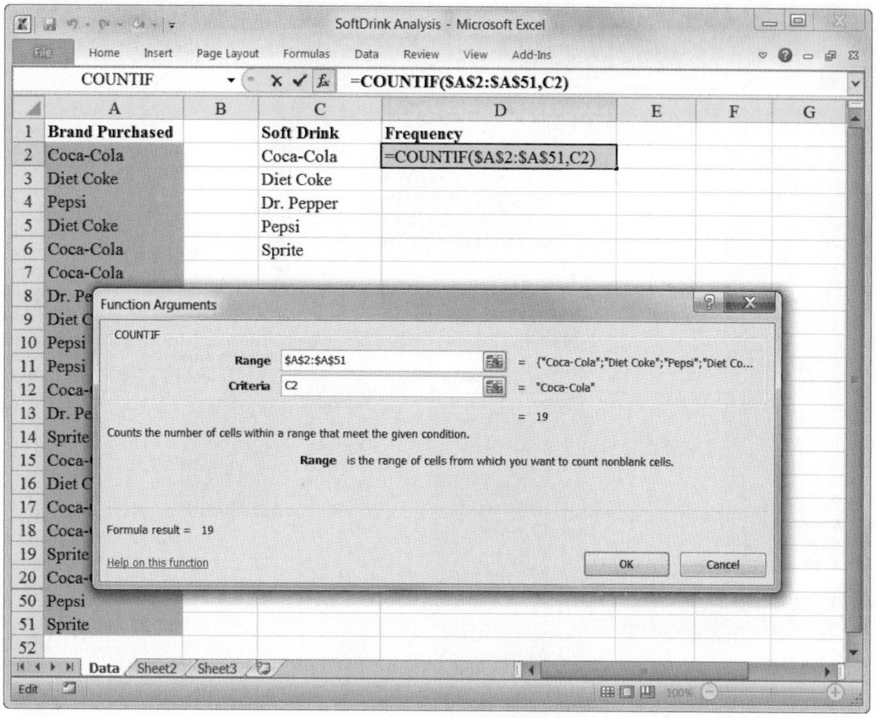

Note: Rows 21–49 are hidden.

Step 4. When the **Function Arguments** dialog box appears (see Figure 10):
Enter $A$2:$A$51 in the **Range** box
Enter C2 in the **Criteria** box (At this point, the value of the function will appear on the next-to-last line of the dialog box. Its value is 19.)
Click **OK**

Step 5. Copy cell D2 to cells D3:D6

The worksheet then appears as in Figure 11. The formula worksheet is in the background; the value worksheet appears in the foreground. The formula worksheet shows that the COUNTIF function was inserted into cells D2:D6. The value worksheet shows the proper class frequencies as computed.

We illustrated the use of Excel's capability to provide assistance in using the COUNTIF function. The procedure is similar for all Excel functions. This capability is especially helpful if you do not know what function to use or forget the proper name and/or inputs required for a function.

Using Excel Add-Ins

Excel's Data Analysis Add-In

Excel's Data Analysis add-in, included with the basic Excel package, is a valuable tool for conducting statistical analysis. Before you can use the Data Analysis add-in it must be installed. To see if the Data Analysis add-in has already been installed, click the Data tab on the Ribbon. In the Analysis group you should see the Data Analysis command. If you do

FIGURE 11 EXCEL WORKSHEET SHOWING THE USE OF EXCEL'S COUNTIF FUNCTION TO CONSTRUCT A FREQUENCY DISTRIBUTION

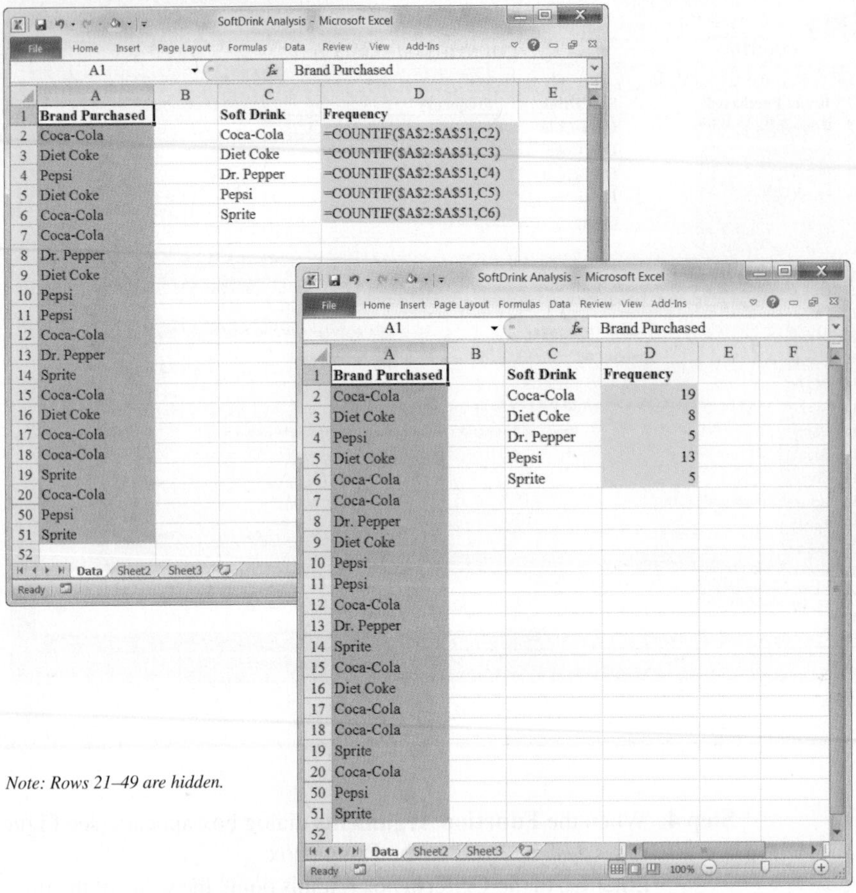

Note: Rows 21–49 are hidden.

not have an Analysis group and/or the Data Analysis command does not appear in the Analysis group, you will need to install the Data Analysis add-in. The steps needed to install the Data Analysis add-in are as follows:

Step 1. Click the **File** tab
Step 2. Click **Options**
Step 3. When the Excel Options dialog box appears:
Select **Add-Ins** from the list of options (on the pane on the left)
In the **Manage** box, select **Excel Add-Ins**
Click **Go**
Step 4. When the Add-Ins dialog box appears:
Select **Analysis ToolPak**
Click **OK**

Outside Vendor Add-Ins

One of the leading companies in the development of Excel add-ins for statistical analysis is Palisade Corporation. In this text we use StatTools, an Excel add-in developed by

Palisade. StatTools provides a powerful statistics toolset that enables users to perform statistical analysis in the familiar Microsoft Office environment.

In the appendix to Chapter 1 we describe how to download and install the StatTools add-in and provide a brief introduction to using the software. In several appendices throughout the text we show how StatTools can be used when no corresponding basic Excel procedure is available or when additional statistical capabilities would be useful.

Typically the add-ins offered with textbooks are designed primarily for classroom use. StatTools, however, was developed for commercial applications. As a result, students who learn how to use StatTools will be able to continue using StatTools throughout their professional career.

Appendix F: Computing p-Values Using Minitab and Excel

Here we describe how Minitab and Excel can be used to compute p-values for the z, t, χ^2, and F statistics that are used in hypothesis tests. As discussed in the text, only approximate p-values for the t, χ^2, and F statistics can be obtained by using tables. This appendix is helpful to a person who has computed the test statistic by hand, or by other means, and wishes to use computer software to compute the exact p-value.

Using Minitab

Minitab can be used to provide the cumulative probability associated with the z, t, χ^2, and F test statistics. So the lower tail p-value is obtained directly. The upper tail p-value is computed by subtracting the lower tail p-value from 1. The two-tailed p-value is obtained by doubling the smaller of the lower and upper tail p-values.

The z test statistic We use the Hilltop Coffee lower tail hypothesis test in Section 9.3 as an illustration; the value of the test statistic is $z = -2.67$. The Minitab steps used to compute the cumulative probability corresponding to $z = -2.67$ follow.

Step 1. Select the **Calc** menu
Step 2. Choose **Probability Distributions**
Step 3. Choose **Normal**
Step 4. When the Normal Distribution dialog box appears:
Select **Cumulative probability**
Enter 0 in the **Mean** box
Enter 1 in the **Standard deviation** box
Select **Input Constant**
Enter -2.67 in the **Input Constant** box
Click **OK**

Minitab provides the cumulative probability of .0038. This cumulative probability is the lower tail p-value used for the Hilltop Coffee hypothesis test.

For an upper tail test, the p-value is computed from the cumulative probability provided by Minitab as follows:

$$p\text{-value} = 1 - \text{cumulative probability}$$

For instance, the upper tail p-value corresponding to a test statistic of $z = -2.67$ is $1 - .0038 = .9962$. The two-tailed p-value corresponding to a test statistic of $z = -2.67$ is 2 times the minimum of the upper and lower tail p-values; that is, the two-tailed p-value corresponding to $z = -2.67$ is $2(.0038) = .0076$.

The t test statistic We use the Heathrow Airport example from Section 9.4 as an illustration; the value of the test statistic is $t = 1.84$ with 59 degrees of freedom. The Minitab steps used to compute the cumulative probability corresponding to $t = 1.84$ follow.

Step 1. Select the **Calc** menu
Step 2. Choose **Probability Distributions**

Step 3. Choose **t**
Step 4. When the t Distribution dialog box appears:
Select **Cumulative probability**
Enter 59 in the **Degrees of freedom** box
Select **Input Constant**
Enter 1.84 in the **Input Constant** box
Click **OK**

Minitab provides a cumulative probability of .9646, and hence the lower tail p-value = .9646. The Heathrow Airport example is an upper tail test; the upper tail p-value is 1 − .9646 = .0354. In the case of a two-tailed test, we would use the minimum of .9646 and .0354 to compute p-value = 2(.0354) = .0708.

The χ^2 test statistic We use the St. Louis Metro Bus example from Section 11.1 as an illustration; the value of the test statistic is $\chi^2 = 28.18$ with 23 degrees of freedom. The Minitab steps used to compute the cumulative probability corresponding to $\chi^2 = 28.18$ follow.

Step 1. Select the **Calc** menu
Step 2. Choose **Probability Distributions**
Step 3. Choose **Chi-Square**
Step 4. When the Chi-Square Distribution dialog box appears:
Select **Cumulative probability**
Enter 23 in the **Degrees of freedom** box
Select **Input Constant**
Enter 28.18 in the **Input Constant** box
Click **OK**

Minitab provides a cumulative probability of .7909, which is the lower tail p-value. The upper tail p-value = 1 − the cumulative probability, or 1 − .7909 = .2091. The two-tailed p-value is 2 times the minimum of the lower and upper tail p-values. Thus, the two-tailed p-value is 2(.2091) = .4182. The St. Louis Metro Bus example involved an upper tail test, so we use p-value = .2091.

The F test statistic We use the Dullus County Schools example from Section 11.2 as an illustration; the test statistic is $F = 2.40$ with 25 numerator degrees of freedom and 15 denominator degrees of freedom. The Minitab steps to compute the cumulative probability corresponding to $F = 2.40$ follow.

Step 1. Select the **Calc** menu
Step 2. Choose **Probability Distributions**
Step 3. Choose **F**
Step 4. When the F Distribution dialog box appears:
Select **Cumulative probability**
Enter 25 in the **Numerator degrees of freedom** box
Enter 15 in the **Denominator degrees of freedom** box
Select **Input Constant**
Enter 2.40 in the **Input Constant** box
Click **OK**

Minitab provides the cumulative probability and hence a lower tail p-value = .9594. The upper tail p-value is 1 − .9594 = .0406. Because the Dullus County Schools example is a two-tailed test, the minimum of .9594 and .0406 is used to compute p-value = 2(.0406) = .0812.

Using Excel

WEB file
p-Value

Excel functions and formulas can be used to compute p-values associated with the z, t, χ^2, and F test statistics. We provide a template in the data file entitled p-Value for use in computing these p-values. Using the template, it is only necessary to enter the value of the test statistic and, if necessary, the appropriate degrees of freedom. Refer to Figure F.1 as we describe how the template is used. For users interested in the Excel functions and formulas being used, just click on the appropriate cell in the template.

The z test statistic We use the Hilltop Coffee lower tail hypothesis test in Section 9.3 as an illustration; the value of the test statistic is $z = -2.67$. To use the p-value template for this hypothesis test, simply enter -2.67 into cell B6 (see Figure F.1). After doing so, p-values for all three types of hypothesis tests will appear. For Hilltop Coffee, we would use the lower tail p-value $= .0038$ in cell B9. For an upper tail test, we would use the p-value in cell B10, and for a two-tailed test we would use the p-value in cell B11.

The t test statistic We use the Heathrow Airport example from Section 9.4 as an illustration; the value of the test statistic is $t = 1.84$ with 59 degrees of freedom. To use the p-value template for this hypothesis test, enter 1.84 into cell E6 and enter 59 into cell E7 (see Figure F.1). After doing so, p-values for all three types of hypothesis tests will appear.

FIGURE F.1 EXCEL WORKSHEET FOR COMPUTING p-VALUES

	A	B	C	D	E	F
1	**Computing *p*-Values**					
2						
3						
4	**Using the Test Statistic z**			**Using the Test Statistic *t***		
5						
6	**Enter *z* -->**	-2.67		**Enter *t* -->**	1.84	
7				***df* -->**	59	
8						
9	***p*-value (Lower Tail)**	0.0038		***p*-value (Lower Tail)**	0.9646	
10	***p*-value (Upper Tail)**	0.9962		***p*-value (Upper Tail)**	0.0354	
11	***p*-value (Two Tail)**	0.0076		***p*-value (Two Tail)**	0.0708	
12						
13						
14						
15						
16	**Using the Test Statistic Chi Square**			**Using the Test Statistic *F***		
17						
18	**Enter Chi Square -->**	28.18		**Enter *F* -->**	2.40	
19	***df* -->**	23		**Numerator *df* -->**	25	
20				**Denominator *df* -->**	15	
21						
22	***p*-value (Lower Tail)**	0.7909		***p*-value (Lower Tail)**	0.9594	
23	***p*-value (Upper Tail)**	0.2091		***p*-value (Upper Tail)**	0.0406	
24	***p*-value (Two Tail)**	0.4181		***p*-value (Two Tail)**	0.0812	
25						

The Heathrow Airport example involves an upper tail test, so we would use the upper tail p-value = .0354 provided in cell E10 for the hypothesis test.

The χ^2 test statistic We use the St. Louis Metro Bus example from Section 11.1 as an illustration; the value of the test statistic is $\chi^2 = 28.18$ with 23 degrees of freedom. To use the p-value template for this hypothesis test, enter 28.18 into cell B18 and enter 23 into cell B19 (see Figure F.1). After doing so, p-values for all three types of hypothesis tests will appear. The St. Louis Metro Bus example involves an upper tail test, so we would use the upper tail p-value = .2091 provided in cell B23 for the hypothesis test.

The F test statistic We use the Dullus County Schools example from Section 11.2 as an illustration; the test statistic is $F = 2.40$ with 25 numerator degrees of freedom and 15 denominator degrees of freedom. To use the p-value template for this hypothesis test, enter 2.40 into cell E18, enter 25 into cell E19, and enter 15 into cell E20 (see Figure F.1). After doing so, p-values for all three types of hypothesis tests will appear. The Dullus County Schools example involves a two-tailed test, so we would use the two-tailed p-value = .0812 provided in cell E24 for the hypothesis test.

Index

Note: Chapters 21 and 22 can be found with the Online Content for this book. Index entries found in these chapters are denoted by the chapter number, hyphen, and page number.

Page numbers followed by a **n** indicate a footnote.

D

E

F

G

H

I

J

K

L

M

N

O

P

Q

R

T

U

V

W

X

Y

Z

UNIVERSITY OF GLAMORGAN
PRIFYSGOL MORGANNWG
Learning Resources Centre

WA 1390365 9

STATISTICS FOR BUSINESS AND ECONOMICS 12e

David R. Anderson
University of Cincinnati

Dennis J. Sweeney
University of Cincinnati

Thomas A. Williams
Rochester Institute of Technology

Jeffrey D. Camm
University of Cincinnati

James J. Cochran
Louisiana Tech University

SOUTH-WESTERN
CENGAGE Learning®

Australia · Brazil · Canada · Mexico · Singapore · Spain · United Kingdom · United States

SOUTH-WESTERN
CENGAGE Learning

Statistics for Business and Economics, Twelfth Edition
David R. Anderson, Dennis J. Sweeney, Thomas A. Williams, Jeffrey D. Camm, James J. Cochran

Senior Vice President, LRS/Acquisitions & Solutions Planning: Jack W. Calhoun

Editorial Director, Business & Economics: Erin Joyner

Editor-In-Chief: Joe Sabatino

Sr. Acquisitions Editor: Charles McCormick, Jr.

Sr. Brand Manager: Kristen Hurd

Developmental Editor: Maggie Kubale

Sr. Content Project Manager: Tamborah Moore

Media Editor: Chris Valentine

Manufacturing Planner: Ron Montgomery

Production Service: MPS Limited

Sr. Art Director: Stacy Jenkins Shirley

Internal Designer: Michael Stratton/cmiller design

Cover Designer: Craig Ramsdell

Cover B/W Image: Getty Images/PhotoDisc/Chad Baker

Cover Color Image: Shutterstock/anyunov

Rights Acquisitions Specialist: Anne Sheroff

Text permissions researcher: Sarah Carey/PMG

Image permissions researcher: Sheeja Mohan/PMG

UNIVERSITY OF GLAMORGAN PRIFYSGOL MORGANNWG
Learning Resources Centre

13903659

Printed in Canada
1 2 3 4 5 6 7 16 15 14 13 12

© 2014, 2012 South-Western, Cengage Learning

ALL RIGHTS RESERVED. No part of this work covered by the copyright herein may be reproduced, transmitted, stored, or used in any form or by any means graphic, electronic, or mechanical, including but not limited to photocopying, recording, scanning, digitizing, taping, web distribution, information networks, or information storage and retrieval systems, except as permitted under Section 107 or 108 of the 1976 United States Copyright Act, without the prior written permission of the publisher.

For product information and technology assistance, contact us at
Cengage Learning Customer & Sales Support, 1-800-354-9706
For permission to use material from this text or product,
submit all requests online at **www.cengage.com/permissions**
Further permissions questions can be emailed to
permissionrequest@cengage.com

Exam*View*® is a registered trademark of eInstruction Corp. Windows is a registered trademark of the Microsoft Corporation used herein under license. Macintosh and Power Macintosh are registered trademarks of Apple Computer, Inc. used herein under license.
© 2008 Cengage Learning. All Rights Reserved.

Microsoft Excel® is a registered trademark of Microsoft Corporation. © 2014 Microsoft.

Library of Congress Control Number: 2012941311

ISBN-13: 978-1-285-17230-9

ISBN-10: 1-285-17230-2

Cengage Learning International Offices

Asia
www.cengageasia.com
tel: (65) 6410 1200

Australia/New Zealand
www.cengage.com.au
tel: (61) 3 9685 4111

Brazil
www.cengage.com.br
tel: (011) 3665 9900

India
www.cengage.co.in
tel: (91) 11 30484837/38

Latin America
www.cengage.com.mx
tel: +52 (55) 1500 6000

UK/Europe/Middle East/Africa
www.cengage.co.uk
tel: (44) 207 067 2500

Represented in Canada by Nelson Education, Ltd.
www.nelson.com
tel: (416) 752 9100 / (800) 668 0671

Cengage Learning is a leading provider of customized learning solutions with office locations around the globe, including Singapore, the United Kingdom, Australia, Mexico, Brazil, and Japan. Locate your local office at: **www.cengage.com/global**

For product information: **www.cengage.com/international**

Visit your local office: **www.cengage.com/global**

Visit our corporate website: **www.cengage.com**